Correlation Guide For

Friedland and Relyea Environmental Science for the AP® Course

Third Edition

Find a complete correlation guide to the AP® Environmental Science curriculum framework at https://www.highschool.bfwpub.com/APES3e

About the Cover

Wildlife Can Thrive in Constructed Habitats

The peregrine falcon *(Falco peregrinus)* is a formidable hunter and typically preys on other birds in flight. Today it is found throughout the world but the peregrine experienced substantial decline in its population beginning in the 1940s, due to the widespread use of the pesticide Dichlorodiphenyltrichloroethane, known commonly as DDT. Because DDT causes egg shells to thin, many fewer birds successfully hatched and this species as well as others declined. DDT was banned in the United States in 1972 and in other countries in the following decades. By the late 1970s the peregrine falcon began to recover and in 1999 it was removed from the endangered species list.

Peregrine falcons perch and nest on cliffs and they often hunt from these positions high on the landscape. Human-made structures such as bridges and tall buildings are also suitable peregrine falcon habitat for hunting and nesting. Cities provide excellent habitat for pigeons, which are prey for the peregrine falcon. Peregrines are now found in cities throughout the United States.

In 2015, a nesting pair of peregrine falcons occupied the balcony of a twenty-eighth floor condominium in Chicago, Illinois. The owner set up a webcam and people all over the world began to follow the activity of the birds. In 2016, a wildlife photographer from the Netherlands spent weeks photographing the birds and documented the hatching and fledging of four peregrine falcon young. The cover photo is of one of the adult falcons and the four chicks on the balcony in Chicago.

The story of the peregrine's comeback is an example of the complexity of environmental science. Yes, humans can disturb natural environments, leading to the decline of a species. Yet human structures, such as bridges and apartment buildings, can also provide habitat for birds such as falcons, and their prey, and create an environment in which they can thrive.

(Luke Massey/naturepl.com)

FRIEDLAND and RELYEA

Environmental Science for the AP® Course

THIRD EDITION

FRIEDLAND and RELYEA

Environmental Science for the AP® Course

THIRD EDITION

Andrew Friedland
Dartmouth College

Rick Relyea
Rensselaer Polytechnic Institute

bedford, freeman & worth
high school publishers
Boston | New York

Senior Vice President, Content Strategy: Charles Linsmeier
Vice President and General Manager, High School Business: Paul Altier
Senior Program Director, High School: Ann Heath
Executive Program Manager, High School STEM: Yolanda Cossio
Editorial Assistant: Carla Duval
Marketing Manager: Thomas Menna
Marketing Assistant: Kelly Noll
Developmental Editor: Rebecca Kohn
Media Editor: Kimberly Morté
Director, Content Management Enhancement: Tracey Kuehn
Senior Managing Editor: Lisa Kinne
Content Project Manager: Pamela Lawson
Media Project Manager: Jodi Isman
Director of Design, Content Management: Diana Blume
Design Services Manager: Natasha Wolfe
Text Designer: Kevin Kall
Art Manager: Matthew McAdams
Illustrations: Joseph BelBruno
Cover Design: John Callahan
Media Permissions Manager: Christine Buese
Photo Research: Donna Ranieri, Lumina Datamatics, Inc.
Senior Workflow Project Supervisor: Susan Wein
Project Management: Lumina Datamatics, Inc.
Composition: Lumina Datamatics, Inc.
Production Supervisor: Lawrence Guerra
Printing and Binding: Transcontinental Printing
Cover Image and Details: Luke Massey/naturepl.com

Library of Congress Control Number: 2018936172
ISBN-13: 978-1-319-11329-2
ISBN-10: 1-319-11329-X

2 3 4 5 6 7 24 23 22 21 20 19

Printed in Canada

Bedford, Freeman & Worth High School Publishers
One New York Plaza
Suite 4600
New York, NY 10004-1562
highschool.bfwpub.com/catalog

To Katie, Jared, and Ethan

—AJF

To Christine, Isabelle, and Wyatt

—RAR

Brief Contents

(Luke Massey/naturepl.com)

Contents

UNIT 3
BIOLOGICAL AND HUMAN POPULATIONS

UNIT 4
EARTH SYSTEMS AND RESOURCES

About the Authors

Nancy Nutile-McMenemy

Brian Mattes

Andrew Friedland is Richard and Jane Pearl Professor in Environmental Studies and former chair of the Environmental Studies Program at Dartmouth College. He was the founding chair of the Advanced Placement® Test Development Committee (College Board®) for Environmental Science. He has a strong interest in high school science education, and in the early years of AP® environmental science he participated in many trainer and teacher workshops. For more than 15 years, Andy has been a guest lecturer at various Advanced Placement® Institutes for Secondary Teachers and in high school APES classrooms. He also served on the College Board AP® Environmental Science Curriculum Development and Assessment Committee.

Andy regularly teaches introductory environmental science and energy courses at Dartmouth and has taught courses in forest biogeochemistry, global change, and soil science, as well as study abroad courses in Kenya. He taught an online introductory environmental science course that is accessible through edX.org and YouTube.

Andy received a BA degree in both biology and environmental studies, and a PhD in earth and environmental science, from the University of Pennsylvania. For three decades, Andy has been investigating the effects of air pollution on the cycling of carbon, nitrogen, and lead in high-elevation forests of New England and the Northeast. More recently, he has examined the impact of increased demand for wood as a fuel, and the subsequent effect on carbon stored deep in forest soils.

Andy has served on panels for the National Science Foundation, USDA Forest Service, and Science Advisory Board of the Environmental Protection Agency. He has authored or coauthored 80 peer-reviewed publications and one book, *Writing Successful Science Proposals,* Third Edition (Yale University Press). In 2015, he was named a Fellow of the American Association for the Advancement of Science.

Andy is passionate about saving energy and at his home has installed a 4 kW photovoltaic tracker that follows the sun during the day.

Rick Relyea is the David Darrin Senior '40 Endowed Chair in Biology and the director of the Darrin Freshwater Institute at Rensselaer Polytechnic Institute. Rick teaches courses in ecology, evolution, and animal behavior at the undergraduate and graduate levels. He received a BS in environmental forest biology from the State University of New York College of Environmental Science and Forestry, an MS in wildlife management from Texas Tech University, and a PhD in ecology and evolution from the University of Michigan.

Rick is recognized throughout the world for his work in the fields of ecology, evolution, animal behavior, and ecotoxicology. He has served on multiple scientific panels for the National Science Foundation and has been an associate editor for the journals of the Ecological Society of America. For two decades, he has conducted research on a wide range of topics, including predator-prey interactions, phenotypic plasticity, eutrophication of aquatic habitats, sexual selection, disease ecology, long-term dynamics of populations and communities across the landscape, and pesticide impacts on aquatic ecosystems. He has authored more than 160 scientific articles and book chapters, and has presented research seminars throughout the world. Rick was a professor at the University of Pittsburgh for 15 years, where he was named the Chancellor's Distinguished Researcher in 2005 and received the Tina and David Bellet Teaching Excellence Award in 2014. In 2014, he moved to Rensselaer Polytechnic Institute to direct The Jefferson Project, which is the most technologically advanced research endeavor to study a freshwater lake.

Rick has a strong interest in high school education. High school science teachers conduct research in his laboratory and he offers summer workshops for high school teachers in the fields of ecology, evolution, and ecotoxicology. Rick also works to bring cutting-edge research experiments into high school classrooms.

Rick's commitment to the environment extends to his personal life. He lives in a home constructed with a passive solar building design and equipped with active solar panels on the roof.

Acknowledgments

We would like to thank the many people at Bedford, Freeman, and Worth/Macmillan who helped guide us through the publication process of this book. They have taught us a great deal and have been crucial to our book becoming greatly appreciated by so many people. We especially want to acknowledge: Ann Heath, Yolanda Cossio, Rebecca Kohn, Joseph BelBruno, Diana Blume, Fred Burns, Christine Buese, Carla Duval, Ellen Irwin, Pamela Lawson, Matt McAdams, Kimberly Morté, Thomas Menna, Kelly Noll, Donna Ranieri, and Natasha Wolfe. Thanks also to all the people who helped with aspects of the program, including Jabin Burnworth, Suzanne Carmody, Leah Kohn, Aaron Stoler, and Erika Yates. We thank Elizabeth Jones and Elisa McCracken for AP® tips and other assistance that only teachers of AP® Environmental Science can offer, and David Courard-Hauri, Ross Jones, and Susan Weisberg for contributions to the first edition of this book.

We also wish to convey our appreciation to the dozens of reviewers who constantly challenged us to write a clear, correct, and philosophically balanced textbook.

From Andy Friedland . . .

I would like to thank all of my teachers, students, and colleagues. Professors Robert Giegengack and Arthur Johnson introduced me to environmental science as an undergraduate and graduate student. My current and previous colleagues in the Environmental Studies Program at Dartmouth and elsewhere have contributed in a variety of ways. I thank Doug Bolger, Michael Cox, Rich Howarth, Anne Kapuscinski, Rosi Kerr, Nick Reo, Bill Roebuck, Jack Shepherd, Chris Sneddon, Ross Virginia, D. G. Webster, and Elizabeth Wilson for all sorts of contributions to my teaching and scholarship and to this book. Graduate students Morgan Peach and Hunter Snyder have also contributed. Madison Sabol and Catherine Rocchi, Dartmouth undergraduates who have taken courses from me, provided excellent editorial, proofreading, and writing assistance. Many other colleagues have had discussions with me or evaluated sections of text including William Schlesinger, Carol Folt, Justin Richardson, Jim Kaste, Kathy Cottingham, and Mark McPeek. Since the time when AP® Environmental Science was just an idea at a College Board® workshop, Beth Nichols, Tom Corley, and many others, especially a large number of teachers, have helped me learn about the world of Advanced Placement® teaching.

I wish to acknowledge Dana Meadows and Ned Perrin, both of whom have since passed away, for contributions during the early stages of this work. Terry Tempest Williams has been a tremendous source of advice and wisdom about topics environmental, scientific, and practical.

I am grateful to Dick Pearl and the late Janie Pearl for friendship and support through the Richard and Jane Pearl Professorship in Environmental Studies. Finally, I thank Katie, Jared, and Ethan Friedland, and my mother Selma.

From Rick Relyea . . .

I would like to thank my family—my wife Christine and my children Isabelle and Wyatt. They continually inspire me.

I am also grateful to the many people at Bedford, Freeman, and Worth who helped guide me and taught me a great deal about the publication process. I would like to especially thank Jerry Correa for convincing me to join the first edition of this book.

Reviewers

High School Focus Group Participants and Reviewers

Our deep appreciation and heartfelt thanks are due to the experienced AP® teachers who participated in focus groups and/or reviewed the manuscript during the development of this book. Their contributions have been invaluable.

Cynthia Ahmed, *Signature School, IN*
Timothy Allen, *Thomas A. Edison Preparatory High School, OK*
Julie Back, *Kecoughtan High School, VA*
Maureen Bagwell, *Collierville High School, TN*
Fredrick Baldwin, *Kendall High School, NY*
Lisa Balzas, *Indian Springs School, AL*
Debra Bell, *Montgomery High School, TX*
Melinda Bell, *Flagstaff Arts and Leadership Academy, AZ*
Karen Benton, *South Brunswick High School, NJ*
Richard Benz, *Wickliffe High School, OH*
Cindy Birkner, *Webber Township High School, IL*
Christine Bouchard, *Milford Public Schools, CT*
Gail Boyarsky, *East Chapel Hill High School, NC*
Rebecca Bricen, *Johnsonburg High School, PA*
Deanna Brunlinger, *Elkhorn Area High School, WI*
Kevin Bryan, *Woodrow Wilson Senior High School, CA*
Tanya Bunch, *Carter High School, TN*

Diane Burrell, *Starr's Mill High School, GA*
Teri Butler, *New Hanover High School, NC*
Charles Campbell, *Russellville High School, AR*
Sande Caton, *Concord High School, DE*
Andrea Charles, *West Side Leadership Academy, IN*
Linda Charpentier, *Xavier High School, CT*
Blanca Ching, *Fort Hamilton High School, NY*
Ashleigh Coe, *Bethesda-Chevy Chase High School, MD*
Bethany Colburn, *Randolph High School, MA*
Jonathan D. Cole, *Holmdel High School, NJ*
Robert Compton, *Walled Lake Northern High School, MI*
Ann Cooper, *Oseola High School, AR*
Thomas Cooper, *The Walker School, GA*
Joyce Corriere, *Hampton High School, VA*
Stephanie Crow, *Milford High School, MI*
Stephen Crowley, *Winooski High School, VT*
Linda D'Apolito, *Trinity School, NY*
Brygida DeRiemaker, *Eisenhower High School, MI*
Chand Desai, *Martin Luther King Magnet High School, TN*
Michael Douglas, *Bronx Prep Charter School, NY*
Nancy Dow, *A. Crawford Mosley High School, FL*
Nat Draper, *Deep Run High School, VA*
Denis DuBay, *Leesville High School, NC*
John Dutton, *Shaw High School, OH*
Heather Earp, *West Johnston High School, NC*
Kim Eife, *Academy of Notre Dame, PA*
Brian Elliot, *San Dimas High School, CA*
Christina Engen, *Crescenta Valley High School, CA*
Mary Anne Evans, *Allendale Columbia School, NY*
Kay Farkas, *Rush-Henrietta High School, NJ*
Tim Fennell, *LASA at LBJ High School, TX*
Michael Finch, *Greene County Tech High School, AR*
Robert Ford, *Fairfield College Preparatory School, CT*
Paul Frisch, *Fox Lane High School, NY*
Bob Furhman, *The Covenant School, VA*
Nivedita (Nita) Ganguly, *Oak Ridge High School, TN*
Mike Gaule, *Ladywood High School, MI*
Billy Goodman, *Passaic Valley High School, NJ*
Amanda Graves, *Mt. Tahoma High School, WA*
Barbara Gray, *Richmond Community High School, VA*
Jack Greene, *Logan High School, UT*
Julie Quinn Kiernan, *Cretin-Durham Hall, NC*
Jeannie Kornfeld, *Hanover High School, NH*
Jen Kotkin, *St. Philip's Academy, NJ*
Pat Kretzer, *Timber Creek High School, FL*
Michelle Krug, *Coral Springs High School, FL*
Jim Kuipers, *Chicago Christian High School, IL*
Claire Kull, *Career Center, NC*
Jay Kurima, *O. D. Wyatt High School, TX*
Tom LaHue, *Aptos High School, CA*
Cathy Larson, *Patuxent High School, MD*
Michael Lauer, *Danville High School, KY*
Sonia Laureni, *West Orange High School, NJ*
Amy Lawson, *Naples High School, FL*
Jim Lehner, *The Taft School, CT*
Dr. Avon Lewis, *Lexington High School, MA*
Marie Lieberman, *Ravenscroft School, NC*
John Ligget, *Conestoga High School, PA*
Ann Linsley, *Bellaire High School, TX*
Mark Little, *Broomfield High School, CO*
Leyana Lloyd, *Washington Senior Academy, GA*
Larry Lollar, *Alice High School, TX*
Stephanie Longfellow, *Deltona High School, FL*
Leslie Lopez, *Round Rock High School, TX*
Sue Ellen Lyons, *Holy Cross School, LA*
Theresa Lyster, *Camden County High School, GA*
John F. Madden, *Ashley Hall School, SC*
Jeremy Magee, *Sandy High School, OR*
Mike Mallon, *James I. O'Neill High School, NY*
Scott Martin, *Deer Creek High School, OK*
Kristi Martinez, *Eastlake High School, WA*
Christeena Mathews, *The Philadelphia High School for Girls, PA*
Courtney Mayer, *Winston Churchill High School, TX*
Monica Maynard, *Schurr High School, CA*
James McAdams, *Center Grove High School, IN*
Kristen McClellen, *Grand Junction High School, CO*
Sandy McDonough, *North Salem Middle/High School, NY*
Diane Medford, *Los Alamos High School, NM*
Andrew Milbauer, *Poudre High School, CO*
Leslie Miller, *Flintridge Sacred Heart Academy, CA*
Lonnie Miller, *El Diamante High School, CA*
Melody Mingus, *Breckinridge County High School, KY*
Myra Morgan, *National Math & Science Initiative, AP® Environmental Consultant*
Tammy Morgan, *Lake Placid High School, NY*
David Moscarelli, *Ponaganset High School, RI*
Terri Mountjoy, *Greene County Career Center, OH*
Bill Mulhearn, *Archmere Academy, DE*
Sharna Murphy, *Millikan High School, CA*
Jeanine Musgrove, *Oakton High School, VA*
Anna Navarro, *Veterans Memorial High School, TX*
Barbara Nealon, *Southern York County School District, PA*
Dara Nix-Stevenson, *American Hebrew Academy, NC*
Bennett O'Connor, *Dallas ISD, TX*
Robert Oddo, *Horace Greeley High School, NY*
Kate Oitzinger, *El Molino High School, CA*
Paul Olson, *Redwood High School, CA*
Janet Ort, *Hoover High School, AL*
Roger Palmer, *Bishop Dunne High School, TX*
Annetta Pasquarello, *Triton Regional High School, NJ*
Lynn Paulsen, *Mayde Creek High School, TX*
Nicole Peffley, *Cinco Ranch High School, TX*
Judy Perrella, *Academy of the Holy Names, FL*
Carolyn Phillips, *Southeastern High School, IL*
Pam Phillips, *Hayden High School, AL*
Alanna Piccillo, *Palisade High School, CO*
Jenny Ramsey, *Charlotte Christian School, NC*
Susan Ramsey, *VASS, VA*
Cristen Rasmussen, *Costa Mesa High School, CA*
Alesa Rehmann, *Coral Shores High School, FL*
Mark Reilly, *Jeffersonville High School, IN*
Kimbell Reitz, *Penn High School, IN*
Cheryl Rice, *Howard High School, MD*
Sharon Riley, *Springfield High School, OH*
Chris Robson, *Ironwood Ridge High School, AZ*
James Rodewald, *Shaker High School, NY*
Kurt Rogers, *Northern Highlands Regional High School, NJ*
Kris Rohrbeck, *Almont High School, MI*
David Rouby, *Hall High School, AR*
Rebecca Rouch, *East Bay High School, FL*
Jennifer Roy, *TrekNorth Junior & Senior High School, MN*
Reva Beth Russell, *Lehi High School, UT*
Sheila Scanlan, *Highland High School, AZ*
Kristi Schertz, *Saugus High School, CA*

Greg Schiller, *James Monroe High School, CA*
Amy Schwartz, *Aragon High School, CA*
Kristin Shapiro, *Plano East Senior High School, TX*
Shashi Sharma, *Henry Snyder High School, NJ*
Tonya Shires, *Edgewood High School, MD*
Pamela Shlachtman, *South Dade Senior High School, FL*
Julie Smiley, *Winchester Community High School, IN*
Amy Snodgrass, *Central High School, AR*
Bill Somerlot, *New Albany High School, OH*
Anne Soos, *Stuart Country Day School of the Sacred Heart, NJ*
Joan Stevens, *Arcadia High School, CA*
Marianne Strickhart, *Henry Snyder High School, NJ*
Timothy Strout, *Jericho High School, NY*
Robert Summers, *A+ College Ready, AL*
Jeff Sutton, *The Harker School, CA*
Dave Szaroleta, *Salesianum School, DE*
Kristen Thomson, *Saratoga High School, CA*
James Timmons, *Carrboro High School, NC*
Thomas Tokarski, *Woodlands High School, NY*
Susan Tully, *Salem Academy Charter School, MA*
Debra Tyson, *Brooke Pointe High School, VA*
Melissa Valentine, *Elizabeth Seton High School, MD*
Dirk Valk, *McKeel Academy, FL*
Gene Vann, *Head-Royce School, CA*
Rebecca Van Tassell, *Herron High School, IN*
Ashley Veenema, *Lebanon High School, NH*
Marc Vermeire, *Friday Harbor High School, WA*
Naomi Volain, *Springfield Central High School, MA*
Betty Walden, *Merritt Island High School, FL*
Craig Wallace, *North Oldham High School, KY*
Abbie Walston, *North Haven High School, CT*
Annette Weeks, *Battle Ground High School, WA*
Pamela Weghorst, *Ardrey Kell High School, NC*
Matthew Wells, *Cypress Lakes High School, TX*
Michelle Whitehurst, *Powhatan High School, VA*
Jane Whitelock, *Easton High School, MD*
Laurie Whitesell, *Eli Whitney Middle School, OK*
Robert Whitney, *Westview High School, CA*
Carol Widegren, *Lincoln Park High School, IL*
Sarrah Williams, *Hamden Hall Country Day School, CT*
Robert Willis, *Lakeside High School, GA*

College Reviewers

We are also indebted to numerous college instructors, many of whom are also involved in AP® Environmental Science, for their insights and suggestions through various stages of development. The content experts who carefully reviewed chapters in their area of expertise are designated with an asterisk (★).

M. Stephen Ailstock, PhD, *Anne Arundel Community College*
Deniz Z. Altin-Ballero, *Georgia Perimeter College*
Daphne Babcock, *Collin County Community College District*
Jay L. Banner, *University of Texas at San Antonio*
James W. Bartolome, *University of California, Berkeley*
Ray Beiersdorfer, *Youngstown State University*
Grady Price Blount, *Texas A&M University, Corpus Christi*
Dr. Edward M. Brecker, *Palm Beach Community College, Boca Raton*
Anne E. Bunnell, *East Carolina University*
Ingrid C. Burke, *Colorado State University*
Anya Butt, *Central Alabama Community College*
John Callewaert, *University of Michigan* ★
Kelly Cartwright, *College of Lake County*
Mary Kay Cassani, *Florida Gulf Coast University*
Young D. Choi, *Purdue University Calumet*
John C. Clausen, *University of Connecticut* ★
Richard K. Clements, *Chattanooga State Technical Community College*
Thomas Cobb, *Bowling Green State University, OH*
Stephen D. Conrad, *Indiana Wesleyan University*
Terence H. Cooper, *University of Minnesota, Saint Mary's Winona Campus*
Douglas Crawford-Brown, *University of North Carolina at Chapel Hill*
Wynn W. Cudmore, *Chemeketa Community College*
Katherine Kao Cushing, *San Jose State University*
Maxine Dakins, *University of Idaho*
Robert Dennison, *Heartland Community College*
Michael Denniston, *Georgia Perimeter College*
Roman Dial, *Alaska Pacific University*
Robert Dill, *Bergen Community College*
Michael L. Draney, *University of Wisconsin, Green Bay*
Anita I. Drever, *University of Wyoming* ★
James Eames, *Loyola University. New Orleans*
Kathy Evans, *Reading Area Community College*
Mark Finley, *Heartland Community College*
Dr. Eric J. Fitch, *Marietta College*
Karen F. Gaines, *Northeastern Illinois University*
James E. Gawel, *University of Washington, Tacoma*
Carri Gerber, *Ohio State University Agricultural Technical Institute*
Julie Grossman, *Saint Mary's University of Minnesota, Saint Mary's Winona Campus*
Lonnie J. Guralnick, *Roger Williams University*
Sue Habeck, *Tacoma Community College*
Hilary Hamann, *Colorado College*
Dr. Sally R. Harms, *Wayne State College*
Floyd Hayes, *Pacific Union College*
Keith R. Hench, *Kirkwood Community College*
William Hopkins, *Virginia Tech* ★
Richard Jensen, *Hofstra University*
Sheryll Jerez, *Stephen F. Austin State University*
Shane Jones, *College of Lake County*
Caroline A. Karp, *Brown University*
Erica Kipp, *Pace University, Pleasantville/Briarcliff*
Christopher McGrory Klyza, *Middlebury College* ★
Frank T. Kuserk, *Moravian College*
Matthew Landis, *Middlebury College* ★
Kimberly Largen, *George Mason University*
Larry L. Lehr, PhD, *Baylor University*
Zhaohui Li, *University of Wisconsin, Parkside*
Thomas R. MacDonald, *University of San Francisco*
Robert Stephen Mahoney, *Johnson & Wales University*
Bryan Mark, *Ohio State University, Columbus Campus*
Paula J.S. Martin, *Juniata College*
Robert J. Mason, *Tennessee Temple University*
Michael R. Mayfield, *Ball State University*
Alan W. McIntosh, *University of Vermont*
Dr. Kendra K. McLauchlan, *Kansas State University* ★
Patricia R. Menchaca, *Mount San Jacinto Community College*
Dr. Dorothy Merritts, *Franklin and Marshall College* ★
Bram Middeldorp, *Minneapolis Community and Technical College*
Tamera Minnick, *Mesa State College*
Mark Mitch, *New England College*
Ronald Mossman, *Miami Dade College, North*

William Nieter, *St. John's University*
Mark Oemke, *Alma College*
Victor Okereke, PhD, PE, *Morrisville State College*
Duke U. Ophori, *Montclair State University*
Chris Paradise, *Davidson College*
Dr. Clayton A. Penniman, *Central Connecticut State University*
Christopher G. Peterson, *Loyola University Chicago*
Craig D. Phelps, *Rutgers, The State University of New Jersey, New Brunswick*
F. X. Phillips, PhD, *McNeese State University*
Rich Poirot, *Vermont Department of Environmental Conservation* ★
Bradley R. Reynolds, *University of Tennessee, Chattanooga*
Amy Rhodes, *Smith College* ★
Marsha Richmond, *Wayne State University*
Sam Riffell, *Mississippi State University*
Jennifer S. Rivers, *Northeastern Illinois University*
Ellison Robinson, *Midlands Technical College*
Bill D. Roebuck, *Dartmouth Medical School* ★
William J. Rogers, *West Texas A&M University*
Thomas Rohrer, *Central Michigan University*
Aldemaro Romero, *Arkansas State University*
William R. Roy, *University of Illinois at Urbana-Champaign*
Steven Rudnick, *University of Massachusetts, Boston*
Heather Rueth, *Grand Valley State University*
Eleanor M. Saboski, *University of New England*
Seema Sah, *Florida International University*
Shamili Ajgaonkar Sandiford, *College of DuPage*
Robert M. Sanford, *University of Southern Maine*
Nan Schmidt, *Pima Community College*
Jeffery A. Schneider, *State University of New York at Oswego*
Bruce A. Schulte, *Georgia Southern University*
Eric Shulenberger, *University of Washington*
Michael Simpson, *Antioch University New England* ★
Annelle Soponis, *Reading Area Community College*
Douglas J. Spieles, *Denison University*
David Steffy, *Jacksonville State University*
Christiane Stidham, *State University of New York at Stony Brook*
Peter F. Strom, *Rutgers, The State University of New Jersey, New Brunswick*
Kathryn P. Sutherland, *University of Georgia*
Christopher M. Swan, *University of Maryland, Baltimore County* ★
Melanie Szulczewski, *University of Mary Washington*
Jamey Thompson, *Hudson Valley Community College*
John A. Tiedemann, *Monmouth University*
Conrad Toepfer, *Brescia University*
Todd Tracy, *Northwestern College*
Steve Trombulak, *Middlebury College*
Zhi Wang, *California State University, Fresno*
Jim White, *University of Colorado, Boulder*
Rich Wolfson, *Middlebury College* ★
C. Wesley Wood, *Auburn University*
David T. Wyatt, *Sacramento City College*

Getting the Most from This Book

Daily life is filled with decisions large and small that affect our environment. From the food we eat, to the cars we drive or choose not to drive, to the chemicals we put into the water, soil, and air. The impact of human activity is wide-ranging and deep. And yet making decisions about the environment is often not easy or straightforward. Is it better for the environment if we purchase a new, energy-efficient hybrid car or should we continue using the older car we already own? Should we remove a dam that provides electricity for 70,000 homes because it interferes with the migration of salmon? Are there alternatives to fossil fuel for heating our homes?

The purpose of this book is to give you a working knowledge of the big ideas of environmental science and help you to prepare for the AP® Environmental Science Exam. The book is designed to provide you with a strong foundation in the scientific fundamentals, to introduce you to the policy issues and conflicts that emerge in the real world, and to offer you an in-depth exploration of all the topics covered on the advanced placement exam in environmental science.

Like the first two editions, *Friedland and Relyea's Environmental Science for the AP® Course*, Third Edition, is organized to closely follow the AP® Environmental Science curriculum framework. Every item on the College Board's curriculum framework is covered thoroughly in the text. The textbook offers comprehensive coverage of all required AP® course topics and will help you succeed on the exam with many unique features including the following.

- **Module format.** Each chapter is divided into manageable modules that will help you organize the material and keep up with the challenging pace of the AP® Environmental Science course.
- **Real-world cases.** The chapter-opening case studies will help you to see how environmental science is grounded in your daily life and in the world around you. "Working Toward Sustainability" at the end of each chapter describes how people are putting the concepts of environmental science into action. Understanding how to apply chapter concepts to actual situations will help motivate learning and prepare you for the kind of analysis the AP® Environmental Science Exam requires.
- **Stunning and informative visuals.** The figures, photographs, graphs, and other visuals in the text will help you understand and remember the big ideas and important concepts that will be on the exam. Many figures contain visual representations of the models you will learn about in the course.
- **Opportunities for AP® practice.** Throughout every chapter you will have many opportunities to practice for the AP® Environmental Science Exam. Look for the following to help sharpen your AP® Environmental Science test-taking skills:
    - AP® Exam tips
    - End-of-module AP® review questions
    - "Do the Math" boxes
    - "Practice Math and Graphing" boxes in the chapter review
    - Chapter AP® practice exams
    - Unit AP® practice exams
    - Two cumulative AP® practice exams at the end of the book

The next few pages offer you a brief tour of the features of this book that have been designed to help you succeed in the course and on the exam.

Use the chapter structure to help you master the material.

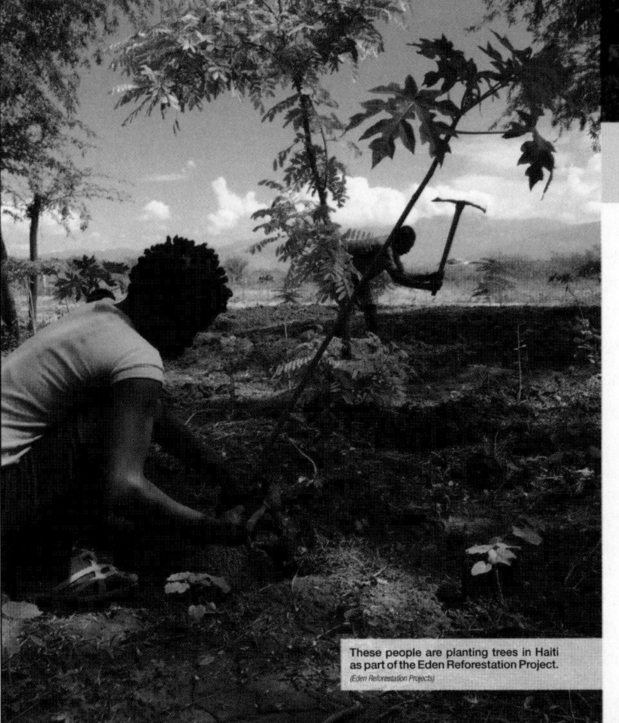

These people are planting trees in Haiti as part of the Eden Reforestation Project.
(Eden Reforestation Projects)

Ecosystem Ecology

CHAPTER 3

MODULE 6 The Movement of Energy
MODULE 7 The Movement of Matter
MODULE 8 Responses to Disturbances

CASE STUDY

Reversing the Deforestation of Haiti

Even before the devastating earthquake of 2010, life in Haiti was hard. On the streets of the capital city, Port-au-Prince, people lined up to buy charcoal to cook their meals. According to the United Nations, 76 percent of Haitians lived on less than $2 a day. Because other forms of cooking fuel, including oil and propane, were too expensive, people turned to the forests, cutting trees to make charcoal from firewood.

The deforestation of Haiti has a long history that goes back to the seventeenth and eighteenth centuries when much of the forest was cleared for lumber, fuel, and agriculture. By 1923, only 60 percent of this mountainous country was covered in forest. However, as agriculture expanded with new plantations of rubber trees in the 1940s and as the population grew with a greater demand for fuel, the amount of forest declined even more. By 2017, with more than 11 million people living in this small nation, only about 30 percent of its land remained forested.

Deforestation disrupts the services that living trees provide in an ecosystem. When Haitian forests are cleared of trees the land becomes much more susceptible to erosion. As the tree roots die, they can no longer stabilize the soil, so the soil is eroded away by the heavy rains of tropical storms and hurricanes. Unimpeded by vegetation, the rainwater runs quickly down the mountainsides, dislodging the topsoil that is so important for forest growth. In addition, this oversaturation of the soil causes massive mudslides that can destroy entire villages.

By 2017, with more than 11 million people living in this small nation, only about 30 percent of its land remained forested.

But the news from Haiti is not all bad. Since the 1980s, the U.S. Agency for International Development, in cooperation with many other groups, has provided funds to plant 60 million trees in Haiti. Unfortunately, simply planting trees is not enough; the local people can't afford to let trees grow when they are in desperate need of firewood

Chapter Opening Case Study

Chapters start with a relevant case study to help you focus your learning. Understanding how to apply chapter concepts to actual situations prepares you for the kind of text analysis the AP® Environmental Science Exam requires.

MODULE 4

Systems and Matter

All questions about the environment involve *matter,* the atoms and molecules that compose materials, and the systems in which the matter circulates. In this module we will look at the basic building blocks of matter and how matter moves among systems. We will look in detail at water, an important component of most environmental systems. Finally, we will explore how matter is conserved in chemical and biological systems.

Learning Goals

After reading this module you should be able to

- describe how matter comprises atoms and molecules that move among different systems.
- explain why water is an vital component of most environmental systems.
- discuss how matter is conserved in chemical and biological systems.

Matter comprises atoms and molecules that move among different systems

What do rocks, water, air, the book in your hands, and the cells in your body have in common? They are all forms of *matter.* **Matter** is anything that occupies space and has *mass.* The **mass** of an object is a measurement of the amount of matter it contains. Note that the words "mass" and "weight" are often used interchangeably, but they are not the same thing. Weight is the force that results from the action of gravity on mass. Your own weight, for example, is determined by the amount of gravity pulling you toward the planet's center. Whatever your weight on Earth, you would weigh less on the Moon where the action of gravity is weaker. In contrast, mass stays the same under any gravitational influence. So although your weight would change on the Moon, your mass would remain the same because the amount of matter you are made of would be the same. In this section we will look at important properties of matter, starting with the building blocks.

Atoms and Molecules

All matter is composed of tiny particles that cannot be broken down into smaller pieces. The basic building blocks of matter are known as *atoms.* An **atom** is the smallest particle that can contain the chemical properties of an *element.* An **element** is a substance composed of atoms that cannot be broken down into smaller, simpler components. At Earth's surface temperatures, elements can occur as solids (such as gold), liquids (such as mercury), or gases (such as helium). Atoms are so small that a single human hair measures about a few hundred thousand carbon atoms across.

Ninety-four elements occur naturally on Earth, and another twenty-four have been produced in laboratories. The **periodic table** lists all of the elements currently known, organized by their properties. The full periodic table is reproduced in an appendix to this book. As you can see, each element is identified by a one- or two-letter symbol; for example, the symbol for carbon is C, and the symbol for sodium is Na. These symbols are used to describe the atomic makeup

Matter Anything that occupies space and has mass.
Mass A measurement of the amount of matter an object contains.
Atom The smallest particle that can contain the chemical properties of an element.
Element A substance composed of atoms that cannot be broken down into smaller, simpler components.
Periodic table A chart of all chemical elements currently known, organized by their properties.

Module Structure

Chapters are divided into short modules to help you tackle concepts in manageable chunks. Each module opens with a brief description of topics to be covered.

Learning Goals

A list of key ideas at the beginning of the module helps keep you focused as you read.

Running Glossary

Knowing and understanding the language of environmental science is critical for success on the AP® Environmental Science Exam. Key terms appear in bold type in the text, and are defined at the bottom of the page. All terms can be found in the Glossary/Glossário at the end of the book.

Math practice makes perfect.

Do the Math

Among the biggest challenges on the AP® Environmental Science Exam are questions that ask you to solve environmental science math problems. "Do the Math" teaches the math skills you need to tackle these problems on the exam.

DO THE MATH How Much Energy Can You Save with a New Refrigerator?

Preparing for the AP® Exam

Your electricity bill shows that you use 600 kWh of electricity each month. Your refrigerator, which is 15 years old, could be responsible for up to 25 percent of this electricity consumption. Newer refrigerators are more efficient, meaning that they use less energy to do the same amount of work. If you wish to conserve electrical energy and save money, should you replace your refrigerator? How can you compare the energy efficiency of your old refrigerator with that of more efficient newer models?

Your refrigerator uses 500 watts when the motor is running. The motor runs for about 30 minutes per hour (or a total of 12 hours per day). How much energy in kilowatt-hours per year will you save by using the best new refrigerator instead of your current one? How long will it take you to recover the cost of the new appliance?

1. Start by calculating the amount of energy your current refrigerator uses.

$$0.5\text{ kW} \times 12\text{ hours/day} = 6\text{ kWh/day}$$

$$6\text{ kWh/day} \times 365\text{ days/year} = 2{,}190\text{ kWh/year}$$

2. How much more efficient is the best new refrigerator compared with your older model?

The best new model uses 400 kWh per year. Your refrigerator uses 2,190 kWh per year.

$$2{,}190\text{ kWh/year} - 400\text{ kWh/year} = 1{,}790\text{ kWh/year}$$

3. Assume that you are paying, on average, $0.10 per kilowatt-hour for electricity. A new refrigerator would cost $550. You will receive a rebate of $50 from your electric company for purchasing an energy-efficient refrigerator. This lowers the cost of the refrigerator to $500. If you replace your refrigerator, how long will it be before your energy savings compensate you for the cost of the new appliance? You will save

$$1{,}790\text{ kWh/year} \times \$.10\text{/kWh} = \$179\text{/year}$$

To determine how long it will take for you to recover the cost of the new refrigerator divide the cost by your savings per year.

$$\$500 \div \$179\text{/year} = 2.79\text{ years (rounding to significant figures)}$$

It will take less than 3 years to recover the cost of the new appliance.

YOUR TURN Environmental scientists must often convert energy units in order to compare various types of energy. For instance, you might want to compare the energy you would save by purchasing an energy-efficient refrigerator with the energy you would save by driving a more fuel-efficient car. Assume that for the amount you would spend on the new refrigerator ($500), you can make repairs to your car engine that would save you 20 gallons (76 liters) of gasoline per month. (Note that 1 L of gasoline contains the energy equivalent of about 10 kWh.) Using this information and Table 5.1 on page 46, convert the quantities of both gasoline and electricity into joules and compare the energy savings. Which decision would save the most energy?

Your Turn

Each "Do the Math" box has a "Your Turn" practice problem to help you review and practice the math skills introduced.

Practice Math and Graphing

At the end of each chapter, "Practice Math and Graphing" gives you an opportunity to work on your data analysis skills with problems like those you will encounter on the AP® Environmental Science Exam.

Practice Math and Graphing

Preparing for the AP® Exam

…nswer the following questions. Be sure to show all your work.

…Practice Math

To get a better sense of how per capita water use varies in households around the world, use the data in the table below to calculate per capita water use in seven different countries. Show your work.

Nation	Total annual water use (m^3)	Population size (millions of people)	Per capita annual water use (m^3/person)
France	111×10^9	59 million	
Iran	102×10^9	63 million	
Brazil	233×10^9	169 million	
Germany	127×10^9	82 million	
India	987×10^9	1007 million	
China	882×10^9	1257 million	
Tanzania	372×10^8	33 million	

Data from Chapagain, A. K., and Hoekstra, A. Y. 2004. *Water Footprints of Nations*, Vol. 1, Main Report, UNESCO-IHE. Research Report Series #16.

2. Practice Graphing

(a) Using the data in the table from "Practice Math, create three bar graphs that compare the following for each country:

1. Total daily water use
2. Population size
3. Per capita water use

(b) Why are there such large differences in per capita daily water use?

Analyze and interpret visual Representations and data.

FIGURE 28.9 Landscaping in the desert. By using plants that are adapted to a desert environment, homeowners in Arizona can greatly reduce the need for irrigation, resulting in a considerable reduction in water use compared with that required for growing grass. *(Bruce C. Murray/Shutterstock)*

Photos and Illustrations

The visual representations in this book are more than just pretty pictures. They have been carefully chosen and developed to help you comprehend learning goals and work with key models.

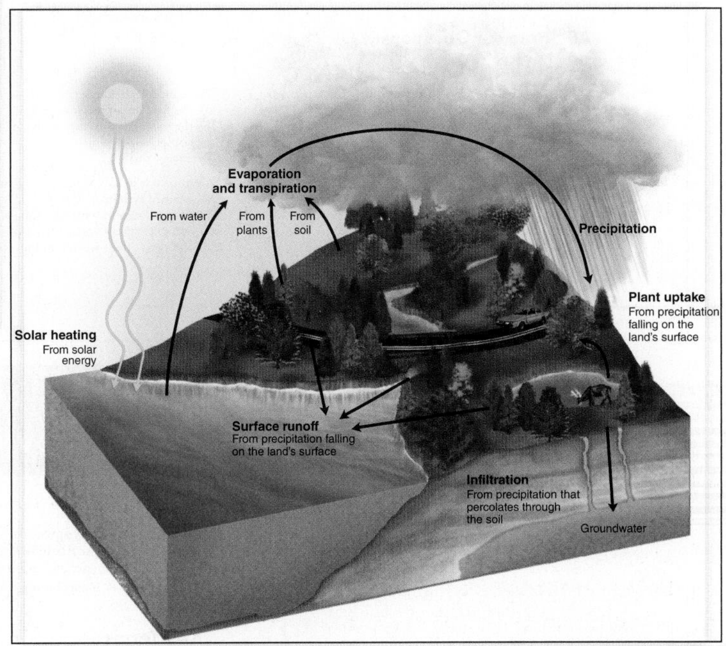

Tables and Graphs

Tables, graphs, and charts have been created to help you understand and analyze data. Working with this material will help you develop the ability to read and interpret data presented in a variety of formats.

FIGURE 22.4 Projected world population growth. Demographers project that the global human population will be between 8 billion and 10 billion by 2050. By 2100, it is projected to be between 7 billion and 11 billion. The dashed lines represent estimated values. *(After Millennium Ecosystem Assessment, 2005; Population Reference Bureau (www.prb.org) 2017.)*

TABLE 5.1 Common units of energy and their conversion into joules

Unit	Definition	Relationship to joules	Co…
calorie	Amount of energy it takes to heat 1 gram of water 1°C	1 calorie = 4.184 J	En… in ecosystems; human food consumption
Calorie (with a capital "C")	Calories in food	1 Calorie = 1,000 calories = 1 kilocalorie (kcal) = 4,184 J	Food labels; human food consumption
British thermal unit (Btu)	Amount of energy it takes to heat 1 pound of water 1°F	1 Btu = 1,055 J	Energy transfer in air conditioners and home water heaters
Kilowatt-hour (kWh)	Amount of energy expended by using 1 kilowatt of electricity for 1 hour	1 kWh = 3,600,000 J = 3.6 megajoules (MJ)	Energy use by electrical appliances; often given in kWh per year

Review to solidify your knowledge.

MODULE 5 AP® Review

Preparing for the AP® Exam

In this module, we have seen that energy is a fundamental component of environmental systems and that there are different forms of energy. The first and second laws of thermodynamics describe energy behavior. Matter and energy flow within and between systems and are subject to feedbacks that regulate and influence the behavior of systems. Natural systems change, sometimes quickly, sometimes slowly, and change can be natural or caused by human beings.

AP® Practice Questions

Choose the best answer for the following.

1. If a solar photovoltaic panel produces 1,000 watts of electrical energy and is active for 12 hours each day, how many kWh of electricity will be produced in a week?
   (a) 63 kWh
   (b) 12 kWh
   (c) 84 kWh
   (d) 70 kWh
2. A car traveling down the highway best represents
   (a) kinetic energy.
   (b) electromagnetic radiation.
   (c) potential energy.
   (d) chemical energy.
3. The concept of energy efficiency is used to quantify
   (a) the first law of thermodynamics.
   (b) the second law of thermodynamics.
   (c) conservation of matter.
   (d) energy quality.
4. A terminal lake like Mono Lake is an example of
   (a) a closed system.
   (b) an open system with only inputs.
   (c) an open system with only outputs.
   (d) an open system with both inputs and outputs.

Module Review

Use the review at the end of each module to check your understanding.

Chapter Review

At the end of each chapter, take time to review the main ideas and key terms.

Chapter 2 Review

Throughout this chapter, we have examined environmental systems. Earth is one large interconnected system. Components of the system follow basic principles of chemistry and biology. Energy is an important component of these systems. Energy conversions are frequently used in systems analysis. Natural systems change over space and time and humans are sometimes major actors in causing system change.

Key Terms

Matter
Mass
Atom
Element
Periodic table
Molecule
Compound
Atomic number
Mass number
Isotopes
Radioactive decay
Half-life
Covalent bond
Ionic bond
Hydrogen bond
Polar molecule
Surface tension
Capillary action
Acid
Base
pH
Ocean acidification
Chemical reaction
Law of conservation of matter
Inorganic compound
Organic compound
Carbohydrate
Protein
Nucleic acid
DNA (deoxyribonucleic acid)
RNA (ribonucleic acid)
Lipid
Cell
Energy
Joule (J)
Power
Electromagnetic radiation
Photon
Potential energy
Chemical energy
Kinetic energy
Temperature
First law of thermodynamics
Second law of thermodynamics
Energy efficiency
Energy quality
Entropy
Open system
Closed system
Input
Output
Systems analysis
Steady state
Negative feedback loop
Positive feedback loop
Adaptive management plan

Learning Goals Revisited

Module 4 Systems and Matter

Describe how matter comprises atoms and molecules that move among different systems.

Matter is composed of atoms, which are made up of protons, neutrons, and electrons. Atoms and molecules can interact in chemical reactions in which the bonds between particular atoms may change.

Explain why water is an important component of most environmental systems.

Water facilitates the transfer of chemical elements and compounds from one system to another. The molecular structure of water gives it unique properties that support the conditions necessary for life on Earth. These properties are essential to physiological functioning of plants and animals and the movement of elements through systems.

Discuss how matter is conserved in chemical and biological systems.

Matter cannot be created or destroyed, but its form can be changed within chemical and biological systems. This is part of the reason we cannot easily dispose of certain chemical compounds, such as hazardous materials.

Module 5 Energy, Flows, and Feedbacks

Distinguish among various forms of energy and understand how they are measured.

Energy can take various forms, including energy that is stored (potential energy) and the energy of motion (kinetic energy). Joules and calories are two important energy units.

Learning Goals Revisited

Check your notes against summaries of the learning goals for each module in the chapter.

Practice for the AP® Environmental Science Exam all year.

AP® Exam Tip

Make sure you can describe the formation of soil. You should be able to distinguish between biological and physical processes that contribute to the formation of the various soil horizons. ●

AP® Exam Tips

Use the AP® Exam Tips throughout each chapter to prepare for the exam by making sure you know key information that is likely to appear on the exam.

Chapter AP® Environmental Science Practice Exam

When you finish a chapter, take the practice exam to check your understanding of chapter concepts. The practice exam will help you become familiar with the style of questions on the AP® Environmental Science Exam.

Multiple-Choice Questions

Each chapter exam begins with multiple-choice questions modeled after those you'll see on the exam. Many of the questions ask you to analyze or interpret tables, graphs, or figures.

Chapter 1 **AP® Environmental Science Practice Exam** Preparing for the AP® Exam

Section 1: Multiple-Choice Questions

Choose the best answer for questions 1–11.

1. Which of the following events has increased the impact of humans on the environment?
I. advances in technology
II. reduced human population growth
III. use of tools for hunting
(a) I only
(b) I and II
(c) II and III
(d) I and III

2. As described in this chapter, environmental indicators
(a) always tell us what is causing an environmental change.
(b) can be used to analyze the health of natural systems.
(c) are useful only when studying large-scale changes.
(d) do not provide information regarding sustainability.

3. Which statement regarding a global environmental indicator is NOT correct?
(a) Concentrations of atmospheric carbon dioxide have been rising quite steadily since the Industrial Revolution.
(b) World grain production has increased fairly steadily since 1950, but worldwide production of grain per capita has decreased dramatically over the same period.
(c) For the past 130 years, average global surface temperatures have shown an overall increase that seems likely to continue.
(d) World population is expected to be between 8 billion and 10 billion by 2050.

4. Figure 2.5 (on page 12) shows atmospheric carbon dioxide concentrations over time. The measured concentration of CO_2 in the atmosphere is an example of
(a) a sample of air from over the Antarctic.
(b) an environmental indicator.
(c) replicate sampling.
(d) how to study seasonal variation in Earth's temperatures.

5. Environmental metrics such as the ecological footprint are most informative when they are considered along with other environmental indicators. Which indicator, when considered in conjunction with the ecological footprint, would provide the most information about environmental impact?
(a) biological diversity
(b) food production
(c) human population
(d) CO_2 concentration

6. In science, which of the following is the most certain?
(a) hypothesis
(b) idea
(c) natural law
(d) observation

7. All of the following would be exclusively caused by anthropogenic activities EXCEPT
(a) combustion of fossil fuels.
(b) overuse of resources such as uranium.
(c) forest clearing for crops.
(d) forest fires.

8. Use Figure 2.3 (on page 11) to calculate the approximate percentage change in world grain production per person between 1950 and 2000.
(a) 10 percent
(b) 20 percent
(c) 30 percent
(d) 40 percent

9. A species of bat nearly went extinct due to the destruction of its forest habitat. However, conservationists preserved a small remnant population of the species in a large forested reserve. The population grew and it appeared that the species would rebound when a disease suddenly wiped out the entire population. What is the likely primary reason for why this occurred?
I. the low genetic diversity of the population
II. the reduced ecosystem diversity of the forest
III. greater contact with humans, who passed on the disease
(a) I only
(b) II only
(c) III only
(d) I and II

10. The fraction of the global human population living in developed nations is currently 17 percent. If each person in the developed world consumes on average 100 kg of meat per year, how much meat will be consumed in the developed world in 1 year?
(a) 102 billion kg
(b) 130 billion kg
(c) 500 billion kg
(d) 630 billion kg

11. Which scenario would be most appropriate to study using a natural experiment?
I. the effects of an herbicide on human health
II. the influence of sleep on exam performance
III. the effects of an earthquake on blood pressure
(a) I only
(b) II only
(c) III only
(d) I and III

Section 2: Free-Response Questions

Write your answer to each part clearly. Support your answers with relevant information and examples. Where calculations are required, show your work.

1. A company conducts an experiment to test whether their coffee or their energy drinks are more likely to keep people up at night. They have five male college students drink a cup of coffee at 5 PM one day, 6 PM the next day, 7 PM on the third day, and 8 PM on the fourth day. They have five other male college students do the same with energy drinks. Each night the researchers record how many minutes it takes for each student to fall asleep. With the results from their experiment, they produce the graphs shown below.

   The researchers conclude that energy drinks are more likely to keep people up at night than coffee.
   (a) Describe THREE problems with the company's experimental design. If you were designing a new experiment, how would you fix TWO of these problems? (5 points)
   (b) Describe ONE problem with the company's graphs. Without repeating anything from part (a) provide at least TWO reasons supporting or opposing the position that these graphs show energy drinks keep people awake later than coffee. (3 points)
   (c) During this study, the researchers do a survey of the students at the university and notice that students who report drinking more coffee also tend to have higher grades. Can any conclusions be drawn from this observation? Why? (2 points)

2. The study of environmental science sometimes involves examining the overuse of environmental resources.
   (a) Identify one general effect of overuse of an environmental resource. (3 points)
   (b) For the effect you listed above, describe a more sustainable strategy for resource utilization. (3 points)
   (c) Describe how a dust bowl, large-scale erosion, or other human-caused land deterioration might be indicative of environmental issues on Earth today. (4 points)

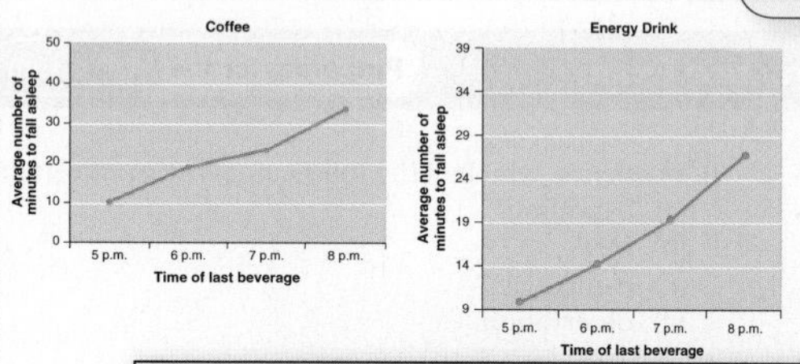

Free-Response Questions

Chapter exams include two or three practice free-response questions. Points are assigned to indicate how a complete, correct answer would be scored on the AP® Environmental Science Exam. The more practice you have in writing answers to practice free-response questions, the better you will do on the exam.

Unit 3 AP® Environmental Science Practice Exam — Preparing for the AP® Exam

Section 1: Multiple-Choice Questions

Choose the best answer for questions 1–25.

1. Which is NOT true regarding limiting resources?
   (a) A limiting resource is an important density-independent factor.
   (b) A limiting resource is common to all members of a population.
   (c) Foraging territory is an example of a limiting resource.
   (d) The number of reproducing individuals in a population is not a limiting resource.
2. Which is most likely to result in a random population distribution?
   (a) clumping of individuals around limiting resources
   (b) competition among individuals
   (c) density-independent factors
   (d) variable carrying capacities among populations within a metapopulation
3. The common reed (*Phragmites australis*) is a perennial grass found throughout temperate and tropical regions of the world. Researchers recently discovered that many North American populations of the reed are composed entirely of genetically ident
   Which is most likely true of these populat
   (a) The age structure is unlikely to va populations.

6. Which of the following is NOT a likely consequence of niche partitioning?
   (a) reduced interactions between different predator species
   (b) clumped species distributions
   (c) genetic variation within species
   (d) reduced competition between species
7. Mycorrhizal fungi are commonly found attached to plant roots. The fungi are more efficient than plants at extracting nutrients from the soil. In exchange for nutrients, plants provide the fungi with sugars made through photosynthesis. However, when nutrients in the soil are abundant, researchers have found that plant growth increases if mycorrhizal fungi are removed. This phenomenon suggests that
   (a) mycorrhizae have switched from being mutualists to parasites.
   (b) mycorrhizae have switched from being mutualists to predators.
   (c) mycorrhizae have switched from being mutualists to commensalists.
   (d) mycorrhizae have ceased to act as a keystone species.

Unit AP® Environmental Science Practice Exam

The textbook is divided into eight major units. At the end of each unit, you are provided with a longer practice exam containing 20 to 25 multiple-choice questions and 2 to 3 practice free-response questions. These exams give you a chance to review material across multiple chapters and to practice your test-taking skills.

Cumulative AP® Environmental Science Practice Exam 1

Section 1: Multiple-Choice Questions

Choose the best answer for questions 1–100.

Question 1 refers to the following table.

Net	Amount of organic matter
1	400 g
2	450 g
3	750 g
4	200 g
5	350 g

1. To estimate the amount of organic matter deposited on the soil within a forest covering 10,000 m^2, researchers set out five nets that can catch any material falling from above. They randomly place the nets throughout the forest. Each net has a surface area of 1 m^2. The results of their collection over 24 hours are listed in the table. Using this data, estimate how much organic matter was deposited throughout the entire forest during the time of collection.
   (a) 3,000 kg
   (b) 4,300 kg
   (c) 10,500 kg
   (d) 21,600 kg

Question 5 refers to the following figure.

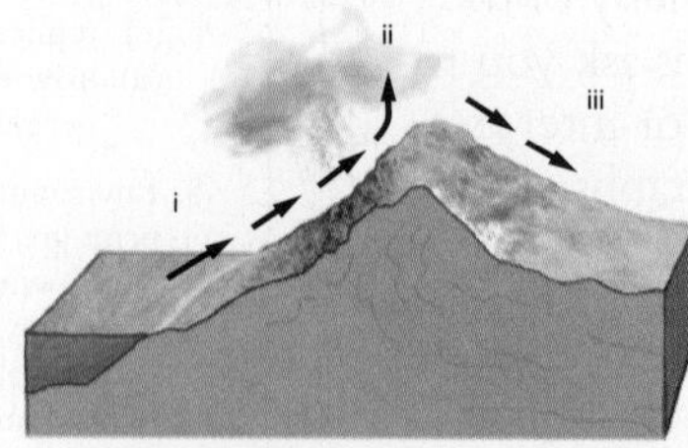

5. Which processes identify the labels associated with i, ii, and iii, respectively?
   (a) evaporation; adiabatic cooling; adiabatic heating
   (b) precipitation; evaporation; adiabatic heating
   (c) evapotranspiration; cloud formation; wind
   (d) evaporation; adiabatic heating; adiabatic cooling
6. One indicator of an improperly functioning septic system might be
   (a) scum that develops on the water's surface in a

Two Model AP® Environmental Science Practice Exams

At the end of the text, two cumulative AP® Environmental Science practice exams match the actual AP® Environmental Science Exam in length and scope. Each exam offers you the opportunity to test your mastery of the content and practice your test-taking skills.

Be inspired to make a difference.

Working Toward Sustainability

Reducing Food Waste

Food is essential to life and food production has substantial impacts on the environment, including most of the indicators we discussed earlier in the chapter. For example, producing food involves clearing land, which impacts biodiversity. It also involves plowing and tilling soils, which leads to more oxygen in soils, which leads to more decomposition and release of carbon dioxide to the atmosphere. And virtually all food production involves fossil fuel energy which, when combusted, also releases carbon dioxide to the atmosphere.

The U.S. Department of Agriculture (USDA)—the primary agency in the United States responsible for overseeing farming, agriculture, forestry, and food—estimates that between 30 and 40 percent of the food supply is wasted. This estimate is based on the percentage of food at the retail and consumer level that is not eaten. It includes food that spoils before it reaches the consumer as well as food put on the plate and not eaten.

Students at the University of Hawaii are part of a Fo[od] Recovery Network that delivers unsold food from th[e] dining facilities to food shelters. *(Courtesy Food Recovery N[etwork])*

Working Toward Sustainability
At the end of each chapter read about environmental solutions that are making a difference.

Critical Thinking Questions
Working Toward Sustainability provides questions that give you a chance to hone your text analysis skills.

Critical Thinking Questions

1. How are the 2013 food waste challenge objectives similar to those in the basic recycling slogan: Reduce, Reuse, Recycle?
2. Why do you think the goal for 50 percent food loss and waste reduction is 2030? Is that a long time from now or a short time?

Learn about science in the real world.

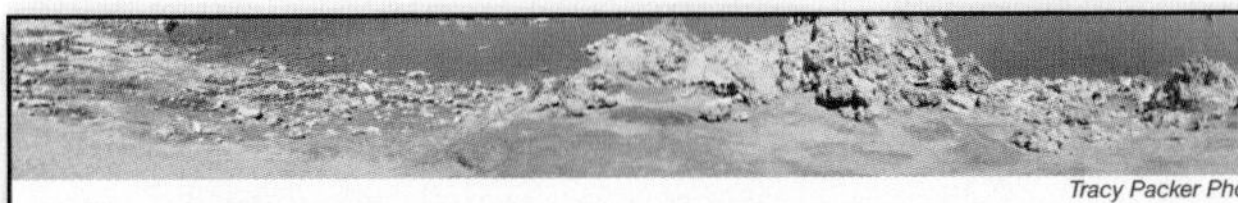

Tracy Packer Photo

Science Applied
At the end of each unit, Science Applied presents a concept application of a current issue in environmental science related to chapter topics.

science applied 1

Where Is the Missing Salt of Mono Lake?

As we discussed at the beginning of Chapter 2, in 1941 the City of Los Angeles redirected water away from Mono Lake in California to meet the needs of its growing population. Before this happened, approximately 120 billion liters of stream water (31 billion gallons) flowed into Mono Lake in an average year. As the city diverted a portion of the lake's incoming water over four decades, the lake dropped to half its prior average depth and the lake's salt concentration increased, which caused a decline in the algae and animals that lived in the lake and other animals that fed on them.

To understand how water diversion alters water depth and salt accumulation, ecosystem scientists had to examine the flows of water and salts into the lake from streams and the loss of water due to evaporation out of the lake (**FIGURE SA1.1**). They made a series of observations and drew conclusions that helped them account for the movement of water, but they could not account for the movement of salt. In short, they discovered that more salt came into the lake than existed in the water of the lake. The question then arose, where is the missing salt?

The balanced inputs and outputs of water

As we mentioned at the beginning of Chapter 2, Mono Lake is categorized as a terminal lake because water flows into the lake from rivers, streams, and from precipitation, but water does not flow out. In a typical year before Los Angeles began diverting water, the lake level did not rise or fall. Given that this lake has no outlets, water outputs must equal water inputs. This equilibrium suggests that the amount of water entering the lake was being balanced by the amount of water leaving the lake by evaporation. If this was not the case, the lake would eventually either dry out or overflow its banks.

The balance inputs and outputs of salt

While the balance of water in the lake is straightforward, the balance of salt in the lake is much more interesting. Because all streams contain some amount of salt water, the stream water that entered Mono Lake also contained salt. Salt concentrations of the stream water flowing into Mono Lake can vary through the seasons, but a typical concentration is 50 mg per liter, which is equivalent to 50 parts per million.

To calculate the total amount of salt that historically entered Mono Lake each year, we can multiply the concentration of salt in the stream water (50 mg per liter) by the number of liters of water flowing into the lake, prior to being diverted by the City of Los Angeles (120 billion liters per year):

$$50 \text{ mg/L salt} \times 120 \text{ billion L/year} = 6 \text{ trillion mg salt/year}$$

Questions

1. How did Los Angeles inadvertently conduct an experiment at Mono Lake?
2. What chemical principle causes terminal lakes to become more salty?
3. What is the reason for the discrepancy between the two calculations of salt content in Mono Lake?

Preparing for the AP® Exam

Practice AP® Free-Response Question

Write your answer to each part clearly. Support your answers with relevant information and examples. Where calculations are required, show your work.

Water that flows into M[ono Lake ...] smaller concentration of sa[lt ...] the lake. This inflowing wa[ter ...] on top of existing water, be[...] that salt water. As salt from [...] the upper layer, nutrients f[...] also rise to the surface. This [...] ical for the growth of algae i[...] research suggests that the r[...]

Practice Free-Response Questions
Science Applied essays include a practice free-response question related to the topic in the article.

Use additional resources for more practice and exam preparation.

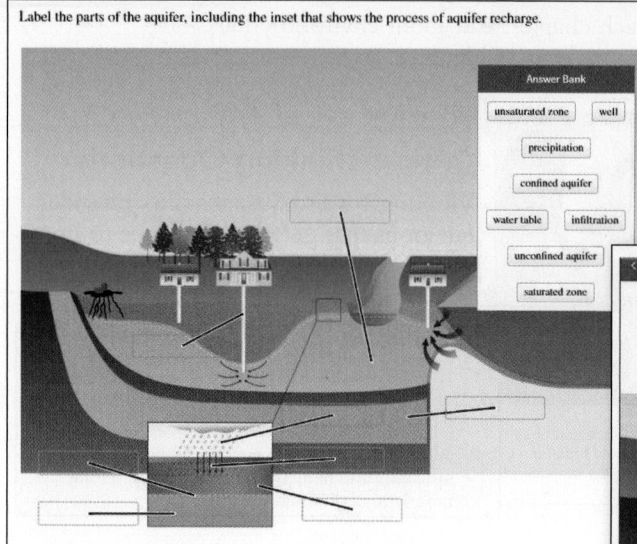

Digital Platform

Created and supported by educators, this program includes an online instructional homework system with an interactive e-book to drive your success. Every homework problem contains hints, answer-specific feedback, and solutions to ensure you get the help you need with course content.

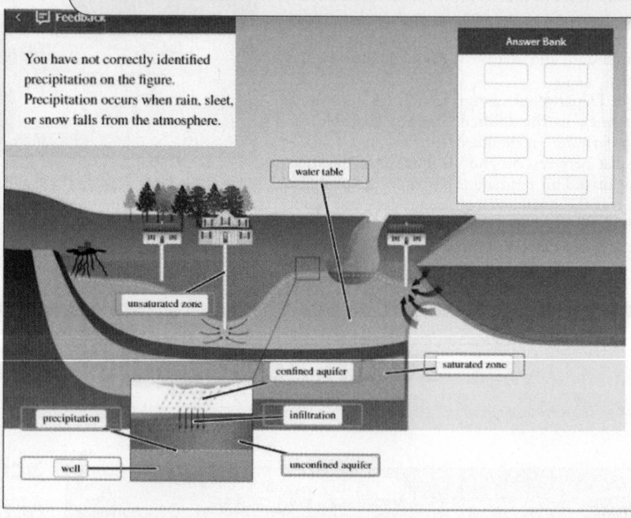

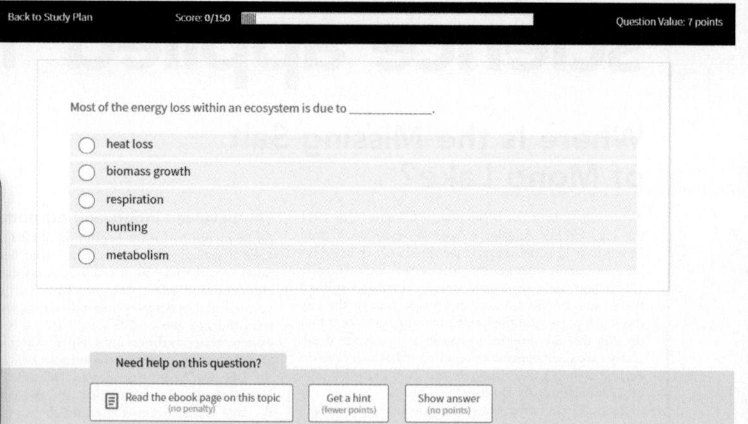

LearningCurve

LearningCurve is a powerful adaptive quiz system with a game-like format that engages you with the course material and adapts to your needs based on performance. You can link to the e-book and view reports on how you are performing in each topic.

Strive for a 5: Preparing for the AP® Environmental Science Examination

Strive for a 5: Preparing for the AP® Environmental Science Examination provides a thorough review of environmental science with essential tips for test preparation. Designed to align with the third edition of *Friedland and Relyea's Environmental Science for the AP® Course,* this workbook gives you the review and practice you need to succeed in the AP® Environmental Science course and on the exam.

FRIEDLAND and RELYEA

Environmental Science
for the AP® Course

THIRD EDITION

Natural gas flares are visible at this fracking site in the Bakken Shale formation, North Dakota. *(Jim West/Alamy Stock Photo)*

Environmental Science: Studying the State of Our Earth

CHAPTER 1

MODULE 1 Environmental Science

MODULE 2 Environmental Indicators and Sustainability

MODULE 3 Scientific Method

CASE STUDY

Earthquakes, Leakage, and Wastewater: Modern-Day Consequences of Energy Production

The United States—like other developed countries—is highly dependent on **fossil fuels** such as coal, oil, and natural gas, which are derived from biological material that became fossilized millions of years ago. The use of these fossil fuels is the cause of many environmental problems including the release of pollutants into air and water and land degradation. Natural gas, also known as methane, is the least harmful producer of air pollution among the fossil fuels; it burns more completely and cleanly than coal or oil, and it contains fewer impurities.

Due to advances in technology, oil and mining companies have recently increased their reliance on *fracking*. **Fracking**, short for hydraulic fracturing, is a method of oil and gas extraction that uses high-pressure fluids to force open existing cracks in rocks deep underground. This technique allows extraction of natural gas from locations that were previously so difficult to reach that extraction was economically unfeasible. As a result, large quantities of natural gas are now available in the United States at a lower cost than before. And fracking has increased our reliance on a domestic energy source and created jobs. Roughly 40 percent of energy in the United States is used to generate electricity and from 1980 through 2000, 50 percent of that energy came from coal and only 15 percent came from natural gas. In 2017, as a result of greater natural gas availability and lower price due to fracking, 33 percent of electricity generated was from natural gas and only 30 percent from coal. Since coal emits more air pollutants—including carbon dioxide—than does natural gas, fracking and the increase in natural gas use for electricity generation initially appeared to be beneficial to the environment.

> More and more studies have begun to suggest that significant quantities of natural gas escape as "leakage" during the fracking and gas extraction process.

However, reports gained attention both in the popular press and in scientific journals about the negative consequences of fracking. Large amounts of water are used in the fracking process, with millions of gallons of water taken out of local streams and rivers and pumped down or injected into each gas well. A portion of this water is later removed from the well and must be properly treated after use to avoid contaminating local water bodies.

A variety of chemicals are added to the fracking fluid to facilitate the release of natural gas. Mining companies are not required to publicly identify all of these chemicals. Environmental scientists and concerned citizens began to wonder if fracking was responsible for chemical contamination of underground water. Some drinking-water wells near fracking sites became contaminated with natural gas, and

Fossil fuel A fuel derived from biological material that became fossilized millions of years ago.

Fracking Hydraulic fracturing, a method of oil and gas extraction that uses high-pressure fluids to force open cracks in rocks deep underground.

homeowners and public health officials asked if fracking was the culprit. Water with high concentrations of natural gas can be flammable, and footage of flames shooting from kitchen faucets after someone ignited the water became popular on YouTube, in documentaries, and in feature films. However, it wasn't clear if fracking caused natural gas to contaminate well water or if some of these wells contained natural gas long before fracking began. Several reputable studies showed that drinking-water wells near some fracking sites were contaminated, with natural gas concentrations in the nearby wells being much higher than in more distant wells. In addition, there has been a higher frequency of earthquakes in the central and southwest United States and the U.S. Geological Survey, the primary government agency responsible for understanding and assessing earthquakes and other geological hazards, reported in 2016 that the injection of waste water from fracking is believed to be the primary cause. These issues need further study, which may take years.

In a direct connection to global climate change, more and more studies have begun to suggest that significant quantities of natural gas escape as "leakage" during the fracking and gas extraction process. It is not known how much of the natural gas extracted each year leaks into the atmosphere, but estimates range from 2 to10 percent. As we will learn in Chapter 19 methane is 25 times more efficient at trapping heat from Earth than carbon dioxide, the *greenhouse gas most* commonly produced by human activity. Due to its high heat trapping ability, the consequences of methane leakage are potentially substantial.

Almost certainly, using natural gas is better for the environment than using coal, though using less fossil fuel—or using no fossil fuel at all—would be even better. However, at present it is difficult to know whether the benefits of using natural gas outweigh the problems that extraction causes. Many years may pass before the extent and nature of harm from fracking is known.

The story of natural gas fracking provides a good introduction to the study of environmental science. It shows us that human activities that are initially perceived as causing little harm to the environment can in fact have adverse effects, and that we may not recognize these effects until we better understand the science surrounding the issue. It also illustrates the difficulty in obtaining absolute answers to questions about the environment and demonstrates that environmental science can be controversial. Finally, it shows us that making assessments and choosing appropriate actions in environmental science are not always as clear-cut as they first appear.

Sources: U.S. Geological Survey Induced Earthquakes Program. https://earthquake.usgs.gov/research/induced/,Drilling down. Multiple authors in 2011 and 2012. *New York Times*, http://www.nytimes.com/interactive/us/DRILLING_DOWN_SERIES.html.

The process of scientific inquiry builds on previous work and careful, sometimes lengthy, investigations. For example, we are still accumulating a body of knowledge on the effects of hydraulic fracturing of natural gas. Until we have this knowledge, we will not be able to make a fully informed decision about the policies of energy extraction. In the meantime, we may need to make interim decisions based on incomplete information. The urgency and lack of complete information are two particularly exciting aspects of environmental science.

To investigate topics such as the extraction and use of fossil fuels, environmental science relies on a number of indicators, methodologies, and tools. This chapter introduces you to the study of the environment and outlines some of the foundations and assumptions you will use throughout your study.

MODULE 1

Environmental Science

Humans are dependent on Earth's air, water, and soil for our existence. However, we have altered the planet in many ways, both large and small. The study of environmental science can help us understand how we have changed the planet and identify ways of responding to those changes.

Learning Goals

After reading this module you should be able to

- define the field of environmental science and discuss its importance.
- identify ways in which humans have altered and continue to alter our environment.

Environmental science offers important insights into our world and how we influence it

Stop reading for a moment and look up to observe your surroundings. Consider the air you breathe, the heating or cooling system that keeps you at a comfortable temperature, and the natural or artificial light that helps you see. Our **environment** is the sum of all the conditions surrounding us that influence life. These conditions include living organisms as well as nonliving components such as soil, temperature, and water. The influence of humans is an important part of the environment as well. The environment we live in determines how healthy we are, how fast we grow, how easy it is to move around, and even how much food we can obtain. One environment may be strikingly different from another—a hot, dry desert versus a cool, humid tropical rainforest, or a coral reef teeming with marine life versus a crowded city street.

We are about to begin the study of **environmental science**, the discipline that examines interactions among human and natural science systems. By *system* we mean any set of interacting components that influence one another by exchanging energy, materials, or information. We have already seen that a change in one part of a system—for example, fracking in a particular geologic formation—can cause changes throughout the entire system, such as in a nearby well that supplies drinking water.

An environmental system may be completely human-made, like a subway system, or it may be natural, like weather. The scope of an environmental scientist's work can vary from looking at a small population of individuals, to multiple populations that make up a species, to a community of interacting species, or to even larger systems, such as the global climate system. Some environmental scientists are interested in regional problems. The case of fracking at a particular location in the United States, for example, is a regional problem. Other environmental scientists work on global issues, such as species extinction and climate change.

Many environmental scientists study a specific type of natural system known as an *ecosystem*. An **ecosystem** is a particular location on Earth with interacting components that include living, or **biotic**, components and nonliving, or **abiotic**, components.

As a student of environmental science, you should recognize that environmental science is different from **environmentalism**, which is a social movement that

Environment The sum of all the conditions surrounding us that influence life.

Environmental science The field of study that looks at interactions among human systems and those found in nature.

Ecosystem A particular location on Earth with interacting biotic and abiotic components.

Biotic Living.

Abiotic Nonliving.

Environmentalism A social movement that seeks to protect the environment through lobbying, activism, and education.

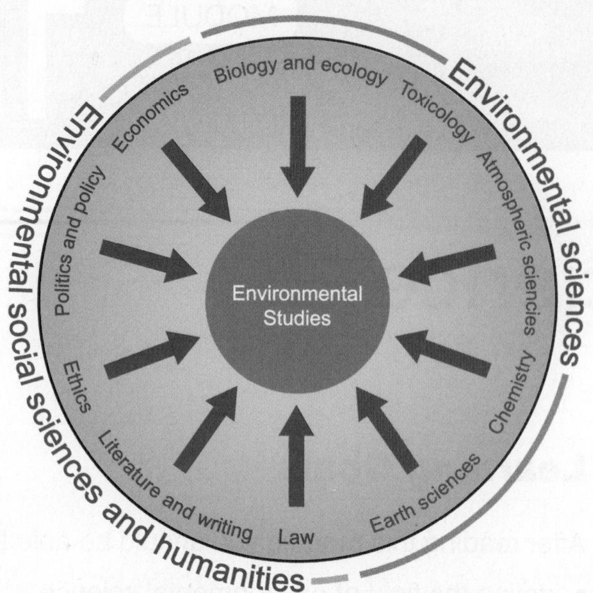

FIGURE 1.1 Environmental studies. The study of environmental science uses knowledge from many disciplines.

seeks to protect the environment through lobbying, activism, and education. An environmentalist is a person who participates in environmentalism. In contrast, an environmental scientist, like any scientist, follows the process of observation, hypothesis testing, and field and laboratory research. We'll learn more about the process of science later in this chapter.

So what does the study of environmental science actually include? As **FIGURE 1.1** shows, environmental science encompasses topics from many scientific disciplines, such as chemistry, biology, and Earth science. Environmental science is itself a subset of the broader field known as **environmental studies**, which includes additional subjects such as environmental policy, economics, literature, and ethics. Throughout the course of this book you will become familiar with these and many other disciplines.

We have seen that environmental science is a deeply interdisciplinary field. It is also a rapidly growing area of study. As human activities continue to affect the environment, environmental science can help us understand the consequences of our interactions with our planet and help us make better decisions about our actions.

Environmental studies The field of study that includes environmental science and additional subjects such as environmental policy, economics, literature, and ethics.

Humans alter natural systems

Think of the last time you walked in a wooded area. Did you notice any dead or fallen trees? Chances are that even if you did, you were not aware that living and nonliving components were interacting all around you. Perhaps an insect pest killed the tree you saw and many others of the same species. Over time, dead trees in a forest lose moisture. The increase in dry wood makes the forest more vulnerable to intense wildfires. But the process doesn't stop there. Wildfires trigger the germination of certain tree seeds, some of which lie dormant until after a fire. And so what began with the activity of insects leads to a transformation of the forest. In this way, biotic factors interact with abiotic factors to influence the future of the forest. All of these factors are part of a system.

Systems can vary in size. A large system may contain many smaller systems within it. **FIGURE 1.2** shows an example of complex, interconnecting systems that operate at multiple space and time scales: the fisheries of the North Atlantic. A physiologist who wants to study how codfish survive in the North Atlantic's coldwaters must consider all the biological adaptations of the cod that enable it to be part of one system. In this case, the fish and its internal organs are the system being studied. In the same environment, a marine biologist might study the predator-prey relationship between cod and herring. That relationship constitutes another system, which includes two fish species and the environment they live in. At an even larger scale, a scientist might examine a system that includes all of these systems as well as people, fishing technology, policy, and law. The global environment is composed of both small-scale and large-scale systems.

Humans manipulate the systems in their environment more profoundly than any other species. We convert land from its natural state into urban, suburban, and agricultural areas. We change the chemistry of our air, water, and soil, both intentionally—for example, by adding fertilizers—and unintentionally—for example, by our activities that generate pollution. Even where we don't manipulate the environment directly, the simple fact that there are so many of us affects our surroundings.

Humans and our direct ancestors (other members of the genus *Homo*) have lived on Earth for about 2.5 million years. During this time, and especially during the last 10,000 to 20,000 years, we have shaped and influenced our environment. As tool-using, social animals, we have continued to develop a capacity to directly alter our environment in substantial ways.

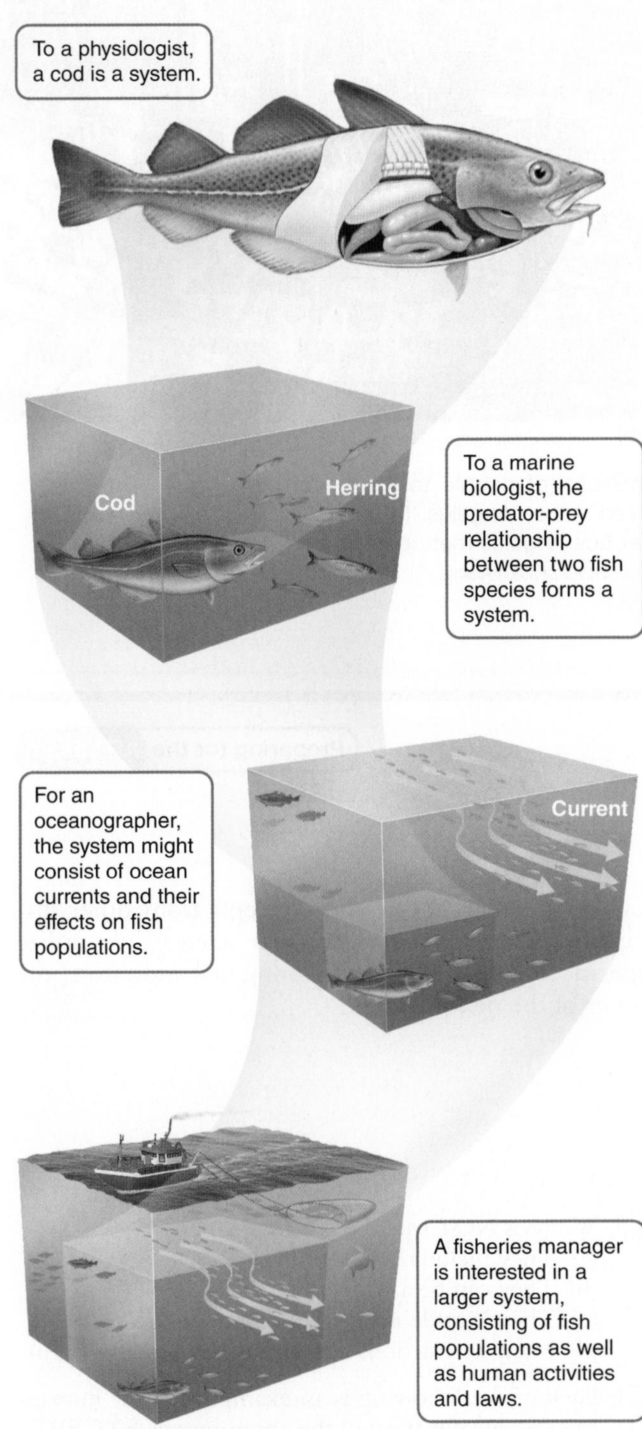

To a physiologist, a cod is a system.
Cod
Herring
To a marine biologist, the predator-prey relationship between two fish species forms a system.
For an oceanographer, the system might consist of ocean currents and their effects on fish populations.
Current
A fisheries manager is interested in a larger system, consisting of fish populations as well as human activities and laws.

FIGURE 1.2 Systems within systems. The boundaries of an environmental system may be defined by the researcher's point of view. Physiologists, marine biologists, oceanographers, and fisheries managers would all describe the North Atlantic Ocean fisheries system differently.

Homo sapiens—genetically modern humans—evolved to be successful hunters; when they entered a new environment, they often hunted large animal species to extinction. In fact, early humans are thought to be responsible for the extinction of mammoths, mastodons, giant ground sloths, and many types of birds. More recently, hunting in North America led to the extinction of the passenger pigeon (*Ectopistes migratorius*) and nearly caused the loss of the American bison (*Bison bison*).

But the picture isn't all bleak. Human activities and manipulation of the natural environment have also created opportunities for certain species to thrive. For example, for thousands of years Native Americans on the Great Plains used fire to capture animals for food. The fires they set kept trees from encroaching on the plains, which in turn created a window for an entire ecosystem to develop. Because of human activity, this ecosystem—the tallgrass prairie—is now home to numerous unique species.

During the last two centuries, the rapid and widespread development of technology, coupled with dramatic human population growth, has substantially increased both the rate and the scale of our global environmental impact. Modern cities with electricity, running water, sewer systems, Internet connections, and public transportation systems have improved human well-being, but they have come at a cost. Because cities cover land that was once undisturbed by people, species that relied on the specific conditions of that land had to adapt or relocate, or they went extinct. Human-induced changes in climate—for example, in patterns of temperature and precipitation—affect the health of natural systems on a global scale. Current changes in land use and climate are rapidly outpacing the rate at which natural systems can evolve. Some species have not "kept up" and can no longer compete in the human-modified environment.

Moreover, as the number of people on the planet has grown, their effect has multiplied. Six thousand people can live in a relatively small area with only minimal effects on the environment. But when roughly 4 million people live in a modern city like Los Angeles, their combined activity will cause environmental damage that will inevitably pollute the water, air, and soil as well as introduce other adverse consequences (**FIGURE 1.3** on page 6).

(a)

(b)

FIGURE 1.3 Human impact on Earth. It is impossible for millions of people to inhabit an area without altering it. (a) In 1880, fewer than 6,000 people lived in Los Angeles. (b) In 2017, Los Angeles had a population of 4 million people, and the greater Los Angeles metropolitan area was home to nearly 13 million people. *(a: The Granger Collection, New York; b: LA/AeroPhotos/Alamy)*

MODULE 1

AP® Review

Preparing for the AP® Exam

In this module we have seen that the study of environmental science helps us understand the role humans have played in the natural environment, and how that role has changed over time. There are specific approaches to the study of environmental science, some of which utilize terms and concepts from other disciplines. To study environmental science, we utilize specific techniques and environmental indicators, the focus of the next module.

AP® Practice Questions

Choose the best answer for the following.

1. Impacts of fracking include
   I. contamination of ground water.
   II. increased use of coal.
   III. lower natural gas prices.
   (a) I only
   (b) I and II
   (c) II and III
   (d) I and III

2. Which is an abiotic component?
   (a) an eagle
   (b) a rock
   (c) a tree
   (d) a human

3. Which is NOT true about ecosystems?
   (a) They include biotic components.
   (b) They can be a wide range of sizes.
   (c) They include no human components.
   (d) Many interactions among species occur in them.

4. Each of the following is an example of how humans have negatively affected the environment EXCEPT
   (a) hunting large mammals.
   (b) conversion of arid land to agricultural use.
   (c) the use of fire to create the Great Plains.
   (d) slash-and-burn forest clearing.

MODULE 2

Environmental Indicators and Sustainability

As we study the way humans have altered the natural world, it is important to have techniques for measuring and quantifying human impact. Environmental indicators allow us to assess the impact of humans on Earth. The use of these indicators helps us determine whether or not the quality of the natural environment is improving and informs discussions on the sustainability of humans on the planet.

Learning Goals

After reading this module you should be able to

- identify key environmental indicators and their trends over time.
- define sustainability and explain how it can be measured using the ecological footprint.

Environmental scientists use key environmental indicators to monitor natural systems for signs of stress

One critical question that environmental scientists investigate is whether the planet's natural life-support systems are being degraded by human-induced changes. Natural environments provide **ecosystem services**—the processes by which life-supporting resources such as clean water, timber, fisheries, and agricultural crops are produced. These are the resources necessary for organisms to survive, such as clean water and air, soil, woody plants, fisheries, and agricultural crops. Ecosystem services may benefit people directly, for example by providing the food that we eat. They may also benefit us indirectly, for example by providing a diversity of conditions, nutrients, and species, all of which make the ecosystem healthier for all organisms, including humans. Although we often take a healthy ecosystem for granted, we notice when an ecosystem is degraded or stressed because it is unable to provide the same services such as water purification or produce the same goods such as fruit or firewood.

To understand the extent of our effect on the environment, we need to be able to measure the health of Earth's ecosystems. Just as body temperature and heart rate can indicate whether a person is healthy or sick, **environmental indicators** describe the current state of an environmental system. These indicators do not always tell us what is causing a change, but they do tell us when we might need to look more deeply into a particular issue. Environmental indicators provide valuable information about natural systems on both small and large scales. Some of these indicators and the chapters in which they are covered are listed in **TABLE 2.1** on page 8.

In this book we will focus on five global-scale environmental indicators:

- Biodiversity
- Food production
- Average global surface temperature and carbon dioxide concentrations in the atmosphere
- Human population
- Resource depletion

Throughout the text we will cover each of these five indicators in greater detail. Here we take a first look.

Ecosystem services The processes by which life-supporting resources such as clean water, timber, fisheries, and agricultural crops are produced.

Environmental indicator An indicator that describes the current state of an environmental system.

TABLE 2.1 Some common environmental indicators		
Environmental indicator	**Unit of measure**	**Chapter**
Human population	Individuals	7
Ecological footprint	Hectares of land	1
Total food production	Metric tons of grain	11
Food production per unit area	Kilograms of grain per hectare of land	11
Per capita food production	Kilograms of grain per person	11
Carbon dioxide	Concentration in air (parts per million)	19
Average global surface temperature	Degrees centigrade	19
Sea level change	Millimeters	19
Annual precipitation	Millimeters	4
Species diversity	Number of species	5, 18
Fish consumption advisories	Present or absent; number of fish allowed per week	17
Water quality (toxic chemicals)	Concentration	14
Water quality (conventional pollutants)	Concentration; presence or absence of bacteria	14
Deposition rates of atmospheric compounds	Milligrams per square meter per year	15
Fish catch or harvest	Kilograms of fish per year or weight of fish per effort extended	11
Extinction rate	Number of species per year	5
Habitat loss rate	Hectares of land cleared or "lost" per year	18
Infant mortality rate	Number of deaths of infants under age 1 per 1,000 live births	7
Life expectancy	Average number of years an infant born today can be expected to live under current conditions	7

Biodiversity

Biodiversity, a combination of the words "biological" and "diversity," is the diversity of life forms in an environment. It exists on three scales: ecosystem, species, and genetic, illustrated in **FIGURE 2.1**. Each level of biodiversity is an important indicator of environmental health and quality.

Genetic Diversity

Genetic diversity is a measure of the genetic variation among individuals in a population. Populations with high genetic diversity are better able to respond to environmental change than populations with lower genetic diversity. For example, if a population of fish possesses high genetic diversity for disease resistance, at least some individuals are likely to survive whatever diseases move through the population. If the population declines in number, however, the amount of genetic diversity it can possess is also reduced, and this reduction increases the likelihood that the population will decline further when exposed to a disease.

Biodiversity The diversity of life forms in an environment.

Genetic diversity A measure of the genetic variation among individuals in a population.

Species A group of organisms that is distinct from other groups in its morphology (body form and structure), behavior, or biochemical properties.

Species diversity The number of species in a region or in a particular ecosystem.

Species Diversity

A **species** is defined as a group of organisms that is distinct from other groups in its morphology (body form and structure), behavior, or biochemical properties. Individuals within a species can breed and produce fertile offspring. Scientists have identified and cataloged approximately 2 million species on Earth. Estimates of the total number of species on Earth range between 5 million and 100 million, with the most common estimate at 10 million. This number includes a large array of organisms with a multitude of sizes, shapes, colors, and roles.

Species diversity indicates the number of species in a region or in a particular ecosystem. Scientists have observed that ecosystems with more species—that is, higher species diversity—are more productive and resilient—that is, better able to recover from disturbances

(a) Ecosystem diversity

(b) Species diversity

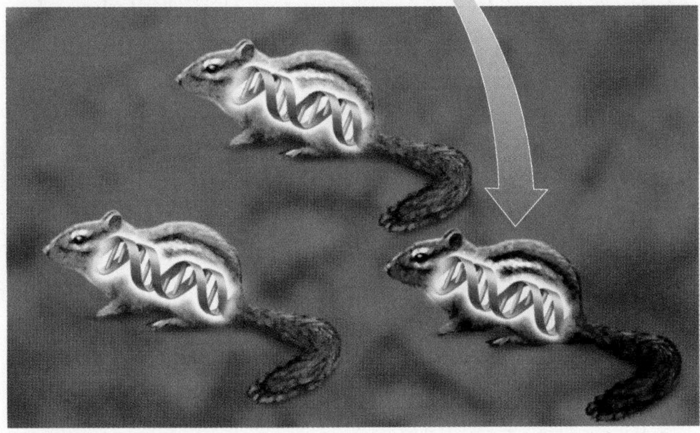

(c) Genetic diversity

FIGURE 2.1 Levels of biodiversity. Biodiversity exists at three scales. (a) Ecosystem diversity is the variety of ecosystems within a region. (b) Species diversity is the variety of species within an ecosystem. (c) Genetic diversity is the variety of genes among individuals of a species.

such as hurricanes or fires. For example, a tropical forest with a large number of plant species growing in the understory is likely to be more productive, and better able to withstand change, than a nearby tropical forest plantation with one crop species growing in the understory.

Environmental scientists often focus on species diversity as a critical environmental indicator. The number of frog species, for example, is used as an indicator of regional environmental health because frogs are exposed to both the water and the air in their ecosystems. A decrease in the number of frog species in a particular ecosystem may be an indicator of environmental problems there. Species losses in several ecosystems can indicate environmental problems on a larger scale. Not all species losses are indicators of environmental problems, however. Species arise and others go extinct as part of the natural evolutionary process. The evolution of new species, known as **speciation**, typically happens very slowly—perhaps on the order of one to three new species per year worldwide. The average rate at which species go extinct over the long term is referred to as the **background extinction rate**. The background extinction rate is also very slow: about one species in a million every year. So with 2 million identified species on Earth, the background extinction rate should be about 2 species per year.

Under conditions of environmental change or biological stress, species may go extinct faster than new ones evolve. Some scientists estimate that more than 1,000 species are currently going extinct each year—which is about 500 times the background rate of extinction. Destruction and degradation of the areas in which species live are the major causes of species extinction today, although climate change, overharvesting, and pressure from introduced species also contribute to species loss. In some situations, human pressures on the natural environment can lead to increased speciation. Human intervention has saved certain species, including the American bison, peregrine falcon (*Falco peregrinus*), bald eagle (*Haliaeetus leucocephalus*), and American alligator (*Alligator mississippiensis*). But other large animal species, such as the Bengal tiger (*Panthera tigris tigris*), snow leopard (*Panthera uncia*), Siberian tiger (*Panthera tigris altaica*), and West Indian manatee (*Trichechus manatus*), remain endangered and may go extinct if present trends are not reversed (**FIGURE 2.2** on page 10). Overall, the number of species has been declining.

Ecosystem Diversity

Ecosystem diversity is a measure of the diversity of ecosystems that exist in a given region. A greater number of healthy and productive ecosystems means a healthier environment overall. As an environmental indicator, the current loss of biodiversity tells us that natural systems are facing strains unlike any in the recent past. We will look at this important topic in greater detail in Chapters 5 and 18.

Some measures of biodiversity are given in terms of land area, so becoming familiar with measurements of

Speciation The evolution of new species.

Background extinction rate The average rate at which species become extinct over the long term.

(a)

(b)

FIGURE 2.2 Species on the brink. Humans have saved some species from the brink of extinction, such as (a) the American bison. Other species, such as (b) the Siberian tiger remain endangered.
(a: David Osborn/Alamy Stock Photo, b: Jean-Louis Klein & Marie-Luce Hubert/Science Source)

land area is important to understanding them. A hectare (ha) is a unit of area used primarily in the measurement of land. It represents 100 meters by 100 meters. In the United States we measure land area in terms of square miles and acres. However, the rest of the world measures land in hectares. "Do the Math: Converting Between Hectares and Acres" shows you how to do the conversion.

AP® Exam Tip

You are required to show all work for math problems on the AP® Environmental Science Exam. You will not get full credit for a correct answer unless you show each step, no matter how simple. Remember to include units. Develop the habit of showing your work from the beginning of the year. Start with the math practice in this chapter. ●

Food Production

The second of our five global indicators is food production: our ability to grow food to nourish the human population. Just as a healthy ecosystem supports a wide range of species, a healthy soil supports abundant and continuous food production. Food grains such as wheat, corn, and rice provide more than half the calories and protein humans consume. Still, the growth of the human population is straining our ability to grow and distribute adequate amounts of food.

In the past, we have used science and technology to increase the amount of food we can produce on a given area of land. World grain production increased fairly steadily between 1950 and 2014 as a result of expanded irrigation, fertilization, new crop varieties, and other innovations. Since 2015, grain production has been approximately level. At the same time, worldwide production of grain *per person* has leveled

DO THE MATH Converting Between Hectares and Acres

Preparing for the AP® Exam

In the metric system, land area is expressed in hectares. A hectare (ha) is 100 meters by 100 meters. In the United States, land area is most commonly expressed in acres. There are 2.47 acres in 1 ha. The conversion from hectares is relatively easy to do without a calculator; rounding to two significant figures gives us 2.5 acres in 1 ha. If a nature preserve is 100 ha, what is its size in acres?

$$100 \text{ ha} \times 2.5 \text{ acres} = 250 \text{ acres}$$

YOUR TURN A particular forest is 10,000 acres. Determine its size in hectares.

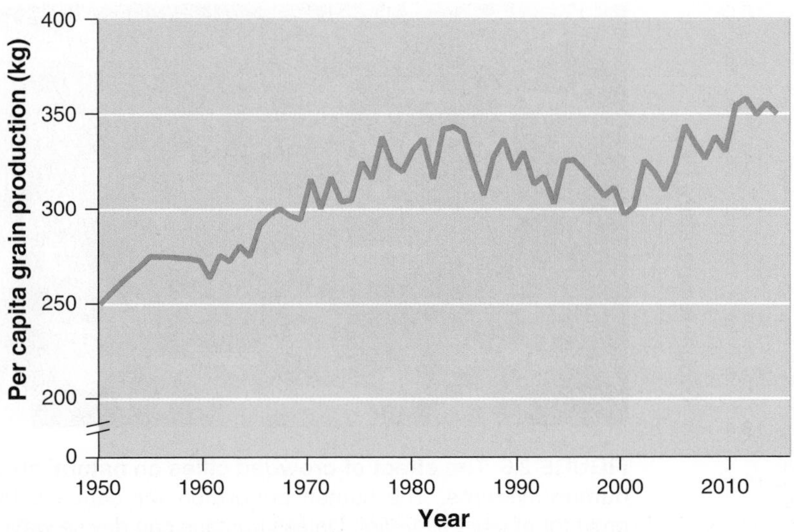

FIGURE 2.3 World grain production per person. Grain production has increased since the 1950s, but it has recently begun to level off. *(Data from: http://www.earth-policy.org/index.php?/indicators/C54)*

off as well. **FIGURE 2.3** shows the variation in per capita grain production through 2017.

In 2008, food shortages around the world led to higher food prices and even riots in some places. Why did this happen? The amount of grain produced worldwide is influenced by many factors. These factors include climatic conditions, the amount and quality of land under cultivation, irrigation, and the human labor and energy required to plant, harvest, and bring the grain to market. Grain production may not be keeping up with population growth because in some areas the productivity of agricultural ecosystems has declined as a result of soil degradation, crop diseases, and unfavorable weather conditions such as drought or flooding. In addition, demand is outpacing supply. While the rate of human population growth has outpaced increases in food production, humans currently use more grain to feed livestock than they consume themselves. Finally, some government policies discourage food production by making it more profitable for land to remain uncultivated or by encouraging farmers to grow crops for fuels such as ethanol and biodiesel instead of food.

Will there be sufficient grain to feed the world's population in the future? In the past, whenever a shortage of food has loomed, humans have discovered and employed technological or biological innovations to increase production. However, these innovations often put a strain on the productivity of the soil. If we continue to overexploit the soil, its ability to sustain food production may decline dramatically. We will take a closer look at soil quality in Chapter 8 and food production in Chapter 11. "Working Toward Sustainability: Reducing Food Waste" on page 27 examines another approach to increasing the availability of food, by reducing the amount of food waste.

Average Global Surface Temperature and Carbon Dioxide Concentrations

We have seen that biodiversity and abundant food production are necessary for life. One of the things that makes them possible is a stable climate. Earth's temperature has been relatively constant since the earliest forms of life began, about 3.5 billion years ago. The temperature of Earth allows the presence of liquid water, which is necessary for life.

What keeps Earth's temperature so constant? As **FIGURE 2.4** shows, our thick planetary atmosphere contains many gases. Some of these atmospheric gases, known as **greenhouse gases**, trap heat near Earth's surface. The gas that contributes most to warming of the atmosphere is carbon dioxide (CO_2). During most of the history of life on Earth, greenhouse gases have been present in the atmosphere at fairly constant concentrations for relatively long periods. They help keep Earth's surface within the range of temperatures at which life can flourish.

In the past 2 centuries, however, the concentrations of CO_2 and other greenhouse gases in the atmosphere have risen. Today, atmospheric CO_2 concentrations are

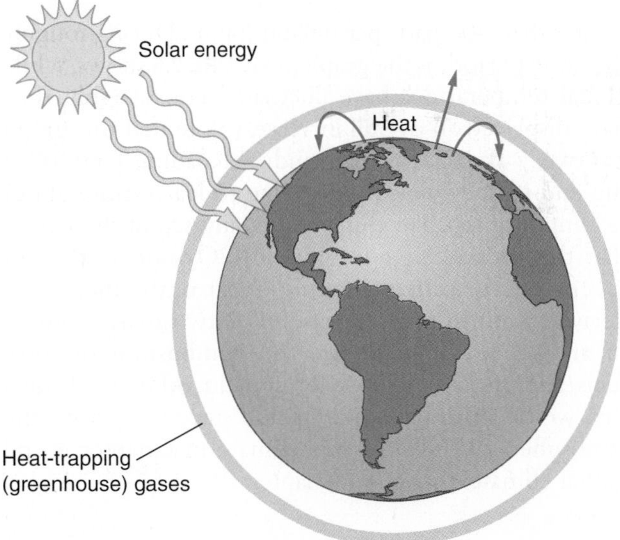

FIGURE 2.4 The Earth-surface energy balance. As Earth's surface is warmed by the Sun, it radiates heat outward. Heat-trapping gases absorb the outgoing heat and reradiate some of it back to Earth. Without these greenhouse gases, Earth would be much cooler.

Greenhouse gases Gases in Earth's atmosphere that trap heat near the surface.

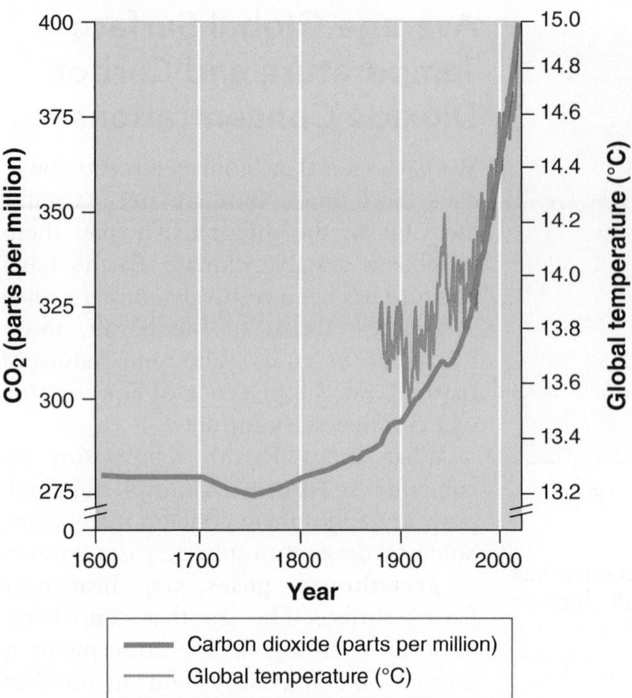

FIGURE 2.5 Changes in average global surface temperature and in atmospheric CO_2 concentrations. Earth's average global surface temperature has increased steadily for at least the past 100 years. Carbon dioxide concentrations in the atmosphere have varied over geologic time, but have risen steadily since 1960. *(Data from: http://data.giss.nasa.gov/gistemp/graphs_v3/and http://www.esrl.noaa.gov/gmd/ccgg/trends/#mlo_full)*

greater than 400 parts per million (ppm). During roughly the same period, as the graph in **FIGURE 2.5** shows, while global temperatures have fluctuated considerably, they have displayed an overall increase. (Note that this graph has two *y* axes. See the appendix "Reading Graphs" at the end of the book if you'd like to learn more about reading a graph like this one.) Many scientists believe that the increase in atmospheric CO_2 during the last 2 centuries is **anthropogenic**—that is, the increase is derived from human activities. The two major sources of anthropogenic CO_2 are the combustion of fossil fuels and the net loss of forests and other land areas that would otherwise take up and store CO_2 from the atmosphere. We will discuss climate in Chapter 4 and global climate change in Chapter 19.

Human Population

In addition to biodiversity, food production, and global surface temperature, the size of the human population can tell us a great deal about the health of our global environment. The human population is currently 7.6 billion and growing. The increasing world

Anthropogenic Derived from human activities.

FIGURE 2.6 The effect of crowded cities on natural and human systems. The human population will continue to grow for at least 50 years. Unless humans can devise ways to live more sustainably, these population increases will put additional strains on natural systems. Here, a street scene in Kolkata. *(Deshakalyan Chowdhury/AFP/Getty Images)*

population places additional demands on natural systems, since each new person requires food, water, and other resources. In any given 24-hour period, 380,000 infants are born and 155,000 people die. The net result is 225,000 new inhabitants on Earth each day, or over a million additional people every 5 days. Although the rate of population growth has been slowing since the 1960s, world population size will still continue to increase for at least another 50 to 100 years. Most population scientists project that the human population will be somewhere between 8 billion and 10 billion in 2050 and will stabilize between 7 billion and 11 billion by 2100.

Can the planet sustain so many people? Even if the human population eventually stops growing, the billions of additional people will create a greater demand on Earth's finite resources, including food, energy, and land (**FIGURE 2.6**). Unless humans work to reduce these pressures, the human population will put a rapidly growing strain on natural systems for at least the first half of this century. We discuss human population issues in Chapter 7.

Resource Depletion

Natural resources provide the energy and materials that support human civilization but, as the human population grows, the resources necessary for our survival become increasingly depleted. In addition, extracting these natural resources can affect the health of our environment in many ways. Pollution and land degradation caused by mining, waste from discarded manufactured products, and air pollution from fossil fuel combustion are just a few of the negative environmental consequences of resource extraction and use.

Some natural resources, such as coal, oil, and uranium, are finite and cannot be renewed or reused. Others,

Rates of Forest Clearing

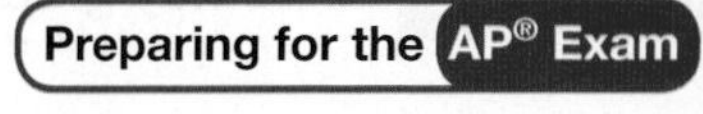

A web search of environmental organizations yielded a range of estimates of the amount of forest clearing that is occurring worldwide:

Estimate 1: 1 acre per second
Estimate 2: 80,000 acres per day
Estimate 3: 32,000 ha per day

Convert Estimate 1 into hectares per year and Estimate 2 into hectares per day.

There are 2.47 acres per hectare. (See "Do the Math: Converting Between Hectares and Acres" on page 10.). Therefore, 1 acre = 0.40 ha.

Estimate 1: 1.0 acre/second × 0.40 ha/acre = 0.40 ha/second
0.40 ha/second × 60 seconds/minute × 60 minutes/hour × 24 hours/day × 365 days/year
= 12,614,400 ha cleared per year

Estimate 2: 80,000 acres/day × 0.40 ha/acre = 32,000 ha cleared per day

YOUR TURN Notice that Estimate 2, when converted to hectares, is identical to Estimate 3. Now convert the estimate of 32,000 ha/day into the amount cleared per year. How much larger is Estimate 1 than Estimate 2? Why might environmental organizations, or anyone else, choose to present similar information in different ways?

such as aluminum or copper, also exist in finite quantities but can be used multiple times through reuse or recycling. Renewable resources, such as timber, can be grown and harvested indefinitely, but in some locations these resources are being used faster than they can be naturally replenished. "Do the Math: Rates of Forest Clearing" provides an opportunity to calculate rates of one type of resource depletion.

Sustaining the global human population requires vast quantities of resources. However, in addition to the total amounts of resources used by humans, we must consider **per capita** resource, which means the amount per each person in a country or other unit of population. Any quantity expressed "per capita" is the total quantity divided by the total number of people in the population unit being described, typically a country. The resulting value is the amount per each person.

Patterns of resource consumption vary enormously among nations depending on their level of development. What exactly do we mean by *development?* **Development** is defined as improvement in human well-being through economic advancement. Development influences personal and collective human lifestyles—things such as automobile use, the amount of meat in the diet, and the availability and use of technologies such as cell phones and personal computers. As economies develop, resource consumption also increases: People drive more automobiles, live in larger homes, and purchase more goods. These increases can often have implications for the natural environment.

According to the United Nations Development Programme, people in developed nations—including the United States, Canada, Australia, most European countries, and Japan—use a disproportionate share of the world's resources. **FIGURE 2.7** shows that the 17 percent of the global population that lives in developed nations

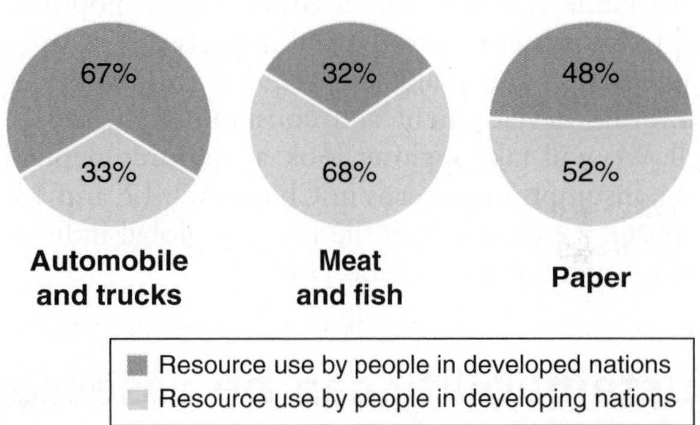

FIGURE 2.7 Fraction of resource use in developed and developing countries. Approximately 17 percent of the world's population lives in developed countries, but that 17 percent uses a disproportionate share of the world's resources. The remaining 83 percent of the population lives in developing countries and uses far fewer resources per capita. *(Data from http://www.oica.net/wp-content/uploads//total-inuse-2013.pdf; http://faostat.fao.org/site/610/DesktopDefault.aspx?PageID=610#ancor; http://www.paperonweb.com/Country.htm)*

Per capita Amount per each person in a country or unit of population.

Development Improvement in human well-being through economic advancement.

TABLE 2.2 Five key global indicators

Indicator	Recent trend	Outlook for the future	Overall impact on environmental quality
Biological diversity	Large number of extinctions, extinction rate increasing	Extinctions will continue	Negative
Food production	Per capita production possibly leveling off	Unclear	May affect the number of people Earth can support
Average global surface temperature and CO_2 concentration	CO_2 concentrations and temperatures increasing	Probably will continue to increase, at least in the short term	Effects are uncertain and varied but probably detrimental
Human population	Still increasing, but growth rate slowing	Population leveling off; resource consumption rates also a factor	Negative
Resource depletion	Many resources being depleted at rapid rate, but human ingenuity develops "new" resources, and efficiency of resource use is increasing in many cases	Unknown	Increased use of most resources has negative effects

owns 67 percent of the world's automobiles and consumes 32 percent of all meat and fish, and 48 percent of all paper. The poorest 20 percent of people in the world consume 5 percent or less of these resources. Thus, even though the number of people in the developing countries is much larger than the number in the developed countries, their total consumption of natural resources is smaller.

So while it is true that a larger human population will have greater environmental impacts than a smaller population, a full evaluation requires that we look at economic development and consumption patterns as well. We will take a closer look at resource depletion and consumption patterns in Chapters 7, 12, and 13.

TABLE 2.2 summarizes the five key global indicators that we will focus on in this book.

Sustainability can be assessed using the ecological footprint

The five key environmental indicators that we have just discussed help us analyze the health of the planet. We can use this information to guide us toward **sustainability**, by which we mean living on Earth in a way that allows us to use its resources without depriving future generations of those resources. Many people maintain that achieving sustainability is the single most important goal for the human species. It is also one of the most challenging tasks we face.

Sustainability Living on Earth in a way that allows humans to use its resources without depriving future generations of those resources.

The Impact of Consumption on the Environment

We have seen that people living in developed nations consume a far greater share of the world's resources than do people in developing countries. What effect does this consumption have on our environment? It is easy to imagine a very small human population living on Earth without degrading its environment because there simply would not be enough people to do significant damage. Today, however, Earth's population is 7.6 billion people and growing.

There are numerous examples of human beings overexploiting forests, fisheries, and other resources. When humans initially settle an area, they typically increase in number in the hospitable environment. They cut down trees to build homes and other structures. Eventually they overuse soil and water resources and possibly fish and wildlife. Without trees to hold the soil in place, erosion can occur, and the loss of soil causes food production to decrease. The classic example of this type of overexploitation occurred on the Great Plains in the midwestern United States in the 1930s. Vast amounts of prairie land were converted to agricultural lands and that land was subject to subsequent droughts and strong winds. During this time, known as the Dust Bowl, dust storms across the entire Midwest carried away large quantities of soil.

Ultimately, consumption in the United States and in the world is related to the size of the human population. While people in some countries use more resources than those in other countries, the population has been growing since 1600. And resource use has grown since then as well. **FIGURE 2.8** shows world population change

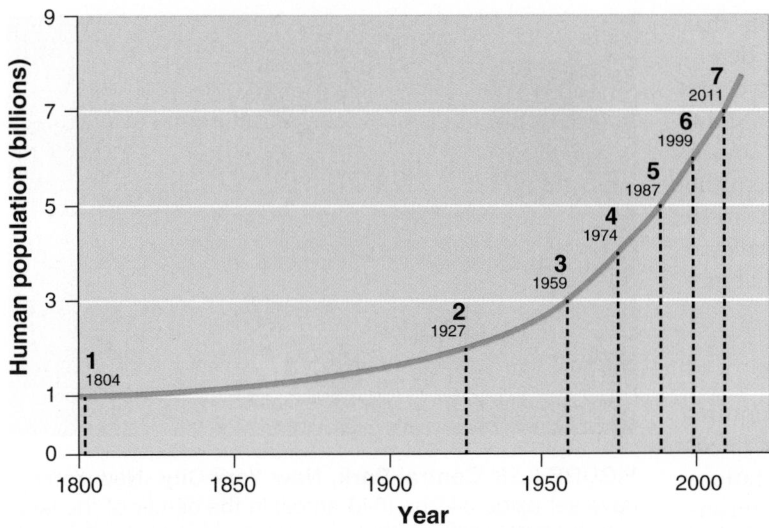

FIGURE 2.8 Adding 1 billion people. The number of years to add an additional billion people on Earth is shown since the first billion were added in 1800. It took 123 years to add 1 billion people from 1804 and only 12 years to add a billion people from 1999. *(Data from: http://www.un.org/esa/population/publications/sixbillion/sixbilpart1.pdf, http://www.pragcap.com/seven-billion-people/)*

since 1800. Each additional billion people on Earth are demanding and using resources, which clearly has an impact on the environment. Many environmental scientists ask how we will be able to continue to produce sufficient food, build needed infrastructure, and process pollution and waste. Our current attempts to sustain the human population have already modified many environmental systems.

Can we continue our current level of resource consumption without jeopardizing the well-being of future generations? Most environmental scientists believe that there are limits to the supply of clean air and water, nutritious foods, and other life-sustaining resources our environment can provide. They also believe there is a point at which Earth will no longer be able to maintain a stable climate. We must meet several requirements in order to live sustainably:

- Environmental systems must not be damaged beyond their ability to recover.
- Renewable resources must not be depleted faster than they can regenerate.
- Nonrenewable resources must be used sparingly.

Sustainable development is development that balances current human well-being and economic advancement with resource management for the benefit of future generations. This is not as easy as it sounds. The issues involved in evaluating sustainability are complex, in part because sustainability depends not only on the number of people using a resource but also on how that resource is used. For example, eating chicken is sustainable when people raise their own chickens and allow them to forage for food on the land. However, if all people, including city dwellers, wanted to eat chicken six times a week, the amount of resources needed to raise that many chickens would probably make the practice of eating chicken unsustainable.

Living sustainably means acting in a way such that activities that are crucial to human society can continue. It includes practices such as conserving and finding alternatives to nonrenewable resources as well as protecting the capacity of the environment to continue to supply renewable resources (**FIGURE 2.9**).

Consider iron, a nonrenewable resource derived from metal ore removed from the ground. Iron is the major constituent of steel, which we use to make many things, including automobiles, bicycles, and strong frames for tall buildings. Historically, our ability to smelt iron for steel limited our use of that resource, but as we have improved steel manufacturing technology, steel has become more readily available and the demand for it has grown. Because of this increased demand, our current use of iron is unsustainable. What would happen if we ran out of iron? While not too long ago the depletion of iron ore might have been a catastrophe, today we have developed materials that can substitute for certain uses of steel—for example, carbon fiber—and we also know how to recycle steel. Developing substitutes and recycling materials are two ways to address the problem of resource depletion and to increase sustainability.

FIGURE 2.9 Living sustainably. Sustainable choices such as bicycling to work or school can help protect the environment and conserve resources for future generations. *(Jim West/The Image Works)*

Sustainable development Development that balances current human well-being and economic advancement with resource management for the benefit of future generations.

The example of iron leads us to a question that environmental scientists often ask: How do we determine the importance of a given resource? If we use up a resource such as iron for which substitutes exist, it is possible that the consequences will not be severe. However, if we are unable to find an alternative to the resource—for example, something to replace fossil fuels—people in the developed nations may have to make significant changes in their consumption habits.

Defining Human Needs

We have seen that sustainable development requires us to determine how we can meet our current needs without compromising the ability of future generations to meet their own needs. Let's look at how environmental science can help us achieve that goal. We will begin by defining needs.

If you have ever experienced an interruption of electricity to your home or school, you know how frustrating it can be. Without the use of lights, computers, televisions, air-conditioning, heating, and refrigeration, many people feel disconnected and uncomfortable. Almost everyone in the developed world would insist that they need—indeed, cannot live without—electricity. In other parts of the world, however, people have never had these modern conveniences. So, when we speak of basic needs, we are referring only to the essentials that sustain human life, including air, water, food, health, and shelter.

But humans also have more complex needs. Many psychologists have argued that we require meaningful human interactions in order to live a satisfying life, and so a community of some sort might be considered a human need. Biologist Edward O. Wilson wrote that humans exhibit **biophilia**—that is, love of life—which is a need to make "the connections that humans subconsciously seek with the rest of life." Thus, our needs for access to natural areas, for beauty, and for social connections can be considered as vital to our well-being as our basic physical needs and must be considered as part of our long-term goal of global sustainability (**FIGURE 2.10**).

FIGURE 2.10 Central Park, New York City. New Yorkers have set aside 341 ha (843 acres) in the center of the largest city in the United States—a testament to the compelling human need for interactions with nature. *(Sylvain Sonnet /Getty Images)*

The Ecological Footprint

We have begun to see the multitude of ways in which human activities affect the environment. As countries prosper, their populations use more resources. Economic development can sometimes improve environmental conditions. For instance, wealthier countries may have the resources to implement pollution controls and invest money to protect native species. So although people in developing countries do not consume the same quantity of resources as those in developed nations, they may be less likely to have the financial resources to implement environmental protections.

How do we determine what lifestyles have the greatest environmental impact? This is an important question for environmental scientists if we are to understand the effects of human activities on the planet and develop sustainable practices. Calculating sustainability, however, is more difficult than one might think because we must consider the impacts of our activities and lifestyles on different aspects of our environment. We use land to grow food and to build structures on, and for parks and recreation. We require water for drinking, for cleaning, and for manufacturing products such as paper, and we need clean air to breathe. Yet these goods and services are all interdependent: Using or protecting one has an effect on the others. For example, using land for conventional agriculture may require water for irrigation, fertilizer to promote plant growth, and pesticides to reduce crop damage. This use of land reduces the amount of water available for human use: The plants consume it and the pesticides pollute it.

One method used to assess whether we are living sustainably is to measure the impact of a person or country on world resources. The tool many environmental scientists use for this purpose, the *ecological footprint,* was developed in 1995 by Professor William E. Rees and his graduate student Mathis Wackernagel. An individual's **ecological footprint** is a measure of how much that person consumes, expressed in area of land. The output from the total amount of land required to support a person's lifestyle represents that person's ecological footprint as shown in **FIGURE 2.11**.

Rees and Wackernagel maintained that if our lifestyle demands more land than is available, then we must be living unsustainably—using up resources more quickly

Biophilia Love of life.

Ecological footprint A measure of how much an individual consumes, expressed in area of land.

than they can be produced, or producing wastes more quickly than they can be processed. For example, each person requires a certain number of food calories every day. We know the number of calories in a given amount of grain or meat. We also know how much farmland or rangeland is needed to grow the grain to feed people or livestock such as sheep, chickens, or cows. If a person eats only grains or plants, the amount of land needed to provide that person with food is simply the amount of land needed to grow the plants they eat. If that person eats meat, however, the amount of land required to feed that person is greater because we must also consider the land required to raise and feed the livestock that ultimately become meat. Thus one factor in the size of a person's ecological footprint is the amount of meat in the diet. Meat consumption is a lifestyle choice, and per capita meat consumption is much greater in developed countries. We can calculate the ecological footprint of the food we eat, the water and energy we use, and even the activities we perform that contribute to climate change. Other metrics for calculating our impact on Earth exist as well.

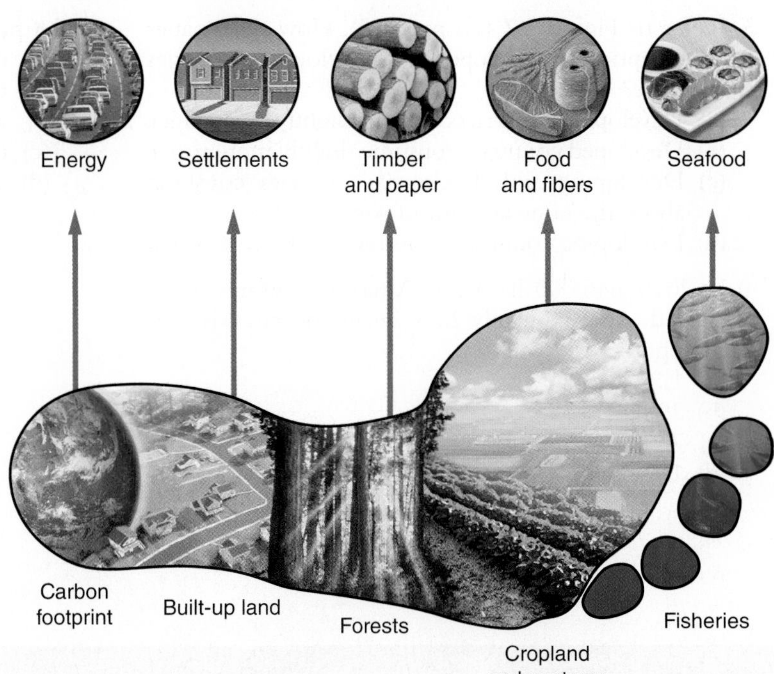

FIGURE 2.11 The ecological footprint. An individual's ecological footprint is a measure of how much land is needed to supply the goods and services that individual uses. Only some of the many factors that go into the calculation of the footprint are shown here. (The actual amount of land used for each resource is not drawn to scale.)

MODULE 2 AP® Review

Preparing for the AP® Exam

In this module we have identified global-scale indicators of environmental health that allow us to monitor specific parameters over time. Ultimately, these indicators contribute to a picture of the sustainability of human activities on Earth. We have identified that biodiversity is decreasing and that food production has leveled off. Atmospheric carbon dioxide concentrations are steadily increasing and global temperatures fluctuate, although the overall change is toward an increase. The human population continues to increase in size but the rate of increase has been declining. One measurement that allows us to assess the sustainability of these different parameters and how they change over time is the ecological footprint. The final module in this chapter introduces us to some of the scientific methods and techniques we will use in order to measure these indicators and make assessments about sustainability.

AP® Practice Questions

Choose the best answer for the following.

1. Common global-scale environmental indicators include all of the following EXCEPT
   (a) atmospheric carbon dioxide concentrations.
   (b) human population.
   (c) natural resource depletion.
   (d) pollution in a local stream.

2. How many hectares of land is a 500-acre park? (1 acre = 0.40 ha)
   (a) 200 ha
   (b) 250 ha
   (c) 500 ha
   (d) 750 ha

3. Refer to Figure 2.7 (on page 13). How does paper consumption in developed and developing countries compare?
   (a) Developing countries consume slightly more paper.
   (b) Developed countries consume slightly more paper.
   (c) Developing and developed countries consume about the same amount of paper.
   (d) Developing countries consume much more paper.

4. In 2015, 640,000 ha of the Amazon rainforest were cleared. Approximately how many hectares is that each hour?
   (a) 1.2 ha
   (b) 29 ha
   (c) 73 ha
   (d) 178 ha

5. A person's ecological footprint is
   (a) the land that a person lives on.
   (b) the amount of carbon dioxide a person contributes to climate change.
   (c) the land required to produce a person's food.
   (d) the land needed to support all of a person's activities.

Scientific Method

Environmental indicators are important for understanding human impacts on Earth systems and the sustainability of those systems. In order to evaluate environmental indicators, we need to use reproducible scientific methods. An understanding of the *scientific method* is essential for environmental science.

Learning Goals

After reading this module you should be able to

- explain the scientific method and its application to the study of environmental problems.
- describe some of the unique challenges and limitations of environmental science.

The scientific method is an important process in environmental science

Humans during the past century have learned a lot about the impact of their activities on the natural world. Scientific inquiry has provided great insights into the challenges we are facing and has suggested ways to address those challenges. For example, a hundred years ago, we did not know how significantly or rapidly we could alter the chemistry of the atmosphere by burning fossil fuels. Nor did we understand the effects of many common materials, such as lead and mercury, on human health. Much of our knowledge comes from the work of researchers who study a particular problem or situation to understand why it occurs and to determine how we can fix or prevent it from occurring. In this section we will look at the process scientists use to ask and answer questions about the environment.

The Scientific Method

To investigate the natural world, scientists, such as those who examined the effects of fracking as described at the beginning of this chapter, have to be as objective and methodical as possible. They must conduct their research in such a way that other researchers can understand how their data were collected and agree on the validity of their findings. To do this, scientists follow a process known as the **scientific method**, which is an objective way to explore the natural world, draw inferences from it, and predict the outcome of certain events, processes, or changes. The scientific method is used in some form by scientists in all parts of the world and is a generally accepted way to conduct science.

As we can see in **FIGURE 3.1**, the scientific method has a number of steps, including observing and questioning, forming hypotheses, collecting data, interpreting results, and disseminating findings.

Observing and Questioning

In some areas where fracking occurred homeowners and scientists noticed that certain household wells contained high methane concentrations. They wanted to know whether fracking might be responsible. Such observing and questioning is where the process of scientific research begins.

Forming Hypotheses

Observing and generating questions lead a scientist to formulate a *hypothesis.* A **hypothesis** is a testable conjecture about how something works. It may be an idea, a proposition, a possible mechanism of interaction, or a statement about an effect. For example, we might hypothesize that when the air temperature rises over time, certain plant species will be more likely to persist, and others will be less likely to persist.

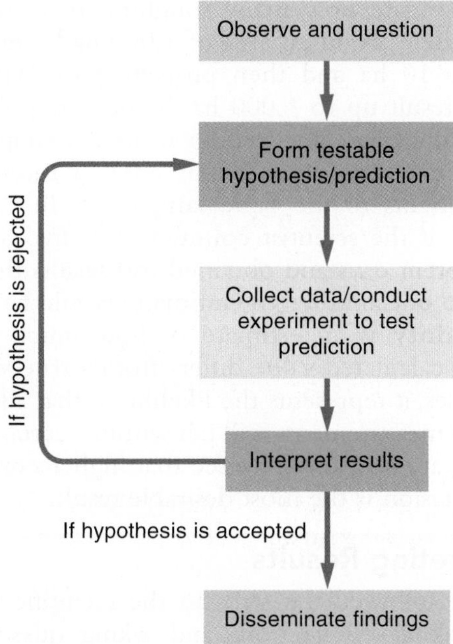

FIGURE 3.1 The scientific method. In an actual investigation, a researcher might reject a hypothesis and investigate further with a new hypothesis, several times if necessary, depending on the results of the experiment.

What makes a hypothesis testable? We can test the idea about the relationship between air temperature and plant species by conducting an experiment. In this case, that would require growing plants in a greenhouse at different temperatures. Consider this statement: "Fish kills are caused by something in the water." The statement is a testable hypothesis because it speculates that there is an interaction between something in the water and the observed dead fish. We can collect water from areas where fish have been found dead and identify harmful chemicals in the water. In a laboratory, we can add those harmful chemicals, one at a time, to fish tanks and determine which, if any, of the chemicals cause harm to fish. These are examples of testing hypotheses.

AP® Exam Tip

Make sure you understand the definition of a testable hypothesis and can give examples. ●

In each case, we are measuring **variables**, which are any categories, conditions, factors or traits that differ in various parts of the natural environment or experimental situation. An **independent variable** is one that is not dependent on factors. Age of an individual or elevation above sea level are examples of independent variables. A **dependent variable** is one that is influenced by other factors. Concentrations of a pollutant or number of deformities in a fish population are dependent variables.

It is almost always easier to prove something wrong than to prove it is true beyond doubt. In this case, scientists use a *null hypothesis.* A **null hypothesis** is a prediction that there is no difference between the groups or treatments that are being compared. The statement "Fish deaths have no relationship to something in the water" is an example of a null hypothesis. Establishing absolute

Scientific method An objective method to explore the natural world, draw inferences from it, and predict the outcome of certain events, processes, or changes.

Hypothesis A testable conjecture about how something works.

Variable Any categories, conditions, factors, or traits that differ in the natural world or in experimental situations.

Independent variable A variable that is not dependent on other factors.

Dependent variable A variable that is dependent on other factors.

Null hypothesis A prediction that there is no difference between the groups or conditions that are being compared.

Range of Electric Vehicles

Preparing for the AP® Exam

All-electric vehicles, which run on batteries only, are becoming more common. The range that a given vehicle can travel without being charged is a function of the efficiency of the vehicle and the storage capacity of the vehicle. The mileage ranges of five common electric vehicles are: 107 miles, 53 miles, 100 miles, 114 miles, and 237 miles. What is the average mileage range of these five electric vehicles? Round to the nearest whole number.

Add the mileage ranges together and divide by the number of electric vehicles:

107 miles + 53 miles + 100 miles + 114 miles + 237 miles = 611 miles

611 miles ÷ 5 vehicles = 122 miles average range per vehicle

YOUR TURN Five common hybrid-electric vehicles (which are powered by an internal combustion engine and batteries) have mileage ranges as follows: 409 miles, 702 miles, 513 miles, 583 miles, and 492 miles. What is the average mileage range of these five hybrid-electric vehicles? Round to the nearest whole number.

proof that a particular chemical killed fish in a river at some time in the past would be very difficult. So by generating a null hypothesis, scientists examine how easy or difficult it is to accept or reject the null hypothesis. If the null hypothesis is accepted, then they conclude that the pollutant, chemical, or process in question did not affect whatever factor they were measuring. They then move on to a different question or concern.

Collecting Data

Scientists typically take several, repeated sets of measurements of what they believe to be the same sample—a procedure called **replication**. The number of times a measurement is replicated is the **sample size** (sometimes referred to as *n*). A sample size that is too small can cause misleading results. For example, if a scientist chose three men out of a crowd at random and found that they all had size 10 shoes, she might conclude that all men have a shoe size of 10. If, however, she chose a larger sample size—100 men—it is very unlikely that all 100 individuals would happen to have the same shoe size. We know that shoe size varies among men. However, if the scientist were studying an unknown population of earthworms, fish, or trees, she may not know whether or not they would have the same length, weight, or height. Therefore, she needs to collect data to obtain this information. Once data are collected, they can be analyzed, which may involve summing or averaging the data. "Do the Math: Range of Electric Vehicles" provides a data averaging exercise.

Proper procedures—testable hypotheses, appropriate sample size, careful measurements—yield results that are accurate and precise. They also help us determine the possible relationship between our measurements or calculations and the true value. **Accuracy** refers to how close a measured value is to the actual or true value. For example, an environmental scientist might estimate how many songbirds of a particular species there are in an area of 1,000 ha by randomly sampling 10 ha and then projecting or extrapolating the result up to 1,000 ha. If the extrapolation is close to the true value, it is an accurate extrapolation. **Precision** is how close to one another the repeated measurements of the same sample are. In the same example, if the scientist counted birds five times on five different days and obtained five results that were similar to one another, the estimates would be precise. **Uncertainty** is an estimate of how much a measured or calculated value differs from a true value. In some cases, it represents the likelihood that additional repeated measurements will fall within a certain range. Looking at **FIGURE 3.2**, we see that high accuracy and high precision is the most desirable result.

Replication The data collection procedure of taking repeated measurements.

Sample size (*n*) The number of times a measurement is replicated in data collection.

Accuracy How close a measured value is to the actual or true value.

Precision How close the repeated measurements of a sample are to one another.

Uncertainty An estimate of how much a measured or calculated value differs from a true value.

Interpreting Results

We have followed the steps in the scientific method from making observations and asking questions, to forming a hypothesis, to collecting data. What happens next? Once results have been obtained, analysis of data begins. A scientist may use a variety of techniques to assist with data analysis, including summaries, graphs, charts, and diagrams.

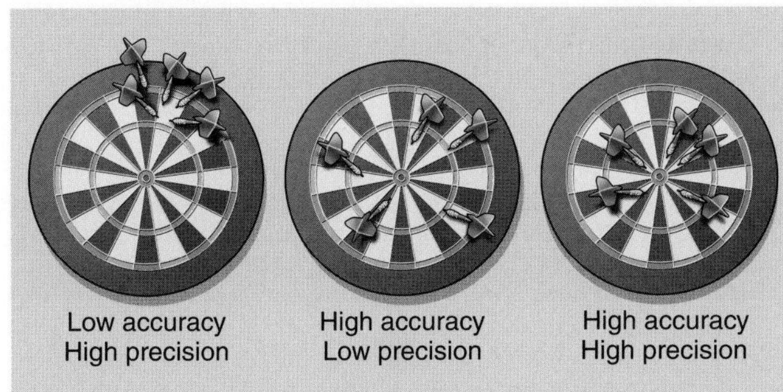

FIGURE 3.2 Accuracy and precision. Accuracy refers to how close a measured value is to the actual or true value. Precision is how close repeated measurements of the same sample are to one another.

As data analysis proceeds, scientists begin to interpret their results. This process normally involves two types of reasoning: inductive and deductive. Inductive reasoning is the process of making general statements from specific facts or examples. If the scientist who sampled a songbird species in the preceding example made a statement about all birds of that species, she would be using inductive reasoning. It might be reasonable to make such a statement if the songbirds that she sampled were representative of the whole population. Deductive reasoning is the process of applying a general statement to specific facts or situations. For example, if we know that, in general, air pollution kills trees, and we see a single, dead tree, we may attribute that death to air pollution. But a conclusion based on a single tree might be incorrect, since the tree could have been killed by something else, such as a parasite or fungus. Without additional observations or measurements, and possibly experimentation, the observer would have no way of knowing the cause of death with any degree of certainty.

The most careful scientists always maintain multiple working hypotheses—that is, they entertain many possible explanations for their results. They accept or reject certain hypotheses based on what the data show or do not show. Eventually, they determine that certain explanations are the most likely, and they begin to generate conclusions based on their results.

Disseminating Findings

A hypothesis is never confirmed by a single experiment. That is why scientists not only repeat their experiments themselves, but also present papers at conferences and publish the results of their investigations. This dissemination of scientific findings allows other scientists to repeat the original experiment and verify or challenge the results. The process of science involves ongoing discussion among scientists, who frequently disagree about hypotheses, experimental conditions, results, and the interpretation of results. Two investigators may even obtain different results from similar measurements and experiments, as happened with investigations of fracking. Only when the same results are obtained over and over by different investigators can we begin to trust that those results are valid. In the meantime, the disagreements and discussion about contradictory findings are a valuable part of the scientific process. They help scientists refine their research to arrive at more consistent, reliable conclusions.

AP® Exam Tip

The free-response section of the AP® Environmental Science Exam regularly asks students to design an experiment that covers each step of the scientific method. Be sure you know how to do a lab write-up using the scientific method. ●

Like any scientist, you should always read reports of "exciting new findings" with a critical eye. Question the source of the information, consider the methods or processes that were used to obtain the information, evaluate the journal or location where the information was published, and draw your own conclusions. This process, essential to all scientific endeavors, is known as critical thinking.

A hypothesis that has been repeatedly tested and confirmed by multiple groups of researchers and has reached wide acceptance becomes a **theory**. Current theories about how plant species distributions change with air temperature, for example, are derived from decades of research and evidence. Notice that this sense of theory is different from the way we might use the term in everyday conversation (such as, "But that's just a theory!"). To be considered a theory, a hypothesis must be consistent with a large body of experimental results. A theory cannot be contradicted by any replicable tests.

Scientists work under the assumption that the world operates according to fixed, knowable laws. We accept this assumption because it has been successful in explaining a vast array of natural phenomena and continues to lead to new discoveries. When the scientific process has generated a theory that has been tested multiple times, we can call that theory a natural law. A natural law is a theory to which there are no known exceptions and which has withstood rigorous testing. Familiar examples include the law of gravity and the laws of thermodynamics, which we will look at in the next chapter. These theories are accepted as facts by the scientific community, but they remain subject to revision if contradictory data are found.

Theory A hypothesis that has been repeatedly tested and confirmed by multiple groups of researchers and has reached wide acceptance.

Scientific Method in Action: The Chlorpyrifos Investigation

Let's look at what we have learned about the scientific method in the context of an actual scientific investigation. In the 1990s, scientists suspected that organophosphates—a group of chemicals commonly used in insecticides—might have serious effects on the human central nervous system. By the early part of the decade, scientists suspected that organophosphates might be linked to problems such as neurological disorders, birth defects, ADHD, and palsy. One of these chemicals, chlorpyrifos (klor-PEER-i-fos), was of particular concern because it is among the most widely used pesticides in the world, with large amounts applied in homes in the United States and elsewhere.

The Hypothesis

The researchers investigating the effects of chlorpyrifos on human health formulated a hypothesis: Chlorpyrifos causes neurological disorders and negatively affects human health. Because this hypothesis would be hard to prove conclusively, the researchers also proposed a null hypothesis: Chlorpyrifos has no observable negative effects on the central nervous system. We can follow the process of their investigation in **FIGURE 3.3**.

Testing the Hypothesis

To test the null hypothesis, the scientists designed experiments using rats. One experiment used two groups of rats, with 10 individuals per group. The first group—the experimental group—was fed small doses of chlorpyrifos for each of the first 4 days of life. In order to measure differences that will be ascribed to the experimental variable, in this case doses of chlorpyrifos, a second group was not fed any chlorpyrifos. That second group was a **control group**: a group that experiences exactly the same conditions as the experimental group, except for the single variable under study. In this experiment, the only difference between the control group and the experimental group was that the control group was not fed any chlorpyrifos. By designating a control group, scientists can determine whether an observed effect is the result of the experimental treatment or of something else in the environment to which all the subjects are exposed. For example, if the control rats—those that were not fed chlorpyrifos—and the experimental rats—those that were exposed to chlorpyrifos—showed no differences in their brain chemistry, researchers could conclude that the chlorpyrifos had no effect. If the control group and experimental group had very different brain chemistry after the experiment, the scientists could conclude that the difference must have been due to the chlorpyrifos.

Control group In a scientific investigation, a group that experiences exactly the same conditions as the experimental group, except for the single variable under study.

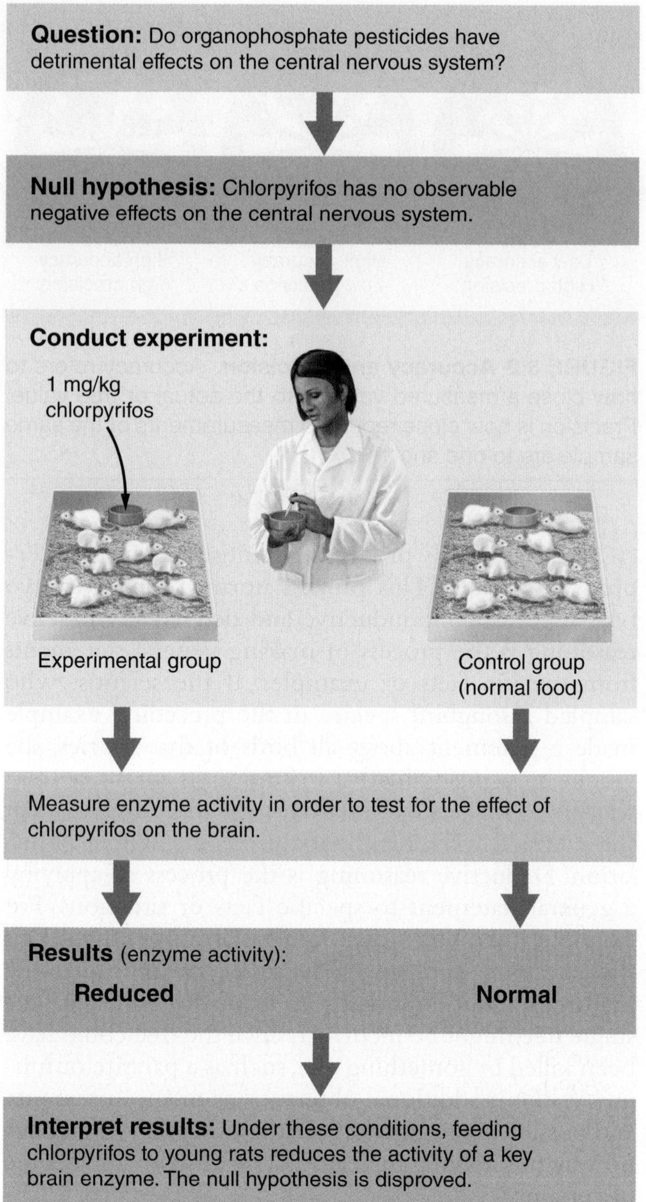

Question: Do organophosphate pesticides have detrimental effects on the central nervous system?
Null hypothesis: Chlorpyrifos has no observable negative effects on the central nervous system.
Conduct experiment:
1 mg/kg chlorpyrifos
Experimental group
Control group (normal food)
Measure enzyme activity in order to test for the effect of chlorpyrifos on the brain.
Results (enzyme activity):
Reduced
Normal
Interpret results: Under these conditions, feeding chlorpyrifos to young rats reduces the activity of a key brain enzyme. The null hypothesis is disproved.

FIGURE 3.3 A typical experimental process. An investigation of the effects of chlorpyrifos on the central nervous system illustrates how the scientific method is used.

The Results

At the end of the experiment, the researchers found that the rats exposed to chlorpyrifos had much lower levels of the enzyme choline acetyltransferase in their brains than the rats in the control group. But without a control group for comparison, the researchers would never have known whether the chlorpyrifos or something else caused the change observed in the experimental group.

The discovery of the relationship between ingesting chlorpyrifos and a single change in brain chemistry

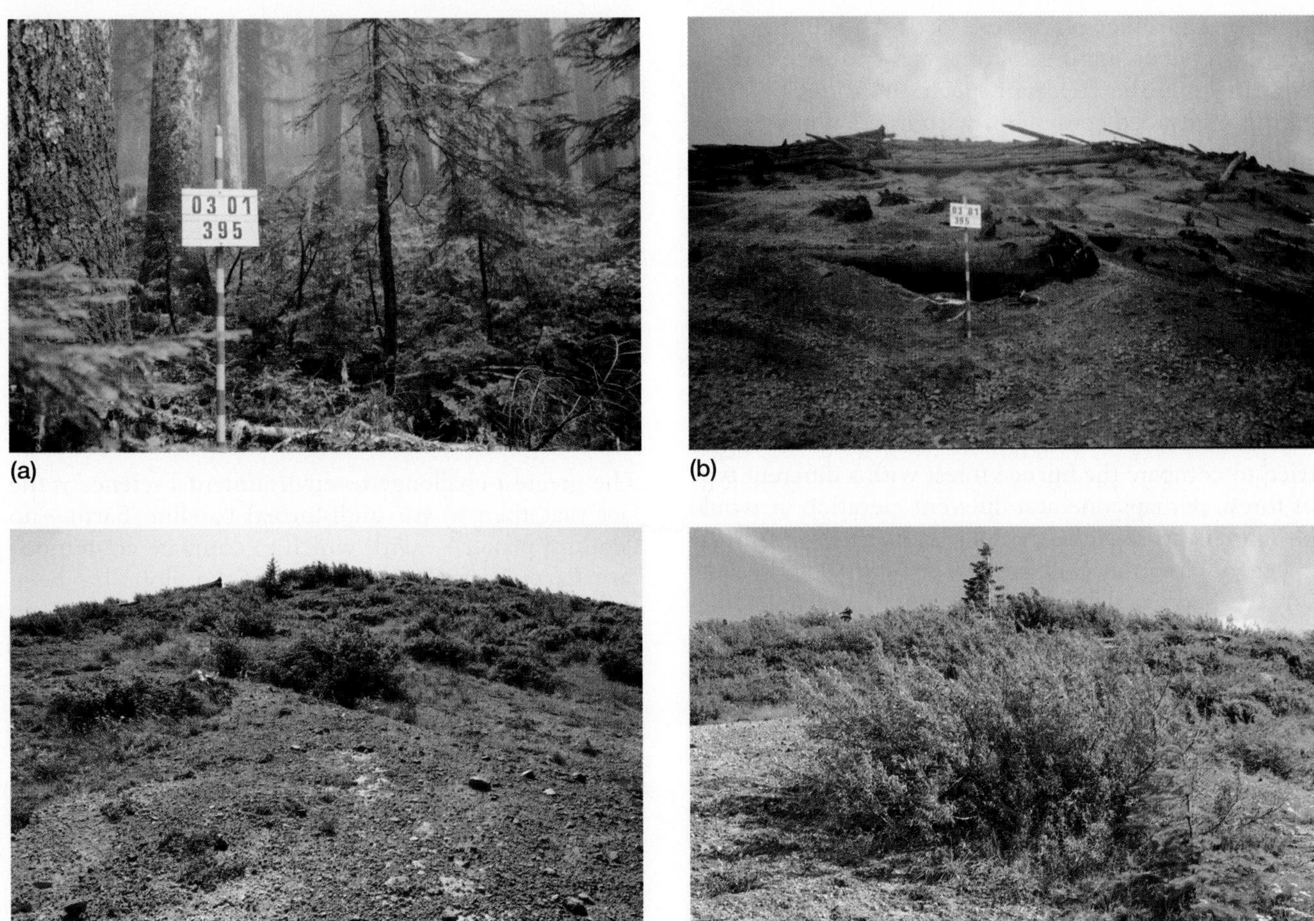

FIGURE 3.4 A natural experiment. The Mount St. Helens eruption in 1980 created a natural experiment for understanding large-scale forest regrowth. (a) A pre-eruption forest near Mount St. Helens in 1979; (b) the same location, post-eruption, in 1982; (c) the same location in 2009 begins to show forest regrowth; (d) the same location in 2014 showing further growth. *(a–d: US Forest Service)*

might seem relatively small. But that is how most scientific research works: Very small steps establish that an effect occurs and, eventually, how it occurs. In this way, we progress toward a more thorough understanding of how the world works. This particular research on chlorpyrifos, combined with numerous other experiments testing specific aspects of the chemical's effect on rat brains, demonstrated that chlorpyrifos was capable of damaging developing rat brains at fairly low doses. The results of this research have been important for our understanding of human health and toxic substances in the environment.

Controlled Experiments and Natural Experiments

The chlorpyrifos experiment we have just described was conducted in the controlled conditions of a laboratory. However, not all experiments can be done under such controlled conditions. For example, it would be difficult to study the interactions of wolves and caribou in a controlled setting because both species need large amounts of land and because their behavior changes in captivity. Other reasons that a controlled laboratory experiment may not be possible include prohibitive costs and ethical concerns.

Under these circumstances, investigators look for a *natural experiment.* A **natural experiment** occurs when a natural event acts as an experimental treatment in an ecosystem. For example, a volcano that destroys thousands of hectares of forest provides a natural experiment for understanding large-scale forest regrowth (**FIGURE 3.4**). We would never destroy that much forest just to study regrowth, but we can study such natural disasters when they occur. Still other cases of natural experiments do not involve disasters. For example, we can study the process of ecological succession by

Natural experiment A natural event that acts as an experimental treatment in an ecosystem.

looking at specific areas where forests have been growing for different amounts of time and comparing them. We can study the effects of species invasions by comparing uninvaded ecosystems with invaded ones.

Because a natural experiment is not controlled, many variables can change at once, and results can be difficult to interpret. Ideally, researchers compare multiple examples of similar systems in order to exclude the influences of different variables. For example, after a forest fire, researchers might not only observe how a burned forest responds to the disturbance but also compare it with a nearby forest that did not burn. In this case, the researchers are comparing similar forests that differ in only one variable, fire. If, however, they tried to compare the burned forest with a different type of forest, perhaps one at a different elevation, it would be difficult to separate the effects of the fire from the effects of elevation. Still, because they may be the only way to obtain vital information, natural experiments are indispensable.

Let us return to the study of chlorpyrifos. Researchers wanted to know if human brains that were exposed to the chemical would react in the same way as rat brains. Because researchers would never feed pesticides to humans to study their effects, for obvious ethical reasons, they conducted a natural experiment. They looked for groups of people who were similar in most ways—for example, income, age, level of education—but who varied in their exposure to chlorpyrifos. To gather data on variation of exposure they looked at how often people in each group used pesticides that contained the chemical, the brand they used, and the frequency and location of use. Researchers found that tissue concentrations of chlorpyrifos were highest in groups that were exposed to the chemical in their jobs and among poor urban families whose exposure to residential pesticides was high. Among these populations, a number of studies connected exposure to chlorpyrifos with low birth weight and other developmental abnormalities.

Science and Progress

The chlorpyrifos experiment is a good example of the process of science. Based on observations, the scientists proposed a hypothesis (that chlorpyrifos causes neurological disorders) and a null hypothesis (that there was no effect from chlorpyrifos). The null hypothesis was tested and rejected. Multiple rounds of additional testing gave researchers confidence in their understanding of the problem. Moreover, as the research progressed, the scientists informed the public, as well as the scientific community, about their results. Finally, in 2000, as a result of the step-by-step scientific investigation of chlorpyrifos, the U.S. Environmental Protection Agency (EPA) decided to prohibit its use for most residential applications. It also prohibited agricultural use on fruits that are eaten without peeling, such as apples and pears, and those that are especially popular with children, such as grapes.

Environmental science presents unique challenges

Environmental science has many things in common with other scientific disciplines. However, it presents a number of challenges and limitations that are not usually found in most other scientific fields. These challenges and limitations are a result of the nature of environmental science and the way research in the field is conducted.

Lack of Baseline Data

The greatest challenge to environmental science is the fact that there is no undisturbed baseline Earth—no "control planet"—with which to compare contemporary Earth. Virtually every part of the globe has been altered by humans in some way (**FIGURE 3.5**). Even though some remote regions appear to be undisturbed, we can still find quantities of lead in the Greenland ice sheet, traces of the anthropogenic compound PCB in the fatty tissue of penguins in Antarctica, and invasive species from many locations carried by ship to remote tropical islands. This situation makes it difficult to know the original levels of contaminants or numbers of species that existed before humans began to alter the planet. Consequently, we can only speculate about how the current conditions deviate from those of pre-human activity.

Subjectivity

A second challenge unique to environmental science lies in the dilemmas raised by subjectivity. For example,

FIGURE 3.5 The global nature of human impacts. The trash that washed up onto the beach of this remote Pacific island vividly demonstrates the difficulty of finding any part of Earth unaffected by human activities. *(Ashley Cooper/Alamy)*

when you go to the grocery store, the bagger may ask, "Paper or plastic?" How can we know for certain which type of bag has the least environmental impact? There are techniques for determining what harm may come from using the petrochemical benzene to make a plastic bag and from using chlorine to make a paper bag. However, different substances tend to affect the environment differently: Benzene may pose more of a risk to people, whereas chlorine may pose a greater risk to organisms in a stream. It is difficult, if not impossible, to decide which is better or worse for the environment overall. There is no single measure of environmental quality. Ultimately, our assessments and our choices involve value judgments and personal opinions. In the case of "Paper or plastic?" many of us would maintain that the best answer is "neither," if the bag is going to be thrown away, but rather a reusable bag made from a sturdier material.

Interactions

A third challenge is the complexity of natural and human-dominated systems. All scientific fields examine interacting systems, but those systems are rarely as complex and as intertwined as they are in environmental science. Because environmental systems have so many interacting parts, the results of a study of one system cannot always be easily applied to similar systems elsewhere.

There are also many examples in which human preferences and behaviors affect environmental systems as much as the natural laws that describe them. For example, many people assume that if more efficient automobiles were available, the overall consumption of gasoline in the United States would decrease. To decrease gas consumption, however, it is necessary not only to build more efficient automobiles, but also to get people to purchase those vehicles and use them in place of less efficient ones. In previous decades, even though there were many fuel-efficient cars available, the majority of buyers in the United States continued to purchase larger, heavier, and less fuel-efficient cars, minivans, light trucks, and sport-utility vehicles. Environmental scientists thought they knew how to reduce gasoline consumption, but they neglected to account for consumer behavior.

Human Well-Being

As we continue our study of environmental science, we will see that many of its topics touch on human well-being. In environmental science, we study how humans impact the biological systems and natural resources of the planet. We also study how changes in natural systems and the supply of natural resources affect humans.

We know that people who are unable to meet their basic needs are less likely to be interested in or able to be concerned about the state of the natural environment. The principle of environmental equity—the fair distribution of Earth's resources—adds a moral issue to questions raised by environmental science. Pollution and environmental degradation are inequitably distributed, with the poor receiving much more than an equal share. Is this a situation that we, as fellow humans, can tolerate? Environmental justice is a social movement and field of study that works toward equal enforcement of environmental laws and the elimination of disparities, whether intended or unintended, in how pollutants and other environmental harms are distributed among the various ethnic and socioeconomic groups within a society (**FIGURE 3.6**).

FIGURE 3.6 Electronic waste recycling. The poor are exposed to a disproportionate amount of pollutants and other hazards. The people shown here, located in a small village on the outskirts of New Delhi, India, are recycling circuit boards from discarded electronics products. *(Aurora Photos/Alamy)*

MODULE 3 AP® Review

Preparing for the AP® Exam

In this module, we have seen how specific aspects of the scientific method are used to conduct field and laboratory evaluations of how human activity affects the natural environment. The scientific method follows a process of observations and questions, testable hypotheses and predictions, and data collection. Results are interpreted and shared with other researchers. Experiments can be either controlled (manipulated) experiments or natural experiments that make use of natural events. There are often challenges in environmental science including the lack of baseline data and the interactions with social factors such as human preferences.

AP® Practice Questions

Choose the best answer for the following.

1. The first step in the scientific process is
   (a) collecting data.
   (b) observations and questions.
   (c) forming a hypothesis.
   (d) forming a theory.

Use the following information for questions 2 and 3:
Two new devices for measuring lead contamination in water are tested for accuracy. Scientists test each device with seven samples of water known to contain 400 ppb of lead. Their data are shown in the table. Concentration is in parts per billion.

Water Sample	1	2	3	4	5	6	7
Device 1	415	417	416	417	415	416	416
Device 2	398	401	400	402	398	400	399

2. The data from device 1 is
   (a) accurate, but not precise.
   (b) precise, but not accurate.
   (c) both accurate and precise.
   (d) neither accurate nor precise.

3. Assuming the devices were used correctly, and assuming we want to choose a device that accurately reflects the true concentration of lead in the water samples, which conclusion does the data support?
   (a) Device 1 is superior to device 2 because it is more precise.
   (b) Device 2 is superior to device 1 because it is more precise.
   (c) Device 1 is superior to device 2 because it is more accurate.
   (d) Device 2 is superior to device 1 because it is more accurate.

4. Challenges in the study of environmental science include all of the following EXCEPT
   (a) dangers of studying natural systems.
   (b) lack of baseline data.
   (c) subjectivity of environmental impacts.
   (d) complexity of natural systems.

5. A control group is
   (a) a group with the same conditions as the experimental group.
   (b) a group with conditions found in nature.
   (c) a group with a randomly assigned population.
   (d) a group with the same conditions as the experimental group except for the study variable.

Working Toward Sustainability

Reducing Food Waste

Food is essential to life and food production has substantial impacts on the environment, including most of the indicators we discussed earlier in the chapter. For example, producing food involves clearing land, which impacts biodiversity. It also involves plowing and tilling soils, which leads to more oxygen in soils, which leads to more decomposition and release of carbon dioxide to the atmosphere. And virtually all food production involves fossil fuel energy which, when combusted, also releases carbon dioxide to the atmosphere.

The U.S. Department of Agriculture (USDA)—the primary agency in the United States responsible for overseeing farming, agriculture, forestry, and food—estimates that between 30 and 40 percent of the food supply is wasted. This estimate is based on the percentage of food at the retail and consumer level that is not eaten. It includes food that spoils before it reaches the consumer as well as food put on the plate and not eaten.

The USDA and the U.S. Environmental Protection Agency (EPA) recently announced a food waste challenge for farms, food processors, grocery stores, restaurants, and even colleges and universities. The stated objectives were to

- reduce food waste with better storage, ordering and cooking methods; and in "all you can eat" dining facilities, encouraging people to take less, or giving them smaller plates.
- recover food waste by connecting food donors to hunger relief organizations such as food banks.
- recycle food waste by feeding it to animals or composting it.

In 2015, the USDA and U.S. EPA established a food loss and waste goal for the United States, and targeted a 50 percent reduction by the 2030 goal. Although not enough time has passed to determine meaningful data on nationwide reductions, there are over 4,000 organizations, both companies and nonprofits, that have begun to help achieve these goals. Many of the nonprofits operate in universities, colleges, and high schools.

The Food Recovery Network was founded by a University of Maryland undergraduate and has recovered over 907,000 kilograms (2,000,000 pounds) of food and delivered it to people in need. The Campus Kitchens Project operates at 60 schools around the country from the University of Florida to Gonzaga College High School in Washington, D.C. They report that during the 2015–2016 academic year, they recovered 590,000 kilograms of food (1.3 million pounds) and prepared 350,000 nutritious meals to almost 20,000 people.

Students at the University of Hawaii are part of a Food Recovery Network that delivers unsold food from their dining facilities to food shelters. *(Courtesy Food Recovery Network)*

Because it takes so much energy to grow, process, and cook food, and because food disposed in the landfill generates methane, a greenhouse gas 25 times more potent than carbon dioxide, reducing, recovering, and recycling food is an excellent way to minimize the impact of people on the natural environment. And it appears to be off to a good start.

Sources: USDA Office of the Chief Economist. Food Waste Challenge viewed 20 July 2017. https://www.usda.gov/oce/foodwaste/foodtank: The Think Tank for Food https://foodtank.com/

Critical Thinking Questions

1. How are the 2013 food waste challenge objectives similar to those in the basic recycling slogan: Reduce, Reuse, Recycle?
2. Why do you think the goal for 50 percent food loss and waste reduction is 2030? Is that a long time from now or a short time?

Throughout this chapter, we have outlined principles, techniques, and methods that will allow us to approach environmental science from an interdisciplinary perspective as we evaluate the current condition of Earth and the ways that human beings have influenced it. We identified that we can use environmental indicators to show the status of specific environmental conditions in the past, at present, and, potentially, into the future. These indicators and other environmental metrics must be measured using the same scientific process used in other fields of science. Environmental science does contain some unique challenges because there is no undisturbed baseline—humans began manipulating Earth long before we have been able to study it.

Key Terms

Fossil fuel
Fracking
Environment
Environmental science
Ecosystem
Biotic
Abiotic
Environmentalism
Environmental studies
Ecosystem services
Environmental indicator
Biodiversity
Genetic diversity
Species
Species diversity
Speciation
Background extinction rate
Greenhouse gases
Anthropogenic
Per capita
Development
Sustainability
Sustainable development
Biophilia
Ecological footprint
Scientific method
Hypothesis
Variable
Independent variable
Dependent variable
Null hypothesis
Replication
Sample size (*n*)
Accuracy
Precision
Uncertainty
Theory
Control group
Natural experiment

Learning Goals Revisited

Module 1 Environmental Science

Define the field of environmental science and discuss its importance.

Environmental science is the study of the interactions among human-dominated systems and natural systems and how those interactions affect environments. Studying environmental science helps us identify, understand, and respond to anthropogenic changes.

Identify ways in which humans have altered and continue to alter our environment.

The impact of humans on natural systems has been significant since early humans hunted some large animal species to extinction. However, technology and population growth have dramatically increased both the rate and the scale of human-induced change.

Module 2 Environmental Indicators and Sustainability

Environmental scientists use key environmental indicators to monitor natural systems for signs of stress.

Five important global-scale environmental indicators are biological diversity, food production, average global surface temperature and atmospheric CO_2 concentrations, human population, and resource depletion. Biological diversity is decreasing as a result of human actions, most notably destruction and degradation of ecosystems. Food production appears to be leveling off and may be decreasing. Carbon dioxide concentrations are steadily increasing as a result of fossil fuel combustion and land conversion. Human population continues to increase and probably will continue to do so throughout this century. Resource depletion for most natural resources continues to increase.

Sustainability can be assessed using the ecological footprint.

Sustainability is the use of Earth's resources to meet our current needs without jeopardizing the ability of future generations to meet their own needs. The ecological footprint is the land area required to support a person's (or a country's) lifestyle. We can use that information to say something about how sustainable that lifestyle would be if it were adopted globally.

Module 3 Scientific Method

The scientific method is an important process in environmental science.

The scientific method is a process of observation, hypothesis generation, data collection, analysis of results, and dissemination of findings. Repetition of measurements or experiments is critical if one is to determine the validity of findings. Hypotheses are tested and often modified before being accepted.

Describe some of the unique challenges and limitations of environmental science.

We lack an undisturbed "control planet" with which to compare conditions on Earth today. Assessments and choices are often subjective because there is no single measure of environmental quality. Environmental systems are so complex that they are poorly understood, and human preferences and policies may affect them as much as do natural laws.

Practice Math and Graphing

Preparing for the AP® Exam

Answer the following questions. Be sure to show all your work.

1. Practice Math

The table shows global average temperature in degrees centigrade from the year 2000 through 2016.

(a) What is the temperature difference between 2000 and 2016?

(b) What is the average temperature for the last 5 years of measurement (from 2012 through 2016)?

2. Practice Graphing

Using the data from the table, plot a graph of global temperature on the *y* axis and time from 2000 through 2016 on the *x* axis.

Year	Temperature °C
2000	14.42
2001	14.55
2002	14.64
2003	14.62
2004	14.55
2005	14.70
2006	14.63
2007	14.66
2008	14.54
2009	14.65
2010	14.72
2011	14.61
2012	14.64
2013	14.66
2014	14.75
2015	14.87
2016	14.99

Section 1: Multiple-Choice Questions

Choose the best answer for questions 1–11.

1. Which of the following events has increased the impact of humans on the environment?
   I. advances in technology
   II. reduced human population growth
   III. use of tools for hunting
   (a) I only
   (b) I and II
   (c) II and III
   (d) I and III

2. As described in this chapter, environmental indicators
   (a) always tell us what is causing an environmental change.
   (b) can be used to analyze the health of natural systems.
   (c) are useful only when studying large-scale changes.
   (d) do not provide information regarding sustainability.

3. Which statement regarding a global environmental indicator is NOT correct?
   (a) Concentrations of atmospheric carbon dioxide have been rising quite steadily since the Industrial Revolution.
   (b) World grain production has increased fairly steadily since 1950, but worldwide production of grain per capita has decreased dramatically over the same period.
   (c) For the past 130 years, average global surface temperatures have shown an overall increase that seems likely to continue.
   (d) World population is expected to be between 8 billion and 10 billion by 2050.

4. Figure 2.5 (on page 12) shows atmospheric carbon dioxide concentrations over time. The measured concentration of CO_2 in the atmosphere is an example of
   (a) a sample of air from over the Antarctic.
   (b) an environmental indicator.
   (c) replicate sampling.
   (d) how to study seasonal variation in Earth's temperatures.

5. Environmental metrics such as the ecological footprint are most informative when they are considered along with other environmental indicators. Which indicator, when considered in conjunction with the ecological footprint, would provide the most information about environmental impact?
   (a) biological diversity
   (b) food production
   (c) human population
   (d) CO_2 concentration

6. In science, which of the following is the most certain?
   (a) hypothesis
   (b) idea
   (c) natural law
   (d) observation

7. All of the following would be exclusively caused by anthropogenic activities EXCEPT
   (a) combustion of fossil fuels.
   (b) overuse of resources such as uranium.
   (c) forest clearing for crops.
   (d) forest fires.

8. Use Figure 2.3 (on page 11) to calculate the approximate percentage change in world grain production per person between 1950 and 2000.
   (a) 10 percent
   (b) 20 percent
   (c) 30 percent
   (d) 40 percent

9. A species of bat nearly went extinct due to the destruction of its forest habitat. However, conservationists preserved a small remnant population of the species in a large forested reserve. The population grew and it appeared that the species would rebound when a disease suddenly wiped out the entire population. What is the likely primary reason for why this occurred?
   I. the low genetic diversity of the population
   II. the reduced ecosystem diversity of the forest
   III. greater contact with humans, who passed on the disease
   (a) I only
   (b) II only
   (c) III only
   (d) I and II

10. The fraction of the global human population living in developed nations is currently 17 percent. If each person in the developed world consumes on average 100 kg of meat per year, how much meat will be consumed in the developed world in 1 year?
   (a) 102 billion kg
   (b) 130 billion kg
   (c) 500 billion kg
   (d) 630 billion kg

11. Which scenario would be most appropriate to study using a natural experiment?
   I. the effects of an herbicide on human health
   II. the influence of sleep on exam performance
   III. the effects of an earthquake on blood pressure
   (a) I only
   (b) II only
   (c) III only
   (d) I and III

Section 2: Free-Response Questions

Write your answer to each part clearly. Support your answers with relevant information and examples. Where calculations are required, show your work.

1. A company conducts an experiment to test whether their coffee or their energy drinks are more likely to keep people up at night. They have five male college students drink a cup of coffee at 5 PM one day, 6 PM the next day, 7 PM on the third day, and 8 PM on the fourth day. They have five other male college students do the same with energy drinks. Each night the researchers record how many minutes it takes for each student to fall asleep. With the results from their experiment, they produce the graphs shown below.

   The researchers conclude that energy drinks are more likely to keep people up at night than coffee.

   (a) Describe THREE problems with the company's experimental design. If you were designing a new experiment, how would you fix TWO of these problems? (5 points)

   (b) Describe ONE problem with the company's graphs. Without repeating anything from part (a) provide at least TWO reasons supporting or opposing the position that these graphs show energy drinks keep people awake later than coffee. (3 points)

   (c) During this study, the researchers do a survey of the students at the university and notice that students who report drinking more coffee also tend to have higher grades. Can any conclusions be drawn from this observation? Why? (2 points)

2. The study of environmental science sometimes involves examining the overuse of environmental resources.

   (a) Identify one general effect of overuse of an environmental resource. (3 points)

   (b) For the effect you listed above, describe a more sustainable strategy for resource utilization. (3 points)

   (c) Describe how a dust bowl, large-scale erosion, or other human-caused land deterioration might be indicative of environmental issues on Earth today. (4 points)

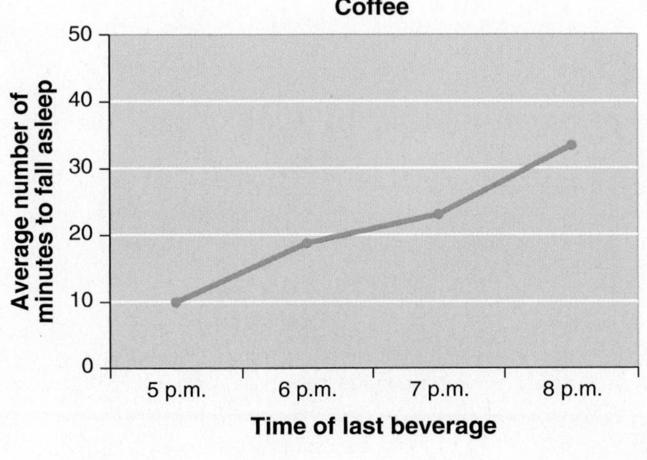

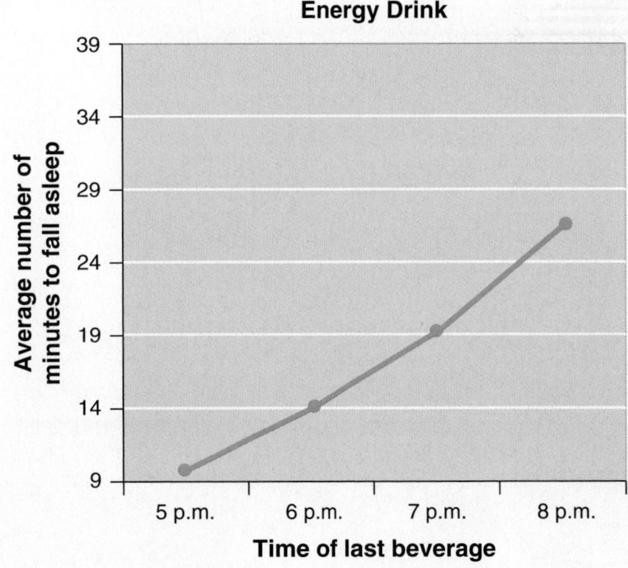

Tufa towers rise out of the salty water of Mono Lake. *(Don Smith/Getty Images)*

Environmental Systems

CHAPTER 2

MODULE 4 Systems and Matter

MODULE 5 Energy, Flows, and Feedbacks

CASE STUDY

A Lake of Salt Water, Dust Storms, and Endangered Species

Located between the deserts of the Great Basin and the mountains of the Sierra Nevada, California's Mono Lake is an unusual site. It is characterized by eerie towers of limestone rock known as tufa, unique animal species, glassy waters, and frequent dust storms. Mono Lake is a terminal lake, which means that water flows into it but does not flow out. As water moves through the mountains and desert soil, it picks up salt and other minerals, which it deposits in the lake. As the water evaporates, these minerals are left behind. Over time, evaporation has caused a buildup of salt concentrations so high that the lake is actually saltier than the ocean, and no fish can survive in the lake's water. Mono Lake is an excellent example of an environmental system.

The Mono brine shrimp (*Artemia monica*) and the larvae of the Mono Lake alkali fly (*Ephydra hians*) are two of only a few animal species that can tolerate the conditions of the lake. The brine shrimp and the fly larvae consume microscopic algae, millions of tons of which grow in the lake each year. In turn, large flocks of migrating birds, such as sandpipers, gulls, and flycatchers, use the lake as a stopover, feeding on the brine shrimp and fly larvae to replenish their energy stores. The lake is an oasis on the migration route for these birds and they have come to depend on its food and water resources. The health of Mono Lake is therefore critical for many species.

In 1913, the City of Los Angeles drew up a controversial plan to redirect water away from Mono Lake and its neighbor, the larger and shallower Owens Lake. Owens Lake was diverted first, via a 359-km (223-mile) aqueduct that drew water away from the springs and streams that kept Owens Lake full. Soon, the lake began to dry up, and by the 1930s only an empty salt flat remained. Today the dry lake bed covers roughly 44,000 ha (109,000 acres). It is one of the nation's largest sources of windblown dust, which lowers visibility in nearby national parks. Even worse, because of the local geology, the dust contains high concentrations of arsenic, which is a threat to human health.

> Just when it appeared that Mono Lake would never recover, circumstances changed.

In 1941, despite the environmental degradation at Owens Lake, Los Angeles extended the aqueduct to draw water from the streams feeding Mono Lake. By 1982, with less fresh water feeding the lake, its depth had decreased by half, to an average of 14 m (45 feet), and the salinity of the water had doubled to more than twice that of the ocean. The salt killed algae in the lake and without the algae to eat, the Mono brine shrimp also died. Most birds stayed away, and newly exposed land bridges allowed coyotes from the desert to prey on the colonies of nesting birds that remained.

However, just when it appeared that Mono Lake would never recover, circumstances changed. In 1994, after years of litigation led by the National Audubon Society and tireless work by environmentalists and environmental scientists, the Los Angeles Department of Water and Power finally agreed to reduce the amount of water it diverted and to allow the lake to refill to about two-thirds of its historical depth. By summer 2009, lake levels had risen to just short of that goal, and the ecosystem was slowly recovering. In 2012, water levels were close to the targeted goals. The brine shrimp were thriving, and many birds returned to Mono Lake. A subsequent drought caused water levels to fall again, bringing concerns that low water levels would adversely affect bird populations and possibly require further restrictions on diversion of water to Los Angeles. Heavy rain and snow in January 2017 averted the potential crisis. The complex Mono Lake system continues to thrive, partly because of weather patterns but also because of close monitoring of the inputs and outputs of the environmental system.

Water is a scarce resource in the Los Angeles area, and demand is particularly high. To decrease the amount of water diverted from Mono Lake, the City of Los Angeles had to reduce its water consumption. The city converted grass lawns requiring a great deal of water to native shrubs that are drought-tolerant, and it imposed new rules requiring low-flow showerheads and water-saving toilets. Through these seemingly small, but effective, measures, the inhabitants of Los Angeles were able to cut their water consumption and, in turn, protect nesting birds, Mono brine shrimp, and algae populations, and restore the Mono Lake ecosystem.

Sources: J. Kay, It's rising and healthy, *San Francisco Chronicle*, July 29, 2006; Mono Lake Committee, *Mono Lake* (2013), http://www.monolake.org/. L. Sahagun, Once teetering, Mono Lake is revived by heavy rains, snow, *Los Angeles Times*, January 10, 2017. http://www.latimes.com/local/california/la-me-mono-lake-storm-20170109-story.html.

The story of Mono Lake shows us that the activities of humans, the lives of other organisms, and abiotic processes in the environment are interconnected. Humans, water, animals, plants, and the desert environment all interact at Mono Lake to create a complex environmental system. The story also demonstrates a key principle of environmental science: a change in any one factor often has both expected and unexpected effects.

In Chapter 1, we learned that a system is a set of interacting components connected in such a way that a change in one part of the system affects one or more other parts of the system. The Mono Lake system is relatively small. Other systems exist on a much larger scale. The largest system that environmental science considers is Earth. Many of the most important current environmental issues—including human population growth and climate change—exist at the global scale. Throughout this book we will define a given system in terms of the environmental issue we are studying and the scale in which we are interested.

Organisms, nonliving matter, and energy all interact in natural systems. Taking a systems approach to an environmental issue decreases the chance of overlooking important components of that issue. Whether investigating ways to reduce pollution, increase food supplies, or find alternatives to fossil fuels, environmental scientists must have a thorough understanding of matter and energy and how these components interact within and across systems. In this chapter, we lay the foundation for the systems approach in environmental science. We will begin by exploring the properties of matter, and we will then discuss the various types of energy and how they influence and limit systems.

MODULE 4

Systems and Matter

All questions about the environment involve *matter,* the atoms and molecules that compose materials, and the systems in which the matter circulates. In this module we will look at the basic building blocks of matter and how matter moves among systems. We will look in detail at water, an important component of most environmental systems. Finally, we will explore how matter is conserved in chemical and biological systems.

Learning Goals

After reading this module you should be able to

- describe how matter comprises atoms and molecules that move among different systems.
- explain why water is a vital component of most environmental systems.
- discuss how matter is conserved in chemical and biological systems.

Matter comprises atoms and molecules that move among different systems

What do rocks, water, air, the book in your hands, and the cells in your body have in common? They are all forms of *matter.* **Matter** is anything that occupies space and has *mass.* The **mass** of an object is a measurement of the amount of matter it contains. Note that the words "mass" and "weight" are often used interchangeably, but they are not the same thing. Weight is the force that results from the action of gravity on mass. Your own weight, for example, is determined by the amount of gravity pulling you toward the planet's center. Whatever your weight on Earth, you would weigh less on the Moon where the action of gravity is weaker. In contrast, mass stays the same under any gravitational influence. So although your weight would change on the Moon, your mass would remain the same because the amount of matter you are made of would be the same. In this section we will look at important properties of matter, starting with the building blocks.

Atoms and Molecules

All matter is composed of tiny particles that cannot be broken down into smaller pieces. The basic building blocks of matter are known as *atoms.* An **atom** is the smallest particle that can contain the chemical properties of an *element.* An **element** is a substance composed of atoms that cannot be broken down into smaller, simpler components. At Earth's surface temperatures, elements can occur as solids (such as gold), liquids (such as mercury), or gases (such as helium). Atoms are so small that a single human hair measures about a few hundred thousand carbon atoms across.

Ninety-four elements occur naturally on Earth, and another twenty-four have been produced in laboratories. The **periodic table** lists all of the elements currently known, organized by their properties. The full periodic table is reproduced in an appendix to this book. As you can see, each element is identified by a one- or two-letter symbol; for example, the symbol for carbon is C, and the symbol for sodium is Na. These symbols are used to describe the atomic makeup

Matter Anything that occupies space and has mass.

Mass A measurement of the amount of matter an object contains.

Atom The smallest particle that can contain the chemical properties of an element.

Element A substance composed of atoms that cannot be broken down into smaller, simpler components.

Periodic table A chart of all chemical elements currently known, organized by their properties.

of **molecules**, which are particles that contain more than one atom. Molecules that contain more than one element are called **compounds**. For example, a carbon dioxide molecule (CO_2) is a compound composed of one carbon atom (C) and two oxygen atoms (O_2). Let's take a closer look at atoms and how they behave.

The Structure of an Atom

FIGURE 4.1 shows a simplified model of single atom of nitrogen. Like the nitrogen atom, every atom consists of a nucleus, or core, surrounded by electrons. The nucleus consists of protons and neutrons. Protons and neutrons have roughly the same mass—both minutely small. Protons have a positive electrical charge, like the "plus" side of a battery. The number of protons in the nucleus of a particular element—called the **atomic number**—is unique to that element. The periodic table lists the atomic number; in the periodic table reproduced in an appendix to this book, it is the whole number next to each element symbol. Neutrons have no electrical charge, but they are critical to the stability of nuclei because they keep the positively charged protons together. Without them, the protons would repel one another and separate.

Turning back to Figure 4.1 you can see that the space around the nucleus is occupied by electrons. Electrons are negatively charged, like the "minus" side of a battery, and have a much smaller mass than protons or neutrons. In the molecular world, opposites always attract, so negatively charged electrons are attracted to positively charged protons. This attraction binds the electrons to the nucleus. In a neutral atom, the numbers of protons and electrons are equal.

The total number of protons and neutrons in an element is known as its **mass number**. Because the mass of an electron is insignificant compared with the mass of a proton or neutron, we do not include electrons in mass number calculations. The periodic table lists the mass number of each element under the element name.

Although the number of protons in a chemical element is constant, atoms of the same element may have different numbers of neutrons and, therefore, different mass numbers. Atoms of the same element with different numbers of neutrons are called **isotopes**. Isotopes of the element carbon, for example, have six protons, but can occur with six, seven, or eight neutrons, which yield mass numbers of 12, 13, or 14, respectively. In nature, carbon occurs as a mixture of carbon isotopes. All carbon isotopes behave the same chemically. However, biological processes sometimes favor one isotope over another. Certain isotopic signatures, or ratios of isotopes, can be left behind by different biological processes. Environmental scientists use these signatures to learn about certain processes or to evaluate environmental conditions such as air pollution. Modern carbon found in wood and fossil carbon found in oil or coal have different carbon isotope signatures. Therefore the proportions of different isotopes in polluted air can tell us whether the pollution comes from a forest fire or combustion of fossil fuels.

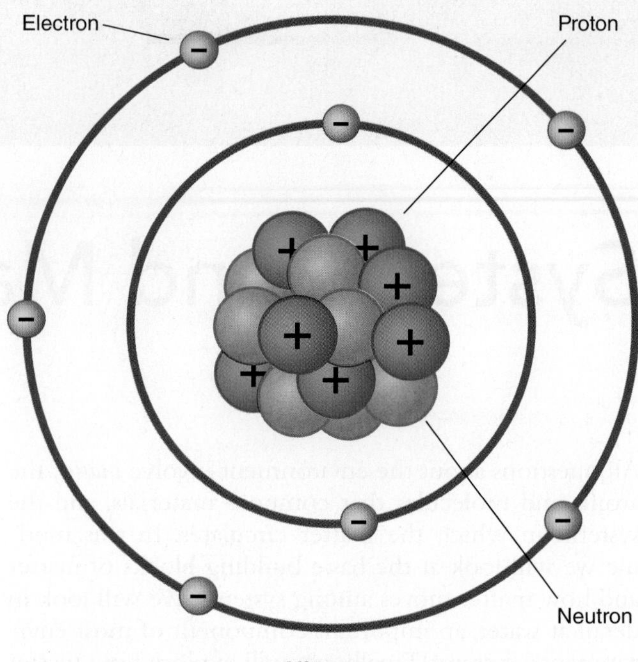

FIGURE 4.1 Structure of the atom. An atom is composed of protons, neutrons, and electrons. Neutrons and positively charged protons make up the nucleus. Negatively charged electrons surround the nucleus.

Molecule A particle that contains more than one atom.

Compound A molecule containing more than one element.

Atomic number The number of protons in the nucleus of a particular element.

Mass number A measurement of the total number of protons and neutrons in an element.

Isotopes Atoms of the same element with different numbers of neutrons.

Radioactive decay The spontaneous release of material from the nucleus of radioactive isotopes.

Radioactivity

The nuclei of isotopes can be stable or unstable, depending on the mass number of the isotope and the number of neutrons it contains. Unstable isotopes are radioactive. Radioactive isotopes undergo **radioactive decay**, the spontaneous release of material from the nucleus. Radioactive decay changes the radioactive element into a different element. For example, uranium-235 (^{235}U) decays to form thorium-231 (^{231}Th). The original atom (uranium) is called the parent and the resulting decay product (thorium) is called the daughter. The radioactive decay of ^{235}U and certain other elements emits a great deal of energy that can be captured as heat. Nuclear power plants use this heat to produce steam that turns turbines to generate electricity.

We measure radioactive decay by recording the average rate of decay of a quantity of a radioactive element. This measurement is commonly stated in terms of the **half-life** of the element, which is the time it takes for one-half of the original radioactive parent atoms to decay. An element's half-life is a useful parameter to know because some elements that undergo radioactive decay emit harmful radiation. Knowledge of the half-life allows scientists to determine the length of time that a particular radioactive element may be dangerous. For example, using the half-life allows scientists to calculate the period of time that people and the environment must be protected from depleted nuclear fuel, like that generated by a nuclear power plant. As it turns out, many of the elements produced during the decay of ^{235}U have half-lives of tens of thousands of years and more. From this we can see why long-term storage of radioactive nuclear waste is so important. We will discuss this further in Chapter 12.

The measurement of isotopes has many applications in environmental science as well as in other scientific fields. For example, carbon in the atmosphere exists in a known ratio of the isotopes carbon-12 (99 percent), carbon-13 (1 percent), and carbon-14 (which occurs in trace amounts, on the order of one part per trillion). Carbon-14 is radioactive and has a half-life of 5,730 years. Carbon-13 and carbon-12 are stable isotopes. Living organisms incorporate carbon into their tissues at roughly the known atmospheric ratio. But after an organism dies, it stops incorporating new carbon into its tissues. Over time, the radioactive carbon-14 in the organism decays to nitrogen-14. By calculating the proportion of carbon-14 in dead biological material—a technique called carbon dating—researchers can determine how many years ago an organism died.

Chemical Bonds

We have seen that matter is composed of atoms, which form molecules or compounds. In order to form molecules or compounds, atoms must be able to interact or join together. This happens by means of chemical bonds of various types. Chemical bonds fall into three categories: *covalent bonds, ionic bonds,* and *hydrogen bonds.*

Covalent Bonds

Elements that do not readily gain or lose electrons form compounds by sharing electrons. Compounds formed by sharing electrons are said to be held together by **covalent bonds**. FIGURE 4.2 illustrates the covalent bonds in a molecule of methane (CH_4, also called natural gas). A methane molecule is made up of one carbon (C) atom surrounded by four hydrogen (H) atoms. Covalent bonds form between the single carbon atom and each hydrogen atom. Covalent bonds also hold the two hydrogen atoms and the oxygen atom in a water molecule together.

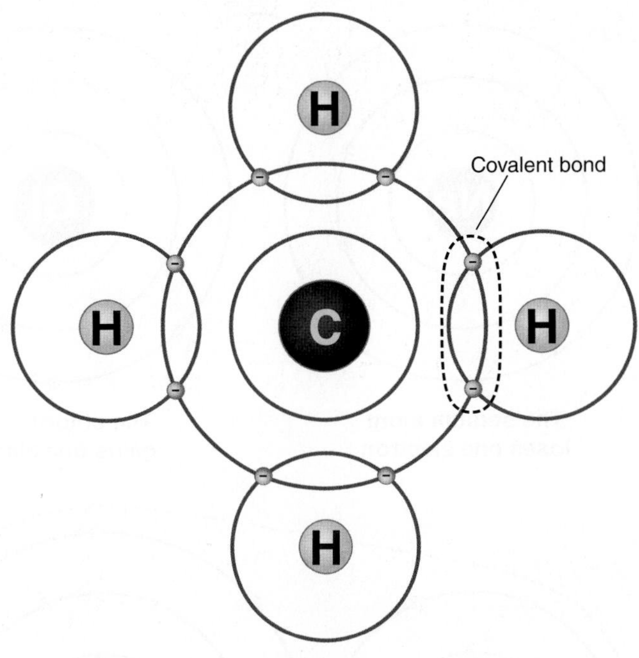

FIGURE 4.2 Covalent bonds. Molecules such as methane (CH_4) are associations of atoms held together by covalent bonds, in which electrons are shared between the atoms. As a result of the four hydrogen atoms sharing electrons with a carbon atom, each atom has a complete set of electrons in its outer shell—two for the hydrogen atoms and eight for the carbon atom.

Ionic Bonds

In a covalent bond, atoms share electrons. Another kind of bond between two atoms involves the transfer of electrons. When such a transfer happens, one atom becomes electron deficient (positively charged), and the other becomes electron rich (negatively charged). The charged atoms are called ions. The charge imbalance holds the two atoms together and the attraction between ions of opposite charges forms a chemical

Half-life The time it takes for one-half of an original radioactive parent atom to decay.

Covalent bond The bond formed when elements share electrons.

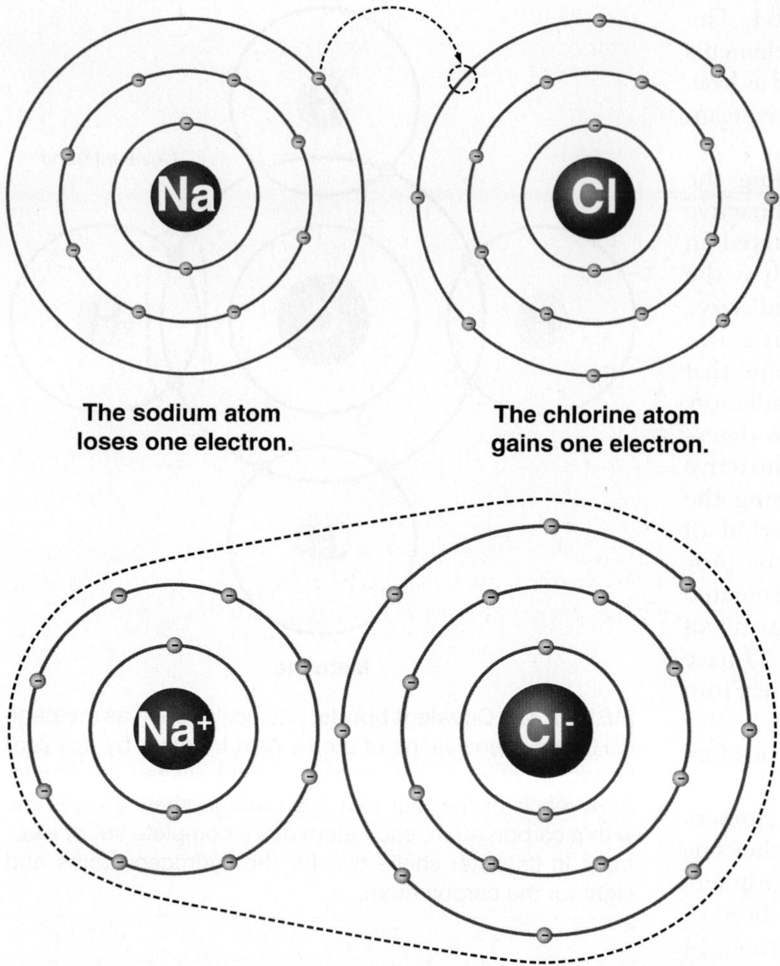
Na
Cl
The sodium atom loses one electron.
The chlorine atom gains one electron.
Na+
Cl-
The sodium ion and the chloride ion form an ionic bond: NaCl.

FIGURE 4.3 Ionic bonds. To form an ionic bond, the sodium atom loses an electron and the chlorine atom gains one. As a result, the sodium atom becomes a positively charged ion (Na^+) and the chlorine atom becomes a negatively charged ion (Cl^-, known as chloride). The attraction between ions of opposite charges—an ionic bond—forms sodium chloride (NaCl), or table salt.

bond known as an **ionic bond**. **FIGURE 4.3** shows an example of this process. Sodium (Na) donates one electron to chlorine (Cl), which gains one electron, to form sodium chloride (NaCl), or table salt.

An ionic bond is not usually as strong as a covalent bond. This means that the compound can readily dissolve. For example, as long as sodium chloride remains in a salt shaker, it remains in solid form. But if you shake some into water, the salt dissolves into sodium and chloride ions (Na^+ and Cl^-).

Ionic bond A chemical bond between two ions of opposite charges.

Hydrogen bond A weak chemical bond that forms when hydrogen atoms that are covalently bonded to one atom are attracted to another atom on another molecule.

Polar molecule A molecule in which one side is more positive and the other side is more negative.

Hydrogen Bonds

The third type of chemical bond is weaker than both covalent bonds and ionic bonds. A **hydrogen bond** is a weak chemical bond that forms when hydrogen atoms that are covalently bonded to one atom are attracted to an atom on another molecule. When atoms of different elements form bonds, their electrons may be shared unequally; that is, shared electrons may be pulled closer to one atom than to the other. In some cases, the strong attraction of the hydrogen electron to other atoms creates a charge imbalance within the covalently bonded molecule.

Looking at **FIGURE 4.4a**, we see that water is an excellent example of this type of unequal electron distribution. Each water molecule as a whole is neutral; that is, it carries neither a positive nor a negative charge. But water has unequal covalent bonds between its two hydrogen atoms and one oxygen atom. Because of these unequal bonds and the angle formed by the H—O—H bonds, water is known as a *polar molecule*. In a **polar molecule**, one side is more positive and the other side is more negative. We can see the result in Figure 4.4b where a hydrogen atom in one water molecule is attracted to the oxygen atom in another nearby water molecule. That attraction forms a hydrogen bond between the two molecules.

Hydrogen bonds also occur in nucleic acids such as DNA, the biological molecule that carries the genetic code for all organisms.

Water is a vital component of most environmental systems

Understanding the chemistry of water is essential in environmental science because so many topics such as air pollution, water pollution, and even photosynthesis involve water chemistry. Because water is often the vehicle for transferring chemical elements and compounds from one system to another, it is also vital for environmental scientists to understand how water behaves. The molecular structure of water gives it unique properties that support the conditions necessary for life on Earth. Among these properties are surface

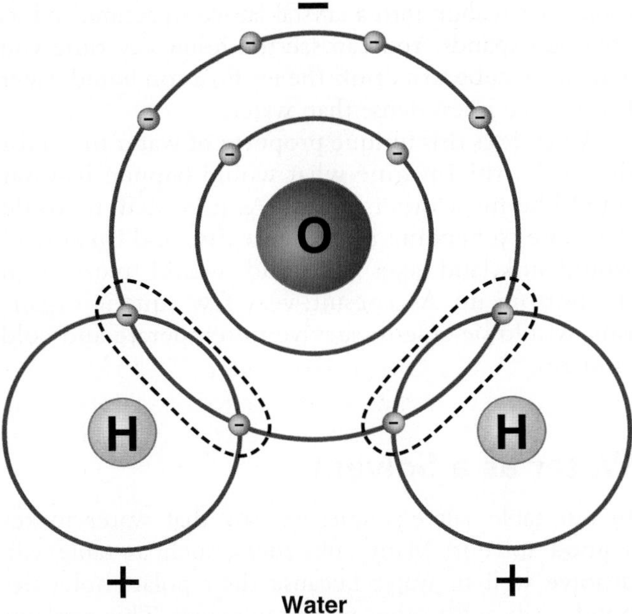

(a) Water molecule

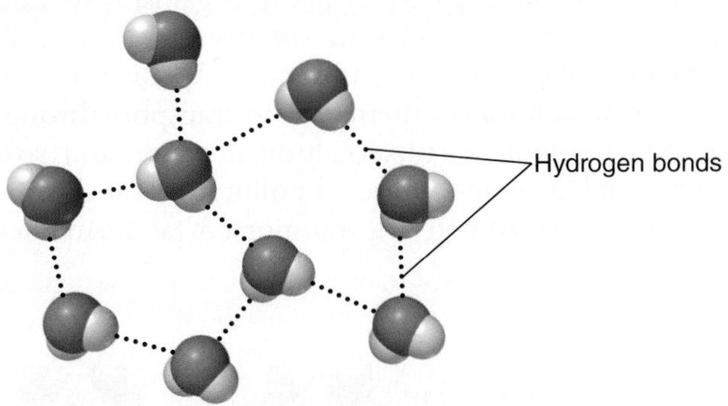

(b) Hydrogen bonds between water molecules

FIGURE 4.4 The polarity of the water molecule. (a) Water (H_2O) consists of two hydrogen atoms covalently bonded to one oxygen atom. Water is a polar molecule because its shared electrons spend more time near the oxygen atom than near the hydrogen atoms. The hydrogen atoms thus have a slightly positive charge, and the oxygen atom has a slightly negative charge. (b) The slightly positive hydrogen atoms are attracted to the slightly negative oxygen atom of another water molecule. The result is a hydrogen bond between the two molecules.

tension, capillary action, a high boiling point, and the ability to dissolve many different substances. Each of these properties is essential to physiological functioning and the movement of elements through systems.

Surface Tension and Capillary Action

Although we don't generally think of water as being sticky, hydrogen bonding makes water molecules stick strongly to one another in an action known as cohesion. Hydrogen bonding also makes water molecules stick

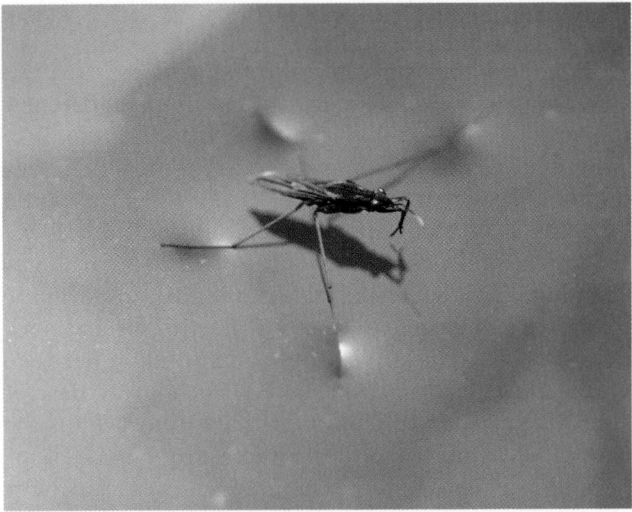

FIGURE 4.5 Surface tension. Hydrogen bonding between water molecules creates the surface tension necessary to support this water strider. *(optimarc/Shutterstock)*

strongly to certain other substances, an action known as adhesion. The ability to cohere or adhere underlies two unusual properties of water: *surface tension* and *capillary action*.

Surface Tension

Surface tension, which results from the cohesion of water molecules at the surface of a body of water, creates a sort of skin on the water's surface. Have you ever seen an aquatic insect, such as a water strider, walk across the surface of the water? This is possible because of surface tension (**FIGURE 4.5**). Surface tension also makes water droplets smooth and more or less spherical as they cling to a water faucet before dropping.

Capillary Action

Capillary action happens when adhesion of water molecules to a surface is stronger than cohesion between the molecules. The absorption of water by a paper towel or a sponge is the result of capillary action. This property is important in thin tubes, such as the water-conducting vessels in tree trunks, and in small pores in soil. It is also important in the transport of underground water, as well as dissolved pollutants, from one location to another.

Surface tension A property of water that results from the cohesion of water molecules at the surface of a body of water and that creates a sort of skin on the water's surface.

Capillary action A property of water that occurs when adhesion of water molecules to a surface is stronger than cohesion between the molecules.

Boiling and Freezing

At the atmospheric pressures found at Earth's surface, water boils (becomes a gas) at 100°C (212°F) and freezes (becomes a solid) at 0°C (32°F). If water behaved like structurally similar compounds such as hydrogen sulfide (H_2S), which boils at −60°C (−76°F), it would be a gas at typical Earth temperatures and life as we know it could not exist. Because of cohesion, however, water can be a solid, a gas, or—most importantly for living organisms—a liquid at Earth's surface temperatures. In addition, the hydrogen bonding between water molecules means that it takes a great deal of energy to change the temperature of water. Thus the water in organisms protects them from wide temperature swings. Hydrogen bonding also explains why geographic areas near large lakes or oceans have moderate climates. The water body holds summer heat, which releases slowly as the atmosphere cools in the fall. Similarly, the water body warms slowly in spring, which prevents the adjacent land area from heating up too quickly.

Water has another unique property: It takes up a larger volume in solid form than it does in liquid form. **FIGURE 4.6** illustrates the difference in molecular structure between liquid water and ice. As liquid water cools, it becomes denser, until it reaches 4°C (39°F), the temperature at which it reaches maximum density. As it cools from 4°C down to freezing at 0°C, its molecules realign into a crystal lattice structure, and its volume expands. You can see the result any time you add an ice cube to a drink: the ice floats on liquid water because ice is less dense than water.

What does this unique property of water mean for life on Earth? Imagine what would happen if water acted like most other liquids. As it cooled, it would continue to become more dense. Its solid form (ice) would sink, and lakes and ponds would freeze from the bottom up. As a result, very few aquatic organisms would be able to survive in temperate and cold climates.

Water as a Solvent

In our table salt example, we saw that water makes a good solvent. Many substances, such as table salt, dissolve well in water because their polar molecules bond easily with other polar molecules. This explains the high concentrations of dissolved ions in seawater as well as the capacity of living organisms to store many types of molecules in solution in their cells. Unfortunately, many toxic substances also dissolve well in water, which makes them easy to transport through the environment. Fertilizers, human waste, and road deicers such as road salt are all pollutants that dissolve easily in water and so are transported far from their sources.

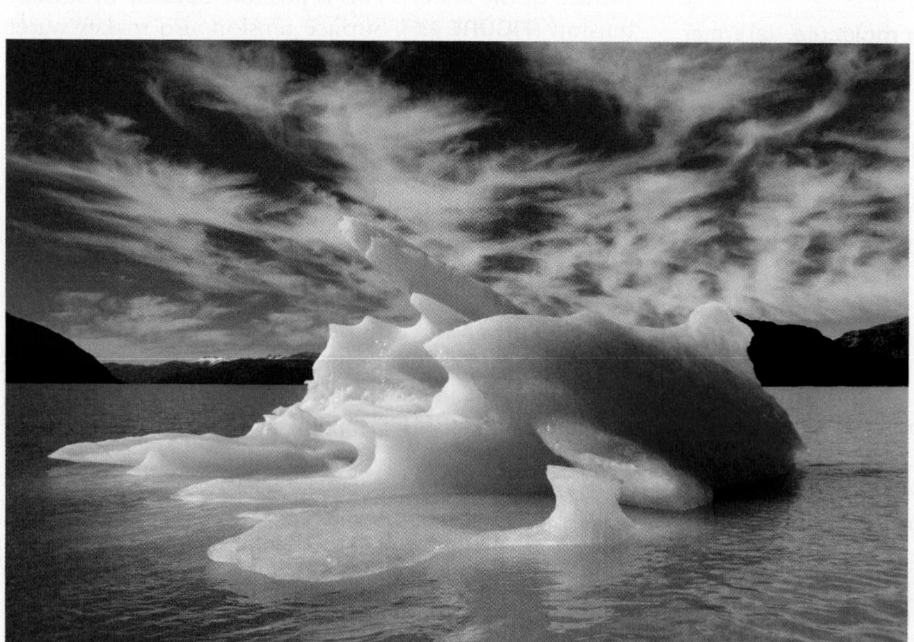

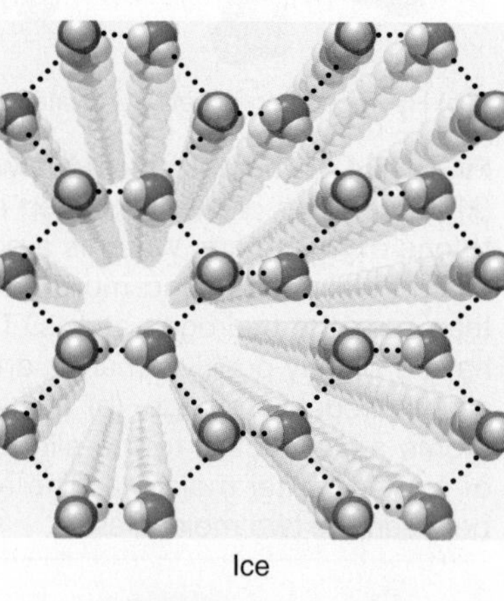

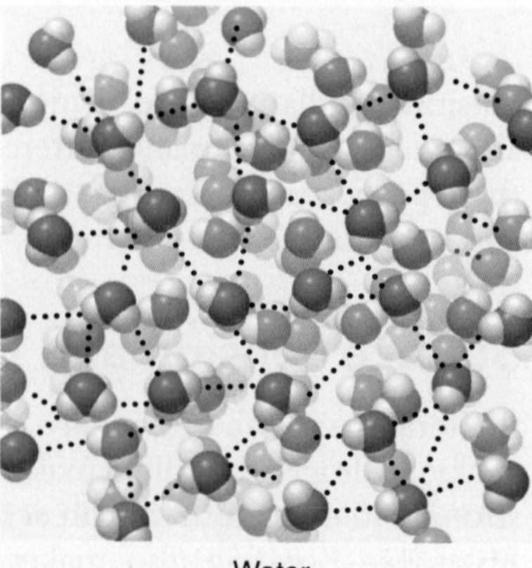

FIGURE 4.6 The structure of water. Below 4°C, water molecules realign into a crystal lattice structure. With its molecules farther apart, solid water (ice) is less dense than liquid water. This property allows ice to float on liquid water. *(Patrick Poendl/Shutterstock)*

Acids, Bases, and pH

Another important property of water is its ability to dissolve hydrogen or hydroxide-containing compounds known as *acids* and *bases.* An **acid** is a substance that contributes hydrogen ions to a solution. A **base** is a substance that contributes hydroxide ions to a solution. Both acids and bases typically dissolve in water.

When an acid is dissolved in water, it dissociates into positively charged hydrogen ions (H^+) and negatively charged ions. Two important acids we will discuss in this book are nitric acid (HNO_3) and sulfuric acid (H_2SO_4), the primary constituents of acid deposition, one form of which is acid rain.

Bases, on the other hand, dissociate into negatively charged hydroxide ions (OH^-) and positively charged ions. Some examples of bases are sodium hydroxide ($NaOH$) and calcium hydroxide ($Ca(OH)_2$), which can be used to neutralize acidic emissions from power plants.

The **pH** indicates the relative strength of acids and bases in a substance. A pH value of 7—the pH of pure water—is neutral, meaning that the number of hydrogen ions is equal to the number of hydroxide ions. Anything above 7 is basic, or alkaline, and anything below 7 is acidic. The lower the number, the stronger the acid, and the higher the number, the more basic the substance. The pH scale is logarithmic, meaning that each number on the scale changes by a factor of 10. For example, a substance with a pH of 5 has 10 times the hydrogen ion concentration of a substance with a pH of 6—it is 10 times more acidic. Water in equilibrium with Earth's atmosphere typically has a pH of 5.65 because carbon dioxide from the atmosphere dissolves in it, which makes it weakly acidic. **FIGURE 4.7** lists the pH of many familiar substances on the pH scale, which ranges from 0 to 14.

AP® Exam Tip

You are expected to understand pH on the AP® Environmental Science Exam and to know the relative acidity of substances shown in Figure 4.7, especially water in various environments. ●

In Chapter 1, we showed that there has been an increase in atmospheric concentrations of carbon dioxide due to human activity. The increase in carbon dioxide in the atmosphere has had an important influence on the pH of ocean water. Carbon dioxide in the atmosphere is constantly dissolving into the ocean, and carbon dioxide in the ocean is constantly coming out of solution into the atmosphere. As atmospheric concentrations of CO_2 have increased, more atmospheric CO_2 has dissolved into the oceans and the oceans have become more acidic, in a phenomenon known as **ocean acidification**.

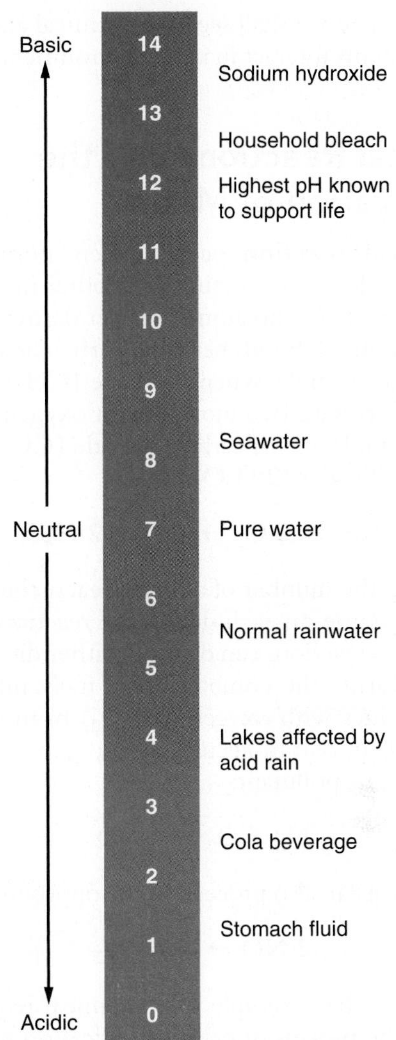

FIGURE 4.7 The pH scale. The pH scale shows how acidic or how basic a solution is.

Environmental systems contain both chemical and biological reactions

The chemical principles that we have described play an important role in many environmental systems through chemical and biological reactions. Although we will look at biological and chemical reactions separately in

Acid A substance that contributes hydrogen ions to a solution.

Base A substance that contributes hydroxide ions to a solution.

pH The number that indicates the relative strength of acids and bases in a substance.

Ocean acidification An increase in the acidity of the oceans.

Chapter 3, here we shall see that chemical and biological components interact in most environmental systems.

Chemical Reactions and the Conservation of Matter

A **chemical reaction** occurs when atoms separate from molecules or recombine with other molecules. In a chemical reaction, no atoms are ever destroyed or created, although the bonds between particular atoms may change. For example, when methane (CH_4) is burned in air, it reacts with two molecules of oxygen ($2\ O_2$) to create one molecule of carbon dioxide (CO_2) and two molecules of water ($2\ H_2O$):

$$CH_4 + 2\ O_2 \rightarrow CO_2 + 2\ H_2O$$

Notice that the number of atoms of each chemical element is the same on each side of the reaction.

Chemical reactions can occur in either direction. For example, during the combustion of fuels, nitrogen gas (N_2) combines with oxygen gas (O_2) from the atmosphere to form two molecules of nitrogen oxide (NO), which is an air pollutant:

$$N_2 + O_2 \rightarrow 2\ NO$$

This reaction can also proceed in the opposite direction:

$$2\ NO \rightarrow N_2 + O_2$$

Here's another example of a chemical reaction. We can show the process of ocean acidification mentioned in the previous section with equations as follows.

First, carbon dioxide and water form carbonic acid:

$$CO_2 + H_2O \rightarrow H_2CO_3$$

Next carbonic acid dissociates or breaks apart to form hydrogen ion and bicarbonate.

$$H_2CO_3 \rightarrow H^+ \text{ and } HCO_3-$$

The free hydrogen ion contributes to the acidity of the ocean.

Finally, bicarbonate dissociates further to form carbonate, which contributes one more hydrogen ion, further increasing ocean acidity:

$$HCO_3- \rightarrow H^+ + CO_3^{2-}$$

This is an excellent example of the chemical connections between two environmental systems: the atmosphere and the oceans.

The observation that no atoms are created or destroyed in a chemical reaction leads us to the **law of conservation of matter**, which states that matter cannot be created or destroyed; it can only change form. For example, when paper burns, it may seem to vanish, but no atoms are lost. In this case, the carbon and hydrogen that make up the paper combine with oxygen in the air to produce carbon dioxide, water vapor, and other materials, which either enter the atmosphere or form ash. Combustion converts most of the solid paper into gases, but all of the original atoms remain. The same process occurs in a forest fire, but on a much larger scale (**FIGURE 4.8**). The only known exception to the law of conservation of matter occurs in nuclear reactions, in which small amounts of matter change into energy.

The law of conservation of matter explains why we cannot easily dispose of hazardous materials. If something is hazardous, it typically will remain hazardous even if we try to convert it through combustion. For example, when we burn material that contains heavy metals, such as an automotive battery, the atoms of the metals in the battery do not disappear. They turn up elsewhere in the environment, where they may harm humans and other organisms. For this and other reasons, understanding the law of conservation of matter is crucial to the study of environmental science.

Biological Molecules and Cells

As we mentioned earlier, chemical and biological reactions happen simultaneously in most environmental systems. We have seen how chemical compounds form and how they respond to various processes such as burning and freezing. To understand biological processes, we must first look at the distinction between compounds that are *inorganic* and those that are *organic*. **Inorganic compounds** are compounds that either do not contain the element carbon or contain carbon that is bound to elements other than hydrogen. Examples include ammonia (NH_3), sodium chloride (NaCl), water (H_2O), and carbon dioxide (CO_2). **Organic compounds** are compounds that have carbon-carbon and carbon-hydrogen bonds. Examples of organic compounds include glucose ($C_6H_{12}O_6$) and fossil fuels, such as natural gas (CH_4).

Organic compounds are the basis of the biological molecules that are important to life: *carbohydrates, proteins, nucleic acids,* and *lipids.* Because these four types of molecules are relatively large, they are also known as macromolecules.

Chemical reaction A reaction that occurs when atoms separate from molecules or recombine with other molecules.

Law of conservation of matter A law of nature stating that matter cannot be created or destroyed; it can only change form.

Inorganic compound A compound that does not contain the element carbon or contains carbon bound to elements other than hydrogen.

Organic compound A compound that contains carbon-carbon and carbon-hydrogen bonds.

FIGURE 4.8 The law of conservation of matter. Even though these trees seem to be disappearing as they burn, all the matter they contain is conserved in the form of water vapor, carbon dioxide, and solid particles. This fire is one of many that broke out during the extremely hot summer and fall of 2017 in northern California. *(JOSH EDELSON/AFP/Getty Images)*

Carbohydrates

Carbohydrates are compounds composed of carbon, hydrogen, and oxygen atoms. Glucose ($C_6H_{12}O_6$) is a simple sugar (a monosaccharide, or single sugar) easily used by plants and animals for quick energy. Sugars can link together in long chains called complex carbohydrates, or polysaccharides (many sugars). For example, plants store energy as starch, which is made up of long chains of covalently bonded glucose molecules. The starch can also be used by animals that eat the plants. Cellulose, a component of plant leaves and stems, is another polysaccharide consisting of long chains of glucose molecules. Cellulose is the raw material for cellulosic ethanol, a type of fuel that has the potential to replace or supplement gasoline.

Proteins

Proteins, a critical component of living organisms, are made up of long chains of nitrogen-containing organic molecules called amino acids. Proteins play a role in structural support, energy storage, internal transport, and defense against foreign substances. Enzymes are proteins that help control the rates of chemical reactions. The antibodies that protect us from infections are also proteins.

Nucleic Acids

Nucleic acids are organic compounds found in all living cells. Long chains of nucleic acids form *DNA* and *RNA*. **DNA (deoxyribonucleic acid)** is the genetic material organisms pass on to their offspring; it contains the code for reproducing the components of the next generation. **RNA (ribonucleic acid)** translates the code stored in DNA, which makes possible the synthesis of proteins.

Carbohydrate A compound composed of carbon, hydrogen, and oxygen atoms.

Protein A critical component of living organisms made up of a long chain of nitrogen-containing organic molecules known as amino acids.

Nucleic acid Organic compounds found in all living cells.

DNA (deoxyribonucleic acid) A nucleic acid, the genetic material that contains the code for reproducing the components of the next generation, and which organisms pass on to their offspring.

RNA (ribonucleic acid) A nucleic acid that translates the code stored in DNA, which makes possible the synthesis of proteins.

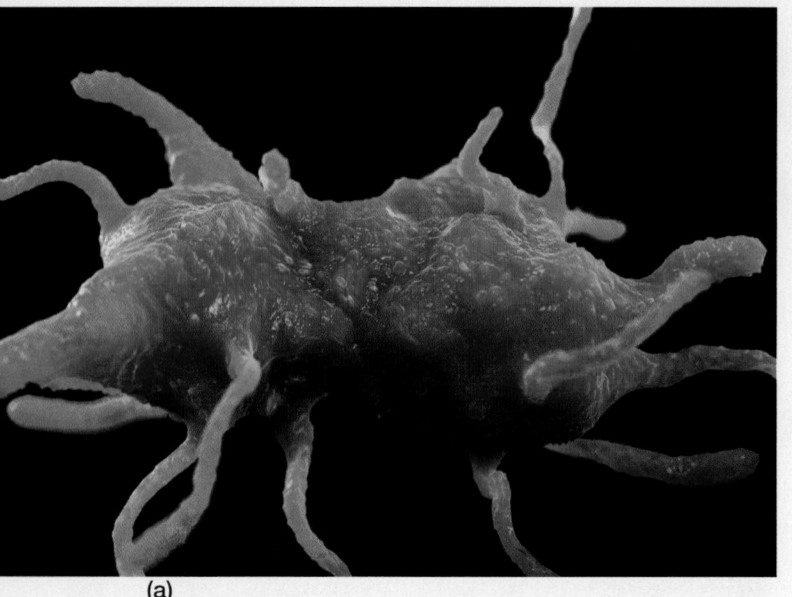

(a)

(b)

FIGURE 4.9 Cellular composition of organisms. (a) Some organisms, such as this amoeba (*A.proteus*), consist of a single cell. (b) More complex organisms, such as this Ana's hummingbird (*Calypte anna*), are made up of millions of cells. *(a: royaltystockphoto.com/Shutterstock; b: Alan Mahood/Alamy Stock Photo)*

Lipids

Lipids are smaller biological molecules that do not mix with water. Fats, waxes, and steroids are all lipids. Lipids form a major part of the membranes that surround cells.

Cells

We have looked at the four types of macromolecules required for life. But how do they work as part of a living organism? The smallest structural and functional component of organisms is known as a *cell*.

A **cell** is a highly organized living entity that consists of the four types of macromolecules and other substances in a watery solution, surrounded by a membrane. Some organisms, such as most bacteria and some algae, consist of a single cell. This cell contains all of the functional structures, or organelles, needed to keep the cell alive and allow it to reproduce (**FIGURE 4.9a**). Larger and more complex organisms, such as Mono Lake's brine shrimp, are multicellular (Figure 4.9b). Throughout this book, we will be examining biological reactions and the resulting effects on the environment. Some effects are at the cellular level, while others are at the organismal level. Each is important to understanding environmental science.

Lipid A smaller organic biological molecule that does not mix with water.

Cell A highly organized living entity that consists of the four types of macromolecules and other substances in a watery solution, surrounded by a membrane.

MODULE 4 AP® Review

Preparing for the AP® Exam

In this module, we saw that all environmental systems consist of matter. Matter has specific properties that can be described and measured. Properties of water include surface tension, capillary action, a high boiling point, and the ability to dissolve many different substances. These properties make water a critical part of most environmental systems. Matter is conserved in chemical reactions and the components of biological reactions. In the next module we will expand our view to look at how the movement of matter in environmental systems is strongly influenced by energy inputs, flows, and outputs.

AP® Practice Questions

Choose the best answer for the following.

1. If two atoms of an element are isotopes, then they have a different
   (a) number of protons.
   (b) number of neutrons.
   (c) number of electrons.
   (d) atomic number.

2. The chemical bond that forms from the attraction of sodium ions and chlorine atoms in table salt (NaCl) is called
   (a) a covalent bond.
   (b) a polar bond.
   (c) a hydrogen bond.
   (d) an ionic bond.

3. Which is NOT a property of water that allows it to support life?
   (a) surface tension
   (b) capillary action
   (c) solvent ability
   (d) high viscosity

4. Which has the highest pH?
   (a) bleach
   (b) cola beverage
   (c) seawater
   (d) acid rain

5. Which is an organic compound?
   (a) CH_4
   (b) NH_3
   (c) NaCl
   (d) CO_2

6. Which is NOT a macromolecule?
   (a) carbohydrates
   (b) nucleic acids
   (c) organelles
   (d) proteins

MODULE 5

Energy, Flows, and Feedbacks

Energy flows within and among systems. Plants and other photosynthetic organisms such as the algae in Mono Lake absorb solar energy and use it in photosynthesis to convert carbon dioxide and water into sugars they need to survive, grow, and reproduce. Animals such as the brine shrimp in Mono Lake eat those plants and the energy is transferred. When migrating gulls use Mono Lake as a stopover, they consume the brine shrimp and transfer the energy and nutrients elsewhere. Some transfers occur more effectively than others. Some transfers occur within a given system while others, like those of migrating gulls, result in transfers of material and energy to another system.

Learning Goals

After reading this module you should be able to

- distinguish among various forms of energy and understand how they are measured.
- identify the first and second laws of thermodynamics and explain how they influence environmental systems.
- explain how scientists keep track of energy and matter inputs, outputs, and changes to environmental systems.

TABLE 5.1 Common units of energy and their conversion into joules

Unit	Definition	Relationship to joules	Common uses
calorie	Amount of energy it takes to heat 1 gram of water 1°C	1 calorie = 4.184 J	Energy expenditure and transfer in ecosystems; human food consumption
Calorie (with a capital "C")	Calories in food	1 Calorie = 1,000 calories = 1 kilocalorie (kcal) = 4,184 J	Food labels; human food consumption
British thermal unit (Btu)	Amount of energy it takes to heat 1 pound of water 1°F	1 Btu = 1,055 J	Energy transfer in air conditioners and home water heaters
Kilowatt-hour (kWh)	Amount of energy expended by using 1 kilowatt of electricity for 1 hour	1 kWh = 3,600,000 J = 3.6 megajoules (MJ)	Energy use by electrical appliances; often given in kWh per year

Energy is a fundamental component of environmental systems

Earth systems cannot function, and organisms cannot survive, without *energy.* **Energy** is the ability to do work, or transfer heat. Water flowing into a lake has energy because it moves and can move other objects in its path.

All living systems absorb energy from their surroundings and use it to organize and reorganize molecules within their cells and to power movement. The sugars in plants are also an important energy source for many animals. Humans, like other animals, absorb the energy they need for cellular respiration from food. This provides the energy for our daily activities, from waking to sleeping to walking, and everything in between. Constructed human systems also utilize energy. If you took mass transportation or an automobile to get to school, you most likely utilized fossil fuel energy that was converted to the energy of motion of your vehicle.

The basic unit of energy in the metric system is the *joule* (J). A **joule** is the amount of energy used when a 1-watt light bulb is turned on for 1 second—a very small amount. Although the joule is the preferred energy unit in scientific study, many other energy units are commonly used. Conversions between these units and joules are given in **TABLE 5.1**.

Although we often use the words "energy" and "power" interchangeably, they are not the same thing. We have seen that energy is the ability to do work. But nothing in the term "energy" explains how quickly or slowly that work gets done. Approximately the same quantity of energy is used to move a pile of rocks up a hill. That energy could be expended slowly, by a human being making multiple trips, or quickly by a tractor in one trip. **Power** is the rate at which work is done. In this example, the tractor can expend more power than the person. This relationship can be expressed as follows:

$$\text{energy} = \text{power} \times \text{time}$$

$$\text{power} = \text{energy} \div \text{time}$$

When we talk about generating electricity, we often hear about kilowatts and kilowatt-hours. The kilowatt (kW) is a unit of power while the kilowatt-hour (kWh) is a unit of energy. Therefore, the capacity of a turbine is given in kW because that measurement refers to the turbine's power. Your monthly home electricity bill reports energy use—the amount of energy from electricity that you have used in your home—in kWh. "Do the Math: How Much Energy Can You Save with a New Refrigerator" gives you an opportunity to practice working with these units.

Energy The ability to do work or transfer heat.

Joule The amount of energy used when a 1-watt electrical device is turned on for 1 second.

Power The rate at which work is done.

AP® Exam Tip

Make sure you can convert units. For example, can you go from millimeters to meters, or from joules to megajoules? These conversions show up frequently in free-response questions and often cause students to lose points. ●

Forms of Energy

Energy exists in different forms and can be converted from one form to another. Potential energy, kinetic energy, light energy, chemical energy, and sound energy are all important energy forms in the environmental sciences.

DO THE MATH How Much Energy Can You Save with a New Refrigerator?

Preparing for the AP® Exam

Your electricity bill shows that you use 600 kWh of electricity each month. Your refrigerator, which is 15 years old, could be responsible for up to 25 percent of this electricity consumption. Newer refrigerators are more efficient, meaning that they use less energy to do the same amount of work. If you wish to conserve electrical energy and save money, should you replace your refrigerator? How can you compare the energy efficiency of your old refrigerator with that of more efficient newer models?

Your refrigerator uses 500 watts when the motor is running. The motor runs for about 30 minutes per hour (or a total of 12 hours per day). How much energy in kilowatt-hours per year will you save by using the best new refrigerator instead of your current one? How long will it take you to recover the cost of the new appliance?

1. Start by calculating the amount of energy your current refrigerator uses.

$$0.5 \text{ kW} \times 12 \text{ hours/day} = 6 \text{ kWh/day}$$

$$6 \text{ kWh/day} \times 365 \text{ days/year} = 2{,}190 \text{ kWh/year}$$

2. How much more efficient is the best new refrigerator compared with your older model?

The best new model uses 400 kWh per year. Your refrigerator uses 2,190 kWh per year.

$$2{,}190 \text{ kWh/year} - 400 \text{ kWh/year} = 1{,}790 \text{ kWh/year}$$

3. Assume that you are paying, on average, \$0.10 per kilowatt-hour for electricity. A new refrigerator would cost \$550. You will receive a rebate of \$50 from your electric company for purchasing an energy-efficient refrigerator. This lowers the cost of the refrigerator to \$500. If you replace your refrigerator, how long will it be before your energy savings compensate you for the cost of the new appliance? You will save

$$1{,}790 \text{ kWh/year} \times \$.10/\text{kWh} = \$179/\text{year}$$

To determine how long it will take for you to recover the cost of the new refrigerator divide the cost by your savings per year.

$$\$500 \div \$179/\text{year} = 2.79 \text{ years (rounding to significant figures)}$$

It will take less than 3 years to recover the cost of the new appliance.

YOUR TURN Environmental scientists must often convert energy units in order to compare various types of energy. For instance, you might want to compare the energy you would save by purchasing an energy-efficient refrigerator with the energy you would save by driving a more fuel-efficient car. Assume that for the amount you would spend on the new refrigerator (\$500), you can make repairs to your car engine that would save you 20 gallons (76 liters) of gasoline per month. (Note that 1 L of gasoline contains the energy equivalent of about 10 kWh.) Using this information and Table 5.1 on page 46, convert the quantities of both gasoline and electricity into joules and compare the energy savings. Which decision would save the most energy?

Electromagnetic Radiation

Ultimately, most energy on Earth derives from the Sun. The Sun emits **electromagnetic radiation**, a form of energy that includes, but is not limited to, visible light, ultraviolet light, and infrared energy, which we perceive as heat. The scale at the top of **FIGURE 5.1** on page 48 shows these and other types of electromagnetic radiation.

Electromagnetic radiation is carried by **photons**, massless packets of energy that travel at the speed of light and can move even through the vacuum of space. The amount of energy contained in a photon depends on its wavelength—the distance between two peaks or troughs in a wave, as shown in the inset in Figure 5.1 on page 48. Photons with long wavelengths, such as radio waves, have very low energy, while those with short wavelengths, such as X-rays, have high energy. Photons of different wavelengths are used by humans for different purposes. For example, high-energy, short-wavelength X-rays are used for diagnostic medical purposes while lower-energy, long-wavelength infrared rays are used to identify heat loss from buildings during an environmental energy audit.

Potential Energy

Many stationary objects possess a large amount of **potential energy**—energy that is stored but has not yet been released. For example, water impounded

Electromagnetic radiation A form of energy emitted by the Sun that includes, but is not limited to, visible light, ultraviolet light, and infrared energy.

Photon A massless packet of energy that carries electromagnetic radiation at the speed of light.

Potential energy Stored energy that has not been released.

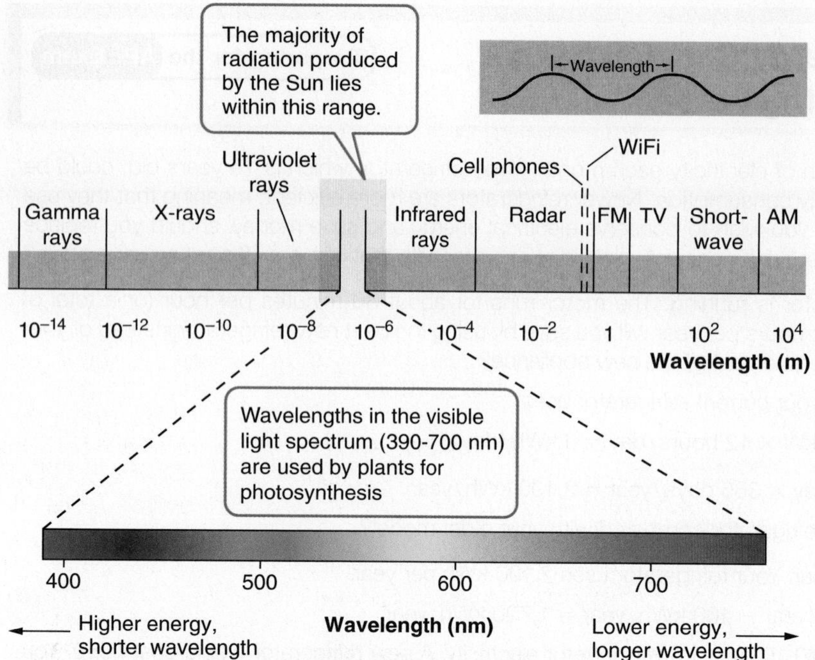

FIGURE 5.1 The electromagnetic spectrum. Electromagnetic radiation can take numerous forms, depending on its wavelength. The Sun releases photons of various wavelengths, but primarily between 250 and 2,500 nanometers (nm).

behind a dam contains a great deal of potential energy. Potential energy stored in chemical bonds is known as **chemical energy**. The energy in food is a familiar example. By breaking down the high-energy bonds in the salad you had for lunch, your body obtains energy to power its activities and functions. Likewise, an automobile engine captures released energy to propel the vehicle—by combusting gasoline in a gasoline engine, or by converting electrical energy in an electric car.

Kinetic Energy

We noted that water impounded behind a dam contains a great deal of potential energy. When the water is released and flows downstream, that potential energy becomes **kinetic energy**, the energy of motion (**FIGURE 5.2**). The kinetic energy of moving water can be captured at a dam and transferred to a turbine and generator, and ultimately to the energy in electricity. Can you think of other common examples of kinetic energy? A car moving down the street, a flying honeybee, and a football traveling through the air all have kinetic energy. Sound also has kinetic energy because it travels in waves through the coordinated motion of atoms. Systems can contain potential energy, kinetic energy, or some of each.

All matter, even the frozen water in the world's ice caps, contains some energy. When we say that energy moves matter, we mean that it is moving the molecules within a substance. The measure of the average kinetic energy of a substance is its **temperature**.

Changes in temperature—and, therefore, in energy—can convert matter from one state to another such as liquid water freezing and becoming ice. At a certain temperature, the molecules in a solid substance start moving so fast that they begin to flow, and the substance melts into a liquid. At an even higher temperature, the molecules in the liquid move still faster, with increasing amounts of energy. Finally, the molecules move with such speed and energy that they overcome the forces holding them together and become gases.

Energy Conversions

Individual organisms rely on a continuous input of energy in order to survive, grow, and reproduce. But interactions beyond the organism can also be seen as a process of converting energy into organized structures such as leaves and branches. Consider a forest ecosystem. Trees absorb water through their roots and carbon dioxide through their leaves. By combining these compounds in the presence of sunlight, they convert water and carbon dioxide into sugars that will provide them with the energy they need. But then a deer grazes on tree leaves, and later a mountain lion eats the deer. At each step, energy is converted by organisms into work. Work is any physical activity to achieve a purpose, such as moving an organism from one place to another.

FIGURE 5.2 Potential and kinetic energy. The water stored behind the Roosevelt Dam in Arizona has potential energy. The potential energy is converted into kinetic energy as the water flows through the dam. *(Richard Kolar/Earth Scenes/Animals Animals)*

Chemical energy Potential energy stored in chemical bonds.

Kinetic energy The energy of motion.

Temperature The measure of the average kinetic energy of a substance.

(a)

(b)

(c)

FIGURE 5.3 The role of energy in a natural system. The amount of energy in a natural system determines which organisms can live in it. (a) A tropical rainforest such as this one in Costa Rica has abundant energy available from the Sun and enough moisture for plants to make use of that energy. (b) Arctic tundra, for example this area in Denali National Park, Alaska, has much less energy available, so plants grow more slowly there and do not reach large sizes. (c) The energy supporting this deep-ocean vent community in the Pacific Ocean comes from chemicals emitted from the vent. Bacteria convert the chemicals into forms of energy that other organisms, such as this giant tube worm (*Riftia pachyptila*), can use. *(a: Steffen Foerster/Shutterstock; b: NancyS/Shutterstock; c: EMORY KRISTOF/National Geographic Creative)*

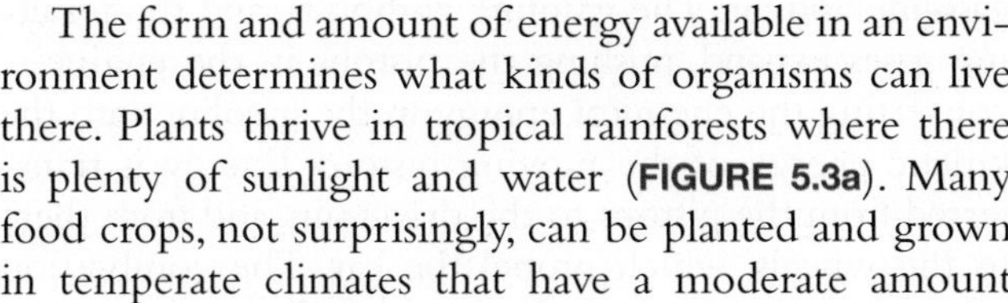

The form and amount of energy available in an environment determines what kinds of organisms can live there. Plants thrive in tropical rainforests where there is plenty of sunlight and water (**FIGURE 5.3a**). Many food crops, not surprisingly, can be planted and grown in temperate climates that have a moderate amount of sunlight. There is less plant productivity and lower species diversity at high latitudes, toward the North and South Poles, where less solar energy is available to organisms (Figure 5.3b). These landscapes are populated mainly by small plants and shrubs, insects, and migrating animals. Plants cannot live at all on the deep

DO THE MATH Comparing Fuel Usage

Preparing for the AP® Exam

Which contains more energy: Fuel usage for a car for 1 year (400 gallons) or the electricity used to power an average U.S. home in the northern part of the United States for 1 year (8,000 kilowatt hours)?

To compare different quantities of energy in different units, in this case gallons and kwh, we convert both to joules.

400 gallons of gasoline

1 gallon gasoline = 132 megajoules energy content

400 ~~gallons~~ × 132 megajoules/~~gallon~~ = 52,800 megajoules of energy content in the gasoline used in 1 year

1 kilowatt-hour = 3.6 megajoules energy content

8,000 ~~kwh~~/year × 3.6 megajoules/~~kwh~~ = 28,800 megajoules in the electricity used in 1 year

Answer: The energy content of the gasoline used in 1 year is greater than the energy content of the electricity used in 1 year.

YOUR TURN Which is greater: the energy contained in 4 metric tons of coal or the energy contained in 1,000 liters of diesel oil?

Note that 1 metric ton of coal = 29,300 megajoules energy equivalent and 1 liter of diesel fuel = 36 megajoules

FIGURE 5.4 Conservation of energy within a system. In an internal combustion gasoline-powered car, the potential energy of gasoline is converted into other forms of energy. Some of that energy leaves the system, but all of it is conserved.

ocean floor, where no solar energy penetrates. The animals that live there, such as eels, anglerfish, and squid, get their energy by feeding on dead organisms that sink from above. Chemical energy, in the form of sulfides emitted from deep-ocean vents, supports a plantless ecosystem that includes sea spiders, 2.4-m (8-foot) tube worms, and bacteria (Figure 5.3c).

The laws of thermodynamics describe how energy behaves

In Chapter 1 we saw that some theories have no known exceptions. Two theories concerning energy fall in this category. The laws of thermodynamics are among the most significant principles in all of science. In environmental science, they describe energy transformation from fuels to useful work, and they dictate why energy transformations result in waste heat and pollution.

First Law of Thermodynamics

Impounded water behind a dam is quiet, still, and unmoving. It doesn't seem like it contains a lot of energy. But in fact that water contains a great deal of potential energy. If you release that water by opening a gate in the dam, the water rushes out. The potential energy of the impounded water becomes the kinetic energy of the water rushing through the gates of the dam. This is an illustration of the **first law of thermodynamics**, which states that energy is neither created nor destroyed but it can change from one form to another.

The first law of thermodynamics dictates that you can't get something from nothing. When an organism needs biologically usable energy, it must convert it from an energy source such as the Sun or food. The potential energy contained in firewood never goes away but is transformed into heat energy permeating a room when the wood is burned in a fireplace. Sometimes it may be difficult to identify where the energy is going, but it is always conserved.

First law of thermodynamics A physical law which states that energy can neither be created nor destroyed but can change from one form to another.

Look at **FIGURE 5.4**, which uses an internal combustion gasoline-powered car to show the first law in action through a series of energy conversions. Think of the car, including its fuel tank, as a system. The potential energy of the fuel (gasoline) is converted into kinetic energy when the battery supplies a spark in the presence of gasoline and air. The gasoline combusts, and the resulting gases expand, pushing the pistons in the engine—converting the chemical energy in the gasoline into the kinetic energy of the moving pistons. Energy is transferred from the pistons to the drivetrain, and from there to the wheels, which propel the car. The combustion of gasoline also produces heat, which dissipates into the environment outside the system. The kinetic energy of the moving car is converted into heat and sound energy as the tires create friction with the road and the body of the automobile moves through the air. When the brakes are applied to stop the car, friction between brake parts releases heat energy. No energy is ever destroyed in this example, but chemical energy is converted into motion, heat, and sound. Notice that some of the energy stays within the system and some energy, for example the heat from burning gasoline, leaves the system.

Second Law of Thermodynamics

We have seen how the potential energy of gasoline is transformed into the kinetic energy of moving pistons in a car engine. But as Figure 5.4 shows, some of that energy is converted into a less usable form—in this case, heat. The heat that is created is called waste heat, meaning that it is not used to do any useful work. And it is inevitable—it is a natural law—that any time there is a conversion of energy from one form to another, some of that energy will be lost as heat. This is one of

(a)

(b)

FIGURE 5.5 Energy efficiency. (a) A modern woodstove, which can heat a room using much less wood, is considerably more energy efficient. (b) The energy efficiency of a traditional fireplace is low because so much heated air can escape through the chimney. *(a: Woodstock Soapstone Co; b: dimetradim/Getty Images)*

the implications of another very important law: The **second law of thermodynamics** tells us that when energy is transformed, the quantity of energy remains the same, but its ability to do work diminishes.

AP® Exam Tip

Make sure you understand the difference between the first law of thermodynamics and the second law of thermodynamics. ●

Energy Efficiency

To quantify the second law of thermodynamics, we use the concept of *energy efficiency*. **Energy efficiency** is the ratio of the amount of energy expended in the desired form to the total amount of energy that is introduced into the system. Two machines or engines that perform the same amount of work, but use different amounts of energy to do that work, have different energy efficiencies. Consider the difference between modern woodstoves and traditional open fireplaces. A woodstove that is 70 percent efficient might use 2 kg of wood to heat a room to a comfortable 20°C (68°F), whereas a fireplace that is 10 percent efficient would require 14 kg of wood to achieve the same temperature—a sevenfold greater energy input (**FIGURE 5.5**).

We can also calculate the energy efficiency of transforming one form of energy into other forms of energy. Let's consider what happens when we convert the chemical energy of coal into the electricity that provides light from a reading lamp and the heat that the lamp releases. **FIGURE 5.6** on page 52 shows the process.

A modern coal-burning power plant can convert 1 metric ton of coal, containing 24,000 megajoules (MJ; 1 MJ = 1 million joules) of chemical energy into about 8,400 MJ of electricity. Since 8,400 is 35 percent of 24,000, this means that the process of turning coal into electricity is about 35 percent efficient. The rest of the energy from the coal—65 percent—is lost as waste heat.

In the electrical transmission lines between the power plant and the house, 10 percent of the electrical energy from the plant is lost as heat and sound, so the transport of energy away from the plant is about 90 percent efficient. We know that the conversion of electrical energy into light in an incandescent bulb is 5 percent efficient; again, the rest of the energy is lost as heat. From beginning to end, we can calculate the energy efficiency of converting coal into incandescent lighting by multiplying all the individual efficiencies:

Calculating energy efficiency:

$$\text{Coal to electricity} \times \text{transport of electricity} \times \text{light bulb efficiency} = \text{overall efficiency}$$

$$0.35 \times 0.90 \times 0.05 = 0.016$$

(1.6% efficiency)

"Do the Math: Comparing the Efficiencies of Two Systems" (see page 52) gives you an opportunity to explore this topic further.

Energy Quality

Most of us have an intuitive sense about the relative effectiveness of various energy sources. For example, we realize that gasoline is a more useful source of

Second law of thermodynamics The physical law stating that when energy is transformed, the quantity of energy remains the same, but its ability to do work diminishes.

Energy efficiency The ratio of the amount of energy expended in the form you want to the total amount of energy that is introduced into the system.

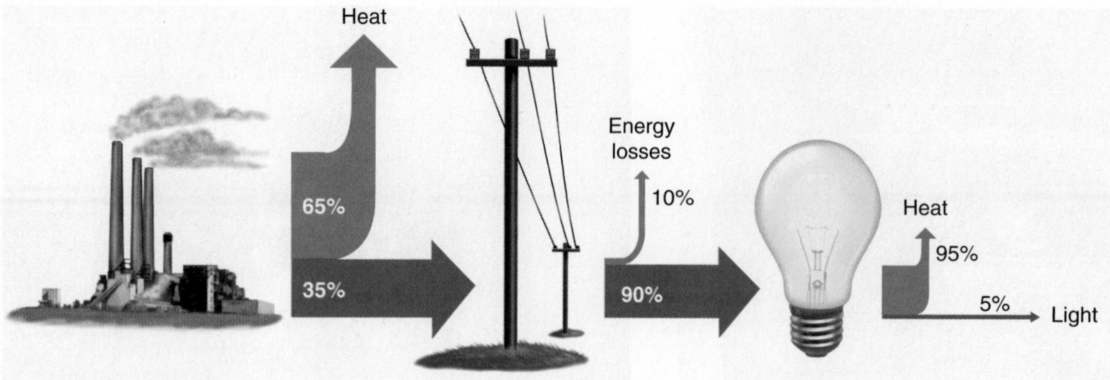

FIGURE 5.6 The second law of thermodynamics. Whenever one form of energy is transformed into another, some of that energy is converted into a less usable form of energy, such as heat. In this example, we see that the conversion of coal into the light of an incandescent bulb is only 1.6 percent efficient.

DO THE MATH Comparing the Efficiencies of Two Systems

Preparing for the AP® Exam

Figure 5.6 illustrates an overall efficiency of 1.6 percent in a system that converts coal to electricity to the light of an incandescent bulb. Calculate the overall efficiencies of the two systems shown in the table.

Power Plant	Transmission	Light bulb
(a) Coal to electricity 35%	90%	Compact fluorescent 20%
(b) Natural gas to electricity 50%	90%	LED (Light emitting diode) 25%

Efficiencies are multiplicative. Rounding to two significant figures:

(a) Coal to electricity: $0.35 \times 0.90 \times 0.20 = 0.06 \times 100\% = 6\%$ efficient

(b) Natural gas to electricity: $0.50 \times 0.90 \times 0.25 = 0.11 \times 100\% = 11\%$ efficient

YOUR TURN The most efficient natural gas to electricity generation plants are 60 percent efficient. The most efficient electric cars are about 60 percent efficient. What is the efficiency of this natural gas to electricity to electric car system?

energy than paper. This difference is a function of each material's **energy quality**. the ease with which an energy source can be used for work. A high-quality energy source has a convenient, concentrated form so that it does not take too much energy to move it from one place to another. Energy quality is one important factor humans must consider when they make energy choices.

Energy quality The ease with which an energy source can be used for work.

Gasoline, for example, is a high-quality energy source because its chemical energy is concentrated (about 44 MJ/kg), and because we have an infrastructure that can conveniently transport it from one location to another. In addition, it is relatively easy to convert gasoline energy into work and heat. Wood, on the other hand, is a lower-quality energy source. It has less than half the energy concentration of gasoline (about 20 MJ/kg) and using it presents more challenges. For example, imagine how difficult it would be to use wood to power an automobile. Clearly, gasoline is a higher-quality energy source than wood.

(a)

(b)

FIGURE 5.7 Energy and entropy. Entropy increases in a system unless an input of energy from outside the system creates order. (a) To reduce the entropy of this messy room, a human must expend energy, which comes from food. (b) A tornado has increased the entropy of this forest system in Wisconsin. *(a: Richard Hutchings/Getty Images; b: AP Photo/The Post Crescent, Dan Powers)*

Entropy

The second law of thermodynamics, which states that all energy conversions result in losses of usable energy, also tells us that all systems move toward randomness rather than toward order. This randomness in a system, called **entropy**, is always increasing unless new energy from outside the system is added to create order.

Think of your bedroom as a system. At the start of the week, your books may be in the bookcase, your clothes may be in the dresser, and your shoes may be lined up in a row in the closet. But what happens if, as the week goes on, you don't expend energy to put your things away (**FIGURE 5.7**)? Unfortunately, your books will not spontaneously line up in the bookcase, your clothes will not fall folded into the dresser, and your shoes will not pair up and arrange themselves in the closet. Unless you bring energy into the system to put things in order, your room will slowly become more and more disorganized.

The energy you use to pick up your room comes from the energy stored in food. Food is a relatively high-quality energy source because the human body easily converts it into usable energy. The molecules of food are ordered rather than random. In other words, food is a low-entropy energy source. Only a small portion of the energy in your digested food is converted into work, however; the rest becomes body heat, which may or may not be needed. This waste heat has a high degree of entropy because heat is the random movement of molecules. Thus, in using food energy to power your body to organize your room, you are decreasing the entropy of the room, but increasing the entropy in the universe by producing waste body heat.

As we know from the Second Law, randomness is always increasing and energy quality is always decreasing. These ideas translate in the real world to the observation that energy always flows from hot to cold. Therefore, a pot of water will never boil without an input of energy, but hot water left alone will gradually cool as its energy dissipates into the surrounding air. This application of the second law is important in many of the global circulation patterns that are powered by the energy of the Sun.

Matter and energy flow in the environment

Why do environmental scientists study whole systems rather than focusing on the individual plants, animals, or substances within a system? Imagine taking apart your cell phone and trying to understand how it works simply by focusing on the microphone. You wouldn't get very far. Similarly, it is important for environmental scientists to look at the whole picture and not just the individual parts of a system in order to understand how that system operates. With a working knowledge of how a system functions, we can predict how changes to any part of the system—for example, fluctuations in the water level at Mono Lake—will change the entire system.

Entropy Randomness in a system.

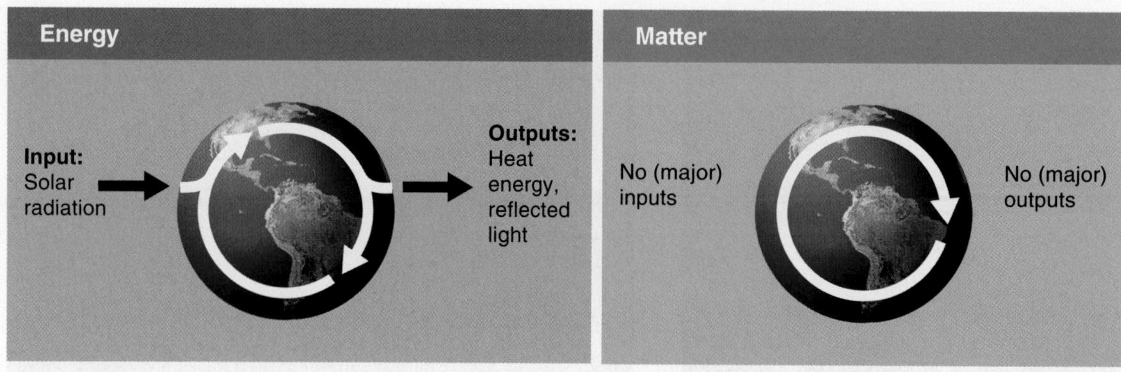

(a) Open system

(b) Closed system

FIGURE 5.8 Open and closed systems. (a) Earth is an open system with respect to energy. Solar radiation enters the Earth system, and energy leaves it in the form of heat and reflected light. (b) Earth is essentially a closed system with respect to matter because very little matter enters or leaves Earth's system. The white arrows indicate the cycling of energy and matter.

Studying systems allows scientists to think about how matter and energy flow in the environment. In this way, researchers can learn about the complex relationships between organisms and the environment. In this section, we will explore system dynamics and changes in systems across space and over time. In each case we will focus on how energy and matter flow in the environment.

System Dynamics

As we suggested at the beginning of this chapter, the Mono Lake ecosystem changes over time. Some years there is more algae growing in the lake and that feeds more brine shrimp. Some years fewer migrating gulls stop over and feed, removing less matter and energy from the system, while in other years, more gulls stop over. Some years there are droughts and in other years, heavy rains. These changing parameters describe the system dynamics of the Mono Lake ecosystem. There are a number of terms used to describe systems and we will present some of them in this section.

Open and Closed Systems

Systems can be either *open* or *closed*. In an **open system**, exchanges of matter or energy occur across system boundaries. Most systems are open. Even at remote Mono Lake, water flows in and birds fly to and from the lake. The ocean is also an open system. Energy from the Sun enters the ocean, warming the waters and providing energy to plants and algae. Energy and matter are transferred from the ocean to the atmosphere as energy from the Sun evaporates the water, giving rise to meteorological events such as tropical storms in which clouds form and send rain back to the ocean surface. Matter, such as sediment and nutrients, enters the ocean from rivers and streams and leaves it through geologic cycles and other processes.

In a **closed system**, matter and energy exchanges do not occur across system boundaries. Closed systems are less common than open systems. Some underground cave systems are almost completely closed systems.

As **FIGURE 5.8** shows, Earth is an open system with respect to energy. Solar radiation enters Earth's atmosphere, and heat and reflected light leave it. But because of its gravitational field, Earth is essentially a closed system with respect to matter. Only an insignificant amount of material enters or leaves the Earth system. All important material exchanges occur within the system.

Inputs and Outputs

By now you have seen numerous examples of both **inputs**, which are additions to a given system, and **outputs**, which are losses from the system. People who study systems often conduct a **systems analysis**, in which they determine inputs, outputs, and changes in the system under various conditions. For instance, researchers studying Mono Lake might quantify the inputs to that system—such as water and salts—and the outputs—such as water that evaporates from the lake and brine shrimp removed by migratory birds. Because no water flows out of the lake, salts are not removed, and even without the aqueduct, Mono Lake, like other terminal lakes, would slowly become saltier.

Steady States

In any given period at Mono Lake, the same amount of water that enters the lake eventually evaporates. In many cases, the most important aspect of conducting

Open system A system in which exchanges of matter or energy occur across system boundaries.

Closed system A system in which matter and energy exchanges do not occur across boundaries.

Input An addition to a system.

Output A loss from a system.

Systems analysis An analysis to determine inputs, outputs, and changes in a system under various conditions.

a systems analysis is determining whether your system is in **steady state**—that is, whether inputs equal outputs, so that the system is not changing over time. This information is particularly useful in the study of environmental science. For example, it allows us to know whether the amount of a valuable resource or a harmful pollutant is increasing, decreasing, or staying the same.

The first step in determining whether a system is in steady state is to measure the amount of matter and energy within it. If the scale of the system allows, we can perform these measurements directly. Consider the leaky bucket shown in **FIGURE 5.9**. We can measure the amount of water going into the bucket and the amount of water flowing out through the holes in the bottom. However, some properties of systems, such as the volume of a lake or the size of an insect population, are difficult to measure directly, so we must calculate or estimate the amount of energy or matter stored in the system. We can then use this information to determine the inputs to and outputs from the system to determine whether it is in steady state.

Many aspects of natural systems, such as the water vapor in the global atmosphere, have been in steady state for at least as long as we have been studying them. The amount of water that enters the atmosphere by evaporation from oceans, rivers, and lakes is roughly equal to the amount that falls from the atmosphere as precipitation. One concern about the effects of global climate change is that some global systems, such as the system that includes water balance in the oceans and atmosphere, may no longer be in steady state. For example, until recently, the oceans have also been in steady state: The amount of water that enters from rivers and streams has been roughly equal to the amount that evaporates into the air. Today, due to an increase in atmospheric temperature, ice on continents is melting more rapidly than new ice is forming, and therefore this system is no longer in steady state.

It's interesting to note that one part of a system can be in steady state while another part is not. Before the Los Angeles Aqueduct was built, the Mono Lake system was in steady state with respect to water but not with respect to salt. The inflow of water equaled the rate of water evaporation but salt was slowly accumulating, as it does in all terminal lakes.

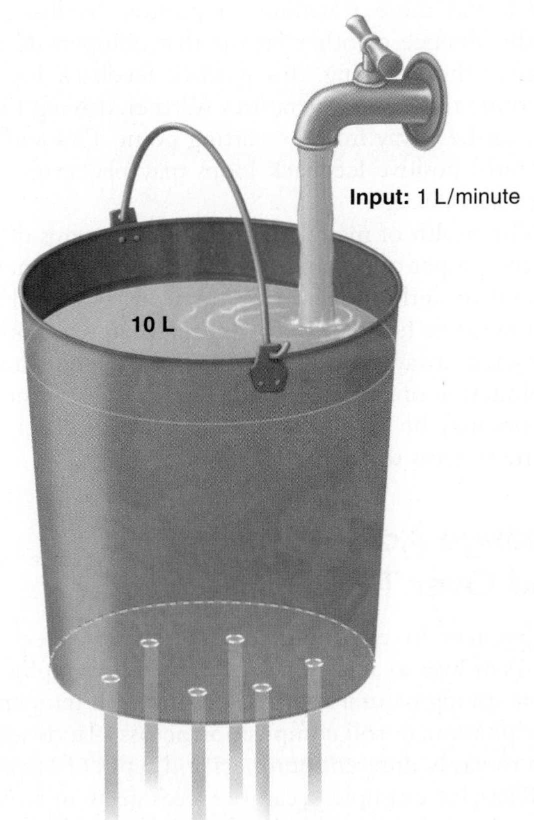

FIGURE 5.9 A system in steady state. In this leaky bucket, inputs equal outputs. As a result, there is no change in the total amount of water in the bucket; the system is in steady state.

Feedbacks

Most natural systems are in steady state. Why? A natural system can respond to changes in its inputs and outputs. For example, during a period of drought, evaporation from a lake will be greater than combined precipitation and stream water flowing into the lake. Therefore, the lake will begin to dry up and the area of the lake will become smaller. Soon there will be less surface water available for evaporation, so the evaporation rate will continue to fall until it matches the new, lower precipitation rate. When this happens, the system returns to steady state, and the lake stops shrinking.

Of course, the opposite is also true. In very wet periods, the size of the lake will grow, and evaporation from the expanded surface area will continue to increase until the system returns to a steady state at which inputs and outputs are equal.

Adjustments in input or output rates caused by changes to a system are called feedbacks; the results of a process feed back into the system to change the rate of that process. Feedbacks, which can be diagrammed as loops or cycles, are found throughout the environment.

Feedback can be either negative or positive. In natural systems, scientists most often observe **negative feedback loops**, in which a system responds to a change by returning to its original state, or by decreasing the

Steady state A state in which inputs equal outputs, so that the system is not changing over time.

Negative feedback loop A feedback loop in which a system responds to a change by returning to its original state, or by decreasing the rate at which the change is occurring.

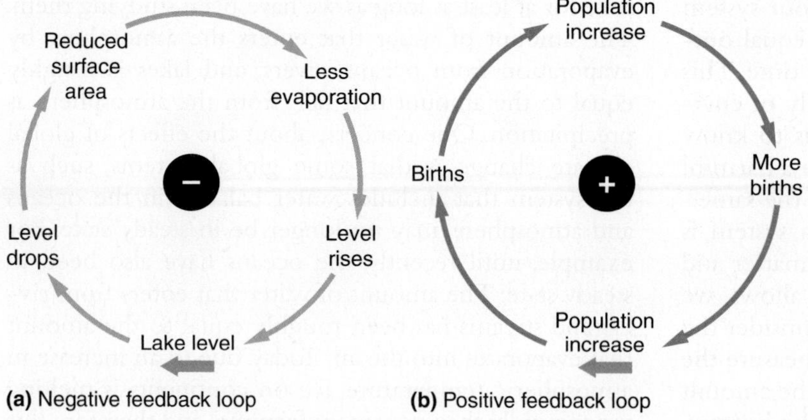

FIGURE 5.10 Negative and positive feedback loops. (a) A negative feedback loop occurs at Mono Lake: When the water level drops, the lake surface area is reduced and evaporation decreases. As a result of the decrease in evaporation, the lake level rises again. (b) Population growth is an example of positive feedback. As members of a species reproduce, they create more offspring that will be able to reproduce in turn, creating a cycle that increases the population size. The green arrow indicates the starting point of each cycle.

rate at which the change is occurring. **FIGURE 5.10a** shows how the negative feedback loop for Mono Lake works: When water levels drop, there is less lake surface area, so evaporation decreases. With less evaporation, the water in the lake slowly returns to its original volume.

Positive feedbacks also occur in the natural world. Figure 5.10b shows an example of how births in a population can give rise to a **positive feedback loop** in which change in a system is amplified. The more members of a species that can reproduce, the more births there will be, creating even more of the species to give birth, and so on.

It's important to note that positive and negative here do not mean good and bad; instead, positive feedback amplifies changes, whereas negative feedback resists changes. People often refer to "the balance of nature." That balance is the logical result of systems reaching a state at which negative feedbacks predominate—although positive feedback loops play important roles in environmental systems as well.

Environmental scientists are especially concerned with the extent to which temperatures on Earth are regulated by feedback loops. Understanding the role of feedback loops in temperature regulation, as well as the types of feedbacks and their scale, can help us make better predictions about climatic changes in the coming decades.

In general, warmer Earth surface temperatures increase the evaporation of water. The additional water vapor that enters the atmosphere by evaporation causes two kinds of clouds to form. Low-altitude clouds reflect sunlight back into space. The result is less heating of Earth's surface, less evaporation, and less warming—a negative feedback loop. High-altitude clouds, on the other hand, absorb terrestrial energy that might otherwise have escaped the atmosphere, leading to higher temperatures near Earth's surface, more evaporation of water, and more warming—a positive feedback loop. In the absence of other factors that compensate for or balance the warming, this positive feedback loop will continue making temperatures warmer, driving the system further away from its starting point. This and other potential positive feedback loops may play critical roles in climate change.

The health of many environmental systems depends on the proper operation of feedback loops. Sometimes, natural or anthropogenic factors lead to a breakdown in a negative feedback loop and drive an environmental system away from its steady state. As you study the exploitation of natural resources, try to determine what factors may be disrupting the negative feedback loops of the systems that provide those resources.

Positive feedback loop A feedback loop in which change in a system is amplified.

Change Across Space and Over Time

Differences in environmental conditions affect what grows or lives in an area, which creates geographic variation among natural systems. Variations in temperature, precipitation, or soil composition across a landscape can lead to vastly different numbers and types of organisms. In Texas, for example, sycamore trees grow in river valleys where there is plenty of water available, whereas pine trees dominate mountain slopes because they can tolerate the cold, dry conditions there. Paying close attention to these natural variations may help us predict the effect of any change in an environment. We know that if the rivers that support the sycamores in Texas dry up, then the trees will probably die.

Natural systems are also affected by the passage of time. Thousands of years ago, when the climate of the Sahara was much wetter than it is today, it supported large populations of Nubian farmers and herders. Small changes in Earth's orbit relative to the Sun, along with a series of other factors, led to the disappearance of monsoon rains in northern Africa. As a result of these changes, the Sahara—now a desert nearly the size of the continental United States—became one of Earth's driest regions (**FIGURE 5.11**). Other, more dramatic changes have occurred on the planet. In the last few million years, Earth has moved in and out of several ice ages; 70 million years ago, central North America was covered by a sea; 240 million years ago, Antarctica was warm enough for 2-meter-long (6.6-foot) salamander-like amphibians to roam its swamps. Natural systems respond to such changes in the global environment with migrations and extinctions of species as well as the evolution of new species.

FIGURE 5.11 The Sahara. The Sahara was once a lush grassland that dried up over time. *(Rachel Carbonell/Getty Images)*

Throughout the history of Earth, small natural changes have had large effects on complex systems, but human activities have increased both the pace and the intensity of these natural environmental changes, as they did at Mono Lake. Studying variations in natural systems over space and time can help scientists learn more about what to expect from the alterations humans are making to the world today.

MODULE 5 AP® Review

Preparing for the AP® Exam

In this module, we have seen that energy is a fundamental component of environmental systems and that there are different forms of energy. The first and second laws of thermodynamics describe energy behavior. Matter and energy flow within and between systems and are subject to feedbacks that regulate and influence the behavior of systems. Natural systems change, sometimes quickly, sometimes slowly, and change can be natural or caused by human beings.

AP® Practice Questions

Choose the best answer for the following.

1. If a solar photovoltaic panel produces 1,000 watts of electrical energy and is active for 12 hours each day, how many kWh of electricity will be produced in a week?
   (a) 63 kWh
   (b) 12 kWh
   (c) 84 kWh
   (d) 70 kWh

2. A car traveling down the highway best represents
   (a) kinetic energy.
   (b) electromagnetic radiation.
   (c) potential energy.
   (d) chemical energy.

3. The concept of energy efficiency is used to quantify
   (a) the first law of thermodynamics.
   (b) the second law of thermodynamics.
   (c) conservation of matter.
   (d) energy quality.

4. A terminal lake like Mono Lake is an example of
   (a) a closed system.
   (b) an open system with only inputs.
   (c) an open system with only outputs.
   (d) an open system with both inputs and outputs.

5. Which of the following will be most likely to return to steady state after a disturbance?
   (a) a system with mostly positive feedback loops
   (b) a system with mostly negative feedback loops
   (c) a system with the same number of positive and negative feedback loops
   (d) an open system with many inputs and outputs

6. Entropy is
   (a) the amount of heat in a system.
   (b) the lowest level of energy quality.
   (c) the ease with which an energy source can be used for work.
   (d) the randomness of a system.

Working Toward Sustainability

Managing Environmental Systems in the Florida Everglades

South Florida's vast Everglades ecosystem extends over 5,000,000 ha (12,400,000 acres). The region, which includes the Everglades and Biscayne Bay national parks and Big Cypress National Preserve, is home to many threatened and endangered bird, mammal, reptile, and plant species, including the Florida panther (*Puma concolor coryi*) and the West Indian manatee (*Trichechus manatus latirostris*). The 400,000 ha (988,000 acre) subtropical wetland area for which the region is best known has been called a "river of grass" because a thin sheet of water flows constantly through it, allowing tall water-tolerant grasses to grow.

A hundred years of rapid human population growth, and the resulting need for water and farmland, have had a dramatic impact on the region. Flood control, dams, irrigation, and the need to provide fresh water to Floridians have led to a 30 percent decline in water flow through the Everglades. Much of the water that does flow through the region is polluted by phosphorus-rich fertilizer and waste from farms and other sources upstream. Cattails thrive on the input of phosphorus, choking out other native plants. The reduction in water flow and water quality is, by most accounts, adversely affecting the Everglades. In 2016, the U.S. Geological Survey (USGS) estimated that 35 percent of the natural habitat in South Florida has been converted to other kinds of land use such as agriculture and urbanization. Altered water flow is the most dramatic effect of human activity. The USGS estimates that water levels have dropped as much as 1.5 to 1.8 m (5 to 6 feet). The Florida Audubon Society, also in 2016, recorded nesting declines in storks, egrets, and herons of between 40 percent and 60 percent compared to the previous decade. Many people have asked if it is possible to maintain this spectacular, natural system while still providing water and other ecosystem services to the people who need it.

River of grass. The subtropical wetland portion of the Florida Everglades has been described as a river of grass because of the tall water-tolerant grasses that cover its surface. *(Panther Media GmbH/Alamy Stock Photo)*

The response of scientists and policy makers has been to treat the Everglades as a set of interacting systems and to manage the inputs and outputs of water and pollutants to those systems. The Comprehensive Everglades Restoration Plan of 2000 is a systems-based approach to the region's problems. It covers 16 counties and 46,600 km^2 (11,500,000 acres) of South Florida. The plan is based on three key steps: increasing water flow into the Everglades, reducing pollutants coming in, and developing strategies for dealing with future problems.

The first step—increasing water flow—will counteract some of the effects of decades of drainage by local communities. Its goal is to provide enough water to support the Everglades' aquatic and marsh organisms. The plan calls for restoring natural water flow as well as natural hydroperiods (seasonal increases and decreases in water flow). Its strategies include removal of over 390 km (240 miles) of inland levees, canals, and water control structures that have blocked this natural water movement.

Water conservation will also be a crucial part of reaching this goal. New water storage facilities and restored wetlands will capture and store water during rainy seasons for use during dry seasons, redirecting much of the 6.4 billion liters (1.7 billion gallons) of fresh water that currently flow to the ocean every day. About 80 percent of this fresh water will be redistributed back into the ecosystem via wetlands and aquifers. The remaining water will be used by cities and farms. The federal and state governments also hope to purchase nearby irrigated cropland and return it to a more natural state. In 2009, for example, the state of Florida purchased 29,000 ha (71,700 acres) of land from the United States Sugar Corporation, the first of a number of actions that could potentially allow engineers to restore the natural flow of water from Lake Okeechobee into the Everglades. Florida has options to purchase even more land from United States Sugar through 2020, but additional purchase agreements have gone through a maze of roadblocks, hurdles, and political debate.

To achieve the second goal—reducing water pollution—local authorities will improve waste treatment facilities and place restrictions on the use of agricultural chemicals. Marshlands are particularly effective at absorbing nutrients and breaking down toxins. Landscape engineers have designed and built more than 21,000 ha (52,000 acres) of artificial marshes upstream of the Everglades to help clean water before it reaches Everglades National Park. Although not all of the region has seen water quality improvements, phosphorus concentrations in runoff from farms south of Lake Okeechobee are lower, meaning that fewer pollutants are reaching the Everglades.

The third goal—to plan for addressing future problems—requires an **adaptive management plan**: a strategy that provides flexibility so that managers can modify it as future changes occur. Adaptive management is an answer to scientific uncertainty. In a highly complex system such as the Everglades, any changes, however well intentioned, may have unexpected consequences. Management strategies must adapt to the actual results of the restoration plan as they occur. In addition, an adaptive management plan can be changed to meet new challenges as they come. One such challenge is global warming. As the climate warms, glaciers melt, and sea levels rise, so much of the Everglades could be inundated by seawater, which would destroy freshwater habitat. Adaptive management essentially means paying attention to what works and adjusting methods accordingly. The Everglades restoration plan has been adjusted numerous times since its inception when results of ongoing observations have led to modifications to the plan. There is a formal mechanism in the plan to ensure that modifications will occur repeatedly over time.

The Everglades plan has its critics. Some people are concerned that control of water flow and pollution will restrict the use of private property and affect economic development, possibly even harming the local economy. Yet other critics fear that the restoration project is underfunded or moving too slowly, and that current farming practices in the region are inconsistent with the goal of restoration.

In spite of its critics, the Everglades restoration plan is, historically speaking, a milestone project, not least because it is based on the concept that the environment is made up of interacting systems.

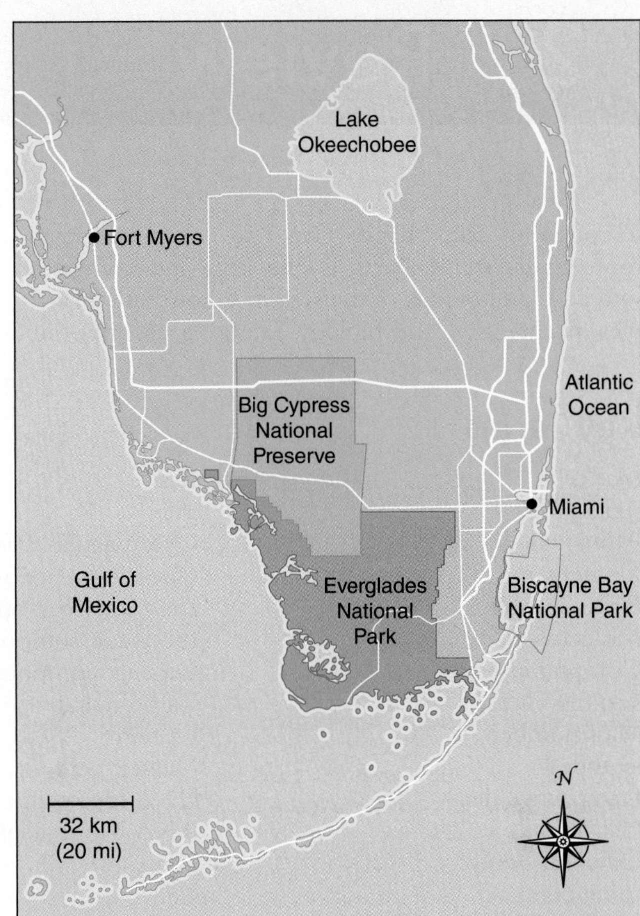

The Florida Everglades Ecosystem. This map shows the locations of Lake Okeechobee and the broader Everglades ecosystem, which includes Everglades and Biscayne Bay national parks and Big Cypress National Preserve.

Critical Thinking Questions

1. Why are the Florida Everglades environmentally significant?
2. How does your understanding of the Florida Everglades change when you think of the Everglades as a set of interacting systems?
3. What are some adaptive management strategies utilized in the Florida Everglades?

References

University of Florida Institute of Food and Agricultural Sciences. "Keeping phosphorus out of the Everglades." ScienceDaily, 24 August 2015. [[www.sciencedaily.com/releases/2015/08/150824102001.htm]].

The Comprehensive Everglades Restoration Plan (CERP) Website. https://www.nps.gov/ever/learn/nature/cerp.htm/. Accessed 15 August 2017.

Adaptive management plan A plan that provides flexibility so that managers can modify it as changes occur.

Chapter

2 Review

Throughout this chapter, we have examined environmental systems. Earth is one large interconnected system. Components of the system follow basic principles of chemistry and biology. Energy is an important component of these systems. Energy conversions are frequently used in systems analysis. Natural systems change over space and time and humans are sometimes major actors in causing system change.

Key Terms

Matter
Mass
Atom
Element
Periodic table
Molecule
Compound
Atomic number
Mass number
Isotopes
Radioactive decay
Half-life
Covalent bond
Ionic bond
Hydrogen bond
Polar molecule
Surface tension
Capillary action
Acid
Base
pH
Ocean acidification
Chemical reaction
Law of conservation of matter
Inorganic compound
Organic compound
Carbohydrate
Protein
Nucleic acid
DNA (deoxyribonucleic acid)
RNA (ribonucleic acid)
Lipid
Cell
Energy
Joule (J)
Power
Electromagnetic radiation
Photon
Potential energy
Chemical energy
Kinetic energy
Temperature
First law of thermodynamics
Second law of thermodynamics
Energy efficiency
Energy quality
Entropy
Open system
Closed system
Input
Output
Systems analysis
Steady state
Negative feedback loop
Positive feedback loop
Adaptive management plan

Learning Goals Revisited

Module 4 Systems and Matter

Describe how matter comprises atoms and molecules that move among different systems.

Matter is composed of atoms, which are made up of protons, neutrons, and electrons. Atoms and molecules can interact in chemical reactions in which the bonds between particular atoms may change.

Explain why water is an important component of most environmental systems.

Water facilitates the transfer of chemical elements and compounds from one system to another. The molecular structure of water gives it unique properties that support the conditions necessary for life on Earth. These properties are essential to physiological functioning of plants and animals and the movement of elements through systems.

Discuss how matter is conserved in chemical and biological systems.

Matter cannot be created or destroyed, but its form can be changed within chemical and biological systems. This is part of the reason we cannot easily dispose of certain chemical compounds, such as hazardous materials.

Module 5 Energy, Flows, and Feedbacks

Distinguish among various forms of energy and understand how they are measured.

Energy can take various forms, including energy that is stored (potential energy) and the energy of motion (kinetic energy). Joules and calories are two important energy units.

Discuss the first and second laws of thermodynamics and explain how they influence environmental systems.

The first law of thermodynamics states that energy cannot be created or destroyed, but it can be converted from one form into another. The second law of thermodynamics states that in any conversion of energy, some energy is converted into unusable waste energy, and the entropy of the universe is increased. The quantities and forms of energy present in various systems influence the types of organisms in those systems.

Explain how scientists keep track of energy and matter inputs, outputs, and changes to environmental systems.

Systems can be open or closed to exchanges of matter, energy, or both. A systems analysis determines what goes into, what comes out of, and what has changed within a given system. Environmental scientists use systems analysis to calculate inputs to and outputs from a system and its rate of change. If there is no overall change, the system is in steady state. Changes in one input or output can affect the entire system.

Practice Math and Graphing

Preparing for the AP® Exam

Answer the following questions. Be sure to show all your work.

1. Practice Math

Wood pellets are a common heating fuel for small wood pellet stoves used in homes. A typical bag of wood pellets weighs 40 pounds and wood pellets typically contain 7,450 BTUs of energy per pound. A typical 18-year old using a rowing machine commonly found in a gym can produce 100 watts while rowing. Compare the energy in a 40-pound bag of wood pellets to the energy produced by rowing for 1 hour.

2. Practice Graphing

Measuring pH allows a researcher to understand the acidity of a water sample. An ocean research vessel cruising in the Pacific Ocean near Hawaii recorded the pH data every year between 1991 and 2015. These data are shown in the table.

(a) Calculate the 5-year average pH values. Take the yearly pH values for each 5-year period and average them. Note this on the chart.

(b) Graph the 5-year averages versus time. Plot pH on y axis and time on x axis. Describe the trend in pH. State whether or not the data from this research vessel support the ocean acidification hypothesis.

Year	pH	5-Year Average
1991	8.16	
1992	8.13	
1993	8.11	
1994	8.09	
1995	**8.09**	
1996	8.08	
1997	8.10	
1998	8.08	
1999	8.07	
2000	**8.09**	
2001	8.08	
2002	8.07	
2003	8.05	
2004	8.07	
2005	**8.08**	
2006	8.09	
2007	8.06	
2008	8.06	
2009	8.07	
2010	**8.06**	
2011	8.06	
2012	8.04	
2013	8.05	
2014	8.06	
2015	**8.03**	

Section 1: Multiple-Choice Questions

Choose the best answer for questions 1–14.

1. Which statement about atoms and molecules is correct?
   (a) The mass number of an element is always less than its atomic number.
   (b) Isotopes are the result of varying numbers of neutrons in atoms of the same element.
   (c) Ionic bonds involve electrons while covalent bonds involve protons.
   (d) Protons and electrons have roughly the same mass.

2. Which does NOT demonstrate the law of conservation of matter?
   (a) $CH_4 + 2\ O_2 \rightarrow CO_2 + 2\ H_2O$
   (b) $NaOH + HCl \rightarrow NaCl + H_2O$
   (c) $2\ NO_2 + H_2O \rightarrow HNO_3 + HNO_2$
   (d) $PbO + C \rightarrow 2\ Pb + CO_2$

3. Pure water has a pH of 7 because
   (a) its surface tension equally attracts acids and bases.
   (b) its polarity results in a molecule with a positive and a negative end.
   (c) its capillary action attracts it to the surfaces of solid substances.
   (d) its H^+ concentration is equal to its OH^- concentration.

4. Which is NOT a type of organic biological molecule?
   (a) lipids
   (b) carbohydrates
   (c) salts
   (d) nucleic acids

5. A wooden log that weighs 1.00 kg is placed in a fireplace. Once lit, it is allowed to burn until there are only traces of ash, weighing 0.04 kg, left. Which best describes the flow of energy?
   (a) The potential energy of the wooden log was converted into the kinetic energy of heat and light.
   (b) The kinetic energy of the wooden log was converted into 0.04 kg of ash.
   (c) The potential energy of the wooden log was converted into 1.00 J of heat.
   (d) The burning of the 1.00 kg wooden log produced 0.96 kg of gases and 0.04 kg of ash.

6. The following table shows the efficiencies of systems providing lighting to three different houses.

	House A (percent efficiency)	House B (percent efficiency)	House C (percent efficiency)
Power plant	55	33	33
Electrical transmission lines	90	90	90
Lighting	5	20	25

   Overall, which represents the most efficient system?
   (a) House A
   (b) House B
   (c) House C
   (d) Houses A and B

7. A researcher conducts an experiment to test how rain of varying acidity will affect the growth of four different plant species, identified as A, B, C, and D. The researcher creates four groups of plants, each with a different treatment. Each group contains 10 individuals of each of the four species. One group is watered daily with a solution of pH 3, a second is watered daily with a solution of pH 4, and a third is watered daily with a solution of pH 6. At the end of 5 days, the growth of each plant is measured, producing the graph below.

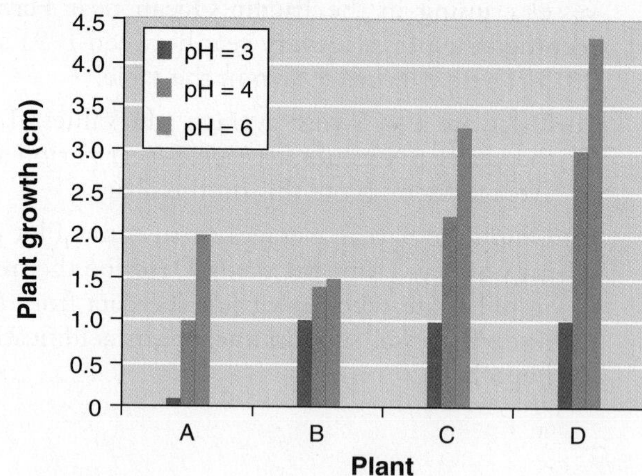

   Which species appears to be least sensitive to the pH of rainfall across the range of exposure?
   (a) species A
   (b) species B
   (c) species C
   (d) species D

8. Sarah is currently spending \$100 per month on electricity. She pays \$0.20 per kWh. Her space heater is responsible for 10 percent of her electricity consumption. How many kilowatt-hours does her space heater use in a month? Assume a month is 30 days.
   (a) 0.07 kWh
   (b) 2 kWh
   (c) 20 kWh
   (d) 50 kWh

9. If the average adult woman consumes approximately 2,000 kcal per day, how long would she need to run in order to utilize 25 percent of her caloric intake, given that the energy requirement for running is 42,000 J per minute? Recall that 1 kcal = 4184 joules.
   (a) 200 minutes
   (b) 50 minutes
   (c) 5 minutes
   (d) 0.05 minutes

10. The National Hurricane Center studies the origins and intensities of hurricanes over the Atlantic and Pacific oceans and it attempts to forecast their tracks, predict where they will make landfall, and assess what damage will result. This systems analysis involves
   (a) changes within a closed system.
   (b) inputs and outputs within a closed system.
   (c) inputs from a closed system and outputs in an open system.
   (d) inputs, outputs, and changes within an open system.

11. Based on the graph below, which is the best interpretation of the data?
   (a) The atmospheric carbon dioxide concentration is in steady state.
   (b) The output of carbon dioxide from the atmosphere is greater than the input into the atmosphere.
   (c) The atmospheric carbon dioxide concentration appears to be decreasing.
   (d) The input of carbon dioxide into the atmosphere is greater than the output from the atmosphere.

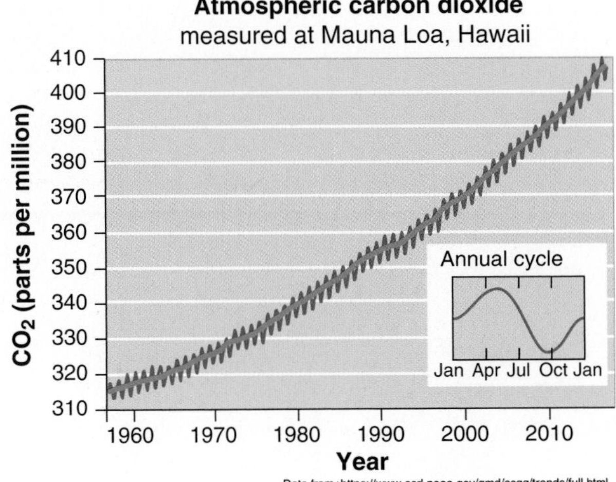

12. Study the diagram below and select the concept it represents.
   (a) a negative feedback loop, because melting of permafrost has a negative effect on the environment by increasing the amounts of carbon dioxide and methane in the atmosphere
   (b) a closed system, because only the concentrations of carbon dioxide and methane in the atmosphere contribute to the permafrost thaw
   (c) a positive feedback loop, because more carbon dioxide and methane in the atmosphere result in greater permafrost thaw, which releases more carbon dioxide and methane into the atmosphere
   (d) an open system that resists change and regulates global temperatures

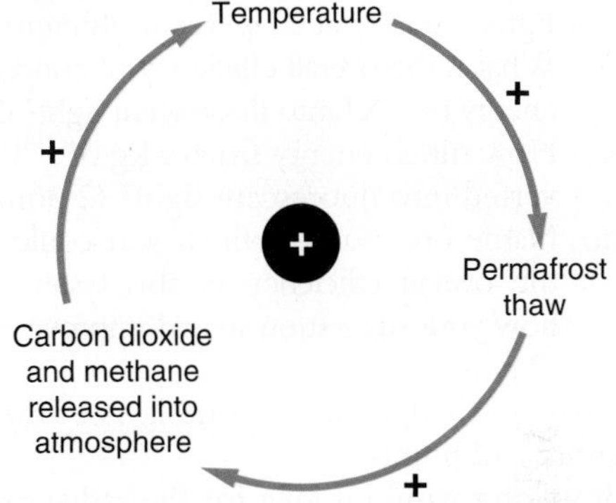

13. Which represents a system in steady state?
   I. The birth of chameleons on the island of Madagascar equals their death rate.
   II. Evaporation from a lake is greater than precipitation and runoff flowing into the lake.
   III. The steady flow of the Colorado River results in more erosion than deposition of rock particles.
   (a) I only
   (b) II only
   (c) III only
   (d) I and II

14. Which statement about the Comprehensive Everglades Restoration Plan is NOT correct?
   (a) Human and natural systems interact because feedback loops lead to adaptations and changes in both systems.
   (b) Improvements in waste treatment facilities and restrictions on agricultural chemicals will reduce the nutrients and toxins in the water that reaches the Everglades.
   (c) Adaptive management will allow for the modification of strategies as changes occur in this complex system.
   (d) The Florida Everglades is a closed system that includes positive and negative feedback loops and is regulated as such.

Section 2: Free-Response Questions

Write your answer to each part clearly. Support your answers with relevant information and examples. Where calculations are required, show your work.

1. The atomic number of uranium-235 is 92, its half-life is 704 million years, and the radioactive decay of 1 kg of ^{235}U releases 6.7×10^{13} J. Radioactive material must be stored in a safe container or buried deep underground until its radiation output drops to a safe level. Generally, it is considered "safe" after 10 half-lives.
   (a) Assume that a nuclear power plant can convert energy from ^{235}U into electricity with an efficiency of 35 percent, the electrical transmission lines operate at 90 percent efficiency, and fluorescent lights operate at 22 percent efficiency.
      (i) What is the overall efficiency of converting the energy of ^{235}U into fluorescent light? (2 points)
      (ii) How much energy from 1 kg of ^{235}U is converted into fluorescent light? (2 points)
      (iii) Name one way in which you could improve the overall efficiency of this system. Explain how your suggestion would improve efficiency. (2 points)
   (b) What are the first and second laws of thermodynamics? (2 points)
   (c) How long would it take for the radiation from a sample of ^{235}U to reach a safe level? (2 points)

2. Because of human activity, there has been an increase in the concentration of carbon dioxide in the atmosphere.
   (a) Explain how the increased carbon dioxide influences ocean chemistry in three steps. (3 points)
   (b) In the following reaction, calcium carbonate, like that used to make the shells of many marine organisms, reacts with hydrogen:

$$CaCO_3 + H^+ \rightarrow Ca^{2+} + HCO_3^-$$

   In that equation, calcium (Ca^{2+}) and carbonate (CO_3^{2-}) are held together by an ionic bond. What is an ionic bond, and how does it affect the solubility (ability to be dissolved) of a compound compared to covalent bonds? (2 points)
   (c) Mussels are marine organisms with a hard shell of calcium carbonate. A scientist notices that the population of a particular species of mussel in Alaska has been declining. What do you hypothesize is happening? (1 point)
   (d) Design an experiment to test your hypothesis from part (c). (3 points)
   (e) Gases are often more soluble (easily dissolved into water) at lower temperatures. If this is the only factor influencing shelled marine organism abundance, how do you think the population of the organism in part (c) will differ off the coast of Hawaii versus Alaska? (1 point)

Tracy Packer Photography/Getty Images

science applied 1

Where Is the Missing Salt of Mono Lake?

As we discussed at the beginning of Chapter 2, in 1941 the City of Los Angeles redirected water away from Mono Lake in California to meet the needs of its growing population. Before this happened, approximately 120 billion liters of stream water (31 billion gallons) flowed into Mono Lake in an average year. As the city diverted a portion of the lake's incoming water over four decades, the lake dropped to half its prior average depth and the lake's salt concentration increased, which caused a decline in the algae and animals that lived in the lake and other animals that fed on them.

To understand how water diversion alters water depth and salt accumulation, ecosystem scientists had to examine the flows of water and salts into the lake from streams and the loss of water due to evaporation out of the lake (**FIGURE SA1.1**). They made a series of observations and drew conclusions that helped them account for the movement of water, but they could not account for the movement of salt. In short, they discovered that more salt came into the lake than existed in the water of the lake. The question then arose, where is the missing salt?

FIGURE SA1.1 Research at Mono Lake. Scientists collect water sample at Mono Lake to monitor the concentrations of salt and nutrients in the water. *(Henry Bortman/NASA)*

The balanced inputs and outputs of water

As we mentioned at the beginning of Chapter 2, Mono Lake is categorized as a terminal lake because water flows into the lake from rivers, streams, and from precipitation, but water does not flow out. In a typical year before Los Angeles began diverting water, the lake level did not rise or fall. Given that this lake has no outlets, water outputs must equal water inputs. This equilibrium suggests that the amount of water entering the lake was being balanced by the amount of water leaving the lake by evaporation. If this was not the case, the lake would eventually either dry out or overflow its banks.

The balanced inputs and outputs of salt

While the balance of water in the lake is straightforward, the balance of salt in the lake is much more interesting. Because all streams contain some amount of salt water, the stream water that entered Mono Lake also contained salt. Salt concentrations of the stream water flowing into Mono Lake can vary through the seasons, but a typical concentration is 50 mg per liter, which is equivalent to 50 parts per million.

To calculate the total amount of salt that historically entered Mono Lake each year, we can multiply the concentration of salt in the stream water (50 mg per liter) by the number of liters of water flowing into the lake, prior to being diverted by the City of Los Angeles (120 billion liters per year):

$$50 \text{ mg/L salt} \times 120 \text{ billion L/year} = 6 \text{ trillion mg salt/year}$$

We can then convert from milligrams to kilograms:

$$6 \text{ trillion mg salt/year} \times \frac{1 \text{ million kg}}{1 \text{ trillion mg}} = 6 \text{ million kg salt/year}$$

This is the annual input of salt to Mono Lake. Scientists believe that no water has flowed out of the Mono Lake basin since it was formed about 120,000 years ago. If we assume that Earth's climate hasn't changed significantly over that time and that water inputs to Mono Lake have not changed drastically over that period, how much salt should be in the water of Mono Lake?

At the historic input rate of 6 million kg salt/year, the total amount of salt in the water of Mono Lake today should be:

$$6 \text{ million kg/year} \times 120{,}000 \text{ years} = 720 \text{ billion kg of dissolved salt}$$

However, the lake today only contains about 285 billion kg of dissolved salt, which is 435 billion kg less salt than predicted. This made scientists wonder about the fate of the huge amount of unaccounted salt that entered the lake over the past 120,000 years.

The lake's tall tufa towers, prominently featured in the photograph at the beginning of Chapter 2, hold the answer. The amount of salt that can remain dissolved in water is limited. As the concentrations of salt (primarily calcium) increase over time, the water reaches this limit. When the water can hold no more dissolved salt, the calcium precipitates out of solution by combining with carbonate ions to form calcium carbonate, which is a form of limestone. Over thousands of years, the precipitating calcium carbonate creates the tufa towers. In this way, the excess salt has been removed from the lake water, but not from the Mono Lake system as a whole. If we account for the amount of salt in the water and in the tufa towers, we can balance the amount of salt coming into the lake and leaving the lake water. **FIGURE SA1.2** summarizes these inputs to and outputs from the Mono Lake system.

Input:
Stream water carries dissolved salts.
Output:
Water evaporates, leaving salts behind.

FIGURE SA1.2 The Mono Lake System. In this terminal lake system, inputs are from stream water while outputs are evaporated water only. In contrast, all salts that enter the lake from the surrounding streams remain in the lake.

The research on Mono Lake is an excellent example of how environmental scientists make observations and develop hypothesis for how the natural world works. By examining the inputs and outputs of both salt and water, they were able to determine how human diversion of essential water has a cascading effect on water levels, salt concentrations, and the consequences to the algae and animals that depend on the lake.

Questions

1. How did Los Angeles inadvertently conduct an experiment at Mono Lake?
2. What chemical principle causes terminal lakes to become more salty?
3. What is the reason for the discrepancy between the two calculations of salt content in Mono Lake?

Preparing for the AP® Exam

Practice AP® Free-Response Question

Write your answer to each part clearly. Support your answers with relevant information and examples. Where calculations are required, show your work.

Water that flows into Mono Lake contains a much smaller concentration of salt than the water already in the lake. This inflowing water tends to stratify, or float on top of existing water, because fresh water is less dense that salt water. As salt from the lower layer dissolves into the upper layer, nutrients from the bottom of the lake also rise to the surface. This exchange of nutrients is critical for the growth of algae in the surface waters. Recent research suggests that the reduction of water diversion from Mono Lake had unexpected results.

In 1995, the reduction of stream diversions from Mono Lake, combined with greater than average quantities of fresh water from snowmelt runoff, led to a rapid rise in lake water level. The large volume of fresh water from streams led to a long-term stratification of the lake, with fresh water on the surface and salt water on the bottom. Relative to data taken before the initial stream diversions in 1941, the current stratification has reduced the rate at which nutrients rise from the bottom of the lake. Long-term projections, based on mathematical models, suggest that the current degree of stratification will persist for decades.

(a) List three potential consequences of reduced lake mixing. (3 points)
(b) Describe two adaptive management strategies that could reduce lake stratification in Mono Lake. (3 points)
(c) What is the chemical property of water that allows salt to dissolve? (2 points)
(d) Why would the mixing of salt water with fresh water be considered an example of increased entropy? (2 points)

Section 1: Multiple-Choice Questions

Choose the best answer for questions 1–20.

1. Which best describes how humans have altered natural systems?
   I. overhunted many large mammals to extinction
   II. created habitat for species to thrive
   III. emitted greenhouse gases
   (a) I only
   (b) II and III only
   (c) I and III only
   (d) I, II, and III

2. Which does NOT describe a benefit of biodiversity?
   (a) Genetic biodiversity improves the ability of a population to cope with environmental change.
   (b) Ecosystems with higher species diversity are more productive.
   (c) Species serve as environmental indicators of problems on a global scale.
   (d) Speciation reduces natural rates of species extinction.

3. Which is NOT a consequence of human population growth?
   (a) depletion of natural resources
   (b) background extinction
   (c) emission of greenhouse gases
   (d) rise in sea level

4. An example of sustainable development is
   (a) harvesting enough crops to provide the basic needs of all humans.
   (b) increasing the price of vegetables.
   (c) reducing the use of all major modes of transportation.
   (d) creating renewable sources of construction material.

5. The ecological footprint of a human is
   (a) a measure of how much a human consumes, expressed in joules.
   (b) a measure of human consumption, expressed in area of land.
   (c) a measure of biodiversity loss stemming from industrial processes.
   (d) a measure of plant biomass removed by a farmer.

6. The greatest value of the scientific method is best stated as:
   (a) The scientific method permits researchers a rapid method of disseminating findings.
   (b) The scientific method removes bias from observation of natural phenomenon.
   (c) The scientific method allows findings to be reproduced and tested.
   (d) The scientific method promotes sustainable development.

7. Researchers conducted an experiment to test the hypothesis that the use of fertilizer near wetlands is associated with increased growth of algae. An appropriate null hypothesis would be:
   (a) Growth of algae in wetlands is never associated with increased fertilizer use.
   (b) Application of fertilizers near wetlands is always associated with increased growth of algae.
   (c) Fertilizer use near wetlands has no association with growth of algae.
   (d) Fertilizer use near wetlands leads to increased growth of algae as a result of elevated nutrient concentrations.

8. Which is an example of a null hypothesis?
   (a) Plants grow faster when exposed to classical music.
   (b) If you smoke several cigarettes a day, you are more likely to get lung cancer.
   (c) The ability to sing in tune is unaffected by age.
   (d) Egg size is influenced by female body mass.

9. Which challenge is unique to environmental science?
   I. lack of baseline data
   II. sample size
   III. objectivity
   (a) I
   (b) II
   (c) III
   (d) I and II

10. Which constitutes baseline data on the effects of humans on natural ecosystems?
    (a) concentrations of atmospheric CO_2 before humans existed
    (b) current rates of species extinction
    (c) global rate of freshwater consumption from 1900 to 2010
    (d) average plant productivity on a remote island uninhabited by humans

11. During radioactive decay
    I. there is a release of material from the nucleus of unstable isotopes.
    II. there is a change in the half-life of an element.
    III. an element is changed into a different element.
    (a) I only
    (b) II only
    (c) I and II
    (d) I and III

12. The mass number of the element selenium is 80 and the atomic number is 34. How many neutrons does selenium have?
    (a) 114
    (b) 80
    (c) 46
    (d) 34

13. The upward movement of water through soil is an example of
    (a) capillary action.
    (b) ionic bonding.
    (c) covalent bonding.
    (d) surface tension.

14. Which list contains only organic material?
    (a) proteins, lipids, salts
    (b) dead trees, decomposing leaves, earthworms
    (c) cellulose, ethanol, calcium chloride
    (d) NH_3, NaOH, NO_2^-

15. A grasshopper can extract energy from ingested food at an efficiency of 10 percent. If the grasshopper consumes 10 Calories of food and uses the energy content of that food during 1 minute, how much energy did it exert per second?
    (a) 70 J
    (b) 700 J
    (c) 7,000 J
    (d) 10,000 calories

16. Which is NOT involved when driving an internal combustion (conventional) car?
    (a) potential energy
    (b) kinetic energy
    (c) waste energy (heat)
    (d) electromagnetic energy

17. When the seed pods of a pea plant dry in the sun, the skin of the pods exert inward pressure on the encased seeds. This provides the seeds with potential energy that is converted to kinetic energy when the pod is ruptured and the seeds shoot far distances. A researcher claims that the seed's potential energy is converted to kinetic energy with 100 percent efficiency. This result would violate
    (a) the law of conservation of matter.
    (b) the law of conservation of energy.
    (c) the first law of thermodynamics.
    (d) the second law of thermodynamics.

18. A researcher team does an experiment to test how watching horror movies might affect heart rate. They attach a heart rate monitor to each of 20 subjects. Each subject's resting heart rate is measured before watching the movie and again while watching a clip from a horror movie in a dark room. The researchers do the same with another 20 subjects, but in a well-lit room. In this experiment, which is the dependent variable?
    (a) human subjects
    (b) heart rate
    (c) horror movies
    (d) darkness

For question 19, refer to the following table, which documents the material inputs and outputs to a 100-m section of forest stream

Type of matter	Headwater inputs	Downstream outputs
Leaf litter	250 g/m^2	100 g/m^2
Woody debris	100 g/m^2	90 g/m^2
Dead insects	1 g/m^2	0.5 g/m^2
Stream sediment	10 g/m^2	3.5 g/m^2
Fish	30 g/m^2	150 g/m^2
Insects	5 g/m^2	52 g/m^2

19. When leaf litter inputs to the stream decrease, the amount of fish and insect biomass leaving the downstream section decreases by a similar amount. This represents
    (a) a negative feedback loop.
    (b) conservation of potential energy.
    (c) a positive feedback loop.
    (d) a decrease in entropy.

20. The reaction of sodium hydroxide, NaOH, and hydrochloric acid, HCl, results in the following reaction: $NaOH + HCl \rightarrow NaCl + H_2O$. This product represents
    (a) an acidic product.
    (b) a basic product.
    (c) a pH-neutral product.
    (d) the formation of both an inorganic and organic compound.

Section 2: Free-Response Questions

Write your answer to each part clearly. Support your answers with relevant information and examples. Where calculations are required, show your work.

1. A housing development is built in the middle of a large forested area that was previously a reserve. The surrounding area contains several lakes and rivers.
   (a) Name TWO potential ecosystem services provided by the area surrounding the development. (2 points)
   (b) Is the area encompassing the development and forest an open or closed system with regards to both energy and matter? Explain your answer. (2 points)
   (c) A team of scientists wants to assess the development's impact on the surrounding ecosystem. Describe TWO environmental indicators they can use to conduct the assessment. (2 points)
   (d) Many of the houses have fireplaces that are 15 percent efficient and use 15 kg of wood to heat a room to 20°C (68°F). Wanting to reduce their ecological footprint, several of the homeowners are considering switching to woodstoves that are 75 percent efficient. If they do change to using a woodstove, how many kilograms of wood will they save per use? (2 points)
   (e) Define "ecological footprint." Identify and describe another action, other than investing in more efficient heating, that the homeowners could utilize to reduce their ecological footprint. (2 points)

2. Approximately 72 billion liters of milk are produced each year in the United States, from 8 million cows. On average, a single cow consumes 13,500 kg of corn feed each year. It requires 40 MJ to produce a kilogram of corn feed, which contains 20 MJ of energy. There are 15 MJ of energy in a single liter of milk.
   (a) Calculate the energy efficiency of growing corn and converting it into milk. (4 points)
   (b) Describe two processes that reduce efficiency of milk production. Consider the entire process of milk production from the growth of cattle feed to the collection of milk. (2 points)
   (c) To increase the energy efficiency of milk production farmers can harvest the fecal waste (manure) from cows and use the gas it produces as a source of energy.
      (i) Name the main chemical in gas produced by cow manure that can be used as a source of energy. (1 point)
      (ii) Explain how energy derived from this compound at the molecular level. (1 point)
      (iii) If 10 percent of the food energy not used by cows could be captured as chemical energy from gas released by manure, what would be the energy efficiency of converting corn into milk? (2 points)

These people are planting trees in Haiti as part of the Eden Reforestation Project.
(Eden Reforestation Projects)

Ecosystem Ecology

CHAPTER 3

CASE STUDY

Reversing the Deforestation of Haiti

Even before the devastating earthquake of 2010, life in Haiti was hard. On the streets of the capital city, Port-au-Prince, people lined up to buy charcoal to cook their meals. According to the United Nations, 76 percent of Haitians lived on less than $2 a day. Because other forms of cooking fuel, including oil and propane, were too expensive, people turned to the forests, cutting trees to make charcoal from firewood.

The deforestation of Haiti has a long history that goes back to the seventeenth and eighteenth centuries when much of the forest was cleared for lumber, fuel, and agriculture. By 1923, only 60 percent of this mountainous country was covered in forest. However, as agriculture expanded with new plantations of rubber trees in the 1940s and as the population grew with a greater demand for fuel, the amount of forest declined even more. By 2017, with more than 11 million people living in this small nation, only about 30 percent of its land remained forested.

Deforestation disrupts the services that living trees provide in an ecosystem. When Haitian forests are cleared of trees the land becomes much more susceptible to erosion. As the tree roots die, they can no longer stabilize the soil, so the soil is eroded away by the heavy rains of tropical storms and hurricanes. Unimpeded by vegetation, the rainwater runs quickly down the mountainsides, dislodging the topsoil that is so important for forest growth. In addition, this oversaturation of the soil causes massive mudslides that can destroy entire villages.

By 2017, with more than 11 million people living in this small nation, only about 30 percent of its land remained forested.

But the news from Haiti is not all bad. Since the 1980s, the U.S. Agency for International Development, in cooperation with many other groups, has provided funds to plant 60 million trees in Haiti. Unfortunately, simply planting trees is not enough; the local people can't afford to let trees grow when they are in desperate need of firewood and charcoal. One solution has been to plant mango trees (*Mangifera indica*). Because a mature mango tree can provide $70 to $150 worth of mangoes annually, this value provides an economic incentive to let the trees grow to maturity. The deforestation problem is also being addressed through efforts to develop alternative fuel sources, such as discarded paper that is processed into dried cakes that can be burned. In 2013, the president of Haiti announced a new program to plant 50 million trees every year, with the goal of reversing deforestation. Unfortunately, hurricanes are common in the Caribbean and in 2016 Hurricane Matthew passed over the island with 225 km/hour (140 mph) winds that devastated the newly planted forests and many of the island's agricultural crops.

Extensive forest removal is a problem in many developing nations. Widespread removal of trees on mountains causes rapid soil erosion and substantial disruptions of the natural cycles of water and soil nutrients. This leads to long-term degradation of the environment. The results not only illustrate the connectedness of ecological systems, but also show how forest ecosystems, like all ecosystems, can be influenced by human decisions.

Sources: "Haitians seek remedies for environmental ruin," National Public Radio, July 15, 2009, http://www.npr.org/templates/story/story.php?storyId=104684950; "Haiti to plant millions of trees to boost forests and help tackle poverty," *The Guardian*, March 28, 2013, http://www.theguardian.com/world/2013/mar/28/haiti-plant-millions-trees-deforestation; "Who will speak for Haiti's trees?" *New York Times*, October 17, 2016, https://www.nytimes.com/2016/10/18/opinion/who-will-speak-for-haitis-trees.html

The story of deforestation in Haiti reminds us that all the components of an ecosystem are interrelated. As we noted in Chapter 1, an ecosystem is a particular location on Earth distinguished by its particular mix of interacting biotic and abiotic components. A forest, for example, contains many interacting biotic components, such as trees, wildflowers, birds, mammals, insects, fungi, and bacteria, that are quite distinct from those found in a grassland. Ecosystems also have abiotic components such as sunlight, temperature, soil, water, pH, and nutrients. The abiotic components of the ecosystem help determine which organisms can live there. In this chapter, we will see that ecosystems control the movement of the energy, water, and nutrients organisms must have to grow and reproduce. By understanding the processes that determine these movements, environmental scientists can better understand how human activities alter these processes and understand how to reduce such negative impacts.

MODULE 6

The Movement of Energy

To understand how an ecosystem functions and the relationship of its biotic and abiotic components, we must first study how energy moves through the ecosystem. In this module we will examine how we delineate ecosystems so that we can examine the movement of energy. We will then consider how photosynthesis and respiration capture and release energy. Finally, we will look at how energy moves through the different components of the ecosystem.

Learning Goals

After reading this module you should be able to

- explain the concept of ecosystem boundaries.
- describe the processes of photosynthesis and respiration.
- distinguish among the trophic levels that exist in food chains and food webs.
- quantify ecosystem productivity.
- explain energy transfer efficiency and trophic pyramids.

Ecosystem boundaries are not clearly defined

The characteristics of any given ecosystem are highly dependent on the climate that exists in that location on Earth. For example, ecosystems in the dry desert of Death Valley, California, where temperatures may reach 50°C (120°F), are very different from those on the continent of Antarctica, where temperatures may drop as low as −85°C (−120°F). Similarly, water can range from being immeasurable in deserts to being the defining feature of the ecosystem in lakes and oceans. On less extreme scales, small differences in precipitation and the ability of the soil to retain water can favor different terrestrial ecosystem types. Regions with greater quantities of water in the soil can support trees, whereas regions with less water in the soil can support only grasses.

The biotic and abiotic components of an ecosystem provide the boundaries that distinguish one ecosystem from another. Some ecosystems have well-defined boundaries, whereas others do not. A cave, for example, is a well-defined ecosystem (**FIGURE 6.1**). It contains identifiable biotic components, such as animals and microorganisms that are specifically adapted to live in a cave environment, as well as distinctive abiotic components, including temperature, salinity, and water that flows through the cave as an underground stream. Roosting bats fly out of the cave each night and consume insects. When the bats return to the cave and defecate, their feces provide energy that passes through the relatively few animal species that live in the cave. In many caves, for example, small invertebrate animals consume bat feces and are in turn consumed by cave salamanders.

The cave ecosystem is relatively easy to study because its boundaries are clear. With the exception of the bats feeding outside the cave, the cave ecosystem is easily defined as everything from the point where the stream enters the cave to the point where it exits. Likewise, many aquatic

FIGURE 6.1 A cave ecosystem. Cave ecosystems, such as this one in Tanzania with emerging bats, typically have distinct boundaries and are home to highly adapted species. *(Michele Menegon/ardea.com)*

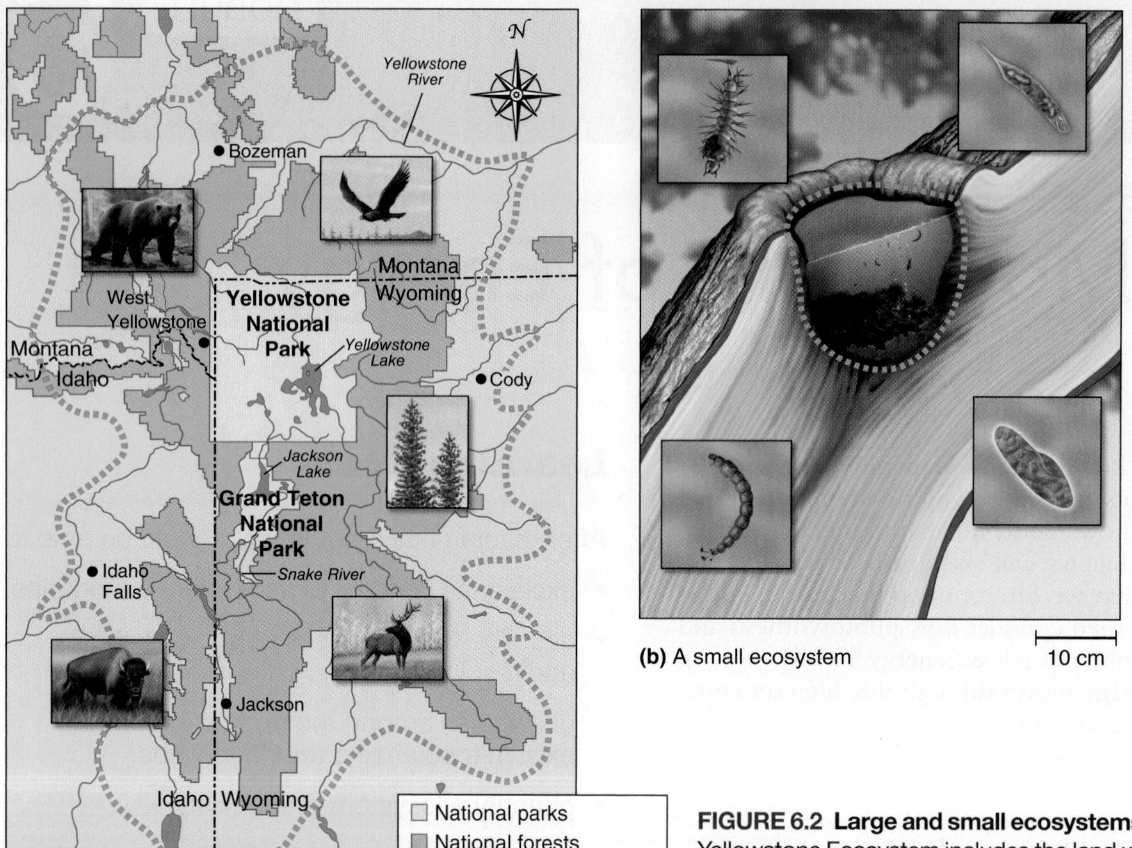

(a) The Greater Yellowstone Ecosystem

(b) A small ecosystem

FIGURE 6.2 Large and small ecosystems. (a) The Greater Yellowstone Ecosystem includes the land within Yellowstone National Park and many adjacent properties. (b) Some ecosystems are very small, such as a rain-filled tree hole that houses a diversity of microbes and aquatic insects.

ecosystems, such as lakes, ponds, and streams, are relatively easy to define because the ecosystem's boundaries correspond to the boundaries between land and water. Knowing the boundaries of an ecosystem makes it easier to identify the system's biotic and abiotic components and to trace the cycling of energy and matter through the system.

In most cases, however, determining where one ecosystem ends and another begins is difficult. For this reason, ecosystem boundaries are often subjective. Environmental scientists might define a terrestrial ecosystem as the range of a particular species of interest, such as the area where wolves roam, or they might define it by using topographic features, such as two mountain ranges enclosing a valley. The boundaries of some managed ecosystems, such as national parks, are set according to administrative rather than scientific criteria. Yellowstone National Park, for example, was once managed as its own ecosystem until scientists began to realize that many species of conservation interest, such as grizzly bears (*Ursus arctos horribilis*), spent time both inside and outside the park, despite the park's massive area of 898,000 ha (2.2 million acres). To manage these species effectively, scientists had to think much more broadly; they had to include nearly 20 million ha (50 million acres) of public and private land outside the park. This larger region, shown in **FIGURE 6.2a**, was named the Greater Yellowstone Ecosystem. As the name suggests, the actual ecosystem extends well beyond the administrative boundaries of the park.

As we saw in Chapter 2, not all ecosystems are as vast as the Greater Yellowstone Ecosystem. Some can be quite small, such as a water-filled hole in a fallen tree trunk, illustrated in Figure 6.2b. Such tiny ecosystems include all the physical and chemical components necessary to support a diverse set of species, although the species are typically very small as well, including microbes, mosquito larvae, and other insects.

Although it is helpful to divide locations on Earth into distinct ecosystems, it is important to remember that each ecosystem interacts with surrounding ecosystems through the exchange of energy and matter. Organisms such as bats—which fly in and out of caves—move across ecosystem boundaries, as do chemical elements, such as carbon or nitrogen dissolved in water. As a result, changes in any one ecosystem can ultimately have far-reaching effects on the global environment.

The combination of all ecosystems on Earth forms the **biosphere**, which is the region of our planet where

Biosphere The region of our planet where life resides, the combination of all ecosystems on Earth.

life resides. The biosphere is a 20-km (12-mile) thick layer around Earth between the deepest ocean bottom and the highest mountain peak.

Photosynthesis captures energy and respiration releases energy

To understand how ecosystems function and how best to protect and manage them, ecosystem ecologists also study the processes that move energy and matter within an ecosystem. To understand energy relationships, we need to look at the way energy flows across an ecosystem.

Consider the Serengeti Plain in East Africa (**FIGURE 6.3**). Plants, such as grasses and acacia trees, absorb energy directly from the Sun. That energy spreads throughout an ecosystem as plants are eaten by animals, such as gazelles, and the animals are subsequently eaten by predators, such as cheetahs. There are millions of herbivores, such as zebras and wildebeests, in the Serengeti ecosystem, but there are far fewer carnivores, such as lions (*Panthera leo*) and cheetahs (*Acinonyx jubatus*), that feed on herbivores. In accordance with the second law of thermodynamics, when one organism consumes another, not all of the energy in the consumed organism is transferred to the consumer. We can think about this using an example of a cheetah catching a gazelle. After the cheetah catches a gazelle, it does not consume a large portion of the total energy available in the gazelle's body, for example the energy in the bones of the gazelle. Of the portion that is consumed, some is not digestible, and is excreted by the cheetah. Of the portion that is digested, some is used to generate heat and the rest is used for the cheetah's growth and reproduction. In the end, only a small portion of the total energy available in the gazelle's body is converted into an increase in the mass of the cheetah's body. Let's trace this energy flow in more detail by looking at the processes of *photosynthesis* and *cellular respiration*.

FIGURE 6.3 The flow of energy in the Serengeti ecosystem of Africa. The Serengeti ecosystem has more plants than herbivores, and more herbivores than carnivores. *(MICHAEL NICHOLS/National Geographic Creative)*

Photosynthesis

Nearly all of the energy that powers ecosystems comes from the Sun as solar energy, which is a form of kinetic energy. Plants, algae, and some bacteria that use the Sun's energy to produce usable forms of energy are called **producers**, or **autotrophs**. As you can see in the top half of **FIGURE 6.4**, these producers use **photosynthesis**, which means they use solar energy to convert carbon dioxide (CO_2) and water (H_2O) into glucose ($C_6H_{12}O_6$). Glucose is a form of potential energy that can be used by a wide range of organisms. The photosynthesis process also produces oxygen (O_2) as a waste product. That is why plants and other producers are beneficial to our atmosphere; they produce the oxygen we need to breathe.

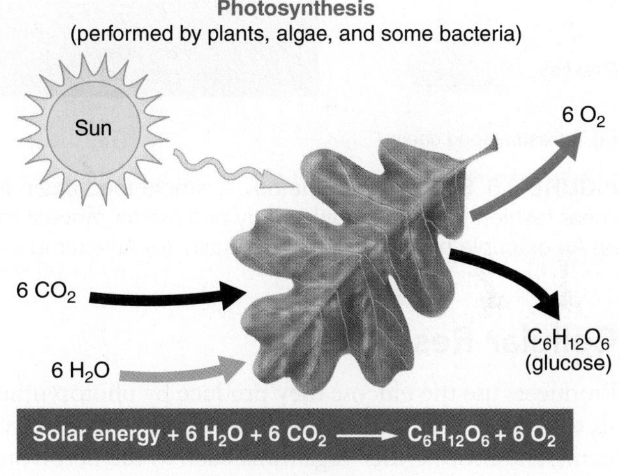

FIGURE 6.4 Photosynthesis and respiration. Photosynthesis is the process by which producers use solar energy to convert carbon dioxide and water into glucose and oxygen. Respiration is the process by which organisms convert glucose and oxygen into water and carbon dioxide, releasing the energy needed to live, grow, and reproduce. All organisms, including producers, perform respiration.

Producer An organism that uses the energy of the Sun to produce usable forms of energy. *Also known as* **Autotroph**.

Photosynthesis The process by which producers use solar energy to convert carbon dioxide and water into glucose.

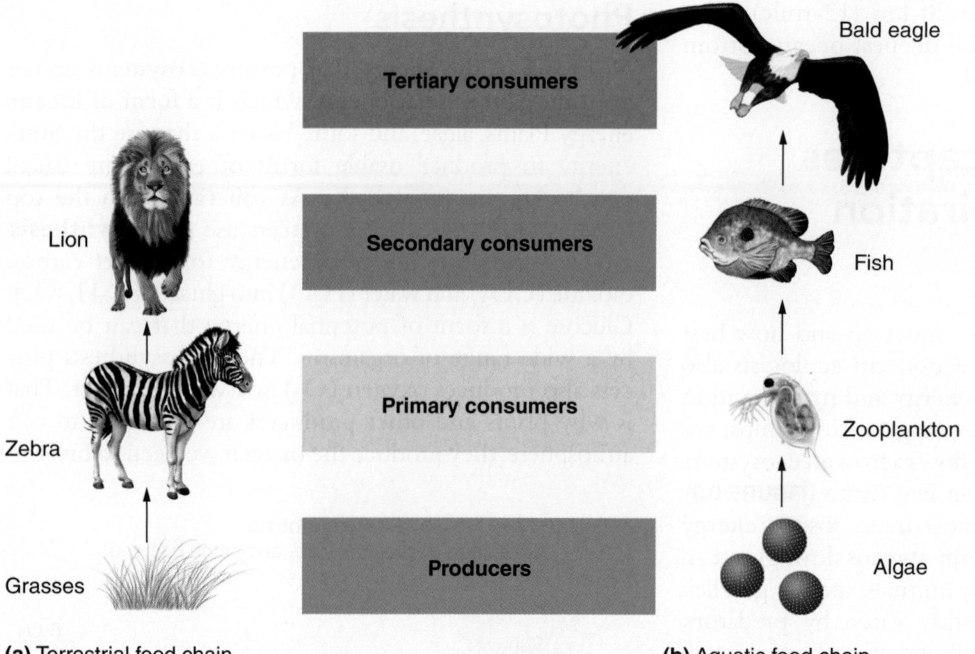

FIGURE 6.5 Simple food chains. A simple food chain that links producers and consumers in a linear fashion illustrates how energy and matter move through the trophic levels of an ecosystem. (a) An example of a terrestrial food chain. (b) An example of an aquatic food chain.

Cellular Respiration

Producers use the glucose they produce by photosynthesis to store energy and to build structures such as leaves, stems, and roots. Other organisms, such as the herbivores on the Serengeti Plain, eat the tissues of producers and gain energy from the chemical energy contained in those tissues. They do this through **cellular respiration**, a process by which cells unlock the energy of chemical compounds. **Aerobic respiration**, which is shown in the bottom half of Figure 6.4, is the opposite of photosynthesis; cells convert glucose and oxygen into energy, carbon dioxide, and water. In essence, organisms conducting aerobic respiration run photosynthesis backward to recover the solar energy stored in glucose. Some organisms, such as bacteria that live in the mud underlying a swamp where oxygen is not available, conduct **anaerobic respiration**, a process by which cells convert glucose into energy in the absence of oxygen. Anaerobic respiration does not provide as much energy as aerobic respiration.

Cellular respiration The process by which cells unlock the energy of chemical compounds.

Aerobic respiration The process by which cells convert glucose and oxygen into energy, carbon dioxide, and water.

Anaerobic respiration The process by which cells convert glucose into energy in the absence of oxygen.

Consumer An organism that is incapable of photosynthesis and must obtain its energy by consuming other organisms. *Also known as* **Heterotroph**.

Herbivore A consumer that eats producers. *Also known as* **Primary consumer**.

Many organisms—including producers—carry out aerobic respiration to fuel their own metabolism and growth. Thus producers both produce and consume oxygen. When the Sun is shining and photosynthesis occurs, producers generate more oxygen through photosynthesis than they consume through respiration. At night, when producers only respire, they consume oxygen without generating it. Overall, producers photosynthesize more than they respire. The net effect is an excess of oxygen released into the air and an excess of carbon stored in the tissues of producers.

AP® Exam Tip

You should know how photosynthesis and cellular respiration work, including the various inputs and outputs of each process. ●

Energy captured by producers moves through many trophic levels

We have seen that producers make their own food. However, **consumers**, or **heterotrophs**, are incapable of photosynthesis and must obtain their energy by consuming other organisms. In **FIGURE 6.5**, we can see that consumers in both terrestrial and aquatic ecosystems fall into different categories. Consumers that eat producers are called **herbivores** or **primary consumers**. Primary consumers include a variety of familiar plant- and alga-eating animals, such as zebras,

FIGURE 6.6 A simplified food web. Food webs are more realistic representations of trophic relationships than simple food chains. They include scavengers, detritivores, and decomposers, and they recognize that some species feed at multiple trophic levels. Arrows indicate the direction of energy movement. This is a real but somewhat simplified food web; in an actual ecosystem, many more organisms are present. In addition, there are many more energy movements.

grasshoppers, tadpoles, and zooplankton. Consumers that eat other consumers are called **carnivores**. Carnivores that eat primary consumers are called **secondary consumers**. Secondary consumers include creatures such as lions, hawks, and rattlesnakes. Carnivores that eat secondary consumers are called **tertiary consumers**. As you can see on the right side of Figure 6.5, animals such as bald eagles can be tertiary consumers. In this food chain, the algae (producers) living in lakes convert sunlight into glucose, zooplankton (primary consumers) eat the algae, fish (secondary consumers) eat the zooplankton, and eagles (tertiary consumers) eat the fish.

The successive levels of organisms consuming one another are known as **trophic levels** (from the Greek word *trophe,* which means "nourishment"). The sequence of consumption from producers through tertiary consumers is known as a **food chain**, where energy moves from one trophic level to the next. A food chain helps us visualize how energy and matter move between trophic levels.

Species in natural ecosystems are rarely connected in such a simple, linear fashion as suggested by a food chain diagram. A more realistic type of model, shown in **FIGURE 6.6**, is known as a *food web*. A **food web** is a complex model of how energy and matter move through trophic levels. Food webs illustrate one of the most important concepts of ecology: All species in an ecosystem are connected to one another.

Not all organisms fit neatly into a single trophic level. Some organisms, called *omnivores,* operate at several

Carnivore A consumer that eats other consumers.

Secondary consumer A carnivore that eats primary consumers.

Tertiary consumer A carnivore that eats secondary consumers.

Trophic levels The successive levels of organisms consuming one another.

Food chain The sequence of consumption from producers through tertiary consumers.

Food web A complex model of how energy and matter move between trophic levels.

trophic levels. Omnivores include grizzly bears, which eat berries and fish, and the Venus flytrap (*Dionaea muscipula*), which can photosynthesize as well as digest insects that become trapped in its leaves.

Each trophic level eventually produces dead individuals and waste products. Three groups of organisms feed on this dead organic matter: *scavengers, detritivores,* and *decomposers*. **Scavengers** are organisms, such as vultures, that consume dead animals. **Detritivores** are organisms, such as dung beetles, that specialize in breaking down dead tissues and waste products (referred to as detritus) into smaller particles. These particles can then be further processed by **decomposers**, which are the fungi and bacteria that complete the breakdown process by converting organic matter into small elements and molecules that can be recycled back into the ecosystem. Without scavengers, detritivores, and decomposers, there would be no way of recycling organic matter and energy, and the world would rapidly fill up with dead plants and animals.

Some ecosystems are more productive than others

The amount of energy available in an ecosystem determines how much life the ecosystem can support. For example, the amount of sunlight that reaches a lake surface determines how much algae can live in the lake. In turn, the amount of algae determines the number of zooplankton the lake can support, and the size of the zooplankton population determines the number of fish the lake can support.

To understand where the energy in an ecosystem comes from and how it is transferred through food webs, environmental scientists measure the ecosystem's productivity. The **gross primary productivity (GPP)** of the ecosystem is a measure of the total amount of solar energy that the producers in the system capture via photosynthesis over a given amount of time. Note that the term *gross,* as used here, indicates the total amount of energy captured by producers. In other words, GPP does not subtract the energy is that lost when the producers respire. The energy captured minus the energy respired by producers is the ecosystem's **net primary productivity (NPP)**:

$$\text{net primary productivity} = \text{gross primary productivity} - \text{respiration by producers}$$

You can think of GPP and NPP in terms of a paycheck: GPP is the total amount your employer pays you whereas NPP is the actual amount you take home after taxes are deducted.

GPP is essentially a measure of how much photosynthesis is occurring over some amount of time. Determining GPP is a challenge for scientists because a plant rarely photosynthesizes without simultaneously respiring. However, if we can determine the rate of photosynthesis and the rate of respiration, we can use this information to calculate GPP.

We can determine the rate of photosynthesis by measuring the compounds that participate in the reaction. So, for example, we can measure the rate at which CO_2 is taken up during photosynthesis and the rate at which CO_2 is produced during respiration. A common approach to measuring GPP is to first measure the production of CO_2 in the dark. Because no photosynthesis occurs in the dark, this measure eliminates CO_2 uptake by photosynthesis. Next, we measure the uptake of CO_2 in sunlight. This measure gives us the net movement of CO_2 when respiration and photosynthesis are both occurring. By adding the amount of CO_2 produced in the dark to the amount of CO_2 taken up in the sunlight, we can determine the gross amount of CO_2 that is taken up during photosynthesis:

$$CO_2 \text{ taken up during photosynthesis} = CO_2 \text{ taken up in sunlight} + CO_2 \text{ produced in the dark}$$

In this way, we can derive the GPP of an ecosystem per day within a given area. We can give our answer in units of kilograms of carbon taken up per square meter per day ($kg\ C/m^2/day$).

Converting sunlight into chemical energy is not an efficient process. As **FIGURE 6.7** shows, only about 1 percent of the total amount of solar energy that reaches the producers in an ecosystem—the sunlight on a pond surface, for example—is converted into chemical energy via photosynthesis. Most of that solar energy is lost from the ecosystem as heat that returns to the atmosphere. Some of the lost energy consists of wavelengths of light that producers cannot absorb. Those wavelengths are either reflected from the surfaces of producers or pass through their tissues.

The NPP of ecosystems ranges from 25 to 50 percent of GPP. Given that only 1 percent of the available solar energy is converted into chemical energy, which is measured as GPP, this means that as little as 0.25 percent of the solar energy striking the planet is represented by NPP. Clearly, it takes a lot of energy to conduct photosynthesis. Let's look at the math. Recall from Figure 6.7 that on average, of the 1 percent of the Sun's energy that is captured by a producer, about 60 percent is used to fuel the producer's respiration. The remaining

Scavenger An organism that consumes dead animals.

Detritivore An organism that specializes in breaking down dead tissues and waste products into smaller particles.

Decomposers Fungi and bacteria that convert organic matter into small elements and molecules that can be recycled back into the ecosystem.

Gross primary productivity (GPP) The total amount of solar energy that producers in an ecosystem capture via photosynthesis over a given amount of time.

Net primary productivity (NPP) The energy captured by producers in an ecosystem minus the energy producers respire.

40 percent can be used to support the producer's growth and reproduction. A forest in North America, for example, might have a GPP of 2.5 kg C/m^2/year and lose 1.5 kg C/m^2/year to respiration by plants. Because NPP = GPP − respiration, the NPP of the forest is 1 kg C/m^2/year (1.8 pounds C/yard^2/year). This means that the plants living in 1 m^2 of forest will add 1 kg of carbon to their tissues every year by means of growth and reproduction. So, in this example, NPP is 40 percent of GPP.

Measurement of NPP allows us to compare the productivity of different ecosystems, as shown in **FIGURE 6.8**. As you can see, producers grow best in ecosystems where they have plenty of sunlight, lots of available water and nutrients, and warm temperatures, such as tropical rainforests and salt marshes, which are the most productive ecosystems on Earth. Conversely, producers grow poorly in the cold regions of the Arctic, dry deserts, and the dark regions of the deep sea. In general, the greater the productivity of an ecosystem, the more primary consumers can be supported.

Measuring NPP is also a useful way to measure change in an ecosystem. For example, after a drastic change alters an ecosystem, such as a hurricane devastating the forests on Haiti, the amount of stored energy (NPP) tells us whether the new system is more or less productive than the previous system.

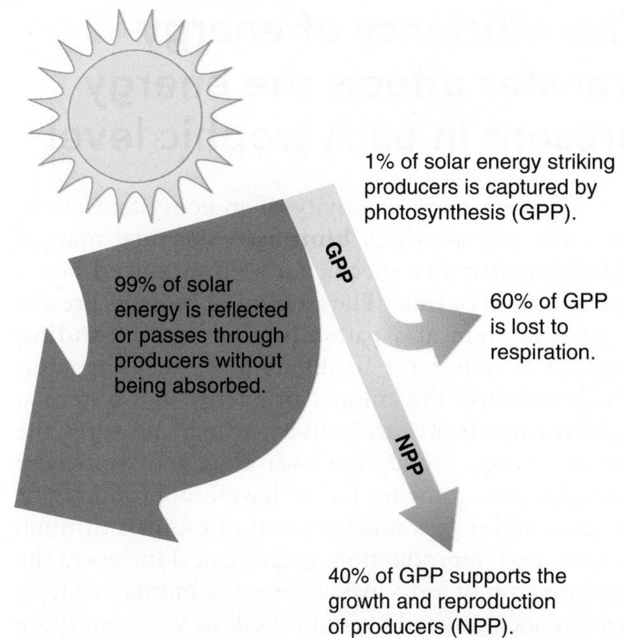

FIGURE 6.7 Gross and net primary productivity. Producers typically capture only about 1 percent of available solar energy via photosynthesis. This is known as gross primary productivity, or GPP. About 60 percent of GPP is typically used for respiration. The remaining 40 percent of GPP is used for the growth and reproduction of the producers. This is known as net primary productivity, or NPP.

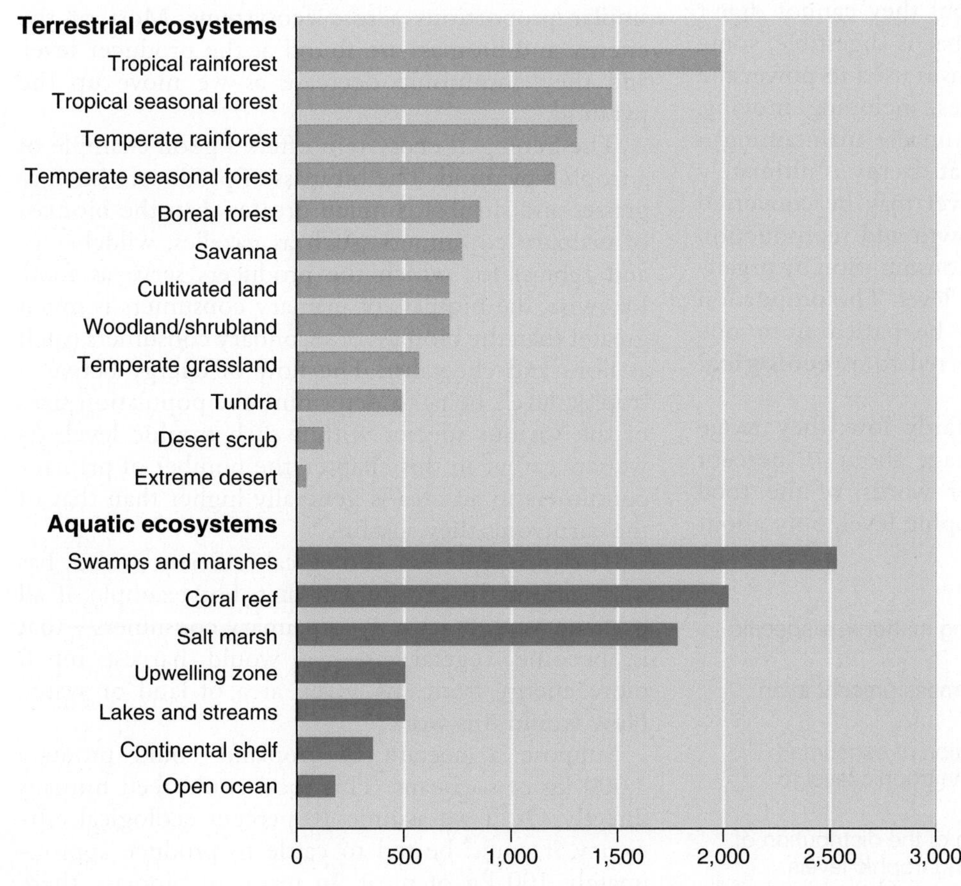

FIGURE 6.8 Net primary productivity. Net primary productivity varies among ecosystems. Productivity is highest where temperatures are warm and water and solar energy are abundant. As a result, NPP varies tremendously among different areas of the world.

The efficiency of energy transfer affects the energy present in each trophic level

The net primary productivity of an ecosystem establishes the rate at which **biomass**—the total mass of all living matter in a specific area—is produced over a given amount of time. The amount of biomass present in an ecosystem at a particular time is its **standing crop**. It is important to differentiate standing crop, which measures the amount of energy in a system at a given time, from productivity, which measures the rate of energy production over a span of time. For example, slow-growing forests have low productivity; the trees add only a small amount of biomass through growth and reproduction each year. However, the standing crop of long-lived trees—the biomass of trees that has accumulated over hundreds of years—is quite high. In contrast, the high growth rates of algae living in the ocean make them extremely productive. But because primary consumers eat these algae so rapidly, the standing crop of algae at any particular time is relatively low.

Not all of the energy contained in a particular trophic level is in a usable form. Some parts of plants are not digestible and are excreted by primary consumers. Secondary consumers such as owls consume the muscles and organs of their prey, but they cannot digest bones and hair. Of the food that is digestible, some fraction of the energy it contains is used to power the consumer's day-to-day activities, including moving, eating, and (for birds and mammals) maintaining a constant body temperature. That energy is ultimately lost as heat. Any energy left over may be converted into consumer biomass for growth and reproduction and thus becomes available for consumption by organisms at the next higher trophic level. The proportion of consumed energy that can be passed from one trophic level to another is referred to as **ecological efficiency**.

Ecological efficiencies are fairly low; they range from 5 to 20 percent and average about 10 percent across all ecosystems. In other words, of the total biomass available at a given trophic level, only about 10 percent can be converted into energy at the next higher trophic level. We can represent the distribution of biomass among trophic levels using a **trophic pyramid**, like the one for the Serengeti ecosystem shown in **FIGURE 6.9**. Trophic pyramids tend to have similar proportions across ecosystems. Most of the energy and biomass are found at the producer level, and they commonly decrease as we move up the pyramid.

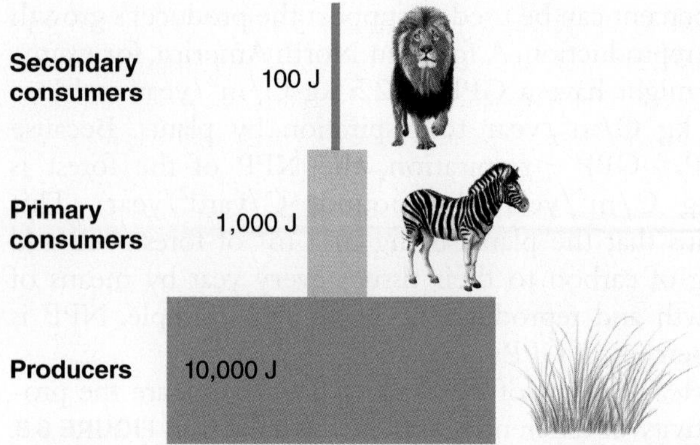

FIGURE 6.9 Trophic pyramid for the Serengeti ecosystem. This trophic pyramid represents the amount of energy that is present at each trophic level, measured in joules (J). While this pyramid assumes 10 percent ecological efficiency, actual ecological efficiencies range from 5 to 20 percent across different ecosystems. For most ecosystems, graphing the numbers of individuals or biomass within each trophic level would produce a similar pyramid.

The Serengeti ecosystem offers a good example of a trophic pyramid. The biomass of producers (such as grasses and shrubs) is much greater than the biomass of primary consumers (such as gazelles, wildebeests, and zebras) for which the producers serve as food. Likewise, the biomass of primary consumers is much greater than the biomass of secondary consumers (such as lions and cheetahs). The flow of energy between trophic levels helps to determine the population sizes of the various species within each trophic level. As we saw earlier in this chapter, the number of primary consumers in an area is generally higher than that of the carnivores they sustain.

The principle of ecological efficiency also has implications for the human diet. For example, if all humans were to act only as primary consumers—that is, become vegetarians—we would harvest much more energy from any given area of land or water. How would this work?

Suppose a hectare of cropland could produce 1,000 kg of soybeans. This food could feed humans directly. Or, if we assume 10 percent ecological efficiency, it could be fed to cattle to produce approximately 100 kg of meat. In terms of biomass, there

Biomass The total mass of all living matter in a specific area.

Standing crop The amount of biomass present in an ecosystem at a particular time.

Ecological efficiency The proportion of consumed energy that can be passed from one trophic level to another.

Trophic pyramid A representation of the distribution of biomass, numbers, or energy among trophic levels.

would be 10 times more food available for humans acting as primary consumers by eating soybeans than for humans acting as secondary consumers by eating beef. However, 1 kg of soybeans actually contains about 2.5 times as many calories as 1 kg of beef. Therefore, 1 ha of land would produce 25 times more calories when used for soybeans than when used for beef. In general, when we act as secondary consumers, the animals we eat require land to support the producers they consume. When we act as primary consumers, we require only the land necessary to support the producers we eat.

MODULE 6 AP® Review

Preparing for the AP® Exam

In this module, we have learned that ecosystems can range widely in size and have boundaries that are either natural or defined by humans. Energy from the Sun is captured by producers during the process of photosynthesis and released during the process of respiration. The energy captured by producers helps to determine the productivity of different ecosystems around the world. This energy moves from producers through the many trophic levels. The efficiency of such energy transfers determines the biomass found in different trophic levels. In the next module we will apply this understanding of how energy moves through an ecosystem to consider how matter cycles around an ecosystem.

AP® Practice Questions

Choose the best answer for the following.

1. Autotrophs
   (a) use photosynthesis.
   (b) are able to survive without oxygen.
   (c) are primary consumers.
   (d) cannot assimilate carbon.

2. A zebra is an example of
   (a) a secondary consumer.
   (b) a detritivore.
   (c) a primary consumer.
   (d) a scavenger.

3. If gross primary productivity in a wetland is 3 kg C/m^2/year and respiration is 1.5 kg C/m^2/year, what is the net primary productivity of the wetland?
   (a) 1.5 kg C/m^2/year
   (b) 2 kg C/m^2/year
   (c) 3 kg C/m^2/year
   (d) 4.5 kg C/m^2/year

4. The average efficiency of energy transfer between trophic levels is approximately
   (a) 1 percent.
   (b) 4 percent.
   (c) 10 percent.
   (d) 40 percent.

5. The gross primary productivity of an ecosystem is
   (a) the total amount of biomass.
   (b) the total energy captured by photosynthesis.
   (c) the energy captured after accounting for respiration.
   (d) the energy available to primary consumers.

6. Ecosystem boundaries are
   (a) based primarily on topographic features.
   (b) rarely based on human-created features.
   (c) are only used for ecosystems smaller than a few square hectares.
   (d) depend on many subjective factors.

MODULE 7

The Movement of Matter

As we saw in Chapter 2, Earth is an open system with respect to energy, but a closed system with respect to matter. In the previous module, we examined how energy flows through the biosphere; it enters as energy from the Sun, moves among the living and nonliving components of ecosystems, and is ultimately emitted back into space. In contrast, matter—such as water, carbon, nitrogen, and phosphorus—does not enter or leave the biosphere, but cycles within the biosphere in a variety of forms. In this module, we will explore the cycling of matter with a focus on the major biotic and abiotic processes that cause this cycling.

The specific chemical forms that elements take determine how they cycle within the biosphere. We will look at the elements that are the most important to the productivity of photosynthetic organisms. We will begin with the hydrologic cycle—the movement of water. Then we will explore the cycles of carbon, nitrogen, and phosphorus. Finally, we will take a brief look at the cycles of calcium, magnesium, potassium, and sulfur.

Learning Goals

After reading this module you should be able to

- describe how water cycles within ecosystems.
- explain how carbon cycles within ecosystems.
- describe how nitrogen cycles within ecosystems.
- explain how phosphorus cycles within ecosystems.
- discuss the movement of calcium, magnesium, potassium, and sulfur within ecosystems.

The hydrologic cycle moves water through the biosphere

Because the movement of matter within and between ecosystems involves cycles of biological, geological, and chemical processes, these cycles are known as **biogeochemical cycles**. To keep track of the movement of matter in biogeochemical cycles, we refer to the components that contain the matter—including air, water, and organisms—as "pools." Processes that move matter between pools are known as "flows." We will begin our study of biogeochemical cycles with water.

Biogeochemical cycle The movements of matter within and between ecosystems.

Hydrologic cycle The movement of water through the biosphere.

Transpiration The release of water from leaves during photosynthesis.

Water is essential to life. It makes up over one-half of a typical mammal's body weight, and no organism can survive without it. Water allows essential molecules to move within and between cells, draws nutrients into the leaves of trees, dissolves and removes toxic materials, and performs many other critical biological functions. On a larger scale, water is the primary agent responsible for dissolving and transporting the chemical elements necessary for living organisms. The movement of water through the biosphere is known as the **hydrologic cycle**.

The Hydrologic Cycle

FIGURE 7.1 shows how the hydrologic cycle works. Heat from the Sun causes water to evaporate from oceans, lakes, and soils. Solar energy also provides the energy for photosynthesis, during which plants release water from their leaves into the atmosphere—a process known as **transpiration**. The water vapor that enters the atmosphere eventually cools and forms clouds, which, in turn, produce precipitation in the form of rain, snow, and hail. Some precipitation falls back into the ocean and some falls on land.

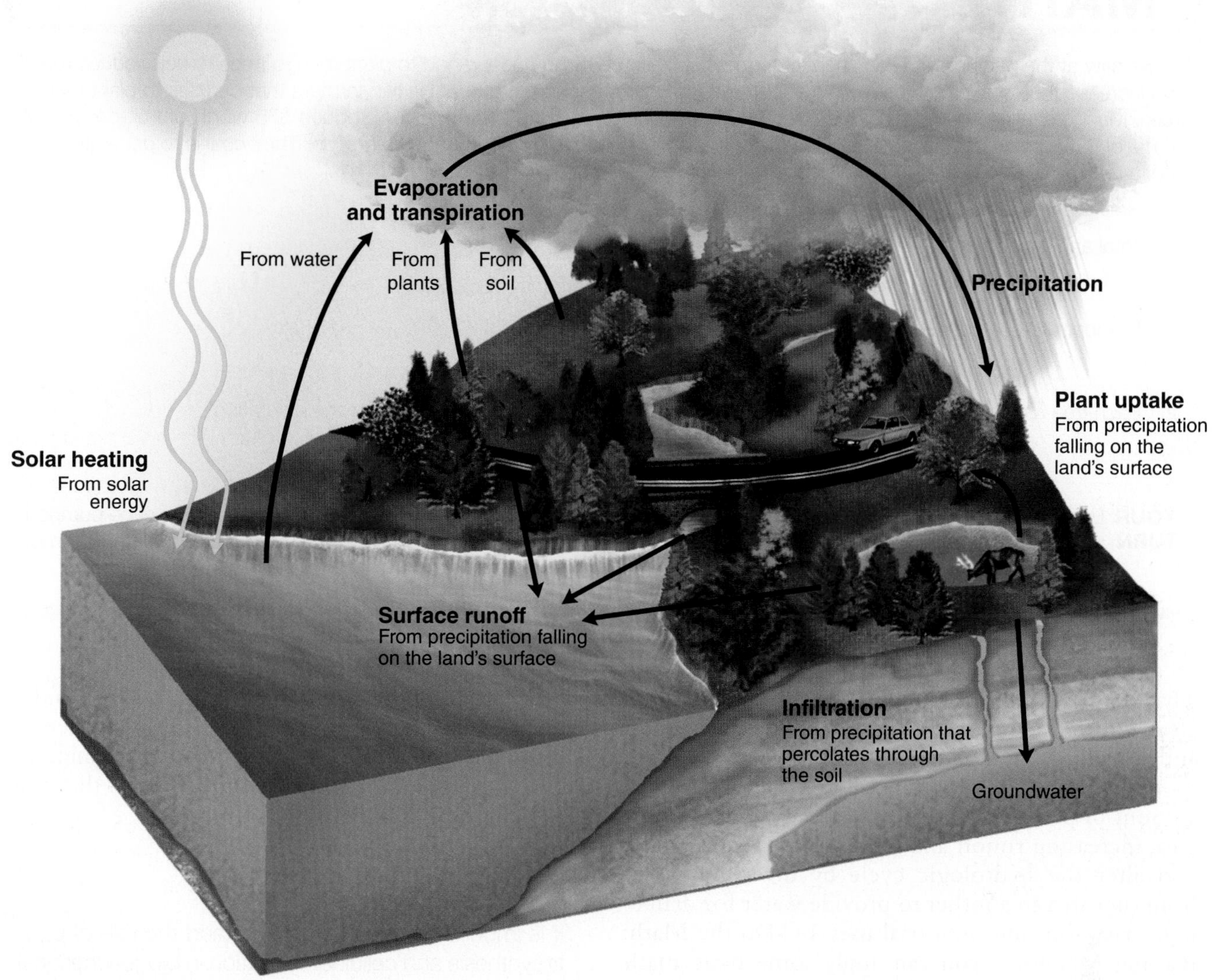

FIGURE 7.1 The hydrologic cycle. Water moves from the atmosphere to Earth's surface and back to the atmosphere.

When water falls on land, it may take one of three distinct routes. First, it may return to the atmosphere by evaporation or, after being taken up by plant roots, by transpiration. The combined amount of evaporation and transpiration is called **evapotranspiration**. Alternatively, water can be absorbed by the soil and percolate down into the groundwater. Finally, water can move as **runoff** across the land surface and into streams and rivers, eventually reaching the ocean, which is the ultimate pool of water on Earth. As water in the ocean evaporates, the cycle begins again.

While the hydrologic cycle describes the movement of water between Earth and the atmosphere, it also plays a key role in moving many elements that are dissolved in the water. As you read about biogeochemical cycles, notice the role that water plays in these processes.

Human Activities and the Hydrologic Cycle

Because Earth is a closed system with respect to matter, water never leaves it. Nevertheless, human activities can alter the hydrologic cycle in a number of ways. For example, harvesting trees from a forest can reduce evapotranspiration by reducing plant biomass. If evapotranspiration decreases, then runoff or percolation will increase. On a moderate or steep slope, most water will leave the land surface as runoff.

Evapotranspiration The combined amount of evaporation and transpiration.

Runoff Water that moves across the land surface and into streams and rivers.

DO THE MATH Raising Mangoes

Preparing for the AP® Exam

As we saw at the beginning of this chapter, farmers in Haiti are being encouraged to plant mango trees to help reduce runoff and increase the uptake of water by the soil and the trees. Consider a group of Haitian farmers that decides to plant mango trees. Mango saplings cost \$10 each. Once the trees become mature, each tree will produce \$75 worth of fruit per year. A village of 225 people decides to pool its resources and set up a community mango plantation. Their goal is to generate a per capita income of \$300 per year for everyone in the village.

1. How many mature trees will the village need to meet the goal?

   Total annual income desired:

   $$\$300/\text{person} \times 225 \text{ persons} = \$67{,}500$$

   Number of trees needed to produce \$67,500 in annual income:

   $$\$67{,}500 \div \$75/\text{tree} = 900 \text{ trees}$$

2. Each tree requires 25 m^2 of space. How many hectares must the village set aside for the plantation?

   $$900 \text{ trees} \times 25 \text{ m}^2/\text{tree} = 22{,}500 \text{ m}^2 = 2.25 \text{ ha}$$

YOUR TURN Each tree requires 20 L of water per day during the 6 hot months of the year (180 days). The water must be pumped to the plantation from a nearby stream. How many liters of water are needed each year to water the plantation of 900 trees?

That is why, as we saw at the opening of this chapter, clear-cutting a mountain slope can lead to erosion and flooding. Similarly, paving over land surfaces to build roads, businesses, and homes reduces the amount of percolation that can take place in a given area, increasing runoff and evaporation. Humans can also alter the hydrologic cycle by diverting water from one area to another to provide water for drinking, irrigation, and industrial uses. In "Do the Math: Raising Mangoes" you can apply some basic math skills to get a sense of the decisions involved in planting trees to reforest Haiti and reduce the impacts of humans on the hydrologic cycle.

The carbon cycle moves carbon between air, water, and land

Carbon is the most important element in living organisms; it makes up about 20 percent of their total body weight. Carbon is the basis of the long chains of organic molecules that form the membranes and walls of cells, constitute the backbones of proteins, and store energy for later use. Other than water, there are few molecules in the bodies of organisms that lack carbon. To understand the **carbon cycle**, which is the movement of carbon around the biosphere, we need to examine the flows that move carbon and the major pools that contain carbon.

Carbon cycle The movement of carbon around the biosphere.

AP® Exam Tip

It is important for you to understand the role of photosynthesis and cellular respiration in biogeochemical cycles, particularly the carbon cycle. ●

The Carbon Cycle

FIGURE 7.2 illustrates the seven processes that drive the carbon cycle: *photosynthesis, respiration, exchange, sedimentation, burial, extraction,* and *combustion*. These processes can be categorized as either fast or slow. The fast part of the cycle involves processes that are associated with living organisms. The slow part of the cycle involves carbon that is held in rocks, in soils, or as petroleum hydrocarbons (the materials we use as fossil fuels). Carbon may be stored in these forms for millions of years.

Photosynthesis and Respiration

Let's take a closer look at Figure 7.2, which depicts the carbon cycle. When producers photosynthesize, whether on land or in the water, they take in CO_2 and incorporate the carbon into their tissues. Some of this carbon is returned as CO_2 when organisms respire.

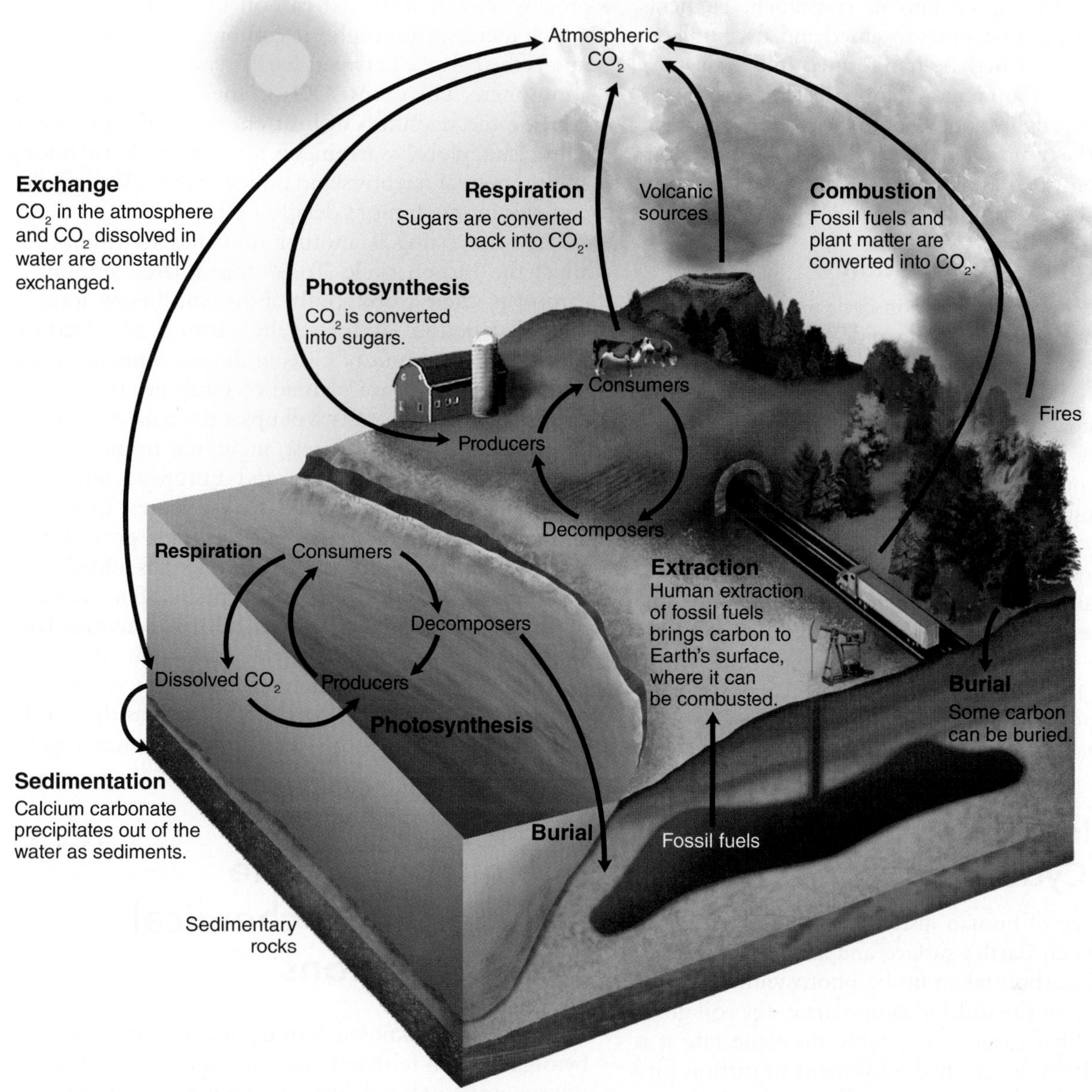

FIGURE 7.2 The carbon cycle. Producers take up carbon from the atmosphere via photosynthesis and pass it on to consumers and decomposers. Some inorganic carbon sediments precipitate out of the water to form sedimentary rock while some organic carbon may be buried and become fossil fuels. Respiration by organisms returns carbon to the atmosphere and water. Combustion of fossil fuels and other organic matter returns carbon to the atmosphere.

It is also returned after organisms die. When organisms die, carbon that was part of the live biomass pool becomes part of the dead biomass pool. Decomposers break down the dead material, which returns CO_2 to the water or air via respiration and continues the cycle.

Exchange, Sedimentation, and Burial

As you can see on the left side of Figure 7.2, carbon is exchanged between the atmosphere and the ocean. The amount of carbon released from the ocean into the atmosphere roughly equals the amount of atmospheric CO_2 that diffuses into ocean water. Some of the CO_2 dissolved in the ocean enters the food web via photosynthesis by algae.

Another portion of the CO_2 dissolved in the ocean combines with calcium ions in the water to form calcium carbonate ($CaCO_3$), a compound that can precipitate out of the water and form limestone and dolomite rock via sedimentation and burial. Although sedimentation is a very slow process, the small amounts of calcium carbonate sediment formed each year have accumulated over millions of years to produce the largest carbon pool in the slow part of the carbon cycle.

A small fraction of the organic carbon in the dead biomass pool is buried and incorporated into ocean sediments before it can decompose into its constituent elements. This organic matter becomes fossilized and, over millions of years, some of it may be transformed into fossil fuels. The amount of carbon removed from the food web by this slow process is roughly equivalent to the amount of carbon returned to the atmosphere by weathering of rocks containing carbon (such as limestone) and by volcanic eruptions, so the slow part of the carbon cycle is in steady state.

Extraction and Combustion

The final processes in the carbon cycle are extraction and combustion, as are shown in the middle of Figure 7.2. The extraction of fossil fuels by humans is a relatively recent phenomenon that began when human society started to rely on coal, oil, and natural gas as energy sources. Extraction by itself does not alter the carbon cycle; it is the subsequent step of combustion that makes the difference. Combustion of fossil fuels by humans and the natural combustion of carbon by fires or volcanoes release carbon into the atmosphere as CO_2 or into the soil as ash.

As you can see, combustion, respiration, and decomposition operate in very similar ways: All three processes cause organic molecules to be broken down to produce CO_2, water, and energy. However, respiration and decomposition are biotic processes, whereas combustion is an abiotic process.

Human Impacts on the Carbon Cycle

In the absence of human disturbance, the exchange of carbon between Earth's surface and atmosphere is in a steady state. Carbon taken up by photosynthesis eventually ends up in the soil. Decomposers in the soil gradually release that carbon at roughly the same rate it is added. Similarly, the gradual movement of carbon into the buried or fossil fuel pools is offset by the slow processes that release it. Before the Industrial Revolution, the atmospheric concentration of carbon dioxide had changed very little for 10,000 years (see Figure 2.5 on page 12). So, until recently, carbon entering any of these pools was balanced by carbon leaving these pools.

Since the Industrial Revolution, however, human activities have had a major influence on carbon cycling. The best-known and most significant human alteration of the carbon cycle is the combustion of fossil fuels. This process releases fossilized carbon into the atmosphere, which increases atmospheric carbon concentrations and upsets the balance between Earth's carbon pools and the atmosphere. The excess CO_2 in the atmosphere acts to increase the retention of heat energy in the biosphere. The result, global warming, is a major concern among environmental scientists and policy makers. We will discuss this in much more detail in Chapter 19.

Tree harvesting is another human activity that can affect the carbon cycle. Trees store a large amount of carbon in their wood, both above and below ground. The destruction of forests by cutting and burning increases the amount of CO_2 in the atmosphere. Unless enough new trees are planted to recapture the carbon, the destruction of forests will upset the balance of CO_2. To date, large areas of forest, including tropical forests as well as North American and European temperate forests, have been converted into pastures, grasslands, and croplands. In addition to destroying a great deal of biodiversity, this destruction of forests has added large amounts of carbon to the atmosphere. The increases in atmospheric carbon due to human activities have been partly offset by an increase in carbon absorption by the ocean. While some regions of the world have experienced increased reforestation, such as the northeastern United States, the loss of forest remains a global concern.

Macronutrient One of six key elements that organisms need in relatively large amounts: nitrogen, phosphorus, potassium, calcium, magnesium, and sulfur.

Limiting nutrient A nutrient required for the growth of an organism but available in a lower quantity than other nutrients.

Nitrogen cycle The movement of nitrogen around the biosphere.

The nitrogen cycle includes many chemical transformations

Six key elements, known as **macronutrients**, are needed by organisms in relatively large amounts: nitrogen, phosphorus, potassium, calcium, magnesium, and sulfur.

Nitrogen is used to form amino acids, the building blocks of proteins, and nucleic acids, the building blocks of DNA and RNA. Because so much of it is required, nitrogen is often a *limiting nutrient* for producers. A **limiting nutrient** is a nutrient required for the growth of an organism but available in a lower quantity than other nutrients. Lack of a limiting nutrient such as nitrogen constrains the growth of an organism. Adding other nutrients, such as water or phosphorus, will not improve plant growth in nitrogen-poor soil.

The Nitrogen Cycle

The **nitrogen cycle** is the movement of nitrogen around the biosphere. As nitrogen moves through an ecosystem, it experiences many chemical transformations. **FIGURE 7.3** shows the five major transformations in the nitrogen cycle: *nitrogen fixation, nitrification, assimilation, mineralization,* and *denitrification.*

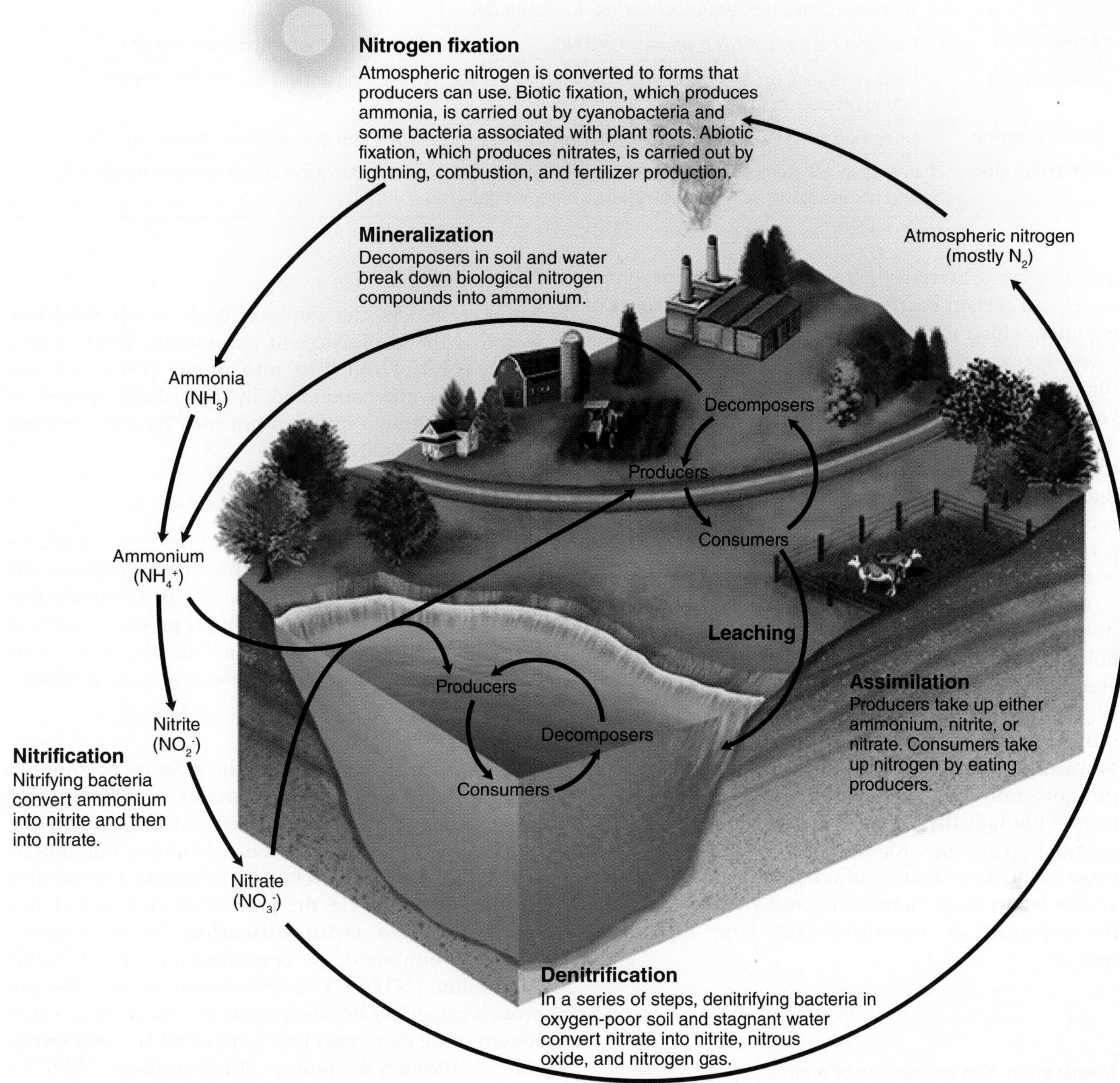
Nitrogen fixation
Atmospheric nitrogen is converted to forms that producers can use. Biotic fixation, which produces ammonia, is carried out by cyanobacteria and some bacteria associated with plant roots. Abiotic fixation, which produces nitrates, is carried out by lightning, combustion, and fertilizer production.
Mineralization
Decomposers in soil and water break down biological nitrogen compounds into ammonium.
Atmospheric nitrogen (mostly N_2)
Ammonia (NH_3)
Decomposers
Producers
Consumers
Ammonium (NH_4^+)
Leaching
Producers
Decomposers
Consumers
Assimilation
Producers take up either ammonium, nitrite, or nitrate. Consumers take up nitrogen by eating producers.
Nitrite (NO_2^-)
Nitrification
Nitrifying bacteria convert ammonium into nitrite and then into nitrate.
Nitrate (NO_3^-)
Denitrification
In a series of steps, denitrifying bacteria in oxygen-poor soil and stagnant water convert nitrate into nitrite, nitrous oxide, and nitrogen gas.

FIGURE 7.3 The nitrogen cycle. The nitrogen cycle moves nitrogen from the atmosphere and into soils through several fixation pathways, including the production of fertilizers by humans. In the soil, nitrogen can exist in several forms. Denitrifying bacteria release nitrogen gas back into the atmosphere.

Nitrogen Fixation

Although Earth's atmosphere is 78 percent nitrogen by volume, the vast majority of that nitrogen is in a gas form that most producers cannot use. **Nitrogen fixation** is the process that converts nitrogen gas in the atmosphere (N_2) into forms of nitrogen that producers can use. As you can see in Figure 7.3, nitrogen fixation can occur through biotic or abiotic processes.

In the biotic process, a few species of bacteria can convert N_2 gas directly into ammonia (NH_3), which is rapidly converted to ammonium (NH_4^+), a form that is readily used by producers. Nitrogen-fixing organisms

Nitrogen fixation The process that converts nitrogen gas in the atmosphere (N_2) into forms of nitrogen that producers can use.

TABLE 7.1	A summary of the five transformations of nitrogen that occur in the nitrogen cycle
Nitrogen fixation	Nitrogen fixation converts N_2 from the atmosphere. Biotic processes convert N_2 to ammonium (NH_4^+), whereas abiotic processes convert N_2 to nitrate (NO_3^-).
Nitrification	Nitrifying bacteria convert ammonium (NH_4^+) into nitrite (NO_2^-) and then into nitrate (NO_3^-).
Assimilation	Producers take up either ammonium (NH_4^+) or nitrate (NO_3^-). Consumers assimilate nitrogen by eating producers.
Mineralization	Decomposers in soil and water break down biological nitrogen compounds into ammonium (NH_4^+).
Denitrification	In a series of steps, denitrifying bacteria in oxygen-poor soil and stagnant water convert nitrate (NO_3^-) into nitrous oxide (N_2O) and eventually nitrogen gas (N_2).

include cyanobacteria (also known as blue-green algae) and certain bacteria that live within the roots of legumes, which include plants such as peas, beans, and a few species of trees. Nitrogen-fixing organisms use the fixed nitrogen to synthesize their own tissues, then excrete any excess. Cyanobacteria, which are primarily aquatic organisms, excrete excess ammonium ions into the water, where they can be taken up by aquatic producers. Nitrogen-fixing bacteria that live within plant roots excrete excess ammonium ions into the plant's root system; the plant, in turn, supplies the bacteria with sugars it produces via photosynthesis.

Nitrogen fixation can also occur through two abiotic pathways. N_2 can be fixed in the atmosphere by lightning or during combustion processes such as fires and the burning of fossil fuels. These processes convert N_2 into nitrate (NO_3^-), which is usable by plants. The nitrate is carried to Earth's surface in precipitation. Humans have developed techniques for nitrogen fixation into ammonia or nitrate to be used in plant fertilizers. Although these processes require a great deal of energy, humans now fix more nitrogen than is fixed in nature. The development of synthetic nitrogen fertilizers has led to large increases in crop yields, particularly for crops such as corn that require large amounts of nitrogen.

Nitrification The conversion of ammonia (NH_4^+) into nitrite (NO_2^-) and then into nitrate (NO_3^-).

Assimilation The process by which producers incorporate elements into their tissues.

Mineralization The process by which fungal and bacterial decomposers break down the organic matter found in dead bodies and waste products and convert it into inorganic compounds.

Ammonification The process by which fungal and bacterial decomposers break down the organic nitrogen found in dead bodies and waste products and convert it into inorganic ammonium (NH_4^+).

Denitrification The conversion of nitrate (NO_3^-) in a series of steps into the gases nitrous oxide (N_2O) and, eventually, nitrogen gas (N_2), which is emitted into the atmosphere.

Nitrification

Another step in the nitrogen cycle is **nitrification**, which is the conversion of ammonium (NH_4^+) into nitrite (NO_2^-) and then into nitrate (NO_3^-). These conversions are conducted by specialized species of bacteria. Although nitrite is not used by most producers, nitrate is readily used.

Assimilation

Once producers take up nitrogen in the form of ammonia, ammonium, nitrite, or nitrate, they incorporate the element into their tissues in a process called **assimilation**. When primary consumers feed on the producers, some of the producer's nitrogen is assimilated into the tissues of the consumers while the rest is eliminated as waste products.

Mineralization

Eventually, organisms die and their tissues decompose. In a process called **mineralization**, fungal and bacterial decomposers break down the organic matter found in dead bodies and waste products and convert organic compounds back into inorganic compounds. In the nitrogen cycle, the process of mineralization is sometime called **ammonification** because organic nitrogen compounds are converted into the inorganic ammonium (NH_4^+). The ammonium produced by this process can either be taken up by producers in the ecosystem or be converted into nitrite (NO_2^-) and nitrate (NO_3^-) through the process of nitrification.

Denitrification

The final step that completes the nitrogen cycle is **denitrification**, which is the conversion of nitrate (NO_3^-) in a series of steps into the gases nitrous oxide (N_2O) and, eventually, nitrogen gas (N_2), which is emitted into the atmosphere. Denitrification is conducted by specialized bacteria that live under anaerobic conditions, such as waterlogged soils or the bottom sediments of oceans, lakes, and swamps.

As you can see, the nitrogen cycle is a fairly complicated cycle because of the many transformations of nitrogen that take place. A summary of these processes can be found in **TABLE 7.1**.

Human Impacts on the Nitrogen Cycle

Nitrogen is a limiting nutrient in most terrestrial ecosystems, so excess inputs of nitrogen can have consequences in these ecosystems. For example, nitrate is readily transported through the soil with water through **leaching**, a process in which dissolved molecules are transported through the soil via groundwater. In addition, adding nitrogen to soils in fertilizers ultimately increases atmospheric concentrations of nitrogen in regions where the fertilizer is applied. This nitrogen can be transported through the atmosphere and deposited by rainfall in natural ecosystems that have adapted over time to a particular level of nitrogen availability. The added nitrogen can alter the distribution or abundance of species in those ecosystems.

In one study of nine different terrestrial ecosystems across the United States, scientists added nitrogen fertilizer to some plots and left other plots unfertilized as controls. They found that adding nitrogen reduced the number of species in a plot by up to 48 percent because some species that could survive under low-nitrogen conditions could no longer compete against larger plants that thrived under high-nitrogen conditions. Other studies have documented cases in which plant communities that have grown on low-nitrogen soils for millennia are now experiencing changes in their species composition. An influx of nitrogen due to human activities has favored colonization by new species that are better adapted to soils with higher fertility.

The phosphorus cycle moves between land and water

Organisms need phosphorus for many biological processes. Phosphorus is a major component of DNA and RNA as well as ATP (adenosine triphosphate), the molecule cells use for energy transfer. Required by both producers and consumers, phosphorus is a limiting nutrient second only to nitrogen in its importance for successful agricultural yields. Thus phosphorus, like nitrogen, is commonly added to soils in the form of fertilizer.

The Phosphorus Cycle

The **phosphorus cycle** is the movement of phosphorus around the biosphere. As shown in **FIGURE 7.4** on page 90, it primarily operates between land and water. There is no gas phase, although phosphorus does enter the atmosphere in very small amounts when dust is dissolved in rainwater or sea spray. Unlike nitrogen, phosphorus rarely changes form; it is typically found in the form of phosphate (PO_4^{3-}).

Assimilation and Mineralization

The biotic processes that affect the phosphorus cycle are not complex. Producers on land and in the water take up inorganic phosphate and assimilate the phosphorus into their tissues as organic phosphorus. The waste products and eventual dead bodies of these organisms are decomposed by fungi and bacteria, which causes the mineralization of organic phosphorus back to inorganic phosphate.

Sedimentation, Geologic Uplift, and Weathering

The abiotic processes of the phosphorus cycle involve movements between the water and the land. In water, phosphorus is not very soluble, so much of it precipitates out of solution in the form of phosphate-laden sediments in the ocean. You can see this process in the left corner of Figure 7.4. Over time, geologic forces can lift these ocean layers up and they become mountains. The phosphate rocks in the mountains are slowly weathered by natural forces including rainfall and this weathering brings phosphorus to terrestrial and aquatic habitats. Phosphorus is tightly held by soils, so it is not easily leached from soils and into water bodies. Because so little phosphorus leaches into water bodies and because much of what enters water precipitates out of solution, very little dissolved phosphorus is naturally available in streams, rivers, and lakes. As a result, phosphorus is a limiting nutrient in many aquatic systems.

Human Impacts on the Phosphorus Cycle

Humans have had a dramatic effect on the phosphorus cycle. For example, humans mine the phosphate sediments from mountains to produce fertilizer. When these fertilizers are applied to lawns, gardens, and agricultural fields, excess phosphorus can leach into water bodies. Because aquatic systems are commonly limited by a low availability of phosphorus, even small inputs of leached phosphorus into these systems can greatly increase the growth of producers. Phosphorus inputs can cause a rapid increase in the algal population of a waterway, known as

Leaching The transportation of dissolved molecules through the soil via groundwater.

Phosphorus cycle The movement of phosphorus around the biosphere.

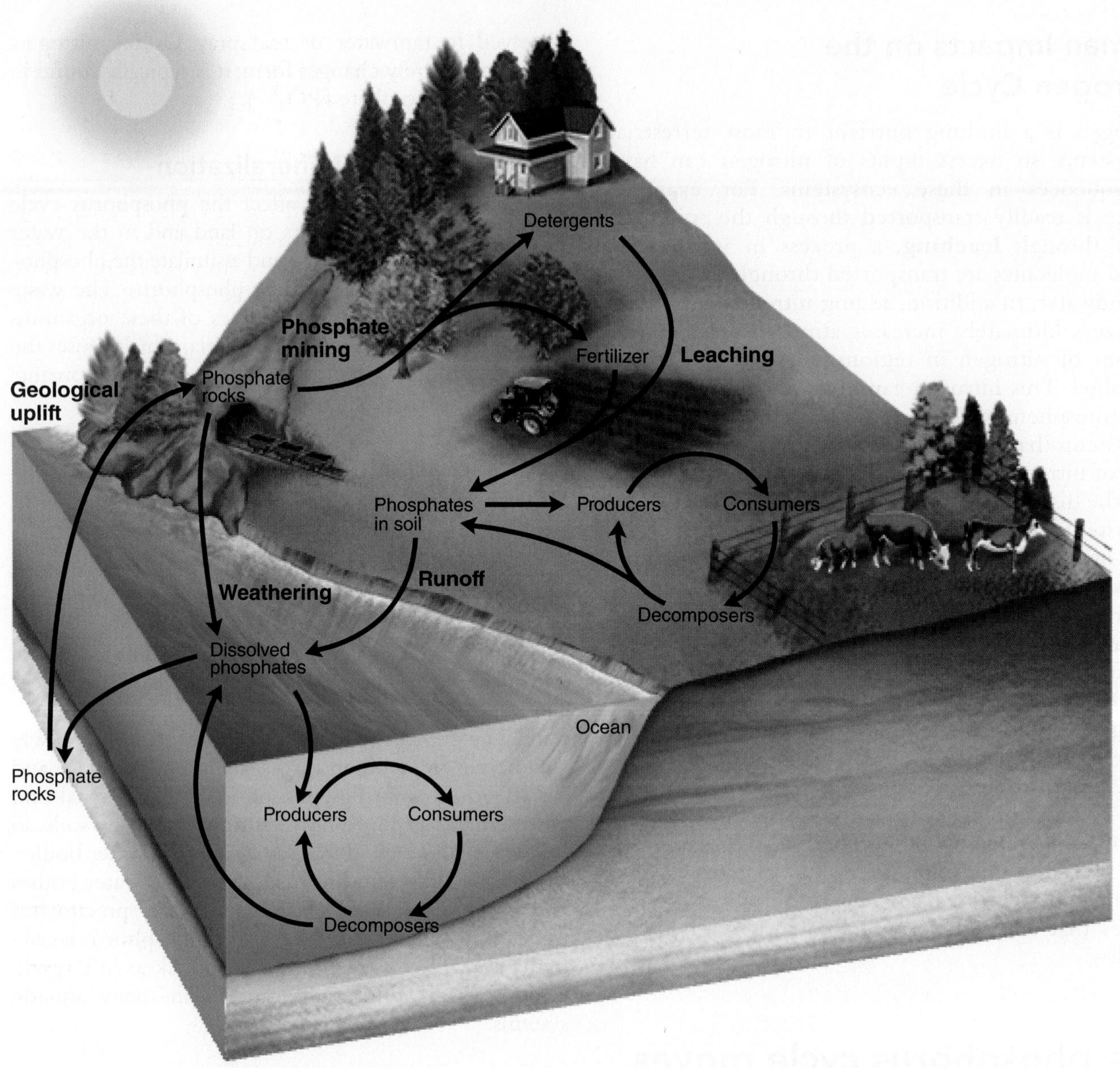

FIGURE 7.4 The phosphorus cycle. The phosphorus cycle begins with the weathering or mining of phosphate rocks and use of phosphate fertilizer, which releases phosphorus into the soil and water. This phosphorus can be used by producers and subsequently moves through the food web. In water, phosphorus can precipitate out of solution and form sediments, which over time are transformed into new phosphate rocks.

an **algal bloom**, that can quickly increase the biomass of algae in the ecosystem (**FIGURE 7.5**). As the algae die, decomposition consumes large amounts of oxygen. As a result, the water becomes low in oxygen, or **hypoxic**. When oxygen concentrations become so low that it kills fish and other aquatic animals, we refer to it as a **dead zone**. Dead zones occur around the world, including where the Mississippi River empties into the Gulf of Mexico.

Algal bloom A rapid increase in the algal population of a waterway.

Hypoxic Low in oxygen.

Dead zone When oxygen concentrations become so low that it kills fish and other aquatic animals.

A second major source of phosphorus in waterways is from the use of household detergents. From the 1940s through the 1990s, laundry detergents in the United States contained phosphates to make clothes cleaner. As a result, the water discharged from washing machines contained phosphorus, which inadvertently fertilized streams, rivers, and lakes. Because ecological dead zones triggered by excess phosphorus cause substantial environmental and economic damage, manufacturers stopped adding phosphates to laundry detergents in 1994 and 17 states have impose partial or total bans on phosphates in dishwashing detergents since 2010. In Europe, the European Union passed a ban on phosphates in laundry detergents starting in 2013 and a ban on phosphates in dishwashing detergents starting in 2017.

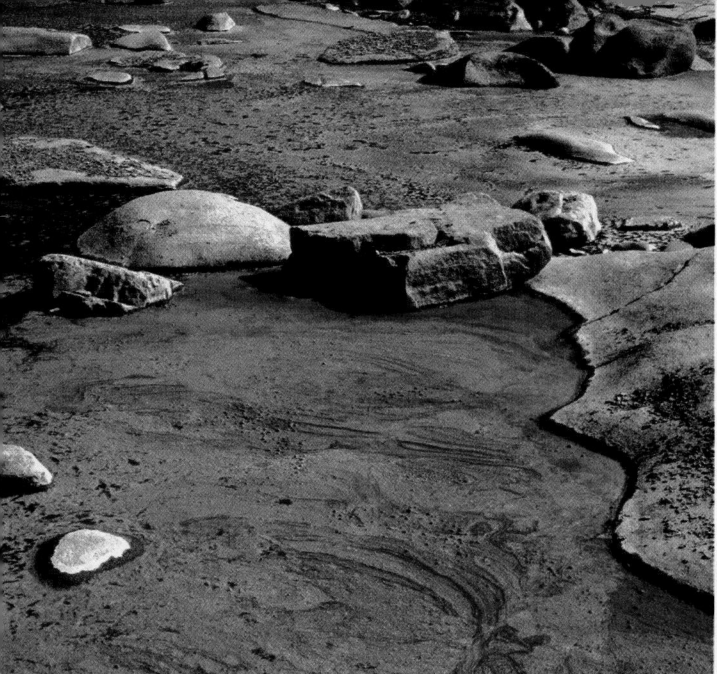

FIGURE 7.5 Algal bloom. When excess phosphorus enters waterways, it can stimulate a sudden and rapid growth of algae that turns the water bright green, like this area along the Susquehanna River in Pennsylvania. The algae eventually die, and the resulting increase in decomposition can reduce dissolved oxygen to levels that are lethal to fish and shellfish. *(Michael P. Gadomski/Science Source)*

In addition to causing algal blooms, increases in phosphorus concentrations can alter plant communities. We have already seen one example in Chapter 2, where we discussed the deterioration of the environment in the Florida Everglades. Because of agricultural expansion in southern Florida, the water that flows through the Everglades has experienced elevated phosphorus concentrations. This change in nutrient cycling has altered the ecosystem of the Everglades. For example, over time, cattails have become more common and sawgrass has declined. Animals that depend on sawgrass are now experiencing reduced food and habitat from this plant.

Calcium, magnesium, potassium, and sulfur also cycle in ecosystems

Calcium, magnesium, and potassium play important roles in regulating cellular processes and in transmitting signals between cells. Like phosphorus, these macronutrients are derived primarily from rocks and decomposed vegetation. All three can be dissolved in water as positively charged ions: Ca^{2+}, Mg^{2+}, and K^{+}. None is

DO THE MATH — Measuring Inputs of Nitrogen and Phosphorus in an Urban Environment

Preparing for the AP® Exam

As we have seen in our discussion, nutrients can come from multiple sources. Scientists recently quantified the various sources of nitrogen and phosphorus that enter the Mississippi River from St. Paul, Minnesota. Using the data below on the amount of nitrogen coming into the river, we can calculate the percentages of each source.

Source	Nitrogen ($kg/km^2/year$)	Phosphorus ($kg/km^2/year$)
Atmospheric deposition	1,300	40
Household pet waste	2,100	280
Residential fertilizer	3,800	0
Weathering	0	10
County compost	200	30

To calculate the percentage of each nitrogen source, we can begin by calculating the sum of all sources:

$$\text{Total nitrogen} = 1{,}300\ kg/km^2/year + 2{,}100\ kg/km^2/year + 3{,}800\ kg/km^2/year + 0\ kg/km^2/year + 200\ kg/km^2/year$$
$$= 7{,}400\ kg/km^2/year$$

We can then calculate each percentage by dividing each source by the total of all nitrogen sources (rounded to two significant digits):

Percent nitrogen from atmospheric deposition = $1{,}300\ kg/km^2/year \div 7{,}400\ kg/km^2/year \times 100\% = 18\%$

Percent nitrogen from household pet waste = $2{,}100\ kg/km^2/year \div 7{,}400\ kg/km^2/year \times 100\% = 28\%$

Percent nitrogen from residential fertilizer = $3{,}800\ kg/km^2/year \div 7{,}400\ kg/km^2/year \times 100\% = 51\%$

Percent nitrogen from weathering = $0\ kg/km^2/year \div 7{,}400\ kg/km^2/year \times 100\% = 0\%$

Percent nitrogen from county compost = $200\ kg/km^2/year \div 7{,}400\ kg/km^2/year \times 100\% = 3\%$

YOUR TURN Using the phosphorus data from the table, calculate the percentage of each phosphorus source that enters the Mississippi from the city of St. Paul.

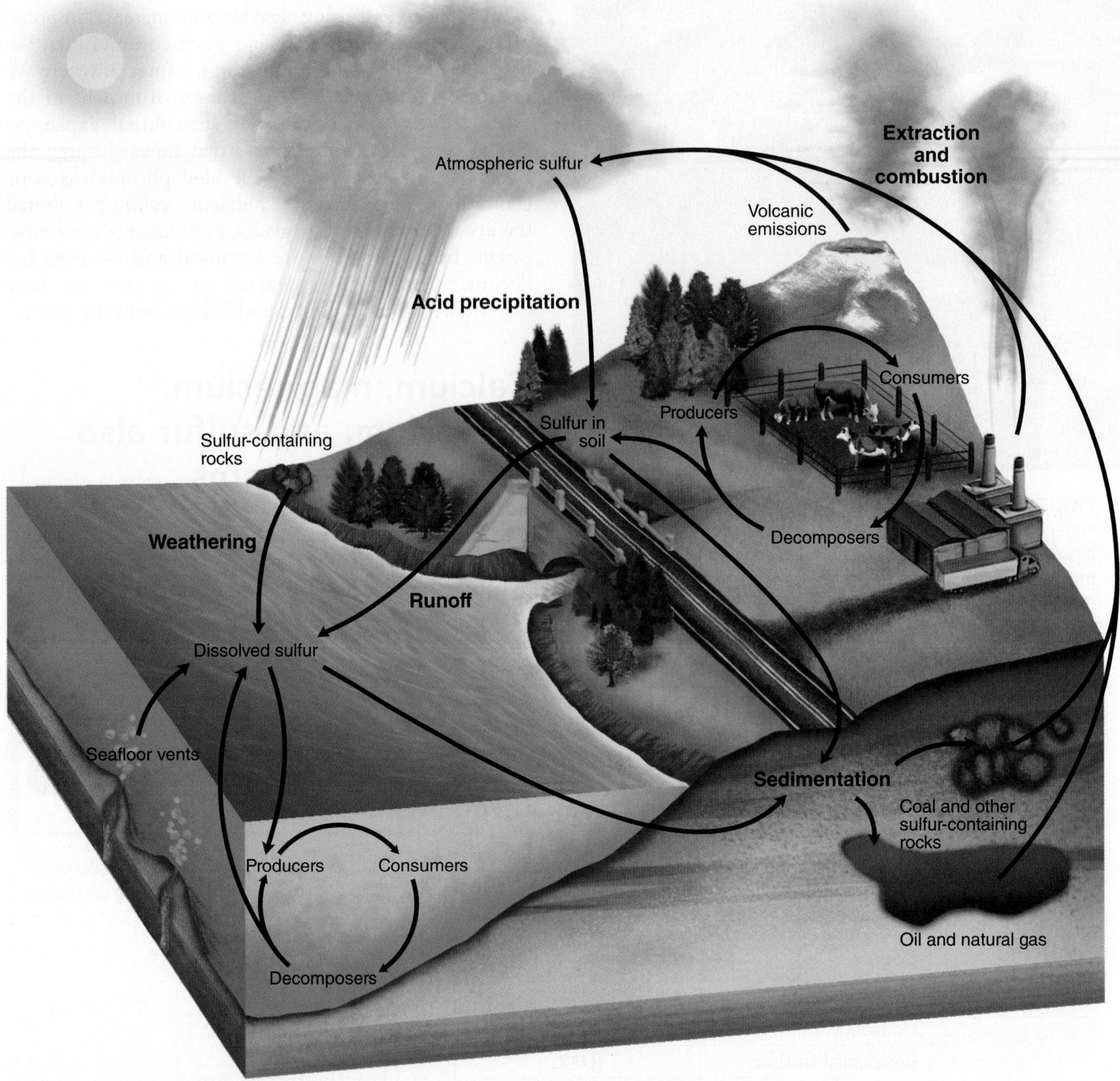

FIGURE 7.6 The sulfur cycle. Most sulfur exists as rocks. As these rocks are weathered over time, they release sulfate ions (SO_4^{2-}) that producers can take up and assimilate. This assimilated sulfur then passes through the food web. Volcanoes, the burning of fossil fuels, and the mining of copper put sulfur dioxide (SO_2) into the atmosphere. In the atmosphere, sulfur dioxide combines with water to form sulfuric acid (H_2SO_4). This sulfuric acid is carried back to Earth when it rains or snows.

present in a gaseous phase, but all can be deposited from the air in small amounts as dust.

Because of their positive charges, calcium, magnesium, and potassium ions are attracted to the negative charges present on the surfaces of most soil particles. Calcium and magnesium occur in high concentrations in limestone and marble. Because Ca^{2+} and Mg^{2+} are strongly attracted to soil particles, they are abundant in many soils overlying these types of rock. In contrast, K^+ is only weakly attracted to soil particles and is therefore more susceptible to being leached away by water moving through the soil. Leaching of potassium can lead to potassium-deficient soils that constrain the growth of plants and animals.

The final macronutrient, sulfur, is a component of proteins and plays an important role in allowing organisms to use oxygen. The **sulfur cycle**, which is the movement of sulfur around the biosphere, is shown in **FIGURE 7.6**. Most sulfur exists in rocks and is released

Sulfur cycle The movement of sulfur around the biosphere.

into soils and water as the rocks weather over time. Producers absorb sulfur through their roots in the form of sulfate ions (SO_4^{2-}), and the sulfur then cycles through the food web. The sulfur cycle also has a gaseous component. Volcanic eruptions are a natural source of atmospheric sulfur in the form of sulfur dioxide (SO_2). Human activities also add sulfur dioxide to the atmosphere, especially the burning of fossil fuels and the mining of metals such as copper. In the atmosphere, SO_2 is converted into sulfuric acid (H_2SO_4) when it mixes with water. The sulfuric acid can then be carried back to Earth when it rains or snows. As humans add more sulfur dioxide to the atmosphere, we cause more acid precipitation, which can negatively affect terrestrial and aquatic ecosystems. Although anthropogenic deposition of sulfur remains an environmental concern, clean air regulations in the United States have significantly lowered these deposits since 1995.

MODULE 7 AP® Review

Preparing for the AP® Exam

In this module, we have learned that water, carbon, and the six macronutrients cycle around ecosystems. These elements become available to organisms in different ways, but once they are obtained by producers from the atmosphere or as ions dissolved in water, the elements cycle through trophic levels in similar ways. Consumers then obtain these elements by eating producers. Finally, decomposers absorb these elements from dead producers and consumers and their waste products. Through the process of decomposition, they convert the elements into forms that are once again available to producers. In the next module, we will explore how ecosystems respond to disturbances in these cycles.

AP® Practice Questions

Choose the best answer for the following.

1. Which one of the following is fixed from the atmosphere by bacteria?
   (a) phosphorus
   (b) sulfur
   (c) nitrogen
   (d) potassium

2. Human construction of buildings and pavement affect the hydrological cycle by
   I. increasing runoff.
   II. increasing evaporation.
   III. increasing percolation.
   (a) I only
   (b) I and II only
   (c) I and III only
   (d) II and III only

3. The largest carbon pool is found in
   (a) oceans.
   (b) the atmosphere.
   (c) living organisms.
   (d) fossil fuels.

4. Phosphorus
   (a) is a limiting nutrient in many aquatic systems.
   (b) has an important gaseous phase.
   (c) is easily lost from soils due to leaching.
   (d) is often produced by volcanic eruptions.

5. Acid rain is associated with which geochemical cycle?
   (a) potassium
   (b) carbon
   (c) phosphorus
   (d) sulfur

6. Which of the following processes is also known as ammonification?
   (a) nitrogen fixation
   (b) nitrification
   (c) mineralization
   (d) denitrification

7. Haitian farmers are buying mango tree saplings for \$10 each, and the full-grown trees produce \$75 of fruit each year. If a farmer wishes to earn \$1,500 per year when the trees are grown, how much will the farmer have to spend on saplings?
   (a) \$100
   (b) \$150
   (c) \$200
   (d) \$750

MODULE 8

Responses to Disturbances

As we have seen in the previous modules, flows of energy and matter in ecosystems are essential to the species that live in them. However, sometimes ecosystems experience major disturbances that alter how they operate. Disturbances can occur over both short- and long-time scales, and ecosystem ecologists are often interested in how disturbances affect the flow of energy and matter through an ecosystem. More specifically, they are interested in whether an ecosystem can resist the impact of a disturbance and whether a disturbed ecosystem can recover its original condition. In this module, we will look at how scientists study disturbances, how ecosystems are affected by disturbances, and how quickly these ecosystems can bounce back to their pre-disturbance condition. Finally, we will apply our knowledge to an important theory about how systems respond to disturbances.

Learning Goals

After reading this module you should be able to

- distinguish between ecosystem resistance and ecosystem resilience.
- explain the insights gained from watershed studies.
- explain the intermediate disturbance hypothesis.

Ecosystems are affected differently by disturbance and in how well they bounce back after the disturbance

When Hurricanes Irma and Maria hit the Caribbean and Florida in 2017, they caused a tremendous amount of damage to homes, businesses, and ecosystems (**FIGURE 8.1**). Hurricanes, ice storms, tsunamis, tornadoes, volcanic eruptions, and forest fires can all be classified as a **disturbance** because they are events caused by physical, chemical, or biological agents that results in changes in population size or community composition in ecosystems. Disturbances also can be due to anthropogenic causes, such as human settlements, agriculture, air pollution, forest clear-cutting, and the removal of entire mountaintops for coal mining.

Disturbance An event, caused by physical, chemical, or biological agents, resulting in changes in population size or community composition.

Resistance A measure of how much a disturbance can affect flows of energy and matter in an ecosystem.

Not every ecosystem disturbance is a disaster. For example, a low-intensity fire might kill some plant species, but at the same time it might benefit fire-adapted species that can use the additional nutrients released from the dead plants. So, although the population of a particular producer species might be diminished or even eliminated, the net primary productivity of all the producers in the ecosystem might remain the same. When this is the case, we say that the productivity of the system is *resistant*. The **resistance** of an ecosystem

FIGURE 8.1 Ecosystem disturbance. Large disturbances can have major effects on ecosystems, such as this area in Puerto Rico which was devastated by Hurricane Maria in 2017. *(RICARDO ARDUENGO/Getty Images)*

(a)

(b)

FIGURE 8.2 Wetland restoration. (a) Once a forested wetland, this property in Maryland was cleared and drained for agricultural use in the 1970s. (b) Beginning in 2003, efforts began to plug the drainage ditches, remove undesirable trees, and plant wetland plants to restore the property to a wetland habitat. *(Rich Mason, USFWS)*

is a measure of how much a disturbance can affect the flows of energy and matter. When a disturbance influences populations and communities, but has no effect on the overall flows of energy and matter, we say that the ecosystem has high resistance.

When the flows of energy and matter of an ecosystem are affected by a disturbance, environmental scientists often ask how quickly and how completely the ecosystem can recover its original condition. The rate at which an ecosystem returns to its original state after a disturbance is termed **resilience**. A highly resilient ecosystem returns to its original state relatively rapidly; a less resilient ecosystem does so more slowly. For example, imagine that a severe drought has eliminated half the species in an area. In a highly resilient ecosystem, the flows of energy and matter might return to normal in the following year. In a less resilient ecosystem, the flows of energy and matter might not return to their pre-drought conditions for many years.

An ecosystem's resilience often depends on specific interactions of the biogeochemical and hydrologic cycles. For example, as human activity has led to an increase in global atmospheric CO_2 concentrations, terrestrial and aquatic ecosystems have increased the amount of carbon they absorb. In this way the carbon cycle as a whole has offset some of the changes that we might expect from increases in atmospheric CO_2 concentrations, including global climate change. Conversely, when a drought occurs, the soil may dry out and harden so much that when it eventually does rain, the soil cannot absorb as much water as it did before the drought. The soil changes in response to the drought, which leads to further drying and intensifies the drought damage. In this case, the hydrologic cycle does not relieve the effects of the drought; instead, a positive feedback in the system makes the situation worse.

Many anthropogenic disturbances—for example, housing developments, clear-cutting, or draining of wetlands—are so large that they eliminate an entire ecosystem. In some cases, however, scientists can work to reverse these effects and restore much of the original function of the ecosystem (**FIGURE 8.2**). Growing interest in restoring damaged ecosystems has led to the creation of a new scientific discipline called **restoration ecology**. Restoration ecologists are currently working on two high-profile ecosystem restoration projects, in the Florida Everglades and in the Chesapeake Bay, to restore water flows and nutrient inputs that are closer to historic levels so that the functions of these ecosystems can be restored.

Watershed studies help us understand how disturbances affect ecosystem processes

Understanding the natural rates and patterns of biogeochemical cycling in an ecosystem provides a basis for determining how a disturbance has changed the system. Because it is difficult to study biogeochemical cycles on a global scale, most of this research takes place on a smaller scale where scientists can

Resilience The rate at which an ecosystem returns to its original state after a disturbance.

Restoration ecology The study and implementation of restoring damaged ecosystems.

measure all of the ecosystem processes. Scientists commonly conduct such studies in a *watershed.* As shown in **FIGURE 8.3**, a **watershed** is all of the land in a given landscape that drains into a particular stream, river, lake, or wetland.

FIGURE 8.3 Watershed. A watershed is the area of land that drains into a particular body of water.

One of the most thorough studies of disturbance at the watershed scale has been ongoing for more than 50 years in the Hubbard Brook ecosystem of New Hampshire. Since 1962, investigators have monitored the hydrological and biogeochemical cycles of six watersheds at Hubbard Brook, ranging in area from 12 to 43 ha (30 to 106 acres). The soil in each watershed is underlain by impenetrable bedrock, so there is no deep percolation of water; all precipitation that falls on the watershed leaves it either by evapotranspiration or by runoff. Scientists measure precipitation throughout each watershed, and a stream gauge at the bottom of the main stream that drains a given watershed allows them to measure the amounts of water and nutrients leaving the system.

Researchers at Hubbard Brook investigated the effects of clear-cutting and subsequent suppression of plant regrowth. The researchers cut down the forest in one watershed and used herbicides to suppress the regrowth of vegetation for several years. A nearby watershed that was not clear-cut served as a control (**FIGURE 8.4**). The concentrations of nitrate in stream water were similar in the two watersheds before the clear-cutting. Within 6 months after the cutting, the clear-cut watershed showed significant increases in stream nitrate concentrations. With this information, the researchers were able to determine that when trees are no longer present to take up nitrate from the soil, nitrate leaches out of the soil and ends up in the stream that drains the watershed. This study and subsequent research have demonstrated the importance of plants in regulating the cycling of nutrients, as well as the consequences of not allowing new vegetation to grow when a forest is cut.

FIGURE 8.4 Studying disturbance at the watershed scale. In the Hubbard Brook ecosystem, researchers clear-cut one watershed to determine the importance of trees in retaining soil nutrients. They compared nutrient runoff in the clear-cut watershed with that in a control watershed that was not clear-cut. (The two other watersheds shown in the photo received other experimental treatments.) *(US Forest Service, Northern Research Station)*

Studies such as the one done at Hubbard Brook allow investigators to learn a great deal about biogeochemical cycles. We now understand that as forests and grasslands

Watershed All land in a given landscape that drains into a particular stream, river, lake, or wetland.

grow, large amounts of nutrients accumulate in the vegetation and in the soil. The growth of forests allows the terrestrial landscape to accumulate nutrients that would otherwise cycle through the system and end up in the ocean. Forests, grasslands, and other terrestrial ecosystems increase the retention of nutrients on land. This is an important way in which ecosystems directly influence their own growing conditions.

Intermediate levels of disturbance favor high species diversity

Ecosystem disturbance can play an important role in affecting species diversity. The **intermediate disturbance hypothesis** states that ecosystems experiencing intermediate levels of disturbance will favor a higher diversity of species than those with high or low disturbance levels. The graph in **FIGURE 8.5a** illustrates this relationship between ecosystem disturbance and species diversity. Ecosystems in which disturbances are rare experience intense competition among species. Because of this, populations of only a few highly competitive species eventually dominate the ecosystem. In places where disturbances are frequent, only a few species with high population growth rates are able to counter the effects of frequent disturbance and prevent species extinction. At intermediate level of disturbance, many more species are capable of persisting.

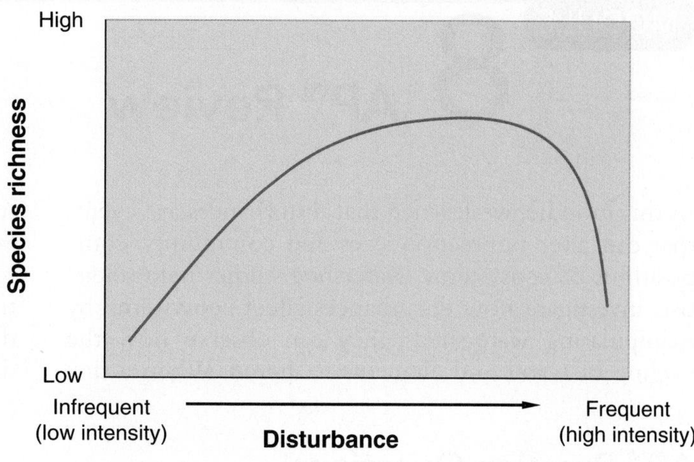

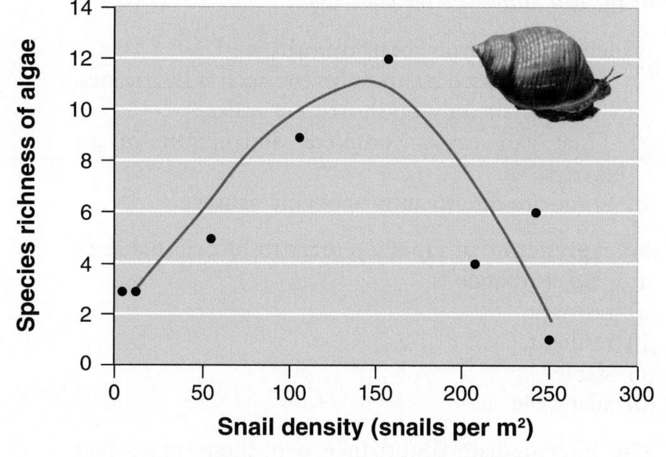

FIGURE 8.5 Intermediate disturbance hypothesis. (a) In general, we expect to see the highest species diversity at intermediate levels of disturbance. Rare disturbances favor the best competitors, which outcompete other species. Frequent disturbances eliminate most species except those that have evolved to live under such conditions. At intermediate levels of disturbance, species from both extremes can persist. (b) An example of the intermediate disturbance in the number of algal species observed in response to different amounts of herbivory by marine snails. When few or many snails are present, there is a low diversity of algal species, but when an intermediate density of snails are consuming algae, the snails cause an intermediate amount of disturbance and a higher diversity of algal species can persist in the ecosystem.

An example of the intermediate disturbance hypothesis can be found in marine algae that spend their lives attached to rocks along the rocky coast of New England. In areas containing low densities of common periwinkle snails (*Littorina littorea*), which eats multiple species of marine algae. When few snails are present, they only eat a small amount of the algae, which represents a small disturbance. Under these conditions, only a few algal species—the most competitive species—dominate the rocks, as shown in Figure 8.5b. In areas containing high densities of sails, however, a lot more algae are consumed, which represents a high disturbance. Under these conditions, only the most unpalatable algal species can persist. However, when snails were present at an intermediate density, this represents an intermediate disturbance to the algae; there is too much consumption by snails for the best competitors to dominate and the other species are never driven extinct. As a result, the intermediate density of snails results in 12 species of algae being able to persist on the rocks. As you can see in this example, the highest diversity of species can occur when ecosystems experience an intermediate frequency of disturbance.

Intermediate disturbance hypothesis The hypothesis that ecosystems experiencing intermediate levels of disturbance are more diverse than those with high or low disturbance levels.

MODULE 8 AP® Review

Preparing for the AP® Exam

In this module, we learned that disturbances are events that can alter population sizes and community compositions of ecosystems. Watershed studies help scientists investigate how disturbances affect ecosystems; by manipulating watersheds, they can observe how the cycling of water and elements is altered. When a disturbance occurs, some ecosystems exhibit high resistance to the disturbance. Other ecosystems show high resilience, which allows them to bounce back quickly to pre-disturbance conditions. Finally, an ecosystem disturbance of intermediate magnitude favors a high diversity of species.

AP® Practice Questions

Choose the best answer for the following.

1. Which is NOT true about disturbances?
   (a) They are caused by natural events such as hurricanes.
   (b) They occur only on short time scales.
   (c) They can cause complete destruction of an ecosystem.
   (d) Some are due to anthropogenic causes.

2. An ecosystem that rapidly returns to its original state after a disturbance is
   (a) resistant.
   (b) resilient.
   (c) stable.
   (d) adaptable.

3. The intermediate disturbance hypothesis states that intermediate levels of disturbance will
   (a) increase runoff.
   (b) decrease primary productivity.
   (c) increase species diversity.
   (d) decrease biomass.

4. The Hubbard Brook experiment showed that
   (a) freshwater aquatic ecosystems are often very resilient.
   (b) river restoration can take many years to complete.
   (c) evapotranspiration increases with more vegetation cover.
   (d) deforestation increases nutrient runoff.

5. Which is a measure of how much a disturbance can affect the flows of energy and matter in an ecosystem?
   (a) diversity
   (b) intensity
   (c) resistance
   (d) resilience

Working Toward Sustainability

The Practice of Precision Agriculture

As we have discussed in this chapter, nutrients such as phosphorus and nitrogen make their way into water bodies and subsequently cause harmful algal blooms and dead zones. Much of this nutrient input comes when agricultural fertilizer is not taken up by crops and washes into local streams, rivers, and lakes when it rains. Not only is this movement of fertilizer harmful to the environment, it also represents a loss to farmers who have purchased the fertilizer and used their time and equipment to distribute it over the fields. As a result, reducing the runoff of excess nutrients from agricultural fields represents a major global challenge.

Precision agriculture. A farmer in Indiana plants soybeans using a GPS-equipped tractor that automatically steers itself. Such tractors can be used to plant crops and tailor the amount of fertilizer or pesticides applied to different areas of a field. *(United Soybean Board)*

One reason nutrients wash off fields is that the fertility can vary across a field so that crops in some portions of a field need more fertilizer than crops in other portions of the same field. However, farmers traditionally spread fertilizer evenly across a field. In a perfect world, a farmer could customize fertilizer applications across the field, adding high amounts only where high amounts are needed. Such an approach should use less fertilizer, improve the growth of the crops, and result in less unused fertilizer washing off the fields and into water bodies.

The practice of precision agriculture began in the 1980s, but today, a customized approach is becoming a reality thanks to two advanced technologies. The first advance has been the ability to assess variation in crop growth across large fields. This is commonly done by using satellite photos that quantify areas of a field where crops are growing well, which is a sign of high fertility, and areas where crops are growing poorly. Using this information, combined with data on how the type of soils varies across the field, a digital map can be created to tell the farmer which areas of the field need more or less fertilizer. The second advance is the use of Global Positioning System (GPS) technology, which is the same technology used on modern smart phones. By combining a digital map of where fertilizer needs to be applied with GPS coordinates of where the tractor is located in the field, farmers today can customize their fertilizer application rates as they move across the field. For very large fields that are fertilized by small planes, the same principles can be used. Such applications can increase the efficiency of nitrogen use by more than 300 percent, which means there is much less fertilizer wasted and headed to streams, rivers, and lakes.

In addition to these two advances, other technological advances have been essential to the development of precision farming. These advances include dramatic improvements in software, huge increases in the ability of computers to analyze large amounts of data from satellite images across thousands of square kilometers, and advancements in soil sensors that are placed in fields to detect variation in nutrients and water. One of the most fascinating advances is the ability to have GPS-equipped tractors automatically steer themselves as a farmer drives across large fields. Such automated steering not only helps control where different amounts of fertilizer are applied, but also helps the tractor minimize overlapping fertilizer applications as it moves back and forth across the field.

Precision agriculture is not only valuable for applying commercial fertilizers, but also for applying manure and irrigating the fields. As animal waste, manure is a natural fertilizer, but it can cause serious problems if it runs off fields and into waterways. Not only does manure runoff add excess nutrients to water bodies, but it can also be a source of disease-causing organisms, as we will discuss further in Chapter 14. Varying the amount of manure to match the needs of the crops reduces this problem. Similarly, because large fields vary in their capacity to retain water, precision agriculture can be used to customize the amount of irrigation applied to different parts of a field.

As you might guess, the use of precision agriculture requires a larger financial investment, but it also results in greater crop production, reduced fertilizer costs, and even reduced fuel costs because of reduced tractor use. In a recent analysis of corn farming in the United States, researchers found that farmers could save up to $62 per ha ($25 per acre) by using these advanced technologies. In short, the investment in precision agriculture has increased crop production, saved farmers money, and helped the environment.

Critical Thinking Questions

1. Given that crop-eating pests can also vary in number across a farm, how might farmers use precision technologies for applying pesticides?
2. With the environmental benefits provided by precision agriculture, how might governments motivate farmers to adopt these technologies?

References

Hedley, C. 2014. The role of precision agriculture for improved nutrient on farms. *Journal of the Science of Food and Agriculture.* 95:12–19.

Cost Savings from Precision Agriculture Technologies on U.S. Corn Farms. May 2, 2016. https://www.ers.usda.gov/amber-waves/2016/may/cost-savings-from-precision-agriculture-technologies-on-us-corn-farms/

Chapter 3 Review

In this chapter, we have learned how ecosystems function by looking at how energy moves through an ecosystem and how matter cycles around an ecosystem. This energy and matter form the basis of the trophic groups that exist in ecosystems and they are responsible for the abundance of each group in nature. The typical movement of energy and matter can be altered by ecosystem disturbances, although the magnitude of the impact depends on the resistance of a particular ecosystem and its resilience after the disturbance.

Key Terms

Biosphere
Producer
Autotroph
Photosynthesis
Cellular respiration
Aerobic respiration
Anaerobic respiration
Consumer
Heterotroph
Herbivore
Primary consumer
Carnivore
Secondary consumer
Tertiary consumer
Trophic levels
Food chain
Food web
Scavenger
Detritivore
Decomposers
Gross primary productivity (GPP)
Net primary productivity (NPP)
Biomass
Standing crop
Ecological efficiency
Trophic pyramid
Biogeochemical cycle
Hydrologic cycle
Transpiration
Evapotranspiration
Runoff
Carbon cycle
Macronutrient
Limiting nutrient
Nitrogen cycle
Nitrogen fixation
Nitrification
Assimilation
Mineralization
Ammonification
Denitrification
Leaching
Phosphorus cycle
Algal bloom
Hypoxic
Dead zone
Sulfur cycle
Disturbance
Resistance
Resilience
Restoration ecology
Watershed
Intermediate disturbance hypothesis

Learning Goals Revisited

Module 6 The Movement of Energy

Explain the concept of ecosystem boundaries.

Ecosystem boundaries distinguish one ecosystem from another. Although boundaries can be well-defined, often they are not. Boundaries are commonly defined either by topographic features, such as mountain ranges, or are subjectively set by administrative criteria rather than biological criteria.

Describe the processes of photosynthesis and respiration.

Photosynthesis captures the energy of the Sun to convert CO_2 and water into carbohydrates. Respiration, whether aerobic or anaerobic, unlocks the chemical energy stored in the cells of organisms.

Distinguish among the trophic levels that exist in food chains and food webs.

The trophic levels consist of producers that convert solar energy into producer biomass through photosynthesis, primary consumers that eat the producers, secondary consumers that eat the primary consumers, and tertiary consumers that eat the secondary consumers. Omnivores eat individuals from more than one trophic group. Trophic groups that eat waste products and dead organisms are scavengers, detritivores, and decomposers.

Quantify ecosystem productivity.

Ecosystem productivity can be quantified by measuring the total amount of solar energy that producers capture, which is gross primary productivity, or by measuring the total amount of solar energy captured minus the amount of energy used for respiration, which is net primary productivity.

Explain energy transfer efficiency and trophic pyramids.

The energy present in one trophic level can be transferred to a higher trophic level and the efficiency of this transfer is generally about 10 percent. Because of this low energy transfer efficiency, the amount of energy present in each trophic level declines as we move to higher trophic levels. We can represent the energy in each trophic level as a rectangular block in a pyramid, with a size that is proportional to the energy found in the trophic level. Low ecological efficiency results in a large biomass of producers, but a much lower biomass of primary consumers and an even lower biomass of secondary consumers.

Module 7 The Movement of Matter

Describe how water cycles within ecosystems.

In the water cycle, water evaporates from water bodies and transpires from plants. The resulting water vapor cools and forms clouds, which ultimately drop water back to Earth in the form of precipitation. When the water falls onto the land, it can evaporate, be taken up by plants and transpired, percolate into the groundwater, or run off along the soil surface and ultimately return to lakes and oceans.

Explain how carbon cycles within ecosystems.

In the carbon cycle, producers take up CO_2 for photosynthesis and transfer the carbon to consumers and decomposers. Some of this carbon is converted back into CO_2 by respiration, while the rest is lost to sedimentation and burial. The extraction and combustion of fossil fuels, as well as the destruction of forests, returns CO_2 to the atmosphere.

Describe how nitrogen cycles within ecosystems.

In the nitrogen cycle, nitrogen fixation by organisms, lightning, or human activities converts nitrogen gas into ammonium (NH_4^+) or nitrate (NO_3^-). Nitrification is a process that converts ammonium into nitrite (NO_2^-) and then into nitrate (NO_3^-). Once producers take up nitrogen as ammonia, ammonium, nitrite, or nitrate, they incorporate it into their tissues in a process called assimilation. Eventually, organisms die and their tissues decompose and are converted to ammonium in a process called mineralization. Finally, denitrification returns nitrogen to the atmosphere.

Explain how phosphorus cycles within ecosystems.

The phosphorus cycle involves a large pool of phosphorus in rock that is formed by the precipitation of phosphate onto the ocean floor. Geologic forces can lift these sediments and form mountains. The phosphorus in the mountains can be made available to producers either by weathering or by mining. Producers assimilate phosphorus from the soil or water and consumers assimilate it when they eat producers. The waste products and dead bodies of organisms experience mineralization, which returns phosphorus to the environment where it can be ultimately transferred back to the ocean.

Discuss the movement of calcium, magnesium, potassium, and sulfur within ecosystems.

Calcium, magnesium, and potassium are derived from rock and can be held by soils. Producers can assimilate these elements, and mineralization of waste products and dead organisms returns the elements back to the environment. Most sulfur exists in the form of rocks and is released through the process of weathering, which makes it available for plant assimilation. Some sulfur exists as a gas in the form of sulfur dioxide (SO_2), which can be produced by volcanic eruptions and the burning of fossil fuels. In the atmosphere, SO_2 is converted into sulfuric acid (H_2SO_4) when it mixes with water. The sulfuric acid can then be carried back to the ground when it rains or snows.

Module 8 Responses to Disturbances

Distinguish between ecosystem resistance and ecosystem resilience.

The resistance of an ecosystem is a measure of how much a disturbance can affect its flows of energy and matter. In contrast, the resilience of an ecosystem is the rate at which an ecosystem returns to its original state after a disturbance has occurred.

Explain the insights gained from watershed studies.

Because watersheds contain all of the land in a given landscape that drains into a particular water body, experimental manipulations such as logging allow scientists to determine how a disturbance to an ecosystem alters the flow of energy and matter.

Explain the intermediate disturbance hypothesis.

The intermediate disturbance hypothesis states that ecosystems experiencing intermediate levels of disturbance are more diverse than those with high or low disturbance levels.

Practice Math and Graphing

Preparing for the AP® Exam

Answer the following questions. Be sure to show all your work.

1. Practice Math

In Cedar Bog Lake in Minnesota, a scientist measured the productivity of producers, primary consumers, and secondary consumers. Using the values in the table below, calculate the percent energy transferred from producers to primary consumers and the percent energy transferred from primary consumers to secondary consumers.

Trophic level	Productivity (cal/cm²/year)
Producers	111.3
Primary consumers	14.8
Secondary consumers	3.1

2. Practice Graphing

As we discussed in this chapter, the availability of nutrients affects the productivity of many ecosystems. To quantify the sensitivity of terrestrial ecosystems to additions of nitrogen and phosphorus, researchers examined studies conducted around the world and determined the average increase in productivity when nitrogen, phosphorus, or both were added to an ecosystem compared to when no extra nutrients were added.

(a) Using the data in the table below, create a bar graph that illustrates the percent increase in productivity among three terrestrial ecosystems.

(b) Based on this graph, which ecosystems appear to be the most sensitive to human inputs of nitrogen, phosphorus, or both?

(c) Why might the combination of added nitrogen and phosphorus cause a larger increase in plant productivity than either nutrient alone?

Ecosystem	Increase in productivity with added nitrogen (percent)	Increase in productivity with added phosphorus (percent)	Increase in productivity with added nitrogen and phosphorus (percent)
Grassland	23	24	60
Forest/shrubland	29	52	5
Tundra	23	1	42

Chapter 3 AP® Environmental Science Practice Exam

Preparing for the AP® Exam

Section 1: Multiple-Choice Questions

Choose the best answer for questions 1–20.

1. Which is NOT an example of an abiotic component of an ecosystem?
   (a) water
   (b) sunlight
   (c) fungi
   (d) air

2. Which is NOT characteristic of most ecosystems?
   (a) biotic components
   (b) abiotic components
   (c) distinct boundaries
   (d) a wide range of sizes

3. Which is produced during photosynthesis?
   I. glucose
   II. sulfur
   III. carbon dioxide
   (a) I only
   (b) III only
   (c) I and III
   (d) II and III

4. Which could be a cause of decreased evapotranspiration?
   (a) increased precipitation
   (b) decreased runoff
   (c) increased percolation
   (d) increased vegetation

For questions 5 and 6, select from the following choices:

(a) producers
(b) primary consumers
(c) secondary consumers
(d) tertiary consumers

5. At which trophic level are eagles that consume fish that consume zooplankton that eat algae?

6. At which trophic level do organisms use a process that produces oxygen as a waste product?

7. Which statement about precision agriculture is false?
   (a) It focuses on areas where crops are growing poorly.
   (b) It uses satellite imagery and GPS technology.
   (c) It prevents fertilizer runoff to local streams and rivers.
   (d) It increases the efficiency of fertilizer application.

8. Which food chain found on the Serengeti Plain is in the correct sequence from lowest trophic level to highest?
   (a) shrubs–gazelles–cheetahs–decomposers
   (b) shrubs–decomposers–cheetahs–gazelles
   (c) gazelles–decomposers–cheetahs–shrubs
   (d) decomposers–cheetahs–shrubs–gazelles

9. Which macronutrient is required by humans in the largest amounts?
   (a) calcium
   (b) nitrogen
   (c) sulfur
   (d) potassium

10. Roughly what percentage of incoming solar energy is converted into chemical energy by producers?
    (a) 99
    (b) 80
    (c) 50
    (d) 1

11. The net primary productivity of an ecosystem is 1 kg C/m^2/year, and the energy needed by the producers for their own respiration is 1.5 kg C/m^2/year. The gross primary productivity of such an ecosystem would be
    (a) 0.5 kg C/m^2/year.
    (b) 1.5 kg C/m^2/year.
    (c) 2.0 kg C/m^2/year.
    (d) 2.5 kg C/m^2/year.

12. A particular plot of land can produce 700 kg of beef per hectare. Beef sells for $4/kg. If that land is converted to producing corn, which sells for $0.15/kg, approximately how much will the farmer make selling corn? Assume 10 percent ecological efficiency.
    (a) $100
    (b) $500
    (c) $1,000
    (d) $3,000

13. Which biogeochemical cycle does NOT have a gaseous component?
    I. potassium
    II. sulfur
    III. phosphorus
    (a) II only
    (b) I and II
    (c) II and III
    (d) I and III

14. Which statement about the carbon cycle is TRUE?
    (a) Carbon transfer from photosynthesis is in steady state with respiration and death.
    (b) The majority of dead biomass is accumulated in sedimentation.
    (c) Combustion of carbon is equivalent in mass to sedimentation.
    (d) Most of the carbon entering the oceans is from terrestrial ecosystems.

15. Human interaction with the nitrogen cycle is primarily due to
    (a) the leaching of nitrates into terrestrial ecosystems.
    (b) the breakdown of ammonium into ammonia for industrial uses.
    (c) the acceleration of the nitrification process in aquatic ecosystems.
    (d) the decreased assimilation of ammonium and nitrates.

16. Research at Hubbard Brook showed that stream nitrate concentrations in two watersheds were _______ before clear-cutting, and that after one watershed was clear-cut, its stream nitrate concentration was _______.
    (a) similar; decreased
    (b) similar; increased
    (c) different; increased
    (d) different; decreased

17. Small inputs of this substance, commonly a limiting factor in aquatic ecosystems, can result in algal blooms and dead zones.
   (a) dissolved carbon dioxide
   (b) sulfur
   (c) dissolved oxygen
   (d) phosphorus

18. After a severe drought, the productivity in an ecosystem took many years to return to pre-drought conditions. This observation indicates that the ecosystem has
   (a) high resilience.
   (b) low resilience.
   (c) high resistance.
   (d) low resistance.

19. Which describes a key difference between photosynthesis and aerobic respiration?
   (a) Photosynthesis generates organic compounds whereas aerobic respiration generates inorganic compounds.
   (b) Photosynthesis releases energy whereas aerobic respiration consumes energy.
   (c) Aerobic respiration has a higher energy efficiency than photosynthesis.
   (d) Photosynthesis consumes chemical energy whereas aerobic respiration generates chemical energy.

20. Ocean acidification represents a key component of the
   (a) nitrogen cycle.
   (b) hydrologic cycle.
   (c) carbon cycle.
   (d) sulfur cycle.

Section 2: Free-Response Questions

Write your answer to each part clearly. Support your answers with relevant information and examples. Where calculations are required, show your work.

1. Nitrogen is crucial for sustaining life in both terrestrial and aquatic ecosystems.
   (a) Draw a fully labeled diagram of the nitrogen cycle. (4 points)
   (b) Describe the following steps in the nitrogen cycle:
      (i) nitrogen fixation (1 point)
      (ii) ammonification (1 point)
      (iii) nitrification (1 point)
      (iv) denitrification (1 point)
   (c) Describe one reason why nitrogen is crucial for sustaining life on Earth. (1 point)
   (d) Describe one way that the nitrogen cycle can be disrupted by human activities. (1 point)

2. A team of scientists wants to test the effects of disturbance on a temperate forest ecosystem. They create three forest plots:
   - Plot A, which is left undisturbed
   - Plot B, which is clear-cut
   - Plot C, where half the trees are cut down throughout the forest

   After 6 months, they measure plant species diversity in each plot, and get the following data:

Plot	Species Diversity
A	17
B	3
C	25

   (a) Explain the pattern in species diversity in Plots A and C versus Plot B. What environmental problems are associated with deforestation? (4 points)
   (b) Describe a possible reason Plot C has the greatest species diversity (2 points).
   (c) The team does a follow-up study in which they apply a nitrogen and phosphorus fertilizer to each of the plots.
      (i) Describe what might happen to species diversity in all the plots. Explain your answer. (2 points)
      (ii) A pond is located near Plot B. Describe what might happen to this pond after the application of the nitrogen and phosphorus fertilizer. (2 points)

3. Read the following article written for a local newspaper and answer the questions below.

   Neighbors Voice Opposition to Proposed Clear-Cut

   A heated discussion took place last night at the monthly meeting of the Fremont Zoning Board. Local landowner Julia Taylor has filed a request that her 150-acre woodland area be rezoned from residential to multiuse in order to allow her to remove all of the timber from the site.

   "This is my land, and I should be able to use it as I see fit," explained Ms. Taylor. "In due course, all of the trees will return and everything will go back to the same as it is now. The birds and the squirrels will still be there in the future. I have to sell the timber because I need the extra revenue to supplement my retirement as I am on a fixed income. I don't see what all the fuss is about," she commented.

   A group of owners of adjacent properties see things very differently. Their spokesperson, Ethan Jared, argued against granting a change in the current zoning. "Ms. Taylor has allowed the community to use these woods for many years, and we thank her for that. But I hope that the local children will be able to hike and explore the woods with their children as I have done with mine. Removing the trees in a clear-cut will damage our community in many ways, and it could lead to contamination of the groundwater and streams and affect many animal and plant species. Like the rest of us property owners, Ms. Taylor gets her drinking water from a well, and I do not think she has really looked at all the ramifications should her plan go through. We strongly oppose the rezoning of this land—it has a right to be left untouched."

   After more than 2 hours of debate between Ms. Taylor and many of the local residents, the chair of the Zoning Board decided to research the points raised by the neighbors and report on his findings at next month's meeting.

   (a) Name and describe the ecosystem value(s) that are being expressed by Ms. Taylor in her proposal to clear-cut the wooded area. (2 points)
   (b) Name and describe the ecosystem value(s) that Mr. Jared is placing on the wooded area. (2 points)
   (c) Provide three realistic suggestions for Ms. Taylor that could provide her with revenue from the property but leave the woods intact. (3 points)
   (d) Identify and then discuss the validity of the environmental concerns that were raised by Mr. Jared. (3 points)

Grapes used for fine wines grow best in only a few regions of the world that have a similar climate. This vineyard is in Napa Valley, California. *(Rob Hainer/Shutterstock.com)*

Global Climates and Biomes

CHAPTER 4

CASE STUDY

Growing Grapes to Make a Fine Wine

Wine making has its origin in the Mediterranean region, in places such as Egypt, Greece, Italy, France, and Spain. As the European nations began to colonize other parts of the world, they brought grape vines with them. Today wine is made throughout the world, but the regions known for the finest wines are the Mediterranean, California, Chile, South Africa, and southwestern Australia. What is it about these regions that favors the production of great wines?

Wine making critically depends on having proper growing conditions for grapes. The best conditions are mild, moist winters and hot, dry summers. Mild winters are important because temperatures that fall below freezing can damage the grape vines. Hot, dry summers are important because they are moderately stressful for the plants, which causes them to create the perfect balance of sugars and acids in the grapes. The dry climate also reduces outbreaks of many grape vine diseases that are more prevalent in humid environments.

The five regions with the best growing conditions are all situated between 30° and 50° latitude, next to the ocean, and typically on the western side of continents. Their similarity in geographic position causes them to have air and water currents that produce comparable climates that provide similar growing conditions. Because of this, these regions also contain plants that look quite similar. Although the plant species in these five locations are not closely related, they consist of drought-tolerant grasses, wildflowers, and shrubs. In short, the plants that grow naturally in these areas, much like grapes, are well adapted to the local climate.

Because wine grapes grow best under a narrow range of climatic conditions, global climate change is causing great concern among winegrowers.

Because wine grapes grow best under a narrow range of climatic conditions, global climate change is causing great concern among winegrowers. For example, scientists in 2013 evaluated the current and future climates in the wine region of California, where 90 percent of U.S. wine is produced. They predicted that many vineyards will have to shift northward to Oregon and Washington by 2040 because the ideal climate for wine growing will shift northward. Similarly, winemakers in France have experienced a decade of exceptionally hot and dry summers, which has made it difficult to grow the unique varieties of French wine grapes. In 2016, scientists reported that warming temperatures in France caused the date of grape harvesting to be 2 weeks earlier than the average harvest date for the past 400 years. In contrast, English wineries, which are located farther north and have not previously had ideal conditions for growing wine grapes, are now producing some of their best wines because the current English climate is starting to resemble the historic French climate. Moreover, new research predicts that temperatures in England will continue to increase by another 2°C by the year 2100. As winemakers face the reality of global warming and changing climates, they will have to decide whether to move their vineyards to more hospitable climates, modify the growing conditions by adding irrigation, or plant varieties of grapes that are more tolerant to the changing climate, and therefore produce different types of wines.

The story of wine grapes illustrates the point that different regions of the world contain distinct climates and that these climates affect the species that can live in each region. It further demonstrates that when these climates change, we can expect changes in the species that live in the regions and changes in the way humans use these ecosystems.

Sources: J. T. Iverson, How global warming could change the winemaking map, *Time*, December 3, 2009, at http://www.time.com/time/specials/packages/article/0,28804,1929071_1929070_1945282,00.html; L. Hannah et al., Climate change, wine and conservation, *Proceedings of the National Academy of Sciences* 110 (2013): 6907–6912; B. Cook, et al., Climate change decouples drought from early wine grape harvests in France, *Nature Climate Change* 6(2016):715–719; S. Chaudhuri, Climate change uncorks British wine production, *Wall Street Journal,* December 1, 2016, at https://www.wsj.com/articles/britains-wine-production-could-be-boosted-by-climate-change-1480619748

As we have seen with wine grapes, annual patterns of temperature and precipitation help to determine the types of plants and animals that can live in aquatic and terrestrial ecosystems around the world. These annual patterns represent a region's **climate**, which is the average *weather* that occurs in a given region over a long period—typically over several decades. We can contrast climate with the concept of **weather**, which is the short-term conditions of the atmosphere in a local area that include temperature, humidity, clouds, precipitation, and wind speed. To understand how climate differences arise around the world, we need to look at the factors that affect the distribution of heat and precipitation across the globe, which include the unequal heating of Earth, air currents, and ocean currents. Once we understand the pattern of climate differences around the world, we can explore how similar climates support similar types of plants. We can also look at characteristics of different aquatic environments.

Climate The average weather that occurs in a given region over a long period of time.

Weather The short-term conditions of the atmosphere in a local area, which include temperature, humidity, clouds, precipitation, and wind speed.

MODULE 9

The Unequal Heating of Earth

The unequal heating of Earth by the Sun is a major driver of the climates found around the world. To understand how this unequal heating occurs, we need to examine the makeup of Earth's atmosphere. We can then investigate how the angle of sunlight striking Earth affects the area over which the Sun's energy is spread, and the ability of different regions of the world to absorb this energy. Finally, we must consider how the tilt of Earth on its axis affects the energy received from one season to another throughout the year.

Learning Goals

After reading this module you should be able to

- identify the five layers of the atmosphere.
- discuss the factors that cause unequal heating of Earth.
- describe how Earth's tilt affects seasonal differences in temperatures.

Earth's atmosphere is composed of layers

As **FIGURE 9.1** shows, Earth's atmosphere consists of five layers of gases. The pull of gravity on the gas molecules keeps these layers of gases in place. Because each layer of gas has mass, the layers closest to Earth have a greater mass of air above them. This causes the layers closest to Earth to have more densely packed molecules, which causes higher air pressure.

The atmospheric layer closest to Earth's surface, the **troposphere**, extends roughly 16 km (10 miles) above Earth. It is the densest layer of the atmosphere and it is the layer where most of the atmosphere's nitrogen, oxygen, and water vapor occur. The troposphere experiences a great deal of circulation of liquids and gases, and it is the layer where Earth's weather occurs. Air temperature in the troposphere decreases with distance from Earth's surface and varies with latitude. Temperatures can fall as low as −52°C (−62°F) near the top of the troposphere.

Above the troposphere is the **stratosphere**, which extends roughly 16 to 50 km (10–31 miles) above Earth's surface. The stratosphere is less dense than the troposphere. Ozone, a pale blue gas composed of molecules made up of three oxygen atoms (O_3), forms a layer within the stratosphere. This ozone layer absorbs most of the Sun's ultraviolet-B (UV-B) radiation and all of its ultraviolet-C (UV-C) radiation. UV radiation can cause DNA damage and cancer in organisms, so the stratospheric ozone layer provides critical protection for our planet. The upper layers in the stratosphere absorb the UV radiation and convert it to infrared radiation, which is released as heat. Because UV radiation from the Sun reaches the higher altitudes of the stratosphere, where much of the UV radiation is absorbed first, there is less UV radiation remaining to be absorbed in the lower stratosphere. As a result, the upper stratosphere is warmer than the lower stratosphere.

Beyond the stratosphere are the mesosphere, the thermosphere, and the exosphere. The atmospheric pressure and density in each of these layers continues to decrease as we move toward the outer layers of the atmosphere. The thermosphere is particularly important to organisms on Earth's surface because of its ability to block harmful X-ray and UV radiation from reaching our planet. The thermosphere is also interesting because it contains charged gas molecules that, when hit by solar energy, begin to glow and produce light, in the same way that a light bulb glows when

Troposphere A layer of the atmosphere closest to the surface of Earth, extending up to approximately 16 km (10 miles).

Stratosphere The layer of the atmosphere above the troposphere, extending roughly 16 to 50 km (10–31 miles) above the surface of Earth.

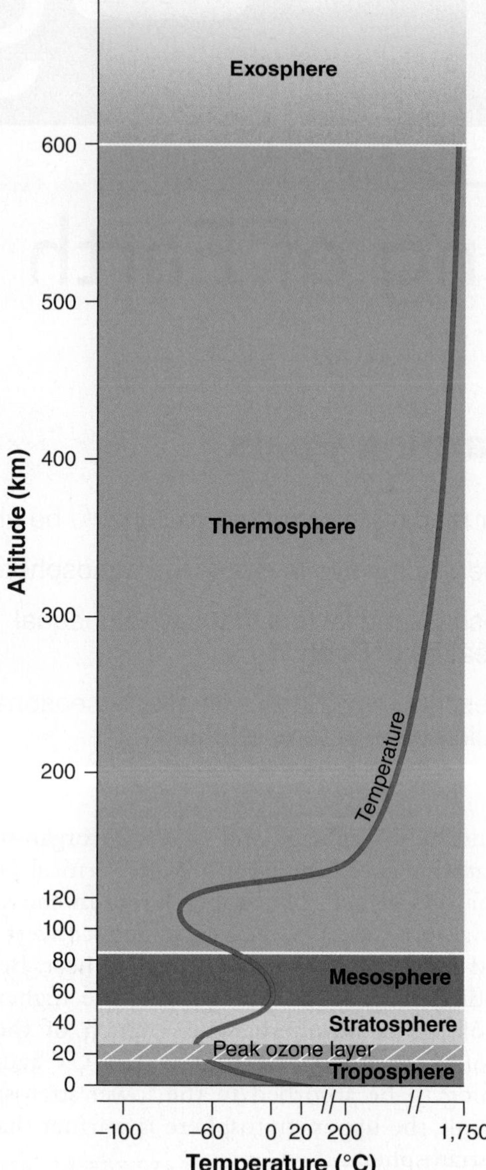

FIGURE 9.1 The layers of Earth's atmosphere. The troposphere is the atmospheric layer closest to Earth. Because the density of air decreases with altitude, the troposphere's temperature also decreases with altitude. Temperature increases with altitude in the stratosphere because the Sun's UV-B and UV-C rays warm the upper part of this layer. Temperatures in the thermosphere can reach 1,750°C (3,182°F).

electricity is applied. Because this interaction between solar energy and gas molecules is driven most intensely by magnetic forces at the North Pole and South Pole, the best places to view the phenomenon are at high latitudes. In the northern United States, Canada, and northern Europe, these glowing gases are known as the northern lights, or aurora borealis. In Australia and southern South America, they are called the southern lights, or aurora australis (**FIGURE 9.2**).

FIGURE 9.2 Northern lights. The glowing, moving lights that are visible at high latitudes in both hemispheres are the product of solar radiation energizing the gases of the thermosphere. *(Stephen Mcsweeny/Shutterstock)*

The amount of solar energy reaching Earth varies with location

Now that we know something about Earth's atmosphere, we can take a closer look at the processes that affect heat and precipitation distribution. As the Sun's energy passes through the atmosphere and strikes land and water, it warms the planet's surface. But this warming does not occur evenly across the planet. This uneven warming pattern has three primary causes.

The first cause of unequal warming is variation in the angle at which the Sun's rays strike Earth. As we can see in **FIGURE 9.3**, in the region nearest to the equator—the tropics—the Sun strikes at a perpendicular, or right, angle. In the mid-latitude and polar regions, the Sun's rays strike at a more oblique angle. As a result, the Sun's rays travel a relatively short distance through the atmosphere to reach Earth's surface in the tropics but they must travel a longer distance through the atmosphere to reach Earth's surface near the poles. Because solar energy is lost as it passes through the atmosphere, more solar energy reaches the equator than the mid-latitude and polar regions.

The second cause of the uneven warming of Earth is variation in the amount of surface area over which the Sun's rays are distributed. As you may know, the Sun's rays strike Earth at different angles in different places on the globe, and the angle can change with the time of year. When the Sun's rays strike near the equator, the solar energy is distributed over a smaller surface area than near the poles. Thus, regions near the equator receive more solar energy per square meter than mid-latitude and polar regions.

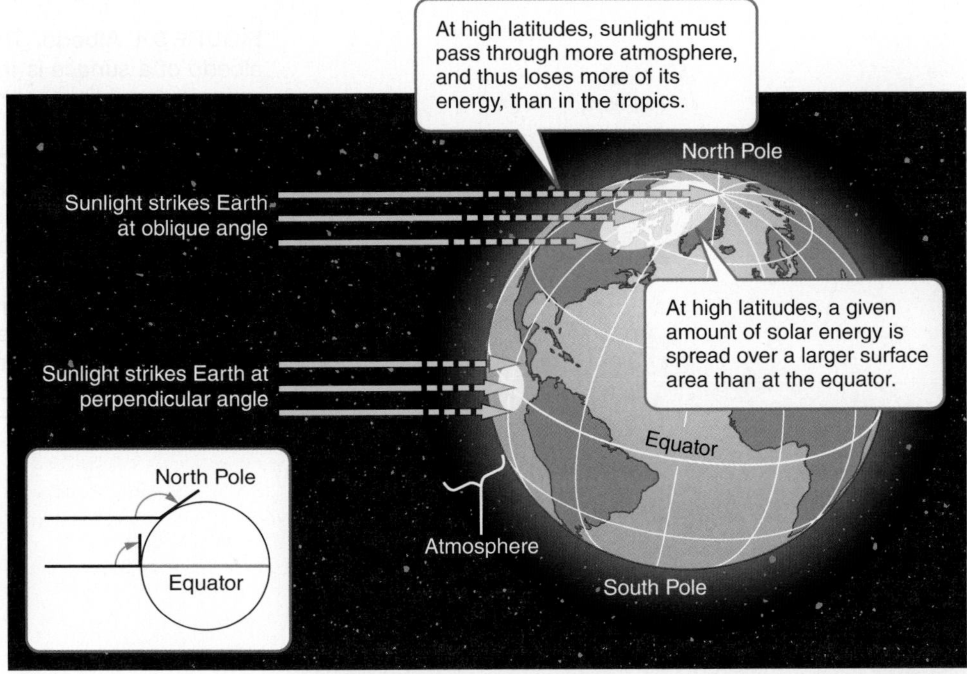
At high latitudes, sunlight must pass through more atmosphere, and thus loses more of its energy, than in the tropics.
North Pole
Sunlight strikes Earth at oblique angle
At high latitudes, a given amount of solar energy is spread over a larger surface area than at the equator.
Sunlight strikes Earth at perpendicular angle
Equator
North Pole
Equator
Atmosphere
South Pole

FIGURE 9.3 Differential heating of Earth. Tropical regions near the equator receive more solar energy than mid-latitude and polar regions, where the Sun's rays strike Earth's surface at an oblique angle.

You can replicate this phenomenon by shining a flashlight onto a round object, such as a basketball, in a dark room. If you shine the light perpendicular to the surface of the ball, you will create a small circle of bright light. If you shine the flashlight at an oblique angle, you will create a large oval pool of dim light because the light is distributed over a larger area.

Finally, some areas of Earth reflect more solar energy than others. The percentage of incoming sunlight that is reflected from a surface is called its **albedo**. The higher the albedo of a surface, the more solar energy it reflects and the less it absorbs. A white surface has a higher albedo than a black surface, so it tends to stay cooler. **FIGURE 9.4** on page 112 shows albedo values for various surfaces on Earth. Although Earth has an average albedo of 30 percent, tropical regions with dense green foliage have albedo values of 10 to 20 percent, whereas the snow-covered polar regions have values of 80 to 95 percent.

Earth's tilt causes seasonal changes in climate

As we saw in the previous section, differences in the amount of solar energy striking various latitudes on Earth depend on the angle of the Sun's rays. For the same reason, the amount of solar energy reaching various latitudes shifts over the course of the year. Because Earth's axis of rotation is tilted 23.5°, Earth's orbit around the Sun causes most regions of the world to experience seasonal changes in temperature and precipitation. Specifically, when the Northern Hemisphere is tilted toward the Sun, the Southern Hemisphere is tilted away from the Sun, and vice versa.

FIGURE 9.5 on page 112 will help us visualize how this works. The Sun's rays strike the equator directly twice a year: during the March equinox, on March 20 or 21, and again during the September equinox, on September 22 or 23. On those days, all regions of Earth except those nearest the poles receive 12 hours of daylight and 12 hours of darkness. For the 6 months between the March and September equinoxes, the Northern Hemisphere tilts toward the Sun, experiencing more hours of daylight than darkness. The opposite is true in the Southern Hemisphere. On June 20 or 21, the Sun is directly above the Tropic of Cancer at 23.5° N latitude. On this day—the June solstice—the Northern Hemisphere experiences more daylight hours than on any other day of the year.

Albedo The percentage of incoming sunlight reflected from a surface.

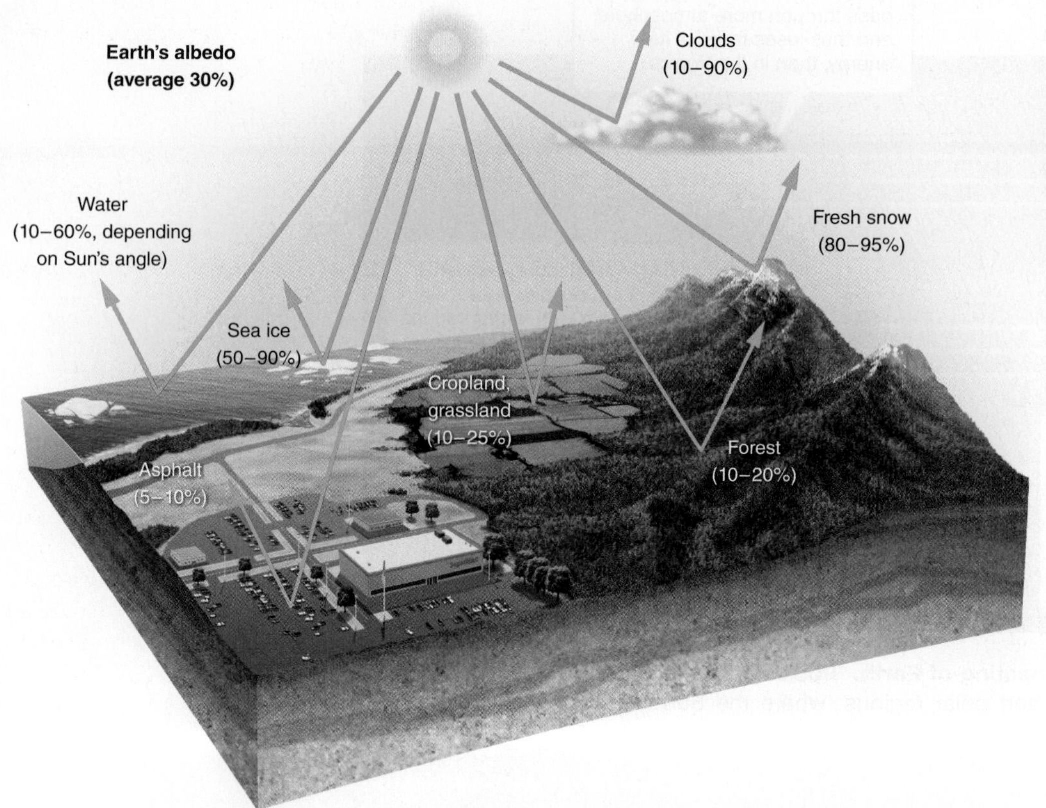

FIGURE 9.4 Albedo. The albedo of a surface is the percentage of the incoming solar energy that it reflects. Snow and ice reflect much of the solar energy that they receive, but darker objects such as forests and asphalt paving reflect very little energy, which means that they absorb most of the solar energy that strikes them.

June solstice
The Northern Hemisphere is maximally tilted toward the Sun and experiences the longest day of the year. Summer begins in the Northern Hemisphere. Winter begins in the Southern Hemisphere.

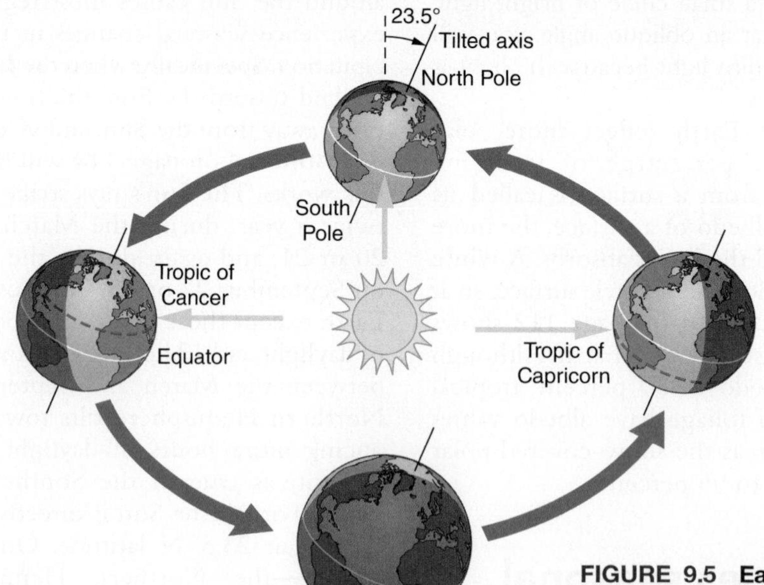

FIGURE 9.5 Earth's seasons. Because Earth's axis of rotation is tilted 23.5°, the latitude that receives the most direct rays of the Sun and the most hours of daylight changes throughout the year as Earth orbits the Sun. Thus Earth's tilt produces predictable seasons. This diagram illustrates the pattern of seasons in the Northern Hemisphere.

For the 6 months between the September and March equinoxes, the Northern Hemisphere tilts away from the Sun, experiencing fewer hours of daylight than darkness. On December 21 or 22—the December solstice—the Sun is directly over the Tropic of Capricorn at 23.5° S latitude. On this day, the Northern Hemisphere experiences its shortest daylight period of the year, and the Southern Hemisphere experiences its longest daylight period of the year.

MODULE 9

Preparing for the AP® Exam

AP® Review

In this module, we have learned that the atmosphere has five layers, each with different properties. These layers absorb most of the UV radiation and X-rays that are emitted toward Earth from the Sun, while much of the Sun's remaining energy passes through the atmosphere to strike Earth's surface. The amount of solar energy that strikes the surface of Earth at any location depends on the angle of the Sun's rays and the albedo. Earth's tilted axis causes the planet's unequal heating to shift throughout the year, which produces seasonal changes in the amount of solar energy that strikes different regions of the world. In the next module, we will examine how this unequal heating of Earth affects the circulation of air currents, which, in turn, help determine global climates.

AP® Practice Questions

Choose the best answer for the following.

1. Which list is in the correct order of atmospheric layers starting from Earth's surface?
   (a) exosphere, troposphere, mesosphere, stratosphere, thermosphere
   (b) mesosphere, stratosphere, thermosphere, exosphere, troposphere
   (c) thermosphere, troposphere, stratosphere, exosphere, mesosphere
   (d) troposphere, stratosphere, mesosphere, thermosphere, exosphere
2. Which contributes to the unequal warming of Earth?
   I. variation in the angle of sunlight that reaches Earth
   II. variation in the amount of surface area over which the Sun's rays are distributed
   III. variation in the amount of sunlight reflected from clouds
   (a) I only
   (b) I and II
   (c) II and III
   (d) III only
3. Which has the highest albedo?
   (a) soil
   (b) snow
   (c) water
   (d) pavement
4. Summer in the Northern Hemisphere is warmer primarily because of
   I. increased atmospheric absorption of solar radiation.
   II. the tilt of Earth's axis.
   III. reduced albedo.
   (a) I only
   (b) I and II
   (c) II only
   (d) II and III
5. Which atmospheric layer contains the protective ozone layer?
   (a) exosphere
   (b) stratosphere
   (c) thermosphere
   (d) troposphere

MODULE 10

Air Currents

The unequal heating of Earth has wide-ranging effects on the climates of our planet not only because it determines regional temperatures, but also because it drives the circulation of air around the planet. These air currents bring warm and cold air to different regions. They also move moisture around the planet, which causes some regions to receive more precipitation than others. In this module, we will explore how air currents have a major impact on the climates of the world. We will begin by examining how the properties of air affect the amount of moisture it can carry. We will then explore how the unequal heating of Earth causes air to circulate along the ground and up into the atmosphere. As we will see, the direction of this air movement is also affected by the rotation of Earth. Finally, we will discuss how air currents that travel over mountain ranges can cause different climates to exist on opposite sides of the mountains.

Learning Goals

After reading this module you should be able to

- explain how the properties of air affect the way it moves in the atmosphere.
- identify the factors that drive atmospheric convection currents.
- describe how Earth's rotation affects the movement of air currents.
- explain how the movement of air currents over mountain ranges affects climates.

Air has several important properties that determine how it circulates in the atmosphere

Air has four properties that determine how it circulates in the atmosphere: density, water vapor capacity, adiabatic heating or cooling, and latent heat release.

The first property is air density, which is the mass of all molecules in the air in a given volume. The density of air determines its movement: Less dense air rises, whereas denser air sinks. At a constant atmospheric pressure, warm air has a lower density than cold air. Because of this density difference, warm air rises, whether in a room in your house or in the atmosphere.

Saturation point The maximum amount of water vapor in the air at a given temperature.

Adiabatic cooling The cooling effect of reduced pressure on air as it rises higher in the atmosphere and expands.

Adiabatic heating The heating effect of increased pressure on air as it sinks toward the surface of Earth and decreases in volume.

The second property that determines how air circulates is its capacity to contain water vapor. For example, warm air is not only less dense than cold air, it also has a higher capacity for water vapor. In regions of the world that receive substantial amounts of precipitation, hot summer days are associated with high humidity because the warm air contains a lot of water vapor. The maximum amount of water vapor that can be in the air at a given temperature is called its **saturation point**. **FIGURE 10.1** shows the relationship between the temperature of air and its saturation point. When the temperature of air falls, its saturation point decreases, water vapor condenses into liquid water, clouds form, and precipitation occurs.

A third property of air is its response to changes in pressure. As air rises higher in the atmosphere, the pressure on it decreases. The lower pressure allows the rising air to expand in volume, and this expansion lowers the temperature of the air. The cooling effect of reduced pressure on air as it rises in the atmosphere and expands is called **adiabatic cooling**. Conversely, when air sinks toward Earth's surface, the pressure on it increases. The higher pressure forces the air to decrease in volume, and this decrease raises the temperature of the air. The heating effect of increased pressure on air as it sinks toward the surface of Earth and decreases in volume is called **adiabatic heating**.

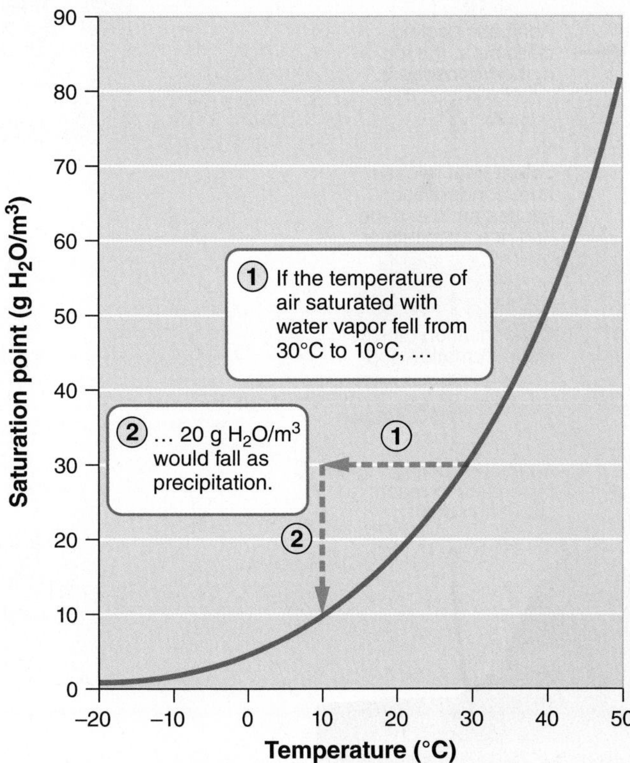

FIGURE 10.1 The saturation point of air. When air cools and its saturation point drops, water vapor condenses into liquid water that forms clouds. These clouds are ultimately the source of precipitation.

The final property of air that determines how it circulates is the production of heat when water vapor condenses from a gas to a liquid. As you may know, the Sun provides the energy necessary to evaporate water on Earth's surface and convert it into water vapor, which enters the atmosphere. In the reverse process, when water vapor in the atmosphere condenses into liquid water, energy is released as heat. The release of energy when water vapor in the atmosphere condenses into liquid is known as **latent heat release**. Because of latent heat release, whenever water vapor in the atmosphere condenses, the air will become warmer and rise in the atmosphere.

Atmospheric convection currents move air and moisture around the globe

A major contributor to the different climates of the world is *atmospheric convection currents.* **Atmospheric convection currents** are global patterns of air movement that are initiated by the unequal heating of Earth. These air currents are called convection currents because they involve the movement of air that absorbs and releases heat as the air moves up into the atmosphere and then back to Earth. Now that we understand the four properties of air, we can look at how these properties contribute to atmospheric convection currents. In the next section we will consider how these atmospheric currents are deflected by Earth's rotation.

We can examine how atmospheric currents develop in **FIGURE 10.2** on page 116. The warming of humid air at Earth's surface decreases the density of the air. As a result, the air begins to rise and, as it does, it begins to experience lower atmospheric pressures and adiabatic cooling. The cooling causes the air to reach its saturation point, which leads to condensation that causes cloud formation and precipitation. Condensation also causes latent heat release, which offsets some of the adiabatic cooling and becomes a strong driving force to make the air expand further and rise higher. Collectively, these processes cause air to rise continuously from Earth's surface near the equator, forming a river of air flowing upward into the troposphere.

As the air reaches the top of the troposphere, it is chilled by adiabatic cooling. This air now contains relatively little water vapor. As warmer air continually rises from below, the cold, dry air at the top of the troposphere is displaced horizontally and eventually begins to sink back to Earth's surface. As the air sinks, it experiences higher atmospheric pressures, and its reduction in volume causes adiabatic heating. By the time the air reaches Earth's surface, it is hot and dry. This hot and dry air circulates back to the starting point and picks up moisture along the way.

Atmospheric convection currents on Earth are found at specific locations. For example, **Hadley cells** are convection currents that cycle between the equator and approximately 30° N and 30° S. In a Hadley cell, illustrated in **FIGURE 10.3** on page 117, the unequal warming of the tropics causes the air near the equator to expand and rise. This rising air cools and the condensed water falls as precipitation over the tropics. After being chilled by adiabatic cooling near the top of the troposphere, the cold dry air is displaced horizontally north and south of the equator. This air descends back to Earth at approximately 30° N and 30° S. As it sinks, adiabatic heating causes the air to warm. As a result, regions at 30° N and 30° S are typically hot, dry deserts. Once the warm, dry air reaches Earth, it flows back toward the equator to complete the cycle.

The circulation of the Hadley cells is driven by the intense solar energy that strikes Earth near the equator. The latitude that receives the most intense sunlight, which causes the ascending branches of the two

Latent heat release The release of energy when water vapor in the atmosphere condenses into liquid water.

Atmospheric convection current Global patterns of air movement that are initiated by the unequal heating of Earth.

Hadley cell A convection current in the atmosphere that cycles between the equator and 30° N and 30° S.

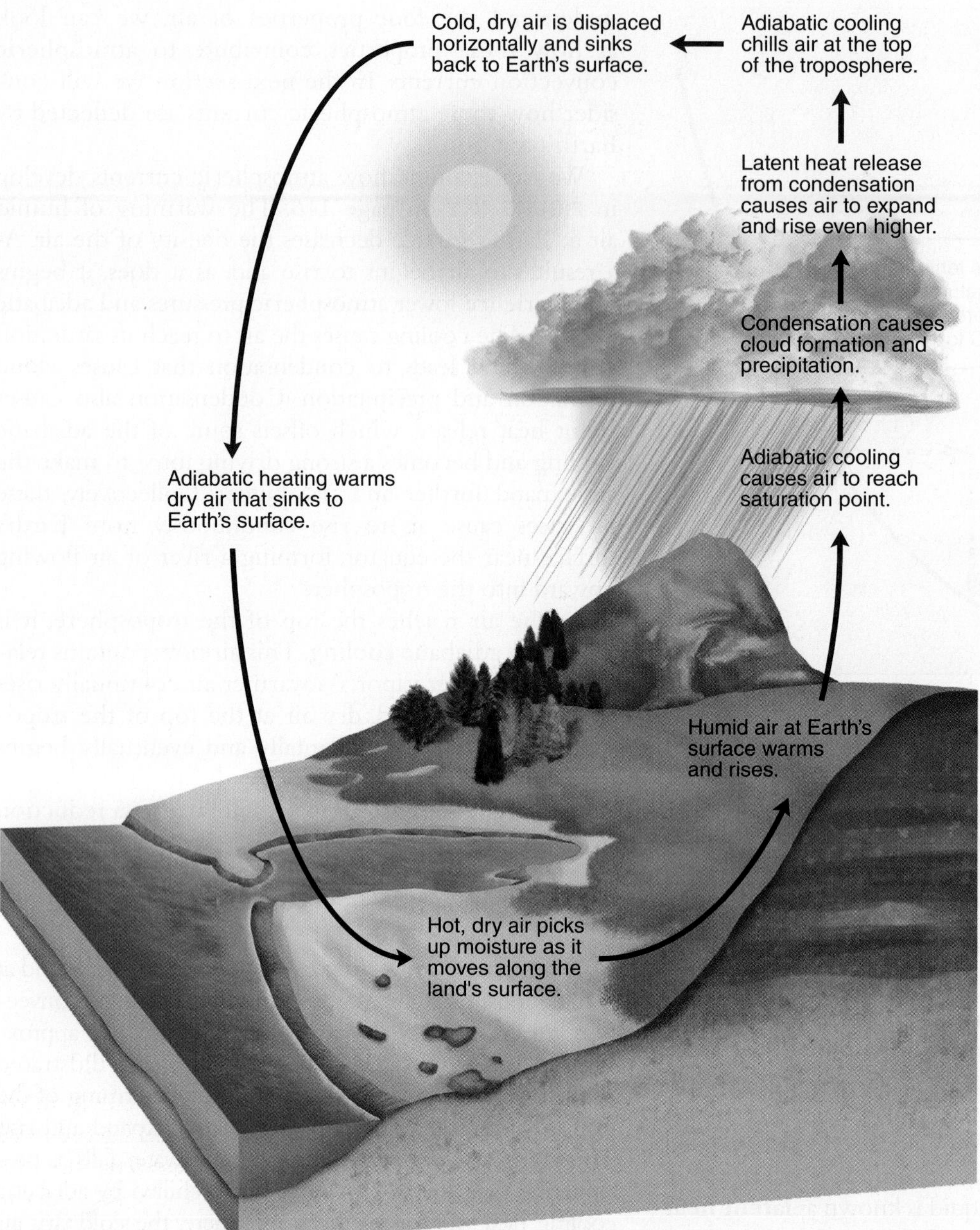
Cold, dry air is displaced horizontally and sinks back to Earth's surface.
Adiabatic cooling chills air at the top of the troposphere.
Latent heat release from condensation causes air to expand and rise even higher.
Condensation causes cloud formation and precipitation.
Adiabatic cooling causes air to reach saturation point.
Adiabatic heating warms dry air that sinks to Earth's surface.
Humid air at Earth's surface warms and rises.
Hot, dry air picks up moisture as it moves along the land's surface.

FIGURE 10.2 Atmospheric currents. Warming at Earth's surface causes air to rise up into the atmosphere where it experiences lower pressures, adiabatic cooling, and latent heat release. The cool air near the top of the atmosphere is then displaced horizontally before it sinks back to Earth. As it sinks, the air experiences adiabatic heating and then moves horizontally along the surface of Earth to complete the cycle.

Hadley cells to converge, is called the **intertropical convergence zone (ITCZ)**. Because humid air rises at this latitude and precipitation falls, the ITCZ is typified by dense clouds and intense thunderstorm activity.

Intertropical convergence zone (ITCZ) The latitude that receives the most intense sunlight, which causes the ascending branches of the two Hadley cells to converge.

Polar cell A convection current in the atmosphere, formed by air that rises at 60° N and 60° S and sinks at the poles, 90° N and 90° S.

The latitude of the ITCZ is not fixed. Because Earth's axis of rotation is tilted, the area receiving the most intense sunlight shifts between approximately 23.5° N and 23.5° S as Earth orbits the Sun. Because the ITCZ occurs at the latitudes that receive the most intense sunlight, the ITCZ moves north and south of the equator over the course of a year. As a result, the tropics experience seasons of high and low precipitation.

Another set of atmospheric convection currents occur near the poles. **Polar cells** are convection currents

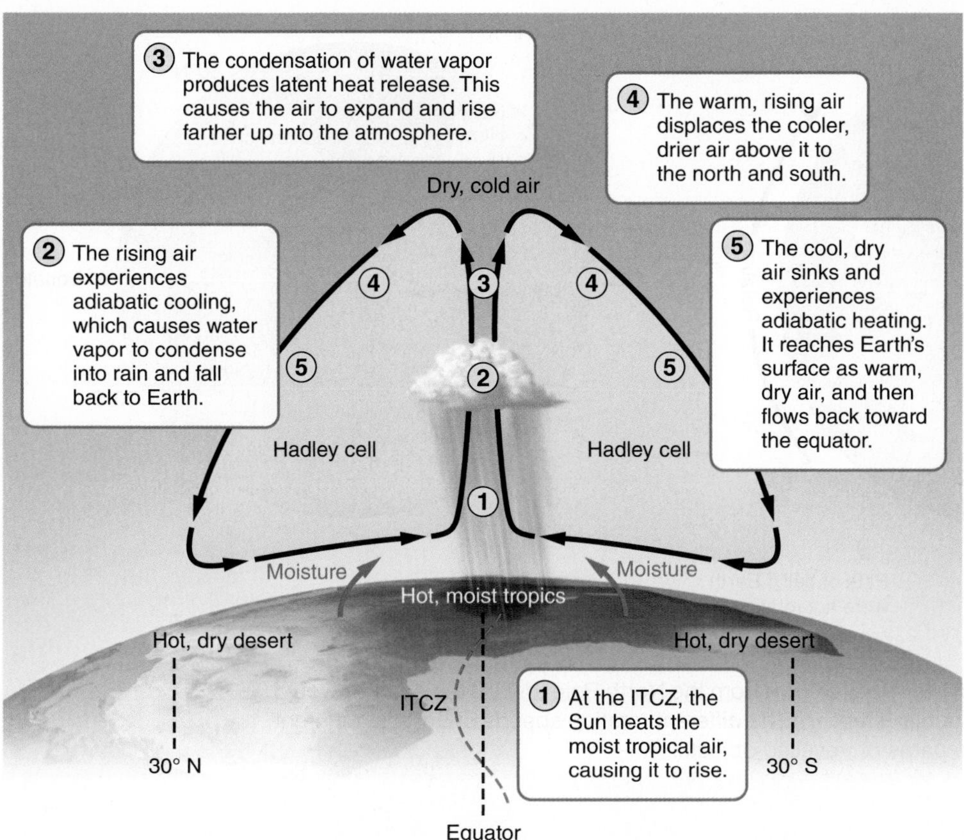

FIGURE 10.3 Hadley cells. Hadley cells are atmospheric convection currents that operate between the equator and 30° N and 30° S. Solar energy warms humid air in the tropics. The warm air rises and eventually cools below its saturation point. The water vapor it contains condenses into clouds and precipitation. The air, which now contains little moisture, sinks to Earth's surface at approximately 30° N and 30° S. As the air descends, it is warmed by adiabatic heating. This descent of hot, dry air causes desert environments to develop at those latitudes.

formed by air that rises at 60° N and 60° S and sinks at the poles (90° N and 90° S). At 60° N and 60° S, the rising air cools and the water vapor condenses into precipitation. The air dries as it moves toward the poles, where it sinks back to Earth's surface. At the poles, the air moves back toward 60° N and 60° S, completing the cycle.

A third convection current, known as **Ferrell cells**, lies between Hadley cells and polar cells. Air currents at these latitudes do not form distinct convection cells, but are driven by the circulation of the neighboring Hadley cells and polar cells. At Earth's surface, some of the warmer air from the Hadley cells moves toward the poles from 30° N and 30° S, and some of the cooler air from the polar cells moves toward the equator from 60° N and 60° S. This movement not only helps to distribute warm air away from the tropics and cold air away from the poles, but also allows a wide range of warm and cold air currents to circulate between 30° and 60°. In this latitudinal range, which includes most of the United States, wind direction can be quite variable, both at Earth's surface and at the top of the troposphere.

Collectively, these convection currents slowly move the warm air of the tropics toward the mid-latitude and polar regions. Because these currents determine the patterns of temperature and precipitation around the world, they are largely responsible for the locations of rainforests, deserts, and grasslands on Earth.

Earth's rotation causes the Coriolis effect

The atmospheric convection currents cause air to move directly north or south. However, the actual air currents at Earth's surface are deflected to the east or west because our planet is rotating. The rotation of Earth causes the deflection of objects that are moving directly north or south, a phenomenon that we call the **Coriolis** (core-ee-oh-lis) **effect**.

Ferrell cell A convection current in the atmosphere that lies between Hadley cells and polar cells.

Coriolis effect The deflection of an object's path due to the rotation of Earth.

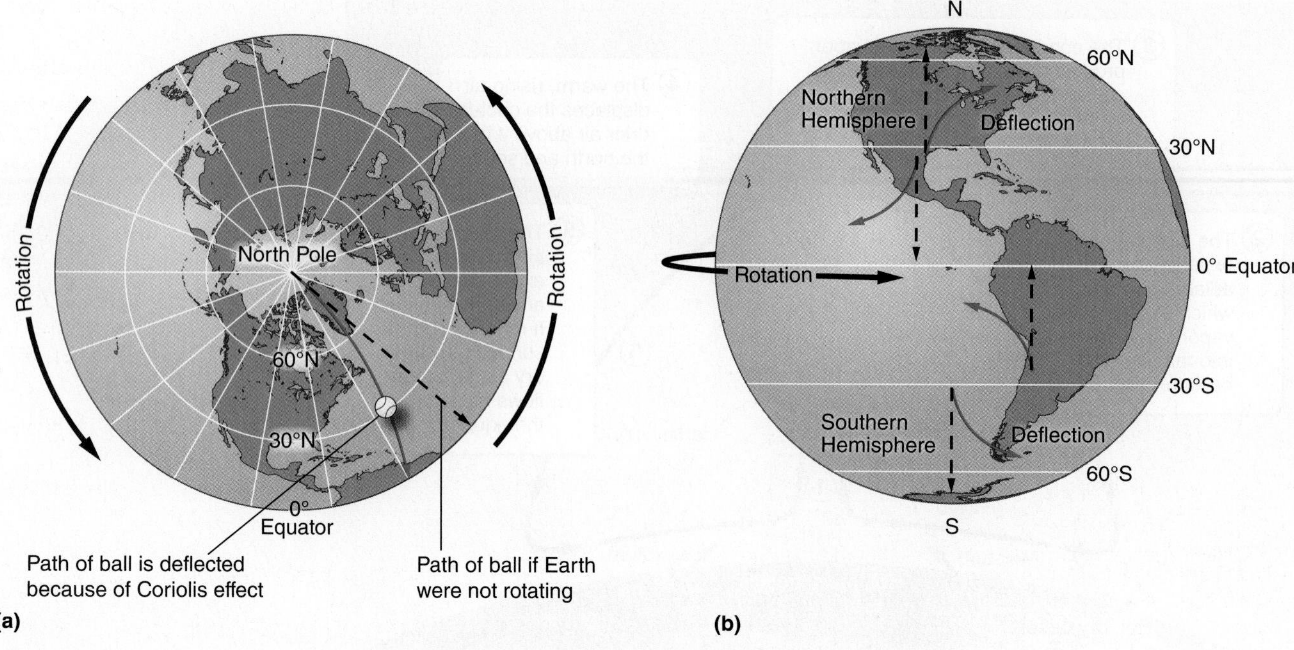

FIGURE 10.4 The Coriolis effect. (a) A ball thrown from the North Pole toward the equator would be deflected to the west by the Coriolis effect. (b) The different rotation speeds of Earth at different latitudes cause a deflection in the paths of traveling objects.

Look at **FIGURE 10.4a** and imagine that you can stand at the North Pole and throw a ball directly south, all the way down to the equator (0° latitude). If Earth did not rotate, the ball would travel south to the equator in a straight line. But because Earth rotates to the east while the ball is traveling through the air, the ball will land in a location that is west of its intended target. This happens because Earth's surface beneath the ball is moving faster and faster as the ball moves toward the equator. In other words, the path of the ball is deflected with respect to a given location on the globe because Earth is rotating. Now imagine you are standing at 30° N or 30° S latitude. As shown in Figure 10.4b, throwing a ball toward the equator results in a deflection of the ball's path to the west. However, throwing a ball toward the nearest pole results in a deflection of the ball's path to the east.

The deflection of objects moving north or south occurs because the planet's surface moves much faster at the equator than at higher latitudes. The planet's circumference is 40,000 km (25,000 miles) at the equator, but decreases to 7,000 km (4,350 miles) at 80° N or 80° S latitude. Now imagine traveling all the way around Earth in 24 hours at the equator versus at 80° N latitude. **FIGURE 10.5** will help you visualize this journey. At the equator, you must travel 40,000 km; at 80° N latitude you must travel only 7,000 km. To make the trips in 24 hours, you must travel much faster at the equator. Indeed, Earth's surface moves at 1,670 km per hour (1,038 miles per hour) at the equator but only 291 km per hour (181 miles per hour) at 80° N or 80° S.

If Earth did not rotate, the air in the convection cells that we discussed earlier would simply move directly north or south and cycle back again. For example, the air in a Hadley cell sinks to Earth's surface at approximately 30° latitude and travels along Earth's surface toward the equator. However, the Coriolis effect produced by Earth's rotation deflects the path of atmospheric convection currents that move toward or

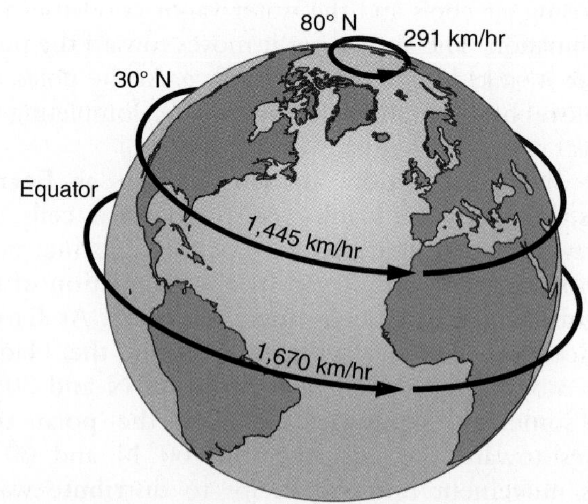

FIGURE 10.5 The speed of Earth's rotation. Because all locations on Earth complete one revolution every 24 hours, and because Earth has a greater circumference near the equator than near the poles, its speed of rotation is much faster at the equator than near the poles.

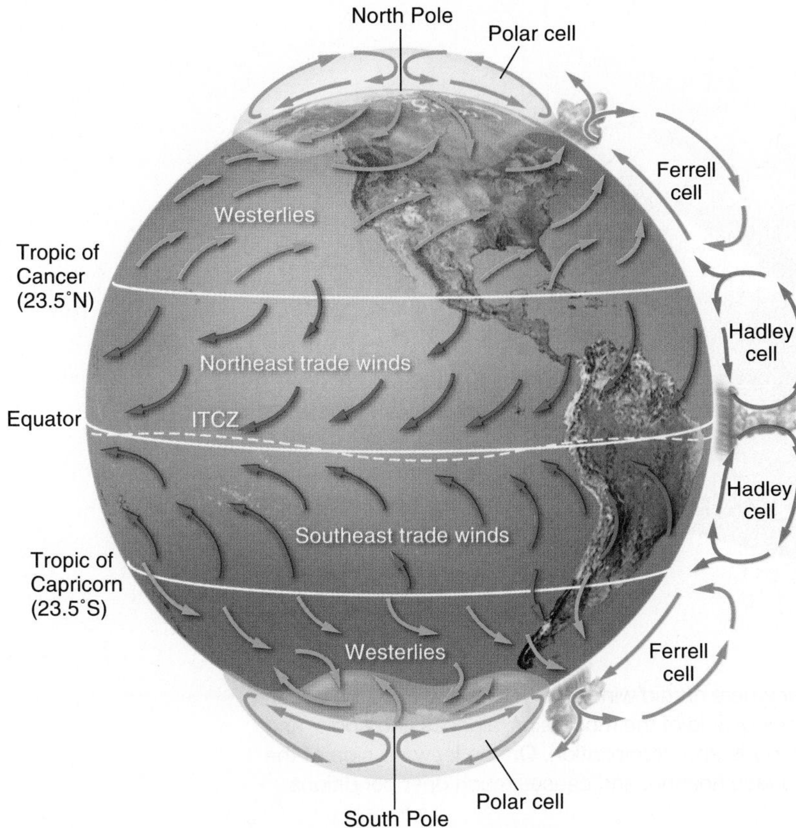

FIGURE 10.6 Prevailing wind patterns. Prevailing wind patterns around the world are produced by a combination of atmospheric convection currents and the Coriolis effect.

away from the equator, as we can see in **FIGURE 10.6**. Because Earth's speed of rotation is faster near the equator, the path of the air current moving toward the equator is deflected to the west, just as we saw in our example of throwing a ball toward the equator. This deflection causes the Hadley cell north of the equator to produce prevailing winds along Earth's surface that come from the northeast (called northeast trade winds), whereas the cell south of the equator produces prevailing winds that come from the southeast (called southeast trade winds).

The Coriolis effect also explains the prevailing wind directions in the mid-latitudes (between 30° and 60°) where we find the Ferrell cells. These winds can be quite variable as a result of the mixing of air currents from the adjacent Hadley cells and polar cells. Closer to 30°, air tends to move along Earth's surface toward the poles. If Earth were not rotating, this air would move straight north in the Northern Hemisphere and straight south in the Southern Hemisphere. Because Earth is rotating faster at 30° than at 60°, this air movement toward the poles is deflected to the east. As a result, the prevailing winds above 30° in the Northern Hemisphere come from the southwest, while in the Southern Hemisphere the prevailing winds above 30° are from the northwest. In both cases, these winds are called westerlies.

Finally, the Coriolis effect helps us understand the prevailing wind directions in the polar regions. Along Earth's surface, the polar cells move air away from the poles and toward 60° latitude. Since Earth is rotating faster at 60° than it is at 90°, the air movement is deflected to the west. As a result, polar winds come out of the northeast in the Northern Hemisphere and out of the southeast in the Southern Hemisphere. These winds are called easterlies.

Rain shadows cause mountains to be dry on one side

Although many processes that affect weather and climate operate on a global scale, local features, such as mountain ranges, can also play a role. Air moving inland from the ocean often contains a large amount of water vapor. As shown in **FIGURE 10.7** on page 120, when this air meets the windward side of a mountain range—the side facing the wind—it rises and begins to experience adiabatic cooling. Because water vapor condenses as air cools, clouds form and precipitation falls. As is the case in Hadley cells, this condensation causes latent heat release, which helps to accelerate the upward movement of the air. Thus the presence of the mountain range causes large amounts of precipitation to fall on its windward side. The cold, dry air then travels to the other side of the mountain range—called the leeward side—where it descends and experiences higher pressures, which cause adiabatic heating. This now warm, dry air produces arid conditions on the leeward side of the range. It is common to see lush vegetation growing on the windward side of a mountain range while very dry conditions prevail on the leeward side. The dry region formed on the leeward side of the mountain as a result of the humid winds from the ocean that cause precipitation on the windward side is known as a **rain shadow**.

Rain shadow A region with dry conditions found on the leeward side of a mountain range as a result of humid winds from the ocean causing precipitation on the windward side.

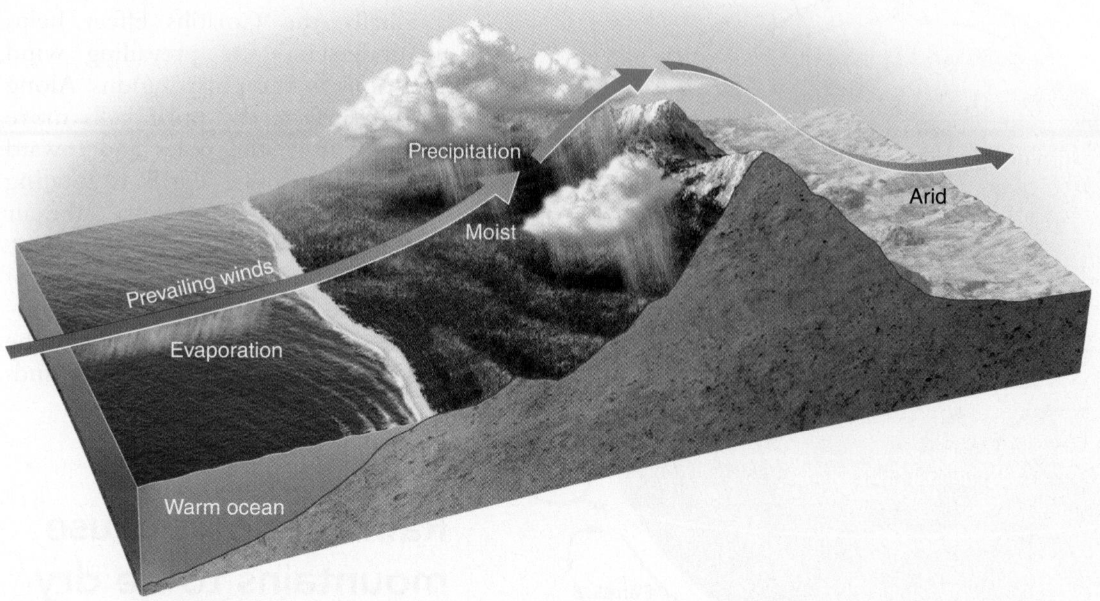

FIGURE 10.7 Rain shadow. Rain shadows occur where humid winds blowing inland from the ocean meet a mountain range. On the windward (wind-facing) side of the mountains, air rises and cools, and large amounts of water vapor condense to form clouds and precipitation. On the leeward side of the mountains, cold, dry air descends, warms via adiabatic heating, and causes much drier conditions.

MODULE 10 AP® Review

Preparing for the AP® Exam

In this module, we have seen that the properties of air affect how it moves in the atmosphere and how much water vapor it contains. These properties help us understand how the intense sunlight at latitudes near the equator drives the formation of Hadley cells, which are responsible for also producing the polar cells and Ferrell cells. These air convection currents, combined with the Coriolis effect, cause the predominant wind directions that exist around the planet. Air convection currents and the effects of rain shadows help to determine the distribution of heat and precipitation around the globe. In the next module, we will see that many of these processes also drive ocean currents, which also affect climates throughout the world.

AP® Practice Questions

Choose the best answer for the following.

1. Temperature change in adiabatic heating occurs due to
   (a) water vapor condensing.
   (b) a pressure increase.
   (c) absorption of sunlight.
   (d) a pressure decrease.

2. Latent heat release
   (a) increases air temperature.
   (b) decreases air temperature.
   (c) increases air pressure.
   (d) decreases air pressure.

3. Which does NOT contribute to atmospheric convection currents?
   (a) Hadley cells
   (b) gyres
   (c) polar cells
   (d) the intertropical convergence zone
4. The maximum amount of water vapor air can hold at a given temperature is called the
   (a) saturation point.
   (b) absolute humidity.
   (c) dew point.
   (d) vapor pressure.
5. The Coriolis effect is responsible for
   (a) the convergence of two ascending Hadley cells.
   (b) the location of atmospheric currents caused by the unequal heating of Earth.
   (c) the cooling effect of air as it rises and expands.
   (d) the deflection of atmospheric currents.
6. Which does NOT play a role in causing a rain shadow?
   (a) cold polar air
   (b) a mountain range
   (c) adiabatic cooling
   (d) humid ocean air
7. Which is the correct order of the convection cells, starting from the equator?
   (a) Ferrell, Hadley, polar
   (b) polar, Hadley, Ferrell
   (c) Hadley, Ferrell, polar
   (d) Hadley, polar, Ferrell

MODULE 11

Ocean Currents

We have seen that the unequal heating of Earth together with the Coriolis effect drives air currents around the world. These two phenomena also affect ocean currents. In this module, we will explore the circulation of ocean waters, both at the surface and in the deep ocean. We will see how ocean circulation affects the transport of heat around the globe, how this heat movement affects the climates of continents, and how disruptions to these circulation patterns can dramatically alter global climates.

Learning Goals

After reading this module you should be able to

- describe the patterns of surface ocean circulation.
- explain the mixing of surface and deep ocean waters from thermohaline circulation.
- identify the causes and consequences of the El Niño–Southern Oscillation.

Surface ocean currents move warm and cold water around the globe

The flow of ocean water is an important factor for global climates because it moves warm and cold waters to different parts of the globe. For example, the Gulf Stream is a current of warm water that emerges from the Gulf of Mexico and warms much of the eastern coast of the United States. In doing so, ocean currents affect the primary productivity found in different ocean regions as well as the climate of the adjacent continents. When warm ocean currents flow from the tropics to higher latitudes, the neighboring continents receive warm air from the

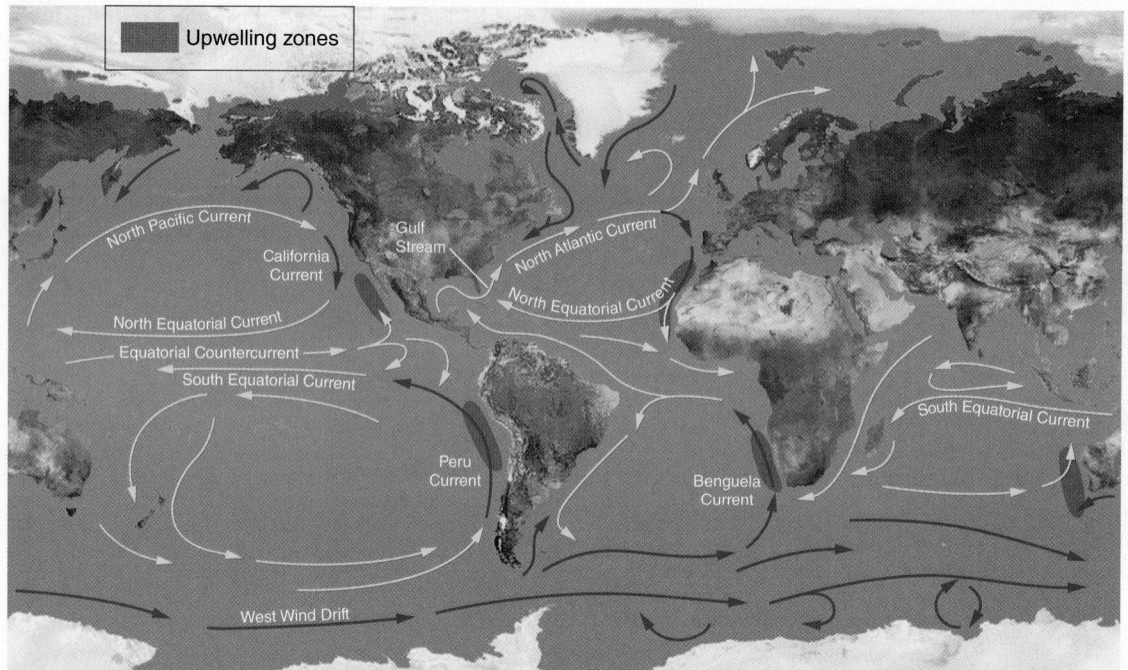

FIGURE 11.1 Oceanic circulation patterns. Oceanic circulation patterns are the result of differential heating, gravity, prevailing winds, the Coriolis effect, and the locations of continents. Each of the five major ocean basins contains a gyre driven by the trade winds in the tropics and the westerlies at mid-latitudes. The result is a clockwise circulation pattern in the Northern Hemisphere and a counterclockwise circulation pattern in the Southern Hemisphere. Along the west coasts of many continents, currents diverge and cause the upwelling of deeper and more fertile water.

ocean. In this section, we will examine the major ocean currents that occur near the oceans and then consider currents that mix the surface water with much deeper water.

Ocean currents are driven by a combination of temperature, gravity, prevailing winds, the Coriolis effect, salinity, and the locations of continents. As we have already observed, the tropics receive the most direct sunlight throughout the year, which generally makes tropical waters warm. Warm water, like warm air, expands and rises. This process raises the tropical water surface about 8 cm (3 inches) higher than mid-latitude waters. While this difference might seem trivial, the slight slope is sufficient for the force of gravity to make water flow away from the equator. As we will see, this movement away from the equator due to gravity combines with other forces to make the surface waters circulate.

Gyres

As the warm tropical waters are pushed away from the equator by gravity, prevailing wind patterns around the globe also play a role in determining the direction in which ocean surface water moves. In the Northern Hemisphere, for example, the trade winds near the equator push water from the northeast to the southwest. However, the Coriolis effect deflects this wind-driven current so that water actually moves from east to west at the equator. Similarly, when winds in northern mid-latitude regions push water from the southwest to the northeast, the Coriolis effect deflects this current so that water actually moves from west to east. **FIGURE 11.1** shows the overall effect: Ocean surface currents between the equator and the mid-latitudes rotate in a clockwise direction in the Northern Hemisphere and in a counterclockwise direction in the Southern Hemisphere. These large-scale patterns of water circulation that move clockwise in the Northern Hemisphere and counterclockwise in the Southern Hemisphere are called **gyres**.

Gyres redistribute heat in the ocean, just as atmospheric convection currents redistribute heat in the atmosphere. Cold water from the polar regions moves along the west coasts of continents, and the cool air immediately above these waters brings cooler temperatures to the adjacent continents. For example, the California Current, which flows south from the North Pacific along the coast of California, causes coastal areas of California to have cooler temperatures than areas at similar latitudes on the east coast of the United States. Similarly, warm water from the tropics moves along the east coasts of continents, and the warm air immediately above these waters causes warmer temperatures on land.

Gyre A large-scale pattern of water circulation that moves clockwise in the Northern Hemisphere and counterclockwise in the Southern Hemisphere.

Upwelling

Ocean currents also help explain why some regions of the ocean support highly productive ecosystems. Along the west coasts of most continents, for

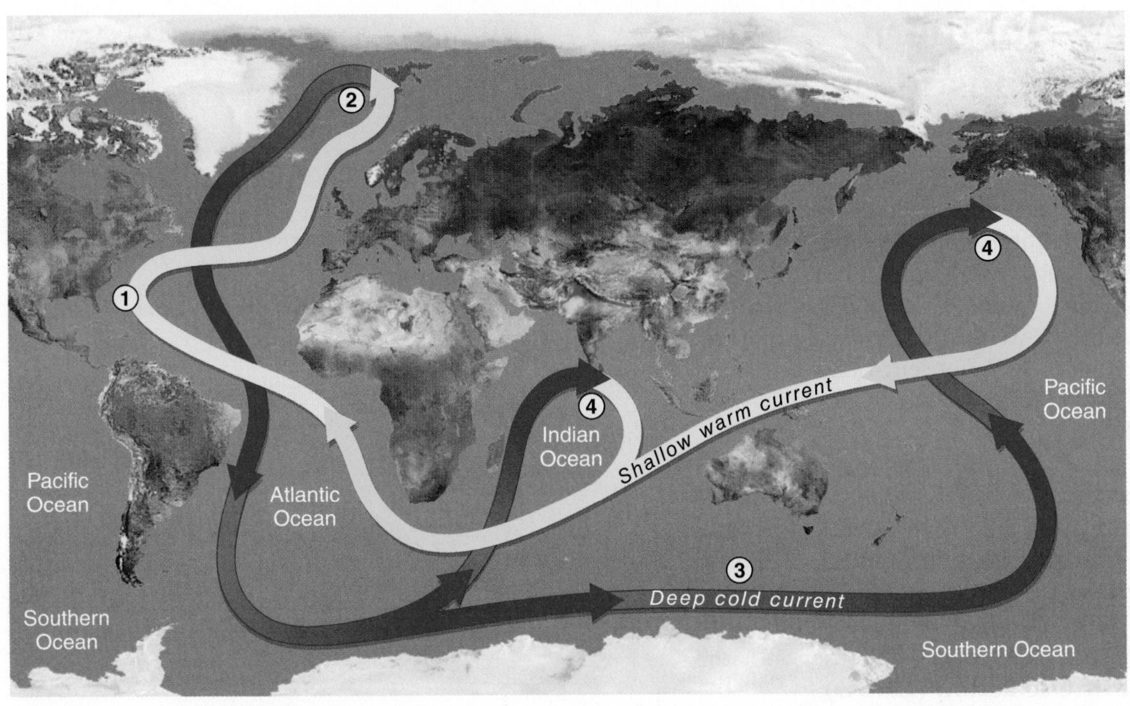

1
2
3
4
Pacific Ocean
Atlantic Ocean
Indian Ocean
Pacific Ocean
Southern Ocean
Southern Ocean
Shallow warm current
Deep cold current
(1) Warm water flows from the Gulf of Mexico to the North Atlantic, where some of it freezes and evaporates.
(2) The remaining water, now saltier and denser, sinks to the ocean bottom.
(3) The cold water travels along the ocean floor, connecting the world's oceans.
(4) The cold, deep water eventually rises to the surface and circulates back to the North Atlantic.

FIGURE 11.2 Thermohaline circulation. The sinking of dense, salty water in the North Atlantic drives a deep, cold current that moves slowly around the world.

example, the surface currents diverge, or separate from one another, causing deeper waters to rise and replace the water that has moved away. This upward movement of water toward the surface, shown as dark blue regions in Figure 11.1, is called **upwelling**. The deep waters bring with them nutrients from the ocean bottom that support large populations of producers. The producers, in turn, support large populations of fish that have long been important to commercial fisheries.

Deep ocean currents circulate ocean water over long time periods

Another oceanic circulation pattern, **thermohaline circulation**, drives the mixing of surface water and deep water. This process is crucial for moving heat and nutrients around the globe. Thermohaline circulation appears to be driven by surface waters that contain unusually large amounts of salt. As Figure 11.1 shows, warm currents flow from the Gulf of Mexico to the very cold North Atlantic. Some of this water freezes or evaporates, and the salt that remains behind increases the salt concentration of the water. This cold, salty water is relatively dense, so it sinks to the bottom of the ocean, mixing with deeper ocean waters. Two processes—the sinking of cold, salty water at high latitudes and the rising of warm water near the equator—create the movement necessary to drive a deep, cold current that slowly moves past Antarctica and northward to the northern Pacific Ocean, where it returns to the surface and then makes its way back to the Gulf of Mexico. This global round trip, shown in **FIGURE 11.2**, can take hundreds of years to complete. As you can see in the figure, thermohaline circulation helps to mix the water of all the oceans.

Ocean currents associated with thermohaline circulation affect the temperature of nearby landmasses much as the major gyres affect surface waters. For example, the ocean current known as the Gulf Stream originates in the tropics near the Gulf of Mexico and flows toward western Europe. This movement of warm tropical waters brings warmer temperatures to latitudes that would otherwise be much colder. For instance, England's average winter temperature is 20°C (36°F) warmer than that of Newfoundland, Canada, which is located at a similar latitude but receives cold ocean currents from the North Atlantic.

Scientists are concerned that global warming could affect thermohaline circulation. If increased air temperatures accelerate the melting of glaciers in the Northern Hemisphere, the waters of the North Atlantic could become less salty and thus less likely to sink. Such a change could potentially shut down thermohaline circulation and stop the transport of warm water to western Europe, making it a much colder place.

Upwelling The upward movement of ocean water toward the surface as a result of diverging currents.

Thermohaline circulation An oceanic circulation pattern that drives the mixing of surface water and deep water.

The El Niño–Southern Oscillation is caused by a shift in ocean currents

Our discussion of ocean currents has focused on the commonly occurring ocean currents. In the southern Pacific Ocean, for example, trade winds from South America and the Coriolis effect cause the equatorial water to flow from east to west. As we have discussed, this current produces an upwelling of cold, nutrient-rich water along the west coast of South America, which produces large populations of fish.

Every 3 to 7 years, however, the tropical current moves in the opposite direction. Because this phenomenon often begins around the December 25 Christmas holiday, it has been named El Niño ("the baby boy"). However, since the cause of the phenomenon is due to a reversal of wind and water currents in the South Pacific, the preferred name for the phenomenon is the **El Niño–Southern Oscillation (ENSO)**.

You should know when El Niño occurs, as well as the implications of this phenomenon. ●

FIGURE 11.3 shows this process in action. First, the trade winds that normally blow from South America and push the surface waters to the west weaken or reverse direction. This change in the wind allows warm equatorial water from the western Pacific to move eastward toward the west coast of South America. The movement of warm water and air toward South America suppresses upwelling off the coast of Peru. The effect can last from a few weeks to a few years.

Because ocean currents are so important to global climates, the ENSO has several consequences. As we have noted, it reduces the upwelling off the South American coast, which decreases productivity and dramatically reduces fish populations. It also has widespread effects around the world, including cooler and wetter conditions in the southeastern United States and unusually dry weather in southern Africa and Southeast Asia. Such changes can have major impacts of crop production, with some ENSO events leading to poor crop yields and thousands of human deaths in developing countries. The ENSO event that occurred during 2015-2016 was the most severe ENSO in 30 years. Scientists estimate that nearly 100 million people experienced food shortages as a result.

El Niño–Southern Oscillation (ENSO) A reversal of wind and water currents in the South Pacific.

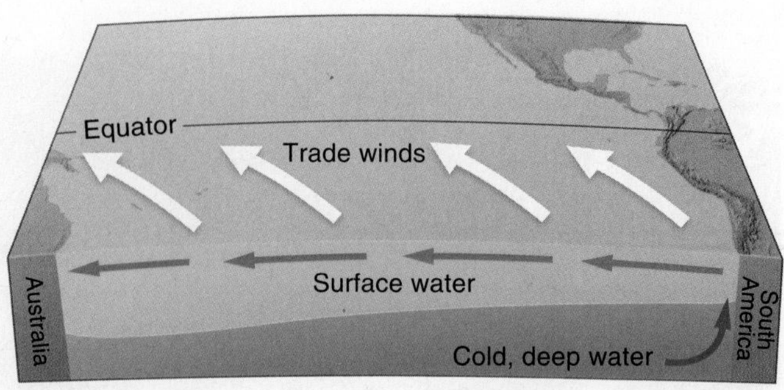

(a) Normal year

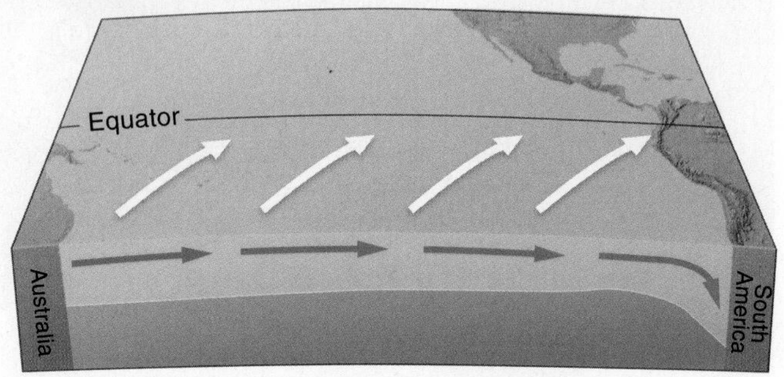

(b) El Niño year

FIGURE 11.3 The El Niño–Southern Oscillation. (a) In a normal year, trade winds push warm surface waters away from the coast of South America and promote the upwelling of water from the ocean bottom. (b) In an El Niño year, trade winds weaken or reverse direction, so warm waters build up along the west coast of Peru.

MODULE 11 AP® Review

Preparing for the AP® Exam

In this module, we have learned that the circulation of surface ocean currents and deep ocean currents is driven by the unequal heating of Earth, air currents, Coriolis forces, and differences in salt concentrations. These currents help to transport heat to different regions of the world and therefore help determine the climates around the globe. The disruption of these currents—due to the ENSO, and potentially because of glacier melting—can have major impacts on these climates. In the next module, we will examine how differences in climate help to determine the plants and animals that live in different parts of the world.

AP® Practice Questions

Choose the best answer for the following.

1. Which does NOT drive ocean currents?
   (a) salinity
   (b) prevailing winds
   (c) temperature
   (d) precipitation

2. High productivity and nutrient availability in the ocean occurs in areas with
   (a) gyres.
   (b) upwelling.
   (c) warm waters.
   (d) prevailing westerlies.

3. Climate change could potentially disrupt which phenomenon?
   (a) El Niño–Southern Oscillation (ENSO)
   (b) thermohaline circulation
   (c) upwelling off the coasts of Australia and New Zealand
   (d) the polar current along Antarctica

4. Which is NOT true of an El Niño–Southern Oscillation (ENSO) event?
   (a) It causes increased upwelling on the west coast of South America.
   (b) It causes wetter conditions in the southern United States.
   (c) It includes a reversal in tropical ocean currents.
   (d) It often begins near Christmas.

5. Gyres do all of the following EXCEPT
   (a) redistribute heat in the ocean.
   (b) result from the Coriolis effect.
   (c) redistribute nutrients from the deep ocean.
   (d) affect the temperatures of coastal areas.

MODULE 12

Terrestrial Biomes

Climate affects the distribution of species around the globe. For example, only species that are well adapted to hot and dry conditions can survive in deserts. A very different set of organisms survives in cold, snowy places. In this module, we will examine how differences in climate help determine the species of plants and animals that live in each region and how scientists categorize these regions.

Learning Goals

After reading this module you should be able to

- explain how we define terrestrial biomes.
- interpret climate diagrams.
- identify the nine terrestrial biomes.

Terrestrial biomes are defined by the dominant plant growth forms

Scientists have long recognized that organisms possess distinct growth forms, many of which represent adaptations to local temperature and precipitation patterns. For example, if we were to examine the plants from all the deserts of the world, we would find many cactus-like species. Plants that look like cacti in North American deserts are indeed members of the cactus family, but those in the Kalahari Desert are members of the euphorb family. These two distantly related families of plants look similar because they have evolved similar adaptations to hot, dry environments—including the ability to store large amounts of water in their tissues and a waxy coating to reduce water loss (**FIGURE 12.1**).

(a)

(b)

FIGURE 12.1 Similar growth forms. Cacti and euphorbs are not closely related, but they have evolved many similar adaptations that allow them to live in hot, dry environments. (a) Organ pipe cactus (*Stenocereus thurberi*) in Arizona. (b) Euphorbia (*E. damarana*) in Namibia. *(a: Dhoxax/Shutterstock; b: Patti Murray/Earth Scenes/Animals Animals)*

Despite these common adaptations to desert conditions, these two plant families have distinctive flowers and spines, and only the euphorb family produces a milky sap. These differences help to confirm that while the two groups may look similar, genetically they are not closely related. Instead, exposure to similar selective pressures over long periods of time has favored the evolution of many similar adaptations. We will discuss this in more detail in Chapter 5. As a result, scientists can categorize terrestrial regions of the world into **terrestrial biomes**, which are geographical regions that each have a particular combination of average annual temperature and precipitation and contain distinctive plant growth forms that are adapted to that climate. In the next module we will examine **aquatic biomes**, which are categorized by particular combinations of salinity, depth, and water flow. **FIGURE 12.2** shows the range of terrestrial biomes on Earth in the context of precipitation and temperature. For example, boreal and tundra biomes have average annual temperatures below 5°C (41°F), whereas temperate biomes have average annual temperatures between 5°C and 20°C (68°F), and tropical biomes have average annual temperatures above 20°C. Within each of these temperature ranges, we can observe a wide range of precipitation. **FIGURE 12.3** shows the distribution of biomes around the world.

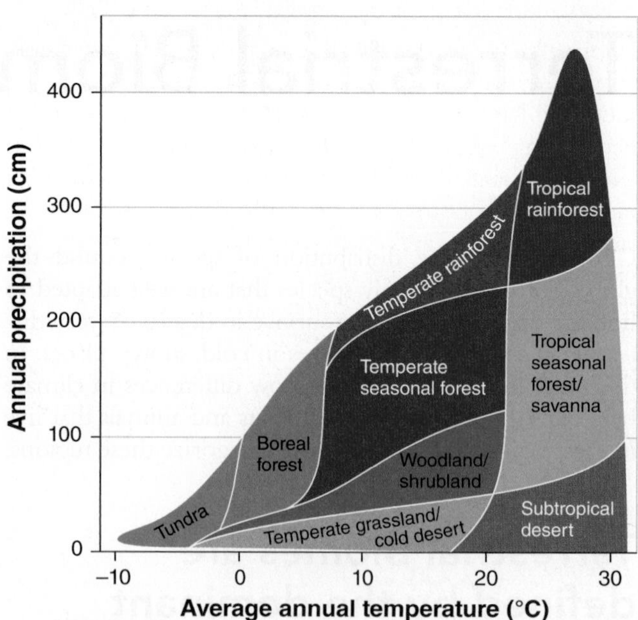

FIGURE 12.2 Locations of the world's biomes. Tundra and boreal biomes generally exist when the average annual temperature is less than 5°C (41°F). In contrast, temperate biomes exist in regions that experience average annual temperatures between 5°C and 20°C (41°F and 68°F). Tropical biomes exist in regions that experience average annual temperatures above 20°C.

Terrestrial biome A geographic region categorized by a particular combination of average annual temperature, annual precipitation, and distinctive plant growth forms on land.

Aquatic biome An aquatic region characterized by a particular combination of salinity, depth, and water flow.

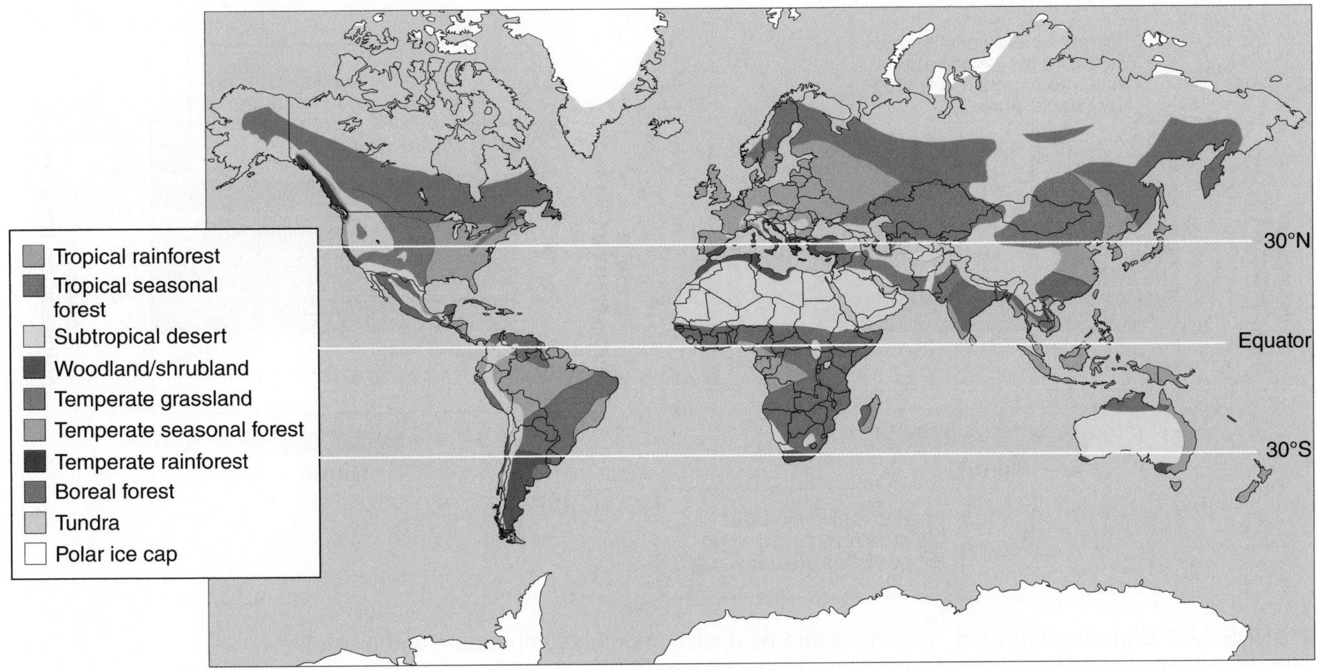

FIGURE 12.3 Biomes. Biomes are categorized by particular combinations of average annual temperature and annual precipitation.

AP® Exam Tip

Students often confuse biomes with habitats, so make sure that you can differentiate between the two. ●

Note that although terrestrial biomes are categorized by plant growth forms, the animal species living in different biomes are often quite distinctive as well. For example, rodents inhabiting deserts around the world have a number of adaptations for hot, dry climates, including highly efficient kidneys that allow very little water loss through urination. For both plants and animals, biomes refer to large geographical areas that are home to large numbers of species. In contrast, a **habitat** is an area where a particular species lives in nature. A habitat is a subset of a biome. For example, the forests of the northeastern United States represent a biome, but this biome contains multiple terrestrial habitats including agricultural fields, abandoned fields, young forests with sapling trees, and mature forests.

Climate diagrams illustrate patterns of annual temperature and precipitation

Climate diagrams, such as those shown in **FIGURE 12.4** on page 128, are a helpful way to visualize regional patterns of temperature and precipitation. By graphing average monthly temperature and precipitation, these diagrams illustrate how the conditions in a biome vary during a typical year. They also indicate when the temperature is warm enough for plants to grow—that is, the months when it is above 0°C (32°F), known as the growing season. In Figure 12.4a, we can see that the growing season is mid-March through mid-October.

In addition to the growing season, climate diagrams can show the relationship between precipitation, temperature, and plant growth. For every 10°C (18°F) temperature increase, plants need 20 mm (0.8 inches) of additional precipitation each month to supply the extra water demand that warmer temperatures cause. As a result, plant growth can be limited either by temperature or by precipitation. In Figure 12.4a, the precipitation line is above the temperature line during all months. This means that water supply of plants exceeds the water demand, so plant growth is more constrained by temperature than by precipitation.

In Figure 12.4b, we see a different scenario. When the precipitation line intersects the temperature line, the amount of precipitation available to plants equals the amount of water lost by plants through evapotranspiration. At any point where the precipitation line is below the temperature line, water demand exceeds supply. In this situation, plant growth will be constrained more by precipitation than by temperature.

Climate diagrams also help us understand how humans use different biomes. For example, areas of the world that have warm temperatures, long growing seasons, and abundant rainfall are generally highly productive and therefore are well suited to growing many crops. Warm regions that have less abundant precipitation are suitable for growing

Habitat An area where a particular species lives in nature.

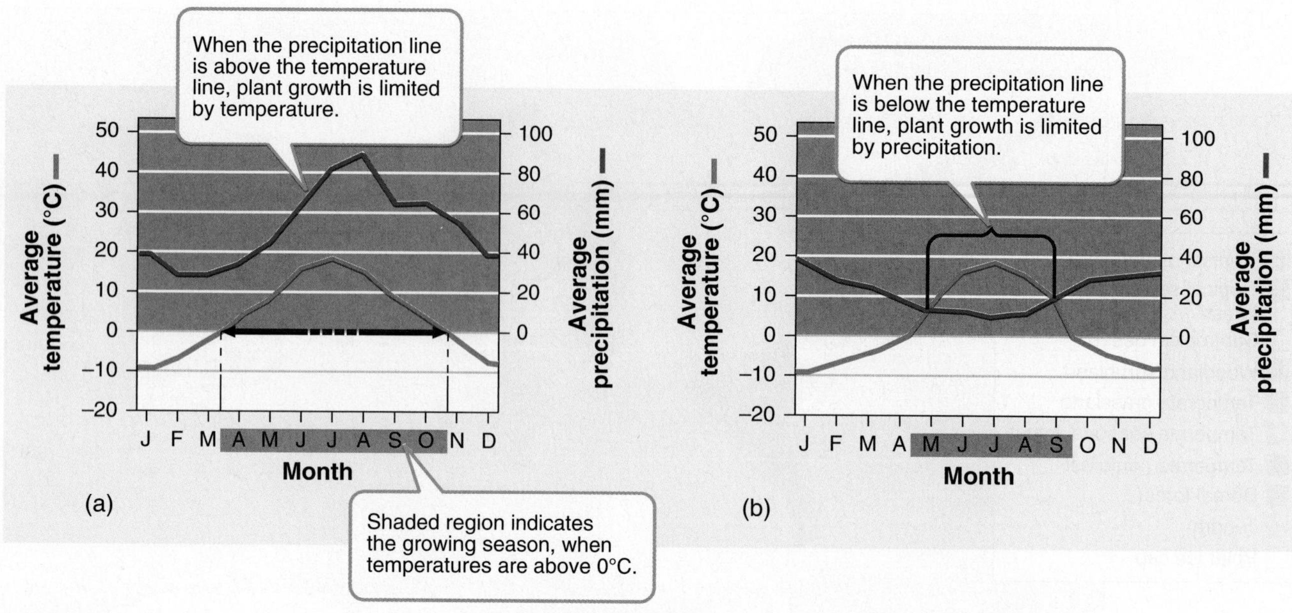

FIGURE 12.4 Climate diagrams. Climate diagrams display monthly temperature and precipitation values, which help determine the productivity of a biome.

grains such as wheat and for grazing domesticated animals, including cattle and sheep. Colder regions are often best used to grow forests that will be harvested for lumber.

Terrestrial biomes range from tundra to tropical forests

We can divide terrestrial biomes into three categories: tundra and boreal forest, temperate, and tropical. These three categories contain a total of nine biomes. We will examine each of these biomes in turn, looking at temperature and precipitation patterns, geographic distribution, and typical plant and animal growth forms.

Tundra

The **tundra**, shown in **FIGURE 12.5**, is a cold and treeless biome, with low-growing vegetation. In winter, the soil is completely frozen. Arctic tundra is found in the northernmost regions of the Northern Hemisphere in Russia, Canada, Scandinavia, and Alaska. Antarctic tundra is found along the edges of Antarctica and on nearby islands. At lower latitudes, alpine tundra can be found on high mountains, where high winds and low temperatures prevent trees from growing.

Tundra A cold and treeless biome with low-growing vegetation.

Permafrost An impermeable, permanently frozen layer of soil.

Boreal forest A forest biome made up primarily of coniferous evergreen trees that can tolerate cold winters and short growing seasons.

The growing season in the tundra is very short, usually only about 4 months during summer, when the polar region is tilted toward the Sun and the days are very long. During this time the upper layer of soil thaws, creating pools of standing water that are ideal habitat for mosquitoes and other insects. The underlying subsoil, known as **permafrost**, is an impermeable, permanently frozen layer that prevents water from draining and roots from penetrating. Permafrost, combined with the cold temperatures and a short growing season, prevents deep-rooted plants such as trees from living in the tundra.

While the tundra receives little precipitation, there is enough to support some plant growth. The characteristic plants of this biome, such as small woody shrubs, mosses, heaths, and lichens, can grow in shallow, waterlogged soil and can survive short growing seasons and bitterly cold winters. At these cold temperatures, chemical reactions occur slowly, and as a result, dead plants and animals decompose slowly. This slow rate of decomposition results in the accumulation of organic matter in the soil over time with relatively low levels of soil nutrients. Common animals in the Arctic tundra include muskoxen (*Ovibos moschatus*), Arctic foxes (*Vulpes lagopus*), and polar bears (*Ursus maritimus*). The major human impact on the tundra biome is the threat of warming global temperatures melting the permafrost, which would then decompose at a faster rate than usual and such decomposition would release additional greenhouse gases. We will discuss this in more detail in Chapter 19.

Boreal Forest

The **boreal forest** biome, shown in **FIGURE 12.6**, consists primarily of coniferous (cone-bearing) evergreen trees that can tolerate cold winters and short growing seasons. Boreal forests are sometimes called taiga.

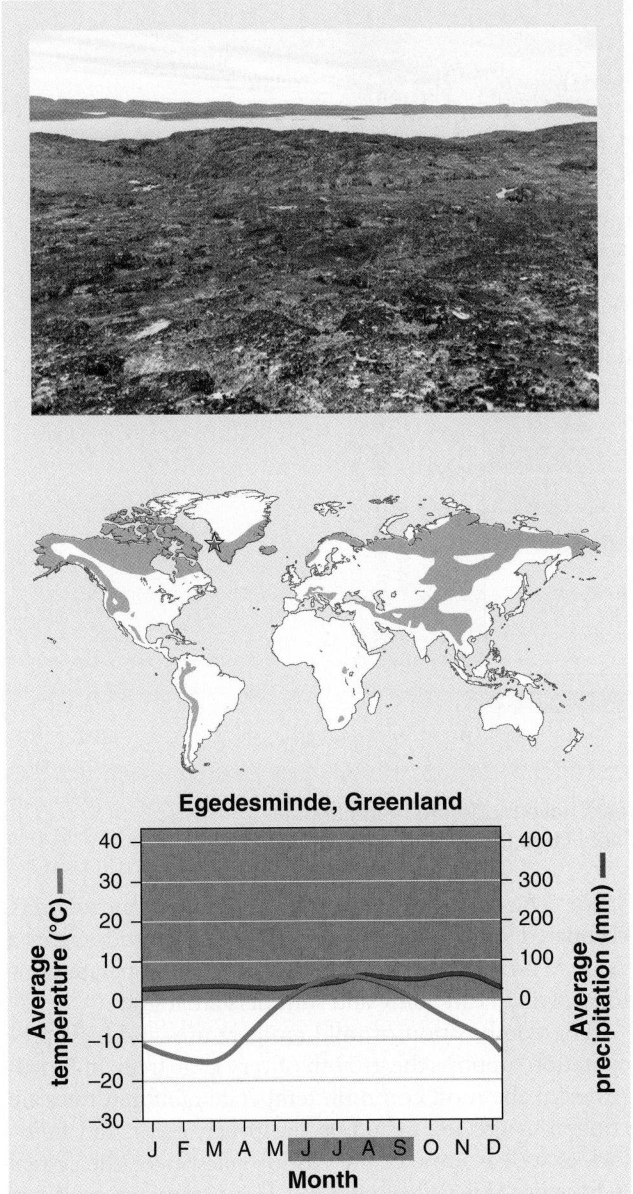

FIGURE 12.5 Tundra biome. The tundra is cold and treeless, with low-growing vegetation. *(Panther Media GmbH/Alamy)*

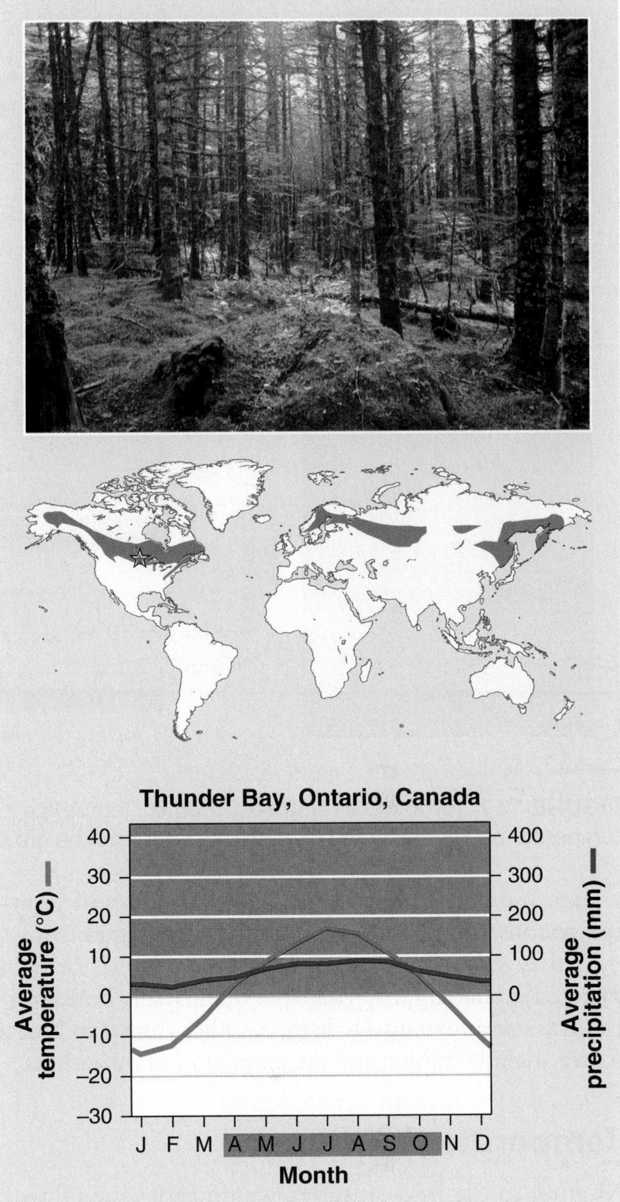

FIGURE 12.6 Boreal forest biome. Boreal forests are made up primarily of coniferous evergreen trees that can tolerate cold winters and short growing seasons. *(Bill Brooks/Alamy)*

Evergreen trees appear green year-round because they drop only a fraction of their needles each year. Boreal forests are found between about 50° N and 60° N in Europe, Russia, and North America. This subarctic biome has a very cold climate, and plant growth is more constrained by temperature than by precipitation.

As in the tundra, cold temperatures and relatively low precipitation make decomposition in boreal forests a slow process. In addition, the waxy needles of evergreen trees contain compounds that are resistant to decomposition. As a result of the slow rate of decomposition and the low nutrient content of the needles, boreal forest soils are covered in a thick layer of organic material, but are poor in nutrients.

The climate characteristics of boreal forests—cold temperatures, low precipitation, and nutrient-poor soil—determine the species of plants that can survive in them. In addition to coniferous trees such as pine, spruce, and fir, some deciduous trees, such as birch, maple, and aspen, can also be found in this biome. The needles of coniferous trees can tolerate below-freezing conditions, but the deciduous trees drop all their leaves in autumn before the subfreezing temperatures of winter can damage them. When the weather warms, the deciduous trees produce new leaves and grow rapidly. Common animal species include beavers (*Castor Canadensis*), brown bears (*Ursus arctos*), and wolverines (*Gulo gulo*).

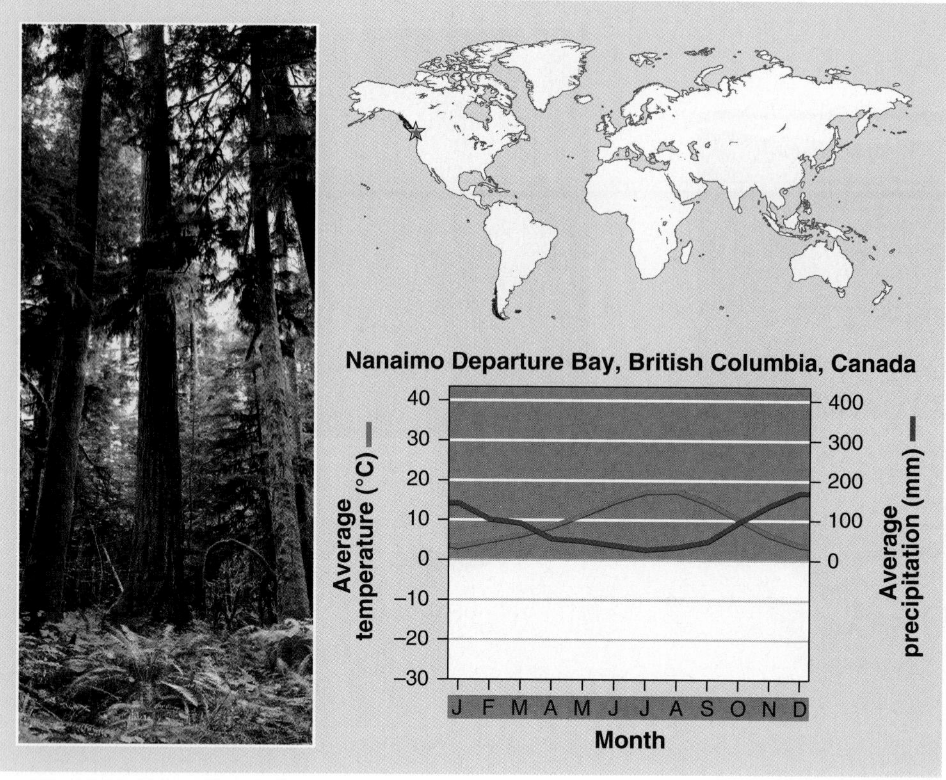

FIGURE 12.7 Temperate rainforest biome. Temperate rainforests have moderate mean annual temperatures and high precipitation that supports the growth of very large trees. *(BGSmith/Shutterstock)*

Because boreal forests have poor soils and short growing seasons, they are poorly suited for agriculture. However, they serve as an important source of trees for pulp, paper, and building materials. As a result, many boreal forests have been extensively logged. Other threats to boreal forests include mining and the extraction of oil and gas.

Temperate Rainforest

Moving to the mid-latitudes, we find that the climate is more temperate, with average annual temperatures between 5°C and 20°C (41°F and 68°F). A range of temperate biomes exists in this area including temperate rainforest, temperate seasonal forest, woodland/shrubland, and temperate grassland/cold desert.

Moderate temperatures and high precipitation typify the **temperate rainforest** biome, shown in **FIGURE 12.7**. The temperate rainforest is a coastal biome and can be found in relatively narrow areas around the world. Temperate rainforests exist along the west coast of North America from northern California to Alaska, in southern Chile, on the east coast of Australia and in neighboring Tasmania, and on the west coast of New Zealand. Ocean currents along these coasts help to moderate temperature fluctuations, and ocean water provides a source of water vapor. The result is relatively mild summers and winters, compared with other biomes at similar latitudes, and a nearly 12-month growing season. In the temperate rainforest, winters are rainy and summers are foggy.

Temperate rainforest A coastal biome typified by moderate temperatures and high precipitation.

The combination of mild temperatures and high precipitation supports the growth of very large trees. In North America, the most common temperate rainforest trees are coniferous species, including fir, spruce, cedar, and hemlock as well as some of the world's tallest trees: the coastal redwoods (*Sequoia sempervirens*). These immense trees can live hundreds to thousands of years and achieve heights of 90 m (295 feet) and diameters of 8 m (26 feet). Because many of these large tree species are attractive sources of lumber, much of this biome has been logged and subsequently converted into single-species tree plantations.

As we have already seen, coniferous trees produce needles that are slow to decompose. The relatively cool temperatures in the temperate rainforest also favor slow decomposition, although it is not nearly as slow as in boreal forest and tundra. The nutrients released are rapidly taken up by the trees or leached down through the soil by the abundant rainfall, which leaves the soil low in nutrients. Ferns and mosses, which can survive in nutrient-poor soil, are commonly found living under the enormous trees. Animals that are characteristic of the temperate rainforest in North America include the black-tailed deer (*Odocoileus hemionus columbianus*), the Pacific giant salamander (*Dicamptodon ensatus*), and the Pacific treefrog (*Pseudacris regilla*).

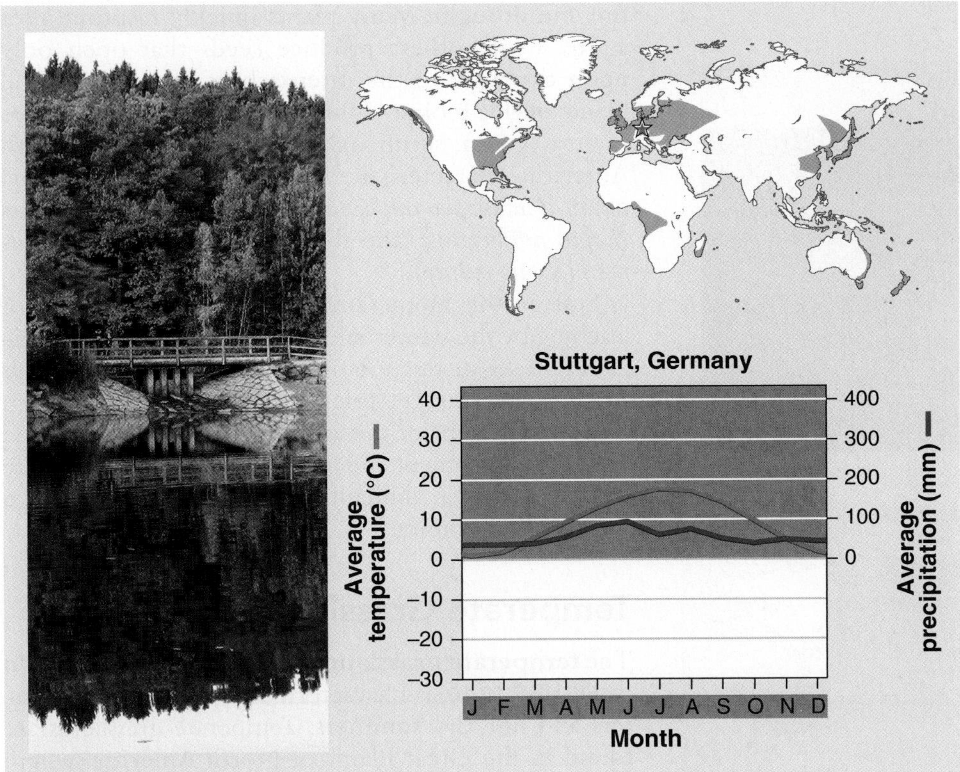

FIGURE 12.8 Temperate seasonal forest biome. Temperate seasonal forest biomes have moderate mean annual temperatures and moderate amounts of precipitation that support broadleaf deciduous trees such as beech, maple, oak, and hickory. *(Ursula Sander/Getty Images)*

Temperate Seasonal Forest

The **temperate seasonal forest** biome, shown in **FIGURE 12.8**, experiences warm summers and cold winters with over 1 m (39 inches) of annual precipitation. It is found in the eastern United States, Japan, China, Europe, Chile, and eastern Australia. Away from the moderating influence of the ocean, these forests experience much warmer summers and colder winters than temperate rainforests. They are dominated by broadleaf deciduous trees such as beech, maple, oak, and hickory, although some coniferous tree species may also be present. Because of the predominance of deciduous trees, these forests are also called temperate deciduous forests.

The warm summer temperatures in temperate seasonal forests favor rapid decomposition. Because the leaves shed by broadleaf trees are more readily decomposed than the needles of coniferous trees, the soils of temperate seasonal forests generally contain more nutrients than those of boreal forests. Their higher soil fertility, combined with their longer growing season, means that temperate seasonal forests have greater plant productivity than boreal forests. Common animal species include white-tailed deer (*Odocoileus virginianus*), red foxes (*Vulpes vulpes*), and gray squirrels (*Sciurus carolinensis*).

Because temperate seasonal forests are so productive, they have historically been one of the first biomes to be converted to agriculture on a large scale. When European settlers arrived in North America, they cleared large areas of the eastern forests for agriculture, although much of this has since grown back.

Woodland/Shrubland

The **woodland/shrubland** biome, illustrated in **FIGURE 12.9**, is characterized by hot, dry summers and mild, rainy winters. This biome is found on the coast of southern California (where it is called chaparral), in southern South America (matorral), in southwestern Australia (mallee), in southern Africa (fynbos), and in a large region surrounding the Mediterranean Sea (maquis). There is a 12-month growing season, but plant growth is constrained by high temperatures and low precipitation in summer and by cool temperatures

Temperate seasonal forest A biome with warm summers and cold winters with over 1 m (39 inches) of precipitation annually.

Woodland/shrubland A biome characterized by hot, dry summers and mild, rainy winters.

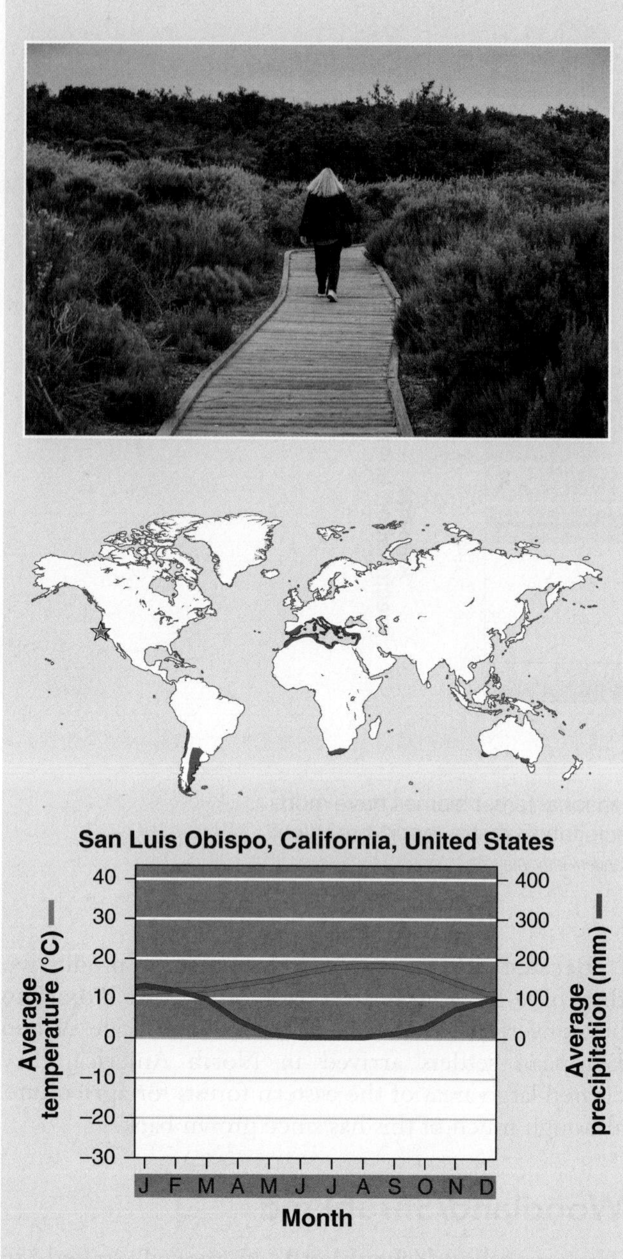

FIGURE 12.9 Woodland/shrubland biome. The woodland/shrubland biome is characterized by hot, dry summers and mild, rainy winters. *(George Rose/Getty Images)*

and higher precipitation in winter. As discussed at the beginning of the chapter, these are the ideal conditions for growing grapes for making wine.

The hot, dry summers of the woodland/shrubland biome favor the natural occurrence of wildfires. Plants of this biome are well adapted to both fire and drought. Many plants quickly resprout after a fire, while others produce seeds that open only upon exposure to the intense heat of a fire. Typical plants of this biome include drought-resistant shrubs such as yucca, scrub oak, and sagebrush. In North America, characteristic animals include California quail (*Callipepla californica*), black-tailed jackrabbits (*Lepus californicus*), and the San Joaquin kit fox (*Vulpes macrotis mutica*).

Soils in this biome are low in nutrients because of leaching by the winter rains. As a result, the major agricultural uses of this biome are for grazing by animals and growing drought-tolerant, deep-rooted crops, such as grapes. The major threats from humans are increased human development and an increased frequency of fire that is making it difficult for many native species to survive in this biome.

Temperate grassland/cold desert A biome characterized by cold, harsh winters, and hot, dry summers.

Temperate Grassland/Cold Desert

The **temperate grassland/cold desert** biome, shown in **FIGURE 12.10**, is characterized by cold, harsh winters and hot, dry summers. Temperate grasslands are found in the Great Plains of North America (where they are called prairies), in South America (pampas), and in central Asia and eastern Europe (steppes). Cold, harsh winters and hot, dry summers characterize this biome. Thus, as in the woodland/shrubland biome, plant growth is constrained by insufficient precipitation in summer and cold temperatures in winter. Fires are common, as the dry and frequently windy conditions fan flames ignited by lightning. Although estimates vary, it is thought that, historically, large wildfires occurred in this biome every few years, sometimes burning as much as 10,000 ha (nearly 25,000 acres) in a single fire.

Typical plants of temperate grasslands include grasses and nonwoody flowering plants. These plants are generally well adapted to wildfires and frequent grazing by animals. Their deep roots store energy to enable quick regrowth. Within this biome, the amount of rainfall determines which plants can survive in a region. In the North American prairies, for example, nearly 1 m (39 inches) of rain falls per year on the eastern edge of the biome, supporting grasses that can grow up to 2.5 m (8 feet) high. Although these tallgrass prairies receive sufficient rainfall for trees to grow, frequent wildfires keep trees from encroaching. In fact, the Native American people are thought to have intentionally kept the eastern prairies free of trees by using controlled burning. To the west, annual precipitation drops to 0.5 m (20 inches), favoring the growth of grasses less than 0.5 m (20 inches) tall. These shortgrass prairies are simply too dry to support trees or tall grasses. Farther west, in the rain shadow of the Rocky Mountains, annual precipitation continues to decline to 0.25 m (10 inches). In this region, the shortgrass prairie gives way to cold desert.

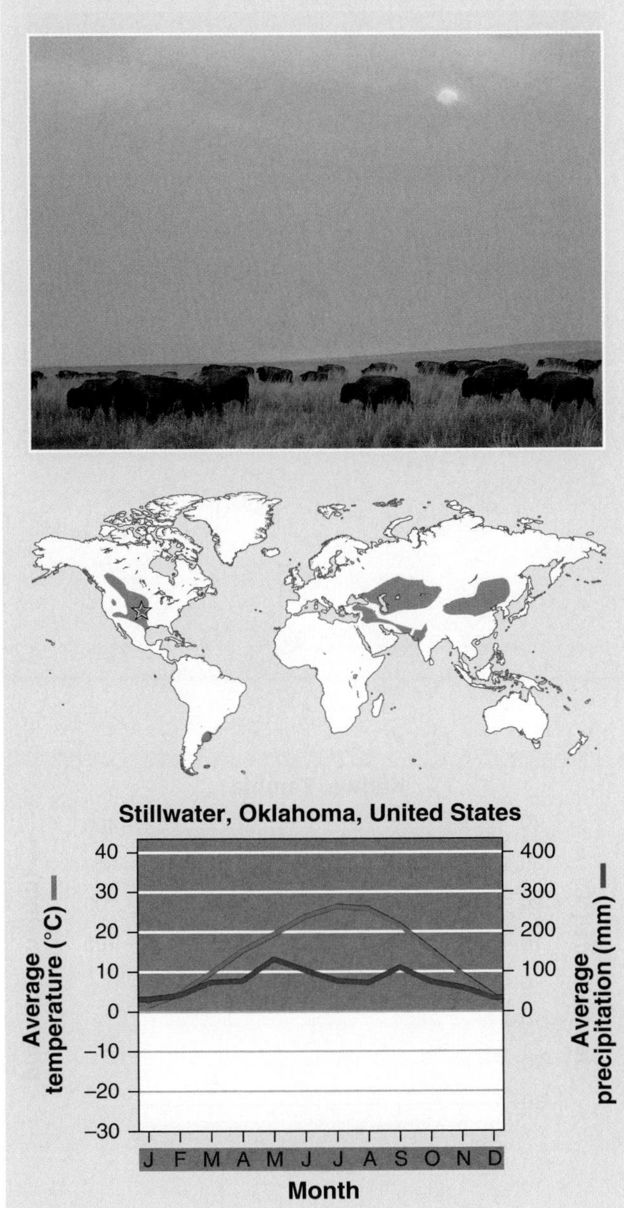

FIGURE 12.10 Temperate grassland/cold desert biome. The temperate grassland/cold desert biome has cold, harsh winters and hot, dry, summers that support grasses and nonwoody flowering plants. *(AP Photo/Brandi Simons)*

Cold deserts, also known as temperate deserts, have even sparser vegetation than shortgrass prairies. Cold deserts are distinct from subtropical deserts in that they have much colder winters and do not support the characteristic plant growth forms of hot deserts, such as cacti and euphorbs. Typical animals of the North American prairie include bison, greater prairie chickens (*Tympanuchus cupido*), and the prairie kingsnake (*Lampropeltis calligaster*).

The combination of a relatively long growing season and rapid decomposition that adds large amounts of nutrients to the soil makes temperate grasslands very productive. More than 98 percent of the tallgrass prairie in the United States has been converted to agriculture, while the less productive shortgrass prairie is predominantly used for growing wheat and grazing cattle.

Tropical Rainforest

In the tropics, average annual temperatures exceed 20°C. Here we find the tropical biomes: tropical rainforests, tropical seasonal forests/savannas, and subtropical deserts.

The **tropical rainforest** biome, shown in **FIGURE 12.11** on page 134, is a warm and wet biome that lies within approximately 20° N and 20° S of the equator. This biome is found in Central and South America, Africa, Southeast Asia, and northeastern Australia. It is also found on large tropical islands, where the oceans provide a constant source of atmospheric water vapor.

Precipitation occurs frequently, although there are seasonal patterns in precipitation that depend on when the ITCZ passes overhead. Because of the warm temperatures and abundant rainfall, productivity is high, and decomposition is extremely rapid. The lush vegetation takes up nutrients quickly, leaving few nutrients to accumulate in the soil. Because of its high productivity, approximately 24,000 ha (59,500 acres) of tropical rainforest are cleared each year for agriculture. However, the high rate of decomposition causes the soils to lose their fertility quickly, so they have relatively little undecomposed organic matter (humus). As a result, farmers growing crops on tropical soils often have to keep moving to newly deforested areas.

Tropical rainforests contain more biodiversity per hectare than any other terrestrial biome. As much as two-thirds of Earth's terrestrial species are found in this biome. These forests have several distinctive layers of vegetation. Large trees form a forest canopy that shades the underlying vegetation. Several layers of successively shorter trees make up the subcanopy, also known as the understory. Attached to the trunks and branches of the trees are epiphytes, plants that hold small pools of water that support small aquatic ecosystems far above the forest floor. Numerous species of woody vines (also called lianas) are rooted in the soil, but climb up the trunks of trees and often into the canopy. Common animals in tropical rainforests around the world include jaguars (*Panthera onca*), orangutans, and red-eyed treefrogs (*Agalychnis callidryas*). The domin-ant human threat to tropical rainforests is deforestation due to expanding forestry and agriculture.

Tropical rainforest A warm and wet biome found between 20° N and 20° S of the equator, with little seasonal temperature variation and high precipitation.

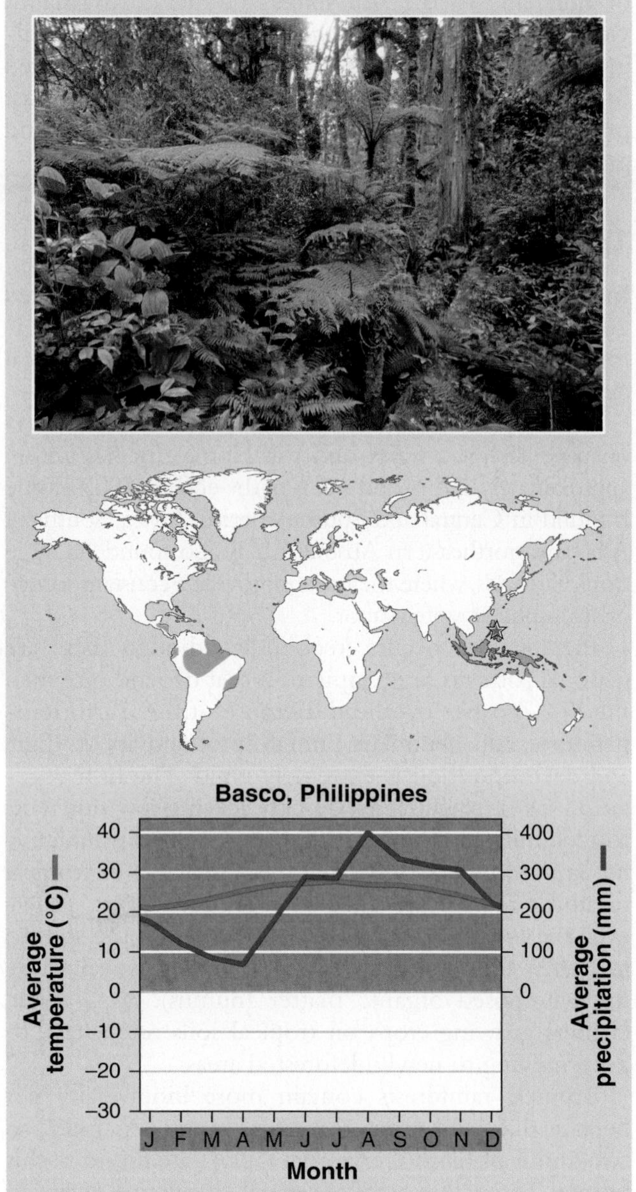

FIGURE 12.11 **Tropical rainforest biome.** Tropical rainforests are warm and wet, with little seasonal temperature variation. These forests are highly productive with several distinctive layers of vegetation. *(Doug Weschler/Earth Scenes/Animals Animals)*

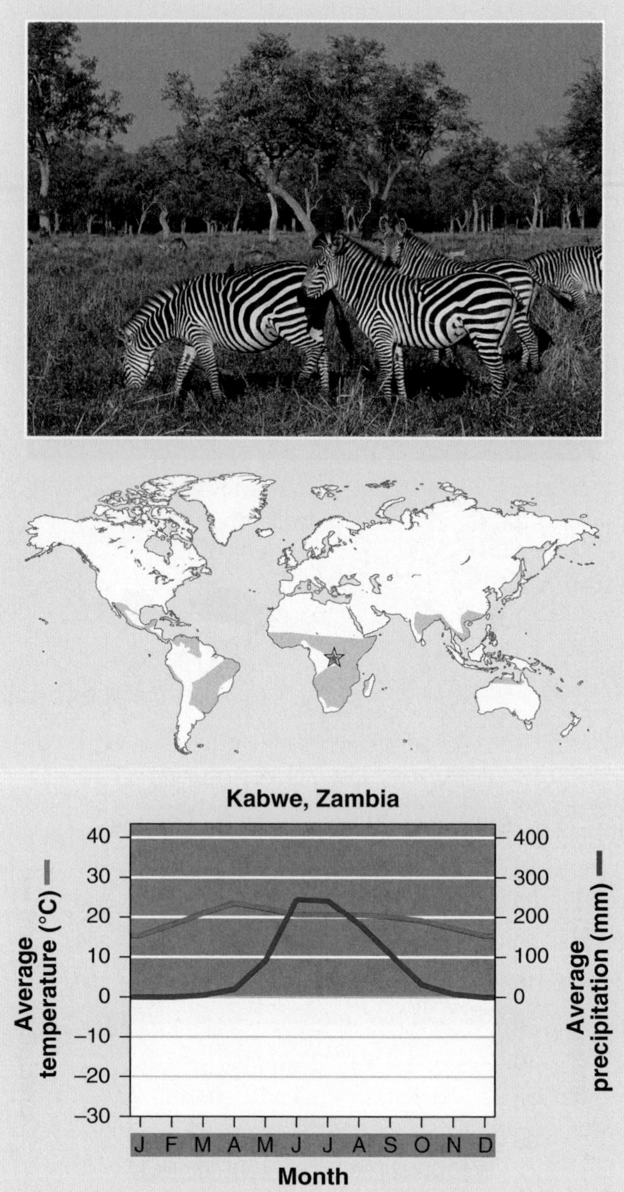

FIGURE 12.12 **Tropical seasonal forest/savanna biome.** Tropical seasonal forest and savannas have warm temperatures and distinct wet and dry seasons. Vegetation ranges from dense stands of shrubs and trees to relatively open landscapes dominated by grasses and scattered deciduous trees. *(John Warburton-Lee/DanitaDelimont.com)*

Tropical Seasonal Forest/Savanna

The **tropical seasonal forests/savanna** biome, shown in **FIGURE 12.12**, is marked by warm temperatures and distinct wet and dry seasons. This seasonal pattern is caused by the seasonal movement of the ITCZ, which, because it tracks the seasonal movement of the most intense sunlight, passes overhead and drops precipitation only during summer. The trees drop their leaves during the dry season as an adaptation to survive the drought conditions and produce new leaves during the wet season. Thus these forests are also called tropical deciduous forests.

Tropical seasonal forest/savanna A biome marked by warm temperatures and distinct wet and dry seasons.

Tropical seasonal forests are common in much of Central America, on the Atlantic coast of South America, in southern Asia, in northwestern Australia, and in sub-Saharan Africa. Areas with moderately long dry seasons support dense stands of shrubs and trees. In areas with the longest dry seasons, the tropical seasonal climate leads to the formation of savannas, relatively open landscapes dominated by grasses and scattered deciduous trees. Common plants in this biome include acacia and baobab trees. Grazing and fire discourage the growth of many smaller woody plants and keep the savanna landscape open. The presence of trees and a warmer average annual temperature distinguish savannas from grasslands.

The warm temperatures of the tropical seasonal forest/savanna biome promote decomposition, but the low amounts of precipitation constrain plants from using the soil nutrients that are released. As a result, the soils of this biome are fairly fertile and can be farmed. Their fertility has resulted in the conversion of large areas of tropical seasonal forest and savanna into agricultural fields and grazing lands. For example, over 99 percent of the tropical seasonal forest of Pacific Central America and the Atlantic coast of South America has been converted to human uses, including agriculture and grazing. Common animals in this biome from around the world include large grazing animals such as several species of gazelles and zebras (*Equus* spp.), and large predators such as African lions and cheetahs.

Subtropical Desert

The **subtropical desert** biome, shown in **FIGURE 12.13**, is located at roughly 30° N and 30° S, and is characterized by hot temperatures, extremely dry conditions, and sparse vegetation. Also known as hot deserts, this biome includes the Mojave Desert in the southwestern United States, the Sahara in Africa, the Arabian Desert of the Middle East, and the Great Victoria Desert of Australia. Cacti, euphorbs, and succulent plants are well adapted to this biome. To prevent water loss, the leaves of desert plants may be small, nonexistent, or modified into spines, and the outer layer of the plant is thick, with few pores for water and air exchange. Most photosynthesis occurs along the plant stem, which stores water so that photosynthesis can continue even during very dry periods. To protect themselves from herbivores, desert plants have developed defense mechanisms such as spines to discourage grazing. Desert-adapted animals around the world include multiple species of tortoises, camels, and roadrunners.

When rain does fall, the desert landscape is transformed. Annual plants—those that live for only a few months, reproduce, and die—grow rapidly during periods of rain. In contrast, perennial plants—those that live for many years—experience spurts of growth when it rains, but then exhibit little growth during the rest of the year. The slow overall growth of perennial plants in subtropical deserts makes them particularly vulnerable to disturbance, and they have long recovery times. The major threats to this biome include climate change and the draining of underground water for human use.

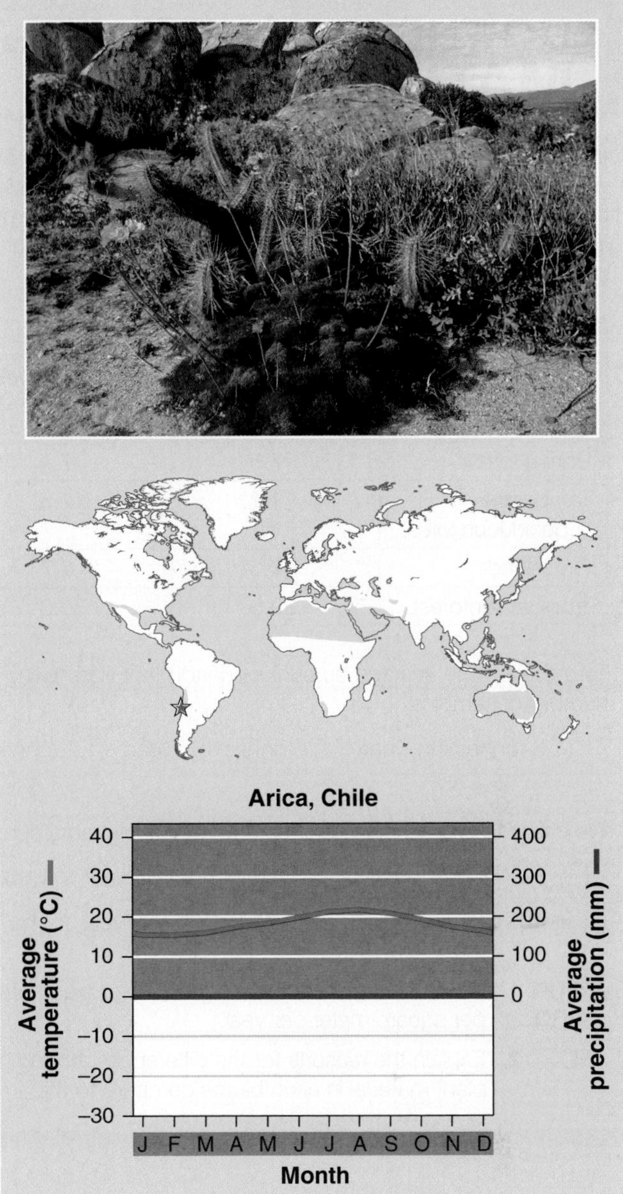

FIGURE 12.13 Subtropical desert biome. Subtropical deserts have hot temperatures, extremely dry conditions, and sparse vegetation. *(Topham/The Image Works)*

Subtropical desert A biome prevailing at approximately 30° N and 30° S, with hot temperatures, extremely dry conditions, and sparse vegetation.

DO THE MATH Organic Matter Inputs Among Terrestrial Biomes

Preparing for the AP® Exam

In our discussion of the terrestrial biomes, we have seen that each biome differs in the productivity of plants and in the amount of organic matter left to decompose in the soil. To get a better understanding of these differences, scientists set out containers in different biomes to determine exactly how much dead plant material actually falls to the ground. Using their data, shown in the table below, we can calculate the average amount of dead plant material falling in each biome in units of metric tonnes per hectare per year.

Biome	Sample 1 (tonnes/ha/year)	Sample 2 (tonnes/ha/year)	Sample 3 (tonnes/ha/year)	Sample 4 (tonnes/ha/year)	Sample 5 (tonnes/ha/year)
Tundra	1.3	1.1	1.7	1.5	1.9
Boreal forest	7.5	7.4	7.6	7.5	7.5
Temperate deciduous forest	11.1	12.5	10.4	9.9	13.2
Grassland	7.3	7.7	7.4	7.6	7.5
Tropical rainforest	29.0	30.0	31.0	35.0	25.0

To calculate the average amount of dead plant litter falling in the tundra biomes, we sum the five values and divide by the number of samples

$$(1.3 \text{ tonnes/ha/year} + 1.1 \text{ tonnes/ha/year} + 1.7 \text{ tonnes/ha/year} + 1.5 \text{ tonnes/ha/year} + 1.9 \text{ tonnes/ha/year}) \div 5 = 1.5 \text{ tonnes/ha/year}$$

We can also express this number in terms of kilograms per square meter per year:

$$1.5 \text{ tonnes/ha/year} \times 1{,}000 \text{ kg/tonnes} = 1{,}500 \text{ kg/ha/year}$$

$$1{,}500 \text{ kg/ha/year} \times 1\text{ha}/10{,}000 \text{ m}^2 = 0.15 \text{ kg/m}^2\text{/year}$$

YOUR TURN

1. Calculate the average amount of dead plant material that fall in the other biomes and convert each into kilograms per square meter per year.
2. Explain the reasons for the differences among biomes. How do these relative differences in the amount of dead plant material in each biome compare to the amount of organic matter in the soil of each biome?

MODULE 12 AP® Review

Preparing for the AP® Exam

In this module, we have learned that terrestrial biomes are categorized by the dominant plant growth forms that exist in a region. These dominant plants coincide with the climates of a region, which we can illustrate graphically using climate diagrams. Based on patterns of climate and plant growth forms, we can categorize nine terrestrial biomes. In the next module, we will see that aquatic biomes can also be categorized, though the criteria are quite different from terrestrial biomes.

AP® Practice Questions

Choose the best answer for the following.

1. In addition to temperature, a terrestrial biome is defined by
   I. annual precipitation.
   II. distinctive animal species.
   III. distinctive plant species.
   (a) I only
   (b) I and II
   (c) I and III
   (d) II and III

2. The precipitation line below the temperature line in a climate diagram shows
   (a) the primary growing season.
   (b) the biome is a desert or tundra.
   (c) when plant growth will be limited by precipitation.
   (d) the seasons in which droughts are most likely to occur.

3. Permafrost is an important factor in which biome?
   I. tundra
   II. boreal forest
   III. cold desert
   (a) I only
   (b) I and II
   (c) II only
   (d) I and III

4. In which biome is plant growth primarily constrained by precipitation?
   (a) boreal forest
   (b) temperate seasonal forest
   (c) temperate grassland
   (d) tropical rainforest

5. Which biome has the highest soil nutrient levels?
   (a) tropical rainforest
   (b) boreal forest
   (c) woodland/shrubland
   (d) temperate seasonal forest

MODULE 13

Aquatic Biomes

Whereas terrestrial biomes are categorized by dominant plant growth forms, aquatic biomes are categorized by physical characteristics such as salinity, depth, and water flow. Temperature is an important factor in determining which species can survive in a particular aquatic habitat, but it is not a factor used to categorize aquatic biomes. Aquatic biomes fall into two broad categories: freshwater and marine. Freshwater biomes include streams, rivers,

Learning Goals

After reading this module you should be able to

- identify the major freshwater biomes.
- identify the major marine biomes.

lakes, and wetlands. Saltwater biomes, also known as marine biomes, include shallow marine areas such as estuaries and coral reefs as well as the open ocean.

Freshwater biomes have low salinity

Freshwater biomes can be categorized as streams and rivers, lakes and ponds, or freshwater wetlands.

Streams and Rivers

Streams and rivers are characterized by flowing fresh water that may originate from underground springs or as runoff from rain or melting snow (**FIGURE 13.1**). Streams (also called creeks) are typically narrow and carry relatively small amounts of water. Rivers are typically wider and carry larger amounts of water. It is not always clear, however, at what point a particular stream, as it combines with other streams, becomes large enough to be called a river.

As water flow changes, biological communities also change. Most streams and many rapidly flowing rivers have few plants or algae to act as producers. Instead, inputs of organic matter from terrestrial biomes, such as fallen leaves, provide the base of the food web. This organic matter is consumed by insect larvae and crustaceans such as crayfish, which then provide food for secondary consumers such as fish. As fast-moving streams combine to form rivers, the water flow typically slows, sediments and organic material settle to the bottom, and rooted plants and algae are better able to grow.

FIGURE 13.1 Streams and rivers. Streams and rivers are freshwater aquatic biomes that are characterized by flowing water. This photo shows Berea Falls on the Rocky River near Cleveland, Ohio. *(Jim West/The Image Works)*

Littoral zone The shallow zone of soil and water in lakes and ponds where most algae and emergent plants grow.

Limnetic zone A zone of open water in lakes and ponds.

Phytoplankton Floating algae.

Profundal zone A region of water where sunlight does not reach, below the limnetic zone in very deep lakes.

Benthic zone The muddy bottom of a lake, pond, or ocean.

Fast-moving streams and rivers typically have stretches of turbulent water called rapids, where water and air are mixed together. This mixing allows large amounts of atmospheric oxygen to dissolve into the water. Such high-oxygen environments support fish species such as trout and salmon that need large amounts of oxygen. Slower-moving rivers experience less mixing of air and water. These lower-oxygen environments favor species such as catfish that can better tolerate low-oxygen conditions. Today, the major threats to streams and rivers are excess nutrients and pollutants.

Lakes and Ponds

Lakes and ponds contain standing water, at least some of which is too deep to support emergent vegetation (plants that are rooted to the bottom and emerge above the water's surface). Lakes are larger than ponds, but as with streams and rivers, there is no clear point at which a pond is considered large enough to be called a lake (**FIGURE 13.2**).

As **FIGURE 13.3** shows, lakes and ponds can be divided into several distinct zones. The **littoral zone** is the shallow area of soil and water near the shore where algae and emergent plants such as cattails grow. Most photosynthesis occurs in this zone. In the open water, or **limnetic zone**, rooted plants can no longer survive; floating algae called **phytoplankton** are the only photosynthetic organisms. The limnetic zone extends as deep as sunlight can penetrate. Very deep lakes have a region of water below the limnetic zone, called the **profundal zone**. Because sunlight does not reach the profundal zone, producers cannot survive there, so nutrients are not easily recycled into the food web. Bacteria decompose the detritus that reaches the profundal zone, but they consume oxygen in the process. As a result, dissolved oxygen concentrations are not sufficient to support many large organisms. The muddy bottom of a lake or pond beneath the limnetic and profundal zones is called the **benthic zone**.

FIGURE 13.2 Lakes and ponds. Lakes, such as Lake George in New York State, are characterized by standing water and a central zone of water that is too deep for emergent vegetation. *(Frank Paul/Alamy)*

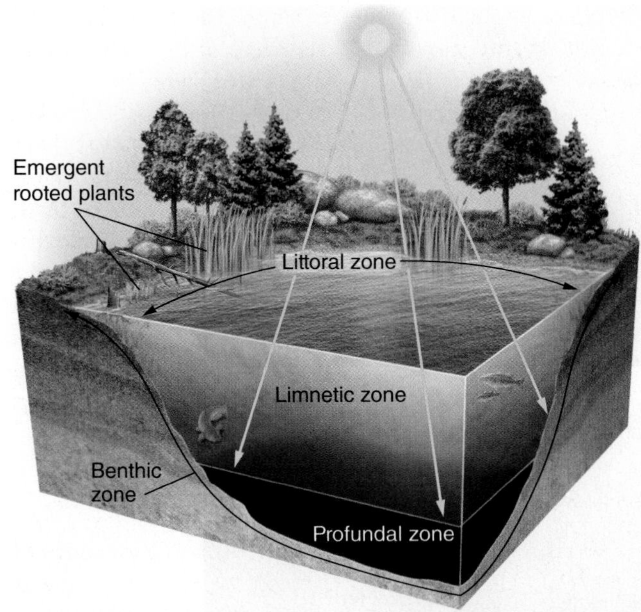

FIGURE 13.3 Lake zones. The littoral zone consists of shallow water with emerging, rooted plants whereas the limnetic zone is the deeper water where plants do not emerge. The deepest water, where oxygen can be limiting because little sunlight penetrates to allow photosynthesis by producers, is the profundal zone. The sediments that lie beneath the littoral, limnetic, and profundal zones constitute the benthic zone.

Lakes are classified by their level of primary productivity. Lakes that have low productivity due to low amounts of nutrients such as phosphorus and nitrogen in the water are called **oligotrophic** lakes. In contrast, lakes with a moderate level of productivity are called **mesotrophic** lakes, and lakes with a high level of productivity are called **eutrophic** lakes. A current concern for many oligotrophic lakes is that human activities are causing increased nutrient inputs, which is causing them to become less clear due to the increased growth of the phytoplankton.

Freshwater Wetlands

Freshwater wetlands are aquatic biomes that are submerged or saturated by water for at least part of each year, but shallow enough to support emergent vegetation. They support species of plants that are specialized to live in submerged or saturated soils.

Freshwater wetlands include swamps, marshes, and bogs. Swamps are wetlands that contain emergent trees, such as the Great Dismal Swamp in Virginia and North Carolina and the Okefenokee Swamp in Georgia and Florida (**FIGURE 13.4a**, page 140). Marshes are wetlands that contain primarily nonwoody vegetation, including cattails and sedges (Figure 13.4b). Bogs, in contrast, are very acidic wetlands that typically contain sphagnum moss and spruce trees (Figure 13.4c).

Freshwater wetlands are among the most productive biomes on the planet, and they provide several critical ecosystem services. For example, wetlands can take in large amounts of rainwater and release it slowly into the groundwater or into nearby streams, thus reducing the severity of floods and droughts. Wetlands also filter pollutants from water, recharging the groundwater with clean water. Many bird species depend on wetlands during migration or breeding. As many as one-third of all endangered bird species in the United States spend some part of their lives in wetlands, even though this biome makes up only 5 percent of the nation's land area. More than half of the freshwater wetland area in the United States has been drained for agriculture or development or to eliminate breeding grounds for mosquitoes and various disease-causing organisms.

Marine biomes have high salinity

Marine biomes contain salt water and can be categorized as salt marshes, mangrove swamps, intertidal zones, coral reefs, and the open ocean.

Salt Marshes

Like freshwater marshes, **salt marshes**—found along the coast in temperate climates—contain nonwoody emergent vegetation (**FIGURE 13.5**, page 140). The salt marsh is one of the most productive biomes in the world. Many salt marshes are found within an **estuary** which is an area along the coast where the fresh water of rivers mixes with salt water from the ocean. Because rivers carry large amounts of nutrient-rich organic material, estuaries are extremely productive places for plants and algae, and the abundant plant life helps filter contaminants out of the water. Salt marshes provide important habitat for spawning fish and shellfish; two-thirds of marine fish and shellfish species spend their larval stages in estuaries. Like many freshwater wetlands, salt marshes have suffered from being filled in, being developed to support growing human populations, and pollutants that arrive from the rivers that supply water.

Oligotrophic Describes a lake with a low level of productivity.

Mesotrophic Describes a lake with a moderate level of productivity.

Eutrophic Describes a lake with a high level of productivity.

Freshwater wetland An aquatic biome that is submerged or saturated by water for at least part of each year, but shallow enough to support emergent vegetation.

Salt marsh A marsh containing nonwoody emergent vegetation, found along the coast in temperate climates.

Estuary An area along the coast where the fresh water of rivers mixes with salt water from the ocean.

FIGURE 13.4 Freshwater wetlands. Freshwater wetlands have soil that is saturated or covered by fresh water for at least part of the year and are characterized by particular plant communities. (a) In this swamp in southern Illinois, bald cypress trees emerge from the water. (b) This marsh in south central Wisconsin is characterized by cattails, sedges, and grasses growing in water that is not acidic. (c) This bog in northern Wisconsin is dominated by sphagnum moss as well as shrubs and trees that are adapted to acidic conditions. *(a–c: Lee Wilcox)*

Mangrove Swamps

Mangrove swamps occur along tropical and subtropical coasts and, like freshwater swamps, contain trees whose roots are submerged in water (**FIGURE 13.6**). Unlike most trees, however, mangrove trees are salt tolerant. They often grow in estuaries, but they can also be found along shallow coastlines that lack inputs of fresh water. The trees help to protect those coastlines from erosion and storm damage. Falling leaves and trapped organic material produce a nutrient-rich environment. Like salt marshes, mangrove

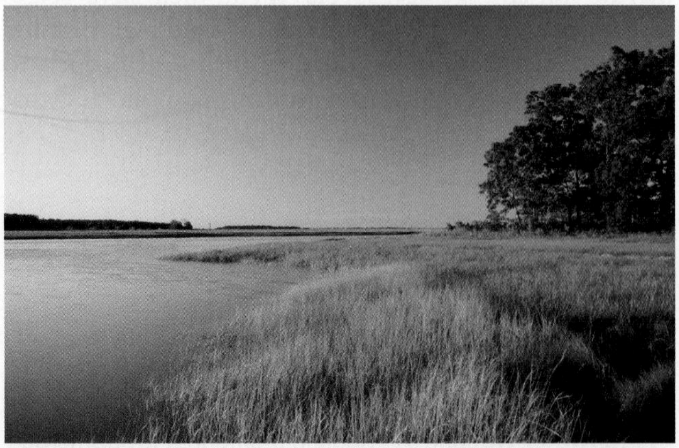

FIGURE 13.5 Salt marsh. The salt marsh is a highly productive biome typically found in temperate regions where fresh water from rivers mixes with salt water from the ocean. This salt marsh is in Plum Island Sound in Massachusetts. *(© 2006, 2007 Jerry and Marcy Monkman/www.ecophotography.com)*

FIGURE 13.6 Mangrove swamp. Salt-tolerant mangrove trees, such as these in Everglades National Park, are important in stabilizing tropical and subtropical coastlines and in providing habitat for marine organisms. *(© T. & S. Allofs/Biosphoto)*

swamps provide sheltered habitat for fish and shellfish. Nearly one-third of the world's mangrove swamps have been destroyed, primarily for human habitation or to make space to grow crops such as rice and rubber trees.

Intertidal Zones

The **intertidal zone** is the narrow band of coastline that exists between the levels of high tide and low tide (**FIGURE 13.7**). Intertidal zones range from steep, rocky areas to broad, sloping mudflats. Environmental conditions in this biome are relatively stable when submerged during high tide. However, conditions can become quite harsh during low tide when organisms are exposed to direct sunlight, high temperatures, and desiccation. Moreover, waves crashing onto shore can make it a challenge for organisms to hold on and not get washed away. Intertidal zones are home to a wide variety of organisms that have adapted to these conditions, including barnacles, sponges, algae, mussels, crabs, and sea stars. The main threats of human impacts are pollution, including trash, chemical pollution, and oil spills.

Coral Reefs

Coral reefs, which are found in warm, shallow waters beyond the shoreline, represent Earth's most diverse marine biome. Corals are tiny animals that secrete a layer of limestone (calcium carbonate) to form an external skeleton. The animal living inside this tiny skeleton is essentially a hollow tube with tentacles that draw in plankton and detritus. Corals live in water that is relatively poor in nutrients and food, which is possible because of their relationship with single-celled algae that live within the tissues of the corals. When a coral digests the food it captures, it releases CO_2 and nutrients. The algae use the CO_2 during photosynthesis to produce sugars and the nutrients stimulate the algae to release their sugars to the coral. The coral gains energy in the form of sugars, and the algae obtain CO_2, nutrients, and a safe place to live within the coral's tiny limestone skeleton. But this association with photosynthetic algae means that corals can live only in shallow waters where light can penetrate.

Although each individual coral is tiny, most corals live in vast colonies. As individual corals die and decompose, their limestone skeletons remain. Over time, these skeletons accumulate and develop into coral reefs, which can become quite massive. The Great Barrier Reef of Australia, for example, covers an area of 2,600 km^2 (1,600 $miles^2$). A tremendous diversity of other organisms, including fish and invertebrates, use the structure of the reef as both a refuge in which to live and a place to find food. At the Great Barrier Reef there are more than 400 species of coral, 1,500 species of tropical fish, and 200 species of birds.

Coral reefs are currently facing a wide range of challenges, including pollutants and sediments that make it difficult for the corals to survive. Coral reefs also face the growing problem of **coral bleaching**, a phenomenon in which the algae inside the corals die (**FIGURE 13.8**). Without the algae, the corals soon die as well, and the reef turns white. Scientists believe that the algae are dying from a combination of disease and environmental changes, including lower ocean pH and abnormally high water temperatures. Coral bleaching is a serious problem: Without the corals, the entire coral reef biome is endangered.

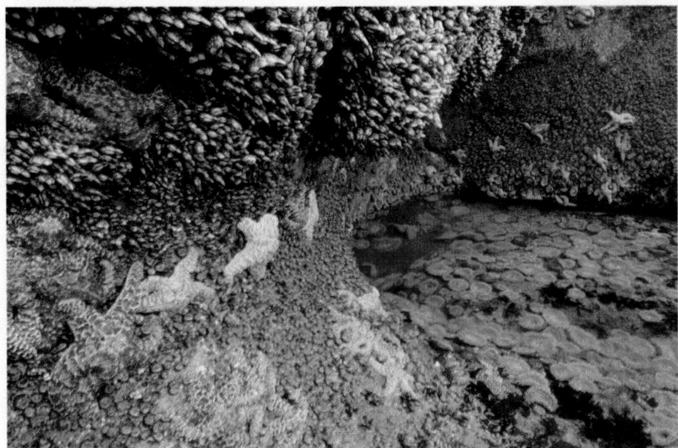

FIGURE 13.7 Intertidal zone. Organisms that live in the area between high and low tide, such as these giant green sea anemones (*Anthopleura xanthogrammica*), goose barnacles (*Lepas anserifera*), and ochre sea stars (*Pisaster ochraceus*), must be highly tolerant of the harsh, desiccating conditions that occur during low tide. This photo was taken at Olympic National Park, Washington. *(Jim Zipp/Science Source)*

The Open Ocean

The **open ocean** contains deep ocean water that is located away from the shoreline where sunlight can no longer reach the ocean bottom. The exact depth of penetration by sunlight depends on a number of factors, including the amounts of sediment and algae suspended in the water, but it generally does not exceed 200 m (approximately 650 feet).

Like a pond or lake, the ocean can be divided into zones. These zones are shown in **FIGURE 13.9** on page 142. The upper layer of ocean water that receives enough sunlight

Mangrove swamp A swamp that occurs along tropical and subtropical coasts, and contains salt-tolerant trees with roots submerged in water.

Intertidal zone The narrow band of coastline between the levels of high tide and low tide.

Coral reef The most diverse marine biome on Earth, found in warm, shallow waters beyond the shoreline.

Coral bleaching A phenomenon in which algae inside corals die, causing the corals to turn white.

Open ocean Deep ocean water, located away from the shoreline where sunlight can no longer reach the ocean bottom.

FIGURE 13.8 **Coral reef.** This coral reef is in American Samoa in the South Pacific. The image on the left was taken prior to coral bleaching in December 2014. The image on the right was taken 2 months later, after coral bleaching occurred. *(The Ocean Agency/XL Catlin Seaview Survey/Richard Vevers)*

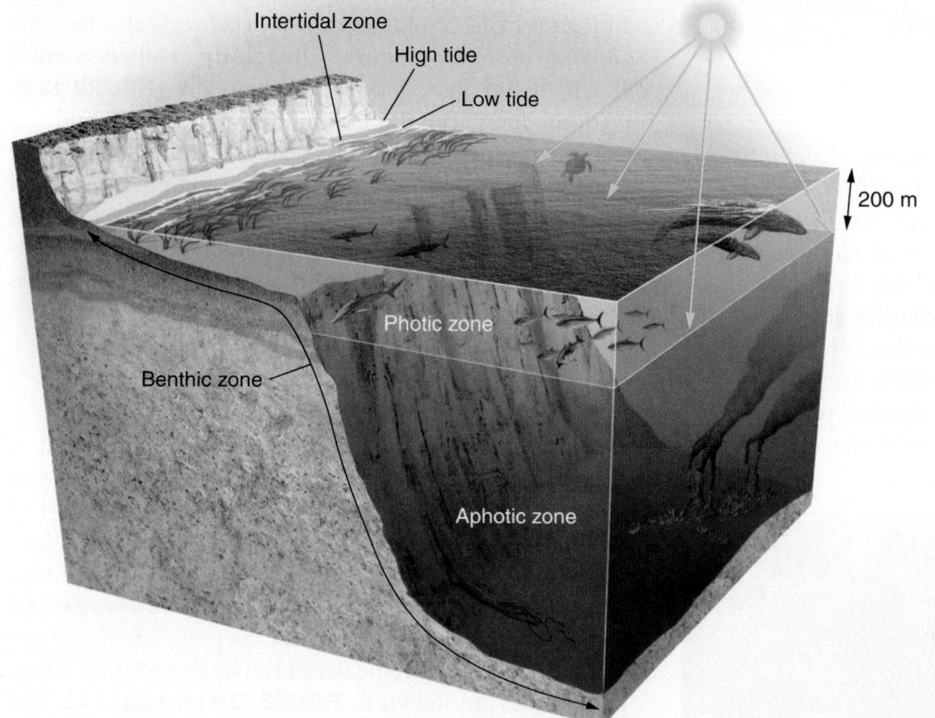

FIGURE 13.9 **The open ocean.** The open ocean can be separated into several distinct zones.

to allow photosynthesis is the **photic zone**, and the deeper layer of water that lacks sufficient sunlight for photosynthesis is the **aphotic zone**. The ocean floor is called the benthic zone.

Photic zone The upper layer of ocean water in the ocean that receives enough sunlight for photosynthesis.

Aphotic zone The deeper layer of ocean water that lacks sufficient sunlight for photosynthesis.

Chemosynthesis A process used by some bacteria in the ocean to generate energy with methane and hydrogen sulfide.

In the photic zone, algae are the major producers. They form the base of a food web that includes tiny zooplankton, fish, and whales. In the aphotic zone, because of the lack of light, there are no photosynthetic producers. However, there are some species of bacteria that can use the energy contained in the bonds of methane and hydrogen sulfide, which are both found in the deep ocean, to generate energy via **chemosynthesis** rather than photosynthesis. These bacteria form the base of a deep-ocean food web that includes animals such as tube worms (see Figure 5.3c on page 49). The aphotic zone also contains a variety of organisms that can generate their own light to help them feed in the dark waters. These organisms include several species of crustaceans, jellyfish, squid, and fish.

MODULE 13 AP® Review

Preparing for the AP® Exam

In this module, we have learned that aquatic biomes are characterized by physical features such as salinity, depth, and water flow. Freshwater biomes include streams and rivers, which have flowing water, and lakes, ponds, and wetlands, which have standing water. Marine biomes contain salt water and include salt marshes, mangrove swamps, intertidal zones, coral reefs, and the open ocean. Differences in water flow, depth, and salinity help us understand why different species of producers and consumers, including commercially important species of fish and shellfish, live in different aquatic regions of the world.

AP® Practice Questions

Choose the best answer for the following.

1. Which ecosystem experiences harsh conditions due to conditions from tides?
   (a) coral reef
   (b) open ocean
   (c) intertidal zone
   (d) ponds and lakes

2. Most of the photosynthesis in lakes and ponds occurs in the
   (a) benthic zone.
   (b) littoral zone.
   (c) limnetic zone.
   (d) profundal zone.

3. Which is NOT an important ecosystem service provided by wetlands?
   (a) flood control
   (b) breeding habitat for birds
   (c) water filtration
   (d) seed dispersal

4. Aquatic biomes are categorized by
   I. dominant plant growth forms.
   II. depth.
   III. salinity.
   (a) I and II
   (b) I and III
   (c) II and III
   (d) I, II, and III

5. Which biome contains the aphotic zone?
   (a) coral reefs
   (b) streams and rivers
   (c) freshwater wetlands
   (d) open ocean

Working Toward Sustainability

Is Your Coffee Made in the Shade?

Around the world, people enjoy drinking coffee. Worldwide, people buy 8.5 billion kilograms (18.7 billion pounds) of coffee beans each year. In the United States, 54 percent of adults drink coffee every day, at an average of 3 cups per day. In 2017, the percentage of Americans who drank coffee continued to increase, with a dramatic 150 percent increase since 2014 for people 13–18 years old. In short, coffee is an important part of many people's lives. But have you ever thought about where your coffee comes from?

Coffee beans come from several species of shrubs that historically grew in Ethiopia under the shade of the tropical rainforest canopy. In the fifteenth century, coffee was brought to the Middle East and eventually spread throughout the world. Because of its popularity, coffee is now farmed in many places around the world, including South America, Africa, and Southeast Asia.

As farmers began cultivating coffee, they grew it like many other crops by clearing large areas of rainforest and planting coffee bushes close together in large open fields. Because the coffee plant's native habitat is a shady forest, coffee farmers found that they had to construct shade over the plants to prevent them from becoming sunburned in the intense tropical sunlight. Over the past several decades, however, plant breeders have developed more sunlight-tolerant plants that can handle intense sunlight and produce many more coffee beans per plant.

As coffee was transformed from a plant that was naturally scattered throughout a diverse, shady rainforest to one that was grown as a single species in large numbers in open

Shade-grown coffee in Ecuador. Coffee grown in the shade requires less pesticide, helps to preserve the plant diversity of the rainforest, and even tastes better. *(Dr Morley Read/Shutterstock.com)*

fields, the coffee fields became attractive targets for insect pests and diseases. In response, farmers have applied a variety of pesticides to combat these pests, which has increased the cost of farming coffee, poisoned workers, and polluted the environment. Given the world's demand for coffee, what other options do coffee farmers have?

Some coffee farmers thought back to the natural environment in which coffee grows and wondered if they could farm coffee under more natural conditions. Such coffee, called shade-grown coffee, is grown in one of three ways: by planting coffee bushes in an intact rainforest, by planting the bushes in a rainforest that has had some of the trees removed, or by planting the bushes in a field alongside trees that produce other marketable products, including fruit. Coffee bushes grown in this way attract fewer pests, so less money is needed to buy and apply pesticide, and there is less risk to workers and the nearby soil and water. Using these methods, coffee can be grown while still preserving some of the plant diversity of the rainforest. And the coffee often tastes better. The density of coffee plants is lower in these more diverse landscapes, however, which means that only about one-third as much coffee is produced per hectare. So, while there are cost savings, the yield is lower. Economically, this means that owners of shade-grown coffee farms need to charge higher prices to match the profits of other farms.

How can farmers producing shade-grown coffee stay in business? A number of environmental groups that want to preserve biodiversity in tropical rainforests have stepped in to help. Researchers found that shade-grown coffee farms provided habitat for approximately 150 species of rainforest birds, whereas open-field coffee farms provided habitat for only 20 to 50 bird species. Not surprisingly, researchers also found that other groups of animals were more diverse on shade-grown coffee farms. In response to these findings, the Smithsonian Migratory Bird Center in Washington, D.C., developed a program to offer a "Bird Friendly" seal of approval to coffee farmers who were producing shade-grown coffee, and doing so without the use of any pesticides. Combined with an advertising campaign that explained the positive effects of shade-grown coffee on biodiversity, this seal of approval alerted consumers to make a conscious choice about the impact that their favorite beverage was having on rainforests. From 2005 to 2011, the sales of the shade-grown coffee with the Smithsonian seal of approval increased by an amazing 250 percent, but it still was only 2 percent of all coffee sales. The Arbor Day Foundation, an environmental organization that promotes the planting of trees, joined the effort by selling its own brand of shade-grown coffee. Over the past 20 years, it has become clear that when consumers are informed about how coffee is grown, many people are willing to choose the shade-grown varieties, even if it requires spending more money to reduce adverse impacts on the tropical rainforest biome.

Today, selling shade-grown coffee to the global market of consumers remains challenging. In fact, a 2014 study found that the total area of land dedicated to shade-grown coffee had declined from 43 to 24 percent over the past 15 years. There are multiple reasons for this decline, including the fact that while shade-grown coffee commands a higher price, the increased price paid to individual farmers does not always compensate for the lower coffee yields and the cost of the certification programs. Thus, it remains clear that shade-grown coffee has environmental benefits, but the decision of individual farmers to grow such coffee depends on their economic conditions.

Critical Thinking Questions

1. If shade-grown coffee produces less coffee per hectare, what economic factor might prevent all coffee from being grown this way?
2. If three times as much coffee can be grown in the Sun than in the shade, what are the trade offs in terms of the amount of land used for growing coffee under these two alternative agricultural practices?

References

Philpott, S. M., et al. 2008. Biodiversity loss in Latin American coffee landscapes: Review of the evidence on ants, birds, and trees. *Conservation Biology* 22:1093–1105.

Smithsonian Migratory Bird Center. *Coffee Drinkers and Bird Lovers.* http://nationalzoo.si.edu/SCBI/MigratoryBirds/Coffee/lover.cfm.

Jha, S., et al. 2014. Shade coffee: Update on a disappearing refuge for biodiversity. *BioScience* 64:416–428.

In this chapter we have examined how global processes such as air and water currents determine regional climates and how these regional climates have a major effect on the types of organisms that can live in different parts of the world. Among the terrestrial biomes, temperature and precipitation affect the rate of decomposition of dead organisms and the productivity of the soil. Understanding these patterns helps us understand how humans have come to use the land in different ways: growing crops in regions with enough water and a sufficient growing season, grazing domesticated animals in drier areas, and harvesting lumber from forests. Among the aquatic biomes, differences in flow, salinity, and depth help to determine the aquatic species that can live in different aquatic regions of the world.

Key Terms

Climate
Weather
Troposphere
Stratosphere
Albedo
Saturation point
Adiabatic cooling
Adiabatic heating
Latent heat release
Atmospheric convection current
Hadley cell
Intertropical convergence zone (ITCZ)
Polar cell
Ferrell cell
Coriolis effect
Rain shadow
Gyres
Upwelling
Thermohaline circulation
El Niño–Southern Oscillation (ENSO)
Terrestrial biome
Aquatic biome
Habitat
Tundra
Permafrost
Boreal forest
Temperate rainforest
Temperate seasonal forest
Woodland/shrubland
Temperate grassland/cold desert
Tropical rainforest
Tropical seasonal forest/savanna
Subtropical desert
Littoral zone
Limnetic zone
Phytoplankton
Profundal zone
Benthic zone
Oligotrophic
Mesotrophic
Eutrophic
Freshwater wetland
Salt marsh
Estuary
Mangrove swamp
Intertidal zone
Coral reef
Coral bleaching
Open ocean
Photic zone
Aphotic zone
Chemosynthesis

Learning Goals Revisited

Module 9 The Unequal Heating of Earth

Identify the five layers of the atmosphere.

Above Earth's surface, the first layer of atmosphere is the troposphere, followed by the stratosphere, mesosphere, thermosphere, and the exosphere.

Discuss the factors that cause unequal heating of Earth.

The unequal heating of Earth is caused by differences in the angle of the Sun's rays that strike Earth, the amount of atmosphere that the Sun's rays must pass through before striking Earth's surface, and how much of the solar energy that reaches Earth is reflected rather than absorbed.

Describe how Earth's tilt affects seasonal differences in temperatures.

Earth's central axis is tilted at 23.5°, which causes seasonal changes in the latitudes that receive the most intense sunlight.

Module 10 Air Currents

Explain how the properties of air affect the way it moves in the atmosphere.

Air rises when it becomes less dense and sinks when it becomes more dense. Warm air has a higher saturation point for water vapor than cold air. Changes in air pressure result in adiabatic cooling or heating; when water condenses it emits heat, which is known as latent heat release.

Identify the factors that drive atmospheric convection currents.

Atmospheric convection currents are driven by the intense sunlight that strikes Earth near the tropics. This solar energy warms the surface of Earth, which causes moist air to rise, cool, and release water as precipitation. As the air continues to rise, it reaches the top of the troposphere. The air, which is now cold and dry, moves toward the poles until it descends at approximately 30° N or 30° S latitude. As it descends back to Earth's surface, the air warms and then moves back toward the equator.

Describe how Earth's rotation affects the movement of air currents.

Because the surface of Earth travels faster near the equator than near the poles, the Coriolis effect causes convection currents traveling north and south to be deflected, thereby creating trade winds, westerlies, and easterlies.

Explain how the movement of air currents over mountain ranges affects climates.

When moist air from the ocean moves up a mountain, the air cools and releases water as precipitation, which results in a moist environment on the windward side. On the other side of the mountain, the cool, dry air descends, which results in a dry environment on the leeward side of the mountain.

Module 11 Ocean Currents

Describe the patterns of surface ocean circulation.

Ocean currents are driven by a combination of temperature, gravity, prevailing winds, the Coriolis effect, and the locations of continents. Together, prevailing winds and ocean currents distribute heat and precipitation around the globe.

Explain the mixing of surface and deep ocean waters from thermohaline circulation.

As ocean water flows from the Gulf of Mexico to the North Atlantic, water evaporates or freezes, and this causes the remaining water to have a high salt concentration and therefore a high density. This dense water sinks to the bottom of the ocean and later comes back to the surface near the equator.

Identify the causes and consequences of the El Niño–Southern Oscillation.

The El Niño–Southern Oscillation occurs when the typical trade winds from South America weaken or reverse, which allows the equatorial current that usually flows from east to west to reverse direction. When this happens, the upwelling along the western coast of South America is impeded, which affects climates around the world.

Module 12 Terrestrial Biomes

Explain how we define terrestrial biomes.

Terrestrial biomes are categorized by the dominant plant forms that exist in a region.

Interpret climate diagrams.

Climate diagrams illustrate monthly patterns of temperature and precipitation during the year. They also illustrate the growing season of a biome and the months during which plants are more constrained by temperature or precipitation.

Identify the nine terrestrial biomes.

The nine terrestrial biomes are tundra, boreal forests, temperate rainforests, temperate seasonal forests, woodland/shrublands, temperate grasslands/cold deserts, tropical rainforests, tropical seasonal forests/savannas, and subtropical deserts.

Module 13 Aquatic Biomes

Identify the major freshwater biomes.

There are three types of freshwater biomes. Streams and rivers have flowing fresh water. Lakes and ponds have standing water, at least some of which is too deep to support emergent vegetation. Freshwater wetlands are submerged or saturated by water for at least part of the year, but shallow enough to support emergent vegetation.

Identify the major marine biomes.

There are five types of marine biomes. Salt marshes are found along the coast in temperate climates and contain nonwoody emergent vegetation. Mangrove swamps occur along tropical and subtropical coasts and contain trees that have roots submerged in the water. The intertidal zone is the narrow band of coastline that exists between the levels of high tide and low tide. Coral reefs are found in warm, shallow waters beyond the shoreline and represent Earth's most diverse marine biome. The open ocean is characterized by deep water where sunlight can no longer reach the ocean bottom.

Practice Math and Graphing

Preparing for the AP® Exam

Answer the following questions. Be sure to show all your work.

1. Practice Math

To create climate graphs, we need to know the typical precipitation for a biome. Scientists typically do this by selecting a location somewhere within a biome and then using weather data from that location to determine the average precipitation over multiple years. The two tables on the next page show data collected for Nashville, Tennessee for 3 years. Use the data in Table 1 to calculate the average precipitation for each month.

TABLE 1 Precipitation

Month	Year 1 (mm)	Year 2 (mm)	Year 3 (mm)
January	91	90	95
February	89	98	101
March	95	99	103
April	91	93	104
May	142	135	134
June	99	93	105
July	87	89	97
August	81	84	87
September	72	101	94
October	82	83	78
November	94	105	105
December	83	125	95

TABLE 2 Temperature

Month	Average temperature (°C)
January	8.3
February	11.1
March	16.1
April	21.7
May	25.6
June	30.0
July	31.7
August	31.7
September	27.8
October	22.2
November	15.6
December	9.4

2. Practice Graphing

(a) Using the average precipitation data from your answers to the "Practice Math" problem above and Table 2 on average monthly temperatures, create a climate diagram.

(b) Based on your climate diagram, what is the growing season for this location?

(c) Based on your climate diagram, is the biome more limited by temperature or precipitation?

Chapter 4 AP® Environmental Science Practice Exam

Preparing for the AP® Exam

Section 1: Multiple-Choice Questions

Choose the best answer for questions 1–20.

1. In which layer of Earth's atmosphere does most weather occur?
   (a) troposphere
   (b) stratosphere
   (c) mesosphere
   (d) thermosphere

2. Which statement best explains why polar regions are colder than tropical regions?
   (a) Polar regions have lower albedo values.
   (b) Polar regions receive less solar energy per unit of surface area.
   (c) Tropical regions receive less direct sunlight throughout the year.
   (d) Sunlight travels through more atmosphere and loses more energy in tropical regions.

3. Which statement about patterns of air convection is NOT correct?
   (a) The air in a Hadley cell rises where sunlight strikes Earth most directly.
   (b) The greatest amount of precipitation occurs at the intertropical convergence zone.
   (c) The air in a Hadley cell descends near 30° N and 30° S, causing the formation of deserts.
   (d) Along Earth's surface, the air of a Hadley cell moves away from the equator.

4. An increase in evaporation near the equator would most likely cause
   (a) increased precipitation at the ITCZ.
   (b) decreased precipitation at the ITCZ.
   (c) increased precipitation in Ferrell cells.
   (d) decreased precipitation in Ferrell cells.

5. The high heat capacity of water causes what effect when combined with ocean circulation?
   (a) the high salinity of deep polar water in the thermohaline cycle
   (b) warm temperatures in continental coastal areas
   (c) the suppression of upwelling during an ENSO event
   (d) the heat transfer between the ocean and atmosphere due to evaporation along the equator

6. Which process is NOT characteristic of oceanic circulation?
   (a) counterclockwise gyres in the Northern Hemisphere
   (b) slow thermohaline circulation of surface and deep ocean waters
   (c) unequal heating of tropical versus polar ocean waters
   (d) El Niño–Southern Oscillation

7. Which statement about rain shadows is correct?
   (a) They occur on the western sides of mountain ranges in the Northern Hemisphere.
   (b) As air rises over a mountain range, water vapor condenses into precipitation.
   (c) They occur on the eastern sides of mountain ranges in the Southern Hemisphere.
   (d) The rain shadow side of a mountain range receives the most rain.

8. Why do scientists use dominant plant growth forms to categorize terrestrial biomes?
   (a) Plants with similar growth forms are always closely related genetically.
   (b) Different plant growth forms indicate climate differences, whereas different animal forms do not.
   (c) Plants from similar climates evolve different adaptations.
   (d) Similar plant growth forms are found in climates with similar temperatures and amounts of precipitation.

9. Which information is NOT found in climate diagrams?
   (a) average annual temperature
   (b) average annual humidity
   (c) the months when plant growth is limited by precipitation
   (d) the length of the growing season

10. Which statement about tundras and boreal forests is correct?
   (a) Both are characterized by slow plant growth, so there is little accumulation of organic matter.
   (b) Tundras are warmer than boreal forests.
   (c) Boreal forests have shorter growing seasons than tundras.
   (d) Boreal forests have larger dominant plant growth forms than tundras.

11. Which statement about temperate biomes is NOT correct?
   (a) Temperate biomes have average annual temperatures above 20°C.
   (b) Temperate rainforests receive the most precipitation, whereas cold deserts receive the least precipitation.
   (c) Temperate rainforests can be found in the northwestern United States.
   (d) Temperate shrublands are adapted to frequent fires.

12. Which statement about tropical biomes is correct?
   (a) Tropical rainforests have the highest precipitation due to the proximity of the ITCZ.
   (b) Savannas are characterized by the densest forests.
   (c) Tropical rainforests have the slowest rates of decomposition due to high rainfall.
   (d) Subtropical deserts have the highest species diversity.

13. Which statement about aquatic biomes is correct?
   (a) They are characterized by dominant plant growth forms.
   (b) They can be categorized by temperature and precipitation.
   (c) Lakes contain littoral zones and intertidal zones.
   (d) Freshwater wetlands have emergent plants in their deepest areas, whereas ponds and lakes do not.

14. Heavy precipitation over the intertropical convergence zone (ITCZ) and the windward side of mountain ranges are both caused by
   (a) the Coriolis effect.
   (b) prevailing winds.
   (c) adiabatic heating.
   (d) reduced air pressure.

Question 15 refers to the graph.

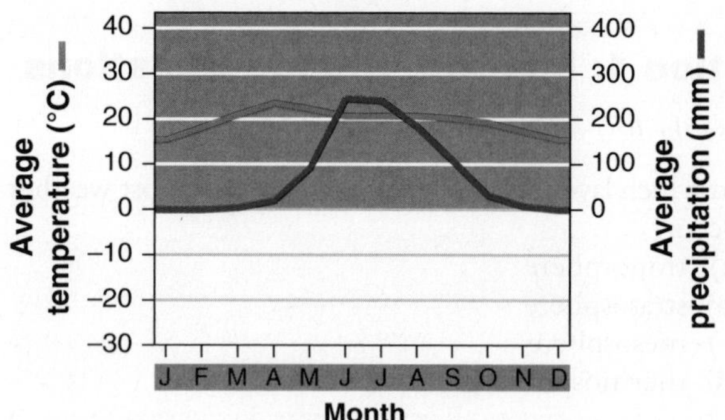

15. Which biome does this graph represent?
   (a) boreal forest
   (b) temperate rainforest
   (c) temperate seasonal rainforest
   (d) tropical seasonal forest

16. Which is NOT true regarding an El Niño–Southern Oscillation (ENSO) event?
(a) An ENSO event relies on thermohaline circulation.
(b) An ENSO event brings warmer air temperatures to South America.
(c) An ENSO event typically starts around Christmas.
(d) ENSO events are driven by a reversal of trade winds.

17. The limnetic zone is defined as the area of a lake where
(a) algae and emergent plants grow.
(b) sunlight does not reach.
(c) the thermocline exists.
(d) phytoplankton are the only photosynthetic organisms.

18. Which statement about the stratosphere is FALSE?
(a) It is where UV radiation is converted to infrared radiation.
(b) It is where Earth's weather occurs.
(c) It is above the troposphere.
(d) It contains a layer of ozone.

19. Which biome provides ecosystem services that include reducing the severity of floods and filtering pollutants from the water?
(a) freshwater wetlands
(b) temperate grassland
(c) coral reef
(d) open ocean

20. Which explains the large amount of precipitation over the tropics?
I. the unequal warming of the tropics
II. adiabatic heating
III. adiabatic cooling
(a) I only
(b) III only
(c) I and II
(d) I and III

Section 2: Free-Response Questions

Write your answer to each part clearly. Support your answers with relevant information and examples. Where calculations are required, show your work.

1. As the greenhouse effect continues to warm the planet slowly, the glaciers of Greenland are melting at a rapid rate. Scientists are concerned that this melting may dilute the salt water in that region of the ocean enough to shut down thermohaline circulation. Use what you know about climate to answer the following questions.
(a) Explain how shutting down thermohaline circulation would affect the temperature of western Europe. (2 points)
(b) Explain the possible consequences for agriculture in western Europe. (2 points)
(c) Why might the populations of fish along the west coasts of most continents increase if thermohaline circulation shuts down? (3 points)
(d) How would shutting down thermohaline circulation affect the transport of nutrients among the oceans of the world? (3 points)

2. A number of Earth's features determine the locations of biomes around the world.
(a) Explain why the tropical rainforests are found in regions of the world that receive the most direct sunlight. (4 points)
(b) Describe how movement of the ITCZ over the year influences the location of seasonal forests in tropical regions. (2 points)
(c) Identify the mechanisms by which albedo and the angle of the Sun's rays cause colder temperatures to occur on Earth near the North and South Poles. (2 points for each mechanism)

3. The climate data displayed in the table is collected from a particular biome.

Month	Average temperature (°C)	Average precipitation (mm)
January	15	717
February	15	499
March	13	640
April	11	585
May	8	641
June	6	440
July	5	418
August	7	427
September	9	523
October	10	688
November	12	522
December	13	648

(a) Using the given data, draw a climate diagram, making sure to label all axes. (3 points)
(b) Identify the growing season for this biome. (1 point)
(c) Is the biome more limited by temperature or precipitation? Explain your reasoning. (2 points)
(d) Is this biome likely tundra, temperate rainforest, cold desert, or temperate seasonal forest? Why? Name one area in the world where this biome is found. (3 points)
(e) Terrestrial biomes are characterized by plant growth forms in addition to temperature and precipitation. Name one factor that characterizes aquatic biomes. (1 point)

The biodiversity of ecological communities, such as this site in the San Juan National Forest in Colorado, is critical to the overall productivity and stability of natural ecosystems.
(Carol Barrington/Getty Images)

Evolution of Biodiversity

CHAPTER 5

CASE STUDY

The Benefits of Biodiversity

Why should we preserve the biodiversity of the world? We have long appreciated that species in nature provide benefits to humans including food, fiber for clothing, lumber, and medicines. We also know that collections of species in ecosystems provide primary productivity in the land and water, the production of clean water, and buffers against natural disasters. The question is, do the important benefits provided by these ecosystems critically depend on maintaining the biodiversity that has historically existed, or can we enjoy the same benefits even when biodiversity declines due to human activities?

For more than two decades, scientists have debated whether greater biodiversity provides greater ecosystem benefits. Researchers conducting experiments in aquatic and terrestrial ecosystems have commonly found that food webs with more species have a higher overall productivity. The ecosystems are also more stable whenever the abiotic environment changes, such as during a drought. While these results have been encouraging, critics have pointed out that because most experiments contain many fewer species than actually exist in nature, we cannot be sure that the results apply to nature. It is also possible that the productivity of natural ecosystems is much more affected by variation in climatic conditions and available nutrients than by the number of species in the food web. If so, the positive results of biodiversity in experiments may be trivial in nature compared to the effects of climate and nutrients.

We have learned that biodiversity is critical to the proper functioning of ecosystems.

To determine whether having higher biodiversity matters in nature, we can examine whether natural sites with a higher number of species function better in terms of productivity. In 2017, researchers compiled data from 67 research studies that measured the number of species and ecosystem productivity from 600,000 sample locations around the world. Across terrestrial, freshwater, and marine biomes, they found that sites with more species did indeed have higher productivity, which was consistent with the results of the earlier experiments. In addition, when the researchers considered how variation in climatic conditions and available nutrients affected the productivity of these ecosystems, they found that biodiversity was a more important factor than climate in half of the studies and a more important factor than nutrients in two-thirds of the studies. In short, we have learned that biodiversity is critical to the proper functioning of ecosystems. Such conclusions make it clear that we need to protect Earth's biodiversity so that we can continue to enjoy the many benefits that it provides.

Source: E. Duffy, et al. Biodiversity effects in the world are common and as strong as key drivers of productivity. *Nature* 549:261–264.

We have seen that biodiversity is an important indicator of environmental health. Conversely, a rapid decline of biodiversity in an ecosystem indicates that it is under stress. The biodiversity on Earth today is the result of evolution and extinction. Knowledge of these processes helps us to understand past and present environmental changes and their effects. In this chapter, we will examine how scientists quantify biodiversity and then look at how the process of evolution creates biodiversity. We will also examine the processes of speciation and extinction and how species have evolved unique ways of life that determine the abiotic and biotic conditions under which they can live.

MODULE 14

The Biodiversity of Earth

As you will recall from Chapter 1, we can think about biodiversity at three different scales (see Figure 2.1 on page 9). Within a given region, for example, the variety of ecosystems is a measure of ecosystem diversity. Within a given ecosystem, the variety of species constitutes species diversity. Within a given species, we can think about the variety of genes as a measure of genetic diversity. Every individual organism is distinguished from every other organism, at the most basic level, by the differences in the information coded by their genes. Because genes form the blueprint for an organism's traits, the diversity of genes on Earth ultimately helps determine the species diversity and ecosystem diversity on Earth. In other words, all three scales of biodiversity contribute to the overall biodiversity of the planet. In this module, we will examine how we estimate the number of species on Earth and how scientists quantify biodiversity. We will then examine how scientists illustrate relatedness among species.

Learning Goals

After reading this module you should be able to

- understand how we estimate the number of species living on Earth.
- quantify biodiversity.
- describe patterns of relatedness among species using a phylogeny.

It is difficult to estimate the number of species on Earth

A short walk through the woods, a corner lot, or a city park makes one thing clear: Life comes in many forms. A small plot of untended land or a tiny pond contains dozens, perhaps hundreds, of different kinds of plants and animals visible to the naked eye as well as thousands of different kinds of microscopic organisms. In contrast, a carefully tended lawn or a commercial timber plantation usually supports only a few types of grasses or trees (**FIGURE 14.1**, page 154). The total number of organisms in the plantation or lawn may be the same as the number in the pond or in the untended plot, but the number of species will be far smaller.

Recall from Chapter 1 that a species is defined as a group of organisms that is distinct from other such groups in terms of size, shape, behavior, or biochemical properties, and that can interbreed with other individuals in its group to produce viable offspring. This last requirement is important because sometimes individuals from different species can mate, but they do not produce offspring that survive.

The number of species in any given place is the most common measure of biodiversity, but estimating the total number of species on Earth is a challenge. Many species are easy to find, such as the birds or small mammals you might see in your neighborhood. Others are not so easy to find. Some species are active only at night, live in inaccessible locations such as the deep ocean, or cannot be seen without a microscope. To date, scientists have named approximately 2 million species, which means the total must be larger than that. While 2 million described species is an impressive number, it is equally impressive that the rate of discovering new species remains very high. In 2016, for example, a staggering 18,000 new species were discovered from around the world. This highlights the fact that we still have a long way to go in knowing the biodiversity of our planet.

Because insects contain more species than most other groups, scientists reason that if we could get a good

FIGURE 14.1 Species diversity and ecosystems. (a) Natural forests contain a high diversity of tree species. (b) In forest plantations, in which a single tree species has been planted for lumber and paper products, species diversity is low. *(a: Ron and Patty Thomas/Getty Images; b: Brent Waltermire/Alamy)*

estimate for the number of insect species in the world, we would have a much better sense of the total number of species. In one study, researchers fumigated the canopies of a single tree species in the tropical rainforest and then collected all the dead insects that fell from the trees onto a tarp on the ground. From this collection, they counted the number of beetle species that fed on only the one tree species they fumigated. By multiplying this number of beetle species by the total number of tropical tree species, they estimated that in the tropics there were perhaps 8 million species of beetles that feed on a single species of tree. Because beetles make up about 40 percent of all insect species, and because insect species in the forest canopy tend to be about twice as numerous as insect species on the forest floor, the researchers suggested that a reasonable estimate for the total number of tropical insect species might be 30 million. More recent work has indicated that this number is probably too high. Current estimates for the total number of species on Earth range between 5 million and 100 million, but most scientists estimate that there are about 10 million species.

Species richness The number of species in a given area.

Species evenness The relative proportion of individuals within the different species in a given area.

We can measure biodiversity in terms of species richness and evenness

Because species are not uniformly distributed, the number of species on Earth is not a useful indicator of how many species live in a particular location. Often we desire to know the number of species at a given location to determine whether a region is being affected by human activities. To measure species diversity at local or regional scales, scientists have developed two measures: *species richness* and *species evenness*.

The number of species in a given area, such as a pond, the canopy of a tree, or a plot of grassland, is known as **species richness**. Species richness is used to give an approximate sense of the biodiversity of a particular place. However, we may also want to know the **species evenness**, which is the relative proportion of individuals within the different species in a location. Species evenness tells us whether a particular ecosystem is numerically dominated by one species or whether all of its species have similar abundances. An ecosystem has high species evenness if its species are all represented by similar numbers of individuals. An ecosystem has low species evenness if one species is represented by many individuals whereas other species are represented by only a few individuals. In this case, there is effectively less diversity.

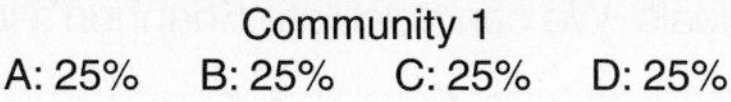

FIGURE 14.2 Measures of species diversity. Species richness and species evenness are two different measures of species diversity. Although both communities contain the same number of species, community 1 has a more even distribution of species and is therefore more diverse than community 2. Percentages of the four tree species are provided below each community.

AP® Exam Tip

The free-response section of the AP® Environmental Science Exam asks students to describe and quantify different processes and systems. It is important to remember that using a term is not enough without including the meaning of the term in your description. For example, if you are asked to describe a measure used to quantify biodiversity, you might answer "species richness." You would then need to follow your answer with a description of species richness, which is the number of species in a given area. ●

Scientists evaluating the biodiversity of an area must often look at both species richness and species evenness. Consider the two forest communities, community 1 and community 2, shown in **FIGURE 14.2**. Both forests contain 20 trees that are distributed among four species. In community 1, each species is represented by 5 individuals. In community 2, one species is represented by 14 individuals and each of the other three species is represented by 2 individuals. Although the species richness of the two forests is identical, the four species are more evenly represented in community 1. That forest therefore has greater species evenness and is considered to be more diverse.

Because species richness or evenness often declines after a human disturbance, knowing the species richness and species evenness of an ecosystem gives environmental scientists a baseline they can use to determine how much that ecosystem has changed. "Do the Math: Measuring Species Diversity" on page 156 demonstrates one common way of calculating species diversity.

The evolutionary relationship among species can be illustrated using a phylogeny

Scientists organize species into categories that indicate how closely related they are to one another. The branching pattern of evolutionary relationships is called a **phylogeny**. Phylogenies can be described with a diagram like the one shown in **FIGURE 14.3** on page 156, called a phylogenetic tree.

Phylogeny The branching pattern of evolutionary relationships.

DO THE MATH Measuring Species Diversity

Preparing for the AP® Exam

Environmental scientists are often interested in evaluating both species richness and species evenness, so they have come up with indices of species diversity that take both measures into account. One commonly used index is Shannon's index of diversity. To calculate this index, we must know the total number of species in a community (*n*) and, for each species, the proportion of the individuals in the community that represent that species (p_i). Once we have this information, we can calculate Shannon's index (*H*) by taking the product of each proportion (p_i) and its natural logarithm [$\ln (p_i)$] and then summing these products, as indicated by the summation symbol ($\sum$):

$$H = -\sum_{i=1}^{n} p_i \ln (p_i)$$

The minus sign makes the index a positive number. Higher values of *H* indicate higher diversity.

Imagine a community of 100 individuals that are evenly divided among four species, so that the proportions (p_i) of the species all equal 0.25. We can calculate Shannon's index as follows:

$$H = -[(0.25 \times \ln 0.25) + (0.25 \times \ln 0.25) + (0.25 \times \ln 0.25) + (0.25 \times \ln 0.25)]$$
$$H = -[(-0.35) + (-0.35) + (-0.35) + (-0.35)]$$
$$H = 1.40$$

Now imagine another community of 100 individuals that also contains four species, but in which one species is represented by 94 individuals and the other three species are each represented by 2 individuals. We can calculate Shannon's index to see how this difference in species evenness affects the value of the index:

$$H = -[(0.94 \times \ln 0.94) + (0.02 \times \ln 0.02) + (0.02 \times \ln 0.02) + (0.02 \times \ln 0.02)]$$
$$H = -[(-0.06) + (-0.08) + (-0.08) + (-0.08)]$$
$$H = 0.30$$

Because this value of *H* is lower than the value we calculated for the first community, we can conclude that the second community has lower diversity. Note that the total number of individuals does not affect Shannon's index of diversity; only the number of species and the proportion of individuals within each species matter.

YOUR TURN Imagine a third community of 100 individuals in which those individuals are distributed evenly among all the species, but there are only two species, not four. Calculate Shannon's index to see how this difference in species richness affects the value of the index.

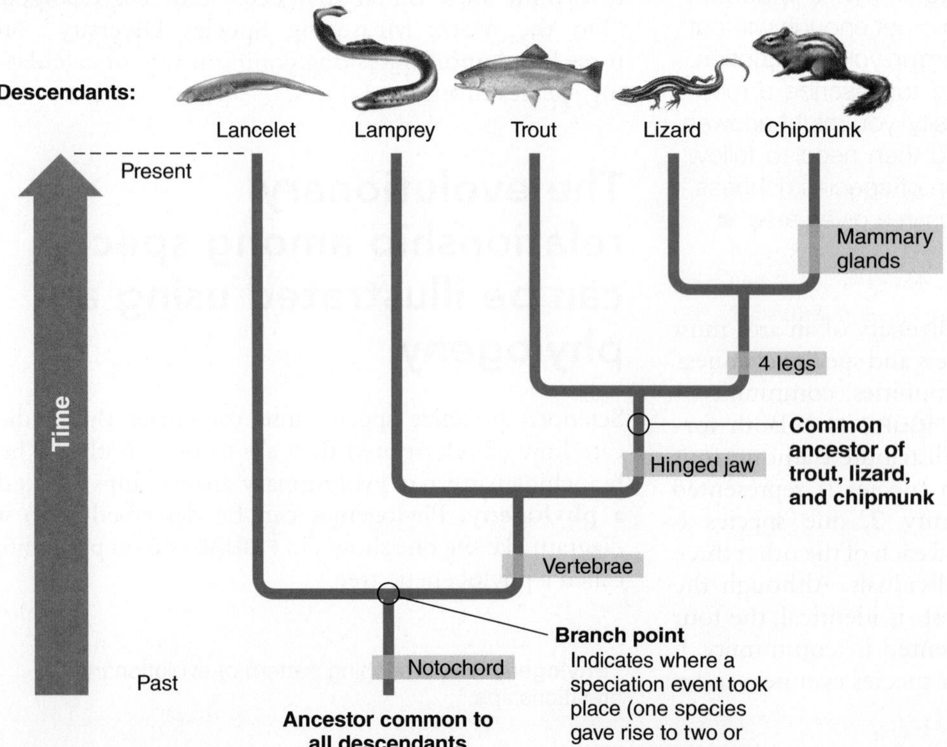

FIGURE 14.3 A phylogenetic tree. Phylogenies are based on the similarity of traits among species. Scientists can assemble phylogenetic trees that indicate how different groups of organisms are related and show where speciation events have occurred. The brown boxes indicate when major morphological changes evolved over evolutionary time.

The relatedness of the species in a phylogeny is determined by similarity of traits: The more similar the traits of two species, the more closely related the two species are assumed to be. Historically, scientists measure similarity using mostly morphological traits, which refers to form and structure. For example, it was common to use bone measurements to determine relationships among species. Today, scientists base phylogenies on a variety of characteristics, including morphology, behavior, and genetics.

MODULE 14

AP® Review

Preparing for the AP® Exam

In this module, we learned that scientists are not able to determine the exact number of species on Earth. Most estimates agree on approximately 10 million species, of which we have identified approximately 2 million. We saw that biodiversity can be quantified in terms of species richness and evenness. Species can be arranged on a phylogenetic tree that illustrates the evolutionary steps that gave rise to current species. In the next module, we will examine how evolution has produced such a large diversity of species.

AP® Practice Questions

Choose the best answer for the following.

1. How many species are estimated to exist on Earth?
   (a) 2 million
   (b) 8 million
   (c) 10 million
   (d) 30 million

2. Two savanna communities both contain 15 plant species. In community A, each of the 15 species is represented by 20 individuals. In community B, 10 of the species are each represented by 12 individuals; the remaining 5 species are each represented by 3 individuals. Which statement best describes the two communities?
   (a) Community A has the same biodiversity as community B.
   (b) Community B has a higher species richness.
   (c) Community A has a higher species evenness.
   (d) Community B has a lower species richness.

3. Phylogeny is
   (a) the number of evolutionarily related species in an ecosystem.
   (b) the branching pattern of evolutionary relationships.
   (c) the process of evolution that creates new species.
   (d) the genetic biodiversity of a species.

4. Which is used to calculate Shannon's index of diversity?
   I. the proportion of individuals in each species
   II. the total number of species
   III. the number of individuals in each species
   (a) I only
   (b) I and II
   (c) II only
   (d) I, II, and III

MODULE 15

How Evolution Creates Biodiversity

Learning Goals

After reading this module you should be able to

- identify the processes that cause genetic diversity.
- explain how evolution can occur through artificial selection.
- explain how evolution can occur through natural selection.
- explain how evolution can occur through random processes.

We have seen the importance of biodiversity. In this module we will look at the processes of *evolution*, which is the source of biodiversity. Because the evolution of biodiversity depends on genetic diversity, we will examine how genetic diversity is created. We begin by exploring the sources of genetic variation and then consider how humans and the natural world select from this variation to favor particular individuals that go on to reproduce in the next generation. Finally, we will examine how several random processes can also cause evolution.

Genetic diversity is created through mutation and recombination

Earth's biodiversity is the product of **evolution**, which can be defined as a change in the genetic composition of a population over time. Evolution can occur at multiple levels. Evolution below the species level, such as the evolution of different varieties of apples or potatoes, is called **microevolution**. In contrast, when genetic changes give rise to new species, or to new genera, families, classes, or phyla—larger categories of organisms into which species are organized—we call the process **macroevolution**. Among these many levels of macroevolution, the term speciation is restricted to the evolution of new species.

To understand how genetic diversity is created, we first need to understand *genes.* **Genes** are physical locations on chromosomes within each cell of an organism. A given gene has DNA that codes for a particular trait, such as body size, but the DNA can take different forms known as alleles. An organism's genes determine the range of possible traits (physical or behavioral characteristics) that it can pass down to its offspring. The complete set of genes in an individual is called its **genotype**. In this section, we will discuss how genotypes help to determine the traits of individuals and the two processes that can create genetic diversity in a population: *mutation* and *recombination*.

Evolution A change in the genetic composition of a population over time.

Microevolution Evolution below the species level.

Macroevolution Evolution that gives rise to new species, genera, families, classes, or phyla.

Gene A physical location on the chromosomes within each cell of an organism.

Genotype The complete set of genes in an individual.

Phenotype A set of traits expressed by an individual.

Genotypes versus Phenotypes

An individual's genotype serves as the blueprint for the complete set of traits that organism may potentially possess. An individual's **phenotype** is the actual set of traits expressed in that individual. Among these traits are the individual's anatomy, physiology, and behavior.

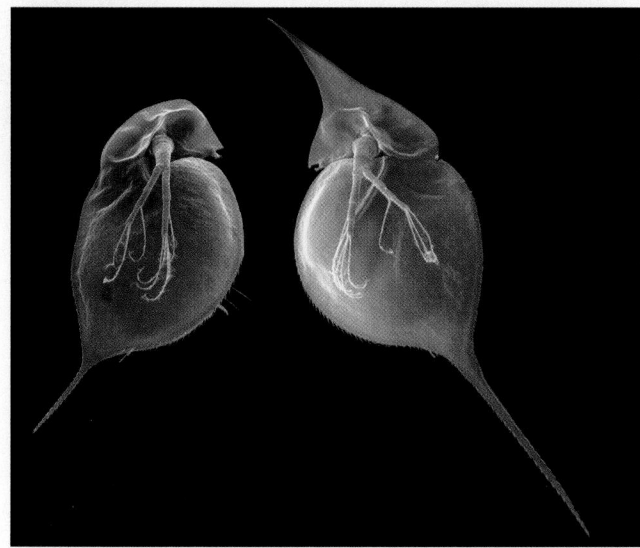

FIGURE 15.1 Environmental effects on phenotype. Water fleas raised in the absence of predators produce relatively small heads and short tail spines (left), whereas individuals raised in the presence of predators produce relatively large heads and long tail spines (right). *(Christian Laforsch/Science Source)*

FIGURE 15.2 Mutations. A mutation in the genetic code of the dusky-headed conure causes these normally green-feathered parrots to develop feathers that appear blue. In nature the mutation makes individuals more conspicuous and prone to predation. *(Howard Voren/Courtesy of George Voren)*

The color of your eyes, for example, is your phenotype, whereas the genes that code for eye color are a part of your genotype. Changes in genotypes can produce important changes in phenotypes.

In some cases, an individual's phenotype is determined almost entirely by genes. For instance, a person who inherits the genes for brown eyes will have brown eyes, regardless of where that person lives. Most phenotypes, however, are the product of an individual's environment as well as its genotype. For example, in many turtle and crocodile species, the temperature of eggs during incubation determines whether the offspring will hatch as males or females. The water flea, a tiny animal that lives in ponds and lakes, offers another interesting example. The body shape of the water flea depends on whether or not a young individual smells predators in its environment (**FIGURE 15.1**). If predators are absent, the water flea develops a relatively small head and tail spine. If predators are present, however, the water flea develops a much larger head and a long tail spine. Although the larger head and longer tail spine help prevent the water flea from being eaten, they come at the cost of slower reproduction. Therefore, it is beneficial that the water flea not produce these defenses unless they are needed. By being able to respond to changing environmental conditions, organisms such as the water flea can improve their ability to survive and reproduce in a variety of environmental conditions.

Mutation

DNA is copied millions of times during an organism's lifetime as cells grow and divide. An occasional mistake in the copying process produces a random change in the genetic code, which is known as a **mutation**. Environmental factors, such as ultraviolet radiation from the Sun, can also cause mutations. When mutations occur in cells responsible for reproduction, such as the eggs and sperm of animals, those mutations can be passed on to the next generation.

Most mutations are detrimental, and many cause the offspring that carry them to die while they are embryos. The effects of some mutations are less severe, but can still be detrimental. For example, some dusky-headed conures (*Aratinga weddellii*) have a mutation that makes these normally green-feathered parrots produce feathers that appear to be blue (**FIGURE 15.2**). In the wild, individuals with this mutation have a poor chance of survival because blue feathers stand out against the green vegetation and make them conspicuous to predators.

Sometimes a mutation improves an organism's chances of survival or reproduction. If such a mutation is passed along to the next generation, it adds new genetic diversity to the population. Some mosquitoes, for example, possess a mutation that makes them less vulnerable to insecticides. In areas that are sprayed with insecticides, this mutation improves an individual mosquito's chance of surviving and reproducing.

Recombination

Genetic diversity can also be created through *recombination*. In plants and animals, genetic **recombination**

Mutation A random change in the genetic code produced by a mistake in the copying process.

Recombination The genetic process by which one chromosome breaks off and attaches to another chromosome during reproductive cell division.

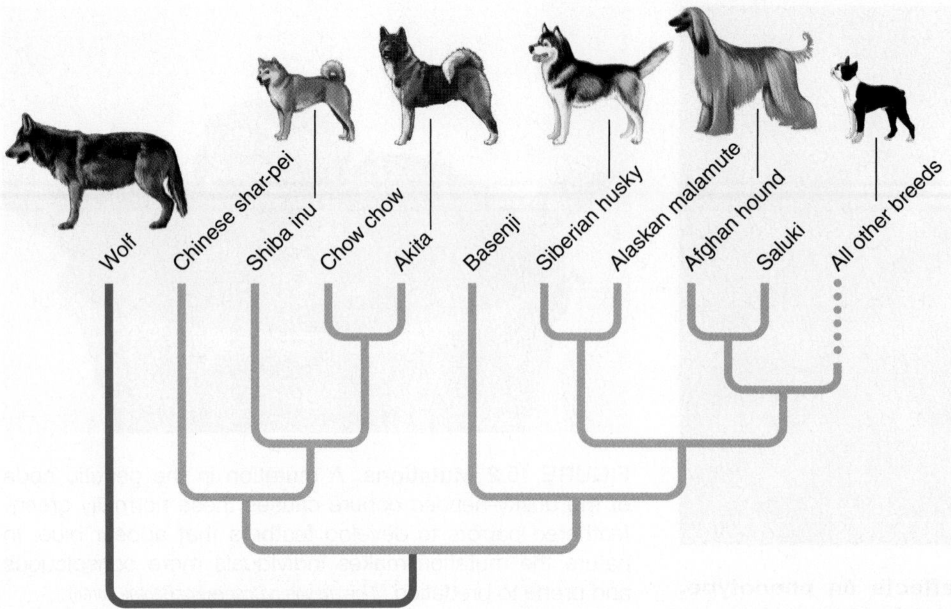

FIGURE 15.3 Artificial selection on animals. The diversity of domesticated dog breeds is the result of artificial selection on wolves. The wolf is the ancestor of the various breeds of dogs. It is illustrated at the same level as the dogs in this phylogeny because it is a species that is still alive today. *(Data from H.G. Parker et al., Science 304 (2004): 1160–1164)*

occurs as chromosomes are duplicated during reproductive cell division and a piece of one chromosome breaks off and attaches to another chromosome. This process does not create new genes, but it does bring together new combinations of alleles on a chromosome and can therefore produce novel traits. For example, the human immune system must battle a large variety of viruses and bacteria that regularly attempt to invade the body. Recombination allows new allele combinations to come together, and this provides new immune defenses that may prove to be effective against the invading organisms.

Evolution can occur through artificial selection

Evolution occurs in three primary ways: *artificial selection, natural selection,* and *random processes*. In this section we will look at artificial selection.

Humans have long influenced evolution by breeding plants and animals for desirable traits. For example, all breeds of domesticated dogs belong to the same species as the gray wolf (*Canis lupus*), yet dogs exist in an amazing variety of sizes and shapes, ranging from toy poodles to Siberian huskies. **FIGURE 15.3** shows the phylogenetic relationships among the wolf and different breeds of domestic dogs that were bred from the wolf by humans. Beginning with the domestication of wolves, dog breeders have selectively bred individuals that had particular qualities they desired, including body size, body shape, and coat color. After many generations of breeding, the selected traits became more and more exaggerated until breeders felt satisfied that the desired characteristics of a new dog breed had been achieved. As a result of this carefully controlled breeding, we have a tremendous variety of dog sizes, shapes, and colors today. Yet dogs remain a single species: All dog breeds can still mate with one another and produce viable offspring.

When humans determine which individuals to breed, typically with a preconceived set of traits in mind, we call the process **evolution by artificial selection**. Artificial selection has produced numerous breeds of horses, cattle, sheep, pigs, and chickens with traits that humans find useful or aesthetically pleasing. Most of our modern agricultural crops are also the result of many years of careful breeding. For example, starting with a single species of wild mustard, *Brassica oleracea,* plant breeders have produced a variety of food crops, including cabbage, cauliflower,

Evolution by artificial selection The process in which humans determine which individuals breed, typically with a preconceived set of traits in mind.

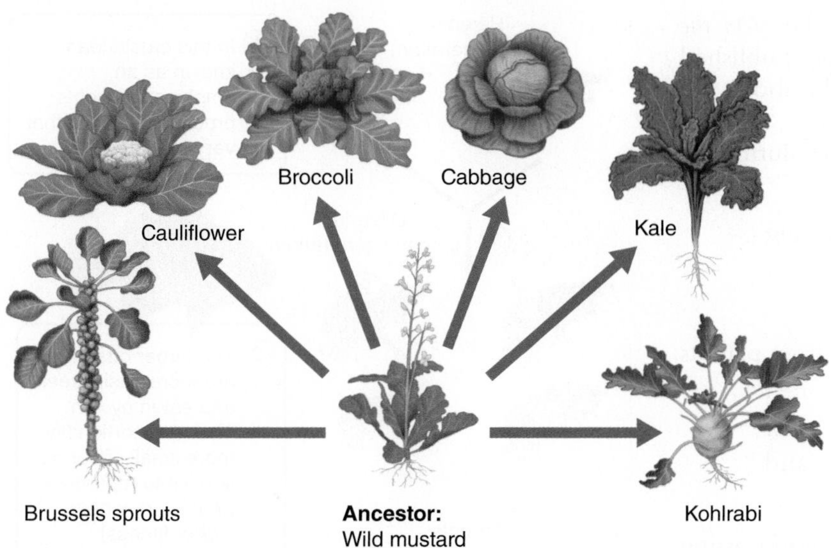

FIGURE 15.4 Artificial selection on plants. Plant breeders have produced a wide range of edible plants from a single species of wild mustard.

broccoli, Brussels sprouts, kale, and kohlrabi, shown in **FIGURE 15.4**.

As useful as artificial selection has been to humans, it can also produce a number of unintended results. For example, farmers often use herbicides to kill weeds. However, as we cover larger and larger areas with herbicides, there is an increasing chance that at least one weed will possess a mutation that allows it to survive the herbicide application. If that one mutant plant passes on its herbicide resistance to its offspring, we will have artificially selected for herbicide resistance in that weed. This process is occurring in many parts of the world where increased use of the popular herbicide Roundup (chemical name: glyphosate) has led to the evolution of several species of Roundup-resistant weeds. A similar process has occurred in hospitals, where the use of antibiotics and antibacterial cleaners has caused artificial selection of harmful drug-resistant bacteria. These examples underscore the importance of understanding the mechanisms of evolution and the ways in which humans can either purposefully or inadvertently direct the evolution of organisms.

Evolution can occur through natural selection

Evolution also takes place through natural mechanisms. In **evolution by natural selection**, the environment determines which individuals survive and reproduce. Members of a population naturally vary in their traits, and certain combinations of those traits make individuals better able to survive and reproduce. As a result, the genes that produce those traits are more common in the next generation.

Prior to the mid-nineteenth century, the idea that species could evolve over time had been suggested by a number of scientists and philosophers. However, the concept of evolution by natural selection did not become synthesized into a unifying theory until two scientists, Alfred Wallace (1823–1913) and Charles Darwin (1809–1882), independently put the various pieces together.

Of the two scientists, Charles Darwin is perhaps the better known. At age 22, he became the naturalist on board HMS *Beagle,* a British survey ship that sailed around the world from 1831 to 1836. During his journey, Darwin made many observations of trait variation across a tremendous variety of species. In addition to observing living organisms, he found fossil evidence of a large number of extinct species. He also recognized that organisms produce many more offspring than are needed to replace the parents, and that most of these offspring do not survive. Darwin questioned why, out of all the species that had once existed on Earth, only a small fraction had survived. Similarly, he wondered why, among all the offspring produced in a population in a given year, only a small fraction survived to the next year. During the decades following his voyage, he

Evolution by natural selection The process in which the environment determines which individuals survive and reproduce.

developed his ideas into a robust theory. His *On the Origin of Species by Means of Natural Selection,* published in 1859, changed the way people thought about the natural world.

The key ideas of Darwin's theory of evolution by natural selection are the following:

- Individuals produce an excess of offspring.
- Not all offspring can survive.
- Individuals differ in their traits.
- Differences in traits can be passed on from parents to offspring.
- Differences in traits are associated with differences in the ability to survive and reproduce.

FIGURE 15.5 shows how this process works using the example of body size in a group of crustaceans known as amphipods. We can begin with parents producing offspring that vary in their body size. The largest offspring are consumed by fish because fish prefer to eat large prey rather than small prey. As a result, the smaller offspring are left to reproduce. Because body size is, in part, determined by an individual's genes, the next generation of amphipods will be smaller. This process can continue over many generations and over time the fish will cause the evolution of smaller body sizes in amphipods.

Both artificial and natural selection begin with the requirement that individuals vary in their traits and that these variations are capable of being passed on to the next generation. In both cases, parents produce more offspring than necessary to replace themselves, and some of these offspring either do not survive or do not reproduce. But in the case of artificial selection, humans decide which individuals will breed, based on those individuals that possess the traits that tend toward some predetermined goal, such as a curly coat or large size. Natural selection does not select for specific traits that tend toward some predetermined goal. Rather, natural selection favors any combination of traits that improves an individual's **fitness**—its ability to survive and reproduce, as we saw in the case of the smallest amphipods surviving predation by fish. Traits that improve an individual's fitness are called **adaptations**.

Natural selection can favor multiple solutions to a particular environmental challenge, as long as each solution improves an individual's ability to survive and reproduce. For example, while all plants living in the desert face the challenge of low water availability in the soil, different species have evolved different solutions to this common challenge. Some species have evolved the ability to store excess water during infrequent rains (**FIGURE 15.6**). Other species have evolved large taproots to draw water from deep in the soil. Still other species have evolved waxy or hairy leaf surfaces that reduce water loss. Each of these very different adaptations allows the plants to survive and reproduce in a desert environment.

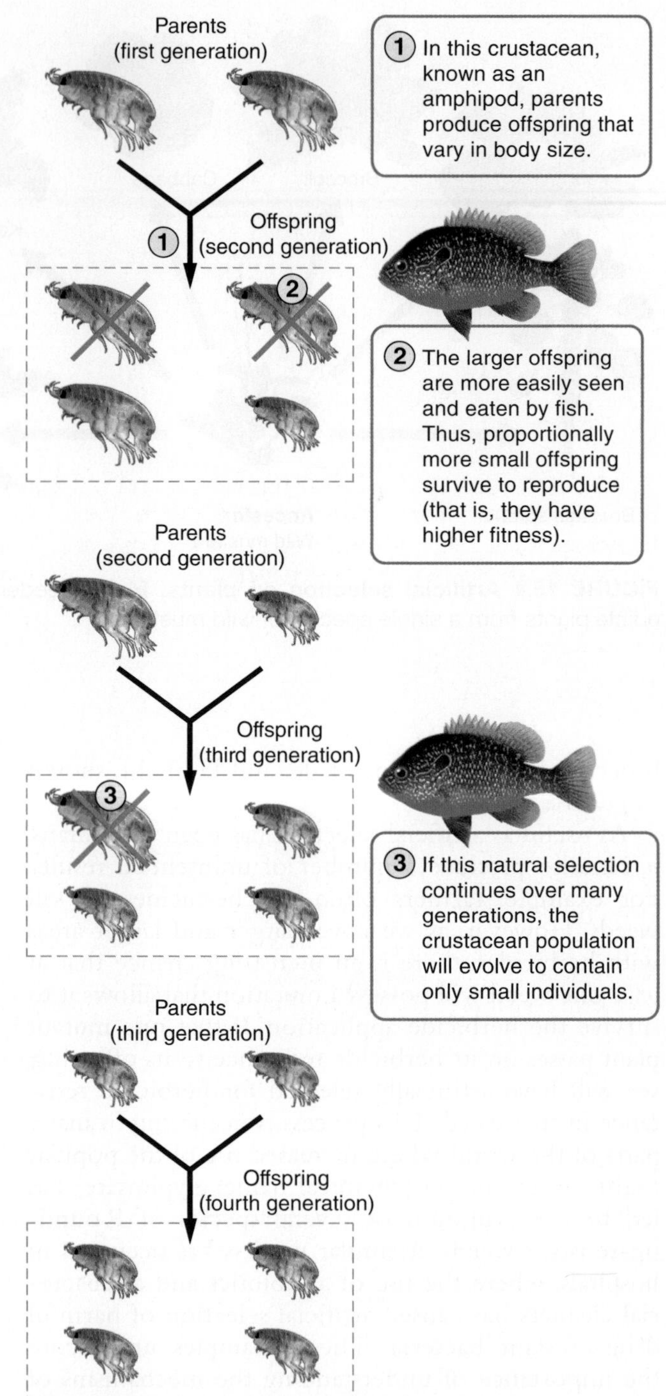

Parents
(first generation)
① In this crustacean, known as an amphipod, parents produce offspring that vary in body size.
①
Offspring
(second generation)
②
② The larger offspring are more easily seen and eaten by fish. Thus, proportionally more small offspring survive to reproduce (that is, they have higher fitness).
Parents
(second generation)
Offspring
(third generation)
③
③ If this natural selection continues over many generations, the crustacean population will evolve to contain only small individuals.
Parents
(third generation)
Offspring
(fourth generation)

FIGURE 15.5 Natural selection. All species produce an excess number of offspring. Only those offspring with the fittest genotypes will pass on their genes to the next generation.

Fitness An individual's ability to survive and reproduce.
Adaptation A trait that improves an individual's fitness.

(a) (b) (c)

FIGURE 15.6 Adaptations. Desert plants have evolved several different adaptations to their desert environment. (a) The wedgeleaf draba (*Draba cuneifolia*) has leaf hairs that reduce water loss. (b) The Leuchtenbergia cactus (*Leuchtenbergia principis*) has a large taproot to draw water from deep in the soil. (c) The waxy outer layers of *Aloe vera* reduce water loss. *(a: FPI/Alamy; b: Ian Nartowicz; c: Scott Green/EyeEm/Getty Images)*

Evolution can also occur through random processes

Artificial and natural selection are important mechanisms of evolution, but evolution can also occur by random, or nonadaptive, processes. In these cases, the genetic composition of a population changes over time, but the changes are not related to differences in fitness among individuals. There are five random processes: *mutation, gene flow, genetic drift, bottleneck effects,* and *founder effects.*

Mutation

If a random mutation is not lethal, it can add to the genetic variation of a population. As shown in **FIGURE 15.7**, the larger the population, the more opportunities there will be for mutations to appear within it. As the number of mutations accumulates in the population over time, evolution occurs.

Gene Flow

Gene flow is the process by which individuals move from one population to another and thereby alter the genetic composition of both populations. Populations can experience an influx of migrating individuals with different alleles. The arrival of these individuals from adjacent populations alters the frequency of alleles in the population. High gene flow between two populations can cause the two populations to become very similar in genetic composition. In a population that is experiencing natural or artificial selection, high gene flow from outside can prevent the population from responding to selection.

Gene flow can be helpful in bringing in genetic variation to a population that lacks it. For example, the Florida panther is a subspecies of panther that once roamed throughout much of the southeastern United States and likely experienced gene flow with other subspecies of panthers. By 1995, the Florida

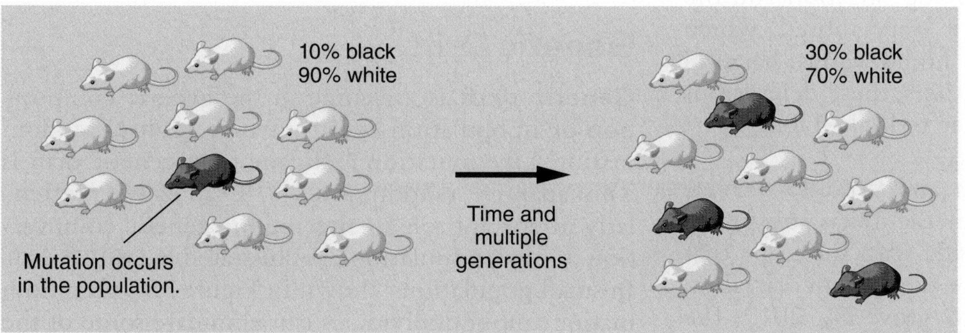

FIGURE 15.7 Evolution by mutation. A mutation can arise in a population and, if it is not lost, it may increase in frequency over time.

Gene flow The process by which individuals move from one population to another and thereby alter the genetic composition of both populations.

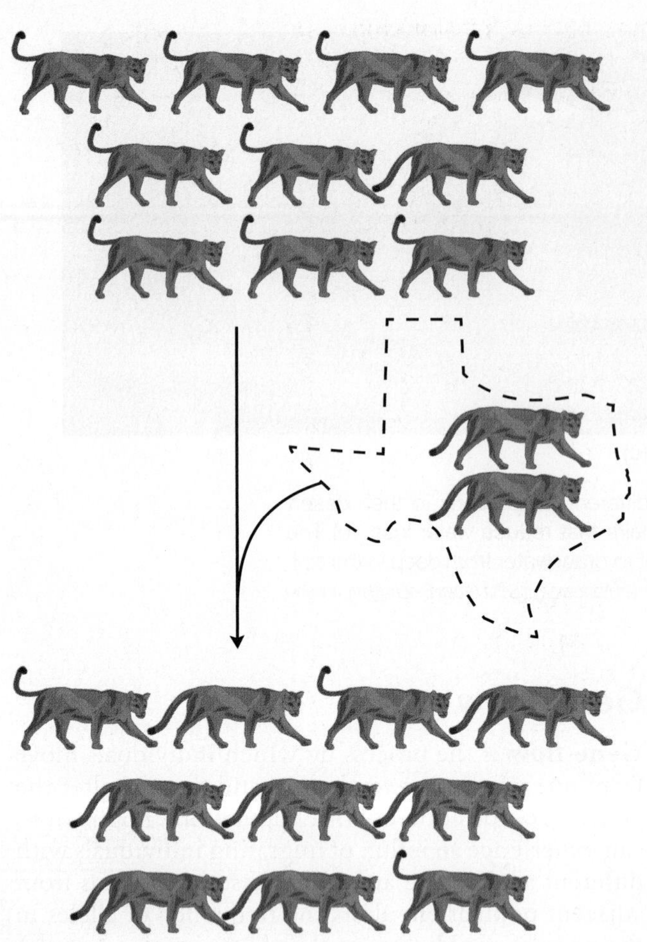

Florida
As the population of Florida panthers declined to very low numbers, the percentage of kinked tails increased to approximately 90 percent.
Texas
Several Texas panthers were brought to Florida.
Florida
After breeding with the Texas panthers, the Florida panther population was less inbred and the percentage of kinked tails declined.

FIGURE 15.8 Evolution by gene flow. As the Florida panther declined in population size, the animals experienced low genetic variation and showed signs of inbreeding, which lead to kinky tails, heart defects, and low sperm counts. With the introduction of eight panthers from Texas, the Florida population experienced a decline in the prevalence of defects and a growth in population from 30 to 160 individuals.

panther only lived in southern Florida, occupying just 5 percent of its original habitat. Moreover, the number of panthers declined to only about 30 individuals, and because the Florida subspecies was isolated from other subspecies, it did not experience gene flow. As a result, shown in **FIGURE 15.8**, the small population had low genetic variation and the remaining individuals became very inbred, which causes individuals to express homozygous, harmful alleles. In the case of the panthers, these deleterious alleles caused a high prevalence of kinked tails, heart defects, and low sperm counts.

In response, the U.S. Fish and Wildlife Service captured eight panthers from the Texas subspecies and introduced them to Florida with the hope that this gene flow would increase the genetic variation and allow the population to grow. By 2017, the Florida panther population had grown to more than 120 individuals and the prevalence of defects previously seen from inbreeding had declined. These improvements in the panther population underscore the importance of gene flow to providing beneficial genetic diversity.

Genetic drift A change in the genetic composition of a population over time as a result of random mating.

Genetic Drift

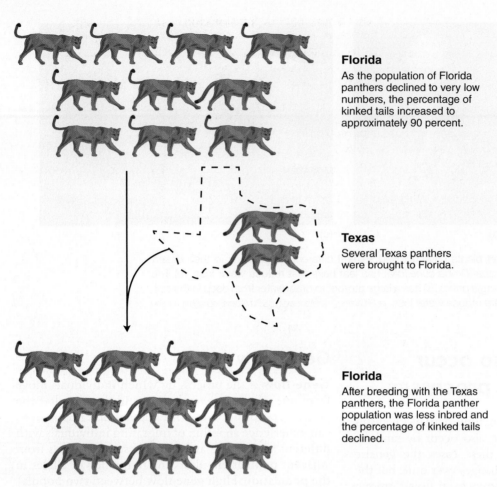

Genetic drift is a change in the genetic composition of a population over time as a result of random mating. Like mutation and gene flow, genetic drift is a nonadaptive, random process. It can have a particularly important role in altering the genetic composition of small populations, as illustrated in **FIGURE 15.9**. In small populations, shown in Figure 15.9a, random mating among individuals can eliminate some of the individuals that carry unique traits simply because they did not find a mate in a given year. For example, imagine a small population of five animals, in which two individuals carry genes that produce black hair

(a) Small population

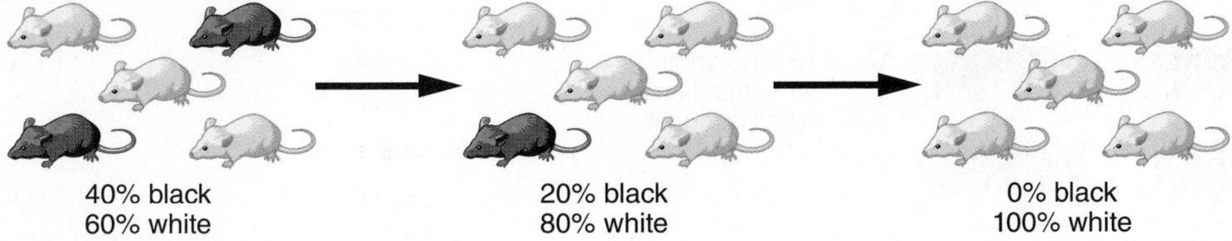

(b) Large population

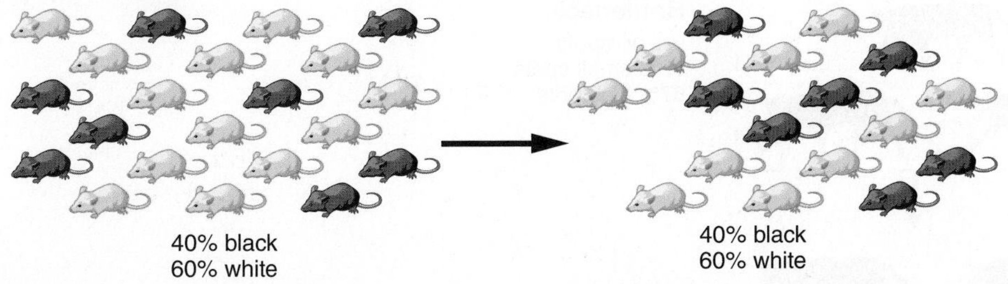

FIGURE 15.9 Evolution by genetic drift. (a) In a small population, some less-common genotypes can be lost by chance as random mating among a small number of individuals can result in the less-common genotype not mating. As a result, the genetic composition can change over time. (b) In a large population, it is more difficult for the less-common genotypes to be lost by chance because the absolute number of these individuals is large. As a result, the genetic composition tends to remain the same over time in larger populations.

and three individuals carry genes that produce white hair. If, by chance, the individuals that carry the genes for black hair fail to find a mate, those genes will not be passed on. The next generation will be entirely white-haired, and the black-haired phenotype will be lost. In this case, the genetic composition of the population has changed, and the population has therefore evolved. The cause underlying this evolution is random; the failure to find a mate has nothing to do with hair color. In contrast, a large population that has the same proportion of black-haired mice, shown in Figure 15.9b, has a greater absolute number of mice. As a result, it is less likely that random mating events will cause all of the black-haired mice to not find a mate, so their genes are passed on to the next generation and genetic drift is less likely to occur.

Bottleneck Effect

A drastic reduction in the size of a population that reduces genetic variation—known as a **bottleneck effect**—is another random process that can change a population's genetic composition. A population might experience a drastic reduction in its numbers for many reasons, including habitat loss, a natural disaster, harvesting by humans, or changes in the environment. **FIGURE 15.10** illustrates the bottleneck effect using the example of cheetahs and spots. When the size of a population is reduced, the amount of genetic variation that can be present in the population is also reduced. With fewer individuals there are fewer unique genotypes remaining in the population.

Low genetic variation in a population can cause several problems, including increased risk of disease and low fertility. In addition, species that have been through a population bottleneck are often less able to adapt to future changes in their environment. In some cases, once a species has been forced through a bottleneck, the resulting low genetic diversity causes it to decline to **extinction**, which occurs when the last member of a species dies. Such declines are thought to be occurring in a number of species today. The cheetah, for example, has relatively little genetic variation due to a bottleneck that appears to have occurred 10,000 years ago.

Founder Effect

Imagine that a few individuals of a particular bird species happen to be blown off their usual migration route and land on a hospitable oceanic island, as illustrated

Bottleneck effect A reduction in the genetic diversity of a population caused by a reduction in its size.

Extinction The death of the last member of a species.

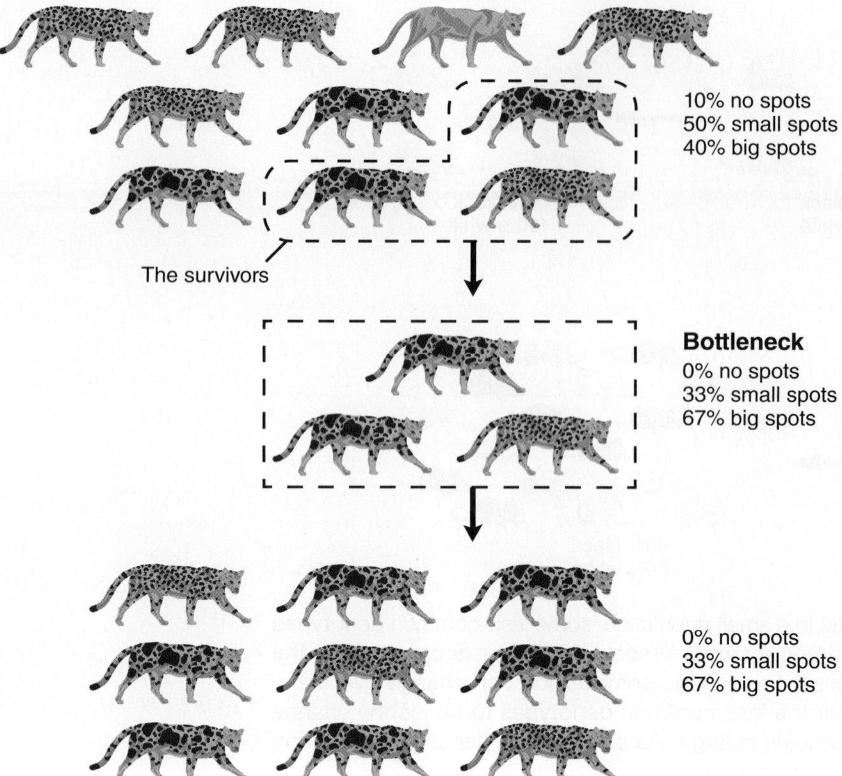

FIGURE 15.10 Evolution by the bottleneck effect. If a population experiences a drastic decrease in size (goes through a "bottleneck"), some genotypes will be lost, and the genetic composition of the survivors will differ from the composition of the original group.

in **FIGURE 15.11**. These two individuals will have been drawn at random from the mainland population, and the genotypes they possess are only a subset of those in the original mainland population. These colonizing individuals, or founders, will give rise to an island population that has a genetic composition very different from that of the original mainland population. A change in the genetic composition of a population as a result of descending from a small number of colonizing individuals is known as the **founder effect**. Like mutation, genetic drift, and the bottleneck effect, the founder effect is a random process that is not based on differences in fitness. "Do the Math: Estimating the Impact of the Founder Effect" provides practice converting the number of each phenotype into percentages of each phenotype.

We can see an example of the founder effect in the Amish communities of Pennsylvania. The Amish population was founded by a relatively small number of individuals—about 200 people from Germany. This group happened to carry a mutation for Ellis-van Creveld syndrome, a condition that causes a variety of malformations including extra fingers. The mutation is rare in humans around the world, but by chance the frequency of the mutation was higher in the early Amish colonists and has remained higher because the population is an isolated group with little gene flow from outside its community.

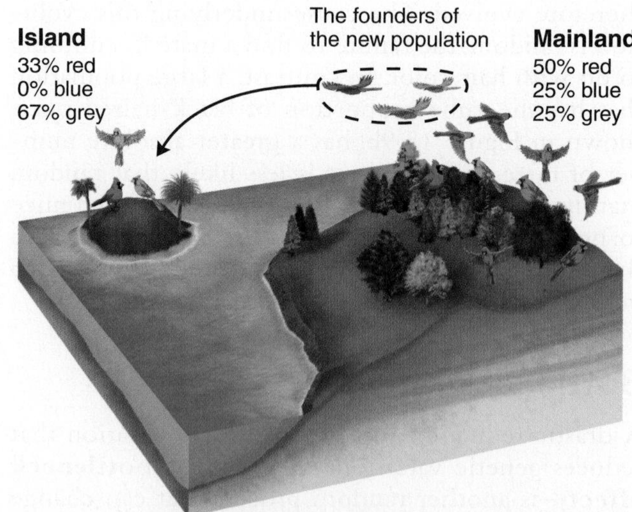

FIGURE 15.11 Evolution by the founder effect. If a few individuals from a mainland population colonize an island, the genotypes on the island will represent only a subset of the genotypes present in the mainland population. As with the bottleneck effect, some genotypes will not be present in the new population.

Founder effect A change in the genetic composition of a population as a result of descending from a small number of colonizing individuals.

DO THE MATH Estimating the Impact of the Founder Effect

Preparing for the AP® Exam

As we have discussed, the founder effect can alter the phenotypes of a population in a newly colonized island compared to the phenotypes in the mainland population. Imagine a mainland population of rats that was comprised of 1,000 individuals with three unique coat colors: brown, black, and white. If there are 550 brown rats, 150 black rats, and 300 white rats, what is the percentage of each phenotype in the population?

To estimate the percentages, we divide the number of each phenotype by the total number of rats:

Percent of brown rats = 550 brown rats ÷ 1,000 rats = 0.55 = 55%

Percent of black rats = 150 black rats ÷ 1,000 rats = 0.15 = 15%

Percent of white rats = 300 white rats ÷ 1,000 rats = 0.30 = 30%

YOUR TURN If the founding population is comprised of 40 percent brown rats, 50 percent black rats, and 10 percent white rats, how many rats of each coat color would you expect in an island population of 200 rats?

MODULE 15 AP® Review

Preparing for the AP® Exam

In this module, we learned that genetic variation helps to determine the traits that individuals express. We also learned that artificial selection, natural selection, and random processes can all cause the evolution of populations. In the next module, we will examine how these evolutionary processes can cause the evolution of new species, which increases biodiversity, and how environmental change that is too rapid for selection can result in extinctions that cause declines in biodiversity.

AP® Practice Questions

Choose the best answer for the following.

1. A phenotype is
   (a) an adaptation that creates a new species.
   (b) the genes of a particular individual.
   (c) a result of genetic recombination.
   (d) the set of traits expressed in an individual.

2. Which process creates genetic diversity in a population?
   I. mutation
   II. allele division
   III. recombination
   (a) I only
   (b) I and II
   (c) I and III
   (d) II and III

3. Evolved resistance to a pesticide is an example of
   (a) a nonadaptive process.
   (b) the bottleneck effect.
   (c) natural selection.
   (d) artificial selection.

4. Which is the best definition of an adaptation?
   (a) a mutation that creates a new species
   (b) a trait that improves an individual's fitness
   (c) a trait that is passed on to the next generation
   (d) a trait created by natural selection

5. The change in the genetic composition of a population over time due to random mating is called
   (a) the bottleneck effect.
   (b) gene flow.
   (c) genetic drift.
   (d) mutation.

6. In a particular zoo the population of spider monkeys has a higher proportion of individuals with light golden brown fur than spider monkeys in the wild. If the monkeys were recently captured from the wild and if fur color is largely determined by genetics, what evolutionary process is at work?
   (a) the founder effect
   (b) the bottleneck effect
   (c) artificial selection
   (d) genetic drift

MODULE 16

Speciation and the Pace of Evolution

Learning Goals

After reading this module you should be able to

- explain the processes of allopatric and sympatric speciation.
- understand the factors that affect the pace of evolution.

Over time, speciation has given rise to the millions of species present on Earth today. Beyond determining how many species exist, environmental scientists are also interested in understanding how quickly existing species can change, how quickly new species can evolve, and how quickly species can go extinct. In this section we will examine the processes that produce new species and the factors that determine how rapidly species can evolve in response to changes in the environment.

Speciation can be allopatric or sympatric

Microevolution is happening all around us, from the breeding of agricultural crops, to the unintentional evolution of drug-resistant bacteria in hospitals, to the bottleneck that reduced genetic variation in the cheetah. But how do we move from the evolution of genetically distinct populations of a species to the evolution of genetically distinct species? That is, how do we move from microevolution to macroevolution? Two common processes are *allopatric speciation* and *sympatric speciation*.

Allopatric Speciation

One common way in which evolution creates new species is through **geographic isolation**, which means the physical separation of a group of individuals from others of the same species. The process of speciation that occurs with geographic isolation is known as **allopatric speciation** (from the Greek *allos,* meaning "other," and *patris,* meaning "fatherland"). As shown in **FIGURE 16.1**, geographic isolation can occur when a subset of individuals from a larger population colonizes a new area of habitat that is physically separated from that larger population. For example, a single large population of field mice might be split into two smaller populations as geographic barriers change over time. For example, a river might change course and divide a large prairie into two halves, a large lake might split into two smaller lakes, or a new mountain range could rise. In such cases, the genetic composition of the isolated populations might diverge over time, either because of random processes or because natural selection favors different adaptations on each side of the barrier.

If the two separated habitats differ in environmental conditions, such as temperature, precipitation, or the occurrence of predators, natural selection will favor different phenotypes in each of the habitats. If individuals cannot move between the populations, then over time the two geographically isolated populations will continue to become more and more genetically distinct. Eventually, the two populations will be separated not only by geographic isolation but also by **reproductive isolation**, which means the two populations of a species have evolved separately to the point that they can no longer interbreed and produce viable offspring. At this point, the two populations will have become distinct species.

Geographic isolation Physical separation of a group of individuals from others of the same species.

Allopatric speciation The process of speciation that occurs with geographic isolation.

Reproductive isolation The result of two populations within a species evolving separately to the point that they can no longer interbreed and produce viable offspring.

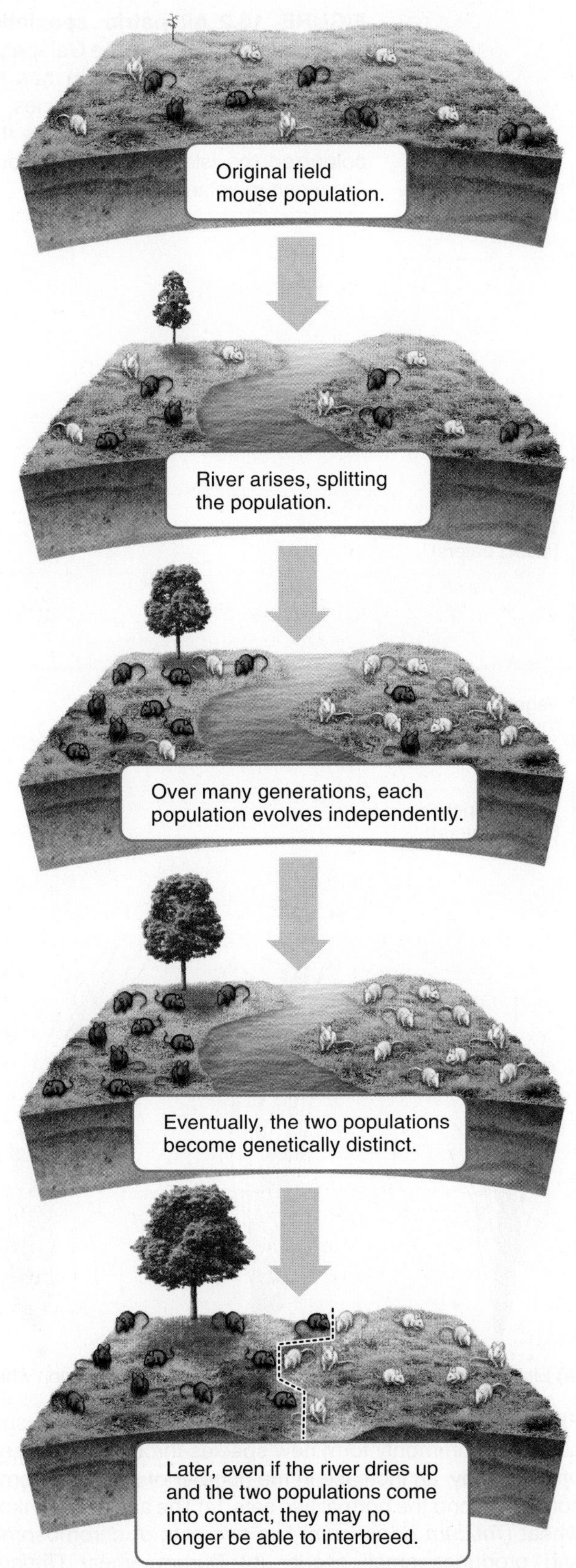

Original field mouse population.
River arises, splitting the population.
Over many generations, each population evolves independently.
Eventually, the two populations become genetically distinct.
Later, even if the river dries up and the two populations come into contact, they may no longer be able to interbreed.

FIGURE 16.1 Allopatric speciation. Geographic barriers can split populations. Natural selection may favor different traits in the environment of each isolated population, resulting in different adaptations. Over time, the two populations may become so genetically distinct that they are no longer capable of interbreeding.

Allopatric speciation is thought to be responsible for the diversity of the group of birds known as Darwin's finches. When Charles Darwin visited the Galápagos Islands, located just west of Ecuador, he noted a large variety of finch species, each of which seemed to live in different habitats or to eat different foods. Research on these birds has demonstrated that they all share a common ancestor that colonized the islands from the mainland long ago. **FIGURE 16.2** on page 170 is a phylogenetic tree for these finches. Over a few million years, as Darwin discovered, the finches that were geographically isolated on different islands became genetically distinct and eventually became reproductively isolated.

Sympatric Speciation

Allopatric speciation is thought to be the most common way in which evolution generates new species. However, it is not the only way. **Sympatric speciation** is the evolution of one species into two species without geographic isolation. It usually happens through a process known as polyploidy. Most organisms are diploid: They have two sets of chromosomes. In polyploidy, the number of chromosomes increases to three, four, or even six sets. Such increases can occur during the division of reproductive cells, either accidentally in nature or as a result of deliberate human actions. Plant breeders, for example, have found several ways to interrupt the normal cell division process. Polyploid organisms include some species of snails and salamanders, 15 percent of all flowering plant species, and a wide variety of agricultural crops such as bananas, strawberries, and wheat. As **FIGURE 16.3** on page 170 shows for wheat, polyploidy often results in larger plants and larger fruits.

The key feature of polyploid organisms is that once they become polyploid, they generally cannot interbreed with their diploid ancestors. At the instant polyploidy occurs, the polyploid and diploid organisms are reproductively isolated from each other and are therefore distinct species, even though they may continue to live in the same place.

The pace of evolution depends on several factors

How long does evolution take? A significant change in a species' genotype and phenotype, such as an adaptation to a completely different food source, can take anywhere from hundreds to millions of years. In this

Sympatric speciation The evolution of one species into two, without geographic isolation.

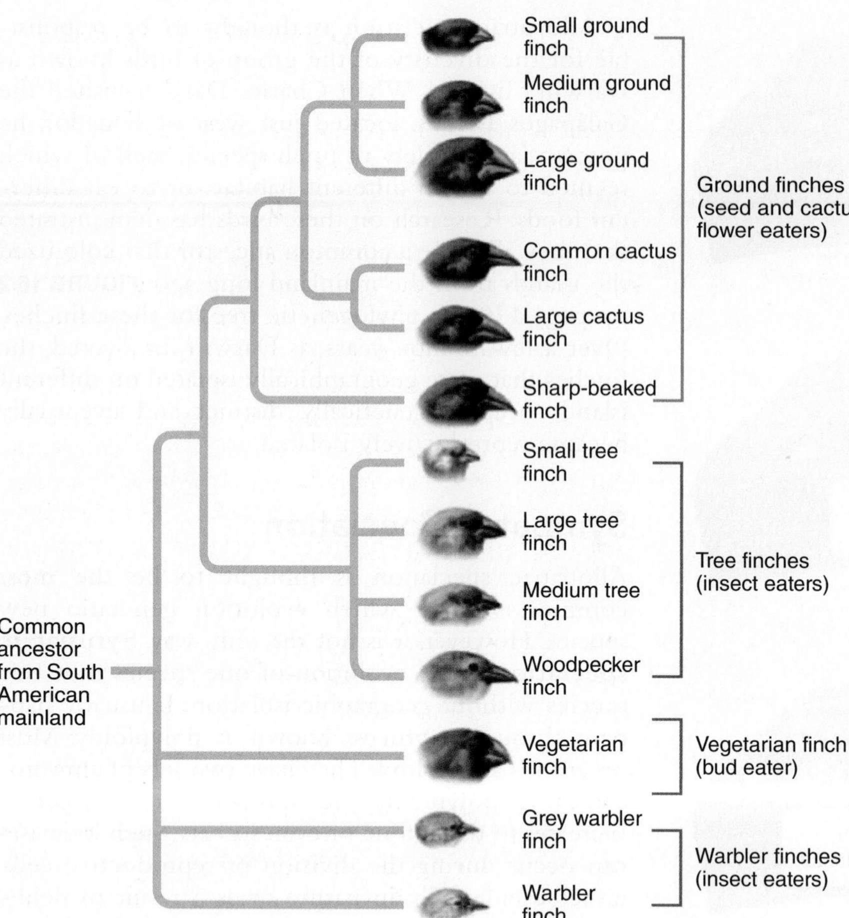

FIGURE 16.2 Allopatric speciation of Darwin's finches. In the Galápagos Islands, allopatric speciation has led to a large variety of finch species, all descended from a single species that colonized the islands from the South American mainland.

section, we will consider examples of rapid evolution by natural selection and very rapid evolution by artificial selection.

Rapid Evolution by Natural Selection

Sometimes evolution can occur rapidly, as in the case of the cichlid fishes of Lake Tanganyika, one of the African Great Lakes. You can see a variety of these species in **FIGURE 16.4**. Evidence indicates that the roughly 200 different species of cichlids in the lake evolved from a single ancestral species over a period of several million years. During this period, some cichlid species specialized to become insect eaters and others to become fish eaters, while still others evolved to eat invertebrates such as snails and clams.

Although the cichlids of Lake Tanganyika evolved quickly in evolutionary terms, the pupfishes of the Death Valley region of California and Nevada evolved even more rapidly. In the 20,000 to 30,000 years since the large lakes of the region were reduced to isolated springs, several species of pupfish have evolved.

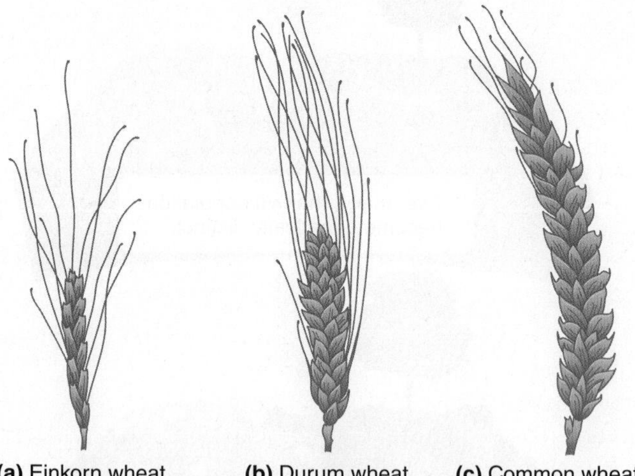

FIGURE 16.3 Sympatric speciation. Flowering plants such as wheat commonly form new species through the process of polyploidy, an increase in the number of sets of chromosomes beyond the normal two sets. (a) The ancestral einkorn wheat (*Triticum boeoticum*) has two sets of chromosomes and produces small seeds. (b) Durum wheat (*Triticum durum*), which is used to make pasta, was bred to have four sets of chromosomes and produces medium-sized seeds. (c) Common wheat (*Triticum aestivum*), which is used mostly for bread, was bred to have six sets of chromosomes and produces the largest seeds.

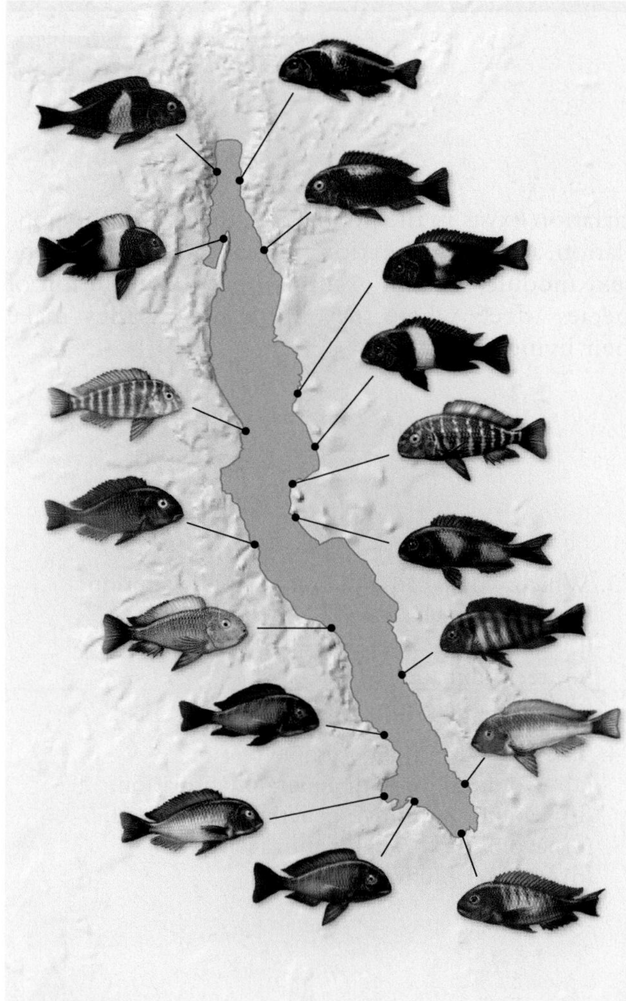

FIGURE 16.4 Rapid evolution. The cichlid fishes of Lake Tanganyika have evolved approximately 200 distinct and colorful species in the relatively short period since the lake formed in eastern Africa. The location where each species can be found in the lake is indicated by a corresponding black dot on the map.

The ability of a species to survive an environmental change depends greatly on how quickly it evolves the adaptations needed to thrive and reproduce under the new conditions. If a species cannot adapt quickly enough, it will go extinct. This can happen when the rate of environmental change is faster than the rate at which evolution can respond. Slow rates of evolution can occur when a population has long generation times or when a population contains low genetic variation on which natural selection can act.

Very Rapid Evolution by Artificial Selection

The pace of evolution by artificial selection can be incredibly fast. Such rapid evolution is occurring in many species of commercially harvested fish, including the Atlantic cod (*Gadus morhua*). Intensive fishing over several decades has targeted the largest adults, selectively removing most of those individuals from the population and, therefore, also removing the genes that produce large adults. Because larger fish tend to reach sexual maturity later, the genes that code for a later onset of sexual maturity have also been removed. As a result, after just a few decades of intensive fishing, the Atlantic cod population has evolved to reach reproductive maturity at a smaller size and a younger age. This evolution of shorter generation times also means that the cod may be able to evolve even faster in the future.

Evolution occurs even more rapidly in populations of *genetically modified organisms*. Using genetic engineering techniques, scientists can now copy genes from a species with some desirable trait, such as rapid growth or disease resistance. Scientists can insert these genes into other species of plants, animals, or microbes to produce a **genetically modified organism (GMO)**. When a GMO reproduces, it passes on the inserted genes to its offspring. For example, scientists have found that a soil bacterium (*Bacillus thuringiensis*) naturally produces an insecticide as a defense against being consumed by insects in the soil. Plant breeders have identified the bacterial genes that are responsible for making the insecticide, copied those genes, and inserted them into the genomes of crop plants. Such crops can now naturally produce their own insecticide, which makes them less attractive to insect herbivores. Common examples include Bt-corn and Bt-cotton, so named because they contain genes from a soil bacterium whose Latin name is *Bacillus thuringiensis*. As you might guess, inserting genes into an organism is a much faster way to produce desired traits than traditional plant and animal breeding, which can only select from the naturally available variation in a population.

Genetically modified organism (GMO) An organism produced by copying genes from a species with a desirable trait and inserting them into another species.

MODULE 16 AP® Review

Preparing for the AP® Exam

In this module, we learned that species evolve through the mechanisms of allopatric and sympatric speciation. We also saw that there are a number of factors that can affect the pace of evolution including how quickly the environment changes, how much genetic variation exists in the population, the size of the population, and the generation time of the species. In the next module, we will examine how the evolution of species affects where they live and how they make their living.

AP® Practice Questions

Choose the best answer for the following.

1. Which contributes to allopatric speciation?
   I. genetic convergence over time
   II. geographic isolation
   III. reproductive isolation
   (a) I and II
   (b) II only
   (c) II and III
   (d) I, II, and III
2. Which is often a cause of sympatric speciation?
   (a) polyploidy
   (b) geographic separation
   (c) artificial selection
   (d) genetic modification
3. Which would cause the most-rapid evolution?
   (a) artificial selection
   (b) geographic isolation
   (c) recombination
   (d) natural selection
4. The Bt-cotton is an example of
   (a) evolution through geographic isolation.
   (b) evolution through genetic modification.
   (c) evolution through natural selection.
   (d) macroevolution.

MODULE 17

Evolution of Niches and Species Distributions

Because evolution alters the traits that species express and because different traits perform well in some environments but not in others, the evolution of species affects where species are able to live on Earth. As environments have changed over millions of years, species have responded by altering their distributions. This response also suggests that environmental changes in the future will continue to affect the distributions of species. Species that can move easily will likely persist whereas species that cannot adjust or move will likely go extinct.

Learning Goals

After reading this module you should be able to

- explain the difference between a fundamental and a realized niche.
- describe how environmental change can alter species distributions.
- discuss how environmental change can cause species extinctions.

Every species has a niche

Every species has an optimal environment in which it performs particularly well. All species have a **range of tolerance**, or limits to the abiotic conditions they can tolerate, such as extremes of temperature, humidity, salinity, and pH. **FIGURE 17.1** illustrates this concept using one environmental factor—temperature. As conditions move further away from the ideal, individuals may be able to survive, and perhaps even to grow, but not be able to reproduce. As conditions continue to move away from the ideal, individuals can only survive. If conditions move beyond the range of tolerance, individuals will die. Because the combination of abiotic conditions in a particular environment fundamentally determines whether a species can persist there, the suite of abiotic conditions under which a species can survive, grow, and reproduce is the **fundamental niche** of the species.

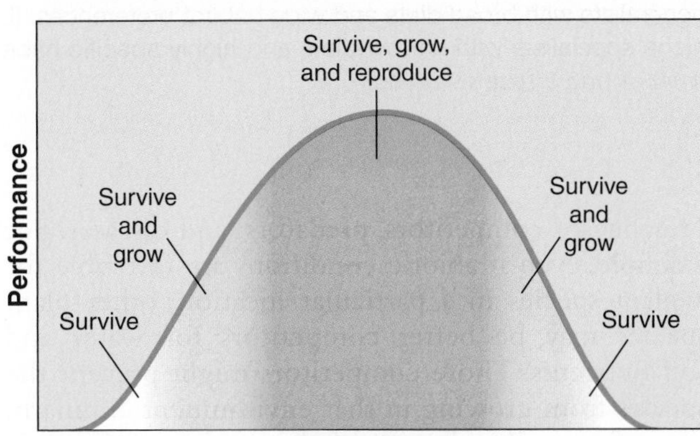

FIGURE 17.1 Range of tolerance. All species have an ideal range of abiotic conditions, such as temperature, under which their members can survive, grow, and reproduce. Under more extreme conditions, their ability to perform these essential functions declines.

AP® Exam Tip

When answering free-response questions, avoid statements such as, "it will kill all the animals and plants." Become familiar with the phrase "range of tolerance" to answer how organisms will respond when conditions change. For example, if ocean temperatures increase, some organisms will not survive because the temperature exceeds their range of tolerance. ●

The fundamental niche establishes the abiotic limits for the persistence of a species. However, biotic factors can further limit the locations where a species can live. Common biotic limitations include the

Range of tolerance The limits to the abiotic conditions that a species can tolerate.

Fundamental niche The suite of abiotic conditions under which a species can survive, grow, and reproduce.

(a)

(b)

FIGURE 17.2 Generalists and specialists. (a) Some organisms, such as gray kangaroos, are niche generalists with broad diets and wide habitat preferences. (b) Other organisms, such as koalas, are niche specialists with narrow diets and highly specific habitat preferences. *(a: Sohns, Juergen & Christine/ Animals Animals; b: Lourens Smak/Alamy)*

presence of competitors, predators, and diseases. For example, even if abiotic conditions are favorable for a plant species in a particular location, other plant species may be better competitors for water and soil nutrients. Those competitors might prevent the species from growing in that environment. Similarly, even if a small rodent can tolerate the temperature and humidity of a tropical forest, a deadly rodent disease might prevent the species from persisting in the forest. Therefore, biotic factors further narrow the fundamental niche that a species actually uses. The range of abiotic and biotic conditions under which a species actually lives is called its **realized niche**. Once we determine what contributes to the realized niche of a species, we have a better understanding of the **distribution** of the species, or the areas of the world in which the species lives.

When we examine the realized niches of species in nature, we see that some species, known as **niche generalists**, can live under a very wide range of abiotic or biotic conditions. For example, some mammals such as gray kangaroos (*Macropus giganteus*) (**FIGURE 17.2a**), feed on numerous plant species. Other species, known as **niche specialists**, are specialized to live under a very narrow range of conditions or feed on a small group of species. For example, the koala (*Phascolarctos cinereus*) (Figure 17.2b) feeds on only the leaves of several species of eucalyptus trees. Niche specialists can persist quite well when environmental conditions remain relatively constant, but they are more vulnerable to extinction if conditions change because the loss of a favored habitat or food source leaves them with few

Realized niche The range of abiotic and biotic conditions under which a species actually lives.

Distribution Areas of the world in which a species lives.

Niche generalist A species that can live under a wide range of abiotic or biotic conditions.

Niche specialist A species that is specialized to live in a specific habitat or to feed on a small group of species.

alternatives for survival. In contrast, niche generalists should fare better under changing conditions because they have a number of alternative habitats and food sources available.

Environmental change can alter the distribution of species

Because species are adapted to particular environmental conditions, we would expect that changes in environmental conditions would alter the distribution of species on Earth. For example, scientists have found evidence for the relationship between environmental change and species distribution in layers of sediments that have accumulated over time at the bottom of modern lakes. Each sediment layer contains pollen from plants that lived in the region when the sediments were deposited. In some cases, this pollen record goes back thousands of years.

In much of northern North America, lakes formed 12,000 years ago at the end of the last Ice Age when temperatures warmed and the glaciers slowly retreated to the north. The retreating glaciers left behind a great deal of barren land, which was quickly colonized by plants, including trees. Some of the pollen produced by these trees fell into lakes and was buried in the lake sediments. Scientists can measure the ages of these sediment layers with carbon dating (see Chapter 2). Furthermore, because each tree species has uniquely shaped pollen, it is possible to determine when a particular tree species arrived near a particular lake and how the entire community of plant species changed over time. **FIGURE 17.3** shows pollen records for three tree species in North America. These pollen records suggest that changes in climatic conditions after the Ice Age produced substantial changes in the distributions of plants over time. Pine trees, spruce trees, and birch trees all moved north with the retreat of the glaciers. As the plants moved north, the animals followed.

AP® Exam Tip

The AP® Environmental Science Exam includes various models such as data tables, graphs, and other data graphics. Practice analyzing various models so that you are comfortable gathering data and making predictions and inferences. ●

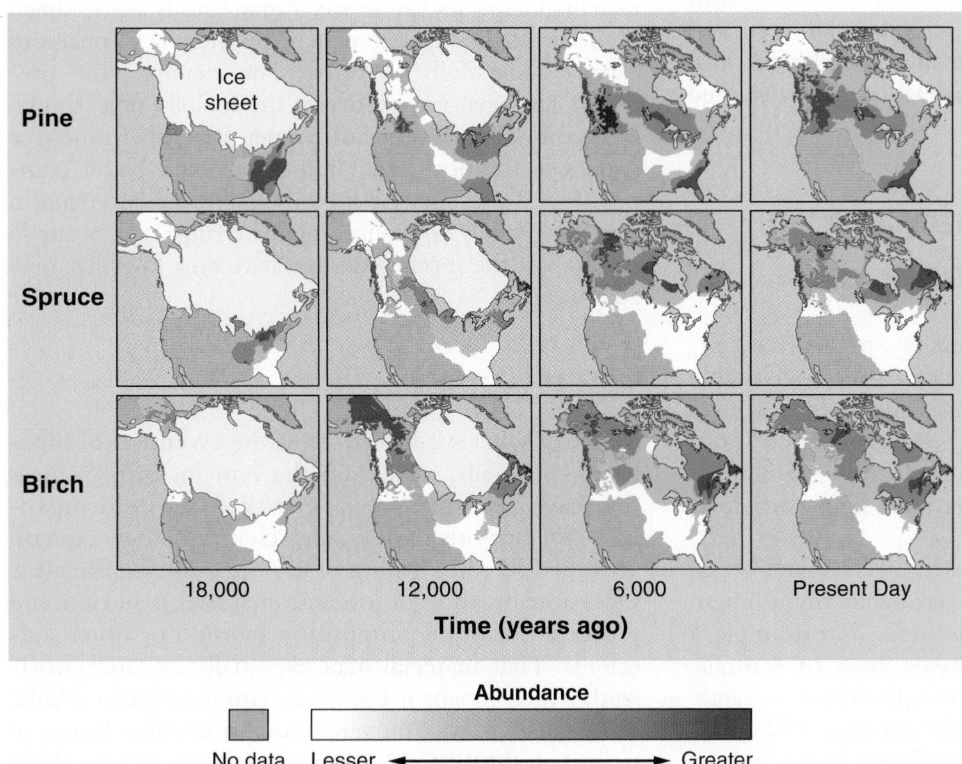

FIGURE 17.3 Changes in tree species distributions over time. Pollen recovered from lake sediments indicates that plant species moved north as temperatures warmed following the retreat of the glaciers, beginning about 12,000 years ago. Areas shown in color or white were sampled for pollen, whereas areas shown in gray were not sampled. *(Data from http://vemages.gsfc.nasa.gov//3453/boreal_model.gif)*

Because past changes in environmental conditions have led to changes in species distributions, it is reasonable to ask whether current and future changes in environmental conditions might also cause changes in species distributions. For example, as the global climate warms, some areas of the world are expected to receive less precipitation, while other areas are predicted to receive more. If our predictions of future environmental changes are correct, and if we have a good understanding of the niche requirements of many species, we should be able to predict how the distribution of species will change in the future. North American trees, for example, are expected to have more northerly distributions with future increases in global temperatures. As **FIGURE 17.4** shows, the loblolly pine (*Pinus taeda*) is expected to move from its far southern distribution and become common throughout the eastern half of the United States.

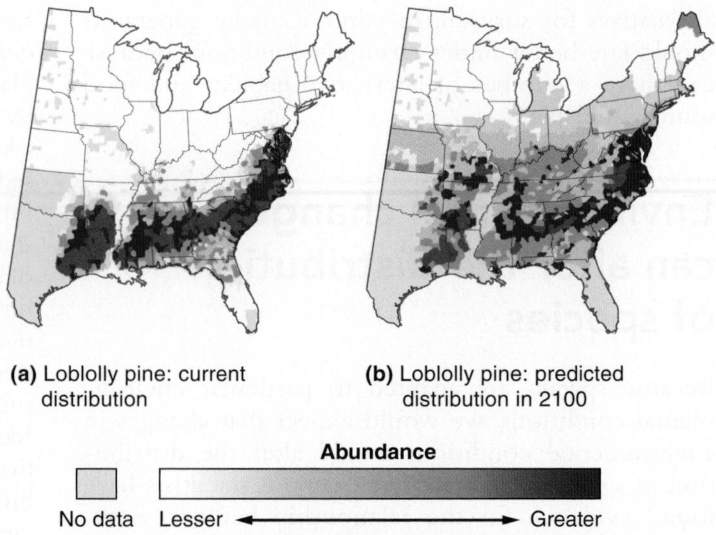

FIGURE 17.4 Predicting future species distributions. Based on our knowledge of the niche requirements of the loblolly pine tree, we can predict how the distributions might change as a result of future changes in environmental conditions. *(Data from: http://www.fs.fed.us/ne/delaware/atlas/web_atlas.html#)*

Species vary in their ability to move physically across the landscape as the environment changes. Some species are highly mobile at particular life stages: Adult birds and wind-dispersed seeds, for example, move across the landscape easily. Other organisms, such as the desert tortoise (*Gopherus agassizii*), are slow movers. Furthermore, the movements of many species may be impeded by obstacles built by humans, including roads and dams. It remains unclear how species that face these challenges will shift their distributions as global climate change occurs.

Environmental change can cause species extinctions

If environmental conditions change, species that cannot adapt to the changes or move to more favorable environments will eventually go extinct. The average life span of a species appears to be only between about 1 million to 10 million years. In fact, 99 percent of the species that have ever lived on Earth are now extinct.

There are several reasons why species might go extinct. First, there may be no favorable environment close enough to which they can move. For example, a small area of land on the southwest coast of Australia experiences a current climate that supports a woodland/shrubland biome (see Figure 12.9 on page 132). This biome is surrounded by a large desert biome to the north and east and an ocean to the south and west. If scientists are right in predicting that the coming century will bring a hotter and drier climate to southwestern Australia, then the unique species of plants that live in this small biome will have nowhere to go where they can survive.

Even if there is an alternative favorable environment to which a species can move, it may already be occupied by other species against which the moving populations cannot successfully compete. For example, the predicted northern movement of the loblolly pine, shown in Figure 17.4b, might not happen if another pine tree species in the northern United States is a better competitor and prevents the loblolly pine from surviving in that area. Finally, an environmental change may occur so rapidly that the species does not have time to evolve new adaptations.

The Fossil Record

Much of what we know about the evolution of life is based on fossils, which are the remains of organisms that have been preserved in rock. Most dead organisms decompose rapidly; the elements they contain are recycled and nothing of the organism is preserved. Occasionally, though, organic material is buried and protected from decomposition by mud or other sediments. That material may eventually become fossilized, which means it has been hardened into rocklike material as it was buried under successive layers of sediment (**FIGURE 17.5** on page 177). When these layers are uncovered, they reveal a record of at least some of the organisms that existed at the time the sediments were deposited. Because of the way layers of sediment are deposited on top of one another over

FIGURE 17.5 Fossils. Fossils, such as this fish discovered in Fossil Butte National Monument in Wyoming, are a record of evolution. *(Danita Delimont/Getty Images)*

time, the oldest fossilized organisms are found in the deepest layers of the fossil record. We can therefore use the fossil record to determine when different species existed on Earth.

The Five Global Mass Extinctions

Throughout Earth's history, individual species have evolved and gone extinct at random intervals. The fossil record has revealed five periods of global **mass extinction**, in which large numbers of species went extinct over relatively short periods of time. The times of these mass extinctions are shown in **FIGURE 17.6**. Note that because species are not always easy to discriminate in the fossil record, scientists count the number of genera, rather than species, that once roamed Earth but are now extinct.

The greatest mass extinction on record took place 251 million years ago when roughly 90 percent of marine species and 70 percent of land vertebrates went extinct. The cause of this mass extinction is not known.

A better-known mass extinction occurred at the end of the Cretaceous period (65 million years ago), when roughly one-half of Earth's species, including the dinosaurs, went extinct. The cause of this mass extinction has been the subject of great debate, but there is now a near-consensus that a large meteorite struck Earth and produced a dust cloud that circled the planet and blocked incoming solar radiation. This resulted in an almost complete halt to photosynthesis, and thus an almost total lack of food at the bottom of the food chain. Among the few species that survived was a small squirrel-sized primate that was the ancestor of humans.

Many scientists view extinctions as the ultimate result of change in the environment. Environmental scientists can learn about the potential effects of both large and small environmental changes by studying historic environmental changes and applying the lessons learned to help predict the effects of the environmental changes that are taking place on Earth today.

The Sixth Mass Extinction

During the last 2 decades, scientists have reached a consensus that we are currently experiencing a sixth global mass extinction of a magnitude within the range of the previous five mass extinctions. Given that we have poor data on the abundance of most species, especially the very small invertebrate animals, estimates of extinction rates vary widely. These estimates are extrapolated from extinction rates observed in a few well-described groups of organisms. For example, the Millennium Ecosystem Assessment, which is conducted by the United Nations, has estimated the extinction rate at 24 species per day. In contrast, The United Nations Convention on Biological Diversity

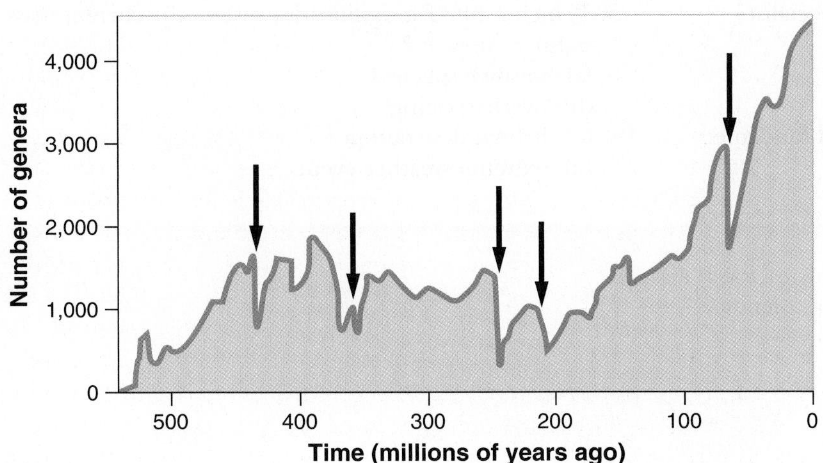

FIGURE 17.6 Mass extinctions. Five global mass extinction events have occurred since the evolution of complex life roughly 500 million years ago. *(Data from GreenSpirit, http://www.greenspirit.org.uk/resources/Timelies.jpg)*

Mass extinction A large extinction of species in a relatively short period of time.

has estimated the extinction rate at 150 species per day. What we know for sure is that there have been about 800 confirmed extinctions during the past 400 years, which represents a rate of about 2 species per year. However, these confirmed extinctions do not include any possible extinctions of species not yet discovered, which is why it is difficult to know the true extinction rate.

Although there is a range of estimates for the correct rate of extinction, there is widespread agreement among scientists that the current mass extinction has human causes. These causes include habitat destruction, overharvesting, introductions of invasive species, climate change, and emerging diseases. We will examine all of these factors in detail in Chapters 18 and 19. Because much of the current environmental change caused by human activities is both dramatic and sudden, environmental scientists contend that many species may not be able to move or adapt in time to avoid extinction.

The recovery of biodiversity from earlier mass extinctions took about 10 million years, an unthinkably long time from a human perspective. Recovery from the present mass extinction could take just as long—500,000 human generations. Much of the current debate among environmental scientists and government officials centers on the true magnitude of this crisis and on the costs of reducing the human impact on extinction rates.

MODULE 17 AP® Review

Preparing for the AP® Exam

In this module, we learned that species differ in their tolerance to abiotic conditions and that the range of abiotic and biotic conditions that allow survival, growth, and reproduction all help to determine where a species can live. Past changes in environmental conditions have altered species distributions and caused extinctions. Future environmental changes are expected to have similar effects.

AP® Practice Questions

Choose the best answer for the following.

1. The abiotic conditions under which a species can survive and reproduce is called its
   (a) range of tolerance.
   (b) realized niche.
   (c) environmental distribution.
   (d) fundamental niche.

2. Which would be expected for a niche specialist?
   (a) many food sources
   (b) a narrow fundamental niche
   (c) adaptability to changing conditions
   (d) no difference between its realized and fundamental niche

3. Which will NOT affect the distribution of a species?
   (a) climate change
   (b) a change in the distribution of the species' food
   (c) a new food source outside its range of tolerance
   (d) the introduction of a predator species

4. How many mass extinction events have there been in Earth's history?
   (a) 2
   (b) 4
   (c) 5
   (d) 7

5. Which is NOT a significant cause of the current mass extinction event?
   (a) invasive species
   (b) overharvesting
   (c) habitat destruction
   (d) extreme weather events

Working Toward Sustainability

Protecting the Oceans When They Cannot Be Bought

For over 50 years, The Nature Conservancy (TNC) has protected biodiversity by using a simple strategy: Buy it. The Conservancy uses grants and donations to purchase privately owned natural areas or to buy development rights to those areas. TNC owns over 0.8 million hectares (2 million acres) of land and has protected over 46 million hectares (115 million acres) of land by buying development rights. As a nonprofit, nongovernmental organization, TNC has great flexibility to use innovative conservation and restoration techniques on natural areas in its possession.

TNC focuses its efforts on areas containing rare species or that have high biodiversity, including the Florida Keys in southern Florida and Santa Cruz Island in California. Recently, it has set its sights on the oceans, including coastal marine ecosystems. Coastal ecosystems have experienced steep declines in the populations of many fish and shellfish, including oysters, clams, and mussels, due to a combination of overharvesting and pollution. By preserving these coastal ecosystems, TNC hopes to create reserves that will serve as breeding grounds for declining populations of overharvested species. In this way, protecting a relatively small area of ocean will benefit much larger unprotected areas, and even benefit the very industries that have led to the population declines.

Shellfish are particularly valuable in many coastal ecosystems because they are filter feeders: They remove tiny organisms, including algae, from large quantities of water, cleaning the water in the process and helping to prevent harmful algal blooms. However, shellfish worldwide have been harvested unsustainably, leading to a cascade of effects throughout many coastal regions. For example, oyster populations in the Chesapeake Bay were once sufficient to filter the water of the entire bay in 3 to 6 days. Now there are so few oysters that it would take a year for them to filter the same amount of water. As a result, the bay has become much murkier, and excessive algae have led to lowered oxygen levels that make the bay less hospitable to fish.

Conserving marine ecosystems is particularly challenging because private ownership is rare. State and federal governments generally do not sell areas of the ocean. Instead, they have allowed industries to lease the harvesting or exploitation rights to marine resources such as oil, shellfish, and physical space for marinas and aquaculture. So how can a conservation group protect coastal ecosystems if it cannot buy an area of the ocean? The Nature Conservancy's strategy is to purchase harvesting and exploitation rights and use them as a conservation tool. In some cases, TNC will not harvest any shellfish in order to allow the populations to rebound. In many cases, the leases require at least some harvesting, and TNC hopes to demonstrate sustainable management practices that will serve as an example of how shellfish harvests can be conducted while restoring the shellfish beds.

From 2002 to 2004, TNC acquired the rights to 5,261 ha (13,000 acres) of oyster beds in New York's Great South Bay, along the southern shore of Long Island. These rights, which were donated to TNC by the Blue Fields Oyster Company, were valued at $2 million. TNC developed restoration strategies, including seeding the area with 7 million clams during 2010–2012 to give the clam population a jump start. While populations of shellfish have shown early signs of recovery, the bay remains challenged

Restoring ocean populations. In areas that have declining populations of shellfish, such as this site in New Hampshire, researchers seed the area with additional individuals to help speed up the recovery of the populations. *(Joe Klementovich)*

by the excess inputs of nutrients that can cause toxic algal blooms. If the nutrient problem can be remedied, TNC hopes to engage in sustainable harvesting over part of this area and conduct research in the rest of it.

TNC has similar projects under way off the coasts of Virginia, North Carolina, and Washington State. In California, TNC has accumulated the rights to 10,000 ha (25,000 acres) of marine fisheries along the coast. Today, it has developed a collaborative program with local commercial fishermen to harvest fish and shellfish that will use sustainable practices, which includes reporting their catches on an iPad app to help the TNC and its partners better assess fish abundance and fishing pressure. These fishermen can also post any information about areas of the ocean where they accidentally catch protected species, which helps other fishing boats avoid these areas. One result is that the amount of unintended species caught, known as "bycatch," has declined from between 15 to 20 percent to only 1 percent. In making these changes, TNC has found effective ways to continue the tradition of fishing that provides jobs to the local communities while working to ensure that the fish populations persist long into the future.

Critical Thinking Questions

1. What are the advantages of working to increase fish populations by leasing areas of the ocean and banning fishing in those areas versus leasing areas and working with local commercial fishing boats to change fishing practices?
2. What are some of the challenges in getting commercial fishing operations to change their practices?

References

California Fishermen, Once Blocked by Conservationists, Now Work with Them. *Market Place,* June 19, 2017. https://www.marketplace.org/2017/06/19/sustainability/california-fishermen-once-blocked-conservationists-now-work-with-them

The Nature Conservancy. 2002. *Leasing and Restoration of Submerged Lands: Strategies for Community-Based, Watershed-Scale Conservation.*

The Nature Conservancy. 2006. *Annual Report.* http://www.nature.org/aboutus/annualreport/file/annualreport2008.pdf.

Partnership Preserves Livelihoods and Fish Stocks. *New York Times,* November 27, 2011. http://www.nytimes.com/2011/11/28/science/earth/nature-conservancy-partners-with-california-fishermen.html?_r=0.

Chapter 5 Review

In this chapter, we have learned about the biodiversity of Earth, how this biodiversity came to be, and how environmental changes can cause it to decline. The estimated number of species on Earth varies widely, but many scientists agree on approximately 10 million. In any given location, we can quantify the diversity of species in terms of both species richness and species evenness. The diversity that exists came about through the process of evolution. We can view patterns of evolution by placing species on a phylogeny. Evolution occurs when there are changes in the genetic composition of a population. This can happen through artificial selection, natural selection, or random processes. The species that evolve through these processes each have a niche that helps to determine their geographic distributions. Historic changes in the environment have altered species distributions and caused many species to go extinct; current and future environmental changes are expected to have similar effects.

Key Terms

Species richness
Species evenness
Phylogeny
Evolution
Microevolution
Macroevolution
Gene
Genotype
Phenotype
Mutation
Recombination
Evolution by artificial selection
Evolution by natural selection
Fitness
Adaptation
Gene flow
Genetic drift
Bottleneck effect
Extinction
Founder effect
Geographic isolation
Allopatric speciation
Reproductive isolation
Sympatric speciation
Genetically modified organism
Range of tolerance
Fundamental niche
Realized niche
Distribution
Niche generalist
Niche specialist
Mass extinction

Learning Goals Revisited

Module 14 The Biodiversity of Earth

Understand how we estimate the number of species living on Earth.

Scientists have estimated the number of species on Earth by collecting samples of diverse groups of organisms, determining the proportion of all known species, and then extrapolating these numbers to other groups to estimate the total number of species.

Quantify biodiversity.

Biodiversity can be quantified using a variety of measurements including species richness, species evenness, or both. Such measurements provide scientists with a baseline they can use to determine how much an ecosystem has been affected by a natural or anthropogenic disturbance.

Describe patterns of relatedness among species using a phylogeny.

Patterns of relatedness are depicted as phylogenies. Phylogenies indicate how species are related to one another and the likely steps in evolution that gave rise to current species.

Module 15 How Evolution Creates Biodiversity

Identify the processes that cause genetic diversity.

Every individual has a genotype that, in combination with the environment, determines its phenotype. In a population, genetic diversity is produced by the processes of mutation and recombination.

Explain how evolution can occur through artificial selection.

Evolution by artificial selection occurs when humans select individuals with a particular phenotypic goal in mind. Such selection has produced various breeds of domesticated animals and numerous varieties of crops. It has also produced harmful outcomes, including selection for pesticide-resistant pests and drug-resistant bacteria.

Explain how evolution can occur through natural selection.

Evolution by natural selection occurs when individuals vary in traits that can be passed on to the next generation and this variation in traits causes different abilities to survive and reproduce in the wild. Natural selection does not target particular traits, but simply favors any trait changes that result in higher survival or reproduction.

Explain how evolution can occur through random processes.

Because evolution is defined as a change in the genetic composition of a population, evolution can also occur when there are mutations within a population or gene flow into or out of a population. It can also occur due to the processes of genetic drift, the bottleneck effect, or the founder effect.

Module 16 Speciation and the Pace of Evolution

Explain the processes of allopatric and sympatric speciation.

Allopatric speciation occurs when a portion of a population experiences geographic isolation from the rest of the population. The composition of isolated populations diverges over time due to random processes or natural selection. Sympatric speciation occurs when one species separates into two species without any geographic isolation. The production of polyploidy individuals is a common mechanism of sympatric speciation.

Understand the factors that affect the pace of evolution.

Evolution can produce adapted populations more easily when environmental changes are slow rather than rapid. The populations are also more likely to adapt when they have high genetic variation, as is often found in large populations. Should there be a beneficial mutation arise, however, the mutation can spread through small populations more rapidly than large populations. Finally, evolution can occur more rapidly in populations that have shorter generation times.

Module 17 Evolution of Niches and Species Distributions

Explain the difference between a fundamental and a realized niche.

Every species has a range of tolerance to the abiotic conditions of the environment. The conditions under which a species can survive, grow, and reproduce is known as its fundamental niche. The portion of the fundamental niche that a species actually occupies due to biotic interactions, including predation, competition, and disease, is known as the realized niche. Some species are niche generalists whereas other species are niche specialists. The niche a species occupies determines its distribution.

Describe how environmental change can alter species distributions.

Because a species niche represents the environmental conditions under which a species can live, environmental change can cause a change in the distribution of a species.

Discuss how environmental change can cause species extinctions.

When environmental changes are too rapid and too extensive to permit evolutionary changes, or if a species is unable to move to more hospitable environments, a species will not be able to persist and will go extinct.

Practice Math and Graphing

Preparing for the AP® Exam

Answer the following questions. Be sure to show all your work.

1. Practice Math

Mosquitoes carry a wide variety of diseases including malaria, so insecticides are often used to reduce mosquito populations. Given that mosquito populations can grow very large, it is common for a few mosquitoes to possess a mutation that confers insecticide resistance. With repeated insecticide spraying, more and more sensitive individuals die from the insecticide while resistant individuals survive. Using the data in the table below, calculate the percentage of the sensitive and resistant phenotypes in the population over a 3-year period.

Years	Number of resistant individuals	Number of tolerant individuals	Percent of resistant individuals	Percent of tolerant individuals
1	995	5		
2	473	527		
3	75	925		

2. Practice Graphing

Use the mosquito data in the table from "Practice Math" to answer the following:

(a) Create a bar graph that illustrates the percent of sensitive and resistant phenotypes of mosquitoes during each year of the insecticide applications.

(b) Based on this graph, how have the insecticide applications affect the population's phenotypes over time?

(c) Explain why the application of the insecticide represents either artificial or natural selection.

Section 1: Multiple-Choice Questions

Choose the best answer for questions 1–18.

1. Which is NOT a measure of biodiversity?
   (a) economic diversity
   (b) ecosystem diversity
   (c) genetic diversity
   (d) species diversity

2. Which statement describes an example of artificial selection?
   (a) Cichlids have diversified into nearly 200 species in Lake Tanganyika.
   (b) Thoroughbred racehorses have been bred for speed.
   (c) Whales have evolved tails that help propel them through water.
   (d) Darwin's finches have beaks adapted to eating different foods.

3. The following table represents the number of individuals of different species that were counted in three forest communities.

Species	Community A	Community B	Community C
Deer	95	20	10
Rabbit	1	20	10
Squirrel	1	20	10
Mouse	1	20	10
Chipmunk	1	20	10
Skunk			10
Opossum			10
Elk			10
Raccoon			10
Porcupine			10

   Which statement best interprets these data?
   (a) Community A has greater species richness than Community B.
   (b) Community B has greater species evenness than Community C.
   (c) Community C has greater species richness than Community A.
   (d) Community A has greater species evenness than Community C.

4. The yellow perch (*Perca flavescens*) is a fish that breeds in spring. A single female can produce up to 40,000 eggs at one time. This species is an example of which of the key ideas of Darwin's theory of evolution by natural selection?
   (a) Individuals produce an excess of offspring.
   (b) Individuals vary in their phenotypes.
   (c) Phenotypic differences in individuals can be inherited.
   (d) Different phenotypes have different abilities to survive and reproduce.

5. In 2002, Peter and B. Rosemary Grant studied a population of Darwin's finches on one of the Galápagos Islands that feeds on seeds of various sizes. After a drought that caused only large seeds to be available to the birds, they found that natural selection favored those birds that had larger beaks and bodies. Once the rains returned and smaller seeds became much more abundant, however, natural selection favored those birds that had smaller beaks and bodies. Which process is the best interpretation of this scenario?
   (a) founder effect
   (b) microevolution
   (c) macroevolution
   (d) bottleneck effect

6. When a population of monkeys migrates to a new habitat across a river and encounters another population of the same species, what evolutionary effect may occur as a result?
   (a) gene flow
   (b) the founder effect
   (c) genetic drift
   (d) recombination

7. The northern elephant seal (*Mirounga angustirostris*) was once hunted to near-extinction. Only 20 animals remained alive in 1890. Then, after the species was protected from hunting, its population grew to nearly 30,000 animals. However, this large population possesses very low genetic variation. Which process is the best interpretation of this scenario?
   (a) evolution by natural selection
   (b) evolution by the founder effect
   (c) evolution by the bottleneck effect
   (d) evolution by genetic drift

8. Which statement is NOT correct?
   (a) Most speciation is thought to occur through allopatric speciation.
   (b) Polyploidy is an example of sympatric speciation.
   (c) Geographic isolation can eventually lead to reproductive isolation.
   (d) Speciation cannot occur without geographic isolation.

9. Which allows more-rapid evolution?
   (a) long generation times
   (b) slow environmental change
   (c) small population sizes
   (d) high genetic variation

10. Which condition does NOT define the fundamental niche of a species?
    (a) humidity
    (b) predators
    (c) temperature
    (d) salinity

11. Some scientists estimate that the current global extinction rate is about 30,000 species per year. If there are currently 10 million species on Earth, how long will it take to destroy all of Earth's biodiversity?
    (a) between 100 and 300 years
    (b) between 300 and 500 years
    (c) between 500 and 700 years
    (d) between 700 and 1,000 years

12. Global climate change can cause the extinction of species due to all of the following EXCEPT
    (a) the inability of individuals to move.
    (b) the absence of a favorable environment nearby.
    (c) the rapid rate of environment change.
    (d) the presence of high genetic variation.

Questions 13 and 14 refer to the following scenario.

A researcher collects samples of aquatic insects from a variety of streams, as shown in the table.

	Stream A	**Stream B**	**Stream C**	**Stream D**
Mayfly	10	5	1	7
Caddisfly	5	2	2	7
Midge			30	7
Stonefly	5			
Dragonfly larvae	7	5	1	
Black fly		3	10	5
Water pennies	8		3	2

13. Which stream has the greatest species richness?
    (a) stream A
    (b) stream B
    (c) stream C
    (d) stream D

14. Scientists often use aquatic insect communities as indicators of water quality. Streams that are polluted can have low species evenness, and can be dominated by pollutant-tolerant taxa like midges and black flies. Based on this information, which of the four streams is most likely to be polluted?
    (a) stream A
    (b) stream B
    (c) stream C
    (d) stream D

15. A species of parakeet inhabits a large area of forest that has been fragmented by housing developments. It is difficult for the birds to move between forest fragments. Which process is likely to be affected by the fragmentation of the parakeet habitat?
    I. gene flow
    II. founder effect
    III. sympatric speciation
    (a) I only
    (b) II only
    (c) III only
    (d) I and II

16. Which is NOT a key idea of Darwin's theory of natural selection?
    (a) Individuals differ in their phenotypes.
    (b) Differences in traits can be passed on from parents to offspring.
    (c) The genotype of an organism determines its phenotype.
    (d) Differences in traits are associated with differences in the ability to survive and reproduce.

17. ______ is a change in the genetic composition of a population as a result of descending from a small number of colonizing individuals.
(a) The founder effect
(b) Inbreeding depression
(c) Gene flow
(d) The bottleneck effect

18. The following table contains the number of different spider species found in a survey of 1-m^2 plots from four gardens. Empty spaces indicate that no spiders of that species were found in the garden.

	Garden A	Garden B	Garden C	Garden D
Grass Spider	2	7	3	2
Orb Weaver	1		3	4
Jumping Spider		8	3	4
Wolf Spider	2	2	2	6
Sac Spider	7		3	2

Which garden has the highest Shannon's Index?
(a) garden A
(b) garden B
(c) garden C
(d) garden D

Section 2: Free-Response Questions

Write your answer to each part clearly. Support your answers with relevant information and examples. Where calculations are required, show your work.

1. Look at the photograph and answer the following questions.

(Radius Images/Alamy)

(a) Explain how this human impact on a forest ecosystem might affect the ability of some species to move to more suitable habitats as Earth's climate changes. (2 points)
(b) Propose and explain one alternative plan that could have preserved this forest ecosystem. (2 points)
(c) Distinguish between the terms microevolution and macroevolution. Explain how the organisms in the forest on the left could evolve into species different from those in the forest on the right. (6 points)

2. On April 26, 1986, a series of power surges occurred during a routine systems test at the Chernobyl nuclear power plant in Russia. These surges resulted in the most catastrophic nuclear accident in history, and released 80 petabecquerel of radioactive material on 200,000 km^2 of surrounding land (1 petabecquerel = 10^{15} becquerel). This created an immediate, severe, and lasting health hazard for humans and wildlife. Although most wildlife died without producing viable offspring, evidence suggests that animals may now be adapting to low levels of chronic radiation.

Exposure to radiation generates health problems through the formation of free radicals that damage DNA and other organic molecules. Twenty-five years after the disaster at Chernobyl, several bird species in the area are producing a greater abundance of antioxidants that can remove free radicals. However, birds that produce these antioxidants demonstrate lower reproduction.

(a) Describe the processes that could have led to changes in antioxidant production in birds around Chernobyl. (2 points)

(b) Define the bottleneck effect, describe how it may apply to wildlife populations around Chernobyl, and discuss potential problems associated with the bottleneck effect. (2 points)

(c) Discuss the importance of having intact bird habitat in areas surrounding Chernobyl that were not exposed to radiation. (2 points)

(d) In general, birds that produce high levels of antioxidants exhibit lower levels of reproduction than birds that produce lower levels of antioxidants. However, researchers hypothesize that birds that produce high levels of antioxidants in the area around Chernobyl are likely to have a higher overall fitness than birds in the area without this trait. Design an experiment to test this hypothesis. In your experiment, define

(i) a null hypothesis. (1 point)

(ii) treatments and controls. (1 point)

(iii) a description of the experimental procedure. (2 points)

3. The graph below shows survival data of two different species of mouse, recorded in a laboratory experiment in which the mice are raised at differing temperatures, with all other conditions kept constant.

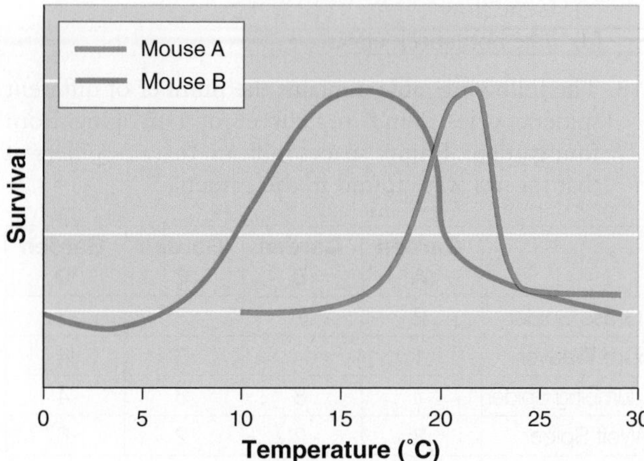

(a) According to the graph, which species is a niche generalist with regard to temperature? Explain your answer. (2 points)

(b) The mice species have overlapping distributions. Scientists predict that the area where they live will become several degrees warmer due to climate change, increasing from 20°C to 22°C.

(i) Based on the graph, determine which species is more likely to survive in that area and explain why. (1 point)

(ii) Identify which species is currently better able to survive in that area. (1 point)

(c) Identify whether the graph shows an aspect of fundamental or realized niche for the two species. Explain your answer and define fundamental niche and realized niche. (2 points)

(d) One mouse species inhabits a large town. Because this species is regarded as a pest, humans put out traps and poison for it. The traps disproportionately catch mice with a better sense of smell, because they are better at smelling the bait in the traps.

(i) Assuming that mice with a better sense of smell continue to be caught in this particular area, describe how the mouse population in this area will evolve over time. (1 point)

(ii) Identify the process this describes. (1 point)

(iii) Explain whether the sense of smell in this scenario is part of an individual's genotype or phenotype. (1 point)

Zeljko Radojko/Shutterstock.com

science applied 2

How Should We Prioritize the Protection of Species Diversity?

As a result of human activities, we have seen a widespread decline in biodiversity across the globe. Many people agree that we should try to slow or even stop this loss. But how do we do this? While ideally we might want to preserve all biodiversity, in reality preserving biodiversity requires compromises. For example, in order to preserve the biodiversity of an area, we might have to set aside land that would otherwise be used for housing developments, shopping malls, or strip mines. If we cannot preserve all biodiversity, how do we decide which species receive our attention? (**FIGURE SA2.1**)

In 1988, Oxford University professor Norman Myers noted that much of the world's biodiversity is concentrated in areas that make up a relatively small fraction of the globe. Part of the reason for this uneven pattern of biodiversity is that so many species are *endemic species*. **Endemic species** are species that live in a very small area of the world and nowhere else, often in isolated locations such as the Hawaiian Islands. Because they are home to so many endemic species, these isolated areas end up containing a high proportion of all the species found on Earth. Myers called these areas **biodiversity hotspots**.

FIGURE SA2.1 The California tiger salamander (*Ambystoma californiense*). This salamander is endemic to the biodiversity hotspot of California and is threatened with extinction due to habitat destruction and the introduction of non-native predators. *(James Gerholdt/Getty Images)*

Scientists originally identified 10 biodiversity hotspots, including Madagascar, western Ecuador, and the Philippines. Myers argued that these 10 areas were in need of immediate conservation attention because human activities there could have disproportionately large negative effects on the world's biodiversity. A year later, the group Conservation International adopted Myers's concept of biodiversity hotspots to guide its conservation priorities. As of 2013, Conservation International had identified the 35 biodiversity hotspots shown in **FIGURE SA2.2**. Although these hotspots collectively represent only 2.3 percent of the world's land area, more than 50 percent of all plant species and 42 percent of all vertebrate species are confined to these areas. As a result of this categorization, major conservation organizations have adjusted their funding priorities and are spending hundreds of millions of dollars to conserve these areas. What does environmental science tell us about the hotspot approach to conserving biodiversity?

What makes a hotspot hot?

Since Norman Myers initiated the idea of biodiversity hotspots, scientists have debated which factors should be considered most important when deciding where to focus conservation efforts. For example, most scientists agree that species richness is an important factor. There are more than 1,300 bird species in the small nation of Ecuador—more

Endemic species A species that lives in a very small area of the world and nowhere else.

Biodiversity hotspot An area that contains a high proportion of all the species found on Earth.

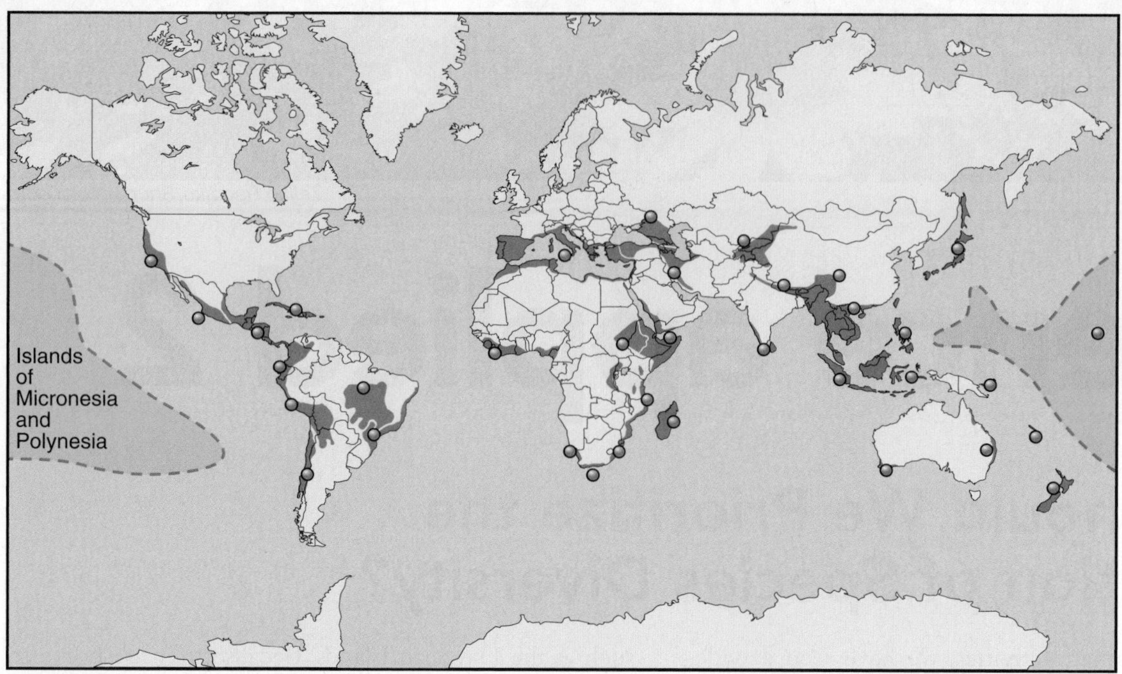

FIGURE SA2.2 Biodiversity hotspots. Conservation International has identified 34 biodiversity hotspots (shown in red) that have at least 1,500 endemic plant species and a loss of at least 70 percent of all vegetation. *(Data from Conservation International at http://www.conservation.org/where/priority_areas/hotspots/Documents/CI_Biodiversity-Hotspots_2013_Map.pdf.)*

than twice the number of bird species living in the United States and Canada. For this reason, protecting a habitat in Ecuador has the potential to save many more bird species than protecting the same amount of habitat in the United States or Canada. From this point of view, the choice to protect areas with a lot of species makes sense.

Identifying biodiversity hotspots is challenging, however, because scientists have not yet discovered and identified all the species on Earth. Because the distribution of plants is typically much better known than that of animals, the most practical way to identify hotspots has been to locate areas containing high numbers of endemic plant species. It is reasonable to expect that areas with high plant diversity will contain high animal diversity as well.

Conservation International considers two criteria when determining whether an area qualifies as a hotspot. First, the area must contain at least 1,500 endemic plant species. By conserving plant diversity, the hope is that we will simultaneously conserve animal diversity, especially for those groups, such as insects, that are poorly cataloged. Second, the area must have lost more than 70 percent of the vegetation that contains those endemic plant species. In this way, high-diversity areas with a high level of habitat loss receive the highest conservation priority. High-diversity areas that are not being degraded receive lower conservation priority.

What else can make a hotspot hot?

The number of endemic species in an area is undoubtedly important in identifying biodiversity hotspots, but other scientists have argued that this criterion alone is not enough. They suggest that we also consider the total number of species in an area or the number of species currently threatened with extinction in an area. Would all three approaches identify similar regions of conservation priority? A recent analysis of birds suggests they would not. When scientists identified bird diversity hotspots using each of the three criteria—endemic species, total species richness, and threatened species—their results, shown in **TABLE SA2.1**, identified very different areas. Scientists using the three criteria to identify hotspots of mammal diversity reached the same conclusion.

As we can see, some areas of the world that have high species richness do not contain high numbers of endemic species. The Amazon, for example, has a high number of bird species, but not a particularly high number of endemic bird species compared with other more-isolated regions of the world, such as the islands of the Caribbean. Similarly, areas with high numbers of threatened or endangered species do not always have high numbers of endemic species. These findings highlight the critical problem of deciding whether conservation efforts should be focused on areas containing the greatest number of species, areas containing the greatest number of threatened species, or areas containing the greatest number of endemic species. All three approaches are reasonable.

In addition to considering species diversity, some scientists have argued that we must also consider the size of the human population in diverse areas. For example, we might expect that natural areas containing more people face a greater probability of being affected by human

TABLE SA2.1 Biodiversity hotspots for birds, identified by three criteria

Rank	Total number of species	Number of endemic species	Number of threatened species
1	Andes	Andes	Andes
2	Amazon Basin	New Guinea and Bismarck Archipelago	Amazon Basin
3	Western Great Rift Valley	Panama and Costa Rica Highlands	Guyana Highlands
4	Eastern Great Rift Valley	Caribbean	Himalayas
5	Himalayas	Lesser Sunda Islands	Atlantic Coastal Forest, Brazil

(Data from: C.D. Orme et al., Global hotspots of species richness are not congruent with endemism or threat, Nature 436 (2005): 1016–1019)

activities. Whereas the world has an average human population density of 42 people per square kilometer, the average hotspot has a human population of 73 people per square kilometer. Such places may be at a higher risk of degradation from human activities. This risk should be considered when determining priority areas for conservation, and it should motivate us to promote development that does not come at the cost of species diversity.

What are the costs and benefits of conserving biodiversity hotspots?

Conservation efforts that focus on regions with large numbers of species place a clear priority on preserving the largest number of species possible. However, people making these efforts do not explicitly consider the likelihood of succeeding in this goal, nor do they necessarily consider the costs associated with the effort. For example, there may be many ways of helping a species persist in an area, including buying habitat, entering into agreements with landowners not to develop their land, or removing threats such as invasive species. In the case of the California tiger salamander, for example, while eliminating invasive predators may not be feasible, it is possible to protect salamander habitat. In 2011, the U.S. Fish and Wildlife Service agreed to designate more than 20,000 ha (50,000 acres) of habitat as being critical for the salamander's persistence and in 2016 the agency announced their plan to help populations of the salamander to recover by working with public and private landowners to protect critical salamander habitat.

What about biodiversity coldspots?

The concept of biodiversity hotspots assumes that our primary goal is to protect the maximum number of species. That goal is admirable, but it could come at the cost of many important ecosystems that do not fall within hotspots. Yellowstone National Park, for example, has a relatively low diversity of species, yet it is one of the few places in the United States that contains remnant populations of large mammals, including wolves, grizzly bears, and bison. Does this mean that places such as Yellowstone National Park should receive decreased conservation attention?

Biodiversity coldspots also provide ecosystem services that humans value at least as much as species diversity. For example, wetlands in the United States are incredibly important for flood control, water purification, wildlife habitat, and recreation. Many wetlands, however, have relatively low plant diversity and, as a result, would not be identified as biodiversity hotspots. It is true that increased species richness leads to improved ecosystem services, but only as we move from very low species richness to moderate species richness. Moving from moderate species richness to high species richness generally does not further improve the functioning of an ecosystem. Since very high species diversity is not expected to provide any substantial improvement in ecosystem function, protecting more and more species produces diminishing returns in terms of protecting ecosystem services. Hence, if our primary goal is to preserve the functioning of the ecosystems that improve our lives, we do not necessarily need to preserve every species in those ecosystems.

How can we reach a resolution?

During the past 2 decades, it has become clear that scientists and policy makers need to set priorities for the conservation of biodiversity. No single criterion may be agreed upon by everyone. However, it is important to appreciate the bias of each approach and to consider the possible unintended consequences of favoring some geographic regions over others. Our investment in conservation cannot be viewed as an all-or-nothing choice of some areas over others. Instead, our decisions must take into account the costs and benefits of alternative conservation strategies and incorporate current and future threats to both species diversity and ecosystem function. In this way, we can strike a balance between our desire to preserve Earth's species diversity and our desire to protect the functioning of Earth's ecosystems.

Questions

1. If you were in the role of a decision maker, how might you balance your desire to protect biodiversity around the world with the high cost of preserving biodiversity in any given location?
2. When identifying hotspots around the world, why is it important to simultaneously consider the biodiversity of different areas and the future human population size in each of these locations?
3. If you had to make a choice between protecting biodiversity hotspots and biodiversity coldspots, how would you make this decision?

Preparing for the AP® Exam

Practice AP® Free-Response Question

Write your answer to each part clearly. Support your answers with relevant information and examples. Where calculations are required, show your work.

Florida Reef is the only living coral barrier reef in the continental United States. Like other coral reefs on the planet, Florida Reef is home to many species of coral, fish, and other aquatic life forms. Also like other coral reefs, its health is at risk due to such factors as coral bleaching. However, Conservation International does not list this as a biodiversity hotspot.

(a) List two reasons why Florida Reef might not be internationally recognized as a hotspot. (2 points)
(b) List and describe two ecosystem services provided by Florida Reef. (2 points)
(c) What is coral bleaching? (2 points)
(d) Researchers recently found that the disturbance generated by hurricanes is important for maintaining the diversity of coral reef ecosystems. However, massive hurricanes often put many coral reef species at risk of local extinction. Generate a hypothesis to explain this pattern. (4 points)

References

Bacchetta, G., et al. 2012. A new method to set conservation priorities in biodiversity hotspots. *Plant Biosystems* 146:638–648.

Joppa, L. N., et al. 2011. Biodiversity hotspots house most undiscovered plant species. *PNAS* 108:13171–13176.

Kareiva, P., and M. Marvier. 2003. Conserving biodiversity coldspots. *American Scientist* 91:344–351.

Marchese, C. 2015. Biodiversity hotspots: A shortcut for a more complicated concept. *Global Ecology and Conservation* 3:297–309.

Myers, N., et al. 2000. Biodiversity hotspots for conservation priorities. *Nature* 403:853–858.

Orme, C. D., et al. 2005. Global hotspots of species richness are not congruent with endemism or threat. *Nature* 436:1016–1019.

Key Terms

Endemic species
Biodiversity hotspots

Unit 2 AP® Environmental Science Practice Exam

Preparing for the AP® Exam

Section 1: Multiple-Choice Questions

Choose the best answer for questions 1–25.

1. A forest is clear-cut and turned into an agricultural field to grow corn. Which of the following will likely occur in that ecosystem?
   (a) Erosion will decrease.
   (b) Albedo will increase.
   (c) Evapotranspiration will decrease.
   (d) NPP will increase.

Question 2 refers to the following diagram:

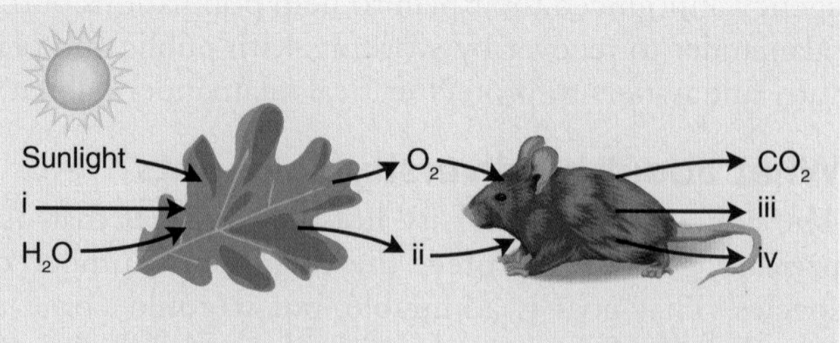

2. By ingesting leaf tissue, herbivores utilize the products of photosynthesis for cellular respiration. The diagram above shows the flow of inputs and outputs for these two processes. Labels i, ii, iii, and iv refer to
   (a) CO_2, $C_6H_{12}O_6$, energy, and water.
   (b) CO_2, CO_2, energy, and excreted waste.
   (c) CO_2, $C_6H_6O_6$, energy, and O_2.
   (d) heat, organic tissue, energy, and O_2.

3. To estimate gross primary production, researchers can use all of the following variable combinations EXCEPT
   (a) net primary production and plant respiration.
   (b) CO_2 taken up by plant in sunlight and CO_2 produced by plants in the dark.
   (c) CO_2 released by plants in sunlight and CO_2 produced by plants in the dark.
   (d) the amount of photosynthesis occurring over time.

4. While collecting data from a lake, a researcher finds that the lake is hypoxic in places and has a shallow limnetic zone. Based on this information, the lake is likely
   (a) oligotrophic.
   (b) autotrophic.
   (c) mesotrophic.
   (d) eutrophic.

5. Which process is NOT part of the carbon cycle?
   (a) growth of algae in the open ocean
   (b) combustion of fossil fuels
   (c) weathering of mineral rock
   (d) decomposition of organic material

6. Production of synthetic fertilizers is an example of ______ and one process that enables plants to use fertilizer is _______.
   (a) nitrogen fixation; nitrification
   (b) nitrification; nitrogen fixation
   (c) nitrification; assimilation
   (d) denitrification; nitrification

7. Which describes a difference between the nitrogen and phosphorus cycles?
   I. The nitrogen cycle has a gas phase, while the phosphorus cycle does not.
   II. The phosphorus cycle involves only abiotic processes, while the nitrogen cycle involves both biotic and abiotic processes.
   III. Phosphorus rarely changes form in its cycle, while the nitrogen cycle involves many chemical transformations.
   (a) II only
   (b) III only
   (c) I and II
   (d) I and III

8. Intertidal communities are frequently disturbed by storms that generate large waves. Following a period of intermediate wave disturbance, a group of researchers examined the species richness of north and south Pacific intertidal communities. Four days after the disturbance, they found that northern communities had returned to their original state whereas southern communities were still recovering. This result suggests that northern intertidal communities
   (a) have greater resilience.
   (b) have greater resistance.
   (c) follow predictions of the intermediate disturbance hypothesis.
   (d) experience more frequent disturbance.

9. Which is NOT a property that determines how air circulates in the atmosphere?
   (a) water vapor capacity
   (b) adiabatic heating/cooling
   (c) density
   (d) latent heat release

10. If Earth's axis of rotation were tilted at 45° instead of 23.5° how many hours of daylight would the Northern Hemisphere receive during the March and September equinoxes?
    (a) 6 hours
    (b) 12 hours
    (c) 18 hours
    (d) 24 hours

11. Which order of processes results in the occurrence of hot and dry deserts at 30° S and 30° N latitudes?
    (a) adiabatic cooling, condensation, latent heat release, adiabatic heating
    (b) adiabatic heating, condensation, latent heat release, adiabatic cooling
    (c) latent heat release, adiabatic cooling, air displacement, condensation
    (d) condensation, adiabatic heating, adiabatic cooling, latent heat release

12. Lush vegetation on the western slopes of the Colorado Rockies may be attributed to
    I. Westerlies generated by the Coriolis effect.
    II. a rain shadow.
    III. adiabatic cooling.
    (a) I only
    (b) III only
    (c) I and III
    (d) I, II, and III

13. The mixing of surface water and deep water in the oceans that occurs because of differences in salinity is known as
    (a) upwelling.
    (b) a gyre.
    (c) the El Niño–Southern Oscillation.
    (d) thermohaline circulation.

14. Which does NOT occur during an El Niño–Southern Oscillation year?
    (a) major changes in weather conditions around the world
    (b) a reversal of trade wind direction
    (c) a buildup of warm water along the western coast of South America
    (d) increased upwelling of water along the western coast of South America

Questions 15 and 16 refer to the following graph:

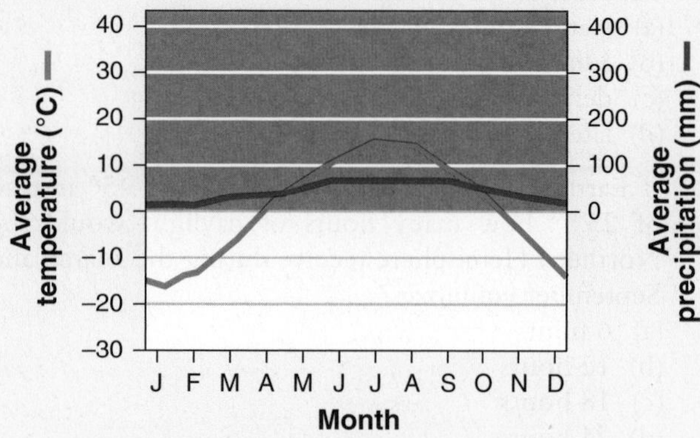

15. For the region represented by the graph, when is the growing season?
    (a) January to December
    (b) January to July
    (c) April to October
    (d) July to December

16. What biome is this graph most likely to represent?
    (a) boreal forest
    (b) temperate seasonal forest
    (c) tropical seasonal forest
    (d) cold desert

17. Which list contains terms for lake classification, from systems with the lowest primary productivity to systems with the highest primary productivity?
    (a) eutrophic, mesotrophic, oligotrophic
    (b) mesotrophic, eutrophic, oligotrophic
    (c) mesotrophic, oligotrophic, eutrophic
    (d) oligotrophic, mesotrophic, eutrophic

18. Iron is a limiting nutrient for algae in the open ocean. After researchers released 5,000 kg of iron into the ocean, they notice an algal bloom in the _______ zone.
    (a) photic
    (b) profundal
    (c) aphotic
    (d) benthic

Question 19 refers to the following table comparing forest communities in different regions of North America.

Total percent state-wide cover			
Species	**Connecticut (CT)**	**Pennsylvania (PA)**	**Georgia (GA)**
Red maple	25	40	20
Black oak	30	20	50
White pine	15	10	10
Eastern hemlock	20	10	0
Black cherry	20	20	30

19. Researchers collected data on tree abundance for a group of forest communities in the three states shown in the table above. Which statement best describes the forest communities within each state?
    (a) PA has the highest species richness; CT has the highest species evenness.
    (b) CT and PA have equal species richness; GA has the highest species evenness.
    (c) GA and PA have equal species richness; GA has the highest species evenness.
    (d) CT and PA have equal species richness; CT has the highest species evenness.

20. Humans have developed varieties of grape vines that produce grapes that make fine wines if grown under particular environmental conditions. Therefore, the flavor of grapes on a given vine depends on
    I. genes.
    II. environments.
    III. artificial selection.
    (a) I only
    (b) III only
    (c) I and II
    (d) I, II, and III

21. Which could cause a dramatic decline in genetic diversity?
    I. mutation
    II. a bottleneck effect
    III. a founder effect
    (a) II only
    (b) III only
    (c) I and II
    (d) II and III

22. Which is an example of sympatric speciation?
    (a) A stream diversion divides a single turtle population into two isolated populations that evolve into two species.
    (b) Forest fragmentation isolates two species of mice.
    (c) Two bird species from separate forests evolve reproductive isolation.
    (d) Two fish species in a single lake evolve from a single ancestor that once lived in the lake.

23. Which is the best definition of a realized niche?
    (a) the range of abiotic and biotic conditions under which a species can potentially live
    (b) the total amount of area in which a species can live
    (c) the range of abiotic and biotic conditions under which a species actually lives
    (d) the total amount of area in which a species actually lives

24. Phosphorus is often limiting in aquatic environments because
    (a) phosphorus is in higher demand by aquatic organisms relative to all other minerals.
    (b) phosphorus is not easily leached from soils and rapidly precipitates in water.
    (c) weathering of rock is an extremely slow process.
    (d) human activity regularly absorbs phosphorus from aquatic environments.

25. Which explains why the distribution of biomass in ecosystems is often structured like a pyramid?
    I. the second law of thermodynamics
    II. low ecological efficiency in natural ecosystems
    III. photosynthesis is more efficient than cellular respiration
    (a) I only
    (b) II only
    (c) I and II
    (d) I and III

Section 2: Free-Response Questions

Write your answer to each part clearly. Support your answers with relevant information and examples. Where calculations are required, show your work.

1. A researcher notices that over several decades, a species of fish appears to be getting smaller and smaller in size, as humans tend to preferentially catch larger fish.
    (a) Explain whether or not this is evolution by natural selection, and why. In your response define the conditions that would need to be met for natural selection to occur. (4 points)
    (b) Define the term "fitness." Describe how a fish's body size could influence fitness. (2 points)
    (c) The researcher finds that the fish species consumes phytoplankton. Name the trophic level of this fish and explain your answer. (2 points)
    (d) The fish is found in an area near an active volcano and a city that burns a large amount of fossil fuels. Describe how each might these affect the pH of the fish's habitat. (2 points)

2. During El Niño–Southern Oscillation (ENSO) years, the buildup of warm water off the coast of South America alters the circulation of the Hadley cell. One consequence of this altered circulation is a dip in the subtropical jet stream that carries storms from west to east across North America. Subsequently, southern California experiences greater precipitation during El Niño years.
    (a) Draw a diagram that shows both Hadley and Ferrell cells, noting latitudinal locations, air direction and areas of condensation, adiabatic heating, and adiabatic cooling. (4 points)
    (b) Why would the western slopes of the southern Colorado Rockies experience greater precipitation than the eastern slopes? (2 points)
    (c) A researcher wants to test the hypothesis that gross primary productivity (GPP) on the western slopes of the Rockies is greater during ENSO years.
        (i) Write an equation that defines gross primary productivity. (1 point)
        (ii) What characteristics of local plants should the researcher measure to test the hypothesis? Provide justification for your responses. (2 points)
        (iii) Provide a null hypothesis for this study. (1 point)

3. In temperate forests, average daily temperatures during winter often drop to 0°C or lower. Although bacteria and fungi can with stand very cold temperatures, they also rely on the insulating capacity of snow, which prevents soil from experiencing temperature extremes. However, during winters when snowfall is less abundant, soil is exposed to freezing temperatures that kill many of the bacteria and fungi residing in the very top layers of soil. When bacteria and fungi die, nutrients and other organic material are easily leached out of cell membranes and these materials enter the soil.
    (a) What are two consequences of reduced winter snowfall on the nitrogen cycle in temperate forests? (4 points)
    (b) What is one consequence of reduced winter snowfall on the carbon cycle in temperate forests? (2 points)
    (c) Describe two ways in which stream and pond food webs might be affected by reduced snowfall during the winter. (2 points)
    (d) Exposure of the soil to extreme temperatures can also kill larger organisms, such as frogs and burrowing insects. For example, wood frogs (*Lithobates sylvaticus*) that burrow too close to the soil surface are likely to die whereas frogs that burrow deeper in the soil are likely to survive. Describe one way in which reduced winter snowfall could alter the local population genetics of wood frogs. (2 points)

A former New England farm is now a forest. *(Erin Paul Donovan/Alamy)*

Population and Community Ecology

CHAPTER 6

MODULE 18 The Abundance and Distribution of Populations

MODULE 19 Population Growth Models

MODULE 20 Community Ecology

MODULE 21 Community Succession

CASE STUDY

New England Forests Come Full Circle

When the Pilgrims arrived in Massachusetts in 1620, they found immense areas of undisturbed temperate seasonal forest containing a variety of tree species, including sugar maple (*Acer saccharum*), American beech (*Fagus grandifolia*), white pine (*Pinus strobus*), and eastern hemlock (*Tsuga canadensis*). Over the next 200 years, settlers cut down most of the trees to clear land for farming and housing. This deforestation peaked in the 1800s, at which point up to 80 percent of all New England forests had been cleared. Between 1850 and 1950, however, many people abandoned their New England farms to take jobs in the growing textile industry. Others moved to the Midwest, where farmland was more fertile and considerably less expensive. The old stone walls that are so common in the New England countryside are the only evidence that this forest was once farmland.

What happened to the former farmland is a testament to the resilience of the forest ecosystem. The transformation began shortly after the farmers left. Seeds of grasses and wildflowers were carried to the abandoned fields by birds or blown there by the wind. Within a year, the fields were carpeted with a large variety of plant species. Eventually, a single group of plants—the goldenrods—came to dominate the fields by growing taller and outcompeting other species of plants for sunlight. The other species remained in the fields, but they were not very abundant. Nevertheless, the dominance of the goldenrods was short-lived.

Goldenrods and other wildflowers play an important part in old-field communities because they support a diverse group of plant-eating insects. Some of these herbivorous insects are generalists that feed on a wide range of plant species, while others specialize on only a small number of plant species. The number of individuals of each insect species varies from year to year, and occasionally some species experience very large population increases, or outbreaks. One such species is a leaf beetle (*Microrhopala vittata*) that specializes in eating goldenrods. Periodic outbreaks of this species in the abandoned fields of New England dramatically reduced goldenrod populations. With fewer goldenrods, other plant species could compete and prosper.

The old stone walls that are so common in the New England countryside are the only evidence that this forest was once farmland.

The complex interactions among populations of goldenrods, insects, and other species created an ever-changing ecosystem. For example, as the leaf beetle population increased in the community, so did the populations of predators and parasites that fed on them. As these predators and parasites reduced the population of leaf beetles, the goldenrod population began to rebound. As the goldenrods surged, they once again caused other plant species to decline in numbers.

Over time, tree seeds arrived and tree seedlings began to grow, which changed the species composition of the old fields once again. One species in particular, the fast-growing white pine, eventually came to dominate. The pine trees cast so much shade that the goldenrods and other sunlight-loving plant species could not survive.

White pines dominated the old-field communities until humans began harvesting them for lumber in the 1900s. Just as the reduction of goldenrod populations made room for other plant species, logging of the white pines made room for broadleaf tree species. Two of these broadleaf species, American beech and sugar maple, are dominant in New England forests today. The New England fields that had been abandoned earlier were slowly transformed into communities that resemble the original forests of centuries ago, with a mix of pines, hemlocks, and broadleaf trees.

The story of the New England forests shows us that populations can increase or decrease dramatically over time. It also illustrates how species interactions within a community can alter species abundance. Finally, it demonstrates how human activity can alter the distribution and diversity of species within an ecosystem.

Sources: M. J. Duveneck, et al., Recovery dynamics and climate change effects to future New England forests, *Landscape Ecology* (2017) 32:1385–1397; J.R. Thompson, et al., Four centuries of change in northeastern United States forests, *PLoS ONE* (2013) 8: e72540; W. P. Carson and R. B. Root, Herbivory and plant species coexistence: Community regulation by an outbreaking phytophagous insect, *Ecological Monographs* 70 (2000): 73–99; T. Wessels, *Reading the Forested Landscape* (Countryman Press, 1997).

A New England forest is a wonderful reminder of the intricate complexity of the natural world. As we saw in this account, there are clear patterns in the distribution and abundance of species over space and time. Understanding the factors that generate these patterns can help us find ways to preserve global biodiversity. These factors include the ways in which populations increase and decrease in size and how species interact in ecological communities. In this chapter, we will examine the factors that help determine the abundance and distribution of populations. We will then look at interactions among species that live within ecological communities and how these interactions further determine whether a species can persist in a particular location on Earth. Finally, we will examine how ecological communities change over time.

MODULE 18

The Abundance and Distribution of Populations

Complexity in the natural world ranges from a single individual to all of the biotic and abiotic components of Earth. In this module, we will begin by broadly examining the different levels of complexity in nature and then narrow our focus to the population level. We will consider the characteristics of populations and then discuss factors that determine how populations increase and decrease over time.

Learning Goals

After reading this module you should be able to

- explain how nature exists at several levels of complexity.
- discuss the characteristics of populations.
- contrast the effects of density-dependent and density-independent factors on population growth.

Nature exists at several levels of complexity

As **FIGURE 18.1** on page 198 shows, the environment around us exists at a series of increasingly complex levels: individuals, populations, communities, ecosystems, and the biosphere. The simplest level is the individual—a single organism. As we saw in Chapter 5, natural selection operates at the level of the individual because it is the individual that must survive and reproduce.

The second level of complexity is a *population*. A **population** is composed of all individuals that belong to the same species and live in a given area at a particular time. Evolution occurs at the level of the population. Scientists who study populations are also interested in the factors that cause the number of individuals to increase or decrease. As we saw in Chapter 1, the boundaries of a population are rarely clear and may be set arbitrarily by scientists. For example, depending on what we want to learn, we might study the entire population of white-tailed deer in North America, or we might focus on the deer that live within a single state, or even within a single forest.

The third level of complexity is the *community*. A **community** incorporates all of the populations of organisms within a given area. Like those of a population, the boundaries of a community may be defined by the state or federal agency responsible for managing it. Scientists who study communities are generally interested in how species interact with one another. In Chapter 4, we saw that terrestrial communities can be grouped into biomes that contain plants with similar growth forms. However, the actual composition of tree species varies from community to community. While the temperate seasonal forests of the eastern United States and Europe experience similar patterns of temperature and precipitation, they contain different tree species.

Communities exist within an ecosystem, which consists of all of the biotic and abiotic components in a particular location. Ecosystem ecologists study flows of energy and matter, such as the cycling of nutrients through the system.

Population The individuals that belong to the same species and live in a given area at a particular time.

Community All of the populations of organisms within a given area.

FIGURE 18.1 **Levels of complexity.** Environmental scientists study nature at several different levels of complexity, ranging from the individual organism to the biosphere. At each level, scientists focus on different processes.

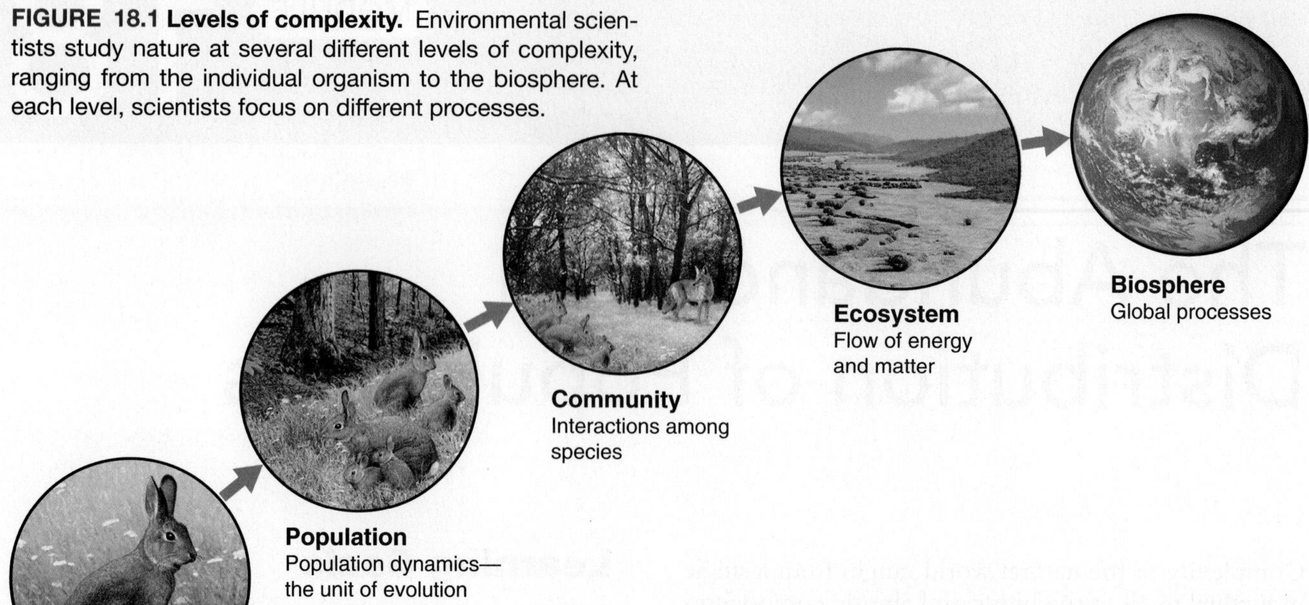

The largest and most complex system environmental scientists study is the biosphere, which incorporates all of Earth's ecosystems. Scientists who study the biosphere are interested in the movement of air, water, and heat around the globe.

Populations have distinctive characteristics

When we study nature at the level of the population, one of the first things that becomes apparent is that populations are dynamic—that is, they are constantly changing. Individuals in the population die, new individuals are produced, and individuals can move from one population to another. The study of factors that cause populations to increase or decrease is the science of **population ecology**.

There are many circumstances in which scientists find it useful to identify the factors that influence population size over time. For example, in the case of endangered species such as the California condor (*Gymnogyps californianus*), knowing the factors that affect its population size has helped us to implement measures that improve its survival and reproduction. Similarly, knowing the factors that influence the population size of a pest species can help us control it. For instance, population ecologists are currently studying the emerald ash borer (*Agrilus planipennis*), an insect from Asia that was accidentally introduced to the American Midwest and is causing the widespread death of ash trees. Once we understand the population ecology of this destructive insect, we can begin to explore and develop strategies to control or eradicate it.

To understand how populations change over time, we must first examine the basic characteristics of populations. These characteristics are *size, density, distribution, sex ratio,* and *age structure*.

Population Size

Population size (*N*) is the total number of individuals within a defined area at a given time. For example, the California condor once ranged throughout California and the southwestern United States (**FIGURE 18.2**). Over the past 2 centuries, however, a combination of poaching, poisoning, and accidents (such as flying into electric power lines) greatly reduced the population's size. By 1987, only 22 birds remained in the wild. Scientists realized that the species was nearing extinction and decided to capture all the wild birds and start a captive breeding program in zoos. As a result of these captive breeding programs and other conservation efforts, the condor population size has increased to more than 440 birds by 2016. This includes birds living in captivity and wild birds that have been reintroduced to California, Arizona, and New Mexico. For much larger populations, we can express the size of the

Population ecology The study of factors that cause populations to increase or decrease.

Population size (*N*) The total number of individuals within a defined area at a given time.

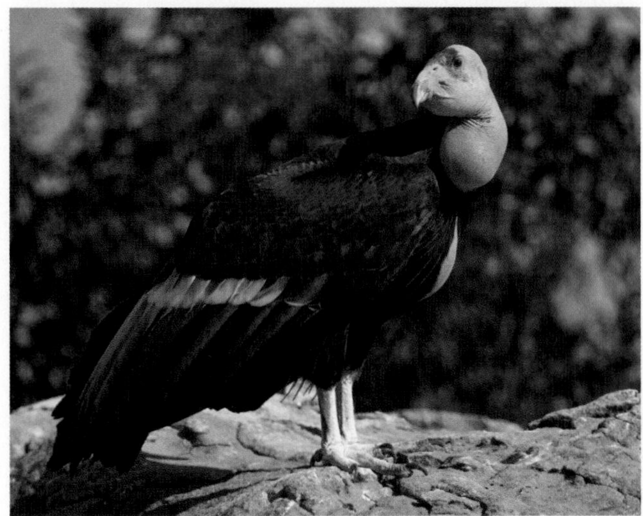

FIGURE 18.2 California Condor. The condor population had declined to only 22 birds left on Earth in 1987. Thanks to conservation efforts, more than 440 birds exist today. *(Steve Johnson/Getty Images)*

population using scientific notations, as described in "Do the Math: Simplifying Numbers with Scientific Notation" on page 200.

Population Density

Population density is the number of individuals per unit area (or volume, in the case of aquatic organisms) at a given time. Knowing a population's density, in addition to its size, can help scientists estimate whether a species is rare or abundant. For example, the density of coyotes (*Canis latrans*) in some parts of Texas might be only 1 per square kilometer, but in other parts of the state it might be as high as 12 per square kilometer. Scientists also study population density to determine whether a population in a particular location is so dense that it might outstrip its food supply.

Population density can be a particularly useful measure for wildlife managers who must set hunting or fishing limits on a species. For example, managers may divide into management zones the entire population of an animal species that is hunted or fished. Management zones may be designated political areas, such as counties, or areas with natural boundaries, such as the major water bodies in a state. Wildlife managers might offer more hunting or fishing permits for zones with high-density populations and fewer permits for zones with low-density populations.

Population Distribution

In addition to population size and density, population ecologists are interested in how a population occupies space. **Population distribution** is a description of how individuals are distributed with respect to one another. **FIGURE 18.3** on page 201 shows three types of population distributions. In some populations, such as a population of trees in a natural forest, the distribution of individuals is random (Figure 18.3a). In other words, there is no pattern to the locations where the individual trees grow.

In other populations, such as a population of trees in a plantation, the distribution of individuals is uniform, or evenly spaced (Figure 18.3b). Uniform distributions are common among territorial animals, such as nesting birds that defend areas of similar sizes around their nests. Uniform distributions are also observed among plants that produce toxic chemicals to prevent other plants of the same species from growing close to them.

In still other populations, the distribution of individuals is *clumped* (Figure 18.3c). Clumped distributions, which are common among schooling fish, flocking birds, and herding mammals, are often observed when living in large groups provides enhanced feeding opportunities or protection from predators.

Population Sex Ratio

The **sex ratio** of a population is the ratio of males to females. In most sexually reproducing species, the sex ratio is usually close to 50:50, although sex ratios can be far from equal in some species. In fig wasps, for example, there may be as many as 20 females for every male. Because the number of offspring produced is primarily a function of how many females there are in the population, knowing a population's sex ratio helps scientists estimate the number of offspring a population will produce in the next generation.

Population Age Structure

Many populations are composed of individuals of varying ages. A population's **age structure** describes how many individuals fit into particular age categories.

Population density The number of individuals per unit area at a given time.

Population distribution A description of how individuals are distributed with respect to one another.

Sex ratio The ratio of males to females in a population.

Age structure A description of how many individuals fit into particular age categories in a population.

DO THE MATH Simplifying Numbers with Scientific Notation

Preparing for the AP® Exam

When we are dealing with very large numbers in science, such as the abundance of deer in a state, it is often easier to express the number using scientific notation. For example, if the number of deer is 2,350,000, we can rewrite this number as 2.35×10^6. The first part of this notation (2.35) is called the digit term while the second part of the notation (10^6) is called the exponential term. When the exponential term is positive, it tells how many places we move the decimal place to the right to return to the original number. In our example, the exponent is a 6, so we can take the 2.35 and move the decimal 6 places to the right to end up at 2,350,000.

Scientific notation also allows us to deal with very small numbers, such as the abundance of a rare species. For example, imagine there were only 25 individuals remaining of an endangered species spanning 10,000 hectares of land. That would translate into a density of 0.0025 animals per hectare. In this case, we can use scientific notation to rewrite the number as 2.5×10^{-4}. When the exponential term is negative, it tells how many places we move the decimal place to the left to return to the original number. In our example, the exponent is a -4, so we can take the 2.5 and move the decimal 4 places to the left to end up at 0.0025.

Here are some additional examples of numbers written in scientific notation:

$$10{,}000 = 1 \times 10^4$$
$$100 = 1 \times 10^2$$
$$1 = 1 \times 10^0$$
$$34{,}267 = 3.4267 \times 10^4$$
$$488 = 4.88 \times 10^2$$
$$0.053 = 5.3 \times 10^{-2}$$

ADDING AND SUBTRACTING EXPONENTS

If we need to add or subtract two numbers that are written using scientific notation, we need to first make sure that both numbers use the same exponent. When they use the same exponent, we can add or subtract the digit terms and simply retain the exponential term. For example:

$$1.23 \times 10^5 + 15.1 \times 10^4 =$$
$$1.23 \times 10^5 + 1.51 \times 10^5 =$$
$$(1.23 + 1.51) \times 10^5 =$$
$$2.74 \times 10^5$$

MULTIPLYING EXPONENTS

If we need to multiply two numbers that are written using scientific notation, we multiply the two digit terms and then add the two exponential terms. For example:

$$(2.4 \times 10^2) \times (3.0 \times 10^4) =$$
$$(2.4 \times 3.0) \times (10^{2+4}) =$$
$$7.2 \times 10^6$$

DIVIDING EXPONENTS

If we need to divide two numbers that are written using scientific notation, we divide the two digit terms and then subtract the two exponential terms. For example:

$$(3.6 \times 10^4) \div (1.2 \times 10^2) =$$
$$(3.6 \div 1.2) \times (10^{4-2}) =$$
$$3.0 \times 10^2$$

YOUR TURN

1. Convert the numbers 1,200,000 and 600,000 into scientific notation.
2. Add the two numbers in scientific notation.
3. Subtract the two numbers in scientific notation.
4. Multiply the two numbers in scientific notation.
5. Divide the two numbers in scientific notation.

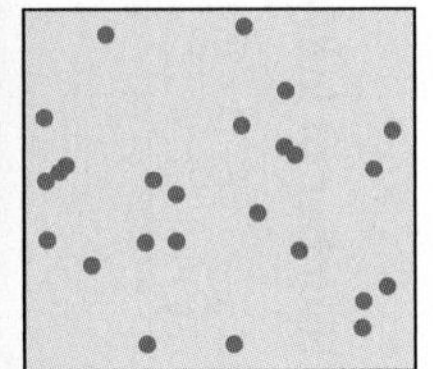

(a) Random distribution

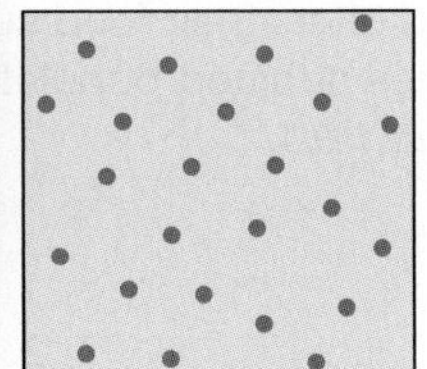

(b) Uniform distribution

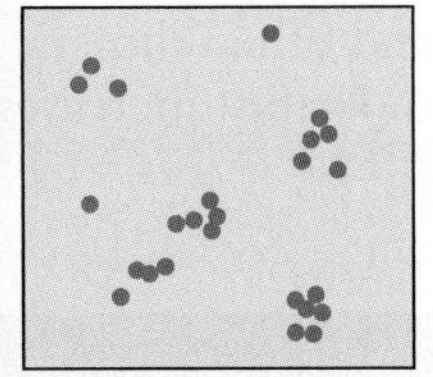

(c) Clumped distribution

FIGURE 18.3 Population distributions. Populations in nature distribute themselves in three ways. (a) Many of the tree species in this New England forest are randomly distributed, with no apparent pattern in the locations of individuals. (b) Territorial nesting birds, such as these Australasian gannets (*Morus serrator*), exhibit a uniform distribution, in which all individuals maintain a similar distance from one another. (c) Many pairs of eyes are better than one at detecting approaching predators. The clumped distribution of these meerkats (*Suricata suricatta*) provides them with extra protection. *(a: David R. Frazier Photolibrary, Inc./Science Source; b: Michael Thompson/Earth Scenes/Animals Animals; c: Clem Haagner/ARDEA)*

Knowing a population's age structure helps ecologists predict how rapidly a population can grow. For instance, a population with a large proportion of old individuals, or with a large proportion of individuals too young to reproduce, will produce far fewer offspring than a population that has a large proportion of individuals of reproductive age.

Population size is affected by density-dependent and density-independent factors

Factors that influence population size can be classified as *density dependent* or *density independent*. We will look at each type in turn.

Density-dependent Factors

In 1932, Russian biologist Georgii Gause conducted a set of experiments that demonstrated that food supply can control population growth. Gause monitored population growth in two species of *Paramecium* (a type of single-celled aquatic organism) living under ideal conditions in test tubes. Each day he added a constant amount of food. As the graph in **FIGURE 18.4a** on page 202 shows, both species of *Paramecium* initially experienced rapid population growth, but over time the rate of growth began to slow. Eventually, the two population sizes reached a plateau and remained there for the rest of the experiment.

Gause suspected that *Paramecium* population growth was limited by food supply. To test this hypothesis, he conducted a second experiment in which he doubled the amount of food he added to the test tubes. Again, both species of *Paramecium* experienced rapid population growth early in the experiment, and the rate of growth, again, slowed over time. However, as Figure 18.4b shows, their maximum population sizes were approximately double those observed in the first experiment.

Gause's results confirmed that food is a *limiting resource* for *Paramecium*. A **limiting resource** is a resource that a population cannot live without and that occurs in quantities lower than the population would require to increase in size. If a limiting resource decreases, so does the size of a population that depends on it. For terrestrial plant populations, water and nutrients such as nitrogen and phosphorus are common limiting resources. For animal populations, food, water, and nest sites are common limiting resources.

Factors such as food that influence an individual's probability of survival and its amount of reproduction in a manner that depends on the size of the population are called **density-dependent factors**. For example,

Limiting resource A resource that a population cannot live without and that occurs in quantities lower than the population would require to increase in size.

Density-dependent factor A factor that influences an individual's probability of survival and reproduction in a manner that depends on the size of the population.

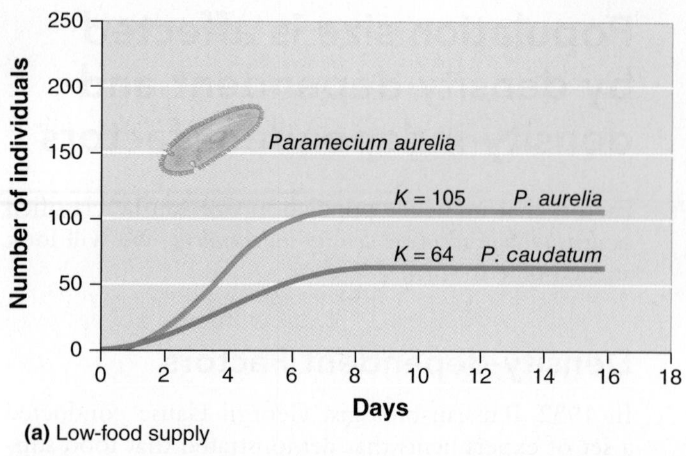

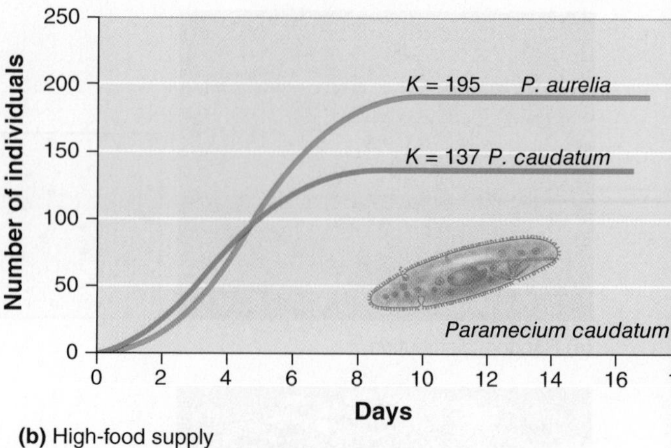

FIGURE 18.4 Gause's experiments. (a) Under low-food conditions, the population sizes of two species of *Paramecium* initially increased rapidly, but then leveled off as their food supply became limiting. (b) When twice as much food was provided, both species attained population sizes that were nearly twice as large, but they again leveled off. *(Data from Gause, 1932)*

because a smaller population requires less total food, food scarcity will have little or no effect on the survival and reproduction of individuals in a small population but it will have large negative effects on a large population.

In Gause's *Paramecium* experiments, population growth slowed as population size increased because there was a limit to how many individuals the food supply could sustain. This limit to the number of individuals that can exist in a population is called the **carrying capacity** of the environment and is denoted as ***K***. Knowing the carrying capacity for a species and its limiting resource helps us predict how many individuals an environment can sustain.

Density-independent Factors

We have seen that density-dependent factors affect an individual's probability of survival and amount of reproduction differently depending on whether the population is large or small. In contrast, **density-independent factors** have the same effect on an individual's probability of survival and amount of reproduction regardless of the population size. A tornado, for example, can uproot and kill a large number of trees in an area (**FIGURE 18.5**). However, a given tree's probability of being killed does not depend on whether it resides in a forest with a high or low density of other trees. Other density-independent factors include hurricanes, floods, fires, volcanic eruptions, and other climatic events. An individual's likelihood of mortality increases during such an event regardless of

Carrying capacity (*K*) The limit of how many individuals in a population the environment can sustain.

Density-independent factor A factor that has the same effect on an individual's probability of survival and the amount of reproduction at any population size.

FIGURE 18.5 Path of Destruction. This forest in Pennsylvania was damaged by a tornado. Natural disasters such as tornadoes can cause large declines in population sizes and therefore regulate plant and animal populations. Because the magnitude of the decline is not related to the density of the population prior to the disaster, these natural disasters represent density-independent factors in population regulation. *(Michael P. Gadomski/Science Source)*

whether the population happens to be at a low or high density.

Bird populations are often regulated by density-independent factors. For example, in the United Kingdom, a particularly cold winter can freeze the surfaces of ponds, making amphibians and fish inaccessible to wading birds such as herons. With their food supply no longer available, herons have an increased risk of starving to death, regardless of whether the heron population is at a low or a high density.

MODULE 18 AP® Review

Preparing for the AP® Exam

In this module, we learned that nature exists at a series of different levels of complexity, which include individuals, populations, communities, and ecosystems. We then examined the level of the population and observed that populations possess a number of characteristics that can be used to describe them, including their abundance and distribution. Finally, we discussed how density-dependent factors can regulate populations more strongly as populations grow whereas density-independent factors can regulate populations at any population size. In the next module, we will see how scientists use mathematical models of populations to obtain insights into how populations change in abundance over time.

AP® Practice Questions

Choose the best answer for the following.

1. Which is the correct order of ecological levels from less complex to more complex?
   (a) individual, population, ecosystem, biosphere, community
   (b) individual, community, ecosystem, population, biosphere
   (c) individual, population, community, ecosystem, biosphere
   (d) individual, population, community, biosphere, ecosystem

2. Population distribution is
   (a) often clumped in response to predation.
   (b) used by wildlife managers when regulating hunting and fishing.
   (c) uniform in most tree species.
   (d) important when estimating the number of offspring expected.

3. Which is true about a population's carrying capacity?
   (a) It is denoted as C.
   (b) It depends on a limiting resource.
   (c) It is controlled by density-independent factors.
   (d) The population of a species cannot exceed it.

4. Which is an example of a density-independent factor?
   (a) limited water for plants
   (b) limited prey for predators
   (c) a fixed amount of grass for cows
   (d) a hurricane

5. Which does NOT have a significant effect on the number of offspring produced by a population?
   (a) population distribution
   (b) age structure
   (c) population size
   (d) carrying capacity

MODULE 19

Population Growth Models

Learning Goals

After reading this module you should be able to

- explain the exponential growth model of populations, which produces a J-shaped curve.
- describe how the logistic growth model incorporates a carrying capacity and produces an S-shaped curve.
- compare the reproductive strategies and survivorship curves of different species.
- explain the dynamics that occur in metapopulations.

Scientists often use models to help them explain how things work and to predict how things might change in the future. Population ecologists use growth models that incorporate density-dependent and density-independent factors to explain and predict changes in population size. These models are important tools for population ecologists, whether they are protecting an endangered condor population, managing a commercially harvested fish species, or controlling an insect pest. In this module we will look at several growth models and other tools for understanding changes in population size.

The exponential growth model describes populations that continuously increase

Population growth models are mathematical equations that can be used to predict population size at any moment in time. In this section, we will examine one commonly used growth model.

As we saw in Gause's experiments, a population can initially grow very rapidly when its growth is not limited by scarce resources. We can define **population growth rate** as the number of offspring an individual can produce in a given time period, minus the deaths of the individual or its offspring during that same period. Under ideal conditions, with unlimited resources available, every population has a particular maximum potential for growth, which is called the **intrinsic growth rate** and denoted as r. When food is abundant, individuals have a tremendous ability to reproduce. For example, domesticated hogs (*Sus domestica*) can have litters of 10 piglets, and American bullfrogs (*Lithobates catesbeianus*) can lay up to 20,000 eggs. Under ideal conditions, the probability of an individual surviving also increases. Together, a high number of births and a low number of deaths produce a high population growth rate. When conditions are less than ideal due to limited resources, a population's growth rate will be lower than its intrinsic growth rate because individuals will produce fewer offspring (or forgo breeding entirely) and the number of deaths will increase.

If we know the intrinsic growth rate of a population (r) and the number of reproducing individuals that are currently in the population (N_0), we can estimate the population's future size (N_t) after some period of time (t) has passed. The formula that allows us to estimate future population is known as the **exponential growth model**

$$N_t = N_0 e^{rt}$$

Population growth models Mathematical equations that can be used to predict population size at any moment in time.

Population growth rate The number of offspring an individual can produce in a given time period, minus the deaths of the individual or its offspring during the same period.

Intrinsic growth rate (r) The maximum potential for growth of a population under ideal conditions with unlimited resources.

Exponential growth model ($N_t = N_0 e^{rt}$) A growth model that estimates a population's future size (N_t) after a period of time (t), based on the intrinsic growth rate (r) and the number of reproducing individuals currently in the population (N_0).

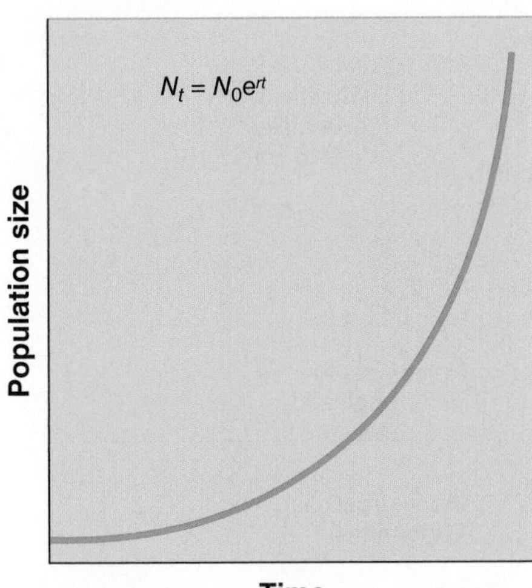

FIGURE 19.1 The exponential growth model. When populations are not limited by resources, their growth can be very rapid. More births occur with each step in time, creating a J-shaped growth curve.

where e is the base of the natural logarithms (the e^x key on your calculator, or 2.72) and t is time. This equation tells us that, under ideal conditions, the future size of the population (N_t) depends on the current size of the population (N_0), the intrinsic growth rate of the population (r), and the amount of time (t) over which the population grows.

When populations are not limited by resources, growth can be very rapid because more births occur with each step in time. When graphed, the exponential growth model produces a **J-shaped curve**, as shown in **FIGURE 19.1**. The J-shape of the curve represents the change in a growing population over time. At first the population is so small that it cannot increase rapidly because there are few individuals present to reproduce. As the population increases, there are more reproducing individuals, and so the growth rate increases.

One way to think about exponential growth in a population is to compare it to the growth of a bank account where N_0 is the account balance and r is the interest rate. Let's say you put $1,000 in a bank account at an annual interest rate of 5 percent. After 1 year the new balance in the account would be:

$$N_t = \$1{,}000 \times e^{(0.05 \times 1)}$$

$$N_t = \$1{,}051.27$$

In the second year, the balance in the account would be:

$$N_t = \$1{,}000 \times e^{(0.05 \times 2)}$$

$$N_t = \$1{,}105.17$$

In the tenth year, the account would grow to a balance of $1,648.72. Moving forward to the twentieth year, the same 5 percent interest rate would produce a balance of $2,718.28.

Applying an annual rate of growth to an increasing amount, whether money in a bank account or a population of organisms, produces rapid growth over time. Exponential growth is density independent because the value will grow by the same percentage every year. "Do the Math: Calculating Exponential Growth on page 206" gives a step-by-step example to show how this principle works in populations.

The exponential growth model is an excellent starting point for understanding population growth. Indeed, there is solid evidence that real populations—even small ones—can grow exponentially, at least initially. However, no population can experience exponential growth indefinitely. In Gause's experiments with *Paramecium,* the two populations initially grew exponentially until they approached the carrying capacity of their test-tube environment, at which point their growth slowed and eventually leveled off to reflect the amount of food that was added daily.

The logistic growth model describes populations that experience a carrying capacity

While the exponential growth model describes a continuously increasing population that grows at a fixed rate, populations do not experience exponential growth indefinitely. For this reason, ecologists have modified the exponential growth model to incorporate environmental limits on population growth, including limiting resources. The **logistic growth model** describes a population whose growth is initially exponential, but slows as the population approaches the carrying capacity of the environment (K). As we can see in **FIGURE 19.2** on page 206, if a population starts out small, its growth can be very rapid. As the population size nears about one-half of the carrying capacity, however, population growth begins to slow. As the population size approaches the carrying capacity, the population stops growing. When graphed, the logistic growth model produces an **S-shaped curve**.

J-shaped curve The curve of the exponential growth model when graphed.

Logistic growth model A growth model that describes a population whose growth is initially exponential, but slows as the population approaches the carrying capacity of the environment.

S-shaped curve The shape of the logistic growth model when graphed.

We observed this pattern in Gause's *Paramecium* experiments: At the carrying capacity, the populations stopped growing and remained at a constant size (see Figure 18.4).

The logistic growth model is used to predict the growth of populations that are subject to density-dependent constraints as the population grows, such as increased competition for food, water, or nest sites. Because density-independent factors such as hurricanes and floods are inherently unpredictable, the logistic growth model does not account for them.

One of the assumptions of the logistic growth model is that the number of offspring produced depends on the current population size and the carrying capacity of the environment. However, many species of mammals mate during the fall or winter, and the number of offspring that develop depends on the food supply at the time of mating. Because these offspring are not actually born until the following spring, there is a risk that food availability will not match the new population size. If there is less food available in the spring than needed

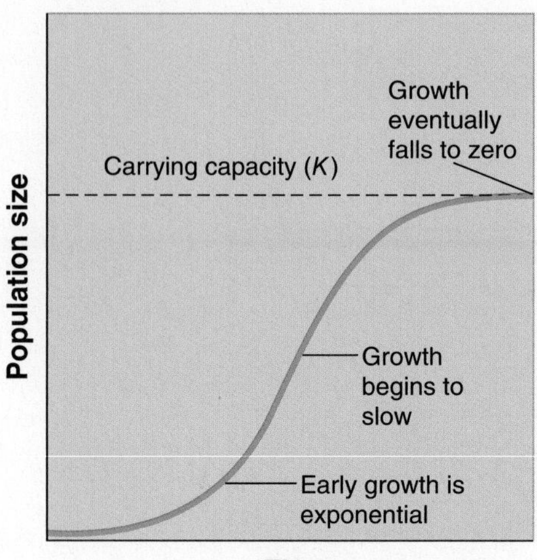

FIGURE 19.2 The logistic growth model. A small population initially experiences exponential growth. As the population becomes larger, however, resources become scarcer, and the growth rate slows. When the population size reaches the carrying capacity of the environment, growth stops. As a result, the pattern of population growth follows an S-shaped curve.

DO THE MATH Calculating Exponential Growth

Preparing for the AP® Exam

Consider a population of rabbits that has an initial population size of 10 individuals ($N_0 = 10$). Let's assume that the intrinsic rate of growth for a rabbit is $r = 0.5$ (or 50 percent), which means that each rabbit produces a net increase of 0.5 rabbits each year. With this information, we can predict the size of the rabbit population 2 years from now:

$$N_t = N_0\, e^{rt}$$
$$N_t = 10 \times e^{0.5 \times 2}$$
$$N_t = 10 \times e^{1}$$
$$N_t = 10 \times (2.72)^{1}$$
$$N_t = 10 \times 2.72$$
$$N_t = 27 \text{ rabbits}$$

We can then ask how large the rabbit population will be after 4 years:

$$N_t = 10 \times e^{0.5 \times 4}$$
$$N_t = 10 \times e^{2}$$
$$N_t = 10 \times 7.4$$
$$N_t = 74 \text{ rabbits}$$

We can also project the size of the rabbit population 10 years from now:

$$N_t = 10 \times e^{0.5 \times 10}$$
$$N_t = 10 \times e^{5}$$
$$N_t = 10 \times 148.4$$
$$N_t = 1{,}484 \text{ rabbits}$$

YOUR TURN Now assume that the intrinsic rate of growth is 1.0 for rabbits. Calculate the predicted size of the rabbit population after 1, 5, and 10 years. Create a graph that shows the growth curves for an intrinsic rate of growth at 0.5, as calculated above, and an intrinsic rate of growth at 1.0. (Note that you will need to use your calculator to complete this problem.)

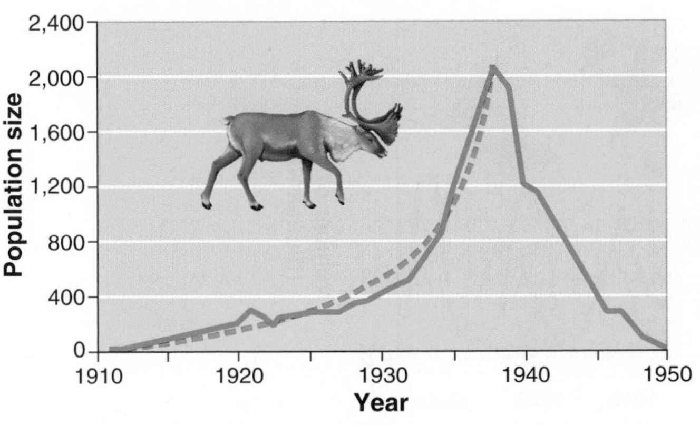

FIGURE 19.3 Growth and decline of a reindeer population. Humans introduced 25 reindeer to St. Paul Island, Alaska, in 1910. The population initially experienced rapid growth (blue line) that approximated a J-shaped exponential growth curve (orange line). In 1938, the population began to crash, probably because the animals exhausted the food supply. *(Data from V. B. Scheffer, The rise and fall of a reindeer herd, Scientific Monthly 73 (1951): 356–362.)*

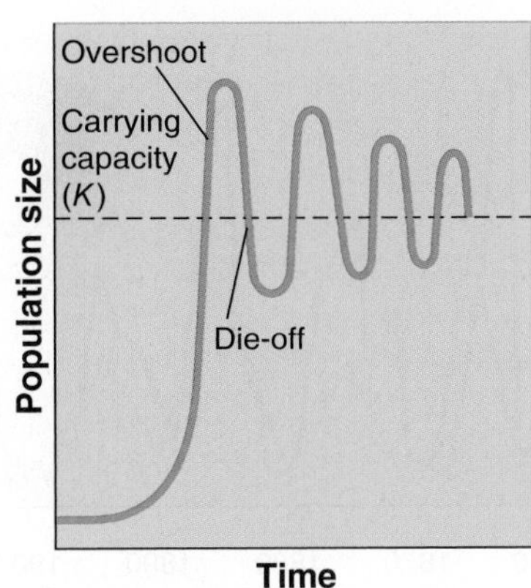

FIGURE 19.4 Population oscillations. Some populations experience recurring cycles of overshoots and die-offs that lead to a pattern of oscillations around the carrying capacity of their environment.

to feed the offspring, the population will experience an **overshoot** by becoming larger than the environment's carrying capacity. As a result of this overshoot, there will not be enough food for all the individuals in the population, and the population will experience a **die-off**, which is a rapid population decline due to death.

The reindeer (*Rangifer tarandus*) population on St. Paul Island in Alaska is a good example of this pattern. As you can see in **FIGURE 19.3**, a small population of 25 reindeer was introduced to the island to serve as a food source for the people living on the island. The island contained abundant vegetation for the reindeer and there were no large predators, so the population grew exponentially until it reached more than 2,000 individuals in 1938. The population subsequently crashed to only 8 animals by 1950, most likely because the reindeer ran out of food. After this die-off and a rebounding of the vegetation, 31 new reindeer were introduced to the island. Today, the St. Paul Tribal Government manages the reindeer population to avoid exponential increases followed by die-offs so that the population provides a stable source of food for the people of the island. In 2017, the reindeer population was approximately 400 animals.

As we have just seen, die-offs can take a population well below the carrying capacity of the environment. In subsequent cycles of reproduction, however, the population may grow large again. **FIGURE 19.4** illustrates the dynamics of a population experiencing a recurring cycle of overshoots and die-offs that causes it to oscillate around the carrying capacity. In many cases, these oscillations decline over time and approach the carrying capacity.

So far we have considered only how populations are limited by resources such as food, water, and the availability of nest sites. Predation may play an important additional role in limiting population growth. A classic example is the relationship between snowshoe hares (*Lepus americanus*) and lynx (*Lynx canadensis*) that prey on them in North America. Trapping records from the Hudson's Bay Company, which purchased hare and lynx pelts for nearly 90 years in Canada, indicate that the populations of both species cycle over time. **FIGURE 19.5** on page 208 shows how this interaction works. The lynx population peaks 1 or 2 years after the hare population peaks. As the hare population increases, it provides more prey for the lynx, and thus the lynx population begins to grow. As the hare population reaches a peak, food for hares becomes scarce, and the hare population dies off. The decline in hares leads to a subsequent decline in the lynx population. Because the low lynx numbers reduce predation, the hare population increases again.

A similar case of predator control of prey populations has been observed on Isle Royale, Michigan, an island in Lake Superior. Wolves (*Canis lupus*) and moose (*Alces alces*) have coexisted on Isle Royale for several decades and their population sizes have been estimated since 1959. Starting in 1981, the wolf population declined sharply, probably as a result of a deadly canine virus. As wolf predation on moose declined, the moose population grew rapidly until it ran out of food and experienced a large die-off that began in 1995. **FIGURE 19.6** on page 208 shows the changes in both populations over a period of more than 50 years. By 2017, only two wolves remained on the island and warmer winter temperatures made it less likely that the water would freeze and allow new wolves to naturally come to the island. As a result, and after a great deal of debate among scientists and the public, the National Park Service announced that they would be introducing 20 to 30 new wolves to the island in the coming years.

Overshoot When a population becomes larger than the environment's carrying capacity.

Die-off A rapid decline in a population due to death.

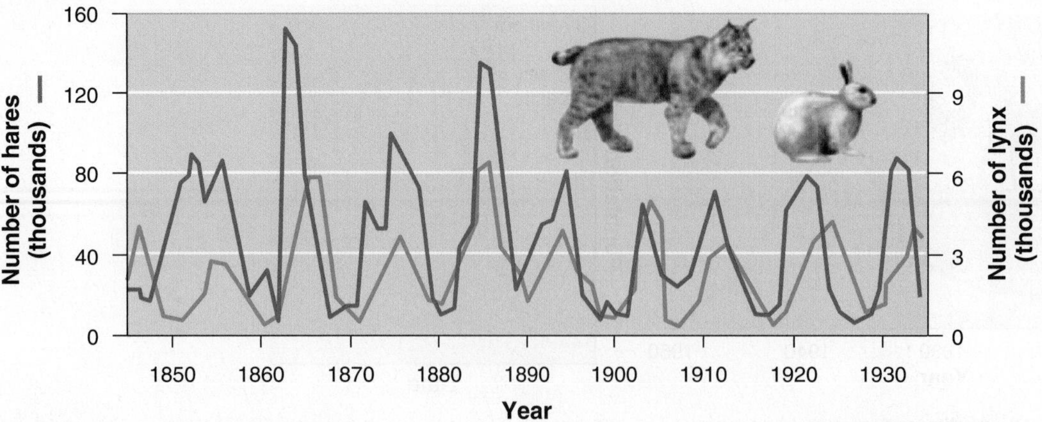

FIGURE 19.5 Population oscillations in lynx and hares. Both lynx and hares exhibit repeated oscillations of abundance, with the lynx population peaking 1 to 2 years after the hare population. When hares are not abundant, there is plenty of food, which allows the hare population to increase. As the hare population increases, there are more hares for lynx to eat, so then the lynx population increases. As the hare population becomes very abundant, they start to run out of food and the hare population dies off. As hares become less abundant, the lynx population subsequently dies off. With less predation and more food once again available, the hare population increases again, and the cycle repeats. *(Data from Hudson's Bay Company.)*

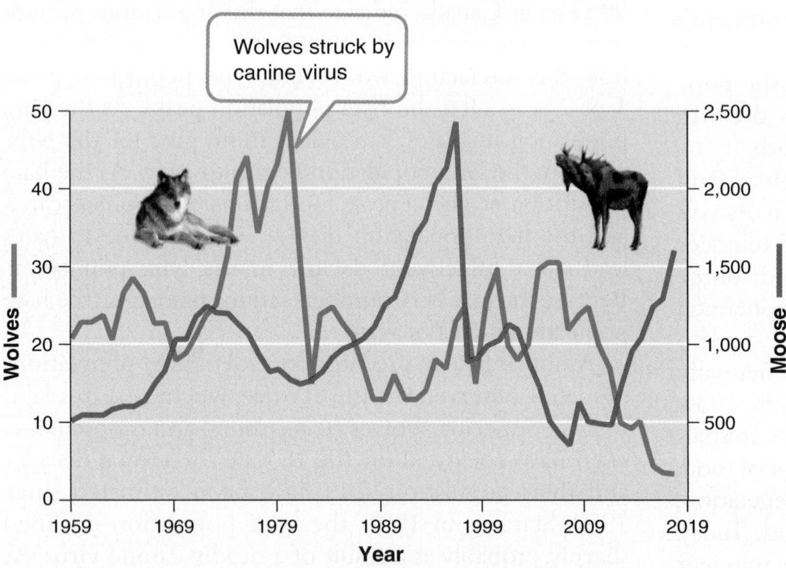

FIGURE 19.6 Predator control of prey populations. As the population of wolves on Isle Royale succumbed to a canine virus, their moose prey experienced a dramatic population increase. The moose population ultimately grew so large that they exceeded their carrying capacity and then experienced a large die-off. In 2017, only two wolves remained and the moose population began to once again increase. *(Data from R. O. Peterson and J. A. Vucetich, Ecological Studies of Wolves on Isle Royale: Annual Report 2016–2017, School of Forest Resources and Environmental Science, Michigan Technological University.)*

Species have different reproductive strategies and distinct survivorship curves

Population size most commonly increases through reproduction. Population ecologists have identified a range of reproductive strategies in nature.

K-selected Species

***K*-selected species** are species that have a low intrinsic growth rate, which causes the population to increase slowly until it reaches the carrying capacity of the environment. The population size of a *K*-selected species is largely determined by the carrying capacity (*K*), and their population fluctuations are small.

K-selected species have certain traits in common. For instance, *K*-selected animals are typically large organisms that reach reproductive maturity relatively late, produce a few, large offspring, and provide substantial parental care. Elephants, for example, have populations that grow slowly; they do not become reproductively mature until they are 13 years old, breed only once every 2 to 4 years, and produce only one calf at a time. However, once an elephant population approaches its carrying capacity, the population remains near the carrying capacity and does not widely fluctuate

***K*-selected species** A species with a low intrinsic growth rate that causes the population to increase slowly until it reaches carrying capacity.

in size. Large mammals and most birds are *K*-selected species. For environmental scientists interested in biodiversity management or protection, the slow growth of *K*-selected species poses a challenge; in practical terms, it means that an endangered *K*-selected species cannot respond quickly to efforts to save it from extinction.

r-selected Species

At the opposite end of the spectrum from *K*-selected species, ***r*-selected species** have a high intrinsic growth rate and their population sizes rarely stay near the carrying capacity. Instead, the populations of *r*-selected species commonly experience large overshoots of the carrying capacity followed by major die-offs. Such species reproduce often and produce large numbers of offspring. It is this ability to have a high intrinsic growth rate that allows *r*-selected species to rapidly surpass their carrying capacity. Once this happens, a major die-off is unavoidable. The name *r*-selected species refers to the fact that the intrinsic growth rate is designated as *r* in population models.

Among animals, *r*-selected species tend to be small organisms with the potential for high population growth rates: They reach reproductive maturity relatively early, reproduce frequently, produce many small offspring, and provide little or no parental care. House mice (*Mus musculus*), for example, become reproductively mature at 6 weeks of age, can breed every 5 weeks, and produce up to a dozen offspring at a time. Other *r*-selected organisms include small fishes, many insect species, and weedy plant species. Many organisms that humans consider to be pests, such as cockroaches, dandelions, and rats, are *r*-selected species.

TABLE 19.1 summarizes the traits of *K*-selected and *r*-selected species. These two categories represent opposite ends of a wide spectrum of reproductive strategies. Most species fall somewhere in between these two extremes, and many exhibit combinations of traits from the two extremes. For example, tuna and redwood trees are both long-lived species that take a long time to reach reproductive maturity. Once they do, however, they produce millions of small offspring that receive no parental care.

TABLE 19.1 Traits of *K*-selected and *r*-selected species

Trait	*K*-selected species	*r*-selected species
Life span	Long	Short
Time to reproductive maturity	Long	Short
Number of reproductive events	Few	Many
Number of offspring	Few	Many
Size of offspring	Large	Small
Parental care	Present	Absent
Population growth rate	Slow	Fast
Population regulation	Density dependent	Density independent
Population dynamics	Stable, near carrying capacity	Highly variable

AP® Exam Tip

You should understand the different ways ecologists describe populations and be able to use these descriptors appropriately when writing for the AP® Environmental Science Exam. For example, be prepared to describe the differences between *K*-selected and *r*-selected reproductive strategies. Table 19.1 provides a concise chart to compare and contrast the traits of these two reproductive strategies a species may employ. ●

Survivorship Curves

In addition to different reproductive strategies, species have distinct patterns of survival over the life span of individuals. These patterns can be plotted on a graph as **survivorship curves**, as shown in **FIGURE 19.7** on page 210. There are three basic types of survivorship curves. A **type I survivorship curve** has high survival throughout most of the life span, but then individuals start to die in large numbers as they approach old age. Species with type I curves include *K*-selected species such as elephants, whales, and humans. In contrast, a **type II survivorship curve** has a relatively constant decline in survivorship throughout most of the life span. Species with a type II curve include corals and squirrels. A **type III survivorship curve** has low survivorship early in life with few individuals reaching adulthood. Species with type III curves include *r*-selected species such as mosquitoes and dandelions.

***r*-selected species** A species that has a high intrinsic growth rate, which often leads to population overshoots and die-offs.

Survivorship curve A graph that represents the distinct patterns of species survival as a function of age.

Type I survivorship curve A pattern of survival over time in which there is high survival throughout most of the life span, but then individuals start to die in large numbers as they approach old age.

Type II survivorship curve A pattern of survival over time in which there is a relatively constant decline in survivorship throughout most of the life span.

Type III survivorship curve A pattern of survival over time in which there is low survivorship early in life with few individuals reaching adulthood.

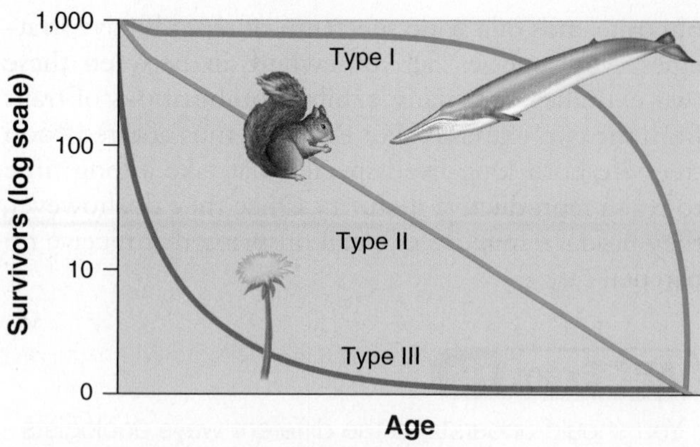

FIGURE 19.7 Survivorship curves. Different species have distinct patterns of survivorship over the life span. Species range from exhibiting excellent survivorship until old age (type I curve) to exhibiting a relatively constant decline in survivorship over time (type II curve) to having very low rates of survivorship early in life (type III curve). *K*-selected species tend to exhibit type I curves, whereas *r*-selected species tend to exhibit type III curves.

Interconnected populations form metapopulations

Cougars (*Puma concolor*)—also called mountain lions or pumas—once lived throughout North America but, because of habitat destruction and overhunting, they are now found primarily in the remote mountain ranges of the western United States. In New Mexico, cougar populations are distributed in patches of mountainous habitat scattered across the desert landscape. These mountain habitats allow the cats to avoid human activities and provide them with reliable sources of water and of prey such as mule deer (*Odocoileus hemionus*).

Because areas of desert separate the cougar's mountain habitats, we can consider the cougars of each mountain range to be a distinct population. Each population has its own dynamics based on local abiotic conditions and prey availability: Large mountain ranges support large cougar populations and smaller mountain ranges support smaller cougar populations. As **FIGURE 19.8** shows, cougars sometimes move between mountain ranges, often using strips of natural habitat that connect the separated populations. Strips of natural habitat that connect populations are known as **corridors** and provide some connectedness among the populations. A group of spatially distinct populations that are connected by occasional movements of individuals between them is called a **metapopulation**.

The connectedness among the populations within a metapopulation is an important part of each population's overall persistence. Small populations are more likely than large ones to go extinct. As we have seen, small populations contain relatively little genetic variation and therefore may not be able to adapt to changing environmental conditions. Small populations can also experience *inbreeding depression*. **Inbreeding depression** occurs when individuals with similar genotypes—typically relatives—breed with each other and produce offspring that have an impaired ability to survive and reproduce. This impaired ability occurs when each parent carries one copy of a harmful mutation in his or her genome. When the parents breed, some of their offspring receive two copies of the harmful mutation and, as a result, have poor chances of survival and successful reproduction.

Small populations are also more vulnerable than large populations to catastrophes such as particularly harsh winters that drive down their populations to critically low numbers. In a metapopulation, occasional immigrants from larger nearby populations can add to the size of a small population and introduce new genetic diversity, both of which help reduce the risk of extinction.

Metapopulations can also provide a species with some protection against threats such as diseases. A disease could cause a population living in a single large habitat patch to go extinct. But if a population living in an isolated habitat patch is part of a much larger metapopulation, then, while a disease could wipe out that isolated population, immigrants from other populations could later recolonize the patch and help the species to persist.

Because many habitats are naturally patchy across the landscape, many species are part of metapopulations. For instance, numerous species of butterflies specialize on plants with patchy distributions. Some amphibians live in isolated wetlands, but occasionally disperse to other wetlands. The number of species that exist as metapopulations is growing because human activities have fragmented habitats, dividing single large populations into several smaller populations. Identifying and managing metapopulations is thus an increasingly important part of protecting biodiversity.

Corridor Strips of natural habitat that connect populations.

Metapopulation A group of spatially distinct populations that are connected by occasional movements of individuals between them.

Inbreeding depression When individuals with similar genotypes—typically relatives—breed with each other and produce offspring that have an impaired ability to survive and reproduce.

FIGURE 19.8 A cougar metapopulation. Populations of cougars live in separate mountain ranges in New Mexico. Occasionally, however, individuals move between mountain ranges. These movements can recolonize mountain ranges with extinct populations and add individuals and genetic diversity to existing populations.

MODULE 19 AP® Review

Preparing for the AP® Exam

In this module, we examined population growth models, which help us understand population increases and decreases. Exponential growth models are the simplest because they assume unlimited resources. Logistic growth models are more realistic because they incorporate a carrying capacity and the associated density-dependent factors that occur in natural populations. By varying assumptions of growth models, we see that a population can overshoot its carrying capacity and experience a die-off. Populations can also oscillate due to predator-prey interactions. Populations can be characterized as either *K*-selected or *r*-selected and have unique types of survivorship curves. Finally, we have seen that a species can exist as a metapopulation composed of multiple, interconnected populations. In the next module, we will move from the population level to the community level and examine how species interactions help to determine which species can persist in natural communities.

AP® Practice Questions

Choose the best answer for the following.

1. The intrinsic growth rate of a population
   (a) occurs at the population's carrying capacity.
   (b) depends on the limiting resources of the population.
   (c) increases as the population size increases.
   (d) only occurs under ideal conditions.

2. Population growth using the exponential growth model
   (a) increases at a constant rate.
   (b) has an increasing intrinsic growth rate.
   (c) represents ideal conditions that rarely occur in natural populations.
   (d) incorporates the carrying capacity of the population.

3. An *r*-selected species characteristically has
   (a) a type I survivorship curve.
   (b) few offspring.
   (c) a population near carrying capacity.
   (d) a fast population growth rate.
4. Which is true of a population overshoot?
   (a) It occurs when reproduction quickly responds to changes in food supply.
   (b) It is followed by a die-off.
   (c) It is most likely to be experienced by *K*-selected species.
   (d) It occurs when a species stops growing after reaching the carrying capacity.
5. Inbreeding depression
   (a) rarely occurs in highly connected metapopulations.
   (b) results from the creation of corridors between populations.
   (c) results in increased rates of reproduction.
   (d) occurs in species that are experiencing overshoot.

MODULE 20

Community Ecology

We have explored the factors that determine population size, as well as ways to model or predict how populations will grow or decline. However, the size of a population tells us nothing about what determines the distribution of populations across the planet.

There are two factors that determine whether a species will persist in a location. As we learned in Chapter 5, if a species is able to disperse to an area, its ability to persist there depends on the fundamental niche of that species, which is the range of abiotic conditions that it can tolerate. The survival of a species in a habitat is also determined by the set of interactions that it has with other species. In this module we will explore these relationships.

Community ecology The study of interactions between species.

Learning Goals

After reading this module you should be able to

- identify species interactions that cause negative effects on one or both species.
- discuss species interactions that cause neutral or positive effects on both species.
- explain the role of keystone species.

Some species interactions cause negative effects on one or both of the species

The science of **community ecology** is the study of interactions among species. Some interactions have negative effects on one or both species. Other interactions involve neutral or positive effects on the two species. As we will see, throughout the world species have **symbiotic relationships**, which means that the two species are living in close association with each other.

Some of the best-known species interactions are those that cause one or both species to be negatively affected by the other species. These interactions include *competition, predation, parasitism,* and *herbivory.*

Competition

In 1934, 2 years after his experiments on logistic growth in populations of *Paramecium,* Georgii Gause studied how different *Paramecium* species affected each other's population growth. **FIGURE 20.1** shows the results of his experiments. When the two species—*P. caudatum* and *P. aurelia*—were grown in separate laboratory cultures, each species thrived and reached a relatively high population size within 10 days. However, when the two species were grown together, *P. aurelia* continued to thrive but *P. caudatum* declined to extinction. **Competition**, the struggle of individuals to obtain a shared limiting resource, caused *P. aurelia* to do better. Gause's observations, combined with additional experiments with other organisms, led researchers to formulate the **competitive exclusion principle**, which states that two species competing for the same limiting resource cannot coexist. Under a given set of environmental conditions, when two species have the same realized niche, one species will perform better and will drive the other species to extinction.

We see competition at work throughout nature. For example, many plant species influence the distribution and abundance of other plant species. Goldenrods are dominant in the old fields of New England because of their superior competitive ability: They can grow taller than other wildflowers and obtain more of the available sunlight. Similarly, the wild oat plant (*Avena fatua*) can outcompete crop plants on the Great Plains of North America because its seeds ripen earlier, permitting oat seedlings to start growing before the other species.

Competition for a limiting resource can lead to **resource partitioning**, in which two species divide a resource based on differences in their behavior or morphology. In evolutionary terms, when competition reduces the ability of individuals to survive and reproduce, natural selection will favor individuals that overlap less with other species in the resources they use. **FIGURE 20.2** on page 214 shows how this process works. Let's imagine two species of birds that eat seeds of different sizes. In species 1, represented by blue, some individuals eat small seeds and others eat medium seeds. In species 2, represented by yellow, some individuals eat medium seeds and others eat large seeds. As a result, some individuals of both species compete for medium seeds. This overlap is represented by green. If species 1 is the better competitor for medium seeds, then individuals of species 2 that compete for medium seeds will have poor

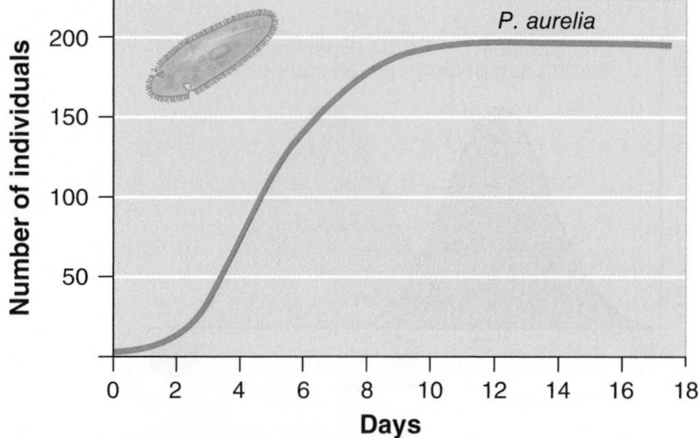

(a) *P. aurelia* grown separately

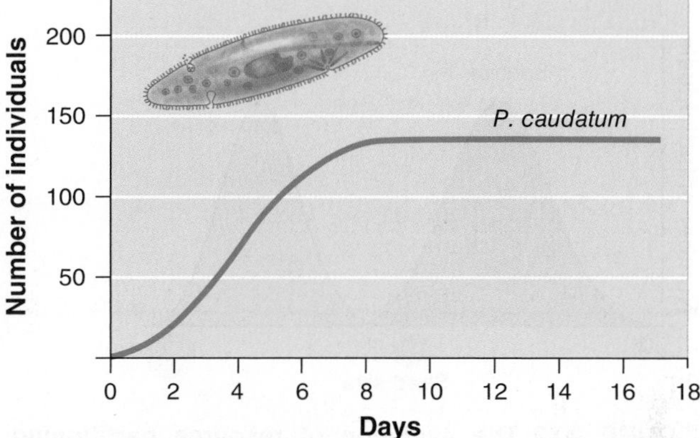

(b) *P. caudatum* grown separately

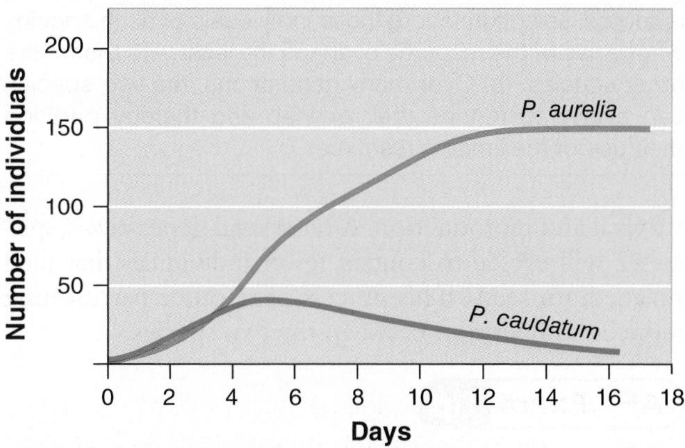

(c) *P. aurelia* and *P. caudatum* grown together

FIGURE 20.1 Competition for a limiting resource. When Gause grew two species of Paramecium separately, both achieved large population sizes. However, when the two species were grown together, *P. aurelia* continued to grow well, while *P. caudatum* declined to extinction. These experiments demonstrated that two species competing for the same limiting resource cannot coexist. *(Data from Gause, 1934.)*

Symbiotic relationship The relationship between two species that live in close association with each other.

Competition The struggle of individuals to obtain a shared limiting resource.

Competitive exclusion principle The principle stating that two species competing for the same limiting resource cannot coexist.

Resource partitioning When two species divide a resource based on differences in their behavior or morphology.

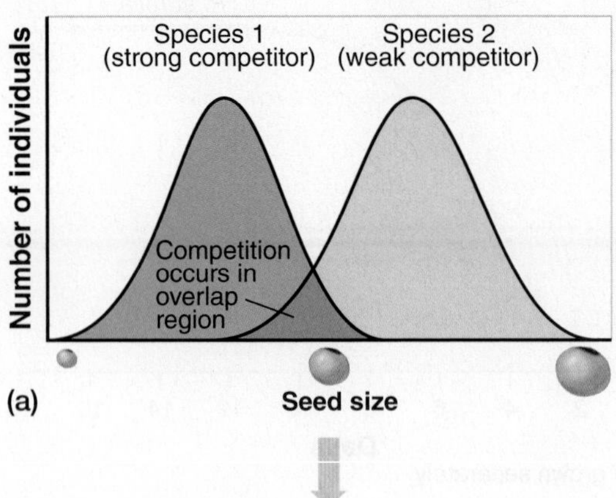

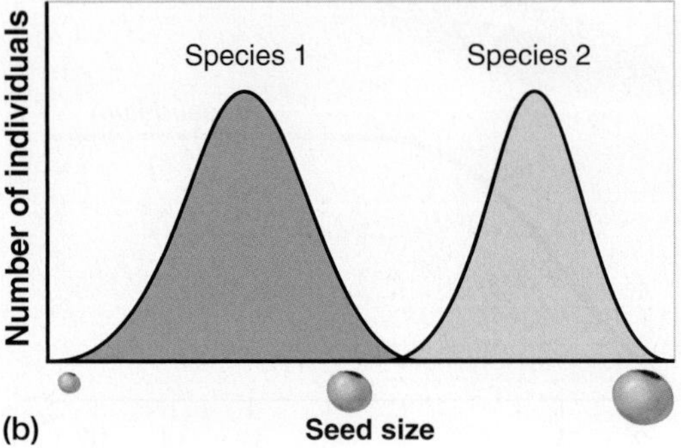

FIGURE 20.2 The evolution of resource partitioning. (a) When two species overlap in their use of a limiting resource, selection favors those individuals of each species whose use of the resource overlaps the least with that of the other species. (b) Over many generations, the two species can evolve to reduce their overlap and thereby partition their use of the limiting resource.

survival and reproduction. After several generations, species 2 will evolve to contain fewer individuals that feed on medium seeds. This process of resource partitioning reduces competition between the two species.

AP® Exam Tip

You may be asked to design an experiment on the AP® Environmental Science exam. To do this you must start with a question, and then determine the variables you would test and measure to answer the question. Create a hypothesis to include your variables. Explain how you will proceed with your experiment. Have you included an appropriate time period for your experiment? Write a conclusion statement using the data collected. Did the data support or reject your hypothesis? Remember data never proves a hypothesis. In your conclusion statement, be sure to mention the value of repeated trials. ●

Predation An interaction in which one animal typically kills and consumes another animal.

Parasitoid A specialized type of predator that lays eggs inside other organisms—referred to as its host.

Species in nature reduce resource overlap in several ways. One strategy is temporal resource partitioning, a process in which two species utilize the same resource but at different times. For example, wolves and coyotes that live in the same territory are active at different times of day to reduce the overlap in the times that they are hunting. Similarly, some plants reduce competition for pollinators by flowering at different times of the year.

If two species reduce competition by using different habitats, they are exhibiting spatial resource partitioning. For example, desert plant species have evolved a variety of different root systems that reduce competition for water and soil nutrients; some have very deep roots while others have shallow roots. For example, black grama grass (*Bouteloua eriopoda*) has shallow roots that extend over a large area to capture rainwater, whereas tarbush (*Flourensia cernua*) sends roots deep into the ground to tap deep sources of water.

A third method of reducing resource overlap is by morphological resource partitioning, which is the evolution of differences in body size or shape. As we saw in Chapter 5, Charles Darwin observed morphological resource partitioning among the finches of the Galápagos Islands (see Figure 16.2 on page 170). The 14 species of finches on the islands descended from a single finch species that colonized the islands from mainland South America. This ancestral finch species would have consumed seeds on the ground. Over time, however, the finches have evolved into 14 species with uniquely shaped beaks that allow each species to eat different foods. The beaks of some species are well suited for crushing seeds, whereas the beaks of others are better suited for catching insects. This morphological resource partitioning reduces competition among the finch species.

Predation

One of the most dramatic species interactions occurs between predators and their prey. **Predation** is an interaction in which one animal typically kills and consumes another animal. Predators include African lions that eat gazelles and great horned owls (*Bubo virginianus*) that eat small rodents. Organisms of all sizes may be predators, and their effects on their prey vary widely.

One specialized type of predator is the group known as **parasitoids**, which are animals that lay eggs inside other organisms—referred to as their host. When the eggs hatch, the parasitoid larvae slowly consume the host from the inside out, eventually leading to the host's death. Parasitoids include certain species of wasps and flies.

As we learned from the lynx-hare interactions in Canada and the wolf-moose interactions on Isle

Royale, predators can play an important role in controlling the abundance of prey. In Sweden, red foxes prey on several species of hares and grouse. In the 1970s and 1980s, this fox population was reduced by a disease called mange. With fewer foxes around, the foxes' prey increased in abundance: Populations of grouse doubled in size, and populations of hares increased sixfold. The predators in all of these examples play a critical role in their ecosystems; they regulate prey populations.

To avoid being eaten or harmed by predators, many prey species have evolved defenses. These defenses may be behavioral, morphological, or chemical, or may simply mimic another species' defense. **FIGURE 20.3** shows two examples of antipredator defenses. Animal prey commonly use behavioral defenses, such as hiding and reduced movement, so as to attract less attention from predators. Other prey species have evolved impressive morphological defenses, including camouflage to help them hide from predators and spines to help deter predators. For example, flounders are a group of fish that lay on the ocean floor with a body color that blends in with their surroundings. Many plants, for example, have evolved spines that deter herbivores from grazing on their leaves and fruits. Similarly, many animals, such as porcupines, stingrays, and puffer fish, have spines that deter predator attack.

Chemical defenses are another common mechanism of protection from predators. Several species of insects, frogs, and plants emit chemicals that are toxic or distasteful to their predators. For example, the poison dart frog shown in **FIGURE 20.4** on page 216 produces a toxin on its skin that is toxic to many species of predators. Many toxic prey are also brightly colored, and predators learn to recognize and avoid consuming them. In some cases, prey have not evolved toxic defenses of their own, but instead have evolved to mimic the physical characteristics of other prey species that do possess chemical defenses. By mimicking toxic species, these nontoxic species can fool predators into not attacking them.

(a)

(b)

FIGURE 20.3 Prey morphological defenses. Predation and herbivory has favored the evolution of fascinating defenses. (a) The camouflage of this stone flounder (*Kareius bicoloratus*) makes it difficult for predators to see it. (b) Sharp spines protect this cactus from predators. *(a: Stephen Frink Collection/Alamy; b: pernsanitfoto/Getty Images)*

Parasitism

Parasitism is an interaction in which one organism lives on or in another organism—referred to as the host. Because parasites typically consume only a small fraction of their host, a single parasite rarely causes the death of its host. Parasites include tapeworms that live in the intestines of animals as well as the protists that live in the bloodstream of animals and cause malaria.

Parasites that cause disease in their host are called **pathogens**. Pathogens include viruses, bacteria, fungi, protists, and wormlike organisms called helminths. These organisms cause many of the most well-known diseases in the world, ranging from the common cold to some forms of cancer.

Herbivory

Herbivory is an interaction in which an animal consumes a producer. They typically eat only a portion of a producer without killing it. The gazelles of the African plains are well-known herbivores, as are the

Parasitism An interaction in which one organism lives on or in another organism.

Pathogen A parasite that causes disease in its host.

Herbivory An interaction in which an animal consumes a producer.

(a)

(b)

FIGURE 20.4 Prey chemical defenses. (a) The poison dart frog (*Epipedobates bilinguis*) has a toxic skin. (b) This nontoxic frog (*Allobates zaparo*) mimics the appearance of the poison dart frog. (a: Morley Read/Alamy; b: Luis Louro/Alamy Stock Photo)

various species of deer, rabbits, goats, and sheep. In aquatic systems, herbivores include sea urchins in the ocean and tadpoles in ponds that consume various species of algae. When herbivores become abundant, they can have dramatic effects on producers. One way to assess the effect of herbivores is by excluding herbivores from an area and seeing how the producers respond. For example, when deer are excluded from an area by putting up a high fence, scientists commonly observe an extraordinary increase in plants, especially in those species of plants that deer prefer to consume (**FIGURE 20.5**). As in the case of predators and prey, many species of producers have evolved defenses against herbivores that include sharp spines and distasteful chemicals.

Mutualism An interaction between two species that increases the chances of survival or reproduction for both species.

FIGURE 20.5 Herbivory. When fences are erected to exclude herbivory by deer, plant growth typically increases dramatically, especially for species that deer prefer to eat. (Jean-Louis Martin RGIS/CEFE-CNRS)

Other species interactions cause neutral or positive effects on one or both species

Some species interactions can be quite beneficial to the participants. In this section, we will example two such positive interactions: *mutualisms* and *commensalisms.*

Mutualisms

One of the most ecologically important interactions is the relationship between plants and their pollinators, which include birds, bats, and insects. The plants depend on the pollinators for their reproduction, and the pollinators depend on the plants for food. In some cases, one pollinator species might visit many species of plants, while in others, a range of pollinator species may visit many plant species. In still other cases, one animal species pollinates only one plant species, and that plant is pollinated only by that animal species. For example, there are about 900 species of fig trees, and almost every one is pollinated by a particular species of fig wasp.

When two interacting species benefit each other by increasing both species' chances of survival or reproduction, we call it a **mutualism**. Each species in a mutualistic interaction is ultimately assisting the other species in order to benefit itself. If the benefit is too small, the interaction will no longer be worth the cost of helping the other species.

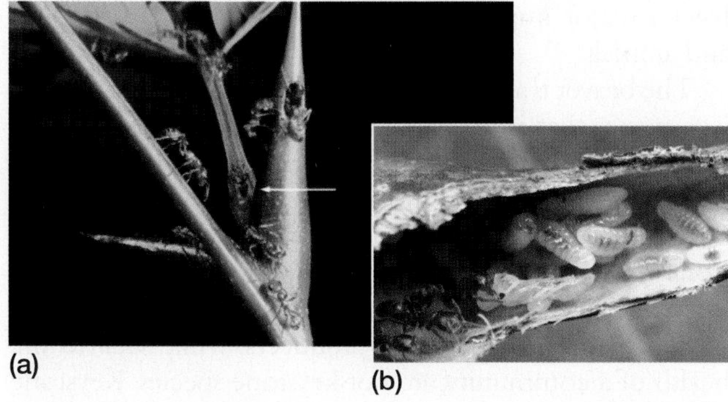

FIGURE 20.6 Mutualism. Acacia trees and *Pseudomyrmex* ants have each evolved adaptations that enhance their mutualistic interaction. The trees' thorns serve as nest sites for the ants (inset) and provide them with food in the form of nectar produced in specialized nectaries (arrow). In return, the ants protect the trees from herbivores and competing plants. *(a: Oxford Scientific/Getty Images; b: Alex Wild Photography)*

Under such conditions, natural selection will favor individuals that no longer engage in the mutualistic interaction.

One well-studied example of plant-animal mutualism is the interaction between acacia trees and several species of *Pseudomyrmex* ants in Central America. The acacia tree supplies the ants with food and shelter, and the ants protect the tree from herbivores and competitors. The ants eat nectar produced by the tree in special structures called nectaries, and they live in the tree's large thorns, which have soft cores that the ants can easily hollow out (**FIGURE 20.6**). The ants protect the tree by attacking anything that falls on it—for example, by stinging intruding insects that might eat the acacia's leaves and destroying vines that might shade the tree's branches. Through natural selection over generations of close association, the ants and the trees have both evolved traits that make this mutualism work. The ants have evolved several behavioral adaptations: living in the hollowed-out thorns of the tree, extensive patrolling of branches and leaves, and attacking all foreign animals and plants regardless of whether they are useful to the ants as food. For its part, the acacia tree has evolved specialized thorns, year-round leaf production, and modified leaves and nectaries that provide food for the ants.

Two other well-known mutualistic relationships are those found in coral reefs and in lichens. As we saw in Chapter 4, algae live within the tiny coral animals that build the coral reefs. Their relationship is critical: If the algae die, the coral will die as well. Many of the lichens that grow on the surfaces of rocks and trees are made up of an alga, one or two species of fungus, and often bacteria living in close association. The fungus provides many of the nutrients the alga needs, and the alga provides carbohydrates for the fungus via photosynthesis (**FIGURE 20.7**).

FIGURE 20.7 Lichens. Lichens, such as this *Hypogymnia physodes*, are composed of one or two fungi and an alga that are tightly linked together in a mutualistic interaction. The fungi provides nutrients to the alga, and the alga provides carbohydrates for the fungus via photosynthesis. *(Duncan Shaw/Science Source)*

Commensalisms

Sometimes species can interact in a way that benefits one species but has no effect on the other species. For example, tree branches can serve as perch locations and nest sites for birds. The birds benefit from the presence of the tree because it provides them perches to search for food and a place to raise their offspring. However, the survival and reproduction of the tree is not affected by the presence of the bird. Species interactions in which one species benefits but the other is neither harmed nor helped are called **commensalisms**. Commensalisms are common in nature. For example, many species of fish use coral reefs as hiding places to avoid predators. The fish receive a major benefit from the presence of the coral, but the species of coral that construct the reefs are typically neither harmed nor benefited by the presence of the fish.

Interactions among species are important in determining which species can live in a community. **TABLE 20.1** on page 218 summarizes these interactions and the effects they have on each of the interacting species, whether positive (+), negative (−), or neutral (0). Competition for a limiting resource has a negative effect on both of the competing species. In contrast, predation, parasitism, and herbivory each has a positive effect on the consumer, but a negative effect on the organism being consumed. Mutualism has positive effects on both interacting species. Commensalism has a positive effect on one species and no effect on the other species.

Commensalism A relationship between species in which one species benefits and the other species is neither harmed nor helped.

TABLE 20.1	Interactions between species and their effects	
Type of interaction	**Species 1**	**Species 2**
Competition	–	–
Predation	+	–
Parasitism	+	–
Herbivory	+	–
Mutualism	+	+
Commensalism	+	0

Keystone species have large effects on communities

Some species that are not abundant can still have very large effects on a community. The beaver is a prime example. Although they make up only a small percentage of the total biomass of the North American forest, beavers play a critical role in the forest community. They build dams that convert narrow streams into large ponds, thereby creating new habitat for pond-adapted plants and animals (**FIGURE 20.8**). These ponds also flood many hectares of forest, causing the trees to die and creating habitat for animals that rely on dead trees. Several species of woodpeckers and some species of ducks make their nests in cavities that are carved into the dead trees. In short, one beaver can have a major impact on the entire community of plants and animals.

FIGURE 20.8 Beavers. Beavers are an important species to the community because of the role they play in creating new pond and wetland habitat. For a species that is not particularly abundant, the beaver has a strong influence on the presence of other organisms in the community. *(a: Thomas & Pat Leeson/Science Source; b: Eric and David Hosking/Getty Images)*

Keystone species A species that is not very abundant but has large effects on an ecological community.

The beaver is an example of a **keystone species**, which is a species that is not very abundant but has large effects on an ecological community. The name *keystone species* is a metaphor that comes from architecture. As **FIGURE 20.9** shows, in a stone arch, the keystone is the single center stone that supports all the other stones. Without the keystone, the arch would collapse. Typically, the most abundant species or the major energy producers, while vital to the health of a community, are not keystone species. Keystone species typically exist in low numbers. They may be predators, sources of food, mutualistic species, or providers of some other essential service.

AP® Exam Tip

You may be asked to describe how a keystone species affects an ecosystem. Most students will answer that a keystone species has a significant role within the ecosystem. While this is correct, the response in incomplete. A complete description should note that keystone species play a more significant role in their ecosystems than their relative abundance would suggest. ●

The role of keystone predators was well demonstrated in a classic experiment conducted in intertidal communities off the coast of Washington State, as shown in **FIGURE 20.10**. These intertidal communities include mobile animal species such as sea stars and snails as well as dozens of other species that make their living attached to the rocky substratum, including mussels, barnacles, and algae. When sea stars are present, they prey on mussels (*Mytilus californianus*). This predation continually clears spaces where other species can attach to the rocks. When researchers removed the sea stars from the community, the mussels were no longer subject to predation by sea stars, and they outcompeted the other species in the community. The mussels became numerically dominant, while 25 other species declined in abundance. Thus the predatory sea stars, while not particularly numerous, played a key role in reducing the abundance of a superior competitor—the mussel—and allowing inferior competitors to persist.

The ability of predators to alter the outcome of competition is common in nature. We observed it in the opening of this chapter, where we described the effects of outbreaks of the leaf beetle that specializes

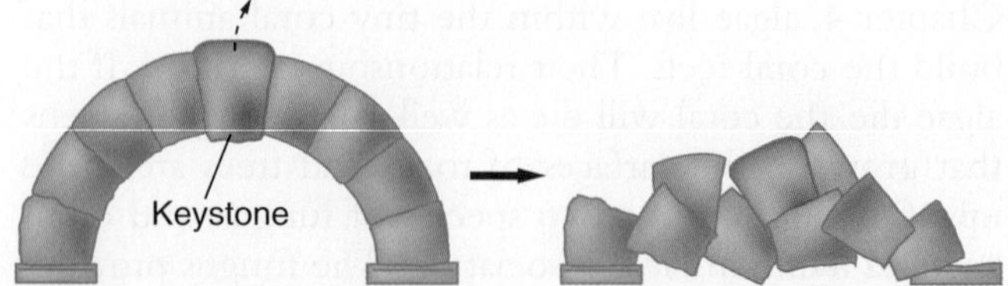

FIGURE 20.9 Keystone. Keystone species get their name from the keystone of an arch. Without the keystone in place, the arch would fall apart.

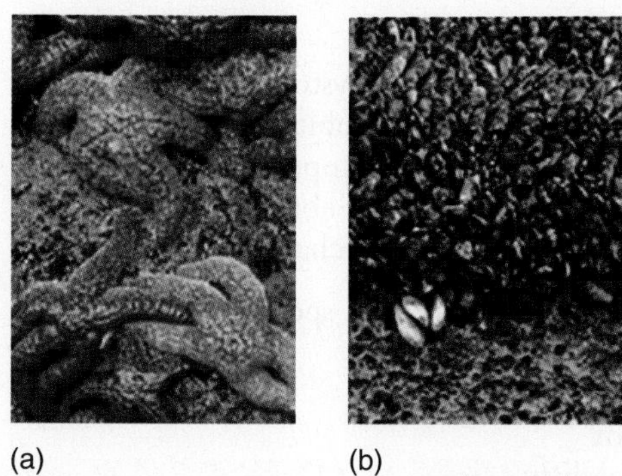

(a) (b)

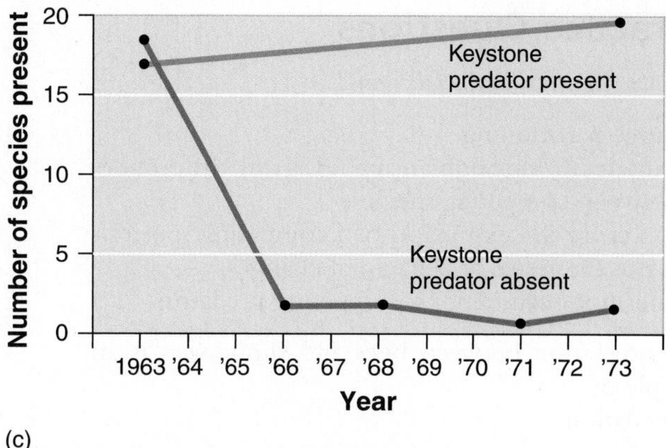

(c)

FIGURE 20.10 Keystone predators. (a) Sea stars are keystone predators in their rocky intertidal communities in Washington State. When sea stars are present, they consume mussels, which are strong competitors for space. This predation creates open spaces that inferior competitors can colonize. As a result, the diversity of species is high. (b) In the absence of sea stars, the mussels dominate the surfaces of the intertidal rocks, and the diversity of species declines dramatically. (c) Evidence of the role sea stars play in intertidal communities was discovered in an experiment in which the sea stars were allowed to be present in intertidal communities or were removed. *(a: blueeyes/Shutterstock; b: Alex L. Fradkin/Getty Images; c: Data from R. T. Paine, "Intertidal community structure: Experimental studies on the relationship between a dominant competitor and its principal predator," Oecologia 15 (1974): 93–120.)*

on goldenrods in the old fields of New England. An outbreak of these herbivorous beetles approximately once per decade dramatically reduces the abundance of goldenrod plants, which are superior competitors, and this allows the abundance of many wildflower species, which are inferior competitors, to increase.

Some species are considered keystone species because of the importance of their mutualistic interactions with other species. For example, most animal pollinators are abundant or provide a service that can be duplicated by other species. Some communities, however, critically rely on relatively rare pollinator species, which makes these pollinators keystone species. On many South Pacific islands, a species of bat known as the flying fox (*Pteropus vampyrus*) is the only pollinator and seed disperser for hundreds of tropical plant species. Flying foxes have been hunted for food to near-extinction. Without the pollinating and seed-dispersing functions of the flying foxes, many plant species may become extinct, dramatically changing these island communities.

Mycorrhizal fungi are another group of mutualists that serve as keystone species. These fungi are found on and in the roots of many plant species, where they increase the plants' ability to extract nutrients from the soil. They play a critical role in the growth of plant species, which in turn provide habitat and resources for other members of a forest or field community.

Finally, a keystone species that creates or maintains habitat for other species is known as an **ecosystem engineer**. We have already seen that beavers are ecosystem engineers. Alligators play a similar role in their communities by digging deep "gator holes" in summer. These holes serve as critical sources of water for many other animals during dry months.

Ecosystem engineer A keystone species that creates or maintains habitat for other species.

MODULE 20 AP® Review

Preparing for the AP® Exam

In this module, we saw that interactions among species affect whether a species can persist. Some interactions such as competition, predation, parasitism, and herbivory have negative effects on one or both species. Other interactions such as mutualisms and commensalisms have neutral or positive effects on both species. Keystone species have particularly large effects on a community; they can dramatically alter the habitat, such as when beavers construct dams. In the next module, we will examine how species in communities change over time as a result of changes in the biotic and abiotic environment.

AP® Practice Questions

Choose the best answer for the following.

1. Resource partitioning
   (a) can occur through morphological differences between competing species.
   (b) can cause the extinction of a competing species.
   (c) is not the result of behavioral changes.
   (d) does not occur among competing predators.

2. The interaction between bees and sunflowers is an example of
   (a) predation.
   (b) parasitism.
   (c) mutualism.
   (d) commensalism.

3. Pathogens are a type of
   (a) mutualist.
   (b) parasite.
   (c) predator.
   (d) herbivore.

4. Which is NOT typical of a keystone species?
   (a) It can shape and maintain habitats for other species.
   (b) It can limit a dominant competitive species.
   (c) It can have a relatively low abundance.
   (d) It is at the top of the food chain.

5. Which interaction harms both species involved?
   (a) competition
   (b) predation
   (c) parasitism
   (d) commensalism

MODULE 21

Community Succession

Learning Goals

After reading this module you should be able to

- explain the process of primary succession.
- explain the process of secondary succession.
- explain the process of aquatic succession.
- describe the factors that determine the species richness of a community.

Even without human activity, natural communities do not stay the same forever. Change in the species composition of communities over time is a perpetual process in nature. In this module we will look at how both terrestrial and aquatic communities change over time.

Primary succession starts with no soil

Virtually every community experiences **ecological succession**, which is the predictable replacement of one group of species by another group of species over time. Depending on the community type, ecological succession can occur over time spans varying from decades to centuries.

Ecological succession The predictable replacement of one group of species by another group of species over time.

Some terrestrial communities begin with bare rock and no soil. For example, a community may begin forming

FIGURE 21.1 Primary succession. Primary succession occurs in areas devoid of soil. Early-arriving plants and algae can colonize bare rock and begin to form soil, making the site more hospitable for other species to colonize later. Over time, a series of distinct communities develops. In this illustration, representing an area in New England, bare rock is initially colonized by lichens and mosses and later by grasses, shrubs, and trees.

on newly exposed rock left behind after a glacial retreat, newly cooled lava after a volcanic eruption, or an abandoned parking lot. When succession begins with bare rock and no soil, we call it **primary succession**.

FIGURE 21.1 shows the process of primary succession in a temperate forest biome in New England. The bare rock can be colonized by organisms such as algae, lichens, and mosses—organisms that can survive with little or no soil. As these early-successional species grow, they excrete acids that allow them to take up nutrients directly from the rock. The resulting chemical alteration of the rock also makes it more susceptible to erosion. When the algae, lichens, and mosses die, they become the organic matter that mixes with minerals eroded from the rock to create new soil.

Over time, soil develops on the bare surface and it becomes a hospitable environment for plants with deep root systems. Mid-successional plants such as grasses and wildflowers are easily dispersed to such areas. These species are typically well adapted to exploiting open, sunny areas and are able to survive in the young, nutrient-poor soil. The lives and deaths of these mid-successional species gradually improve the quality of the soil by increasing its ability to retain nutrients and water. As a result, new species colonize the area and outcompete the mid-successional species.

The type of community that eventually develops is determined by the temperature and rainfall of the region. In the United States, succession produces forest communities in the East, grassland communities in the Midwest, and shrubland communities in the Southwest. In some areas, the number of species increases as succession proceeds. In others, late-successional communities have fewer species than early-successional communities.

Secondary succession starts with soil

Secondary succession occurs in areas that have been disturbed but have not lost their soil. Secondary succession follows an event, such as a forest fire or hurricane, that removes vegetation but leaves the soil mostly intact.

Primary succession Ecological succession occurring on surfaces that are initially devoid of soil.

Secondary succession The succession of plant life that occurs in areas that have been disturbed but have not lost their soil.

FIGURE 21.2 Secondary succession. Secondary succession occurs where soil is present, but all plants have been removed. Early-arriving plants set these areas on a path of secondary succession. Secondary succession in a New England forest begins with grasses and wildflowers, which are later replaced by trees.

Secondary succession also occurs on abandoned agricultural fields, such as the New England farms we discussed at the beginning of the chapter.

FIGURE 21.2 shows the process of secondary succession, again for a typical forest in New England. It usually begins with rapid colonization by plants that can easily disperse to the disturbed area. The first to arrive are typically such plants as grasses and wildflowers that have light, wind-borne seeds. As in primary succession, these species are eventually replaced by species that are better competitors for sunlight, water, and soil nutrients. In regions that receive sufficient rainfall, trees replace grasses and flowers. The first tree species to colonize the area are those that can disperse easily and grow rapidly. For example, the seeds of aspen trees are carried on the wind, and cherry tree seeds are carried by birds that consume the fruit and excrete the seeds on the ground. Trees such as aspen and cherry are often called **pioneer species** because of their ability to colonize new areas rapidly and grow well in full sunshine. As the pioneer trees increase in number and grow larger, however, they cause an increased amount of shade on the ground. Because pioneer trees need full sunshine to grow, new seedlings of these tree species cannot persist. In contrast, species that are more shade tolerant, including many species of beech and maple, survive and grow well in the shade of the pioneer tree species. These shade-tolerant species grow up through the pioneer canopy and eventually outcompete the pioneer trees and dominate the forest community.

The process of secondary succession happens on many spatial scales, from small-scale disturbances such as a single treefall to huge areas cleared by natural disasters such as hurricanes or wildfires. The process is similar in both cases. For example, if a large tree falls and creates an opening for light to reach the forest floor, the seedlings of trees that grow rapidly in full sunlight outcompete other seedlings, and secondary succession begins again.

Historically, ecologists described succession as having a final stage known as a **climax community**. In forests, for example, ecologists considered the oldest forests to be climax forests. However, it is now recognized that, because natural disturbances such as fire, wind, and outbreaks of insect herbivores are a regular part of most communities, a late-successional stage is typically not final because at any moment it can be reset to an earlier stage.

Pioneer species A species that can colonize new areas rapidly and grow well in full sunshine.

Climax community Historically described as the final stage of succession.

Succession occurs in a variety of aquatic ecosystems

Disturbances also create opportunities for succession in aquatic environments. For example, the rocky intertidal zone along the Pacific coast of North America is exposed to the air during low tide and lies under water during high tide. From time to time, major storms turn over rocks or clear their surfaces of living things. These bare rocks can then be colonized through the process of primary succession. Diatoms and short-lived species of red and green algae are usually the first to arrive. Not long afterward, the rocks are colonized by barnacles and several species of long-lived red algae. This mid-successional stage develops rather quickly—in about 15 months. Finally, if the rock is not disturbed again for 3 years or more, the area may become dominated by several species of attached barnacles and mussels.

FIGURE 21.3 shows the pattern of succession in freshwater lakes. A glacier may carve out a lake basin, scouring it of sediments and vegetation. Over time, algae and aquatic plants colonize the lake. The growth of plants and algae and the erosion of surrounding rock and soil slowly fill the basin with sediment and organic matter, making it increasingly shallow. This process, which may take hundreds or even thousands of years, will eventually fill in the lake completely and make it a terrestrial habitat.

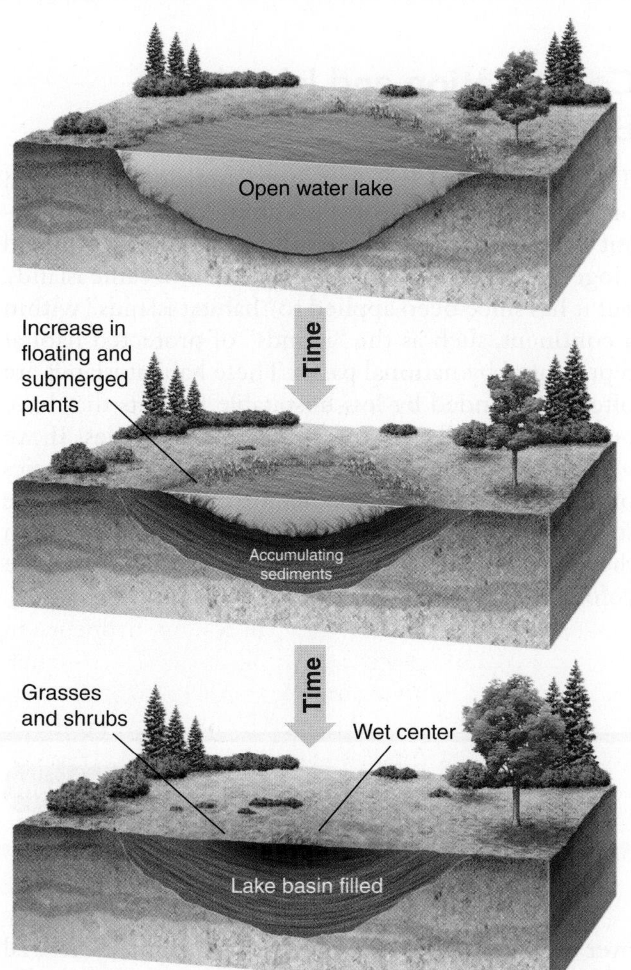

FIGURE 21.3 Succession in lakes. Over a time span of hundreds to thousands of years, lakes are filled with sediments and slowly become terrestrial habitats.

The species richness of a community is influenced by many factors

As we have seen, species are not distributed evenly on Earth. They are organized into biomes by global climate patterns and into communities whose composition changes regularly as species interact. In a given region within a biome, the number and types of species present are determined by three basic processes: colonization of the area by new species, speciation within the area, and losses from the area by extinction. The relative importance of these processes varies from region to region and is influenced by four factors: *latitude, time, habitat size,* and *distance from other communities.*

Latitude

As we move from the equator toward the North or South Pole, the number of species declines. For example, the southern latitudes of the United States support more than 12,000 species of plants, whereas a similar-sized area in northern Canada supports only about 1,700 species. This latitudinal pattern is also observed among birds, reptiles, amphibians, and insects. For more than a century, scientists have sought to understand the reasons for this pattern, yet those reasons remain unclear.

Time

Patterns of species richness are also regulated by time. The longer a habitat exists, the more colonization, speciation, and extinction can occur in that habitat. For example, Lake Baikal in Siberia—more than 25 million years old—is one of the oldest lakes in the world. Its benthic zone is home to over 580 species of invertebrates. By contrast, only 4 species of invertebrates inhabit the benthic zone of the Great Slave Lake in northern Canada—a lake that is similar in size and latitude to Lake Baikal, but is only a few tens of thousands of years old. This difference suggests that older communities have had more opportunities for speciation.

Habitat Size and Distance from a Source of Species

The final two factors that influence species richness are the size of the habitat and the distance of that habitat from a source of colonizing species. These factors are the basis for the **theory of island biogeography**, which demonstrates the dual importance of habitat size and distance from a mainland in determining species richness.

Larger habitats typically contain more species. In **FIGURE 21.4**, for example, we can see that larger islands of reed habitat in Hungary contain a greater number of bird species than smaller islands. There are three reasons for this pattern. First, dispersing species are more likely to find larger habitats than smaller habitats, particularly when those habitats are islands. Second, at any given latitude, larger habitats can support more species than smaller habitats. Larger habitats are capable of supporting larger populations of any given species, and larger populations are less prone to extinction. Third, larger habitats often contain a wider range of environmental conditions, which in turn provide more niches that support a larger number of species. A wider range of environmental conditions also provides greater opportunities for speciation over time.

The distance between a habitat and a source of colonizing species is the second factor that affects the species richness of communities. For example, oceanic islands that are more distant from continents generally have fewer species than islands that are closer to continents. Distance matters because, while many species can disperse short distances, only a few can disperse long distances. In other words, if two islands are the same size and contain the same resources, the nearer island should accumulate more species than the farther island because it has a higher rate of immigration by new species.

Theory of island biogeography A theory that demonstrates the dual importance of habitat size and distance in determining species richness.

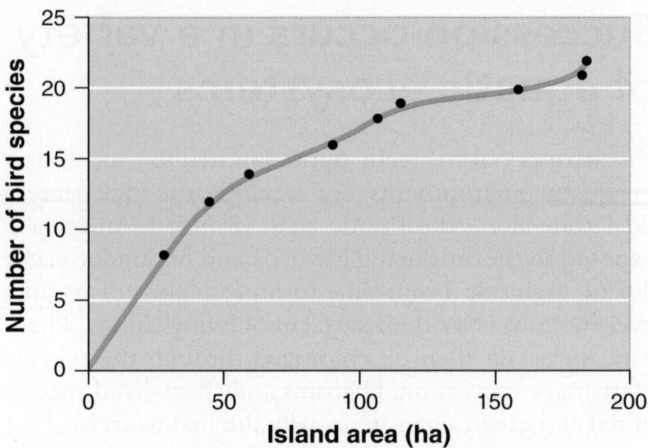

FIGURE 21.4 Habitat size and species richness. Species richness increases as the size of the habitat increases. In this example, researchers counted the number of bird species that inhabited reed islands in Lake Velence, Hungary. As island area increased, the number of bird species initially rose quickly and then began to slow. *(Data from A. Baldi and T. Kisbenedek, "Bird species numbers in an archipelago of reeds at Lake Velence, Hungary," Global Ecology and Biogeography 9 (2000): 451–461.)*

Conservation and Island Biogeography

The effects of colonization, speciation, and extinction on the species richness of communities have important implications for conservation. The theory of island biogeography was originally applied to oceanic islands, but it has since been applied to "habitat islands" within a continent, such as the "islands" of protected habitat represented by national parks. These habitat islands are often surrounded by less hospitable habitats that have been dramatically altered by human activities. If we wish to set aside natural habitat for a given species or a group of species, we need to consider both the size of the protected area and the distance between the protected area and other areas that could provide colonists.

MODULE 21 AP® Review

Preparing for the AP® Exam

In this module, we learned that communities change over time in a process known as succession. In terrestrial communities, an area that begins with no soil undergoes primary succession whereas an area that begins with soil undergoes secondary succession. In aquatic communities, lakes and ponds can slowly fill in with sediments over thousands of years to eventually become terrestrial habitats. We also learned that the species richness of a community is affected by the latitude of a habitat, the length of time that the habitat has been present, the size of the habitat, and the distance from the habitat to other habitats that serve as sources of species.

AP® Practice Questions

Choose the best answer for the following.

1. Which is true of primary succession?
   (a) It starts with bare soil.
   (b) As it progresses the number of species initially decreases.
   (c) It occurs after forest fires.
   (d) It begins with colonization by algae, lichens, and mosses.

2. The process of succession in lakes
   (a) results in a terrestrial ecosystem.
   (b) occurs rarely because disturbances are rare.
   (c) progresses fastest in very deep lakes.
   (d) has no climax species.

3. Which tree species is a pioneer species in North American forests?
   (a) beech
   (b) fir
   (c) aspen
   (d) maple

4. Which factor does NOT affect species richness?
   (a) latitude
   (b) survivorship curves
   (c) time
   (d) habitat size

5. Which is NOT a factor in the theory of island biogeography?
   (a) Dispersing species are more likely to find a large habitat.
   (b) Large populations are less likely to go extinct.
   (c) Larger habitats contain a larger range of environmental conditions.
   (d) Islands farther from the continent have more species due to increased speciation.

Working Toward Sustainability

Bringing Back the Black-footed Ferret

Throughout the western United States, the Great Plains were once covered with large congregations of prairie dogs. Spanning several different species, these prairie dog "towns" consisted of networks of underground tunnels. They were great attractions for out-of-town tourists, but the local ranchers who had to live with the prairie dogs were not nearly as fond of the little rodents. Prairie dogs are effective herbivores that consume a great deal of plant biomass, including agricultural crops. The ranchers viewed prairie dogs as competitors for their crops and sought a variety of ways, such as poisoning, to eradicate them. An unintended consequence of the ranchers' victory against the prairie dogs was the near-extinction of another species, the black-footed ferret (*Mustela nigripes*).

The black-footed ferret is a member of the weasel family and the only species of ferret native to North America. It lives in burrows and preys on prairie dogs. In fact, a single ferret can consume 125 to 150 prairie dogs in a year. Although prairie dogs had been the ferrets' main source of food for millions of years, things started to change when settlers began plowing the Great Plains for agriculture. This destroyed many prairie dog towns, while other prairie dog towns were poisoned. Together, the plowing and poisoning have reduced the prairie dog population by 98 percent.

As we would expect in a density-dependent scenario, the reduction in the prairie dog population reduced the carrying capacity of the environment for the black-footed ferret. The poisoning campaign poisoned ferrets as well as prairie dogs. Because both small and large populations of ferrets were poisoned, and a large proportion of ferrets died in both cases, the poisoning had a density-independent effect on the ferret population.

In 1967, the black-footed ferret was officially listed as an endangered species, and the last known population died out in 1974. People feared that the black-footed ferret was extinct. In 1981, however, a small population of 130 ferrets was discovered in Wyoming. Conservation efforts began, but a highly lethal disease known as canine distemper passed through the population, reducing the entire species to a mere 18 animals.

As part of a collaboration between federal biologists, private landowners, and several zoos, all 18 ferrets were immediately brought into captivity in the hope that a captive breeding program might be able to restore their numbers. The black-footed ferret is a *K*-selected species; it breeds once a year and has 3 to 4 offspring, for which the parents provide a great deal of parental care. As a result, population increases could not occur rapidly. Nevertheless, the captive population grew to 120 animals by 1989.

At this point, the captive breeding program was considered successful enough to allow reintroduction of the species into the wild. However, the biologists first had to consider exactly where they wanted to reintroduce the ferrets. They understood the risks of introducing all the animals into a single site because another round of disease could kill the entire population. To manage the reintroduction process, the biologists decided to make use of the metapopulation concept. They reintroduced the ferrets in several places across the Great Plains, choosing locations with healthy prairie dog populations that had a low risk of extermination.

From 1991 to 2013, black-footed ferrets were reintroduced at 19 sites, from northern Mexico all the way up to southern Canada. These efforts are clearly paying off. Today, more than 1,000 ferrets live in the wild, and hundreds more are part of the ongoing captive breeding program. But the rescue effort is not over. During years of high rainfall, outbreaks of plague, a bacterial disease carried by fleas, kill many ferrets. To combat plague, biologists have been dusting prairie dog towns with an insecticide powder that kills fleas. They are also working on a new vaccine against plague that they hope to distribute to prairie dog towns using drones. As of 2017, the ferrets have been introduced to 28 sites throughout the United States, Mexico, and Canada and their total population size numbers in the hundreds. However, they continue to face the challenges of a reduced carrying capacity during drought years and higher populations of plague-carrying fleas during wet years.

Black-footed ferret. Once critically endangered, populations of the black-footed ferret are rebounding as a result of collaborative conservation efforts. *(USFWS/National Black-footed Ferret Conservation Center)*

Critical Thinking Questions

1. How can humans balance the interests of ranchers who want to control prairie dog populations with the interests of conservationists who want to prevent the extinction of the black-footed ferret?
2. How has the metapopulation concept helped conservationists manage the recovery of the black-footed ferret?

References

Black-footed Ferret Recovery Implementation Team. *The Black-footed Ferret Recovery Program*. http://www.blackfootedferret.org

Robbins, J. 2008. Efforts on 2 fronts to save a population of ferrets. *New York Times*, July 15.

Rogers, N. 2016. Black-footed ferret recovery comes full circle. http://wildlife.org/black-footed-ferret-recovery-comesfull-circle/.

Chapter 6 Review

Community ecology examines how species interactions help to determine the species that are present in a community. Different characteristics of populations affect their abundance and distribution including density-dependent and density-independent factors. Population growth models help us understand how populations increase and decrease over time. Species also have distinctive reproductive strategies and growth curves, which affect population size and characteristics. At the community level, major types of species interactions include competition, predation, parasitism, mutualism, and commensalism. Communities experience ecological succession. The species richness of a given community depends on latitude, elapsed time, habitat size, and habitat distance to other sources of species.

Key Terms

Population
Community
Population ecology
Population size (N)
Population density
Population distribution
Sex ratio
Age structure
Limiting resource
Density-dependent factor
Carrying capacity (K)
Density-independent factor
Population growth model
Population growth rate
Intrinsic growth rate (r)
Exponential growth model ($N_t = N_0e^{rt}$)
J-shaped curve
Logistic growth model
S-shaped curve
Overshoot
Die-off
K-selected species
r-selected species
Survivorship curve
Type I survivorship curve
Type II survivorship curve
Type III survivorship curve
Corridor
Metapopulation
Inbreeding depression
Community ecology
Symbiotic relationship
Competition
Competitive exclusion principle
Resource partitioning
Predation
Parasitoid
Parasitism
Pathogen
Herbivory
Mutualism
Commensalism
Keystone species
Ecosystem engineer
Ecological succession
Primary succession
Secondary succession
Pioneer species
Climax community
Theory of island biogeography

Learning Goals Revisited

Module 18 The Abundance and Distribution of Populations

Explain how nature exists at several levels of complexity.

Nature exists at several levels of complexity: individuals, populations, communities, ecosystems, and the biosphere.

Discuss the characteristics of populations.

Populations can have distinct population sizes, densities, distributions, sex ratios, and age structures.

Contrast the effects of density-dependent and density-independent factors on population growth.

Density-dependent factors influence an individual's probability of survival and reproduction in a manner that is related to the size of the population. Density-independent factors have the same effect on an individual's probability of survival and reproduction in populations of any size.

Module 19 Population Growth Models

Explain the exponential growth model of populations, which produces a J-shaped curve.

The exponential growth model describes rapid growth under ideal conditions when resources are not limited. The J-shaped curve occurs because the population initially grows slowly when few individuals are present to reproduce but then grows rapidly as the number of reproducing individuals increases.

Describe how the logistic growth model incorporates a carrying capacity and produces an S-shaped curve.

The logistic growth model incorporates density-dependent factors that allow rapid initial growth but then cause population growth to slow down as populations approach their carrying capacity. When we allow lag times between when resources change in abundance and when populations produce new offspring, we can observe population overshoots and die-offs.

Compare the reproductive strategies and survivorship curves of different species.

Organisms have a range of reproductive strategies. At the extremes are r-selected species, which experience rapid population growth rates, and K-selected species, which experience high survivorship and slow population growth rates. Patterns of survivorship over the life span can be graphically represented as type I, II, and III survivorship curves.

Explain the dynamics that occur in metapopulations.

Metapopulations are groups of spatially distinct populations that are connected by occasional movements of individuals. These movements reduce the probability of any of the populations going extinct.

Module 20 Community Ecology

Identify species interactions that cause negative effects on one or both species.

Competition is an interaction between two species that share a limiting resource. Over time, competition for a resource can cause natural selection to favor those individuals that have reduced overlap in resource use and this can lead to spatial, temporal, or morphological resource partitioning. Predation is an interaction in which animals partially or entirely consume another animal. Predators can affect the abundance of prey populations and cause the evolution of antipredator defenses in prey populations. Parasitism is an interaction in which one organism lives on or in another organism. Those parasites that can cause disease in their hosts are known as pathogens. Herbivory is an interaction in which animals consume producers. In some cases, herbivores can have dramatic effects on plant and algal communities by removing the most palatable species.

Discuss species interactions that cause neutral or positive effects on both species.

Mutualisms are interactions that benefit two interacting species by increasing the chances of survival or reproduction for both. One of the most common mutualisms is the interaction between flowering plants and their pollinators. A second well-known mutualism is between acacia trees and the ants that defend the trees in exchange for food and a place to live. Commensalisms are interactions in which one species benefits but the other is neither harmed nor helped. Examples include birds perching on trees and marine fish using coral reefs for protection from predators.

Explain the role of keystone species.

Keystone species play a role in the community that is far more important than their relative abundance might suggest. Common examples include predatory sea stars that alter the outcome of competition in intertidal communities and beavers that create large ponds by constructing dams on streams.

Module 21 Community Succession

Explain the process of primary succession.

Primary succession occurs on surfaces that are initially devoid of soil, such as bare rock that is exposed after the retreat of glaciers or the cooled lava from a volcanic eruption. Over time, plants and animals arrive at the site and modify the environment, making it more favorable for other species to arrive and persist.

Explain the process of secondary succession.

Secondary succession occurs in areas that have been disturbed but have not lost their soil. A common example is the bare soil left behind when farmers stop planting crops in a field. Over time, plants and animals colonize the site, alter the environmental conditions, and favor the persistence of other species.

Explain the process of aquatic succession.

Lakes and ponds experience sedimentation over long periods of time and this slowly fills in the basin. Over thousands of years, the lakes and ponds can be slowly converted into terrestrial habitats.

Describe the factors that determine the species richness of a community.

The species richness of a community is typically higher at latitudes that are closer to the equator. Richness is also higher in older sites where evolution has been producing new species for longer periods of time. Finally, more species exist in larger habitats and habitats that are closer to sources of new species, as is the case for oceanic islands that are located close to continents.

Practice Math and Graphing

Preparing for the AP® Exam

Answer the following questions. Be sure to show all your work.

1. Practice Math

As we have seen, black-footed ferrets live in prairie dog towns and the prairie dogs often are parasitized by ticks. Using the abundances below and the size of the area that was sampled, calculate the density of each species and then convert each density into scientific notation.

Species	Number of individuals	Area	Density	Density in scientific notation
Ferrets	7	1,000 m^2		
Prairie dogs	85	1,000 m^2		
Ticks	950	1,000 m^2		

2. Practice Graphing

(a) Use the data in the table to create a line graph that illustrates the growth of a population over time.

(b) Why might the population growth slow down over time?

(c) Explain whether these data represent density-dependent or density-independent growth.

Time	Population size
1	10
2	11
3	14
4	19
5	26
6	31
7	34
8	35
9	35
10	35

Section 1: Multiple-Choice Questions

Choose the best answer for questions 1–20.

1. Which is NOT an example of a density-independent factor?
   (a) competition
   (b) forest fire
   (c) hurricane
   (d) flood

2. As the size of a white-tailed deer population increases,
   (a) the carrying capacity of the environment for white-tailed deer will be reduced.
   (b) a volcanic eruption will have a greater proportional effect than it would on a smaller population.
   (c) the effect of limiting resources will decrease.
   (d) the number of gray wolves, a natural predator of white-tailed deer, will increase.

3. The following graph shows population growth of Canada geese in Ohio between 1955 and 2002.

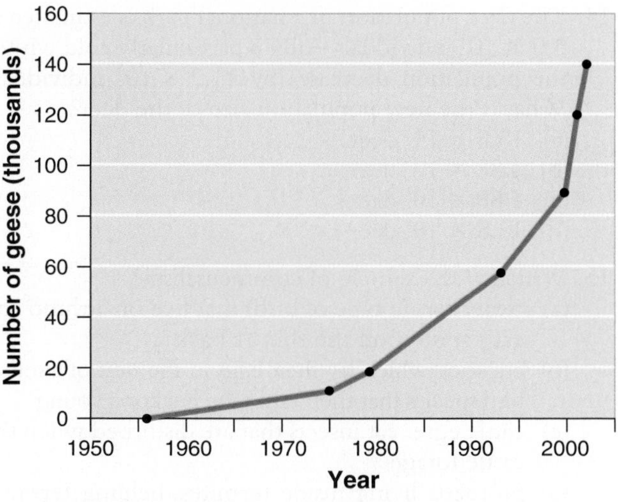

   This graph can best be described as:
   (a) an exponential growth curve.
   (b) a logistic growth curve.
   (c) a stochastic growth curve.
   (d) oscillation between overshoot and die-off.

4. Which is NOT a true statement based on the logistic growth model?
   (a) Population growth is limited by density-dependent factors.
   (b) Future population growth cannot be predicted mathematically.
   (c) Population growth slows as the number of individuals approaches the carrying capacity.
   (d) A graph of population growth produces an S-shaped growth curve over time.

5. Which characteristic is typical of *r*-selected species?
   I. They produce many offspring in a short period of time.
   II. They have very low survivorship early in life.
   III. They take a long time to reach reproductive maturity.
   (a) I only
   (b) III only
   (c) I and II
   (d) II and III

6. A high intrinsic growth rate would most likely be characteristic of
   (a) a *K*-selected species such as elephants.
   (b) an *r*-selected species such as the American bullfrog.
   (c) a *K*-selected species that lives close to its carrying capacity.
   (d) a species with a low reproductive rate that takes a long time to reach reproductive maturity.

7. The following graph presents three survivorship curves.

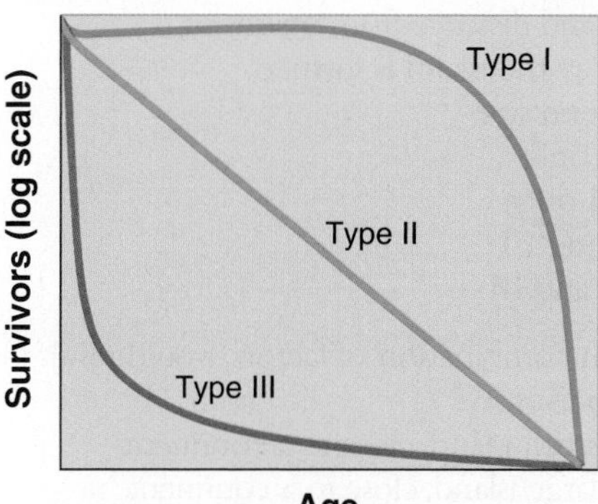

   Choose the description that best describes the graph.
   (a) Type I could represent the house mouse, which is a typical *r*-selected species.
   (b) Type III could represent elephants, which are typical *K*-selected species.
   (c) Type I could represent an oak tree species that experiences very low survivorship early and late in life.
   (d) Type II could represent a coral species that has a constant decline in survivorship throughout its life.

8. In the coniferous forests of Oregon, eight species of woodpeckers coexist. Four species select their nesting sites based on tree diameter, while the fifth species nests only in fir trees that have been dead for at least 10 years. The sixth species also nests in fir trees, but only in live or recently dead trees. The two remaining species nest in pine trees, but each selects trees of different sizes. This pattern is an example of
   (a) resource partitioning.
   (b) commensalism.
   (c) predator-mediated competition.
   (d) a keystone species.

9. Which statement about ecological succession is correct?
   (a) Secondary succession begins in a community lacking soil.
   (b) Succession is influenced by competition for limiting resources such as available soil, moisture, and nutrients.
   (c) In forest succession, less shade-tolerant trees replace more shade-tolerant trees.
   (d) Forest fires and hurricanes lead to primary succession because a soil base still exists.

10. Which sequence of secondary succession would be likely to occur in abandoned farmland in the eastern United States?
   (a) bare soil, lichens, mosses, grasses, deciduous trees
   (b) bare rock, lichens, mosses, grasses, shrubs, mixed shade- and sunlight-tolerant trees
   (c) bare soil, grasses and wildflowers, shrubs, sunlight-tolerant trees, shade-tolerant trees
   (d) bare rock, grasses and wildflowers, lichens, mosses, shrubs, coniferous trees

11. The theory of island biogeography suggests that species richness is affected by which of the following factors?
   I. island distance from mainland
   II. how the island is formed
   III. island size
   (a) II only
   (b) III only
   (c) I and III
   (d) II and III

12. Which combination of factors would result in the highest biodiversity?
   (a) a small island, close to a continent
   (b) a large island, close to a continent
   (c) a small island, distant from a continent
   (d) a large island, distant from a continent

13. Scientists track mouse density in a particular forest over time and produce the following graph.

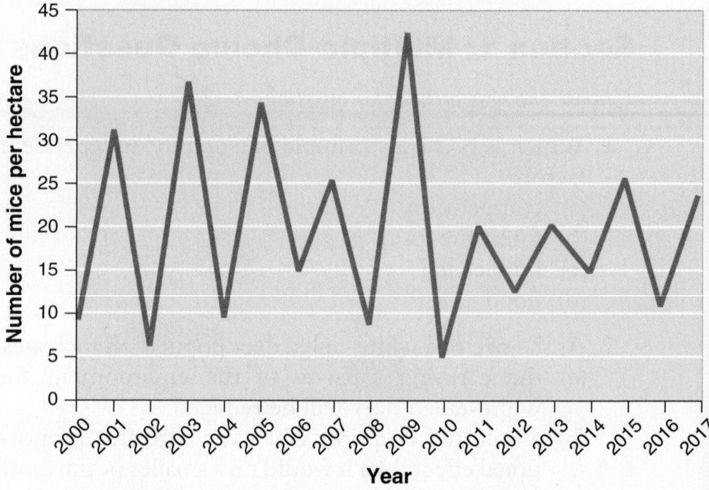

   What could be causing the pattern in the figure?
   (a) depletion of food resources
   (b) predation
   (c) variation in food availability and predation
   (d) all of the above

14. The deer population in a national park is estimated as 3.0×10^4 individuals. After a particularly cold winter, the population decreases by 11.2×10^3 individuals. What is the new population size of the deer?
   (a) 18.8×10^1 deer
   (b) 1.88×10^1 deer
   (c) 1.88×10^4 deer
   (d) 18.8×10^4 deer

15. Which is an example of commensalism?
   (a) oxpeckers (a type of bird) that live on a rhino and eat parasites off the rhino's back
   (b) cuckoos, which lay their eggs in the nest of another bird species that then raises the cuckoo's young
   (c) cattle egret eat insects that are disturbed when the cattle forage
   (d) protozoa living inside termites, helping termites to digest wood

16. Which is least likely to affect the carrying capacity of a habitat for a particular species?
   (a) limited food
   (b) density-dependent factors
   (c) abundance of competing species
   (d) population distribution

17. Suppose that a population of squirrels can increase by 70 percent each year under ideal conditions. For a population of 30 squirrels living in a temperate forest, what is the most likely number of squirrels after 3 years of population growth?
   (a) $30 \times e^{0.5\times3}$
   (b) $30 \times e^{0.7\times3}$
   (c) $70 \times e^{30\times3}$
   (d) $0.7 \times e^{30\times3}$

18. Which is a requirement for the existence of a metapopulation?
    I. a system of habitat corridors that link individual populations
    II. the presence of both *r*- and *K*-selected species
    III. variation in the size of individual populations
    (a) I only
    (b) II only
    (c) I and II
    (d) II and III

19. Which is NOT true regarding keystone species?
    (a) Keystone species can be mutualists with other species.
    (b) Keystone species are often rare individuals in a community.
    (c) Keystone species are most likely to be commensalists.
    (d) Some keystone species are habitat engineers.

20. During the last Ice Age, glaciers covered much of the higher latitudes in North America. Considering the factors that influence the diversity of a community, which of the following is probably NOT true?
    (a) At northern latitudes, large lakes contain more fish species than small lakes.
    (b) At northern latitudes, large lakes have larger populations of each species than small lakes.
    (c) At northern latitudes, large lakes have experienced fewer species extinctions than small lakes.
    (d) At northern latitudes, small lakes were colonized by new species faster than large lakes.

Section 2: Free-Response Questions

Write your answer to each part clearly. Support your answers with relevant information and examples. Where calculations are required, show your work.

1. The California Department of Fish and Wildlife is developing a plan to connect mountain "habitat islands" that are separated by open areas of flat, arid land in the deserts of southeastern California. These mountain areas are habitats for desert bighorn sheep (*Ovis canadensis*), which move extensively among the islands through habitat corridors. The habitat corridors provide opportunities for recolonization, seasonal migration, and maintenance of genetic variation among the metapopulation of desert bighorn sheep.
    (a) Explain what is meant by a metapopulation and how it relates to the desert bighorn sheep. (1 point)
    (b) Identify two density-dependent factors and one density-independent factor that could affect the populations of desert bighorn sheep. (3 points)
    (c) Explain the consequences to the desert bighorn sheep population if the plan to connect the mountain habitat islands is not implemented. (2 points)
    (d) Explain how the theory of island biogeography applies to the mountainous areas of southeastern California. (4 points)

2. *Read the following information, which was posted in the great apes exhibit of the Fremont Zoo, and answer the questions that follow.*

    The western lowland gorilla (*Gorilla gorilla*) lives in the moist tropical rainforests of western Africa. The gorillas' primary diet consists of fruits, leaves, foliage, and sometimes ants and termites. Occasionally they venture onto farms and feed on crops. Their only natural enemy is the leopard, the only animal other than humans that can successfully kill an adult gorilla. However, disease, particularly the Ebola virus, is threatening to decimate large populations of these gorillas in Congo.

    The western lowland gorilla has a life span of about 40 years and produces one offspring every 4 years. Males usually mate when they are 15 years old. Females reach sexual maturity at about 8 years of age, but rarely mate before they are 10 years old.

    (a) Based on the information above, is the western lowland gorilla an *r*-selected or a *K*-selected species? Provide evidence to support your answer. (3 points)
    (b) Identify and explain four community interactions that involve the western lowland gorilla. (4 points)
    (c) Explain what is meant by secondary succession and describe how it may be initiated in a tropical forest. (3 points)

3. A species of woodpecker inhabits a reserve adjacent to a suburban housing development. The species nests in cavities that it excavates in large trees, and rarely crosses roads or areas that do not provide at least some vegetative cover. The population inside the reserve is small and relatively isolated from other larger populations in the region.
    (a) Describe ONE density-dependent and ONE density-independent factor that could affect the woodpecker population in the reserve. (2 points)
    (b) Describe TWO potential problems with the isolation and lack of gene flow the population experiences due to the housing development. (2 points)
    (c) The particular species of woodpecker is an *r*-selected species. Explain what this is, including THREE typical characteristics of an *r*-selected species. (4 points)
    (d) After creating and nesting in a tree cavity, the woodpeckers abandon the cavity, allowing it to be used by several other species of cavity-nesting birds that are unable to excavate their own cavities. Based on this description, identify the role the woodpecker species plays in its ecosystem. (2 points)

The rapidly growing middle class in China has created a large increase in the consumption of consumer goods and other materials that impact the environment. *(Lucas Schifres/Getty Images)*

The Human Population

CHAPTER 7

MODULE 22 Human Population Numbers

MODULE 23 Economic Development, Consumption, and Sustainability

CASE STUDY

The Environmental Implications of China's Large Population

Human population size, affluence, and resource consumption all have interrelated impacts on the environment. The example of China is striking. With more than 1.3 billion people—almost 20 percent of the human population on Earth—China is the world's most populous nation. Because of its rapid economic development, it may one day become the world's largest economy. Once-scarce consumer goods such as automobiles and refrigerators are becoming increasingly commonplace in China. Although the United States, with more than 325 million people, has historically been the world's largest consumer of resources and the greatest producer of many pollutants, China is rapidly surpassing the United States in both consumption and pollution. It is already the largest emitter of carbon dioxide and sulfur dioxide, and it consumes one-third of commercial fish and seafood. The Chinese are facing considerable environmental challenges as their affluence increases.

To manage the presence of humans on Earth in a sustainable way, we must address both population growth and resource consumption. China has already taken dramatic steps to limit its population growth. For over 3 decades, China had a "one-child" policy. Couples that restricted themselves to a single child were rewarded financially, while those with three or more children faced sanctions, such as a 10 percent salary reduction. Chinese officials used numerous tools—many controversial—to meet population targets, including abortions, sterilizations, and the designation of certain pregnancies as "illegal."

The Chinese are facing considerable environmental challenges as their affluence increases.

China is one of only a few countries where government-mandated population control measures contributed to reducing population growth. After decades of having one of the world's highest fertility rates, China now has a fertility rate of 1.6 births per woman. In addition to the one-child policy, economic prosperity has also contributed to slowing population growth. If China's current population dynamics continue, its population will reach a maximum by approximately 2030 and begin declining shortly after that.

Population decline is only one part of the picture, however. Even if China's population were to stop growing today, the country's resource consumption would continue to increase as standards of living improve. The middle class in China is approaching 500 million people, which is larger than the total population of every country in the world other than China and India. Greater numbers of Chinese people are purchasing cars, home appliances, and other material goods that people in Western nations commonly own. All of these products require resources to produce and use. Manufacturing a refrigerator requires mining and processing raw materials such as steel and copper, producing plastic from oil, and using large quantities of electricity. Having a refrigerator in the home increases daily electricity demand. All of these processes generate carbon dioxide, air and water pollution, and other waste products.

A look at a typical Chinese city street is evidence of the country's growing affluence. Between 1985 and 2002, China's population increased 30 percent, but the number of motor vehicles used in

China grew by over 500 percent, from 3 million to 20 million. According to current Chinese government estimates, China now has over 170 million vehicles. China is the second largest consumer of petroleum (after the United States), and concentrations of urban air pollutants, such as carbon monoxide and photochemical smog, are high relative to many other countries. In recent years, Chinese cities have often been among the top 10 for levels of various pollutants. A study in the *Proceedings of the National Academy of Sciences of the United States of America* documented that air pollution from China travels across the Pacific Ocean and has resulted in declining air quality in the western United States.

There is some good news, however. China already has higher fuel efficiency standards for cars than does the United States, and it is a leader in the manufacturing of renewable energy technologies. In 2017, China announced its intention to ban gasoline- and diesel-powered automobiles at some point in the future, but they have not released a target date yet.

China's influence on the environment is dramatic because of its size and its increasing industrial activity. However, increasing industrial activity is occurring in many other parts of the world as human populations and consumption increase. In fact, the population of India is expected to surpass that of China by 2025. Even today, the middle class in India may be larger than that in China. What will be the environmental impact of humans in 2025 and beyond, and what can we do to reduce it?

Sources: H. Kan, B. Chen, and C. Hong, Health impact of outdoor air pollution in China: Current knowledge and future research needs, *Environmental Health Perspectives* 117 (2009): A187; J. Lin, et al., China's international trade and air pollution in the United States, *Proceedings of the National Academy of Sciences of the United States of America* 111(2014) 1736–1741.

Human population growth and associated resource consumption have large impacts on the environment. In 2019, Earth's human population was 7.6 billion and increasing by 225,000 people per day. At the same time, people in many parts of the world are using more resources than ever before. Almost all aspects of environmental science are affected by the numbers of people on Earth and their activities. This chapter describes how human populations grow, what limits human population numbers, and the relative impacts of human population size and human consumption behavior.

MODULE 22

Human Population Numbers

Although it may seem surprising, environmental scientists do not know the human carrying capacity of Earth. However, we are able to discuss the factors that contribute to the carrying capacity of humans on Earth and what drives human population growth. We can also look at patterns of past human population growth to gain an understanding of what might occur in future decades.

Learning Goals

After reading this module you should be able to

- explain factors that may potentially limit the carrying capacity of humans on Earth.
- describe the drivers of human population growth.
- read and interpret an age structure diagram.

Scientists disagree on Earth's carrying capacity

Every 5 days, the global human population increases by roughly 1.3 million lives: 2.1 million infants are born and 800,000 people die. The human population has not always grown at this rate, however. As **FIGURE 22.1** shows, until a few hundred years ago the human population was relatively stable: Deaths and births occurred in roughly equal numbers. This situation changed about 400 years ago, when agricultural output increased and sanitation began to improve. Better living conditions caused death rates to fall, but birth rates remained relatively high. This was the beginning of a period of rapid population growth that has brought us to the current human population of 7.6 billion people.

As we saw in Chapter 6, under ideal conditions all populations grow exponentially. In most cases, exponential growth slows or stops when an environmental limit is reached. The limiting factor can be a scarcity of resources such as food or water or an increase in predators, parasites, or diseases. Limiting factors determine the carrying capacity of a habitat. Are human populations constrained by limiting factors?

Environmental scientists have differing opinions on Earth's carrying capacity for humans. Some scientists believe we have already outgrown, or eventually will outgrow, the available supply of food, water, timber, fuel, and other resources on which humans rely. One of the first proponents of the notion that the human population could exceed Earth's carrying capacity was English clergyman and professor Thomas Malthus. In 1798, Malthus observed that the human population was growing

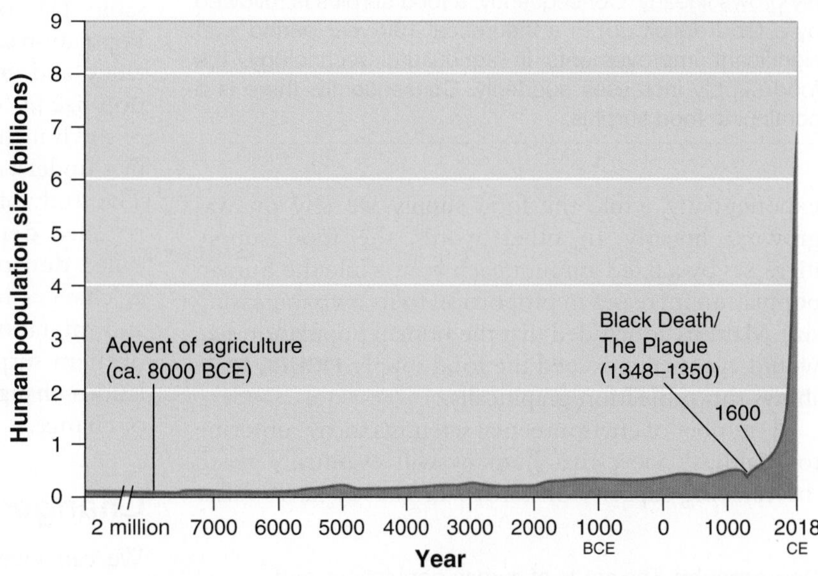

FIGURE 22.1 Human population growth. The global human population has grown more rapidly in the last 400 years than at any other time in history.

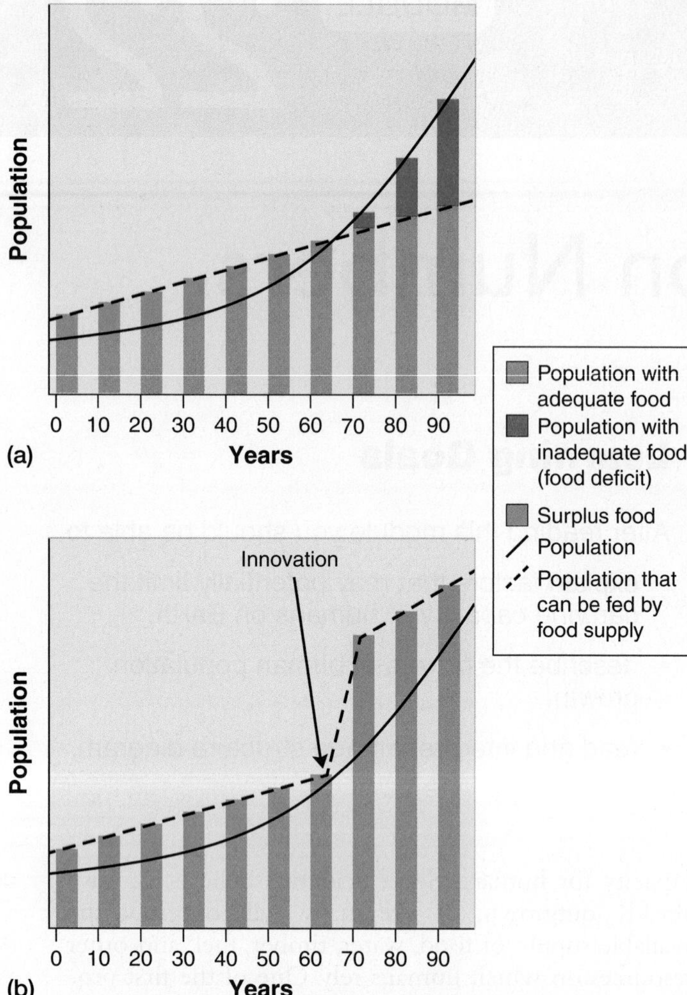

FIGURE 22.2 A theoretical model of food supply and population size. (a) In a theoretical 100-year period *without* significant improvements in agricultural technology, the human population grows exponentially, while the food supply grows linearly. Consequently, a food surplus is followed by a food deficit. (b) In a theoretical 100-year period *with* significant improvements in agricultural technology, the food supply increases suddenly. Consequently, there is a continuing food surplus.

exponentially while the food supply we rely on was growing linearly. In other words, the food supply increases by a fixed amount each year, while the human population increases in proportion to its own increasing size. Malthus concluded that the human population size would eventually exceed the food supply. **FIGURE 22.2a** shows this projection graphically.

A number of environmental scientists today subscribe to Malthus's view that humans will eventually reach the carrying capacity of Earth—or already have—after which the rate of population growth will decline. Other scientists do not believe that Earth has a fixed carrying capacity for humans. They argue that the growing population of humans provides an increasing supply of intellect that leads to increasing amounts of innovation. Humans can alter Earth's carrying capacity by employing creativity—one of the fundamental ways in which humans differ from most other species on Earth.

For example, in the past whenever the food supply seemed small enough to limit the human population, major technological advances increased food production. This progression began thousands of years ago. The development of arrows made hunting more efficient, which allowed hunters to feed a larger number of people. Early farmers increased crop yields with hand plows and later with oxen- or horse-drawn plows. More recently, mechanical harvesters made farming even more efficient. Each of these inventions increased the planet's carrying capacity for humans, as Figure 22.2b shows.

The ability of humans to innovate in the face of challenges has led some scientists to expect that we will continue to make technological advances indefinitely. This expectation is reasonable, but questions remain. Based on our history, should we assume that humans will continue to find ways to feed a growing population? Are there other limits to human population growth? And how do we know if we have exceeded Earth's carrying capacity?

Demography The study of human populations and population trends.

Demographer A scientist in the field of demography.

Many factors drive human population growth

As we know from our study of biological populations, a variety of factors influence the growth, reproduction, and success of plant and animal species. Many of these same factors influence human populations as well. Population size, birth and death rates, fertility, life expectancy, and migration are factors that influence population size in countries. In order to understand the impact of the human population on the environment, we must first understand what drives human population growth. The study of human populations and population trends is called **demography**, and scientists in this field are called **demographers**. By analyzing specific data such as changes in population size, fertility, life expectancy, and migration, demographers can offer insights—some of them surprising—into how and why human populations change and what can be done to influence rates of change.

Changes in Population Size

We can view the human population as a system with inputs and outputs, like all biological systems. If there are more births than deaths, the inputs are greater than the outputs, and the system expands. For most of

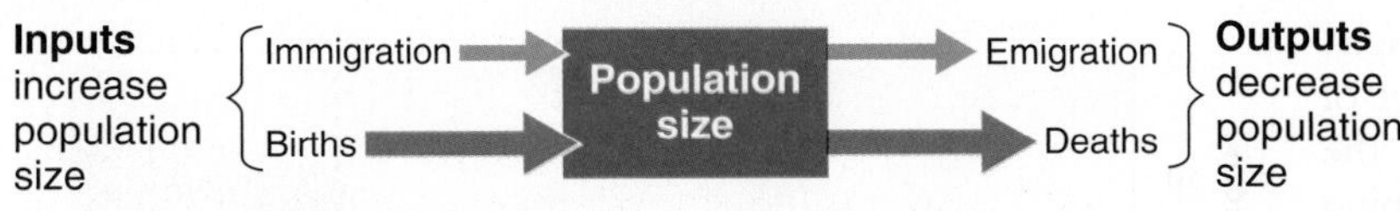

FIGURE 22.3 The human population as a system. We can think of the human population as a system, with births and immigration as inputs and deaths and emigration as outputs.

human history, total births slightly outnumbered total deaths, resulting in very slow population growth. If the reverse had been true, the human population would have decreased and would have eventually become extinct.

When demographers look at population trends in individual countries, they take into account inputs and outputs. As **FIGURE 22.3** shows, inputs include both births and **immigration**, which is the movement of people into a country or region from another country or region. Outputs, or decreases, include deaths and **emigration**, which is the movement of people out of a country or region. When inputs to the population are greater than outputs, the growth rate is positive. Conversely, if outputs are greater than inputs, the growth rate is negative.

Demographers use specific measurements to determine yearly birth and death rates. The **crude birth rate (CBR)** is the number of births per 1,000 individuals per year. The **crude death rate (CDR)** is the number of deaths per 1,000 individuals per year. Worldwide, there were 20 births and 8 deaths per 1,000 people in 2017. We do not factor in migration for the global population because, even though people move from place to place, they do not leave Earth. Thus, in 2017, the global population increased by 12 people per 1,000 people. This rate can be expressed mathematically as a percentage:

$$\text{Global population growth rate in percent} = \frac{[\text{CBR} - \text{CDR}]}{10} = \frac{[20 - 8]}{10} = \frac{[12]}{10} = 1.2\%$$

To calculate the population growth rate for a single nation, we take immigration and emigration into account:

$$\text{National population \% growth rate} = \frac{[(\text{CBR} + \text{immigration}) - (\text{CDR} + \text{emigration})]}{10}$$

If we know the growth rate of a population and assume that growth rate is constant, we can calculate the number of years it takes for a population to double, which is known as its **doubling time**. As a population grows rapidly, the doubling time gives us a better sense of the magnitude of the change than the growth rate alone. Because growth rates may change in future years, we can never determine a country's doubling time with certainty. Therefore, we say that a population will double in a certain number of years if the growth rate remains constant.

The doubling time can be approximated mathematically using a formula called the rule of 70:

$$\text{Doubling time (years)} = \frac{70}{\text{growth rate (expressed in \%)}}$$

Therefore, a population growing at 2 percent per year will double every 35 years:

$$\frac{70}{2} = 35 \text{ years}$$

Note that this is true of any population growing at 2 percent per year, regardless of the size of that population. At a 2 percent growth rate, a population of 50,000 people will increase by 50,000 in 35 years, and a population of 50 million people will increase by 50 million in 35 years.

As we saw in Figure 22.1, Earth's population has doubled several times since 1600. It is almost certain, however, that Earth's population will not double again. **FIGURE 22.4** on page 238 shows the current projections through the year 2100. Most demographers believe that the human population will be somewhere between 8 billion and 10 billion in 2050 and will stabilize between 7 billion and 11 billion by roughly 2100.

AP® Exam Tip

You will be expected to know how to work with the rule of 70 on the AP® Environmental Science exam. Make sure you are familiar with the formula and know how to use it. ●

Fertility

To understand more about the role births play in population growth, demographers look at the **total fertility rate (TFR)**, an estimate of the average number of children that

Immigration The movement of people into a country or region, from another country or region.

Emigration The movement of people out of a country or region.

Crude birth rate (CBR) The number of births per 1,000 individuals per year.

Crude death rate (CDR) The number of deaths per 1,000 individuals per year.

Doubling time The number of years it takes a population to double.

Total fertility rate (TFR) An estimate of the average number of children that each woman in a population will bear throughout her childbearing years.

each woman in a population will bear throughout her childbearing years (between the onset of puberty and menopause). For example, in the United States in 2017, the TFR was 1.8, meaning that, on average, each woman of childbearing age gave birth to just under two children. Note that, unlike crude birth rate and crude death rate, TFR is not calculated per 1,000 people. Instead, it is a measure of births per woman over her lifetime.

To gauge changes in population size, demographers also calculate **replacement-level fertility**, the TFR required to offset the average number of deaths in a population so that the current population size remains stable. Typically, replacement-level fertility is just over two children. In theory, exactly two children will replace the two parents who conceived them. In practice, the replacement level fertility—the number of children needed to replace two parents—is higher because some children die before they are able to have children of their own, and some people do not have children.

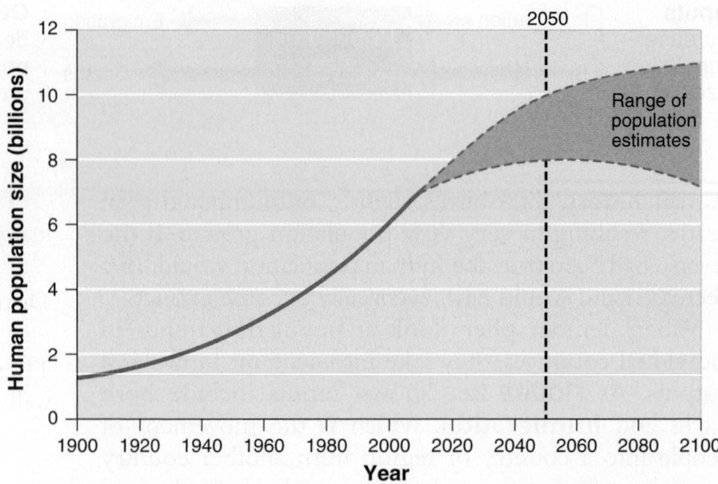

FIGURE 22.4 Projected world population growth. Demographers project that the global human population will be between 8 billion and 10 billion by 2050. By 2100, it is projected to be between 7 billion and 11 billion. The dashed lines represent estimated values. *(After Millennium Ecosystem Assessment, 2005; Population Reference Bureau (www.prb.org) 2017.)*

DO THE MATH Comparing Population Growth in Two Countries

Preparing for the AP® Exam

Hungary and Sweden are two European countries each with approximately 10 million people. When will the population of each country double? Hungary has a CBR of 10 and a CDR of 13. Sweden has a CBR of 12 and a CDR of 9. Ignore migration in and out of each country and calculate the national population growth rate of each country.

1. Hungary

$$(\text{CBR} - \text{CDR}) \div 10 = \frac{10 - 13}{10} = -0.3$$

So the growth rate is -0.3%.

The doubling time is $\frac{70}{-0.3} = -233$. Because this is a negative number, it means the time for the population to double will never occur.

At current birth and death rates, Hungary will never double in population.

2. Sweden

$$(\text{CBR} - \text{CDR}) \div 10 = \frac{12 - 9}{10} = 0.3$$

So the growth rate is 0.3%

The doubling time is $\frac{70}{0.3} = 233$ years.

At current birth and death rates, Sweden will double in size in 233 years.

YOUR TURN The Dominican Republic has 10.7 million people and a national population growth rate of 1.5 percent. Paraguay has 6.8 million people and a national population growth rate of 1.5 percent. In how many years will each country double in size?

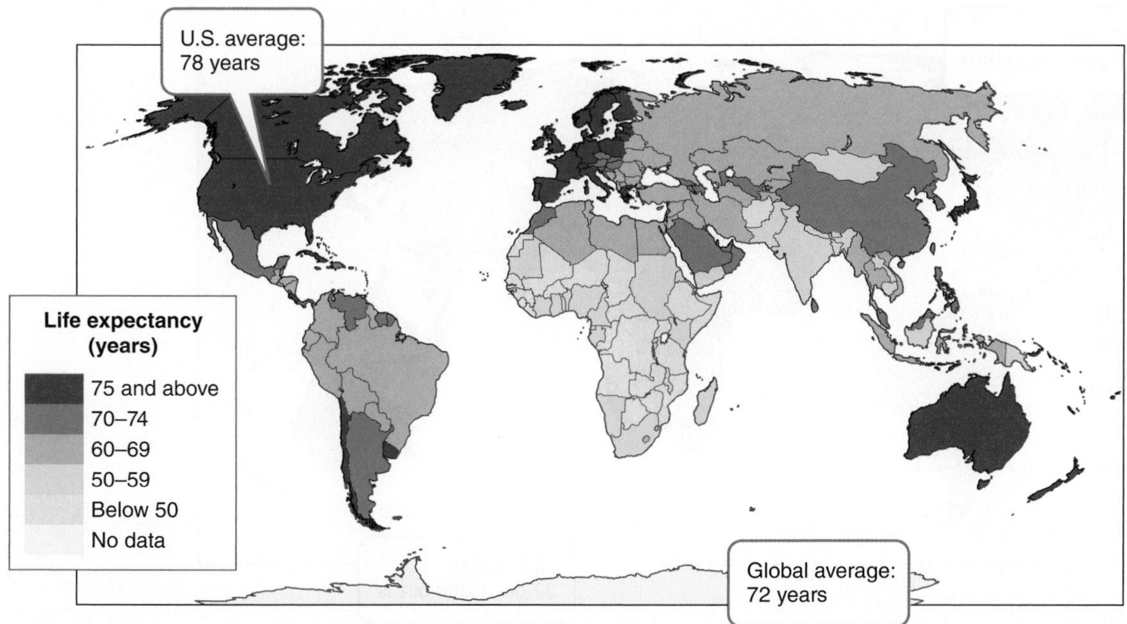

FIGURE 22.5 Average life expectancies around the world. Life expectancy varies significantly by continent and in some cases by country. *(Data from: https://www.americashealthrankings.org https://www.americashealthrankings .org/learn/reports/2015-annual-report/comparison-nations-2)*

As we shall see, the rate of death among children depends on a number of factors including the economic status of each country.

In **developed countries**—countries with relatively high levels of industrialization and income—we typically see a replacement-level fertility of about 2. In **developing countries**—those with relatively low levels of industrialization and incomes of less than $3 per person per day—a TFR of greater than 2.1 is needed to achieve replacement-level fertility. Replacement level fertility is higher in developing countries because mortality among young people tends to be higher.

In a country where TFR is equal to replacement-level fertility, and where immigration and emigration are equal, the country's population is stable. A country with a TFR of less than 2.1 and no net increase from immigration is likely to experience a population decrease because that country's TFR is below replacement-level fertility. In contrast, a developed country with a TFR of more than 2.1 and no net decrease from emigration is likely to experience population growth because that country's TFR is above replacement-level fertility.

Life Expectancy

To understand more about the outputs in a human population system, demographers study the human life span. **Life expectancy** is the average number of years that an infant born in a particular year in a particular country can be expected to live, given the current average life span and death rate in that country. Life expectancy is generally higher in countries with better health care. A high life expectancy also tends to be a good predictor of high resource consumption rates and environmental impacts. **FIGURE 22.5** shows life expectancies around the world.

Life expectancy is often reported in three different ways: for the overall population of a country, for males only, and for females only. For example, in 2017, global life expectancy was 72 years overall, 70 years for men, and 74 years for women. In the United States, life expectancy was 78 years overall, 76 years for men, and 81 years for women. In general, human males have higher death rates than human females, leading to a shorter life expectancy for men. In addition to biological factors, men have historically tended to face greater dangers in the workplace, made more hazardous lifestyle choices, and been more likely to die in wars. Cultures have changed over time, however, and as more and more women enter the workforce and the armed forces, the life expectancy gap between men and women will probably decrease.

Replacement-level fertility The total fertility rate required to offset the average number of deaths in a population in order to maintain the current population size.

Developed country A country with relatively high levels of industrialization and income.

Developing country A country with relatively low levels of industrialization and income.

Life expectancy The average number of years that an infant born in a particular year in a particular country can be expected to live, given the current average life span and death rate in that country.

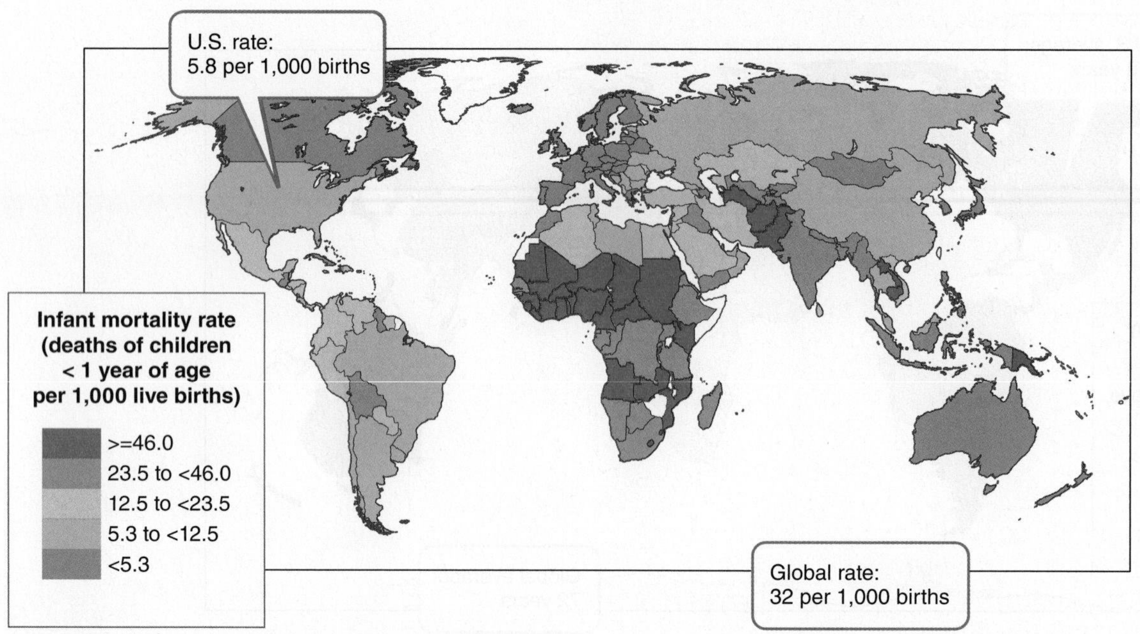

FIGURE 22.6 Infant mortality around the world. Infant mortality rates are lower in developed countries and higher in developing countries. *(Data from: https://www.americashealthrankings.org/learn/reports/2015-annual-report/comparison-nations-2.)*

Infant and Child Mortality

The availability of health care, access to good nutrition, and exposure to pollutants are all factors in life expectancy, *infant mortality,* and *child mortality*. The **infant mortality** rate is defined as the number of deaths of children under 1 year of age per 1,000 live births. The **child mortality** rate is defined as the number of deaths of children under age 5 per 1,000 live births. **FIGURE 22.6** shows infant mortality rates around the world.

If a country's life expectancy is relatively high and its infant mortality rate is relatively low, it is likely that the country has a high level of available health care, an adequate food supply, potable drinking water, good sanitation, and a moderate level of pollution. Conversely, if its life expectancy is relatively low and its infant mortality rate is relatively high, it is likely that the country's population does not have sufficient health care or sanitation and that potable drinking water and food are in limited supply. Pollution and exposure to other environmental hazards may also be high. In 2017, the global infant mortality rate was 32. In the United States, the infant mortality rate was 5.8. In other developed countries, such as Sweden (2.5) and France (3.5), the infant mortality rate was even lower. Availability of prenatal care is an important predictor of the infant mortality rate. For example, the infant mortality rate is 54 in Liberia and 39 in Bolivia, both countries where many women do not have adequate access to prenatal care.

Infant mortality The number of deaths of children under 1 year of age per 1,000 live births.

Child mortality The number of deaths of children under age 5 per 1,000 live births.

Sometimes, life expectancy and infant mortality in a given sector of a country's population differ widely from life expectancy and infant mortality in the country as a whole. In this case, even when the overall numbers seem to indicate a high level of health care throughout the country, the reality may be starkly different for a portion of its population. For example, whereas the infant mortality rate for the U.S. population as a whole is 5.8, in 2015, it was 12.0 for African Americans, 7.6 for Native Americans, and 5.3 for Caucasians. This variation in infant mortality rates is probably related to socioeconomic status and varying degrees of access to adequate nutrition, prenatal care, and overall health care. These differences are often issues of environmental justice, a topic we discuss in more detail in Chapter 20.

Aging and Disease

Even with a high life expectancy and a low infant mortality rate, a country may have a high crude death rate, in part because it has a large number of older individuals. The United States, for example, has a higher standard of living than Mexico, which is consistent with the higher life expectancy and lower infant mortality rate in the United States. At the same time, the United States has a much higher CDR, at 8 deaths per 1,000 people on average, than Mexico, which has 5 deaths per 1,000 people on average. This higher CDR results from the much larger elderly population in the United States, with 15 percent of its population aged 65 years or older, compared with the 6 percent of the population aged 65 or older in Mexico.

Disease is an important regulator of human populations. According to the World Health Organization, infectious diseases—those caused by microbes that are transmissible from one person to another—are the second biggest killer worldwide after heart disease. In the past, tuberculosis and malaria were two of the infectious diseases responsible for the greatest number of human deaths. Today, the human immunodeficiency virus (HIV), which causes acquired immune deficiency syndrome (AIDS), is responsible for more deaths annually than either tuberculosis or malaria. In 2016, there were 1 million AIDS-related deaths. Between 1990 and 2015, AIDS-related illnesses killed more than 30 million adults and children. Because HIV disproportionately infects people aged 15 to 49—the most productive years in a person's life span—HIV has had a more disruptive effect on society than other illnesses that affect the very young and the very old.

HIV has a significant effect on infant mortality, child mortality, population growth, and life expectancy. In Lesotho, in southern Africa, where 23 percent of the adult population was infected with HIV, life expectancy fell from 63 years in 1995 to 40 years in 2009. It has since been increasing and is now approximately 53.

As **FIGURE 22.7** shows, approximately 37 million people were living with HIV in 2016, with 26.5 million of them in Africa, south of the Sahara. The annual number of deaths due to AIDS reached a peak of 2.1 million in 2005 but has decreased since then. We will talk more about AIDS and other infectious diseases in Chapter 17.

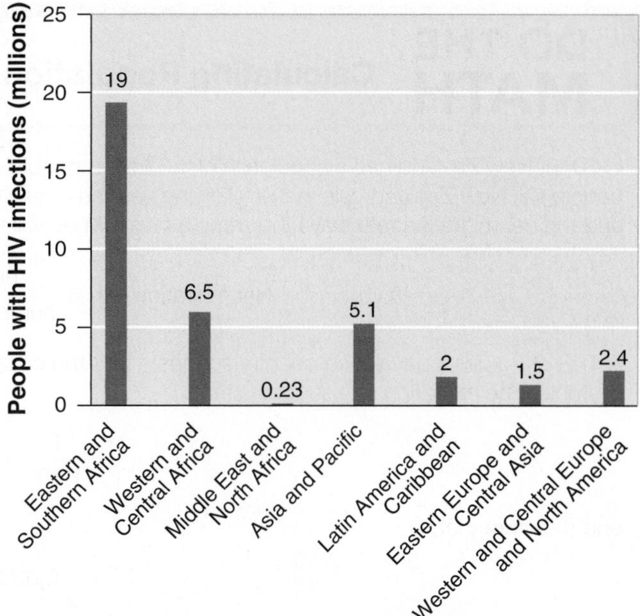

FIGURE 22.7 HIV infection worldwide. Worldwide, about 37 million people are living with HIV, two-thirds of them in sub-Saharan Africa, which includes all of Africa south of the Sahara (essentially all of Africa except North Africa). *(Data from: http://www.unaids.org/sites/default/files/media_asset/2016-AIDS-data_en.pdf.)*

Migration

Regardless of its birth and death rates, a country may experience population growth, stability, or decline as a result of migration. **Net migration rate** is the difference between immigration and emigration in a given year per 1,000 people in a country. A positive net migration rate means there is more immigration than emigration, and a negative net migration rate means the opposite. For example, approximately 1 million people immigrate to the United States each year, and only a small number emigrate. With a U.S. population of 325 million, these rates are equal to 3 immigrants per 1,000 people.

A country with a relatively low CBR but a high immigration rate may still experience population growth. For example, the United States has a TFR of 1.8, which is below replacement-level fertility, but it has a high net migration. As a result, the U.S. population will probably increase by 20 percent by 2050. Canada has a net migration rate of 9 per 1,000 and a TFR of 1.6, which is well below replacement-level fertility. The current population in Canada of 36.7 million will probably increase by 28 percent by 2050. Therefore, both the United States and Canada will experience net population growth over the next few decades, but the growth will come from immigration rather than from births originating within the existing population. "Do the Math: Calculating Population Growth" on page 242 shows how this can happen.

AP® Exam Tip

You are expected to know the approximate population sizes of the world and the three most populous countries, China, India, and the United States. AP® Environmental Science Exam problems may ask you to do a calculation that assumes you know this information. ●

Net migration rate The difference between immigration and emigration in a given year per 1,000 people in a country.

DO THE MATH Calculating Population Growth

Preparing for the AP® Exam

In 2012, New Zealand had a population of 4.3 million people, a TFR of 2.1, and a net migration rate of 2 per 1,000. How many people will New Zealand gain in the following year as a result of immigration? If the TFR stays the same for the next century, and the net migration rate stays the same as well, when will the population of New Zealand double?

$$\text{Net migration rate} = \frac{\text{number of immigrants/year}}{\text{number of people in the population}}$$

A TFR of 2.1 for a developed country suggests that the country is at replacement-level fertility and, therefore, the population is stable. The migration rate suggests that

$$\frac{2}{1{,}000} = \frac{x}{4{,}300{,}000}$$

and therefore that

$$x = 8{,}600 \text{ people/year}$$

So, 8,600 people are added to New Zealand each year by migration. If there is no growth due to biological replacement, then the rate of increase is

$$\frac{8{,}600 \text{ people/year}}{4{,}300{,}000 \text{ people/year}} = 0.002 \times 100\%/\text{year} = 0.2\%/\text{year}$$

YOUR TURN How many years will it take for the New Zealand population to double if the population increases due to migration only? Recall that to calculate doubling time, we use the rule of 70:

$$\text{Doubling time (years)} = \frac{70}{\text{growth rate}}$$

In countries with a negative net migration rate and a low TFR, the population actually decreases over time. Very few countries fit this model. One is the country of Georgia, in western Asia. In 2015, it had a growth rate of 0.2 percent, a TFR of 1.7, and a net migration rate of −5 per 1,000. Georgia is projected to have a 20 percent population decrease by 2050.

We have noted that although the movement of people around the world does not affect the total number of people on the planet, migration is still an important issue in environmental science. The movement of people displaced because of disease, natural disasters, environmental problems, or conflict can create crowded, unsanitary conditions, and shortages of food and water. In some cases people are moved into refugee camps where they have little opportunity to improve their conditions through employment or emigration. All of these situations can easily become humanitarian and environmental health issues. The movement of people from developing countries to developed countries tends to increase the ecological footprint of those people because, over time, immigrants typically adopt the lifestyle and consumption habits of their new country. A person who migrates from Mexico to the United States, for example, is likely to use more resources as a U.S. resident than as a resident of Mexico because the United States typically has a more affluent lifestyle.

Age structure diagram A visual representation of the number of individuals within specific age groups for a country, typically expressed for males and females.

Age structure diagrams describe how populations are distributed across age ranges

Demographers use data on age to predict how rapidly a population will increase and what its size will be in the future. The age structure of a population describes how its members are distributed across age ranges, usually in 5-year increments. **Age structure diagrams**, examples of which are shown in **FIGURE 22.8**, are visual

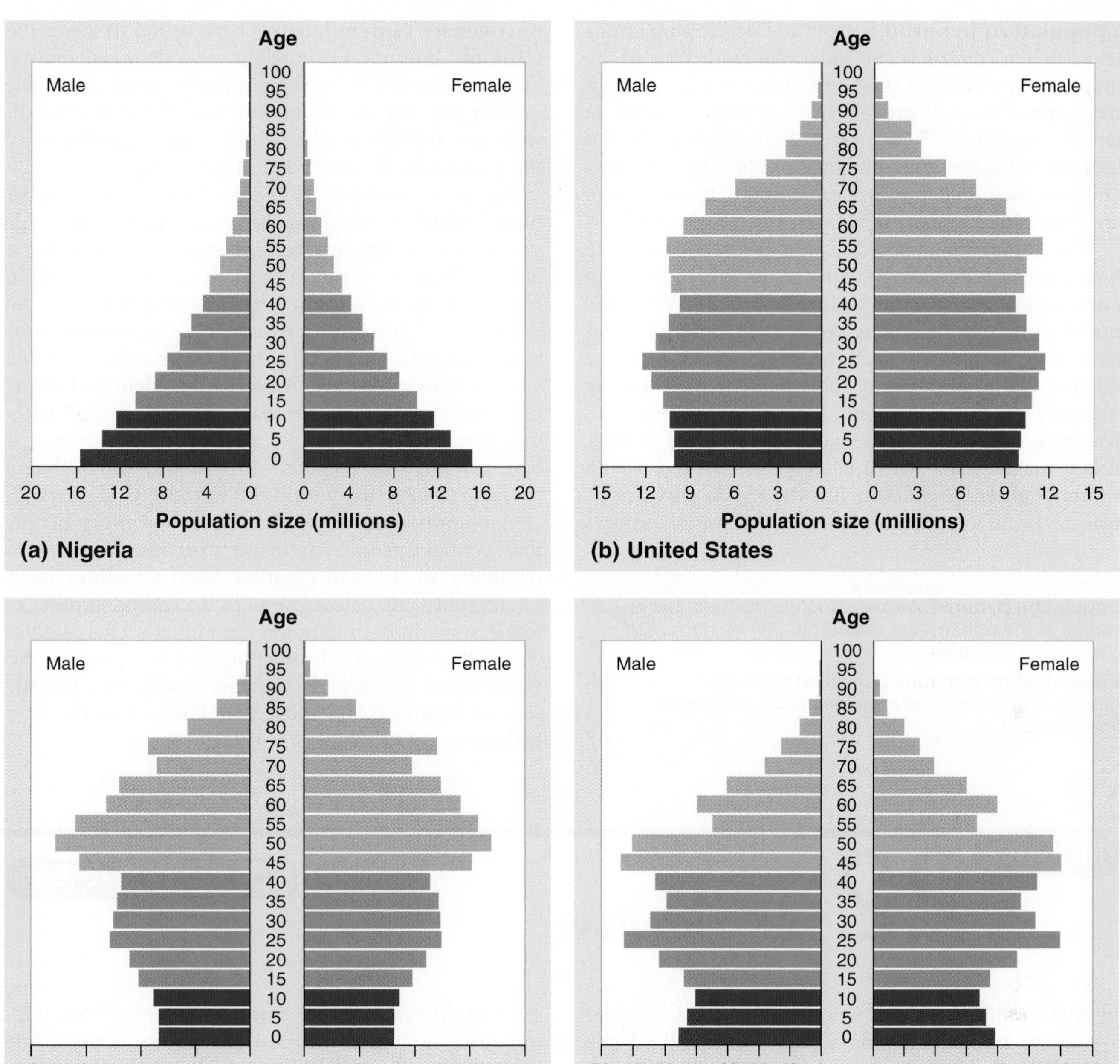

FIGURE 22.8 Age structure diagrams for four countries in 2017. The horizontal axis of the age structure diagram shows the population size in millions for males and females in each 5-year age group shown on the vertical axis. (a) A population pyramid illustrates a rapidly growing population (Nigeria). (b) A column-shaped age structure diagram indicates population stability (United States). (c) In some developed countries, the population is declining (Germany). (d) China's population control measures will eventually lead to a population decline. *(Data from: https://www.census.gov/population/international/data/idb.)*

representations of the number of individuals within specific age groups for a country, typically expressed for males and females. Each horizontal bar of the diagram represents a 5-year age group. The total area of all the bars in the diagram represents the size of the whole population.

While every nation has a unique age structure, we can group countries very broadly into three categories. A country with many more younger people than older people has an age structure diagram that is widest at the bottom and narrowest at the top, as shown in Figure 22.8a. This type of age structure diagram, called a **population pyramid**, is typical of developing countries, such as Venezuela and India. The wide base of the graph compared with the levels above it indicates that the population will grow because a large number of females aged 0 to 15 have yet to bear children. Even if each one of these future potential mothers has only two children, the population will grow simply because there are increasing numbers of women able to give birth.

The population pyramid can also be used to illustrate how long a time it takes for changes to affect a growing population. **Population momentum** is continued population growth after growth reduction measures have been implemented. It occurs because there are relatively large numbers of individuals at reproductive maturity in the population. Population momentum has been compared with the momentum of a long, heavy freight train, which takes longer to stop than a shorter, lighter freight train. It is the reason why a population keeps on growing after birth control policies or voluntary birth reductions have begun to lower the CBR of a country. Eventually, over several generations, those actions will bring the population to a more stable growth rate but the momentum of all the individuals who have recently reached child-bearing age will carry the population forward for a number of years.

Population pyramid An age structure diagram that is widest at the bottom and smallest at the top, typical of developing countries.

Population momentum Continued population growth after growth reduction measures have been implemented.

A country with little difference between the number of individuals in younger age groups and in older age groups has an age structure diagram that looks more like a column from age 0 through age 50, as shown in Figure 22.8b. If a country has few individuals in the younger age classes, we can deduce that it has slow population growth or is approaching no growth at all. The United States, Canada, Australia, Sweden, and many other developed countries have this type of age structure diagram. A number of developing countries that have recently lowered their growth rates should begin to show this pattern within the next 10 to 15 years.

A country with a greater number of older people than younger people has an age structure diagram that resembles an inverted pyramid. Such a country has a total fertility rate below 2.1 and a decreasing number of females within each younger age range. Such a population will continue to shrink. Italy, Germany, Russia, and a few other developed countries display this pattern, seen in Figures 22.8c and 22.8d. China is in the very early stages of showing this pattern.

MODULE 22 AP® Review

Preparing for the AP® Exam

In this module we have seen that scientists disagree about the human carrying capacity of Earth. While some believe that we have exceeded human carrying capacity on Earth, others suggest that human ingenuity and innovation will allow an almost infinite increase in the carrying capacity for human existence. Many factors drive human population growth, including crude birth rate, crude death rate, and fertility rates. The age structure diagram is an important tool for evaluating the age composition of populations.

AP® Practice Questions

Choose the best answer for the following.

1. Thomas Malthus introduced
   (a) the study of human demographics.
   (b) the hypothesis that humans could exceed Earth's carrying capacity.
   (c) the concept of variation in replacement-level fertility due to differences in industrialization.
   (d) the age structure diagram.

2. Doubling time can be approximated mathematically by
   (a) $\frac{70}{\text{growth rate}}$.
   (b) $\frac{10\%}{\text{growth rate}}$.
   (c) $2\% \times$ growth rate.
   (d) $\frac{\text{CBR} - \text{CDR}}{10}$.

3. Replacement-level fertility is
   (a) just under 2 children in the United States.
   (b) the number of births per 1,000 women.
   (c) always greater than 2.
   (d) the base level of 1 child to replace the mother.
4. A country has a net immigration rate of 3 per 1,000, a CBR of 9 per 1,000, and a CDR of 11 per 1,000. What is the growth rate of this country?
   (a) 0.1 percent
   (b) 0.2 percent
   (c) 0.5 percent
   (d) 1 percent
5. An age structure diagram with roughly similar numbers of individuals in each group is representative of which country?
   (a) India
   (b) Mexico
   (c) Sweden
   (d) Venezuela

MODULE 23

Economic Development, Consumption, and Sustainability

Human populations undergo change as a variety of natural and societal conditions in those populations change over time. Some of these changes occur with specific patterns that can be described and explained. In this module we will look at demographic transitions, the effect of these transitions on the environment, and the relationship between economic development and sustainability.

Learning Goals

After reading this module you should be able to

- describe how demographic transition follows economic development.
- explain how relationships among population size, economic development, and resource consumption influence the environment.
- describe why sustainable development is a common but elusive goal.

Demographic transition follows economic development

As we saw in Module 22, populations in different countries change over time and the population of Earth has also fluctuated. Demographers are interested in understanding the reasons behind fluctuations in population growth in the past and whether they apply to contemporary or future demographic issues. We will begin this module by looking at an important theory

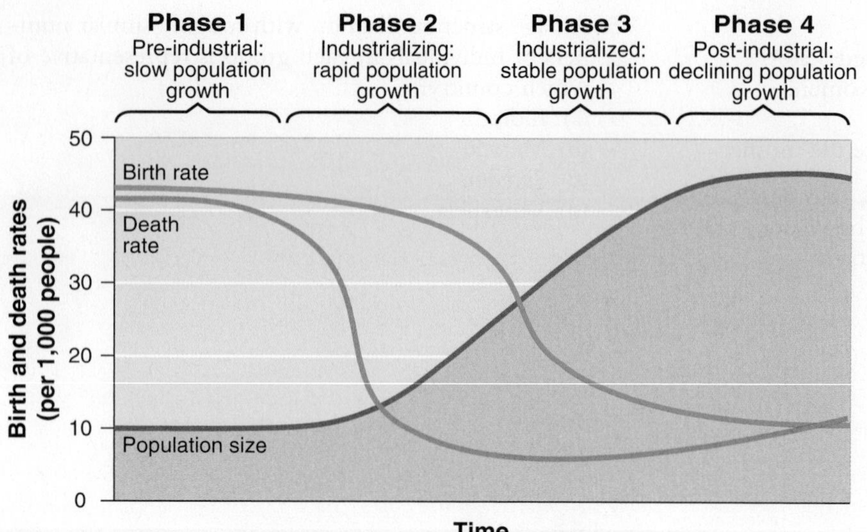

FIGURE 23.1 Demographic transition. The theory of demographic transition models the way that birth, death, and growth rates for a nation change with economic development. Phase 1 is a preindustrial period characterized by high birth rates and high death rates. In phase 2, as the society begins to industrialize, death rates drop rapidly, but birth rates do not change. Population growth is greatest at this point. In phase 3, birth rates decline for a variety of reasons. In phase 4, the population stops growing and sometimes begins to decline as birth rates drop below death rates.

of how populations change. We will also look at family planning because it is an important component of demographic transition.

The Theory of Demographic Transition

Historically, nations that have gone through similar processes of economic development have experienced similar patterns of population growth. Scientists who studied the population growth patterns of European countries in the early 1900s described a four-phase process they referred to as a demographic transition. The **theory of demographic transition** says that as a country moves from a subsistence economy to industrialization and increased affluence, it undergoes a predictable shift in population growth. The four phases of a demographic transition are shown in **FIGURE 23.1**.

The theory of demographic transition, while helpful as a learning tool, does not adequately describe the population growth patterns of some developing countries either today or during the last quarter-century. Both birth and death rates have declined rapidly in a number of developing countries because of a variety of factors that are not yet entirely understood. In some developing countries governments have taken measures to improve health care and sanitation and promote birth control, in spite of the country's poverty.

Despite the limitations of the theory of demographic transition, it is worth examining in more detail because it allows us to visualize a representation of the way some countries influence the environment as they undergo growth and development.

Phase 1: Slow Population Growth

Phase 1 represents a population that is nearly at steady state. The size of the population will not change very quickly because high birth rates and high death rates offset one another. In other words, CBR equals CDR. This pattern is typical of countries before they begin to modernize and was typical of the entire world until a few centuries ago. In these countries, life expectancy for adults is relatively short due to difficult and often dangerous living or working conditions. The infant mortality rate is also high because of disease, lack of health care, and poor sanitation. In a subsistence economy, where most people are farmers, having numerous children is an asset. Children can do jobs such as collecting firewood, tending crops, watching livestock, and caring for younger siblings. With no social security system, parents also count on having many children to care for them when they become old.

AP® Exam Tip

You should know the characteristics of each phase of the demographic transition model and be prepared to give examples. ●

Western Europe and the United States were in phase 1 before the Industrial Revolution, which began in the late eighteenth century. Today, crude birth rates exceed crude death rates in almost every country in the world, so even the poorest nations have moved beyond phase 1. However, an increase in crude death

Theory of demographic transition The theory that as a country moves from a subsistence economy to industrialization and increased affluence it undergoes a predictable shift in population growth.

rates due to war, famine, and diseases such as AIDS has pushed some countries back in the direction of phase 1. For example, more than a decade ago, Lesotho (CBR = 26, CDR = 28 in 2005) moved back into phase 1 as a result of its high death rate.

Phase 2: Rapid Population Growth

In phase 2, death rates decline while birth rates remain high and, as a result, the population grows rapidly. As a country modernizes, better sanitation, clean drinking water, increased access to food and goods, and access to health care, including childhood vaccinations, all reduce the infant mortality rate and CDR. However, the CBR does not markedly decline. Couples continue to have large families because it takes at least one generation, if not more, for people to notice the decline in infant mortality and adjust to it. This is another example of population momentum. It also takes time to implement educational systems and birth control measures.

A phase 2 country is in a state of imbalance: Births outnumber deaths. India is in phase 2 today. The U.S. population exhibited a phase 2 population structure in the early twentieth century when there were high birth rates, low death rates, and a high total fertility rate.

Phase 3: Stable Population Growth

A country enters phase 3 as its economy and educational system improve. In general, as family income increases, people have fewer children, as **FIGURE 23.2** shows. As a result, the CBR begins to fall. Phase 3 is typical of many developed countries, including the United States and Canada.

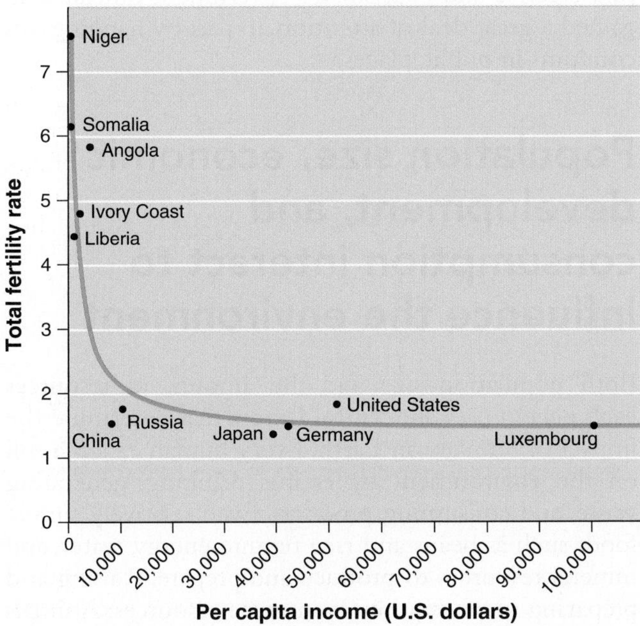

FIGURE 23.2 Total fertility rate and per capita income. Wealthier nations tend to have lower total fertility rates. *(Data from The Economist Pocket World in Figures, 2017, and World Bank.)*

FIGURE 23.3 Elderly populations. Some countries have very large elderly populations. Here, men and women gather at a senior residence home in San Diego, California. *(Stockbroker/Alamy)*

As societies transition from subsistence farming to more complex economic specializations, having large numbers of children may become a financial burden rather than an economic benefit. **Affluence** is the state of having plentiful wealth including the possession of money, goods, or property. When people are relatively affluent they spend more time pursuing education and they are more likely to have access to birth control and to choose to have smaller families. However, cultural, societal, and religious norms may also play a role in birth rates.

As birth rates and death rates decrease in phase 3, the system returns to a steady state. Population growth levels off during this phase, and population size does not change very quickly, because low birth rates and low death rates are roughly equal.

Phase 4: Declining Population Growth

Phase 4 is characterized by declining population size and often by a relatively high level of affluence and economic development. Japan, Germany, Portugal, and Italy are phase 4 countries, with the CBR below the CDR.

The declining population in phase 4 means a country will have fewer young people and a higher proportion of elderly people (**FIGURE 23.3**). This demographic shift can have important social and economic effects. With fewer people in the labor force and more people retired or working part-time, the ratio of dependent elderly to wage earners increases, and the costs of pension programs and social security services will increase the tax burden on

Affluence The state of having plentiful wealth including the possession of money, goods, or property.

each wage earner. There may be a shortage of health care workers to care for an aging population. Governments may encourage immigration as a source of additional workers. In some countries, such as Japan, the government provides economic incentives to encourage families to have more children in order to offset the demographic shift.

Recent studies on demographic shifts in highly developed countries suggest that the TFR actually increases after reaching a low point between 1.2 and 1.5. The reasons for this increase are unclear, but it appears that when a population becomes affluent and well educated, it becomes somewhat easier for families to raise children, and they choose to do so in slightly greater numbers. Such a pattern is occurring in Norway, Italy, the United States, and other developed nations.

Family Planning

We have already observed that as family income increases, people tend to have fewer children. In fact, there is a link between higher levels of education and affluence among females, in particular, and lower birth rates. As the educational levels of women increase and women enter the workplace, fertility generally decreases. Even in developed countries where the TFR has increased slightly after hitting its apparent low point, women have fewer children than those in developing countries. Educated and working women tend to have fewer children than other women, and many delay having children because of the demands of school and work. Having a first child at an older age means that a woman is likely to have fewer children in her lifetime.

Women with more education and income also tend to have more access to information about methods of birth control, they are more likely to interact with their partners as equals, and they may choose to practice *family planning* with or without the consent of their partners. **Family planning** is regulation of the number or spacing of offspring through the use of birth control. When women have the option to use family planning, crude birth rates tend to drop. **FIGURE 23.4** shows how female education levels correlate with crude birth rates. In Ethiopia, for instance, women with a secondary school education or higher have a TFR of 2.0, whereas the TFR among uneducated women is 4.2.

There have been many examples of effective family planning campaigns in the last few decades. In the 1980s, Kenya had one of the highest population growth rates in the world, and its TFR was almost 8. By 1990, its TFR was about 4—one-half of the previous rate. Kenya's government achieved these dramatic results by implementing an active family planning campaign. The campaign, which began in the 1970s, encouraged

Family planning The practice of regulating the number or spacing of offspring through the use of birth control.

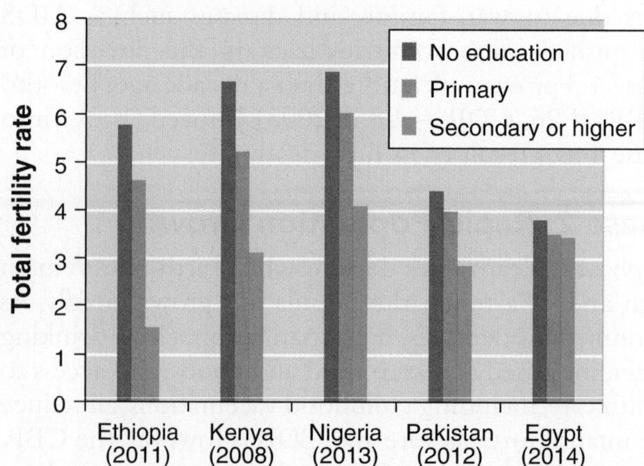

FIGURE 23.4 Total fertility rates for educated and uneducated women in five countries. Fertility is strongly related to female education in many developing countries. *(Data from: www.theglobalist.com.)*

smaller families. Advertising directed toward both men and women emphasized that overpopulation led to both unemployment and harm to the natural environment. The ads also promoted condom use.

Thailand also successfully used family planning campaigns to lower its growth rate and TFR. Beginning in 1971, national population policy encouraged married couples to use birth control. Contraceptive use increased from 15 to 70 percent, and within 15 years the population growth rate fell from 3.2 to 1.6 percent. Today, Thailand's growth rate is 0.4 percent, among the lowest in Southeast Asia. Some of the credit for this hugely successful reduction in the growth rate is given to a creative and charismatic government official who gained a great deal of attention, in part by handing out condoms in public places.

Population size, economic development, and consumption interact to influence the environment

Both population size and the amount of resources each person uses are critical factors that determine the impact of humans on Earth. Every human exacts a toll on the environment by eating, drinking, generating waste, and consuming products. Even relatively simple foods such as beans and rice require energy, water, and mineral resources to produce and prepare. Raising and preparing meat requires even more resources: Animals require crops to feed them as well as water and energy resources. Building homes, manufacturing cars, and making clothing and consumer products all require energy, water, wood, steel, and other resources. These

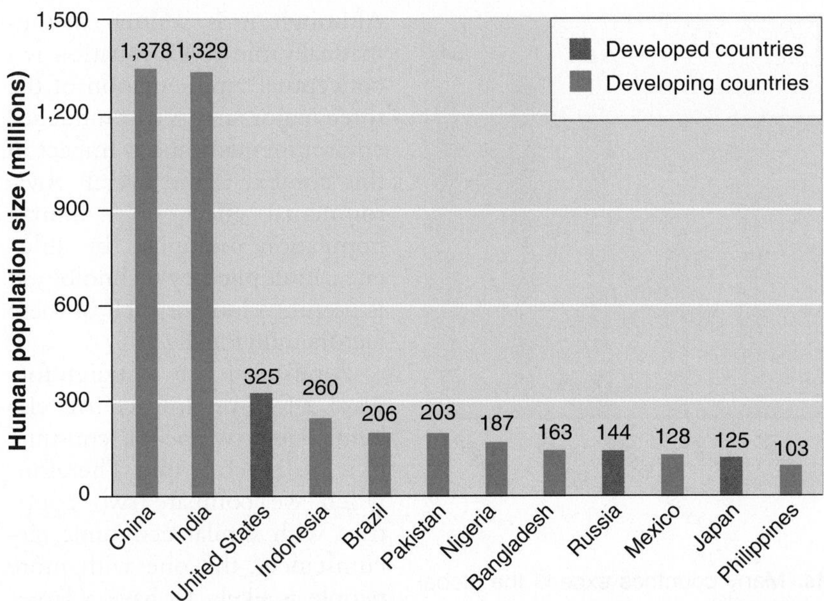

FIGURE 23.5 The 12 most populous countries in the world. China and India are by far the largest nations in the world. Only 3 of the 12 most populous countries are developed nations. *(Data from Population Reference Bureau www.prb.org.)*

and many other human activities contribute to environmental degradation.

Both population and economic development contribute to the consumption of resources and to human impact on the environment. In this section we will examine how relationships among population size, economic development, and resource consumption influence the environment. We will then look more closely at local, urban, and global impacts.

Resource Use

In Module 22 we described developing countries as those with relatively low levels of industrialization and incomes of less than $3 per person per day. In contrast, developed nations have relatively high levels of both industrialization and income. Of Earth's 7.6 billion human inhabitants, roughly 6.3 billion live in developing countries, and 1.3 billion live in developed countries. As **FIGURE 23.5** shows, 9 of the 12 most populous nations on Earth are developing countries. **FIGURE 23.6** charts the relationship between economic development and population growth rate for developing nations. Populations in developing parts of the world have continued to grow relatively rapidly, at an average rate of 1.5 percent per year. At the same time, populations in the developed world have almost leveled off, with an average growth rate of 0.1 percent per year. Impoverished countries are increasing their populations more rapidly than are affluent countries.

Differences in resource use are striking in terms of how population and wealth affect the environment. Calculating the per capita ecological footprint for a country provides a way to measure the effect of affluence—the state of having plentiful wealth that includes the possession of money, goods, or property—and consumption on the planet. Although affluence tends to be associated with higher consumption, it is possible to be affluent without having a large ecological footprint. **FIGURE 23.7** on page 250 shows some examples of ecological footprints for selected countries for the latest year available. The world average ecological footprint is 1.7 ha (4.2 acres) per capita. The United States has one of the largest ecological footprint of any nation, at 8.6 ha (21 acres) per capita. China's footprint is 3.6 ha (8.9 acres) per capita and has been increasing rapidly. Haiti, the poorest country in the Western Hemisphere, has one of the smallest footprints in the world, 0.6 ha (1.5 acres) per capita. In other words, a person living in the United States has almost 2.5 times the environmental impact of a person living in China and 14 times that of a person living in Haiti.

In addition to looking at data on a per capita basis, it is useful to examine the footprints of entire countries. We can do this by multiplying the per capita ecological footprint of a country by the number of people in the country. We find that the United States has a footprint of 2,795 million hectares (6,900 million acres). China's footprint is 4,968 million hectares

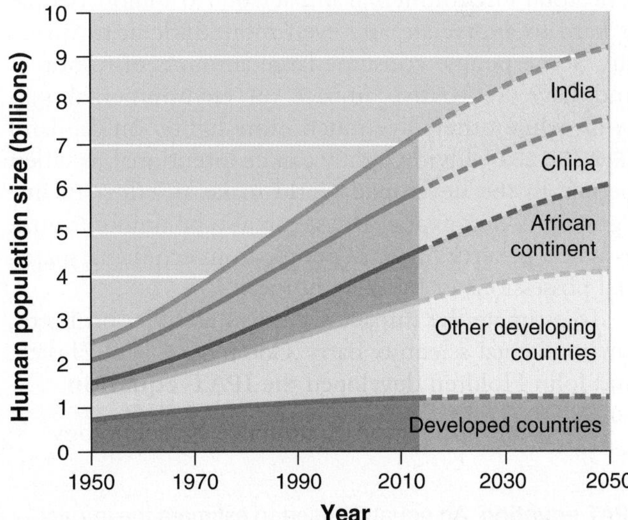

FIGURE 23.6 Population growth past and future. Population growth in developed countries has mostly ceased, while that in developing countries is slowing, but is expected to continue beyond 2050. *(Data from United Nations Population Division.)*

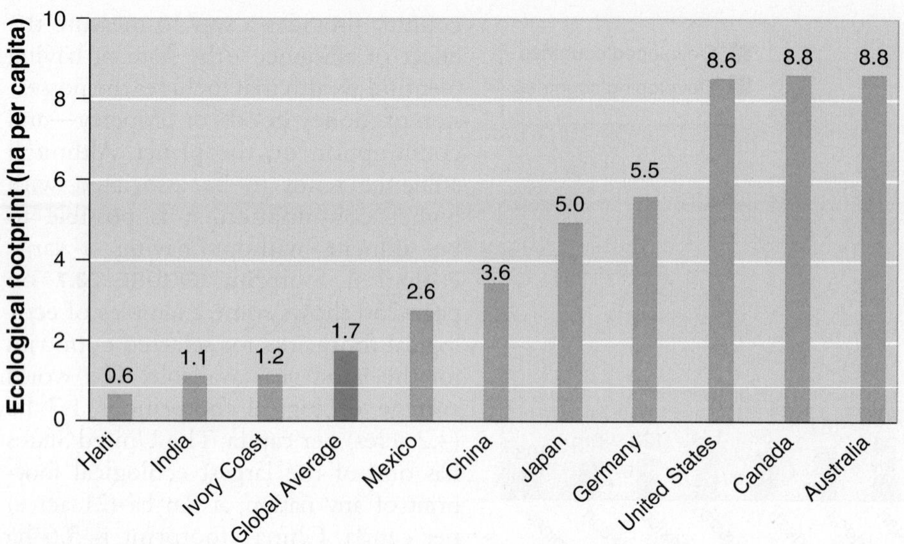

FIGURE 23.7 Per capita ecological footprints. Many countries exceed the global average footprint of 1.7 ha per capita. *(Data from Global Footprint Network, 2013, published in 2017: http://www.footprintnetwork.org/content/documents/ecological_footprint_nations/ecological_per_capita.html.)*

(12,270 million acres), and Haiti has a footprint of 6.6 million hectares (16.5 million acres). Viewed another way, the United States, with roughly one-fourth the population of China, has an ecological footprint that is almost one-half that of China because of the relatively higher levels of consumption in the United States. However, China's rapid development, as described at the beginning of this chapter, means that its ecological footprint is likely to continue to grow in the coming years.

The IPAT Equation

The total environmental impact of 7.6 billion people is hard to appreciate and even more difficult to quantify. Some people consume large amounts of resources and have a negative impact on environmental systems, while others live much more lightly on the land (**FIGURE 23.8**). Living lightly can be intentional, as when people in the developed world make an effort to live "green," or sustainably. But it can also be unintentional, as when poverty prevents people from acquiring material possessions or building homes.

To estimate the impact of human lifestyles on Earth, environmental scientists Barry Commoner, Paul Ehrlich, and John Holdren developed the **IPAT equation**

$$\text{impact} = \text{population} \times \text{affluence} \times \text{technology}$$

IPAT equation An equation used to estimate the impact of the human lifestyle on the environment: impact = population × affluence × technology.

Gross domestic product (GDP) A measure of the value of all products and services produced in 1 year in one country.

Although it is written mathematically, the IPAT equation is a conceptual representation of the three major factors that influence environmental impact. Impact in this context is the overall environmental effect of a human population multiplied by affluence, multiplied by technology. It is useful to look at each of these factors individually.

Population has a straightforward effect on impact. All else being equal, two people consume twice as much as one. Therefore, when we compare two countries with similar economic circumstances, the one with more people is likely to have a larger impact on the environment.

Affluence is created by economic opportunity and does not have as simple a relationship to impact as population does. One person in a developed country can have a greater impact than two or more people in a developing country. A family of four in the United States that owns two large sport utility vehicles and lives in a spacious home with a lawn and swimming pool uses a much larger share of Earth's resources than a Bangladeshi family of four living in a two-room apartment and traveling on bicycles and buses. The more affluent a society or individual is, the higher the environmental impact.

The effect of technology is even more complicated. Technology can both degrade the environment and create solutions to minimize our impact on the environment. For example, the manufacturing of chlorofluorocarbons (CFCs) resulted in safe and effective refrigeration and air conditioning that was beneficial to human health, yet CFCs led to ozone destruction in the stratosphere. In contrast, the hybrid electric car helps to reduce the impact of the automobile on the environment because it has greater fuel efficiency than a conventional internal combustion vehicle of the same size and weight. The IPAT equation originally used the term *technology,* but some scientists now use the term *destructive technology* to differentiate it from beneficial technologies such as the hybrid electric car.

The Impact of Affluence

To help gauge a country's wealth and its potential impact on the environment, environmental scientists often turn to the most commonly used measure of a nation's wealth. **Gross domestic product (GDP)** is the value of all products and services produced in 1 year in one country. GDP is made up of four types

(a)

(b)

FIGURE 23.8 Material possessions. Most families in developing countries have few possessions compared with their counterparts in developed countries. In each of these photos, members of a family are shown outside their homes with all their possessions. (a) This family lives in a rural region of Tibet. (b) This individual lives in an urban part of China, where more development has occurred. *(a: Huang Quingjun/http://www.huangqinglun.com/; b: Huang Qingjun/http://www.huangqingjun.com/)*

of economic activity: consumer spending, investments, government spending, and exports minus imports. A country's per capita GDP often correlates with its pollution levels. At very low levels of per capita GDP, industrial activity is too low to produce much pollution; the country uses very little fossil fuel and generates relatively little waste. Many developing countries fit this pattern.

As GDP increases, a nation begins to be able to afford to burn fossil fuels, especially coal, which, although relatively inexpensive, emits a substantial amount of pollution. The country may also rely on rudimentary, inefficient equipment that emits large amounts of pollutants. It is at this point in its development that a country emits pollution at the highest levels. The United States fit this pattern during the twentieth century. China, which is going through a similar rapid industrialization, currently relies on coal as its primary energy source. Many people view this shift as a trade off that occurs as a country's GDP increases: The risk posed to human health by breathing dirty air is different from the risk of poverty. Although it is not always a more desirable risk, many people would choose dirty air as a byproduct of economic advancement over poverty.

As a nation's GDP increases further, it may reach a turning point. It can afford to purchase equipment that burns fossil fuels more efficiently and cleanly, which helps to reduce the amounts and types of pollution generated. People may also be willing to expend resources and support government efforts to regulate polluting industries. Wealthier societies are also able to afford better policing and enforcement mechanisms that ensure environmental regulations are being followed. Many Western European countries and the United States and Canada fit this scenario today. We will explore this turning point in more depth in Chapter 20.

Some environmental scientists argue that increasing the GDP of developing nations is the best way to save the environment, for at least two reasons. First, as we have seen, rising income generally correlates with falling birth rates, and a reduced population size should lead to a reduction in environmental impact. Second, wealthier countries can afford to make environmental improvements and increase their efficiency of resource use.

Local versus Global Impacts, and Urban Impacts

Impacts on the environment may occur locally—within the borders of a region, city, or country—or they may be global in scale. The scale of impact depends on the nature of the economy and the degree to which the society has developed. For example, a person may create a local impact by using products created within the country's borders. That same person may create a global impact by using materials that are imported. In addition to local and global impacts, some impacts are specific to people who live in urban environments.

Local versus Global

In general, highly localized impacts are typical of rural, agriculturally based societies. Most of the materials consumed in developing countries are produced locally. While this may benefit the local economy, it can lead to regional overuse of resources and environmental degradation. Chapter 3 described deforestation in Haiti as an example of such overuse.

Two commonly overused local resources are the land itself and woody biomass from trees and other plants. A growing population requires increasing amounts of food. In developing countries that do not import their food, local demand for agricultural land increases with population size. To put more land into cultivation, farmers may convert forests or natural grasslands into cropland.

Brazil provides an example of land overuse in a developing nation. The United Nations Food and Agriculture Organization estimates that in Brazil, approximately 3 million hectares (7.4 million acres) were cleared per year during the peak years between 2000 and 2005. Some of this land was used for small-scale agriculture and some for industrial production of soybeans, sugarcane, and corn. The local environmental impacts of converting land to agricultural use include erosion, soil degradation, and habitat loss (**FIGURE 23.9**). Agriculture has global as well as local impacts, whether it occurs in developed or developing countries. Conversion of land to agriculture reduces the total amount of atmospheric carbon dioxide uptake by plants, which affects the global carbon cycle. In addition, an increase in the use of fertilizers made from fossil fuels increases the release of greenhouse gases into the atmosphere.

Global impacts are more common in affluent or urban societies because they tend to specialize production in the industrial and high technology sectors. For example, more than half the ecological footprint of the United States comes from its use of fossil fuels, of which approximately one-third are imported. China's ecological footprint has more than doubled since 1970, and its ratio of local to global impact has shifted. Most of its ecological footprint was previously driven by demand for food, fiber crops such as cotton, hemp, and flax, and woody biomass. However, in recent years, China's demand for fossil fuels has increased dramatically.

Families in suburban areas of developed countries such as the United States consume far fewer local resources than rural families in developing countries, but they have a much greater impact on the global environment. In general, populations with large global impacts tend to deplete more environmental resources. Much of the impact comes from consumption of imported energy sources such as oil and other imported resources such as food. When people are affluent, they are more likely to purchase imported bananas, fish, and coffee from other countries, drive long distances in automobiles that were manufactured in factories hundreds or thousands of miles away, and live in homes surrounded by lawns that require large quantities of water, fertilizer, and pesticides.

Urban Impacts

Urban populations represent one-half of the human population but consume three-fourths of Earth's resources. While definitions vary by country, an **urban area**, according to the U.S. Census Bureau, contains more than 386 people per square kilometer (1,000 people per square mile). New York City is the most densely populated city in the United States, with 10,400 people per square kilometer (27,000 people per square mile). Mumbai, India, is the most densely populated city in the world, with 23,000 people per square kilometer (60,000 people per square mile).

Urban area An area that contains more than 386 people per square kilometer (1,000 people per square mile).

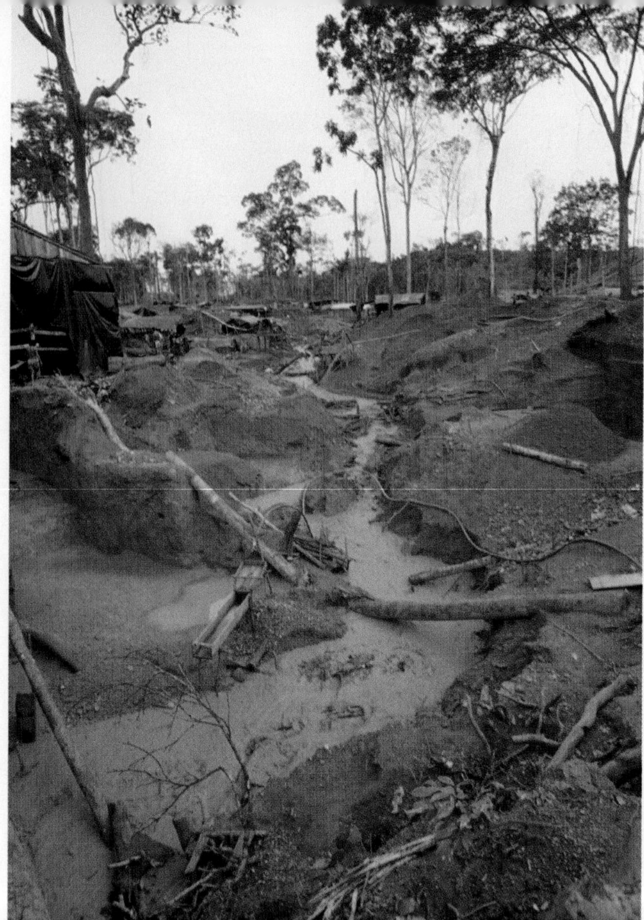

FIGURE 23.9 Deforested land in Brazil. Clearing forests and grasslands for agriculture can lead to erosion, soil degradation, and habitat loss. This photo shows erosion caused by clear-cutting in Brazil. *(Publiphoto/Science Source)*

More than 75 percent of people in developed countries live in urban areas, as **FIGURE 23.10** shows, and that number is expected to increase slightly over the next 15 years. In developing countries, 48 percent of people live in urban areas. This number will probably increase to 56 percent by 2030. **TABLE 23.1** shows that, of the 20 largest cities in the world, 16 are in developing countries. Worldwide, almost 5 billion people are expected to live in urban areas by 2030.

Urban living in both developed and developing countries presents environmental challenges. Most developed countries employ city planning. As urban areas expand, experts design and install public transportation facilities, water and sewer lines, and other municipal services. In addition, while urban areas produce greater amounts of solid waste, pollution, and carbon dioxide emissions than suburban or rural areas, they tend to have smaller per capita ecological footprints.

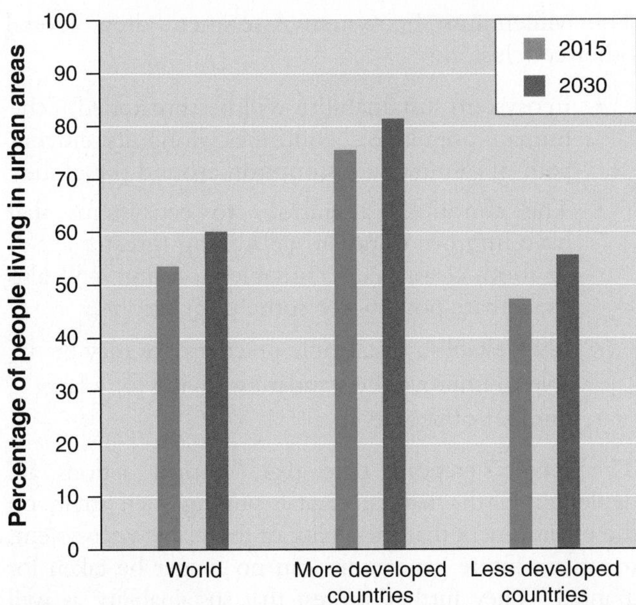

FIGURE 23.10 Urban growth. More than one-half of the world's population will live in urban settings by 2030. *(Data from U.N. POP Fund.)*

FIGURE 23.11 Informal settlements. Informal collections of dwellings such as this squatter settlement on the outskirts of Lima, Peru, expand outward from the urban center. *(Reuters/Mariana Bazo)*

TABLE 23.1 The largest 20 urban areas in the world

Rank	City, Country	Population (millions)
1	Tokyo, Japan	38.1
2	Delhi, India	26.5
3	Shanghai, China	24.5
4	Mumbai, India	21.4
5	São Paulo, Brazil	21.3
6	Beijing, China	21.2
7	Mexico City, Mexico	21.2
8	Osaka-Kobe, Japan	20.3
9	Cairo, Egypt	19.1
10	New York-Newark, United States	18.6
11	Dhaka, Bangladesh	18.2
12	Karachi, Pakistan	17.1
13	Buenos Aires, Argentina	15.3
14	Kolkata, India	15.0
15	Istanbul, Turkey	14.4
16	Chongqing, China	13.7
17	Lagos, Nigeria	13.7
18	Manila, Philippines	13.1
19	Guangzhou, Guangdong, China	13.1
20	Rio de Janeiro, Brazil	13.0

Source: Data from United Nations Population Division.

Note: Data are from 2016 and contain the areas defined by the United Nations as "urban agglomerations." Other agencies agglomerate urban areas differently and obtain slightly different results.

There are many reasons for this difference, including greater access to public transportation and services that are nearby, such as shopping.

In developing countries, the relatively affluent portions of urban areas have safe drinking water, sewage treatment systems, and systems for disposal of household solid waste to minimize their impact on the surrounding environment. However, many less-affluent urban residents have no access to these services. Rapid urbanization in the developing world often results in an influx of the very poor who often cannot afford permanent housing and instead construct temporary shelters with whatever materials they find available, including mud, cardboard, or plastic. Whether they are squatter settlements, shantytowns, or slums, these overcrowded and underserved living situations are a common fact of life in most cities in the developing world. The United Nations organization UN-HABITAT estimates that 1 billion people live in squatter settlements and other similar areas throughout the world. Most residents live in housing structures without flooring, safe walls and ceilings, or such basic amenities as water, sanitation, or health care (**FIGURE 23.11**).

Sustainable development is a common, if elusive, goal

We have seen that economic development and advancement improve human well-being, but they also have a strong influence on the environmental impact of a society. While many people believe that we cannot have both economic development and environmental protection, a growing number of social and natural scientists maintain that, in fact, sustainable economic development is possible.

As we saw in Chapter 1, sustainable development goes beyond economic development to meet the essential needs of people in the present without compromising the ability of future generations to meet their needs. In other words, sustainable development strives to improve standards of living—which involves greater expenditures of energy and resources—without causing additional environmental harm. Many cities in Scandinavia have achieved something close to sustainable development. For example, Övertorneå, a Swedish town of 5,600 people, recently revived the local economy by focusing its economic activities on renewable energy, free public transportation, organic agriculture, and land preservation. The city still has an ecological footprint, of course, but it is much smaller, and thus much more sustainable, than it was previously.

How can sustainable development be achieved? There are no simple answers to that question, nor is there a single path that all people must follow. The Millennium Ecosystem Assessment project, completed in 2005, offers some insights. This project's reports constitute a global analysis of the effects of the human population on ecosystem services such as clean water, forest products, and natural resources. They are also a blueprint for sustainable development. The reports, prepared at the request of the United Nations, concluded that human demand for food, water, lumber, fiber, and fuel has led to a large and irreversible loss of biodiversity. The Millennium Ecosystem Assessment drew several other conclusions:

- Ecosystem sustainability will be threatened if the human population continues along its current path of resource consumption around the globe.
- The continued alterations to ecosystems that have improved human well-being (greater access to food, clean water, suitable housing) will also exacerbate poverty for some populations.
- If we establish sustainable practices, we may be able to improve the standard of living for a large number of people.

The project's reports state that "human actions are depleting Earth's natural capital, putting such strain on the environment that the ability of the planet's ecosystems to sustain future generations can no longer be taken for granted." They further suggest that sustainability, as well as sustainable development, will be achieved only with a broader and accelerated understanding of the connections between human systems and natural systems. This means that governments, nongovernmental organizations, and communities of people will have to work together to raise standards of living while understanding the impacts of those improvements on the local, regional, and global environments.

MODULE 23 AP® Review

Preparing for the AP® Exam

In this module we examined the theory of the demographic transition, which states that in some countries, the process of industrialization causes a reduction in the population growth rate. Slower population growth is typical of many countries that have industrialized and become wealthier. Some nations that have slowed in population growth have not industrialized. Those countries that have industrialized typically increase their consumption of resources, with adverse impacts on the environment. Impacts can occur locally and globally. Some locations are beginning to take on the challenge of fostering industrialization and economic development in a sustainable manner.

AP® Practice Questions

Choose the best answer for the following.

1. According to the theory of demographic transition, countries move through growth phases in which order?
   (a) stable growth, rapid growth, slow growth, declining growth
   (b) rapid growth, slow growth, stable growth, declining growth
   (c) slow growth, declining growth, rapid growth, stable growth
   (d) slow growth, rapid growth, stable growth, declining growth

2. Which has contributed to increased family planning worldwide?
   I. women's education
   II. increased income
   III. advertising campaigns
   (a) I and II
   (b) I and III
   (c) II and III
   (d) I, II, and III

3. Which is NOT a factor in gross domestic product?
   (a) consumer spending
   (b) cost of environmental damage
   (c) government spending
   (d) imports and exports
4. The IPAT equation was developed to estimate
   (a) the affluence of a country's population.
   (b) the rate of demographic transition.
   (c) the impact of human lifestyles on Earth.
   (d) the gross domestic product of a country.
5. Approximately what percentage of the population of developed countries lives in urban areas?
   (a) 44 percent
   (b) 50 percent
   (c) 75 percent
   (d) 86 percent

Working Toward Sustainability

Gender Equity and Population Control in Kerala

India is the world's second most populous country and will probably overtake China as the most populous by 2025. Since the middle of last century, India has tried various methods of population control, steadily reducing its growth rate to the current 1.5 percent. Although this growth rate would suggest below-replacement-level fertility, India's population will continue to increase for some time because of its large number of young people. The Population Reference Bureau estimates that India's population will not stabilize until after 2050.

Decades ago, India attempted to enforce nationwide population control through sterilization, but massive protests against this coercive approach led to changes in policy. Since the 1970s, India has emphasized family planning and reproductive health, although each Indian state chooses its own approach to population stabilization. Some states have had tremendous success in lowering growth rates. Kerala, in southwestern India, is one of those states.

The state of Kerala is about twice the size of Connecticut. But Connecticut has a population of roughly 3.5 million people while Kerala has a population of more than 30 million. Kerala's population density is about 820 people per square kilometer (2,100 people per square mile), almost 3 times that of India as a whole, and 25 times that of the United States. At one point, Kerala's population was growing even faster than the population in the rest of India, in large part because the state government had implemented an effective health care system that decreased infant and adult mortality rates.

However, for the last 40 years, Kerala's birth rate, like its mortality rate, has fallen to levels similar to those in North America and other industrialized countries. The

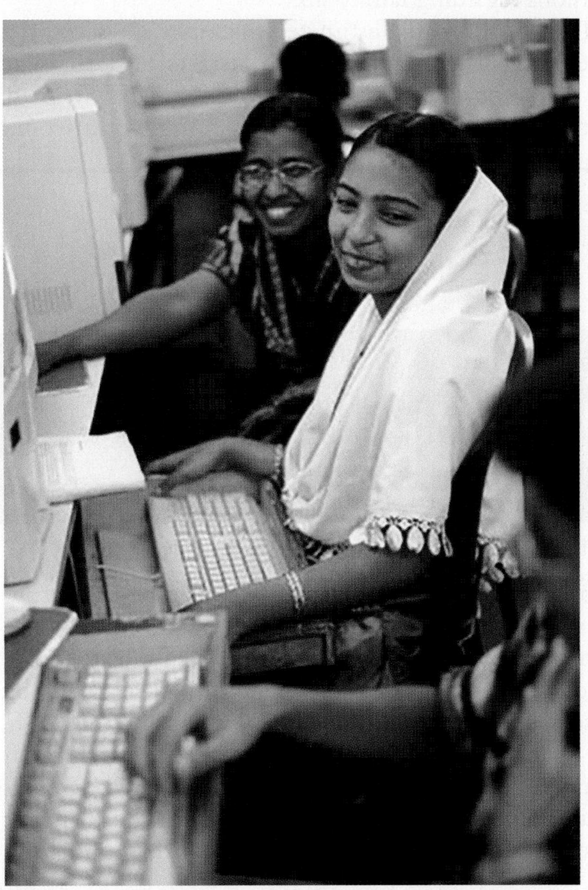

Computer students in Kerala, India. Literacy rates similar to those in developed countries may be a factor in the population stabilization that has occurred in the state of Kerala, India. *(Owen Franken/Getty Images)*

current total fertility rate in Kerala is about 1.7—lower than the U.S. TFR of 1.8 and much lower than the TFR of 2.3 for India as a whole. As a result, Kerala's population has stabilized. How did the state's population stabilize despite its poverty?

A special combination of social and cultural factors seems to be responsible for Kerala's sustainable population growth. Kerala has emphasized "the three Es" in its approach: education, employment, and equality. In addition to its accessible health care system, Kerala has good schools that support a literacy rate of over 90 percent, the highest in India. Furthermore, unlike rates in the rest of India, male and female literacy rates in Kerala are almost identical. In other parts of India and in most other developing countries, women attend school, on average, only half as long as men do. As we have learned, higher education levels for women lead to greater female empowerment and increased use of family planning. In Kerala, 63 percent of women use contraceptives, compared with 48 percent in the rest of India, and many women delay childbirth and join the workforce before deciding to start families. Kerala has a strong matriarchal tradition in which women are highly valued and their education is encouraged. Education and empowerment allow these women to be in a better position to make decisions regarding family size.

The World Bank estimates that if the rest of the developing world had followed Kerala's lead 30 years ago and equalized education for men and women, current TFR throughout the developing world would be close to replacement-level fertility. Evidence for this view can be found in the reduced fertility rates of other relatively poor countries where women and men have equal status, such as Sri Lanka and Cuba. In fact, gender equality is an important component of a number of United Nations Programs today.

Critical Thinking Questions

1. Name some typical ways countries have tried to lower birth rates.
2. What are some other innovative ways to lower birth rates?
3. Can achieving replacement-level fertility sometimes cause difficulties? If so, describe some problems that might occur.
4. What is preventing the model used to promote slower population growth in Kerala from being used elsewhere in the world?

References

Franke, R. W., and B. H. Chasin. 2005. Kerala: Radical reform as development in an Indian state. In *The Anthropology of Development and Globalization: From Classical Political Economy to Contemporary Neoliberalism,* ed. M. Edelman and A. Haugerud. Blackwell.

Weeks, J. R. 2016. *Population: An Introduction to Concepts and Issues.* 12th ed. Cengage.

Chapter 7 Review

Throughout this chapter, we have examined the ways in which the human population changes in response to natural biological factors and in response to other factors that are specifically human in nature. Together, these factors determine the carrying capacity of Earth, the total number of human beings that the world can support. People in the developed world are currently exhibiting very different behavior from those in the developing world. In the developed world, population growth has almost stopped while consumption of resources is still quite high. In the developing world, growth rates, while slowing, are still high, and consumption of resources is still rather low. As the world becomes more urban, and the impacts of development and consumption from larger numbers of people increase, the challenge is to achieve sustainable growth while also promoting the improvement of living conditions in the developing world.

Key Terms

Demography	Replacement-level fertility	Population pyramid
Demographer	Developed country	Population momentum
Immigration	Developing country	Theory of demographic transition
Emigration	Life expectancy	Affluence
Crude birth rate (CBR)	Infant mortality	Family planning
Crude death rate (CDR)	Child mortality	IPAT equation
Doubling time	Net migration rate	Gross domestic product (GDP)
Total fertility rate (TFR)	Age structure diagram	Urban area

Learning Goals Revisited

Module 22 Human Population Numbers

Explain factors that may potentially limit the carrying capacity of humans on Earth.

Scientists disagree about the size of Earth's carrying capacity for humans. Some scientists believe we have already exceeded that carrying capacity. Others believe that innovative approaches and new technologies will allow the human population to continue to grow beyond the environmental limits currently imposed by factors such as the supply of food, water, and natural resources.

Describe the drivers of human population growth.

The human population is currently 7.6 billion people, and it is growing at a rate of about 1.3 million people every 5 days. If we think of the human population—for the entire globe or in individual countries—as a system, there are more inputs (births and immigration) than outputs (deaths and emigration). To understand changes in population size, demographers measure crude birth rate, crude death rate, total fertility rate, replacement-level fertility, life expectancy, infant and child mortality, age structure, and net migration rate.

Read and interpret an age structure diagram

Age structure diagrams are visual representations of age structures for males and females. Each horizontal bar in the diagram represents an age group. The total area of all the bars in the diagram represents the size of the whole population. A country with more younger people than older people has an age structure diagram that is widest at the bottom and narrowest at the top. This is often called a population pyramid and is typical of developing countries. Developed countries tend to have closer to an even age distribution and their age structure diagrams are more vertical or column-like.

Module 23 Economic Development, Consumption, and Sustainability

Describe how demographic transition follows economic development.

A number of countries have undergone a demographic transition as their economies have modernized. Economic development generally leads to increased affluence, increased education, less need for children to help their families generate subsistence income, and increased family planning. These factors have reduced the average size of families in developed countries, which leads to slower population growth. Eventually, population size may even decline.

Explain how relationships among population size, economic development, and resource consumption influence the environment.

Most population growth today is occurring in developing countries. Only one-sixth of the global population lives in developed countries, but those countries consume more than half the world's energy and resources. The IPAT equation states that the environmental impact of a population is a result of population size, affluence, and technology. A relatively small population can have a high environmental impact if its affluence leads to high consumption and extensive use of destructive technology. However, an affluent nation can more easily take measures to reduce its environmental impact through the use of technology that counters pollution and increases the efficiency of resource use.

Describe why sustainable development is a common but elusive goal.

Sustainable development attempts to raise standards of living without increasing environmental impact. However, this is a challenging goal that is very difficult to achieve. Raising standards of living almost always involves greater resource use, with subsequent environmental degradation. The Millennium Ecosystem Assessment is a blueprint for sustainable development.

Practice Math and Graphing

Preparing for the AP® Exam

Answer the following questions. Be sure to show all your work.

1. Practice Math

At the beginning of 2016, the United States had a CBR of 12, a CDR of 8, and a population of 320 million. Based on births and deaths, and ignoring migration, how many people were expected to be in the United States at the end of 2016?

2. Practice Graphing

The U.S. Census Bureau reports the following total population values for the United States over the past 6 decades (with some rounding).

Year	Population
1960	180 million
1970	200 million
1980	225 million
1990	250 million
2000	280 million
2010	310 million

(a) Graph the U.S. population over time with millions of people on the *y* axis and year on the *x* axis.

(b) Calculate the percentage increase (relative to the 1960 population) in the U.S. population between 1960 and 2010. Describe how that compares to the 2016 percentage population increase given in "Practice Math".

Chapter 7 AP® Environmental Science Practice Exam

Preparing for the AP® Exam

Section 1: Multiple-Choice Questions

Choose the best answer for questions 1–20.

1. Which does NOT support the theory that humans can devise ways to expand their carrying capacity on Earth?
   (a) the development of CFCs for use in refrigeration
   (b) the development of hydraulic fracturing to reach natural gas reserves
   (c) the use of arrows for hunting animals
   (d) the use of waste methane from landfills for electricity generation

2. A metropolitan region of 100,000 people has 2,000 births, 500 deaths, 200 emigrants, and 100 immigrants over a 1-year period. Its population growth rate is
   (a) 1.2 percent.
   (b) 1.4 percent.
   (c) 1.6 percent.
   (d) 1.8 percent.

3. Which pair of indicators best reflects the availability of health care in a country?
   (a) crude death rate and growth rate
   (b) growth rate and life expectancy
   (c) infant mortality rate and crude death rate
   (d) infant mortality rate and life expectancy

4. In 2019, the population of Earth was about _______ billion, with about _______ billion living in China.
(a) 6.1; 1.1
(b) 6.1; 1.2
(c) 6.3; 1.4
(d) 7.6; 1.3

Use the following age structure diagram to answer question 5.

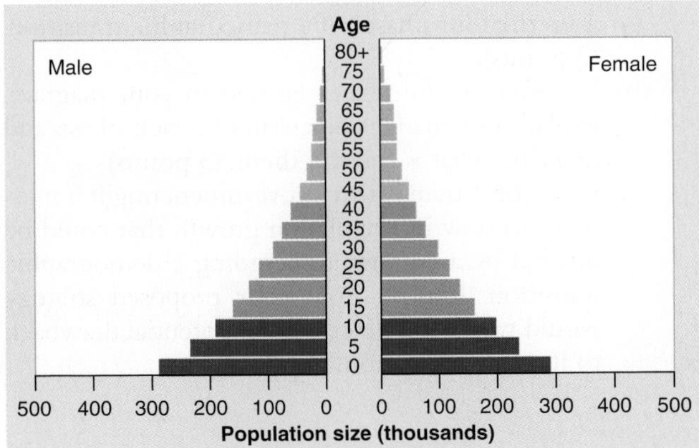

5. A country with an age structure diagram like the one shown is most likely experiencing
(a) a high life expectancy.
(b) slow population growth.
(c) a short doubling time.
(d) a low infant mortality rate.

6. Which statement about total fertility rate is correct?
(a) TFR is equal to the crude birth rate minus the crude death rate.
(b) TFR is the average number of children each woman must have to replace the current population.
(c) TFR is generally higher in developed countries than in developing countries.
(d) TFR is the average number of children each woman will give birth to during her childbearing years.

7. Even if a country reduces its birth rate and maintains replacement-level fertility, its population will still continue to grow for several decades because of
(a) lower death rates.
(b) increased income.
(c) population momentum.
(d) increased life expectancy.

8. At current growth rates, which country will probably be the most populous in the world after 2050?
(a) China
(b) Brazil
(c) India
(d) Indonesia

9. Which is typical of developed countries?
I. high technology use
II. low GDP
III. small-scale sustainable agriculture
(a) I only
(b) II only
(c) I and III
(d) II and III

10. What percentage of the world's population lives in developing countries?
(a) 34 percent
(b) 50 percent
(c) 66 percent
(d) 83 percent

11. Which country best exemplifies phase 4 of a demographic transition?
(a) Argentina
(b) China
(c) India
(d) Japan

12. As Brazil has become more developed and industrialized, its population growth has stabilized. At the same time, the use of technology and raw materials has increased to meet the demands of a wealthier and more prosperous population. This increased consumption is described by
(a) the work of Thomas Malthus.
(b) the Millennium Ecosystem Assessment.
(c) the theory of demographic transition.
(d) the IPAT equation.

13. In 2012, there were 20 births per 1,000 people and 5 deaths per 1,000 people in Peru. What is the approximate doubling time of Peru's population? Assume equal rates of immigration and emigration.
(a) 5 years
(b) 15 years
(c) 47 years
(d) 72 years

14. Which is most likely to be the replacement-level fertility for a developing nation in phase 3 of demographic transition?
(a) 1.8
(b) 2.0
(c) 2.2
(d) 3.0

15. Which is most likely to result in an increase in pollution?
(a) an increase in GDP
(b) switching from inland to offshore oil drilling
(c) a shift from subsistence energy sources to natural gas
(d) an increase in affordable health care

16. A population with an age distribution shaped like a pyramid
   I. exhibits a stable population size.
   II. is likely to exhibit a high population momentum.
   III. has a high crude birth rate.
   (a) III only
   (b) I and II
   (c) I and III
   (d) II and III

17. The four factors represented in the IPAT equation include
   (a) impact on the environment, population size, affluence, and total fertility rate.
   (b) infant mortality rate, population size, affluence, and total fertility rate.
   (c) impact on the environment, population size, affluence, and technology.
   (d) industry, population growth, architecture, and technology.

Questions 18 and 19 refer to the following scenario:

The town of Fremont has a population of 2,000. In 2019, the town has 50 deaths, 20 immigrants, and 30 emigrants. The town's population growth rate is 4.5 percent.

18. What was Fremont's crude birth rate in 2019?
   (a) 70
   (b) 75
   (c) 95
   (d) 105

19. What is the approximate doubling time of Fremont?
   (a) 16 years
   (b) 22 years
   (c) 45 years
   (d) 88 years

20. Which is part of the equation used to estimate the impact of humans on the environment?
   I. affluence
   II. population
   III. CBR
   (a) I only
   (b) II only
   (c) III only
   (d) I and II

Section 2: Free-Response Questions

Write your answer to each part clearly. Support your answers with relevant information and examples. Where calculations are required, show your work.

1. Answer the following questions about the theory of demographic transition.
   (a) Draw a fully labeled diagram that shows how birth and death rates change as a country undergoes the four phases of a demographic transition. (3 points)
   (b) For each of the phases labeled in your diagram, explain the changes occurring in each phase and describe what is causing them. (3 points)
   (c) Describe a strategy that a government might implement to slow its population growth that could be utilized by a country undergoing a demographic transition. Explain how your proposed strategy would work, and describe one potential drawback to its implementation. (4 points)

2. Look at the age structure diagrams for country A and country B below and answer the questions that follow.

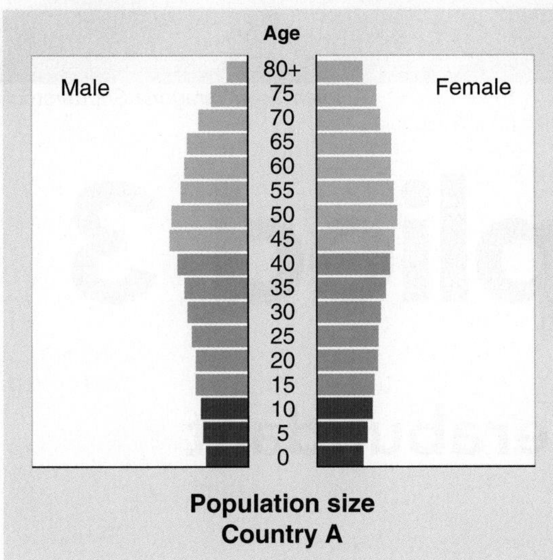

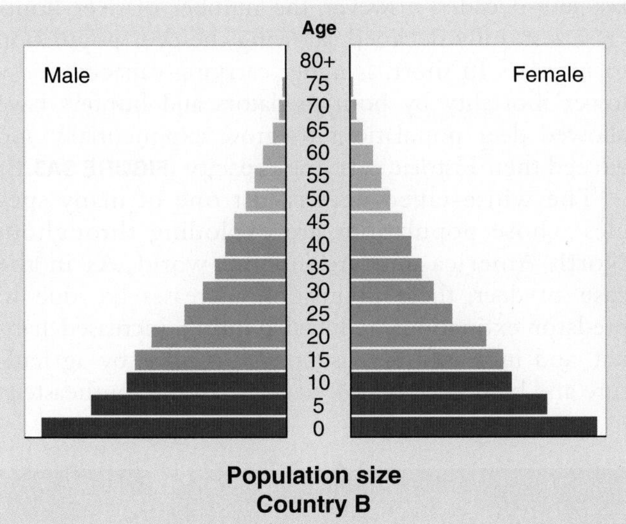

(a) What observations and educated predictions can you make about the following characteristics of country A?
   (i) the age structure of its population (1 point)
   (ii) the total fertility rate of the country (1 point)
   (iii) the life expectancy of the population (1 point)
   (iv) the growth rate and doubling time of the population (2 points)

(b) Describe one socioeconomic feature of country A. (2 points)

(c) Explain how country B differs from country A in terms of
   (i) the age structure of its population. (1 point)
   (ii) its infant mortality rate. (1 point)
   (iii) its rate of population growth. (1 point)

3. Country Y and country Z are two nations in different phases of demographic transition. Country Y is in stage two, while Country Z is in phase four.

(a) Draw two labeled age structure diagrams, one for each country.

(b) Explain the differences between country Y and Z in terms of the following.
   (i) CBR, CDR, and the rate of population growth (2 points)
   (ii) the age structure of the population (1 point)

(c) Explain the social and economic implications of a high immigration rate into country Z. (2 points)

(d) Give one example of an actual country that could be represented by country Y and one example that could be represented by country Z. (1 point)

Anitham Raju Yaragorla/Shutterstock.com

science applied 3

How Can We Control Overabundant Animal Populations?

The growing human population has caused population declines in many animal species. At the same time, human alteration of the landscape has caused other species to flourish. Increases in the population of these species have had many consequences. How have these increases happened, and what, if anything, should we do about them?

What causes animal population explosions?

The white-tailed deer is a high-profile example of dramatic animal population growth in the United States. Before European colonists arrived, its densities were approximately 4 to 6 deer per square kilometer (10 to 15 deer per square mile). In many areas today, however, deer densities are up to 20 times higher. In New York State, for example, unregulated hunting and the clearing of preferred deer habitats for farming had nearly eliminated the deer population by 1900. The population has since grown exponentially to 1 million deer.

This population growth can be explained by processes we discussed in Chapters 6 and 7. For example, white-tailed deer prefer to live in early-successional fields containing wildflowers and shrubs. Abandoned farms that have undergone secondary succession during the twentieth century, such as the old fields of New England that we described in Chapter 6, are prime deer habitat. Such increases in habitat and food supply have increased the carrying capacity of the environment for deer. With a higher carrying capacity, density-dependent controls on the population are weaker, and reproduction increases.

In addition to providing deer with habitat and food, humans have eliminated many of the deer's major predators, including wolves and cougars. During much of the twentieth century, regulated hunting held deer populations in check, taking the place of natural predators that had largely been eliminated. Over the last few decades, however, the number of deer hunters has steadily declined, allowing the deer population to increase. In short, a higher carrying capacity and a lower mortality by both predators and hunters have allowed deer populations to grow exponentially and exceed their historic carrying capacity (**FIGURE SA3.1**).

The white-tailed deer is just one of many species whose populations are exploding throughout North America and around the world. As in the case of deer, these population increases are due to predator extinctions, reduced hunting, increased habitat, and increased food supplies provided by agriculture and by people who feed them. In the southeastern

FIGURE SA3.1 Deer overpopulation. As humans have provided increased food and habitat for white-tailed deer and simultaneously removed their major predators, deer populations have increased to the point that they exceed the historic carrying capacity. This photo shows deer in a yard of a home in Hemlock Farms, Pennsylvania. *(Raymond Gehman/Getty Images)*

United States, some varieties of domesticated European and Russian hogs (*Sus scrufa*) have been allowed to be free ranging and this has allowed feral populations to evolve and spread. Over the past 3 decades, feral hogs have spread to 39 states. Wild horses (*Equus ferus*) are another species that is not native to the United States but has a growing population that is a perpetual challenge to control in the American West (**FIGURE SA3.2**).

Large herbivorous mammals are not the only animals that experience overpopulation as the result of human activity. Florida's 500,000 monk parakeets (*Myiopsitta monachus*) are the offspring of individuals that have escaped from the pet trade since the 1960s. Populations of Canada geese (*Branta canadensis*) have grown along with the number of agricultural fields and golf courses on which they can feed. Other former pets, including dogs and cats, quickly establish feral populations when released into the wild.

What effects do overabundant species have on communities?

Populations that exceed their historic carrying capacity can cause widespread problems. Large populations of herbivore species such as deer, hogs, and horses can overgraze the plants in their community, leaving little food for other herbivorous species. Hogs are also causing major problems with rooting up the soil. In Australia, grazing by the overabundant kangaroos removes many of the plants needed by other herbivores, including several species of endangered animals that depend on these plants for food and habitat.

These large animal populations also have direct effects on humans. Herbivorous mammals consume agricultural crops, landscape plants, and valuable wild

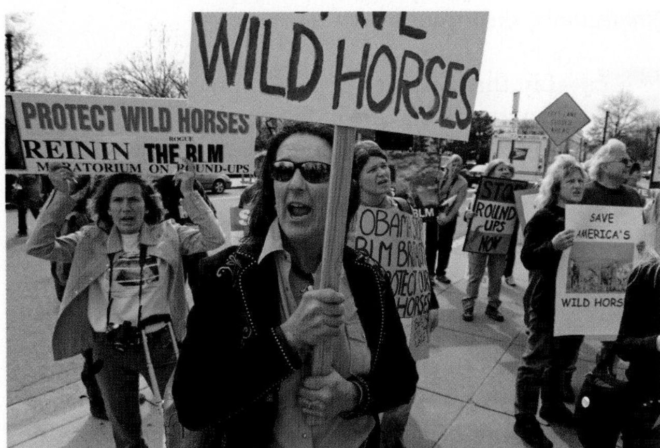

FIGURE SA3.2 Overabundant horses. Wild horses in the American West were introduced more than a century ago. Without major predators, their population has increased to a point where they are impacting the habitat needed by native species. How to best control horse populations is a controversial issue with many different ethical and scientific perspectives. *(Cliff Owen/AP Images)*

plant species. In the United States, for example, the deer population so greatly exceeds the land's natural carrying capacity that it is now difficult to regenerate several valuable forest tree species, including northern red oak (*Quercus rubra*) and sugar maple. Deer also consume suburban landscape plants; in New York State, for example, it has been estimated that deer damage $60 million worth of agricultural crops each year. Wild hogs across the United States cause $1.5 billion in crop damage each year.

Wildlife overabundance increases wildlife-human encounters, such as deadly collisions of vehicles with deer or kangaroos. For example, white-tailed deer in the United States cause more than 1.2 million vehicle collisions, with an average cost of about $4,000 per accident. More than 200 people die each year from these collisions.

Over-abundant animal populations can also have less appreciated impacts. For example, most people are surprised to know that even house cats can have a major impact on native animal populations. For example, researchers in 2014 estimated that domestic cats annually kill 1.4 to 3.7 billion birds and 6.9 to 20.7 billion mammals.

How can we control overabundant species?

Controlling the population of any species involves issues that are ecological, political, moral, and ethical. From an ecological perspective, it is clear how to control a wildlife population: reduce the available food and habitat to lower the carrying capacity, compensate for the missing predators by killing individuals in the population, or slow the population's ability to reproduce. On a local scale, it can be helpful to discourage people from feeding wildlife. On a larger scale, however, reducing food and habitat is often the least viable option because it would typically require restricting the access of large herbivores to millions of hectares of land.

Compensating for absent predators by culling animal herds through hunting can be an effective approach, but it is often met with resistance from members of the public. For example, in 2009 the Australian government ordered the army to shoot 6,000 kangaroos to reduce an over-abundant population living on a large military compound. The objective was to reduce the amount of grazing done by the kangaroo herd and therefore protect rare plants and the insects that live on those plants. This decision was opposed by animal rights activists and by many Australians who consider the kangaroo a national symbol that should be protected. In other cases, such as overabundant deer in suburban areas of the United States, shooting the animals is not a viable approach for reasons of safety. In short, culling a herd through shooting is an increasingly unpopular strategy. Moreover, most populations that are reduced in number rapidly bounce back because of a sudden increase in the abundance of their food.

FIGURE SA3.3 Administering contraceptives to wild animals. Researchers deliver the PZP contraceptive drug to large herbivores, such as wild horses, using a dart gun. *(Anne Sherwood/The New York Times/Redux)*

A strategy that has more public support is the use of birth control. This approach is common in cities, where many feral cats are trapped, neutered, and returned to the wild. Researchers are also investigating ways to give animals contraceptive drugs to reduce their populations over time. The drugs being developed differ in their modes of action, ranging from chemicals that prevent animals from making critical reproductive hormones to chemicals that prevent sperm from fertilizing an egg. Contraceptive strategies are currently being tested on a variety of species, including deer, kangaroos, elephants, and pigeons (**FIGURE SA3.3**).

Reducing animal overpopulation with contraceptives has widespread public appeal because it eliminates the need to kill animals. Getting the contraceptive drugs into the animals, however, presents substantial challenges. Drugs can be administered to some animals via dart guns. For others, such as monk parakeets, researchers are working on ways to administer medications in food. Some contraceptive drugs last for a single breeding season and require a new dose each season, which can be difficult and expensive. The cost of giving an animal a single dose of a contraceptive drug ranges from $100 to $500 per animal. However, recent tests with a new contraceptive drug, known as PZP (porcine zona pellucida), suggest that the birth control effects on deer may last for up to 5 years, which may make the approach more economically viable. At the same time, researchers are developing less labor-intensive methods of delivering the drugs to animals, including setting up feeders with facial recognition software to ensure that the drug is given to the targeted species and not to other species.

The issue of overabundant wildlife highlights a number of other environmental issues. The underlying causes of animal overpopulation are typically associated with human population increases. These human-caused wildlife population explosions can, in turn, have substantial negative effects on humans and their environment. Although there are several potential solutions to the problem of animal overpopulation, successful solutions must take into account the public's affection for wildlife, its opposition to killing animals, and its frequent lack of knowledge about the negative effects of overabundant species.

References

Foderaro, L. 2013. A kinder, gentler way to thin the deer herd. *New York Times*, July 5. (http://www.nytimes.com/2013/07/06/nyregion/providing-birth-control-to-deer-in-an-overrun-village.html)

Nordstrum, A. 2014. Can wild pigs ravaging the U.S. be stopped? *Scientific American*, October 21. https://www.scientificamerican.com/article/can-wild-pigs-ravagingthe-u-s-be-stopped/.

Klein, A. 2016. Cont-roo-ception: Hormone implants bring kangaroos under control. *New Scientist*, June 20. https://www.newscientist.com/article/2094401-cont-roo-ceptionhormone-implants-bring-kangaroosunder-control/.

Masters, B. 2017. Can fertility control keep wild horse herds in check? *National Geographic*, February 8. https://www.nationalgeographic.com/adventure/features/environment/wild-horses-part-three/

Questions

1. If humans are responsible for environmental changes that cause increases in the populations of native species, under what conditions should we work to reverse these population increases?
2. How might the options to control overabundant populations differ in rural versus urban areas?

Preparing for the AP® Exam

Practice AP® Free-Response Question

Write your answer to each part clearly. Support your answers with relevant information and examples. Where calculations are required, show your work.

(a) List three density-dependent factors and three density-independent factors that can influence the population size of white-tailed deer. (2 points)

(b) Describe two ways that deer populations at low abundance can *accelerate* the process of ecological succession in a forest toward a climax stage. (4 points)

(c) Consider a region of temperate forest that contains 100,000 deer with a sex ratio of 3 females to 1 male, an infant mortality rate of 10, a crude birth rate of 40, and a crude death rate of 20. Only 10 percent of the females are able to reproduce.

(i) Calculate population growth rate for white-tailed deer in this region. (2 points)

(ii) Calculate the number of deer that hunters would need to shoot in order to keep the deer population stable. (2 points)

Section 1: Multiple-Choice Questions

Choose the best answer for questions 1–25.

1. Which is NOT true regarding limiting resources?
   (a) A limiting resource is an important density-independent factor.
   (b) A limiting resource is common to all members of a population.
   (c) Foraging territory is an example of a limiting resource.
   (d) The number of reproducing individuals in a population is not a limiting resource.

2. Which is most likely to result in a random population distribution?
   (a) clumping of individuals around limiting resources
   (b) competition among individuals
   (c) density-independent factors
   (d) variable carrying capacities among populations within a metapopulation

3. The common reed (*Phragmites australis*) is a perennial grass found throughout temperate and tropical regions of the world. Researchers recently discovered that many North American populations of the reed are composed entirely of genetically identical clones. Which is most likely true of these populations?
   (a) The age structure is unlikely to vary among populations.
   (b) Evolution by natural selection occurs very slowly in the absence of seed dispersal among populations.
   (c) Density-dependent factors will be more important than density-independent factors in regulating population densities.
   (d) Density-dependent factors will be less important than density-independent factors in regulating population densities.

4. Species with a type III survivorship curve are likely to exhibit
   I. fast population growth rate.
   II. large amounts of parental care.
   III. density-independent population regulation.
   (a) I only
   (b) II only
   (c) III only
   (d) I and III

5. In an experimental grassland plot, researchers planted 20 seeds of a wildflower species that exhibits exponential growth with an intrinsic growth rate of 0.60. Approximate the maximum population size after 10 years.
   (a) 20×0.60
   (b) $20 \div 0.60$
   (c) $e^{0.60 \times 10}$
   (d) $e^{0.60 \times 11}$

6. Which of the following is NOT a likely consequence of niche partitioning?
   (a) reduced interactions between different predator species
   (b) clumped species distributions
   (c) genetic variation within species
   (d) reduced competition between species

7. Mycorrhizal fungi are commonly found attached to plant roots. The fungi are more efficient than plants at extracting nutrients from the soil. In exchange for nutrients, plants provide the fungi with sugars made through photosynthesis. However, when nutrients in the soil are abundant, researchers have found that plant growth increases if mycorrhizal fungi are removed. This phenomenon suggests that
   (a) mycorrhizae have switched from being mutualists to parasites.
   (b) mycorrhizae have switched from being mutualists to predators.
   (c) mycorrhizae have switched from being mutualists to commensalists.
   (d) mycorrhizae have ceased to act as a keystone species.

8. Given two islands of the same size, the theory of island biogeography suggests that the _______ island should have a higher _______.
   (a) farther; extinction rate
   (b) closer; speciation rate
   (c) farther; species diversity
   (d) closer; species richness

9. Ecological succession
   (a) begins with primary succession, followed by secondary succession, and ends with a climax stage.
   (b) begins with either primary or secondary succession and ends with a climax stage.
   (c) cannot occur if there are no species initially present at a site.
   (d) does not end with a climax stage.

10. The theory of island biogeography does not apply to
    (a) protected wildlife areas.
    (b) islands very far away from the mainland.
    (c) metapopulations.
    (d) mountaintop communities.

11. Which may lead to an increase in the carrying capacity of the global human population?
    I. consuming food at a lower trophic level
    II. technological advancement
    III. reduction in per capita consumption rate
    (a) I only
    (b) II only
    (c) I and II
    (d) I, II, and III

12. In Vietnam, there were 16 births and 6 deaths out of every 100 people in the year 2011. Using the rule of 70, estimate the doubling time for Vietnam's population.
(a) 0.7 years
(b) 7 years
(c) 0.14 years
(d) 1.4 years

13. Which is NOT likely to describe a population in a developing country?
(a) Replacement level fertility is higher than 2.1.
(b) Infant mortality is at 100 per 1,000 live births or more.
(c) Immigration is relatively low or nonexistent.
(d) The age structure will be shaped like an inverted pyramid.

14. World Bank records indicate that in 2010, the total fertility rate in Puerto Rico was 1.6 and population size was 4,000,000. If net migration rate was −1.0, then the best estimate of the population in 2012 is
(a) 2,500,000.
(b) 3,700,000.
(c) 4,000,000.
(d) 4,500,000.

15. A developing country with a total fertility rate of 4.0 and a net migration rate of −3.5 is likely to have an age structure
(a) shaped like a pyramid.
(b) shaped like an inverted pyramid.
(c) shaped like a column.
(d) that is variable depending on the demographic.

16. Which is NOT a factor that determines population growth under the theory of demographic transition?
(a) Improved sanitation and health care leads to rapid population growth.
(b) Population momentum holds birth rates stable while death rates decline.
(c) Density independent factors slow economic growth.
(d) Education reduces a population growth rate.

17. Increasing the gross domestic product of a nation can be associated with
I. an increase in pollution.
II. a decrease in pollution.
III. an increase in population size.
(a) I only
(b) II only
(c) III only
(d) I, II, and III

Questions 18 and 19 refer to the following table:

Country	Ecological footprint (km^2)	Average monthly salary	Number of coal burning plants	Population growth rate (%)
Argentina	1,000,000	1,100	1	0.98
Colombia	800,000	700	4	1.10
Philippines	1,100,000	300	9	1.84
Brazil	5,500,000	800	7	0.83
Panama	100,000	800	1	1.38

Data from: Global Energy Observatory, http://www.globalenergyobservatory.org; United Nations' International Labour Organization, http://www.ilo.org; World Bank, http://data.worldbank.org.

18. Based solely on the IPAT equation, ______ is likely to have the greatest population size and ______ is likely to have smallest population size.
(a) Argentina; Colombia
(b) Colombia; Philippines
(c) Philippines; Panama
(d) Brazil; Panama

19. Which country is most likely to be in the second phase of demographic transition?
(a) Argentina
(b) Colombia
(c) Philippines
(d) Brazil

20. If everyone moved to urban areas, it is likely that
(a) the global ecological footprint would decrease.
(b) the global ecological footprint would increase.
(c) the global ecological footprint would remain the same.
(d) global sustainability would be achieved.

21. Which is an example of a population?
(a) a family of bears living in an area of a national park
(b) mammalian species inhabiting the area around a river
(c) all red-eyed tree frogs living in a particular nature reserve
(d) the flora and fauna of Glacier National Park

Questions 22 and 23 refer to a species of bird, the North Island robin of New Zealand.

22. The North Island robin is very territorial, and individuals often remain in and defend the same patch of forest throughout their lives. The distribution of this species within a particular forest is likely to be
(a) clumped.
(b) uniform.
(c) random.
(d) scattered.

23. Robin territories are much larger when invasive predators such as rats and weasels are present than when such predators are absent. These predators often disproportionately kill females because the females are vulnerable while incubating eggs on the nest. If rats and weasels are introduced to a reserve where they were previously absent, robin density would most likely_______ and the male:female sex ratio would become _______.
    (a) increase; even
    (b) decrease; skewed female
    (c) increase; skewed female
    (d) decrease; skewed male

24. A city has a CBR of 15 and a CDR of 17. Ignoring immigration and emigration, when will the population of this city double?
    (a) 20 years
    (b) 35 years
    (c) 200 years
    (d) 350 years

25. As a country moves into phase 2 of the demographic transition, you would expect replacement level fertility to _______ and crude death rate to _______.
    (a) increase; decrease
    (b) decrease; increase
    (c) increase; increase
    (d) decrease; decrease

Section 2: Free-Response Questions

Write your answer to each part clearly. Support your answers with relevant information and examples. Where calculations are required, show your work.

1. *Wolbachia* is a genus of bacteria and one of the world's most common parasitic microbes. It infects an enormous number of species, including many insects. Estimates indicate that as many as 70 percent of all insect species are infected with *Wolbachia,* including mosquitoes, spiders, and mites. The bacteria reproduce inside a host and are transmitted directly from mother to offspring or when infected males mate with uninfected females. After infection, *Wolbachia* consume host resources, which leaves the host with fewer resources for reproduction. As a result, infected females often have fewer offspring.
    (a) What is a possible limiting resource for *Wolbachia,* aside from the amount of resources within the host? (2 points)
    (b) Describe the population trajectory of *Wolbachia* inside of a host. (2 points)
    (c) Describe the population trajectory of *Wolbachia* in a population of hosts. (2 points)
    (d) Based on your answer for the previous question, what is one possible way that natural selection is altering the interaction between *Wolbachia* and its hosts? (2 points)
    (e) Discuss how the presence of nearby host populations might alter the spread of *Wolbachia.* (2 points)

2. By 2012, China's one-child policy had drastically reduced total fertility rate to 1.4 and led to a reduction in population growth rate to 0.50 percent. One problem associated with this is the development of an inverted population pyramid. Care for the elderly requires a substantial amount of energy, resources, and manpower. Chinese elderly primarily rely on their families for health care.
    (a) Given China's one-child policy, provide two examples of how China's growing elderly population will affect future population growth. (3 points)
    (b) Describe one potential benefit and one potential problem associated with the emigration of elderly individuals from rural to urban areas. (2 points)
    (c) The total population size of China was 1,400,000,000 in 2012, with a crude birth rate of 12 births per 1,000 individuals. The number of individuals that emigrated from China was 700,000 and the number of individuals that immigrated to China was 200,000. Calculate the crude death rate. (3 points)
    (d) What would the total fertility rate be if all families in China obeyed the one-child policy rule? (1 point)

3. A researcher is examining the communities of two islands. Island A has an area of 53×10^5 m^2 and is 2 km from the mainland. Island B is 4.5×10^6 m^2 and is 3 km from the mainland.
    (a) Which island would you expect to have a greater number of species, and why? Use the theory of island biogeography in your answer. (3 points)
    (b) The researcher notices that Island A has two species of parakeets, which seem to look and behave very similarly, except that one species forages on the forest floor and the other high in the forest canopy. Describe the phenomenon that is occurring here. (2 points)
    (c) Give an example of ONE density-dependent factor and ONE density-independent factor that could influence the parakeet species on the island. (2 points)
    (d) A species of deer that lives on Island B and feeds on a particular type of tree is exterminated from the island due to hunting. Within several years, plant species evenness on that island drops dramatically as a particular type of tree dominates the ecosystem. What is a likely explanation for this pattern, and what specific role do the deer play in this? (3 points)

Lithium, lanthanum, and other metals are vital components of environmentally friendly hybrid-electric cars, but mining these metals has adverse environmental consequences. This open-pit mine is in Mountain Pass, California.

(Bloomberg/Getty Images)

Earth Systems

CHAPTER

8

MODULE 24 Mineral Resources and Geology

MODULE 25 Weathering and Soil Science

CASE STUDY

Are Hybrid Electric Vehicles as Environmentally Friendly as We Think?

In 2017, there were approximately 2 million all-electric vehicles sold worldwide including half a million in the United States. At that time, hybrid electric vehicle (HEV) sales numbers throughout the world totaled over 11 million. These sales numbers are quite small compared to the total number of roughly 1.2 billion automobiles worldwide. However, many people in the environmental science community believe that the latest improvements to HEV and all-electric vehicles are some of the most exciting environmental advancements of the last decade. Cars that run on electricity or on a combination of electricity and gasoline use energy much more efficiently than similarly sized internal combustion (IC) automobiles. Some of these cars use no gasoline at all, while others are able to travel twice the distance as a conventional IC car on the same amount of gasoline.

Although HEV and all-electric vehicles reduce our consumption of liquid fossil fuels, and perhaps reduce overall energy use as well, they do come with environmental trade offs. The manufacturing of HEV vehicles uses scarce *metals*, including neodymium, lithium, and lanthanum. Neodymium is needed to form the magnets used in the electric motors, and lithium and lanthanum are used in the compact high-performance batteries the vehicles require. At present, there appears to be enough of these metals to meet demand and the availability of rare metals will probably not be the limiting factor in HEV and all-electric vehicle production. However, there are environmental costs to the increased demand for metals for HEV and all-electric cars.

Although HEV and all-electric vehicles reduce our consumption of liquid fossil fuels, they do come with environmental tradeoffs.

The mining of metals occurs around the globe. Toyota currently obtains its lanthanum from China. There are also supplies of lanthanum in various geologic deposits in California, Australia, Bolivia, Canada, and elsewhere. Neodymium is mined in China, the United States, Brazil, and India. Lithium is mined in Australia, Chile, Argentina, China, and the United States.

Wherever mining occurs, it has a number of environmental consequences. Material extraction leaves a landscape fragmented by holes—some quite large. Road construction necessary for access to the mining site further alters the habitat. Air pollution, erosion, and water contamination are also common results of mining. And the rock and soil debris that remains after the metals are extracted can persist and cause harm to the landscape for a very long time.

A typical Toyota Prius HEV contains approximately 1 kg (2.2 pounds) of neodymium and 10 kg (22 pounds) of lanthanum. Mining these elements involves pumping acids into deep boreholes to dissolve the surrounding rock and then removing the acids and resulting mineral slurry. Lithium is extracted from certain rocks, and lithium carbonate is extracted from brine pools and mineral springs adjacent to or under salt flats. Both extraction procedures are types of surface mining, which can have substantial environmental impacts. The holes, open pits, and ground disturbance created by mining these minerals provide the opportunity for air and water to react with other minerals in the rock, such as sulfur, to form an acidic slurry. The slurry, a combination of drainage water and material, flows over the land or underground toward rivers and streams and dissolves metals and other elements. As a result, water near surface mining operations is highly acidic—sometimes with a pH of 2.5 or lower. It may also contain harmful levels of dissolved metals and minerals.

Current HEV technology does represent a step forward in our search to reduce energy use, particularly the use of liquid fossil fuels such as gasoline and diesel. However, to make good decisions about the use of resources, we must have a full understanding of the environmental and other costs as well as the benefits.

Sources: M. Armand and J.-M. Tarascon, Building better batteries, *Nature* 451 (2008): 652–657; A. Lovins, Clean energy and rare earths: Why not to worry. *Bull. of Atomic Scientists* http://thebulletin.org/clean-energy-and-rare-earths-why-not-worry10785.

As we saw in our account of hybrid and all-electric cars, decisions about which resources to use are not always obvious. Why are some of Earth's mineral resources relatively limited? Why do certain elements occur in some locations but not in others? What processes create minerals and other Earth materials, and what are the consequences of extracting them? Understanding the answers to these questions helps environmental scientists make informed decisions about the environmental and economic costs and benefits of resource use.

In this chapter we will explore the subjects of resources, geology, and soil science. We will look inside Earth to explore its structure, its formation, and the ongoing processes that affect the composition and availability of elements and minerals on our planet. We will then turn to the surface of Earth to explore how the rock cycle distributes these resources. With this foundation, we will examine the formation of soil, which is essential for so many biological activities on Earth. Finally, we will look at the problems we face in extracting resources from Earth—an issue of great concern to environmental scientists and others.

MODULE 24

Mineral Resources and Geology

Almost all of the mineral resources on Earth accumulated when the planet formed 4.6 billion years ago. But Earth is a dynamic planet. Earth's geologic processes form and break down rocks and minerals, drive volcanic eruptions and earthquakes, determine the distribution of scarce mineral resources, and create the soil in which plants grow. In this module we will examine the distribution of Earth's mineral resources and some of the geological processes that continue to affect this distribution.

Learning Goals

After reading this module you should be able to

- describe the formation of Earth and the distribution of critical elements on Earth.
- define the theory of plate tectonics and discuss its relevance to the study of the environment.
- describe the rock cycle and discuss its importance in environmental science.

The availability of Earth's resources was determined when the planet was formed

Earth's history is measured using the geologic time scale, shown in **FIGURE 24.1** on page 272. It's hard to believe that events that took place 4.6 billion years ago continue to have such an influence on humans and their interactions with Earth. The distribution of chemicals, minerals, and ores around the world is in part a function of the processes that occurred during the formation of Earth.

The Formation and Structure of Earth

Nearly all the elements found on Earth today are as old as the planet itself. **FIGURE 24.2** on page 273 illustrates how Earth formed roughly 4.6 billion years ago from cosmic dust in the solar system. The early Earth was a hot, molten sphere. For a period of time, additional debris from the formation of the Sun bombarded Earth. As this molten material slowly cooled, the elements within it separated into layers according to their mass. Heavier elements such as iron sank toward Earth's center, and lighter elements such as silica floated toward its surface. Some gaseous elements left the solid planet and became part of Earth's atmosphere. Although asteroids occasionally strike Earth today, the bombardment phase of planet formation has largely ceased and the elemental composition of Earth has stabilized. In other words, the elements and minerals that were present when the planet formed—and which are distributed unevenly around the globe—are all that we have today.

Some minerals such as silicon dioxide—the primary component of sand and glass—are readily available worldwide on beaches and in shallow marine and glacial deposits. Others such as diamonds—which are formed from carbon that has been subjected to intense pressure—are found in relatively few isolated locations. Over the course of human history, this uneven geographic distribution has driven many economic and political conflicts.

Because Earth's elements settled into place according to their mass, the planet is characterized by distinct vertical zonation. If we could slice into Earth as shown in **FIGURE 24.3** on page 273 we would see concentric layers composed of various materials. The innermost zone is the planet's **core**, over 3,000 km (1,860 miles) below

Core The innermost zone of Earth's interior, composed mostly of iron and nickel. It includes a liquid outer layer and a solid inner layer.

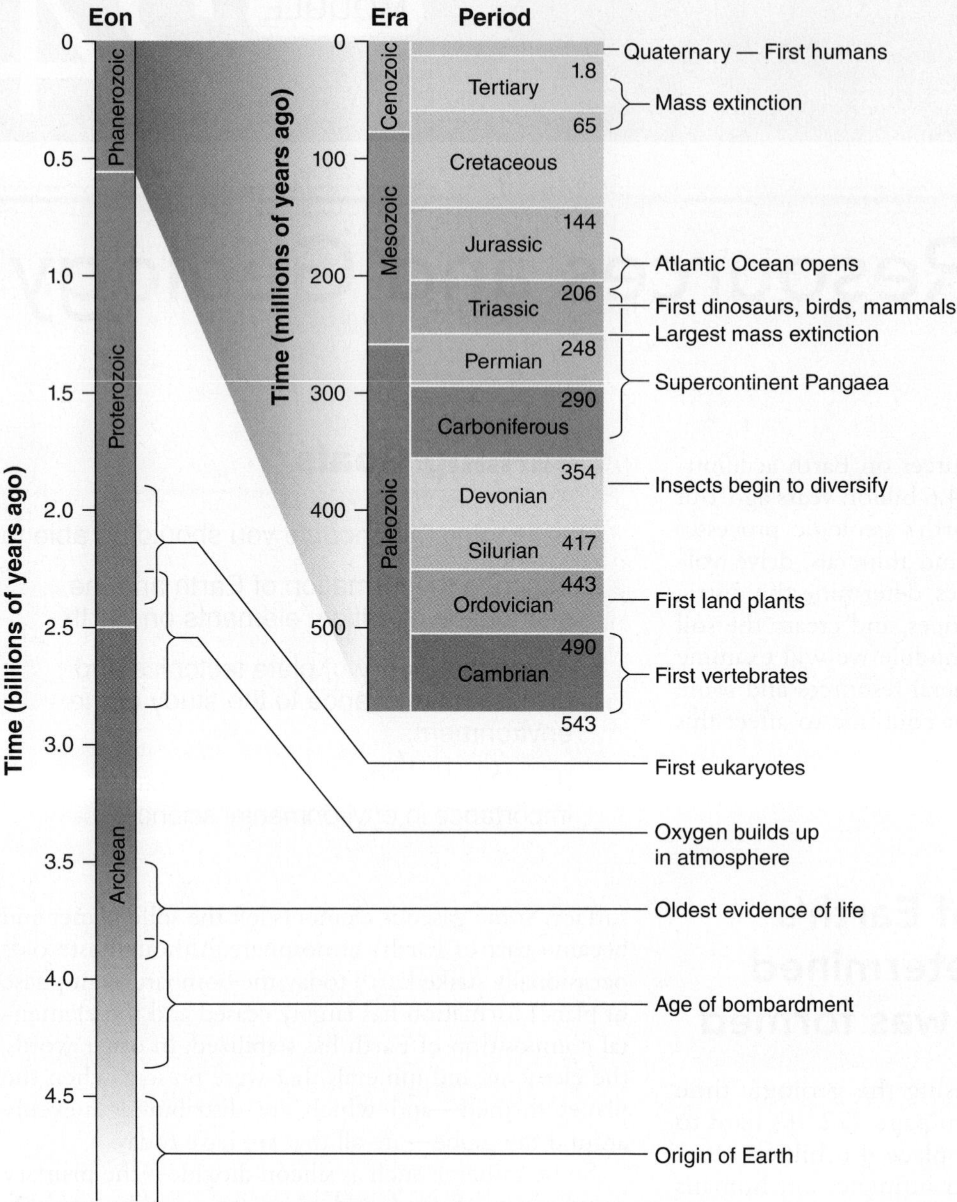

FIGURE 24.1 The geologic time scale. Time since the origin of Earth through the present is divided into three eons. The most recent eon, the Phanerozoic, is broken down into three eras. The Phanerozoic can also be divided into 11 periods, the most recent of which is the Quaternary.

Earth's surface. The core is a dense mass largely made of nickel and some iron. The inner core is solid, and the outer core is liquid. Above the core is the **mantle**, containing molten rock, or **magma**, that slowly circulates in convection cells, much as the atmosphere does. The **asthenosphere**, located in the outer part of the mantle, is composed of semi-molten, ductile (flexible) rock. The brittle outermost layer of the planet, called the **lithosphere** (from the Greek word *lithos,* which means "rock"), is approximately 100 km (60 miles) thick. It includes the solid upper mantle as well as the **crust**, the chemically distinct outermost layer of the lithosphere. It is important to recognize that these regions overlap: The lowest part of the lithosphere is also the uppermost portion of the mantle.

The lithosphere is made up of several large and numerous smaller plates, which overlie the convection cells within the asthenosphere. Over the crust lies the

Mantle The layer of Earth above the core, containing magma.

Magma Molten rock.

Asthenosphere The layer of Earth located in the outer part of the mantle, composed of semi-molten rock.

Lithosphere The outermost layer of Earth, including the mantle and crust.

Crust In geology, the chemically distinct outermost layer of the lithosphere.

FIGURE 24.2 Formation of Earth and the solar system. The processes that formed Earth 4.6 billion years ago determined the distribution and abundance of elements and minerals today.

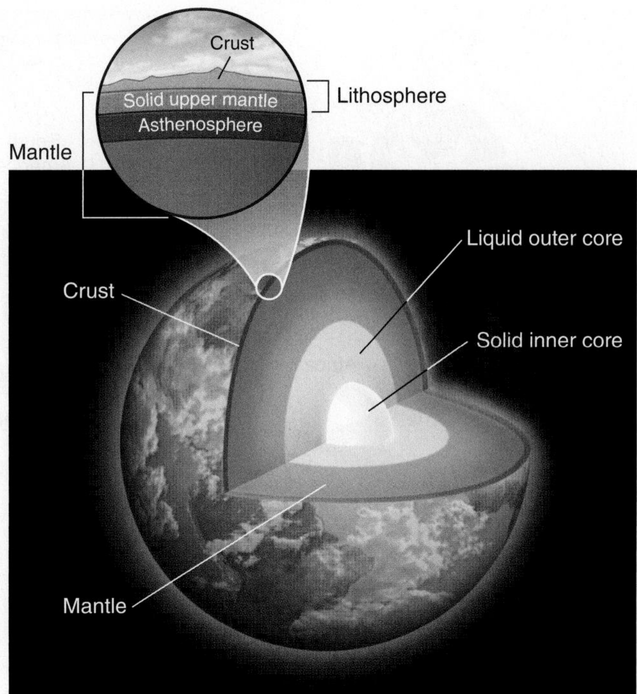

(a) Earth's vertical zonation

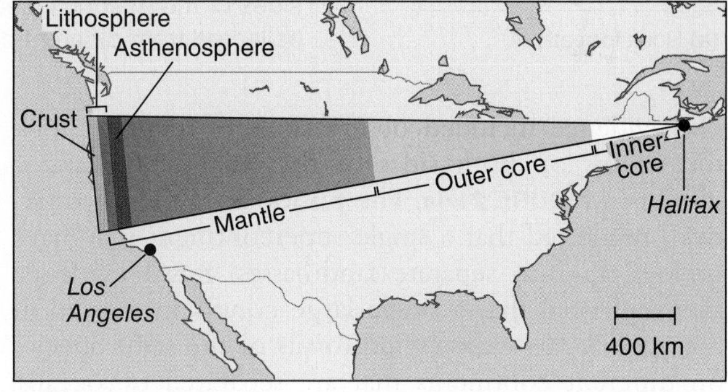

(b) Scale of Earth's layers

FIGURE 24.3 Earth's layers. (a) Earth is composed of concentric layers. (b) If we were to slice a wedge from Earth, it would cover the width of the United States.

thin layer of soil that allows life to exist on the planet. The crust and overlying soil provide most of the chemical elements that make up life.

Because Earth contains only a finite supply of mineral resources, we will not be able to extract resources from the planet indefinitely. In addition, once we have mined the deposits of resources that are most easily obtained, we must use more energy to extract the remaining resources. Both of these realities provide an incentive for us to minimize our use of mineral resources and to reuse and recycle them whenever possible.

Hot Spots

One of the critical consequences of Earth's formation and elemental composition is that the planet remains very hot at its center. The high temperature of Earth's outer core and mantle is thought to be the result of the radioactive decay of various isotopes of elements such as potassium, uranium, and thorium, which release heat. The heat causes plumes of hot magma to well upward from the mantle. These plumes produce **hot spots**: places where molten material from the mantle reaches the lithosphere. As we shall see in the following pages, hot spots are an important component of the surface dynamics of Earth.

The theory of plate tectonics describes the movement of the lithosphere

Prior to the 1900s, scientists believed that the major features of Earth—such as continents and oceans—were fixed in place. In 1912, a German meteorologist named Alfred Wegener published a revolutionary hypothesis proposing that the world's continents had once been joined in a single landmass, which he called "Pangaea."

Hot spot In geology, a place where molten material from Earth's mantle reaches the lithosphere.

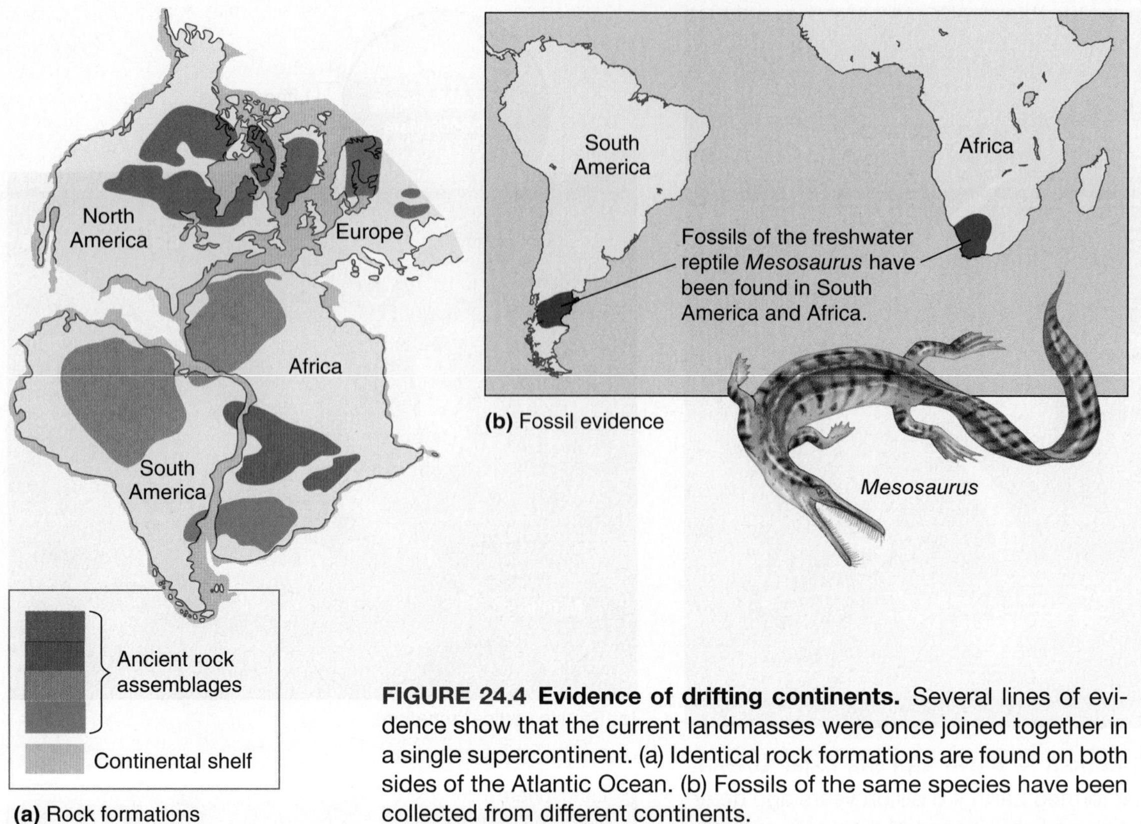

FIGURE 24.4 Evidence of drifting continents. Several lines of evidence show that the current landmasses were once joined together in a single supercontinent. (a) Identical rock formations are found on both sides of the Atlantic Ocean. (b) Fossils of the same species have been collected from different continents.

His evidence included observations of identical rock formations on both sides of the Atlantic Ocean, as shown in **FIGURE 24.4a**. The positions of these formations suggested that a single supercontinent may have broken up into separate landmasses. Fossil evidence also suggested that a single large continent existed in the past. Today, we can find fossils of the same species on different continents that are separated by oceans; Figure 24.4b shows one example.

Many resisted the idea that Earth's lithosphere could move laterally. However, following the publication of Wegener's hypothesis, scientists found additional evidence that Earth's landmasses had existed in several different configurations over time. This led to the theory of **plate tectonics**, which states that Earth's lithosphere is divided into plates, most of which are in constant motion. The **tectonic cycle** is the sum of the processes that build up and break down the lithosphere.

The theory of plate tectonics is often called a unifying theory in geology and earth sciences because it relates to so many different aspects of the earth sciences.

Plate tectonics The theory that the lithosphere of Earth is divided into plates, most of which are in constant motion.

Tectonic cycle The sum of the processes that build up and break down the lithosphere.

Subduction The process of one crustal plate passing under another.

Plate Movement

We now know that the lithosphere consists of a number of plates. Oceanic plates lie primarily beneath the oceans, whereas continental plates lie beneath landmasses. The crust of oceanic plates is dense and rich in iron, while the crust of continental plates generally contains more silicon dioxide, which is much less dense than iron. The continental plates are therefore lighter and typically rise above the oceanic plates. **FIGURE 24.5** identifies Earth's major plates.

Oceanic and continental plates "float" on top of the denser material beneath them. Their slow movements are driven by convection cells in Earth's mantle. The heat from Earth's core creates these convection cells, which are similar to those in the atmosphere. (See Figure 10.6 on page 119.) Mantle convection drives continuous change: the creation and renewal of Earth materials in some locations of the lithosphere and destruction and removal of Earth material in other locations. As oceanic plates move apart, rising magma forms new oceanic crust on the seafloor at the boundaries between those plates. This process, called seafloor spreading, is shown in the center of **FIGURE 24.6**. Where oceanic plates meet continental plates, old oceanic crust is pulled downward beneath the continental lithosphere, removed from the ocean bottom, and pushed toward the center of Earth. In this process, the heavier oceanic plate slides underneath the lighter continental plate. This is best seen on the left side of Figure 24.6. This process of one crustal plate passing under another is known as **subduction**.

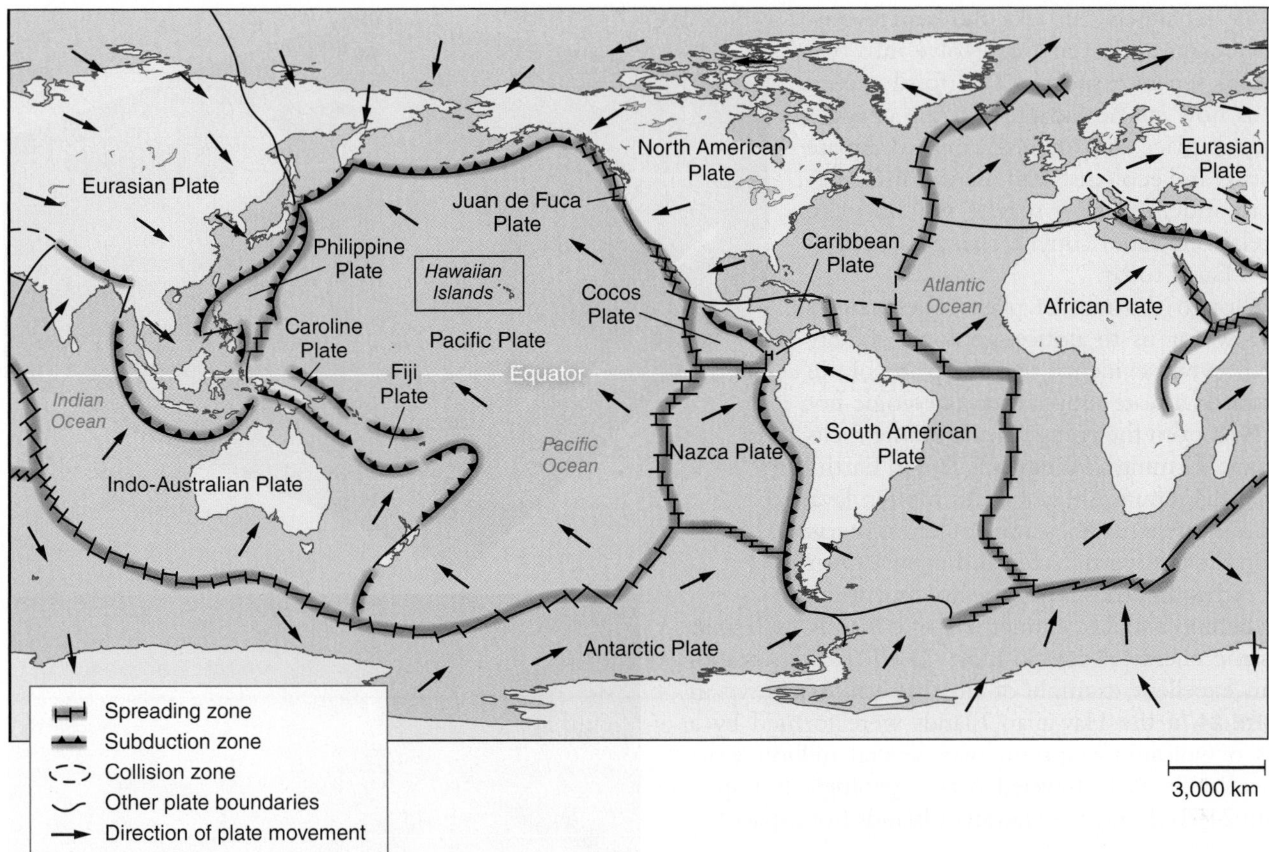

FIGURE 24.5 Tectonic plates. Earth is covered with tectonic plates, most of which are in constant motion. The arrows indicate the direction of plate movement. New lithosphere is added at spreading zones and older lithosphere is recycled into the mantle at subduction zones.

Consequences of Plate Movement

Because the plates move, continents on those plates slowly drift over the surface of Earth. As the continents have drifted, their climates have changed and geographic barriers were formed or removed and, as a result, species evolved and adapted, or slowly or rapidly went extinct. In some places, as the plates moved a continent that straddled two plates broke apart, creating two separate smaller continents or islands in different climatic regions. As you may recall from our discussion of allopatric speciation in Chapter 5, species that

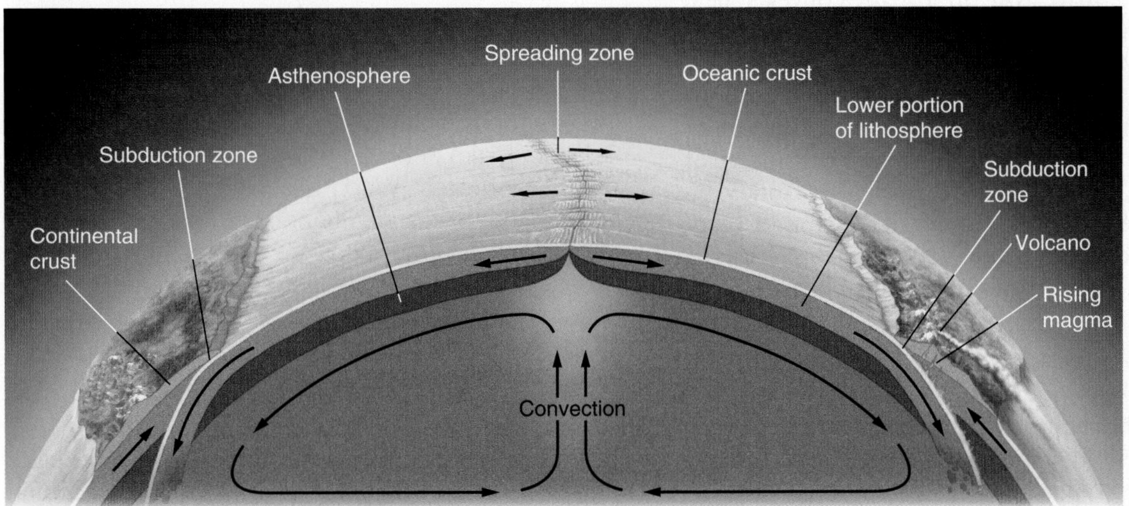

FIGURE 24.6 Convection and plate movement. Convection in the mantle causes oceanic plates to spread apart as new rock rises to the surface at spreading zones. Where oceanic and continental plate margins come together, older oceanic crust is subducted.

become separated can take different evolutionary paths and over time evolve into two or more separate species. The fossil record tells us how species adapted to the changes that took place over geologic time. Climate scientists and ecologists can use this information to anticipate how species will adapt to the relatively rapid climate changes happening on Earth today.

Although the rate of plate movement is too slow for us to notice, geologic activity provides vivid evidence that the plates are in motion. As a plate moves over a geologic hot spot, heat from the rising mantle plume melts the crust, forming a **volcano**: a vent in Earth's surface that emits ash, gases, and molten lava. Volcanoes are a natural source of atmospheric carbon dioxide, particulates, and metals. Over time, as the plate moves past the hot spot, it can leave behind a trail of extinct volcanic islands, each with the same chemical composition. The Hawaiian Islands are an excellent example of this pattern. As shown in **FIGURE 24.7a** the Hawaiian Islands were formed by a series of volcanic eruptions over several million years as the Pacific Plate traveled over a geologic hot spot. Figure 24.7b shows the Hawaiian Islands from space.

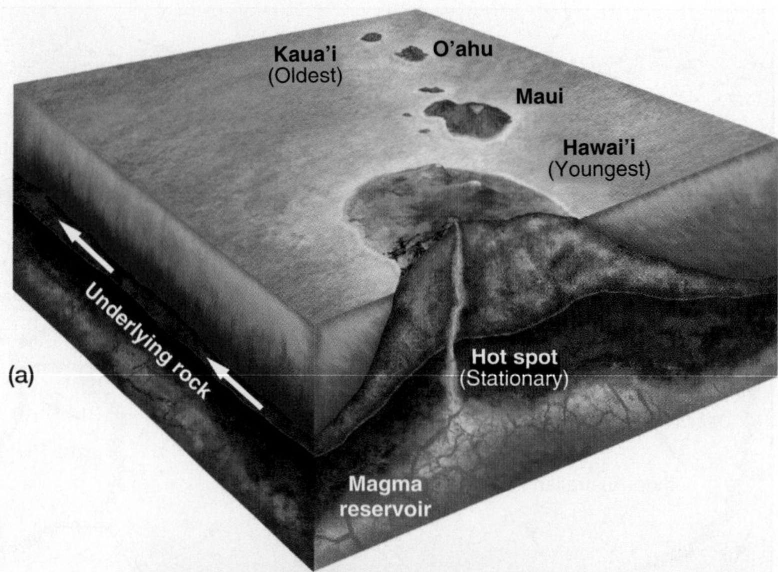

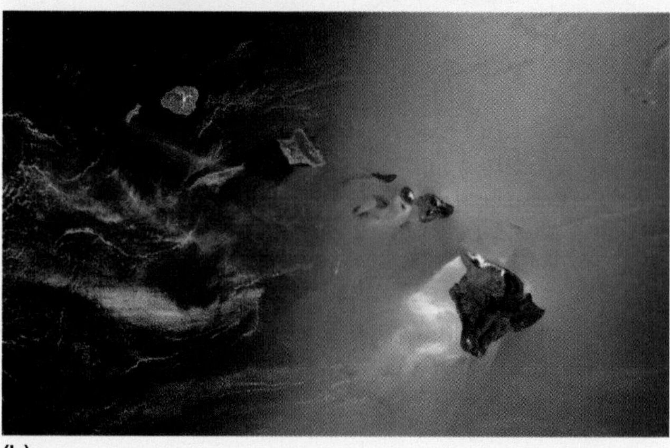

(b)

FIGURE 24.7 Plate movement over a hot spot. (a) As the Pacific Plate moves over a hot spot, a series of volcanic eruptions that occurred over several million years led to formation of the Hawaiian Islands. (b) This photo shows the Hawaiian Islands from space. *(Jacques Descloitres, MODIS Rapid Response Team, NASA/GSFC)*

Types of Plate Contact

Many other geologic events occur at the zones of contact that result from the movements of plates relative to one another. These zones of plate contact can be classified into three types: *divergent plate boundaries, convergent plate boundaries,* and *transform fault boundaries*.

Beneath the oceans, areas in which tectonic plates move away from each other are known as **divergent plate boundaries**. Divergent plate boundaries are illustrated in **FIGURE 24.8a**. At these boundaries, oceanic plates move apart as if on a giant conveyer belt. As magma from the mantle reaches Earth's surface and pushes upward and outward, new rock is formed, a phenomenon called **seafloor spreading**. Seafloor spreading creates new lithosphere and brings important elements such as copper, lead, and silver to the surface of Earth. However, this new rock typically lies under the deep ocean. Over tens to hundreds of millions of years, as the tectonic cycle continues, some of that material forms new land that contains these valuable resources.

AP® Exam Tip

Be specific in your language when you answer free-response questions on the AP® Environmental Science Exam. For example, plate tectonics is a dynamic process. When describing plate moment, indicate that plates are currently in motion. Qualifying terms such as time, position, or direction are often necessary to earn points. For example, earthquakes occur when plates move past each other quickly or suddenly. For full points, the qualifying term must be included in the answer. ●

Volcano A vent in the surface of Earth that emits ash, gases, or molten lava.

Divergent plate boundary An area beneath the ocean where tectonic plates move away from each other.

Seafloor spreading The formation of new ocean crust as a result of magma pushing upward and outward from Earth's mantle to the surface.

Convergent plate boundary An area where plates move toward one another and collide.

Clearly, if tectonic plates are diverging in one place, and if the surface of Earth has a finite area, the plates must be moving together somewhere else. **Convergent plate boundaries** form where plates move toward one another and collide, as shown in Figure 24.8b. The plates generate a great deal of pressure as they push against one another.

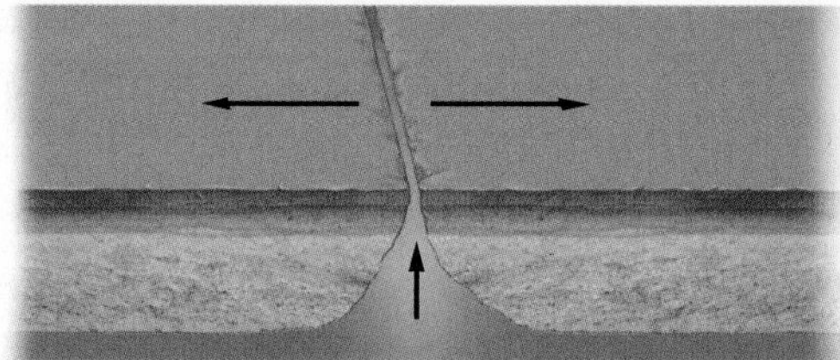
(**a**) Divergent plate boundary

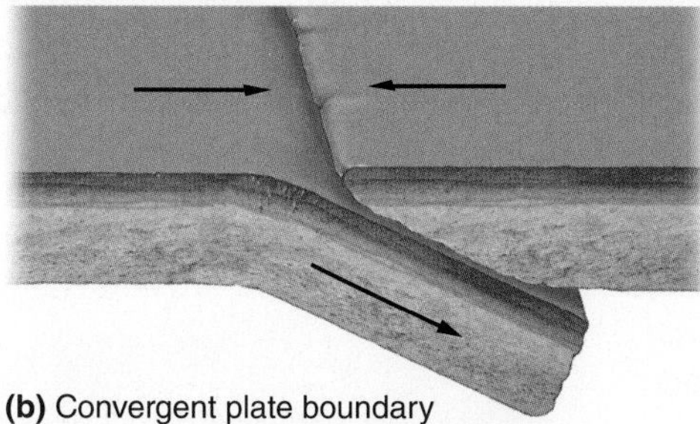
(**b**) Convergent plate boundary

(**c**) Transform fault boundary

FIGURE 24.8 Types of plate boundaries. (a) At divergent plate boundaries, plates move apart. (b) At convergent plate boundaries, plates collide. (c) At transform fault boundaries, plates slide past each other.

When plates move sideways past each other, the result is a **transform fault boundary**, shown in Figure 24.8c. A **fault** is a fracture in rock caused by movement in Earth's crust. Where this occurs, it is said that there is a high level of **seismic activity**, which is the frequency and intensity of earthquakes experienced over time. A **fault zone** is a large expanse of rock where a fault has occurred. Fault zones—also called areas of high seismic activity—form in the brittle upper lithosphere where two plates meet or slide past one another. In these large expanses of rock where movement has occurred, the rock near the plate margins becomes fractured and deformed from the immense pressures exerted by plate movement. We'll discuss faults and earthquakes in more detail in the next section.

If two continental plates meet, both plate margins may be lifted, forming a mid-continental mountain range such as the Himalayas in Asia (**FIGURE 24.9a**). As shown in Figure 24.9b, when both plates are composed of material of equal density, one does not get subducted under the other. Instead, they are forced into one another, and the force of the plate movement pushes material upward in one of the processes that leads to the formation of mountains.

Most plates and continents move at about the same rate as your fingernails grow: roughly 36 mm, or 1.4 inches, per year. While this movement is far too slow to notice on a daily basis, the two plates underlying the Atlantic Ocean have spread apart and come together twice over the past 500 million years, causing Europe and Africa to collide with North America and South America and separate from them again. "Do the Math: Plate Movement" on page 278 shows how we can calculate the time it takes for plates to move.

(**a**)

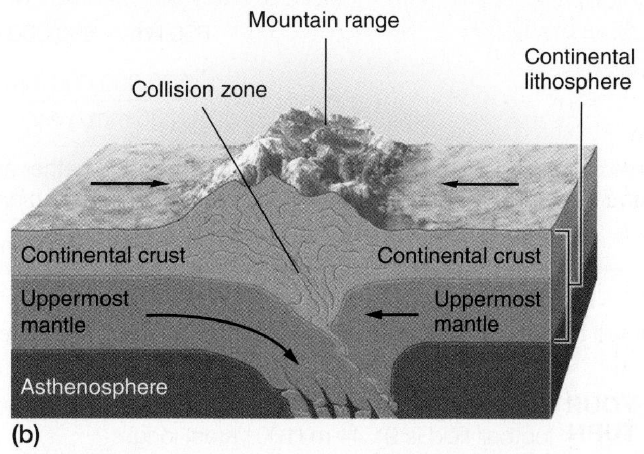

(**b**)

FIGURE 24.9 Collision of two continental plates. (a) This photo shows Mount Everest and Makalu in the Himalayas from space. (b) The Himalayan mountain range, which include the highest mountains on Earth, was formed when the collision of two continental plates forced the margins of both plates upward. *(NASA)*

Faults, Earthquakes, and Volcanoes

Although the plates are always in motion, their movement, while generally slow, is not necessarily smooth. Imagine rubbing two rough and jagged rocks past each other. The rocks would resist that movement and get stuck together.

Transform fault boundary An area where tectonic plates move sideways past each other.

Fault A fracture in rock caused by a movement of Earth's crust.

Seismic activity The frequency and intensity of earthquakes experienced over time.

Fault zone A large expanse of rock where a fault has occurred.

DO THE MATH Plate Movement

Preparing for the AP® Exam

If two cities lie on different tectonic plates, and those plates are moving so that the cities are approaching each other, how many years will it take for the two cities to be situated adjacent to each other?

Los Angeles is 630 km (380 miles) southeast of San Francisco. The plate under Los Angeles is moving northward at about 36 mm per year relative to the plate under San Francisco. Given this average rate of plate movement, how long will it take for Los Angeles to be located next to San Francisco?

The distance traveled is 630 km, and the net distance moved is 36 mm per year. We can use this formula to determine the answer

$$\text{time} = \text{distance} \div \text{rate}$$

$$630 \text{ km} = 630{,}000 \text{ m} = 630{,}000{,}000 \text{ mm}$$

$$\frac{630{,}000{,}000 \text{ mm}}{36 \text{ mm/year}} = 17{,}500{,}000 \text{ years}$$

We could also put these dimensional relationships together as follows and then cancel units that occur in both the numerator and denominator. We are left with an answer in numbers of years

$$630 \text{ km}\left(\frac{1{,}000 \text{ m}}{1 \text{ km}}\right)\left(\frac{1{,}000 \text{ mm}}{1 \text{ m}}\right)\left(\frac{1 \text{ year}}{36 \text{ mm}}\right) = 17{,}500{,}000 \text{ years}$$

It will take about 18 million years for Los Angeles to be located alongside San Francisco.

YOUR TURN How long will it take for a plate that moves at 20 mm per year to travel the distance of one football field? Note that a football field is 91.44 m (100 yards) long.

The rock along a fault is also jagged and thus resists movement, but the mounting pressure eventually overcomes the resistance and the plates give way, slipping quickly. This is an **earthquake**, which is a sudden movement of Earth's crust caused by the release of potential energy along a fault, causing vibration or movement at the surface. Earthquakes occur when the rocks of the lithosphere rupture unexpectedly along a fault. The plates can move up to several meters in just a few seconds. The **epicenter** of an earthquake is the exact point on the surface of Earth directly above the location where the rock ruptures. Earthquakes are common in fault zones, which are areas of seismic activity. **FIGURE 24.10** shows one such area, the San Andreas Fault in California, which is a transform fault. In this widely studied fault, both plates are moving north and west. But the Pacific Plate is moving at a faster rate of speed than the North American Plate, which is one cause of seismic activity. In addition, the North American Plate is moving toward the Pacific Plate, which is another reason for periodic seismic activity.

Earthquake The sudden movement of Earth's crust caused by a release of potential energy along a geologic fault and usually causing a vibration or trembling at Earth's surface.

Epicenter The exact point on the surface of Earth directly above the location where rock ruptures during an earthquake.

Richter scale A scale that measures the largest ground movement that occurs during an earthquake.

Earthquakes are a direct result of the movement of plates and their contact with each other. Volcanic eruptions happen when molten magma beneath the crust is released to the atmosphere. Sometimes the two events are observed together, most often along plate boundaries where tectonic activity is high. **FIGURE 24.11** shows one example, in which earthquake locations and volcanoes form a circle of tectonic activity, called the "Ring of Fire," around the Pacific Ocean.

The Environmental and Human Toll of Earthquakes and Volcanoes

Plate movements, volcanic eruptions, seafloor spreading, and other tectonic processes bring molten rock from deep beneath Earth's crust to the surface, and subduction sends surface crust deep into the mantle. This tectonic cycle of surfacing and sinking is a continuous Earth process. When humans live in close proximity to areas of seismic or volcanic activity, however, the results can be dramatic and devastating.

Earthquakes occur many times a day throughout the world, but most are so small that humans do not feel them. The magnitude of an earthquake is reported on the **Richter scale**, a measure of the largest ground movement that occurs during an earthquake. The Richter

(b)

FIGURE 24.10 A fault zone. (a) The San Andreas Fault in California is an example of a transform fault. Although plates are moving north and west, the Pacific Plate is moving at a greater speed than the North America Plate. The North America Plate is also moving toward the Pacific Plate. (b) In this photo of the San Andreas Fault you can see the results of seismic activity. *(DEA/PUBBLI AER FOTO/Getty Images)*

scale, like the pH scale described in Chapter 2, is logarithmic. On a logarithmic scale, a value increases by a factor of 10 for each unit increase. Thus a magnitude 7.0 earthquake, which causes serious damage, is 10 times greater than a magnitude 6.0 earthquake and 1,000 (10^3) times greater than a magnitude 4.0 earthquake, which only some people can feel or notice. Worldwide, there may be as many as 800,000 small earthquakes of magnitude 2.0 or less per year, but an earthquake of magnitude 8.0 occurs approximately once every year.

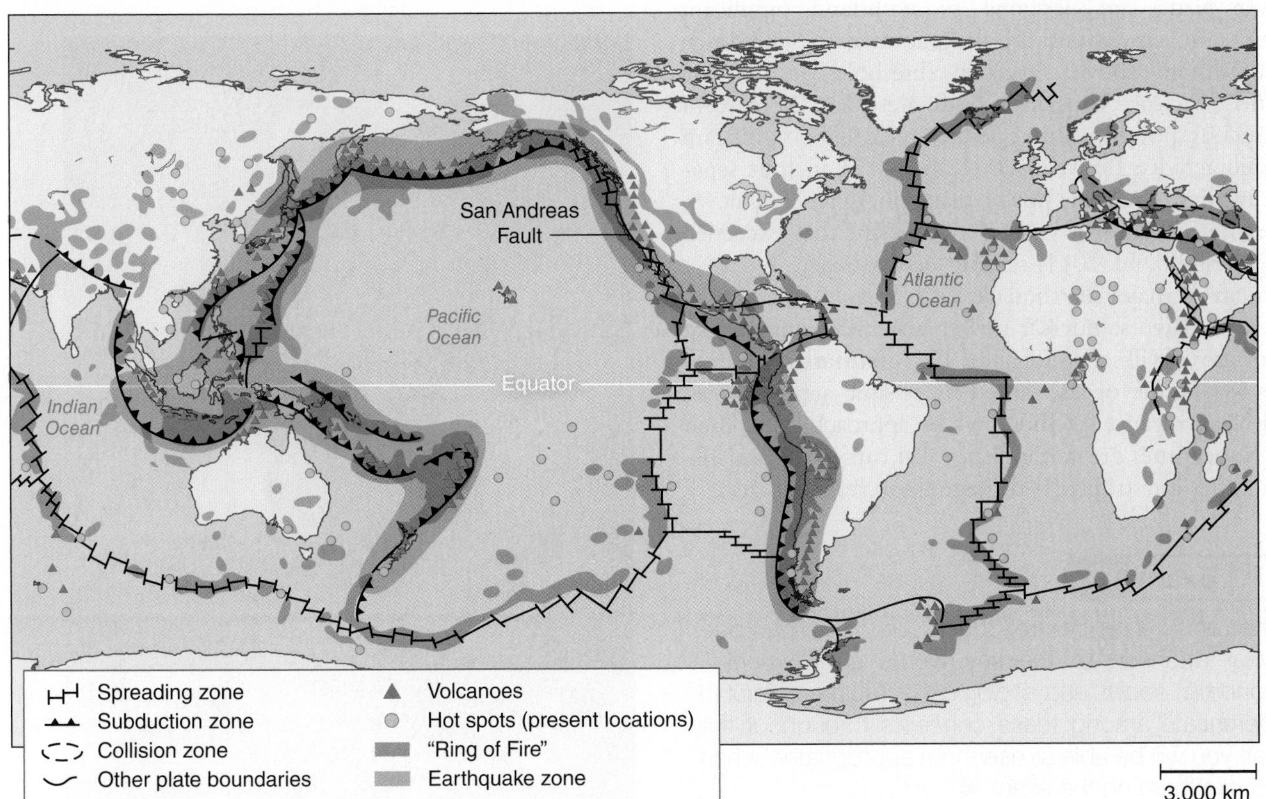

FIGURE 24.11 Locations of earthquakes and volcanoes. A "Ring of Fire" circles the Pacific Ocean along plate boundaries. Other zones of seismic and volcanic activity, including hot spots, are also shown on this map.

Even a moderate amount of Earth movement can be disastrous. Moderate earthquakes (defined as magnitudes 5.0 to 5.9) lead to collapsed structures and buildings, fires, contaminated water supplies, ruptured dams, and deaths. Loss of life is more often a result of the proximity of large population centers to the epicenter than of the magnitude of the earthquake itself. The quality of building construction in the affected area is also an important factor in the amount of damage that occurs. In 2008, a magnitude 7.9 earthquake in the southwestern region of Sichuan Province, China, killed more than 69,000 people. The epicenter was near a populated area where many buildings were probably not built to withstand a large earthquake. In 2010, a magnitude 7.0 earthquake in Haiti killed more than 200,000 people. In 2017, two earthquakes of magnitudes 8.1 and 7.1 occurred within two weeks in Mexico. A series of smaller earthquakes continued for days afterwards and over 300 people were killed. Many of the victims were trapped under collapsed buildings (**FIGURE 24.12**).

FIGURE 24.12 Earthquake damage in Mexico. This photo shows earthquake damage in September 2017 in Mexico City, Mexico, after an earthquake of magnitude 7.1 on the Richter scale. *(MARIO VAZQUEZ/Getty Images)*

Extra safety precautions are needed when dangerous materials are used in areas of seismic activity. Nuclear power plants are designed to withstand significant ground movement and are programmed to shut down if movement above a certain threshold occurs. The World Nuclear Association estimates that 20 percent of nuclear power plants operate in areas of significant seismic activity. Between 2004 and 2009, in four separate incidents, nuclear power plants in Japan shut down operation because of ground movement that exceeded the threshold. In 2011, seismic activity there led to a devastating major earthquake and *tsunami* that caused the second-worst nuclear power plant accident ever to occur, as we will see in Chapter 13. A **tsunami** is a series of waves in the ocean caused by seismic activity or an undersea volcano. As those waves approach land, their height becomes greater and they can cause substantial—sometimes catastrophic—damage upon reaching land.

AP® Exam Tip

When answering free-response questions, pay particular attention to the key words *environmental, economic, social,* and *ecological*. If you are careful to differentiate among these concepts throughout the year, you will be able to use them appropriately when you see them on the exam. ●

Tsunami A series of waves in the ocean caused by seismic activity or an undersea volcano.

Volcanoes, when active, can be equally disruptive and harmful to human life. Active volcanoes are not distributed randomly over Earth's surface; 85 percent of them occur along plate boundaries. As we have seen, volcanoes can also occur over hot spots. Depending on the type of volcano, an eruption may eject cinders, ash, dust, rock, or lava into the air (**FIGURE 24.13**).

FIGURE 24.13 A volcanic eruption. This eruption of the Etna volcano in Italy occurred in 2001. *(John Beatty/Science Source)*

(a)

(b)

(c)

FIGURE 24.14 Some common minerals. (a) Pyrite (FeS_2), also called "fool's gold." (b) Graphite, a form of carbon (C). (c) Halite, or table salt (NaCl). *(a: John Grotzinger/Ramón Rivera-Moret/Harvard Mineralogical Museum, b: John Grotzinger/Ramón Rivera-Moret/Harvard Mineralogical Museum, c: The Natural History Museum/Alamy)*

Volcanoes can result in loss of life, habitat destruction and alteration, reduction in air quality, and many other environmental consequences.

The world gained a new awareness of the impact of volcanoes when eruptions from a volcano in Iceland disrupted air travel to and from Europe in April 2010. Ash from the eruption entered the atmosphere in a large cloud and prevailing winds spread it over a wide area. The ash contained small particles of silicon dioxide, which have the potential to damage airplane engines. Air travel was suspended in many parts of Europe, and millions of travelers were stranded in what may have been the greatest travel disruption ever to have been caused by a volcano.

The rock cycle recycles scarce minerals and elements

The second part of the geologic cycle is the **rock cycle**, which governs the constant formation, alteration, and destruction of rock material that results from tectonics, weathering, and erosion, among other processes. The rock cycle is the slowest of all of Earth's cycles. Environmental scientists are most often interested in the part of the rock cycle that occurs at or near Earth's surface.

AP® Exam Tip

This is a good time to review all the biogeochemical cycles. You should know these cycles and review them regularly. ●

Rock, the substance of the lithosphere, is composed of one or more minerals (**FIGURE 24.14**). Minerals are solid chemical substances with uniform (often crystalline) structures that form under specific temperatures and pressures. They are usually compounds, but may be composed of a single element such as silver or gold.

Formation of Rocks and Minerals

FIGURE 24.15 on page 282 shows the processes of the rock cycle. Rock forms when magma from Earth's interior reaches the surface, cools, and hardens. Once at Earth's surface, rock masses are broken up, moved, and deposited in new locations by processes such as weathering and erosion. New rock may be formed from the deposited material. Eventually, the rock is subducted into the mantle, where it melts and becomes magma again. The rock cycle slowly but continuously breaks down rock and forms new rock.

While magma is the original source of all rock, there are three major ways in which the rocks we see at Earth's surface can form: directly from molten magma; by compression of sediments; and by exposure of rocks and other Earth materials to high temperatures and pressures. These three modes of formation lead to three distinct rock types: *igneous, sedimentary,* and *metamorphic.*

Igneous Rocks

Igneous rocks form directly from magma. They are classified by their chemical composition as basaltic or granitic, and by their mode of formation as intrusive or extrusive.

Basaltic rock is dark-colored rock that contains minerals with high concentrations of iron, magnesium, and calcium. It is the dominant rock type in the crust of oceanic plates. Granitic rock is lighter-colored rock made up of the minerals feldspar, mica, and quartz, and contains elements such as silicon, aluminum, potassium, and calcium. It is the dominant rock type in the crust of continental plates. When granitic rock breaks down due to weathering, it forms sand. Soils that develop from granitic rock tend to be more permeable than

Rock cycle The geologic cycle governing the constant formation, alteration, and destruction of rock material that results from tectonics, weathering, and erosion, among other processes.

Igneous rock Rock formed directly from magma.

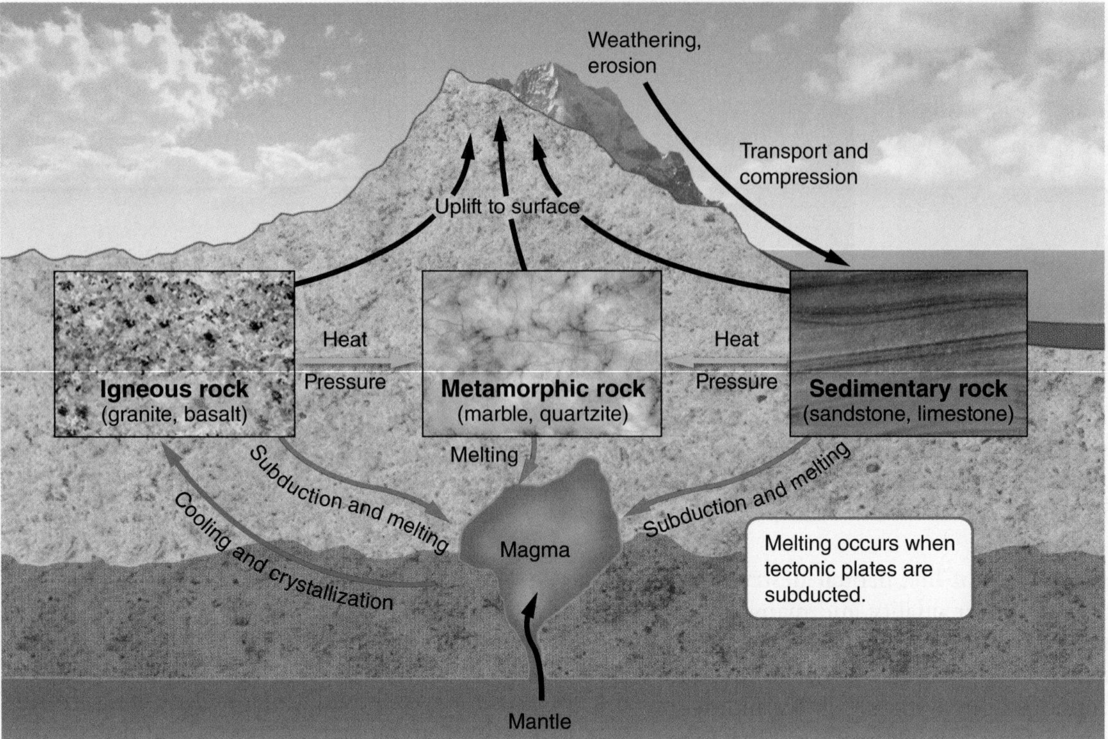

FIGURE 24.15 The rock cycle. The rock cycle slowly but continuously forms new rock and breaks down old rock. Three types of rock are created in the rock cycle: Igneous rock is formed from magma; sedimentary rock is formed by the compression of sedimentary materials; and metamorphic rock is created when rocks are subjected to high temperatures and pressures.

those that develop from basaltic rock, but both types of rock can form fertile soil.

Intrusive igneous rocks form within Earth as magma rises up and cools in place underground. **Extrusive igneous rocks** form when magma cools above Earth's surface, such as when it is ejected from a volcano or released by seafloor spreading. Extrusive rocks cool rapidly, so their minerals have little time to expand into large individual crystals. The result is fine-grained, smooth types of rock such as obsidian. Both extrusive and intrusive rocks can be either granitic or basaltic in composition.

The formation of igneous rock often brings to the surface rare elements and metals that humans find economically valuable, such as the lanthanum described at the beginning of this chapter. When rock cools, it is subject to stresses that cause it to break. Cracks that occur when rock cools, known as **fractures**, can occur in any kind of rock. Water from the surface of Earth running through fractures may dissolve valuable metals, which may precipitate out in the fractures to form concentrated deposits called *veins*. These deposits are important sources of gold- and silver-bearing ores as well as rare metals such as tantalum, which is used to manufacture electronic components of cell phones.

Intrusive igneous rock Igneous rock that forms when magma rises up and cools in a place underground.

Extrusive igneous rock Rock that forms when magma cools above the surface of Earth.

Fracture In geology, a crack that occurs in rock as it cools.

Sedimentary rock Rock that forms when sediments such as muds, sands, or gravels are compressed by overlying sediments.

Sedimentary Rocks

Sedimentary rocks form when sediments such as muds, sands, or gravels are compressed by overlying sediments. Sedimentary rock formation occurs over long periods when environments such as sand dunes, mudflats, lake beds, or areas prone to landslides are buried and the overlying materials create pressure on the materials below. The resulting rocks may be uniform in composition, such as sandstones and mudstones that formed from ancient oceanic or lake environments. Alternatively, they may be highly heterogeneous, such as conglomerate rocks formed from mixed cobbles, gravels, and sands.

Sedimentary rocks hold the fossil record that provides a window into our past. When layers of sediment containing plant or animal remains are compressed over eons, those organic materials may be preserved, as described in Chapter 5.

Metamorphic Rocks

Metamorphic rocks form when sedimentary rocks, igneous rocks, or other metamorphic rocks are subjected to high temperatures and pressures. The pressures that form metamorphic rock cause profound physical and chemical changes in the rock. These pressures can be exerted by overlying rock layers or by tectonic processes such as continental collisions, which cause extreme horizontal pressure and distortion. Metamorphic rocks include stones such as slate and marble as well as anthracite, a type of coal. Metamorphic rocks have long been important as building materials in human civilizations because they are structurally strong and visually attractive.

Metamorphic rock Rock that forms when sedimentary rock, igneous rock, or other metamorphic rock is subjected to high temperature and pressure.

MODULE 24 AP® Review

Preparing for the AP® Exam

In this module, we looked at the formation of Earth and the distribution of minerals. After Earth formed, heavier elements sank to the core, which accounts for the distribution and abundance of elements we see at the surface of Earth. The plates that overlay Earth and move at different rates further distribute elements. Plates are in constant motion around the globe and the resulting seismic activity contributes to various environmental hazards and has led to the creation of different landforms such as mountain ranges. Minerals and rocks of various compositions result from the chemical composition of the material that forms them, and the geologic conditions under which they form. When rocks and minerals break down as a result of various environmental conditions, they release chemical elements and the precursors of soils. In the next module we will examine the conditions under which rocks break down and how that contributes to the variety of soils that form around the world.

AP® Practice Questions

Choose the best answer for the following.

1. Which layer of Earth is composed primarily of iron and nickel?
   (a) the core
   (b) the crust
   (c) the asthenosphere
   (d) the mantle

2. Subduction
   (a) is the result of a hot spot moving near a plate boundary.
   (b) occurs when one plate passes under another.
   (c) occurs when oceanic plates diverge and form volcanoes.
   (d) is the process in transform boundaries that results in earthquakes.

3. The Hawaiian Islands were formed
   (a) at a divergent plate boundary.
   (b) at a hot spot.
   (c) at a convergent plate boundary.
   (d) at a transform fault.

4. How far will a plate travel in 60,000 years if it moves at net rate of 25 mm/year?
   (a) 24 m
   (b) 1,500 m
   (c) 3,000 m
   (d) 4,800 m

5. Which rock is formed at high temperatures and pressures?
   (a) extrusive igneous
   (b) intrusive igneous
   (c) sedimentary
   (d) metamorphic

6. An earthquake of magnitude 7.0 on the Richter scale is how much greater than an earthquake of magnitude 5.0?
   (a) 2
   (b) 5
   (c) 10
   (d) 100

MODULE 25

Weathering and Soil Science

Soil is a combination of geologic and organic material that forms a dynamic membrane over much of the surface of Earth. It is a nonrenewable resource that can take tens of thousands of years or longer to fully develop. A variety of processes that occur in soil connect the overlying biology with the underlying geology. In this module we will explore the weathering of rocks that leads to the formation of soil and the development of specific soil horizons. We will discuss physical, chemical, and biological processes that take place in soils. Finally, we will examine human activities that degrade soils, including the process of mining.

Learning Goals

After reading this module, you should be able to

- understand how weathering and erosion occur and how they contribute to element cycling and soil formation.
- explain how soil forms and describe its characteristics.
- describe how humans extract elements and minerals and the social and environmental consequences of these activities.

The processes of weathering and erosion contribute to the recycling of elements

We have seen that rock forms beneath Earth's surface under intense heat, pressure, or both heat and pressure. When rock is exposed at Earth's surface, it begins to break down through the processes of weathering and erosion. These processes are components of the rock cycle, returning chemical elements and rock fragments to the crust by depositing them as sediments through the hydrologic cycle. This physical breakdown and chemical alteration of rock begins the cycle all over again, as shown in Figure 24.15. Without this part of the rock cycle, elements would never be recycled and the precursors of soils would not be present.

Weathering

Weathering occurs when rock is exposed to air, water, certain chemical compounds, or biological agents such as plant roots, lichens, and burrowing animals. There are two major categories of weathering—*physical* and *chemical*—that work in combination to degrade rocks.

Physical weathering is the mechanical breakdown of rocks and minerals, shown in **FIGURE 25.1**. Physical weathering can be caused by water, wind, or variations in temperature such as seasonal freeze-thaw cycles. When water works its way into cracks or fissures in rock, it can remove loose material and widen the cracks, as illustrated in Figure 25.1a. When water freezes in the cracks, the water expands, and the pressure of its expansion can force rock to break. Different responses to temperature can cause two minerals within a rock to expand and contract differently, which also results in splitting or cracking. Coarse-grained rock formed by slow cooling or metamorphism tends to weather more quickly than fine-grained rock formed by rapid cooling or metamorphism.

Biological agents can also cause physical weathering. Plant roots can work their way into small cracks in rocks and pry them apart, as illustrated in Figure 25.1b. Burrowing animals may also contribute to the breakdown of rock material, although their contributions are usually minor. However it occurs, physical weathering exposes more surface area and makes rock more vulnerable to further degradation. By producing more surface area for weathering processes to act on, physical weathering increases the rate of *chemical weathering*. **Chemical weathering** is the breakdown of rocks and minerals by chemical reactions, the dissolving of chemical elements from rocks, or both these processes.

Physical weathering The mechanical breakdown of rocks and minerals.

Chemical weathering The breakdown of rocks and minerals by chemical reactions, the dissolving of chemical elements from rocks, or both.

(a)

(b)

FIGURE 25.1 Physical weathering. (a) Water can work its way into cracks in rock, where it can wash away loose material. When the water freezes and expands, it can widen the cracks. (b) Growing plant roots can force rock sections apart. *(a: Walt Anderson/Getty Images, b: Bruce Hamms/Alamy)*

It releases essential nutrients from rocks, making them available for use by plants and other organisms.

Chemical weathering occurs most rapidly on newly exposed minerals, known as primary minerals. It alters primary minerals to form secondary minerals and the ionic forms of their constituent chemical elements. For example, when feldspar—a mineral found in granitic rock—is exposed to natural acids in rain, it forms clay particles and releases ions such as potassium, an essential nutrient for plants. Lichens can break down rock in a similar way by producing weak acids. Their effects can commonly be seen on soft gravestones and masonry. Rocks that contain compounds that dissolve easily, such as calcium carbonate, tend to weather quickly. Rocks that contain compounds that do not dissolve readily are often the most resistant to chemical weathering. In examining element cycles in Chapter 3, we noted that weathering of rocks is an important part of the phosphorus cycle.

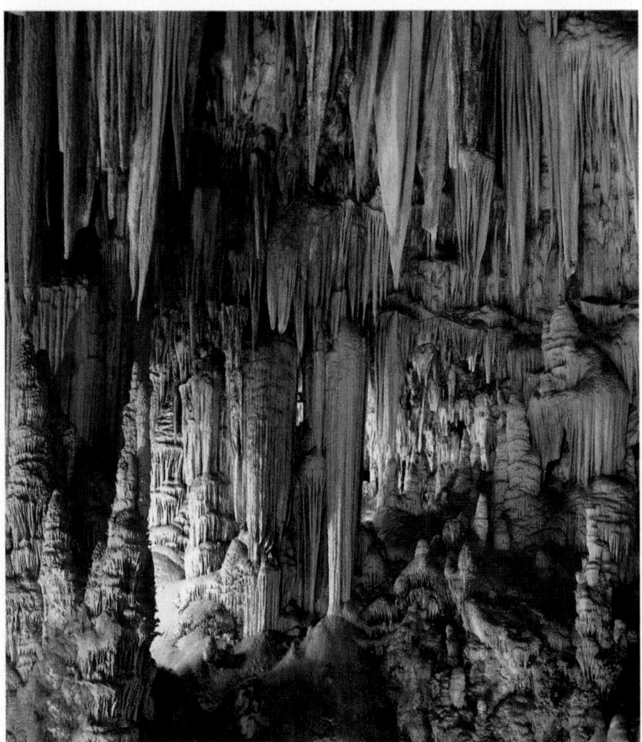

FIGURE 25.2 Chemical weathering. Water that contains carbonic acid wears away limestone, sometimes forming spectacular caves. *(mauritius images/Thonig)*

Recall from Chapter 2 that solutions can be basic or acidic. Depending on the starting chemical composition of rock and the pH of the water that comes in contact with it, hundreds of different chemical reactions can take place. For example, as we saw in Chapter 2, carbon dioxide in the atmosphere dissolves in water vapor to create a weak acid, called carbonic acid. When waters containing carbonic acid flow into geologic regions that are rich in limestone, they dissolve the limestone (which is composed of calcium carbonate) and create spectacular cave systems (**FIGURE 25.2**).

Some chemical weathering is the result of human activities. For example, sulfur emitted into the atmosphere from fossil fuel combustion combines with oxygen to form sulfur dioxide. That sulfur dioxide reacts with water vapor in the atmosphere to form sulfuric acid, which then causes *acid precipitation.* **Acid precipitation**, also called **acid rain**, is precipitation high in sulfuric acid and nitric acid from reactions between water vapor and sulfur and nitrogen oxides in the atmosphere. Acid precipitation is responsible for the rapid degradation of certain old statues and gravestones and other limestone and marble structures. When acid precipitation falls on soil, it can promote chemical weathering of certain minerals in the soil, releasing elements that may then be taken up by plants or leached from the soil into groundwater and streams.

Acid precipitation Precipitation high in sulfuric acid and nitric acid from reactions between water vapor and sulfur and nitrogen oxides in the atmosphere. *Also known as* **Acid rain**.

FIGURE 25.3 Erosion. Some erosion, such as the erosion that created these formations in the Badlands of South Dakota, occurs naturally as a result of the effects of water, glaciers, or wind. The Badlands are the result of the erosion of softer sedimentary rock types, such as shales and clays. Harder rocks, including many types of metamorphic and igneous rocks, are more resistant to erosion. *(welcomia/Shutterstock)*

Chemical weathering, due to either natural processes or acid precipitation, can contribute additional elements to an ecosystem. Knowing the rate of weathering helps researchers assess how rapidly soil fertility can be renewed in an ecosystem. In addition, because the chemical reactions involved in the weathering of certain granitic rocks consume carbon dioxide from the atmosphere, weathering can actually reduce atmospheric carbon dioxide concentrations.

Erosion

We have seen that physical and chemical weathering results in the breakdown and chemical alteration of rock. **Erosion** is the physical removal of rock fragments (sediment, soil, rock, and other particles) from a landscape or ecosystem. Erosion is usually the result of two mechanisms. In one, wind, water, and ice move soil and other materials by downslope creep under the force of gravity. In the other, living organisms, such as animals that burrow under the soil, cause erosion. After eroded material has traveled a certain distance from its source, it accumulates. Deposition is the accumulation or depositing of eroded material such as sediment, rock fragments, or soil.

Erosion is a natural process: Streams, glaciers, and wind-borne sediments continually carve, grind, and scour rock surfaces (**FIGURE 25.3**). In many places, however, human land use contributes substantially to the rate of erosion. Poor land use practices such as deforestation, overgrazing, unmanaged construction activity, and road building can create and accelerate erosion problems. Furthermore, erosion usually leads to deposition of the eroded material somewhere else, which may cause additional environmental problems. We discuss human-caused erosion further in Chapters 10 and 11.

Erosion The physical removal of rock fragments from a landscape or ecosystem.

Soil links the rock cycle and the biosphere

Soil has a number of functions that benefit organisms and ecosystems. As you can see in **FIGURE 25.4**, soil is a medium for plant growth. It also serves as the primary filter of water as water moves from the atmosphere into rivers, streams, and groundwater. Soil contributes greatly to biodiversity by providing habitat for a wide variety of living organisms—from bacteria, algae, and fungi to insects and other animals. Soil and the organisms within it filter chemical compounds deposited by air pollution and by household sewage systems; some of these materials remain in the soil and some are released to the atmosphere or into groundwater.

In this section we will look at the formation and properties of soil.

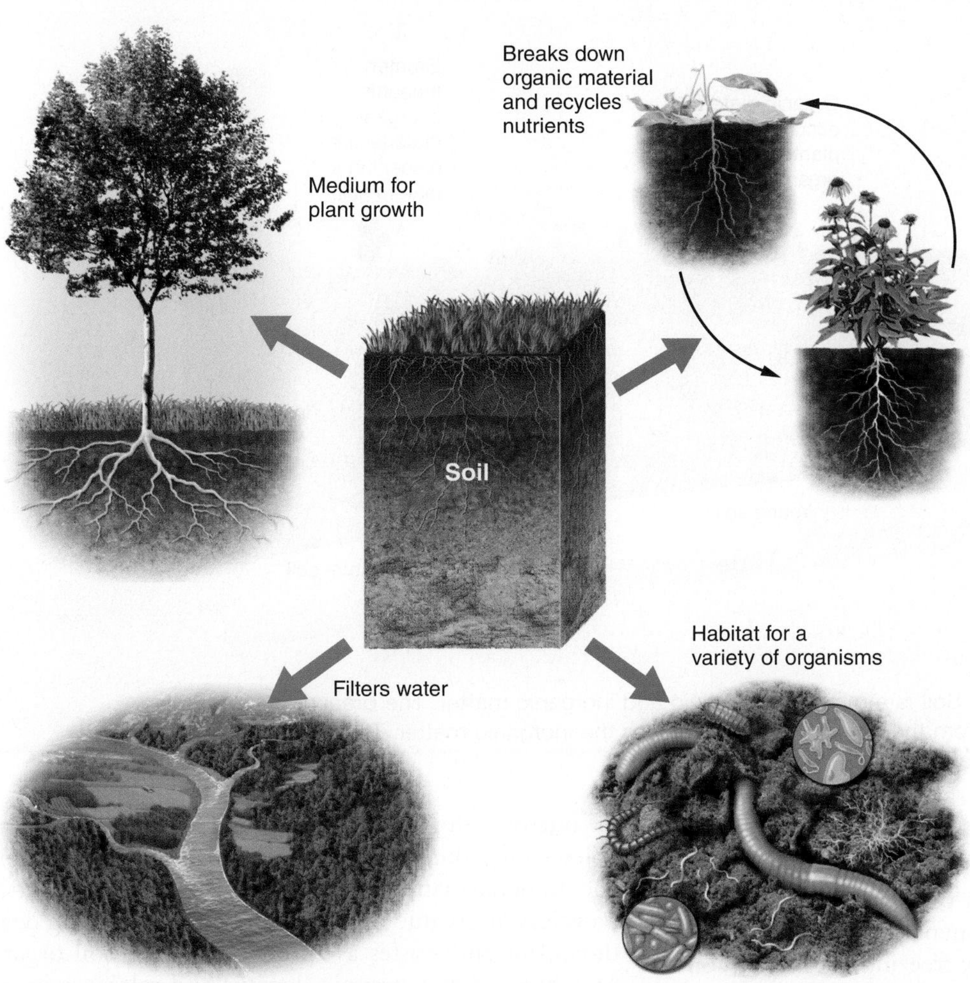

FIGURE 25.4 Ecosystem services provided by soil. Soil serves as a medium for plant growth, as a habitat for other organisms, and as a recycling system for organic wastes. Soil also helps to filter and purify water.

AP® Exam Tip

Make sure you can describe the formation of soil. You should be able to distinguish between biological and physical processes that contribute to the formation of the various soil horizons. ●

The Formation of Soil

In order to appreciate the role of soil in ecosystems, we need to understand how and why soil forms and what happens to soil when humans alter it. It takes hundreds to thousands of years for soil to form. Soil is the result of physical and chemical weathering of rocks and the gradual accumulation of detritus from the biosphere. We can determine the specific properties of a soil if we know its parent rock type, the amount of time it has been forming, and its associated biotic and abiotic components.

FIGURE 25.5 on page 288 shows the stages of soil development from rock to mature soil. The processes that form soil work in two directions simultaneously. The breakdown of rocks and primary minerals by weathering provides the raw material for soil from below. The deposition of organic matter from organisms and their wastes contributes to soil formation from above. What we normally think of as "soil" is a mix of these mineral and organic components. A poorly developed (young) soil has substantially less organic matter and fewer nutrients than a more developed (mature) soil. Very old soils may also be nutrient poor because over time plants remove many essential nutrients and water leaches away others. Five factors simultaneously determine the properties of soils: *parent material,* climate, topography, organisms, and time.

Parent Material

A soil's **parent material** is the underlying rock material from which its inorganic components are derived. Different soil types arise from different parent materials. For example, a quartz sand (made up of silicon dioxide) parent material will give rise to a soil that is nutrient poor, such as those along the Atlantic coast of the United States. By contrast, a soil that has calcium carbonate as its parent material will contain an abundant supply of calcium, have a high pH, and may also support high agricultural productivity. Such soils are found in the area surrounding Lake Champlain in Vermont and northern New York, as well as in many other locations.

Parent material The underlying rock material from which the inorganic components of a soil are derived.

Parent rock is weathered and fragments move upward.
Organic material accumulates as plants and other organisms die.
Greater amounts of organic material are present in a mature soil.
Immature soil
Young soil
Time
Mature soil

FIGURE 25.5 Soil formation. Soil is a mixture of organic and inorganic matter. The breakdown of rock and primary minerals from the parent material provides the inorganic matter. The organic matter comes from organisms and their wastes.

Climate

Climate influences soil formation in a number of ways. The long-term accumulation of temperature, humidity, and water affect soil development. Soils do not develop when temperatures are below freezing because decomposition of organic matter and water movement are both extremely slow in frozen or nearly frozen soils. Therefore, soils at high latitudes of the Northern Hemisphere are composed largely of organic material in an undecomposed state, as we saw in Chapter 4. In contrast, soil development in the humid tropics is accelerated by rapid weathering of rock and soil minerals, leaching of nutrients, and decomposition of organic detritus. Climate also has an indirect effect on soil formation because it affects the type of vegetation that develops, and therefore the type of detritus left after the vegetation dies.

Topography

Topography—the surface slope and arrangement of a landscape—is another factor in soil formation. Soils that form on steep slopes are constantly subjected to erosion and, on occasion, to more drastic mass movements of material as happens in landslides. In contrast, soils that form at the bottoms of steep slopes may continually accumulate material from higher elevations and become quite deep.

Organisms

Many organisms influence soil formation. Plants remove nutrients from soil and excrete organic acids that speed chemical weathering. Animals that tunnel or burrow—for example, earthworms, gophers, and voles—mix the soil, uniformly distributing organic and mineral matter. Collectively, soil organisms act as recyclers of organic matter. In the process of using dead organisms and wastes as an energy source, soil organisms break down organic detritus and release mineral nutrients and other materials that benefit plants.

Human activity has dramatic effects on soils. For centuries, the use and overuse of land for agriculture, forestry, and other human activities has led to significant **soil degradation**: the loss of some or all of the ability of soils to support plant growth. One of the major causes of soil degradation is soil erosion, which occurs when topsoil is disturbed—for example, by plowing—or when vegetation is removed. As we saw in Chapter 3, these activities lead to erosion by water or wind (**FIGURE 25.6**). While topsoil loss

FIGURE 25.6 Erosion from human activity. Erosion in this cornfield in Tennessee is obvious after a brief rainstorm. *(Tim McCabe/USDA Natural Resources Conservation Service)*

Soil degradation The loss of some or all of a soil's ability to support plant growth.

can happen rapidly—in as little as a single growing season—it takes centuries for the lost topsoil to be replaced. Compaction of soil by machines, humans, and livestock can alter its properties and reduce its ability to retain moisture. Compaction and drying of soil can, in turn, reduce the amount of vegetation that grows in the soil and thereby increase erosion. Intensive agricultural use and irrigation can deplete soil nutrients, and the application of agricultural pesticides can pollute the soil. In Chapters 11 and 14 we will return to the ways in which human activity affects the soil.

Time

The final factor that determines the properties of a soil is the amount of time during which the soil has developed. As soils age, they develop a variety of characteristics. The grassland soils that support much of the food crop and livestock feed production in the United States are relatively old soils. Because they have had continual inputs of organic matter for hundreds of thousands of years from the grassland and prairie vegetation growing above them, they have become deep and fertile. Other soils that are equally old, but with less productive communities above them and perhaps greater quantities of water moving through them, can become relatively infertile.

O horizon: Organic matter in various stages of decomposition
A horizon (topsoil): Zone of overlying organic material mixed with underlying mineral material
E horizon: Zone of leaching of metals and nutrients; occurs in some soils beneath either the O horizon or the A horizon
B horizon (subsoil): Zone of accumulation of metals and nutrients
C horizon (subsoil): Least-weathered portion of the soil profile, similar to the parent material
Horizons
O
A
E
B
C
0
<.1 m
.3 m
1 m
1.3 m

FIGURE 25.7 Soil horizons. All soils have horizons, or layers, which vary depending on soil-forming factors such as climate, organisms, and parent material. Most soils have either an O or A horizon and usually not both. Some soils that have an O horizon also have an E horizon.

Soil Horizons

As soils form, they develop characteristic **horizons**, which are horizontal layers with distinct physical features such as color or texture, shown in **FIGURE 25.7**. The specific composition of those horizons depends largely on climate, vegetation, and parent material. At the surface of many soils is a layer known as the **O horizon**, composed of organic detritus such as leaves, needles, twigs, and even animal bodies, all in various stages of decomposition. The O horizon is most pronounced in forest soils and is also found in some grasslands. Organic matter is sometimes called *humus* (pronounced "hu-mus," but often mispronounced by beginning students as hum-mus, the delicious food made from chickpeas, olive oil, and garlic). **Humus** is the most fully decomposed organic matter in the lowest layer of the O horizon. Unlike forms of decomposing organic matter higher up in the O horizon, humus does not contain recognizable plant or animal components. Humus and "humic material" may also occur in smaller concentrations in the mineral soil horizons below the O horizon.

In a soil that is mixed, either naturally or by human agricultural practices, the top layer is the **A horizon**, also known as **topsoil**, a zone of organic material (including humus) and minerals that have been mixed together. In some acidic soils, an **E horizon**—a zone of leaching, or eluviation—forms under the O horizon or, less often, the A horizon. When an E horizon is present, iron, aluminum, and dissolved organic acids from the overlying horizons are transported through and removed from the E horizon and then deposited in the B horizon, where they accumulate. When an E horizon is present, it always occurs above the *B horizon*. The **B horizon**, commonly known as subsoil, is composed primarily of mineral material with very small amounts of organic matter, including humus. If nutrients are present in the subsoil, they will be in the B horizon. The **C horizon**—the least weathered soil horizon—occurs beneath the B horizon and is similar to the parent material.

Horizon A horizontal layer in a soil defined by distinctive physical features such as texture and color.

O horizon The organic horizon at the surface of many soils, composed of organic detritus in various stages of decomposition.

Humus The most fully decomposed organic matter in the lowest section of the O horizon.

A horizon Frequently the top layer of soil, a zone of organic material and minerals that have been mixed together. *Also known as* **Topsoil**.

E horizon A zone of leaching, or eluviation, found in some acidic soils under the O horizon or, less often, the A horizon.

B horizon A soil horizon composed primarily of mineral material with very little organic matter.

C horizon The least-weathered soil horizon, which always occurs beneath the B horizon and is similar to the parent material.

Properties of Soil

Soils with different properties serve different functions for humans. For example, some soil types are good for growing crops and others are more suited for building a housing development. Therefore, to understand and classify soil types, we need to understand the physical, chemical, and biological properties of soils.

Physical Properties of Soil

The physical properties of soil refer to physical characteristics such as size and weight. Sand, silt, and clay are mineral particles of different sizes. The texture of a soil is determined by the percentages of sand, silt, and clay it contains. **FIGURE 25.8a** plots those percentages on a triangle-shaped diagram that allows us to identify and compare soil types. Each location on the diagram has three determinants: the percentages of sand, of silt, and of clay. A point in the middle of the "loam" category (approximately at the "a" in "Loam" in Figure 25.8) represents a soil that contains 40 percent sand, 40 percent silt, and 20 percent clay. We can determine this by following the lines from that point to the scales on each of the three sides of the triangle. If you are not certain which line to follow from a given point, always follow the line that leads to the lower value. For example, if you want to determine the percentage of sand in a sample represented by the red dot in Figure 25.8, you might follow the line out to either 60 percent sand or 40 percent sand; however, since you should always follow the line to the lower value, in this case the percentage of sand is 40 percent. The sum of sand, silt, and clay will always be 100 percent. Conversely, in the laboratory, a soil scientist can determine the percentage of each component in a soil sample and then plot the results. The name for the soil, for example "silty clay loam," follows from the percentage of the components in the soil.

AP® Exam Tip

You should know how to read and interpret a soil properties chart, like the one in Figure 25.8. Questions about this chart appear on every AP® Environmental Science Exam. ●

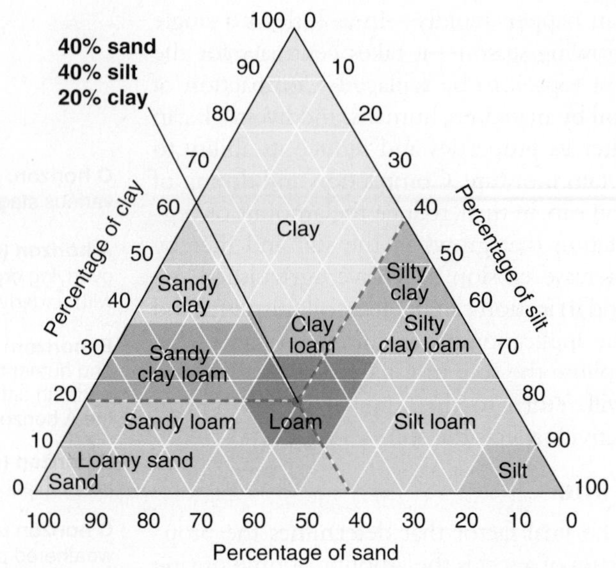

(a) Soil texture chart

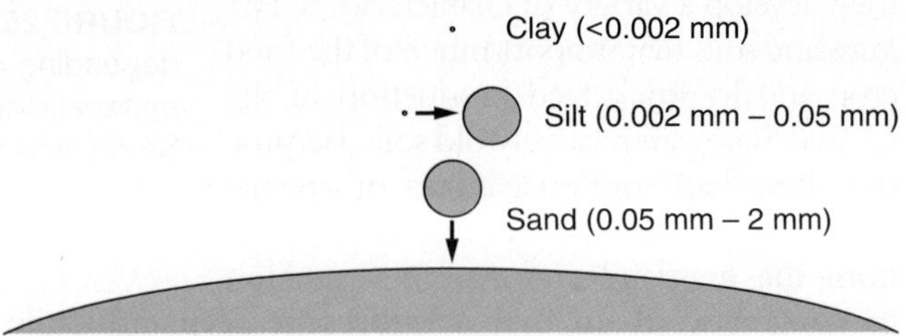

(b) Relative soil particle sizes (magnified approximately 100 times)

FIGURE 25.8 Soil properties. (a) Soils consist of a mixture of clay, silt, and sand. The relative proportions of these particles determine the texture of the soil. (b) The relative sizes of sand, silt, and clay.

The permeability of soil—how quickly soil drains—depends on its texture, shown in **FIGURE 25.9**. Sand particles—the largest of the three components—pack together loosely. Water can move easily between the particles, making sand quick to drain and quick to dry out. Soils with a high proportion of sand are also easy for roots to penetrate, making sandy soil somewhat advantageous for growing plants such as carrots and potatoes. Clay particles—the smallest of the three components—pack together much more tightly than sand particles. As a result, there is less

FIGURE 25.9 Soil permeability. The permeability of soil depends on its texture. Sand, with its large, loosely packed particles, drains quickly. Clay drains much more slowly.

pore space in a soil dominated by clay, and water and roots cannot easily move through it. Silt particles are intermediate both in size and in their ability to drain or retain water.

The best agricultural soil is a mixture of sand, silt, and clay that would be characterized as a loam in Figure 25.8a. This mixture promotes balanced water drainage and retention. In natural ecosystems, however, various herbaceous and woody plants have adapted to growing in wet, intermediate, and dry environments, so there are plants that thrive in soils of virtually all textures.

Soil texture can have a strong influence on how the physical environment responds to environmental pollution. For example, the ground water of western Long Island in New York State has been contaminated over the years by toxic chemicals discharged from local industries. One major reason for the contamination is that Long Island is dominated by sandy soils that readily allow surface water to drain into the groundwater. While soil usually serves as a filter that removes pollutants from the water moving through it, sandy soils are so permeable that pollutants move through them quickly and therefore are not filtered effectively.

Clay is particularly useful where a potential contaminant needs to be contained. Many modern landfills are lined with clay, which helps keep the contaminants in solid waste from leaching into the soil and groundwater beneath the landfill.

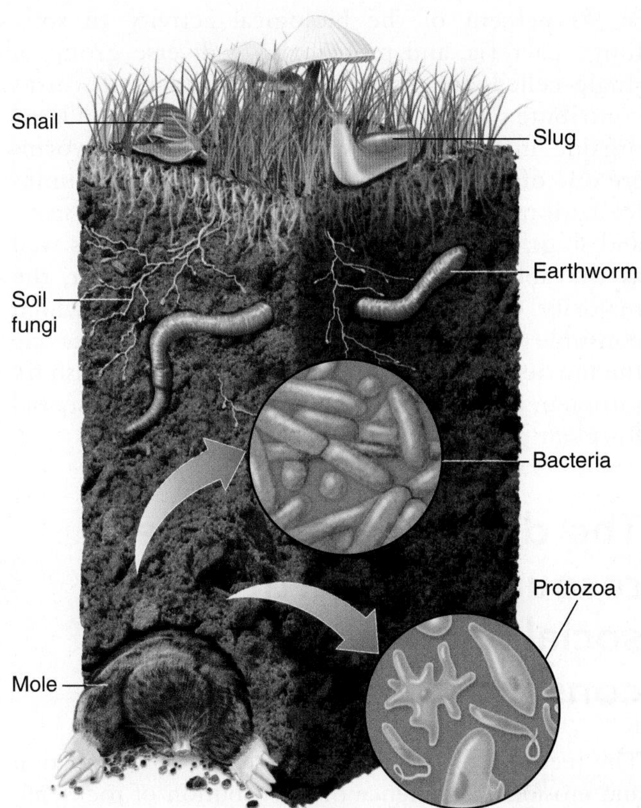

FIGURE 25.10 Soil organisms. Bacteria, fungi, and protozoans account for 80 to 90 percent of soil organisms. Also present are snails, slugs, insects, earthworms, and rodents.

Chemical Properties of Soil

Chemical properties are also important in determining how a soil functions. Clay particles contribute the most to the chemical properties of a soil because of their ability to attract positively charged mineral ions, referred to as cations. Because clay particles have a negative electrical charge, cations are adsorbed—held on the surface—by the particles. The cations can be subsequently released from the particles and used as nutrients by plants.

The ability of a particular soil to adsorb and release cations is called its **cation exchange capacity (CEC)**, sometimes referred to as the nutrient holding capacity. The overall CEC of a soil is a function of the amount and types of clay particles present. Soils with high CECs have the potential to provide essential cations to plants and therefore are desirable for agriculture. If a soil is more than 20 percent clay, however, its water retention becomes too great for most crops as well as many other types of plants. In such waterlogged soils, plant roots are deprived of oxygen. Thus there is a trade off between CEC and permeability.

The relationship between soil bases and soil acids is another important soil chemical property. Calcium, magnesium, potassium, and sodium are collectively called soil bases because they can neutralize or counteract soil acids such as aluminum and hydrogen. Soil acids are generally detrimental to plant nutrition, while soil bases tend to promote plant growth. With the exception of sodium, all the soil bases are essential for plant nutrition. **Base saturation** is the proportion of soil bases to soil acids, expressed as a percentage.

Because of the way they affect nutrient availability to plants, CEC and base saturation are important determinants of overall ecosystem productivity. If a soil has a high CEC, it can retain and release plant nutrients. If it has a relatively high base saturation, its clay particles will hold important plant nutrients such as calcium, magnesium, and potassium. A soil with both high CEC and high base saturation is likely to support high productivity.

Biological Properties of Soil

As we have seen, a diverse group of organisms populates the soil. **FIGURE 25.10** shows a representative sample. Three groups of organisms account for 80

Cation exchange capacity (CEC) The ability of a particular soil to adsorb and release cations.

Base saturation The proportion of soil bases to soil acids, expressed as a percentage.

to 90 percent of the biological activity in soils: fungi, bacteria, and protozoans (a diverse group of single-celled organisms). Rodents and earthworms contribute to soil mixing and the breakdown of large organic materials into smaller pieces. Earthworms are one of many organisms that contribute to humus formation in soils. Some soil organisms, such as snails and slugs, are herbivores that eat plant roots as well as the aboveground parts of plants. However, the majority of soil organisms are detritivores, which consume dead plant and animal tissues and recycle the nutrients they contain. Some soil bacteria also fix nitrogen, which, as we saw in Chapter 3, is essential for plant growth.

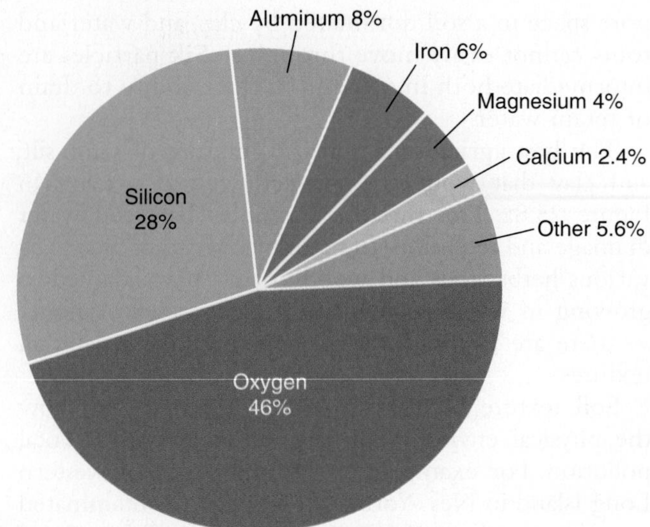

FIGURE 25.11 Elemental composition of Earth's crust. Oxygen is the most abundant element in the crust. Silicon, aluminum, and iron are the next three most abundant elements.

The distribution of mineral resources on Earth has social and environmental consequences

The tectonic cycle, the rock cycle, and soil formation and erosion all influence the distribution of rocks and minerals on Earth. These resources, along with fossil fuels, exist in finite quantities, but are vital to modern human life. In this section we will discuss some important nonfuel mineral resources and how humans obtain them; we will discuss fuel resources in Chapters 12 and 13. Some of these resources are abundant, whereas others are rare and extremely valuable.

Abundance of Ores and Metals

As we saw at the beginning of this chapter, early Earth cooled and differentiated into distinct vertical zones. Heavy elements sank toward the core, and lighter elements rose toward the crust. **Crustal abundance** is the average concentration of an element in Earth's crust. Looking at **FIGURE 25.11**, we can see that four elements—oxygen, silicon, aluminum, and iron—constitute over 88 percent of the crust. However, the chemical composition of the crust is highly variable from one location to another.

Environmental scientists and geologists study the distribution and types of mineral resources around the planet in order to locate them and to manage their extraction or conservation. **Ores** are concentrated accumulations of minerals from which economically valuable materials can be extracted. Ores are typically characterized by the presence of valuable metals, but accumulations of other valuable materials, such as salt or sand, can also be considered ores. **Metals** are elements with properties that allow them to conduct electricity and heat energy and to perform other important functions. Copper, nickel, and aluminum are common examples of metals. They exist in varying concentrations in rock, usually in association with elements such as sulfur, oxygen, and silicon. Some metals, such as gold, exist naturally in a pure form. In the chapter opener we discussed how the manufacturing of HEV vehicles uses scarce metals, including neodymium, lithium, and lanthanum. As the number of vehicles that contain scarce metals increases, so will the demand for these metals. "Do the Math: What Fraction of Automobiles Are in the Developed World?" gives you an opportunity to use scientific notation to explore the number of automobiles in the developing and developed worlds and what percentage of those vehicles are HEV and all-electric.

Ores are formed by a variety of geologic processes. Some ores form when magma comes into contact with water, heating the water and creating a solution from which metals precipitate, while others form after the deposition of igneous rock. Some ores occur in relatively small areas of high concentration, such as veins, and others, called disseminated deposits, occur in much larger areas of rock, although often in lower concentrations. Still other ores, such as copper, can be deposited both throughout a large area and in veins. Nonmetallic

Crustal abundance The average concentration of an element in Earth's crust.

Ore A concentrated accumulation of minerals from which economically valuable materials can be extracted.

Metal An element with properties that allow it to conduct electricity and heat energy, and to perform other important functions.

DO THE MATH What Fraction of Automobiles Are in the Developing World?

Preparing for the AP® Exam

Global mining of metals is influenced by the number of cars in the world. The following table shows the number of automobiles in the developed and developing worlds. Use scientific notation to calculate answers to the questions that follow.

Level of development	Number of automobiles
Developed countries	667,000,000
Developing countries	533,000,000

Express the number of automobiles in the developed and developing worlds in scientific notation. What is the total number of automobiles in the world? What percentage is in the developing world?

$$\text{Developed} = 6.67 \times 10^8$$
$$\text{Developing} = 5.33 \times 10^8$$

To find the total number of automobiles in world, you need to add the developed and developing world numbers together:

$$6.67 \times 10^8 + 5.33 \times 10^8$$
$$= 1.2 \times 10^9$$

To determine the fraction of automobiles in the developing world, we divide the number of automobiles in the developing world by total automobiles in the world:

$$(5.33 \times 10^8) \div (1.2 \times 10^9) = 0.44$$

To determine the value expressed as a percentage, multiply the decimal fraction by 100.

$0.44 \times 100\% = 44\%$ of all automobiles in the world are in the developing world.

YOUR TURN If there are roughly 1.3×10^7 HEV and all-electric vehicles in the world, what percentage is this of all vehicles?

mineral resources, such as clay, sand, salt, limestone, and phosphate, typically occur in concentrated deposits. These deposits occur as a result of their chemical or physical separation from other materials by water, in conjunction with the tectonic and rock cycles. Some ores, such as bauxite—the ore in which aluminum is most commonly found—are formed by intense chemical weathering in tropical regions.

The global supply of mineral resources is difficult to quantify. Because private companies hold the rights to extract certain mineral resources, information about the exact quantities of resources is not always available to the public. The publicly known estimate of how much of a particular resource is available is based on its **reserve**: the known quantity of the resource that can be economically recovered. A resource is considered economically recoverable (or extractable) if the concentration in the host rock is high enough for it to be profitably mined. **TABLE 25.1** on page 294 lists the estimated number of years of remaining supplies for some of the most important metal resources commonly used in the United States, assuming that rates of use do not change. The increased reliance on recycling of many metals will change the time estimates shown. So will innovations that allow substitutions and replacement materials. Some important metals, such as tantalum, have never existed in the United States. The United States has consumed all of the economically extractable reserves of certain metals, such as manganese, and must now import those metals from other countries.

Mining Techniques

Mineral resources are extracted from Earth by mining the ore and separating any other minerals, elements, or residual rock away from the sought-after element or mineral. As illustrated in **FIGURE 25.12** on page 294, two kinds of mining take place on land: surface mining and *subsurface mining*. Each method has different benefits and costs in terms of environmental, human, and social perspectives.

Reserve In resource management, the known quantity of a resource that can be economically recovered.

TABLE 25.1 Approximate supplies of metal reserves remaining

Metal	Global reserves remaining (years)	U.S. reserves remaining (years)
Aluminum (Al)	330	2
Copper (Cu)	40	25
Lead (Pb)	20	15
Zinc (Zn)	20	15
Gold (Au)	20	15
Nickel (Ni)	35	5
Cobalt (Co)	50	30
Manganese (Mn)	40	0
Chromium (Cr)	20	0

Sources: S. Marshak, *Earth: Portrait of a Planet*, 3rd ed. (W. W. Norton, 2007); U.S. Geological Survey Mineral Commodity Summaries 2017, https://minerals.usgs.gov/minerals/pubs/mcs/2017/mcs2017.pdf.

Surface Mining

A variety of surface mining techniques can be used to remove a mineral or ore deposit that is close to the surface of Earth. **Strip mining**, or the removal of "strips" of soil and rock to expose the underlying ore, is used when the ore is relatively close to Earth's surface and runs parallel to it, which is often the case for deposits of sedimentary materials such as coal and sand. In these situations, miners remove a large volume of material, extract the resource, and return the unwanted waste material, called **mine tailings**, to the hole created during the mining. Mine tailings are the mineral and other residues that are left behind after the desired metal or ore is removed. A variety of strategies can be used to restore the affected area to something close to its original condition.

FIGURE 25.12 Surface and subsurface mining. Surface mining methods include strip, open-pit, mountaintop removal, and placer mining.

Open-pit mining, a mining technique that creates a large visible pit or hole in the ground, is used when the resource is close to the surface but extends beneath the surface both horizontally and vertically. Copper mines are usually open-pit mines. One of the largest open-pit mines in the world is the Kennecott Bingham Canyon mine near Salt Lake City, Utah (**FIGURE 25.13**). This copper mine is 4.4 km (2.7 miles) across and 1.1 km (0.7 miles) deep.

In **mountaintop removal**, miners remove the entire top of a mountain with explosives. Large earth-moving equipment removes the resource and deposits the tailings in lower-elevation regions nearby, often in or near rivers and streams.

Placer mining is the process of looking for metals and precious stones in river sediments. Miners use the river water to separate heavier items, such as diamonds, tantalum, and gold, from lighter items, such as sand and mud. The prospectors in the California gold rush in the mid-1800s were placer miners, and the technique is still used today.

Subsurface Mining

When the desired resource is more than 100 m (328 feet) below Earth's surface, miners must turn to **subsurface mining**, which is mining that occurs below the surface of Earth. Typically, a subsurface mine begins with a horizontal tunnel dug into the side of a mountain or other feature containing the resource. From this horizontal tunnel, vertical shafts are drilled, and elevators are used to bring miners down to the resource and back to the surface. The deepest mines on Earth are up to 3.5 km (2.2 miles) deep. Coal, diamonds, and gold are some of the resources removed by subsurface mining.

The Environment and Safety

The extraction of mineral resources from Earth's crust has a variety of environmental impacts on water, soil, biodiversity, and other areas. In addition, mineral resource extraction can have human health consequences that affect miners and others.

Mining and the Environment

As you can see in **TABLE 25.2** on page 296, all forms of mining affect the environment. Mining almost always requires the construction of roads, which can result in soil erosion, damage to waterways, and habitat fragmentation. In addition, all types of mining produce mine tailings. Depending on the location and the amount of rainfall in a given area, mine tailings may contaminate land and water with acids and metals. Acid mine drainage is the general term for water that passes through mine tailings. This water reacts chemically with mine tailings and becomes acidic and may contain metals that are leached from the tailings.

In mountaintop removal, the mine tailings are typically deposited in the adjacent valleys, sometimes blocking or changing the flow of rivers. Mountaintop removal is used primarily in coal mining and is safer for workers than subsurface mining. In environmental terms, mining companies do sometimes make efforts to restore the mountain to its original shape. However, there is considerable disagreement about whether these reclamation efforts are effective. Damage to streams and nearby groundwater during mountaintop removal cannot be completely rectified by the reclamation process.

Placer mining can also contaminate large portions of rivers, and the areas adjacent to the rivers, with sediment and chemicals. In certain parts of the world, the toxic metal mercury is used in placer mining of gold and silver. Mercury is a highly volatile metal; that is, it moves easily among air, soil, and water. Mercury is harmful to plants and animals and can damage the central nervous system in humans; children are especially sensitive to its effects.

The environmental impacts of subsurface mining may be less apparent than the visible scars left behind by surface mining. One of these impacts is acid mine

FIGURE 25.13 Open-pit mining. The Bingham Canyon Copper Mine, near Salt Lake City, Utah, is the largest open-pit mine in the world. *(Lee Prince/Shutterstock)*

Strip mining The removal of strips of soil and rock to expose ore.

Mine tailings Unwanted waste material created during mining including mineral and other residues that are left behind after the desired metal or ore is removed.

Open-pit mining A mining technique that creates a large visible pit or hole in the ground.

Mountaintop removal A mining technique in which the entire top of a mountain is removed with explosives.

Placer mining The process of looking for minerals, metals, and precious stones in river sediments.

Subsurface mining Mining techniques used when the desired resource is more than 100 m (328 feet) below the surface of Earth.

TABLE 25.2 Types of mining operations and their effects

Type of operation	Effects on air	Effects on water	Effects on soil	Effects on biodiversity	Effects on humans
Surface mining	Significant dust from earth-moving equipment	Contamination of water that percolates through tailings	Most soil removed from site; may be replaced if reclamation occurs	Habitat alteration and destruction over the surface areas that are mined	Minimal in the mining process, but air quality and water quality can be adversely affected near the mining operation
Subsurface mining	Minimal dust at the mining site, but emissions from fossil fuels used to power mining equipment can be significant	Acid mine drainage as well as contamination of water that percolates through tailings		Road construction to mines fragments habitat	Occupational hazards in mine; possibility of death or chronic respiratory diseases such as black lung disease

drainage. To keep underground mines from flooding, pumps must continually remove underground water from the mine. These pumps create a form of acid mine drainage. This drainage water lowers the pH of nearby soils and streams and can cause damage to ecosystems.

Mining Safety and Legislation

Subsurface mining is a dangerous occupation. Hazards to miners include accidental burial, explosions, and fires. In addition, the inhalation of gases and particles over long periods can lead to a number of occupational respiratory diseases, including black lung disease and asbestosis, a form of lung cancer. In the United States, in the twentieth century, more than 11,000 coal miners died in underground coal mine explosions and fires. A much larger number died from respiratory diseases. Today, there are relatively few deaths per year in coal mines in the United States, in part because of improved work safety standards and in part because there is much less subsurface mining. In other countries, especially China, mining accidents remain fairly common.

As human populations grow and developing nations continue to industrialize, the demand for mineral resources continues to increase. But as the most easily mined mineral resources are depleted, extraction efforts become more expensive and environmentally destructive. The ores that are easiest to reach and least expensive to remove are always recovered first. When these sources are exhausted, mining companies must turn to deposits that are more difficult to reach. These extraction efforts result in greater amounts of mining tailings and more of the environmental problems we have already noted. Learning to use and reuse limited mineral resources more efficiently will help protect the environment as well as human health and safety.

Governments have sought to regulate the mining process for many years. Early mining legislation was primarily focused on promoting economic development, but later legislation became concerned with worker safety as well as environmental protection. The effectiveness of these mining laws has varied.

Congress passed the Mining Law of 1872 to regulate the mining of silver, copper, and gold ores as well as fuels, including natural gas and oil, on federal lands. This law, also known as the General Mining Act, allowed individuals and companies to recover ores or fuels from federal lands. The law was written primarily to encourage development and settlement in the western United States and, as a result, it contains very few provisions for environmental protection.

The Surface Mining Control and Reclamation Act of 1977 regulates surface mining of coal and the surface effects of subsurface coal mining. The act mandates that land be minimally disturbed during the mining process and be reclaimed after mining is completed. Mining legislation does not regulate all of the mining practices that can have harmful effects on air, water, and land. In later chapters we will learn about other U.S. legislation that does, to some extent, address these issues, including the Clean Air Act, Clean Water Act, and Superfund Act.

MODULE 25 AP® Review

Preparing for the AP® Exam

In this module, we saw that rocks and minerals undergo physical and chemical weathering and become products that are precursors for soil. Weathered materials are subject to erosion, which is a natural process that can be enhanced by human activity. Erosion also influences the precursors to soil. Soil forms from geologic material as well as biological material. Soil properties result from physical, chemical, and biological processes and are influenced by five soil forming factors. Concentrated accumulations of elements and minerals in and below soils that are economically valuable are called ores. When ores are extracted, a variety of consequences affect humans and the environment.

AP® Practice Questions

Choose the best answer for the following.

1. Acid precipitation directly causes
   I. erosion.
   II. physical weathering.
   III. chemical weathering.
   (a) I only
   (b) I and II
   (c) I and III
   (d) III only

2. What are the five primary soil formation factors?
   (a) climate, parent material, pH, organisms, topography
   (b) parent material, topography, organisms, time, latitude
   (c) parent material, climate, pH, latitude, altitude
   (d) climate, parent material, topography, organisms, time

3. Which is the correct order of soil horizons starting at the surface?
   (a) A, B, C, E, O
   (b) O, A, B, C, E
   (c) O, E, A, B, C
   (d) O, A, E, B, C

4. What type of soil would be best to add to the bottom of a constructed pond, where the goal is to have as little leakage of water as possible?
   (a) mostly clay
   (b) mostly silt
   (c) mostly sand
   (d) equal amounts of sand, silt, and clay

5. Which of the following, if added to soil, would lower the base saturation?
   (a) sodium
   (b) potassium
   (c) magnesium
   (d) aluminum

6. Tailings are
   (a) magma ore resulting from seafloor spreading.
   (b) the remaining supply of metals on Earth.
   (c) the nutrients that leach downward in soil.
   (d) the waste material from mining.

Working Toward Sustainability

Mine Reclamation and Biodiversity

One of the environmental impacts of surface mining is the amount of land surface it disturbs. Once a mining operation is finished, the mining company may try to restore it to its original condition. In the United States, the Surface Mining Control and Reclamation Act of 1977 requires coal mining companies to restore the lands they have mined. Regulations also require other types of mining operations to do some level of restoration.

A disturbed ecosystem can return to a state similar to its original condition only if the original physical, chemical, and biological properties of the land are recreated. Therefore, reclamation after mining involves several steps. First, the mining company must fill in the hole or depression it has created in the landscape. Second, the fill material must be shaped to follow the original contours of the land that existed before the mining began.

Before a mining operation begins, the mining company usually scrapes off the topsoil that was on the land and puts it aside. This topsoil will be returned and spread over the landscape after mining is completed. Finally,

(a)

(b)

Mining and reclamation. These photos show a strip mine site reclaimed by the Pennsylvania Abandoned Mine Land Project. (a) In this photo the reclamation contractor is removing mine tailings and other materials that can contribute to acid mine drainage. (b) After reclamation, a 2.5 acre wildlife habitat pond dominates the formerly-mined area. *(Pennsylvania Department of Environmental Protection, Bureau of Abandoned Mine Reclamation.)*

the land must be replanted. In order to re-create the communities of organisms that inhabited the area before mining, the vegetation planted on the site must be native to the area and foster the process of natural succession. Properly completed reclamation makes the soils physically stable so that erosion does not occur and water infiltration and retention can proceed as they did before mining. The materials used in the reclamation must be relatively free of metals, acids, and other compounds that could potentially leach into nearby bodies of water.

Many former mining areas have not been reclaimed properly. However, there are an increasing number of reclamation efforts that have achieved conditions that equal those that existed prior to the mining operation. The Trapper Mine in Craig, Colorado, and other mines like it illustrate reclamation success stories.

The Trapper Mine produced approximately 1.6 million metric tons (1.8 million tons) of coal in 2016, all of which was sold to a nearby electricity generation plant. Although all coal mining operations are required to save excavated rock and topsoil, managers at the Trapper Mine have stated that they are meticulous about saving all the topsoil they remove. The rock they save is used to fill the lower portions of excavated holes. Workers then install drainage pipes and other devices to ensure proper drainage of water. The topsoil that has been set aside is spread over the top of the restored ground and contoured as it was before mining. Trapper Mine reclamation staff then replant the site with a variety of native species of grasses and shrubs, including the native sagebrush commonly found in the high plateaus of northwestern Colorado.

Government officials, and even the Colorado branch of the Sierra Club, have expressed approval of the Trapper Mine reclamation process. But perhaps the strongest evidence of its success is the wildlife that now inhabits the formerly mined areas. The Columbian sharp-tailed grouse (*Tympanuchus phasianellus columbianus*), a threatened bird species, has had higher annual survival and fertility rates on the reclaimed mine land than it has on native habitat in other parts of Colorado. Populations of elk, mule deer, and antelope have all increased on reclaimed mine property.

Reclamation issues must be addressed for each particular area where mining occurs. As we have seen, subsurface and surface water runoff in certain locations can contain high concentrations of toxic metals and acids. In other situations, although certain native species may increase in abundance after reclamation, other native species may decrease. Nevertheless, with supervision, skill, and enough money to pay for the proper reclamation techniques, a reclaimed mining area can become a satisfactory or even an improved habitat for many species.

Critical Thinking Questions

1. If you were on a committee asked to determine if a mine reclamation project had been successful, what three measures or conditions would you want to consider in your evaluation?
2. Identify two environmental problems that might occur in the area where you live if a mining site was not reclaimed.

References

Web, Dennis. 2017. Colorado coal production drops almost 40% in 2016. *Daily Sentinel* (Grand Junction, Colorado), June 8. http://www.gjsentinel.com/news/articles/colorado-coal-production-drops-almost-40-percent-i

Raabe, S. 2002. Nature's Comeback: Trapper Mine reclamation attracts wildlife, wins praise. *Denver Post*, November 12, p. C01.

Chapter 8 Review

In this chapter, we have examined how geologic processes such as plate tectonics, earthquakes, and volcanism have led to the differential distribution of elements and minerals on the surface of Earth. These geologic processes, which have occurred at different rates over long periods of time, have led to the formation of different rocks and minerals on or near the surface of Earth. Rocks and minerals have undergone weathering at different rates and have been eroded and deposited elsewhere on Earth. This has been one of the contributors to soil formation. Soils are a membrane that covers much of the land surface on Earth and these soils contain a mixture of geologic material from below and organic material from plants and animals from above. We have also examined how the actions involved in removal of valuable mineral resources from the surface and below the surface of Earth have affected a number of environmental processes.

Key Terms

Core
Mantle
Magma
Asthenosphere
Lithosphere
Crust
Hot spot
Plate tectonics
Tectonic cycle
Subduction
Volcano
Divergent plate boundary
Seafloor spreading
Convergent plate boundary
Transform fault boundary
Fault
Seismic activity
Fault zone
Earthquake
Epicenter
Richter scale
Tsunami
Rock cycle
Igneous rock
Intrusive igneous rock
Extrusive igneous rock
Fracture
Sedimentary rock
Metamorphic rock
Physical weathering
Chemical weathering
Acid precipitation
Acid rain
Erosion
Parent material
Soil degradation
Horizon
O horizon
Humus
A horizon
Topsoil
E horizon
B horizon
C horizon
Cation exchange capacity (CEC)
Base saturation
Crustal abundance
Ore
Metal
Reserve
Strip mining
Mine tailings
Open-pit mining
Mountaintop removal
Placer mining
Subsurface mining

Learning Goals Revisited

Module 24 Mineral Resources and Geology

Describe the formation of Earth and the distribution of critical elements on Earth.

Earth formed from cosmic dust in the solar system. As it cooled, heavier elements, such as iron, sank toward the core, while lighter elements, such as silica, floated toward the surface. These processes have led to an uneven distribution of elements and minerals throughout the planet.

Define the theory of plate tectonics and discuss its relevance to the study of the environment.

Earth is overlain by a series of plates that move at rates of a few millimeters per year. Plates can move away from each other, move toward each other, or slide past each other. One plate can be subducted under another. These tectonic processes create mountains, earthquakes, and volcanoes.

Describe the rock cycle and discuss its importance in environmental science.

Rocks are made up of minerals, which are formed from the various chemical elements in Earth's crust. The processes of the rock cycle lead to the formation, breakdown, and recycling of rocks.

Module 25 Weathering and Soil Science

Understand how weathering and erosion occur and how they contribute to element cycling and soil formation.

Physical weathering is the mechanical breakdown of rocks and minerals while chemical weathering is a result of chemical reactions. Both occur as a result of natural processes and can be accelerated by human activities. Erosion is the physical removal of rock fragments and weathering products that are subsequently deposited elsewhere.

Explain how soil forms and describe its characteristics.

Soil forms as the result of physical and chemical weathering of rocks and the gradual accumulation of organic detritus from the biosphere. The factors that determine soil properties are parent material, climate, topography, soil organisms, and time. The relative abundances of sand, silt, and clay in a soil determine its texture.

Describe how humans extract elements and minerals and the social and environmental consequences of these activities.

Concentrated accumulations of minerals from which economically valuable materials can be extracted are called ores. Ores are removed by surface or subsurface mining operations. Surface mining generally results in greater environmental impacts, whereas subsurface mining is more dangerous to miners. With the exception of coal mining, legislation directly related to mining does not address most environmental considerations.

Practice Math and Graphing

Preparing for the AP® Exam

Answer the following questions. Be sure to show all your work.

1. Practice Math

(a) Earth formed 4.6 billion years ago. Express this value in scientific notation.

(b) There is evidence of life on Earth in the fossil record as far back as approximately 3.8 billion years ago. How many years did the Earth exist before evidence of life appears in the fossil record? Express this in scientific notation.

2. Practice Graphing

(a) Using the data in the table, create a bar chart graph that shows the number of deaths per year related to coal mining in China.

Year	Number of fatalities
2001	2,613
2002	4,251
2003	4,899
2004	3,603
2005	3,925
2006	1,673
2007	1,368
2008	1,086

Data from: Ming-Xiao, Wang, et al. 2011. Analysis of national coal-mining accident data in China, 2001–2008. *Public Health Reports* 126.2: 270–275.

(b) Describe the pattern or trend in coal mining deaths. Can you propose a hypothesis or suggest a possible explanation for this pattern? How would you go about testing your hypothesis?

Section 1: Multiple-Choice Questions

Choose the best answer for questions 1–20.

1. As Earth slowly cooled
   (a) lighter elements sank to the core and heavier elements moved to the surface.
   (b) lighter elements mixed with heavier elements and sank to the core.
   (c) lighter elements mixed with heavier elements and moved to the surface.
   (d) lighter elements moved to the surface and heavier elements sank to the core.

2. The correct vertical zonation of Earth above the core is
   (a) asthenosphere-mantle-soil-lithosphere.
   (b) asthenosphere-lithosphere-mantle-soil.
   (c) mantle-asthenosphere-lithosphere-soil.
   (d) mantle-lithosphere-soil-asthenosphere.

3. Evidence for the theory of plate tectonics includes
   I. deposits of copper ore around the globe.
   II. identical rock formations on both sides of the Atlantic.
   III. fossils of the same species on distant continents.
   (a) I and III
   (b) II and III
   (c) I and II
   (d) I, II, and III

4. Which type of mining is usually most directly harmful to miners?
   (a) mountaintop removal
   (b) open-pit mining
   (c) placer mining
   (d) subsurface mining

5. Measured on the Richter scale, an earthquake with a magnitude of 7.0 is ____ times greater than an earthquake with a magnitude of 2.0.
   (a) 100
   (b) 1,000
   (c) 10,000
   (d) 100,000

For questions 6 to 9, select from the following choices:

(a) divergent plate boundary
(b) convergent plate boundary
(c) transform fault boundary
(d) epicenter

6. At which type of boundary do tectonic plates move sideways past each other?

7. At which type of boundary does subduction occur?

8. At which type of boundary does seafloor spreading occur?

9. Which term refers to the point on Earth's surface directly above an earthquake?

10. Fossil records are found in
    (a) intrusive rock.
    (b) igneous rock.
    (c) metamorphic rock.
    (d) sedimentary rock.

11. Where would you expect to find extrusive igneous rock?
    (a) along a transform fault
    (b) along the ocean floor
    (c) on the coast of a continent
    (d) at a continental subduction zone

12. Which statement about soil is NOT correct?
    (a) Soil is fairly static and does not change.
    (b) Soil is the medium for plant growth.
    (c) Soil is a primary filter of water.
    (d) A wide variety of organisms live in soil.

13. The soil horizon commonly known as subsoil is the
    (a) A horizon.
    (b) O horizon.
    (c) B horizon.
    (d) C horizon.

14. Which statement about earthquakes is NOT true?
    (a) An earthquake of magnitude 8.0 on the Richter scale occurs about once every year.
    (b) Earthquakes occur when rocks of the asthenosphere rupture along a fault.
    (c) The Ring of Fire describes a geological circle on Earth where earthquakes are likely to occur.
    (d) An earthquake of magnitude 4 on the Richter scale is 100 times greater than an earthquake of magnitude 2.

15. Which contributes to the physical weathering of rocks?
    (a) acid rain
    (b) burial in sediment
    (c) volcanic activity
    (d) growth of plant roots

16. A field containing soil that is 60 percent clay can hold a tremendous amount of nutrients but often has poor crop growth. What is the most likely explanation for this phenomenon?
   (a) Excessive concentrations of nutrients in soil can become toxic to many types of plants.
   (b) Clay can hold heavy metals that are toxic to plants.
   (c) Plant roots have difficulty penetrating a soil that is 60 percent clay.
   (d) The base saturation of clay soil is too high for most plants.

17. Which type of mining is potentially the most harmful to human health?
   (a) open-pit mining
   (b) mountaintop removal
   (c) strip mining
   (d) subsurface mining

18. Which rock is formed by high temperature and pressure like that caused by continental collisions?
   I. igneous
   II. metamorphic
   III. sedimentary
   (a) I only
   (b) II only
   (c) III only
   (d) II and III

19. New Zealand is a country with two main islands located on the boundary of two tectonic plates—the Australian and the Pacific—which are moving in different directions. The city of Christchurch, located on the South Island, is moving away from the city of Auckland, located on the North Island, at a rate of 40 mm per year. How many years will it take for Christchurch to be 500 km farther away from Auckland than it is today?
   (a) 12.5×10^5 years
   (b) 12.5×10^6 years
   (c) 20.0×10^5 years
   (d) 20.0×10^6 years

20. Which statement about oceanic plates is not correct?
   (a) The crust of oceanic plates is rich in iron.
   (b) The movement of oceanic plates away from each other results in seafloor spreading.
   (c) The dominant rock type in the crust of oceanic plates is basaltic rock.
   (d) Continental crust is subducted where oceanic and continental plate margins meet.

Section 2: Free-Response Questions

Write your answer to each part clearly. Support your answers with relevant information and examples. Where calculations are required, show your work.

1. The rock cycle plays an important role in the recycling of Earth's limited amounts of mineral resources.
   (a) The mineral composition of rock depends on how it is formed. Name the *three* distinct rock types, explain how each rock type is formed, and give one specific example of a rock of each type. (6 points)
   (b) Explain either the physical *or* the chemical weathering process that leads to the breakdown of rocks. (2 points)
   (c) Describe the natural processes that can lead to the formation of soil, and discuss how human activities can accelerate the loss of soil. (2 points)
2. Soils play a vital role in ecosystems, including acting as a medium for plant growth, filtering water, and providing habitat for a variety of organisms.
   (a) Describe TWO of the factors that determine the properties of soils. (2 points)
   (b) Name and describe the properties of the four main soil horizons (excluding the E horizon), going from the top layer to the bottom layer. (4 points)
   (c) Describe at least TWO differences between a soil primarily composed of sand and a soil primarily composed of clay. (2 points)
   (d) Describe the role of cation exchange and base saturation in soil productivity. (2 points)
3. *Read the following two paragraphs and answer the questions that follow.*

   Cattle can be a destructive force on a landscape or they can benefit the soil, depending on how they are managed. Researchers compared the soil on six pasture-based ranches in Minnesota to soil from neighboring farms that produced corn, soybean, oats, or hay. The researchers monitored the soil at these locations for 4 years. At the end of this time they found that the soil and related ecosystem health on the managed grazed land was better than the soil on the crop land in the following ways:

   - 53 percent greater soil stability
   - 131 percent more earthworms in soil
   - substantially more organic matter in soil
   - less nitrate pollution of groundwater
   - improved stream quality at adjacent locations
   - better habitat for grassland birds and other wildlife

   The United States is currently losing as much as 3 billion tons of topsoil each year. Much of this loss can be attributed to growing corn and soy for animal feed. As shown in the figure, using fields for pasture instead of conventional row crops can substantially reduce soil loss—by as much as 93 percent.

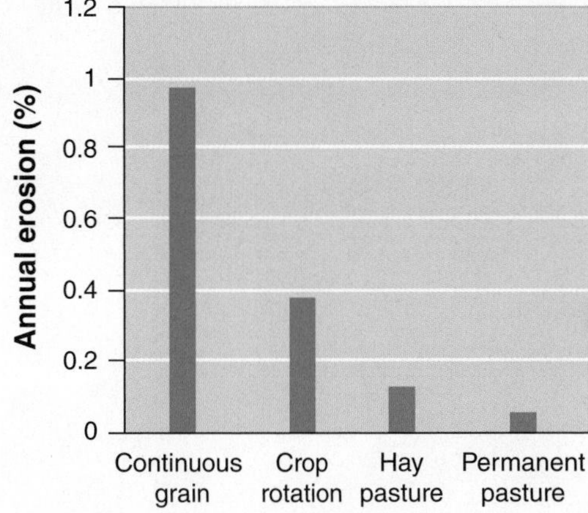

   (a) Explain the consequences of the observations described and comment on the long-term sustainability of each agricultural practice.
      (i) What roles do earthworms play in maintaining soil stability? (1 point)
      (ii) How does the presence of organic matter benefit the soil? (1 point)
      (iii) Suggest why there is less nitrate contamination of groundwater from the permanent pastures. (1 point)
      (iv) In what ways could the stream quality have improved? (1 point)
   (b) Discuss one negative consequence of grazing cattle on pastureland instead of feeding them grain. (3 points)
   (c) Describe one potential negative effect on rivers and streams of grazing cattle on pastureland. (3 points)

The beautiful Klamath River flows from southern Oregon and down through northern California. *(Sarina Finkelstein/Redux)*

Water Resources and Water Use

CHAPTER 9

MODULE 26 The Availability of Water

MODULE 27 Human Alteration of Water Availability

MODULE 28 Human Use of Water Now and in the Future

CASE STUDY

Dams and Salmon on the Klamath River

The Klamath River is a beautiful stretch of water that runs for 400 km (250 miles) from southern Oregon through northern California, where it empties into the ocean. The Klamath River was once a spectacular habitat for salmon and was the third largest salmon fishery on the West Coast. During the past 100 years, however, the Klamath River has seen many changes. It has been dammed for electricity, and farmers have settled the land and diverted water to irrigate their crops. The salmon population has been greatly reduced, and local Native American tribes and the commercial fishing industry have suffered as a result.

The story of the Klamath River began thousands of years ago, when six tribes of Native Americans were the sole inhabitants of the region. For more than 300 generations, these tribes relied on salmon as an important part of their diet. Historically, salmon eggs hatched in the upstream portion of the river. Young fish migrated down the river into the ocean, where they spent several years growing into mature fish. Once mature, they migrated back up the river, returning to lay eggs in the same place where they had hatched. Several species of salmon were abundant in the river. An estimated 800,000 mature Chinook salmon (*Oncorhynchus tshawytscha*)—which can grow as large as 45 kg (100 pounds)—migrated up the river each year.

This migratory behavior worked well for thousands of years, until the region began to change during the twentieth century. In the early 1900s, the U.S. Bureau of Reclamation decided to build canals to drain two large lakes that were the source of the river. The exposed lake bottoms were turned into farmland, and the water that drained into the canals provided a constant source of irrigation for this new farmland. In addition, four hydroelectric dams were built on the river between 1908 and 1962. These dams generated enough electricity to power 70,000 homes. Because hydroelectric dams generate less pollution than other sources of electricity, these dams provided a substantial benefit to the region. But their presence imposed a barrier to the natural migration of salmon. In addition, the formation of large pools behind the dams caused a rise in water temperature that is not favorable to salmon.

> The Klamath River was once a spectacular habitat for salmon.

As the decades passed, the multiple uses of the Klamath River began to come into conflict. In addition, a gradual warming of temperatures in the region has caused less snow to fall in the mountains. Although less snow means less water for the river, agricultural use of the river water continued, which reduced the river flow to low levels that created an additional threat to the salmon. In 1997, the coho salmon (*Oncorhynchus kisutch*) was given protection under the U.S. Endangered Species Act. To protect the salmon, water withdrawals could not reduce the river's flow below the minimal level in which the salmon could thrive. That restriction prevented farmers from receiving enough water to irrigate their crops and threatened their livelihoods.

The conflict among the various interest groups intensified. Native Americans, the hydroelectric company, farmers, the commercial fishing industry, conservationists, and multiple government agencies all debated regional water management policy from different points of view.

In 2002, the migrating salmon experienced a massive die-off in the Klamath River. This catastrophe was caused by unusually warm water, the growth of toxic algae in the dammed waters, and diseases. The sight of 30,000 large, dead chinook salmon brought worldwide attention to the river. Nearly 30 interest groups began meeting in order to find a solution to the river's complex problems.

In 2009, the parties reached a decision. The hydroelectric company, which realized it would soon have to spend $460 million on dam renovations, agreed to remove the four dams. The farmers agreed to conserve water by planting crops that required less water and by updating their irrigation technology. These changes are expected to improve river flow, thereby improving conditions for the salmon and those who depend on them. In 2016, a formal plan was signed by the states of California and Oregon, the federal government, the hydroelectric company, and several Native American tribes. At a cost of $450 million ($250 million from the state of California and $200 from the utility's customers), this endeavor represents the largest dam removal project in history. In 2017, the Trump Administration agreed to allow the plan to move forward. When completed, salmon will have 676 km (420 miles) of uninterrupted river for the first time in a century. However, more work remains to be done regarding how the available water will be shared among the Native American tribes, farmers, and other users.

Conflicts over the use of water resources are found throughout the world as human populations grow and place ever-increasing demands on limited water resources. The compromise reached on the Klamath has the potential to serve as a model for resolving such conflicts in the future.

Sources: R. Rymer, Reuniting a river, *National Geographic,* December 2008, http://ngm.nationalgeographic.com/2008/12/klamath-river/rymer-text/1; C. Sullivan, Landmark agreement to remove four Klamath River dams, *New York Times,* September 30, 2009, http://www.nytimes.com/gwire/2009/09/30/30greenwire-landmark-agreement-to-remove-4-klamath-river-d-72992.html; L. Zuckerman, Interior Department recommends removal of dams on Klamath River to aid salmon, *Reuters*, April 4, 2013, http://www.reuters.com/article/2013/04/05/us-usa-environment-dams-idUSBRE93402O20130405; S. Gilman, This will be the biggest dam-removal project in history, *National Geographic*, April 11, 2016, https://news.nationalgeographic.com/2016/04/160411-klamath-glen-canyon-dam-removal-video-anniversary/; W. Houston, Feds won't oppose Klamath River dam removal, official says, Eureka*Times Standard*, September 29, 2017. http://www.times-standard.com/article/NJ/20170929/NEWS/170929773&template=printart.

Water is a critical resource for all organisms but it can be challenging to obtain. As we saw in Chapter 4, climate variation causes some regions of the world to possess abundant supplies of water, whereas other regions have very little water. This not only affects the plants and animals that live in different regions of the world but also affects growing human populations that have limited water availability. In this chapter, we will examine the many sources of fresh water, which is the type of water that humans can drink. We will then consider how humans have developed ways of altering the availability of water on Earth. We will also examine how humans use water and the prospects for the availability of water in the future. In Chapter 14, we will discuss the issue of water pollution.

MODULE 26

The Availability of Water

A very small amount of Earth's aboveground water is found in the atmosphere and in the form of water bodies such as streams, rivers, wetlands, and lakes. In this module, we will examine the major sources of fresh water on Earth. We will also consider the effects of unusually high and low amounts of precipitation.

Learning Goals

After reading this module you should be able to

- describe the major sources of groundwater.
- identify some of the largest sources of fresh surface water.
- explain the effects of unusually high and low amounts of precipitation.

Groundwater can be extracted for human use

Nearly 70 percent of Earth's surface is covered by water. As **FIGURE 26.1** shows, this water is found in five main repositories, not including the water vapor and precipitation contained in the atmosphere. The vast majority of Earth's water—more than 97 percent—is found in the oceans as salt water. The remainder—less than 3 percent—is fresh water, the type of water that can be consumed by humans. Of this small percentage of Earth's water that is fresh water, nearly three-fourths is aboveground, though mostly in the form of ice and glaciers, and therefore generally not available for human consumption. Only one-fourth resides underground in the form of groundwater. This means that available freshwater is less than 1 percent of all water on Earth.

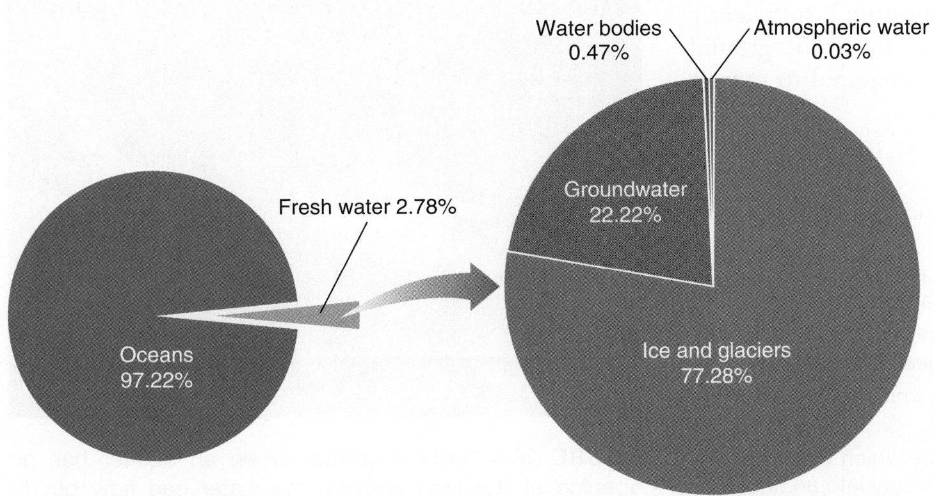

FIGURE 26.1 Distribution of water on Earth. Fresh water represents less than 3 percent of all water on Earth, and only about three-fourths of that fresh water is surface water. Most of that surface water is frozen as ice and in glaciers. Therefore, less than 1 percent of all water on the planet is accessible for use by humans. *(Data from: R.W. Christopherson, Geosystems, 7th ed. Pearson/Prentice Hall, 2009)*

Groundwater exists in the multitude of small spaces found within permeable layers of rock and sediment called **aquifers**. **FIGURE 26.2** shows how different types of aquifers are situated. Water can easily flow in and out of an **unconfined aquifer**, which is made of porous rock covered by soil. In contrast, **confined aquifers** are surrounded by a layer of impermeable rock or clay, which impedes water flow to or from the aquifer. The uppermost level at which the groundwater in a given area fully saturates the rock or soil is called the **water table**.

The process by which water from precipitation percolates through the soil and works its way into the groundwater is known as **groundwater recharge**. If water falls on land that contains a confined aquifer, however, it cannot penetrate the impermeable layer of rock. A confined aquifer can only be recharged if the impermeable layer of rock has a surface opening that can serve as a recharge area.

Aquifers serve as important sources of fresh water for many organisms. Plant roots can access the groundwater in an aquifer and draw down the water table. Water from some aquifers naturally percolates up to the ground surface as **springs** (**FIGURE 26.3**). Springs serve as a natural source of water for freshwater aquatic biomes, and they can be used directly by humans as sources of drinking water. Humans discovered centuries ago that water can also be obtained from aquifers by digging a well—essentially a hole in the ground. Most modern wells are very deep and water is pumped to the surface against the force of gravity. In some confined aquifers, however, the water is under tremendous pressure from the impermeable layer of rock that surrounds it. Drilling a hole into a confined aquifer releases the pressure, which allows the water to burst out of the aquifer and rise up in the well, as shown in Figure 26.2. A well created by drilling a hole into a confined aquifer is called an **artesian well**. If the pressure is sufficiently great, the water can rise all the way up to the ground surface, in which case no pump is required to extract the water from the ground.

Aquifer A permeable layer of rock and sediment that contains groundwater.

Unconfined aquifer An aquifer made of porous rock covered by soil out of which water can easily flow.

Confined aquifer An aquifer surrounded by a layer of impermeable rock or clay that impedes water flow.

Water table The uppermost level at which the water in a given area fully saturates rock or soil.

Groundwater recharge A process by which water percolates through the soil and works its way into an aquifer.

Spring A natural source of water formed when water from an aquifer percolates up to the ground surface.

Artesian well A well created by drilling a hole into a confined aquifer.

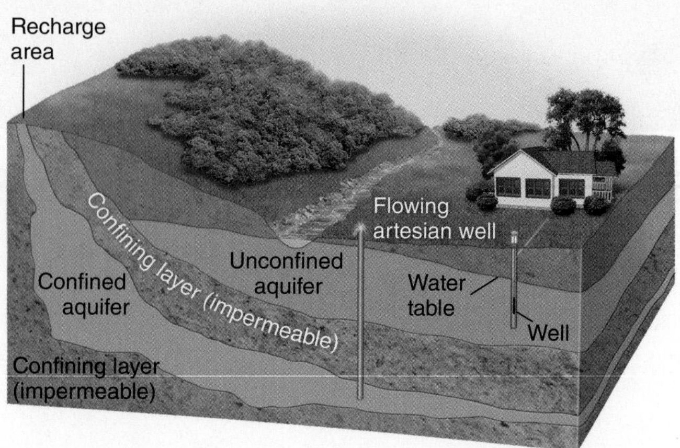

FIGURE 26.2 Aquifers. Aquifers are sources of usable groundwater. Unconfined aquifers are rapidly recharged by water that percolates downward from the land surface. Confined aquifers are surrounded by an impermeable layer of rock or clay, which can cause water pressure to build up underground. Artesian wells are formed when a well is drilled into a confined aquifer and the natural pressure causes water to rise toward the ground surface.

The age of water in aquifers varies, as does the rate at which aquifers are recharged. The water in an unconfined aquifer may originate from water that fell to the ground last year or even last week. This direct and rapid connection with the surface is one of the reasons water from unconfined aquifers is much more likely to be contaminated with chemicals released by human activities. Confined aquifers, on the other hand, are generally recharged very slowly, perhaps over 10,000 to 20,000 years. For this reason, water from a confined aquifer is usually much older and less likely

FIGURE 26.3 Natural spring. When an aquifer has an opening at the land surface, the water can flow out to form a spring. Springs, such as this one in New Mexico, can be important sources of water for organisms and serve as the initial water source for many streams and rivers. *(Jerry Moorman/iStockphoto.com)*

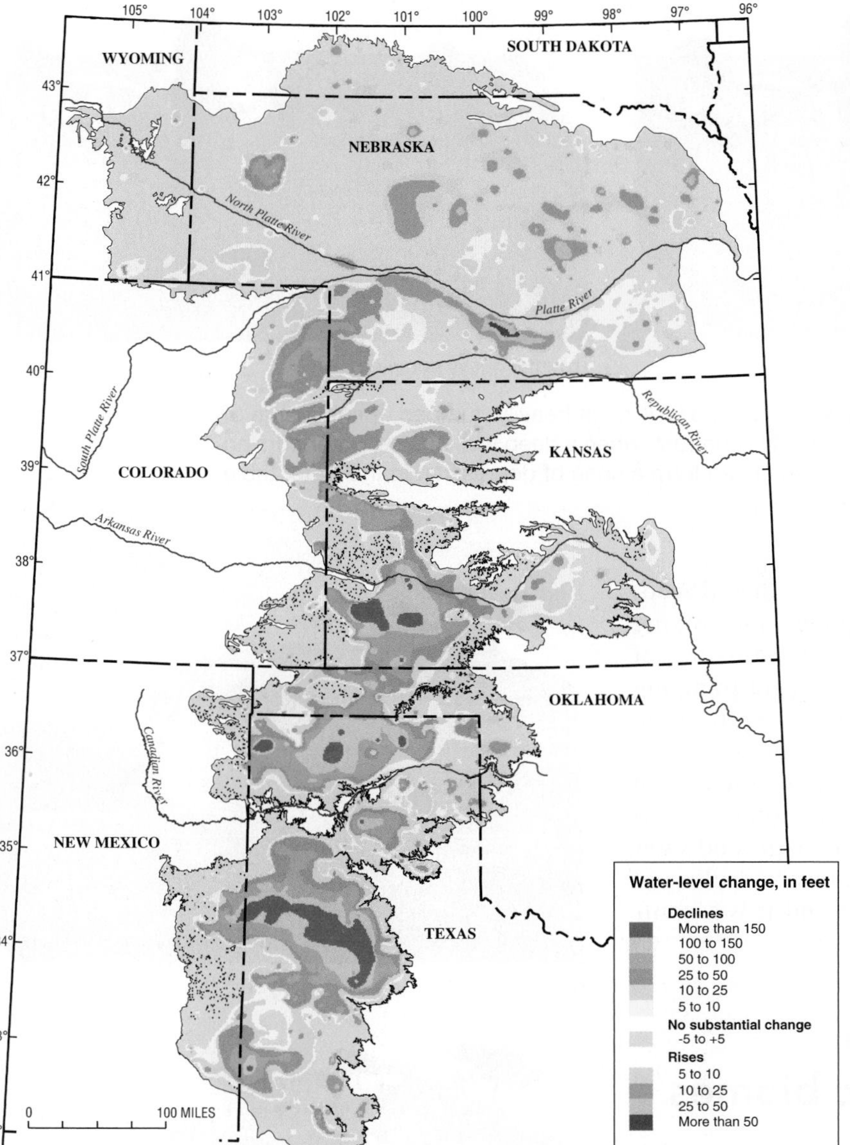

FIGURE 26.4 The Ogallala aquifer. The Ogallala aquifer, also called the High Plains aquifer, is the largest in the United States, with a surface area of about 450,000 km^2 (175,000 miles^2). The aquifer has declined by about 5 m (16 feet) from 1950 to 2015, mostly due to withdrawals for irrigation that have exceeded the aquifer's rate of recharge. *(Data from U.S. Geological Survey 2017.)*

to be contaminated by anthropogenic chemicals than is water from an unconfined aquifer.

Large-scale use of water from a confined aquifer is unsustainable because the withdrawal of water is not balanced by recharge. The largest aquifer in the United States is the massive Ogallala aquifer in the Great Plains, which is located under portions of eight western states from Texas to South Dakota. Large amounts of water have been withdrawn from this aquifer for household, agricultural, and industrial uses. Unfortunately the slow rate of recharge is not keeping pace with the fast rate of water withdrawal, as **FIGURE 26.4** shows. As a result, the depth of the water has declined by about 5 m (16 feet) from 1950 to 2015 and the Great Plains region could run out of water during this century.

FIGURE 26.5 on page 310 shows what happens when more water is withdrawn from an aquifer than enters the aquifer. As the water table drops farther from the ground surface, springs that once bubbled up to the surface no longer emerge, and spring-fed streams dry up. In addition, some shallow wells no longer reach the water table. When water is rapidly withdrawn from a well, it can create an area of that contains no groundwater around the well, which is known as a **cone of depression**. In other words, rapid pumping of a deep well can cause adjacent, shallower wells to go dry.

In some situations, water quality is compromised when fresh water is pumped out of wells faster than the aquifer can be recharged. **FIGURE 26.6** on page 310 shows how this can happen when wells are drilled near coastlines. Some coastal regions have an abundance of fresh water underground. This groundwater

Cone of depression An area lacking groundwater due to rapid withdrawal by a well.

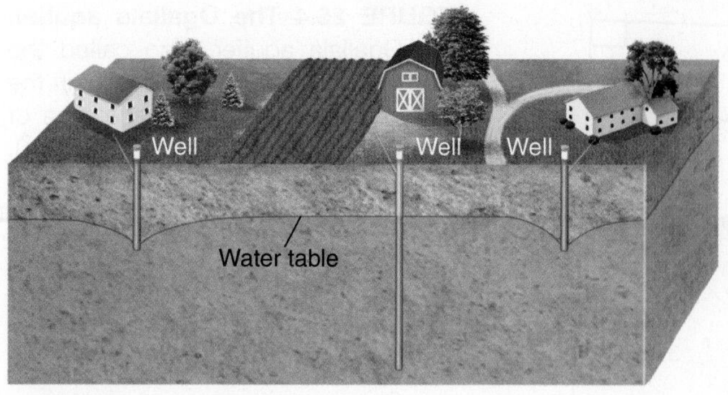

(a) Before heavy pumping

(b) After heavy pumping

FIGURE 26.5 Cone of depression. (a) When a deep well is not heavily pumped, the recharge of the water table keeps up with the pumping. (b) In contrast, when a deep well pumps water from an aquifer more rapidly than it can be recharged, it can form a cone of depression in the water table and cause nearby shallow wells to go dry.

exerts downward pressure that prevents the adjacent ocean water from flowing into aquifers and mixing with the groundwater. As humans drill thousands of wells along these coastlines, however, rapid pumping draws down the water table, reduces the depth of the groundwater, and thereby lessens that pressure. The adjacent salt water is then able to infiltrate the area of rapid pumping, making the water in the wells salty. An infiltration of salt water in an area where groundwater pressure has been reduced from extensive drilling of wells is called **saltwater intrusion** and it is a common problem in coastal areas.

(a)

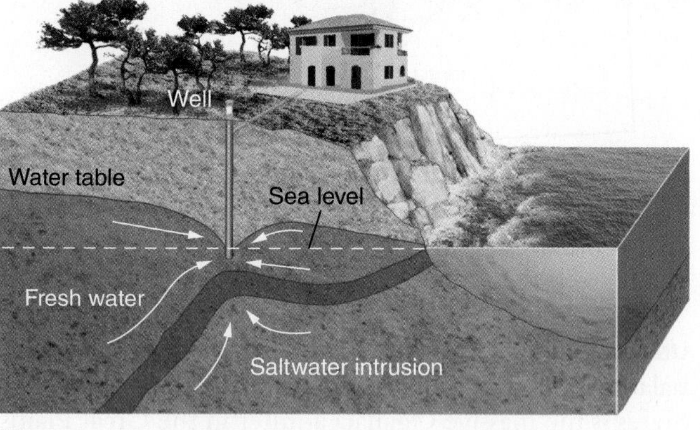

(b)

FIGURE 26.6 Saltwater intrusion. (a) When there are few wells along a coastline, the water table remains high and the resulting pressure prevents salt water from intruding. (b) Rapid pumping of wells drilled in aquifers along a coastline can lower the water table. Lowering the water table reduces water pressure in the aquifer, allowing the nearby salt water to move into the aquifer and contaminate the well water with salt.

Surface water is the collection of aquatic biomes

In contrast to groundwater, surface water is the fresh water that exists aboveground and includes streams, rivers, ponds, lakes, and wetlands. As mentioned in Chapter 4, these waters typically represent freshwater biomes that perform a number of important functions.

The world's three largest rivers, as measured by the volume of water they carry, are the Amazon in South America, the Congo in Africa, and the Yangtze in China. Early human civilizations typically settled along major rivers not only because these waterways served as important means of transportation but also because the land surrounding rivers is often highly fertile. Most streams and rivers naturally overflow their banks during periods of spring snowmelt or heavy rainfall. This excess water spreads onto the land adjacent to the river, called the **floodplain**, where it deposits nutrient-rich sediment that improves the fertility of the soil. The ancient Egyptians built a large and prosperous civilization by using the Nile River for commerce and its fertile floodplains for agriculture. The use of rivers as

Saltwater intrusion An infiltration of salt water in an area where groundwater pressure has been reduced from extensive drilling of wells.

Floodplain The land adjacent to a river.

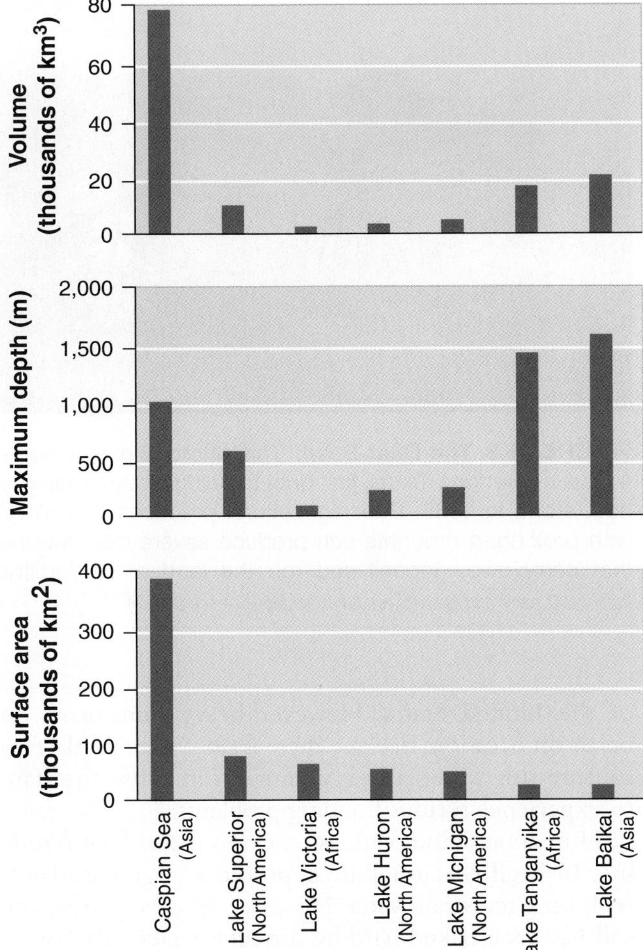

FIGURE 26.7 The world's largest lakes. Measurements of surface area, depth, and volume are shown for each lake. *(Data from http://wldb.ilec.or.jp/LakeDB2/)*

a means of transport continues today. Many large cities in North America, including Pittsburgh, New Orleans, and St. Louis, developed on the banks of major rivers.

Ponds and lakes typically form in depressions in the landscape that are filled by precipitation, runoff that is not absorbed by the surrounding landscape, and groundwater that flows into the depression. As you can see in **FIGURE 26.7**, we can compare the size of Earth's lakes according to surface area, depth, or volume. Some lakes, such as Lake Victoria in Africa, have a fairly large surface area but are not especially deep. Other lakes, such as Lake Baikal in Asia, do not have a large surface area, but are extremely deep and therefore have a large volume.

AP® Exam Tip

Begin making connections between older content and new information to deepen your understanding and prepare for the free-response portion of the AP® Environmental Science Exam. Review the aquatic biomes presented in Chapter 4, Module 13. What new information have you learned about these water resources? ●

Lakes can be created by a variety of processes, including tectonic activity and glaciation. When tectonic activity causes areas of land to rise up, it can isolate a region of the ocean. Because lakes formed in this way were once part of the ocean, they have higher salinities than other lakes. The Caspian Sea in western Asia is an excellent example of a salty lake that was formed by tectonic uplift. Tectonic activity can also cause land to split open into long, deep fissures that subsequently fill with water. Some of the world's largest lakes have formed this way, including Lake Victoria and Lake Tanganyika in Africa and Lake Baikal in Asia. Another way that lakes are formed is through the movement of glaciers. Over thousands of years, glacial movement scrapes large depressions in the land that subsequently fill with water. This process is thought to have played a major role in the creation of the Great Lakes in North America.

As we saw in Chapter 4, freshwater wetlands are essential to water distribution and regulation. During periods of heavy rainfall that might otherwise lead to flooding, these freshwater wetlands—as well as salt marshes and mangrove swamps—can absorb and store the excess water and release it slowly, thereby reducing the likelihood of a flood.

Wetlands are measured by surface area. There is some debate about which freshwater wetland is the world's largest, but most researchers consider the Pantanal in South America to be the largest wetland and the Florida Everglades to be the second largest.

Atmospheric water produces precipitation

Although the atmosphere contains only a very small percentage of the water on Earth, that atmospheric water is essential to global water distribution. People living in arid regions rely heavily on precipitation, in the form of rain and snow, for their water needs. In certain regions, such as East Africa, annual rainfall follows a fairly regular pattern, with April, May, October, and November normally rainy months. However, even areas with predictable rainfall patterns can experience unexpected droughts. In 2004, the worst drought in more than a decade adversely affected many parts of southern Africa, destroying crops, killing cattle, and causing millions of people to go hungry. In 2017, the worst drought in more than a century occurred in South Africa, due to a strong El Niño year. Reduced precipitation and a growing human population caused some reservoirs to decline to only 14 percent of their capacity.

In addition to the direct losses of human lives, livestock, and crops that droughts cause, they also have long-term effects on soil. Because the cycling of many nutrients important to ecosystem productivity, such as nitrogen and phosphorus, depends on the movement of water, droughts affect nutrient cycling and thus soil fertility. Furthermore, prolonged droughts can dry out the soil to such an extent that the fertile upper layer—the topsoil—blows away in the wind. As a result, the land may become useless for agriculture for decades or longer. Finally, soil that has become severely parched may harden and become impermeable. When the rains finally arrive, the water runs off over the hardened land surface rather than soaking in, and this causes the topsoil to erode.

Human activities can contribute to the negative effects of droughts. In the midwestern United States in the 1920s and 1930s, for example, the amount of conversion from native grasslands to wheat fields increased dramatically. Because wheat fields are more susceptible to soil erosion by wind than are native grasslands, this conversion produced many more hectares of land susceptible to erosion. In the early 1930s, a severe and prolonged drought caused a decade of crop failures. With few crop plants remaining alive to hold the soil in place, the drought led to massive dust storms as winds picked up the parched topsoil and blew it away (**FIGURE 26.8**). One dust storm, on April 14, 1935, was so severe that the Sun was entirely blocked out in the southern Great Plains, leading to the name "Black Sunday." Topsoil from these dust storms traveled as far as Washington, D.C., and earned the southern Great Plains the nickname of "the Dust Bowl." The decade of dust storms and the resulting losses of topsoil led to large-scale human migration out of the region and impacted the economy of the entire nation. Today, improved farming practices have reduced the susceptibility of soil to erosion by wind, making major dust storms in the Great Plains and elsewhere much less likely.

Flooding occurs when water input exceeds the ability of an area to absorb that water. Many drought-prone areas of the world rarely experience high amounts of rainfall, but when it does happen, severe flooding can occur. California, for example, is a relatively dry region

Impermeable surface Pavement or buildings that do not allow water penetration.

FIGURE 26.8 The Dust Bowl. This photo shows a farmhouse in Stratford, Texas, just prior to being hit by a massive dust storm in 1935. Poor agricultural practices combined with prolonged droughts can produce severe dust storms that carry away topsoil and rob the land of its fertility. *(National Oceanic and Atmospheric Administration/Science Source)*

of the United States. However heavy rainstorms do occur on occasion. Because there is no system in place to capture this water, a heavy rainfall can cause the state to experience serious flooding problems.

Human activities can also worsen the risk of flooding. In healthy natural areas, porous soil and wetlands soak up excess rainwater. However, in areas where the soil has been baked hard by drought, water that comes from heavy rainfalls cannot soak into the ground. This is also true in urban and suburban areas with large areas of **impermeable surfaces**—pavement or buildings that do not allow water to penetrate the soil. Instead, storm waters run off over impermeable soil or paved surfaces into storm sewers or nearby streams. Excess water that the ground does not absorb fills streams and rivers, which can overflow their banks and may flood lowland areas. Like droughts, floods can lead to crop and property damage as well as losses of animal and human lives.

MODULE 26 AP® Review

Preparing for the AP® Exam

In this module, we learned that fresh water makes up a tiny fraction of all water on Earth and that most of this fresh water is locked up in the form of ice and glaciers. Groundwater is a type of available freshwater and it can be found as either confined or unconfined aquifers that may experience fast or slow recharge. In some parts of the world the rate of water withdrawal for human use is faster than the rate of aquifer recharge. Surface waters such as streams, rivers, wetlands, and lakes are important sources of freshwater, and wetlands can be important for flood control. Finally, we learned that the climatic differences in the availability of precipitation can interact with human land use to cause dramatic floods and droughts in different parts of the world. In the next module, we will examine how humans have worked to alter the natural distribution of water in ways that better suit human needs.

AP® Practice Questions

Choose the best answer for the following.

1. Which is one of the top three largest rivers by volume?
   (a) the Nile
   (b) the Congo
   (c) the Mississippi
   (d) the Yellow River

2. When deep wells are heavily pumped, one result can be
   (a) decreased groundwater recharge.
   (b) spring formation.
   (c) a cone of depression.
   (d) increased groundwater recharge.

3. Which is true of floodplains?
   (a) They have increased fertility.
   (b) Humans are unable to use them.
   (c) They occur near ponds and lakes.
   (d) They are the result of glaciation.

4. Groundwater recharge
   (a) is the result of precipitation.
   (b) sometimes occurs as a spring.
   (c) occurs rapidly in confined aquifers.
   (d) can cause saltwater intrusions.

MODULE 27

Human Alteration of Water Availability

We have seen that the availability of water around the world can be unpredictable. In the Klamath River, for example, years of low precipitation caused an insufficient water supply for human needs. As a result, humans have learned to live with variations in water availability in several ways. We can channel the flow of flood waters with *levees* and *dikes*, block the flow of rivers with *dams* to store water, divert water from rivers and lakes and transport it to distant locations, and even obtain fresh water by removing the salt from salt water. In this module, we will look at each of these water distribution methods and examine their costs and benefits.

Learning Goals

After reading this module you should be able to

- compare and contrast the roles of levees and dikes.
- explain the benefits and costs of building dams.
- explain the benefits and costs of building aqueducts.
- describe the processes used to convert salt water into fresh water.

Levees and dikes are built to prevent flooding

Before the intervention of humans, most rivers periodically overflowed their banks. These occasional overflows of nutrient-rich water made their floodplains particularly fertile. While throughout history, humans have taken advantage of these fertile floodplains for agriculture, in more recent times, they have sought ways to prevent flooding so floodplain land could be developed for residential and commercial use. One way to prevent flooding is by constructing a **levee**, an enlarged bank built up on each side of the river. The Mississippi River has the largest system of levees in the world. This river is enclosed by more than 2,400 km (1,500 miles) of levees that offer flood protection to more than 6 million hectares (15 million acres) of floodplains.

The use of levees has produced several major challenges. First, the fertility of these lands is reduced because natural floodwaters no longer add fertility to floodplains by depositing sediments there. Second, because the sediments do not leave the river, they are carried farther downstream and settle out where the river enters the ocean. Third, levees may prevent flooding at one location, but by doing so they force floodwater farther downstream where it can cause even worse flooding. Finally, the building of levees encourages development in floodplains, although these areas will still occasionally flood. This practice raises the question of whether development should be allowed in areas where flooding remains a high risk.

When floodwaters become too high, levees can either collapse due to the tremendous pressure of the water, or water can come over the top and quickly erode a large hole in the levee. Both events result in massive flooding (**FIGURE 27.1**). One of the most famous levee failures occurred in New Orleans in 2005 as the result of Hurricane Katrina. The levees could not contain the storm surge and heavy rainfall associated with the hurricane. The water overtopped nearly 50 levees, quickly washing out the banks of dirt and flooding many of the neighborhoods that the levees were built to protect. The hurricane and the devastating floodwaters caused more than 1,800 deaths and over $80 billion in damage to homes and businesses.

Dikes are structures built to prevent ocean waters from flooding adjacent land. Functionally, dikes are similar to levees in that they are built to keep water from flooding onto the land. Dikes are common in northern Europe, where large areas of farmland lie below sea level. Perhaps the best-known dikes are those of the Netherlands, where dikes have been used for nearly 2,000 years. Approximately 27 percent of the land in the Netherlands is below sea level. The dikes, combined with pumps that move any intruding water back out to the ocean, have allowed the country to inhabit and farm areas that would otherwise be under water. The original pumps were powered by windmills, which is why we often associate the Netherlands with wind energy. Today, however, the water is pumped with electric and diesel pumps.

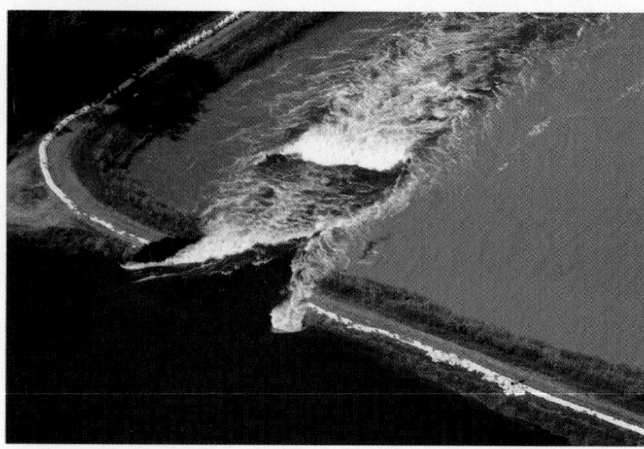

FIGURE 27.1 Levees. Levees are built to prevent rivers from flowing over their banks and onto the floodplain. In 2008, this levee, just north of St. Louis, Missouri, collapsed and allowed the floodwaters to spread over the surrounding fields. *(Anthony Souffle/Chicago Tribune Archives)*

Levee An enlarged bank built up on each side of a river.

Dike A structure built to prevent ocean waters from flooding adjacent land.

Dam A barrier that runs across a river or stream to control the flow of water.

Reservoir The water body created by damming a river or stream.

Dams are built to restrict the flow of streams and rivers

A **dam** is a barrier that runs across a river or stream to control the flow of water. The water body created by damming a river or stream is called a **reservoir**. Dams hold water for a wide variety of purposes, including human consumption, generation of electricity, flood control, and recreation. In 2013, there were 845,000 dams in the world and more than 84,000 dams in the United States. The largest reservoir system in the United States is along the Missouri River. With six dams and reservoirs in Montana, North Dakota, South Dakota, and Nebraska, this system stores nearly 90 trillion liters (24 trillion gallons) of water.

For centuries, humans have used dams to do work, from turning waterwheels that operated grain mills to powering modern turbines that generate electricity in hydroelectric plants. Although hydroelectric dams (discussed further in Chapter 13) are some of the largest dams in the world, they represent only 3 percent of all dams in the United States. In contrast, 18 percent of U.S. dams were built with the primary purpose of flood control—to reduce or prevent flooding farther down the river. Another 38 percent of U.S. dams were constructed primarily for recreation. Dams have also been built to create scenic lakes for housing developments.

Dams provide substantial benefits to humans, but they come with financial, societal, and environmental costs. The world's largest dam, the Three Gorges Dam

FIGURE 27.2 Dams. The photo shows the world's largest dam, the Three Gorges Dam, on the Yangtze River in China. Dams serve a wide variety of purposes, including flood control and electricity generation. *(AP Photo/Xinhua, Du Huaju)*

built across the Yangtze River in China (**FIGURE 27.2**), illustrates these costs and benefits well. This massive structure is 2 km (1.3 miles) wide and 185 m (610 feet) high, and has created a reservoir 660 km (410 miles) long behind it. The Three Gorges Dam, completed in 2006, took 13 years to construct. The reservoir required another 2 years to fill with water.

The benefits of the Three Gorges Dam include the large amount of hydroelectric power it generates, which reduces the extraction and use of fossil fuels. In addition, the dam helps prevent the seasonal flooding that damaged downstream cities and villages and killed more than 1 million people over the past 100 years.

The costs of building dams, both to people and the environment, are substantial. Building a dam, like other large-scale construction, uses large amounts of energy and materials. Projects on such a large scale may displace many people. For example, the Three Gorges Dam and the reservoir it created flooded 13 cities, 140 towns, and 1,350 villages; more than 1.3 million people were forced to relocate.

AP® Exam Tip

You should be familiar with other dams that have large environmental, social and economic impacts such as China's Three Gorges Dam, the James Bay Project in northern Quebec, the Aswan Dam, and the damming of the Colorado River. You may be asked to apply the pros and cons of dams to any of these case studies on the AP® Environmental Science Exam. ●

One of the most common environmental problems associated with dams is the interruption of the natural flow of water to which many organisms are adapted. For migrating fish such as salmon, dams represent an insurmountable obstacle to breeding. The loss of these fish can have a cascading effect on other organisms that depend on them, such as bears that feed on migrating salmon. To alleviate this problem, **fish ladders**, which are built like a set of stairs with water flowing over them, have been added to some dams. Migrating fish can swim up the fish ladders and reach their traditional breeding grounds (**FIGURE 27.3**). Fish are also impacted when water is released from a dam to generate electricity. Passing millions of gallons of water through a spinning turbine kills millions of fish each year.

FIGURE 27.3 Fish ladder. Because dams are an impediment to fish such as salmon that migrate upstream to breed, fish ladders like this one on the Snake River at Little Goose Lock and Dam in Washington State have been designed to allow the fish to get around the dam and continue on their traditional path of migration. *(Theodore Clutter/Science Source)*

Using dams to prevent seasonal flooding has other consequences for the ecology of an area. Seasonal flooding represents a natural disturbance that scours out pools and shorelines, and this disturbance favors colonization by certain plants and animals. Some of the species that are highly dependent on such disturbances are very rare. In recent years, managers have begun to experiment with releasing large amounts of water from reservoirs to simulate seasonal flooding; the results have been very promising. In some areas, dams have been removed where they are no longer needed, as described at the opening of this chapter. To date, nearly 1,300 dams have been removed in the United States. In such cases, the natural flow of water has been restored and much of the ecology has quickly rebounded to its natural state.

Fish ladder A stair-like structure that allows migrating fish to get around a dam.

Aqueducts carry water from one location to another

While dams are designed to hold water back and store it, we often also need to move water from one place to another. **Aqueducts** are canals, ditches, or pipes used to carry water from one location to another. Typically, aqueducts remove water from a lake or river to a place where it is needed. Although aqueducts are famously associated with the Roman Empire, their use dates back to at least the seventh century BCE in Greece. The earliest aqueducts were made of limestone, but modern aqueducts include concrete canals and pressurized steel pipes laid above or under the ground. These structures are more efficient water carriers than the ditches and stone-lined canals of historic times. Older aqueducts, many of which are still in use, can lose as much as 55 percent of the water they carry through leakage or evaporation. These losses can be particularly troublesome in such arid regions as Israel and Jordan where water is already scarce.

In the United States, two of the country's largest cities, New York and Los Angeles, depend on aqueducts to meet their daily water needs. The Catskill Aqueduct brings clean, fresh water over 200 km (120 miles) from the streams and lakes of the Catskill Mountains to New York City. The Colorado River Aqueduct is a canal that carries water 400 km (250 miles) from the Colorado River to Los Angeles (**FIGURE 27.4**).

FIGURE 27.4 Aqueducts. Aqueducts are canals, ditches, or pipes that deliver water from places where it is abundant to places where it is needed. This section of the Colorado River Aqueduct, which diverts water from the Colorado River to Los Angeles, passes through the Mojave Desert. *(iofoto/Shutterstock)*

Using aqueducts to transport water supplies comes with costs and benefits. Although bringing water to cities from pristine areas may ensure a clean supply of water, the construction of aqueducts is expensive and disturbs natural habitats. Aboveground aqueducts can fragment an environment. Even if an aqueduct is buried as an underground pipeline, a great deal of disruption occurs during construction. In **FIGURE 27.5**, you can see an example of the massive underground pipes that are typically used. Often water is diverted from a natural river where it has flowed for millennia. Some major rivers in the United States, including the Colorado River and the Rio Grande, now lose so much water in multiple locations that during certain periods the rivers go dry before they reach the ocean.

Some water diversion projects have international impacts. India and Bangladesh, for example, share nearly 250 rivers that originate in the Himalayas of India and flow into Bangladesh. In 2016, India initiated a large-scale water diversion project, involving more than 50 rivers, to move water from the north where it is abundant to other parts of the country where it is scarce. By diverting water from these rivers, India stands to gain 170 billion liters (55 billion gallons) of water each year for agricultural and household use. However, Bangladesh is situated downstream of many of these river diversion projects and Bangladeshi officials worry that the project will end up dramatically reducing the flow of river water into their country and affect local fish populations and river navigability for commerce. Even farther downstream, the project may reduce the flow of fresh water into estuaries, allowing ocean water to move farther up into the river. This influx of seawater raises the salinity of the

FIGURE 27.5 Aqueduct pipes. Massive water pipes, such as this one for an aqueduct from the Catskill Mountain reservoirs to New York City, are buried deep underground. *(Allyse Pulliam)*

Aqueduct A canal, ditch, or pipe used to carry water from one location to another.

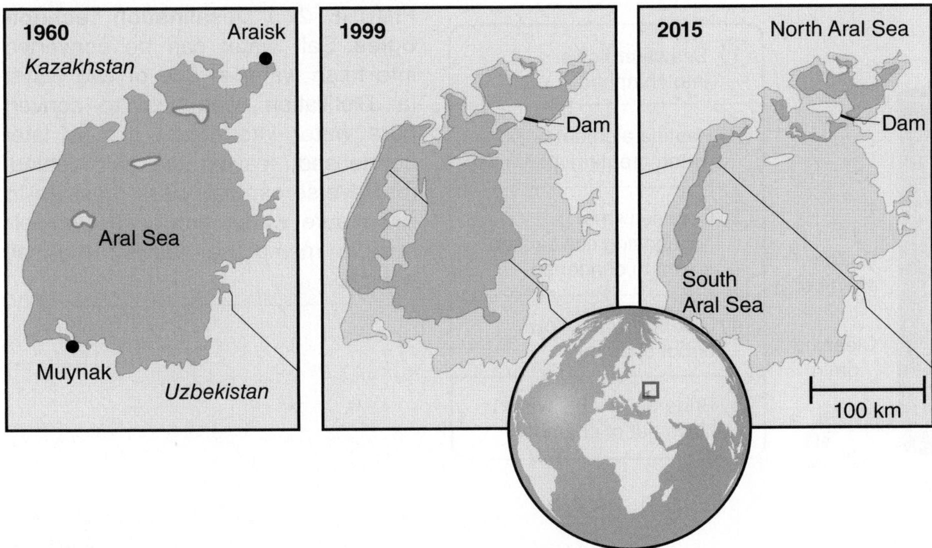

FIGURE 27.6 Consequences of river diversion. Diverting river water can have devastating impacts downstream. The Aral Sea, on the border of Kazakhstan and Uzbekistan, was once the world's fourth largest lake. Since the two rivers that fed the lake were diverted, its surface area has declined by 60 percent and the lake has split into two parts: the North and South Aral seas.

estuary and harms aquatic life that is not adapted to such salty water.

At the same time that India was diverting water that would naturally flow into Bangladesh, neighboring China began the construction of dams on the Yarlung-Zangbo River, which flows from China into India. The first dam was completed in 2015 and additional downstream dams are scheduled to follow. The Chinese dams are primarily used to generate electricity, but there is concern that the water behind the dams could later be diverted to agricultural use in China, which could result in substantially less water flowing through India and Bangladesh. These conflicts over water rights have led many people to call for treaties among the countries in the region to reach a mutually agreeable decision regarding water use.

The most infamous river diversion project happened in the 1950s, when the Soviet Union diverted two rivers that fed the Aral Sea in Central Asia. Diversion of the rivers dramatically decreased freshwater input into the Aral Sea, increased the salinity of the remaining lake water, and destroyed the local fish populations. Although the Aral Sea was originally the fourth largest lake in the world in terms of surface area, the diversion project reduced its surface area by more than 60 percent, as shown in **FIGURE 27.6**. The withdrawals of water caused the Aral Sea to split into two parts: the North Aral Sea and the South Aral Sea. Dust storms have eroded soil, salt, and pesticide residues from the dry lake bed and carried particles of these substances into the atmosphere. The reduction in the size of the lake has also affected the local climate. Without the moderating effect of a large body of water on climate, summers in the region are now hotter and winters are colder. While efforts have been made to save the North Aral Sea by installing a dam and increasing river flows into the sea, saving the South Aral Sea has been deemed too expensive and all efforts have been abandoned. The drying of the South Aral Sea continues today; the shallower eastern half of the South Aral Sea is predicted to be completely dry in 10 years.

Desalination converts salt water into fresh water

Today, some water-poor countries are able to obtain fresh water by removing the salt from salt water, a process called **desalination**, or **desalinization** (**FIGURE 27.7** on page 318). The salt water usually comes from the ocean, but it can also come from salty inland lakes. Currently, the countries of the Middle East produce 50 percent of the world's desalinated water. During the past decade, a number of technological advances have been made and this has greatly reduced the cost of desalination.

Desalination The process of removing the salt from salt water. *Also known as* **Desalinization**.

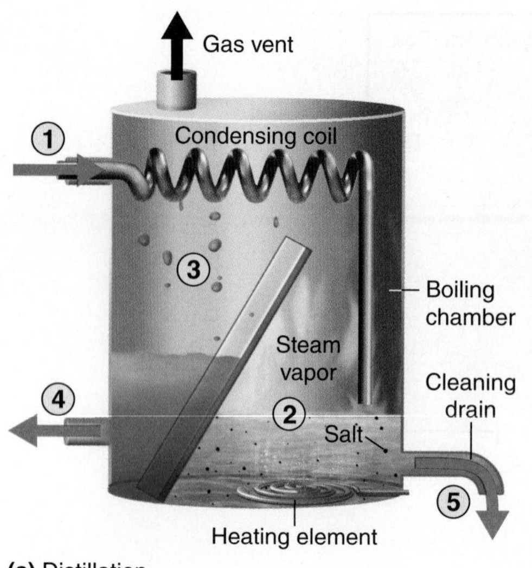
Gas vent
Condensing coil
Boiling chamber
Steam vapor
Cleaning drain
Salt
Heating element
1 Seawater flows into chamber.
2 Heating element boils water, creating steam.
3 Cool seawater in condensing coil causes steam to condense.
4 Salt-free water flows out of chamber.
5 Brine (very salty water) flows out of chamber.

(a) Distillation

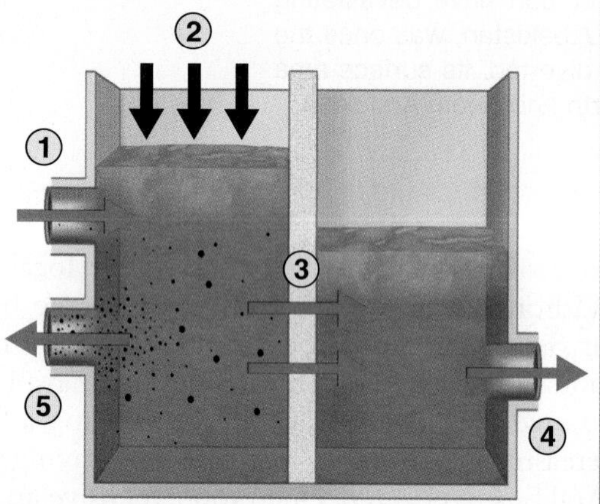
1 Seawater flows into chamber.
2 Pressure is applied to the water.
3 Under pressure, water is pushed through a semipermeable membrane but salt is not.
4 Salt-free water flows out of chamber.
5 Brine (very salty water) flows out of chamber.

(b) Reverse osmosis

FIGURE 27.7 Desalination technologies. Salt water can be converted into fresh water in one of two ways. (a) Distillation uses heat to convert pure water into steam that is later condensed, leaving the salt behind. (b) Reverse osmosis uses pressure to force pure water through a semipermeable membrane, leaving the salt behind.

The two most common desalination technologies are *distillation* and *reverse osmosis*. **Distillation** (Figure 27.7a) is a process of desalination in which water is boiled and the resulting steam is captured and condensed to yield pure water. As the water is converted into steam, salt is left behind. A great deal of energy is required to boil the water and then condense it, so distillation can be a monetarily and environmentally expensive process.

Reverse osmosis (Figure 27.7b) is a process of desalination in which water is forced through a thin semipermeable membrane at high pressure. Water can pass through the membrane, but salt cannot. Reverse osmosis is a newer technology that is more efficient and often less costly than distillation. However, the liquid that remains, called brine, has a very high salt concentration. It cannot be deposited on land because its high salt content would contaminate the soil, potentially harming plant and animal life. If dumped into a bay or coastal area, it can harm fish and other aquatic life. Typically, brine is returned to the open ocean, though its high salt concentrations can still cause harm to ocean life in areas where it is dumped.

Ultimately, all water management systems require a large investment to build, maintain, and repair. This means that many water-poor countries are not able to implement large-scale methods of water management to provide for their needs. Water availability varies greatly around the world, as **FIGURE 27.8** shows. Middle Eastern and North African countries represent about 5 percent of the world's population, yet they have less than 1 percent of the fresh water available for drinking. The United Nations estimates that nearly 1.2 billion people live in regions that have a scarcity of water.

Distillation A process of desalination in which water is boiled and the resulting steam is captured and condensed to yield pure water.

Reverse osmosis A process of desalination in which water is forced through a thin semipermeable membrane at high pressure.

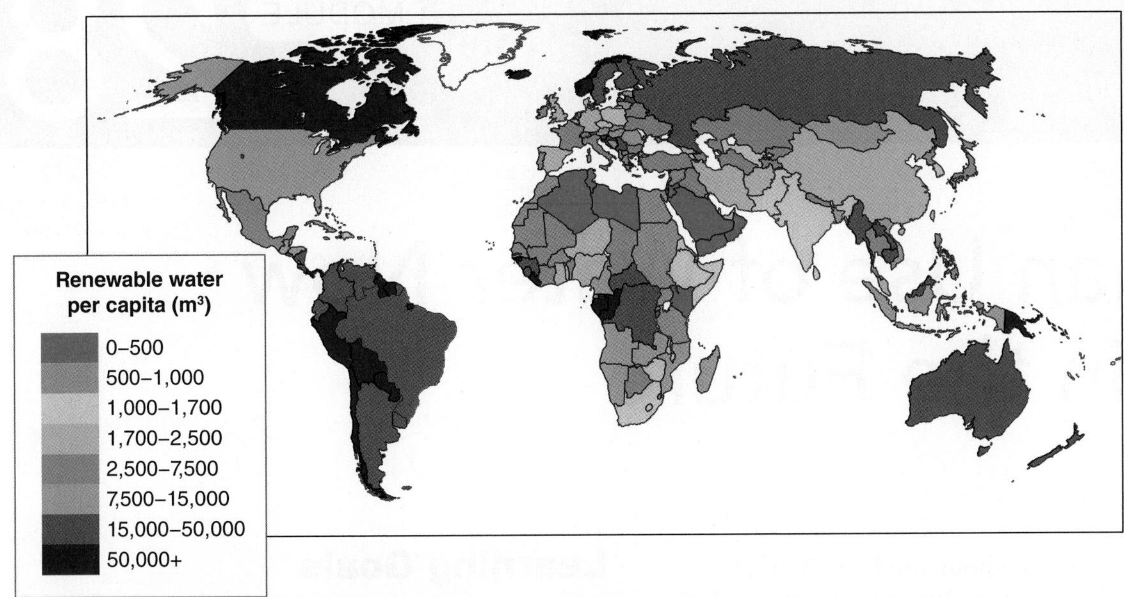

FIGURE 27.8 Water availability per capita. The amount of water available per person varies tremendously around the world. North Africa and the Middle East are the regions with the lowest amounts of available fresh water. (*Data from: United Nations World Water Development Report 2015 http://www.unesco.org/fileadmin/MULTIMEDIA/HQ/SC/images/WWDR2015_03.pdf*)

MODULE 27

AP® Review

Preparing for the AP® Exam

In this module, we have seen that humans have developed numerous methods of altering the distribution of water including the use of levees, dikes, dams, and aqueducts. We also learned that desalination technologies can be used to convert undrinkable salt water into drinkable fresh water. In the next module, we will examine how humans consume all of this water for agriculture, industry, and home use.

AP® Practice Questions

Choose the best answer for the following.

1. Which is NOT true of levees?
   (a) They limit the benefits of floodplains.
   (b) They can lead to increased flooding downstream.
   (c) They discourage the use of floodplains.
   (d) They increase sediment deposition in the ocean.

2. Which is NOT a primary use of dams?
   (a) recreation
   (b) electricity generation
   (c) flood control
   (d) habitat restoration

3. Which cannot reduce the effects of flooding?
   (a) levees
   (b) dams
   (c) freshwater wetlands
   (d) impermeable surfaces

4. Water diversion has significantly reduced the size of
   (a) Lake Erie.
   (b) the Aral Sea.
   (c) Lake Baikal.
   (d) the Baltic Sea.

5. Which is true of desalination?
   (a) The technique of distillation requires less energy than reverse osmosis.
   (b) Desalination is primarily used in North Africa.
   (c) The technique of reverse osmosis leaves brine.
   (d) Brine can be dumped onto land without affecting plant and animal life.

MODULE 28

Human Use of Water Now and in the Future

A human can survive without food for 3 weeks or more, but cannot survive without water for more than a few days. Water is also essential for agriculture and industry. In fact, 70 percent of the world's freshwater consumption is used for agriculture. The remaining 30 percent is split between industrial and household uses, the proportion of which varies from country to country. On average, experts estimate that about 20 percent of the world's freshwater use is for industry and about 10 percent is for household use. In this module, we will examine the major uses of water by humans in the areas of agriculture, industry, and households. We will then investigate how water ownership and conservation will determine the availability of water in the future.

Learning Goals

After reading this module you should be able to

- compare and contrast the four methods of agricultural irrigation.
- describe the major industrial and household uses of water.
- discuss how water ownership and water conservation are important in determining future water availability.

Water is used for agriculture

As we have seen with so many other resources, the per capita daily use of water varies dramatically among the nations of the world. **FIGURE 28.1** shows total daily per capita use of fresh water for a number of countries, which is known as the **water footprint** of a nation. This water use reflects the total water use by a country for agriculture, industry, and residences divided by the population of that country. This allows us to compare water use among nations. For example, a person living in the United States, Spain, or Canada uses about three times more water than a person living in Kenya or China. "Do the Math: Calculating Per Capita Water Use" will help you understand how we obtain these estimates.

As we have just noted, the largest use of water worldwide is for agriculture. During the last 50 years, as agricultural output has grown along with the human population, the amount of water used for irrigation throughout the world has more than doubled. Indeed, producing a metric ton of grain (1,000 kg, or 2,200 pounds) requires more than 1 million liters of water (264,000 gallons). Together, India, China, the United States, and Pakistan account for more than half the irrigated land in the world. In the United States, approximately one-third of all freshwater use is for irrigation. Raising livestock for meat also requires vast quantities of water. For example, producing 1 kg (2.2 pounds) of beef in the United States requires about 11 times more water than producing 1 kg of wheat. In this section, we will consider some of the technological advances for irrigating crops that make efficient use of water.

Water footprint The total daily per capita use of fresh water.

Irrigation

Since agriculture is the greatest consumer of fresh water throughout the world, agriculture poses a large potential for conserving water. One way to conserve water in agriculture is by changing irrigation practices. There are four major techniques for irrigating crops:

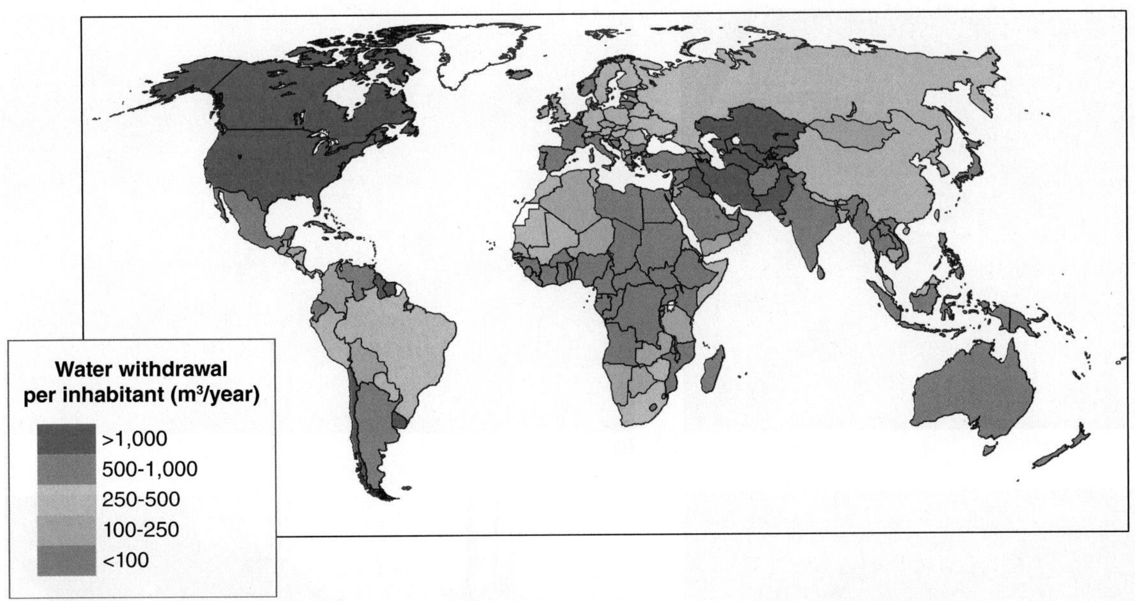

FIGURE 28.1 Total per capita water use per day. The total water use per person for agriculture, industry, and households varies tremendously by country. *(Data from Food and Agriculture Association of the United Nation http://www.fao.org/nr/water/aquastat/maps/WithT.Cap_eng.pdf)*

furrow irrigation, flood irrigation, spray irrigation, and drip irrigation (**FIGURE 28.2** on page 322).

The oldest technique is furrow irrigation (Figure 28.2a), which is easy and inexpensive. The farmer digs trenches, or furrows, along the crop rows and fills them with water, which seeps into the ground and provides moisture to plant roots. Furrow irrigation is about 65 percent efficient; 65 percent of the water is accessible by the plants and the other 35 percent either runs off the field or evaporates.

Flood irrigation (Figure 28.2b) involves flooding an entire field with water and letting the water soak in evenly. This technique is generally more disruptive to plant growth than furrow irrigation, but is also slightly more efficient, ranging from 70 to 80 percent efficiency. In spray irrigation (Figure 28.2c), water is pumped from

DO THE MATH Calculating Per Capita Water Use

Preparing for the AP® Exam

The water footprint of a country is determined by knowing the total water use of the country and the population of the country. When we divide the total water use by the population size, we determine how much water is used by the average person in the country. As we have seen, patterns of per capita water use among countries can be very different from patterns in total water use among countries. This is due to differences among countries in water availability and the need for water in agriculture and industry.

Let's calculate the per capita water use for the United States.

$$\text{Total annual water use in the United States} = 5.14 \times 10^{14}\ \text{L}$$

$$\text{Population of the United States} = 3.25 \times 10^{8}\ \text{people}$$

$$\begin{aligned}\text{Per capita annual water use in the United States} &= (5.14 \times 10^{14})\ \text{L} \div (3.25 \times 10^{8})\ \text{people} \\ &= (5.14 \div 3.25) \times 10^{(14-8)}\ \text{L per person} \\ &= 1.58 \times 10^{6}\ \text{L per person}\end{aligned}$$

YOUR TURN The total daily water use in Canada is 3.59×10^{13} L and the Canadian population is 35 million people. What is the per capita daily water use for Canada?

FIGURE 28.2 Irrigation techniques. Several techniques are used for irrigating agricultural crops, each with its own set of costs and benefits. (a) Furrow irrigation is easy and inexpensive, but has a low water use efficiency; only 65 percent of the water is accessible by the plants. (b) Flood irrigation is easy and inexpensive, with 70 to 80 percent water use efficiency (c) Spray irrigation is more expensive, but water use efficiency is 75 to 95 percent. (d) Drip irrigation is useful for perennial crops because the hoses do not have to be moved for plowing, and it has a water use efficiency of over 95 percent. *(a: Jenny E. Ross/Getty Images, Du Huaju, b: Jeff Vanuga/NRCS/USDA, c: Australian Scenics/Getty Images, d: Lynn Betts/NRCS/USDA)*

a well into an apparatus that contains a series of spray nozzles that spray water across the field, like giant lawn sprinklers. The advantage of spray irrigation is that it is 75 to 95 percent efficient, but it is also more expensive than furrow or flood irrigation and uses a fair amount of energy.

The most efficient method of irrigation, drip irrigation (Figure 28.2d), uses a slowly dripping hose that is either laid on the ground or buried beneath the soil. Drip irrigation using buried hoses is over 95 percent efficient. It has the added benefit of reducing weed growth because the surface soil remains dry, which discourages weed germination. Drip irrigation systems are particularly useful in fields containing perennial crops such as orchard trees, where the hoses do not have to be moved each year in order to plow the field.

Efficient irrigation technology benefits the environment by reducing both water consumption and the amount of energy needed to deliver the water. Many new technologies are being developed to more carefully control when plants are irrigated. As with all human activities, the costs and benefits of each irrigation technique need to be weighed to determine the best solution for each situation.

Hydroponic agriculture The cultivation of plants in greenhouse conditions by immersing roots in a nutrient-rich solution.

Hydroponic Agriculture

For some crops, *hydroponic agriculture* is an alternative to traditional irrigation. **Hydroponic agriculture** is the cultivation of crop plants under greenhouse conditions with their roots immersed in a nutrient-rich solution, but

FIGURE 28.3 **Hydroponic agriculture.** These greenhouse strawberries are growing with their roots immersed in a solution of water and nutrients. By recycling the water, hydroponic operations can use up to 95 percent less water than traditional farms. *(Mawardibahar/Getty Images)*

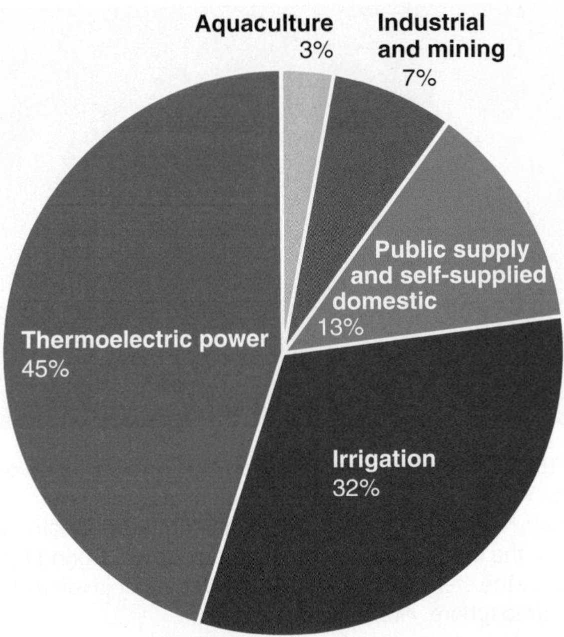

FIGURE 28.4 **Water use in the United States.** Nearly three-quarters of water used in the United States is used for agriculture and thermoelectric power. The remainder is for households, industry, and aquaculture. *(Data from Maupin et al. 2010. Estimated Use of Water in the United States. U.S. Geological Survey Circular 1405.)*

with no soil (**FIGURE 28.3**). Water not taken up by the plants can be reused, so this method uses up to 95 percent less water than traditional irrigation techniques. Hydroponic operations can produce more crops per hectare than traditional farms. They can also grow crops under ideal conditions, grow the crops during every season of the year, and often grow the crops with little or no use of pesticides. Hydroponic agriculture is growing in popularity, especially for vegetables; hydroponic tomatoes, for example, have won awards for their flavor. Although the cost of hydroponic agriculture is higher, consumers are willing to pay more for hydroponically grown vegetables and a number of businesses have been successful. One company in Georgia now grows hydroponic vegetables in a huge facility that covers 129 ha (318 acres).

Water is also used for industrial processes and households

While most water is used for agriculture, as shown in **FIGURE 28.4**, the remainder is used for industrial purposes and households.

Industrial Water Use

Water is required for many industrial processes, such as generating electricity, cooling machinery, and refining metals and paper. In the United States, approximately one-half of all water used goes toward generating electricity. It is important to distinguish between water that is withdrawn from and then returned to its source, and water that is withdrawn and consumed. For example, water that passes through a turbine at a hydroelectric dam is withdrawn from a reservoir to generate electricity, but that water is not consumed. Rather, it passes from the reservoir through the turbine and back to the river flowing away from the dam. However, the creation of the reservoir still has harmful effects on the environment, as we discussed earlier in this chapter.

Some processes that generate electricity do consume water. This means that some of the water used is not returned to the source from which it was removed, but instead enters the atmosphere as water vapor. Thermoelectric power plants, including the many plants that generate heat using coal or nuclear reactors, are large consumers of water. These plants use heat to convert water into steam that is used to turn turbines. The steam needs to be cooled and condensed before it can be returned to its source. In many plants, this cooling is accomplished by using massive cooling towers. If you have ever seen a nuclear power plant, even from a distance, you may have seen the large plumes of water vapor rising up for thousands of meters from the cooling towers. The towers allow much of the steam from the plant to condense and cool into liquid water, but a large fraction of it is lost to the atmosphere. This water vapor represents the water that is consumed by nuclear reactors (**FIGURE 28.5**).

Industrial processes such as refining metals and making paper also require large amounts of water. Copper, used extensively for electrical wiring, requires 440 L (116 gallons) of water per kilogram of refined copper.

FIGURE 28.5 Water consumption in nuclear power plants. When nuclear reactors, such as this one in Germany, heat water to make the steam that turns electrical turbines, the steam must be cooled back down. During the cooling of the steam, a great deal of water vapor is released to the atmosphere. *(RelaxFoto.de/Getty Images)*

Aluminum, used in products as diverse as automobiles and aluminum foil for cooking, requires 410 L (108 gallons) per kilogram of refined aluminum. Steel, used to manufacture home appliances, cars, buildings, and other products, requires 260 L (68 gallons) per kilogram. When we use paper, we indirectly consume water. A kilogram of paper requires 125 L (33 gallons) of water to manufacture. As we will see later in this chapter, there are opportunities to reduce the use of water during these processes.

Household Water Use

According to the U.S. Geological Survey, household use accounts for approximately 10 percent of all water used in the United States. The quantity of water used in households depends on the types of infrastructure available. Households in less developed countries generally do not have the appliances and bathroom fixtures that are common in more developed countries. As a result of such disparities, per capita household water use varies dramatically among nations. For example, on average, an individual in the United States uses 595 L (157 gallons) per day, whereas an average individual in Kenya uses only 41 L (11 gallons) per day. **FIGURE 28.6** shows per capita daily household water use for the United States and 10 other countries.

AP® Exam Tip

Review the definition of "per capita" from Chapter 1 and be sure you understand what this term means when reading graphs and working math problems.

Indoor use of water is quite similar across the United States since households across the country are typically equipped with bathrooms, washing machines, and cooking appliances. **FIGURE 28.7** shows the fraction of indoor water use that goes to each of these functions. Based on data from the Water Research Foundation in 2016, of all household water that is used indoors, 24 percent is used for flushing toilets, 23 percent for bathing, and 16 percent for laundry.

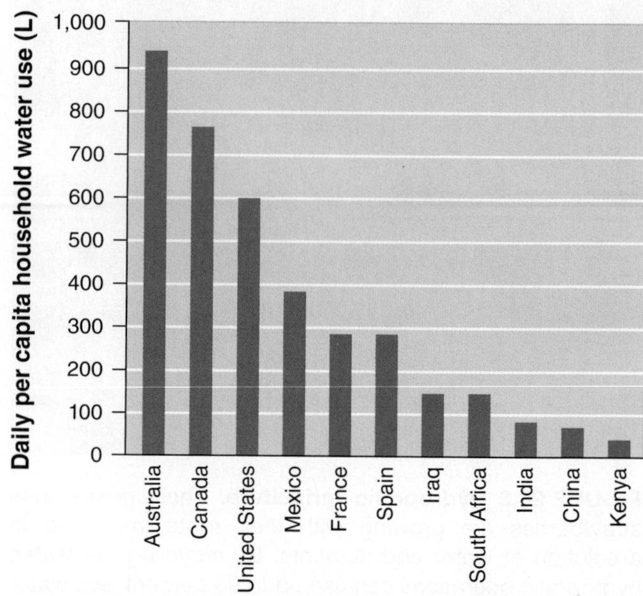

FIGURE 28.6 Comparing household per capita water use per day among several nations. The amount of household water use per capita differs a great deal between developed and developing nations. *(Data from: Chapagain, A.K., and Hoekstra, A.Y. 2004.* Water Footprints of Nations, *Vol. 1, Main Report, UNESCO-IHE. Research Report Series #16.)*

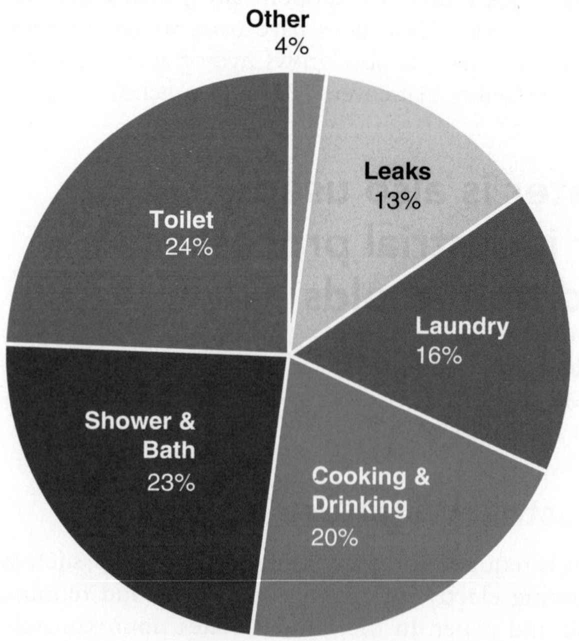

FIGURE 28.7 Indoor household water use. Most water used indoors is used in the bathroom. *(Data from: Water Research Foundation. 2016.* Residential End Uses of Water, *Version 2.)*

Outdoor water use—for watering lawns, washing cars, and filling swimming pools—varies tremendously across the United States by region. In California, for example, the typical household uses about six times more water outdoors than does the typical Pennsylvania family.

Although drinking water represents a relatively small percentage of household water use, it is particularly important. If you live in a developed country, you may not have given much thought to the availability and safety of drinking water. However, more than 1 billion people—nearly 15 percent of the world's population—lack access to clean drinking water. Every year, 1.8 million people die of diarrheal diseases related to contaminated water, and 90 percent of those people are children under 5 years of age. This means that 5,000 people in the world die each day in large part because they do not have access to clean water. As we will see in Chapter 14, developed countries typically have modern sanitation systems, stronger environmental laws, and the technologies to remove harmful contaminants and waterborne pathogens from public water supplies. Poor conditions in many developing countries do not provide safeguards that ensure the availability of clean drinking water.

The future availability of water depends on water ownership and water conservation

The future of water availability will depend on many things, including how we resolve issues of water ownership, how we improve water conservation, and—as world population grows—how we develop new water-saving technologies.

Water Ownership

Water is an essential resource, but who actually owns it? This is a rather complex question. In a particular area, such as the Klamath River region described at the beginning of the chapter, it is clear that multiple interest groups can claim a right to use and consume the water. However, having a right to use the water is not the same as owning it. For example, regional and national governments often set priorities for water distribution, but of course they have no control over whether or not a particular year will bring an abundance of rain and snow. In California, the state government promises specific amounts of water for cities, suburbs, farmers, and fish, but these promises can exceed the actual amount of water that is available in many years.

Throughout the world, the issues of water rights and ownership have created many conflicts. Earlier in this chapter we discussed India's plan to divert water from rivers that flow from the Himalayas in India into Bangladesh (**FIGURE 28.8**). Since the water originates in India, does India own the water? Does Bangladesh have any legitimate claim to some of this water that its people have relied on for millennia? Currently the two countries do not have any water agreements, but they are discussing the possibility for water agreements in the future. In the water-poor Middle East, water rights have long contributed to political tensions. In both the 1967 Arab-Israeli War and the 1980s Iran-Iraq War, disputes over water use added to the conflict. Water experts predict that as populations in these arid regions continue to grow, conflicts over water will increase.

One solution that has been proposed by economists is to allow all interested parties to openly compete for water and let market forces determine its price. In this way, they argue, water could be owned, but the true value of water would be realized and paid for. In 1981, a free-market system of water distribution was initiated in Chile. While water is public property, Chile allows individuals and corporations to buy and sell the water. The idea behind this approach is that having to buy water will encourage more efficient use. While market forces can be useful in determining the appropriate distribution of water among competing needs, government oversight helps to ensure that the needs of the people and the environment are balanced with the needs of agriculture and industry. In the case of Chile, government regulation of water distribution was weak, so allowing market forces to determine who gets the water has allowed large companies to buy and hoard water

FIGURE 28.8 A river in the Himalayas. The government of India is proposing to divert water from some of the rivers flowing out of the Himalayas to provide more water for its citizens. However, rivers, such as this one flowing down from Mera Peak in Nepal, subsequently flow into Bangladesh, raising critical questions regarding who owns the water and how it should be allocated between the two countries. *(robertharding/Alamy)*

while many indigenous groups have found themselves without sufficient water. In the American West, water rights are tied to historic agreements that are often more than 150 years old, when people in the region were scarce and the primary needs were for agriculture. Selling water across state lines is commonly prohibited and if western farmers and ranchers don't use all of their allocated water, their future allocations can be reduced. As a result of these agreements, it has been difficult to have a free-market system of water distribution that would favor more efficient water use. As we will see in Science Applied 4: "Is There a Way to Resolve the California Water Wars?" competing interests are now attempting to implement a more balanced approach.

Another approach to water allocation is the use of a tiered water-pricing system. **Tiered water-pricing systems** charge rates that increase with the amount of water consumed. There is a baseline amount of water that has the lowest cost to ensure that everyone has enough water to meet their basic needs, regardless of household income. Water use above this level is then charged at a higher rate to discourage wasting water. The income generated from those using more water helps offset the reduced cost of providing the baseline amount of water to all users.

Water Conservation and Technologies

Ultimately, there is a finite amount of water that we all must share. In some regions, such as much of the northeastern United States, water is abundant. In regions where water is scarce, such as the southwestern United States, water conservation becomes more of a necessity.

In recent years, many countries have begun to find ways to use water more efficiently through technological improvements in water fixtures, faucets, and washing machines. In 1994, new federal standards were issued for toilets and showerheads in the United States. For example, a toilet manufactured before 1994 typically uses 27 L (7 gallons) per flush, but toilets manufactured after January 1994 must use 6 L (1.6 gallons) or less per flush, representing a 78 percent reduction in water use. Australia and some countries in Europe and Asia have moved to dual-flush toilets. First invented by an Australian company in 1980, this type of toilet allows the user to push one button for a normal 6 L flush to remove solid waste, but another button produces a much more efficient 3 L (0.8 gallon) flush to remove liquid waste. These dual-flush toilets are increasingly used in the United States. Consumers also have embraced improved efficiencies in showerheads and washing machines. For example, a 10-minute shower with an older showerhead might use 150 L (40 gallons) of water, but revised federal standards for new reduced-flow showerheads call for a 10-minute shower to use no more than 95 L (25 gallons)—a 37 percent reduction in water use. Washing machines are not subject to the new federal standards, but newer, more efficient front-loading machines are now available to consumers, although these more efficient models cost nearly twice as much to purchase. "Do the Math: Selecting the Best Washing Machine" looks at the costs and benefits of these newer, more energy-efficient washing machines.

Tiered water-pricing systems A water allocation system that charges rates that increase with the amount of water consumed.

Xeriscaping A style of landscaping that removes water-intensive vegetation from lawns and replaces it with more water-efficient native landscaping.

FIGURE 28.9 Landscaping in the desert. By using plants that are adapted to a desert environment, homeowners in Arizona can greatly reduce the need for irrigation, resulting in a considerable reduction in water use compared with that required for growing grass. *(Bruce C. Murray/Shutterstock)*

In countries where a percentage of the population can be considered wealthy, household water uses can include watering lawns and filling swimming pools, both of which require large amounts of water. In some regions of the United States, homeowners have been encouraged, or even required, to plant vegetation that is appropriate to the local habitat. For example, the city of Las Vegas, Nevada, paid homeowners to use **xeriscaping**, which is a style of landscaping that removes water-intensive vegetation from lawns and replaces it with more water-efficient native landscaping. Changing the vegetation in this manner can result in a savings of 2,000 L of water per square meter (520 gallons per 10 square feet) of lawn per year (**FIGURE 28.9**).

One of the best ways to reduce industrial water consumption is by producing more efficient manufacturing equipment. In the United States, businesses and factories have achieved more sustainable water use in the last 15 years, mainly through the introduction of equipment that either uses less water or reuses water. For example, industries that need water for cooling machinery have switched from once-through

DO THE MATH Selecting the Best Washing Machine

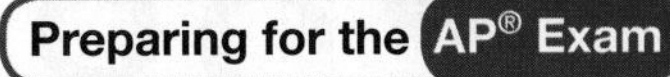

Suppose you move into a new house and need to purchase a washing machine. You've heard that the new front-loading machines use less water, but that they are more expensive. A traditional washing machine costs $300, whereas the water-efficient washing machine costs $900. So how do you choose?

If the average family washes 300 loads of laundry per year, and traditional washing machines use 100 L more water per load than water-efficient machines, how many liters would an efficient washing machine save per year?

$$300 \text{ loads/year} \times 100 \text{ L/load} = 30{,}000 \text{ L/year}$$

If the average cost of water in the United States is $0.50 for every 1,000 L, how much money would you save annually by using the more efficient washing machine?

$$30{,}000 \text{ L/year} \times \$0.50/1{,}000 \text{ L} = \$15/\text{year}$$

Given your calculated annual savings, how many years would it take for your water savings to pay for the more expensive, high-efficiency washer?

$$(\$900 - \$300) \div \$15/\text{year} = 40 \text{ years}$$

YOUR TURN If the average cost of water in the United States doubled to $1.00 for every 1,000 L, how much money would you save annually by using the more efficient washing machine and how many years would it take to pay for the more expensive, high-efficiency washer?

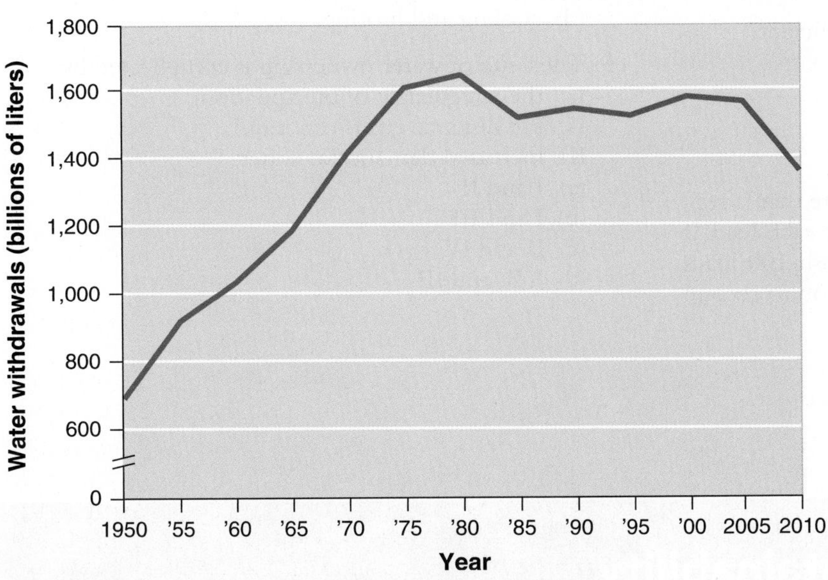

FIGURE 28.10 Water withdrawals in the United States from 1950 to 2010. Although the U.S. population continues to increase, the country's water use has leveled off due to the use of water-efficient technologies. *(Data from: UGS https://water.usgs.gov/watuse/wutrends.html)*

water systems—systems that bring in water for cooling and then pump the heated water back into the environment—to recirculating water systems.

Some simple ways to conserve water can be used throughout the world. For example, the impervious surfaces of buildings represent a potential water-collecting surface. A gutter system can collect rainwater and channel it into rain barrels or, for greater capacity, a large underground water tank. As we will see in Chapter 14, some countries are now using wastewater for irrigation after sending it through a sewage treatment process. There is also great interest in developing monitoring technologies that will detect leaks in the pipes that distribute water over large areas.

The world's growing population and the associated expansion of irrigated agriculture have increased global water withdrawals more than fivefold in the last 100 years. Global water use is expected to continue to grow along with the human population through the early part of this century. However, as **FIGURE 28.10** shows, despite the growing population in the United States, water withdrawals have leveled off since they peaked in 1980. This is largely a result of greater efficiency in the use of water for agricultural irrigation, electricity generation, and household appliances. Reductions in water use are projected to continue until at least 2020. Of course, the continued development of more efficient technologies would allow additional reductions in water use.

MODULE 28

AP® Review

Preparing for the AP® Exam

In this module, we learned that about 70 percent of water that is consumed goes to agriculture, 20 percent goes to industry, and 10 percent goes to households. Much of the agricultural use of water is for irrigation, which includes low-efficiency furrow and flood irrigation, moderate-efficiency spray irrigation, and high-efficiency drip irrigation. An alternative method of growing crops with highly efficient water use is hydroponic agriculture. Nearly half of the water used in industry is for generating electricity. Some of this water is returned to its source and some is released to the atmosphere as water vapor. Industries that manufacture paper and refine metals also use large amounts of water. Among households, those in developed countries often have more water-consuming appliances, swimming pools, and irrigated lawns, and therefore use more water per person than households in developing countries. The future availability of water depends on countries' reaching agreements on water ownership and taking steps to reduce the amount of water used in agriculture, industry, and households.

AP® Practice Questions

Choose the best answer for the following.

1. In the United States, most water is used for
   (a) agriculture.
   (b) households.
   (c) industry.
   (d) hydroelectric plants.

2. Which irrigation method is the most efficient?
   (a) furrow irrigation
   (b) spray irrigation
   (c) flood irrigation
   (d) drip irrigation

3. A water-efficient washer costs $400 more than a regular washer and saves 100 L of water for each load. If water costs $5 per 1,000 L and you wash 100 loads each year, how many years will it take you to recoup the extra cost of the machine?
   (a) 10
   (b) 5
   (c) 4
   (d) 8

4. The primary use of industrial water in the United States is
   (a) mining.
   (b) paper products.
   (c) electricity generation.
   (d) cooling and heating.

5. The issue of water ownership is complicated by
   I. the irregularity of precipitation.
   II. the distance rivers can travel.
   III. increased demand for water.
   (a) I and II
   (b) I and III
   (c) II and III
   (d) I, II, and III

Working Toward Sustainability

From Laundry to Landscape

In certain parts of the world, such as the United States, sanitation regulations impose such high standards on household wastewater that we classify relatively clean water from bathtubs and washing machines as contaminated. This water must then be treated as sewage. On the other hand, we use clean, drinkable water to flush our toilets and water our lawns. Can we combine these two observations to come up with a way to save water? One idea that is gaining popularity is to reuse some of the relatively clean water from bathtubs and washing machines that we normally discard as sewage.

This idea has led creative homeowners and plumbers to distinguish between two categories of wastewater in the home: *gray water* and *contaminated water.* **Gray water** is defined as the wastewater from baths, showers, bathroom sinks, and washing machines. Although no one would want to drink it, gray water is perfectly suitable for watering lawns, washing cars, and flushing toilets. In contrast, **contaminated water** is defined as the wastewater from toilets, kitchen sinks, and dishwashers; such water contains a good deal of waste and contaminants and should therefore be disposed of as sewage.

Many cities in Australia have considered the use of gray water as a way to reduce both the consumption of fresh water and the volume of contaminated water that is sent to sewage treatment plants. The city of Sydney estimates that 70 percent of the water used in the metropolitan area is used in households, and that perhaps 60 percent of that household water becomes gray water. The Sydney Water Corporation estimates that the use of gray water for outdoor purposes could save up to 50,000 L (13,000 gallons) per household per year.

Unfortunately, many local and state regulations in the United States and around the world do not allow households to use gray water, mostly due to concerns about human exposure to bacteria that could be found in the gray water as a result of its initial use for bathing and washing hands and dishes in sinks. Arizona, a state in the arid Southwest, has some of the least restrictive regulations. In 2010, in the face of a severe water shortage, California reversed earlier restrictions on gray water use and agreed to allow gray water to be used outdoors, but there are a few restrictions. For example, gray water cannot be sprayed over a lawn, but it can be delivered to lawns through underground drip irrigation systems to avoid potential bacterial contamination. It can be used to water crops, but not root crops such as potatoes and carrots since the edible part of the plant will have come into direct contact with the bacteria-laden water. Since California has allowed the use of gray water, there has been a surge of interest in designing gray water systems. Ranging from $1,000 to $30,000 per home, these systems perform a variety of functions, including filtering systems to remove any harmful bacteria. Given that the typical household in the United States produces 227,000 L (60,000 gallons) of gray water annually, using gray water for irrigation presents a major opportunity for water conservation.

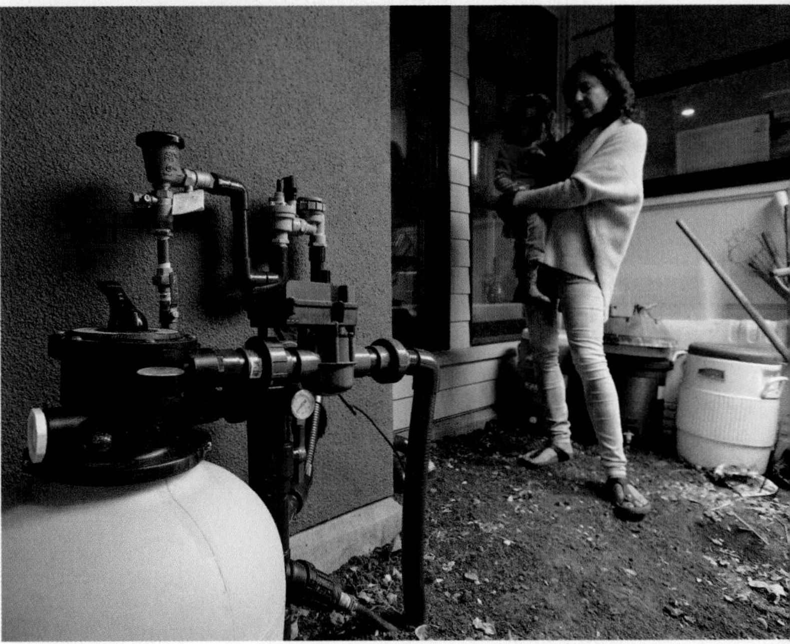

Reusing gray water. In states like California, residents are being allowed to reuse their gray water to irrigate the landscaping of their homes. This home gray water filtering system is located in Berkeley, California. *(Eric Risberg/AP Images)*

Critical Thinking Questions

1. Given that gray water can contain more bacteria than clean water, how should we balance a desire to irrigate with clean water against the need to conserve limited water supplies?
2. Describe how you could formulate an argument to convince a town council to permit the use of gray water for watering lawns. Use the two references below to support your points.

References

California drought spurring "grey water" recycling at home. Associated Press, June 5, 2015, http://www.dailynews.com/2015/06/05/california-drought-spurring-grey-water-recycling-at-home/

California Gray Water Guide. http://www.thegreywaterguide.com/california.html

Grey Water Action. http://greywateraction.org/content/about-greywater-reuse

Gray water Wastewater from baths, showers, bathroom sinks, and washing machines.

Contaminated water Wastewater from toilets, kitchen sinks, and dishwashers.

Chapter 9 Review

In this chapter, we have learned that fresh water is a critical, scarce resource. The major sources of fresh water are groundwater, surface water, and atmospheric water. When groundwater is extracted faster than it is replaced, we can experience dry water wells and saltwater intrusion. Humans alter the availability of water by constructing levees, dikes, and dams to constrain water movement. We transport water from one place to another using aqueducts and we are able to convert salt water into fresh water through distillation or reverse osmosis. The use of water among countries can be determined by calculating the water footprint of a country; developed countries typically have a larger water footprint than developing countries. Most water is used for agriculture, followed by industrial and household use. Recent changes in each of these sectors have improved water use and reuse efficiency and conservation.

Key Terms

Aquifer
Unconfined aquifer
Confined aquifer
Water table
Groundwater recharge
Spring
Artesian well
Cone of depression
Saltwater intrusion
Floodplain
Impermeable surface
Levee
Dike
Dam
Reservoir
Fish ladder
Aqueduct
Desalination
Desalinization
Distillation
Reverse osmosis
Water footprint
Hydroponic agriculture
Tiered water-pricing system
Xeriscaping
Gray water
Contaminated water

Learning Goals Revisited

Module 26 The Availability of water

Describe the major sources of groundwater.

Groundwater can exist as either confined or unconfined aquifers. Unconfined aquifers can be rapidly recharged, but they are susceptible to contamination from chemicals released by human activities. Confined aquifers are typically recharged slowly and are less likely to be contaminated by anthropogenic chemicals. Large withdrawals of water from an aquifer can cause cones of depression that result in wells going dry. In coastal areas, large withdrawals for fresh water can cause saltwater intrusion.

Identify some of the largest sources of fresh surface water.

Sources of fresh surface water include streams, rivers, wetlands, and lakes. Streams and rivers have traditionally overflowed their banks and deposited nutrient-rich sediments along floodplains. Rivers have also been an important resource for transportation. Lakes are formed through a variety of processes including tectonic activity and glaciation.

Explain the effects of unusually high and low amounts of precipitation.

The abundance of precipitation varies with climate around the world and this affects the availability of water for ecosystems and humans. Periods of drought can interact with human activities to alter the cycling of nutrients, cause the topsoil to be blown away, and cause the soil surface to become impermeable to future precipitation.

Module 27 Human Alteration of Water Availability

Compare and contrast the roles of levees and dikes.

Levees and dikes both use barriers to prevent flood water, but levees are built along rivers to prevent rising rivers from spilling over onto the floodplain whereas dikes are built near the ocean to prevent ocean waters from flooding the adjacent land.

Explain the benefits and costs of building dams.

Dams can provide multiple benefits including electric generation, flood control, and recreation. However, dams require a large amount of resources to construct, they often displace people, and they have ecological effects including interference with fish migration and prevention of seasonal flooding that some species require to persist.

Explain the benefits and costs of building aqueducts.

Aqueducts are effective at moving water from places that have water to places that need water. Aqueducts built as open canals can lose a large fraction of water through leaks and evaporation. Newer aqueducts are commonly built as pipelines to reduce water loss, but they still require a large amount of resources to construct and their construction can have negative impacts on the habitat during the construction process.

Describe the processes used to convert salt water into fresh water.

The process of converting salt water into fresh water is known as desalination or desalinization. It can be accomplished either through distillation, which boils water and then condenses the salt-free water vapor, or through reverse osmosis, which pushes salt water against a membrane through which water can pass but salt cannot.

Module 28 Human Use of Water Now and in the Future

Compare and contrast the four methods of agricultural irrigation.

Furrow irrigation involves flooding the furrows between plant rows; it is the easiest and least expensive method, but it also is the least efficient. Flood irrigation involves flooding an entire field with water; it is only slightly more efficient than furrow irrigation. Spray irrigation involves distributing water through spray nozzles that resemble giant lawn sprinklers; it can be 75 to 95 percent efficient. Drip irrigation involves the use of a hose that slowly drips water on or below the ground surface; its efficiency is over 95 percent.

Describe the major industrial and household uses of water.

About half of the water used for industry is used for generating electricity. Of the remainder, major users include refiners of paper and metal. Of all indoor water use in U.S. households, we use 41 percent for flushing toilets, 33 percent for bathing, 21 percent for laundry, and 5 percent for cooking and drinking. Outdoor water use varies tremendously by region as a result of differences in climate and in the use of water for swimming pools and watering lawns.

Discuss how water ownership and water conservation are important in determining future water availability.

Water ownership is a complicated issue because multiple groups can have a right to use the water but it is often unclear who actually owns the water. Among nations, water rights are often determined through treaties. In some regions of the world, people compete for water by purchasing it in a free-market system. The demand from different users helps to determine the price while government oversight can help ensure that there is a proper balance between the needs of industry, agriculture, households, and the environment.

Practice Math and Graphing

Preparing for the AP® Exam

Answer the following questions. Be sure to show all your work.

1. Practice Math

To get a better sense of how per capita water use varies in households around the world, use the data in the table below to calculate per capita water use in seven different countries. Show your work.

Nation	Total annual water use (m^3)	Population size (millions of people)	Per capita annual water use (m^3/ person)
France	111×10^9	59 million	
Iran	102×10^9	63 million	
Brazil	233×10^9	169 million	
Germany	127×10^9	82 million	
India	987×10^9	1007 million	
China	882×10^9	1257 million	
Tanzania	372×10^8	33 million	

Data from Chapagain, A. K., and Hoekstra, A. Y. 2004. *Water Footprints of Nations*, Vol. 1, Main Report, UNESCO-IHE. Research Report Series #16.

2. Practice Graphing

(a) Using the data in the table from "Practice Math, create three bar graphs that compare the following for each country:

1. Total daily water use
2. Population size
3. Per capita water use

(b) Why are there such large differences in per capita daily water use?

Section 1: Multiple-Choice Questions

Choose the best answer for questions 1–16.

1. What percentage of Earth's water is fresh water?
   (a) 3 percent
   (b) 10 percent
   (c) 50 percent
   (d) 90 percent

2. Which contrast between confined and unconfined aquifers is correct?
   (a) Confined aquifers are more rapidly recharged.
   (b) Only confined aquifers can produce artesian wells.
   (c) Only unconfined aquifers are overlain by a layer of impermeable rock.
   (d) Only unconfined aquifers can be drilled for wells to extract water.

3. Which statement about surface waters is NOT correct?
   (a) Historically, most rivers regularly spilled over their banks.
   (b) Levees are used to make reservoirs.
   (c) Dikes are human-made structures that keep ocean water from moving inland.
   (d) Wetlands play an important role in reducing the likelihood of flooding.

4. Which statement about dams is NOT correct?
   (a) Dams are used to reduce the risk of flooding.
   (b) Dams can cause increased water temperatures.
   (c) Fish ladders allow migrating fish to move past dams.
   (d) Most dams are built to generate electricity.

5. Which statement about aqueducts is correct?
   (a) Aqueducts designed as open canals can lose a lot of water through evaporation.
   (b) Aqueducts are a modern invention.
   (c) Aqueducts do not affect the amount of water remaining in rivers.
   (d) Aqueducts move water from locations where the demand for water is high.

6. Which statement about desalination is correct?
   (a) Distillation requires more energy than reverse osmosis.
   (b) The brine left over from desalination is not harmful when returned to the ocean.
   (c) Large-scale desalination of water is affordable to all nations.
   (d) Most desalination occurs in North America.

7. Which list of household water uses is in the correct order, from highest to lowest?
   (a) toilet, laundry, cooking, drinking, bathing
   (b) toilet, bathing, laundry, cooking, drinking
   (c) bathing, toilet, laundry, cooking, drinking
   (d) bathing, toilet, cooking, drinking, laundry

8. Which statement about the industrial use of water is NOT correct?
   (a) It is used to create steam.
   (b) It is important in generating electricity.
   (c) It plays a role in making paper products.
   (d) Its use is becoming less efficient.

9. Which list of agricultural irrigation techniques is in the correct order, from least efficient to most efficient?
   (a) drip irrigation, furrow irrigation, flood irrigation, spray irrigation
   (b) spray irrigation, furrow irrigation, flood irrigation, drip irrigation
   (c) furrow irrigation, flood irrigation, spray irrigation, drip irrigation
   (d) furrow irrigation, spray irrigation, drip irrigation, flood irrigation

10. Which is NOT a water conservation technique?
   (a) flood irrigation
   (b) reduced-flow showerheads
   (c) dual-flush toilets
   (d) front-loading washing machines

11. Drought conditions
   (a) can result in increased nutrient cycling.
   (b) decrease the severity of flooding.
   (c) are rarely affected by human activities.
   (d) cause increased erosion.

12. Saltwater intrusion is likely to occur when
   (a) a cone of depression is created near a coastal ecosystem.
   (b) humans drill into a confined aquifer near the coast.
   (c) farmers irrigate their crops with an excess of saline water.
   (d) an artesian well is created near the coast.

13. The Dead Sea, which lies between Jordan and Israel, is one of the saltiest lakes on Earth. Suppose you are a geologist exploring the area. What geological factors would you expect to see that might explain this salinity?
   I. high concentrations of uranium in or around the Dead Sea
   II. lack of outlets for water to escape
   III. relatively impermeable soil at the bottom of the lake
   (a) II only
   (b) III only
   (c) I and II
   (d) II and III

14. Which is NOT an advantage of hydroponic agricultural operations compared to traditional farms and farming practices?
   (a) They can grow crops any season of the year.
   (b) They are much less expensive to run.
   (c) They can produce more crops per hectare.
   (d) They use up to 95 percent less water.

15. During an extreme drought, water use increases rapidly in a coastal city supported by a local aquifer. After several months, however, the water coming from that aquifer becomes undrinkable. Which is a likely explanation for the deterioration of the water?
    (a) contamination by wastewater
    (b) saltwater intrusion
    (c) the invasion of gray water
    (d) surface impermeability

16. A dam is constructed to prevent seasonal flooding that typically occurs in an area. What might be expected to occur along the shorelines downstream of the dam?
    (a) Species richness will decrease.
    (b) Fish numbers will increase.
    (c) Salination will increase.
    (d) Groundwater recharge will accelerate.

Section 2: Free-Response Questions

Write your answer to each part clearly. Support your answers with relevant information and examples. Where calculations are required, show your work.

1. An important source of fresh water is groundwater. Answer the following questions about human use of groundwater.
    (a) Using a cross section of the ground, draw the water table, an unconfined aquifer, and a confined aquifer. (4 points)
    (b) Explain the factors that create an artesian well. (2 points)
    (c) Describe the concerns that scientists have regarding the pumping of water from confined aquifers. (2 points)
    (d) Explain how cones of depression near the coasts of continents can lead to saltwater intrusion of water wells. (2 points)

2. Answer the following questions about the world's approximately 845,000 dams.
    (a) What are the benefits that dams provide to humans? (4 points)
    (b) What are the costs of dams to human society? (2 points)
    (c) What are the negative impacts of dams on the environment? (2 points)
    (d) If dams have both costs and benefits, how should we decide when and where a dam should be built? (2 points)

3. The situation in the Klamath River in California and Oregon exemplifies the conflict that can occur when many stakeholders rely on a water resource for various purposes.
    (a) Describe the conflict over the river in terms of:
        (i) at least THREE stakeholders. (2 points)
        (ii) at least TWO environmental problems that occurred. (2 points)
        (iii) the compromise that was reached. (2 points)
    (b) Describe the function of a levee, and THREE problems that levees can produce. (4 points)

PhilAugustavo/Getty Images

science applied 4

Is There a Way to Resolve the California Water Wars?

Pick up any tomato, broccoli spear, or carrot in the grocery store and chances are it came from California. California's $54 billion agricultural industry is more than twice the size of the agricultural industry of any other state, and it accounts for nearly one-half of all the fruits, nuts, and vegetables grown in the United States. California's temperatures are ideal for many crops, and its mild climate permits a long growing season. However, water is a scarce resource and most of California's growing regions lack the water that is necessary for farming. To remedy this situation, the state has developed an extensive irrigation system that moves water from areas where it is abundant to areas where it is scarce, as you can see in **FIGURE SA4.1**.

Although agriculture in California has been economically successful, it has increased competition among the state's residents for water. Nearly two-thirds of California's

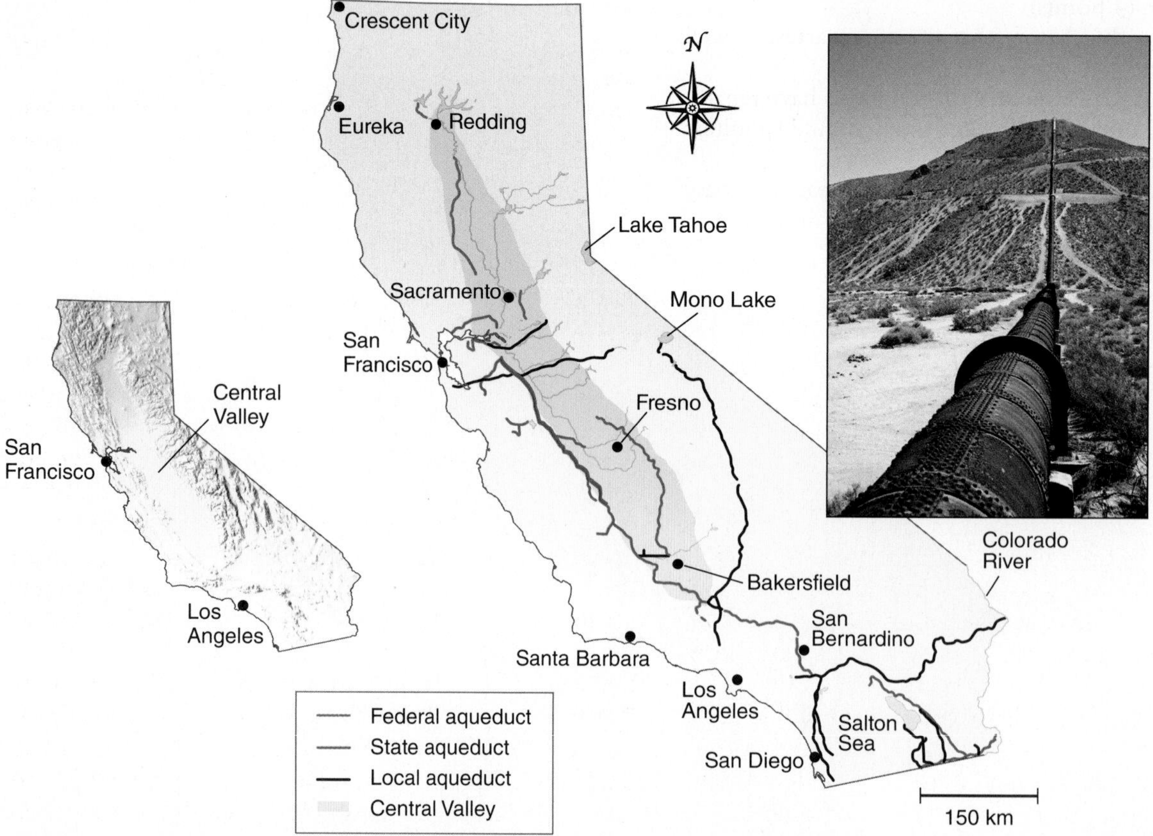

FIGURE SA4.1 Water distribution in California. To supply water to the agricultural areas in the Central Valley and to supply drinking water to municipalities, the state has developed an elaborate system of aqueducts. These aqueducts consist of both open canals and pipes laid above or below the ground. *Jim West/The Image Works*

population lives in the southern part of the state, where residential and business demand for water has long exceeded the locally available supply. Los Angeles, for example, has been importing water from other areas of the state for more than 100 years. Demand for water has continued to grow with the state's population, which has been increasing by about 1 percent per year.

Wildlife also depends on a readily available supply of fresh water. In many regions of California, diverting water away from streams and rivers threatens the habitats of fish and other aquatic organisms, including several endangered species. Salmon, for instance, require cold, clean water to lay their eggs. In the Sacramento-San Joaquin River Delta, which is where the San Joaquin and Sacramento rivers come together just east of San Francisco, so much water has been pumped out for use in the southern part of the state that the salmon in the region's rivers have experienced dramatic declines.

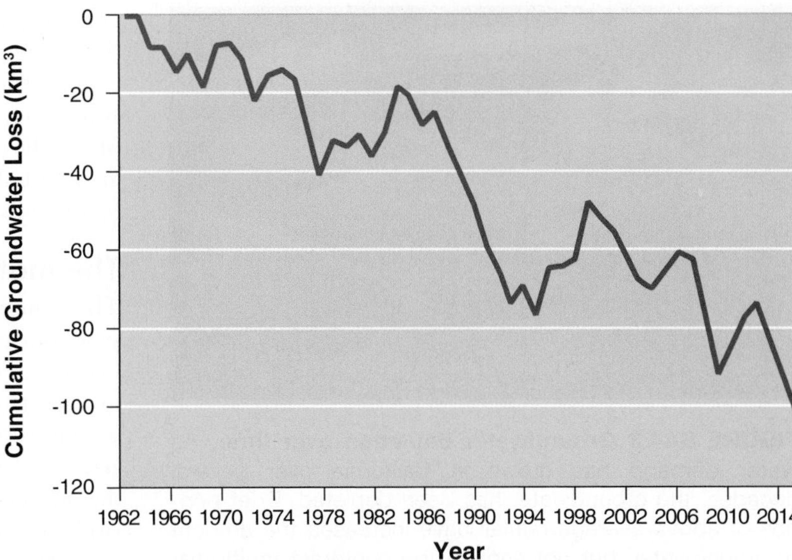

FIGURE SA4.2 Irrigating crops. Crops such as potatoes, tomatoes, and broccoli use less water and have a higher resale value than crops such as alfalfa. Growing such crops would reduce the water needed for agriculture in California. *(Data from Cooley, H., Donnelly, K., Phurisambam, R. and Subramanian, M. (2015)* Impacts of California's Ongoing Drought: Agriculture. *Oakland, California: Pacific Institute.)*

What historical factors have led to California's water wars?

At 144 trillion liters (38 trillion gallons) of fresh water per day, California's water use is more than twice that of any other state. The vast majority of that water is used in the Central Valley region, which produces more than 80 percent of California's farm income. In 1935, the U.S. Department of Reclamation began an immense endeavor, called the Central Valley Project, to divert 8.6 trillion liters (2.3 trillion gallons) of water per year from rivers and lakes throughout the state for both agricultural and municipal uses.

While the project made agriculture possible in many of southern California's desert regions, it also locked into place long-term contracts that have subsidized several decades of cheap water for farmers, who pay rates that are only 5 to 10 percent of those paid by southern California towns and cities. Interestingly, northern California also has farms, but much of its abundant water is sent to the farms and municipalities of southern California. In addition to moving water, farmers also pump a good deal of groundwater. Over the past years, this pumping of groundwater has exceeded the amount of recharged groundwater, particularly during droughts, as you can see in **FIGURE SA4.2**.

In 2005, the federal government renewed nearly 200 large water supply contracts with farmers. These contracts were for 25 years, with an option available to extend them another 25 years, at prices so low that the payments did not even cover the cost of the electricity used to transport the water. Organizations ranging from the Cato Institute to the Natural Resources Defense Council protested this decision. The controversy raised the question of whether farmers should continue to be given cheap access to water to protect the existing agricultural industry or be required to pay the same price for water as everyone else.

How might farmers respond to increased water prices?

What would happen if the federal government raised the price of water for California farmers? If water were more expensive, farmers would face four choices: raise the prices of their crops, switch to less water-intensive crops, use more efficient irrigation methods, or sell their farms. Each of these options would have different effects on the state's environment and economy.

One obvious way to offset higher water prices would be for farmers to charge more for the produce they grow. For example, if California broccoli growers had to pay the same price for water as municipalities paid, that cost would add about 13 cents to the price of broccoli per pound—an increase of less than 10 percent. This action would not change the environmental threat of water depletion, but would have a small negative effect on the demand for broccoli and on the consumer's wallet.

Other farmers faced with higher water prices might decide to switch to crops that require less water. For example, alfalfa, which uses 25 percent of California's irrigation water and has a low resale value compared with other crops, would no longer be a profitable crop. A farm using 296 million liters (78 million gallons) of water can produce about $60,000 worth of alfalfa per year, according to the Natural Resources Defense Council. If farmers had to purchase water at the unsubsidized rate, that water would cost them $168,000; if the alfalfa continued to sell for only $60,000, the farmers would lose more than $100,000 a year growing alfalfa. Tomatoes, broccoli, and potatoes use about

FIGURE SA4.3 Groundwater depletion over time. As water demand has grown in California over several decades, the groundwater has been depleted. Brief periods of above-average rainfall have increased the amount of groundwater, but not enough to counteract multi-year droughts. Data from USGS Professional Paper 1766. Satellite data courtesy of NASA and the National Center for Atmospheric Research. *(Pgiam/Getty Images)*

half as much water per acre as alfalfa and have a higher resale value (**FIGURE SA4.3**). Just by replacing alfalfa with these crops, California would reduce its statewide water use by 20 percent. This action would have a positive environmental impact, since less overall water would be needed for agriculture in the state.

Investment in new irrigation technology that uses less water would also reduce a farmer's costs. As we saw earlier in this chapter, the four major irrigation techniques differ in cost and efficiency. If farmers had to purchase water at the unsubsidized rate, converting to more-efficient irrigation techniques could rapidly pay for itself since the initial investment in equipment would be offset by savings on water. As long as the cost of water continues to be subsidized by the government, farmers have little incentive to upgrade their irrigation technology.

If water prices rise, not all farmers will be able to afford the new costs. This recognition brings us to the final option: Some farmers could be driven out of business and forced to sell their land. This outcome would probably reduce the amount of food grown in the region. Whether conversion from farming to other uses would reduce water demand would depend on how the land was used. For example, if it became a housing development in which residents practiced water conservation, then the residents would use much less water per acre than the amount used for agriculture. However, if the residents chose to plant and water large lawns and were not careful about their use of water indoors, they could use even more water than was used for agriculture.

Because charging more for water could force some farmers out of business, many politicians are reluctant to support this change. In response, some economists argue that farmers who possess long-term subsidized water contracts should be allowed to sell their water allotment to the highest bidder, something that is currently happening with some farms in California. In this scenario, a farmer can buy water from the government at the contracted low cost and then sell that water to a city at a substantial profit rather than using it on the farm. Many critics argue, however, that the government should not be giving preferential water rates to a few users who can then make a profit on the water.

The motivation provided by drought

The competition for water came to a head in recent years during a multi-year drought. From 2013 to 2016, California experienced the worst drought ever recorded for the state, and the state had no choice but respond in substantial ways. In 2015, nearly 200,00 hectares (500,000 acres) of farm fields were allowed to go fallow to reduce the amount of water used. A large portion of this fallow land were rice fields that have to be flooded with water. In addition, farmers spent nearly $300 million per year to pump water out of wells because the aqueducts were not delivering enough. Many farmers also survived by planting crops that are better suited to the local climate, and that require less water. They also added water-saving devices, such as drip irrigation, where feasible. Farmers who grew water-intensive crops such as hay and rice were hard hit. However, overall farm revenues increased during this time because the shortage of produce caused prices to rise. Those farmers who were able to continue crop production by pumping water to irrigate crops did well financially. Many scientists have expressed concern that the short-term gain from pumping groundwater might cause future problems since the groundwater may be slow to recharge.

How have the water wars played out in the political arena?

The California water wars have developed because of the large number of competing interests involved, including northern California farmers and municipalities that are losing water, southern California farmers and municipalities that need more water, endangered species that are declining due to the diversion of water, and beautiful natural areas that have experienced reduced water capacity, seen increased salinity, or gone completely dry.

Competition for California's water has been marked by political maneuvering and numerous lawsuits. In the 1960s, for example, there was a large push to build dams that would create reservoirs to collect water. Today there are about 1,200 dams in the state. Because of the money and time needed to construct new dams and the environmental impact of dams, many Californians do not consider more dams to be a viable solution to today's water problems.

Over the past 2 decades there have been lawsuits that produced very interesting court decisions. For example, the Natural Resources Defense Council won a lawsuit regarding the 2005 water contracts for farmers in the eastern San Joaquin Valley, who use water diverted from the San Joaquin River. The judge ruled that the contracts were illegal because they did not consider how diverting the water might affect endangered species that rely on the river.

In 2007, a federal court was asked to consider how the large, powerful pumps used to divert water from the Sacramento-San Joaquin River Delta affected local fish populations—particularly the delta smelt (*Hypomesus transpacificus*), a federally protected fish whose population in the delta region was declining. The court ruled that pumping from the delta region—the source of much of the state's fresh water—must be reduced. Because this decision came during a multi-year drought, its impact on southern California was even larger than it otherwise would have been. A number of California congressmen accused the state and federal governments of caring more about fish than people.

When a drought hit California in 2008, the spring of 2008 was the driest in 88 years. The California legislature passed a bill to cut water use by 20 percent. But the legislature put most of the responsibility for reducing water use on residential users, despite the fact that agriculture uses a much larger portion of the state's water. Some municipalities started rationing water, and a number of municipalities began enforcing a 2001 state law that requires new building developments to show plans that ensure a 20-year water supply.

As the drought continued into 2009, additional ideas to promote water conservation were proposed. In June 2009, for example, the Los Angeles Department of Water and Power announced that customers who replaced turfgrass with drought-resistant plants or mulch would be credited $11 per square meter.

In 2009, the California legislature passed a series of bills to address the water crisis. The $40 billion package called for the restoration of the San Joaquin–Sacramento River Delta ecosystem as well as the building of new dams and a series of water conservation efforts to reduce water demand. A few days later, the governor signed the bill into law.

In 2014, California passed the Sustainable Groundwater Management Act, which is intended to balance the amount of water pumped out of the ground with the amount of aquifer recharge. It is the first statewide legislation to regulate how groundwater is being used. However, it allows local government agencies to develop their own plans for sustainable water use in their region by 2020.

What can be done?

Given the scarcity of water in California, it stands to reason that its sustainable use is an important topic with many environmental, economic, and political implications. Economists generally argue that all users should be charged the same amount for a resource so that only those who value it most will pay to use it. Doing so would provide increased motivation for more efficient water use by farms, residences, and businesses. Increased water costs could be helpful in reducing the demand for water, but we cannot forget the simultaneous need for all users to better conserve water.

Scenarios similar to the California water wars are being played out in many other parts of the world. Competition for water exists not only among farmers and municipalities, but also among neighboring countries. Indeed, disagreements over water have turned into real wars between nations. If the competing interests in California can devise creative solutions to the state's water problems, those solutions can serve as a model for the rest of the world so that we can continue to have economic prosperity without severely harming the environment.

Questions

1. What are the challenges of balancing the need for water in agriculture with the need for water to protect endangered species?
2. How might a California farmer who benefits from water subsidies argue that the current water distribution rules are beneficial not only to farmers but also to society?

Preparing for the AP® Exam

Practice AP® Free-Response Question

Write your answer to each part clearly. Support your answers with relevant information and examples. Where calculations are required, show your work.

The Los Angeles aqueduct spans 375 km (233 miles) from the Owens Valley through the Mojave Desert to the San Fernando Valley. The aqueduct collects snowmelt from the mountains and is capable of carrying 13.7 m^3 of water per second. The water flows the entire distance by gravity. In the years after the aqueduct was completed in 1913, the area in and around Owens Valley has become increasingly dry and farmers have suffered a tremendous loss of productivity.

(a) Surveys taken from the Los Angeles Department of Water and Power claim that the water table in the Owens Valley has not changed, yet residents in the Owens Valley claim that wells are running dry. Why might this be? (3 points)

(b) Aside from moving out of the area, list three possible ways that farmers may mitigate the loss of water supply and continue farming. (3 points)

(c) Los Angeles is situated along the shore of the Pacific Ocean. Describe two methods of desalination that Los Angeles could use to obtain fresh water. List two costs associated with each method. (2 points)

(d) Considering the position of Owens Valley on the windward side of the Sierra Nevada mountain range, how might water diversion through the aqueduct affect air quality in the Owens Valley? (2 points)

References

Martin, G. 2005. The California water wars: Water flowing to farms, not fish. *San Francisco Chronicle*, October 23.

Steinhauer, J. 2008. Governor declares drought in California. *New York Times*, June 5.

Obegi, D. 2011. California water wars: It isn't fish vs. farmers. *Los Angeles Times*, August 10.

Kasler, D. 2016. Drought costs California farms $600 million, but impact eases. *Sacramento Bee*, August 15, 2016.

Weiser, M. 2016. Despite drought, California farming prospered. *News Deeply*, August 1.

Section 1: Multiple-Choice Questions

Choose the best answer for questions 1–22.

1. Which is the order of layers in Earth's structure, from innermost to outermost?
   (a) core, lithosphere, asthenosphere, mantle
   (b) core, asthenosphere, mantle, lithosphere
   (c) core, mantle, asthenosphere, lithosphere
   (d) core, mantle, lithosphere, asthenosphere

2. Which does NOT provide evidence for the theory of plate tectonics?
   (a) plumes of magma from the mantle that reach the crust
   (b) similar rock formations on both sides of the Atlantic Ocean
   (c) discovery of similar fossils in Nigeria and Brazil
   (d) mountain ranges along the northern boundaries of India

3. An earthquake occurs as a direct result of
   (a) volcanic eruption.
   (b) conversion of kinetic energy to tectonic plate movement.
   (c) magma convection in the mantle.
   (d) potential energy release along a fault.

4. The dominant type of rock in the Hawaiian Islands is most likely
   (a) metamorphic.
   (b) sedimentary.
   (c) basaltic.
   (d) granitic.

Question 5 refers to the following diagram:

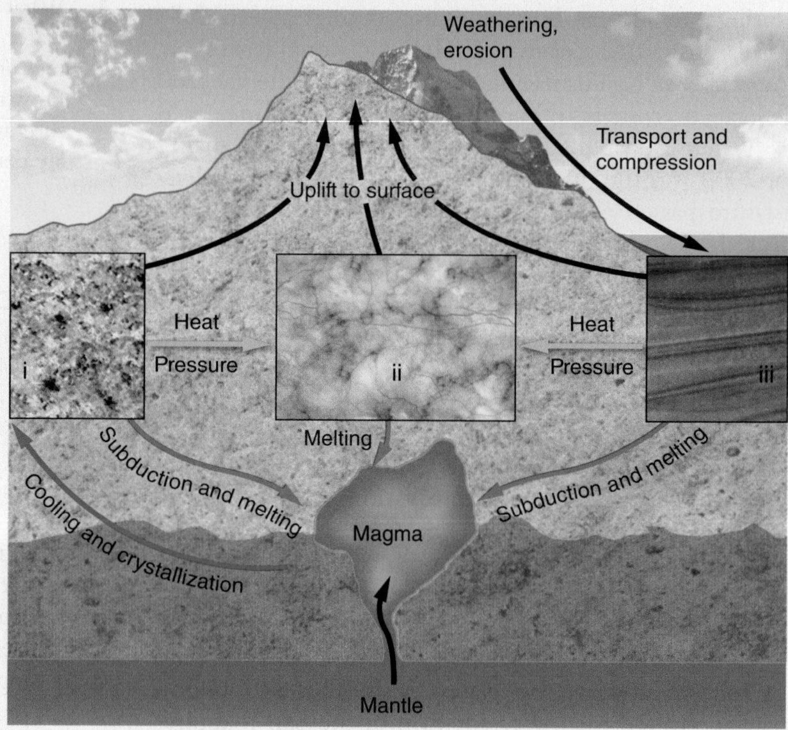

5. The rocks represented by i, ii, and iii are
   (a) igneous, sedimentary, and metamorphic.
   (b) granite, marble, and limestone.
   (c) sedimentary, marble, and limestone.
   (d) basalt, marble, and quartz.

6. Which is true?
   I. Physical weathering of rocks can increase rates of chemical weathering.
   II. High rates of physical weathering can lead to higher primary production.
   III. Weathering by acid precipitation occurs faster than physical weathering.
   (a) I only
   (b) II only
   (c) I and II
   (d) I and III

7. What creates mountain ranges like the Himalayas?
   (a) convergent continental plates
   (b) divergent oceanic plates
   (c) transform fault boundaries
   (d) seismic active zones

8. The ______ soil horizon is the most diagnostic feature of forests, whereas the ______ soil horizon is the most diagnostic feature of agricultural lands.
   (a) O;A
   (b) O;E
   (c) E;A
   (d) B;A

9. A 6.2 earthquake is how many times greater than a magnitude 3.2 earthquake?
   (a) 100
   (b) 300
   (c) 1,000
   (d) 3,000

10. Which best describes the differences between the cation exchange capacity (CEC) and base saturation of a soil?
   (a) CEC is the proportion of nutrients to clay particles; base saturation is the proportion of soil bases to soil acids.
   (b) CEC is the proportion of nutrients to clay particles; base saturation is the proportion of soil acids to bases.
   (c) CEC is the ability of a soil to retain beneficial nutrients for plant production; base saturation is the proportion of beneficial minerals to detrimental minerals in soil.
   (d) CEC is the ability of a soil to adsorb and release nutrients; base saturation is the proportion of soil bases to soil bases plus soil acids.

11. Several types of minerals are mined from locations around subduction zones. As subduction occurs, hot magma is pushed upward into the subsurface. The heat of magma causes groundwater to rise and cool. As the groundwater moves, it extracts minerals from the subsurface and deposits them near the surface in veins that flow in all directions. The most efficient way of mining for these minerals is
   (a) open-pit mining.
   (b) mountaintop removal.
   (c) subsurface mining.
   (d) strip mining.

12. The Surface Mining Control and Reclamation Act of 1977
   (a) protects workers from the hazards of subsurface and surface mining.
   (b) mandates that the land cannot be disturbed during the mining process.
   (c) requires that mined land be fully restored to its original condition.
   (d) regulates all mining practices than can impact the environment.

13. A cone of depression in an unconfined aquifer can be caused by
   I. excessive pumping of water from a well.
   II. low rainfall.
   III. saltwater intrusion.
   (a) I only
   (b) II only
   (c) I and II
   (d) II and III

14. Which is NOT a likely consequence of severe drought?
   (a) reduced rates of nutrient cycling
   (b) severe erosion of soil during subsequent rains
   (c) starvation
   (d) slower recharge of confined aquifers

15. All of the following may be consequences of dam construction EXCEPT
   (a) reduced upstream migration of fish.
   (b) less frequent and less severe inundation of downstream floodplains.
   (c) higher risk of levee failure downstream of the dam.
   (d) lower fertility of floodplains downstream.

16. Desalination by reverse osmosis
   (a) is less efficient than distillation.
   (b) produces a brine that is saltier than seawater.
   (c) has less effect on local wildlife than other methods of desalination.
   (d) is primarily a chemical process.

17. Which is NOT considered when determining a nation's water footprint?
   I. the daily per capita use of fresh water
   II. the amount of water used to produce exported crops
   III. the daily per capita amount of precipitation
   (a) I only
   (b) II only
   (c) I and II
   (d) II and III

18. Common methods of irrigation, in order from least efficient to most efficient, include
   (a) flood, furrow, drip, spray, and hydroponic irrigation.
   (b) furrow, drip, spray, and flood irrigation.
   (c) furrow, spray, drip, and hydroponic irrigation.
   (d) hydroponic, flood, and drip irrigation.

19. Which is NOT a likely consequence of the free-market system of water distribution?
   (a) The free-market system will encourage more efficient use of water.
   (b) The free-market system will encourage illegal diversion of water resources.
   (c) The price of water for homeowners will go down.
   (d) Farmers will begin planting crops that are more water efficient.

20. What is one possible method to increase water use efficiency?
(a) discouraging free-market distribution of water
(b) building new levees
(c) replacing furrow irrigation systems with drip irrigation systems
(d) monetarily subsidizing the water use of farmers

21. What kind of rock would have been formed by the eruption of Mount Vesuvius?
(a) metamorphic
(b) intrusive igneous
(c) extrusive igneous
(d) sedimentary

Question 22 refers to the following diagram:

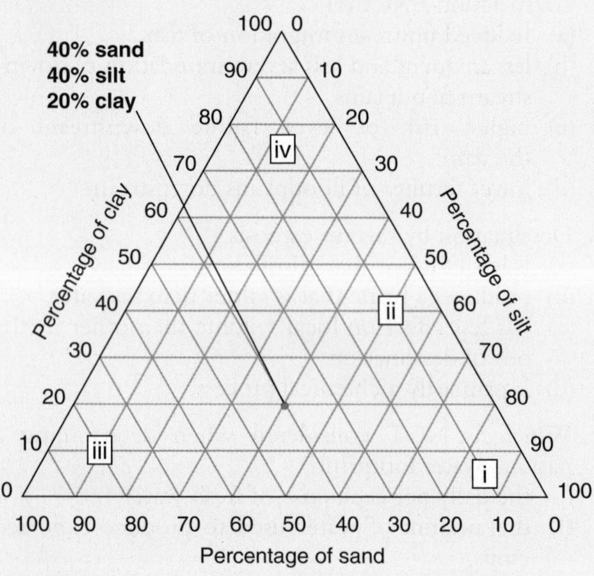

22. Which soil is likely to have the greatest CEC?
(a) i
(b) ii
(c) iii
(d) iv

Section 2: Free-Response Questions

Write your answer to each part clearly. Support your answers with relevant information and examples. Where calculations are required, show your work.

1. Mountaintop mining is the process of removing 50–200 vertical meters of mountaintop to extract coal from underlying seams. Following the removal of topsoil, material overlying coal seams is removed with explosives. This process leaves behind a tremendous amount of soil, rock, and coal tailings. The tailings are often pushed into adjacent valleys that contain headwater streams and floodplains. Much of the coal tailings become oxidized, and they can release sulfuric acid after a rainfall. When coal extraction is complete, the mountains are not rebuilt. However, mining companies are required to contour the landscape so that rainfall and snowmelt drain from the mountain along similar paths as in the pre-mining landscape. Companies are also required to add topsoil to the mountain and seed the soil to facilitate plant growth.
(a) Topsoil is often added by large tractors that compact the soil. How does soil compaction alter the flow of water down the mountain? (2 points)
(b) List two ways in which soil compaction may alter plant growth on the mountain following all reclamation activities. (2 points)
(c) List two environmental consequences that are likely to result from filling valleys with mountaintop tailings, and two consequences that might arise downstream of the mining site. (4 points)
(d) What other coal mining practice(s) could be used to remove the coal from mountains with less environmental impact than mountaintop mining? (2 points)

2. Predicting earthquakes could save many lives and prevent costly damage. Currently, we are only able to predict what areas are likely to experience earthquakes but not when an actual event will occur.
   (a) Describe how an earthquake occurs. In your answer, describe the location of the epicenter. (2 points)
   (b) We may be able to predict earthquakes by measuring the electric charge of minerals in the ground. When pressure is applied to many crystals, they emit a small electric charge that increases with pressure, a phenomenon known as "piezoelectricity." Describe how we might use piezoelectricity to predict the future occurrence of an earthquake. (4 points)
   (c) The amount of ground movement is directly proportional to the magnitude of the earthquake. The amount of ground movement produced by an earthquake with a magnitude of 4.0 on the Richter scale is equal to 10 millimeters. What is the amount of ground movement that would occur during an earthquake of magnitude 6.0 on the Richter scale? (2 points)
   (d) If an area of land experiences an earthquake of magnitude 6.0 on Richter scale every 40 years, how long will it take for that area to move 10 meters? (2 points)

3. A dam is constructed in a river next to a small town, creating a reservoir.
   (a) Name THREE reasons for which that dam may have been built. (3 points)
   (b) Shortly after the dam's construction, the salmon population in the river begins to decline. Propose a hypothesis for why the population declines. Describe a potential solution to this problem. (3 points)
   (c) Following the dam's construction, the density of farms increases along the river. Describe how the construction of a dam might lead to increased farm density in the area. (2 points)
   (d) The reservoir initially acts as a habitat for perch and other fish species. However, it eventually becomes hypoxic and the fish die off. Describe how the changes in the reservoir might be related to an increase in farming intensity. (2 points)

Environmentalists show support for Tim DeChristopher on the opening day of his federal trial in Salt Lake City, Utah.
(Jim Urquhart/AP Images)

Land, Public and Private

CHAPTER 10

MODULE 29 Land Use Concepts and Classification

MODULE 30 Land Management Practices

CASE STUDY

Who Can Use Our Public Lands? And for What Purpose?

There are many categories of public lands, a term for land that is held by a local or federal entity "in trust" for the citizens of the United States. And there have been many struggles over the best way to use and steward lands, both public and private. What are the rules regulating public lands? Who decides how they are used and how competing interests for the same land are evaluated? For over a century, the federal government has awarded the right to use federal lands to private individuals and corporations. The government awards or sells leases that permit a variety of activities including harvesting trees, mining copper, grazing cattle, and extracting oil and gas.

In 2008, a University of Utah student who objected to the designated use of federal lands in Utah attended a Bureau of Land Management (BLM) oil and gas lease auction and was assigned bidder number 70. Tim DeChristopher, a 27-year-old undergraduate and former wilderness guide, successfully bid on thousands of acres of land. When federal officials realized that he had no intention of paying the $1.8 million lease fee, or extracting oil and gas from that land, he was arrested and charged with two felonies that carried a potential sentence of 10 years. He spent almost 2 years in federal prison for trying to obstruct a process that legally allocated federal land to be used for the extraction of fossil fuels that lie beneath that land.

Some supporters argued that he was practicing a form of civil disobedience. Oil rigs, drilling platforms, and other evidence of resource extraction would have been visible from national parks in southern Utah. And certainly the extraction of oil and gas would have consequences both locally and globally. Many environmentalists viewed the BLM auction as a hasty attempt to develop as much land as possible for oil and gas exploration before environmental reviews could occur. Critics of DeChristopher maintained that the leasing of land for a variety of purposes has occurred for more than a century, and this auction was no different from any other. They stated that the government is obligated to allow certain lands to be used for a variety of purposes, most of which benefit some or many people in the United States. After his arrest and during his trial Tim DeChristopher continued to advocate for protecting public lands from drilling and expressed moral and legal objections to leasing public land.

> The conflict over use of public lands raises many questions.

Others have objected in less dramatic ways. In 2016, environmental writers and activists Terry Tempest and Brooke Williams tried to purchase oil and gas leases for federal land in Utah with the express intention of developing other energy resources from the land such as solar and wind. Their lease application was denied. They argue that they have been treated differently from other lease purchasers because they expressly stated they wanted to do something other than extract oil and gas. They are currently suing the federal government for the right to lease the land and use it for purposes other than oil and gas exploration. Their case is pending.

There have also been protests concerning public lands involving biological resources. Advocates for the protection of wolves called for the removal of cattle that had been allowed to graze on public lands in Colville National Forest, Washington. Cattle ranchers, who have grazed on public lands in northeastern Washington for over a hundred years, defended the right to graze their cattle. They also have supported actions by the Washington Department of Fish and Wildlife to protect their cattle, including shooting wolves when there was evidence that they had attacked or simply were in the vicinity of cattle. In 2016, Washington State officials killed seven wolves on public land at a cost of $135,000. The wolves were believed to have killed and eaten six cattle from nearby ranches. In 2017, two wolves were killed by the state. Strong sentiment both pro and con over wolf killing and cattle grazing on public lands has continued to increase. So far, the courts have allowed killings, as long as the agencies involved have studied the situation, collected and analyzed data, and have scientific or wildlife management support for their actions.

The conflict over use of public lands raises many questions. What obligations does a government have when it holds land "in trust" for its citizens? Do citizens of the United States have the right to use public lands for purposes other than those designated by the government? Is Tim DeChristopher a hero or simply a lawbreaker? Are cattle ranchers entitled to use public lands and entitled to federal protection for their cattle? These are challenging questions. From an environmental science perspective, we might begin to evaluate these situations by asking what land designations and uses can minimize harm to lands and increase sustainability.

Sources: Gage, Beth, *Bidder 70* (Film Documentary). First Run Features, 73 minutes. 2012. Loomis, Brandon, DeChristopher Sentenced to Prison, 26 Protesters Arrested, *Salt Lake Tribune* (Salt Lake City, Utah), July 27, 2011. Mapes, Lynda, With Cattle in Washington's Wolf Country, Ranchers Work and Worry, *Seattle Times*, July 15, 2017.

Issues involving land use are complex, and there are no easy solutions to many of the conflicts that arise. In this chapter we will explore how our use of land affects the environment, and how specific land-management practices can be implemented to minimize negative impacts. We will look at land use and management on both public and private land, and we will explore the topic of sustainable land use practices. We will also look at how land use has changed with shifting and growing populations.

MODULE 29

Land Use Concepts and Classification

Humans value land for what it can provide: food, shelter, resources, and intrinsic beauty. There are also various classifications of land that are made for a variety of scientific and political purposes. However, human use of land for any purpose can create problems that may be addressed and possibly prevented by regulations. In this module we will look at how human land use affects the environment and examine land use concepts and classification.

Learning Goals

After reading this module, you should be able to

- explain how human land use affects the environment.
- describe the various categories of public land used globally and in the United States.

Human land use affects the environment in many ways

Agriculture, housing, recreation, industry, mining, and waste disposal are all land uses that benefit humans. But, as we have seen in previous chapters, these activities also have negative consequences. Extensive logging may lead to mudslides, and deforestation of large areas contributes to climate change and many other environmental problems. Change to the landscape is the single largest cause of species extinctions today. Paving over land surfaces reroutes water runoff. These paved surfaces also absorb heat from the sun and reradiate it, creating urban "heat islands." Overuse of farmland can lead to soil degradation and water pollution and, in some cases, mudslides (**FIGURE 29.1**).

Every human use of land alters it in some way. Furthermore, individual activities on any parcel of land can have wide-ranging effects on other land. For this reason, communities around the world use a number of methods, including laws and informal norms, to influence or regulate both private and public land use.

As we saw at the beginning of this chapter, people do not always agree on land use and management priorities. Do we protect a large tract of open land, or do we extract resources from it in order to gain benefits in the form of energy resources, jobs, profits, and economic development? To understand the issues involved in the use of land, we will look at three concepts essential to understanding land use and land conflicts: the *tragedy of the commons, externalities,* and *maximum sustainable yield.*

The Tragedy of the Commons

In virtually every society throughout history, at least some land was viewed as a common resource: Anyone could

FIGURE 29.1 Mudslide. This home in Orinda, California, was damaged by a mudslide in 2017. *(Marcio Jose Sanchez/AP Images)*

Use of the commons is below the carrying capacity of the land. All users benefit.
If one or more users increase the use of the commons beyond its carrying capacity, the commons becomes degraded. The cost of the degradation is incurred by all users.
Unless environmental costs are accounted for and addressed in land use practices, eventually the land will be unable to support the activity.

FIGURE 29.2 The tragedy of the commons. If the use of common land is not regulated in some way—by the users or by a government agency—the land can easily be degraded to the point at which it can no longer support that use.

use certain spaces for foraging, growing crops, felling trees, hunting, or mining. But as populations increased, common lands tended to become degraded—overgrazed, overharvested, and deforested. In 1968, ecologist Garrett Hardin brought the issue of overuse of common resources to the attention of the broader scientific community. Because people often act from self-interest for short-term gain, Hardin observed that when many individuals share a common resource without agreement on or regulation of its use, it is likely to become overused and depleted very quickly. The tendency for a shared, limited resource to become depleted if it is not regulated in some way is known as the **tragedy of the commons**. Regulation might be managed by the users or by a governmental agency.

Consider land as an example. Imagine a communal pasture on which many farmers graze their sheep, like the one shown in **FIGURE 29.2**. At first, no single farmer appears to have too many sheep. But because an individual farmer gains from raising as many sheep as possible, each farmer may be tempted to add sheep to the pasture. However, if the total number of sheep owned by all the farmers continues to grow, the number of sheep will soon exceed the carrying capacity of the land. The sheep will overgraze the pasture to the point at which plants will not have a chance to recover. The common land will be degraded and the sheep will no longer have an adequate source of nourishment. Over a longer period, the entire community will suffer. When the farmers make decisions that benefit only their own short-term gain and do not consider the common good, eventually, everybody loses. That is why it is called a "tragedy"—there are no winners.

Tragedy of the commons The tendency of a shared, limited resource to become depleted if it is not regulated in some way.

Externality The cost or benefit of a good or service that is not included in the purchase price of that good or service or otherwise accounted for.

The tragedy of the commons applies not only to agriculture, but to any publicly available resource that is not regulated, including land, air, and water. For example, the use of global fisheries as commons has led to the overexploitation and rapid decline of many commercially harvested fish species, and has upset the balance of entire marine ecosystems. The release of carbon dioxide to the atmosphere or sewage to an estuary are other examples of the tragedy of the commons (**FIGURE 29.3** on page 347).

Externalities

The tragedy of the commons is the result of an economic phenomenon called a *negative externality.* More generally, an **externality** is a cost or benefit of a good or service that is not included in the purchase price of that good or service, or otherwise not accounted for. For example, if a bakery moves into the building next to you and you wake up every morning to the delicious smell of freshly baked bread, you are benefiting from a positive externality. On the other hand, if the bakers arrive at three in the morning and make so much noise that they interrupt your sleep, making you less productive at your job later that day, you are suffering from a negative externality.

In environmental science, we are especially concerned with negative externalities because they so often lead to serious environmental damage for which no one is held legally, financially, or morally responsible. For example, if one farmer grazes more sheep than is beneficial for the community in a common pasture, his action will ultimately result in more total harm than total benefit. But, as long as the land continues to support grazing, the individual farmer will not have to pay for his overuse.

FIGURE 29.3 Global fisheries. Global fisheries are one of the commons. These sardines were photographed in the North Atlantic Ocean. *(Jens Kuhfs/Getty Images)*

Ultimately, though, other farmers will bear the burden of his overuse because the land will be depleted and they will have to find somewhere else to graze their sheep, or go out of business. But what if the farmer responsible for the extra sheep had to pay an extra fee for using more of the common resource? He probably would not graze the extra sheep on the commons because the cost of doing so would exceed the benefit. From this example, we can see that in order to calculate the true cost of using a resource, we must always include any negative externalities. In other words, we must account for any potential external harm that comes from the use of that resource.

Some economists maintain that private ownership can prevent the tragedy of the commons. After all, a landowner is much less likely to overgraze his own land than he would common land. Governmental regulation is another approach. For example, a local government could prevent overuse of a common pasture by passing an ordinance that permits only a certain number of sheep to graze there.

Challenging the idea that government regulation is necessary, the late Professor Elinor Ostrom (1933–2012) of Indiana University showed that many commonly held resources can be managed effectively at the community level or by user institutions. Ostrom's work, for which she was awarded the 2009 Nobel Prize in Economics, has shown that self-regulation by resource users can prevent the tragedy of the commons. Professor Ostrom also challenged the idea that private property rights can address all environmental problems.

Maximum Sustainable Yield

When we want to obtain the maximum amount of a biological resource, we need to know how much of a given plant or animal population can be harvested without harming the resource as a whole.

Imagine a situation in which deer hunting in a public forest is unregulated, with each hunter free to harvest as many deer as possible. As a result of unlimited hunting, the deer population could be depleted to the point of endangerment. This, in turn, would disrupt the functioning of the forest ecosystem. On the other hand, if hunting were prohibited entirely, in the absence of natural predators, the deer herd might grow so large that there would not be enough food in the forests and fields to support the herd. In extreme cases, such as that of the reindeer of St. Paul Island in Alaska (see Figure 19.3 on page 207), the population could grow unchecked until it crashed due to starvation.

Some intermediate amount of hunting will leave enough adult deer to reproduce at a rate that will maintain the population, but not leave so many that there is too much competition for food. This intermediate harvest is called the *maximum sustainable yield*. Specifically, the **maximum sustainable yield (MSY)** of a renewable resource is the maximum amount that can be harvested without compromising the future availability of that resource. In other words, it is the maximum harvest that will be adequately replaced by population growth.

MSY varies case by case. A reasonable starting point is to assume that population growth is the fastest at about one-half the carrying capacity of the environment, as shown on the S-shaped curve in **FIGURE 29.4** on page 348.

AP® Exam Tip

Make sure that you understand the concept of maximum sustainable yield, as well as its application to natural resources. MSY often appears on the AP® Environmental Science Exam. ●

Looking at the graph, we can see that at a small population size, the growth curve is shallow and growth is relatively slow. As the population increases in size, the slope of the curve is steeper, indicating a faster growth rate. As the population size approaches the carrying capacity, the growth rate slows. The MSY is the amount of harvest that allows the resource population to be maintained at roughly one-half the carrying capacity of the environment.

Forest trees, like animal populations, have a maximum sustainable yield. Loggers may remove a particular percentage of the trees at a site in order to allow a certain amount of light to penetrate to the forest floor and reach younger trees. If they cut too many trees, however, an excess of sunlight will penetrate and dry the forest soil. This drying may create conditions inhospitable to tree germination and growth, thus inhibiting adequate regeneration of the forest. In theory, harvesting the maximum sustainable yield should permit an indefinite use without depletion of the resource. In reality, though, there are two problems with using MSY in environmental policy management.

First, it is very difficult to calculate MSY with certainty because in a natural ecosystem, it is difficult to

Maximum sustainable yield (MSY) The maximum amount of a renewable resource that can be harvested without compromising the future availability of that resource.

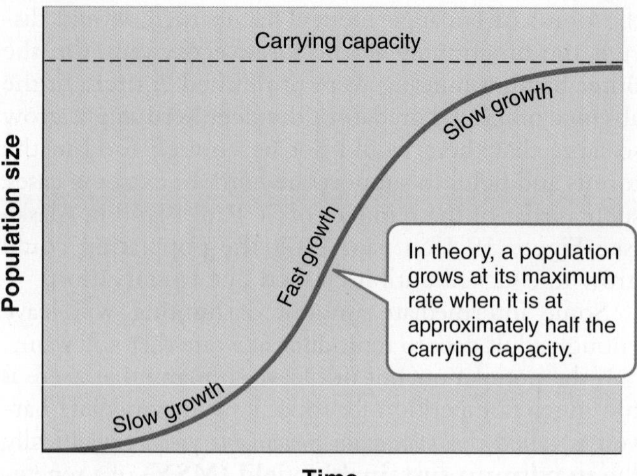

FIGURE 29.4 Maximum sustainable yield. Every population has a point at which a maximum number of individuals can be harvested sustainably. That point, the maximum sustainable yield, is often reached when the population size is about one-half the carrying capacity.

obtain necessary information such as birth rates, death rates, and the carrying capacity of the system. Once an MSY calculation is made, we cannot know if a yield is truly sustainable until years later when we can evaluate the effect of the harvest on reproduction. By that time, if the harvest rate has been too great, it is too late to prevent harm to the population.

In addition to the calculation difficulties, a sizeable group of environmental scientists argue that MSY is not an adequate guide to use for environmental protection. They believe that unmeasured externalities and harm to ecosystems can occur when using MSY as a harvest guideline, and that the best way to protect the environment is to harvest considerably less than the MSY.

Public lands are classified according to their use

All countries have public lands designated for a variety of purposes, including environmental protection. The use of protected areas may vary from excluding almost all human activities to permitting harvest of biological and other resources for human benefit. The 2014 United Nations List of Protected Areas—the most recent global study of protected areas—contains more than 209,000 protected areas around the world covering more than 3.2 billion hectares (7.9 billion acres). Roughly 14 percent of terrestrial and 3.4 percent of marine areas of the surface of Earth are protected in one way or another. These areas are shown on the map in **FIGURE 29.5**. In the following section we examine both international and national categories of land protection. "Do the Math: Changes in the Amount of Protected Lands over Time" gives you the opportunity to explore some of the data associated with protected lands.

International Categories of Public Lands

The 2014 United Nations List of Protected Areas classifies protected public lands into six categories according to how they are used: national parks, managed resource protected areas, habitat/species management areas, strict nature reserves and wilderness areas, protected landscapes and seascapes, and national monuments. The specific regulations concerning protection and use within each protected area classification varies by country: however, we can make some generalizations.

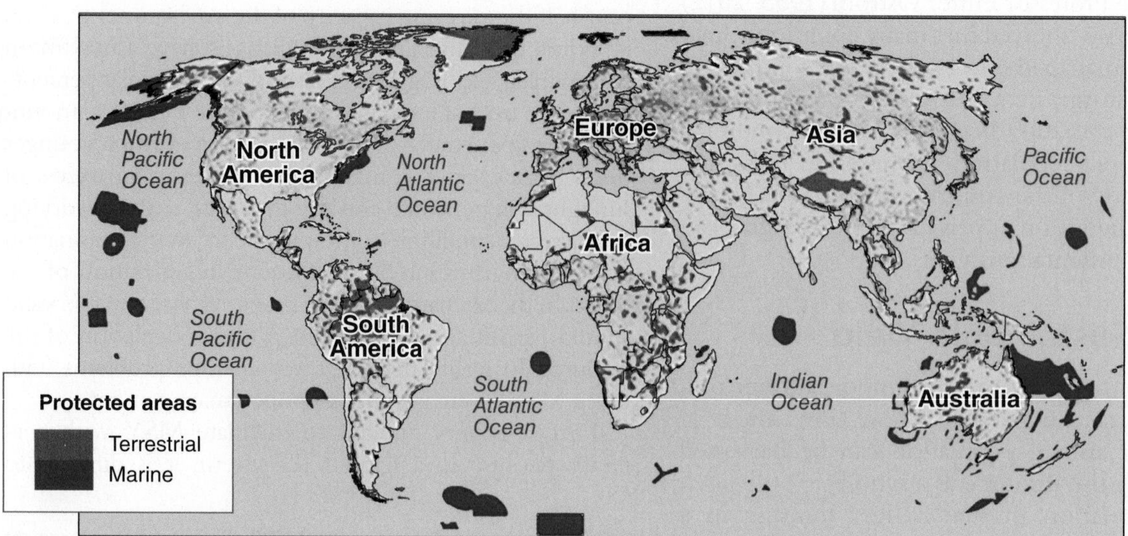

FIGURE 29.5 Protected land and marine areas of the world. Protected areas are distributed around the globe. *(Data from http://wdpa.s3.amazonaws.com/Files_pp_net/Global_map_template_Dec2016.png)*

FIGURE 29.6 National parks. Majestic Grand Teton National Park in Wyoming is one of 58 national parks in the United States, valued for their natural beauty and recreational opportunities. *(John Kercher/Getty Images)*

National Parks

National parks are managed for scientific, educational, and recreational use, and sometimes for their beauty or unique landforms. In most cases, they are not used for the extraction of resources such as timber or ore. Some of the most famous national parks in the United States are Yellowstone, Yosemite, Grand Canyon, Acadia, and Great Smoky Mountains (**FIGURE 29.6**).

Famous national parks are found on every continent. Some of the best known are in Africa and include Amboseli National Park in Kenya, Serengeti National Park in Tanzania, and Kruger National Park in South

DO THE MATH Changes in the Amount of Protected Lands Over Time

Preparing for the AP® Exam

In recent years the United Nations has reported a significant increase in the total area protected worldwide. In 2004 the UN listed total area protected as 18,800,000 square kilometers. Let's convert this value in square kilometers to hectares, express it in scientific notation, and compare that amount to the current 2014 values indicated earlier in the chapter.

$$18{,}800{,}000\ \cancel{\text{km}^2} \times (100\ \text{ha}/1\ \cancel{\text{km}^2}) = 1{,}880{,}000{,}000\ \text{ha}$$

In scientific notation this value = 1.88×10^9 ha protected area in 2004

Protected area in 2014 = 3.2 billion ha = 3.2×10^9 ha

2004 protected area = 1.88×10^9 ha

2014 protected area = 3.20×10^9 ha

(3.20×10^9) ha − (1.88×10^9) ha = 1.32×10^9 ha = change in protected area 2004 to 2014

Change over original 2004 value = $(1.32 \times 10^9\ \text{ha}) \div (1.88 \times 10^9\ \text{ha}) = 0.70 \times 100\% = 70\%$

Protected areas increased 70 percent between 2004 and 2014.

YOUR TURN There is approximately 8,750,000 km^2 of forest in the tropics of Central and South America; 3,500,000 km^2 is in protected areas.

1. Convert to hectares and express in scientific notation for each.
2. Identify the percent of forest in protected areas.

Africa. These parks generally exist to protect animal species such as elephants, rhinoceroses, and lions, as well as areas of great natural beauty. They also generate tourism, which can provide revenue to local and national economies.

Even though national parks in the United States are not used for resource extraction, there still may be some level of management permitted or desired. For example, controlled burns may be deliberately introduced to remove deadwood in the understory of a forest and thereby reduce the risk of a catastrophic wildfire.

Managed Resource Protected Areas

Managed resource protection areas are open for the sustained use of biological, mineral, and recreational resources. In most countries, these areas are managed for multiple uses. This means that mining, logging, and other activities are allowed and sometimes encouraged. In the United States, national forests are one example of this kind of protected area. Often, in order to remove trees or other resources, roads need to be constructed. Road construction causes habitat alteration and fragmentation, which can lead to controversies about the use of the area.

Habitat or Species Management Areas

Habitat or species management areas are actively managed for a variety of purposes. Management may occur in order to maintain biological communities, for example through provision of predator control or fire prevention, or introduction of fire. Introduction of fire may be beneficial to species that have adapted to reproduce following periodic fires. For example, certain species of pine trees produce a type of seed cone (called serotinous cones) that has a sticky resin covering the seeds. The resin prevents the cone from opening and the seeds from being released until a fire of a certain level of intensity occurs. For the species to propagate, the forest must experience periodic fires.

Karelia, in northwest Russia and bordering areas of Finland, has one of the highest proportions of protected areas in Europe: 5 percent of its total area. Of this total, more than one-half consists of habitat or species management areas that are actively managed for hunting and conservation.

Strict Nature Reserves and Wilderness Areas

Strict nature reserves and wilderness areas are established to protect species and ecosystems. There are approximately 6,000 such sites worldwide, covering more than 200 million hectares (490 million acres). For example, the Chang Tang Reserve on the Tibetan Plateau in China was set aside to protect a number of species—including the declining population of wild yak—from hunting, habitat destruction, and hybridization with domesticated animals.

Protected Landscapes and Seascapes

Protected landscapes and seascapes combine the nondestructive use of natural resources with opportunities for tourism and recreation. Orchards, villages, beaches, and other such areas make up the 6,500 such sites worldwide, which cover more than 100 million hectares (250 million acres). Among these protected areas is the Batanes Protected Landscape and Seascape in the northernmost islands of the Philippines, home to several endemic plant and animal species as well as important marine habitats.

National Monuments

National monuments are set aside to protect unique sites of special natural or cultural interest. There are thousands of national monuments and landmarks around the world. Most of these are established to protect historical landmarks, such as the Arc de Triomphe in Paris, France, or to identify areas of natural beauty and historical and cultural importance, such as Bears Ears National Monument in Utah in the United States (**FIGURE 29.7**).

Designation of protected areas typically generates great benefits for the natural environment and plant and animal species. On the negative side, in order to create and maintain protected areas, governments have sometimes evicted and excluded indigenous human populations from the land, creating humanitarian crises. For example, in the winter of 2009, a series of evictions from the Mau Forest in the Rift Valley of Kenya led to the displacement of 20,000 families. One of the leading Kenyan newspapers reported that 2,300 families had not been resettled by 2015. Such programs continue to generate controversy in Kenya and other countries.

Public Lands in the United States

In the United States, publicly held land may be owned by federal, state, or local governments. Of the nation's land area, 42 percent is publicly held—a larger

FIGURE 29.7 National monuments. One of the newest national monuments is Bears Ears National Monument in southern Utah that was established in 2016. It is named for a pair of isolated hills that rise up relative to the surrounding terrain. *(The Washington Post/Getty Images)*

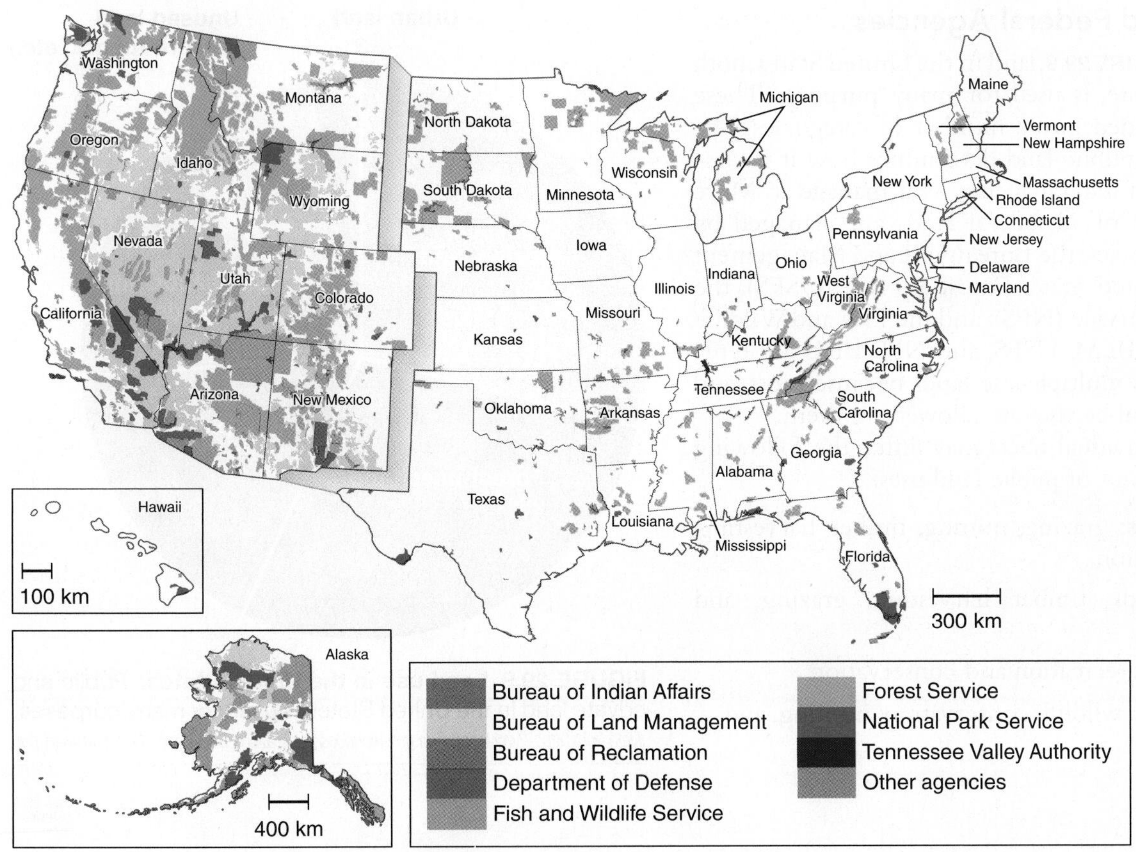

FIGURE 29.8 Federal lands in the United States. Approximately 42 percent of the land in the United States is publicly owned, with 25 percent of the nation's land owned by the federal government. *(Data from http://nationalatlas.gov.)*

percentage than in any other nation. As you can see in **FIGURE 29.8**, the federal government is by far the largest single landowner in the United States. It owns 259 million hectares (640 million acres), or roughly 28 percent of the country. Only 0.3 percent of the land in Connecticut and Iowa are federally owned. But 46 percent of the 11 western continental states is federally owned, as is 61 percent of the land in Alaska. Less than 10 percent of federal land is located in the Midwest and on the East Coast.

Public Land Classifications

Public lands in the United States include rangelands, national forests, national parks, national wildlife refuges, and wilderness areas. Since the founding of the nation, many different individuals and groups have expressed interest in using these public lands. However, most environmental policies, laws, and management plans have been based, at least partially, on the **resource conservation ethic**, which states that people should maximize resource use based on the greatest good for everyone. The resource conservation ethic calls for policy makers to consider the resource or monetary value of nature. In conservation and land use terms, it has meant that areas are preserved and managed for economic, scientific, recreational, and aesthetic purposes.

Of course, the purposes can be in conflict. In order to manage competing interests, the U.S. government has, for decades, adopted the principle of *multiple use* in managing its public resources. Public lands that are classified as **multiple-use lands** might be used for a variety of purposes at the same time, including recreation, grazing, timber harvesting, and mineral extraction. Other areas are designated as protected lands in order to maintain a watershed, preserve wildlife and fish populations, or maintain sites of scenic, scientific, and historical value.

Resource conservation ethic The belief that people should maximize use of resources, based on the greatest good for everyone.

Multiple-use lands A U.S. classification used to designate lands that may be used for recreation, grazing, timber harvesting, and mineral extraction.

Land Use and Federal Agencies

As shown in **FIGURE 29.9**, land in the United States, both public and private, is used for many purposes. These uses can be divided into a number of categories. The probable use of public land determines how it is classified and which federal agency will manage it. More than 95 percent of all federal lands are managed by four federal agencies: the Bureau of Land Management (BLM), the United States Forest Service (USFS), the National Park Service (NPS), and the Fish and Wildlife Service (FWS). BLM, USFS, and NPS lands are typically classified as multiple-use lands because most, and sometimes all, public uses are allowed on them.

Although individual tracts may differ, the following are typical divisions of public land uses:

- BLM lands: grazing, mining, timber harvesting, and recreation
- USFS lands: timber harvesting, grazing, and recreation
- NPS lands: recreation and conservation
- FWS lands: wildlife conservation, hunting, and recreation

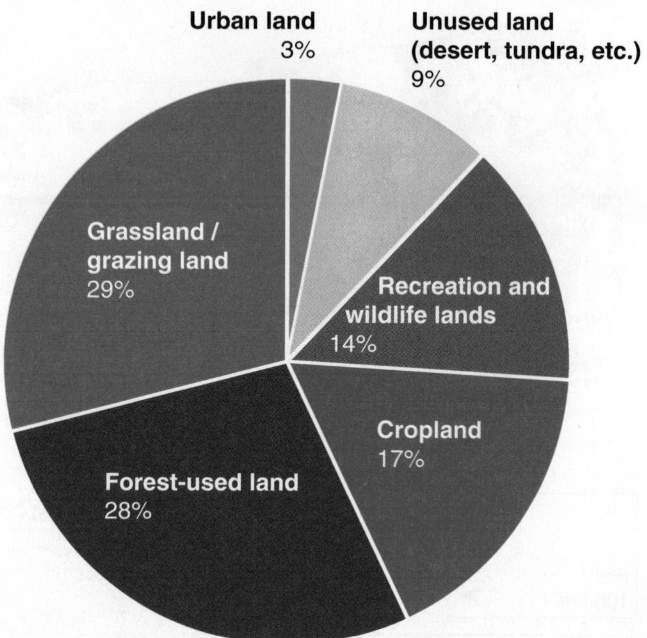

FIGURE 29.9 Land use in the United States. Public and private land in the United States is used for many purposes. *(Data from https://www.ers.usda.gov/webdocs/publications/84880/eib-178_summary.pdf)*

MODULE **29** AP® Review

Preparing for the AP® Exam

In this module, we have seen that the tragedy of the commons describes the tendency of shared, common resources to be over-exploited. Externalities are costs or benefits not included in the price of a good or service, and the tragedy of the commons can be the result of negative externalities. Maximum sustainable yield is a management technique that allows for the greatest amount of a resource that can be harvested from land without compromising the sustainability of that land and the resource. Public lands are classified and designated for various uses differently in different countries. In the United States, public lands fall under the jurisdiction of numerous federal agencies and can be managed for multiple uses. In the next module we will look in more detail at land management practices.

AP® Practice Questions

Choose the best answer for the following.

1. The tragedy of the commons can be prevented by
   (a) externalities.
   (b) fisheries in international waters.
   (c) the use of harvest permits.
   (d) the use of public land for grazing.

2. Which is NOT an example of an externality?
   (a) the use of national forests by private companies
   (b) the use of the atmosphere for air pollution
   (c) the use of forested watersheds for aquifer purification
   (d) roads maintained with a tax on gasoline

3. The maximum sustainable yield of a population usually occurs
   (a) at the maximum population size.
   (b) at the minimum growth rate.
   (c) at half the carrying capacity.
   (d) at approximately 90 percent of the carrying capacity.

4. Which type of protected land might allow limited mining?
   (a) a nature reserve
   (b) a national monument
   (c) a habitat management area
   (d) a managed resource protected area

5. Which federal agency typically classifies federal lands under its jurisdiction as multiple-use?
   I. the Bureau of Land Management
   II. the United States Forest Service
   III. the Fish and Wildlife Service
   (a) I only
   (b) I and II
   (c) I and III
   (d) I, II, and III

MODULE 30

Land Management Practices

We have seen that resources can be managed or they can be over-exploited, depending on a number of factors related to economics, human behavior, and specific management practices. Proper resource management also depends on specific aspects of the type of land involved as well as the agency and the guidelines under which it operates. In this module, we will examine land management practices in rangelands and forests, with a focus on timber management practices and fire management. We will also examine trends in residential land use and the causes and consequences of urban sprawl.

Learning Goals

After reading this module, you should be able to

- explain specific land management practices for rangelands and forests.
- describe contemporary problems in residential land use and some potential solutions.

Land management practices vary according to land use

Some of the most challenging environmental science issues are related to land management theory and practice. Perhaps the most important work in the last century on this topic has come from Aldo Leopold, a University of Wisconsin professor, ecologist, writer, and environmentalist. His book *A Sand County Almanac*, published posthumously in 1949, characterizes land management, conservation, ecology, and interactions with people that in a way that was pioneering when published, and is still timely and important today. Leopold wrote about a "land ethic," which he described as a moral responsibility of humans to the natural world. The moral responsibility to protect land is not typically a topic in environmental science, but many contemporary discussions about the conservation and protection of public and private lands inevitably reference Aldo Leopold.

Now that we have a basic picture of how public land is classified and of the relationship between public land use and management agencies in the United States, let's turn to some of the specific issues involved in managing different types of public lands. Note that many of the management issues we discuss here apply to private lands as well.

Rangelands

Rangelands are dry, open grasslands primarily used for grazing cattle, which is the most common use of land in the United States. Rangelands are semiarid ecosystems and are therefore particularly susceptible to fires and other environmental disturbances. When overused, rangelands can easily lose biodiversity.

Like most human activities, grazing livestock has mixed environmental effects. Grazing livestock on open lands uses much less fossil fuel energy than raising them in feedlots. Furthermore, grazing domesticated animals for short, intense periods of time can mimic the pressure on grasslands that have been experienced over evolutionary time by migrating herds of bison and other herbivores. Therefore, the grazing of livestock can help maintain grasslands and prevent less desirable species from becoming dominant. In addition, grazing introduces animal waste and aerates the soil, both of which are beneficial to ecosystems. However, improperly managed livestock can damage stream banks and pollute surface waters. Grazing too many animals in a finite area can quickly denude a region of vegetation (**FIGURE 30.1**). Overgrazed land is exposed to wind erosion which makes it difficult for soils to absorb and retain water and nutrients when it rains.

Many environmental scientists argue that rangeland ecosystems are too fragile for multiple uses. Certain environmental organizations have suggested that as much as 55 percent of U.S. rangeland soils are in poor or very poor condition, due in large part to overgrazing. However, the BLM, which manages most public rangelands in the United States, has maintained that the percentage of poor or very poor soils is not nearly that high. Reconciling this difference is a challenge because of the many factors that influence how soil condition is determined.

The Taylor Grazing Act of 1934 was passed to halt overgrazing. It converted federal rangelands from a commons into a permit-based grazing system. The goal of a permit-based system is to limit the number of animals grazing in a particular area and thereby avoid a tragedy of the commons situation. However, critics maintain that the low cost of the permits continues to encourage overgrazing. For decades, the federal government has spent seven times more money managing its rangelands than it has received in permit fees. Thus, in effect, grazing is subsidized with federal funds.

FIGURE 30.1 Overgrazed rangeland. Overgrazing can rapidly strip an area of vegetation, as seen in this site in the Mojave Desert, California. *(Jeff Foott/Getty Images)*

The BLM focuses on mitigating the damage caused by grazing and considers "rangeland health" when it sets grazing guidelines. For example, state and regional rangeland managers are required to ensure healthy watersheds, maintain ecological processes such as nutrient cycles and energy flow, preserve water quality, maintain or restore habitats, and protect endangered species. However, because these managers are not given detailed guidance, and the BLM regulations do not require the involvement of environmental scientists, the managers have wide latitude to set their own guidelines and standards. As a result, BLM managers are not consistently successful in preserving vulnerable rangeland ecosystems.

Forests

Forests are dominated by trees and other woody vegetation. Roughly three-quarters of the forests used for commercial timber operations in the United States are privately owned. Many national forests were originally established to ensure a steady and reliable source of timber, and commercial logging companies are currently allowed to harvest U.S. national forests, usually in exchange for a royalty that is a percentage of their revenues. The federal government typically spends more money managing the timber program and building and maintaining logging roads than it receives from these royalties, essentially subsidizing the timber program as it does grazing on rangelands.

Rangeland A dry open grassland primarily used for grazing cattle.

Forest Land dominated by trees and other woody vegetation and sometimes used for commercial logging.

Timber Harvest Practices

The two most common ways in which trees are harvested for timber production are *clear-cutting* and *selective cutting,* both of which are illustrated in **FIGURE 30.2**. **Clear-cutting** involves removing all, or almost all, the trees within an area; it is the easiest harvesting method and, in most cases, the most economical. When a stand, or cluster, of trees has been clear-cut, foresters sometimes replant or reseed the area. Often the entire area will be replanted at the same time, so all the resulting trees will be the same age. Because they are exposed to full sunlight, clear-cut tracts of land are ideal for fast-growing tree species that achieve their maximum growth rates with large amounts of direct sunlight. Because other species may not be so successful in these conditions, there can be an overall reduction in biodiversity.

If a commercially valuable tree species constitutes only a small fraction of a stand of trees, it may not be economically efficient to clear-cut the entire stand. This is particularly true in many tropical forests, where economically valuable species constitute only a small percentage of the trees and are mixed in with many other less valuable species.

Regrowth

(a) Clear-cutting

Regrowth

(b) Selective cutting

FIGURE 30.2 Timber harvest practices. (a) Clear-cutting removes most, if not all, trees from an area and is often coupled with replanting. The resulting trees are then all the same age. (b) In selective cutting, single trees or small numbers of trees are harvested. The resulting forest consists of trees of varying ages.

AP® Exam Tip

Questions about tree-harvesting practices have appeared on both the multiple-choice and free-response sections of the AP® Environmental Science Exam. Make sure that you understand tree harvest techniques and the environmental costs and benefits of each method. ●

Clear-cutting, especially on slopes, increases wind and water erosion, which causes the loss of soil and nutrients (**FIGURE 30.3** on page 356). Erosion also adds silt and sediment to nearby streams, which harms aquatic populations. In addition, the denuded slopes are prone to dangerous mudslides, like the one shown in Figure 29.1 on page 345. Clear-cutting also increases the amount of sunlight that reaches rivers and streams, which raises water temperatures that can adversely affect certain aquatic species. Even the replanting process can have negative environmental consequences. Timber companies often use fire or herbicides to remove bushy vegetation before a clear-cut is replanted. These practices reduce soil quality, and herbicides may contaminate water that runs off into streams and rivers. Many environmental scientists identify clear-cutting as a cause of habitat alteration and destruction as well as forest fragmentation. These effects decrease biodiversity and sometimes lower the aesthetic value of the affected forest. However, in heavily forested regions of the country, such as the northern New England states of Maine, New Hampshire, and Vermont, where forests can cover up to 85 percent of land area, a case can be made that clear-cuts actually increase habitat diversity and, as a result, species diversity.

Selective cutting removes single trees or a relatively small number of trees from the larger forest. This method creates many small openings in a stand

Clear-cutting A method of harvesting trees that involves removing all or almost all of the trees within an area.

Selective cutting The method of harvesting trees that involves the removal of single trees or a relatively small number of trees from the larger forest.

FIGURE 30.3 Clear-cut forest in western Washington state. A clear-cut on a steep slope increases the likelihood of erosion and delayed regeneration of vegetation. *(PhilAugustavo/Getty Images)*

FIGURE 30.4 Sustainable forestry. Logging without the use of fossil fuels, as is done in this forest in Oregon, further enhances the sustainability of a forestry project. *(Michael Thompson/Earth Scenes/Animals Animals)*

where trees can reseed or young trees can be planted, so the regenerated stand contains trees of different ages. Because seedlings and young trees must grow next to larger, older trees, selective cutting produces optimum growth only among shade-tolerant tree species.

The environmental impact of selective cutting is less extensive than that of clear-cutting. However, many of the negative environmental impacts associated with logging remain the same. For example, whether a company uses clear-cutting or selective cutting, it will need to construct logging roads to carry equipment and workers into the area that will be harvested. These roads fragment the forest habitat, which affects species diversity, and compact the soil, which causes nutrient loss and reductions in water infiltration.

A third approach to logging—**ecologically sustainable forestry**—removes trees from the forest in ways that do not unduly affect the viability of other, noncommercial tree species. This approach has a goal of maintaining both plants and animals in as close to a natural state as possible. Some loggers have even returned to using animals such as horses to pull logs in an attempt to reduce soil compaction. However, the costs of such methods make it difficult to compete economically with mechanized logging practices (**FIGURE 30.4**).

Logging, Deforestation, and Reforestation

Almost 30 percent of all commercial timber in the world is produced in the United States and Canada. Compared with South America and Africa, deforestation and land loss and destruction in these two major timber-producing countries have been relatively small over the last several decades. Still, timber production presents important ecological challenges in the United States and Canada.

Throughout this chapter we have seen examples of the conflicts over land use created by competing interests and values. Perhaps nowhere is this conflict so clear as in the case of logging. For example, while timber production has always been a part of the mission of the United States Forest Service, maintaining biodiversity is an equally important goal.

All of the activities associated with logging disrupt habitat and usually have an effect, either negative or positive, on plant and animal species. One such species is the marbled murrelet (*Brachyramphus marmoratus*). This bird spends most of its life along the coastal waters of the Pacific Northwest, but it nests in coastal redwood forests. With the intensive logging of these forests, the marbled murrelet has become scarce and is now listed as a federally threatened species and as an endangered species by the state of California.

Logging often replaces complex forest ecosystems with **tree plantations**, which are large areas typically planted with a single rapidly growing tree species. These same-aged stands can be easily clear-cut for commercial purposes, such as pulp and wood, and then replanted. Because of this cycle of planting and harvesting, tree plantations never develop into mature, ecologically diverse forests. If too many planting and harvesting cycles occur, the soil may become depleted of important nutrients such as calcium.

Since 1982, federal regulations have required the USFS to provide appropriate habitat for plant and animal communities while at the same time meeting multiple-use goals. However, because these regulations

Ecologically sustainable forestry An approach to removing trees from forests in ways that do not unduly affect the viability of other noncommercial tree species.

Tree plantation A large area typically planted with a single rapidly growing tree species.

fail to specify how biodiversity protection should be achieved or how the results should be quantified, the USFS has had to choose its own approach to biodiversity management. Critics charge that the USFS is not adequately protecting biodiversity and forest ecosystems, but the USFS maintains that it is doing the best it can at meeting many different objectives.

Fire Management

In many ecosystems, fire is a natural process that is important for nutrient cycling and regeneration. As discussed briefly in Chapter 3, when fires periodically move through an ecosystem, they liberate nutrients tied up in dead biomass. In addition, areas where vegetation is killed by fire provide openings for early-successional species.

Humans have followed a variety of management policies with respect to fire. For many years, managers of forest ecosystems, including the USFS, worked to suppress fires. This strategy led to the accumulation of large quantities of dead biomass on the forest floor, which built up until a large fire became inevitable. One method for reducing the accumulation of dead biomass is a **prescribed burn**, in which a fire is deliberately set under controlled conditions in order to reduce the accumulation of dead biomass on a forest floor. Prescribed burns help reduce the risk of uncontrolled natural fires and provide some of the other benefits of fire. More recently, forest managers have allowed certain fires that have occurred naturally to burn. This policy appears to have been accepted in many parts of the United States, as long as human life and property are not threatened.

There are also many instances where human behavior and mismanagement can lead to large fires on public or private land. Deliberate and accidental fire setting by individuals is certainly a cause of fires. But poor equipment management may also lead to fires. The 2017 northern California wildfires attained great attention because of the number, extent, and location in and near Sonoma County, famous for its wine making. Downed power lines are the suspected cause of some of the 2017 California fires.

Probably the best-known forest fires in the United States are those that occurred in Yellowstone National Park in the summer of 1988, the driest year on record at the park where a combination of human activity and lightning set off multiple fires. Over 25,000 people participated in fighting the fires. When the fires were out, more than one-third of Yellowstone National Park had burned, as shown in **FIGURE 30.5a**. Initially, many people were outraged that the NPS and others had "allowed" the park to burn. However, within a few years, it became clear that the fires had created new, nutrient-rich habitat for early-successional plant species that attracted elk and other herbivores. Ultimately, researchers and forest managers concluded that the

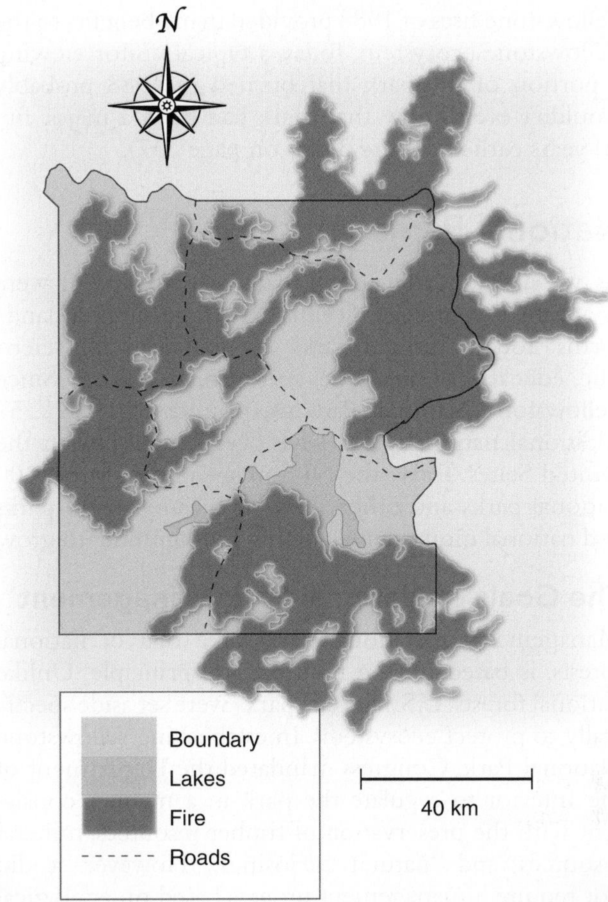

(a)

(b)

FIGURE 30.5 Yellowstone fires of 1988. (a) As can be seen from the map, extensive areas of the park were burned in this exceptionally hot and dry year. (b) Thirty years later, a casual observer might not know that the forest had burned, except for the sign in the foreground. *(Peter Haigh/Alamy)*

Prescribed burn A fire deliberately set under controlled conditions in order to reduce the accumulation of dead biomass on a forest floor.

Yellowstone fires of 1988 provided many benefits to the Yellowstone ecosystem. Today, a typical visitor viewing a portion of the park that burned in 1988 probably wouldn't even know that there had been a major fire 30 years earlier (Figure 30.5b on page 357).

National Parks

As we have already noted, many national parks were established to preserve scenic views and unusual landforms. Today, national parks are managed for scientific, educational, aesthetic, and recreational use. Since Yellowstone National Park was founded in 1872, 57 additional national parks have been established in the United States. Today the NPS manages more than 400 national parks and other areas, such as historical parks and national monuments, and the list continues to grow.

The Goals of National Park Management

Management of national parks, like that of national forests, is based on the multiple-use principle. Unlike national forests, U.S. national parks were set aside specifically to protect ecosystems. In establishing Yellowstone National Park, Congress mandated the Department of the Interior to regulate the park in a manner consistent with the preservation of timber resources, mineral resources, and "natural curiosities." However, it did not require a management process based on ecological principles.

It was not until the 1960s that ecology became a focus of national park management. In 1963, an advisory board on wildlife management presented the Leopold Report (authored by A. Starker Leopold, the oldest son of Aldo) to Secretary of the Interior Stewart Udall. This report established the guiding principles of national park management that are followed today. It proposed that the primary purpose of NPS should be to maintain the parks in the same biotic condition in which they were first found by European settlers. To this end, the authors believed that NPS should focus its efforts on conservation and on protection of wildlife species and their habitats. The report claimed that human activity had severely affected normal ecological processes in the parks and that active intervention was required to achieve the goal of a return to a more "natural" state.

Today, NPS applies knowledge gained from environmental science to maintain biodiversity and ecosystem function in all national parks. As we saw in the example of the Yellowstone fires, NPS policies continue to be controversial. Each national park adapts U.S. policy to its specific needs. In parks with high levels of endemic biodiversity, for example, management focuses on conserving endemic species. The Channel Islands National Park off the coast of southern California, for instance, is an important breeding ground for many seabirds, including a pelican species rarely found elsewhere in the western United States. Because of this, the Channel Islands National Park is managed primarily to conserve its biodiversity. Other national parks balance biodiversity protection with recreational use.

National Parks and Human Activities

Reducing the impact of human activities both outside and inside park borders is the primary challenge of most national parks throughout the world. Air and water pollution from distant sources can reduce biodiversity, recreational value, and economic opportunities. Ongoing development adjacent to park boundaries can be particularly problematic. In many locations, national parks have become islands of biodiversity amid increasing human development, but even the protected status of the parks cannot defend them completely from invasive species and other problems, such as pollution. These external threats require large-scale evaluation, planning, and management that extend beyond the borders of individual parks.

National parks are also victims of their popularity. Although the park system was established in part to make areas of great beauty accessible to people, human overuse can harm the very environment that attracts visitors. For example, all-terrain vehicles (ATVs) are a major cause of air and noise pollution in national parks, as well as a direct cause of habitat destruction (**FIGURE 30.6**). Noise pollution can harm humans and disturb animals. Today, many parks strictly limit or even ban the use of ATVs. The use and impact of snow machines (snowmobiles) for recreation is also a frequent topic of scientific analysis, public policy, and public sentiment in many of the parks in northern climates.

Still, park managers grapple with how best to determine appropriate limits on human activity. In many cases, there is no easy answer to the trade off between short-term recreational uses and long-term protection of biodiversity.

FIGURE 30.6 ATV-caused damage near Olympic National Park in Washington State. Although today many national parks limit ATV use, the conflict between those who want to use the parks for this form of recreation and those who wish to protect biodiversity remains. *(Transtock/Superstock)*

Wildlife Refuges and Wilderness Areas

National wildlife refuges are the only federal public lands managed for the primary purpose of protecting wildlife. The Fish and Wildlife Service manages more than 560 national wildlife refuges and 38 wetland management areas on 61 million hectares (150 million acres) of publicly owned land.

National wilderness areas are set aside with the intent of preserving large tracts of intact ecosystems or landscapes. Sometimes only a portion of an ecosystem is included. Wilderness areas are created from other public lands, usually national forests or rangelands, and are managed by the same federal agency that managed them prior to their designation as wilderness areas.

These areas allow only limited human use and are designated as roadless. Although logging, road building, and mining are banned in national wilderness areas, roads that existed before the designation sometimes remain in use, and activities such as mining that were previously permitted on the land are allowed to continue. More than 43 million hectares (106 million acres) of federal land in 44 states are classified as wilderness.

Federal Regulation of Land Use

Government regulation can influence the use of private as well as public lands. The 1969 **National Environmental Policy Act (NEPA)** mandates an environmental assessment of all projects involving federal money or federal permits. Along with other major laws of the 1960s and 1970s, such as the Clean Air Act, the Clean Water Act, and the *Endangered Species Act,* NEPA creates an environmental regulatory process designed to ensure protection of the nation's resources.

Before a project can begin, NEPA rules require the project's developers to file an *environmental impact statement.* An **environmental impact statement (EIS)** typically outlines the scope and purpose of the project, describes the environmental context, suggests alternative approaches to the project, and analyzes the environmental impact of each alternative. NEPA does not require that developers proceed in the way that will have the least environmental impact. However, in some situations, NEPA rules may stipulate that building permits or government funds be withheld until the developer submits an **environmental mitigation plan** stating how it will address the project's environmental impact. In addition, preparation of the EIS sometimes uncovers the presence of endangered species in the area under consideration. When this occurs, managers must apply the protection measures of the **Endangered Species Act**, a 1973 law designed to protect species from extinction.

Members of the public are entitled to comment on the environmental assessment and decision makers are required to respond. Although developers are not obligated to act in accordance with public wishes, in practice public concern often improves the project's outcome. For this reason, attending informational sessions and providing input is a good way for concerned citizens to learn more about local land use decisions and to help reduce the environmental impact of land development.

Residential land use is expanding

Roughly a century ago, many people in the United States began to move from rural areas to large cities. The last 50 years have seen an increased movement from those cities to the surrounding areas, which has resulted in the growth of two classes of communities: *suburban* and *exurban.* **Suburbs** are areas that surround metropolitan centers and have low population densities compared with urban areas. **Exurbs** are similar to suburbs, but are not connected to any central city or densely populated area. Since 1950, more than 90 percent of the population growth in metropolitan areas has occurred in suburbs, and two out of three people now live in suburban or exurban communities.

The U.S. census bureau currently categorizes a population as urban if the area contains more than 2,500 people. **FIGURE 30.7** on page 360 shows U.S. urban and rural populations between 1910 and 2012. The rural population in the United States has been roughly the same since 1910, and a century later it makes up roughly a sixth of the total U.S. population. Although this population shift from rural to urban areas has brought with

National wildlife refuge A federal public land managed for the primary purpose of protecting wildlife.

National wilderness area An area set aside with the intent of preserving a large tract of intact ecosystem or landscape.

National Environmental Policy Act (NEPA) A 1969 U.S. federal act that mandates an environmental assessment of all projects involving federal money or federal permits.

Environmental impact statement (EIS) A document outlining the scope and purpose of a development project, describing the environmental context, suggesting alternative approaches to the project, and analyzing the environmental impact of each alternative.

Environmental mitigation plan A plan that outlines how a developer will address concerns raised by a project's impact on the environment.

Endangered Species Act A 1973 U.S. act designed to protect species from extinction.

Suburb An area surrounding a metropolitan center, with a comparatively low population density.

Exurb An area similar to a suburb, but unconnected to any central city or densely populated area.

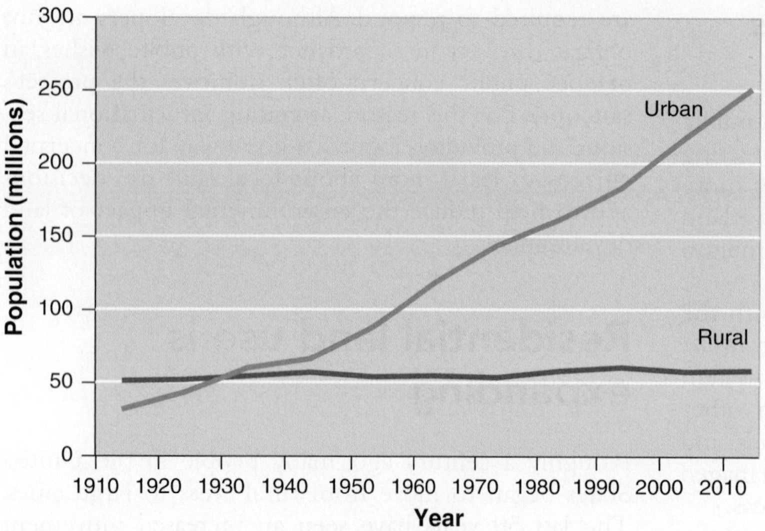

FIGURE 30.7 Distribution of urban and rural populations in the United States between 1910 and 2012. This graph shows a dramatic shift in the population from rural to urban areas. *(Data from http:www.census.gov/geo/reference/ua/urban-rural-2012.html)*

it a new set of environmental problems associated with *urban sprawl,* attempts to find creative solutions have been increasingly successful.

Causes and Consequences of Urban Sprawl

If you have ever been to a strip mall, you are familiar with the phenomenon known as **urban sprawl**—urbanized areas that spread into rural areas and remove clear boundaries between the two. The landscape in these areas is characterized by clusters of housing, retail shops, and office parks, which are separated by miles of road. Large feeder roads and parking lots that separate "big box" and other retail stores from the road discourage pedestrian traffic.

Urban sprawl has had a dramatic environmental impact. Dependence on the automobile causes suburban residents to drive more than twice as much as people who live in cities. Between 1950 and 2000, the number of vehicle miles traveled per person in U.S. suburban areas tripled. Because suburban house lots tend to be significantly larger than their urban counterparts, suburban communities also use more than twice as much land per person as urban communities. Urban sprawl tends to occur at the edge of a city, often replacing farmland and increasing the distance between farms and consumers. In its most recent survey, the U.S. Department of Agriculture estimated that between 1982 and 2007, U.S. farmland was being converted to residential uses at a rate of 405,000 ha (1 million acres) per year. There are four main causes of urban sprawl in the United States: automobiles and highway construction, living costs, urban blight, and government policies.

Urban sprawl Urbanized areas that spread into rural areas, removing clear boundaries between the two.

Automobiles and Highway Construction

Before automobiles and highway systems existed, transportation into and out of cities was difficult: Horses were slow, roads were bad, and trolley services rarely went far beyond the city limits. In those days, if you wanted to take advantage of the many amenities of city life, such as job opportunities, cultural institutions, shopping, and social activities, you had to live within a few miles of the city center.

The advent of the automobile, and the subsequent development of the interstate highway system in the 1950s and 1960s, changed everything. Today we think nothing of working in the city during the day and commuting home to the suburbs at night. And if you live in the suburbs but want to get into the city on a Saturday night for dinner, a concert, or a movie, that's no problem either. With rapid, comfortable transportation by automobile between urban and suburban areas, it became possible to work or play in the city and then return to a large home in a quiet neighborhood. For the first time in history, people could enjoy the best of both worlds.

Living Costs

Land in the suburbs, because it is readily available, is relatively inexpensive when compared with land in the city. For the cost of a tiny one-bedroom condominium in a desirable section of a city, you may be able to purchase a five-bedroom house with a big yard in the suburbs. In addition, because suburban governments often provide fewer public services than cities, tax rates in the suburbs are likely to be lower. For these reasons, moving to the suburbs allows some people to enjoy a higher standard of living than they could afford in the city.

Although single-family homes in the suburbs are likely to be less expensive than desirable housing in the city, they are still financially unobtainable for many people. And because single-family homes are usually the only housing option available in the suburbs, those with lower incomes are excluded. Most suburban communities have little, if any, low-income or "affordable" housing. Furthermore, in many suburban locations, it is difficult to commute to the city without a car, which compounds the difficulties for lower-income individuals. Even when public transportation is available, commuting costs can be high.

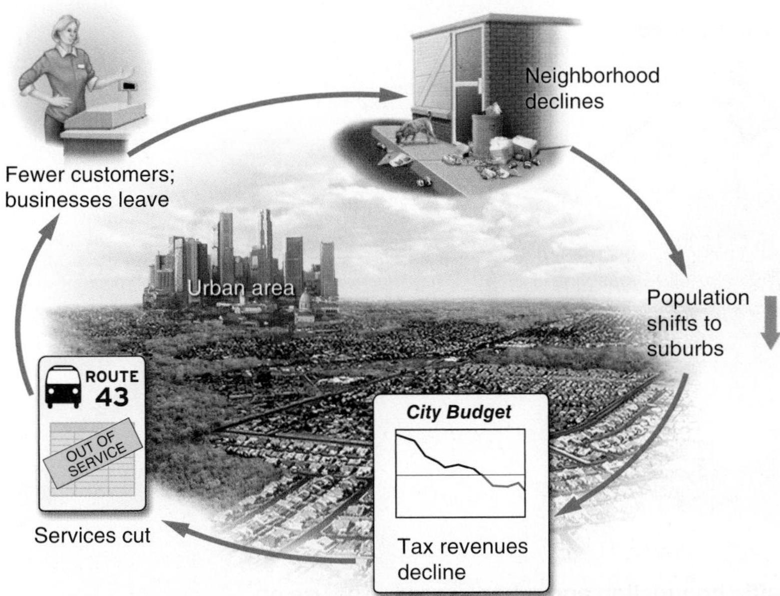

FIGURE 30.8 Urban blight. As people move away from a city to suburbs and exurbs, the city often deteriorates, which causes yet more people to leave. This cycle is an example of a positive feedback system. The green arrow indicates the starting point of the cycle.

Urban Blight

We have seen that urban sprawl encourages population shifts to the suburbs. As people leave the city, city revenue from sources such as property, sales, and service taxes begins to shrink. However, the cost of maintaining urban services, including public transportation, police and fire protection, and social services, does not decrease. Faced with declining tax receipts, cities are forced to reduce services, raise tax rates, or both. As services decline, crime rates may increase, either because police resources are stretched thin or because conditions for lower-income residents decline even further. Infrastructure also deteriorates, leading to a decline in the quality of the built environment. These problems, combined with higher taxes, have made cities less attractive places to live than in previous decades, and those who can afford to move away have often done so.

In addition, as the population shifts to the suburbs, jobs and services follow. Suburbanization has spawned suburban office parks, which have led to an increase in suburb-to-suburb commuting. Commuting patterns develop around cities rather than into and out of them, and these new traffic patterns make it more difficult to provide public transportation to the spreading region. As wealthy and middle-income people leave cities, urban retail stores lose customers and are forced to close, giving people even fewer reasons to go to the city to shop. This further decreases the customer base for the remaining stores. This cascade of effects leads to the positive feedback loop shown in **FIGURE 30.8**. This loop creates **urban blight**—the degradation of the built and social environments of the city that often accompanies and accelerates migration to the suburbs.

Historically, urban blight has contributed to racial segregation. In the 1950s and 1960s, when migration to the suburbs began in earnest, those leaving the cities for the suburbs were predominantly middle- or upper-income Caucasians. This so-called "white flight" resulted in highly concentrated minority populations in city centers and almost entirely white populations in the suburbs. Over time, the disparity of opportunity increased because tax revenues in suburban communities often allowed for better schools. In a number of U.S. cities, some of this segregation has begun to decline as Caucasians return to the cities and more minorities have accumulated enough wealth to move out to the suburbs. In addition, many young professionals are making the decision to work and live in cities. Nevertheless, racial disparities remain.

Government Policies

Urban sprawl has also been enhanced by federal and local laws and policies, including the Highway Trust Fund, zoning laws, and subsidized mortgages.

The **Highway Trust Fund**, begun by the Highway Revenue Act of 1956 and funded by a federal gasoline

Urban blight The degradation of the built and social environments of the city that often accompanies and accelerates migration to the suburbs.

Highway Trust Fund A U.S. federal fund that pays for the construction and maintenance of roads and highways.

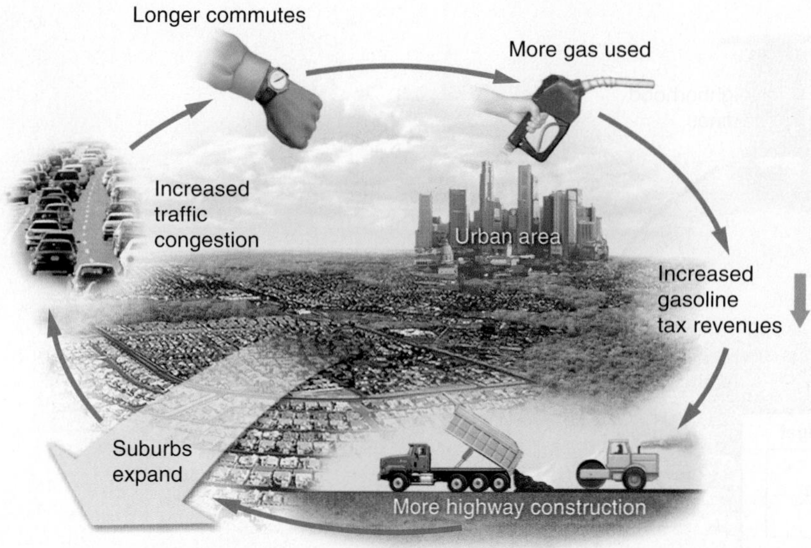

FIGURE 30.9 Induced demand as a cause of traffic congestion and urban sprawl. The use of gasoline tax money to build highways leads to the development of suburbs and traffic congestion, at which point yet more money is spent on highways to alleviate the congestion. The green arrow indicates the starting point of the cycle.

tax, pays for the construction and maintenance of roads and highways. We have already seen that highways allow people to live farther from where they work. **FIGURE 30.9** shows the resulting positive feedback loop. More highways mean more driving and more gasoline purchases, which lead to more gasoline tax receipts that pay for more roads, and so on. As people move farther away from their jobs, traffic congestion increases, and roads are expanded. But the new, larger roads encourage even more people to live farther away from work. This cycle exemplifies a phenomenon known as **induced demand**, in which an increase in the supply of a good causes demand to grow.

Governments may use *zoning* to address issues of traffic congestion, urban sprawl, and urban blight. **Zoning** is a planning tool developed in the 1920s and used to separate industry and business from residential neighborhoods, thereby creating quieter, safer communities. Governments that use zoning can classify land areas into "zones" in which certain land uses are restricted. For instance, zoning ordinances might prohibit developers from building a factory or a strip mall in a residential area or a multi-dwelling apartment building in a single-home neighborhood. Nearly all metropolitan governments across the United States have adopted zoning.

Zoning often regulates much more than land use. The number of parking spaces a building must have, how far from the street a building must be placed, or even the size and location of a home's driveway are among the development features that zoning may stipulate. Zoning laws have been helpful in addressing issues of safety and sometimes in minimizing environmental damage caused by new construction. One negative aspect of zoning, however, is that it generally prohibits suburban neighborhoods from developing a traditional "Main Street" with shops, apartments, houses, and businesses clustered together. Many communities are now attempting to incorporate **multi-use zoning**, which allows retail and high-density residential development to coexist in the same area. However, most zoning in the United States continues to promote automobile-dependent development.

The federal government has also played a large part in encouraging the growth of suburbs through the Federal Housing Administration (FHA). Congress established the FHA during the Great Depression of the 1930s, in part to jump-start the economy by creating more demand for new housing. Through the FHA, people could apply for federally subsidized mortgages that offered low interest rates, which allowed many people who otherwise could not afford a house to purchase one. However, the FHA financed mortgages only for homes in financially "low risk" areas that were almost always the newly built, low-density suburbs. At the end of World War II, the GI Bill extended generous credit terms to war veterans, and the suburban housing

Induced demand The phenomenon in which an increase in the supply of a good causes demand to grow.

Zoning A planning tool used to separate industry and business from residential neighborhoods.

Multi-use zoning A zoning classification that allows retail and high-density residential development to coexist in the same area.

boom continued. Suburbs became even more attractive as cars became more affordable and mass-produced housing allowed developers to sell suburban homes at affordable prices.

Smart Growth

People are beginning to recognize and address the problems of urban sprawl. One approach, **smart growth**, focuses on strategies that encourage the development of sustainable, healthy communities. The Environmental Protection Agency lists 10 basic principles of smart growth:

1. *Create mixed land uses.* Smart growth mixes residential, retail, educational, recreational, and business land uses. Mixed-use development allows people to walk or bicycle to various destinations and encourages pedestrians to be in a neighborhood at all times of the day, increasing safety and interpersonal interactions.
2. *Create a range of housing opportunities and choices.* By providing housing for people of all income levels, smart growth counters the concentration of poverty in failing urban neighborhoods. Mixed housing also allows more people to find jobs near where they live, improves schools, and generates strong support for neighborhood transit stops, commercial centers, and other services.
3. *Create walkable neighborhoods.* Walkable neighborhoods are created by mixing land uses, reducing the speed of traffic, encouraging businesses to build stores directly up to the sidewalk, and placing parking lots behind buildings. In neighborhoods that encourage walking, people use their cars less, reducing fossil fuel use and traffic congestion and providing health benefits. Communities with more pedestrians tend to see more interaction among neighbors because people stop to talk with each other. This, in turn, creates opportunities for civic engagement. When people interact, the environment usually benefits.
4. *Encourage community and stakeholder collaboration in development decisions.* There is no single "right" way to build a neighborhood; residents and **stakeholders**—people with an interest in a particular place or issue—need to work together to determine how their neighborhoods will appear and be structured.
5. *Take advantage of compact building design.* Smart growth incorporates multistory buildings and parking garages—as opposed to sprawling parking lots—to reduce a neighborhood's environmental footprint and protect more open space. Ideally, shops, cafés, and small businesses should be easily accessible to pedestrian traffic on the ground floor near sidewalks, with two or three stories of apartments and offices above.

FIGURE 30.10 The French Quarter of New Orleans. One principle of smart growth is to foster communities with a strong sense of place. The French Quarter of New Orleans, Louisiana, is known for its architecture, food, and, especially, music. *(Design Pics Inc/Alamy)*

6. *Foster distinctive, attractive communities with a strong sense of place.* A **sense of place** is the feeling that an area has a distinct and meaningful character. Many cities have such unique neighborhoods. For example, the French Quarter of New Orleans (**FIGURE 30.10**) has a sense of place that adds to the quality of life for people living there. Smart growth attempts to foster this sense of place through development that fits into the neighborhood.
7. *Preserve open space, farmland, natural beauty, and critical environmental areas.* Working farmland is a source of fresh local produce and other goods. Open space provides opportunities for recreation and enjoyment. Land protected from development through restricted growth also provides habitats for a variety of species.
8. *Provide a variety of transportation choices.* **Transit-oriented development (TOD)** attempts to focus dense residential and retail development around public transportation stops, giving people convenient alternatives to driving. Bicycle racks and safe roads, pleasant sidewalks for walking, frequent bus service, and light rail service can all aid in this goal. Car-sharing networks such as Zipcar can provide easy access to a fleet of rental automobiles. This increasingly popular service

Smart growth A set of principles for community planning that focuses on strategies to encourage the development of sustainable, healthy communities.

Stakeholder A person or organization with an interest in a particular place or issue.

Sense of place The feeling that an area has a distinct and meaningful character.

Transit-oriented development (TOD) Development that attempts to focus dense residential and retail development around stops for public transportation, a component of smart growth.

reduces the need for private car ownership where public transportation is not available. There is also a trend of developing recreational bicycle and walking trails on abandoned railroad corridors (**FIGURE 30.11**).

9. *Strengthen and direct development toward existing communities.* Development strategies can help to reinvigorate urban neighborhoods that are caught in a vicious cycle of depopulation and blight and can protect rural lands from sprawl. Development that fills in vacant lots within existing communities, rather than expanding into new land outside the city, is known as **infill**. Some cities, such as Portland, Oregon, have had success with **urban growth boundaries**, which place restrictions on development outside a designated area.
10. *Make development decisions predictable, fair, and cost-effective.* Suburban developments throughout the country often look the same because standardized designs allow developers to move rapidly through the permitting process. A streamlined approval process that encourages smart growth could increase the number of individualized plans rather than promoting cookie-cutter development.

Of course, no individual new development or neighborhood plan is likely to incorporate all of these ideals, but as a guide to thinking about how to build communities, the smart growth concept has been quite successful.

AP® Exam Tip

You should be familiar with all the factors that facilitate smart growth and have an example or two in mind for the AP® Environmental Science Exam. ●

Infill Development that fills in vacant lots within existing communities rather than expanding into new land outside the city.

Urban growth boundary A restriction on development outside a designated area.

FIGURE 30.11 Railtrail. Because old railroad beds, like this one in northern Virginia, are typically flat and extend for long distances, they make excellent paths for walking, running, and biking. *(SCPhotos/Alamy Stock Photo)*

Smart growth can have important environmental benefits. For example, compact development reduces the amount of impervious surface, and allows room for more trees to be planted, both of which reduce runoff and flooding downstream. A 2000 study found that smart growth in New Jersey would reduce water pollution by 40 percent compared with the more common, dispersed growth pattern. By mixing uses and providing transportation options, smart growth also reduces fossil fuel consumption. A 2005 study in Seattle found that residents of neighborhoods incorporating just a few of the techniques to make non-auto travel more convenient traveled 26 percent fewer vehicle miles than residents of more dispersed, less-connected neighborhoods. A 2012 San Francisco State University study found that similar smart growth practices led to a 20 percent reduction in miles traveled in 18 urban areas across the United States.

MODULE 30 AP® Review

Preparing for the AP® Exam

In this module, we saw that land management varies by land type. Rangelands are managed to maintain ecological processes such as nutrient cycles and to preserve water quality. Forests may be clear-cut or selectively cut. Clear-cutting may have greater environmental consequences than selective cutting. Tree plantations are sometimes established as replacement forest after clear-cutting. Management of forests may include the use of fire. Controls on residential land use are expanding and urban and suburban sprawl are major environmental problems. Smart growth is a response to the problems of sprawl and includes many innovative strategies that are being implemented around the United States.

AP® Practice Questions

Choose the best answer for the following.

1. Which is NOT a problem in rangeland management?
   (a) fires
   (b) overgrazing
   (c) poor soil conditions
   (d) the high cost of grazing permits

2. Which is used to reduce the impact of timber harvesting?
   (a) tree plantations
   (b) selective cutting
   (c) prescribed burns
   (d) mechanized logging

3. The use of environmental impact statements was mandated by
   (a) the Endangered Species Act of 1973.
   (b) the Leopold Report in 1963.
   (c) the Clean Air Act of 1963.
   (d) the National Environmental Policy Act of 1969.

4. Which is NOT a cause of urban sprawl?
   (a) automobile and highway construction
   (b) mixed land use
   (c) living costs
   (d) urban blight

5. The creation of walkable neighborhoods is
   (a) a result of urban sprawl.
   (b) one goal of the Federal Housing Administration.
   (c) discouraged by multi-use zoning.
   (d) a principal of smart growth.

Working Toward Sustainability

What Are the Ingredients for a Successful Neighborhood?

The Dudley Street area of Roxbury and North Dorchester in Boston was once a prime example of urban blight. In the 1980s, years of urban decay and the loss of large numbers of primarily Caucasian families to the suburbs had left 21 percent of the Dudley Street neighborhood vacant—amounting to 1,300 abandoned parcels. Almost 30 percent of neighborhood residents had an income below the federal poverty level, making the neighborhood one of the poorest in Boston. Fires were a particular problem; in some cases, arsonists attempted to gain insurance money on homes that could not be sold. The residents felt that they were being ignored by City Hall. Some suggested that this was because 96 percent of the neighborhood's residents were members of ethnic minority groups. Tony Hernandez remembers this landscape from his childhood when he played on vacant land in the Dudley Street neighborhood. Today he owns a home in the same neighborhood, one of more than 200 units of permanently affordable housing. His community contains parks and gardens, a town commons, a charter school, and working farms.

Tony and his neighbors enjoy their successful neighborhood in part because almost 3 decades ago, in response to the urban decay, residents banded together to turn their neighborhood around. They formed the Dudley Street Neighborhood Initiative (DSNI), designed to allow the residents to move toward a common vision for a sustainable neighborhood. The DSNI still exists today. Perhaps a key to its success is that it obtained something no other neighborhood organization had at that time: the power of *eminent domain*. **Eminent domain** allows a government to acquire property at fair market value even if the owner does not wish to sell it. It is frequently used to acquire land for highway projects, but also has been used recently, and controversially, in urban redevelopment.

By the late 1980s, DSNI had worked with community members to develop a comprehensive revitalization plan, which has been periodically updated since then and still exists today. The main goals of the DSNI plan are

- to rehabilitate existing housing
- to construct homes that are affordable according to criteria set by residents
- to assemble parcels of vacant land for redevelopment, using the power of eminent domain if necessary

Eminent domain A principle that grants government the power to acquire a property at fair market value even if the owner does not wish to sell it.

- to plan environmentally sound, affordable development that is physically attractive
- to convert some vacant properties into safe play areas, gardens, and facilities that the entire community can enjoy
- to increase both the economic and political power of residents, who are the stakeholders

A Dudley Street neighborhood event. This neighborhood in Boston was once a symbol of urban decay. Today it is a thriving urban community that has adopted many of the principles of smart growth. *(Dudley Street Neighborhood Initiative)*

Whether in the Dudley Street area of Boston or elsewhere, stakeholders with the help of local and national agencies have been using the principles of smart growth to turn urban areas into vibrant communities with shops, recreation, housing, and schools. The environmental impact of living in these revitalized urban areas is much less than that of a typical suburban community where people commute to work and school and drive to shop.

Critical Thinking Questions

1. What are some of the features of a community that will make it sustainable?
2. List examples of how the Dudley Street neighborhood engaged in "smart growth."
3. List and describe three ideas for "smart growth" in your own community.

References

Dudley Street Neighborhood Initiative. http://www.dsni.org/

Loh, Penn. January 28, 2015. How one Boston neighborhood stopped gentrification in its tracks. http://www.yesmagazine.org/issues.

Chapter 10 Review

In this chapter we have looked at issues involving human use of land. The tragedy of the commons suggests that common-use resources may be over-exploited. Maximum sustainable yield is a management technique that should allow for the greatest amount of harvest from a resource. Public lands are classified in a variety of ways around the world and in the United States. Rangeland and forest land are managed differently. In forests, clear-cutting and selective cutting are two ways to manage tree removal. Fire is also a tool used to manage forests. An increasingly greater percentage of the U.S. population lives in or near cities, a fact that has created a number of environmental problems. Some of these problems are being addressed by the principals of smart growth.

Key Terms

Tragedy of the commons
Externality
Maximum sustainable yield (MSY)
Resource conservation ethic
Multiple-use lands
Rangeland
Forest
Clear-cutting
Selective cutting
Ecologically sustainable forestry
Tree plantation
Prescribed burn
National wildlife refuge
National wilderness area
National Environmental Policy Act (NEPA)
Environmental impact statement (EIS)
Environmental mitigation plan
Endangered Species Act
Suburb
Exurb
Urban sprawl
Urban blight
Highway Trust Fund
Induced demand
Zoning
Multi-use zoning
Smart growth
Stakeholder
Sense of place
Transit-oriented development (TOD)
Infill
Urban growth boundary
Eminent domain

Learning Goals Revisited

Module 29 Land Use Concepts and Classification

Explain how human land use affects the environment.

Individuals have no incentive to conserve common resources when they do not bear the cost of using those resources. A cost or benefit not included in the price of a good is an externality. The lack of incentive to conserve common resources leads to the overuse of these resources, which may be degraded if their use is not regulated. The maximum sustainable yield (MSY) is the largest amount of a renewable resource that can be harvested indefinitely. Harvesting at the MSY keeps the resource population at about one-half the carrying capacity of the environment.

Describe the various categories of public land used globally and in the United States.

In the United States, public land is managed for multiple uses, including grazing, timber harvesting, recreation, and wildlife conservation. The Bureau of Land Management manages rangeland, which is used for grazing. The United States Forest Service manages national forests, which are used for timber harvesting, recreation, and other uses.

Module 30 Land Management Practices

Explain specific land management practices for rangelands and forests.

Rangelands are grazed by cattle and sheep and need to be managed to prevent overgrazing. Timber can be harvested by clear-cutting or selective cutting, both of which have environmental impacts, or by ecologically sustainable forestry methods.

Describe contemporary problems in residential land use and some potential solutions.

Causes of urban sprawl include the development of the automobile, construction of highways, less expensive land at the urban fringe, and urban blight. Government institutions and policies, such as the federal Highway Trust Fund, zoning, and subsidized mortgages, have also contributed to the problem. The result of urban sprawl is automobile dependence, traffic congestion, and social isolation, including less involvement in community affairs. Smart growth is one possible response to urban sprawl. It advocates more compact, mixed-use development that encourages people to walk, bicycle, or use public transportation. Smart growth consumes less land and offers numerous other environmental benefits.

Practice Math and Graphing

Preparing for the AP® Exam

Answer the following questions. Be sure to show all your work.

1. Practice Math

(a) If U.S. farmland is being converted to residential uses at a rate of 405,000 ha (1 million acres) per year, how much land will be converted in 10 years?

(b) If there is roughly 369 million ha (911 million acres) of farmland in the United States, what percentage is lost each decade?

2. Practice Graphing

(a) Draw a maximum sustainable yield graph based on the data shown in the table. The x axis should show the fish population in metric tons. The y axis should show the annual removal of fish in tons of fish per year.

(b) Determine the MSY for a fishery based on this graph.

Fish population (metric tons)	Annual removal of fish (tons per year)
100	30
200	60
400	95
600	95
800	60
900	20

Chapter 10 AP® Environmental Science Practice Exam

Preparing for the AP® Exam

Section 1: Multiple-Choice Questions

Choose the best answer for questions 1–20.

1. Which is NOT an example of the tragedy of the commons?
   (a) overgrazing by sheep on community-owned pastures
   (b) depletion of fish stocks in international waters
   (c) automobile congestion in Yellowstone National Park
   (d) depletion of soil minerals by farmers on private land

Use the following graph to answer Question 2.

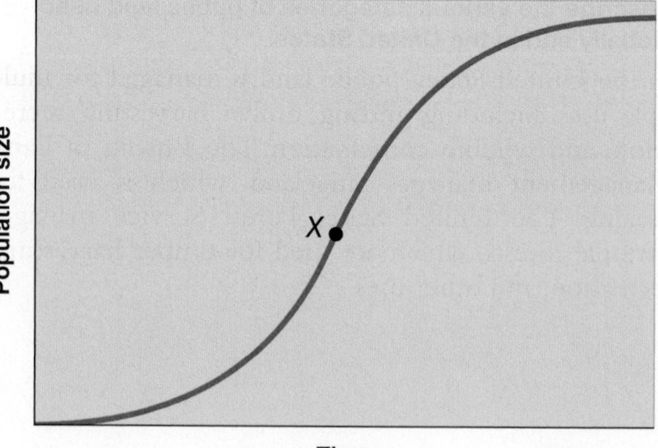

2. This graph shows the population growth of the common pheasant, one of the world's most hunted birds. On the graph, X represents
   (a) carrying capacity.
   (b) maximum sustainable yield.
   (c) resource depletion.
   (d) population overshoot.

3. Under the provisions of the National Environmental Policy Act (NEPA), which of the following would require the preparation of an environmental impact statement (EIS)?
   (a) construction of a house on privately owned land
   (b) paving a parking lot for a local business
   (c) expansion of an interstate highway
   (d) planting trees in front of City Hall
4. Federally owned land in the United States can best be described as
   (a) 28 percent of all U.S. land, with the majority of it in the west.
   (b) 42 percent of all U.S. land, with the majority of it in the east.
   (c) 25 percent of all U.S. land, with most of it in Texas.
   (d) 20 percent of all U.S. land, with 10 percent of it in the west.
5. Which type of protected area is least likely to support tourism and recreation?
   (a) protected landscapes and seascapes
   (b) national parks
   (c) managed resource protected areas
   (d) habitat/species management areas
6. The four major public land management agencies in the United States operate under the principle of multiple use. Which use is common to lands managed by all four agencies?
   (a) mining
   (b) grazing
   (c) timber harvesting
   (d) recreation
7. For many years, forest fires were suppressed to protect lives and property. This policy has led to
   (a) a buildup of dead biomass that can fuel larger fires.
   (b) many forest species being able to live without having their habitats destroyed.
   (c) increased solar radiation in most ecosystems.
   (d) soil erosion on steep slopes.
8. When we purchase an item, we are charged for the labor and supply costs of producing that item. However, we are not charged for the costs of any environmental damage that occurred in manufacturing that item. Those costs are known as
   (a) externalities.
   (b) environmental mitigation.
   (c) infill.
   (d) marginal costs.
9. Which is NOT an environmental consequence of clear-cutting?
   (a) increased soil erosion and sedimentation in nearby streams
   (b) decreased biodiversity due to habitat fragmentation
   (c) increased fish populations due to the influx of nutrients into streams
   (d) stands of same-aged trees
10. Which are environmental impacts of urban sprawl?
    I. greater reliance on the automobile and increased fossil fuel consumption
    II. increased consumption of land for housing and highway construction
    III. loss of valuable farmlands
    (a) I only
    (b) II only
    (c) I and II
    (d) I, II, and III
11. Which has contributed to urban sprawl over the past 50 years?
    (a) migration of people from suburban areas to rural areas
    (b) increased availability of public transportation
    (c) lower property taxes in urban areas
    (d) use of the federal gasoline tax to construct and maintain highways
12. Which is NOT an environmental benefit of smart growth?
    (a) reduced flooding
    (b) increased impervious surfaces
    (c) reduced fossil fuel consumption
    (d) increased open space
13. Which is NOT likely to lead to a tragedy of the commons?
    (a) power plants that emit air pollution
    (b) farmers who siphon water from a county aqueduct for irrigation
    (c) office staff who use a common coffee machine
    (d) a ranch that charges farmers to board their horses

Question 14 refers to the following graph.

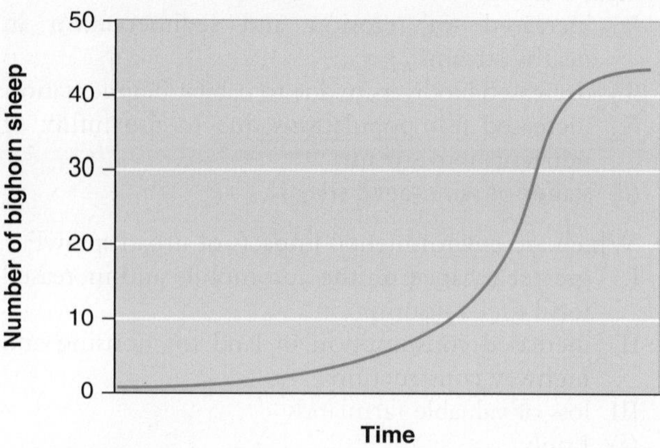

14. The graph shows the population growth of bighorn sheep through time for a 10-km^2 plot without predators. In a more realistic scenario with both sheep and wolf predators, how many bighorn sheep can hunters harvest without compromising the future of the sheep population?
    (a) 0
    (b) 10
    (c) 30
    (d) 35

15. Which is NOT an outcome of a prescribed forest burn?
    (a) Maximum carbon storage in the forest is achieved.
    (b) Nutrients in dead material are liberated for primary producers.
    (c) The risk of forest fires is minimized.
    (d) Habitat is provided for early-succession species.

16. Exurban communities are similar to
    (a) urban communities with low population densities.
    (b) suburban communities with low population densities.
    (c) suburban communities not connected to any central city.
    (d) urban communities whose services are no longer funded by a major city's government.

17. One U.S. city recently held a town hall meeting to decide on the best way to address growing problems of urban sprawl. Several suggestions were considered. Which suggestion does NOT follow the EPA's principles of smart growth?
    (a) fund individual communities within the city to develop themes that allow each community to become distinct from adjacent neighborhoods
    (b) provide public transportation for routes that circle the city in order to encourage driving of personal cars within the city
    (c) involve neighborhood residents in making decisions regarding regulation of local services and development of new infrastructure
    (d) develop apartments and offices in the same areas

18. The Kingdom of Bhutan has an area of approximately 38,000 km^2, about 50 percent of which is protected. How many hectares of Bhutan are protected?
    (a) 1.9×10^4 ha
    (b) 1.9×10^5 ha
    (c) 7.6×10^4 ha
    (d) 7.6×10^5 ha

19. Which is NOT an example of an externality?
    (a) air pollution from motor vehicles
    (b) increased crop pollination around a managed honeybee colony
    (c) increased housing development in coastal areas
    (d) antibiotic resistance caused by increased use of antibiotics

20. The Galápagos, which is actively maintained to protect native flora and fauna, is an example of a
    (a) habitat/species management area.
    (b) strict nature reserve.
    (c) national park.
    (d) protected landscape/seascape.

Section 2: Free-Response Questions

Write your answer to each part clearly. Support your answers with relevant information and examples. Where calculations are required, show your work.

1. The property pictured is the Farm Barn at Shelburne Farms, a National Historic Landmark, nonprofit environmental education center, and 1,400-acre working farm on the shores of Lake Champlain in Vermont. However, for the sake of this exercise, let's assume that the property pictured below belongs to the federal government.

(Farrell Grehan/Getty Images)

   (a) Identify and explain which of the four public land management agencies would be involved in managing this public land. (2 points)
   (b) Applying any three of the basic principles of smart growth, explain how the private land surrounding this federally owned property might be developed to minimize environmental impacts. (4 points)
   (c) Define *environmental impact statement* and describe one condition under which an EIS might be required for the use of either the privately owned or the federally owned lands associated with this tract. (4 points)

2. The town of Fremont met recently to discuss the pros and cons of protecting prairie dogs. Prairie dogs are burrowing rodents the size of rabbits that live in colonies underground in grasslands and prairies. Their numbers have been greatly reduced over the last few decades. Dr. Masser, a local biologist, pointed out that prairie dogs are an important part of the prairie food web, as they are prey for many birds and mammals. Without federal protection from both the Bureau of Land Management and the U.S. Fish and Wildlife Service, they could become extinct in a few years. Dr. Masser also explained that two of the five species of prairie dogs are already listed as either threatened or endangered. Local ranchers disagreed with Dr. Masser. Mr. Smith, a Fremont area rancher, stated that he will continue to poison or shoot the prairie dogs because they destroy the grasses that are needed by his livestock, and he encouraged the BLM to do the same on public lands.
   (a) Explain the tragedy of the commons in general terms. Using the information you just read about the prairie dog and any other relevant information, incorporate the town of Fremont's discussion into your explanation. (4 points)
   (b) Identify and discuss ONE argument in favor of preserving western grasslands as habitat for prairie dogs and ONE argument in favor of maintaining those grasslands for the grazing of livestock. (3 points)
   (c) Identify ONE action that the Bureau of Land Management and one action that the U.S. Fish and Wildlife Service could take to resolve this land use conflict. (3 points)

3. A species of salmon is reintroduced to a river from which it had previously been overfished to extinction. Each year scientists measure the population, until it reaches carrying capacity. The table shows the results.

Year	**Salmon population (metric tons)**
2008	1000
2009	1100
2010	1500
2011	2000
2012	3000
2013	5000
2014	6000
2015	6800
2016	7200
2017	7250

   (a) Plot the data in a graph, and draw an arrow indicating at what population size the maximum sustainable yield (MSY) should likely be. Explain why you chose that point. (6 points)
   (b) Describe TWO problems that can occur when calculating or using MSY. (2 points)
   (c) Explain the tragedy of the commons, and give an example using the salmon fishery. (2 points)

Joel Salatin's Polyface Farm has become a showcase for agricultural techniques that are consistent with ecological principles. *(Greg Kahn Photography)*

Feeding the World

CHAPTER 11

MODULE 31 Human Nutritional Needs

MODULE 32 Modern Large-Scale Farming Methods

MODULE 33 Alternatives to Industrial Farming Methods

CASE STUDY

A Farm Where Animals Do Most of the Work

It is common for a farmer growing juicy tomatoes or sweet apples to become a local hero, but how often does a farmer become known throughout an entire country?

Joel Salatin of Swoope, Virginia, has been featured in a best-selling book and two major motion pictures. He is the author of numerous popular books, including *Holy Cows and Hog Heaven: The Food Buyer's Guide to Farm Friendly Food* and *Everything I Want to Do Is Illegal: War Stories from the Local Food Front.* His farming has drawn the attention of authors, filmmakers, and many others because he raises vegetables and livestock in a sustainable, organic way. In 2015, Joel and Lucille Salatin and their farm were the main subject of the documentary *Polyfaces: A World of Many Choices*. The film followed Polyface Farm over 4 years and explained their philosophy and farming techniques. Thousands of people visit Polyface Farm every year. Polyface Farm also offers a series of self-study and home study courses. So what's so special about this farm and what is Salatin teaching?

Joel Salatin understands that farming, done properly, has to replicate many of the processes that exist in nature. Salatin calls himself a "grass farmer" because grass is at the foundation of the trophic feeding pyramid. On his Polyface Farm, he grows grass that is harvested, dried by the Sun, and turned into hay, which he feeds to cattle. The cattle are processed into beef and consumed by humans. But that is only one piece of the intricate food chain that Salatin has created on his farm—all without chemical pesticides or synthetic fertilizers.

Salatin understands that farming, done properly, has to replicate many of the processes that exist in nature.

As cattle graze a field, they leave behind large piles of manure that become the energy source for large numbers of grubs and fly larvae. Before the larvae can hatch into flies and cause a problem on the farm, Salatin brings in his "sanitation crew"—not people spraying pesticides, but chickens. As the chickens eat the larvae, they spread the nitrogen-rich cow manure and deposit some of their own. They continue to lay eggs, which are tasty and nutritious. Salatin explains that his method is similar to a process in nature in which birds typically follow herbivores to take advantage of a wide range of food offerings they generate by attracting insects and leaving in their path tissue, blood, and organic waste.

Salatin uses other animals to take care of the buildup of cow manure during winter when his cattle remain in a barn. Some farmers push that manure out the doorway and end up with huge manure piles in the spring, but Salatin continuously layers straw, which consists of dried stalks of plants, and wood chips on top of the manure. He allows the floor of his barn to rise higher and higher during winter. Occasionally he throws down some corn. During the winter, the manure-straw-corn layers decompose and produce heat, which allows the cattle to use less energy and their food to stay warm. In spring, when the cattle go out to pasture, he brings pigs into the barn. Pigs have an excellent sense of smell, and they dig up the layers of manure and straw with their snouts to get at the fermented corn that they find especially delectable. Within a few weeks, the pigs provide Salatin with a barn full of thoroughly mixed compost that required no machinery to make: His animals mixed the manure and straw for him as they gained nutrients and calories.

These are just two examples of the ways one farmer uses natural processes and knowledge of plants, animals, and the natural world to grow food for human consumption. Joel Salatin and other farmers interested in sustainable farming are finding ways to feed the human population by working with, not against, the natural world.

Sources: M. Pollan, *The Omnivore's Dilemma,* Penguin, 2007; I. Doherty and L. Heenan, *Polyfaces: A World of Many Choices* (Documentary Film), Regrarians Media, 2015.

Most agricultural methods, especially modern ones practiced in the developed countries of the world, tend to fight the basic laws of ecology in order to provide large amounts of food for humans. These modern methods also consume substantial quantities of fossil fuel in every step of the process, from the use of machinery in plowing and harvesting crops, to the production of fertilizers and pesticides to the transportation and processing of the food. However, there are ways to grow food efficiently without using as much fossil fuel, manufactured fertilizers, and pesticides as generally used today. In this chapter, we will explore the current state of human nutrition around the world, and examine conventional, large-scale methods of agriculture. We will also look at alternatives to conventional agriculture including sustainable and organic agriculture, which have the potential to reduce adverse impacts of agriculture on the environment.

MODULE 31

Human Nutritional Needs

Learning Goals

After reading this module, you should be able to

- describe human nutritional requirements.
- explain why nutritional requirements are not being met in various parts of the world.

In both the developed and developing worlds, human nutritional requirements are being met to varying degrees. Since agriculture first started to be practiced more than 10,000 years ago, farmers have been able to increase agricultural output through improved technology and more efficient use of resources. While these improvements have been beneficial to the overall output of food resources, they have not necessarily been beneficial to overall human well-being. In this module, we will look at basic human nutritional needs, how those needs are met, and the reasons why they are not always met.

Human nutritional requirements are not always satisfied

Advancements in agricultural methods are believed to have greatly improved the human diet over the last 10,000 years. In particular, tremendous gains in agricultural productivity and food distribution were made in the twentieth century. But despite these advances, many people throughout the world do not receive adequate nutrition. As **FIGURE 31.1** on page 376 shows, approximately 815 million people worldwide lack access to adequate amounts of food. Although that number had been declining for a decade, it has increased in the last few years and is still much higher than the target numbers set by various world agencies. Currently, as many as 24,000 people starve to death each day—8.8 million people each year. In this section we will explore human nutritional requirements in relation to food production and other factors.

Chronic hunger, or **undernutrition**, means not consuming enough calories to be healthy. Food calories are converted into usable energy by the human body. Not receiving enough food calories leads to an energy deficit. An average person needs approximately 2,000 kilocalories per day, though this amount varies with gender, age, and weight. A long-term food deficit of only 100 to 400 kilocalories per day—in other words, an intake of 100 to 400 kilocalories less than one's daily need—deprives a person of the energy needed to perform daily activities and makes the person more susceptible to disease. In children, undernutrition can lead to improper brain development and lower IQ.

In addition, a person lacking sufficient food calories is probably lacking sufficient protein and other nutrients. According to the Global Nutrition Report 2017, roughly 2.5 billion people in the world—roughly one-third—are **malnourished**; that is, regardless of the number of calories they consume, their diets lack the correct balance of proteins, carbohydrates, vitamins, and minerals. These people experience malnutrition.

The Food and Agriculture Organization of the United Nations (FAO) defines **food security** as the condition in which people have access to sufficient, safe, and nutritious food that meets their dietary needs for an active and healthy life. Access refers to the economic, social, and physical availability of food. **Food insecurity** refers to the condition in which people do not have adequate access to food.

Undernutrition The condition in which not enough calories are ingested to maintain health.

Malnourished Having a diet that lacks the correct balance of proteins, carbohydrates, vitamins, and minerals.

Food security A condition in which people have access to sufficient, safe, and nutritious food that meets their dietary needs for an active and healthy life.

Food insecurity A condition in which people do not have adequate access to food.

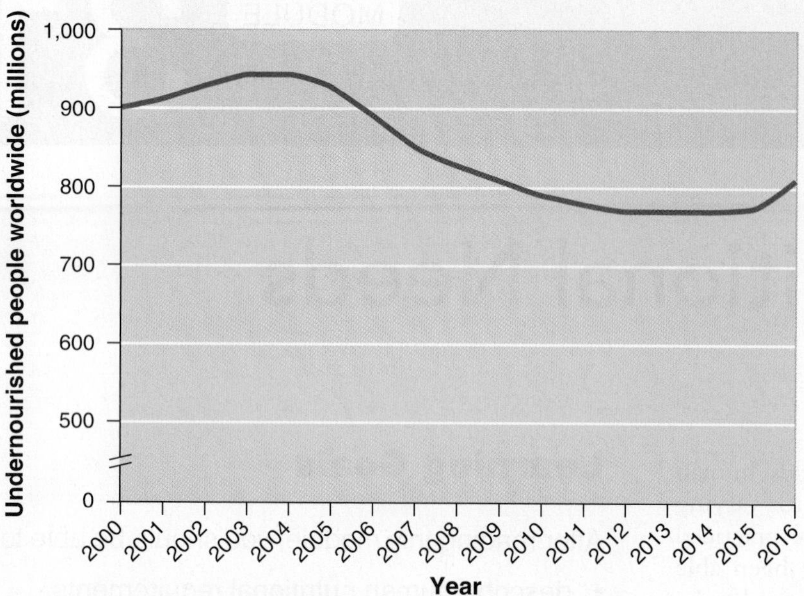

FIGURE 31.1 Global undernutrition. The number of undernourished people in the world declined from 2005 through 2013 but has been increasing in the last few years. *(Data from Food and Agriculture Organization. 2017.* The State of Food Security and Nutrition in the World.*)*

Famine is a condition in which food insecurity is so extreme that large numbers of deaths occur in a given area over a relatively short period. Actual definitions of famine vary widely depending on the agency using the term. One relief agency defines a famine as an event in which there are more than five deaths per day per 10,000 people due to lack of food. According to this definition, there is an annual mortality rate of 18 percent during a famine. Famines are often the result of crop failures, sometimes due to drought, although famines can also have social and political causes.

Even when people have access to sufficient food, a deficit in just one essential vitamin or mineral can have drastic consequences. The WHO estimates that more than 250,000 children per year become blind due to a vitamin A deficiency. Iron deficiency, known as **anemia**, is the most widespread nutritional deficiency in the world. The WHO estimates that there are 2 billion anemic people in the world—many in developing countries, but some in developed countries as well. Lack of dietary iron is the major cause of anemia, but there are other causes, including malaria, AIDS, and parasite infestations. An increase in the ingestion of iron-rich foods, such as certain grains, herbs, vegetables, and meats, can reduce anemia.

In the last few decades, one other form of malnutrition has been increasing. **Overnutrition**, the ingestion of too many calories combined with a lack of proper balance in foods and nutrients, causes a person to become both overweight and malnourished. The WHO estimates that there are almost 2 billion adults in the world who are overweight, and that roughly 600 million of these people are obese, meaning they are more than 20 percent above their ideal weight. Overnutrition is a type of malnutrition that puts people at risk for a variety of diseases, including type 2 diabetes, hypertension, heart disease, and stroke. While overnutrition is common in developed countries such as the United States, it can also coexist with malnutrition in developing countries. Childhood obesity is a related condition that now occurs in greater numbers as well. Overnutrition is often a function of the availability and affordability of certain kinds of foods. For example, overnutrition in the United States has been attributed in part to the easy availability and low cost of processed foods containing ingredients such as high-fructose corn syrup, which are often high in calories but lack other nutritional content.

Humans eat a variety of foods, but grains—the seedlike fruits of corn (maize), rice, wheat, and rye, among others, also referred to as cereals, make up the largest component of the human diet. Worldwide, there are roughly 50,000 edible plant species, but just three of them—corn, rice, and wheat—constitute half of human energy intake. **Meat**, the second largest component of the human diet, is usually defined as livestock (beef, veal, pork, and lamb) and poultry (chicken, turkey, and duck) consumed as food. As income increases with economic growth, people tend to add more meat to their diet. However, in relatively affluent countries, concerns about health consequences of eating "too much" meat or perhaps concerns about the environmental consequences of eating meat sometimes results in a reduction in per capita meat consumption. As **FIGURE 31.2** shows, meat consumption has been increasing globally, although it has decreased slightly in the United States in recent years. Nevertheless, the United States remains as one of the highest per capita consumers of meat in the world.

Famine The condition in which food insecurity is so extreme that large numbers of deaths occur in a given area over a relatively short period.

Anemia A deficiency of iron.

Overnutrition Ingestion of too many calories and a lack of balance of foods and nutrients.

Meat Livestock or poultry consumed as food.

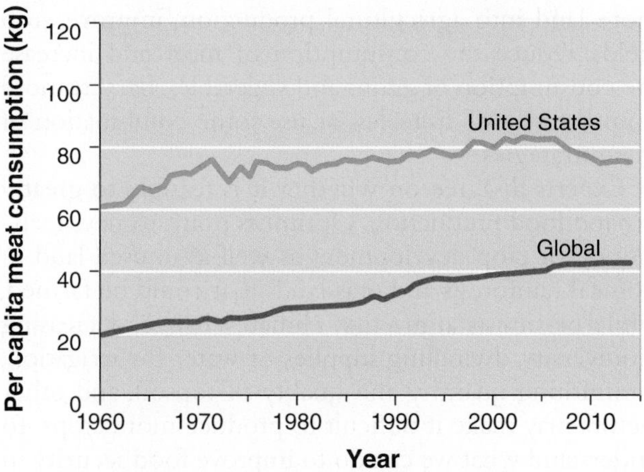

FIGURE 31.2 Per capita meat consumption. Per capita meat consumption has begun to decrease in the United States. It has been increasing worldwide. *(Data from ERS, USDA, and FAO, https://www.ers.usda.gov/data-products/food-availability-per-capita-data-system/.)*

Undernutrition and malnutrition occur primarily because of poverty

Currently, the world's farmers grow enough grain to feed at least 8 billion people, which would appear to be more than enough for the world's population of 7.6 billion. And grain is only a little more than one-half of the food we produce, in terms of caloric content or biomass. Why, then, do we find that undernutrition and malnutrition are common in so many countries of the world? There are many ways to examine this question, and the answers are complex.

The primary reason for undernutrition and malnutrition is poverty: the lack of resources that allows a person to have access to food. According to many food experts, starvation on a global scale is the result of unequal distribution of food rather than absolute scarcity of food. In other words, the food exists but not everyone has access to it. This may mean that people cannot afford to buy the food they need, which is a problem that cannot be solved just by producing more grain.

Political and economic factors can be both the cause and effect of undernutrition and malnutrition. There are numerous examples in modern history where the lack of an adequate food supply has led to political unrest because people without the means to feed themselves or their families may resort to political protest or violence in an attempt to improve their situation. A rise in food prices in 2008 led to food riots in Haiti, Egypt, Ivory Coast, Cameroon, Yemen, and elsewhere. Poor governance can create food shortages. For example, food shortages caused by government agricultural policies led to rioting and deaths in Venezuela in 2017. Political unrest has frequently caused inadequate food supplies. Refugee populations that flee their homes due to war or political unrest typically do not have access to food that they grew and stored but had to leave behind when they became refugees.

Food researchers have also observed that large amounts of agricultural resources are diverted to feed livestock and poultry rather than people. In fact, roughly 40 percent of the grain grown in the world is used to feed livestock. In the United States, the two largest agricultural crops, corn and soybeans, are grown more for animal feed than for people. (Although more corn is used for ethanol than for either livestock or people.) When these foods are fed to livestock, the low efficiency of energy transfer causes much of the energy they contain to be lost from the system, as we saw in Chapter 3. Ultimately, perhaps only 10 to 15 percent of the calories in grain or soybeans fed to cattle are converted into calories in beef. If people ate producers, such as grains and soybeans, rather than primary consumers, such as cattle, it is possible that more food would be available. This concept is referred to as "eating lower on the food chain." It was made popular in the United States in a book called *Diet for A Small Planet* by Frances Moore Lappé in 1971. The main premise is that human beings can utilize fewer calories from agriculture and obtain sufficient nutrition (protein, vitamins, and calories) by eating primary producers (plants) rather than primary consumers (animals). This process is called "eating lower on the food chain" because primary producers are one level lower on the trophic pyramid than primary consumers such as chickens, pigs, and cows. (See Figure 6.9 on page 80.) The book also publicized the fact that with proper attention to food choices, one can obtain sufficient protein from plants.

As we saw in Chapter 1, grain production per capita has increased in the past 2 decades and leveled off recently. Since the population has been growing during this time period, this means that grain production must have increased in order for per capita production to be level. **FIGURE 31.3** on page 378 shows global grain production since 1950. A large number of factors influence grain production, including the amount of land under cultivation, global weather and precipitation patterns, world prices for grain, and the productivity of the land on which grain is being grown. Global grain production has been increasing, although there have been some years within the last 2 decades with no increase in production.

In Chapter 7 we learned that the human population is expected to be between 8 and 10 billion by 2050. To feed everyone in the future, we will have to reduce food waste, reduce caloric consumption, place

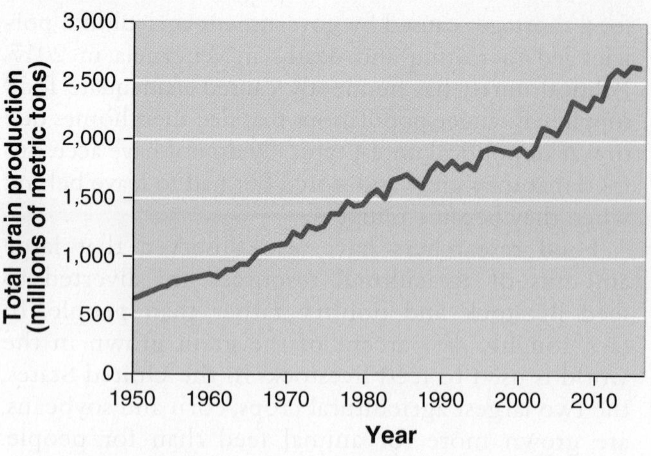

FIGURE 31.3 Global grain production, 1950–2017. Global grain production has increased from 1950 through the present. *(Data from FAO.)*

more land into agricultural production, improve crop yields, reduce the consumption of meat and increase the consumption of grains and vegetables, harvest more from the world's fisheries, or use some combination of these strategies.

Experts disagree on whether it is feasible to greatly expand food production. Optimists point to new techniques for crop development as well as unused land in tropical rainforests and grasslands that could be farmed, while pessimists argue that climate change, decreasing biodiversity, dwindling supplies of water for irrigation, diminishing quantity and quality of topsoil, and other factors may make it difficult to produce more crops. To understand what we can do to improve food security in all nations of the world and reduce the environmental consequences of growing food, we will need to examine the farming methods we use today and the trade offs involved in using them.

MODULE 31 AP® Review

Preparing for the AP® Exam

In this module, we have seen that the advent of agriculture 10,000 years ago has brought about many positive changes to the human population. Nevertheless, there are approximately 815 million people around the world who do not receive adequate amounts of food each day. Nutritional requirements are met by meat consumption to a greater extent in the developed world than in the developing world. Undernutrition, overnutrition, and malnutrition remain issues of concern around the globe.

AP® Practice Questions

Choose the best answer for the following.

1. Approximately what percent of the population is malnourished worldwide?
(a) 18 percent
(b) 25 percent
(c) 33 percent
(d) 45 percent

2. Undernutrition means
(a) lacking the proper balance of proteins, vitamins, and minerals.
(b) not consuming a sufficient number of calories.
(c) lacking a specific nutrient such as vitamin A or iron.
(d) the ingestion of too many poor-quality calories.

3. Anemia, the most widespread nutritional deficiency, is the result of insufficient
(a) vitamin A.
(b) protein.
(c) iron.
(d) vitamin D.

4. The primary reason for malnutrition is
(a) poverty.
(b) political unrest.
(c) insufficient food production.
(d) diversion of food to livestock.

5. Which nutritional trend is TRUE?
(a) Three species of plants contribute roughly 50 percent of human energy intake.
(b) Meat consumption is decreasing globally.
(c) Growing more grain is the only way to feed the world's population.
(d) Overnutrition has remained constant over the last century worldwide.

MODULE 32

Modern Large-Scale Farming Methods

Modern agricultural methods are widespread throughout the developed world. They involve efficient, large-scale practices that produce the greatest amount of food at the lowest economic price, but they do not necessarily result in the least environmental impact. It is important to understand these methods, their benefits, and their costs before examining alternatives.

Learning Objectives

After reading this module you should be able to

- describe modern, large-scale agricultural methods.
- explain the benefits and consequences of genetically modified organisms.
- discuss the large-scale raising of meat and fish.

Modern industrial farming methods have transformed agriculture

The advent of agriculture roughly 10,000 years ago enabled people to move beyond a subsistence level of existence. However, cultivation of food initiated a level of environmental degradation never before experienced on Earth. In addition, the increased abundance of food contributed to the exponential growth of the human population. As we have seen, this growing human population has created even greater stresses on Earth's resources. An increase in food production leads to a positive feedback loop: With more food come more people, who require even more food production.

In the twentieth century, farming became more mechanized, and the use of fossil fuel energy in food production increased. **Industrial agriculture**, or **agribusiness**, applies the techniques of the Industrial Revolution—mechanization and standardization—to the production of food. Today's modern agribusinesses are quite different from the small family farms that dominated agriculture only a few decades ago. In this section we will look at some of the changes that accompanied the era of modern agriculture, including energy dependence and the techniques of the *Green Revolution*.

The Energy Subsidy in Agriculture

The activities associated with growing, harvesting, processing, and preparing food require a great deal of energy input beyond solar energy. The fossil fuel energy and human energy input per calorie of food produced, beyond that which is provided by the sun, is called the **energy subsidy**. In other words, if we use 5 calories of human and fossil fuel energy to produce food, and we receive 1 calorie of energy when we eat that food, then the food has an energy subsidy of 5. We can think of this in another way if we use mass of inputs and outputs as a substitute measure for energy. If it takes 20 kg (44 pounds) of grain to feed cattle to produce 1 kg (2.2 pounds) of beef, the energy subsidy is 20. If it takes 2.8 kg (6 pounds) of grain to feed chickens to produce 1 kg (2.2 pounds) of chicken meat, the energy subsidy is 2.8, which is considerably smaller than the energy subsidy for beef. The comparison is not perfect, because the energy content of 1 kg of beef is greater

Industrial agriculture Agriculture that applies the techniques of mechanization and standardization to the production of food. *Also known as* **agribusiness.**

Energy subsidy The fossil fuel energy and human energy input per calorie of food produced.

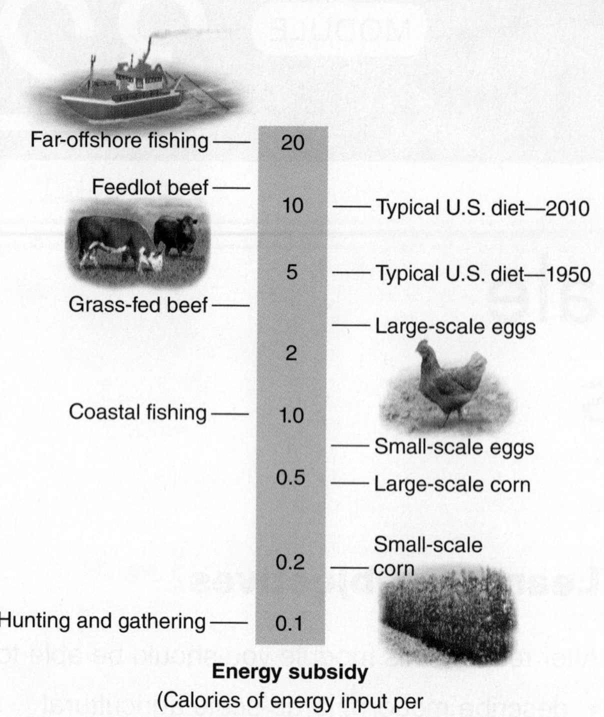

FIGURE 32.1 Energy subsidies for various methods of food production and diets. Additional energy (beyond that provided by the sun) input per calorie of food obtained is greater for modern agricultural practices than for traditional agriculture. Energy inputs for hunting and gathering and for small-scale food production are mostly in the form of human energy, whereas fossil fuel energy is the primary energy subsidy for large-scale modern food production. All values are approximate, and for any given method there is a large range of values.

than that of 1 kg of grain. Nevertheless, it should help you to understand the relationship between energy inputs and outputs in producing food.

As **FIGURE 32.1** shows, traditional small-scale agriculture requires a relatively small energy subsidy beyond the energy provided by the sun: It uses few energy inputs per calorie of food produced. By contrast, food writer and journalism professor Michael Pollan estimates that if you eat the average modern U.S. diet, which contains foods mostly produced by modern agricultural methods, there is a 10-calorie energy input for every calorie you eat. Most of that energy input is fossil fuel energy. Therefore, food choices are also energy choices. "Do the Math: Land Needed for Food" shows you how to calculate the different land requirements for obtaining sufficient calories from corn and from beef.

Green Revolution A shift in agricultural practices in the twentieth century that included new management techniques, mechanization, fertilization, irrigation, and improved crop varieties, that resulted in increased food output.

Most of the energy subsidies in modern agriculture are in the form of fossil fuels, which are used to produce fertilizers and pesticides, to operate tractors, to pump water for irrigation, and to harvest food and prepare it for transport. Other energy subsidies take place off the farm. Although the average food item travels roughly 2,000 km (1,240 miles) from the farm to your plate, transportation accounts for only about 10 percent of the energy subsidy. The Department of Energy reports that in the United States, 17 percent of total commercial energy use goes into growing, processing, transporting, and cooking food. The largest fraction of the energy subsidy goes to food production at the farm and food processing in the factory and in the home. Those of us eating a supermarket diet in the developed world are highly dependent on fossil fuel for our food; the modern agricultural system would not work without it.

The Green Revolution

How did we get to our current levels of energy use for food production? In the twentieth century, the agricultural system was dramatically transformed from a system of small farms relying mainly on human labor with relatively low fossil fuel inputs to a system of large industrial operations with fewer people but much more machinery. This shift in farming methods, known as the **Green Revolution**, involved new management techniques and mechanization as well as the triad of fertilization, irrigation, and improved crop varieties. These changes increased food production dramatically, and farmers were able to feed many more people.

The Green Revolution began with the work of crop scientists, particularly the American scientist Norman Borlaug (1914–2009), who won the Nobel Peace Prize for his contribution to increasing the world food supply. Through intensive breeding, Borlaug and other agricultural researchers developed strains of wheat that were disease resistant and produced higher yields. They also used fertilizers and irrigation to improve yields. In the 1940s, researchers brought these techniques to developing nations such as Mexico and the Philippines to help increase their agricultural output and feed their growing populations. From the 1950s through the 1970s, many countries, particularly those in the developing world, underwent similar shifts in the way they farmed.

The upward trend in world grain production since 1950 (see Figure 31.3 on page 378) is the result of the Green Revolution. From the mid-1960s through the mid-1980s, world grain production doubled. By the 1990s, there were at least 18 organizations around the world promoting and developing Green Revolution techniques, including mechanization, irrigation, use of fertilizer, *monocropping*, and use of pesticides. However, the Green Revolution has also had negative environmental impacts. Let's examine some Green Revolution practices in more detail.

DO THE MATH Land Needed for Food

Preparing for the AP® Exam

We have seen that raising beef requires more resources than growing corn. Let's look at some of the actual numbers.

On farms in the midwestern United States, a hectare of land yields roughly 370 bushels of corn (equivalent to 150 bushels per acre). A bushel consists of 1,250 ears of corn, and each ear typically contains 80 kilocalories. Assume that a person eats only corn and requires 2,000 kilocalories per day. Although this assumption is not very realistic, it allows an approximation of how much land it would take to feed that person.

The person's food requirement is

$$2{,}000 \text{ kilocalories/day} \times 365 \text{ days/year}$$
$$= 730{,}000 \text{ kilocalories/year}$$

A hectare of corn produces

$$370 \text{ bushels/hectare} \times 1{,}250 \text{ ears/bushel} \times 80 \text{ kilocalories/ear}$$
$$= 37{,}000{,}000 \text{ kilocalories/hectare}$$
$$730{,}000 \text{ kilocalories/year} \div 37{,}000{,}000 \text{ kilocalories/hectare} = 0.02 \text{ ha (0.05 acres)}$$
of land to feed one person for a year

Thus, one person eating only corn can obtain sufficient calories in a year from 0.02 ha (0.05 acres) of land. What if that person ate only beef? We have seen that it takes 20 kg of grain to produce 1 kg of beef. So it would take 20 times as much land, or 0.4 ha (1 acre), to feed a person who ate only beef.

What happens if we extend this analysis to a global scale? If Earth has about 1.5 billion ha (3.7 billion acres) of land suitable for growing food, is there sufficient land to feed all 6.8 billion inhabitants of the planet if they all eat a diet of only beef?

$$6.8 \text{ billion people} \times 0.4 \text{ ha/person} = 2{,}720{,}000{,}000 \text{ ha}$$

So 2.72 billion ha (6.72 billion acres) would be needed, and the answer is no.

YOUR TURN How many people eating a beef-only diet can Earth support?

Mechanization

Farming involves many kinds of work. Fields must be plowed, planted, irrigated, weeded, protected from pests, harvested, and prepared for the next season. After harvesting, crops must be dried, sorted, cleaned, and prepared for market in almost as many different ways as there are different crops. Machines do not necessarily do this work better than humans or animals, but it can be economically advantageous to replace humans or animals with machinery, particularly if fossil fuels are abundant, fuel prices are relatively low, and labor prices are relatively high. In developed countries, where wages are relatively high, less than 5 percent of the workforce works in agriculture. In developing countries, where wages tend to be much lower, 40 to 75 percent of the working population is employed in agriculture.

Since the advent of mechanization, large farms producing staple crops such as beans or corn have generally been more profitable than small farms. Size matters because of **economies of scale**, which means that the average costs of production fall as output increases. For example, a new combine harvester—a machine that harvests the crop and separates out the grain or seed for transport—costs between $150,000 and $400,000. This large up-front expenditure is a good investment for a large farm, where the cost of the machine is justified by the profits on the increased production that comes from using it. A small farm, however, does not have enough land to be able to recoup the cost of such expensive equipment. Because of economies of scale, profits tend to increase with size, and large agricultural operations generally outcompete small ones. As a result, between 1950 and 2000, the average farm size in Iowa more than doubled, from about 70 ha (173 acres) to over 140 ha (346 acres).

Mechanization also means that single-crop farms are generally more efficient than farms that grow many crops. Because mechanized crop planting and picking

Economies of scale The observation that average costs of production fall as output increases.

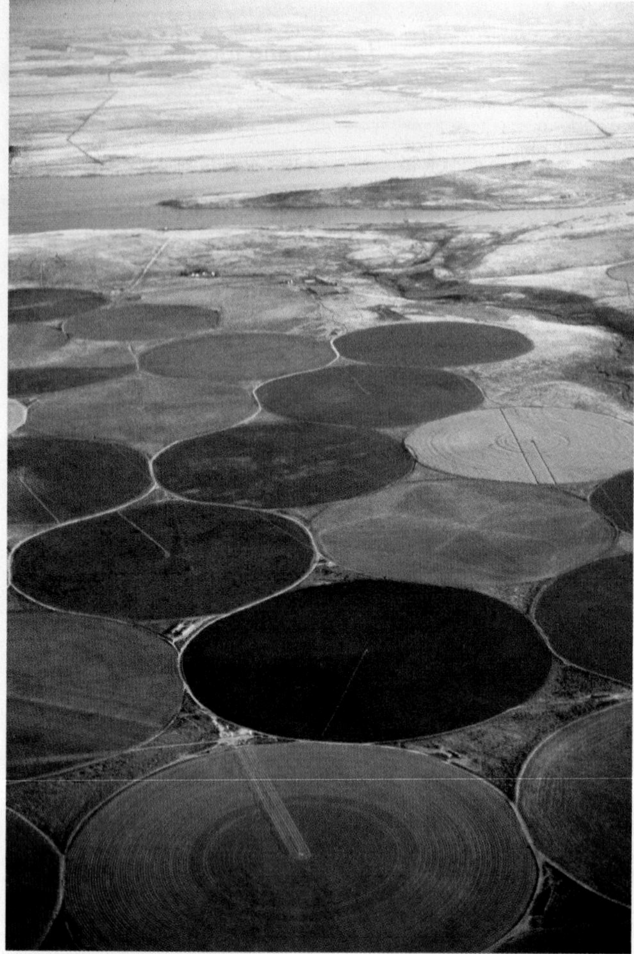

FIGURE 32.2 Irrigation circles. The green circles in this aerial photograph from Oregon are obvious evidence of irrigation. *(Doug Wilson/ARS/USDA)*

require specialized, expensive equipment specific to each crop, planting and harvesting only a single type of crop reduces equipment costs.

Irrigation

Irrigation systems like those described in Chapter 9 can increase crop growth rates or even enable crops to grow where they could not otherwise be grown (**FIGURE 32.2**). For example, irrigation has transformed approximately 400,000 ha (1 million acres) of former desert in the Imperial Valley of southeastern California into a major producer of fruits and vegetables. In other situations, irrigation can allow productive land to become extremely productive land. One estimate suggests that the 16 percent of the world's agricultural land that is irrigated produces 40 percent of the world's food.

While irrigation has many benefits, including more efficient use of water, it can also have a number of negative consequences over time. As we saw in Chapter 9, it can deplete groundwater, draw down aquifers, and cause saltwater intrusion into freshwater wells. It can also contribute to soil degradation through *waterlogging* and *salinization,* shown in **FIGURE 32.3**. **Waterlogging**, a form of soil degradation that occurs when soil remains under water for prolonged periods, impairs root growth because roots cannot get oxygen. **Salinization** occurs when the small amounts of salts in irrigation water become highly concentrated on the soil surface through evaporation. These salts can eventually reach toxic concentrations and impede plant growth.

Waterlogging A form of soil degradation that occurs when soil remains under water for prolonged periods.

Salinization A form of soil degradation that occurs when the small amount of salts in irrigation water becomes highly concentrated on the soil surface through evaporation.

Organic fertilizer Fertilizer composed of organic matter from plants and animals.

Fertilizers

Growing crops and exporting them from the farm system to humans for consumption removes organic matter and nutrients from soil. If these materials are not replenished, they can be quickly depleted. Industrial agriculture, because it keeps soil in constant production, requires large amounts of fertilizers to replace lost organic matter and nutrients. Fertilizers contain essential nutrients for plants—primarily nitrogen, phosphorus, and potassium—and they foster plant growth where one or more of these nutrients is lacking. There are two types of fertilizers used in agriculture: *organic* and *synthetic.*

As their name suggests, **organic fertilizers** are composed of organic matter from plants and animals.

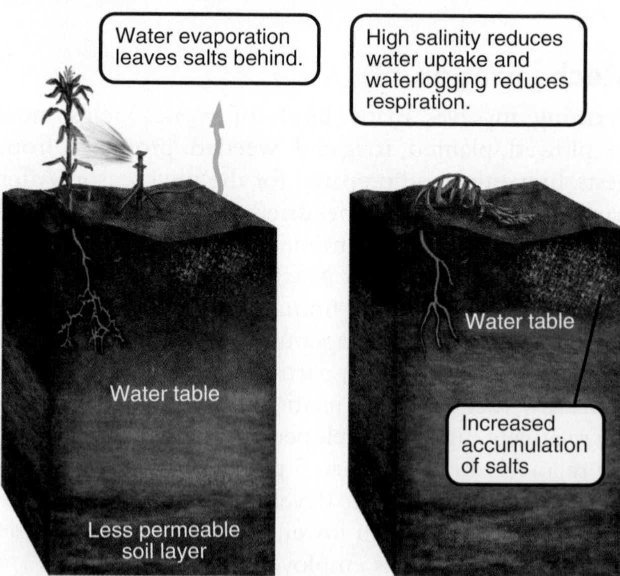

FIGURE 32.3 Irrigation-induced salinization and waterlogging. Over time, irrigation can degrade soil by leaving a layer of highly concentrated salts at the soil surface and waterlogged soil below.

They are typically made up of animal manure and crop residues that have been allowed to decompose. Traditional farmers often spread animal manure and crop wastes onto fields to return some of the nutrients that were removed from those fields when crops were harvested.

Synthetic fertilizers, or **inorganic fertilizers**, are produced commercially, normally with the use of fossil fuels. Nitrogen fertilizers are often produced by combusting natural gas, which allows nitrogen from the atmosphere to be fixed and captured in fertilizer. Fertilizers produced in this way are highly concentrated, and their widespread use has increased crop yields tremendously since the Green Revolution began. Synthetic fertilizers have many advantages over organic fertilizers. They are designed for easy application, their nutrient content can be targeted to the needs of a particular crop or soil, and plants can easily absorb them, even in poor soils. Worldwide, synthetic fertilizer use increased from 20 million metric tons in 1960 to nearly 200 million metric tons in 2018. It can be argued that without synthetic fertilizers, we could not feed all the people in the world.

Despite these advantages, synthetic fertilizers can have several adverse effects on the environment. The process of manufacturing synthetic fertilizers uses large quantities of fossil fuel energy. Producing nitrogen fertilizer is an especially energy-intensive process. Whereas synthetic fertilizers are more readily available for plant uptake than are organic fertilizers, they are also more likely to be carried by runoff into adjacent waterways and aquifers. In surface waters, this nutrient runoff can cause algae and other organisms to proliferate. After these organisms die, they decompose and reduce oxygen levels in the water, a process we will learn more about when we look at water pollution in Chapter 14. Finally, synthetic fertilizers do not add organic matter to the soil, as organic fertilizers do. As we saw in Chapter 8, organic matter contributes many beneficial properties to soil, including increased cation exchange capacity and water retention abilities.

The United States uses somewhat less fertilizer and consequently experiences less nutrient runoff than other nations with similar agricultural output. Still, large amounts of nitrogen and other nutrients run into waterways in intensively farmed regions such as California's Central Valley, farming regions along the East Coast, and the Mississippi River watershed (**FIGURE 32.4**).

FIGURE 32.4 Fertilizer runoff after fertilization. The proximity of this drainage ditch to an agricultural field may lead to runoff of fertilizer during a heavy rainstorm. *(Lynn Betts/USDA Natural Resources Conservation Service)*

Monocropping

Both the mechanization of agriculture and the use of synthetic fertilizers encourage large plantings of a single species or variety, a practice known as **monocropping** (**FIGURE 32.5**, page 384). Monocropping is the dominant agricultural practice in the United States, where corn, soybean, wheat, and cotton are frequently grown in monocrops of 405 ha (1,000 acres) or more.

Monocropping has greatly improved agricultural productivity. This technique allows large expanses of land to be planted at the same time and later harvested at the same time. With the use of large machinery, the harvest can be done easily and efficiently. If fertilizer or pesticide treatments are required, those treatments can also be applied uniformly over large fields, which, because they are planted with the same crop, have the same pesticide or nutritional needs.

Despite the benefits of increased efficiency and productivity, monocropping can lead to environmental degradation. First, soil erosion can become a problem. Because fields that are monocropped are readied for planting or harvesting all at once, soil will be exposed over many hectares at the same time. On a 405-ha (1,000-acre) field that has not yet been planted, wind can blow for over 1.6 km (1 mile) without encountering anything but bare soil. Under these circumstances, the wind can gain enough speed to carry dry soil away from the field. Certain farmland in the United States loses an average of 1 metric ton of topsoil per hectare (2.5 metric tons per acre) per year to wind erosion. As we saw in Chapter 8, this topsoil contains important nutrients and its loss can reduce productivity.

Synthetic fertilizer Fertilizer produced commercially, normally with the use of fossil fuels. *Also known as* **inorganic fertilizer.**

Monocropping An agricultural method that utilizes large plantings of a single species or variety.

FIGURE 32.5 Monocropping. This large field in Oregon contains only one crop species, soybeans. There are both advantages and disadvantages to monocropping. *(Scott Sinklier/Alamy)*

Monocropping also makes crops more vulnerable to attack by pests. Large expanses of a single plant species represent a vast food supply for any pests that specialize on that particular plant. Such pests will establish themselves in the monocrop and reproduce rapidly. Their populations may experience exponential growth similar to what we saw in Georgii Gause's *Paramecium* populations that were supplied with unlimited food (see Figure 18.4 on page 200). Natural predators may not be able to respond rapidly to the exploding pest population. Many predators of crop pests, such as ladybugs and parasitic wasps, are attracted to the pests that feed on monocrops. But these predators also rely on non-crop plants for habitat, which monocropping removes. Therefore, predators that might otherwise control the pest population are largely absent.

Pesticide A substance, either natural or synthetic, that kills or controls organisms that people consider pests.

Insecticide A pesticide that targets species of insects and other invertebrates that consume crops.

Herbicide A pesticide that targets plant species that compete with crops.

Broad-spectrum pesticide A pesticide that kills many different types of pest.

Selective pesticide A pesticide that targets a narrow range of organisms. *Also known as* **narrow-spectrum pesticide.**

Persistent pesticide A pesticide that remains in the environment for a long time.

Pesticides

Pesticides are substances, either natural or synthetic, that kill or control organisms that people consider pests. The use of pesticides has become routine and widespread in modern industrial agriculture. In the United States, the Environmental Protection Agency reports on the manufacture and use of pesticides. In 2012, the latest year for which data are available, 385 million kilograms (850 million pounds) of pesticides were applied in the United States. Almost 90 percent were applied for agricultural purposes. The United States accounts for about one-fifth of worldwide pesticide use.

AP® Exam Tip

Many past free-response questions on the AP® Environmental Science Exam have asked about pesticide use. You should know the different categories of pesticides and understand the uses of each: insecticide, herbicide, broad-spectrum pesticide, and selective pesticide. ●

Insecticides target species of insects and other invertebrates that consume crops, and **herbicides** target plant species that compete with crops. Some pesticides are **broad-spectrum pesticides**, meaning that they kill many different types of pest, and some are **selective pesticides** or **narrow-spectrum pesticides** that target a narrow range of organisms. The broad-spectrum insecticide dimethoate, for example, kills almost any insect or mite—relatives of ticks and spiders—while the more selective narrow-spectrum acequinocyl kills only mites.

The application of pesticides allows a farmer to quickly and relatively easily respond to an infestation of pests on an agricultural crop. In many cases, a single application can significantly reduce a pest population. By preventing crop damage, pesticide application can result in greater crop yields on less land, thereby reducing the area disturbed by agriculture and making agriculture more efficient.

But the application of pesticides, like many other industrial agricultural practices, presents some environmental problems. For example, pesticides often injure or kill more than their intended targets. Some pesticides, such as dichlorodiphenyltrichloroethane, also known as DDT, are **persistent pesticides**, meaning that they remain in the environment for years to decades. In 1972, DDT was banned in the United States, in part because it was found to accumulate in the fatty tissues of animals, such as eagles and pelicans. The increasing concentrations of DDT in these animals caused them to lay eggs with thin shells that easily cracked during incubation by the parents. We will discuss this process in more detail in Chapter 17.

Crop infested with pests
Develop a new pesticide
Apply pesticide.
Most pests die, but some resistant individuals survive.
The proportion of resistant individuals in the pest population grows.
Apply pesticide again, with little result.

FIGURE 32.6 The pesticide treadmill. Over time, pest populations evolve resistance to pesticides, which requires farmers to use higher doses or to develop new pesticides.

Other pesticides, such as the herbicide glyphosate, known by the trade name Roundup, are **nonpersistent pesticides**, meaning that they break down relatively rapidly, usually in weeks to months. Nonpersistent pesticides have fewer long-term effects, but because they must be applied more often, their overall environmental impact is not always lower than that of persistent pesticides.

Another disadvantage of pesticide use is that pest populations may evolve resistance to pesticides over time, as described in Chapter 5. Pest populations are usually large and thus contain significant genetic diversity. In those vast gene pools, there are usually a few individuals that are not as susceptible to a pesticide as others. Individuals that are exposed to a particular pesticide and survive are said to have **pesticide resistance** to that pesticide. If the pesticide is successful in reducing the pest population, the next generation will contain a larger fraction of pesticide resistant individuals in the population. As time goes by, resistant individuals will make up a larger and larger portion of the population. Often the resistance becomes more effective, which makes the pesticide significantly less useful. At this point, crop scientists and farmers must search for a new pesticide. The cycle of pesticide development and pest resistance, illustrated in **FIGURE 32.6**, is known as the **pesticide treadmill**. The pesticide treadmill is an example of a positive feedback system.

Pesticides can cause even wider environmental effects. They may kill organisms that benefit farmers, such as predatory insects that eat crop pests, pollinator insects that pollinate crop plants, and plants that fix nitrogen and improve soil fertility. Furthermore, chemical pesticides, like fertilizers, can run off into surrounding surface waters and pollute groundwater, a problem we will look at in detail in Chapter 14. The toxicity of pesticides to farmworkers has been well documented. The risk to humans who ingest food that has been treated with pesticides is a subject of debate, which will be covered in Chapter 17.

Genetic engineering is revolutionizing agriculture

As noted in Chapter 5, humans have modified plants and animals by artificial selection for thousands of years. The modern techniques of genetic engineering,

Nonpersistent pesticide A pesticide that breaks down rapidly, usually in weeks or months.

Pesticide resistance A trait possessed by certain individuals that are exposed to a pesticide and survive.

Pesticide treadmill A cycle of pesticide development, followed by pest resistance, followed by new pesticide development.

however, go far beyond traditional practices. Scientists today can isolate a specific gene from one organism and transfer it into the genetic material of another, often very different, organism to produce a genetically modified organism, or GMO. By manipulating specific genes, agricultural scientists can rapidly produce organisms with desirable traits that may be impossible to develop with traditional breeding techniques. Genetically modified organisms present both benefits and drawbacks.

The Benefits of Genetic Engineering

Genetically modified crops and livestock offer the possibility of greater yield and food quality, reductions in pesticide use, and higher profits for the agribusinesses that use them. They are also seen as a way to help reduce world hunger by increasing food production and reducing losses to pests and varying environmental conditions.

Increased Crop Yield and Quantity

Genetic engineering can increase food production in several ways. It can create strains of organisms that are resistant to pests and harsh environmental conditions such as drought or high salinity. In addition, agricultural scientists have begun to engineer plants that produce essential nutrients for humans. For example, they have inserted a gene for the production of vitamin A into rice plants, creating new seeds known as golden rice (**FIGURE 32.7**). Although golden rice is still an experimental product, some scientists hope that it will help reduce the incidence of blindness resulting from vitamin A deficiency. Crop plants, animals, and bacteria have also been modified to produce pharmaceuticals and other compounds, a process that can make these products far less expensive to manufacture.

FIGURE 32.7 White rice and golden rice. Crop scientists have inserted a gene that synthesizes a precursor to vitamin A in white rice. The resulting genetically modified rice is called golden rice. *(Reuters/Erik De Castro)*

A number of projects are under way to create genetically modified animals for food production, including a salmon that grows to its full size of 3.6 kg (8 pounds) in 18 months—half the growing time of an unmodified fish. In 2017, the first commercial batch of genetically engineered salmon was sold by a Massachusetts company to an undisclosed recipient in Canada.

Potential Changes in Pesticide Use

Genetic engineering for resistance to pests could reduce the need for pesticides. Corn, for example, is subject to attacks from the bollworm, European corn borer (*Ostrinia nubilalis*), and lepidopteran (butterfly and moth) larvae. *Bacillus thuringiensis* is a natural soil bacterium that produces a toxin that can kill lepidopterans. The insecticidal gene of this bacterium, known as Bt, has been inserted into the genetic material of corn plants, resulting in a genetically modified plant that produces a natural insecticide in its leaves.

A similar technique has been used to create crop plants that are resistant to the herbicide Roundup. The "Roundup Ready" gene allows growers to spray the herbicide on their fields to control the growth of weeds without harming the crop plants. These genes are now widely used in corn, soybean, and cotton plants and are more generally referred to as herbicide tolerant (HT). The success of *no-till agriculture*, which we will discuss later in this chapter, rests largely on the use of herbicides and the introduction of herbicide-resistant crops.

In 2017, roughly 80 percent of the land area planted with corn in the United States was planted with Bt and/or HT corn. Growers of Bt and HT corn have been able to reduce the amount of synthetic pesticide and herbicide used on their crops.

Increased Profits

GMO seed crops can increase farm profits in two ways. Although GMO crop seeds do cost more to purchase than conventional crop seeds, they reduce pesticide use, which can be a significant savings. GMO crops can also raise revenues by producing greater yields. Both of these changes can lead to higher incomes for farmers, lower food prices for consumers, or both.

Concerns About Genetically Modified Organisms

Industrial agriculture relies more heavily on genetically modified crops each year. In 2017, 92 percent of the corn, 94 percent of the soybeans, and 96 percent of the cotton planted in the United States came from genetically modified seeds. However, many European countries, as well as a number of people in the United States, question the safety of GMOs. Genetically modified crops and livestock are the source of considerable controversy, and concerns have been raised about their safety for human consumption and their effects on biodiversity. Regulation of GMOs is also an issue, both in the United States and abroad.

Safety for Human Consumption

Some people are concerned that the ingestion of genetically modified foods may be harmful to humans, although so far there is little evidence to support these concerns. However, researchers are studying the possibility that GMOs may cause allergic reactions when people eat a food containing genes transferred from another food to which they are allergic.

Effects on Biodiversity

There is some concern that if genetically modified crop plants are able to breed with their wild relatives—as many domesticated crop plants are—the newly added genes will spread to the wild plants. The spread of such genes might then alter or eliminate natural plant varieties. Examples of GMOs crossing with wild relatives do exist. Because of these concerns, attempts have been made to introduce buffer zones around genetically modified crops.

The use of genetically modified seeds is contributing to a loss of genetic diversity among food crops. As with any reduction in biodiversity, we cannot know what beneficial genetic traits might be lost. For example, researchers recently discovered that one variety of sorghum, a cereal grass grown for its sweet juice extract, has a natural genetic variation that gives it resistance to a pest called the greenbug. This particular variety has since been used extensively to confer greenbug resistance on the U.S. sorghum crop. If growers had been growing only one or two genetically modified varieties of sorghum, this naturally resistant variety might have been eliminated before its beneficial trait was discovered.

Regulation of Genetically Modified Organisms

In 2016, Congress passed a National Bioengineered Food Disclosure Standard, also known as the GMO labeling law, and President Barack Obama signed it. The USDA is in the process of implementing the law, which requires disclosure and labeling if a food contains a GMO product. Labeling is regarded as a measure of transparency, allowing a consumer to understand how a food has been grown or processed. The standard does not restrict the use of GMOs in plants. The European Union, in contrast, allows very few genetically modified foods. In France, Germany, and Italy, almost all GMOs are banned. Labeling opponents argued that labeling of foods containing GMOs might suggest to consumers that there is something wrong with GMOs. They also argued that such labeling would be too difficult because small amounts of GMO materials are found throughout the U.S. agricultural system. Those who want to avoid consuming GMOs can purchase organic food; the federal definition of "organic" excludes genetically modified foods. (See "Science Applied 5: How Do We Define Organic Food?" on page 410).

As noted earlier, there are companies working on genetically modified animals and one company in Massachusetts has developed a salmon that it sold to a company in Canada. There are currently applications for a number of genetically modified animals under consideration by the Food and Drug Administration. However the U.S. government has not yet approved any genetically modified animals for market in the United States.

Modern agribusiness includes farming meat and fish

We have seen that in order to remain economically viable and feed large numbers of people, modern agriculture in the United States has had to become larger and more mechanized. While many of the practices described in the preceding sections have remained in place as agricultural activities moved to a larger scale, some have been abandoned. In particular, as agriculture developed to supply meat and poultry to large numbers of people at a low cost, the primary objective became implementing methods that promoted the faster growth of animals. Here we will examine some of the modern methods in large-scale production of farm animals and fish.

High-Density Animal Farming

In 2016, according to the U.S. Department of Agriculture, roughly 150 million animals were slaughtered for beef, pork, and lamb, along with billions of chickens, turkeys, and ducks. Many of these animals were raised in feed-lots, or **concentrated animal feeding operations (CAFOs)**, which are large indoor or outdoor structures designed for maximum output (**FIGURE 32.8** on page 388). This type of high-density animal farming is used for beef cattle, dairy cows, hogs, and poultry, all of which are confined or allowed very little room for movement during all or part of their life cycle. A CAFO may contain as many as 2,500 hogs or 50,000 turkeys in a single building. By keeping animals confined, farmers minimize land costs, improve feeding efficiency, and increase the fraction of food energy that goes into the production of animal body mass. Keeping animals confined to a small space, which is criticized

Concentrated animal feeding operation (CAFO) A large indoor or outdoor structure designed for maximum output.

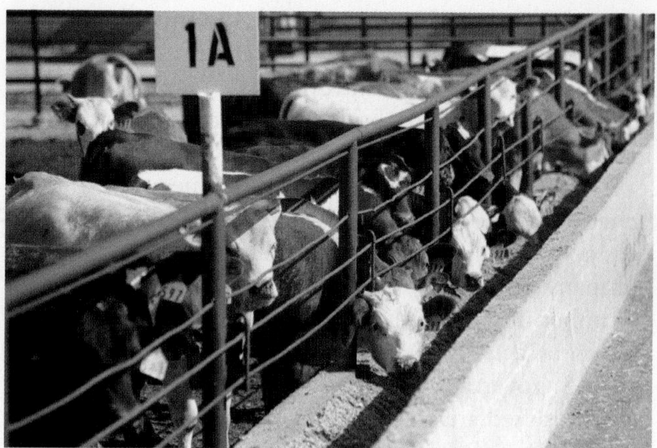

FIGURE 32.8 Cattle in a concentrated animal feeding operation in Kansas. CAFOs are large indoor or outdoor structures that allocate a very small amount of space to each animal. *(ERIC GAY/AP Images)*

by some on ethical grounds, ensures that less energy is expended by the animal in activities such as moving around and the increased respiration that will result. The animals are given antibiotics and nutrient supplements to reduce the risk of adverse health effects and diseases, which would normally be high in such highly concentrated animal populations.

AP® Exam Tip

Make sure that you understand the advantage of a diet with little meat and know examples of other protein sources. ●

High-density animal farming has many environmental and health consequences. There is evidence that antibiotics given to confined animals are contributing to an increase in antibiotic-resistant strains of microorganisms that affect humans. Waste disposal is another serious problem. An average CAFO produces over 2,000 tons of manure annually, or about as much as a town of 5,000 people would produce. The waste is usually used to fertilize nearby agricultural fields, but if over-applied, it can cause the same nutrient runoff problems as synthetic fertilizer. Sometimes animal wastes are stored in lagoons adjacent to feedlots but, during heavy rainstorms, runoff from these lagoons can contaminate nearby waterways. Animal wastes have also been dumped, either inadvertently or intentionally, into natural waters. The U.S. Environmental Protection Agency has concluded that chicken, hog, and cattle waste has caused pollution along 56,000 km (35,000 miles) of rivers in 22 states and has caused some degree of groundwater contamination in 17 states.

Harvesting Fish and Shellfish

Fish is the third major source of food for humans, after grain and meat. In many coastal areas, particularly in Asia and Africa, fish accounts for nearly all animal protein that some people consume. As **FIGURE 32.9** shows, the global production of fish has increased more than 30 percent since 1980. This increase masks two divergent trends: a decrease in wild fish caught in the world's oceans and a relatively rapid increase in **aquaculture**, which is the farming of aquatic organisms such as fish, shellfish, and seaweeds.

A **fishery** is a commercially harvestable population of fish within a particular ecological region. The tragedy of the commons, which we learned about in Chapter 10, is particularly applicable to ocean fisheries. No country has an incentive to protect fish stocks or to attempt to replenish them because fish in the ocean do not belong to any one nation. Since most fish do not live their entire lives within one national border, one country working alone cannot solve the problem of fishery depletion. In fact, a voluntary limit by one or a few countries will give other countries more opportunities to overfish. In many parts of the oceans, the continuing competition for fish has led to a precipitous decline in fish populations.

In 2003, the journal *Nature* reported a dramatic decline in the number of large predatory ocean fish caught over the past 50 years, even though both the number of fishing vessels and the amount of time spent fishing had increased over that period. Fishers were working harder, but catching fewer fish. A study in 2006 found that 30 percent of fisheries worldwide had experienced a 90 percent decline in fish populations. The decline of a fish population by 90 percent or more is referred to as **fishery collapse**. Many studies on this subject focus on large predatory fish. However, a 2011 study in the *Proceedings of the National Academy of Sciences* noted an unexpected increase in fishery collapse among smaller species as well.

Ocean harvests used to be limited by the difficulty of finding fish in a vast ocean as well as by lack of capacity of small boats and nets. Neither limitation is an obstacle today. Current fishing methods make it easy to catch large numbers of fish. Factory ships can stay at sea for months at a time, processing and freezing their harvest without having to return to port. Most marine fish are now caught either by large nets pulled

Aquaculture Farming aquatic organisms such as fish, shellfish, and seaweeds.

Fishery A commercially harvestable population of fish within a particular ecological region.

Fishery collapse The decline of a fish population by 90 percent or more.

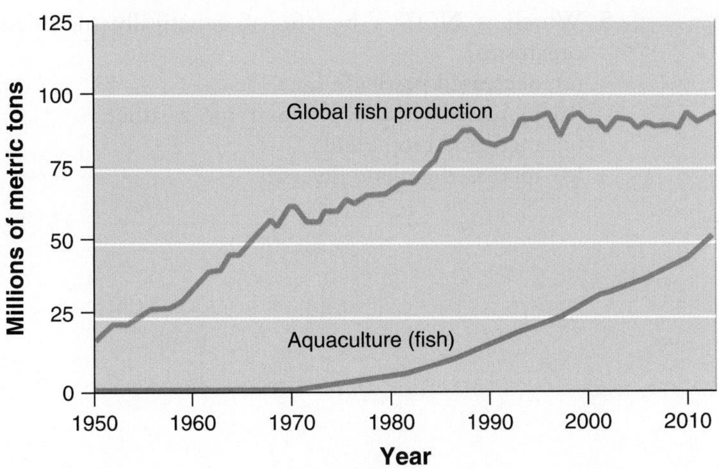

FIGURE 32.9 Global fish production. Global fish production has increased by more than 30 percent since 1980, primarily as a result of the large increase in aquaculture. The graph shows data for aquaculture-raised fish (blue) and wild-caught fish production (orange). *(Data from The State of World Fisheries and Aquaculture, FAO. 2016, http://www.fao.org/3/a-i5555e.pdf.)*

behind one of these ships or by very long fishing lines bearing hundreds or even thousands of baited hooks. Fishers in pursuit of high-value species such as tuna use spotter planes and sonar to locate schools. They encircle schools with nets that can capture up to 3,000 adult tuna at a time—weighing perhaps 450,000 kg (almost 1 million pounds). Fish species that live on or close to the ocean bottom, as well as many shellfish, are caught in dragnets, which are weighted so that they can be pulled across the ocean floor.

Large-scale, high-tech fishing can adversely affect both target and nontarget species. Dragnets can damage ocean-bottom habitats by scouring them of coral, sea sponges, and plants. Many commercially important fish are keystone species, so a decline or loss of their populations can have cascading effects on other marine species. Intensive fishing leads to the loss of juvenile fish of the target species as well as to the loss of noncommercial species that are accidentally caught by nets and lines. This unintentional catch of nontarget species, referred to as **bycatch**, has significantly reduced populations of fish species such as sharks and has endangered other organisms such as sea turtles. Some countries now require the use of technology that minimizes bycatch or harm to endangered species.

Bycatch The unintentional catch of nontarget species while fishing.

MODULE 32 AP® Review

Preparing for the AP® Exam

In this module, we have seen that modern, large-scale farming methods have transformed agriculture. Large amounts of fossil fuel energy and fertilizer are common inputs to most agriculture in the developed world. The Green Revolution introduced a number of changes to agriculture in the 1950s and 1960s that continue to result in increased agricultural yield today. Genetic engineering has also changed agriculture, but it remains controversial. Animals are typically raised in concentrated feeding areas that allow for maximum profit but can result in a variety of environmental concerns. Fishery production is also dependent on modern mechanized systems that have resulted in increased yield at the expense of the wild populations.

AP® Practice Questions

Choose the best answer for the following.

1. If 12,000,000 kilocalories of chicken can be produced per hectare per year, how much land is needed to provide a person with 2,000 kilocalories of chicken/day for a year?
   (a) 0.01 ha
   (b) 0.02 ha
   (c) 0.06 ha
   (d) 0.2 ha

2. The Green Revolution
   (a) discouraged the mechanization of agriculture.
   (b) decreased the energy subsidy of most food.
   (c) encouraged the use of monocropping.
   (d) pertained to leafy green plants only.

3. Broad-spectrum pesticides
   (a) do not cause increased pesticide resistance.
   (b) are almost always nonpersistent.
   (c) are banned in the United States.
   (d) are likely to kill beneficial insects.

4. Bycatch
   (a) is a common problem with increased pesticide use.
   (b) is a management technique in CAFOs.
   (c) is a cause of fishery collapse.
   (d) is common in large-scale fishing.

5. Which is NOT a benefit of genetically modified organisms?
   (a) decreased pesticide use
   (b) increased resistance to extreme weather
   (c) increased crop yield
   (d) increased genetic diversity

MODULE 33

Alternatives to Industrial Farming Methods

As problems with industrial agriculture become more apparent, alternative techniques and methods are gaining more attention. Some of the newer techniques have been used in the developing world for thousands of years. In this module, we will examine some of these alternative farming practices. We will also look at the large-scale farming of meat and fish.

Learning Goals

After reading this module, you should be able to

- describe alternatives to conventional farming methods.
- explain alternative techniques used in farming animals and in fishing and aquaculture.

Alternatives to industrial farming methods are gaining more attention

Industrial agriculture has been so successful and widespread that it has come to be known as conventional agriculture. In the developing world, where labor is relatively inexpensive and farm machinery and fossil fuel are much more costly, small-scale farming is common. In these countries, there are still many farmers who grow crops on small plots of land with mostly manual labor. Traditional farming methods that differ from those of industrial agriculture include *shifting agriculture* and *nomadic grazing,* which are not always sustainable, and more sustainable methods such as *intercropping* and *agroforestry*.

FIGURE 33.1 Shifting agriculture. This forest in Panama has been cleared for agriculture. Clearing land often involves burning it, which may make nutrients susceptible to leaching and soils vulnerable to erosion. *(Oyvind Martinsen Documentary Collection/Alamy Stock Photo)*

Shifting Agriculture and Nomadic Grazing

In locations with a moderately warm climate and relatively nutrient-poor soils, such as the rainforests of Central and South America, a large percentage of the nutrients is contained within the vegetation. In these places, farmers sometimes use **shifting agriculture**, in which the land is cleared, the vegetation is burned, and the land is used for only a few years until the soil is depleted of nutrients. This traditional method of agriculture uses a technique sometimes called "slash-and-burn," in which existing trees and vegetation are cut down, placed in piles, and burned (**FIGURE 33.1**). The resulting ash is rich in potassium, calcium, and magnesium, which make the soil more fertile, although these nutrients are usually depleted quickly. If the deforestation occurs in an area of heavy rainfall, nutrients may be washed away, along with some of the soil, which further reduces the availability of nutrients. After a few years, the farmer usually moves on to another plot and repeats the process.

If a plot is used for a few years and then abandoned for a number of decades, over time the soil may recover its organic content and nutrient supplies and the vegetation may have a chance to regrow. However, population pressures may cause the land to be used too frequently without enough time to allow for its full recovery. One additional consequence is **soil compaction**, a process where repeated trampling by humans, machinery, or animals causes a compaction of soil and a reduction in pore space. This can result in a decrease in permeability. When this happens, soil productivity can decrease rapidly, leaving the land suitable only for animal grazing. In addition, the burning process oxidizes carbon, meaning that it converts it into the oxide compounds carbon monoxide (CO) and carbon dioxide (CO_2). In this way, carbon from the vegetation and the soil is released into the atmosphere and ultimately contributes to higher atmospheric CO_2 concentrations.

In semiarid environments, dry, nutrient-poor soils can be easily degraded by agriculture to the point at which they are no longer viable for any production at all. Irrigation can cause salinization, and topsoil is eroded away because the shallow roots of annual crops fail to hold the soil in place. The transformation of arable, productive land to desert or unproductive land due to climate change or destructive land use is known as **desertification**. The world map in **FIGURE 33.2** on page 392 shows the parts of the world that are most vulnerable to desertification. Today, desertification is occurring most rapidly in Africa, where parts of the Sahara are expanding at a rate of up to 50 km (31 miles) per year. Unsustainable farming practices in northern China are also leading to rapid desertification.

The only sustainable way for people to use soil types with very low productivity is **nomadic grazing**, in which they move herds of animals, often over long distances, to seasonally productive feeding grounds. If grazing animals move from region to region without lingering in any one place for too long, the vegetation can usually regenerate.

Shifting agriculture and nomadic grazing worked well under the conditions in which they were first developed—namely, low human population densities and subsistence farming—but as populations increase, both these forms of traditional agriculture become less sustainable. Sometimes the relocation of people for political or other reasons can cause traditionally sustainable agricultural techniques to become unsustainable. For example, in the early part of the twentieth century, many subsistence farmers in Central America were relocated away from rich floodplains to mountainous areas where the plowing methods that had worked in the flat areas caused severe erosion in the mountains and could not be sustained.

Shifting agriculture An agricultural method in which land is cleared and used for a few years until the soil is depleted of nutrients.

Soil compaction A process where repeated trampling by humans, machinery, or animals causes a compaction of soil and a reduction in pore space.

Desertification The transformation of arable, productive land to desert or unproductive land due to climate change or destructive land use.

Nomadic grazing The feeding of herds of animals by moving them to seasonally productive feeding grounds, often over long distances.

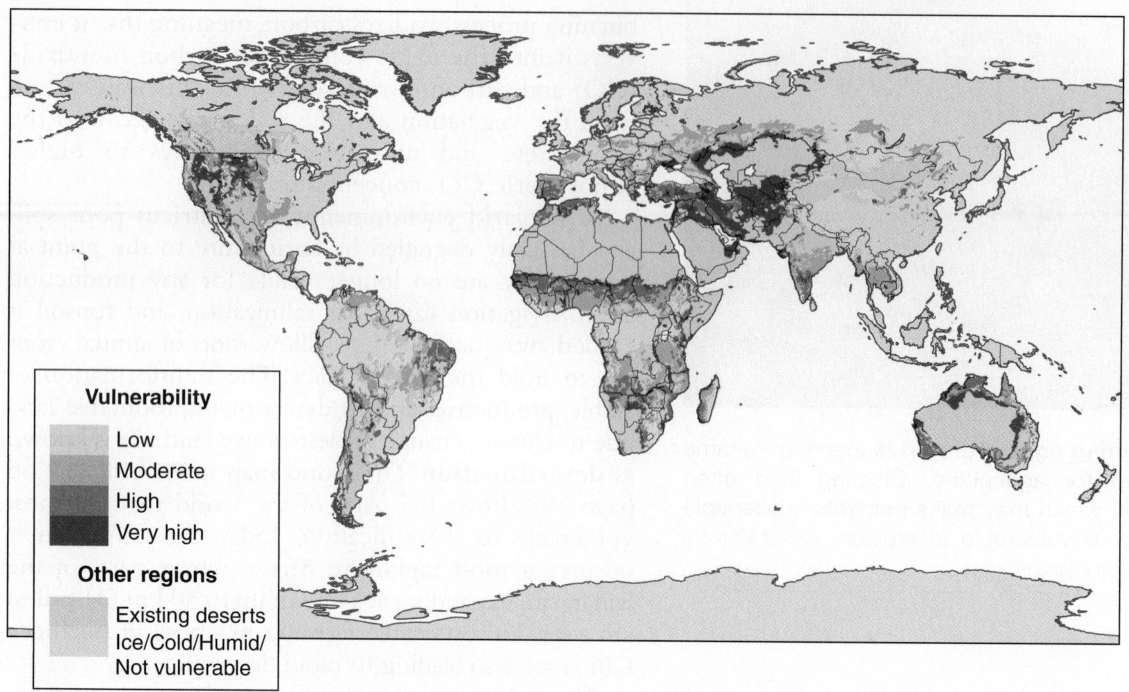

FIGURE 33.2 Vulnerability to desertification. Certain regions of the world are much more vulnerable to desertification than others. *(Data from FAO.)*

Sustainable Agriculture

Is it possible to produce enough food to feed the world's population without destroying the land, polluting the environment, and reducing biodiversity? **Sustainable agriculture** fulfills the need for food and fiber while enhancing the quality of the soil, minimizing the use of nonrenewable resources, and allowing economic viability for the farmer. It emphasizes the ability to continue agriculture on a given piece of land indefinitely through conservation and soil improvement. Sustainable agriculture often requires more labor than industrial agriculture, which makes it more expensive in places where labor costs are high. But practitioners of sustainable agriculture consider the improved long-term productivity of the land to be worth this extra cost.

Sustainable agriculture Agriculture that fulfills the need for food and fiber while enhancing the quality of the soil, minimizing the use of nonrenewable resources, and allowing economic viability for the farmer.

Intercropping An agricultural method in which two or more crop species are planted in the same field at the same time to promote a synergistic interaction.

Crop rotation An agricultural technique in which crop species in a field are rotated from season to season.

Agroforestry An agricultural technique in which trees and vegetables are intercropped.

Contour plowing An agricultural technique in which plowing and harvesting are done parallel to the topographic contours of the land.

Many of the practices used in sustainable agriculture are traditional farming methods (**FIGURE 33.3**). Subsistence farmers in India, Kenya, and Thailand typically use animal and plant wastes as fertilizer because they cannot obtain or afford synthetic fertilizers. Such traditional farmers may also practice **intercropping** (Figure 33.3a), in which two or more crop species are planted in the same field at the same time to promote a synergistic interaction between them. For example, corn, which requires a great deal of nitrogen, can be planted along with peas, a nitrogen-fixing crop. **Crop rotation** achieves the same effect by rotating the crop species in a field from season to season. For example, peas can be planted in a field for 1 year, resulting in increased nitrogen concentrations in the soil at the end of the year. This increase in nitrogen will enhance the growth of the corn crop that is planted there in the following year.

Intercropping trees with vegetables—a practice that is sometimes called **agroforestry**—allows vegetation of different heights, including trees, to act as windbreaks and to catch soil that might otherwise be blown away, greatly reducing erosion (Figure 33.3b). The trees not only protect the vegetable crops and the soil but also provide fruit and firewood.

Alternative methods of land preparation and use can also help to conserve soil and prevent erosion. For instance, **contour plowing**—plowing and harvesting parallel to the topographic contours of the land—helps prevent erosion by water while still allowing for the practical advantages of plowing (Figure 33.3c).

(a)

(b)

(c)

FIGURE 33.3 Sustainable farming methods. A variety of farming methods can be used to improve agricultural yield and retain soil and nutrients, including (a) intercropping such as this corn grown in a peach orchard in the state of Washington, (b) contour plowing, such as this farm in Iowa growing alfalfa and corn, and (c) agroforestry such as this shade-grown coffee in Mexico. *(a: Inga Spense/Getty Images; b: Tim McCabe/USDA National Resources Conservation Service; c: Chris R. Sharp/Science Source)*

Some farmers plant an autumn crop, such as winter wheat, that will sprout before frost sets in, so that the land does not remain bare between regular plantings.

No-Till Agriculture

Perennial plants live for multiple years, and there is usually no need to disturb the soil. In contrast, **annual plants**, such as wheat and corn, live only one season and must be replanted each year. Conventional agriculture of annual plants therefore relies on plowing and tilling, processes that physically turn the soil upside down and push crop residues under the topsoil, thereby killing weeds and insect pupae. Plowing and tilling exert a number of negative impacts on soils. As we learned in Chapter 8, soils may take hundreds or even thousands of years to develop as organic matter accumulates and soil horizons form. Every time soil is plowed or tilled, soil particles that were attached to other soil particles or to plant roots are disturbed and broken apart and become more susceptible to erosion. Soil below the plow zone may become susceptible to compaction. In addition, repeated plowing increases the exposure of organic matter deep in the soil to oxygen. This exposure leads to oxidation of organic matter, a reduction in the organic matter content of the soil, and an increase in atmospheric CO_2 concentrations. Plowing and tilling, in addition to irrigation and overuse of cropland, have led to severe soil degradation in many parts of the world. The world map in **FIGURE 33.4** on page 394 indicates areas of severe soil degradation.

Perennial plant A plant that lives for multiple years.

Annual plant A plant that lives only one season.

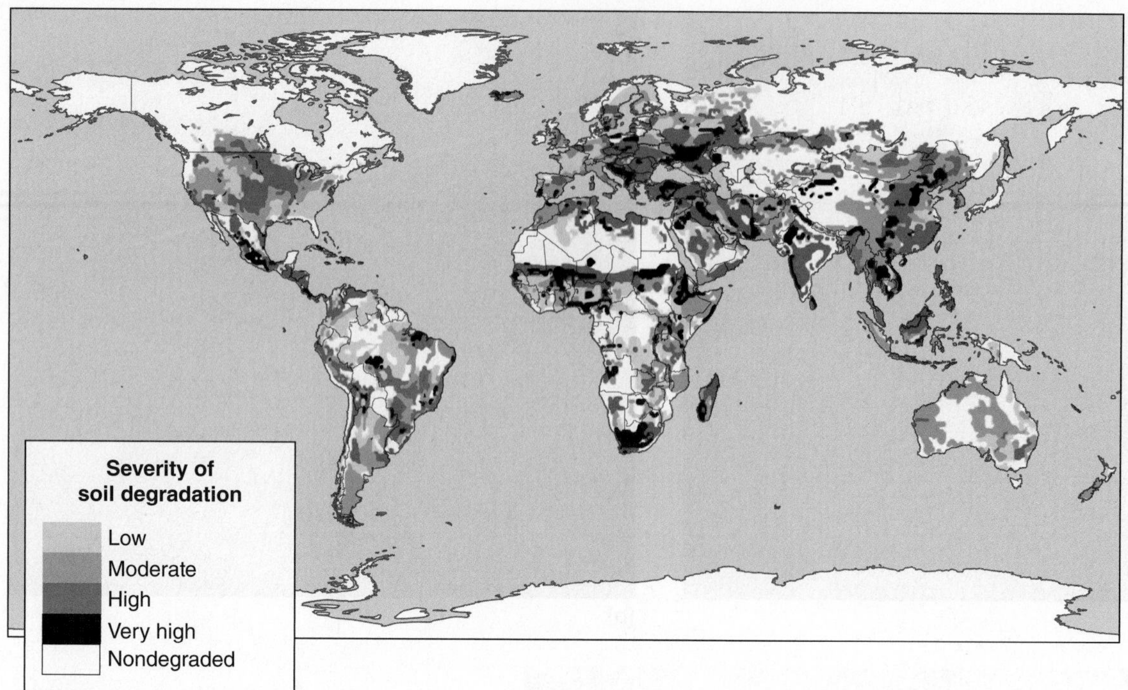

FIGURE 33.4 Global distribution of soil degradation. Soil degradation is a global problem caused by overgrazing and deforestation as well as agricultural mismanagement. *(Data from United Nations Environment Programme.)*

AP® Exam Tip

Soil qualities and properties come up often on the AP® Environmental Science Exam. Be sure that you understand the various soil qualities and properties that are needed for growing crops. ●

No-till agriculture is an agricultural method in which farmers do not turn the soil between seasons and is used as a means of reducing topsoil erosion. No-till agriculture is designed to avoid the soil degradation that comes with conventional agricultural techniques. Farmers using this method leave crop residues in the field between seasons (**FIGURE 33.5**). The intact roots hold the soil in place, reducing both wind and water erosion, and the undisturbed soil is able to regenerate natural soil horizons. No-till agriculture also reduces emissions of CO_2 because the intact soil undergoes less oxidation and carbon stored in the soil is more likely to remain there. "Do the Math: Comparing Carbon Accumulation" allows you to calculate the amounts of carbon in land that has been tilled versus land that has not been tilled.

No-till agriculture An agricultural method in which farmers do not turn the soil between seasons as a means of reducing topsoil erosion.

In many cases, however, in order for no-till agriculture to be successful, farmers must apply herbicides to the fields before, and sometimes after, planting so that weeds do not compete with the crops. Therefore, the downside of no-till methods is an increase in the use of herbicides.

FIGURE 33.5 No-till agriculture. Rows of soybeans emerge between the residues of a corn crop left over from the previous season. *(Lynn Betts/USDA Natural Resources Conservation Service)*

DO THE MATH — Comparing Carbon Accumulation in Tilled and No-Till Agricultural Soils

Preparing for the AP® Exam

Measuring soil carbon content is one way to determine the amount of organic matter in a soil. In theory, the practice of no-till agriculture should result in more soil carbon compared to a similar farm that practices conventional tilled agriculture. A native grassland that has not been farmed should be relatively undisturbed and therefore contain more organic matter and a greater carbon content then a plot of land tilled conventionally or a no-till plot.

A study in Nebraska measured the carbon content in soil on three adjacent areas of land: native grassland, tilled farmland, and no-till farmland. The investigators reported the findings for the first 5 cm of soil and then for the horizon from 5 cm to 20 cm. The study produced the following findings.

Treatment	Soil depth	Carbon content (g C m^{-2})	Nitrogen content (g N m^{-2})
Native grassland	0–5 cm	1,437	131
Native grassland	5–20 cm	2,653	265
Tilled farmland	0–5 cm	699	82
Tilled farmland	5–20 cm	2,208	230
No-till farmland	0–5 cm	1,129	108
No-till farmland	5–20 cm	2,299	366

Data from: J. Six et al. *Soil Sci. Soc. Am. J.* 62:1367–1377. 1998.

1. Add the 0–5 cm and 5–20 cm horizon for each treatment and compare the results of the three treatments.

The carbon content in g C m^{-2} for the 0–20 cm depth can be calculated by adding the 0–5 cm and 5–20 cm horizons:

Native grassland: 1437 g C m^{-2} + 2653 g C m^{-2} = 4090 g C m^{-2}

Tilled farmland: 699 g C m^{-2} + 2208 g C m^{-2} = 2907 g C m^{-2}

No-till farmland: 1129 g C m^{-2} + 2299 g C m^{-2} = 3428 g C m^{-2}

2. Describe the two treatments as a percentage of the native grassland carbon content.

Tilled farmland as a percentage of native grassland = (2907 g C m^{-2}) ÷ (4090 g C m^{-2}) = 0.7108 × 100% = 71.08%

No-till farmland as a percentage of native grassland = (3428 g C m^{-2}) ÷ (4090 g C m^{-2}) = 0.8381 × 100% = 83.81%

The no-till farmland contains 83.81 percent of the carbon contained in the native grassland while the tilled farmland contains only 71.08 percent of the native grassland carbon content.

3. Draw a conclusion about the effects of tilled agriculture versus no-till agriculture compared to native grassland.

This study supports the idea that there is more carbon retained in a no-till agricultural farm then in a conventionally tilled farm, relative to native grassland.

YOUR TURN Conduct a similar analysis for nitrogen.

1. Add the 0–5 cm and 5–20 cm horizon for each treatment and compare the results of the three treatments.
2. Describe the two treatments as a percentage of the native grassland nitrogen content.
3. Draw a conclusion about the effects of tilled agriculture versus no-till agriculture compared to native grassland.

A variation on the idea of no-till agriculture is the effort to produce perennial crops. Because annual crops die at the end of each growing season, there is a need to plow or till and replant at the beginning of the following growing season. Proponents of the breeding and development of perennial crops maintain that the single most beneficial new development in agriculture would be the development of food crops that do not need to be replanted every year.

Plant researchers are exploring a variety of ways to develop perennial crops. Using a combination of conventional selective breeding and technology, they are attempting to convert annual species such as wheat, sorghum, sunflowers, and corn into perennials. At the same time, they are collecting wild perennials such as wheatgrass, then domesticating them and selecting for higher seed yield, size, and quality. Through these efforts, researchers hope to assemble communities of perennial plants, animals, fungi, and microorganisms that will be stable, productive, and resistant to insect pests and diseases. In the process, the need for plowing and tilling will be eliminated.

Integrated Pest Management

Another alternative agricultural practice, **integrated pest management (IPM)**, uses a variety of techniques designed to minimize pesticide inputs. These techniques include crop rotation and intercropping, the use of pest-resistant crop varieties, the creation of habitats for predators of pests, and limited use of pesticides.

Crop rotation and the use of pest-resistant crop varieties prevent pest infestations. Crop rotation can reduce the success of insect pests that are specific to one crop and may have laid eggs in the soil. When the eggs hatch, if a different crop is present, the crop-specific pest will not have a suitable host and will die. In the same way, crop rotation can also hinder crop-specific diseases that may survive on infected plant material from the previous season. Intercropping, as stated earlier, also makes it harder for specialized pests that succeed best with only one crop present to establish themselves. Farmers can also provide habitat for species that prey on crop pests (**FIGURE 33.6**). For example, agroforestry encourages the presence of insect-eating birds (although birds can also damage some crops). Many herbs and flowers attract beneficial insects and are often used in gardens and croplands.

Although IPM practitioners do use pesticides, they limit applications through very careful observation. Farmers regularly inspect their crops for the presence of insect pests and other potential crop hazards to catch them early so they can be treated using natural controls or smaller doses of pesticides than would be needed at later stages of an infestation. These more-targeted methods of pest control can result in significant cash savings on pesticides as well as improved yields. **FIGURE 33.7**—a case study from IPM training in Indonesia—shows the difference IPM can make. Farmers who learned how to determine whether a pesticide application was warranted were able to cut their pesticide applications, and their expenditures on pesticides, in half. Yields also improved after farmers learned and implemented IPM methods.

When farmers take time to inspect their fields carefully, as required by IPM methods, they often notice other crop needs, and this additional attention improves overall crop management. The trade off for these benefits is that farmers must be trained in IPM methods and must spend more time inspecting their crops. But once farmers are trained, the extra income and reduced economic costs associated with IPM often outweigh the

Integrated pest management (IPM) An agricultural practice that uses a variety of techniques designed to minimize pesticide inputs.

FIGURE 33.6 Beneficial insect habitat. Practitioners of integrated pest management often provide habitat for insects that prey on crop pests. This wasp is laying eggs in a caterpillar, which it has paralyzed. *(USDA/Nature Source/Science Source)*

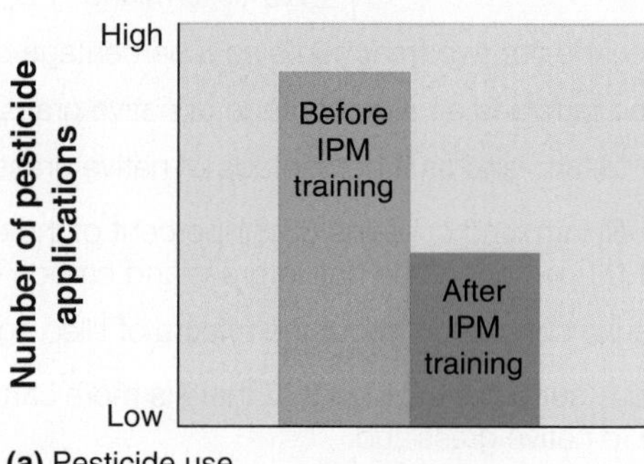

(a) Pesticide use

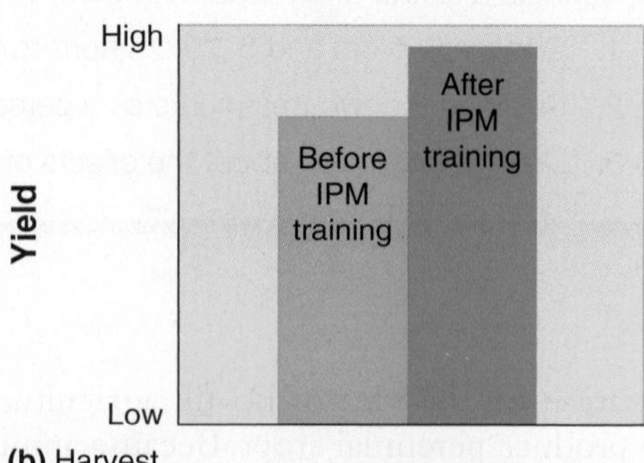

(b) Harvest

FIGURE 33.7 Effects of IPM training. (a) IPM training of farmers in Indonesia led to a significant reduction in pesticide applications. (b) Yield improvements also occurred after the training because of the additional attention the farmers gave to their crops. *(Data from FAO.)*

extra time they must spend in the field. IPM has been especially successful in many parts of the developing world where the high-input industrial farming model is not viable because labor costs are low and farmers lack financial resources.

Organic Agriculture

Organic agriculture is the production of crops in a way that sustains or improves the soil, without the use of synthetic pesticides or fertilizers. Organic agriculture follows several basic principles:

- Use ecological principles and work with natural systems rather than dominating those systems.
- Maintain the soil by increasing soil mass, biological activity, and beneficial chemical properties.
- Keep as much organic matter and as many nutrients in the soil and on the farm as possible.
- Avoid the use of synthetic fertilizers and synthetic pesticides.
- Reduce the adverse environmental effects of agriculture.

In the developed world, organic farming has increased in popularity over the past 3 decades. The U.S. Organic Foods Production Act (OFPA) was enacted as part of the 1990 farm bill to establish uniform national standards for the production and handling of foods labeled organic. "Science Applied 5: How Do We Define Organic Food?" on page 410 discusses organic food labeling in more detail.

For a long time, all organic farms were small, but today this is not necessarily the case. However, because most organic farmers plant diverse crops and encourage beneficial insects, it is common for them to keep their farms relatively small. Organic farmers also must manage the soil carefully because if they lose soil nutrients and see a decline in the health of their crops, they have fewer options than conventional farmers. These practices usually increase labor costs significantly. However, farmers can recoup extra labor costs by selling their harvest at a premium price to consumers who prefer to buy organic food and are sometimes willing to pay more for it (**FIGURE 33.8**).

Organic agriculture is not without some adverse environmental consequences. Because organic farmers typically do not use herbicides, they are less likely than conventional farmers to be able to use no-till methods successfully. And alternative pest control methods are not always environmentally friendly. For example, in order to keep crops such as carrots free of weeds, organic farmers may treat the soil with a flame fueled by propane before planting. While this technique protects carrots without the use of herbicides, it does use propane, a fossil fuel, and causes the release of carbon dioxide to the atmosphere.

FIGURE 33.8 Organic farming. Many customers are willing to pay higher prices for organically grown food, like the produce at this organic farm in Goleta, California. *(UniversalImagesGroup/Getty Images)*

Alternative techniques for farming animals and fish are becoming more popular

Animals and fish are in general more energy intensive to raise and prepare for eating than vegetable crops, but there are methods for animal and fish production that have a lower environmental impact. This section covers some of those processes.

Not all meat comes from CAFOs. Free-range chicken and beef are becoming increasingly popular in the United States. Some people find it more ethically acceptable to eat a chicken or cow that has wandered free than one that has spent its entire life confined in a small space. Free-range meat, if properly produced, is more likely to be sustainable than meat produced in CAFOs. Because these free-range animals are not as likely to spread disease as those that are kept in close quarters, the use of antibiotics and other medications can be reduced or eliminated. The animals graze or feed on the NPP of the land, with little or no supplemental feeding, so less fossil fuel goes into the raising of free-range meat. Finally, manure and urine are dispersed over the range area and are naturally processed by detritivores and decomposers in the soil. As a result, there is no need to treat and dispose of massive quantities of manure. On the negative side, free-range

Organic agriculture The production of crops in a way that sustains or improves the soil, without the use of synthetic pesticides or fertilizers.

operations use more land than CAFOs do, and the cost of meat produced using these techniques is usually significantly higher.

More Sustainable Fishing

In the interest of creating and supporting sustainable fisheries, many countries around the world have developed fishery management plans, often in cooperation with one another. International cooperation is particularly important because fish migrate across national borders, some marine ecosystems span national borders, and many of the world's most important fisheries lie in international waters.

The northwestern Atlantic fisheries, for example, comprise several continental shelf ecosystems that exist along the coast of the northeastern United States and eastern Canada. Historically, these fisheries were among the most productive in the world. However, overfishing by international fleets of factory ships led to a substantial and long-lasting depletion of fish stocks, particularly of cod and pollock, by the early 1990s, as **FIGURE 33.9** shows. The fisheries were forced to close because of the depleted stocks, and the Canadian and U.S. governments imposed a moratorium on bottom fishing in the area. In recent years, a recovery has occurred in some of the fisheries, seemingly in response to the closing of the fisheries. The majority of the Atlantic Canadian fisheries are still closed today. Some fisheries off the coast of the United States are now open, at least for part of the year.

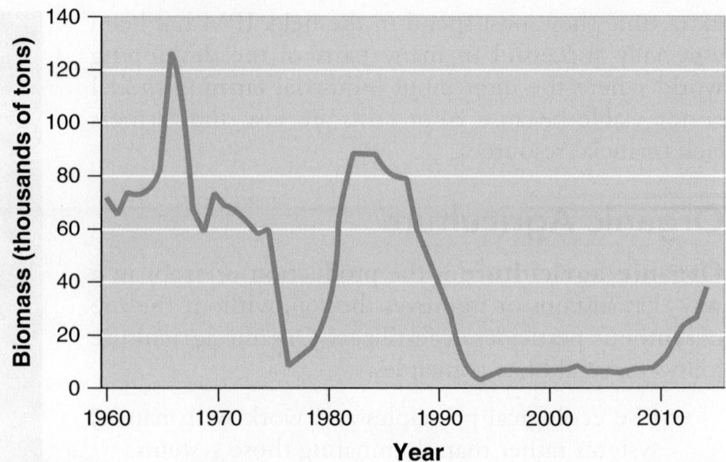

FIGURE 33.9 Fishery collapse in the northwestern Atlantic Ocean. Cod biomass in the Grand Bank, Northwest Atlantic Ocean. A decline beginning in 1985 lasted through 2010. Since then, these fisheries have been recovering. *(Data from FAO, Fisheries and Resources Monitoring System, http://firms.fao.org/firms/resource/10315/en.)*

In response to the fishery collapse, and in order to restore the depleted stocks and manage the ecosystem as a whole, the U.S. Congress passed the Sustainable Fisheries Act in 1996. This act shifted fisheries management from a focus on economic sustainability to an approach that increasingly stressed conservation and the sustainability of species. The act calls for the protection of critical marine habitat, which is important for both commercial fish species and nontarget species. For many commercial species considered to be in danger, such as cod, a sustainable fishery means that no fishing will be permitted until populations recover.

One successful fishery management plan was developed in Alaska, where the commercial salmon fishery declined rapidly between 1940 and 1970. Managers first tried to increase salmon populations by limiting the fishing season, but by 1970, when the season was restricted to less than a week, so many fishers participated that populations continued to drop.

In 1973, fishery managers introduced a system of **individual transferable quotas (ITQs)**, a fishery management program in which individual fishers are given a total allowable catch of fish for a season that they can either catch themselves or sell to others. Before the start of each salmon season, fishery managers establish a total allowable catch and distribute or sell quotas to individual fishers or fishing companies, favoring those with long-term histories in the fishery. Fishers with ITQs have a secure right to catch their quota so they have no need to spend money on bigger boats and better equipment in order to outcompete others. If fishers cannot catch enough salmon to remain economically viable, they can sell all or part of their quota to another fisher. **FIGURE 33.10** shows the results: Since the beginning

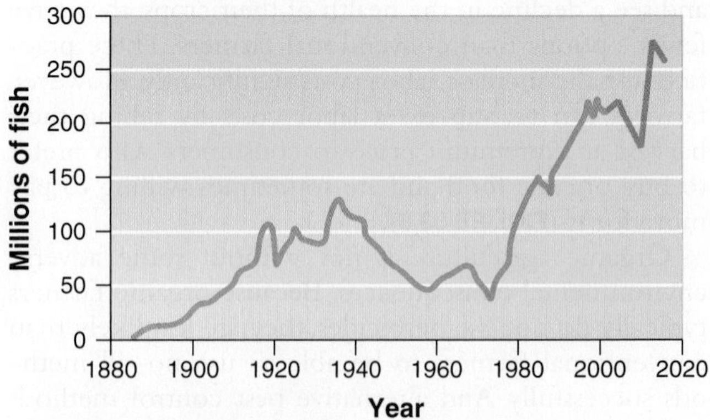

FIGURE 33.10 Commercial salmon harvest in the Alaska fishery. After a peak harvest in 1940, overfishing led to a decline in the number of fish caught. In 1973, fishery managers introduced a system of individual transferable quotas. By 1980, the fishery had rebounded. *(Data from: http://www.adfg.alaska.gov/index.cfm?adfg=wildlifenews.view_article&articles_id=775.)*

Individual transferable quota (ITQ) A fishery management program in which individual fishers are given a total allowable catch of fish in a season that they can either catch or sell.

of the ITQ program, the salmon population and harvest have increased—at times very rapidly—and the adverse effects on fishers and fishing companies have been reduced.

In Alaska, ITQs are sold primarily to small family-run fishing operations. In New Zealand, however, ITQs have been used effectively to control overfishing by large fishing companies. The ITQ system is being used successfully in many other fisheries around the world.

Not all fisheries are declining, but it is often difficult for consumers to know which fish are being overharvested and which are not. To help consumers choose more sustainable fish, the Environmental Defense Fund and other organizations have compiled lists of popular food fish, dividing them into three categories, depending on how sustainable their stocks are. "Best" choices include wild Alaskan salmon and farmed rainbow trout. "Worst" choices include shark and Chilean sea bass. The Monterey Bay Aquarium has a seafood watch app that classifies seafood into three categories—"Best Choice," "Good Alternative," or "Avoid"—depending on its status.

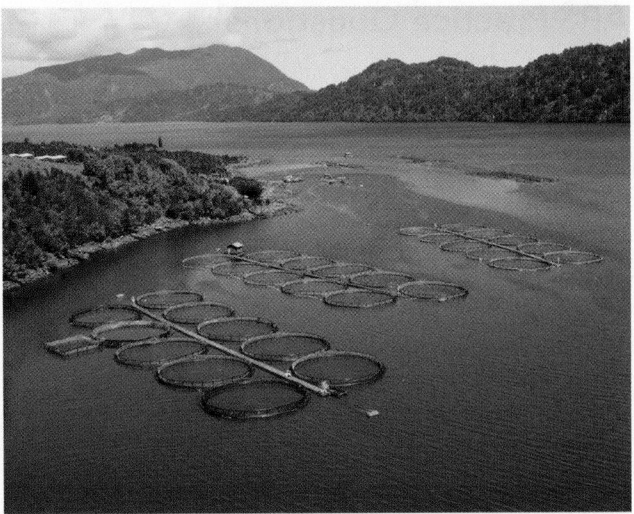

FIGURE 33.11 A salmon farming operation in Chile. Uneaten food and waste released from salmon farms can cause significant nutrient input into natural marine ecosystems. *(MARCELODLT/Shutterstock)*

Aquaculture

The demand for fish has increased even as wild fish catches have been falling. In response, many scientists, government officials, and entrepreneurs have been developing ways to increase the production of seafood through aquaculture. Aquaculture involves constructing an aquatic ecosystem by stocking the organisms, feeding them, and protecting them from diseases and predators. It usually requires keeping the organisms in enclosures (**FIGURE 33.11**), and it may require providing them with food and antibiotics. Most of the catfish, shrimp, and salmon eaten in the United States are produced by aquaculture.

Proponents of aquaculture believe it can alleviate some of the human-caused pressure on overexploited fisheries while providing much-needed protein for the more than 1 billion undernourished people in the world. Aquaculture also has the potential to boost the economies of many developing countries.

Critics of aquaculture, though, point out that it can create many environmental problems. In a typical aquaculture facility, clean water is pumped in at one end of a pond or marine enclosure, and wastewater containing feces, uneaten food, and antibiotics is pumped back into the river or ocean at the other end. The wastewater may also contain bacteria, viruses, and pests that thrive in the high-density habitat of aquaculture facilities and this can infect wild fish and shellfish populations outside the facility. In addition, fish that escape from aquaculture facilities may harm wild fish populations by competing with them, interbreeding with them, or spreading diseases and parasites. Overall, however, aquaculture has many promising characteristics as a means of sustainable food production.

MODULE 33 AP® Review

Preparing for the AP® Exam

Alternatives to modern, industrial agriculture feature a number of characteristics that tend to emulate natural processes. Intercropping of different plants is similar to multi-species associations of plants that exist in nature. Crop rotation is a practice that simulates the natural shifts in species that occur. Integrated pest management uses a variety of techniques to control pests with a minimum of pesticide use. Organic agriculture is the production of crops without the use of synthetic pesticides or fertilizers.

AP® Practice Questions

Choose the best answer for the following.

1. Which agricultural method is most similar to intercropping?
   (a) contour plowing
   (b) agroforestry
   (c) no-till agriculture
   (d) integrated pest management

2. Which agricultural method is used to prevent erosion?
   (a) intercropping
   (b) integrated pest management
   (c) contour plowing
   (d) CAFOs

3. Which is NOT a principle of organic farming?
   (a) avoiding synthetic fertilizers
   (b) keeping as much organic matter as possible on the farm
   (c) increasing soil mass and biological activity
   (d) avoiding the use of fossil fuels

4. Which is a part of integrated pest management?
   (a) no-till
   (b) increased application of pesticides
   (c) contour plowing
   (d) crop rotation

5. Individual transferable quotas
   (a) are used in organic pest management.
   (b) are a part of sustainable fishing.
   (c) are an alternative method used for CAFOs.
   (d) are used to increase yields in communal farms.

Working Toward Sustainability

Urban Agriculture

Urban farms are a small percentage of farmland by area, and they represent a small fraction of total food production, but they have the potential to introduce substantial benefits to human civilization and the natural environment. Remember that more than half the people in the world live in cities. Around the United States and elsewhere in the world, an increasing number of urban gardeners are providing food for themselves and their neighbors. In doing so, they are enhancing ecosystem services by improving urban land that had been degraded, increasing habitat for plants and beneficial insects, and building up soil nutrients and storing atmospheric carbon. They are also providing knowledge about and access to nutritious, healthy food. Beyond environmental benefits, urban farms are a potential source of community revitalization, promoting jobs and helping redevelop neighborhoods that have been on the decline.

The USDA defines urban agriculture as "backyard, rooftop and balcony gardening" as well as farming and gardening in vacant lots and parks in urban and suburban areas. A group called the Detroit Black Community Food Security Network (DBCFSN) has taken advantage of unfortunate circumstances and urban blight to revitalize and energize residents in the city of Detroit, Michigan, with urban agriculture. In 2007, after a series of financial and political crises, one-third of the city's land area was designated as "vacant." It also had no national food chain grocery stores until one relatively expensive food store arrived in 2013. Forty percent of the population was living below the national poverty level, and although there is no subway system in Detroit and only a limited streetcar system, almost one-third of its residents did not own cars. In addition to unemployment, crime and unpleasant living conditions, obtaining affordable, diverse, quality food was a challenge for many Detroit residents.

Created in 2006, DBCFSN started a 0.1 hectare (quarter-acre) organic farm called D-Town Farm that eventually grew to 3 hectares (7 acres). D-Town Farm was established as a model farm to demonstrate farming techniques to school groups and adults that can be used all over the city. The farm practices organic agriculture, so the thin, nutrient-depleted soil they started with has been improved over time, leading to storage of carbon and an increase in soil nutrients. The farm grows more than 30 crop species allowing for an increase in biodiversity of the area.

A community garden. Local residents in a community garden in Detroit are growing both vegetables and flowers. Many gardens in Detroit are located in what used to be abandoned fields and lots. *(Jim West/Alamy)*

There is greater water absorption since the farm was established, now that a rich soil and vegetation cover the land. They also keep bees at D-Town Farm that provide an ecosystem service to the area. From this model farm and others that started over the past 2 decades, there are now over 350 urban farms in Detroit. These farms grow produce that is sold at farmers markets, start seedlings that are distributed to family and school gardens, and educate children and adults about urban agriculture and environmental science. Some of the gardens supply food for a variety of food rescue organizations that gather food that is donated or might otherwise be wasted. They provide this food to people in need. Other gardens compost food that is no longer edible. Both of these actions reduce food waste and prevent spoiled food from ending up in the landfill, where it will decompose anaerobically and produce methane.

One writer estimated that urban farms and gardens in Detroit grew 181,000 kg (400,000 pounds) of produce in 2014. This is a small fraction of the produce consumed in Detroit, but it's a beginning to what could become a larger food contribution from urban areas. The FAO estimates that 800 million people practice urban gardening and farming worldwide. In doing so, they are utilizing what could be called under-utilized space such as backyards, abandoned lots, and rooftops and increasing biodiversity, ecosystem services, and productivity of urban areas while also increasing carbon storage. There's another service provided by urban farming: People in urban areas are often unaware of their food sources. The writer Wendell Berry is famous for saying that "Eating is an agricultural act." Urban gardeners are able to do more than just eat their food—in some case, they are responsible for growing it. If that results in improvements in health, economic and environmental science knowledge, and reductions in food waste, there are many reasons to encourage more urban gardening.

Critical Thinking Questions

1. What are some of the ecosystem services provided by urban agriculture?
2. How would you improve the USDA definition of urban agriculture?

References

Guzmán, M. 2016. Black farmers in Detroit are growing their own food. https://www.pri.org/stories/2016-03-30/black-farmers-detroit-are-growing-their-own-food-theyre-having-trouble-owning

Bonfiglio, O. 2008. Growing green in Detroit. *Christian Science Monitor*, August 21.

In this chapter, we examined human nutritional requirements and contemporary problems of undernutrition and overnutrition. Modern industrial agriculture in the developed world uses large amounts of fossil fuel inputs and relies on other practices such as irrigation, fertilization, and pesticide application. Genetically modified organisms are also a component of modern agriculture in many parts of the world. Raising meat and fish have specific environmental impacts related to concentrated feeding areas for meat and removal of large numbers of fish from the oceans. Alternatives to conventional modern agriculture exist and often make use of natural patterns observed in nature such as crop rotation and interplanting different species. Organic agriculture is an agricultural practice that does not use chemical pesticides or fertilizers. A variety of sustainable organic and other alternative agricultural practices are being utilized in a variety of locations around the world.

Key Terms

Undernutrition
Malnourished
Food security
Food insecurity
Famine
Anemia
Overnutrition
Meat
Industrial agriculture
Agribusiness
Energy subsidy
Green Revolution
Economies of scale
Waterlogging
Salinization
Organic fertilizer
Synthetic fertilizer
Inorganic fertilizer
Monocropping
Pesticides
Insecticide
Herbicide
Broad-spectrum pesticide
Selective pesticide
Narrow-spectrum pesticide
Persistent pesticide
Nonpersistent pesticide
Pesticide resistance
Pesticide treadmill
Concentrated animal feeding operation (CAFO)
Aquaculture
Fishery
Fishery collapse
Bycatch
Shifting agriculture
Soil compaction
Desertification
Nomadic grazing
Sustainable agriculture
Intercropping
Crop rotation
Agroforestry
Contour plowing
Perennial plant
Annual plant
No-till agriculture
Integrated pest management (IPM)
Organic agriculture
Individual transferable quota (ITQ)

Learning Goals Revisited

Module 31 Human Nutritional Needs

Describe human nutritional requirements.

Humans require both a certain number of calories each day and a balance of proteins and other nutrients. Undernutrition is a condition in which not enough food calories are ingested each day. Malnutrition is a condition in which the diet lacks the correct balance of proteins, carbohydrates, vitamins, and minerals.

Explain why nutritional requirements are not being met in various parts of the world.

Malnutrition occurs due to poverty and political situations that prohibit the efficient and equitable distribution of food. In addition, some grains that could be fed to humans are used to feed livestock and are not available for direct human consumption.

Module 32 Modern Large-Scale Farming Methods

Describe modern, large-scale agricultural methods.

In the twentieth century, a variety of modern agricultural innovations including the Green Revolution transformed agriculture from a system of small farms relying mainly on human labor into a system of large industrial operations that use machinery run by fossil fuels. This shift has resulted in larger farms and monocropping. Irrigation can increase crop yields dramatically, but it can also draw down aquifers and lead to soil degradation through waterlogging and salinization. Fertilizers also increase crop yields dramatically, but can run off into surface waters and cause damage to ecosystems. The negative environmental consequences of pesticides include loss of beneficial nontarget organisms, human health problems, and surface water contamination. When pests become resistant to pesticides over time, a pesticide treadmill can develop.

Explain the benefits and consequences of genetically modified organisms.

Agricultural scientists are using genetic engineering to produce genetically modified organisms with desirable traits. GMOs can increase yields and reduce the use of pesticides. Concerns about GMOs include safety for human consumption and their effects on biodiversity.

Discuss the large-scale raising of meat and fish.

Concentrated animal feeding operations and aquaculture facilities enable efficient animal growth and economically inexpensive food production. However, the disposal of concentrated animal waste presents a problem and the environment can be harmed by runoff containing antibiotics and other waste products.

Module 33 Alternatives to Industrial Farming Methods

Describe alternatives to conventional farming methods.

Traditional farming techniques such as intercropping, crop rotation, agroforestry, and contour plowing can sometimes improve agricultural yields and conserve soil and other resources. No-till agriculture is another way to reduce soil erosion and degradation. Integrated pest management reduces the use of pesticides, thus saving money and reducing environmental damage. IPM requires more labor, however, and practitioners must be trained to identify potential hazards to their crops. Organic agriculture focuses on maintaining the soil and avoids the use of synthetic fertilizers and pesticides. This approach often results in more labor-intensive and smaller farms, where many alternative agricultural techniques must be applied.

Explain alternative techniques used in farming animals and in fishing and aquaculture.

Free-range animals can be raised with much lower environmental cost than those raised in feedlots. Wild-caught fish, when caught in appropriate numbers, can be managed sustainably. Some fish species can be raised on fish farms with minimal environmental harm.

Practice Math and Graphing

Preparing for the AP® Exam

Answer the following questions. Be sure to show your work.

1. Practice Math

In 2015, the people in the United States consumed 90 pounds of chicken per capita, 55 pounds of beef per capita, and 50 pounds of pork per capita. Convert each value to kilograms and determine the total meat (chicken + beef + pork) consumption per capita in the United States in 2015.

2. Practice Graphing

(a) Using the data shown in the table, create a line graph that shows the per capita consumption of chicken and beef in the United States between 1990 and 2015. Plot year on the x axis and per capita chicken and beef consumption (in pounds) on the y axis.

Year	Beef, pounds per person	Chicken, pounds per person
1990	68	60
1995	66	68
2000	66	76
2005	64	85
2010	59	82
2015	55	90

(b) Describe the trends in chicken and beef consumption in the United States between 1990 and 2015.

Section 1: Multiple-Choice Questions

Choose the best answer for questions 1–20.

1. Which does NOT explain the rise of the modern farming system?
   (a) Small farms are usually more profitable than large farms.
   (b) Irrigation contributes to greater crop yields.
   (c) Fertilizers improve crop yields and are easy to apply.
   (d) Mechanization facilitates monocropping and improves profits.

2. Irrigation can result in which of the following environmental problems?
   I. reduction of evaporation rates
   II. accumulation of salts in soil
   III. waterlogging of soil and plant roots
   (a) I only
   (b) II only
   (c) I and II
   (d) II and III

3. The use of synthetic fertilizers increases crop yields, but also
   (a) destroys the nitrifying bacteria in the soil.
   (b) increases fish populations in nearby streams.
   (c) decreases phosphorus concentrations in the atmosphere.
   (d) increases nutrient runoff into bordering surface waters.

4. Which statement best describes the pesticide treadmill?
   (a) Broad-spectrum pesticides degrade into selective pesticides, thereby killing a wide range of insect pests over a long period.
   (b) Pesticides accumulate in the fatty tissues of consumers and increase in concentration as they move up the food chain.
   (c) Some pest populations evolve resistance to pesticides, which become less effective over time so that new pesticides must be developed.
   (d) Testing of the toxicity of pesticides to humans cannot keep pace with the discovery and production of new pesticides.

5. How did the Green Revolution increase food production?
   I. through the development of disease-resistant and high-yielding crop plants
   II. through monocropping and the widespread use of machinery
   III. through the application of fertilizers and the use of irrigation techniques
   (a) I only
   (b) II only
   (c) III only
   (d) I, II, and III

6. Which is NOT a traditional farming technique that is used in sustainable agriculture?
   (a) nomadic herding
   (b) intercropping
   (c) crop rotation
   (d) agroforestry

7. Which is an environmental advantage of no-till agriculture?
   (a) The use of herbicides improves the stability of the soil.
   (b) Migratory bird populations are reduced.
   (c) The undisturbed soil is less susceptible to erosion.
   (d) The crop residues reduce the soil profile.

8. Which practice is NOT a part of integrated pest management?
   (a) crop rotation
   (b) elimination of pesticides
   (c) use of pest-resistant crops
   (d) frequent inspection of crops

9. Farmers who practice organic agriculture have less of an impact on the environment than farmers who practice industrial agriculture because they
   (a) use no-till agriculture exclusively.
   (b) import soil to maintain soil fertility.
   (c) maintain large farms with a single crop.
   (d) avoid pesticides and synthetic fertilizers.

10. Critics of genetically modified organisms as food crops are concerned about
    I. introduction of new allergens into the food supply.
    II. loss of genetic diversity in food crops.
    III. decreases in food production worldwide.
    (a) I only
    (b) II only
    (c) III only
    (d) I and II

11. Concentrated animal feeding operations (CAFOs) can best be described as
    (a) facilities where a large number of animals are housed and fed in a confined space.
    (b) a method of producing more meat at a higher cost.
    (c) a means of producing great quantities of manure to fertilize fields organically.
    (d) the storing and compacting of grain for use as a nutrient supplement for cattle.

12. Which is NOT an environmental or health problem that has been associated with CAFOs?
    (a) the increase of antibiotic-resistant bacteria potentially harmful to humans
    (b) the overgrazing of large tracts of land
    (c) the runoff of animal wastes into natural waters
    (d) the production of huge quantities of manure, creating a waste disposal problem

Use the following graphs to answer Question 13.

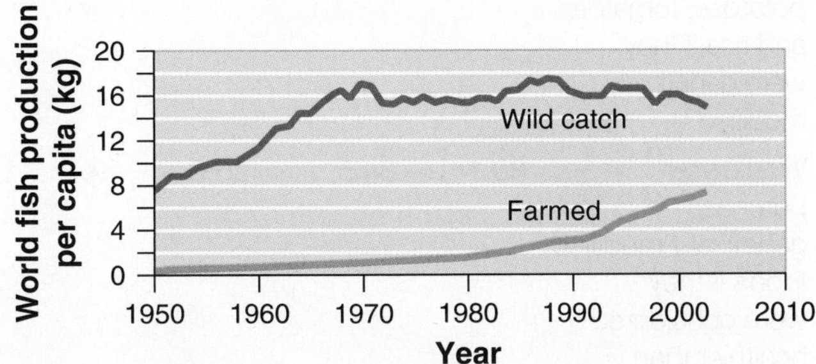

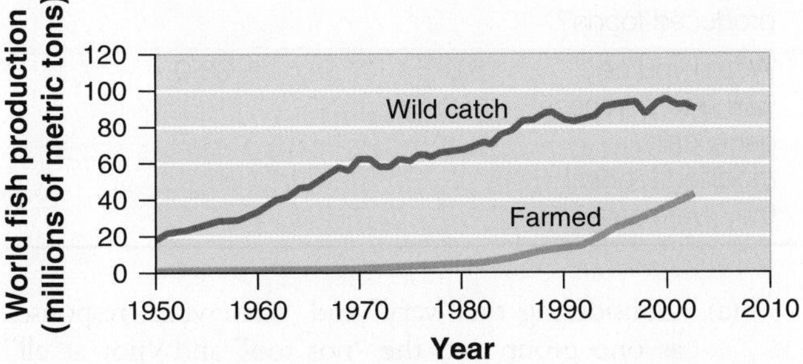

13. The data presented in the graphs, compiled by the FAO, describe world fish production from 1950 to 2003. Based on these data, which of the following statements best describes the global trend in fish production?
    (a) The use of individual transferrable quotas (ITQs) has led to the overfishing of wild species.
    (b) Fish production through aquaculture exceeds fish production in the ocean.
    (c) As the human population continues to increase, the per capita wild catch also continues to increase.
    (d) As the human population continues to increase, per capita farmed fish production also continues to increase.

14. Which is an environmental effect of bycatch?
    (a) Juvenile fish are too small and slip through the nets, thereby ensuring the next generation of commercial fish.
    (b) Predators caught in the nets feed on the commercially important fish also caught, thereby reducing the total catch.
    (c) Nontarget fish populations have declined.
    (d) Cod populations have increased due to the removal of competitor species.

15. It has been projected that aquaculture could supply about one-half the global demand for seafood by 2025. However, future production through aquaculture could be limited because
    (a) concentrated waste from aquaculture facilities contaminates rivers and oceans.
    (b) raising fish in a protected environment could lead to a fish population overshoot.
    (c) the economies of developing countries would be negatively affected.
    (d) ocean fishing operations would go out of business.

16. What is the primary benefit of perennial plants?
    (a) increased crop yields
    (b) decreased pesticide use
    (c) increased nitrogen fixing
    (d) decreased erosion

17. Which describes how intercropping alters the ecology of a farm?
    I. Intercropping increases diversity.
    II. Intercropping promotes plant mutualisms.
    III. Relative to monocropping, intercropping increases the prevalence of insect infestations among crops.
    (a) I only
    (b) II only
    (c) III only
    (d) I and II

18. Which is TRUE about shifting agriculture?
    I. It can result in decreased permeability from soil compaction.
    II. It is often practiced in relatively nutrient-poor soils, like rainforests.
    III. The ash from the burned vegetation decreases soil productivity.
    (a) I only
    (b) III only
    (c) I and II
    (d) I, II, and III

19. Suppose it takes 12 kg of grain to produce 1 kg of lamb and that a person eating only corn can obtain sufficient calories in a year from 0.02 ha of land. How much land would be required to support a person who only eats lamb?
    (a) 0.12 ha
    (b) 0.24 ha
    (c) 12.0 ha
    (d) 24 ha

20. Reasons for global malnutrition include all of the following EXCEPT
    (a) insufficient food production.
    (b) poverty.
    (c) political unrest.
    (d) natural disasters.

Section 2: Free-Response Questions

Write your answer to each part clearly. Support your answers with relevant information and examples. Where calculations are required, show your work.

1. Maintaining dairy cattle in a CAFO requires large quantities of water and produces vast quantities of wastewater and manure. The increasing number of CAFO dairies in eastern New Mexico and west Texas is contributing to significant groundwater contamination and the depletion of the Ogallala aquifer. Consider a CAFO with 1,000 dairy cows and answer the following questions.
   (a) According to the U.S. Department of Agriculture, the average dairy cow can consume up to 200 L of water daily. An additional 120 L per cow per day is required to wash the milking equipment and milking area. How many liters of water are required to operate this dairy daily? (2 points)
   (b) According to the U.S. Environmental Protection Agency, the average dairy cow produces 55 kg of wet manure daily. How many metric tons would this dairy produce each day (1,000 kg = 1 metric ton)? (2 points)

2. The table shows the results of a survey of 113 people who were asked questions about foods that were staples in their diet. Use the table to answer the questions at the top of the next column.

Questions	Very likely (%)	Somewhat likely (%)	Not too likely (%)	Not at all likely (%)
Would you be willing to purchase corn oil, potatoes, tomatoes, and rice if they were genetically modified?	15.9	31.9	38.9	7.1
Would you purchase genetically modified foods if they were considered healthier than conventionally produced foods?	30.1	34.5	25.6	9.7
Would you be willing to purchase genetically modified foods if they were safe?	15.9	31.9	38.9	7.1

   (a) Considering the "very" and "somewhat" responses as one group and the "not too" and "not at all" responses as another group, what percentage of the survey respondents would probably purchase GM foods? What percentage would not? (1 point)
   (b) Referring to the second survey question, identify and explain two possible benefits of the inclusion of GM foods in the diet that might convince people to purchase these products. (4 points)
   (c) Referring to the third survey question, identify and explain two possible dangers of the use of GM foods that might convince people not to purchase these products. (4 points)
   (d) Identify one U.S. federal agency that is responsible for the regulation of GMOs. (1 point)

3. In the twentieth century the Green Revolution transformed agricultural systems from small farms to large industrial operations.
   (a) Green Revolution practices include better irrigation systems and the use of synthetic fertilizers. These have both improved agricultural productivity, but also have contributed to environmental degradation. Describe ONE advantage and ONE disadvantage of each practice. (4 points)
   (b) Explain how mechanization and use of synthetic fertilizers led to monocropping. (2 points)
   (c) Describe TWO sustainable alternatives to monocropping and their advantages. (4 points)

Marlee90/Getty Images

science applied 5

How Do We Define Organic Food?

If you've ever spent time roaming through the aisles of a grocery store, you have undoubtedly seen food labeled "organic." You may even have purchased organic food, despite the fact that it is generally more expensive than conventionally grown food. Some people prefer organic food because they believe it is healthier and tastes better. Others buy it because they think organic food is safer to eat. Still others believe organic food is produced in a way that is healthier for the environment and safer for farmworkers. Are these beliefs accurate? What exactly does the organic food label mean?

How did the organic food movement begin?

The notion of organic food emerged in the 1940s, when farmers first started using synthetic pesticides in agriculture. Three decades later, interest in organic food increased as concerns about environmental contaminants became more widespread. Organic food enthusiasts wanted food that was free of chemicals, including pesticides. At the same time, the popularity of food grown locally on small farms was increasing. The push for organic produce in the 1970s was part of a counterculture movement against the farming practices of large-scale commercial agriculture. Organic food was perceived as the food of people rebelling against the mainstream culture.

As we saw in Chapter 11, the organic farmer is fundamentally concerned with the health of the soil. He or she will work hard to ensure that the organic matter content, base saturation, and cation exchange capacity of the soil will increase each year. Actions that will degrade the soil or promote erosion are carefully avoided (**FIGURE SA5.1**).

Today, organic food is much more common and appeals to people from all walks of life. Indeed, organic farming has increased rapidly during the past decade.

FIGURE SA5.1 An organic farm. This strawberry farm in Florida is a certified organic producer. *(ZUMA Press Inc/Alamy)*

According to the U.S. Department of Agriculture, the number of acres of organic farmland increased more than fivefold from 1992 to 2016. As you can see in **FIGURE SA5.2**, in 2016, there were about 2 million ha (5 million acres) of organic cropland and pastureland in the United States. The states with the most land devoted to organic farmland are California, Montana, New York, and Wisconsin. Although organic food has been the fastest-growing category of food in the nation—with a current value of $50 billion—it still represents only 5 percent of all food sales. The most popular products are fruits and vegetable (14 percent of all sales) and milk (8 percent of all sales).

What does it mean to be organic?

The rise in popularity of organic products has brought increased pressure to define what exactly it means to be organic. Without government guidelines, anyone could claim to be selling organic food. In 1990,

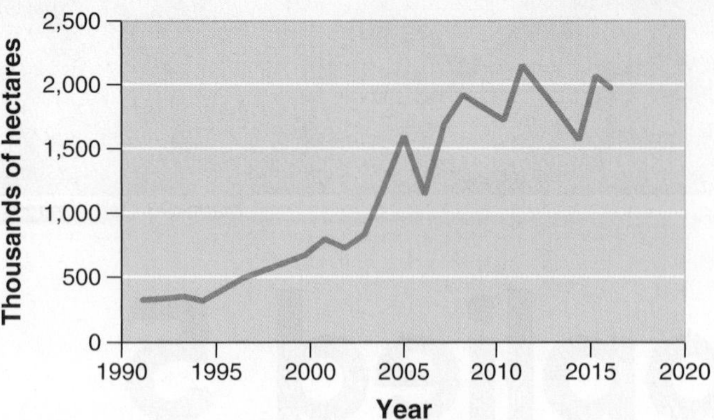

FIGURE SA5.2 Land devoted to organic farming. From 1992 to 2016, the amount of land dedicated to organic farm has increased more than fivefold. *(Data from USDA.)*

the U.S. Department of Agriculture (USDA) began to develop guidelines for organic food. Its goal was to set national organic certification standards that would assure consumers that the food was produced and processed with minimal use of chemicals. Food that met these standards could carry the USDA Organic seal (**FIGURE SA5.3**).

In 1995, the USDA determined that "Organic agriculture is an ecological production management system that promotes and enhances biodiversity, biological cycles, and soil biological activity. It is based on minimal use of off-farm inputs and on management practices that restore, maintain, and enhance ecological harmony." What does this statement mean in practice? First, it means that to be considered organic, food must be produced and processed with minimal use of pesticides. It also means that organic meat, eggs, and dairy products come from animals that have not been given antibiotics or growth hormones. Finally, it means that food labeled organic cannot have been fertilized with synthetic fertilizers or sewage sludge. Rather than relying on chemicals that come from outside the farm, organic farmers must use nonchemical methods of pest control and fertilization, as discussed in Chapter 11. It also means

FIGURE SA5.3 The USDA organic seal. Food that meets USDA organic certification standards can carry the USDA organic seal. *(USDA)*

that the food cannot come from genetically modified organisms.

Pest control is perhaps one of the greatest challenges in organic farming. Agricultural pests include weeds that compete with the crops, insects that eat the crops, and fungal pathogens that can kill the crops. Organic farmers are required to try to control pests without synthetic pesticides. Such efforts can include mechanically destroying weeds and using plant-derived insecticides or the release of insect enemies to combat the pests. If these tactics fail, organic farmers are permitted to use a synthetic pesticide from a relatively small list of approved chemicals. It came as a shock to consumers, however, when reports arose in 2017 that imported organic fruits and vegetables were being fumigated with pesticides when they entered the United States to ensure that they were not introducing any pests from other regions of the world.

Having standards for organic food is effective only if those standards are enforced. One of the important functions of the USDA's National Organic Program is to conduct inspections of organic food producers and processors to ensure that organic food meets the national standards. Since 1990, the government has allowed inspections to be conducted by independent companies that are hired by producers and processors to inspect organic farms and factories to ensure that the organic standards are being met.

As part of this effort, the certifying companies were required to conduct spot checks. In 2010, the USDA announced that it had failed to enforce the spot check requirement and that, as a result, many of the largest independent certifiers had been conducting spot checks on organic food only if they suspected a problem. When problems were not suspected, producers could operate for years without any spot checks. In fact, some growers had been selling nonorganic food under the USDA Organic seal and obtaining a higher price for their products. In response, the federal government announced in 2010 that the funding and staff of the National Organic Program would nearly double to ensure that the certifying companies were conducting the required spot checks. As of 2017, these inspections are commonly conducted with advance notice to the farms and the farms are allowed to select which certifying company can inspect them.

Does organic food mean family farms?

Most organic food production was originally conducted on small family farms. As we saw in Chapter 11, organic farming methods can require more labor, time, and money than conventional methods, so the price of organic food can be significantly higher. As farmers have learned, however, many consumers are willing to pay a premium for what they believe to be a premium product. This consumer behavior has attracted the attention of large factory farms owned by corporations, which have begun to conduct large-scale organic agriculture.

Although organic farms have a history of being small farms, there is no inherent reason why large farms cannot grow organic food (**FIGURE SA5.4**). An increasing number of corporations have decided to enter the business of organic food production. One of the most significant events to drive this movement occurred in 2006 when Wal-Mart, the largest grocery retailer in the United States, announced that it would begin selling a substantial number of organic foods in its stores at prices only slightly higher than those of nonorganic foods. The expansion into this market by Wal-Mart, Costco, Whole Foods, and Kroger has brought organic food to a much larger number of consumers. As a result, demand for organic food has increased, and more agribusinesses are producing it. Increased production of organic food by larger, more efficient farms has driven down the cost of organic food.

As more factory farms entered the organic food business, there was increasing pressure to permit the use of some nonorganic and synthetic chemicals during the production stage, when the food from the field is turned into a form that can be sold in the grocery store. Proponents of the rule change, including agribusinesses, argued that many processed foods, such as frozen meals, could not be made in an organic version unless the rules about additives were changed. Opponents argued that if processed foods could not be made without additives, then processed foods should not be certified as organic. In 2002, the USDA designated several nonorganic and synthetic additives as permissible in organic processed foods. Permitted nonorganic additives include citric acid, pectin, and cornstarch. Synthetic chemicals permitted in the processing of organic foods include alcohol (used for disinfecting machinery) and aspirin (used for medical treatment of animals). The number of nonorganic additives has increased from 77 additives in 2002 to 250 additives in 2013.

FIGURE SA5.4 Large-scale organic farming. Although organic farms traditionally have been small-scale family farms, some organic farms, such as this organic dairy farm in California, operate at much larger scales. *(inga spence/Alamy)*

What does the organic label not mean?

The USDA definition of "organic" does not match what many people think or hope "organic" should mean. For example, many people associate organic food with small family farms, but as we have seen, organic food can be grown on farms of all sizes. The USDA organic seal therefore tells us nothing about the size of the farm that grew the food. Nor does the seal tell us anything about the safety of the food, including whether it contains fewer dangerous food-related pathogens than nonorganic food. Nor does it indicate whether eggs, meat, and dairy products came from animals that were treated humanely or were allowed to roam freely outdoors (known as "free range"). It tells us nothing about where the food was grown, or about labor conditions, such as compensation for the farmworkers who produced it. Furthermore, there are currently no standards for certifying nonagricultural products, such as seafood or cotton, as organic. Although you may see these products labeled "organic," there is currently no agreed-upon definition for such products. Similarly, companies sometimes label their products "natural," a word that seems to imply "organic," but in fact has no standard definition.

In response to public demand, farmers sometimes use food labels that go beyond the federal requirements for organic certification. They may include information about humane treatment of animals, the conditions and wages of the farmworkers, and whether genetically modified organisms were used. Such detailed labels have seen growing popularity.

Although the USDA organic certification has a somewhat narrow definition, it does offer a number of important benefits. In studies reported in 2014 and 2016, researchers examined hundreds of studies on organic versus traditional food and concluded that organic food is healthier; organic meat and dairy contains more omega-3 fatty acids (which combat heart disease), and vegetables are higher in antioxidants (which provide a variety of health benefits). In addition, the certification program provides a standardized set of requirements so that informed consumers understand what they are buying. Given the reduction in synthetic pesticide use, the program also allows consumers to support farming practices that are much better for soils and ecosystems and better for farmworkers—who face decreased risks from applying pesticides—than conventional farming practices.

References

Whoriskey, P. 2017. Audit finds how fake "organic" foods reach U.S. *Washington Post,* September 19.

Aubrey, A. 2016. Is organic more nutritious? A new study adds to the evidence. *Northeast Public Radio,* February 18. https://www.npr.org/sections/thesalt/2016/02/18/467136329/is-organic-more-nutritious-new-study-adds-to-the-evidence.

Chang, K. 2012. Organic food vs. conventional food. *New York Times,* September 4. http://well.blogs.nytimes.com/2012/09/04/organic-food-vs-conventional-food/.

Strom, S. 2012. Has organic been oversized? *New York Times,* July 7. http://www.nytimes.com/2012/07/08/business/organic-food-purists-worry-about-big-companies-influence.html.

U.S. Department of Agriculture. *What is organic production?* http://www.nal.usda.gov/afsic/pubs/ofp/ofp.shtml.

Questions

1. Why do many people think that organic food comes from small farms?
2. Why might large agricultural companies lobby the federal government to permit many more synthetic chemicals to be used in processing organic food?

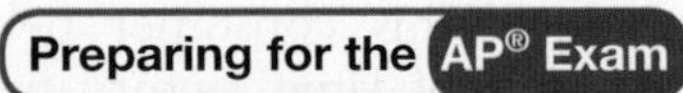

Practice AP® Free-Response Question

Write your answer to each part clearly. Support your answers with relevant information and examples. Where calculations are required, show your work.

For many people, organic farming involves far more than the reduction of pesticide and fertilizer use. Organic farming is also a method of caring for the land in a manner that will benefit landscape ecology and sustain crop production for many years.

(a) Succession is the ecological process by which plants replace each other on a landscape through competition and facilitation. How can the principle of succession be used to generate a sustainable organic farm? (3 points)

(b) Describe three methods of reducing soil erosion. (3 points)

(c) What is integrated pest management, and how might it be applied to organic farming? (2 points)

(d) Before a farm can become certified as organic, it must undergo a 3-year transition period. During this period, no prohibited fertilizers or pesticides can be used on the land. Due to the high cost of farming without the aid of synthetic chemicals, transitioning to organic farming represents a financial risk. Describe two ways that this risk can be reduced. (2 points)

Section 1: Multiple-Choice Questions

Choose the best answer for questions 1–20.

1. Which is an example of a positive externality?
   (a) Individual farmers exploit a common grazing area.
   (b) Deforestation causes the loss of valuable ecosystem services.
   (c) A change in local zoning causes a rise in property value.
   (d) Selective logging occurs in a nationally managed forest.

2. Which of the following are possible methods to prevent the tragedy of the commons?
   I. private ownership of land
   II. government regulation of public land
   III. self-regulation by communities and stakeholders
   (a) I only
   (b) II only
   (c) III only
   (d) I, II, and III

3. Determination of a maximum sustainable yield of plant harvest is most likely to be useful for which classification of land?
   (a) national park
   (b) protected seascape
   (c) nature reserve
   (d) habitat management area

4. Which is NOT a consideration that guides the regulation of public land use in the United States?
   (a) revenue from tourism
   (b) profits to private companies
   (c) resource conservation ethic
   (d) value of ecosystem services

5. The Taylor Grazing Act of 1934
   (a) has resulted in a positive net revenue for the federal government.
   (b) has prevented a tragedy of the commons situation for rangelands.
   (c) converted federal rangelands from a permit-based to common-use grazing system.
   (d) subsidizes grazing activities with federal funds.

6. For several decades, managers of a forest in West Virginia have selectively logged oak trees in a mixed forest that contains primarily oak and hickory. Which of the following is likely to happen over time?
   (a) The abundance of mature oak trees in the forest will increase.
   (b) Large patches of clear-cut forest will become visible from high above the ground.
   (c) There will be an increase in the abundance of shade-tolerant species.
   (d) The abundance of older-aged hickory trees will increase.

7. A section of a heavily forested area in a rural part of the northeastern United States is clear-cut and the trees are removed. Which of the following would you NOT expect as a result?
   (a) greater risk of fire
   (b) increased erosion
   (c) increased habitat diversity
   (d) higher stream temperature

8. Which is likely to have the greatest negative environmental impact?
   (a) greenways established in cities and suburbs
   (b) concentration of the population into cities
   (c) population movement into exurbs
   (d) development of rapid transit between cities and suburbs

9. Which makes up the largest fraction of the energy subsidy in U.S. agriculture?
   I. transportation of the food from farm to plate
   II. food production at the farm
   III. food processing in the factory
   (a) I only
   (b) II only
   (c) I and II
   (d) II and III

10. All of the following are possible strategies for smart growth EXCEPT
    (a) fostering a sense of place.
    (b) transit-oriented development.
    (c) increasing exurban areas.
    (d) infill.

11. The 1969 National Environmental Policy Act (NEPA) requires developers to
    (a) file an environmental impact statement.
    (b) proceed in a way that will have the least environmental impact.
    (c) ensure that developed land allows multiple use.
    (d) adhere to the resource conservation ethic.

12. On average, 1 kg of corn contains 3,000 kilocalories, and a single hectare can produce 8,000 kg of corn per year. Approximately how many humans eating 2,000 calories per day can a single hectare of corn feed in a year?
(a) 12
(b) 21
(c) 25
(d) 33

13. A single ear of corn contains 80 kilocalories. To produce a single ear of corn requires 40 kilocalories of fuel. What is the energy subsidy of corn production?
(a) 0.1
(b) 0.5
(c) 1.0
(d) 5.0

14. Due to the economies of scale, it is economically beneficial for
(a) an owner of a small farm to purchase industrial farming equipment.
(b) farm owners to seek the best purchasing price for their produce.
(c) an owner of a large farm to lease sections of his farm to neighbors.
(d) an owner of a small farm to share equipment with neighboring farms.

15. Monocropping generally requires more ________ than intercropping.
I. fertilization
II. irrigation
III. pesticide application
(a) I only
(b) II only
(c) I and II
(d) I and III

16. Which of the following is NOT true regarding genetically modified organisms (GMOs) in crops?
(a) Use of GMO crops has reduced pesticide use.
(b) Use of GMO crops has eliminated the pesticide treadmill.
(c) Use of GMO crops has increased livestock production.
(d) Use of GMO crops can be combined with the practices of intercropping.

17. Which factor has NOT contributed to the collapse of ocean fisheries?
(a) tragedy of the commons in open waters
(b) increase in offshore aquaculture development
(c) bycatch harvest
(d) use of international transferable quotas

18. No-till agriculture and contour plowing are methods that typically result in a decrease in
(a) pesticide use.
(b) energy subsidies.
(c) loss of topsoil.
(d) pest outbreaks.

19. For produce to be labeled as organic in the United States, the food must have been grown
I. without the use of any pesticide.
II. without the use of synthetic fertilizer.
III. on small, sustainable farms.
(a) II only
(b) III only
(c) I and II
(d) I, II, and III

20. Which set of parameters is most likely to determine the ecological footprint of a free-range livestock ranch?
(a) number of cattle, method of tilling crop lands, amount of land allocated for cattle
(b) diversity of livestock, type of grain fed to cattle, method of waste disposal
(c) diversity of livestock, public interest in the ranch, amount of land allocated for cattle
(d) diversity of livestock, number of cattle, amount of land allocated for cattle

Section 2: Free-Response Questions

Write your answer to each part clearly. Support your answers with relevant information and examples. Where calculations are required, show your work.

1. Detroit, once a bustling U.S. city, now has many fewer residents. For over 3 decades, the city has experienced the departure of many Caucasian residents, along with socioeconomic turmoil. In 2009, two of the city's major employers, Chrysler and General Motors, filed for bankruptcy and eliminated a large number of jobs. The population of the city has declined by 60 percent from what it once was. Many residential areas contain abandoned housing and urban businesses have declined.
(a) Describe two principles related to the strategy of smart growth that might be applied to reinvigorate the city of Detroit. (4 points)
(b) Describe the positive feedback loop of urban blight. (2 points)
(c) There has been a recent proposal to convert several residential neighborhoods with numerous abandoned buildings into urban parks. Describe ONE positive externality and ONE negative externality of this proposal. (2 points)
(d) Detroit is located halfway between Lake Erie and Lake Huron. Some officials have suggested that allowing increased fishing activities could attract people to the area. What are four parameters that state officials would need to measure in order to determine the maximum sustainable yield of Great Lakes fisheries? (2 points)

2. Bt-corn, a genetically modified plant, is now widely used in the United States because it greatly reduces the need to spray insecticides on corn fields. More recently, scientists have begun inserting Bt genes into other plant types, such as rice, canola, and cotton. Consumers are wary of the potential effects these genetically modified organisms may have on human health. However, many environmental scientists are more concerned with the potential effects of the plants on the ecology of land surrounding the farms. Consider this recent report from a study examining the Bt-crops:

   We observed slower decomposition of leaf litter and stems from Bt-positive transgenic strains of rice, tobacco, canola, cotton, and potato relative to decomposition of litter from Bt-negative strains. The amount of carbon dioxide released by soil bacteria and fungi surrounding Bt-positive litter was also less, indicating less microbial activity and slower growth of microbial populations. These results may be caused by the production of certain enzymes by the foliage of Bt-positive crops that deters herbivory while the plant is living and leaches into the soil as the plant decomposes.

   (a) Why is the decomposition of plant litter important on farmlands? (2 points)
   (b) Discuss how and why the use of Bt-positive crop strains might influence future crop production on a farm. (2 points)
   (c) How might the use of Bt-positive crops influence the evolution of pest species? (2 points)
   (d) On average, one kilogram of potato contains 850 kilocalories. Suppose that growing a kilogram of nontransgenic potatoes on a conventional farm requires 1,100 kilocalories of fuel to plant the potatoes, 700 kilocalories of fuel to spray pesticides on the potatoes, and 200 kilocalories to harvest the potatoes. If a farmer switches to planting Bt-potatoes, then the energy investment in spraying pesticides is no longer needed. What is the difference in energy subsidies between nontransgenic and transgenic potatoes? (4 points)

3. Alternatives to industrial agriculture are now gaining more attention in some countries as people strive to practice agriculture more sustainably.
   (a) Organic and no-till agriculture are both alternative methods to industrial agriculture. Describe each of these techniques. (2 points)
   (b) Name TWO advantages and ONE disadvantage of organic agriculture and no-till agriculture. (6 points)
   (c) On the livestock side of agriculture, free-range meat is an alternative to concentrated animal feeding operations (CAFO). Describe ONE advantage and ONE disadvantage of free-range meat. (2 points)

In 2010, an explosion at the British Petroleum Deepwater Horizon oil extraction platform in the Gulf of Mexico killed 11 workers on the drilling platform, injured 17 others, and released more than 206 million gallons of oil into the ocean. *(Gerald Herbert/AP Images)*

Nonrenewable Energy Resources

CHAPTER 12

MODULE 34 Patterns of Energy Use

MODULE 35 Fossil Fuel Resources

MODULE 36 Nuclear Energy Resources

CASE STUDY

All Energy Use Has Consequences

The modern world is dependent on fossil and nuclear fuels for energy. Many of the benefits of our modern society—health care, comfortable living conditions, easy travel, abundant food—rely on readily accessible and relatively affordable fossil and nuclear fuels. But the costs to society are high.

Fossil fuels are named appropriately: They are fuels containing fossilized remains of animals and plants. This means that the carbon released into the atmosphere when fossil fuels are burned is carbon that had been buried deep underground for millions of years. Therefore, fossil fuel combustion contributes additional carbon to the atmosphere. Furthermore, fossil fuel combustion contributes other air pollutants that produce haze and smog in the atmosphere, harm human beings when inhaled, and alter terrestrial and aquatic ecosystems.

In addition to the problems that occur while fossil fuel is used, extraction can cause significant ecosystem and property damage and human injury and death. For example, on April 20, 2010, an explosion and fire occurred at the British Petroleum Deepwater Horizon oil extraction platform in the Gulf of Mexico. Oil gushed from the platform until it was capped 87 days later, on July 15. The BP accident killed 11 workers on the drilling platform, injured 17 others, and released more than 780 million liters (206 million gallons) of oil into the gulf. Scientists from the Center for Biological Diversity later estimated that at least 6,000 sea turtles, 26,000 marine mammals, and more than 82,000 birds were killed as a result of the spill. The oil spread through the Gulf of Mexico and washed up on the shores of Louisiana, Mississippi, Alabama, and Florida.

The modern world is dependent on fossil and nuclear fuels for energy but the costs to society are high.

Before the Deepwater Horizon incident, the largest spill in U.S. waters had been the March 1989 Exxon Valdez accident, when the Valdez, a supertanker carrying 200 million liters (53 million gallons) of oil, crashed into a reef in Prince William Sound, Alaska. Roughly 42 million liters (11 million gallons) of oil spilled into the sound, much of which washed up on shore, contaminating the coastline and killing perhaps half a million birds and thousands of marine mammals. The number of dead animals was so much greater in the Exxon Valdez accident because the spill occurred in a relatively enclosed sound rather than in open waters.

Oil spills remain an ongoing environmental problem in the modern world. Some spills are caused by leaks or explosions at wells, while others occur when oil is being transported by pipeline or tanker. Accidents can happen after oil is extracted and transported to a refinery. In 2005, 15 workers died in an explosion at a British Petroleum oil refinery in Texas.

Other fossil fuels pose similar risks. In the early 1900s, there were hundreds of accidental coal-mining deaths each year in the United States. While safety measures have improved since then, mining is still a dangerous business. In April 2010, the worst U.S. coal mine explosion in 40 years killed 29 miners in West Virginia. Long after they leave the mines, tens of thousands of coal miners develop black lung disease and other respiratory ailments that lead to disability or death.

Natural gas is often considered to be "clean" because its combustion produces lower amounts of particulates, sulfur dioxide, and carbon dioxide than does oil or coal. However, combustion still results in emission of carbon dioxide, which is the major greenhouse gas produced by human activity. The production of natural gas also has negative consequences. "Thumper trucks," which generate seismic vibrations to identify natural gas deposits underground, can disturb soil and alter groundwater flow, causing some areas to flood and some wells to go dry. Drilling for natural gas can also contaminate drinking water. Construction of natural gas pipelines is disruptive to the environment and is often opposed by those who live in the affected communities. And a growing number of studies suggest that an unknown but possibly substantial percentage of natural gas is lost to the atmosphere during extraction and transport.

Nuclear energy has contributed to a few energy catastrophes as well. A 2011 earthquake off the coast of Japan resulted in a tsunami that damaged the nuclear reactors at the Fukushima nuclear power plant, leading to the release of radioactive gases. In 1986, a fire and release of radioactivity at the Chernobyl nuclear power plant in Ukraine resulted in 31 deaths during the event and additional deaths later. There was substantial release of radioactive material and 116,000 people were evacuated from a 30 km (19 mile) radius from the plant. A larger area, and more people, were evacuated later and this area remains uninhabited by humans now, more than 30 years after the accident. Even when operating properly, a nuclear power plant generates radioactive nuclear waste. Storage of this waste is currently an unresolved environmental challenge.

As you can see, there are consequences to all energy choices; every choice is problematic in one way or another. The study of environmental science allows us to identify and articulate the advantages and disadvantages of each energy choice made by an individual or society.

Sources: L. Margonelli, *Oil on the Brain: Petroleum's Long Strange Trip to Your Tank* (Broadway Books, 2008); M.B. McElroy, *Energy and Climate: Vision for the Future* (Oxford, 2016).

We use energy in all aspects of our daily lives: heating and cooling, cooking, lighting, communications, and travel. In these activities, humans convert stored energy resources such as natural gas and oil into useful forms of energy such as motion, heat, and electricity, with varying degrees of efficiency and environmental effects. Each energy choice we make has both positive and negative consequences. In a society like the United States, where each person averages 10,000 watts of energy use continuously—24 hours per day, 365 days per year—there are a lot of consequences to understand and evaluate.

In this chapter, we will look at the supplies of fossil fuel and nuclear fuel that we currently use. Chapter 13 addresses the renewable energy resources that we use to some extent now and that many people expect we will use even more in the future.

MODULE 34

Patterns of Energy Use

In this module we begin our study of *nonrenewable* energy sources by looking at patterns of energy use throughout the world and in the United States. We will see how evaluating energy efficiency can help us determine the best application for different energy sources. Finally, because electricity accounts for such a large percentage of our overall energy use, we will examine the ways in which electricity is generated.

Learning Goals

After reading this module, you should be able to

- describe the use of nonrenewable energy in the world and in the United States.
- explain why different forms of energy are best suited for certain purposes.
- understand the primary ways that electricity is generated in the United States.

Nonrenewable energy is used worldwide and in the United States

Fossil fuels are fuels derived from biological material that became fossilized millions of years ago. Fuels from this source provide most of the energy used in both developed and developing countries. The vast majority of the fossil fuels we use—coal, oil, and natural gas—come from deposits of organic matter that were formed 50 million to 350 million years ago. As we saw in Chapter 3 (see Figure 7.2 on page 85), when organisms die, decomposers break down most of the dead biomass aerobically, and it quickly reenters the food web. However, in an anaerobic environment—for example in places such as swamps, river deltas, and the ocean floor—a large amount of detritus may build up quickly. Under these conditions, decomposers cannot break down all of the detritus. As this material is buried under succeeding layers of sediment and exposed to heat and pressure, the organic compounds within it are chemically transformed into high-energy solid, liquid, and gaseous components that are easily combusted. Because fossil fuel cannot be replenished once it is used up, it is known as a **nonrenewable energy resource**. **Nuclear fuel**, derived from radioactive materials that give off energy, is another major source of nonrenewable energy on which we depend. The supply of this fuel is finite, although at current rates of consumption it will last for a very long time.

Every country in the world uses energy at different rates and relies on different energy resources. Factors that determine the rate at which energy is used include the resources that are available and affordable. In recent decades, people have also begun to consider environmental impacts in some energy-use decisions. In this section we will look at patterns of energy use worldwide and in the United States.

To talk about quantities of energy used, it is helpful to use specific measures. Recall from Chapter 2 that the basic unit of energy is the joule (J). One gigajoule (GJ) is 1 billion (1×10^9) joules, or about as much energy as is contained in 30 L (8 gallons) of gasoline. One exajoule (EJ) is 1 billion (1×10^9) gigajoules. In some figures, we also present the quad, a unit of energy used by the U.S. government to report energy consumption. The quad is 1 quadrillion, or 1×10^{15}, British thermal units, or Btu. One quad is equal to 1.055 EJ.

Fossil fuel A fuel derived from biological material that became fossilized millions of years ago.

Nonrenewable energy resource An energy source with a finite supply, primarily the fossil fuels and nuclear fuels.

Nuclear fuel Fuel derived from radioactive materials that give off energy.

AP® Exam Tip

Remember that a prefix in front of a unit of measure indicates a multiple or fraction of that unit. For example, the prefix *kilo* (k) means 1,000, or 10^3, so a kilogram (kg) is equal to 1,000 g. The prefix *mega* (M) means 1,000,000 or 10^6, and the prefix *giga* (G) means one billion or 10^9. You will encounter many prefixes for units of energy in this chapter and on the AP® Environmental Science Exam. ●

Worldwide Patterns of Energy Use

As **FIGURE 34.1** shows, in 2015 total world energy consumption was approximately 570 EJ per year. This number amounts to roughly 75 GJ per person per year. Oil, coal, and natural gas were the three largest energy sources. Peat, a precursor to coal, is sometimes combined with coal for reporting purposes in certain countries, mostly in the developing world.

Energy use is not evenly distributed throughout the world. **FIGURE 34.2** shows that energy consumption in the United States was 320 GJ per person per year, almost five times greater than the world average. In fact, although only 20 percent of the world's population lives in developed countries, people in those countries use roughly half of the world's energy each year. Note that of the countries shown in Figure 34.2, China has the greatest total energy consumption, whereas the United States has the greatest per capita energy consumption. At 0.17 EJ per year, Tanzania has the lowest annual energy consumption of the countries shown; annual per capita energy consumption in Tanzania is approximately 3 GJ per person per year.

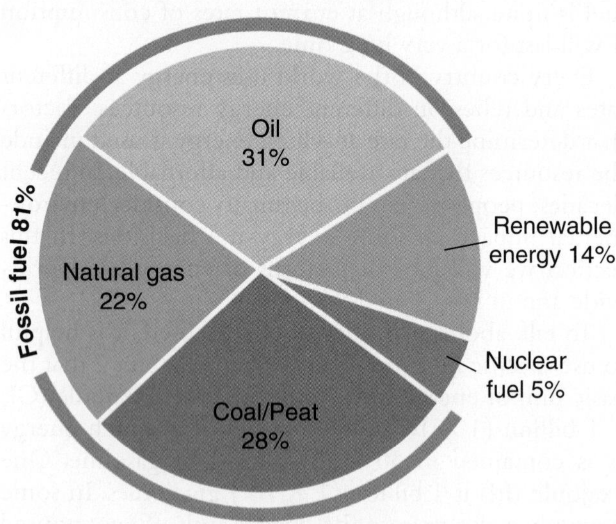

FIGURE 34.1 Worldwide annual energy consumption, by resource, in 2015. Oil, coal and peat, and natural gas are the major sources of energy for the world. *(Data from the International Energy Agency, 2017.)*

Commercial energy source An energy source that is bought and sold.

Subsistence energy source An energy source gathered by individuals for their own immediate needs.

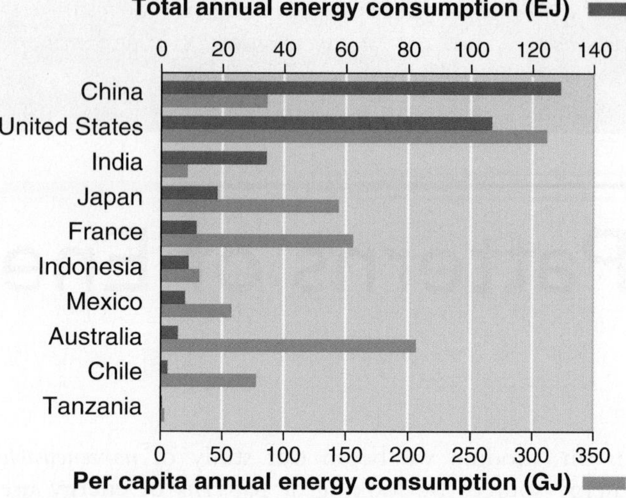

FIGURE 34.2 Global variation in total annual energy consumption and per capita energy consumption. The 10 countries shown are among the largest and the smallest energy users in the world. *(Data from Global Energy Statistical Yearbook 2017 https://yearbook.enerdata.net/ and U.S. Department of Energy, Energy Information Administration, 2017.)*

There are a variety of reasons for the patterns we see in Figure 34.2. In developed countries and in the urban areas of some developing countries, individuals are likely to use fossil fuels such as coal, oil, and natural gas—either directly, or indirectly through the use of electricity that is generated by burning those fuels. However, people living in rural areas of developing countries primarily still utilize such fuels as wood, charcoal, or animal waste. These differences lead us to distinguish between *commercial* and *subsistence* energy sources. **Commercial energy sources** are those that are bought and sold, such as coal, oil, and natural gas, although sometimes wood, charcoal, and animal waste are also sold commercially. **Subsistence energy sources** are those gathered by individuals for their own immediate needs and include straw, sticks, and animal dung. There is much greater use of subsistence energy sources in the developing world, especially in rural areas.

Changes in energy demand generally reflect the level of industrialization in a country or region. As energy demand increases, societies change the types of fuels they use. Today, we see the same patterns of changing energy use in developing countries that have been observed historically in the United States. For example, as more people own automobiles, demand for gasoline and diesel fuel increases. As industries develop and factories are built, demand for electricity and nuclear fuel increases. Although worldwide energy use varies considerably, the United States is a particularly large energy consumer.

Patterns of Energy Use in the United States

FIGURE 34.3 shows the history of energy use in the United States. Wood was the predominant energy source until about 1875, when coal came into wider use. Starting in the early 1900s, oil and natural gas joined coal as the primary sources of energy. By 1950, electricity generated by nuclear energy became part of the mix, and hydroelectricity became more prominent. The 1970s saw a decline of oil and a resurgence of coal. These changes were the result of political, economic, and environmental factors that will continue to shape energy use into the future. Today, the three resources that supply the majority of the energy used in the United States—in order of importance—are oil, natural gas, and coal. The recent increase in the consumption of natural gas and reduction in coal consumption shown in Figure 34.3 is a result of the increase in availability of natural gas, largely because of hydraulic fracturing (fracking), which has reduced the cost.

Energy use in the United States today is the result of the inputs and outputs of an enormous system. The boundaries of the system are political and technological as well as physical. For example, oil inputs enter the U.S. energy system from both domestic production and imports from other countries. Hydroelectric energy comes from water that flows within the physical boundaries of the country, as well as from neighboring Canada, but it is not an energy input until we move it into a technological system, such as a hydroelectric dam. One major output from the system is work—the end use of the energy, such as in transportation, residential, commercial, and industrial. The other major output is waste: heat, CO_2, and other pollutants that are released as energy is converted and entropy increases.

As you can see in **FIGURE 34.4**, U.S. energy consumption in 2016 was just over 100 EJ (1.0×10^{18} J) per year. The energy mix of U.S. consumption is 81 percent fossil fuel, 9 percent nuclear fuel, and 10 percent

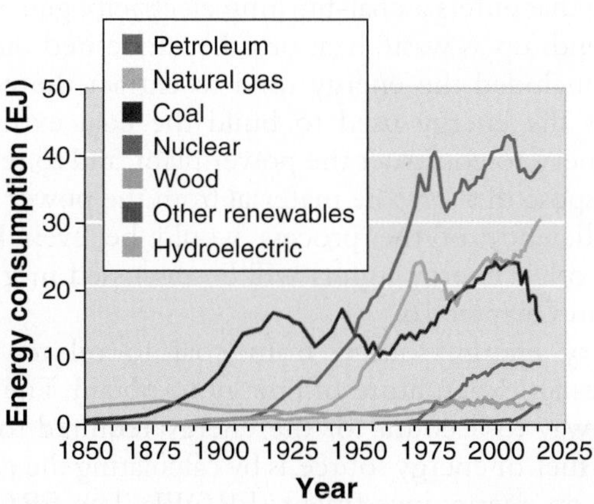

FIGURE 34.3 Energy consumption in the United States from 1850 through 2016. Wood and then coal once dominated our energy supply. Today a mix of three fossil fuels accounts for most of our energy use. The recent increase in natural gas and decrease in oil and coal is quite evident. *(Data from U.S. Department of Energy, Energy Information Administration, 2017.)*

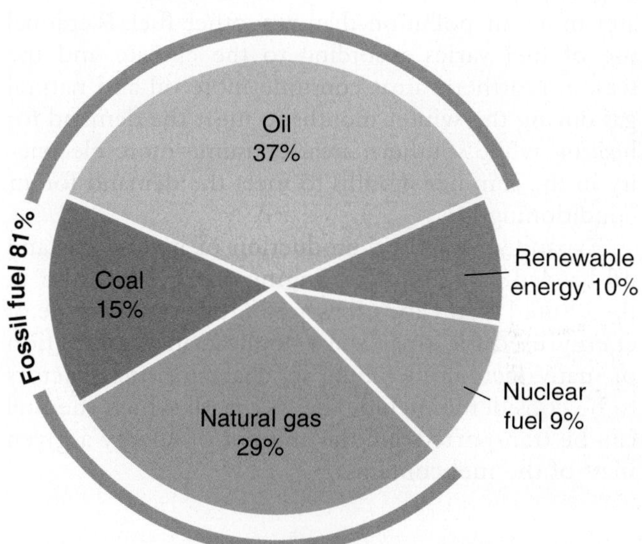

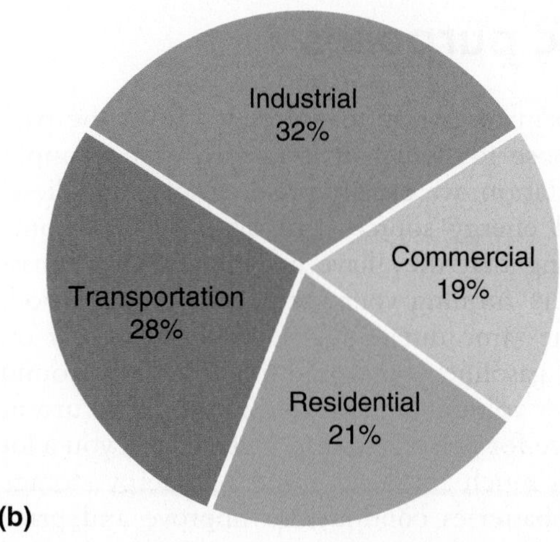

FIGURE 34.4 United States annual energy consumption by resource and end use in 2016. These graphs show energy consumption and end use in the United States. (a) United States annual energy consumption by fuel type in 2016. (b) United States end use energy sectors in 2016. Commercial includes businesses and schools. *(Data from U.S. Department of Energy, Energy Information Administration, 2017.)*

renewable energy resources. The United States produces 85 percent of the energy it needs, while the remainder—roughly 15 percent—comes from other countries, primarily in the form of petroleum imports. The percentage of energy imports has been decreasing as domestic production of natural gas and oil increases; these trends are expected to continue. Industry uses the most energy, followed by the transportation sector.

Energy use varies regionally and seasonally. In the midwestern and southeastern states, coal is the primary fuel burned for electricity generation. The western and northeastern states generate electricity using a mix of nuclear fuels, natural gas, and hydroelectric dams. Highly populated areas tend to use less coal, which creates more air pollution than any other fuel. Regional use of fuel varies according to the climate and the season. Northern areas consume more oil and natural gas during the winter months to meet the demand for heating while southern areas consume more electricity in the summer months to meet the demand for air conditioning.

Expanded domestic production of natural gas and oil has led to a rapidly changing energy portfolio in the United States and elsewhere. However, the type of energy used for a particular application is a function of many factors, including its characteristics. Factors to be considered include the ease with which the fuel can be transported and the amount of energy a given mass of the fuel contains.

Different energy forms are best suited for specific purposes

The best form of energy to use depends on the particular purpose for which it is needed. For example, for transportation, we usually prefer gasoline or diesel fuel—liquid energy sources that are relatively compact, meaning that they have a high energy-to-mass ratio. Imagine running your car on coal or firewood: To travel the same distance as a conventional car on one tank of gasoline, you would have to carry around a much larger volume of material. Gasoline, the current fuel of choice for personal transportation, gets you a lot farther on a much smaller volume. As energy storage capacity in batteries continues to improve and price continues to decrease, electricity may become the fuel of choice for personal transportation.

Energy-to-mass ratio is not the only consideration for fuel choice, however. A wood or coal fire starts relatively slowly and if used in an automobile, would not allow the vehicle to accelerate quickly. Gasoline and diesel are ideal fuels for vehicles because they can provide energy quickly and can be shut off quickly. Unfortunately, compared with other energy sources, such as natural gas or hydroelectricity, gasoline produces large amounts of air pollution per joule of energy released. Furthermore, unlike coal or wood, gasoline requires a good deal of refining—chemical processing—to produce. So there are many factors to consider when determining the most suitable energy source for a particular application.

Quantifying Energy Efficiency

Although all conventional nonrenewable energy sources have environmental impacts, energy efficiency is an important factor in making energy-use decisions. Energy efficiency refers to both the efficiency of the process we use to obtain the fuel, and the efficiency of the process that converts the fuel into the work that is needed.

In Chapter 2 we discussed energy efficiency as well as energy quality, a measure of the ease with which stored energy can be converted into useful work. The second law of thermodynamics dictates that when energy is transformed, its ability to do work diminishes because some energy is lost during each conversion. In addition to these losses, there is an expenditure of energy involved in obtaining almost every fuel that we use.

FIGURE 34.5 outlines the process of energy use from extraction of a resource to electricity generation and disposal of waste products from the power plant. The red arrows indicate that there are many opportunities for energy loss, each of which reduces energy efficiency. You may recall that the efficiency of converting coal into electricity is approximately 35 percent (see Figure 5.6, on page 52). In other words, about two-thirds of the energy that enters a coal-burning electricity generation plant ends up as waste heat or other undesired outputs. If we included the energy used to extract the coal as well as the energy used to build the coal extraction machinery, to construct the power plant, and to remove and dispose of the waste material from the power plant, the efficiency of the process would be even lower. These other energy inputs will be discussed in greater detail in Chapter 16.

Every energy source, from coal to oil to wind, requires an expenditure of energy to obtain. The most direct way to account for the energy required to produce a fuel, or energy source, is by calculating the energy return on energy investment (EROEI). The EROEI is the amount of energy we get out of an energy source for every unit of energy expended on its production. EROEI is calculated as follows:

$$\text{EROEI} = \frac{\text{Energy obtained from the fuel}}{\text{Energy invested to obtain the fuel}}$$

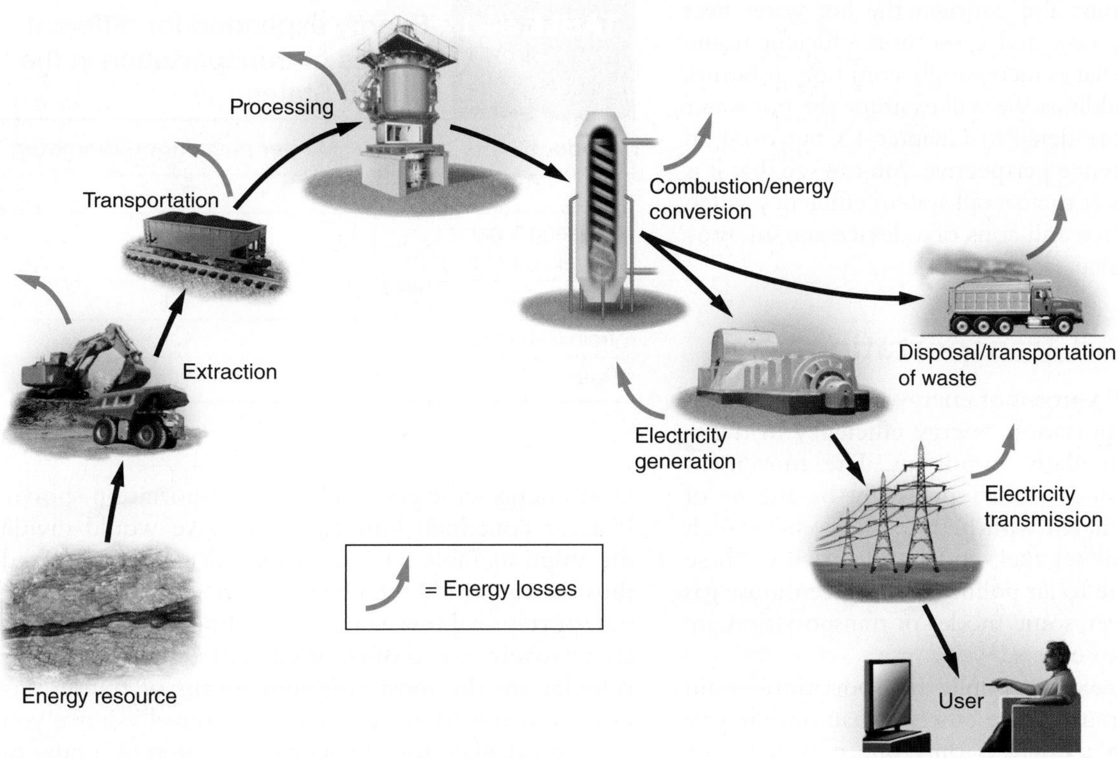

FIGURE 34.5 Inefficiencies in energy extraction and use. Coal provides an example of inefficiencies in energy extraction and use. Energy is lost at each stage of the process, from extraction, processing, and transport of the fuel to the disposal of waste products.

For example, in order to obtain 100 J of coal from a surface coal mine, 5 J of energy is expended. Therefore,

$$100\text{ J EROEI} = \frac{100\text{ J}}{5\text{ J}} = 20$$

As you might expect, a larger value for EROEI suggests a more efficient and more desirable process. "Science Applied 6: Should Corn Become Fuel?" in the following Chapter 13 discusses the EROEI for ethanol, a fuel made from corn.

Finding the Right Energy Source for the Job

When deciding between two energy sources for a given job, it is essential to consider the overall system efficiency. Sometimes the trade-offs are not immediately apparent. The home hot water heater is an excellent illustration of this principle.

Electric hot water heaters are often described as being highly efficient. Even though it is not possible to convert an energy supply entirely to its intended purpose and obtain 100 percent efficiency, converting electricity into hot water in a water heater comes very close. Electric hot water heaters contain a resistance coil that generates heat inside the tank of water. The heat is transferred to the surrounding water, which increases in temperature. Any heat energy that is "lost" to the surrounding environment is actually captured by the water in the tank. The efficiency of this process is 99 percent. In contrast, a typical natural gas water heater transfers energy to water with a flame below the tank. Waste heat and by-products of combustion warm the surrounding environment, as well as the tank of water, and are vented to the outside. This type of water heater is about 60 percent efficient.

However, to compare the total system efficiency of each water heater we need to consider the fuel each uses. The energy expended to extract, process, and deliver natural gas to the home can be considerable. If a coal-fired power plant is the source of the electricity that fuels the electric water heater, we have to consider that conversion of fossil fuel into electricity is only about 35 percent efficient. This means that even though an electric water heater has a higher direct efficiency than a natural gas water heater, the overall efficiency of the electric water heating system is lower—35 percent for the electric water heater compared with something close to 60 percent for the gas water heater. There may be many situations in which it is a better choice to heat water with electricity rather than with natural gas. If the source of the electricity is renewable, wind or solar for example, it probably makes sense to use electricity. You must also consider the hot water heat pump, a relatively new and even more efficient home hot water heater that is increasingly common in homes and apartment buildings. We will examine the hot water heat pump in more detail in Chapter 13, but from an environmental science perspective, you can see that it is important to look at the overall system efficiency when considering the pros and cons of a device and an associated energy choice.

Efficiency and Transportation

Because nearly 30 percent of energy use in the United States is for transportation, energy efficiency in transportation is particularly important. The movement of people and goods occurs primarily by means of vehicles that are fueled with petroleum products, such as gasoline and diesel fuel, and by electricity. These vehicles contribute to air pollution and greenhouse gas emissions. However, some modes of transportation are more efficient than others.

As you might expect, public transportation—train or bus travel—is much more efficient than traveling by car, especially when there is only one person in the car. And public ground transportation is usually more efficient than air travel. **TABLE 34.1** shows the efficiencies of different modes of transportation. Note that the energy values report only energy consumed, in megajoules (MJ, 10^6 J), per passenger-kilometer traveled and do not include the embodied energy used to build the different vehicles. Trains and motorcycles are the most energy-efficient modes of transportation shown. If a car contained four passengers, we would divide the value in Table 34.1 for a lone driver by four, and thus the car would then be the most efficient means of transportation. However, cars traveling with four riders are relatively rare in the United States; single-occupant vehicles are the most common means of transportation. "Do the Math: Efficiency of Travel" shows you how to calculate the efficiencies of different modes of transportation.

TABLE 34.1 Energy expended for different modes of transportation in the United States

Mode	MJ per passenger-kilometer
Air	2.1
Passenger car (driver alone)	3.6
Motorcycle	1.1
Train (Amtrak)	1.1
Bus	1.7

Transportation efficiency calculations do not take into account convenience, comfort, or style. Many people in the developed world are quite particular about how they get from place to place and want the independence of a personal vehicle. They also tend to have strong feelings about what type of personal

DO THE MATH Efficiency of Travel

Preparing for the AP® Exam

Imagine that you had unlimited time and you needed to get from Washington, D.C., to Cleveland, Ohio. The distance is roughly 600 km (370 miles). For each mode of transportation, calculate how many megajoules of energy you would use.

Using the data in Table 34.1, we can determine the following:

Air 2.1 MJ/passenger-kilometer × 600 km/trip = 1,260 MJ/passenger-trip

Car 3.6 MJ/passenger-kilometer × 600 km/trip = 2,160 MJ/passenger-trip

Train 1.1 MJ/passenger-kilometer × 600 km/trip = 660 MJ/passenger-trip

Bus 1.7 MJ/passenger-kilometer × 600 km/trip = 1,020 MJ/passenger-trip

If a gallon of gasoline contains 120 MJ, how many gallons of gasoline does it take to make the trip by car?

$$\frac{2{,}160 \text{ MJ/passenger-trip}}{120 \text{ MJ/gallon}} = 18 \text{ gallons/passenger-trip}$$

Examining total joules expended makes it clear that the train is the most energy-efficient means of travel. Driving alone in a car is the least energy-efficient.

YOUR TURN If you could carpool with three other people who needed to make the same trip, what would the energy expenditure be for each person?

vehicle is most desirable. In the United States, light trucks—a category that usually includes sport-utility vehicles (SUVs), minivans, and pickup trucks—account for roughly one-half of total automobile sales, while hybrid electric vehicles account for roughly 3 percent of total sales.

Light trucks are comparatively heavy vehicles and so generally have mileage ratings of less than 8.5 km per liter, or 20 miles per gallon (mpg). Because they are exempt from certain vehicle emission standards, they emit more of certain air pollutants per liter of fuel combusted than passenger cars. Smaller cars with standard internal combustion engines can travel up to 19 km per liter (45 mpg) on the highway. Some hybrid passenger cars, which use a gasoline engine, electric motors, and special braking systems, obtain closer to 21 km per liter (50 mpg). Electric cars and plug-in hybrid electric cars, which are becoming more common in the United States, obtain even better fuel efficiency. Unfortunately, the technology that powers self-driving cars uses a fair amount of energy and may decrease the fuel efficiency of a given vehicle by 5 to 10 percent. However, a self-driving car can be programmed to drive as fuel efficiently as possible and use less fuel than the most careful and energy-mindful human driver. And presumably, self-driving cars will eventually make fewer navigation and other errors. So over time, self-driving cars may allow fuel efficiency gains to continue.

Despite the availability of fuel-efficient vehicles, many people drive vehicles that yield relatively low fuel efficiencies. As **FIGURE 34.6** shows, the overall fuel efficiency of the U.S. personal vehicle fleet declined from 1985 through 2005 as people chose light trucks and SUVs over cars. In the decade beyond 2005, vehicle choice has changed somewhat and vehicle efficiency has slowly increased. Recently, the EPA instituted a requirement for a gradual increase in the average fuel efficiency of new cars and light trucks sold each year so as to deliver a combined fleet average of 21.5 km per liter (50 mpg) by 2025, although there is a discussion about whether or not increases in efficiency will required through 2022 or 2025.

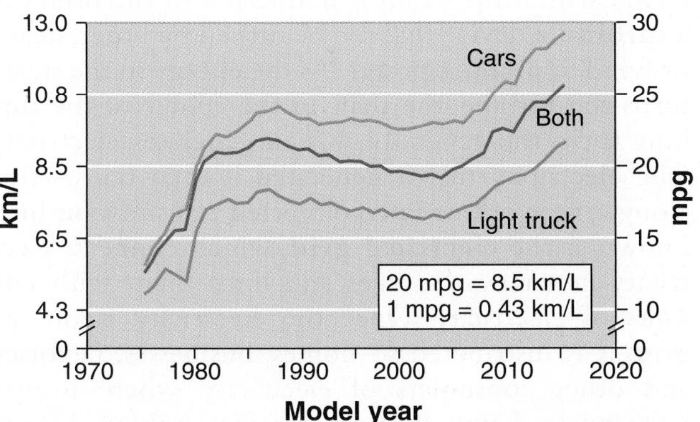

FIGURE 34.6 Overall fuel efficiency of U.S. automobiles from 1975 through 2016. As more buyers moved from cars to light trucks (a category that includes pickup trucks, minivans, and SUVs) for their personal vehicles, the fuel economy of the total U.S. fleet declined. It has recently begun to increase. *(Data from U.S. EPA, 2017.)*

Energy efficiency is an important consideration when making fuel and technology choices, but it is not the only factor we must consider. Fuel choice is also important. Determining the best fuel for a particular activity is not always easy, and it involves trade offs among convenience, ease of use, safety, cost, and pollution.

Electricity accounts for 40 percent of our energy use

Because electricity can be generated from many different sources, including fossil fuels, wind, water, and the Sun, it is a form of energy in its own category. Coal, oil, and natural gas are primary sources of energy. Electricity is a secondary source of energy, meaning that we obtain it from the conversion of a primary source. As a secondary source, electricity is an **energy carrier**—something that can move and deliver energy in a convenient, usable form to end users.

Approximately 40 percent of the energy consumed in the United States is used to generate electricity. But because of conversion losses during the electricity generation process, of that 40 percent, only 13 percent of this energy is available for end uses. In this section we will look at some of the basic concepts and issues related to generating electricity from fossil fuels. Chapter 13 discusses electricity generation from renewable energy sources.

The Process of Electricity Generation

Electricity is produced by conversion of primary sources of energy such as coal, natural gas, or wind. Electricity is clean at the point of use; no pollutants are emitted in your home when you use a light bulb or computer. When electricity is produced by combustion of fossil fuels, however, pollutants are released at the location of its production. And, as we have seen, the transfer of energy from a fuel to electricity is only about 35 percent efficient. Therefore, although electricity is highly convenient, from the standpoint of efficiency of the overall system and the total amount of pollution released, it is usually more desirable to transfer heat directly to a home with wood or oil combustion, for example, than via electricity

Energy carrier Something that can move and deliver energy in a convenient, usable form to end users.

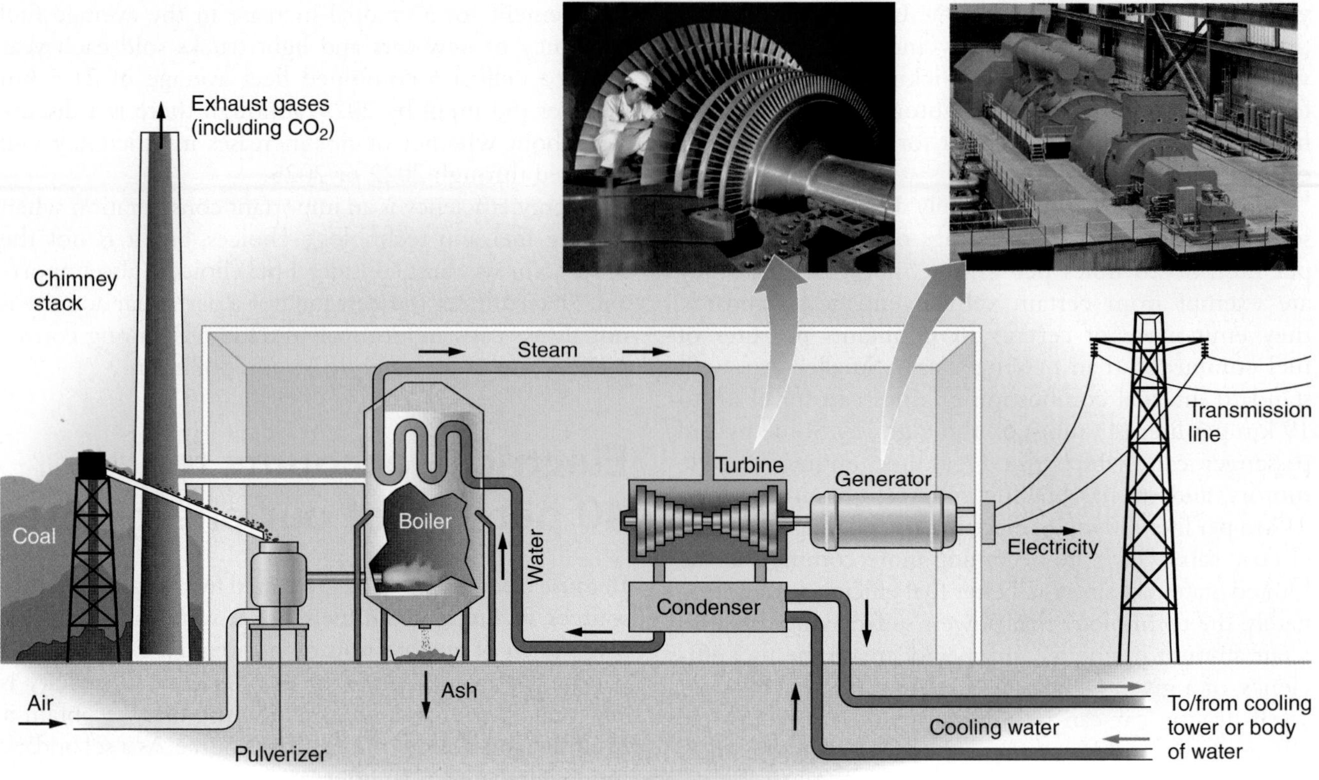

FIGURE 34.7 A coal-fired electricity generation plant. Energy from coal combustion converts water into steam, which turns a turbine. The turbine turns a generator, which produces electricity.
(Peter Bowater/Alamy; Peter Bowater/Science Source)

generated from the same materials. The energy source that entails the fewest conversions from its original form to the end use is likely to be the most efficient.

Many types of fossil fuels, as well as nuclear fuels, can be used to generate electricity. Regardless of which fuel is used, all thermal power plants work in the same basic way—they convert the potential energy of a fuel into electricity.

AP® Exam Tip

Recall that potential energy is energy that has been stored and not yet released, while kinetic energy is the energy of motion. Also make sure you know the first law of thermodynamics, which states that energy can change from one form to another. These concepts are critical for the AP® Environmental Science Exam and will help you understand what happens in a coal-fired electricity generation plant. ●

Turbine A device that can be turned by water, steam, or wind to produce power.

Electrical grid A network of interconnected transmission lines that joins power plants together and links them with end users of electricity.

FIGURE 34.7 illustrates the major features of a typical coal-burning power plant. Fuel—in this case, coal—is delivered to a boiler, where it is burned. The energy contained within the combusted fuel is transferred to water, which becomes steam. The kinetic energy contained within the steam is transferred to the blades of a **turbine**, a device that can be turned by water, steam, or wind to produce power. As the energy in the steam turns the turbine, the shaft in the center of the turbine turns the generator, which generates electricity. The electricity that is generated is then transported along a network of interconnected transmission lines known as the **electrical grid**, which connects electricity generation sources and links them with end users of electricity. Once the electricity is on the grid, it is distributed to homes, businesses, factories, and other consumers of electricity, where it may be converted into heat energy for cooking, kinetic energy in motors, or radiant energy in lights or used to operate electronic and electrical devices. After the steam passes through the turbine, it is condensed back into water. Sometimes the water is cooled in a cooling tower or discharged into a nearby body of water. Once-through use of water for thermal electricity generation is responsible for a substantial amount of water consumption in the United States.

Efficiency of Electricity Generation

Whereas a typical coal-burning power plant has an efficiency of about 35 percent, newer coal-burning power plants may have slightly higher efficiencies of approximately 40 percent. Power plants using other fossil fuels can be even more efficient. An improvement in gas combustion technology has led to the **combined cycle** natural gas–fired power plant, which uses both exhaust gases and steam turbines to generate electricity. Natural gas is combusted, and the combustion products turn a gas turbine. In addition, the waste heat from this process boils water, which turns a conventional steam turbine. For this reason, a combined cycle plant can achieve efficiencies of up to 60 percent.

Capacity

The **capacity** of a power plant is its maximum electrical output. A typical power plant in the United States might have a capacity of 500 megawatts (MW). This means that when the plant is operating, it generates 500 MW of electricity continuously. (Recall from Chapter 2 that a watt is a joule per second, so a power plant with a 500 MW capacity generates 500 MW continuously.) If the plant operated for one day, it would generate 500 MW × 24 hours = 12,000 megawatt-hours (MWh). Most home electricity is measured in kilowatt-hours (kWh). So the typical power plant we've just described would generate 12,000 MWh × 1,000 kWh/MWh = 12,000,000 kWh in a day. If it operated for 365 days per year, it would generate 365 times that daily amount.

Most power plants, however, do not operate every day of the year. They must be shut down for maintenance, refueling, or repairs. Other factors may also influence whether a plant is shut down for a time, including the demands of users on the grid and electricity production by other power plants on the grid. For example, in recent years, many coal-burning power plants in the northeastern United States operated less than 20 percent of the time because utilities have been voluntarily shutting them down. Why? Because it is less expensive for them to operate nearby natural–gas burning power plants and it is economically preferable to leave the coal plant off-line. Therefore, to determine a plant's output, it is useful to report the amount of time it actually operates in a year. This number—the fraction of time a plant is operating in a year—is known as its **capacity factor**. Most thermal power plants have capacity factors of 0.9 or greater, except when they are voluntarily shut down by the operators due to low demand for electricity from that plant or that type of fuel. Nuclear fueled power plants have a capacity factor of 0.9 or greater as do coal-burning power plants in parts of the United States where coal is an abundant and preferred fuel. As we will see in Chapter 13, power plants using some forms of renewable energy, such as wind and solar, may have a capacity factor of only about 0.25. "Do the Math: Calculating Energy Supply" on page 426 shows you how to calculate the amount of energy a power plant can supply.

When it is time to start up a power plant, both nuclear and coal-fired plants may take a number of hours, or even

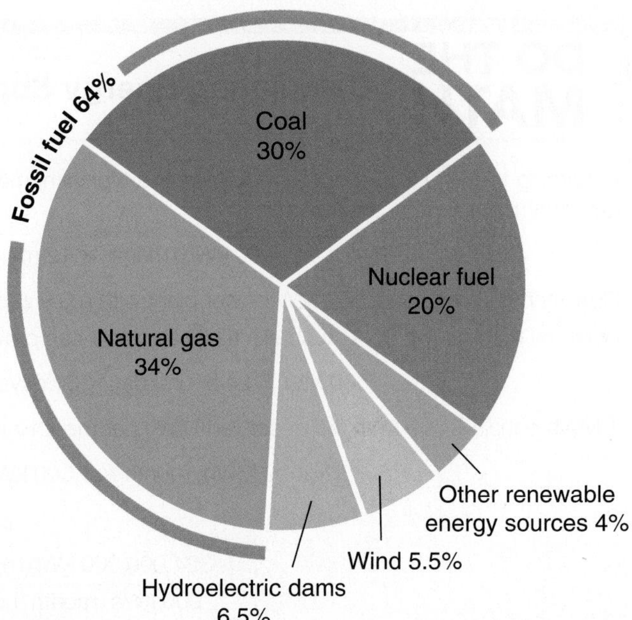

FIGURE 34.8 Fuels used for electricity generation in the United States in 2016. Natural gas is the fuel most commonly used for electricity generation. The electricity fuel mix in the United States has changed considerably in the last decade due to the increased availability and decreasing price of natural gas. *(Data from US Department of Energy, Energy Information Agency, 2017.)*

a full day, to come up to full generating capacity. Because of the time it takes for them to become operational, electric companies tend to keep nuclear, natural gas, and some coal-fired plants running at all times. As demand for electricity changes during the day or week, plants that are more easily powered up, such as those that use water (hydro), are used.

Cogeneration

The use of a fuel to generate electricity and produce steam heat is known as **cogeneration**, also called **combined heat and power**. Cogeneration is a method employed by certain steam power plants for obtaining greater efficiencies. If steam used for industrial purposes or to heat buildings is diverted to turn a turbine first, the user will achieve greater overall efficiency than by generating heat and electricity separately. Cogeneration efficiencies can be as high as 90 percent, whereas steam heating alone might be 75 percent efficient, and electricity generation alone might be 35 percent efficient.

There are roughly 7,000 power plants in the United States with generation capacity of 1 MW or greater. In 2016, they generated approximately 4 billion MWh (4 trillion kWh). **FIGURE 34.8** shows the

Combined cycle A power plant that uses both exhaust gases and steam turbines to generate electricity.

Capacity In reference to an electricity-generating plant, the maximum electrical output.

Capacity factor The fraction of time a power plant operates in a year.

Cogeneration The use of a fuel to generate electricity and produce heat. *Also known as* **Combined heat and power.**

DO THE MATH Calculating Energy Supply

Preparing for the AP® Exam

According to the U.S. Department of Energy, a typical home in the United States uses approximately 900 kWh of electricity per month. On an annual basis, this is

$$900 \text{ kWh/month} \times 12 \text{ months/year} = 10{,}800 \text{ kWh/year}$$

How many homes can a 500 MW power plant with a 0.9 capacity factor support?

Begin by determining how much electricity the plant can provide per month:

$$500 \text{ MW} \times 24 \text{ hours/day} \times 30 \text{ days/month} \times 0.9 = 324{,}000 \text{ MWh/month}$$

1 MWh equals 1,000 kWh, so to convert MWh per month into kWh per month, we multiply by 1,000:

$$324{,}000 \text{ MWh/month} \times 1{,}000 \text{ kWh/MWh} = 324{,}000{,}000 \text{ kWh/month}$$

So,

$$\frac{324{,}000{,}000 \text{ kWh/month}}{900 \text{ kWh/month/home}} = 360{,}000 \text{ homes}$$

On average, a 500 MW power plant can supply roughly 360,000 homes with electricity.

YOUR TURN During summer months, in hot regions of the United States, some homes run air conditioners continuously. How many homes can the same power plant support if average electricity usage increases to 1,200 kWh/month during summer months?

fuels that were used to generate this electricity. As we can see, natural gas–fired power plants are the largest component of electricity generation in the United States, responsible for 34 percent of all electricity produced. Coal, at 30 percent, and nuclear energy, at 20 percent, account for most of the remainder of the generating capacity. Together, these three sources, plus a very small amount of oil, account for 84 percent of electricity generation in the United States. Water and other renewable energy resources such as wind and solar energy are the source of the remainder of electricity generated in the country. As we discussed earlier, expanded domestic production of natural gas has increased the fraction of electricity generated from natural gas while the fraction of electricity generated from coal has decreased. In 2005, coal accounted for 50 percent of electricity generation.

MODULE 34 AP® Review

Preparing for the AP® Exam

In this module, we have seen that fossil fuels are the most common fuels used in the United States and in other countries throughout the world. In the developed world, per capita energy use is many times higher than it is in the developing world. Inefficiencies during energy extraction and use account for losses of usable energy throughout the energy cycle. The fuels that have been consumed in the United States have varied over time and according to the end use. Oil and natural gas together make up 64 percent of overall energy use in the United States. Different fuels are suitable for different tasks, depending on a variety of factors including their energy content and whether they are liquid or solid. Coal and natural gas account for two-thirds of electricity generation. In the next module, we will consider the major fossil fuel resources used around the world as well as in the United States.

AP® Practice Questions

1. Which energy source does not originate from the Sun?
   (a) coal
   (b) solar
   (c) oil
   (d) nuclear

2. Traveling alone in a car uses 3.6 MJ of energy per kilometer. If four people go on a trip of 400 miles, what is the MJ used per person?
   (a) 200 MJ
   (b) 320 MJ
   (c) 580 MJ
   (d) 860 MJ

3. The major source of energy in the United States is
   (a) natural gas.
   (b) coal.
   (c) oil.
   (d) nuclear.

4. Which is a secondary energy source?
   (a) solar
   (b) coal
   (c) electricity
   (d) nuclear

5. Cogeneration is
   (a) the use of two or more energy sources to generate electricity.
   (b) the use of two separate turbines to generate electricity.
   (c) a method of electricity generation that includes renewable energy.
   (d) the use of a fuel to generate steam heat and electricity.

Fossil Fuel Resources

We have seen that fossil fuels provide most of the commercial energy in the world. These fuels are also responsible for a good deal of the pollution that occurs, a topic that will be discussed in Chapter 15. Because fossil fuels are such an important energy source, environmental scientists must study them closely. In this module, we will examine the major fossil fuels—coal, petroleum, and natural gas—and discuss their advantages and disadvantages. We will also consider the future use of fossil fuels.

Learning Goals

After reading this module, you should be able to

- discuss the uses of coal and its consequences.
- discuss the uses of petroleum and its consequences.
- discuss the uses of natural gas and its consequences.
- discuss the uses of oil sands and liquefied coal and their consequences.
- describe future prospects for fossil fuel use.

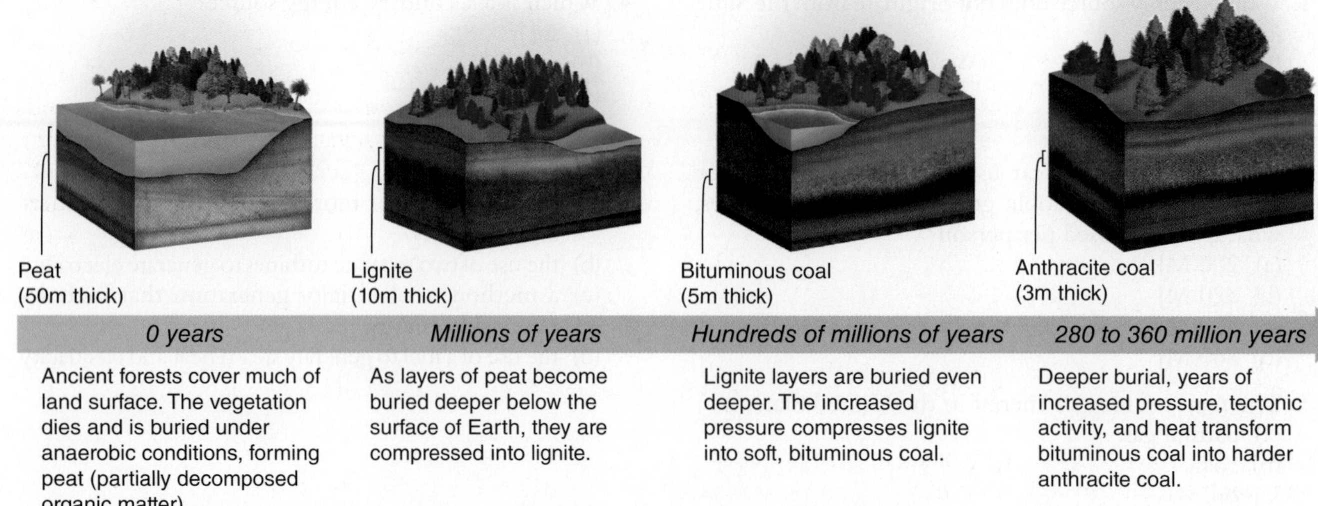

Peat
(50m thick)
Lignite
(10m thick)
Bituminous coal
(5m thick)
Anthracite coal
(3m thick)
0 years
Millions of years
Hundreds of millions of years
280 to 360 million years
Ancient forests cover much of land surface. The vegetation dies and is buried under anaerobic conditions, forming peat (partially decomposed organic matter).
As layers of peat become buried deeper below the surface of Earth, they are compressed into lignite.
Lignite layers are buried even deeper. The increased pressure compresses lignite into soft, bituminous coal.
Deeper burial, years of increased pressure, tectonic activity, and heat transform bituminous coal into harder anthracite coal.

FIGURE 35.1 The coal formation process. Peat is the raw material from which coal is formed. Over millions of years and under increasing pressure due to burial under more and more layers of rock and sediment, various types of coal are formed.

Coal is the most abundant and dirtiest of the fossil fuels

Coal is a solid fuel formed primarily from the remains of trees, ferns, and other plant materials that were preserved 280 million to 360 million years ago. It is available in many areas of the world and often is relatively easy to extract, handle, and process. We have seen that coal is the second-most commonly used fuel for electricity generation in the United States. There are three types of coal, ranked from lesser to greater age, exposure to pressure, and energy content; they are lignite, bituminous, and anthracite. A precursor to coal, called peat, is made up of partly decomposed organic material, including mosses. The formation of coal takes hundreds of millions of years. **FIGURE 35.1** represents factors involved in the formation of the different types of coal. Starting with an organic material such as peat, increasing time and pressure produce successively denser coal with more carbon molecules, and more potential energy, per kilogram.

The largest coal reserves are found in the United States, Russia, China, Australia, and India. The countries that are currently producing the greatest amounts of coal are China, the United States, India, and Australia.

Coal A solid fuel formed primarily from the remains of trees, ferns, and other plant materials preserved 280 million to 360 million years ago.

Advantages of Coal

Because it is energy-dense and plentiful, coal is used to generate electricity and in industrial processes such as making steel. In many parts of the world, coal reserves are relatively easy to exploit by surface mining. The technological demands of surface mining are relatively small and the economic costs are low. Once coal is extracted from the ground, it is relatively easy to handle and needs little refining before it is burned. It can be transported to power plants and factories by train, barge, or truck. All of these factors make coal a relatively easy fuel to use, regardless of technological development or infrastructure.

Disadvantages of Coal

Although coal is a relatively inexpensive fossil fuel, its use does have several disadvantages. As we saw in Chapter 8, the environmental consequences of the tailings from surface mining are significant. As surface coal is used up and becomes harder to find, however, subsurface mining becomes necessary. With subsurface mining, the technological demands and costs increase, as do the consequences for human health.

Coal contains a number of impurities, including sulfur, that are released into the atmosphere when the coal is burned. The sulfur content of coal typically ranges from 0.4 to 4 percent by weight. Lignite and anthracite have relatively low sulfur contents, whereas the sulfur content for bituminous coal is often much higher. Trace metals such as mercury,

FIGURE 35.2 Results of a coal ash spill. This Tennessee home was buried in ash when an ash holding pond at a nearby power plant gave way. *(J. Miles Cary/Knoxville News Sentinel/AP Images)*

lead, and arsenic are also found in coal. Combustion of coal results in the release of these elements, which leads to an increase of sulfur dioxide and other air pollutants, such as particulates, in the atmosphere, as we will see in Chapter 15. Compounds that are not released into the atmosphere remain behind in the resulting ash.

In order to reduce the chemical compounds released into the air, coal companies wash their coal in a variety of organic compounds. In some cases, these compounds have not been widely tested for toxicity or their effect on humans and ecosystems. In 2014, a chemical storage tank containing one such coal cleaning compound leaked in West Virginia. Residents nearby immediately noticed a sweet odor and ultimately over 300,000 residents were advised not to drink their water for a number of days. Unfortunately, chemical spills are not the only accidents related to coal combustion. The residual ash from coal combustion can also be a problem.

According to the U.S. Department of Energy, there are approximately 700 coal-producing mines in the United States and they produced roughly 660 million metric tons of coal in 2015, although this is many fewer mines and much less coal production than 5 or 10 years earlier. Most of this coal is burned in the United States, with anywhere from 3 to 20 percent of it remaining behind as ash. Large deposits of this ash are often stored near coal-burning power plants. One such ash deposit—a mixture of ash and water—was kept in a holding pond at a power plant near Knoxville, Tennessee. In December 2008, the retaining wall that contained the ash gave way and spilled 4.1 billion liters (1.1 billion gallons) of ash (**FIGURE 35.2**). Three houses were destroyed by the flow of muddy ash, which covered over 121 ha (300 acres) of land. The event was the largest of its kind in U.S. history. It took 5 years and over $1 billion to clean up the bulk of the ash.

Finally, coal is a significant source of air pollution. Coal is 60 to 80 percent carbon. When it is burned, most of that carbon is converted into CO_2 and energy is released in the process. Coal produces far more CO_2 per unit of energy released than either oil or natural gas and it contributes to the increasing atmospheric concentrations of CO_2.

Petroleum is cleaner than coal

Petroleum, another widely used fossil fuel, is a liquid mixture of hydrocarbons (such as oil, gasoline, and kerosene), water, and sulfur that occurs in underground deposits. Liquid petroleum that is removed from the ground is known as **crude oil**. The word petroleum is typically used synonymously with oil, although technically petroleum is a compound that contains many hydrocarbons and can be distilled into oil, kerosene, and gasoline. The U.S. Department of Energy refers to oil, crude oil, and petroleum as equivalent substances, and we will do the same in this chapter. While coal is ideal for stationary combustion applications such as those in power plants and industry, the fluid nature of petroleum products such as oil and gasoline makes them more suitable for mobile combustion applications, such as in vehicles.

Petroleum is formed from the remains of ocean-dwelling phytoplankton (microscopic algae) that died 50 million to 150 million years ago. Deposits of phytoplankton are found in locations where porous sedimentary rocks, such as sandstone, are capped by nonporous rocks. Petroleum forms over millions of years and fills the pore spaces in the rock. Geologic events related to the tectonic cycle we discussed in Chapter 8 may deform the rock layers so that they form a dome, as **FIGURE 35.3** on page 430 shows. The petroleum is less dense than the rock, so over time it migrates upward toward the highest point in the porous rock, where it is trapped by the nonporous rock. The natural gas often associated with petroleum is even less dense, so it migrates to the absolute highest point in the dome, above the petroleum. In certain locations, petroleum

Petroleum A widely-used fossil fuel that occurs in underground deposits, composed of a liquid mixture of hydrocarbons, water, and sulfur.

Crude oil Liquid petroleum removed from the ground.

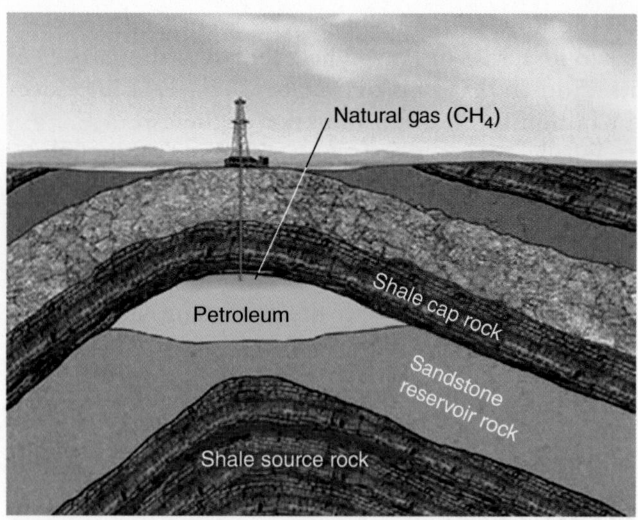

FIGURE 35.3 Petroleum accumulation underground. Petroleum migrates to the highest point in a formation of porous rock and accumulates there. Such accumulations of petroleum can be removed by drilling a well.

flows out under pressure the way water flows from an artesian well, as described in Chapter 9. But usually petroleum producers must drill wells into a deposit and extract the petroleum with pumps. After extraction, the petroleum must be transported to a petroleum refinery, by pipeline if the well is on land, or by supertanker if the well is underwater.

Petroleum contains natural gas. As we can see from Figure 35.3, some of this gas separates out naturally. If you have ever seen a burning flame in photographs of oil wells, it is a gas flare created when oil workers flare, or burn off, the natural gas, which is done under controlled conditions to prevent an explosion. As we will see in the next section, some of the gas is also extracted for use as fuel.

Crude oil can be further refined into a variety of compounds. These compounds, including tar and asphalt, gasoline, diesel, and kerosene, are distinguished by the temperature at which they boil and can therefore be separated by heating the petroleum. This process takes place in an oil refinery, a large factory-like structure. The refining process is complex and dangerous, and requires a major financial investment. There are roughly 140 oil refineries in the United States; some of the larger ones can refine over 80 million liters (21 million gallons) per day. Oil production and sales are measured in barrels of oil; one barrel equals 160 L (42 gallons).

As we saw in Figure 34.4a, the United States uses more petroleum than any other fuel—roughly 3.1 billion liters (816 million gallons) of petroleum products per day. Gasoline accounts for roughly one-half of that amount. While the primary use of petroleum products is for transportation, petroleum is also the raw material for petrochemicals, such as plastics, and for lubricants, pharmaceuticals, and cleaning solvents. Worldwide petroleum consumption is almost 15 billion liters (4 billion gallons) per day, of which the United States is responsible for about 21 percent.

The top petroleum-producing countries are Saudi Arabia, Russia, the United States, Iran, China, Canada, and Iraq, in roughly this order. These seven countries account for approximately one-half of worldwide oil production.

Advantages of Petroleum

Because petroleum is a liquid, it is extremely convenient to transport and use. It is relatively energy-dense and is cleaner-burning than coal. For these reasons, it is an ideal fuel for mobile combustion engines such as those found in automobiles, trucks, and airplanes. Because it is a fossil fuel, it releases CO_2 when burned, although for every joule of energy released, oil produces only about 85 percent as much CO_2 as coal.

Disadvantages of Petroleum

Oil, like coal, contains sulfur and trace metals such as mercury, lead, and arsenic, which are released into the atmosphere when it is burned. Some sulfur can be removed during the refining process, so it is possible, though more expensive, to obtain low-sulfur oil.

As we have seen, oil must be extracted from under the ground or beneath the ocean. Whenever oil is extracted and transported, there is the potential for oil to leak from the wellhead or to be spilled from a pipeline or tanker. Some oil naturally escapes from the rock in which it was stored and seeps into water or out onto land. However, commercial oil extraction has greatly increased the number of leakage and spillage events and the amount of oil that has been lost to land and water around the world. As noted at the beginning of the chapter, the largest oil spill in the United States until 2010 was the *Exxon Valdez* oil tanker accident in 1989. As noted in the chapter-opening case, the blowout of the BP Deepwater Horizon oil well, drilled 81 km (50 miles) off the coast of Louisiana in 1,524 m (5,000 feet) of water, led to a spill of well over 780 million liters (206 million gallons) of oil.

Larger oil spills have occurred elsewhere in the world. For example, during the 1991 Persian Gulf War, approximately 912 million liters (240 million gallons) of petroleum were spilled when wellheads were deliberately sabotaged or destroyed by the Iraqi army in Kuwait.

Oil is spilled into the natural environment in various ways. A 2003 National Academy of Sciences study found that oil extraction and transportation

FIGURE 35.4 The Lac-Mégantic accident. In 2013, a railroad train carrying oil derailed and exploded in Quebec, Canada, near the Maine border. *(Paul Chiasson/AP Images)*

were responsible for a relatively small fraction of the oil spilled into marine waters worldwide. It found that roughly 85 percent of the oil entering marine waterways came from runoff from land and rivers, airplanes, and small boats and personal watercraft, including both deliberate and accidental releases of waste oil.

In the United States, debates continue over the trade off between domestic oil extraction and the consequences for habitat and species living near oil wells or pipelines. For example, when a 1,300-km (800-mile) pipeline was constructed to transport oil overland from the North Slope of Alaska to tankers that would carry it south to the contiguous United States, wildlife biologists predicted that the pipeline might melt permafrost and interfere with the calving grounds of caribou. Scientists continue to monitor the pipeline, but so far have come to no conclusions about its environmental impact. The transportation of petroleum products by means other than pipeline can have significant consequences as well. There have been a number of serious railway accidents in recent years as a result of increased domestic drilling for oil in the United States. In a 2013 accident in Quebec, Canada, near the Maine border, a railroad train carrying oil derailed and exploded in a small town in the early morning hours, killing 47 people and destroying many buildings (**FIGURE 35.4**).

The debate about the environmental effects of land-based oil extraction continues with periodic proposals to allow oil exploration in the Arctic National Wildlife Refuge (ANWR), a 7.7 million hectare (19 million acre) tract of land in northeastern Alaska, shown in **FIGURE 35.5**. Proponents of exploration suggest that ANWR might yield 95 billion liters (25 billion gallons) to 1.4 trillion liters (378 billion gallons) of oil and substantial quantities of natural gas. Opponents maintain that opening ANWR to exploration and petroleum extraction will harm pristine habitat for many species as well as adversely affect people living in the area.

Humans, as well as wildlife, have been harmed by oil extraction. In Nigeria and many other developing countries, oil fields are adjacent to villages. Thick, gelatinous crude oil covers the ground where people walk, sometimes in bare feet. Oil flaring—the burning off of excess natural gas—takes place close to homes. Concerns about the effects of oil extraction on health, human rights, and environmental justice have led to violent political protests against oil companies in Nigeria and elsewhere.

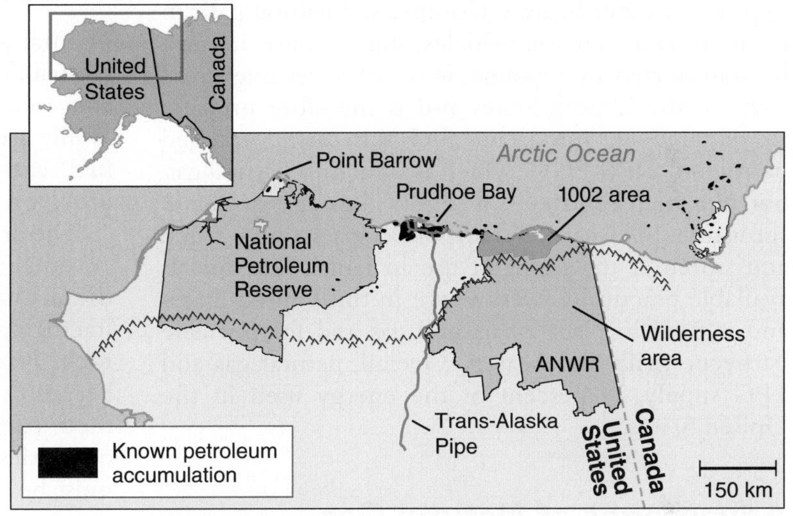

(a)

(b)

FIGURE 35.5 The Arctic National Wildlife Refuge (ANWR). (a) This map shows ANWR and adjacent areas on the North Slope of Alaska. The area in orange is the coastal plain, also called the 1002 area (pronounced as "ten-o-two"), which is under consideration for petroleum exploration and extraction. (b) Caribou are among the many species that live in ANWR. *(Jim Goldstein/DanitaDelimont.com)*

Natural gas is the cleanest of the fossil fuels

We have already mentioned natural gas in connection with petroleum, since it exists as a component of petroleum in the ground as well as in gaseous deposits found separately from petroleum. Natural gas is 80 to 95 percent methane (CH_4) and 5 to 20 percent ethane, propane, and butane. Because natural gas is lighter than oil, it lies above oil in petroleum deposits (see Figure 35.3). Natural gas is generally extracted in association with petroleum; only recently has exploration specifically for natural gas been conducted.

The two largest uses of natural gas in the United States are for electricity generation and industrial processes. Natural gas is also used to manufacture nitrogen fertilizer, and it is used in homes as an efficient fuel for cooking, heating, and operating clothes dryers and water heaters. Compressed natural gas can be used as a fuel for vehicles, but because it must be transported by pipeline, it is not accessible in all parts of the United States and is therefore unlikely to become an important fuel for cars. Liquefied petroleum gas (LPG)—which is similar to natural gas, but in a liquid form—is a slightly less energy-dense substitute. LPG can be transported via train or truck and stored at the point of use in tanks. This fuel is available practically everywhere in the United States and is used in place of natural gas and for portable barbecue grills and heaters. Overall, natural gas and LPG supply 29 percent of the energy used in the United States.

Advantages of Natural Gas

Because of the extensive natural gas pipeline system in many parts of the United States, roughly one-half of homes use natural gas for heating. Compared with coal and oil, natural gas contains fewer impurities and therefore emits almost no sulfur dioxide or particulates during combustion. And for every joule of energy released during combustion, natural gas emits only 60 percent as much CO_2 as coal. So natural gas is the cleanest of the fossil fuels and as long as it can be supplied by pipeline, it is a very convenient and desirable fossil fuel. In some locations where natural gas pipelines are not present, LPG is used although it is slightly less convenient.

Disadvantages of Natural Gas

While natural gas when combusted releases the least carbon dioxide of all the fossil fuels, unburned natural gas—methane—that escapes into the atmosphere is itself a potent greenhouse gas that is 25 times more efficient at absorbing infrared energy than CO_2.

FIGURE 35.6 Natural gas field in Wyoming. Even though natural gas is relatively clean compared with other fossil fuels, its extraction impacts large amounts of land. *(Jim Havey/Alamy stock photo)*

Natural gas that leaks after extraction is a suspected contributor to the steep rise in atmospheric methane concentrations that was observed in the 1990s. While natural gas is referred to as the "clean" fossil fuel, extraction and use still lead to environmental problems (**FIGURE 35.6**).

The process of exploring for natural gas involves the "thumper trucks" already mentioned. As described at the beginning of Chapter 1, the process of hydraulic fracturing (fracking) releases natural gas from host rock. Fracking uses large amounts of chemicals and because those chemicals do not need to be named, their effects are unknown. Fracking also uses large quantities of water that can become contaminated and must be disposed of afterward. There can be groundwater contamination resulting from the drilling of natural gas wells. As hydraulic fracturing has grown more widespread, a number of scientists and gas extraction experts have attempted to quantify exactly how much natural gas escapes during the extraction and transportation process. Estimating the amount of escaped natural gas has become a controversial subject due to the large uncertainties in the natural gas extraction process, measurement difficulties, and industry secrecy. Presently, estimates of escaped gas range from 2 to 9 percent of the total amount of gas extracted.

Oil sands and liquefied coal are also fossil fuels

Other types of fossil fuel deposits contain a great deal of energy, but are not readily available. When we consider using these energy resources, it is important to determine the energy return on energy investment. Two of the less readily available fossil fuels are oil sands and liquid coal.

Oil Sands

Not all petroleum is easily extractable in conventional oil wells. **Oil sands** are slow-flowing, viscous deposits of *bitumen* mixed with sand, water, and clay. **Bitumen** is a degraded type of petroleum that forms when a petroleum deposit is not capped with nonporous rock. The petroleum migrates close to the surface, where bacteria metabolize some of the light hydrocarbons while others evaporate. The remaining mix no longer flows at ambient temperatures and pressures, but it can be extracted by surface mining.

Although oil sand exploitation promises to extend our petroleum supply, it could have serious negative environmental impacts. The mining of bitumen is much more energy-intensive than conventional drilling for crude oil. As we saw in Chapter 8, surface mining creates large open pits. Extraction of the bitumen from the other material contaminates roughly 2 to 3 L of water for every liter of bitumen obtained, and many oil sands are located in areas where water is not an abundant resource. In addition, because oil sands require so much energy before they arrive at the refinery, the overall system efficiency is lower, and the resulting CO_2 release greater, than for conventional oil production.

AP® Exam Tip

Past AP® Environmental Science Exams have asked about oil sands. Make sure that you know how oil sands are mined and processed, and that you understand the environmental consequences of this process.

Liquid Coal

Whereas the availability of petroleum may become much more limited in this century as supplies diminish, the United States and China both have immense coal reserves. The technology to convert solid coal into a liquid fuel—a process known as **CTL**, short for "coal to liquid"—has been available for decades. CTL was widely used by the German military during World War II and has been used by other countries since then. But producing liquefied coal is relatively expensive and has many of the same drawbacks found in the exploitation of oil sands. In terms of total energy content, there is over 1,000 times more energy in the world's coal reserves than in the world's petroleum reserves. Because there is so much coal in the United States, CTL has the potential to eliminate U.S. dependence on foreign oil. On the other hand, the U.S. Environmental Protection Agency estimates that total greenhouse gas emissions from liquefied coal are more than twice those from conventionally produced oil, and as we have seen, the environmental impacts of coal mining can be severe. For these reasons, liquid coal is not highly promising as an important future energy source.

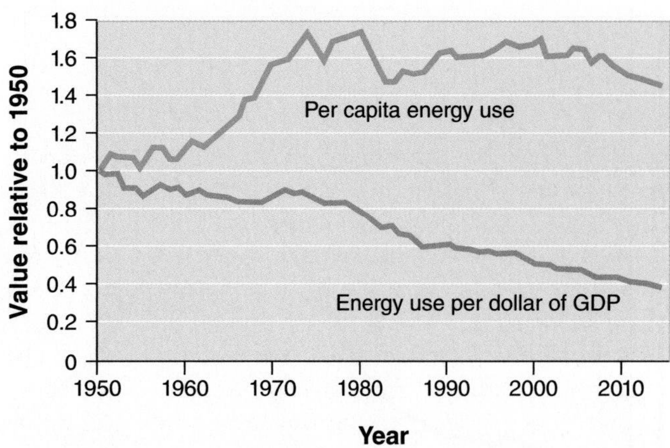

FIGURE 35.7 U.S. energy use per capita and energy intensity. Energy use per capita in the United States was roughly level between 1990 and 2005 and has been dropping in recent years. Our energy intensity, or energy use per dollar of GDP, has been decreasing steadily since 1980. *(Data from the U.S. Department of Energy, Energy Information Administration, 2017.)*

Fossil fuels are a finite resource

Although we know that the supply of fossil fuels is finite, there is some discussion within the environmental science community on whether or not that matters. Recall our discussion in Chapter 7 about human creativity. Many people believe that we will apply human ingenuity to develop new energy sources. In the meantime, total energy use in the United States has leveled off, with small variations from year to year, and energy use per person has been decreasing slightly. An important measure of energy use and productivity is **energy intensity**, which is the energy use per unit of gross domestic product (GDP). In recent decades, energy intensity in the United States has been steadily decreasing, as **FIGURE 35.7** shows. In other words, we are using energy more efficiently. But because the U.S. population has grown, and we are doing more things that use

Oil sands Slow-flowing, viscous deposits of bitumen mixed with sand, water, and clay.

Bitumen A degraded petroleum that forms when petroleum migrates to the surface of Earth and is modified by bacteria.

CTL (coal to liquid) The technology to convert solid coal into liquid fuel.

Energy intensity The energy use per unit of gross domestic product.

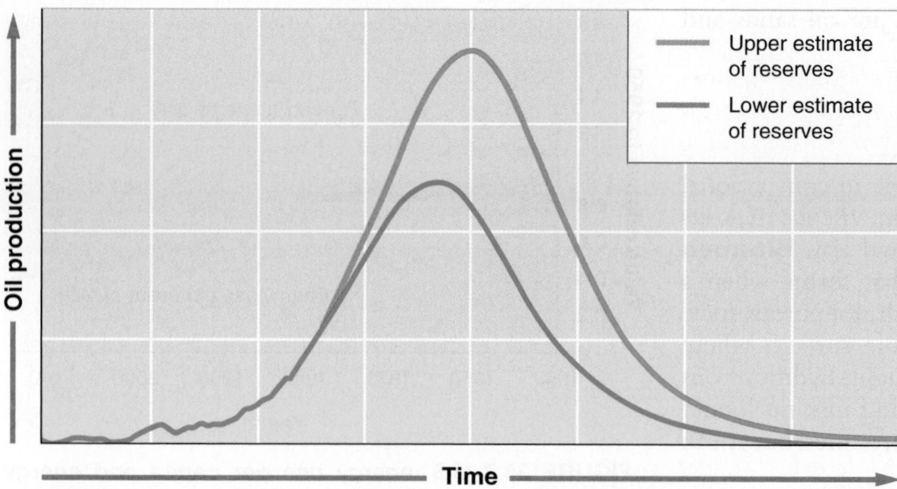

FIGURE 35.8 A generalized version of the Hubbert curve. Regardless of whether an upper estimate or a lower estimate of total petroleum reserves is used, the date by which petroleum reserves will be depleted does not change substantially. A growing body of environmental scientists believes that the consequences of using fossil fuels will require a switch to other fuels long before supplies become limited.

energy, our overall energy use has leveled off but not decreased. Prior to that, it had been growing steadily for the previous 2 centuries, as shown in Figure 34.3 on page 419. You can better understand why there isn't a decrease if you think about how many electronic devices you have, then ask your parents or grandparents how many electronic devices they had when they were high school students. We will begin this section with a survey of ideas about the availability of fuel and its use over time, and then we will look at possible patterns of fossil fuel energy use in the future.

The Hubbert Curve

We have seen that fossil fuels, our primary source of energy, are finite. Because fossil fuels take millions of years to form, they are nonrenewable resources, at least on a human timescale. By definition, then, the use of fossil fuels is not sustainable because there is no way to limit our consumption to the rate at which these fuels are being formed. Energy scientists and others have debated whether our economy will at some point be limited by the availability of this energy resource, although in recent years, concerns have shifted away from the actual supply of fossil fuels to the consequences of fossil fuel combustion, particularly the release of CO_2 and its contribution to global climate change. Many environmental scientists believe that these consequences will manifest themselves in adverse ways long before we run out of fossil fuels. However, there is still a group of energy analysts deeply concerned about "the end of oil."

Hubbert curve A bell-shaped curve representing oil use and projecting both when world oil production will reach a maximum and when the world will run out of oil.

Peak oil The point at which half the total known oil supply is used up.

In 1969, M. King Hubbert, a geophysicist and oil company employee, published a graph showing a bell-shaped curve representing oil use. Shown in **FIGURE 35.8**, the **Hubbert curve** represents oil use and projects both when world oil production will reach a maximum and when world oil will finally be depleted. Hubbert used two estimates of total world petroleum reserves: an upper estimate and a lower estimate. He found that the total reserves did not greatly influence the time it would take to use up all of the oil in known reserves. He predicted that oil extraction and use would increase steadily until roughly half the supply had been used up, a point known as **peak oil**. Hubbert predicted that at peak oil, extraction and use would begin to decline. Some oil experts believe we have already reached peak oil, while others maintain that we may reach it very soon. Back in 1969, Hubbert predicted that 80 percent of the world's total oil supply would be used up in roughly 60 years.

Although there have been discoveries of large oil fields since Hubbert did his work, the conclusion he drew from his model still holds. Regardless of the exact amount of the total reserves, the total number of years we use petroleum will fall within a relatively narrow time window. When we identify a fuel source, we tend to use it until we come upon a better fuel source. As a number of energy experts are fond of saying, "We did not move on from the Stone Age because we ran out of stones." In a similar vein, many people believe that ingenuity and technological advances in the renewable energy sector will one day render oil, and other fossil fuels, much less desirable. A growing number of environmental scientists maintain that wondering about how much oil remains and whether or not we have reached peak oil are no longer meaningful questions to ask. As fossil fuels become less desirable, many believe a more important question is, "What fuel or energy sources will we use next?"

The Future of Fossil Fuel Use

If current global use patterns continue, and no significant additional oil supplies are discovered, we may run out of conventional oil supplies in less than 50 years. Natural gas supplies will last longer. Coal supplies will last for at least 200 years, and probably much longer. While these predictions assume that we will continue our current use patterns, technological advances, a shift to nonfossil fuels, and changes in social choices and population patterns are already altering these usage patterns and rates.

As more people and nations have come to accept the concept that anthropogenic increases in atmospheric greenhouse gas concentrations are causing global climate change, a large number of researchers have suggested that we should turn our attention to how we might transition from fossil fuels before their use causes further problems.

MODULE 35 AP® Review

Preparing for the AP® Exam

In this module, we have seen that fossil fuels have significant advantages as well as significant disadvantages. Coal is an abundant fuel that requires minimal processing and is easily transported. However, it leads to significant human health problems when mined below ground, while surface mining causes environmental damage. Coal emits more carbon dioxide per unit of energy released than do other fossil fuels. Petroleum is less abundant than coal but requires more processing. The possibility of spills is another disadvantage of petroleum. Natural gas is easily transported via pipeline and emits less carbon dioxide per unit of energy released than other fossil fuels. Its use has increased substantially in recent years for electricity generation and other purposes, mainly because of increased availability and decreased price due to fracking. In the last 50 years, estimates of fossil fuel availability and the rate at which it will be depleted have been replaced by concerns about the pollution from fossil fuels and questions about future fossil fuel energy use patterns. As we will see in the next module, there is one other conventional, nonrenewable fuel that offers a very attractive feature: It does not contribute significantly to the addition of greenhouse gases to the atmosphere. That fuel is uranium.

AP® Practice Questions

1. Which type of coal has the highest energy density?
   (a) lignite
   (b) bituminous
   (c) anthracite
   (d) subbituminous

2. What makes petroleum convenient to use as fuel for transportation?
   I. high energy density
   II. clean burning
   III. its liquid state
   (a) I only
   (b) I and III
   (c) II and III
   (d) I, II, and III

3. Natural gas is primarily
   (a) ethane.
   (b) propane.
   (c) butane.
   (d) methane.

4. Bitumen is
   (a) a form of liquid coal.
   (b) a degraded type of petroleum.
   (c) a by-product of natural gas extraction.
   (d) a fast-forming fossil fuel.

5. The Hubbert curve predicts that
   (a) peak oil will occur once half of the supply is used up.
   (b) finding additional reserves could greatly increase the time to peak oil.
   (c) natural gas will be cheaper than oil by the early twenty-first century.
   (d) the cost of oil is independent of the world's oil supply.

MODULE 36

Nuclear Energy Resources

Because the combustion of fossil fuels releases large quantities of CO_2 into the atmosphere, nuclear energy has received renewed interest. Concerns about nuclear energy include radioactivity, the proliferation of radioactive fuels that could be used in weapons, and the potential for accidents. Recently, however, nuclear energy has received positive attention, even from self-proclaimed environmentalists, because of its relatively low emissions of CO_2. In this module we will examine how nuclear energy works and the advantages and disadvantages of nuclear energy.

Learning Goals

After reading this module, you should be able to

- describe how nuclear energy is used to generate electricity.
- discuss the advantages and disadvantages of using nuclear fuels to generate electricity.

Nuclear reactors use fission to generate electricity

Electricity generation from nuclear energy uses the same basic process as electricity generation from fossil fuels: Steam turns a turbine that turns a generator that generates electricity. The difference is that a nuclear power plant uses a radioactive isotope, uranium-235 (^{235}U), as its fuel source. We presented the concepts of isotopes, radioactive decay, and half-lives in Chapter 2. Radioactive decay occurs when a parent radioactive isotope emits alpha or beta particles or gamma rays. Here, we need to introduce one more concept before being able to fully describe a nuclear power plant. The naturally occurring isotope ^{235}U, as well as other radioactive isotopes, undergoes a process called *fission*.

Fission, depicted in **FIGURE 36.1**, is a nuclear reaction in which a neutron strikes a relatively large atomic nucleus, which then splits into two or more parts, releasing additional neutrons and energy in the form of heat. The additional neutrons can, in turn, promote additional fission reactions, which leads to a chain reaction of nuclear fission that gives off an immense amount of heat energy. In a nuclear power plant, that heat energy is used to produce steam, just as in any

Fission A nuclear reaction in which a neutron strikes a relatively large atomic nucleus, which then splits into two or more parts, releasing additional neutrons and energy in the form of heat.

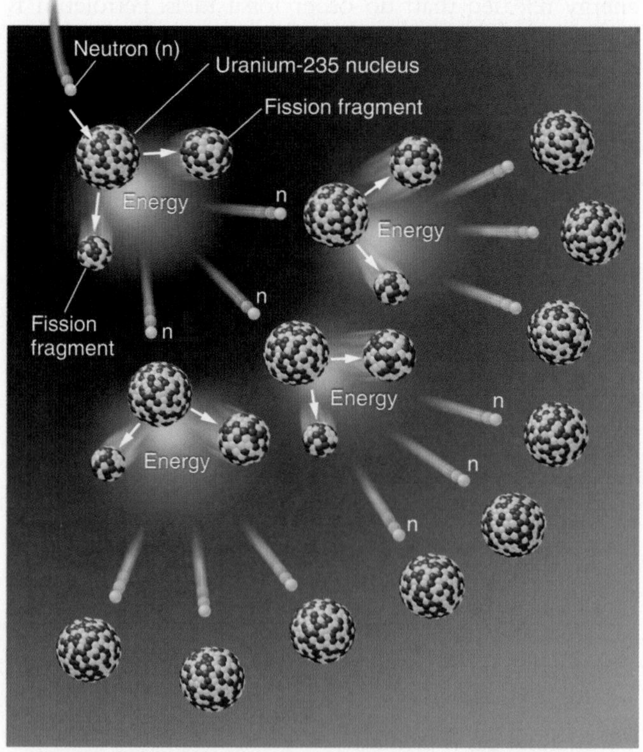

FIGURE 36.1 Nuclear fission. Energy is released when a neutron strikes a large atomic nucleus, which then splits into two or more parts.

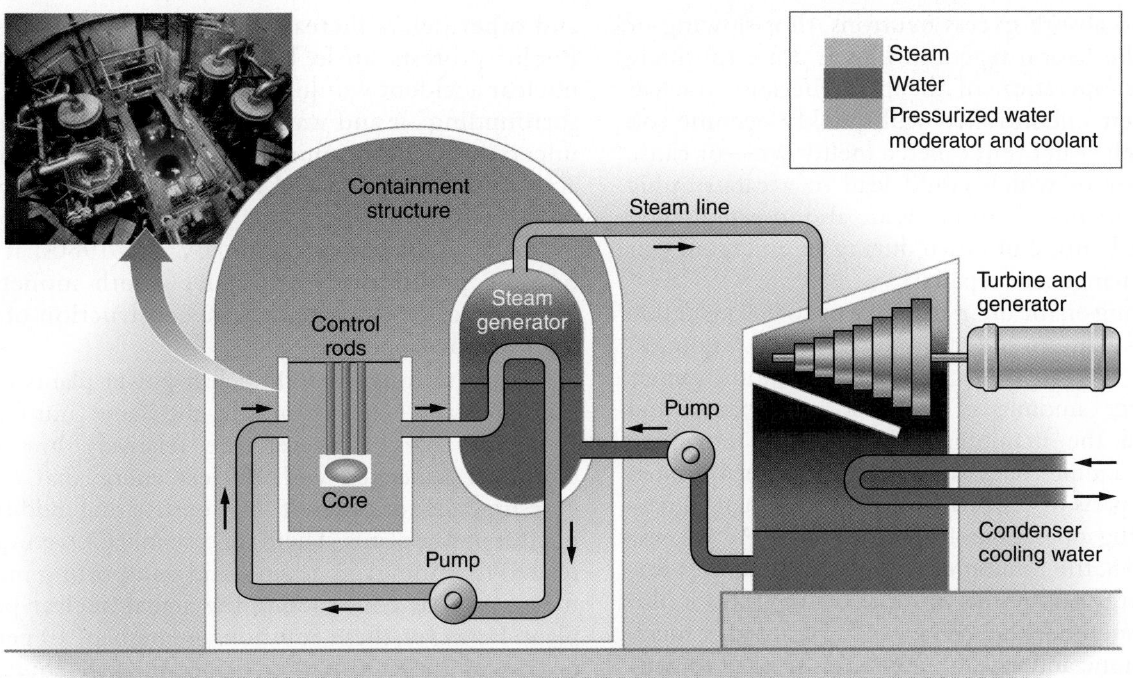

FIGURE 36.2 A nuclear reactor. This schematic shows the basic features of the light-water reactor, the type of reactor found in the United States. *(US Department of Energy/Science Source)*

other thermal power plant. However, 1 g of ^{235}U contains 2 million to 3 million times the energy of 1 g of coal.

Uranium-235 is one of the more easily fissionable isotopes, which makes it ideal for use in a nuclear reactor. A neutron colliding with ^{235}U splits the uranium atom into smaller atoms, such as barium and krypton, and three neutrons. The reaction is as follows:

$$1 \text{ neutron} + {}^{235}U \rightarrow {}^{142}Ba + {}^{91}Kr + 3 \text{ neutrons in motion (kinetic energy)}$$

Many other radioactive daughter products are released as well. A properly designed nuclear reactor will harness the kinetic energy from the three neutrons in motion to produce a self-sustaining chain reaction of nuclear fission. The by-products of the nuclear reaction include radioactive waste that remains hazardous for many half-lives—that is, hundreds of thousands of years or longer.

FIGURE 36.2 shows how a nuclear reactor works. The containment structure encloses the nuclear fuel—which is contained within cylindrical tubes called **fuel rods**—and the steam generator. Uranium fuel is processed into pellets, which are then put into the fuel rods. A typical nuclear reactor might contain hundreds of bundles of fuel rods in the reactor core within the containment structure.

Heat from nuclear fission is used to heat water within the containment structure, which circulates in a loop. This loop passes close to another loop of water, and heat is transferred from one loop to the other. In the process, steam is produced, which turns a turbine, which turns a generator, just as in most other thermal power plants. The nuclear power plant shown in Figure 36.2 is a light-water reactor, the only type of reactor used in the United States and the most common type used elsewhere in the world.

A nuclear power plant is designed to harness heat energy from fission to make steam. But the plant must be able to slow the fission reaction to allow collisions to take place at the appropriate speed. To do this, nuclear reactors contain a moderator, such as water, to slow down the neutrons so that they can effectively trigger the next chain reaction. Because there is also a risk that the reaction will run out of control, nuclear reactors contain **control rods**, cylindrical devices that can be inserted between the

Fuel rod A cylindrical tube that encloses nuclear fuel within a nuclear reactor.

Control rod A cylindrical device inserted between the fuel rods in a nuclear reactor to absorb excess neutrons and slow or stop the fission reaction.

fuel rods to absorb excess neutrons, thus slowing or stopping the fission reaction. This is done routinely during the operation of the plant because nuclear fuel rods left uncontrolled will quickly become too hot and melt—an event called a meltdown—or cause a fire, either of which could lead to a catastrophic nuclear accident. Control rods are also inserted when the plant is being shut down during an emergency or for maintenance and repairs.

Depending on the ore, it may take up to 900 kg (2,000 pounds) of uranium ore to produce 3 kg (6.6 pounds) of nuclear fuel. In order to obtain uranium, miners remove large amounts of the host rock, extract and concentrate the uranium, and leave the remaining material in tailings piles. Australia, the western United States, and parts of Canada have large commercial uranium mining operations for nuclear fuel. As we saw in Chapter 8, the mining of any material requires fossil fuel energy and results in mine tailings. This is also true for uranium, although, as we have noted, a much smaller volume and mass of uranium is needed to generate a similar quantity of electricity than would be the case with coal.

Nuclear power plants rely on ^{235}U as their fuel. However, most uranium ores contain as much as 99 percent ^{238}U, another isotope of uranium that occurs with ^{235}U but does not fission as easily. Therefore, when uranium ore is mined, it must be chemically enriched—a process to increase its concentration—in ^{235}U to be useful as a fuel. Typically, suitable nuclear fuel contains 3 percent or greater ^{235}U.

Nuclear energy has advantages and disadvantages

Nuclear power plants do not produce air pollution during their operation, so proponents of nuclear energy consider it "clean" energy. In countries with limited fossil fuel resources, nuclear energy is one way to achieve independence from imported oil. Nuclear energy generates 70 percent of electricity in France, and it is widely used in Spain, the United Kingdom, China, Canada, Ukraine, and South Korea, as well as in other countries.

As we saw in Figure 34.8, 20 percent of the electricity generated in the United States comes from nuclear energy. Early proponents of nuclear energy in the 1950s and 1960s claimed that it would be "too cheap to meter," meaning that it would be so inexpensive that there would be no point in trying to figure out how much each customer used. However, construction of new nuclear power plants became more and more expensive in the United States, in part because public protests, legal battles, and other delays increased the cost of construction. Public protests arose because of concerns that a nuclear accident would release radioactivity into the surrounding air and water. Other concerns included uncertainty about appropriate locations for radioactive waste disposal and fear that radioactive waste could fall into the hands of individuals seeking to make a nuclear weapon. By the 1980s, it had become prohibitively expensive—both monetarily and politically—to attempt the construction of new nuclear power plants.

There are roughly 100 nuclear power plants in the United States—approximately the same number as 2 decades ago. However, the relatively low CO_2 emissions associated with nuclear energy has caused a resurgence of interest in constructing additional nuclear power plants. There are certainly CO_2 emissions related to mining, processing, and transporting nuclear fuel as well as constructing the actual nuclear power plant. However, these emissions are perhaps 10 percent or less of those related to generating an equivalent amount of electricity from coal. Two major issues of environmental concern remain, however: the possibility of accidents and the disposal of radioactive waste. We will also consider nuclear fusion and summarize the advantages and disadvantages of nuclear power relative to other energy sources.

Nuclear Accidents

Two accidents contributed to the global protests against nuclear energy in the 1980s and 1990s. On March 28, 1979, at the Three Mile Island nuclear power plant in Pennsylvania, operators did not notice that a cooling water valve had been closed the previous day. This oversight led to a lack of cooling water around the reactor core, which overheated and suffered a partial meltdown. The reactor core was severely damaged, and a large part of the containment structure became highly radioactive with an unknown amount of radiation released to the outside environment. A few thousand schoolchildren and pregnant women were evacuated from the area surrounding the plant by order of the governor of Pennsylvania, although an estimated 200,000 other people chose to evacuate as well.

In the days following the accident, residents of the area experienced a great deal of anxiety and fear, especially as reports of a potentially explosive gas bubble in the containment structure were evaluated. Although there has been no documented increase in local health problems as a result of this accident, concerns remain and several investigators maintain that infant mortality rates and cancer rates increased in the following years. The nuclear reactor in which the incident occurred, one of two at the Three Mile Island nuclear facility, has not been used since the accident.

The Three Mile Island event, compounded by the coincidental release of the film *The China Syndrome* about safety violations and a near-catastrophic accident at a nuclear power plant, led to widespread public fear and anger not only in Pennsylvania but elsewhere as well.

A much more serious accident occurred on April 26, 1986, at a nuclear power plant in Chernobyl, Ukraine. The accident occurred during a test of the plant when, in violation of safety regulations, operators deliberately disconnected emergency cooling systems and removed the control rods. With no control rods and no coolant, the nuclear reactions continued without control and the plant overheated. These "runaway" reactions led to an explosion and fire that damaged the plant beyond use. At the time of the accident, 31 plant workers and firefighters died from acute radiation exposure and burns, and many more died later of causes related to exposure.

After the accident, winds blew radiation from the plant across much of Europe, where it contaminated crops and milk from cows grazing on contaminated grass. More than a hundred thousand people were evacuated from the area around Chernobyl. Estimates of health effects vary widely, in part because of the paucity of information provided by the Soviet government, but a U.S. National Academy of Sciences panel estimated that 4,000 additional cancer deaths (over and above the average number of expected deaths) would occur over the next 50 years among people who lived near the plant or worked on the cleanup. There have been approximately 5,000 cases of thyroid cancer, most of them nonfatal, among children who were younger than 18 at the time of the accident and lived near the Chernobyl plant. Thyroid cancer may be caused by the absorption of radioactive iodine, one of the radioactive elements emitted during the accident.

The most recent nuclear power plant accident occurred in Japan on March 11, 2011. A magnitude 9.0 earthquake off the country's coast generated a tsunami that was 15 m high. This caused flooding in northeastern Japan, where the Fukushima nuclear power plant is located. The resulting structural damage to the plant and interruption of the regional electrical supply led to fires, hydrogen gas explosions, and the release of radioactive gases from nuclear reactors within the containment structure. Ultimately, four out of a total of six reactors were damaged beyond repair, radioactive gases were released into the surrounding environment, and over 100,000 people were evacuated from their homes. Approximately 20,000 people, including three workers at the Fukushima plant, were killed by the earthquake and tsunami. However, at present, there have been no deaths attributed to the release of radioactivity from the reactors. After the Fukushima accident, Japan shut down all 54 of its nuclear reactors. As of mid-2018, only five had been restarted but the Japanese government is tentatively planning to restart more reactors in the coming years.

Radioactive Waste

When nuclear fuel is no longer useful in a power plant, yet continues to emit radioactivity, it is considered **radioactive waste**. Long after nuclear fuel can produce enough heat to be useful in a power plant, it continues to emit radioactivity. Because radioactivity can be extremely damaging to living organisms, radioactive materials—the nuclear waste—must be stored in special, highly secure locations.

The use of nuclear fuels produces three kinds of radioactive waste: high-level waste in the form of used fuel rods; low-level waste in the form of contaminated protective clothing, tools, rags, and other items used in routine plant maintenance; and uranium mine tailings, the residue left after uranium ore is mined and enriched. High-level radioactive waste releases ionizing radiation in doses large enough to potentially cause immediate damage to the human body including organ failure, radiation sickness, and death. At lower exposure to high-level waste, and exposure to low-level radioactive waste, humans can experience damage to DNA, which may ultimately lead to cancer or tumors. There are also many other sublethal harmful effects to the eyes, brain, and immune and reproductive systems. Unborn children, children, and adolescents are most vulnerable.

Disposal of all three types of radioactive waste is regulated by the government, but because it has the greatest potential impact on the environment, we will focus here on high-level radioactive waste.

After a time, nuclear fuel rods become spent, meaning they are not sufficiently radioactive to generate electricity efficiently. As we discussed in Chapter 2, each radioactive isotope has a specific half-life: the time it takes for one-half of its radioactive atoms to decay. Uranium-235 has a half-life of 704,000,000 years, which means that 704 million years from today, a sample of ^{235}U will be only half as radioactive as it is today. In another 704 million years it will lose another half of its radioactivity, so it will be one-fourth as radioactive as it is today.

Radioactive waste Nuclear fuel that can no longer produce enough heat to be useful in a power plant but continues to emit radioactivity.

DO THE MATH Calculating Half-Lives

Preparing for the AP® Exam

Strontium-90 is a radioactive waste product from nuclear reactors. It has a half-life of 29 years. How many years will it take for a quantity of strontium-90 to decay to $\frac{1}{16}$ of its original mass?

It will take 29 years to decay to $\frac{1}{2}$ its original mass; another 29 years to $\frac{1}{4}$; another 29 years to $\frac{1}{8}$; and another 29 years to $\frac{1}{16}$:

$$29 + 29 + 29 + 29 = 116 \text{ years}$$

YOUR TURN You have 180 g of a radioactive substance. It has a half-life of 265 years. After 1,325 years, what mass remains?

Radiation can be measured with a variety of units. A **becquerel (Bq)** measures the rate at which a sample of radioactive material decays; 1 Bq is equal to the decay of one atom per second. A **curie**, another unit of measure for radiation, is 37 billion decays per second. If a material has a radioactivity level of 100 curies and has a half-life of 50 years, the radioactivity level in 200 years will be

$$\frac{200 \text{ years}}{50 \text{ years/half-life}} = 4 \text{ half-lives}$$

100 curies → 50 curies (one half-life) → 25 curies (two half-lives) → 12.5 curies (three half-lives) → 6.3 curies (four half-lives)

You can gain more experience in working with these calculations in "Do the Math: Calculating Half-Lives."

Because spent fuel rods remain a threat to human health for 10 or more half-lives, they must be stored until they are no longer dangerous. At present, nuclear power plants are required to store spent fuel rods at the plant itself. Initially, all fuel rods were stored in pools of water at least 6 m (20 feet) deep. The water acts as a shield from radiation emitted by the rods. Currently, more than 100 sites around the country are storing spent fuel rods. However, some facilities have run out of pool storage space and are now storing them in lead-lined dry containers on land (**FIGURE 36.3**). Eventually, all of this material will need to be moved to a permanent radioactive waste disposal facility.

Becquerel (Bq) Unit that measures the rate at which a sample of radioactive material decays; 1 Bq = decay of 1 atom or nucleus per second.

Curie A unit of measure for radiation; 1 curie = 37 billion decays per second.

Disposing of radioactive waste is a challenge. This waste cannot be incinerated, safely destroyed using chemicals, shot into space, dumped on the ocean floor, or buried in an ocean trench because all of these options involve the potential for large amounts of radioactivity to enter the oceans or atmosphere. Therefore, at present, the only solution is to store it safely somewhere on Earth indefinitely. The physical nature of the storage site must ensure that the waste will not leach into the groundwater or otherwise escape into the environment. It must be far from human habitation in case of any accidents and be secure against terrorist attack. In addition, the waste has to be transported to the storage site in a way that minimizes the risk of accidents or theft by terrorists.

In 1978, the U.S. Department of Energy began examining a site at Yucca Mountain, Nevada, 145 km (90 miles) northwest of Las Vegas, as a possible long-term repository for the country's spent nuclear fuel. The Yucca Mountain proposal has generated enormous protest and controversy. In 2006, the Department of Energy released a report confirming the soundness of the research supporting the Yucca Mountain site. However, a few years later, the Yucca Mountain project was canceled. Every few years, interest in the project is generated in Congress or elsewhere but as of 2017, there are no specific plans to proceed with Yucca Mountain as a national nuclear waste repository.

Nuclear energy raises a unique question of sustainability. It is a source of electricity that releases much less CO_2 than coal, and none during the actual generation of electricity. But some argue that the use of a fuel that consistently generates large quantities of high-level radioactive waste is not a sustainable practice. Many critics of nuclear energy maintain that there will never be a place safe enough to store high-level radioactive waste.

FIGURE 36.3 Storage of nuclear waste. Nuclear waste is often stored in metal containers at the power plant where it is produced. *(Toby Talbot/AP Images)*

Fusion Power

As we have seen, nuclear fission occurs when the nuclei of radioactive atoms are broken apart into smaller, lighter nuclei. **Nuclear fusion**, the reaction that powers the Sun and other stars, occurs when lighter nuclei are forced together to produce heavier nuclei. In the process, a great deal of heat is generated. The nuclear fusion reaction that is most promising for electricity generation is that of two hydrogen isotopes fusing together into a helium atom. As the reaction occurs, a small amount of mass is lost and an immense amount of energy is liberated.

Nuclear fusion seems to promise a seemingly unlimited source of energy that requires only hydrogen as an input and produces relatively small amounts of radioactive waste. Unfortunately, creating fusion on Earth requires a reactor that will heat material to temperatures 10 times those in the core of the Sun. These temperatures make containment extremely difficult. So far, the most promising techniques have involved suspending superhot material in a magnetic field, but the amount of energy required is greater than the energy output. Most experts believe that it will be several decades, or perhaps longer, before the promise of fusion power can be realized.

Summary of Nuclear Power and Comparison with Other Fuels

At present, our reliance on nuclear energy in the United States is subject to speculation and projections, but it is hard to know whether the positive aspects of nuclear energy will come to outweigh the risks of accidents and containment of radioactive waste. Indeed, in 2006, applications for new nuclear power plants in the United States were filed with the Nuclear Regulatory Commission for the first time in more than 2 decades. In 2012, licenses were approved for two nuclear power plants. In 2018, there were two plants under construction, but then construction was halted. It is possible—but not a certainty—that up to five new nuclear power plants may come online in the United States in the next decade.

TABLE 36.1 on page 442 summarizes the major benefits and consequences of the conventional fuels we have discussed in this chapter.

Nuclear fusion A reaction that occurs when lighter nuclei are forced together to produce heavier nuclei.

TABLE 36.1 Comparison of nonrenewable energy fuels

Energy Type	Advantages	Disadvantages	Pollutant and greenhouse gas emissions	Electricity (cents/kWh)	Energy return on energy investment*
Oil/gasoline	• Ideal for mobile combustion (high energy/mass ratio) • Quick ignition/turn-off capability • Cleaner burning than coal	• Significant refining required • Oil spill potential effect on habitats near drilling sites • Significant dust and emissions from fossil fuels used to power earth-moving equipment • Human rights/ environmental justice issues in developing countries that export oil • Will probably be much less available in the next 40 years or so	• Second highest emitter of CO_2 among fossil fuels • Hydrocarbons • Hydrogen sulfide	• Relatively little electricity is generated from oil 20 cents/kWh	4.0 (gasoline) 5.7 (diesel)
Coal	• Energy-dense and abundant—U.S. resources will last at least 200 years • No refining necessary • Easy, safe to transport • Economic backbone of some small towns	• Mining practices frequently risk human lives and dramatically alter natural landscapes • Coal power plants are slow to reach full operating capacity • A large contributing factor to acid rain in the United States	• Highest emitter of CO_2 among energy sources • Sulfur • Trace amounts of toxic metals such as mercury	5 cents/kWh	14
Natural Gas	• Cogeneration power plants can have efficiencies up to 60% • Efficient for cooking, home heating, etc. • Fewer impurities than coal or oil	• Risk of leaks/ explosions • Twenty-five times more effective as a greenhouse gas than CO_2 • Not available everywhere because it is transported by pipelines	• Methane • Hydrocarbons • Hydrogen sulfide	3–4 cents/kWh	8
Nuclear Energy	• Emits no CO_2 once plant is operational • Offers independence from imported oil • High energy density, ample supply	• Very unpopular; generates protests • Plants are very expensive to build because of legal challenges • Meltdown could be catastrophic • Possible target for terrorist attacks	• Radioactive waste is dangerous for hundreds of thousands of years • No long-term plan currently in place to manage radioactive waste • No air pollution during production	12–15 cents/kWh	8

*Estimates vary widely.

MODULE

36 AP® Review

Preparing for the AP® Exam

In this module, we have seen that the use of nuclear fuels for generating electricity has significant advantages and disadvantages. Nuclear energy is a relatively clean means of electricity generation, although fossil fuels are used in constructing nuclear power plants and mining and processing the uranium fuels. The possibility of accidents during plant operation and the difficulty of radioactive nuclear waste disposal are major environmental hazards of nuclear energy used for the generation of electricity. Neither of these issues has been satisfactorily resolved at present. Throughout the nuclear electricity generation cycle, carbon dioxide emissions are much less than during electricity generation from fossil fuels.

AP® Practice Questions

1. One g of 235^{u} produces approximately how much more energy than 1 g of coal?
   (a) 19,000 times
   (b) 80,000 times
   (c) 400,000 times
   (d) 2,000,000 times

2. Control rods slow nuclear reactions by
   (a) reducing the amount of fuel available.
   (b) absorbing the heat produced.
   (c) increasing the rate heat transfers to the water.
   (d) absorbing excess neutrons.

3. The process of fusion
   (a) splits atoms.
   (b) requires extremely high temperatures.
   (c) uses plutonium instead of uranium.
   (d) requires several radioactive elements.

4. Interest in nuclear power has recently increased because of
   (a) low energy costs.
   (b) lack of significant accidents.
   (c) low carbon dioxide emissions.
   (d) new solutions for waste disposal.

5. For a sample of Ti-44 with a half-life of 63 years, how long until $\frac{1}{16}$ of the original amount is left?
   (a) 126 years
   (b) 189 years
   (c) 252 years
   (d) 315 years

Working Toward Sustainability

Meet Sense, the Home Energy Monitor

It is generally accepted that peer pressure influences behavior. When hotels placed small signs in bathrooms telling their guests that other guests were reusing their towels rather than insisting on receiving fresh clean towels each day, the need for towel laundering dropped drastically. Many studies have shown that when people receive feedback about their electricity consumption, they will often respond just as the hotel guests did. A number of electric companies have experimented with mailings that show homeowners how much electricity they use in comparison with neighbors in similar-size homes. However, if homeowners are using less electricity than their neighbors, they may not be inclined to reduce their use further.

For decades, environmental scientists have imagined having a device that would provide homeowners with an instantaneous reading of actual electricity use in the home. Ideally, this device would also allow the homeowner to know which devices were using the most electricity and perhaps this would encourage people to curtail electricity usage. Today, there are a number of such devices readily available. One example is Sense: The Home Energy Monitor, which provides a readout of your home electricity use on your smartphone.

Sense and other such electricity monitoring devices provide an instantaneous readout of electricity use in your home. Suppose you are getting ready to head out for the evening. The refrigerator is plugged in, but is quiet, meaning that the compressor is not running at that exact moment. Before you turn out the last light and leave, you glance at Sense and see that your house is drawing 500 watts. Wait a minute, that's not right! Then you remember that you left your Playstation game console and 42-inch flat screen TV on in the other room. You run over and turn them both off. Sense now reads that only 45 watts are being used in your home. But if everything is turned off, why is your house using 45 watts? It's probably due to the phantom load—unnecessary standby electricity—drawn by battery chargers, instant-on features on televisions, computers in sleep mode, and other electrical devices that are on even when you think they are off. Because they are now more aware of phantom load, some homeowners have installed power strips that allow them to truly turn off an appliance. As a result they have seen their phantom loads drop.

An installed home energy monitor. When Sense is installed to monitor home electricity use, people become more aware of phantom loads. *(Sense.com)*

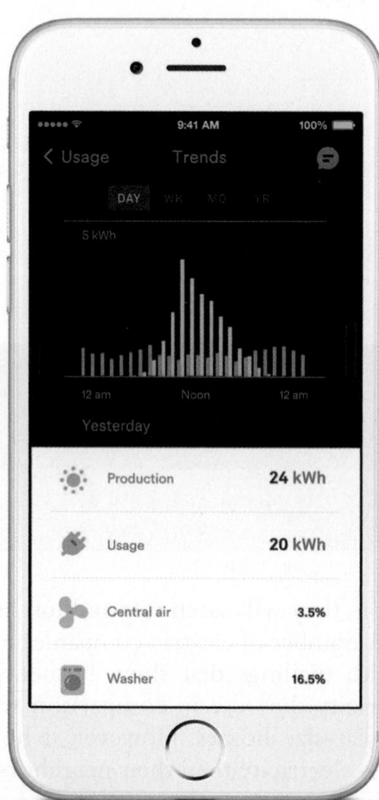

Home energy monitor. Sense allows a user to instantaneously monitor electricity use in a home. *(Sense.com)*

Since we know that all electricity use has environmental implications, a reduction in electricity use by any means is beneficial. The reductions that come from simple changes in behavior are some of the easiest to achieve.

Critical Thinking Questions

1. In what way is a home energy meter important in energy awareness?
2. What is a phantom load and what are some major contributors to phantom load in a home?

References

Laskey, A. 2013. *How behavioral science can lower your energy bill.* TED. https://www.ted.com/talks/alex_laskey_how_behavioral_science_can_lower_your_energy_bill

Tutar, T. 2011. *Time to Fight the Rising Plug Load Monster.* Rocky Mountain Institute. https://www.rmi.org/news/blog_time_to_fight_the_rising_plug_load_monster/

Chapter 12 Review

In this chapter, we have examined the nonrenewable fossil fuels that are used around the world and in the United States. We observed that energy use varies widely in different countries. Fossil fuel resources are used for different purposes and electricity draws 40 percent of U.S. energy demand. In the United States, oil, natural gas, and coal, in that order, meet 81 percent of the nation's energy requirements. Each fossil fuel has its own advantages and disadvantages. Nuclear fuels, which generate 20 percent of the electricity in the United States, have their own set of advantages and disadvantages. One reasonable conclusion is that all energy choices have adverse consequences.

Key Terms

Fossil fuel
Nonrenewable energy resource
Nuclear fuel
Commercial energy source
Subsistence energy source
Energy carrier
Turbine
Electrical grid
Combined cycle
Capacity
Capacity factor
Cogeneration
Combined heat and power
Coal
Petroleum
Crude oil
Oil sands
Bitumen
CTL (coal to liquid)
Energy intensity
Hubbert curve
Peak oil
Fission
Fuel rod
Control rod
Radioactive waste
Becquerel (Bq)
Curie
Nuclear fusion

Learning Goals Revisited

Module 34 Patterns of Energy Use

Describe the use of nonrenewable energy in the world and in the United States.

Energy use has changed over time along with the level of industrial development in a country. The United States and the rest of the developed world have moved from a heavy reliance on wood and coal to other fossil fuels and nuclear energy. The developing world still relies largely on wood, charcoal, and animal waste.

Explain why different forms of energy are best suited for certain purposes.

Each source of energy is best suited for certain activities, and less well suited for others. Energy efficiency is an important consideration in determining the environmental impacts of energy use. In general, the energy source that entails the fewest conversions from its original form to its end use is likely to be the most efficient.

Understand the primary ways that electricity is generated in the United States.

Electricity generation plants convert the chemical energy of fuel into electricity. Natural gas, coal, and nuclear fuels are the energy sources most commonly used for generating electricity. The electrical grid is a network of interconnected transmission lines that tie power plants together and link them with end users of electricity.

Module 35 Fossil Fuel Resources

Discuss the uses of coal and its consequences.

Coal is an energy-dense fossil fuel that is a major energy source in the generation of electricity. Coal combustion, however, is a major source of air pollution and greenhouse gas emissions.

Discuss the uses of petroleum and its consequences.

Petroleum includes both crude oil and natural gas. The United States uses more petroleum than any other fuel, primarily for transportation. Petroleum produces air pollution as well as greenhouse gas emissions. Oil spills are a major hazard to organisms and habitat.

Discuss the uses of natural gas and its consequences.

Natural gas is a relatively clean fossil fuel. The presence of natural gas pipelines makes natural gas a convenient fuel for electricity generation, home heating, and manufacturing processes as well as fertilizer production. Hydraulic fracturing has greatly increased the availability of natural gas in the United States.

Discuss the uses of oil sands and liquefied coal and their consequences.

Oil sands and liquefied coals have the potential to become more important in the U.S. energy portfolio but both require relatively large energy inputs to be obtained and processed into a usable form.

Describe future prospects for fossil fuel use.

Most environmental scientists believe that oil production will begin to decline sometime in the next few decades. The transition away from oil will have important environmental consequences, depending upon how quickly it occurs and whether we move to renewable energy resources or alternative fossil fuels.

Module 36 Nuclear Energy Resources

Describe how nuclear energy is used to generate electricity.

In nuclear fission, a neutron strikes a relatively large atom such as uranium, and two or more smaller parts split off, releasing a great deal of energy. This energy is used to convert water into steam, which turns a turbine, which then turns a generator. A small amount of nuclear fuel can release a great deal of energy and generate a large quantity of electricity.

Discuss the advantages and disadvantages of using nuclear fuels to generate electricity.

All forms of energy have advantages and disadvantages. Considering only emissions during electricity generation, nuclear energy is relatively clean. However, the construction of nuclear power plants and the mining of uranium both use fossil fuels. The major environmental hazards of electricity generation from nuclear energy are the potential for accidents during plant operation and the challenges of radioactive waste disposal.

Practice Math and Graphing

Preparing for the AP® Exam

Answer the following questions. Be sure to show all your work.

1. Practice Math

Coal contains a number of impurities, including sulfur, that are released into the atmosphere when the coal is burned. The sulfur content of coal typically ranges from 0.4 to 4 percent by weight. Assume that a power plant burns one metric ton (1,000 kg) of coal with a 1 percent sulfur content to generate electricity. A ton of coal contains the energy equivalent of 7,700 kWh of electricity.

(a) Assuming typical efficiency, how much electricity will be generated?

(b) How much sulfur will be released?

2. Practice Graphing

The daily maximum 1-hour average concentration of sulfur dioxide (measured in parts per billion or ppb) in the United States has been decreasing over the past few decades.

(a) Use the data in the table to graph sulfur dioxide concentrations from 2000 to 2016.

(b) State the percentage decrease in 2016 relative to the 2000 concentration.

Year	SO_2 concentration
2000	79.5
2001	79.7
2002	71.4
2003	73.2
2004	69.9
2005	70.5
2006	66.8
2007	63.7
2008	57.6
2009	49.1
2010	45.4
2011	38.3
2012	36.8
2013	30.7
2014	29.7
2015	25.7
2016	22.4

Section 1: Multiple-Choice Questions

Choose the best answer for questions 1–17.

1. Which is NOT a nonrenewable energy resource?
(a) coal
(b) natural gas
(c) wind
(d) nuclear fuels

2. The fact that global transfer of energy from fuels to electricity is about 35 percent efficient is mostly a consequence of
(a) the Hubbert curve.
(b) the law of conservation of matter.
(c) the first law of thermodynamics.
(d) the second law of thermodynamics.

3. Which is the most fuel-efficient mode of transportation in terms of joules per passenger-kilometer?
(a) train
(b) bus
(c) airplane
(d) car with one passenger

4. Which is NOT associated with the surface extraction of coal?
(a) low death rates among miners
(b) relatively high economic costs
(c) large piles of mine tailings
(d) underground tunnels and shafts

5. Which statement regarding petroleum is CORRECT?
I. It is formed from the decay of woody plants.
II. It contains natural gas as well as oil.
III. It migrates through pore spaces in rocks.
(a) I, II, and III
(b) I and III
(c) I and II
(d) II and III

6. Which is a disadvantage of natural gas?
(a) high sulfur emissions
(b) groundwater contamination
(c) high carbon dioxide emissions
(d) significant waste disposal

7. Nuclear power plants produce electricity using energy from the radioactive decay of
(a) uranium-235.
(b) uranium-238.
(c) plutonium-235.
(d) plutonium-238.

8. Currently, most high-level radioactive waste from nuclear reactors in the United States is
(a) stored in deep ocean trenches.
(b) buried in Yucca Mountain.
(c) reprocessed into new fuel pellets.
(d) stored at the power plant that produced it.

9. A radioactive isotope has a half-life of 40 years and a radioactivity level of 4 curies. How many years will it take for the radioactivity level to become 0.25 curies?
(a) 80
(b) 120
(c) 160
(d) 200

10. Which energy source is responsible for the largest fraction of electricity generation in the United States?
(a) natural gas
(b) coal
(c) uranium
(d) oil

11. In 1969, M. King Hubbert published a graph known as the Hubbert curve. This graph shows
(a) the amount of nuclear fuel available in North America.
(b) the amount of nuclear fuel available in the world.
(c) the point at which world oil production will reach a maximum and the point at which we will run out of oil.
(d) the point at which world oil production will increase.

12. Which is involved in generating electricity through nuclear fission?
I. a positive feedback loop
II. creation of chemical energy
III. release of kinetic energy
(a) I only
(b) II only
(c) III only
(d) I and III

13. Calculate the energy efficiency of converting coal into heated water if the efficiency of turning coal into electricity is 35 percent, the efficiency of transporting the electricity is 90 percent, and the efficiency of hot water heaters is 90 percent.
(a) 10 percent
(b) 17 percent
(c) 28 percent
(d) 32 percent

Question 14 refers to the following table.

Fuel source	EROEI
Liquid petroleum	20
Ethanol from corn	2
Ethanol from sugarcane	5
Natural gas	10

14. All of the fuel sources listed in the table can be used in personal vehicles. Assuming that each vehicle achieves approximately the same MJ per passenger-mile, which type of fuel is likely to contribute most to overall operation efficiency?
(a) liquid petroleum
(b) ethanol from corn
(c) ethanol from sugarcane
(d) natural gas

15. Which of is NOT one of the top seven petroleum-producing countries?
(a) United States
(b) Canada
(c) Australia
(d) Iran

Question 16 refers to the following table.

Mode	MJ per passenger-kilometer
Air	2.1
Passenger car (driver alone)	3.6
Motorcycle	1.1
Train (Amtrak)	1.1
Bus	1.7

16. Nicole wants to take a trip to visit one of her friends. She can fly 1,200 kilometers to visit Aaria, drive 500 kilometers to visit Sofia, take a bus 1,000 kilometers to visit Callum or take a train 1,500 kilometers to visit Saaid. Use the table to determine which trip will result in the smallest amount of energy expended.
(a) Aaria
(b) Sofia
(c) Callum
(d) Saaid

17. In recent decades in the United States, _______ has/have decreased, while _______ has/have leveled off but not decreased.
(a) fracking; nuclear energy
(b) coal usage; natural gas
(c) energy intensity; overall energy use
(d) subsistence energy sources; commercial energy sources

Section 2: Free-Response Questions

Write your answer to each part clearly. Support your answers with relevant information and examples. Where calculations are required, show your work.

1. Many college students have a mini fridge in their dorm room. A standard mini fridge costs roughly \$100, uses about 100 watts of electricity, and can be expected to last for 5 years. The refrigerator is plugged into an electrical socket 24 hours a day, but it demands electricity for about 12 hours a day. Assume that electricity costs \$0.10/kWh.
(a) Calculate the lifetime monetary cost of owning and operating the refrigerator. (2 points)
(b) Assume that the electricity used to power the refrigerator comes from a coal-burning power plant. One metric ton of coal contains 29.3 GJ (8,140 kWh) of energy. Because of the inefficiency of electricity generation and transmission, only one-third of the energy in coal reaches the refrigerator. How many tons of coal are used to power the refrigerator during its lifetime? (2 points)
(c) Assume that 15 percent of the mass of the coal burned in the power plant ends up as coal ash, a potentially toxic mixture that contains mercury and arsenic. How many tons of coal ash are produced as a result of the refrigerator's electricity use over its lifetime? (2 points)
(d) What externalities does your answer from part (a) not include? Describe one social and one environmental cost associated with using this appliance. (2 points)
(e) Describe two ways a college student could reduce the electricity use associated with having a mini fridge in his or her dorm room. (2 points)

2. A number of U.S. electric companies have filed applications with the Nuclear Regulatory Commission for permits to build new nuclear power plants to meet future electricity demands.
   (a) Explain the process by which electricity is generated by a nuclear power plant. (2 points)
   (b) Describe the two nuclear accidents that occurred in 1979 and 1986, respectively, that led to widespread concern about the safety of nuclear power plants. (2 points)
   (c) Discuss the environmental benefits of generating electricity from nuclear energy rather than coal. (2 points)
   (d) Describe the three types of radioactive waste produced by nuclear power plants and explain the threats they pose to humans. (2 points)
   (e) Discuss the problems associated with the disposal of radioactive waste and outline the U.S. Department of Energy's proposal for its long-term storage. (2 points)

3. While much of the discussion about climate change focuses on the use of oil, natural gas and coal are also commonly used energy fuels.
   (a) Name TWO disadvantages of natural gas and of coal. (4 points)
   (b) Explain how electricity is generated, and describe how cogeneration can allow a power plant to increase efficiency. (3 points)
   (c) Electric water heaters are 99 percent efficient in terms of the conversion of electricity to hot water, while a natural gas water heater is about 60 percent efficient. However, electric water heaters are not always more efficient overall. What other factors might be considered when looking at the total system efficiency of each water heater? (3 points)

William Kamkwamba's cousin climbs one of the wind turbines that William built. *(Lucas Oleniuk/Getty Images)*

Achieving Energy Sustainability

CASE STUDY

Energy from the Wind

In a small village in the African nation of Malawi, a 14-year-old boy named William Kamkwamba and his family did not have enough to eat because of a famine. His family could not afford the required school tax; in many parts of Africa a child whose parents cannot pay the school tax cannot attend school. So instead of attending school, he spent his days in a public library funded by the U.S. government, trying to teach himself. In the library, he studied one book over and over—a textbook titled *Using Energy*. The cover of the book featured a series of windmills. Although William had never seen a windmill, within months he was building his own from abandoned bicycles and old parts he found in scrap heaps. William used the fundamentals of physics he learned from the book and his inherent skills at tinkering and fixing things. He did not have any teachers or mentors but he did rely on assistance from some of his friends. He worked hard and made many attempts to construct something that in his world was seemingly impossible. At first his neighbors thought he was mentally disturbed or was practicing magic. But when he was able to illuminate a small light bulb at the top of what they had called his "junk" tower, people rushed from great distances to see it, and he became a local hero. William had generated electricity without any conventional fuel and far from the nearest power plant. Because there were no visible inputs like fuel and no waste piles or pollution outputs, in many ways it did seem like magic. William used the electricity he generated from wind to light his house and charge cell phones, and eventually to irrigate his family's crops.

> Although William had never seen a windmill, within months he was building his own from abandoned bicycles and old parts he found in scrap heaps.

The use of windmills, also known as wind turbines, to generate electricity is growing in both the developing and developed worlds. The mechanics of how to build a windmill are widely discussed in online sources including YouTube, Wikipedia, and "how to build it" instructional videos. However William had only one book and no access to the Internet. A few years after he built his first windmill, William exclaimed to Jon Stewart on *The Daily Show*, "Where was this Internet when I needed it?" William recently graduated from Dartmouth College, where he majored in environmental studies. He is coauthor of the book *The Boy Who Harnessed the Wind*, which has sold thousands of copies and has been adopted as summer reading in high schools and colleges around the United States and elsewhere in the world. The book was also the basis of a BBC feature film that was released in 2018. Today, William is a designer/coordinator at Widernet, a nonprofit organization in Chapel Hill, North Carolina, that strives to improve digital communications around the world, particularly in developing countries.

Sources: W. Kamkwamba and B. Mealer, *The Boy Who Harnessed the Wind* (Harper, 2009); R. Wolfson, *Energy, Environment and Climate*, 3rd ed. (Norton, 2017).

Throughout this book we have discussed sustainability as the foundation of the environmental health of our planet. Sustainability is particularly important to consider with respect to energy because energy is a resource that humans cannot live without and energy use often has many consequences for the environment. Currently, only a very small fraction of the energy we use, particularly in the developed world, comes from renewable resources. While expanding renewable energy resources is an important step, achieving energy sustainability will require us to rely as much, or more, on reducing the amount of energy that we use.

In Chapter 12 we discussed the finite nature of traditional energy resources such as fossil fuels and the environmental consequences of their use. In this chapter we outline the components of a sustainable energy strategy, beginning with ways to reduce our use of energy through conservation and increased efficiency. We define renewable forms of energy and discuss an important carbon-based energy resource, biomass, as well as energy that is obtained from flowing water. We then describe innovations in obtaining energy from non-carbon-based resources such as the Sun, wind, internal heat from Earth, and hydrogen. We conclude the chapter with a discussion of our energy future.

MODULE 37

Conservation, Efficiency, and Renewable Energy

In any discussion of energy use—whether renewable or nonrenewable—energy conservation and increased energy efficiency rank among the most crucial factors to consider. We begin our discussion of renewable energy with a look at conservation and efficiency. We will then explore the range of renewable energy resources that are available.

Learning Goals

After reading this module, you should be able to

- describe strategies to conserve energy and increase energy efficiency.
- explain differences among the various renewable energy resources.

We can use less energy through conservation and increased efficiency

A truly sustainable approach to energy use must incorporate both *energy conservation* and energy efficiency. Conservation and efficiency efforts save energy that can then be used later, just as you might save money in a bank account to use later when the need arises. In this sense, conservation and efficiency are sustainable energy "sources."

Energy conservation and energy efficiency are the least expensive and most environmentally sound options for maximizing our energy resources. In many cases, they are also the easiest approaches to implement because they often require fairly simple changes to existing systems rather than a switch to a completely new technology. In this section we will examine ways to achieve both objectives.

Energy Conservation and Efficiency

Energy conservation means finding and implementing ways to use less energy. As we saw in Chapters 2 and 12, increasing energy efficiency means obtaining the same work from a smaller amount of energy. Energy conservation and energy efficiency are closely linked. One can conserve energy by not using an electrical appliance; doing so results in less energy consumption. But one can also conserve energy by using a more efficient appliance—one that does the same work but uses less energy.

Conservation

TABLE 37.1 on page 454 lists some of the ways that an individual might conserve energy, including lowering the household thermostat during cold months, consolidating errands in order to drive fewer miles, or turning off a computer when it is not being used. On a larger scale, a government might implement energy conservation measures that encourage or even require individuals to adopt strategies or habits that use less energy. One such top-down approach is to improve the availability of public transportation (**FIGURE 37.1** on page 454). Governments can also facilitate energy conservation by taxing electricity, oil, and natural gas, since higher taxes discourage use. Alternatively, governments might offer rebates or tax credits for retrofitting a

Energy conservation Finding and implementing ways to use less energy.

TABLE 37.1 Reducing energy use

Home	Transportation	Electrical and electronic devices
• Weatherize; insulate, seal gaps • Turn thermostat down in winter, up in summer • Reduce use of hot water; do laundry in cold water; take shorter showers	• Walk or ride a bike • Take public transportation • Carpool • Consolidate trips	• Buy Energy Star devices and appliances • Unplug when possible or use a power strip • Use a laptop rather than a desktop computer

home or business so it will operate on less energy. Some electric companies bill customers with a **tiered rate system** in which customers pay a low rate for the first increment of electricity they use and pay higher rates as use goes up. All of these practices encourage people to reduce the amount of electricity they use.

As we saw in Chapter 12, the demand for energy varies with time of day, season, and weather. When electricity-generating plants are unable to handle the demand during high-use periods, brownouts or blackouts may occur. To avoid this problem, electric companies must be able to provide enough energy to satisfy **peak demand**, the greatest quantity of energy used at any one time. Peak demand may be several times the overall average demand, which means that substantially more energy must be available than is needed under average conditions. To meet peak demand for electricity, electric companies often keep backup generators of electricity available—typically fossil fuel–fired generators, and in some cases, batteries.

Therefore, an important aspect of energy conservation is the reduction of peak demand, which would make it less likely that electric companies will have to build excess generating capacity that is used only sporadically. One way of reducing peak demand is to establish a variable price structure under which customers pay less to use electricity when demand is lowest (typically in the middle of the night and on weekends) and more when demand is highest. This approach helps even out the use of electricity, which both reduces the burden on the generating capacity of the utility and rewards the electric consumer at the same time.

The second law of thermodynamics tells us that whenever energy is converted from one form into another, some energy is lost as unusable heat. In a typical thermal fossil fuel or nuclear power plant, only about one-third of the energy consumed goes to its intended purpose; the rest is lost during energy conversions. We need to consider these losses in order to fully account for all energy conservation savings. So, the amount of energy we save is the sum of both the energy we did not use together with the energy that would have been lost in converting that energy into the form in which we would have used it. For example, if we can reduce our electricity use by 100 kWh, we may actually be conserving 300 kWh of an energy resource such as coal, since we save both the 100 kWh that we decide not to use and the 200 kWh that would have been lost during the conversion process to make the 100 kWh available to us.

Tiered rate system A billing system used by some electric companies in which customers pay higher rates as their use goes up.

Peak demand The greatest quantity of energy used at any one time.

Efficiency

Modern changes in electric lighting are a good example of how steadily increasing energy efficiency results in overall energy conservation. Compact fluorescent light bulbs use one-fourth as much energy to provide the same amount of light as incandescent bulbs. LED (light-emitting diode) light bulbs are even more efficient; they use one-sixth as much energy as incandescent bulbs. Over time, the widespread adoption of these efficient bulbs has resulted in substantially less energy used to provide the same amount of lighting.

Another way in which consumers can increase energy efficiency is by switching to products that meet the efficiency standards of the Energy Star program set by the

FIGURE 37.1 Reducing energy use. There are many ways individuals can reduce their energy use in and outside the home. Taking public transportation rather than traveling by personal vehicle is one method of energy conservation. These students are taking the Washington, D.C., Metro. *(Chicago Tribune/Tribune News Service/Getty Images)*

DO THE MATH Energy Star

Preparing for the AP® Exam

Thinking clearly about energy efficiency and energy conservation can save you a lot of money in some surprising ways. If you are saving money on your electric bill, you are also saving energy and reducing the emission of pollutants. Consider the purchase of an air conditioner. Suppose you have a choice: an Energy Star unit for \$300 or a standard unit for \$200. The two units have the same cooling capacity but the Energy Star unit costs 5 cents per hour less to run. If you buy the Energy Star unit and run it 12 hours per day for 6 months of the year, how long does it take to recover the \$100 extra cost?

You would save

$$\$0.05/\text{hour} \times 12\ \text{hours/day} = \$0.60/\text{day}$$

Six months is about 180 days, so in the first year you would save

$$\$0.60/\text{day} \times 180\ \text{days} = \$108$$

Spending the extra \$100 for the Energy Star unit actually saves you \$8 in just 1 year of use. In 3 years of use, the savings will more than pay for the entire initial cost of the unit (3 × \$108 = \$324), and after that you pay only for the operating costs.

YOUR TURN You are about to invest in a 66-inch flat screen TV. These TVs come in both Energy Star and non–Energy Star models. The cost of electricity is \$0.15 per kilowatt-hour, and you expect to watch TV an average of 4 hours per day.

1. The non–Energy Star model uses 0.5 kW (half a kilowatt). How much will it cost you per year for electricity to run this model?
2. If the Energy Star model uses only 40 percent of the amount of electricity used by the non–Energy Star model, how much money would you save on your electric bill over 5 years by buying the efficient model?

U.S. Environmental Protection Agency. For example, an Energy Star air conditioner may use 0.2 kWh (200 watt-hours) less electricity per hour than a non–Energy Star unit. In terms of cost, a single consumer may save only 2 to 5 cents per hour by switching to an Energy Star unit. However, if 100,000 households in a city switched to Energy Star air conditioners, the city would reduce its energy use by 20 MW, or 4 percent of the output of a typical power plant. "Do the Math: Energy Star" shows you how to calculate Energy Star savings.

Sustainable Design

Sustainable design can improve the efficiency of the buildings and communities in which we live and work. **FIGURE 37.2** on page 456 shows some key features of sustainable design applied to a single-family dwelling. Insulating foundation walls and basement floors, orienting a house properly in relation to the Sun, and planting shade trees in warm climates are all appropriate design features. As we saw in Chapter 10, good community planning also conserves energy. Building houses close to where residents work reduces reliance on fossil fuels used for transportation, which in turn reduces the amount of pollution and carbon dioxide released into the atmosphere.

Buildings consume a great deal of energy for cooling, heating, and lighting. Many sustainable building strategies rely on **passive solar design**, a construction technique designed to take advantage of solar radiation without the use of active technology. **FIGURE 37.3** on page 456 illustrates key features of passive solar design. Passive solar design stabilizes indoor temperatures without the need for pumps or other mechanical devices. For example, in the Northern Hemisphere, constructing a house with windows along a south-facing wall allows the Sun's rays to penetrate and warm the house, especially in winter when the Sun is more prominent in the southern sky. Double-paned windows insulate while still allowing incoming solar radiation to warm the house. Carefully placed windows also allow the maximum amount of light into a building and reduce the need for artificial lighting. Dark materials on the roof or exterior walls of a building absorb more solar energy than light-colored materials, further warming the structure. Conversely, using light-colored materials on a roof reflects heat away from the building, which keeps it cooler. In summer, when the Sun is high in the sky for much of the day, an overhanging roof helps block out sunlight during the hottest period, which makes the indoor temperature cooler and reduces the need for ventilation fans or air conditioning.

Passive solar design Construction designed to take advantage of solar radiation without active technology.

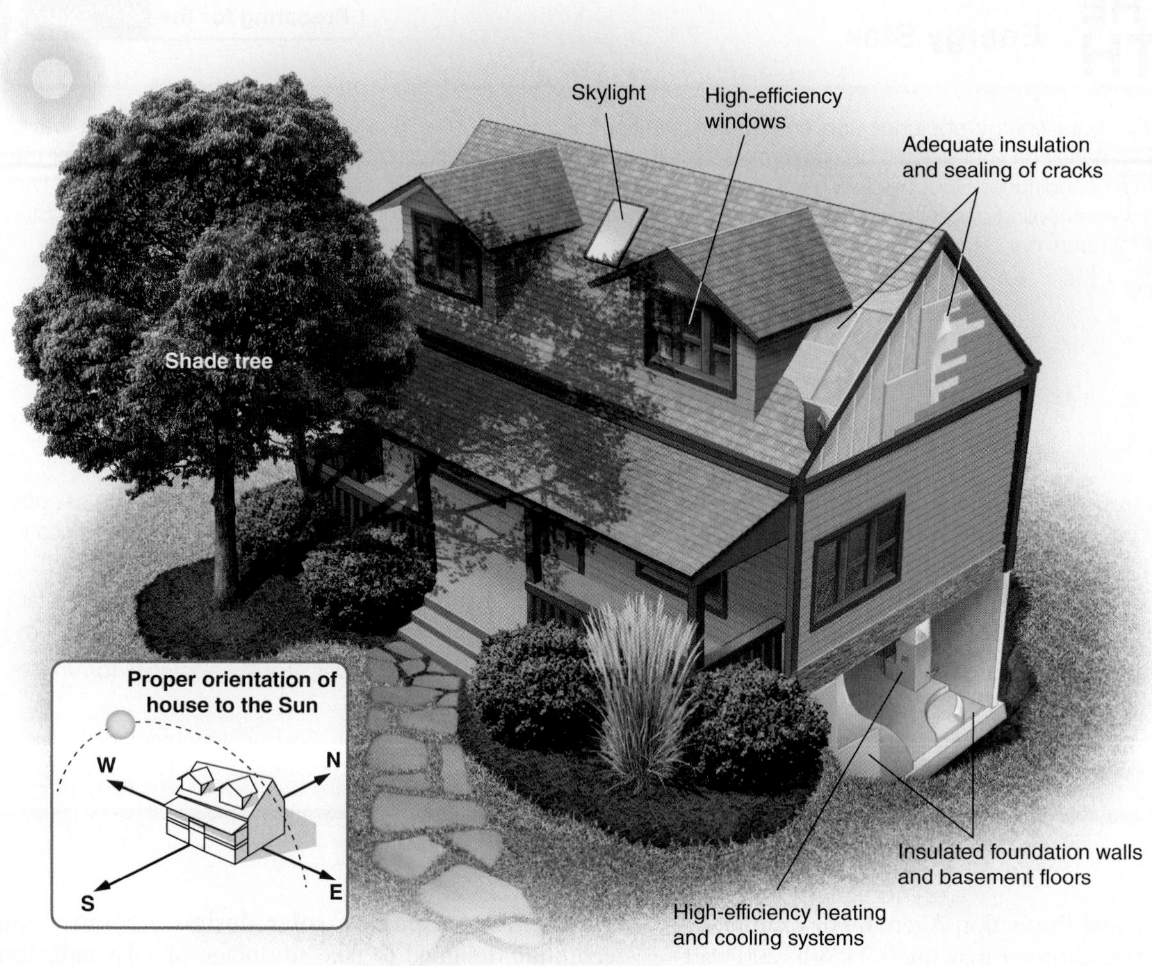

FIGURE 37.2 An energy-efficient home. A sustainable building design incorporates proper solar orientation and landscaping as well as insulated windows, walls, and floors. In the Northern Hemisphere, a southern exposure allows the house to receive more direct rays from the Sun in winter when the path of the Sun is in the southern sky.

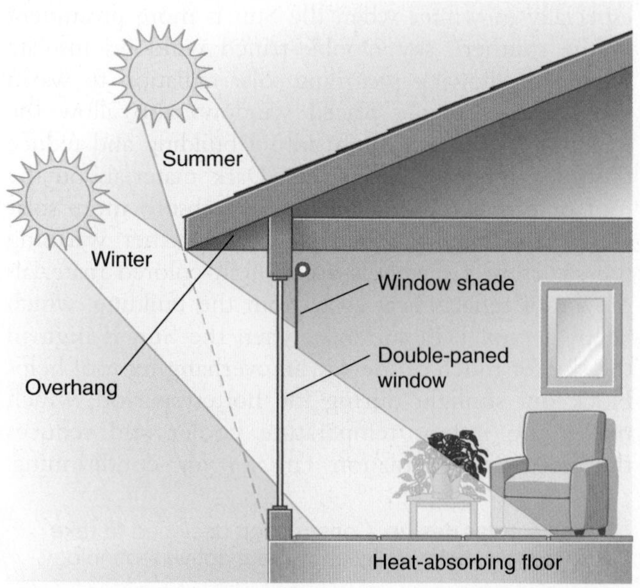

FIGURE 37.3 Passive solar design. Passive solar design uses solar radiation to maintain indoor temperature. Roof overhangs make use of seasonal changes in the Sun's position to reduce energy demand for heating and cooling. In winter, when the Sun is low in the sky, it shines directly into the window and heats the house. In summer, when the Sun is higher in the sky, the overhang blocks incoming sunlight and the room stays cool. High-efficiency windows and building materials with high thermal inertia are also components of passive solar design.

FIGURE 37.4 The California Academy of Sciences. The sustainable design of this San Francisco research institution maximizes the use of natural light and ventilation. The building generates much of its own electricity with solar panels on its roof and captures water in its rooftop garden. *(Nancy Hoyt Belcher/Alamy)*

Window shades can also reduce solar energy entering the house.

To reduce demand for heating at night and for cooling during the day, builders can use construction materials that have high *thermal mass.* **Thermal mass** is a property of a building material that allows it to retain heat or cold. Materials with high thermal mass stay hot once they have been heated and cool once they have been cooled. Stone and concrete have high thermal mass, whereas wood and glass do not; think of how a cement sidewalk stays warm longer than a wooden boardwalk after a hot day. A south-facing room with stone walls and a stone floor will heat up on sunny winter days and retain that heat long after the Sun has set.

Although building a house into the side of a hill or roofing a building with soil and plants are less-common approaches, these measures also provide insulation and reduce the need for both heating and cooling. While "green roofs"—roofs with soil and growing plants—are somewhat unusual in the United States, many European cities, such as Berlin, have them on new or rebuilt structures. They are especially common on high-rise buildings in downtown areas that have little natural plant cover. These green roofs cool and shade the buildings and the surrounding environment. And the addition of plants to an urban environment also improves overall air quality.

The use of recycled building materials is another method of energy conservation. Recycling reduces the need for new construction materials, which reduces the amount of energy required to produce the components of the building. For example, many buildings now use recycled denim insulation in the walls and ceilings, and fly ash (a byproduct recovered from coal-fired power plants) in the foundation.

Homes constructed today may incorporate some or all of these sustainable design strategies, but it is possible to achieve energy efficiency even in very large buildings. The building that houses the California Academy of Sciences in Golden Gate Park in San Francisco is a showcase for several of these sustainable design techniques (**FIGURE 37.4**). This structure, which incorporates a combination of passive solar design, radiant heating, solar panels, and skylights, actually uses 30 percent

Thermal mass A property of a building material that allows it to maintain heat or cold.

less energy than the amount permitted under national building energy requirements. Natural light fills 90 percent of the office space and many of the public areas. Windows, blinds, and skylights open as needed to allow air to circulate, capturing the ocean breezes and ventilating the building. Recycled denim insulation in the walls and a soil-covered rooftop garden provide insulation that reduces heating and cooling costs. As an added benefit, the living green roof grows native plants and captures 13.6 million liters (3.6 million gallons) of rainwater per year, which is then used to recharge groundwater stores.

In addition to these passive techniques, the designers of this building incorporated active technologies that further reduce its use of energy. An efficient radiant heating system carries warm water through tubes embedded in the concrete floor, using a fraction of the energy required by a standard forced-air heating system. To produce some of the electricity used in the building directly, the designers added 60,000 photovoltaic solar cells to the roof. These solar panels convert energy from the Sun into 213,000 kWh of electricity per year and reduce greenhouse gas emissions by about 200 metric tons per year.

The California Academy of Sciences took an innovative approach to meeting its energy needs through a combination of energy efficiency and use of renewable energy resources. However, many, if not all, of these approaches will have to become commonplace if we are going to use energy in a sustainable way.

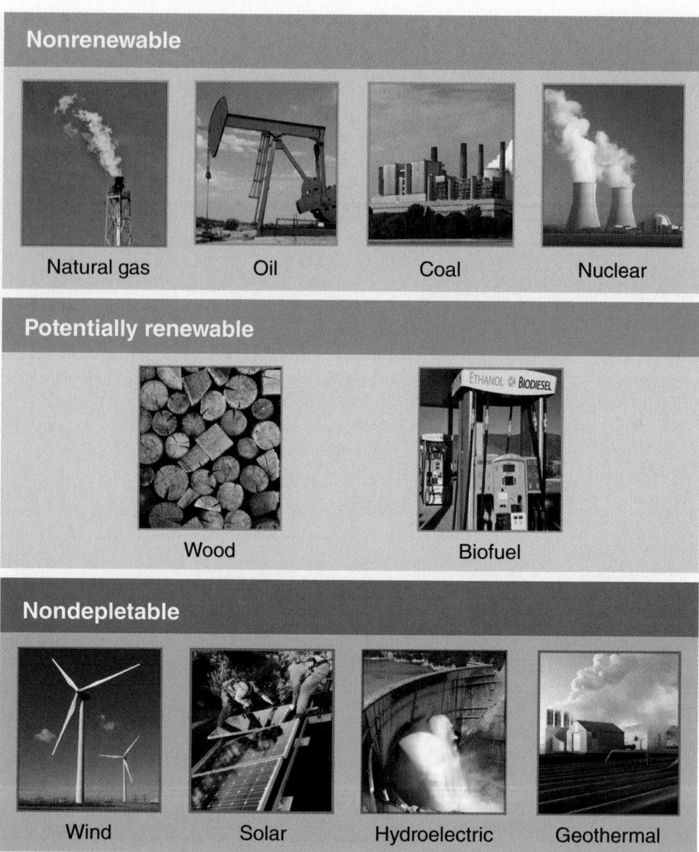

FIGURE 37.5 Renewable and nonrenewable energy resources. Fossil fuels and nuclear fuels are nonrenewable energy resources. Renewable energy resources include potentially renewable energy sources such as biomass, which is renewable as long as humans do not use it faster than it can be replenished, and nondepletable energy sources, such as solar radiation and wind. *(Ian Hamilton/iStockphoto.com, babyblueut/Getty Images, MichaelUtech/Getty Images, RelaxFoto.de/Getty Images, BerndLang/GettyImages, IngaSpence/ScienceSource, acilo/GettyImages, BlendImages-DonMason/Getty Images, Kyodo News/Getty Images, Rhoberazzi/Getty Images)*

Renewable energy is either potentially renewable or nondepletable

As fossil fuels become less available and more expensive, what will take their place? Probably it will be a mix of energy efficiency strategies, energy conservation, and new energy sources. In the rest of this chapter we will explore our renewable energy options.

In Chapter 12 we learned that conventional energy resources, such as petroleum, natural gas, coal, and uranium ore, are nonrenewable. From a systems analysis perspective, fossil fuels constitute an energy reservoir we are depleting much faster than it can ever be replenished. Similarly, we have a finite amount of uranium ore available to use as fuel in nuclear reactors.

In contrast, some other sources of energy can be regenerated rapidly. Biomass energy resources are **potentially renewable** because those resources can be regenerated indefinitely as long as we do not consume them more quickly than they can be replenished. There are still other energy resources that cannot be depleted no matter how much we use them. Solar, wind, geothermal, hydroelectric, and tidal energy are essentially **nondepletable** in the span of human time; no matter how much we use there will always be more. The amount of a nondepletable resource available tomorrow does not depend on how much we use today. In this book we refer to potentially renewable and nondepletable energy resources together as **renewable** energy resources. **FIGURE 37.5** illustrates the categories of energy resources.

Potentially renewable An energy source that can be regenerated indefinitely as long as it is not overharvested.

Nondepletable An energy source that cannot be used up.

Renewable In energy management, an energy source that is either potentially renewable or nondepletable.

Many renewable energy resources have been used by humans for thousands of years. In fact, before humans began using fossil fuels, the only available energy sources were wood and plants, animal manure, and fish or animal oils. Today, in parts of the developing world where there is little access to fossil fuels, people still rely on local biomass energy sources such as manure and wood for cooking and heating—sometimes to such an extent that they overuse the resource. For example, according to the International Energy Agency, biomass is currently the source of 65 percent of the energy consumed in sub-Saharan Africa (excluding South Africa) and much of it is not harvested sustainably.

As **FIGURE 37.6** shows, renewable energy resources account for approximately 14 percent of the energy used worldwide, most of which is in the form of biomass. In the United States, which depends heavily on fossil fuels, renewable energy resources provide only about 10 percent of the energy used. That 10 percent, shown in detail in **FIGURE 37.7**, comes primarily from biomass, hydroelectricity, and wind.

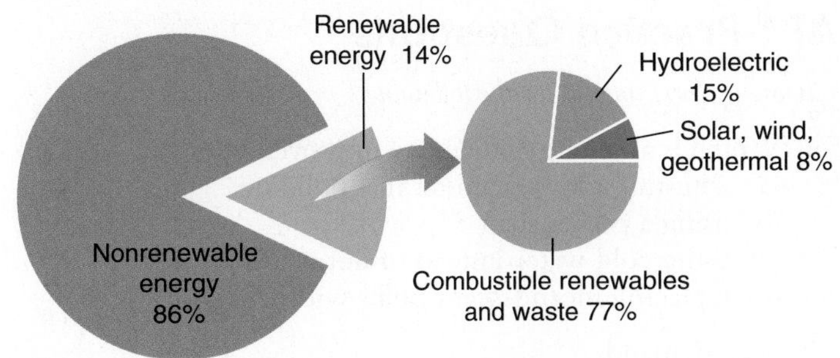

FIGURE 37.6 Global energy use. Renewable energy resources provide about 14 percent of energy worldwide. *(Data from International Energy Agency for the 2015 year, Key World Energy Statistics, 2017.)*

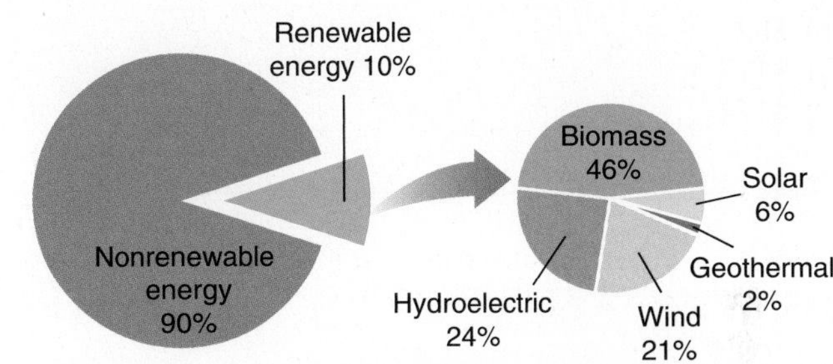

FIGURE 37.7 Energy use in the United States. Roughly 10 percent of the energy used in the United States comes from renewable energy resources. Note that the sum of components does not equal 100 percent due to rounding. *(Data from www.eia.doe.gov.)*

MODULE 37 AP® Review

Preparing for the AP® Exam

In this module, we have seen that to achieve energy sustainability we should begin with conserving energy and increasing energy efficiency. Energy conservation refers to finding ways to use less energy. Increasing energy efficiency means achieving the same amount of work from a smaller quantity of energy. Sustainable design of buildings and communities can decrease energy use. We also looked at different categories of renewable energy resources. Potentially renewable energy sources can regenerate if we do not consume them more quickly than they can be replaced. Nondepletable energy resources such as the wind cannot be depleted no matter how much we use. In the next module, we examine two important renewable resources, biomass and water that is used for generating electricity.

AP® Practice Questions

Choose the best answer for the following.

1. Which is an energy efficiency improvement?
   (a) adjusting the thermostat in a building
   (b) using a power strip
   (c) using cold water instead of hot water
   (d) replacing incandescent bulbs with LEDs

2. Peak demand
   (a) decreases electricity-generating capacity.
   (b) is caused by higher electricity prices.
   (c) occurs equally in all seasons.
   (d) can be managed using a variable price structure.

3. If an Energy Star refrigerator costs 2 cents less per hour to run, and if it runs for 16 hours a day, how much will it save in a year?
   (a) \$44
   (b) \$82
   (c) \$117
   (d) \$174

4. Which is LEAST likely to be used in sustainable design in North America?
   (a) recycled building materials
   (b) an overhanging roof
   (c) windows on the northern wall
   (d) building materials with high thermal mass

5. Which energy source is NOT nondepletable?
   (a) wind
   (b) biomass
   (c) tidal
   (d) solar

MODULE 38

Biomass and Water

As we discussed in Chapter 12, the Sun is the ultimate source of fossil fuels. Fossil fuels are created from dead plants and animals that are buried deep in sediments and that are slowly transformed into petroleum or coal. Most types of renewable energy are also derived from the Sun and cycles driven by the Sun, including solar, wind, and hydroelectric energy as well as plant biomass such as wood. In this module, we present two important renewable energy sources: biomass and water.

Learning Goals

After reading this module, you should be able to

- describe the various forms of biomass.
- explain how energy is harnessed from water.

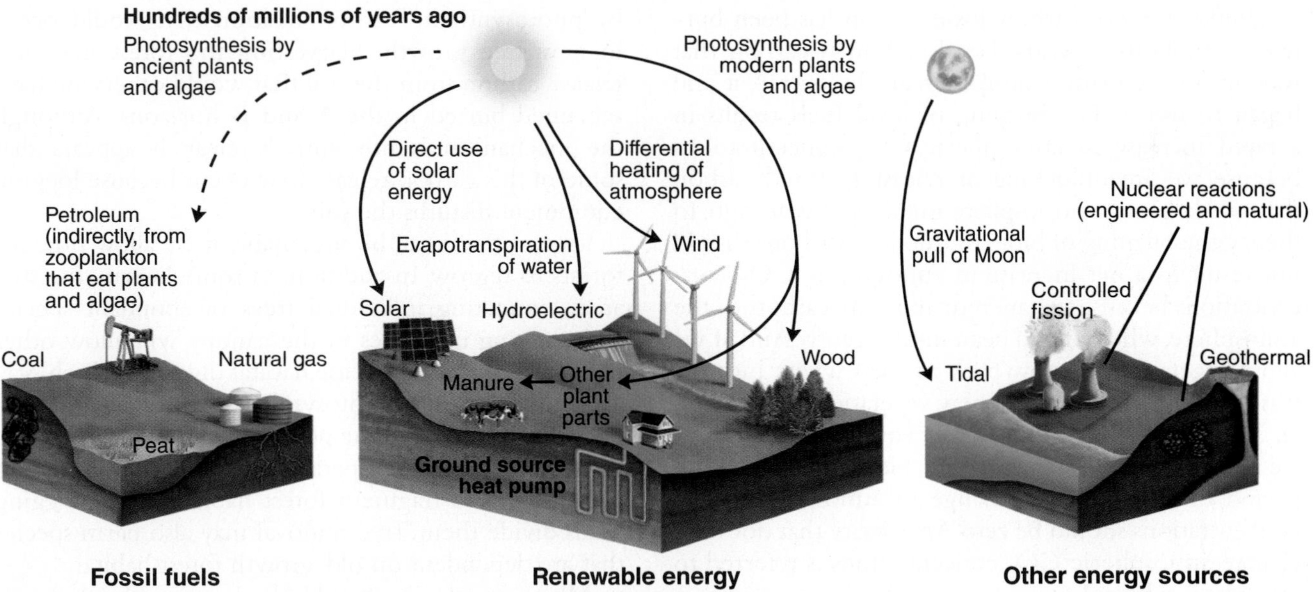

FIGURE 38.1 Energy from the Sun. The Sun is the ultimate source of almost all types of energy.

Biomass is energy from the Sun

As **FIGURE 38.1** shows, all fossil fuel and most renewable energy sources ultimately come from the Sun. Biomass energy resources encompass a large class of fuel types that include wood and charcoal, animal products and manure, plant remains, and municipal solid waste (MSW), as well as liquid fuels such as ethanol and biodiesel. Many forms of biomass used directly as fuel, such as wood and manure, are readily available all over the world. Because these materials are inexpensive and abundant, they account for more than 10 percent of world energy consumption (see Figure 37.6 on page 459), with a much higher percentage in many developing countries. Biomass can also be processed or refined into liquid fuels such as ethanol and biodiesel, known collectively as **biofuels**. These fuels are used in more limited quantities due to the technological demands associated with their use. For example, it is easier to burn a log in a fire than it is to develop the technology to produce a compound such as ethanol.

Biomass—including ethanol and biodiesel—accounts for roughly one-half of the renewable energy and approximately 4.5 percent of all the energy consumed in the United States today (see Figure 37.7 on page 459). However, the mix of biomass used in the United States differs from that found in the developing world. A little less than half of the biomass energy used in the United States comes from wood, and a similar amount comes from biofuels. Only a small amount (less than 5 percent) comes from MSW, while in the developing world, a larger percentage of biomass energy comes from wood and animal manure.

Modern Carbon versus Fossil Carbon

Like fossil fuels, biomass contains a great deal of carbon, and burning it releases that carbon into the atmosphere. Given the fact that both fossil fuels and biomass raise atmospheric carbon concentrations, is it really better for the environment to replace fossil fuels with biomass? The answer depends on how the material is harvested and processed and on how the land is treated during and after harvest. It also depends on how long the carbon has been stored. The carbon found in plants growing today was in the atmosphere in the form of carbon dioxide until recently when it was incorporated into the bodies of the plants through photosynthesis. Depending on the type of plant it comes from, the carbon in biomass fuels may have been captured through photosynthesis as recently as a few months ago, as in the case of a corn plant, or perhaps up to several hundred years ago, as in the case of wood from a large tree. We call the carbon in biomass **modern carbon**, in contrast to the carbon in fossil fuels, which we call **fossil carbon**.

Biofuel Liquid fuel created from processed or refined biomass.

Modern carbon Carbon in biomass that was recently in the atmosphere.

Fossil carbon Carbon in fossil fuels.

Unlike modern carbon, fossil carbon has been buried for millions of years. Fossil carbon is carbon that was out of circulation until humans discovered it and began to use it. The burning of fossil fuels results in a rapid increase in atmospheric CO_2 concentrations because we are unlocking or releasing stored carbon that was last in the atmosphere millions of years ago. In theory, the burning of biomass (modern carbon) should not result in a net increase in atmospheric CO_2 concentrations because we are returning the carbon to the atmosphere, where it had been until recently. And, if we allow vegetation to grow back in areas where biomass was recently harvested, that new vegetation will take up an amount of CO_2 more or less equal to the amount we released earlier by burning the biomass. Over a long period of time, the net change in atmospheric CO_2 concentrations should be zero. An activity that does not change atmospheric CO_2 concentrations is referred to as **carbon neutral**.

Whether using biomass is truly carbon neutral, however, is an important question that is currently being discussed by scientists and policy makers. Sometimes the use of modern carbon ends up releasing CO_2 into the atmosphere that would otherwise have remained in the soil.

Solid Biomass: Wood, Charcoal, and Manure

Throughout the world, 2 billion to 3 billion people rely on wood for heating or cooking. In the United States, approximately 3 million homes use wood as the primary heating fuel, and more than 20 million homes use wood for energy at least some of the time. In addition, the pulp and paper industries, power plants, and other industries use wood waste and by-products for energy. In theory, cutting trees for fuel is sustainable if forest growth keeps up with forest removal. Unfortunately, many forests, like those in Indonesia and parts of Africa and South America, are cut intensively, allowing little chance for regrowth.

Removing more timber than is replaced by growth, or **net removal** of forest, is an unsustainable practice that will eventually lead to deforestation. This net removal of forest together with the burning of wood results in a net increase in atmospheric CO_2: The CO_2 released from the burned wood is not balanced by photosynthetic carbon fixation that would occur in new tree growth. Harvesting the forest may also release carbon from the soil that would otherwise have remained buried in the A and B horizons. Although the mechanism is not entirely clear, it appears that some of this carbon release may occur because logging equipment disturbs the soil.

Tree removal can be sustainable if we allow time for forests to regrow. In addition, in some heavily forested areas, extracting individual trees of abundant species and opening up spaces in the canopy will allow other plants to grow and increase habitat diversity, which may even increase total photosynthesis. More often, however, tree removal has the potential to cause soil erosion, to increase water temperatures in nearby rivers and streams, and to fragment forest habitats when logging roads divide them. Tree removal may also harm species that are dependent on old-growth forest habitat.

Many people in the developing world use wood to make charcoal, which is a superior fuel for many reasons. Charcoal is lighter than wood and contains approximately twice as much energy per unit of weight. A charcoal fire produces much less smoke than wood and does not need to be tended constantly, as does a wood fire. Although it is more expensive than wood, charcoal is a fuel of choice in urban areas of the developing world and for families who can afford it. However, harvesters who clear an area of land for charcoal production often leave it almost completely devoid of trees (**FIGURE 38.2**).

In regions where wood is scarce, such as parts of Africa and India, people often use dried animal manure as a fuel for indoor heating and cooking. Burning manure can be beneficial because it removes harmful microorganisms from surrounding areas, which reduces the risk of disease transmission. However, burning manure also releases particulate matter and other pollutants that cause a variety of respiratory illnesses, from emphysema to cancer. The problem is exacerbated when the manure is burned indoors in poorly ventilated rooms, a common situation in many developing countries. The World Health Organization estimates that indoor air pollution is responsible for 3 million deaths annually. Chapters 15 and 17 cover indoor air pollution and human health in more detail.

Whether indoors or out, burning biomass fuels produces a variety of air pollutants, including particulate matter, carbon monoxide (CO), and nitrogen oxides (NO_x), which are important components of air pollution (**FIGURE 38.3**).

Biofuels: Ethanol and Biodiesel

The liquid biofuels—ethanol and biodiesel—can be used as substitutes for gasoline and diesel, respectively. **Ethanol** is an alcohol made by converting starches and sugars from plant material into alcohol and carbon

Carbon neutral An activity that does not change atmospheric CO_2 concentrations.

Net removal The process of removing more than is replaced by growth, typically used when referring to carbon.

Ethanol Alcohol made by converting starches and sugars from plant material into alcohol and CO_2.

(a)

(b)

FIGURE 38.2 Charcoal as fuel in the developing world. (a) Many people in developing countries rely on charcoal for cooking and heating. This photo shows a charcoal market in the Philippines. (b) Charcoal production can strip the land of all trees *(a: JAY DIRECTO/Getty Images, b: Eduardo Martino/Panos Pictures)*

FIGURE 38.3 Particulate emissions from burning biomass fuels. Burning biomass fuels contributes to air pollution. This photo shows smog and decreased visibility in Montreal, Canada, caused by emissions of particulate matter due to the extensive use of woodstoves. *(alank/Getty Images)*

dioxide. More than 90 percent of the ethanol produced in the United States comes from corn and corn by-products, although ethanol can also be produced from sugarcane, wood chips, crop waste, or switchgrass. **Biodiesel**, a diesel substitute produced by extracting and chemically altering oil from plants, is a substitute for regular petroleum diesel. It is usually produced by extracting oil from algae and plants such as soybean and palm. In the United States, many policy makers are encouraging the production of ethanol and biodiesel as a way to reduce the need to import foreign oil while also supporting U.S. farmers and declining rural economies.

Ethanol

The United States is the world leader in ethanol production, manufacturing roughly 59 billion liters (15.5 billion gallons) in 2017. Brazil, the world's second largest ethanol producer, is making biofuels a major part of its sustainable energy strategy. Brazil manufactures ethanol from sugarcane, which is easily grown in its tropical climate. Unlike corn, which must be replanted

Biodiesel A diesel substitute produced by extracting and chemically altering oil from plants.

every year, sugarcane is replanted every 6 years and is sometimes harvested by hand, factors that reduce the amount of fossil fuel energy needed to grow it.

Ethanol is usually mixed with gasoline, most commonly at a ratio of one part ethanol to nine parts gasoline. The result is gasohol, a fuel that is 10 percent ethanol. Gasohol has a higher oxygen content than gasoline alone and produces less of some air pollutants when combusted. In certain parts of the midwestern United States, especially in corn-growing states, a fuel called E-85 (85 percent ethanol, 15 percent gasoline) is available. **Flex-fuel vehicles** can run on either gasoline or E-85. However, a study by General Motors several years ago revealed that most of the owners of the 7 million flex-fuel vehicles in use at that time did not know that their cars could run on E-85.

Proponents of ethanol claim that it is a more environmentally friendly fuel than gasoline, although opponents dispute that claim. Ethanol does have disadvantages. The carbon bonds in alcohol have a lower energy content than those in gasoline, which means that a 90 percent gasoline/10 percent ethanol mix reduces gas mileage by 2 to 3 percent when compared with 100 percent gasoline fuel. As a result, a vehicle needs more gasohol to go the same distance it could go on gasoline alone. Furthermore, growing corn to produce ethanol uses a significant amount of fossil fuel energy, as well as land that can otherwise be devoted to growing food. Ethanol production has led to concern among economic analysts that this production periodically contributes to short-term food shortages. Furthermore, some scientists argue that using ethanol actually creates a net increase in atmospheric CO_2 concentrations. The benefits and drawbacks of using ethanol as a fuel are discussed in more detail in "Science Applied 6: Should Corn Become Fuel?" that follows this chapter.

Research is under way to find viable alternatives to corn as sources for U.S. ethanol production. Switchgrass is one possibility. It is a perennial crop, which means that farmers can harvest it without replanting, minimizing soil disturbance and erosion. Furthermore, switchgrass does not require as much fossil fuel input as corn to produce. However, crops such as corn and sugarcane produce ethanol more readily due to their high sugar content because sugars are readily and rapidly converted into ethanol. In contrast, switchgrass and other alternative materials, such as wood chips, are composed primarily of cellulose—the material that constitutes plant cell walls—which must be broken down into sugars before it can be used in ethanol production. Scientists have not yet developed an efficient breakdown process for large-scale ethanol production from switchgrass, although such a process would increase the energy and carbon advantages of ethanol.

Flex-fuel vehicle A vehicle that runs on either gasoline or a gasoline/ethanol mixture.

Biodiesel

Biodiesel is a direct substitute for petroleum-based diesel fuel. It is usually more expensive than petroleum diesel, although the difference varies depending on market conditions and the price of petroleum. Biodiesel is typically diluted to "B-20," a mixture of 80 percent petroleum diesel and 20 percent biodiesel, and is available at some gas stations scattered around the United States. It can be used in any diesel engine without modification.

Because biodiesel tends to solidify into a gel at low temperatures, higher concentrations of biodiesel work effectively only in modified engines. However, with a kit sold commercially, a skilled individual or automobile mechanic can modify any diesel vehicle to run on 100 percent straight vegetable oil (SVO), typically obtained as a waste product from restaurants and filtered for use as fuel. Groups of students in the United States have driven buses around the country almost exclusively on SVO, and some municipalities have community-based SVO recycling facilities (**FIGURE 38.4**). Although there is unlikely to be enough waste vegetable oil to significantly reduce fossil fuel consumption, SVO is nevertheless a potential transition fuel that may temporarily reduce our use of petroleum.

In the United States, most biodiesel comes from soybean oil or processed vegetable oil. However, scientists are working on ways to produce large quantities of biodiesel directly from wood or other forms of cellulose—especially waste wood from logging and sawmills. In addition, some species of algae appear to have great potential for producing biodiesel. Algae are photosynthetic microorganisms that can be grown almost anywhere and, of all biodiesel options, produce the greatest yield of fuel per hectare of land area per year and utilize the least amount of energy and fertilizer per quantity of

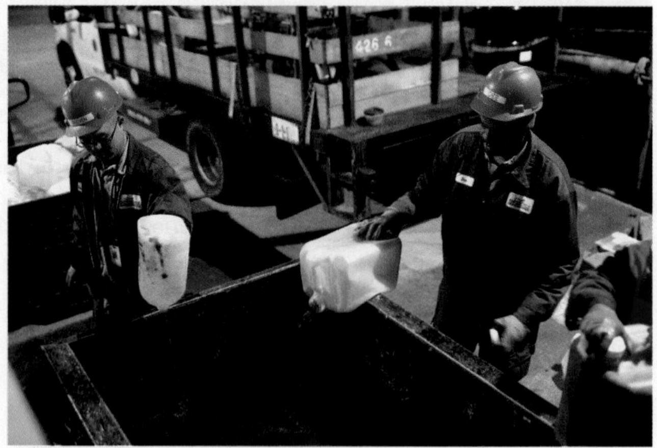

FIGURE 38.4 A cooking oil recycling program. Used cooking oil collected from restaurants in San Francisco, California, is transferred to dumpsters where it will be recycled into biodiesel, reducing demand for fossil fuels and keeping the oil from entering the sewer system. *(Justin Sullivan/Getty Images)*

fuel. One study reported that algae produce 15 to 300 times more fuel per area used than did conventional crops. Algae can be grown on marginal lands, in brackish water, on rooftops, and in other places that are not traditionally thought of as agricultural space.

Emissions of carbon monoxide from combustion of biodiesel are lower than those from petroleum diesel. Since it contains modern carbon rather than fossil carbon, biodiesel should be carbon neutral, although, as with ethanol, some critics question whether biodiesel is truly carbon neutral. For instance, producing biodiesel from soybeans requires less fossil fuel input per liter of fuel than producing ethanol from corn, but soybeans require more cropland, and so they may actually transfer more carbon from the soil to the atmosphere. In contrast, producing biodiesel from wood waste or algae may require very little or no cropland and a minimal amount of other land.

AP® Exam Tip

Make sure that you can differentiate between the impact of biodiesel fuel and fossil fuels on atmospheric CO_2 concentrations.

The kinetic energy of water can generate electricity

Hydroelectricity is electricity generated by the kinetic energy of moving water. It is the second most commonly used form of renewable energy in the United States and in the world, and it is the form most widely used for electricity generation. As we saw in Chapter 12 (Figure 34.8 on page 425), hydroelectricity accounts for approximately 6.5 percent of the electricity generated in the United States. More than one-half of that hydroelectricity is generated in five states: Washington, California, New York, Oregon, and Alabama. Worldwide, 17 percent of all electricity comes from hydroelectric power plants, with China the leading producer, followed by Brazil, Canada, the United States, India, and Russia. In this section we will look at ways in which hydroelectricity is generated, and we will consider whether hydroelectricity is sustainable.

Methods of Generating Hydroelectricity

Moving water, either falling over a vertical distance or flowing with a river or tide, contains kinetic energy. A hydroelectric power plant captures this kinetic energy and uses it to turn a turbine in the same way that the kinetic energy of steam turns a turbine in a coal-fired power plant. The turbine, in turn, transforms the kinetic energy of water or steam into electricity, which is then exported to the electrical grid via transmission lines.

The amount of electricity that can be generated at any particular hydroelectric power plant depends on the flow rate, the vertical distance the water falls, or both. Where falling water is the source of the energy, the amount of electricity that can be generated depends on the vertical distance the water falls; the greater the distance, the more potential energy the water has, and the more electricity it can generate (see Figure 5.2 on page 48). The amount of electricity a hydroelectric power plant can generate also depends on the flow rate: the amount of water that flows past a certain point per unit of time. The higher the flow rate, the more kinetic energy is present, and the more electricity can be generated.

Run-of-the-River Systems

In **run-of-the-river** hydroelectricity generation, water is retained behind a low dam and runs through a channel before returning to the river. Run-of-the-river systems do not store water in a reservoir. These systems have several advantages that reduce their environmental impact: Relatively little flooding occurs upstream, and seasonal changes in river flow are not disrupted. However, run-of-the-river systems are generally small and, because they rely on natural water flows, electricity generation can be intermittent. Heavy spring runoff from rains or snowmelt cannot be stored, and the system cannot generate any electricity in hot, dry periods when the flow of water is low.

Water Impoundment Systems

Storing water in a reservoir behind a dam is known as **water impoundment**. **FIGURE 38.5** on page 466 illustrates the various features of a water impoundment system. By managing the opening and closing of the gates, the dam operators control the flow rate of the water that turns the turbine—and in turn the generator—and thereby influence the amount of electricity produced.

Water impoundment is the most common method of hydroelectricity generation because it usually allows for the generation of electricity on demand. The largest hydroelectric water impoundment dam in the United States is the Grand Coulee Dam in Washington State, which generates 6,800 MW at peak capacity. The Three Gorges Dam on the Yangtze River in China (see Figure 27.2 on page 315) is the largest dam in the world. It has a capacity of 22,500 MW and can generate almost

Hydroelectricity Electricity generated by the kinetic energy of moving water.

Run-of-the-river Hydroelectricity generation in which water is retained behind a low dam or no dam.

Water impoundment The storage of water in a reservoir behind a dam.

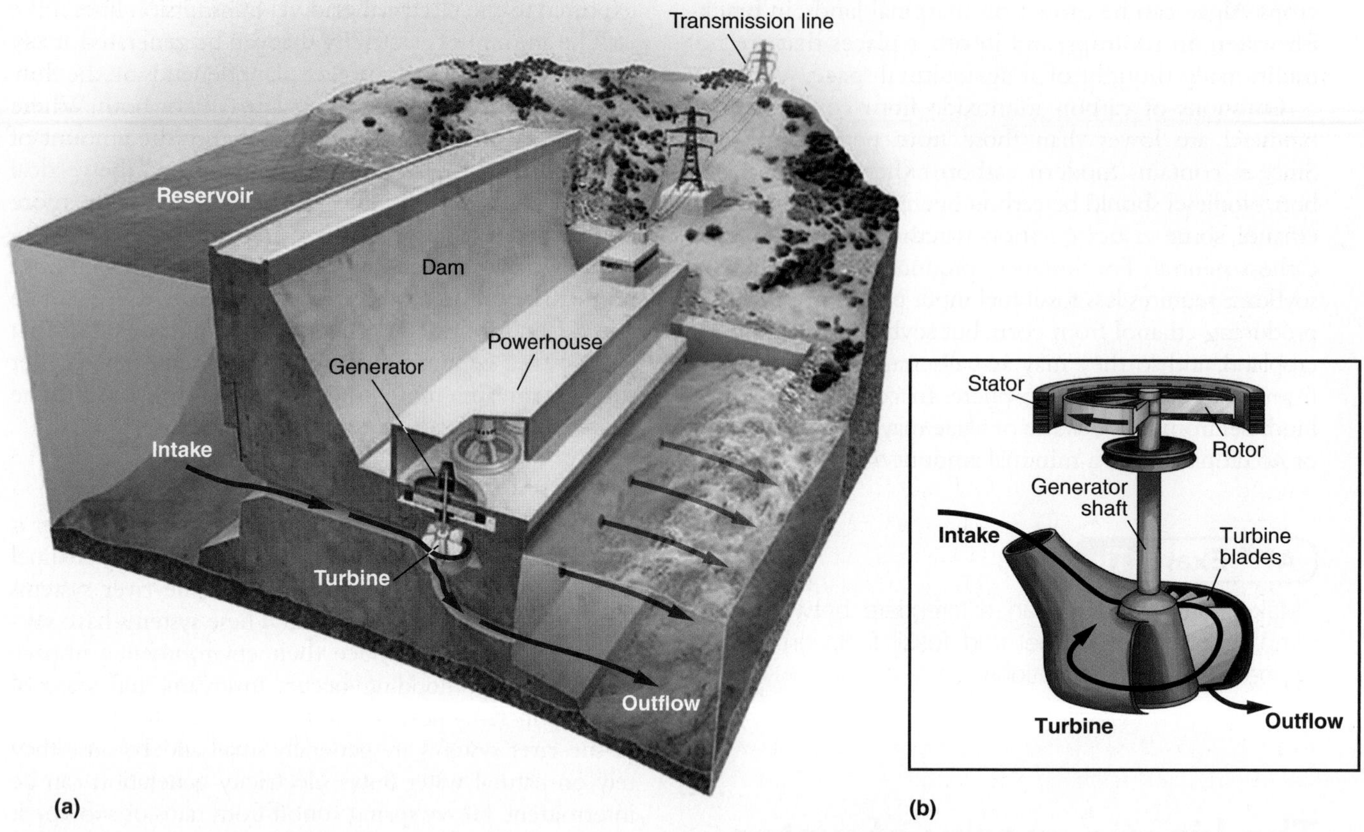

FIGURE 38.5 A water impoundment hydroelectric dam. (a) Water impoundment, a common method of hydroelectricity generation, allows for electricity generation on demand. Dam operators control the rate of water flow by opening and closing the gates. Increased water flow results in increased rotation of the turbines (b) resulting in a greater amount of electricity generated. The arrows indicate the path of water flow.

85 billion kilowatt-hours per year, approximately 11 percent of China's total electricity demand.

Tidal Systems

Tidal energy also comes from the movement of water, although the movement in this case is driven by the gravitational pull of the Moon. Tidal energy systems use gates and turbines similar to those used in run-of-the-river and water impoundment systems to capture the kinetic energy of water flowing through estuaries, rivers, and bays and convert this energy into electricity.

Although tidal power plants are operating in many parts of the world, including France, Korea, and Canada, tidal energy does not have the potential to become a major energy source. In many locations around the world, the difference in water level between high and low tides is not great enough to provide sufficient kinetic energy to generate a large amount of electricity. In addition, to transfer the energy generated, transmission lines must be constructed on or near a coastline or estuary. This infrastructure may have a disruptive effect on coastal, shoreline, and marine ecology as well as on tourism that relies on the aesthetics of a coastal region.

Tidal energy Energy that comes from the movement of water driven by the gravitational pull of the Moon.

Hydroelectricity and Sustainability

Major hydroelectric dam projects have brought renewable energy to large numbers of rural residents in many countries, including the United States, Canada, India, China, Brazil, and Egypt. Although hydroelectric dams are expensive to build, once built, they require a minimal amount of fossil fuel for operation. In general, the benefits of water impoundment hydroelectric systems are great: They generate large quantities of electricity without creating air pollution, waste products, or CO_2 emissions. Electricity from hydroelectric power plants is usually less expensive for the consumer than electricity generated using nuclear fuels or natural gas. In the United States, the price of hydroelectricity ranges from 5 cents to 11 cents per kilowatt-hour.

In addition, the reservoir behind a hydroelectric dam can provide recreational and economic opportunities as well as downstream flood control for flood-prone areas. For example, Lake Powell, the reservoir impounded by the hydroelectric Glen Canyon Dam, draws more than 3 million visitors to the Glen Canyon National Recreation Area each year and generates more than $400 million annually for the local and regional economies in Arizona and Utah (**FIGURE 38.6**). In China, the Three Gorges Dam provides flood control and protection to many millions of people.

Water impoundment, however, does have negative environmental consequences. In order to form an impoundment, a free-flowing river must be held back. The resulting reservoir may flood hundreds or thousands of hectares of prime agricultural land or canyons with great aesthetic or archeological value. It may also force people to relocate. As we saw in Chapter 9, the construction of the Three Gorges Dam displaced more than 1.3 million people and submerged ancient cultural and archaeological sites as well as large areas of farmland. Impounding a river in this way may also make it unsuitable for organisms or recreational activities that depend on a free-flowing river. Large reservoirs of standing water hold more heat and contain less oxygen than do free-flowing rivers, thereby affecting which species can survive in the waters. Certain human parasites also become more abundant in impounded waters in tropical regions.

By regulating water flow and flooding, dams also alter the dynamics of the river ecosystem downstream. Some rivers, for example, have sandbars created during periods of low flow that follow periods of flooding. Some plant species, such as cottonwood trees, cannot reproduce in the absence of these sandbars. The life cycles of certain aquatic species, such as salmon, certain trout species, and freshwater clams and mussels, also depend on seasonal variations in water flow. Impoundment systems disrupt these life cycles by controlling the flow of water so it is consistently plentiful for hydroelectricity generation.

It is possible to address some of these problems. For example, as we saw in Chapter 9, the installation of a fish ladder (see Figure 27.3 on page 315) may allow fish to travel upstream around a dam. Such solutions are not always optimal, however; some fish species fail to utilize them and some predators learn to monitor the fish ladders for their prey.

Other environmental consequences of water impoundment systems include the release of greenhouse gases to the atmosphere, both during dam construction and after filling the reservoir. Production of cement—a major component of dams—is responsible for approximately 5 percent of global anthropogenic CO_2 emissions to the atmosphere. Once the dam is completed, the impounded water usually covers forests or grasslands. The dead plants and organic materials in the flooded soils decompose anaerobically and release methane, a potent greenhouse gas. Some researchers assert that in tropical regions, the methane released from a hydroelectric water impoundment contributes more to climate warming than a coal-fired power plant with about the same electricity generation output.

The accumulation of sediments in reservoirs has negative consequences not only for the environment but also for the electricity-generating capacity of hydroelectric dams. A fast-moving river carries sediments

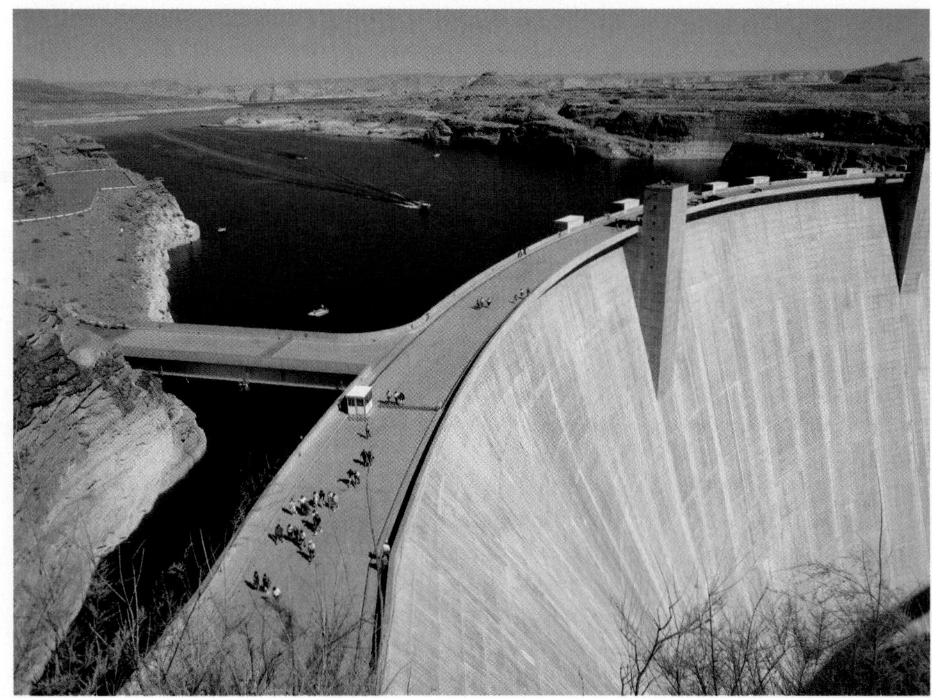

FIGURE 38.6 A recreation area created by water impoundment. Lake Powell at the Glen Canyon National Recreation Area was created by water impoundment at the hydroelectric Glen Canyon Dam in northern Arizona. *(Sal Maimone/Sal Maimone/Superstock)*

(a)

(b)

FIGURE 38.7 Dam removal and environmental restoration. In recent years, a number of dams have been dismantled due to environmental concerns or heavy siltation. (a) This photo shows Oregon's Marmot Dam, on the Sandy River, before removal. (b) The same stretch of the Sandy River after dam removal in 2007 shows how the natural landscape has been restored. *(a and b: Portland General Electric)*

that settle out when the river feeds into the reservoir. The accumulation of these sediments on the bottom of the reservoir is known as **siltation**. Over time, as the reservoir fills up with sediments, the amount of water that can be impounded, and thus the generating capacity and life span of the dam, is reduced. This process may take hundreds of years or only decades, depending on the geology of the area. The only way to reverse the siltation process is by dredging, removal of the sediment, usually with machinery that runs on fossil fuels.

Because of either environmental concerns or heavy siltation, a number of hydroelectric dams are being dismantled, as we saw at the beginning of Chapter 9. In 1999, the Edwards Dam was removed from the Kennebec River in Maine. More than a decade later, native fishes such as bass and alewives have returned to the waters and are flourishing. In 2007, the Marmot Dam on Oregon's Sandy River was removed using explosives. The restored river now hosts migrating salmon and steelhead trout (*Oncorhynchus mykiss*) for the first time since 1912 (**FIGURE 38.7**).

Siltation The accumulation of sediments, primarily silt, on the bottom of a reservoir.

AP® Exam Tip

You should be able to explain the processes by which alternative sources of energy are converted to useable forms of energy.

MODULE

38 AP® Review

Preparing for the AP® Exam

In this module, we have seen that biomass and water are two important renewable energy sources. Biomass is a modern source of carbon, which was formed between a few years ago and hundreds of years ago, as opposed to fossil fuels, which contain carbon formed millions of years ago. Solid biomass includes wood, charcoal, and animal manure, all of which are relatively low-quality sources of energy and release a fair amount of particulates and other pollutants when burned. Liquid fuels include ethanol, an alcohol derived from plant material such as corn, and biodiesel, produced from vegetable oils such as soybean. Algae is another source of oil for biodiesel. Hydroelectricity is generated from the energy in water. The largest hydroelectric projects come from impounding water behind a large dam and releasing it periodically when electricity is needed. The impounded water behind a dam can promote recreational and economic opportunities but can also have numerous impacts on the environment. Water availability in a region may be somewhat variable. In the next module we will examine two continuous, nondepletable sources of renewable energy: the Sun and wind.

AP® Practice Questions

Choose the best answer for the following.

1. Which is NOT a form of biomass?
   (a) coal
   (b) charcoal
   (c) municipal solid waste
   (d) ethanol

2. A hydroelectric power plant's rate of electricity generation depends on
   I. the flow rate of the water.
   II. the vertical distance the water falls.
   III. the amount of water behind the dam.
   (a) I only
   (b) I and II
   (c) II only
   (d) II and III

3. Which is true of solid biofuels?
   (a) Charcoal is the primary replacement when wood is scarce.
   (b) Indoor air pollution from them results in millions of deaths annually.
   (c) Switchgrass is a newly developed replacement for wood.
   (d) They are carbon neutral due to the net removal of forests.

4. Cellulosic ethanol is produced from
   (a) corn.
   (b) beets.
   (c) sugarcane.
   (d) switchgrass.

5. The most common method of hydroelectric generation is
   (a) run-of-the-river.
   (b) tidal.
   (c) water impoundment.
   (d) gorge dams.

Solar, Wind, Geothermal, and Hydrogen

After biomass and water, the most important forms of renewable energy come from the Sun and wind. These nondepletable sources of renewable energy represent the fastest growing forms of energy development throughout the world.

Learning Goals

After reading this module, you should be able to

- list the different forms of solar energy and their application.
- describe how wind energy is harnessed and its contemporary uses.
- discuss the methods of harnessing the internal energy from Earth.
- explain the advantages and disadvantages of energy from hydrogen.

The energy of the Sun can be captured directly

In addition to driving the natural cycles of water and air movement that we can tap as energy resources, the Sun also provides energy directly. Every day, Earth is

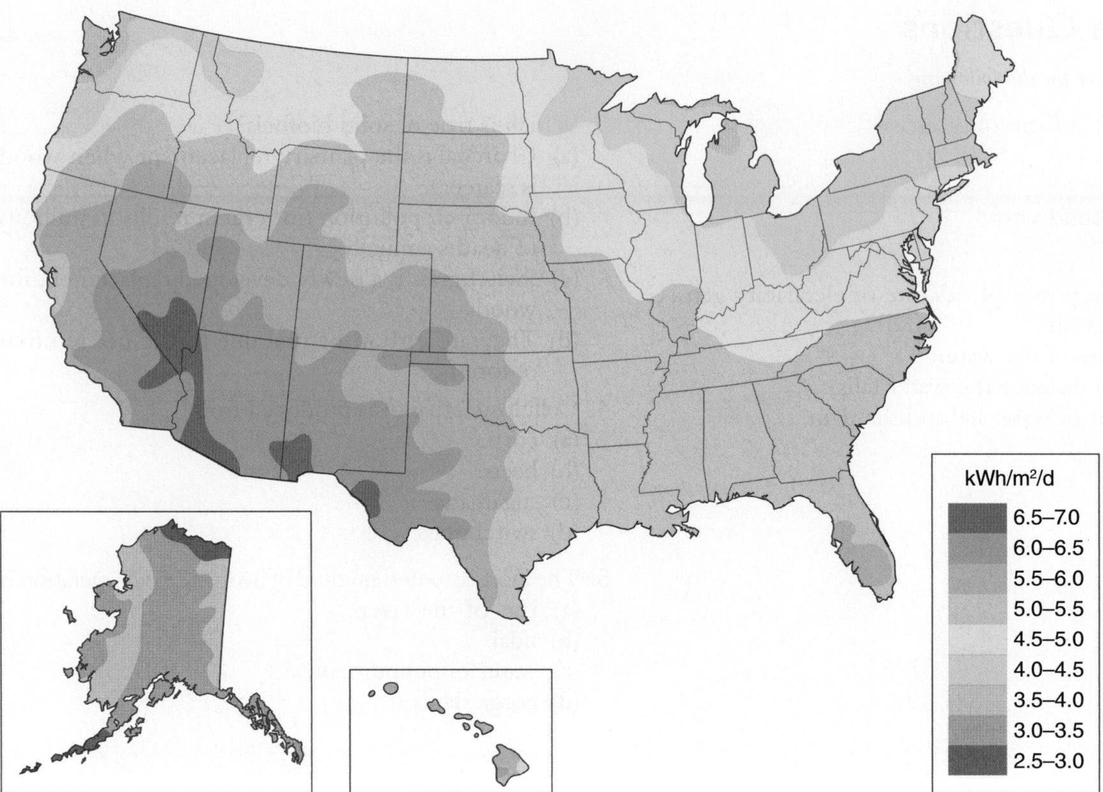

FIGURE 39.1 Geographic variation in solar radiation in the United States. This map shows the amount of solar energy available to a flat photovoltaic solar panel in kilowatt-hours per square meter per day, averaged over a year. *(Data from National Renewable Energy Laboratory, U.S. Department of Energy.)*

bathed in solar radiation, an almost limitless source of energy. The amount of solar energy available in a particular place varies with amount of cloudiness, time of day, and season. The average amount of solar energy available varies geographically. As **FIGURE 39.1** shows, average daily solar radiation in the continental United States ranges from 3 kWh of energy per square meter in the Pacific Northwest to almost 7 kWh per square meter in parts of the Southwest.

Passive Solar Heating

We have already seen several applications of passive solar heating, including positioning windows on south-facing walls to admit solar radiation in winter, covering buildings with dark roofing material in order to absorb the maximum amount of heat, and building homes into the side of a hill. None of these strategies relies on intermediate pumps or technology to supply heat. Solar ovens are another practical application of passive solar heating. For instance, a simple "box cooker" concentrates sunlight as it strikes a reflector on the top of the oven. Inside the box, the solar energy is absorbed by a dark base and a cooking pot and is converted into heat energy. The heat is distributed throughout the box by reflective material lining the interior walls and is kept from escaping by a glass top. On sunny days, such box cookers can maintain temperatures of 175°C (350°F), heat several liters of water to boiling in under an hour, or cook traditional dishes of rice, beans, or chicken in 2 to 5 hours.

Solar ovens have both environmental and social benefits. The use of solar ovens in place of firewood reduces deforestation and, in areas unsafe for travel, having a solar oven means not having to leave the relative safety of home to seek firewood. For example, over 10,000 solar ovens have been distributed in refugee camps in the Darfur region of western Sudan in Africa, where leaving the camps to find cooking fuel would put women at risk of attack (**FIGURE 39.2**).

Active Solar Energy Technologies

In contrast to passive solar design, **active solar energy** technologies capture the energy of sunlight with the use of technologies that include small-scale solar water heating systems, photovoltaic solar cells, and large-scale concentrating solar thermal systems for electricity generation.

Active solar energy Energy captured from sunlight with advanced technologies.

FIGURE 39.2 Solar cookers. This woman in a Mali village in West Africa is preparing food with a solar cooker. *(Joerg Boethling/Alamy)*

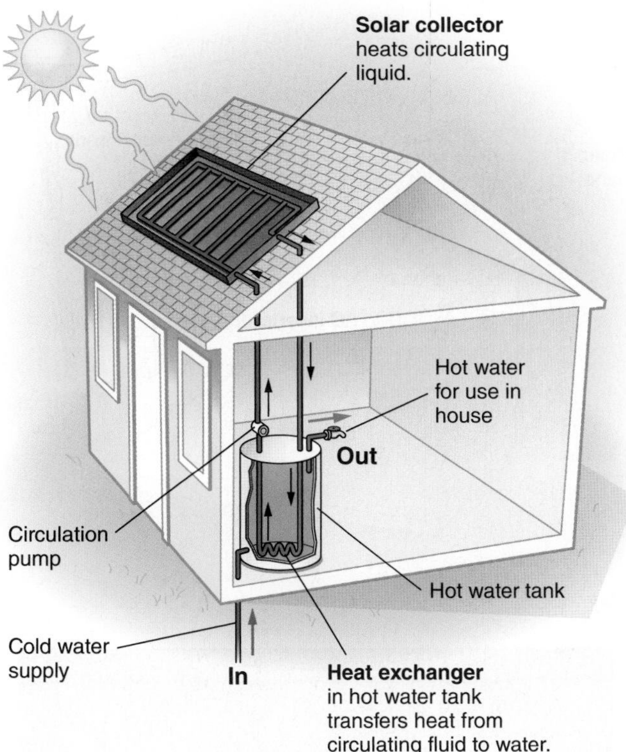

FIGURE 39.3 A solar domestic hot water system. When a solar hot water system is used to heat a house, a nonfreezing liquid is circulated by an electric pump through a closed loop of pipes. This circulating liquid moves from a water storage tank to a solar collector on the roof, where it is heated, and then sent back to the tank, where a heat exchanger transfers the heat to water.

Solar Water Heating Systems

Solar water heating applications range from providing domestic hot water and heating swimming pools to a variety of heating purposes for business and home. In the United States, heating swimming pools is the most common application of solar water heating, and it is also the one that pays for itself the most quickly.

A household solar water heating system, like the domestic hot water system shown in **FIGURE 39.3**, allows heat energy from the Sun to be transferred directly to water or another liquid, which is then circulated to a hot water heating system. The circulation of the liquid is driven either by a pump (in active systems) or by natural convection (in passive systems). In both cases, cold liquid is heated as it moves through a solar collector mounted on the roof or wall of a building or situated on the ground.

The simplest solar water heating systems pump cold water directly to the collector to be heated; the heated water then flows back to an insulated storage tank. In areas that are sunny but experience temperatures below freezing, the water is kept in the storage tank and a "working" liquid containing nontoxic antifreeze circulates in pipes between the storage tank and the solar collector. The nonfreezing circulating liquid is heated by the Sun in the solar collector, then returned to the storage tank where it flows through a heat exchanger that transfers its heat to the water. The energy needed to run the pump is usually much less than the energy gained from using the system, especially if the pump runs on electricity from the Sun. Solar water heating systems typically include a backup energy source, such as an electric heating element or a connection to a fossil fuel–based central heating system, so that hot water is available even when it is cloudy or very cold.

Photovoltaic Systems

In contrast to solar water heating systems, **photovoltaic solar cells** capture energy from the Sun as light, not heat, and convert it directly into electricity.

Photovoltaic solar cell A system of capturing energy from sunlight and converting it directly into electricity.

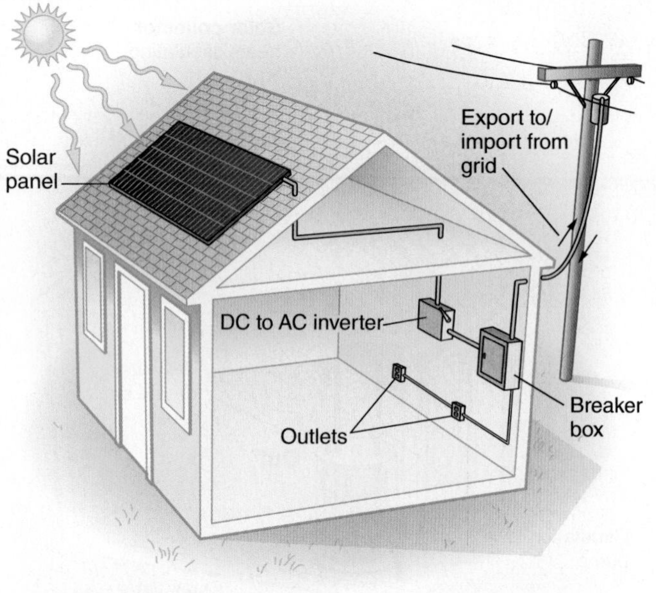

FIGURE 39.4 Photovoltaic solar energy. (a) In this domestic photovoltaic system, photovoltaic solar panels convert sunlight into direct current (DC). An inverter converts DC into alternating current (AC), which supplies electricity to the house. Any electricity not used in the house is exported to the electrical grid. (b) Photovoltaic panels on the roof of this house provide nearly all of the electricity this family uses. *(altrendo images/Getty Images)*

FIGURE 39.4a shows how a photovoltaic system, also referred to as PV, delivers electricity to a house. A photovoltaic solar cell makes use of the fact that certain semiconductors—very thin, ultraclean layers of material—generate a low-voltage electric current when they are exposed to direct sunlight (Figure 39.4b). The low-voltage direct current is usually converted into higher-voltage alternating current for use in homes or businesses. Typically, photovoltaic solar cells are 12 to 20 percent efficient in converting the energy of sunlight into electricity. "Do the Math: Calculating Home PV Generation and Capacity Factor" shows you how to calculate electricity production and dollar savings from a home photovoltaic solar system.

Electricity produced by photovoltaic systems can be used in several ways. Solar panels—arrays of photovoltaic solar cells—on a roof can be used to supply electricity to appliances or lights directly, or they can be used to charge batteries. The vast majority of photovoltaic systems are tied to the electrical grid, meaning that any extra electricity generated and not needed is sent to the electric utility, which buys it or gives the customer credit toward the cost of future electricity use. Homes that are "off the grid" may rely on photovoltaic solar cells as their only source of electricity, using batteries to store the electricity until it is needed. Photovoltaic solar cells have other uses in locations far from the grid where a small amount of electricity is needed on a regular basis. For example, small photovoltaic solar cells charge the batteries that keep highway emergency telephones working. In several U.S. cities, photovoltaic solar cells provide electricity for streetside trash compactors and for new "smart" parking meter systems that have replaced aging coin-operated parking meters.

Concentrating Solar Thermal Electricity Generation

Concentrating solar thermal (CST) systems are a large-scale application of solar energy to electricity generation. CST systems use lenses or mirrors and tracking systems to focus the sunlight falling on a large area into a small beam, in the same way you might use a magnifying glass to focus energy from the Sun and perhaps burn a hole in a piece of paper. In this case, however, the heat of the concentrated beam is used to evaporate water and produce steam that turns a turbine to generate electricity. CST power plants operate much like conventional thermal power plants; the only difference is that the energy to produce the steam comes directly from the Sun, rather than from fossil fuels. The arrays of lenses and mirrors required are large, so CST power plants are best constructed in desert areas where there is consistent sunshine and plenty of open space (**FIGURE 39.5**).

Although CST systems have existed for 10 years or more, they are now becoming more common. In the United States, several plants are under development in California and in the Southwest. For example, the Ivanpah plant in California contains over 250 MW of capacity on 3,500 acres (1,420 ha). These plants, though, have drawbacks that include the large amount of land required and their inability to generate electricity at night.

AP® Exam Tip

Be sure that you understand the different types of solar energy systems, their respective benefits and costs, and their use and design.

DO THE MATH Calculating Home PV Generation and Capacity Factor

Preparing for the AP® Exam

One of this book's authors, Andy Friedland, has a 4,000 watt photovoltaic solar array along the side of the driveway to his house in Vermont. It consists of 16 250-watt solar panels. In 2017, this array generated 5,300 kwh of electricity. Electricity in Vermont costs \$0.15/kwh. What is the capacity factor of this system and how much did Professor Friedland offset in electricity costs?

If 4,000 watts were generating continuously, it would generate this much electricity in a year:

$$4{,}000 \text{ watts} \times 1 \text{ kw}/1{,}000 \text{ watts} = 4 \text{ kw}$$

4 kw × 24 hours per day × 365 days per year = 35,000 kwh per year (after rounding to two significant figures) if the solar panels were generating 24 hours per day.

In actuality, if the system generated 5,300 kwh of electricity, the capacity factor is:

$$5{,}300 \text{ kwh/year} \div 35{,}000 \text{ kwh/year} = 0.15 \times 100\% = 15\% \text{ capacity factor}$$

The capacity factor or percentage of time that the photovoltaic cells were generating electricity at full capacity was 15 percent.

To determine the dollar amount of the offset with this system, multiply the amount generated by the cost of electricity:

5,300 kwh × \$0.15/kwh = \$795. Professor Friedland offset, or avoided paying, \$795 to the electrical utility in 2017.

YOUR TURN A 3 MW wind turbine was installed in an on-land location where the capacity factor was measured to be 22 percent.

1. How much electricity will this wind turbine generate in a year?
2. How much revenue would a utility receive if they were paid \$0.05/kwh for this electricity?

FIGURE 39.5 A concentrating solar thermal power plant. Mirrors and reflectors concentrate the energy of the Sun onto a "power tower," which uses the sunlight to heat water and make steam for electricity generation. *(Lowell Georgia/Science Source)*

Benefits and Drawbacks of Active Solar Energy Systems

Active solar energy systems offer many benefits such as generating hot water or electricity without producing CO_2 or polluting the air or water during operation. In addition, photovoltaic solar cells and CST power plants can produce electricity when it is needed most: on hot, sunny days when demand for electricity is high, primarily for air conditioning. By producing electricity during peak demand hours, these systems can help reduce the need to build new fossil fuel–power plants.

In many areas, small-scale solar energy systems are economically feasible. For a new home located miles away from the grid, installing a photovoltaic system may be much less expensive than running electrical transmission lines to the home site. When a house is near the grid, the initial cost of a photovoltaic system may take 5 to 20 years for payback; once the initial cost is paid back, however, the electricity it generates is almost free.

Despite these advantages, a number of drawbacks have inhibited the growth of solar energy use in the United States. Photovoltaic solar panels are expensive to manufacture and install. Although the technology is changing rapidly as industrial engineers and scientists seek better, cheaper photovoltaic materials and systems, the initial cost to install a photovoltaic system can be daunting and the payback period is a long one. In parts of the country where the average daily solar radiation is low, the payback period can be even longer. Some countries, such as Germany, have made solar energy a part of their sustainable energy agenda by subsidizing their solar industry. In the United States, recent tax breaks, rebates, and funding packages instituted by various states and the federal government have made solar electricity and water heating more affordable for consumers and businesses.

The use of photovoltaic solar cells has environmental as well as financial costs. Manufacturing photovoltaic solar cells requires a great deal of energy and water and involves a variety of toxic metals and industrial chemicals that can be released into the environment during the manufacturing process, although newer types of these solar cells may reduce reliance on toxic materials. For systems that use batteries for energy storage, there are environmental costs associated with manufacturing, disposing of, or recycling the batteries, as well as energy losses during charging, storage, and recovery of electricity in batteries. The end-of-life reclamation and recycling of photovoltaic solar cells is another potential source of environmental contamination, particularly if the cells are not recycled properly. However, solar energy advocates, and even most critics, agree that the energy expended to manufacture photovoltaic solar cells is usually recovered within a few years of their operation, and that if the life span of photovoltaic solar cells can be increased to between 30 and 50 years, they will be a very promising source of renewable energy.

Wind energy is the most rapidly growing source of electricity

The wind is another important source of nondepletable, renewable energy. **Wind energy** is energy generated from the kinetic energy of moving air. As discussed in Chapter 4, winds are the result of the unequal heating of the surface of Earth by the Sun. Warmer air rises and

Wind energy Energy generated from the kinetic energy of moving air.

Wind turbine A turbine that converts wind energy into electricity.

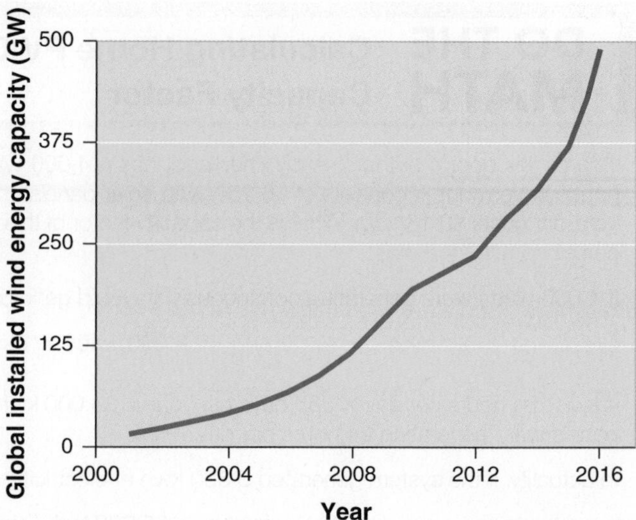

FIGURE 39.6 Global growth of installed wind energy capacity. Worldwide, installed wind energy capacity is now almost 500 gigawatts (GW). *(Data from Global Wind Energy Council.)*

cooler, denser air sinks, creating circulation patterns similar to those in a pot of boiling water. Ultimately, the Sun is the source of all winds—it is solar radiation and ground surface heating that drive air circulation.

Before the electrical grid reached rural areas of the United States in the 1920s, windmills dotted the landscape. Today, wind energy is the fastest-growing major source of electricity in the world. As **FIGURE 39.6** shows, global installed wind energy capacity has risen from less than 24 gigawatts in 2001 to almost 500 gigawatts today. **FIGURE 39.7** shows installed wind energy generating capacity and the percentage of electricity generated by wind for a number of countries. China has the largest wind energy generating capacity in the world, followed by the United States, Germany, India, and Spain.

Despite its large generating capacity, the United States obtains less than 6 percent of its electricity from wind. The largest amounts are generated in Texas, Oklahoma, Iowa, California, and Kansas, although more than 40 U.S. states produce at least some wind-generated electricity. Denmark, a country of 5.8 million people, generates about 37 percent of its electricity from wind and hopes to increase this figure to 50 percent soon. Although the United States currently obtains only a small percent of its electricity from wind, it is the fastest growing source of electricity in the country.

Generating Electricity from Wind

A **wind turbine** converts the kinetic energy of moving air into electricity in much the same way that a hydroelectric turbine harnesses the kinetic energy of moving water. As you can see in **FIGURE 39.8**, wind

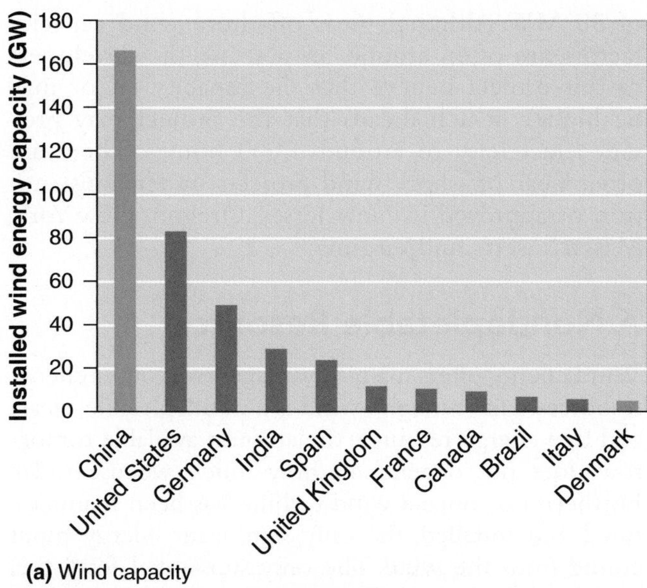

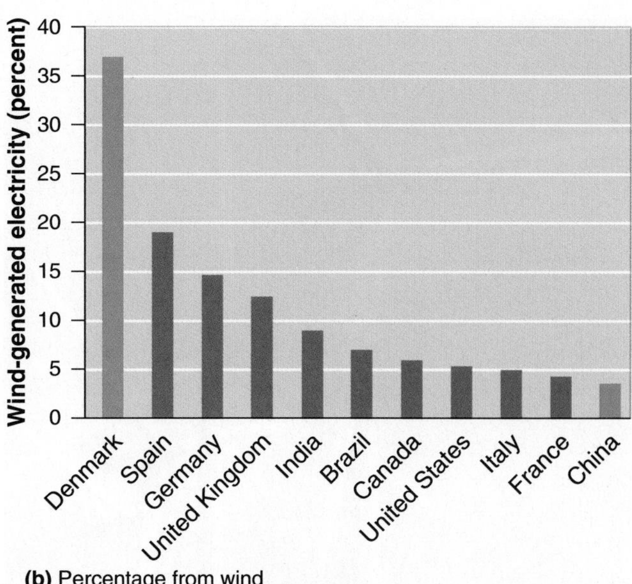

FIGURE 39.7 Installed wind energy capacity by country. (a) China generates more electricity from wind energy than any other country. (b) However, some relatively small countries, such as Denmark, generate a much higher percentage of their electricity from wind. *(Data from Global Wind Energy Council.)*

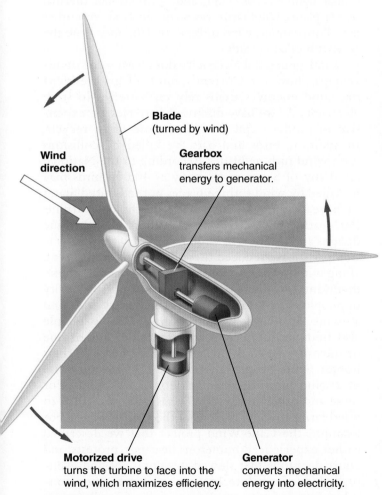

Blade (turned by wind)

Wind direction

Gearbox transfers mechanical energy to generator.

Motorized drive turns the turbine to face into the wind, which maximizes efficiency.

Generator converts mechanical energy into electricity.

FIGURE 39.8 Generating electricity with a wind turbine. The wind turns the blade, which is connected to the generator, which generates electricity.

turns the blades of the wind turbine and the blades transfer energy to the gear box that in turn transfers energy to the generator that generates electricity. A modern wind turbine, like the one shown, may sit on a tower as tall as 100 m (330 feet) and have blades 40 to 75 m (130–250 feet) long. Under average wind conditions, a wind turbine on land might have a capacity factor of 25 percent. While it is spinning, it might generate between 2,000 and 3,000 kW (2–3 MW), and in a year it might produce more than 4.4 million kilowatt-hours of electricity, enough to supply more than 400 homes. Offshore wind conditions are even more desirable for electricity generation with capacity factors of 35 percent to 50 percent, and turbines can be made even larger in an offshore environment.

Wind turbines on land are typically installed in rural locations, away from buildings and population centers. However, they must also be close to electrical transmission lines with enough capacity to transport the electricity they generate to users. For these reasons, as well as for political and regulatory reasons and to facilitate servicing the equipment, the usual practice is to group wind turbines into wind farms or wind parks.

The number of wind farms is increasing in the United States and around the world. Wind farms are often placed on land in locations where the capacity factor can be as high as 25 percent. However, near-offshore coastal locations are even more desirable because the capacity factor can be up to 50 percent. Offshore wind parks, which are clusters of wind turbines, are often located in the ocean within a few miles of the coastline (**FIGURE 39.9** on page 476). Such parks are operating in Denmark, the Netherlands, the United Kingdom, Sweden, and elsewhere.

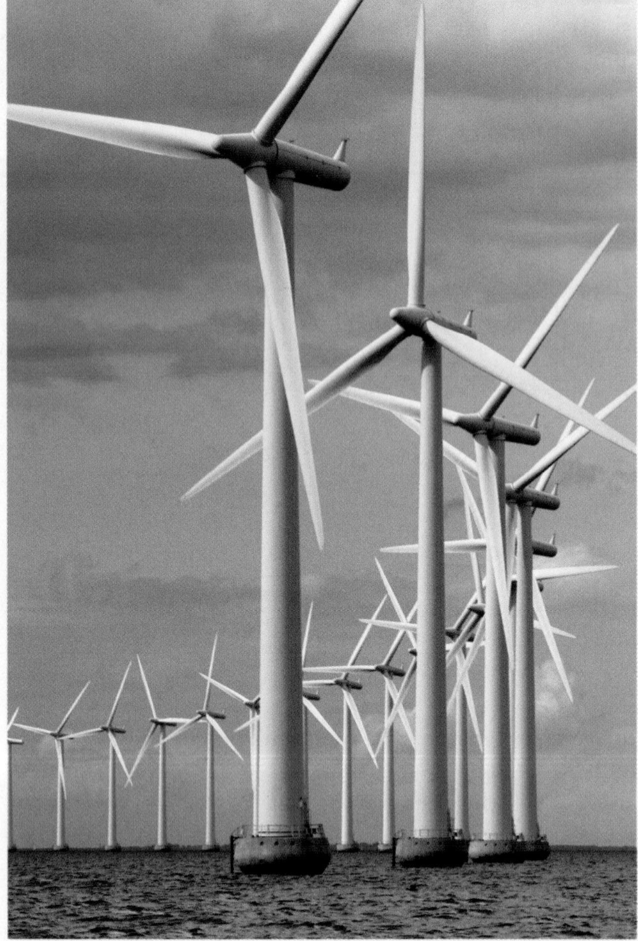

FIGURE 39.9 Offshore wind parks. Capacity factors at near-offshore locations like this one in Denmark are generally higher than on land. *(Max Mudie/Alamy)*

The United States lags behind many other countries in the development of near-offshore wind. A proposed project located off Cape Cod, Massachusetts, in Nantucket Sound languished for almost 2 decades and was never built. Called Cape Wind, it was to become the first offshore wind farm in the United States and would have featured 130 wind turbines with the potential to produce up to 420 MW of electricity, or up to 75 percent of the electricity used by Cape Cod and the nearby islands. In late 2016, its developer declared the project officially canceled. It failed for a variety of reasons. There was a lack of clarity about which U.S. agencies regulate off-shore wind development. In addition, the project was close to sensitive and scenic waterways near Cape Cod. Finally, some local landowners and legislators did not want to see a wind farm close to Nantucket Sound.

In 2016, the Block Island Wind Farm project went online in the near-offshore environment of Rhode Island. Officially the first near-offshore wind project to be completed in the United States, it contains five turbines, each 6 MW, for a total capacity of 30 MW. Although near off-shore wind capacity factors are often around 35 percent, the developers of this project believe that the capacity factor may be higher, which means that the project may provide electricity to around 4,000 homes. There are other near off-shore wind projects under construction or approved in New Jersey, Oregon, New York, Massachusetts, and Virginia.

A Nondepletable Resource

Wind energy offers many advantages over other energy resources. Like sunlight, wind is a nondepletable, clean, and free energy resource; the amount available tomorrow does not depend on how much we use today. Furthermore, once a wind turbine has been manufactured and installed, the only significant energy input comes from the wind. The only substantial fossil fuel input required, once the turbines are installed, is the fuel workers need to travel to the wind farm to maintain the equipment. Thus, wind-generated electricity produces no pollution and no greenhouse gases. Finally, unlike hydroelectric, CST, and conventional thermal power plants, wind farms can share the land with other uses. For example, wind turbines on land may share the area with grazing cattle.

Wind-generated electricity does have some disadvantages, however. Currently, most off-grid residential wind energy systems rely on batteries to store electricity. As we have discussed, batteries are expensive to produce and hard to dispose of or recycle. In addition, birds and bats are killed by collisions with wind turbine blades. According to the National Academy of Sciences, as many as 40,000 birds may be killed by wind turbine blades in the United States each year—approximately four deaths per turbine. Bat deaths are not as well quantified. New turbine designs and location of wind farms away from migration paths have reduced these deaths to some extent, along with mitigating some of the noise and aesthetic disadvantages. There are a small but vocal minority of people in certain regions of the country who find a wind farm visually objectionable. Some people also find the sound of a wind turbine bothersome or intrusive, especially when it can be heard in their homes. Some people feel that the background noise of a wind turbine causes anxiety or irritability. For these and other reasons, there has been resistance to wind farms in some regions of the United States. For example, the Cape Wind project that we described earlier experienced numerous hearings, protests, and court decisions that slowed development and eventually led to its failure. In Vermont, a state often considered to be very environmentally friendly, more and more towns and individuals have argued against the installation of commercial wind projects on or near ridgelines, citing habitat fragmentation and alteration, noise, and aesthetics, among other reasons. Other states have slowed wind development by resisting the construction of above-ground electrical transmission lines, which also fragment habitat but are needed to move renewable electricity through forested areas to large numbers of users, who are usually in or close to metropolitan areas.

Earth's internal heat is a source of nondepletable energy

Unlike most forms of renewable energy, *geothermal energy* does not come from the Sun. **Geothermal energy** is heat that comes from the natural radioactive decay of elements deep within Earth. As we saw in Chapter 8, convection currents in Earth's mantle bring hot magma toward the surface of Earth. Wherever magma comes close enough to groundwater, that groundwater is heated. The pressure of the hot groundwater sometimes drives it to the surface, where it visibly manifests itself as geysers and hot springs, like those in Yellowstone National Park. Where hot groundwater does not naturally rise to the surface, humans may be able to reach it by drilling.

Many countries obtain clean, renewable energy from geothermal resources. The United States, China, and Iceland, all of which have substantial geothermal resources, are the largest geothermal energy producers.

Harvesting Geothermal Energy

Geothermal energy can be used directly as a source of heat. Hot groundwater can be piped directly into household radiators to heat a home. In other cases, heat exchangers can collect heat by circulating cool liquid underground, where heat from the ground flows to the cool circulating liquid, and then returns to the surface. Iceland, a small nation with vast geothermal resources, heats 85 percent of its homes this way.

Geothermal energy can also be used to generate electricity. The electricity-generating process is much the same as that in a conventional thermal power plant although, in this case, the steam to run the turbine comes from water evaporated by Earth's internal heat instead of by burning fossil fuels.

The heat released by decaying radioactive elements deep within Earth is essentially nondepletable in the span of human time. However, the groundwater that so often carries that heat to Earth's surface can be depleted. As we learned in Chapter 9, groundwater, if used sustainably, is a renewable resource. Unfortunately, long periods of harvesting groundwater from a site may deplete it to the point at which the site is no longer a viable source of geothermal energy. Returning the water to the ground to be reheated is one way to use geothermal energy sustainably.

Iceland currently produces about 25 percent of its electricity using geothermal energy. In the United States, geothermal energy accounts for less than 1 percent of the renewable energy used. Geothermal power plants are currently in operation in many states including California, Nevada, New Mexico, Oregon, Hawaii, and Utah. Geothermal energy has less growth potential than wind or solar energy because it is not easily accessible everywhere. Hazardous gases and steam may also escape from geothermal power plants, another drawback of geothermal energy.

Ground Source Heat Pumps

Another approach to tapping Earth's thermal resources is the use of **ground source heat pumps**, a technology that transfers heat from the ground to a building. Ground source heat pumps take advantage of the high thermal mass of the ground. Earth's temperature about 3 m (10 feet) underground remains fairly constant year-round, at 10°C to 15°C (50°F–60°F), because the ground retains the Sun's heat more effectively than does the ambient air. We can take advantage of this fact to heat and cool residential and commercial buildings. Although the heat tapped by ground source heat pumps is often referred to informally as "geothermal," it comes not from geothermal energy but from solar energy.

FIGURE 39.10 on page 478 shows how a ground source heat pump transfers heat from the ground to a house. In contrast to the geothermal systems just described, ground source heat pumps do not remove steam or hot water from the ground. In much the same way that a solar water heating system works, a ground source heat pump cycles fluid through pipes buried underground. In winter, this fluid absorbs heat from underground. The slightly warmed fluid is compressed in the heat pump to increase its temperature even more, and the heat is distributed throughout the house. The fluid is then allowed to expand, which causes it to cool and run through the cycle again, picking up more heat from the ground. In summer, when the underground temperature is lower than the ambient air temperature, the fluid is cooled underground and then pulls heat from the house as it circulates, resulting in a cooler house as heat is transferred underground.

Ground source heat pumps can be installed anywhere in the world, regardless of whether there is geothermal

Geothermal energy Heat energy that comes from the natural radioactive decay of elements deep within Earth.

Ground source heat pump A technology that transfers heat from the ground to a building.

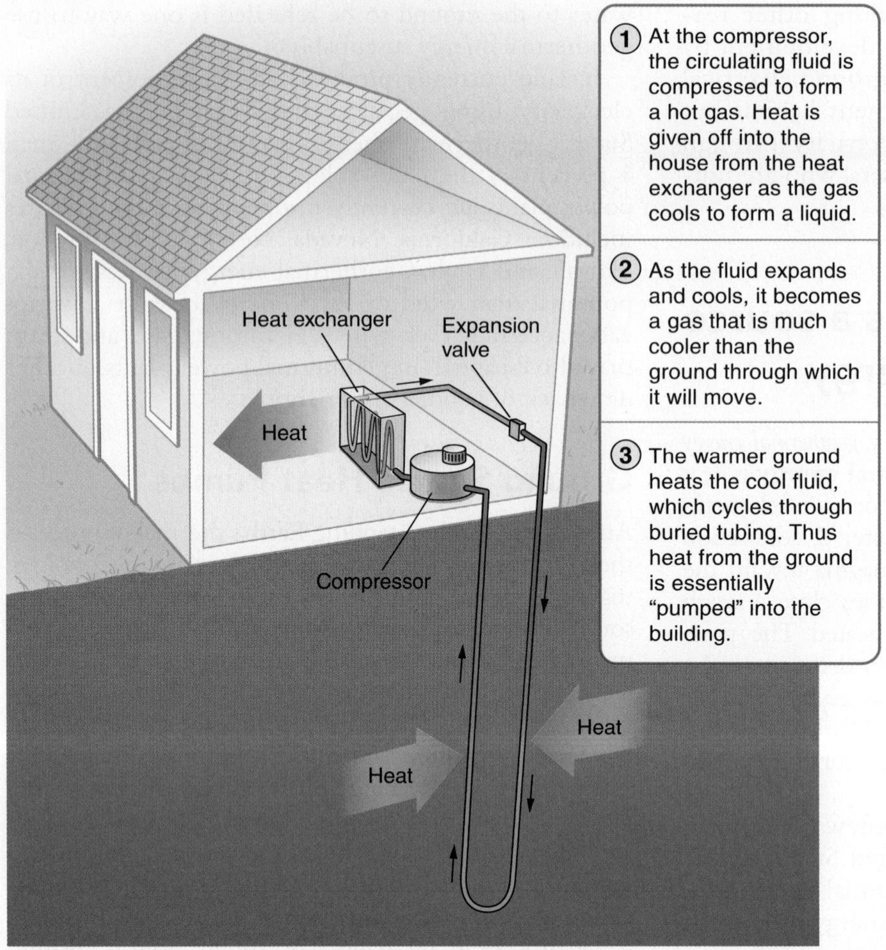

1 At the compressor, the circulating fluid is compressed to form a hot gas. Heat is given off into the house from the heat exchanger as the gas cools to form a liquid.
2 As the fluid expands and cools, it becomes a gas that is much cooler than the ground through which it will move.
3 The warmer ground heats the cool fluid, which cycles through buried tubing. Thus heat from the ground is essentially "pumped" into the building.
Heat exchanger
Expansion valve
Heat
Compressor
Heat
Heat

FIGURE 39.10 Heating and cooling with a ground source heat pump. By exchanging heat with the ground, a ground source heat pump can heat and cool a building using 30 to 70 percent less energy than traditional furnaces and air conditioners.

energy accessible in the vicinity. The operation of the pump requires some energy, but in most cases the system uses 30 to 70 percent less energy to heat and cool a building than a standard furnace or air conditioner.

Hot Water Heat Pumps

The hot water heat pump, a variation of the ground source heat pump, extracts heat from the air in a garage or basement and transfers it to water in a domestic hot water tank. This water is then used for household activities such as washing dishes and taking showers. Hot water heat pump systems are similar to those found in air conditioners and refrigerators as well as the ground source heat pump we just described. A refrigerator extracts heat from the inside of an insulated box—the refrigerator—thereby lowering the temperature of the food and air inside, while discharging heat to the outside—the kitchen. A hot water heat pump works in reverse: It extracts heat from the surrounding air in the basement or garage and pumps it into a tank filled with water destined for the kitchen sink or bathroom shower. We have said that a resistance coil water heater is 99 percent efficient, meaning that 99 percent of the energy in the electricity is transferred to the water. In a hot water heat pump, between 200 and 250 percent of the amount of energy in the electricity used to run the hot water heat pump is transferred to the water in the tank. You might be wondering if this violates the first law of thermodynamics. How can we obtain more energy from the system than the amount of energy put into the system? Remember our efficiency equation introduced in Chapter 2: Energy efficiency is the ratio of the amount of energy obtained in the desired form to the total amount of energy introduced into the system. The heat pump technology utilizes the energy in the electricity and also extracts heat energy from the surrounding air. The amount of electrical energy required to run the heat pump plus the energy extracted from the air in the room, is greater than the electrical energy used to run the heat pump. So this yields a number greater than 100 percent of the electricity used to run the pump. In the case of a home hot water heat pump, the overall energy gain in the hot water tank can be as much as 200 to 250 percent of the electrical energy put into the system. For this reason, hot water heat pumps are becoming more and more popular in homes and rental properties in the United States and can be found in all plumbing supply and big box home improvement stores.

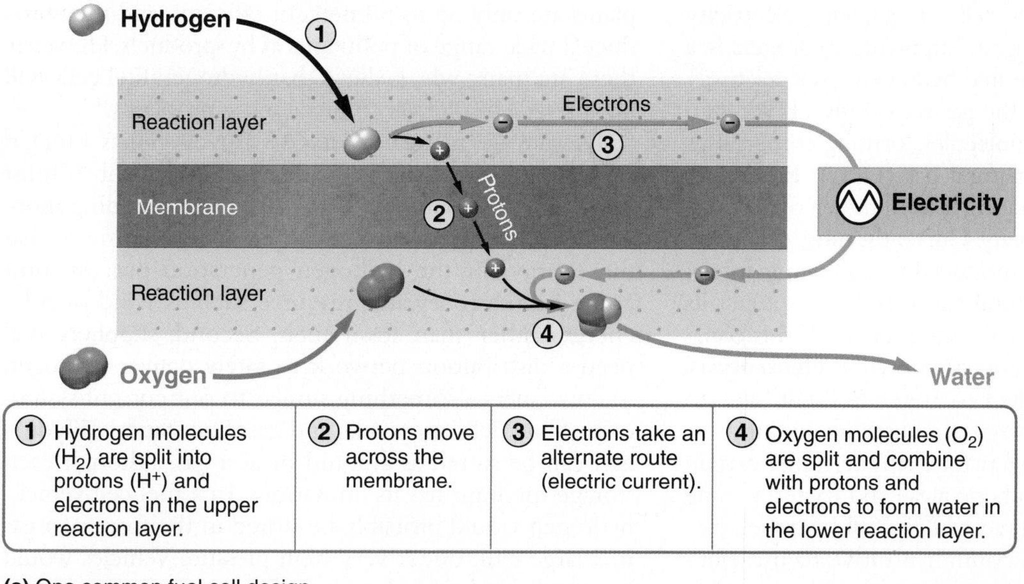

Hydrogen
Reaction layer
Electrons
Membrane
Protons
Electricity
Reaction layer
Oxygen
Water
1
2
3
4
(1) Hydrogen molecules (H_2) are split into protons (H^+) and electrons in the upper reaction layer.
(2) Protons move across the membrane.
(3) Electrons take an alternate route (electric current).
(4) Oxygen molecules (O_2) are split and combine with protons and electrons to form water in the lower reaction layer.

(a) One common fuel cell design

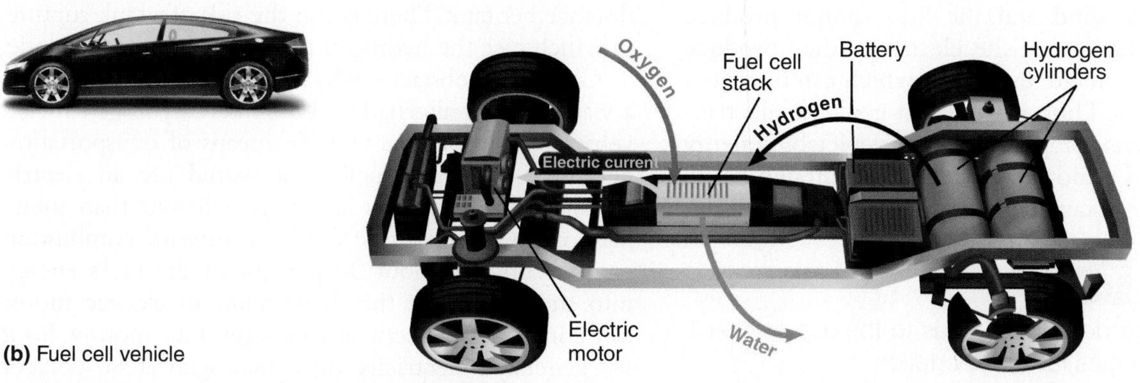

Oxygen
Fuel cell stack
Battery
Hydrogen cylinders
Hydrogen
Electric current
Electric motor
Water

(b) Fuel cell vehicle

FIGURE 39.11 Power from a hydrogen fuel cell. (a) Hydrogen gas enters the cell from an external source. Protons from the hydrogen molecules pass through a membrane, while electrons flow around it, producing an electric current. Water is the only waste product of the reaction. (b) In a fuel cell vehicle, hydrogen is the fuel that reacts with oxygen to provide electricity to run the motor.

Hydrogen fuel cells have many potential applications

We end our coverage of sustainable energy types with one additional energy technology that has received a great deal of attention for many years: hydrogen fuel cells.

The Basic Process in a Fuel Cell

A **fuel cell** is an electrical-chemical device that converts fuel, such as hydrogen, into an electrical current. A fuel cell operates much like a common battery, but with one key difference. In a battery, electricity is generated by a reaction between two chemical reactants, such as nickel and cadmium. This reaction happens in a closed container to which no additional materials can be added; eventually the reactants are used up and the battery goes dead. In a fuel cell, however, the reactants are added continuously to the cell, so the cell produces electricity for as long as it continues to receive fuel.

FIGURE 39.11 shows how hydrogen functions as one of the reactants in a hydrogen fuel cell. Electricity is generated by the reaction of hydrogen with oxygen, which forms water:

$$2H_2 + O_2 \rightarrow \text{energy} + 2\,H_2O$$

Although there are many types of hydrogen fuel cells, the basic process forces protons from hydrogen gas through a membrane, while the electrons take a different pathway. The movement of protons in one direction and electrons in another direction generates an electric current.

Fuel cell An electrical-chemical device that converts fuel, such as hydrogen, into an electrical current.

Using a hydrogen fuel cell to generate electricity requires a supply of hydrogen. Supplying hydrogen is a challenge, however, because free hydrogen gas is relatively rare in nature and because the gas is explosive. Hydrogen tends to bond with other molecules, forming compounds such as water (H_2O) or natural gas (CH_4). Producing hydrogen gas requires separating it from these compounds using either heat or electricity. Currently, most commercially available hydrogen is produced by an energy-intensive process of burning natural gas in order to extract its hydrogen; carbon dioxide is a waste product of this combustion. In an alternative process, known as **electrolysis**, an electric current is applied to water to "split" it into hydrogen and oxygen. Energy scientists are looking for other ways to obtain hydrogen; for example, under certain conditions, some photosynthetic algae and bacteria, using sunlight as their energy source, can give off hydrogen gas.

Although it may seem counterintuitive to use electricity to create electricity, the advantage of hydrogen is that it can act as an energy carrier. Renewable energy sources such as wind and the Sun cannot produce electricity constantly, but the electricity they produce can be used to generate hydrogen, which can be stored until it is needed. Thus, if we could generate electricity for electrolysis using a clean, nondepletable energy resource such as wind or solar energy, hydrogen could potentially be a sustainable energy carrier.

AP® Exam Tip

Be prepared to describe hurdles to the use of alternative technologies for transportation.

The Viability of Hydrogen

Some policy makers consider hydrogen fuel cells to be the future of energy and the solution to many of the world's energy problems. Hydrogen fuel cells are 80 percent efficient in converting the potential energy of hydrogen and oxygen into electricity, with water as their only by-product. In contrast, thermal fossil fuel power plants are only 35 to 50 percent efficient, and they produce a wide range of pollutants as by-products. However, there are many who believe that hydrogen fuel cells will not provide a solution to our energy problems.

Despite the many advantages of hydrogen as a fuel, it also has a number of disadvantages. First, scientists must learn how to obtain hydrogen without expending more fossil fuel energy than its use would save. This means that the energy for the hydrogen generation process must come from a renewable resource such as wind or solar energy rather than fossil fuels. Second, suppliers will need a distribution network to safely deliver hydrogen to consumers—something similar to our current system of gasoline delivery trucks and gasoline stations. Hydrogen can be stored as a liquid or as a gas, although each storage medium has its limitations. In a fuel cell vehicle, hydrogen would probably be stored in the form of a gas in a large tank under very high pressure. Vehicles would have to be redesigned with fuel tanks much larger than current gasoline tanks to achieve an equivalent travel distance per tank. There is also the risk of a tank rupture, in which case the hydrogen might catch fire or explode.

Given these obstacles, why is hydrogen even considered a viable energy alternative? Ultimately, hydrogen-fueled vehicles could be a sustainable means of transportation because a hydrogen-fueled car would use an electric motor. Electric motors are more efficient than internal combustion engines: While an internal combustion engine converts about 20 percent of the fuel's energy into the motion of the drive train, an electric motor can convert 60 percent of its energy into motion. So if we generated electricity from hydrogen at 80 percent efficiency, and used an electric motor to convert that electricity into vehicular motion at 60 percent efficiency, we would have a vehicle that is much more efficient than one with an internal combustion engine. Thus, even if we obtained hydrogen by burning natural gas, the total amount of energy used to move an electric vehicle using hydrogen might still be substantially less than the total amount needed to move a car fueled by gasoline. Using solar or wind energy to produce the hydrogen would lower the environmental cost even more, and the energy supply would be renewable. In those circumstances, an automobile could be fueled by a truly renewable source of energy that is both carbon neutral and pollution free.

Electrolysis The application of an electric current to water molecules to split them into hydrogen and oxygen.

MODULE 39 AP® Review

Preparing for the AP® Exam

In this module, we have seen that the Sun and wind provide viable sources of renewable energy in many locations. Solar energy can be used both passively, such as locating a building in a particular direction, as well as actively, such as using photovoltaic cells. Wind energy is harnessed through the use of a wind turbine, which converts the energy of moving air into electricity. Wind is the fastest growing form of new electricity generation in the world. Harnessing geothermal energy also is a good source of energy in certain locations. Hydrogen has great potential that has not yet been realized. We can use all of these energy forms in appropriate locations under the proper circumstances, which is the focus of the next module.

AP® Practice Questions

Choose the best answer for the following.

1. Which is an application of passive solar technology?
   (a) concentrating solar thermal
   (b) photovoltaic cells
   (c) solar water heating
   (d) solar ovens
2. On average, what percentage of time does a land-based wind turbine generate electricity?
   (a) 60 percent
   (b) 45 percent
   (c) 30 percent
   (d) 25 percent
3. Which is NOT true about geothermal energy?
   (a) It is only available in limited areas.
   (b) It cannot be used for cooling.
   (c) It can be locally depleted due to heavy use.
   (d) Ground source heat pumps require an additional source of energy.
4. A hydrogen fuel cell is most similar to
   (a) an engine.
   (b) a photovoltaic cell.
   (c) a source of coal.
   (d) a battery.
5. Which is NOT a benefit of solar energy systems?
   (a) They typically produce electricity during peak demand.
   (b) They require very little maintenance.
   (c) They produce electricity continuously.
   (d) They do not produce pollution while generating electricity.

MODULE 40

Planning Our Energy Future

Although renewable energy is a more sustainable energy choice than nonrenewable energy, using any form of energy has an impact on the environment. Biomass, for instance, is a renewable resource only if it is used sustainably. Overharvesting wood leads to deforestation and degradation of the land, as we saw in the description of Haiti in Chapter 3. Wind turbines can kill birds and bats, and hydroelectric turbines kill millions of fish. Manufacturing photovoltaic solar panels requires heavy metals and a great deal of water. Because all energy

Learning Goals

After reading this module, you should be able to

- discuss the environmental and economic options we must assess in planning our energy future.
- consider the challenges of a renewable energy strategy.

choices have environmental consequences, minimizing energy use through conservation and efficiency is the best approach. After we achieve that, we must make energy choices wisely, depending on a variety of environmental, economic, and convenience factors.

Our energy future depends on efficiency, conservation, and the development of renewable and nonrenewable energy resources

Each of the renewable energy resources we have discussed in this chapter has unique advantages. None of these resources, however, is a perfect solution to our energy needs. **TABLE 40.1** lists some of the advantages and limitations of each of these resources. In short, no single energy resource that we are currently aware of can replace nonrenewable energy resources in a way that is completely renewable, nonpolluting, and free of impacts on the environment. A sustainable energy strategy, therefore, must combine energy efficiency, energy conservation, and the development of renewable and nonrenewable energy resources, taking into account the costs, benefits, and limitations of each. Convenience and reliability are also important factors. Finally, logistical considerations, such as where an energy source is located and how we transport the energy from that source to users, are also important. This is particularly important with the generation of electricity from renewable sources in remote regions, which requires an electrical transmission grid to get it to users.

A renewable energy strategy presents many challenges

Energy expert Amory Lovins suggests that innovation and technological advances, not the depletion of a resource, have provided the driving force for moving from one energy technology to the next. Extending this concept to the present, one can argue that we will develop new energy technologies before we run out of the fuels on which we currently depend.

Despite their tremendous potential, however, renewable energy resources are unlikely to replace fossil fuels completely in the immediate future unless nations commit to supporting their development and use through direct funding and financial incentives such as tax cuts and consumer rebates. In fact, the U.S. Department of Energy predicts that fossil fuel consumption will continue to increase in the United States well into the middle of the twenty-first century.

TABLE 40.1 Comparison of renewable energy resources

Energy resource	Advantages
Liquid biofuels	• Potentially renewable • Can reduce our dependence on fossil fuels • Reduce trade deficit • Possibly more environmentally friendly than fossil fuels
Solid biomass	• Potentially renewable • Eliminates waste from environment • Available to everyone • Minimal technology required
Photovoltaic solar cells	• Nondepletable resource • After initial investment, no cost to harvest energy
Solar water heating systems	• Nondepletable resource • After initial investment, no cost to harvest energy
Hydroelectricity	• Nondepletable resource • Low cost to run • Flood control • Recreation
Tidal energy	• Nondepletable resource • After initial investment, no cost to harvest energy
Geothermal energy	• Nondepletable resource • After initial investment, no cost to harvest energy • Can be installed anywhere (ground source heat pump)
Wind energy	• Nondepletable resource • After initial investment, no cost to harvest energy • Low up-front cost
Hydrogen fuel cell	• Efficient • Zero Pollution

	Disadvantages	Emissions (pollutants and greenhouse gases)	Electricity cost ($/kWh)	Energy return on energy investment
	• Loss of agricultural land • Higher food costs • Lower gas mileage • Possible net increase in greenhouse gas emissions	CO_2 and methane		1.3 (from corn) 8 (from sugar cane)
	• Deforestation • Erosion • Indoor and outdoor air pollution • Possible net increase in greenhouse gas emissions	• Carbon monoxide • Particulate matter • Nitrogen oxides • Possible toxic metals from MSW • Danger of indoor air pollutants		
	• Manufacturing materials requires high input of metals and water • No plan in place to recycle solar panels • Geographically limited • High initial costs • Storage batteries required for off-grid systems	• None during operation • Some pollution generated during manufacturing of panels	0.1	8
	• Manufacturing materials requires high input of metals and water • After initial investment, no cost to harvest energy and water • No plan in place to recycle solar panels • Geographically limited • High initial costs	• None during operation • Some pollution generated during manufacturing of panels		
	• Limited amount can be installed in any given area • High construction costs • Threats to river ecosystems • Loss of habitat, agricultural land, and cultural heritage; displacement of people • Siltation	• Methane from decaying flooded vegetation	.05–.11	12
	• Potential disruptive effect on some marine organisms • Geographically limited	• None during operation		15
	• Emits hazardous gases and steam • Geographically limited	• None during operation	.05–.30	8
	• Turbine noise • Deaths of birds and bats • Geographically limited to windy areas near transmission lines • Aesthetically displeasing to some • Storage batteries required for off-grid systems	• None during operation	.04–.06	18
	• Producing hydrogen is an energy-intensive process • Lack of distribution network • Hydrogen storage challenges	• None during operation		8

In spite of their extremely rapid growth, wind and solar energy still account for far less than 1 percent of all the energy produced in the United States. Government funding or other sources of capital are needed to support research to overcome the current limitations of many renewable energy resources. One limitation that is already evident relates to the transmission of renewable electricity over the electrical distribution network. Other limitations to consider are energy cost and storage.

Improving the Electrical Grid

An increased reliance on renewable energy means that energy will be obtained in many locations and will need to be delivered to other locations. Delivery can be particularly problematic when electricity for an urban area is generated at a remote location. The electrical distribution system—the grid that we described in Chapter 12—was not originally designed for this purpose. So in addition to investing in new energy sources, the United States will have to upgrade its existing electrical infrastructure—its power plants, storage capacity, and distribution networks. Approximately 40 percent of the energy used in the United States is used to generate electricity. The U.S. electricity distribution system is outdated and subject to overloads and outages, which cost the U.S. economy over $100 billion per year. There are regions of the country that cannot supply enough generating capacity to meet local needs, while in other locations the electrical infrastructure cannot accommodate all the electricity that is generated, including from small generators such as household PV systems. Furthermore, the current system requires that electricity be moved long distances from power plants to consumers. Approximately 5 to 10 percent of the electricity generated is lost as it is transported along electrical transmission lines, and the greater the distance, the more that is lost. While the storage capacity of batteries improves each year, batteries are probably not a sustainable solution for this problem of energy loss. Many people are focusing their attention on improving the electrical grid to make it as efficient as possible at moving electricity from one location to another, thereby reducing the need for storage capacity.

An energy economy based on nondepletable energy sources requires reliable electricity storage and affordable—or at least effective and efficient—distribution networks. U.S. energy scientists maintain that because we currently do not have a cost-effective, reliable means of storing energy, we should not depend on intermittent sources such as wind and solar energy for more than about 20 percent of our total electricity production since it could lead to risky instability in the grid. However, a number of European countries now have more than 20 percent renewables on their electricity grids, which may lead scientists to re-evaluate this recommendation.

One solution currently being implemented is the **smart grid**, an efficient, self-regulating electricity distribution network that accepts any source of electricity and distributes it automatically to end users. A smart grid uses computer programs and the Internet to tell electricity generators when electricity is needed and electricity users when there is excess capacity on the grid. In this way, it coordinates electricity use with electricity availability. Since 2008, government and industry contributions have brought total investment in the smart grid to almost $8 billion, which has been used to fund more than 100 smart grid projects around the country.

How does a smart grid work? **FIGURE 40.1** shows one example. With "smart" appliances plugged into a smart grid, at bedtime a consumer could set an appliance such as a dishwasher to operate overnight. A computer on the dishwasher would be programmed to run the appliance anytime between midnight and 5:00 AM, depending on when there is a surplus of electricity. The dishwasher's computer would query the smart grid and determine the optimal time, in terms of electricity availability, to turn on the appliance. The smart grid could also help manage electricity demand so that peak loads do not become too great. For example, smart grid technology could delay the onset of the cooling cycle in a large supermarket freezer. Perhaps the freezer units would delay by 15 minutes the time at which they would initiate a cycle. This might result in a very small increase in temperature in the freezers, but it wouldn't be large enough to adversely affect the food. And it might delay or stagger electricity demand at a time when demand is high. We cannot control the timing of all electricity demand, but by improving consumer awareness of electricity abundance and shortages, using smart appliances, and setting variable pricing for electricity, we can make electricity use much more regular, and thus more sustainable.

Our current electrical infrastructure relies on a system of large electricity producers—regional electricity generation plants. When one plant goes off-line or shuts down, the reduction in available generating capacity puts greater demands on the rest of the system. Some energy experts maintain that a better system would consist of a large number of small-scale electricity generation "parks" that rely on a mix of fossil fuel and renewable energy sources. These experts maintain that a system of decentralized energy parks would be the least expensive and most reliable electrical infrastructure to meet our future needs. Small, local energy parks would save money and energy by transporting electricity a shorter distance. Such decentralized generators would also be less likely to suffer breakdowns or sabotage. Since each small energy park might serve only a few thousand people, widespread outages would be much less likely.

Smart grid An efficient, self-regulating electricity distribution network that accepts any source of electricity and distributes it automatically to end users.

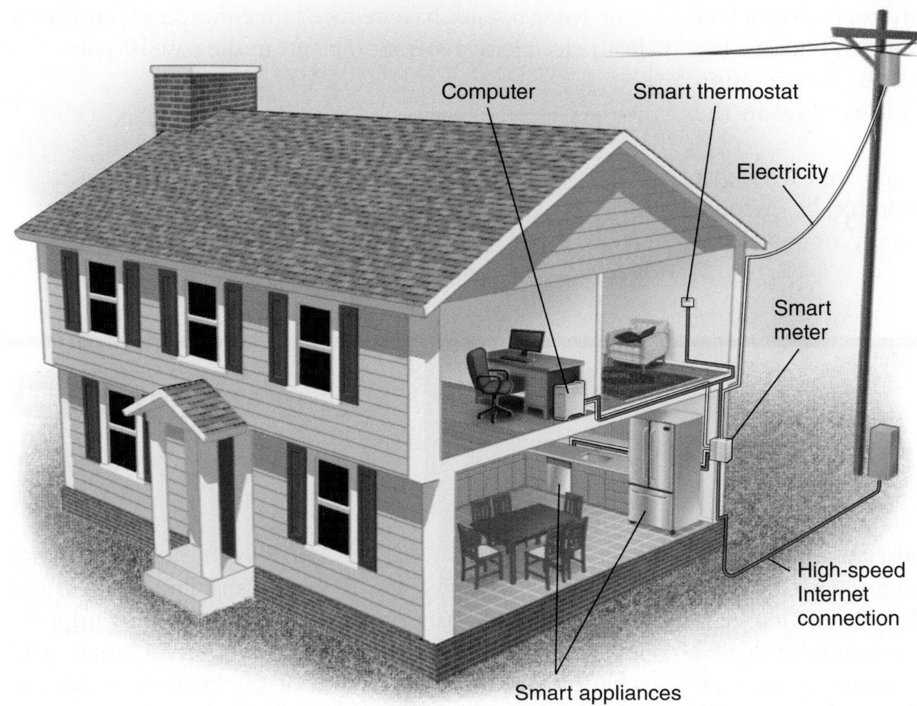

FIGURE 40.1 Using a smart grid. A smart grid optimizes the use of energy in a home by continuously coordinating energy use with energy generation.

Addressing Energy Cost and Storage

The major impediments to widespread use of wind, solar, and tidal energy—the forms of renewable energy with the least environmental impact—are cost and the limitations of energy storage technology. Fortunately, the cost of renewable energy has been falling. For example, in some markets wind energy is now cost-competitive with natural gas and coal. Throughout this book we have seen that the efficiency of production improves with technological advances and experience. In general, as we produce more of something, and get experience from making it, we learn to produce it less expensively. Production processes become dramatically more efficient, more companies enter the market, and developing new technologies has a clear payoff. For the consumer, this technological advancement also has the benefit of lowering prices: For electricity generation from solar, wind, and natural gas, we have seen that costs tend to decline in a fairly regular way as installed capacity grows.

What are the implications of this relationship between experience and efficiency? In general, any technology that has been in widespread use has an advantage over a newer technology because it is familiar and because the less expensive something is, the more people will buy it, leading to further reductions in its price. State and federal subsidies and tax incentives also help to lower the price of a technology. Tax credits and rebates have been instrumental in reducing the cost of solar and wind energy systems for consumers.

Similarly, in time, researchers will develop solutions to the problem of creating efficient energy storage systems, which might reduce the need to transport electricity over long distances. One very simple and effective approach is to use the excess capacity during off-peak hours to pump water uphill with electricity to a reservoir. Then, during hours of peak demand, operators can release the water through a turbine to generate the necessary electricity—cleanly and efficiently. Research into battery technology and hydrogen fuel cell technology continues. Battery capacity and charge/discharge efficiencies are improving rapidly and cost per kilowatt hour of battery storage has been going down. Tesla is probably the best-known company making advances in this area, with products such as the Tesla Powerwall 2, which is a lithium ion battery that can be hung on a wall in a garage or basement.

Progress on these and other technologies may accelerate with government intervention, taxes on industries that emit carbon dioxide, or a market in which consumers are willing to pay more for technologies that have minimal environmental impacts. In the immediate future, we are more likely to move toward a sustainable energy mix if nonrenewable energy becomes more expensive. Consumers have shown more willingness to convert in large numbers to renewable energy sources, or to engage in further energy conservation, when fossil fuel prices increase. We have already seen instances of this shift in behavior. In 2008, energy conservation increased when oil prices rose rapidly to almost

$150 per barrel and gasoline in most of the United States cost more than $4 per gallon. People used public transportation more often, drove more fuel-efficient vehicles, and carpooled more than they did before the price spike. As gasoline prices decrease, conservation measures tend to decrease as well, and we saw in the latter half of 2008.

Other ways to spur conservation are initiatives that regulate the energy mix itself—for example, by encouraging that a certain fraction of electricity be generated using renewable energy sources. One such initiative is the Regional Greenhouse Gas Initiative (RGGI), whereby nine eastern states have reduced greenhouse gas emissions from electricity generation plants in the past 10 years.

AP® Exam Tip

You should be able to explain the different methods of producing electricity.

MODULE 40 AP® Review

Preparing for the AP® Exam

In this module, we have seen that conservation and efficiency need to be considered simultaneously with a wide variety of renewable energy sources. Environmental, economic, and convenience considerations must be evaluated when comparing renewable energy options against each other and with nonrenewable fuels. The path we take in our energy future will depend on the potential for a given energy source to meet the needs of the country, along with the availability and issues of reliability, storage, and accessibility of that source. Challenges include improving the electrical grid and the adoption of smart grid technologies to address some of the challenges that are specific to renewable and sometimes intermittent energy supplies.

AP® Practice Questions

Choose the best answer for the following.

1. Which is NOT a disadvantage of liquid biofuels?
   (a) They are associated with lower gas mileage.
   (b) They create more carbon monoxide than fossil fuels.
   (c) They can contribute to a loss of agricultural land.
   (d) They can increase food costs.

2. Which aspect of renewable energy electricity generation requires updating the electricity transmission grid?
   I. Electricity generators are located in numerous, remote locations.
   II. There is a need to transport electricity long distances.
   III. There are storage problems due to the unpredictable nature of some renewables.
   (a) I and II
   (b) I and III
   (c) II and III
   (d) I, II, and III

3. The smart grid does NOT
   (a) use the Internet to coordinate energy use and energy availability.
   (b) reduce the variability in electricity demand.
   (c) have the potential to provide a cheap way to store electricity.
   (d) increase the need for variable pricing of electricity.

4. Which renewable energy source has become cost-competitive with fossil fuels?
   (a) tidal
   (b) geothermal
   (c) wind
   (d) solar photovoltaic

5. Which will NOT increase adoption of renewable technologies?
   (a) increased cost of fossil fuels
   (b) a carbon dioxide emissions tax
   (c) cheaper energy storage
   (d) decreased government subsidies

Working Toward Sustainability

Building an Alternative Energy Society in Iceland

The people of Iceland use more energy per capita than the people of any other nation, including the United States. However, the energy Icelanders use is almost all in the form of local, renewable resources that do not pollute or contribute greenhouse gases to the environment.

This isolated European island nation has had to learn to be self-sufficient in energy or suffer the high cost of importing fuel. When the Vikings first came to Iceland over a thousand years ago, they relied on biomass, in the form of birch wood and peat, for fuel. The resulting deforestation, and the slow regrowth of forests in Iceland's cold temperatures and limited sunlight, restricted human population growth and economic development for the next thousand years. With the beginning of the Industrial Revolution in the eighteenth century, Iceland began to supplement its biomass fuel with imported coal, but the expense of importing coal also limited its economic growth.

In the late nineteenth and early twentieth centuries, Iceland began to look to its own resources for energy. It began by tapping its abundant freshwater resources to generate hydroelectricity, which became the country's major energy source for residential and industrial use. This transition led to the general electrification and economic modernization of the country and greatly reduced its dependence on imported fossil fuels. Iceland did not stop with hydroelectricity, however, but sought ways to utilize its other major renewable energy source: the thousands of geysers and hot springs on this volcanic island that would provide ready access to geothermal energy.

Geothermal energy is now the primary energy source for home heating in Iceland, and geothermal and hydroelectric resources provide energy for nearly all electricity generation in the country. Even so, the potential of these resources remains relatively underutilized. Iceland has harvested less than 20 percent of its hydroelectric potential, and there is even more potential in its geothermal resources.

Iceland continues to take advantage of local resources and global technology to develop clean, sustainable energy sources. Despite its commitment to renewable energy, Iceland is still dependent on imported fossil fuels to run its cars, trucks, buses, and fishing vessels. In 2000, Iceland embarked on an ambitious project to wean the country from fossil fuels by 2050. The goal was to use its sustainably generated electricity to split water and obtain hydrogen, then use the hydrogen as fuel. In April 2003, Iceland opened one of the world's first filling stations for hydrogen-fueled vehicles. In August of that year, hydrogen-fueled buses for public transportation were introduced in Reykjavik, the capital of Iceland. Hydrogen-fueled rental cars were also available. The financial crisis of 2008 through 2011 delayed and even reversed some of the achievements. The original hydrogen fueling station was dismantled in 2012, but in 2018, Iceland renewed its commitment to hydrogen with the construction of three hydrogen filling stations. The country is small enough that 80 percent of the Icelandic population is within easy reach of a fueling station—the highest percentage in the world.

A geothermal energy power plant in Iceland. This thermal power station in west Iceland uses geothermal energy from Earth rather than a fossil fuel to produce electricity and hot water. *(Feifei Cui-Paoluzzo/Getty Images)*

Critical Thinking Questions

1. Can the knowledge and experience gained in Iceland be applied to many other parts of the world? Why or why not?
2. What challenges would there be if hydrogen fuel was to become the major source of energy for automobiles in the United States?

References

Blanchette, S. 2008. A hydrogen economy and its impact on the world as we know it. *Energy Policy* 36:522–530.

Veal, L. 2017. Hydrogen, Iceland and the Future of Transport. International Press Syndicate. Published on April 14, 2017. https://www.indepthnews.net/index.php/sustainability/affordable-clean-energy/1074-hydrogen-iceland-and-the-future-of-transport

Chapter 13 Review

In this chapter, we have examined the role of conservation as well as increased energy efficiency in reducing the demand for energy. We have described the different categories of renewable energy and examined the two most prominent renewable energy sources: biomass and energy from flowing and standing water. Biomass energy contains modern carbon and can be obtained from wood, charcoal, and animal wastes. Energy can be harnessed from both standing water and free flowing water, typically to generate electricity. Solar energy can be harnessed both passively and actively. The most prominent active collection of solar energy comes from photovoltaic cells that convert sunlight into electricity. Wind energy is harnessed directly and a wind turbine is very similar to the turbines used to generate electricity from fossil fuels. Geothermal energy from Earth can be used in specific locations. Hydrogen is a fuel that has much promise but is not likely to be used widely anytime soon. Each renewable energy resource has its advantages and disadvantages and these can be considered from both environmental and economic perspectives.

Key Terms

Energy conservation
Tiered rate system
Peak demand
Passive solar design
Thermal mass
Potentially renewable
Nondepletable
Renewable
Biofuel
Modern carbon
Fossil carbon
Carbon neutral
Net removal
Ethanol
Biodiesel
Flex-fuel vehicle
Hydroelectricity
Run-of-the-river
Water impoundment
Tidal energy
Siltation
Active solar energy
Photovoltaic solar cell
Wind energy
Wind turbine
Geothermal energy
Ground source heat pump
Fuel cell
Electrolysis
Smart grid

Learning Goals Revisited

Module 37 Conservation, Efficiency, and Renewable Energy

Describe strategies to conserve energy and increase energy efficiency.

Turning down the thermostat and driving fewer miles are examples of steps individuals can take to conserve energy. Buying appliances that use less energy and switching to compact fluorescent light bulbs are examples of steps individuals can take to increase energy efficiency. Buildings that are carefully designed for energy efficiency can save both energy resources and money. Reducing the demand for energy can be an equally effective or a more effective means of achieving energy sustainability than developing additional sources of energy.

Explain differences among the various renewable energy resources.

Renewable energy resources include nondepletable energy resources, such as the Sun, wind, and moving water, and potentially renewable energy resources, such as biomass. Potentially renewable energy resources will be available to us as long as we use them sustainably.

Module 38 Biomass and Water

Describe the various forms of biomass.

Biomass is one of the most common sources of energy in the developing world, but biomass energy is also used in developed countries. In theory, biomass energy is carbon neutral; that is, the carbon produced by combustion of biomass should not add to atmospheric carbon concentrations because it comes from modern, rather than fossil, carbon sources. Wood is a potentially renewable resource because, if harvests are managed correctly, it can be a continuous source of biomass energy. Ethanol and biodiesel have the potential to supply large amounts of renewable energy, but growing and processing these fuels makes demands on land and energy resources.

Explain how energy is harnessed from water.

Most hydroelectric systems use the energy of water impounded behind a dam to generate electricity. Run-of-the-river hydroelectric systems impound little or no water and have fewer environmental impacts, although they often produce less electricity.

Module 39 Solar, Wind, Geothermal, and Hydrogen

List the different forms of solar energy and their application.

Passive solar energy takes advantage of relatively inexpensive strategies such as the direction windows are facing in a building. Active solar technologies use technology to obtain heat or electrical energy from the Sun and have high initial costs but can potentially supply relatively large amounts of energy. Active solar applications can be small, such as those that fit on a rooftop or in a field, or they can be extremely large, on an industrial scale.

Describe how wind energy is harnessed and its contemporary uses.

Wind turbines can be located on land or in the near-offshore environment. Frequently, a number of wind turbines are grouped together in wind farms. Wind is the most rapidly growing source of renewable electricity. It is a clean, nondepletable energy resource, but objections to wind farms are increasing because of aesthetics, sound, and hazards the turbines pose to birds and bats.

Discuss the methods of harnessing the internal energy from Earth.

Geothermal energy from underground can heat buildings directly or can generate electricity. However, geothermal power plants must be located in places where geothermal energy is accessible.

Explain the advantages and disadvantages of energy from hydrogen.

The only waste product from a hydrogen fuel cell is water, but obtaining hydrogen gas for use in fuel cells is an energy-intensive process. If hydrogen could be obtained using renewable energy sources, it could become a truly renewable source of energy.

Module 40 Our Energy Future

Discuss the environmental and economic options we must assess in planning our energy future.

Many scenarios have been predicted for the world's energy future. However, conserving energy, increasing energy efficiency, relying more on renewable energy sources, and improving energy distribution and storage will all be necessary to achieve energy sustainability. Fossil fuel use continues throughout the world today and it does not appear that it will decrease any time soon.

Consider the challenges of a renewable energy strategy.

Improving the electrical grid in the United States is vital if we are to increase reliance on renewable forms of electricity. However, because the grid is so widespread, expanding and maintaining its geographic spread and electrical capacity are expensive. Even with an expanded grid, there are numerous obstacles to increasing renewable electricity generation. The high economic cost—at least initially—for renewable forms of electricity is a challenge. Also, the difficulty and economic cost of storing electricity have no easy solutions at present.

Practice Math and Graphing

Preparing for the AP® Exam

Answer the following questions. Be sure to show all your work.

1. Practice Math

(a) Wind capacity in the United States has undergone a rapid increase over the past 2 decades. Total capacity of wind turbines in the United States was 4,000 MW in 2000 and 85,000 MW in 2017. What was the percentage increase per year over this 17-year period? Round to the appropriate number of significant figures.

(b) There was approximately 3,600 MW of installed geothermal capacity in the United States in 2016. Geothermal electricity generation in the United States went from 14 MWH in 2000 to 17 MWH in 2016. What was the percentage increase over this 16-year period? Round to the appropriate number of significant figures.

2. Practice Graphing

Smart meters have two-way communication between a home or business and an electrical utility. They range from recording hourly electricity usage to lowering electricity demand at specific times of peak demand on the grid. In 2007, there were approximately 1 million smart meters installed in the United States. By 2013, there were more than 50 million. Create the following two graphs with year on the x axis and number of meters in millions on the y axis.

(a) Show the pattern of smart meter growth in the United States, assuming the growth was roughly steady from 2007–2013.

(b) Show the pattern of smart meter growth in the United States assuming that most of the growth actually happened between 2007 and 2013, as was actually observed.

Section 1: Multiple-Choice Questions

Choose the best answer for questions 1–20.

1. Which is NOT an example of a potentially renewable or nondepletable energy source?
   (a) hydroelectricity
   (b) solar energy
   (c) nuclear energy
   (d) wind energy

2. Renewable energy sources are best described as
   (a) those that are the most cost-effective and support the largest job market.
   (b) those that are, or can be, perpetually available.
   (c) those that are dependent on increasing public demand and decreasing supply.
   (d) those that are being depleted at a faster rate than they are being replenished.

3. An energy-efficient building might include all of the following EXCEPT
   (a) building materials with low thermal mass.
   (b) a green roof.
   (c) southern exposure with large double-paned windows.
   (d) reused or recycled construction materials.

4. Which source of energy is NOT (ultimately) solar-based?
   (a) wind
   (b) biomass
   (c) tides
   (d) coal

5. Which demonstrates the use of passive solar energy?
   I. a south-facing room with stone walls and floors
   II. photovoltaic solar cells for the generation of electricity
   III. a solar oven
   (a) I only
   (b) II only
   (c) III only
   (d) I and III

Question 6 refers to the following graphs.

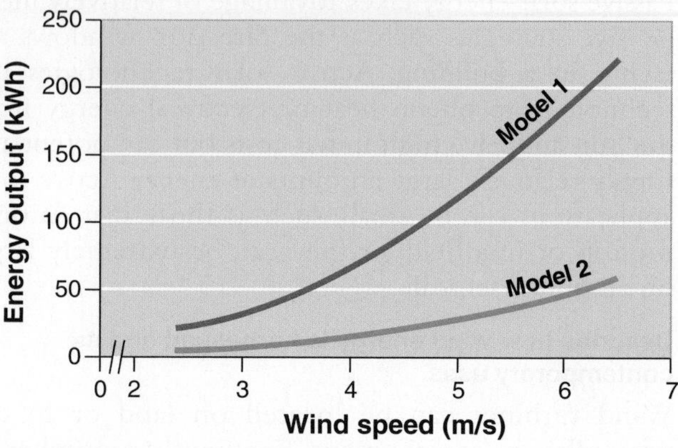

(a) Weekly energy output

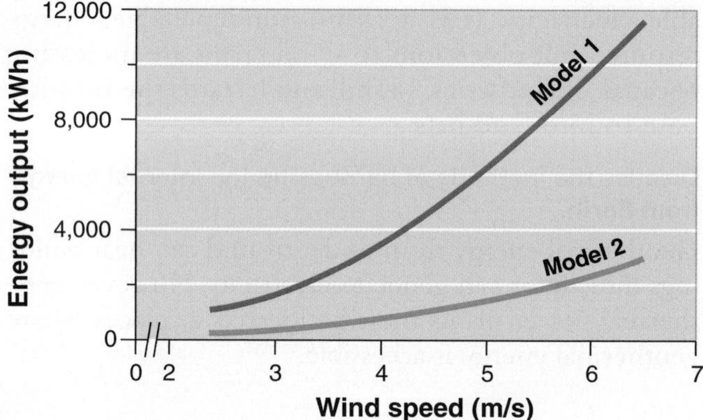

(b) Annual energy output

6. A study of small wind turbines in the Netherlands tested the energy output of two models, shown in the graphs. Which statement can be inferred from these data?
   (a) As wind speed increases, energy output decreases.
   (b) The annual energy output of model 1 can exceed 6,000 kWh.
   (c) As energy output surpasses 50 kWh per week, noise pollution increases.
   (d) Model 2 is likely to cause more bird and bat deaths.

7. The primary sources of renewable energy in the United States are
   (a) solar and wind energy.
   (b) hydroelectricity and tidal energy.
   (c) biomass and hydroelectricity.
   (d) geothermal and tidal energy.

Questions 8 and 9 use the following diagram, which represents annual U.S. energy consumption by source and sector for 2007.

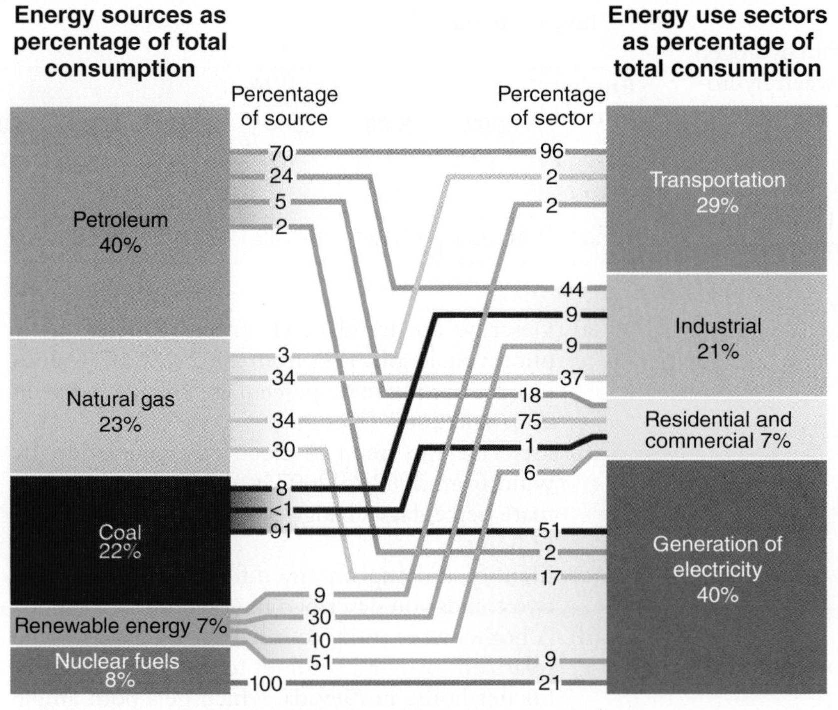

8. Which statement best describes the sources of energy in U.S. energy consumption patterns?
   (a) Most of the renewable energy is used in the industrial, residential, and commercial sectors.
   (b) Most of the electricity generated in the United States comes from nuclear energy.
   (c) The industrial sector is heavily dependent on coal and renewable energy.
   (d) Fossil fuels continue to be the major energy source for all sectors.

9. Which statement best describes U.S. energy use?
   (a) Transportation is the largest end use of energy in the United States.
   (b) Electricity generation is the largest end use of energy in the United States.
   (c) Electricity generation is powered mainly by nuclear energy.
   (d) Industry is the largest end use of energy in the United States.

10. Which statement best describes the role of renewable energy in the United States?
    (a) It is the dominant source of energy.
    (b) It is the largest contributor of greenhouse gases.
    (c) It is a large contributor to the transportation sector.
    (d) Its largest contribution is to the electricity generation sector.

11. The environmental impacts of cutting down a forest to obtain wood as fuel for heating and cooking could include
    I. deforestation and subsequent soil erosion.
    II. release of particulate matter into the air.
    III. a large net rise in atmospheric concentrations of sulfur dioxide.
    (a) I only
    (b) II only
    (c) III only
    (d) I and II

12. What is the fuel source for a flex-fuel vehicle?
    (a) electricity
    (b) biodiesel
    (c) E-85
    (d) solar

13. Which strategy will best help humans to achieve energy sustainability?
    I. building large, centralized power plants
    II. improving energy efficiency
    III. developing new energy technologies
    (a) II only
    (b) III only
    (c) I and II
    (d) II and III

14. Biomass is created through the conversion of _______ energy into _______ energy, which can then be used to generate electricity. In contrast, tidal energy involves the conversion of _______ energy into electricity.
    (a) chemical; potential; potential
    (b) solar; kinetic; potential
    (c) chemical; kinetic; kinetic
    (d) solar; chemical; kinetic

15. Concentrated solar thermal systems implement
    (a) active solar technology.
    (b) photovoltaic cell technology.
    (c) passive solar technology.
    (d) smart grid technology.

16. Which factor should NOT be considered when determining the EROEI of hydrogen fuel cell vehicles?
   (a) efficiency of isolating hydrogen from water or natural gas
   (b) by-products of generating electricity from hydrogen
   (c) energy produced by the reaction between hydrogen and oxygen
   (d) efficiency of electric motors

17. The primary purpose of a smart grid is to
   (a) improve the efficiency of electricity production.
   (b) improve the efficiency of energy transportation through power lines.
   (c) improve the capacity factor of power plants.
   (d) coordinate electricity use with electricity availability.

18. Which is a method of generating electricity from the movement of water?
   I. tidal energy
   II. run-of-the-river
   III. water impoundment
   (a) I only
   (b) II only
   (c) III only
   (d) I, II and III

19. Which is NOT an example of energy conservation?
   (a) consolidating trips
   (b) installing a high efficiency heating system
   (c) taking public transport
   (d) turning the thermostat down in winter

20. Maha wants to buy a dehumidifier. An Energy Star version of this appliance costs $300, while a standard unit costs $150. She expects to run it for 6 hours per day all year round. If the Energy Star unit costs 5 cents less per hour to run, approximately how much money will Maha save in 5 years if she spends the extra $150 for the Energy Star unit?
   (a) $150
   (b) $200
   (c) $350
   (d) $400

Section 2: Free-Response Questions

Write your answer to each part clearly. Support your answers with relevant information and examples. Where calculations are required, show your work.

1. Hydroelectricity provides about 7 percent of the electricity generated in the United States.
   (a) Explain how a hydroelectric power plant converts energy stored in water into electricity. (2 points)
   (b) Identify TWO factors that determine the amount of electricity that can be generated by an individual hydroelectric power plant. (2 points)
   (c) Describe the TWO main types of land-based hydroelectric power plants. (2 points)
   (d) Describe TWO economic advantages and TWO environmental disadvantages of hydroelectricity. (4 points)

2. The following table shows the amounts of electricity generated by photovoltaic solar cells and by wind in the United States from 2002 to 2007 (in thousands of megawatt-hours).

Energy Source	Year					
	2002	2003	2004	2005	2006	2007
PV solar cells	555	534	575	550	508	606
Wind	10,354	11,187	14,143	17,810	26,589	32,143

   (a) Describe the trend in electricity generation by photovoltaic solar cells from 2002 to 2007. Calculate the approximate percentage change between 2002 and 2007. (2 points)
   (b) Describe the trend in electricity generation by wind from 2002 to 2007. Calculate the approximate percentage change between 2002 and 2007. (2 points)
   (c) Identify and explain any difference between the two trends you described in (a) and (b). (2 points)
   (d) A homeowner wants to install either photovoltaic solar cells or wind turbines to provide electricity for her home in Nevada, which gets both ample sunlight and wind. Provide two arguments in favor of installing one of these technologies, and explain two reasons for not choosing the other. (2 points)
   (e) Would the installation of either PV solar cells or wind turbines be considered an application of energy conservation or of energy efficiency? Explain. (2 points)

3. Biomass accounts for approximately one-half of the renewable energy produced in the United States today.
   (a) Biomass is considered potentially renewable. Explain what this means, and contrast it with nondepletable resources (2 points)
   (b) Burning biomass releases carbon into the atmosphere. Compare the environmental effect of burning biomass to using fossil fuels. Is it better for the environment to replace fossil fuels with biomass? In your answer include the terms modern carbon, fossil carbon, and carbon neutral (4 points).
   (c) Biomass can also be processed to create biofuels, including ethanol and biodiesel. Explain the difference between ethanol and biodiesel, both in terms of how they are generated and how they are used. (4 points)

Fedor Selivanov/Shutterstock

science applied 6

Should Corn Become Fuel?

Corn-based ethanol is big business—so big, in fact, that to offset demand for petroleum, U.S. policy has required an increase in annual ethanol production from 34 billion liters (9 billion gallons) in 2008 to 136 billion liters (36 billion gallons) by 2022. By 2017, production had already risen to nearly 60 billion liters (16 billion gallons).

Ethanol proponents maintain that substituting ethanol for gasoline decreases air pollution, greenhouse gas emissions, and our dependence on foreign oil. Opponents counter that when we consider all of the inputs used to grow and process corn into ethanol, it increases air pollutants and greenhouse gases. Moreover, opponents claim that growing corn and converting it into ethanol uses more energy than we obtain when we burn ethanol for fuel and that the impact of ethanol on reducing our import of foreign oil is very small. What does the science tell us?

Does ethanol reduce air pollution?

Ethanol (C_2H_6O) and gasoline (a mixture of several compounds, including heptane, C_7H_{16}) are both hydrocarbons. Under ideal conditions, in the presence of enough oxygen, burning hydrocarbons produces only water and carbon dioxide. In reality, however, gasoline-only vehicles always produce some carbon monoxide (CO) because there can be insufficient oxygen during combustion. Carbon monoxide has direct effects on human health and also contributes to the formation of photochemical smog (see Chapter 15 for more on CO and air pollution).

Because ethanol is an **oxygenated fuel**—a fuel with oxygen as part of the molecule—adding ethanol to the fuel mix of a car should ensure that more oxygen is present and that combustion is more complete, which would reduce the production of CO. When it comes to combustion in vehicles, it is true that ethanol produces lower amounts of air pollutants than gasoline. However, when we compare the entire life cycle of growing, harvesting, and processing the corn for ethanol, it turns out that ethanol produces more of many air pollutants than gasoline.

Is ethanol neutral in the production of greenhouse gases?

Biofuels are modern carbon, not fossil carbon. As a result, burning ethanol should not introduce additional carbon into the atmospheric reservoir because the carbon captured in growing the corn kernels and the carbon released in burning the ethanol should cancel each other out. That is, ethanol should be neutral in terms of the amount of carbon produced. However, when we once again consider the entire life cycle of ethanol, we are reminded that corn production requires fossil fuels for driving a tractor, fertilizer production, and processing the corn kernels to make ethanol. These are all sources of fossil carbon, which means that ethanol production causes a net increase in the amount of greenhouse gases being produced.

Various aspects of corn production, such as plowing and tilling, may release additional CO_2 into the atmosphere from organic matter that otherwise would have remained undisturbed in the A and B horizons of the soil. Furthermore, greater demand for corn will increase pressure to convert land that is forest, grassland, or pasture into cropland. There is increasing evidence that these conversions result in a net transfer of carbon from the soil to the atmosphere and lead to additional increases in atmospheric CO_2 concentrations. Moreover, in the United States, the ethanol production process currently uses more coal than natural gas. Because coal emits nearly twice as much CO_2 per joule of energy as natural gas (see Chapter 12), producing the ethanol may reverse many of the benefits of replacing gasoline's fossil carbon with ethanol's modern carbon. Quite possibly, producing ethanol with

Oxygenated fuel A fuel with oxygen as part of the molecule.

FIGURE SA6.1 An ethanol-producing factory. Ethanol-producing factories, such as this one in South Dakota, convert corn into ethanol to produce a fuel that can be mixed with gasoline to power automobiles. *(Jim Parkin/Alamy)*

coal releases as much carbon into the atmosphere as simply burning gasoline in the first place. This means that when we consider the entire life cycle of ethanol production, it is not carbon neutral (**FIGURE SA6.1**).

Does ethanol provide a substantial return on energy investment?

We can also ask how much ethanol production provides a return on the investment, which is how much energy we get out of ethanol for every unit of energy we put in. Scientists at the U.S. Department of Agriculture have analyzed this problem, examining the energy it takes to grow corn and convert it into ethanol (the inputs) and the return on this energy investment (the outputs). The energy inputs include the energy to run farm machinery, to produce chemicals (especially nitrogen fertilizer), dry the corn, transport the corn, convert it into ethanol, and ship it. The primary output is the ethanol, although several by-products are produced, including distiller's grains, corn gluten, and corn oil. Each by-product would have required energy to produce if it had been manufactured independently of the ethanol manufacturing process, and so energy "credit" is assigned to these by-products. As **FIGURE SA6.2** shows, there is a slight gain of usable energy when corn is converted into ethanol: For every unit of energy we put in, we obtain about 1.3 units of output. For comparison, for every unit of fossil fuel we invest in producing gasoline, we obtain about 15 units of output.

Does ethanol reduce our dependence on gasoline?

If there is a positive energy return on energy investment, then using ethanol should reduce the amount of gasoline we use and therefore the amount of foreign oil we must import. Our current production of 60 billion liters (16 billion gallons) of ethanol translates to about 4 percent of gasoline consumption in the United States. If our goal is to reduce gasoline consumption, much larger gains could be accomplished by having higher fuel mileage standards for cars and trucks.

What are the unintended impacts of ethanol production?

If we create greater demand for a crop that until now has been primarily used as a food source, there are many unintended impacts. For example, increased ethanol production has led to the large-scale conversion of cropland from food to fuel production. Even if we converted every acre of potential cropland to ethanol production, we could not produce enough ethanol to replace 20 percent of U.S. annual gasoline consumption. Furthermore, if all cropland in the United States were devoted to ethanol production, all agricultural products destined for the dinner table would have to be imported from other countries. Clearly this is not a practical solution.

What seems more likely is that we will be able to replace some smaller fraction of gasoline consumption

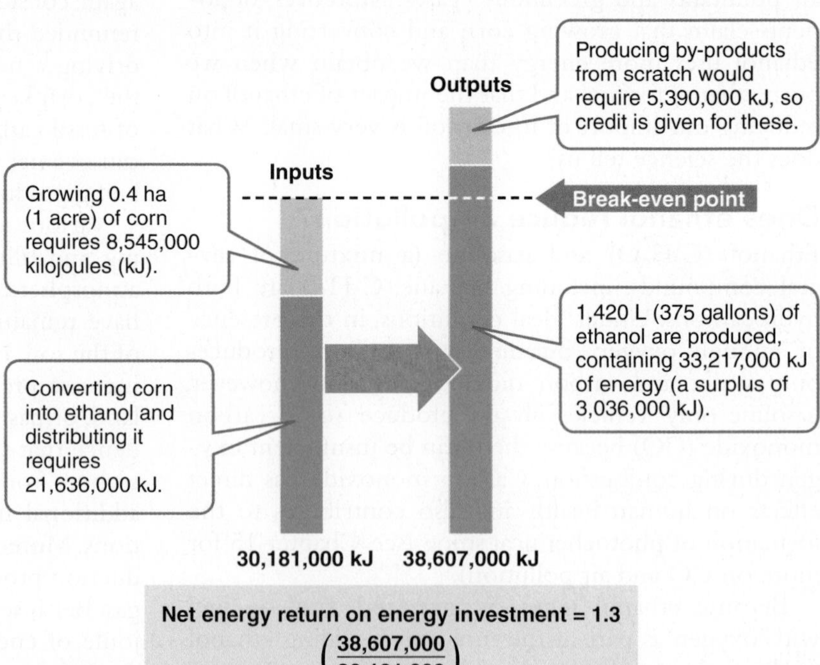

FIGURE SA6.2 Energy required to produce ethanol. An analysis of the energy costs of growing and converting 0.4 ha (1 acre) of corn into ethanol shows a slight gain of usable energy when corn is converted into ethanol. For every 1 unit of energy input, we obtain 1.3 units of energy output.

with biofuels. The Earth Policy Institute points out, however, that the 10 bushels of corn that it takes to produce enough ethanol to fill a 95-L (25-gallon) fuel tank in a vehicle contain the number of calories needed to feed a person for about a year. Higher ethanol demand would increase corn and other grain prices and would thus make it harder for lower-income people around the world to afford food.

Indeed, in the summer of 2007, corn prices in the United States rose to $4 per bushel, roughly double the price in prior years, primarily because of the increased demand for ethanol. In more recent years, prices have stayed above $3 per bushel and at times have gone above $8 per bushel. People in numerous countries have had difficulty obtaining food because of these higher grain prices. In a number of years, most notably 2008 and 2015, there were food riots in many countries around the globe.

FIGURE SA6.3 A potential source of cellulosic ethanol. Miscanthus, a fast-growing tall grass, may be a source of cellulosic ethanol in the future. *(Frank Dohleman)*

Are there alternatives to corn ethanol?

Stimulating demand for corn ethanol may spur the development of another ethanol technology—**cellulosic ethanol**, an ethanol derived from cellulose. Cellulose is the material that makes up the cell walls of plants: Grasses, trees, and plant stalks are made primarily of cellulose. If we were able to produce large quantities of ethanol from cellulose, we could replace fossil fuels with fuel made from a number of sources. Ethanol could be manufactured from fast-growing grasses such as switchgrass (*Panicum virgatum*) or Miscanthus grass (**FIGURE SA6.3**), tree species that require minimal energy input, discarded paper and agricultural waste, including corn stalks that are left behind when corn kernels are harvested for conventional ethanol production. It is also possible that algae could be used as the primary material for ethanol.

Producing cellulosic ethanol requires breaking cellulose into its component sugars before distillation. This is a difficult and expensive task because the bonds between the sugar molecules are very strong. One method of breaking down cellulose is to mix it with enzymes that sever these bonds. In 2007, the first commercial cellulosic ethanol plant was built in Iowa. At the moment, however, cellulosic ethanol is more expensive to produce than corn ethanol.

How much land would it take to produce significant amounts of cellulosic ethanol? Some scientists suggest that the impact of extensive cellulosic ethanol production would be very large, while others have calculated that, with foreseeable technological improvements, we could replace all of our current gasoline consumption without large increases in land under cultivation or significant losses in food production. Because the technology is so new, it is not yet clear who is correct. There will still be other considerations, such as the impact on biodiversity whenever land is dedicated to growing biofuels.

The good news is that many of the raw materials for cellulosic ethanol are perennial crops such as grasses. These crops do not require the high energy, fertilizer, and water inputs commonly used to grow annual plants such as corn. Furthermore, the land used to grow grass would not need to be plowed every year. Fertilizers and pesticides would also be unnecessary, eliminating the large inputs of energy needed to produce and apply them. Algae may be an even more attractive raw material for cellulosic ethanol because its production would not need to use land that could otherwise be used for growing food crops.

What's the bottom line?

When we compare the full life cycle of ethanol production, it is clear that ethanol results in a greater production of many air pollutants and it is not neutral in terms of greenhouse production. However, its greenhouse gas production is less than gasoline. The return on the energy invested to create ethanol is quite low compared to producing gasoline. Moreover, corn-based ethanol has the potential to replace only a small amount of total U.S. gasoline consumption. Increasing corn ethanol consumption to the levels suggested by some policy makers may require troublesome trade offs between driving vehicles and feeding the world. Cellulosic ethanol shows the potential to have a significant effect on fossil fuel use, at least in part because of lower energy inputs required to obtain the raw material and convert it into a fuel.

Questions

1. Legislation requiring that ethanol be added to gasoline was passed to improve air quality and reduce dependence on foreign oil. Do you think the latter goal is achieved by adding ethanol to gasoline?
2. What are some of the other environmental impacts of using corn-based ethanol that are not illustrated by energy calculations shown here?

Cellulosic ethanol An ethanol derived from cellulose, the cell wall material in plants.

Preparing for the AP® Exam

Practice AP® Free-Response Question

Write your answer to each part clearly. Support your answers with relevant information and examples. Where calculations are required, show your work.

The production of ethanol from corn is much like the production of any alcoholic beverage. Corn is harvested, ground, and cooked to form a "mash." Added yeast use the sugar to grow and reproduce. As they grow, yeast respire CO_2 and produce ethanol as a waste product through fermentation. Since ethanol has a lower boiling point than water, the fermented product can be boiled to evaporate the ethanol, which is a process known as distillation. Evaporated ethanol is then collected through condensation.

(a) Identify SIX energy inputs in the process of producing corn ethanol. (2 points)
(b) Using the answers provided in question (a), write an equation that would calculate the energy return on energy investment (EROEI) for ethanol production. (2 points)
(c) The leftover mash consists of water and corn. Ethanol producers will often dry this mash, collect the corn, and sell it as livestock feed. Why might this livestock feed be less beneficial for livestock than unfermented corn? (2 points)
(d) If the price of food increases around the world as a result of corn ethanol production, poorer communities may revert to subsistence energy production. Explain how this reversion could impact the total environmental benefit of ethanol production. (2 points)
(e) Cellulosic ethanol production is similar to corn ethanol production, except producers make use of enzymes that increase the sugar content of the mash, which in turn increases the energy available for yeast to grow. Describe two benefits and two disadvantages of cellulosic ethanol production relative to corn ethanol production.

References

Conca, J, 2014. It's final: Corn ethanol is of no use. *Forbes*, April 20, 2014. https://www.forbes.com/sites/jamesconca/2014/04/20/its-final-corn-ethanol-is-of-no-use/#6385d53867d3

Runge, C. F. 2016. The case against more ethanol: It's simply bad for the environment. Yale Environment 360. http://e360.yale.edu/features/the_case_against_ethanol_bad_for_environment

Searchinger, T. D., et al. 2009. Fixing a critical climate accounting error. *Science* 326:527–528.

Service, R. F. 2013. Battle for the barrel. *Science* 339:1374–1379.

Key Terms

Oxygenated fuel
Cellulosic ethanol

Unit 6 AP® Environmental Science Practice Exam

Preparing for the AP® Exam

Section 1: Multiple-Choice Questions

Choose the best answer for questions 1–25.

1. Within a developing nation, an increase in the use of subsistence energy sources would most likely be caused by
(a) a decrease in the availability of straw, sticks, animal dung, and other local sources of fuel.
(b) an increase in the availability of oil.
(c) an increase in the cost of oil.
(d) the loss of forested land.

Questions 2 and 3 refer to following table:

Energy type	Energy return on energy investment	MJ per kilogram of fuel
Biodiesel	1	40
Coal	80	24
Ethanol from corn	1	30
Ethanol from sugarcane	5	30

2. Which fuel is most likely the least expensive fuel to produce per MJ of fuel?
(a) biodiesel
(b) coal
(c) ethanol from corn
(d) ethanol from sugarcane

3. How many kilograms of fuel will be consumed if someone travels 200 km in a car that uses 3 MJ of biodiesel per kilometer?
   (a) 7.5 kg
   (b) 15 kg
   (c) 22 kg
   (d) 66 kg

4. Which is likely to reduce the efficiency of a power plant?
   (a) Increasing the capacity factor of the plant
   (b) Using anthracite coal instead of bituminous coal
   (c) Shutting off a turbine driven by exhaust gases
   (d) Pumping cold water from a nearby stream into the condenser

5. Which is a benefit of using petroleum instead of coal for energy?
   I. Petroleum releases about 15 percent more CO_2 than coal.
   II. Petroleum flows more easily than coal.
   III. Mining and transport of petroleum is less harmful to the environment.
   (a) I only
   (b) II only
   (c) III only
   (d) I and II

6. Which is true regarding natural gas?
   (a) Extraction and combustion of natural gas have less effect on global warming than extraction and combustion of coal.
   (b) Contamination of water during the extraction process is of little concern.
   (c) Liquefied petroleum gas is slightly more energy-dense than natural gas.
   (d) Pipelines are the primary means of transporting natural gas.

7. In 1969, M. King Hubbert calculated lower and upper estimates for the volume of total world petroleum reserves. Regardless of which estimate was used, he predicted that 80 percent of the total reserves would be used up in approximately 60 years. Which of the following likely explains why his predictions were the same for both estimates?
   I. The lower and upper estimates were not different enough to cause a major significant shift in his predictions.
   II. Per capita energy use will increase with available energy.
   III. Availability and use of petroleum is inversely correlated with cost.
   (a) I only
   (b) II only
   (c) III only
   (d) II and III

8. In a nuclear power plant, control rods are used to
   (a) control the placement of fuel rods.
   (b) increase the efficiency of nuclear reactions.
   (c) transfer heat energy from the fuel rods into water.
   (d) absorb excess neutrons emitted by fuel rods.

9. One particularly large nuclear power plant produces about 400 kilocuries of krypton per year. Krypton has a half-life of 10 years. After 30 years, what will be the radioactivity of the krypton waste generated in a single year?
   (a) 50 kilocuries
   (b) 100 kilocuries
   (c) 200 kilocuries
   (d) 800 kilocuries

10. Which reduces the capacity factor of nuclear power plants?
    (a) the need for long-term storage of radioactive waste
    (b) government regulation of low-level radioactive waste
    (c) competition with plants that use coal to produce energy
    (d) the need to periodically replace fuel rods

11. Which is a renewable energy source that does not originate from solar radiation?
    (a) biomass
    (b) geothermal
    (c) nuclear
    (d) wind

12. A homeowner in the Northern Hemisphere wants to save money and energy by using a passive solar design for heating. The homeowner could consider
    (a) opening window blinds on southern exposure sunny winter days.
    (b) using only Energy Star appliances.
    (c) placing photovoltaic cells on the roof.
    (d) installing a ground source heat pump.

13. Suppose you own a diesel car and the current price of diesel gasoline is $3.00 per gallon. You pay $750 for parts that allow the engine to be converted so it can operate on straight vegetable oil (SVO), which costs $1.50 per gallon. The car gets 50 miles per gallon when running on either type of fuel. After how many miles will you recoup the cost of engine conversion?
    (a) 1,500
    (b) 5,000
    (c) 10,000
    (d) 25,000

14. Relative to burning fossil carbon, the use of modern carbon for energy
    (a) rarely contributes to the removal of vegetation.
    (b) does not use energy that originates from the Sun.
    (c) is cheaper and more efficient.
    (d) is more likely to be carbon-neutral.

15. Which consequence of water impoundment for hydroelectric energy production is most likely to release greenhouse gases?
    (a) flooding of forests and grasslands
    (b) siltation
    (c) the generation of electricity by use of turbines
    (d) transfer of energy to power lines

Question 16 refers to the following diagram.

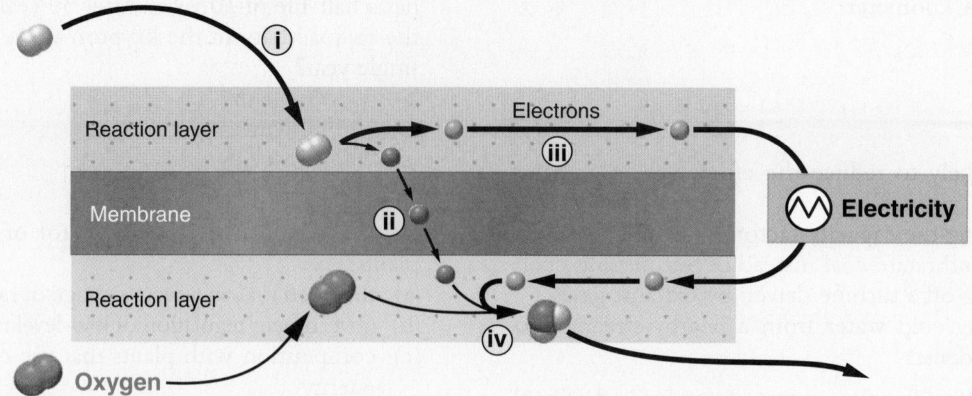

16. In this depiction of a hydrogen powered fuel cell, components i, ii, iii, and iv refer to
    (a) hydrogen, protons, neutrons, water.
    (b) hydrogen, protons, hydrogen ions, water.
    (c) hydrogen, protons, electrons, water.
    (d) hydrogen, electrons, protons, water.

17. The combined use of concentrated solar thermal (CST) and fossil fuel energy generation can provide for greatest grid reliability when
    (a) CST and fossil fuel plants are built near each other.
    (b) fossil fuels are able to be used when CST is unavailable.
    (c) CST and fossil fuel energy are equally distributed to the power grid.
    (d) fossil fuel plants are built in desert areas where there is consistent sunshine and plenty of open space.

18. Which sources of electricity employ the use of turbines?
    (a) photovoltaic cells, windmills, run-of-the-river hydroelectric plants
    (b) photovoltaic cells, windmills, water impoundment hydroelectric dams
    (c) concentrated solar thermal plants, nuclear plants, windmills
    (d) concentrated solar thermal plants, water impoundment hydroelectric dams, fuel cells

19. Which is most likely to increase the efficiency of energy use in the United States?
    (a) government subsidies for active solar construction designs
    (b) decreasing the cost of fossil fuels
    (c) replacement of large power plants with several smaller plants
    (d) substitution of incandescent lighting with LEDs

20. A smart grid system
    I. regulates electrical energy usage according to electrical energy availability.
    II. requires more energy-efficient electricity production.
    III. will reduce the cost of energy production.
    (a) I only
    (b) II only
    (c) III only
    (d) I and II

21. The ________ projects when world oil production will peak and when world oil will be depleted.
    (a) Hubbert curve
    (b) Hadley cell
    (c) petroleum development index
    (d) Kyoto Protocol

22. Which is a subsistence energy source?
    (a) ethanol
    (b) bitumen
    (c) wood
    (d) wind

23. While natural gas is considered to be cleaner than coal or oil, it has disadvantages, including
    I. the release of methane.
    II. increased sulfur dioxide emissions.
    III. groundwater contamination.
    (a) I only
    (b) II only
    (c) III only
    (d) I and III

24. At a particular wind farm, 3 J of energy is expended in order to obtain 45 J of energy, and its capacity factor is 0.25. What is the EROEI of this wind farm?
    (a) 3.75
    (b) 15
    (c) 34
    (d) 60

25. While only 20 percent of the world's population lives in developed countries, people in those countries use approximately _________ of the world's energy each year.
   (a) 30 percent
   (b) 50 percent
   (c) 70 percent
   (d) 80 percent

Section 2: Free-Response Questions

Write your answer to each part clearly. Support your answers with relevant information and examples. Where calculations are required, show your work.

1. Although hybrid and electric vehicles release less carbon dioxide per distance traveled than internal combustion engine—gasoline-powered vehicles, there remains some debate over whether hybrid vehicles are better for the environment and whether their overall efficiency is greater.
   (a) Provide TWO reasons why hybrid and electric vehicles might be less energy-efficient than gas-powered cars. (3 points)
   (b) How might the average lifetime of a hybrid or electric vehicle versus a conventional internal combustion engine vehicle alter an analysis of the total ecological footprint of the vehicles? (2 points)
   (c) Suppose a new all-electric vehicle costs $40,000 and requires 1.5 MJ per passenger-mile. You trade in your old gasoline-powered vehicle, which required 4 MJ per passenger-mile, for $2,000.
      (i) Supposing that a gallon of gasoline contains 20 MJ and costs $4.00, how many miles must you drive before your purchase becomes cost effective? (3 points)
      (ii) How might the use of "smart grid" technology reduce the environmental footprint of an electric vehicle? (2 points)

2. Electricity accounts for 40 percent of overall energy use in the United States. It is a secondary source of energy since we obtain it through the conversion of a primary source of energy, like coal or wind.
   (a) How is electricity generated in a typical power plant? (2 points)
   (b) Define capacity and capacity factor. Explain how capacity factors may differ between plants powered by nonrenewable fuels such as coal, and those powered by wind and solar. (4 points)
   (c) Energy demands can vary depending on season, weather, and time of day. What problem can occur due to this varying demand for energy? What could an electric company do to deal with this issue and increase energy conservation? (2 points)
   (d) If an average home in a city uses 12,000 kWh per year, how many homes can a 700 MW power plant with a 0.5 capacity factor support? (2 points)

The Chesapeake Bay receives water from a very large geographic area that carries sediments, nutrients, and chemicals from a variety of urban, suburban, and agricultural areas. This photo shows the James River estuary near Williamsburg, Virginia. *(De Agostini Picture Library/Getty Images)*

Aquatic Pollution

CHAPTER 14

MODULE 41 Wastewater from Humans and Livestock

MODULE 42 Heavy Metals and Other Chemicals

MODULE 43 Oil Pollution

MODULE 44 Nonchemical Water Pollution

MODULE 45 Water Pollution Laws

CASE STUDY

The Chesapeake Bay

The Chesapeake Bay, located between the states of Virginia and Maryland, is the largest estuary in the United States. It also faces some of the biggest environmental challenges. The Chesapeake Bay receives fresh water from numerous rivers and streams. This fresh water mixes with the salt water of the ocean to produce an extremely productive estuary. Many of the rivers and streams that dump into the bay travel long distances and drain water from a large watershed of urban, suburban, and agricultural areas. Indeed, the watershed that supplies water to the bay extends from northern Virginia all the way up into central New York State. One of the consequences of receiving water from such a large watershed is that the water coming into the bay contains an abundance of nutrients, sediments, and chemicals.

Among the major nutrients that enter the bay each year are 272 million kg (600 million pounds) of nitrogen and 14 million kg (30 million pounds) of phosphorus. The nitrogen and phosphorus come from three major sources. The first source is water discharged from sewage treatment facilities; it carries high amounts of nutrients from human waste. The second source is animal waste produced by concentrated animal feeding operations; these operations generate large amounts of manure that can make their way into nearby streams and rivers. The third source is fertilizer that is spread on both agricultural fields and suburban lawns; much of this fertilizer leaches out of the soil and into local streams that eventually flow into the Chesapeake Bay. When these nutrients reach the bay, the algae in the bay experience an explosive population growth that is known as an algal bloom.

> Estrogen from sewage in the water was the most likely reason that 23 percent of male largemouth bass and 82 percent of male smallmouth bass developed into hermaphrodites.

Increased sediments are also an issue in the Chesapeake Bay. Sediments are soils washed away from fields and forests as well as soils washed away from the banks of streams and the ocean shoreline. According to current estimates, 8.2 billion kilograms (18.7 billion pounds) of sediments come into the bay each year. The tiniest soil particles stay suspended in the water, make the water cloudy, and prevent sunlight from reaching the grasses that historically have been abundant in the bay. These grasses are important because they serve as a habitat for fish and blue crabs (*Callinectes sapidus*). Over the past several decades, the blue crab population has experienced substantial declines.

Inputs of anthropogenic chemicals, including pesticides and pharmaceutical drugs, are also prevalent and commonly bound to the sediments. Pesticides arrive in the bay by direct application over water, by running off the surface of the land when it rains, or by being carried by the wind immediately after application. Pharmaceuticals pass through the human body and enter sewage treatment plants that eventually discharge into the streams and rivers that feed the bay. In 2009, the U.S. Fish and Wildlife Service announced that estrogen from sewage in the water was the most likely reason that 23 percent of male largemouth bass and 82 percent of male smallmouth bass developed into hermaphrodites; their male sex organs grow eggs, a function that normally only occurs in the sex organs of female fish. This impact not only concerns fish, but also is a concern for humans who share many similarities in their reproductive systems and may consume water contaminated with estrogens.

Because the Chesapeake Bay watershed is so large, cleaning it up has required a sustained and monumental effort. In 2000, the states surrounding the Chesapeake Bay formed a partnership along with multiple federal departments to develop the Chesapeake Bay Action Plan. This plan outlines a series of goals to reduce the impacts of nutrients, sediments, and chemicals coming into the bay. An assessment in 2010 concluded that many of the Action Plan's goals were being met, including a reduction in nitrogen, an increase in water clarity, and an increase in the number of blue crabs. Overall, the improved water quality, combined with limits in the blue crab harvest, allowed the blue crab population to increase dramatically. In 2010, the blue crab population in the bay reached its highest density in nearly 20 years. For a few years after this, the population declined, but they have since rebounded with continual increases from 2014 to 2017. Much more work still needs to be done, such as resolving the problem of anthropogenic chemicals that cause hermaphroditic smallmouth bass.

The story of the Chesapeake Bay serves as an excellent example of the wide variety of pollutants that can impact aquatic ecosystems and of effective and substantial efforts that can only be made when all parties work together toward a common goal.

Sources: Chesapeake Bay Program http://www.chesapeakebay.net; Maryland Department of Natural Resources. 2017 blue crab winter dredge survey. 2017, http://dnr.maryland.gov/fisheries/Pages/blue-crab/dredge.aspx; U.S. Fish and Wildlife Service, Intersex fish: Endocrine disruption in smallmouth bass. https://www.fws.gov/chesapeakebay/pdf/endocrine.pdf.

Water is a key resource for life on Earth. All organisms, including humans, require water to live, but increases in human populations combined with industrialization have led to the contamination of water supplies. **Water pollution** is generally defined as the contamination of streams, rivers, lakes, oceans, or groundwater with substances produced through human activities. As we will see, water pollution occurs from a wide range of substances from various sources. The prevalence and impact of pollutants in water also varies. Although this chapter focuses on the contaminants that exist in water and their effects on aquatic organisms and humans, because there are many ecological connections between aquatic and terrestrial ecosystems, water pollution can also affect terrestrial organisms.

The broad range of pollutants that can be found in water includes human and animal waste, inorganic substances, organic compounds, synthetic organic compounds, and nonchemical pollutants. In this chapter we will consider the origin of each category of pollutant, its negative effects on humans and the environment, and what can be done to reduce the pollutant.

Water pollution The contamination of streams, rivers, lakes, oceans, or groundwater with substances produced through human activities.

MODULE 41

Wastewater from Humans and Livestock

A major type of water pollution is **wastewater**, which is the water produced by livestock operations and human activities, including human sewage from toilets and gray water from bathing and washing clothes and dishes. In this module, we will review the effects that wastewater pollution can have on natural water bodies and then discuss the technologies that have been developed to treat wastewater and minimize its effects.

Learning Goals

After reading this module you should be able to

- discuss the three major problems caused by wastewater pollution.
- explain the modern technologies used to treat wastewater.

Wastewater from humans and livestock poses multiple problems

Regardless of the specific contaminant, pollution can come from either *point sources* or *nonpoint sources* (**FIGURE 41.1** on page 504). A **point source** is a distinct location from which pollution is directly produced. Examples of point sources include a particular factory that pumps its waste directly into a nearby stream or a sewage treatment plant that discharges its wastewater from a pipe into the ocean. **A nonpoint source** is a more diffuse area that produces pollution, such as an entire farming region, a suburban community with many lawns and septic systems, or storm runoff from parking lots. The distinction between the two sources can help us control pollutant inputs to waterways. For example, if a municipality can determine one or two point sources for most of its water pollution, it can target those specific sources to reduce its pollution output. Controlling pollution from nonpoint sources is more challenging because it represents a negative externality. As we saw in Chapter 10, it is difficult to address negative externalities since they are not typically reflected in the cost of making the products that generate them.

Environmental scientists are concerned about wastewater as a pollutant for three major reasons. First, wastewater dumped into bodies of water naturally undergoes decomposition by bacteria, which creates a large demand for oxygen in the water. Second, the nutrients that are released from wastewater decomposition can make the water more fertile. Third, wastewater can carry a wide variety of disease-causing organisms.

Oxygen Demand

When organic matter enters a body of water, microbes that are decomposers feed on it. Because these microbes require oxygen to decompose the waste, the more waste that enters the water, the more the microbes grow and the more oxygen they demand. The amount of oxygen a quantity of water uses over a period of time at a specific temperature is its **biochemical oxygen demand (BOD)**. Lower BOD values indicate that a water body is less polluted by wastewater, and higher BOD values indicate that a water body is more polluted by wastewater. If we were to test the BOD of natural waters over a 5-day

Wastewater Water produced by livestock operations and human activities, including human sewage from toilets and gray water from bathing and washing of clothes and dishes.

Point source A distinct location from which pollution is directly produced.

Nonpoint source A diffuse area that produces pollution.

Biochemical oxygen demand (BOD) The amount of oxygen a quantity of water uses over a period of time at a specific temperature.

(a)

(b)

FIGURE 41.1 Two types of pollution sources. Pollution can enter water bodies either from (a) point sources, as in sewage pipes, or from (b) nonpoint sources, as in rainwater that runs off hundreds of square kilometers of agricultural fields and into streams. *(a: age fotostock/Superstock, b: The Irish Image Collection/Superstock)*

period in a liter of water, we might find a BOD of 5 to 20 mg of oxygen coming from the decomposition of leaves, twigs, and perhaps a few dead organisms. In contrast, wastewater might have a BOD of 200 mg of oxygen.

When microbial decomposition uses a large amount of oxygen in a body of water, the amount of oxygen remaining for other organisms can be very low. Low oxygen concentrations are lethal to many organisms including fish. Low oxygen can also be lethal to organisms that cannot move, such as many plants and shellfish. Some areas, known as dead zones, become so depleted of oxygen that little life survives. Dead zones can be self-perpetuating; organisms that die from lack of oxygen decompose and cause microbes to use even more oxygen.

Nutrient Release

When wastewater decomposes it releases nitrogen and phosphorus. As you may recall from Chapter 3, nitrogen and phosphorus are generally the two most important elements for limiting the abundance of producers in aquatic ecosystems. The decomposition of wastewater releases nitrogen and phosphorus, which in turn provide an abundance of fertility to a water body, a phenomenon known as **eutrophication**. As noted at the beginning of this chapter, the Chesapeake Bay experiences eutrophication from wastewater decomposition as well as from nutrients leached from fertilized agricultural lands during precipitation. When a body of water experiences an increase in fertility due to anthropogenic inputs of nutrients, it is called **cultural eutrophication**.

Eutrophication initially causes a rapid growth of algae, known as an algal bloom. As we discussed in Chapter 3, this enormous amount of algae eventually dies and microbes then rapidly begin digesting the dead algae; the increase in microbes consumes most of the oxygen in the water (see Figure 7.5 on page 91). In short, the release of nutrients from wastewater initiates a chain of events that eventually leads to a lack of oxygen and the creation of dead zones once again. One of the most impressive dead zones in the world occurs where the Mississippi River dumps into the Gulf of Mexico, shown in **FIGURE 41.2a**. The Mississippi River receives water from 41 percent of the land of the continental United States. Each summer there is an influx of wastewater and fertilizer that causes large algal blooms followed by substantial decreases in oxygenated water and massive die-offs of fish (Figure 41.2b). While some dead zones arise naturally, most are caused by human activities. In 1910, 4 dead zones were known around the world. This number increased to 87 dead zones in the 1980s and more than 400 dead zones by 2008, which is the most recent estimate.

Eutrophication A phenomenon in which a body of water becomes rich in nutrients.

Cultural eutrophication An increase in fertility in a body of water, the result of anthropogenic inputs of nutrients.

AP® Exam Tip

Make sure you can explain how fertilizers are transported from their application site into the environment through runoff or groundwater filtration. ●

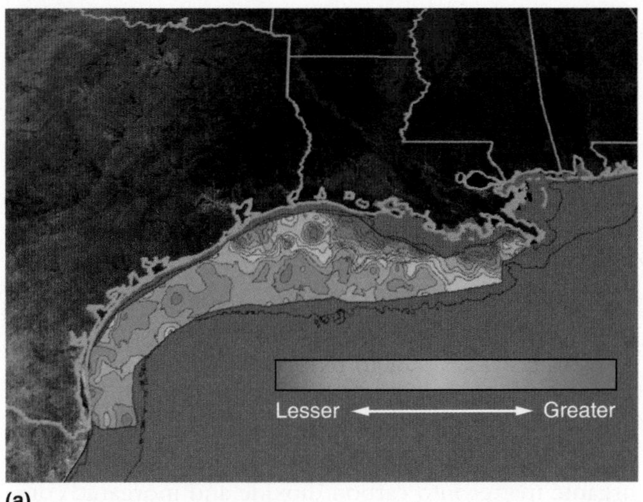

(a)

(b)

FIGURE 41.2 Dead zone. When raw sewage is dumped directly into bodies of water, subsequent decomposition by microbes can consume nearly all of the oxygen in the water and cause a dead zone to develop. Shown here are (a) oxygen concentrations in Gulf Coast waters and (b) a massive fish die-off that occurred as a result of low oxygen conditions in Lake Trafford, Florida *(a: NASA/Goddard Space Flight Center Scientific Visualization Studio; b: Steven David Miller/Nature Picture Library)*

Disease-Causing Organisms

For centuries, humans have faced the challenge of keeping wastewater from contaminating drinking water. This can be difficult because throughout the world many people routinely use the same water source for drinking, bathing, washing, and disposing of sewage (**FIGURE 41.3**). Wastewater can carry a variety of pathogens including viruses, bacteria, and protists. Pathogens in wastewater are responsible for a number of diseases that can be contracted by humans or other organisms coming in contact with or ingesting the water. Such pathogens cause cholera, typhoid fever, various types of stomach flu, and diarrhea.

Worldwide, the major waterborne diseases are cholera, hepatitis, and gastrointestinal illness. Cholera, which claims thousands of lives annually in developing countries (**FIGURE 41.4**), is not common in the United States. However, hepatitis A is appearing more frequently in the United States, usually originating in restaurants that lack adequate sanitation practices. According to the Centers for Disease Control and Prevention, there were nearly 20,000 cases of hepatitis A in the United States annually in the mid-1990s. Today, that number has declined to approximately 1,000 cases. The bacterium *Cryptosporidium* has caused a number of outbreaks of gastrointestinal illness in this country.

FIGURE 41.3 Washing clothes in the Tuo River of China. Using water for bathing, washing, and disposing of sewage without contaminating sources of drinking water is a long-standing challenge. *(Bamboosil/AGE Fotostock)*

FIGURE 41.4 Cholera is prevalent in raw sewage. Children who play in water contaminated by raw sewage, such as this girl in Cambodia, face a high risk of contracting the cholera pathogen. *(imageBROKER/Alamy)*

Large-scale disease outbreaks from municipal water systems are relatively rare in the United States, but they do occasionally occur. They are relatively common in many regions of the developing world.

The World Health Organization estimates that 2.1 billion people, nearly one-fourth of the world's population, do not have access to sufficient supplies of safe drinking water. Approximately 2.3 billion people lack access to proper sanitation and over half of these people live in China and India. In sub-Saharan Africa, only 40 percent of the population has access to proper sanitation. It is estimated that 361,000 children around the world die from diarrhea each year because they do not have safe drinking water, proper sanitation, and proper hygiene.

Because of the risk that pathogens pose, we need the ability to easily test for the presence of pathogens in our drinking water. It is not feasible to test for all of the many different pathogens that can exist in drinking water. Instead, scientists have settled on using an **indicator species**, a species that indicates whether or not disease-causing pathogens are likely to be present. The best indicators for water that is potentially harmful are **fecal coliform bacteria**, a group of generally harmless microorganisms that live in the intestines of human beings and other animals. One of the most common species of fecal coliform bacteria is *Escherichia coli,* abbreviated *E. coli*. Most strains of *E. coli* live naturally in humans and are not harmful, although there are strains that can be deadly to people who are very young, very old, or possess weak immune systems. Given that *E. coli* is commonly found in human intestines, detecting *E. coli* in a body of water is a reliable indicator that human waste has entered the water. This does not necessarily mean that the water is harmful to drink, but the presence of *E. coli* does indicate an increased risk that wastewater pathogens are in the water.

Public water supplies, such as drinking water sources and swimming pools, are routinely tested for *E. coli*. Homeowners with a single-family well might test their water less frequently or not at all. Public health authorities recommend declaring water unsuitable for human consumption if any bacteria are present. For safe swimming and fishing, the acceptable level of *E. coli* is higher; for example, swimming at a public beach or in a river is considered safe as long as the fecal coliform bacteria levels are less than 500 to 10,000 colonies per 100 megaliters of water. A pool, beach, or campground with contaminated water likely would be posted with a sign such as "The Department of Health has closed this water supply because of the presence of fecal coliform bacteria."

Indicator species A species that indicates whether or not disease-causing pathogens are likely to be present.

Fecal coliform bacteria A group of generally harmless microorganisms in human intestines that can serve as an indicator species for potentially harmful microorganisms associated with contaminated sewage.

Septic system A relatively small and simple sewage treatment system, made up of a septic tank and a leach field, often used for homes in rural areas.

Septic tank A large container that receives wastewater from a house as part of a septic system.

We have technologies to treat wastewater

Because proper sanitation is so important, we need ways of treating wastewater to reduce the risk of waterborne pathogens. Humans have devised a number of ways to handle wastewater. The various solutions all have the same basic approach: Bacteria are used to break down the organic matter into carbon dioxide and inorganic compounds, which include nitrogen and phosphorus, and the harmful pathogens are outcompeted by nonharmful pathogens. In this section we will look at the two most widespread systems for treating human sewage: *septic systems* and *sewage treatment plants*. We will also describe the treatment of waste from large livestock operations.

Septic Systems

In rural areas, houses often treat their own sewage with a **septic system**, which is a relatively small and simple sewage treatment system that is made up of a *septic tank* and a *leach field*. Septic systems are often used for homes in rural areas. As shown in **FIGURE 41.5**, the **septic tank** is a large container that receives wastewater from the house. Having a capacity of 1,900 to 4,700 L (500–1,250 gallons), the septic tank is buried underground adjacent to the house. Wastewater from the house flows into the tank at one end and leaves the tank at the other end. After the tank has been operating for some time, three layers develop. Anything

FIGURE 41.5 A septic system. Wastewater from a house is held in a large septic tank where solids settle to the bottom and bacteria break down the sewage. The liquid moves through a pipe at the top of the tank and passes through perforated pipes that distribute the water through a leach field.

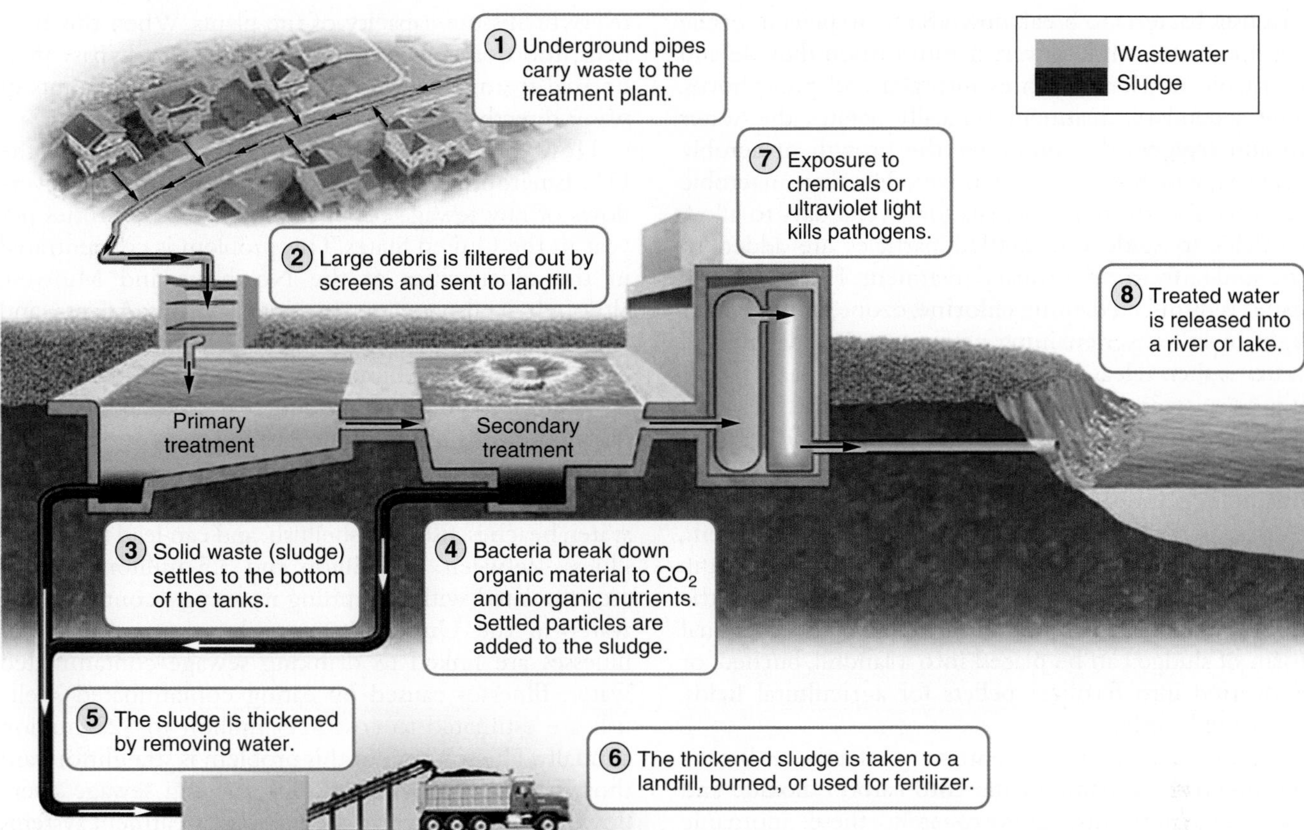

FIGURE 41.6 A sewage treatment plant. In large municipalities, great volumes of wastewater are handled by separating the sludge from the water and then using bacteria to break down both components.

that floats will rise to the top of the tank and form a scum layer. Anything heavier than water, including many pathogens, will sink to the bottom of the tank and form the **sludge** layer, which we define as the solid waste material from wastewater. In the middle is a fairly clear water layer called **septage**. The septage contains large quantities of bacteria and may also contain some pathogens and inorganic nutrients such as nitrogen and phosphorus.

Through gravity, septage moves out of the septic tank into several underground pipes laid out across a lawn below the surface. The combination of pipes and lawn makes up the **leach field**. The pipes contain small perforations so the water can slowly seep out and spread across the leach field. The septage that seeps out of the pipes is slowly absorbed and filtered by the surrounding soil. The harmful pathogens in the septage can be outcompeted by other microorganisms in the septic tank and therefore diminish in abundance, or they can be degraded by soil microorganisms after the septage seeps out into the leach field. The organic matter found in the septage is broken down into carbon dioxide and inorganic nutrients. Eventually, the water and nutrients seeping out into the leach field are taken up by plants or enter a nearby stream or aquifer.

There are significant environmental advantages to septic systems. Because most septic systems rely on gravity—water from the house flows downhill to the septic tank, and water from the septic tank flows downhill to the leach field—no electricity is needed to run a septic system. However, sludge from the septic tank must be pumped out periodically (every 5 to 10 years) and taken to a sewage treatment plant.

Sewage Treatment Plants

Although household septic systems work well for rural areas in which each house has sufficient land for a leach field, they are not feasible for more developed areas with greater population densities and little open land. In developed countries, municipalities use centralized sewage treatment plants that receive the wastewater from hundreds or thousands of households via a network of underground pipes. In a traditional sewage treatment plant, wastewater is handled using a primary treatment followed by a secondary treatment.

As shown in **FIGURE 41.6**, the primary treatment allows the solid waste material to settle into a sludge layer. The remaining wastewater undergoes a secondary treatment

Sludge Solid waste material from wastewater.

Septage A layer of fairly clear water found in the middle of a septic tank.

Leach field A component of a septic system, made up of underground pipes laid out below the surface of the ground.

that uses bacteria to break down 85 to 90 percent of the organic matter and convert it into carbon dioxide and inorganic nutrients such as nitrogen and phosphorus. The secondary treatment typically aerates the water to add oxygen; this promotes the growth of aerobic bacteria, which emit less offensive odors than anaerobic bacteria. The treated water sits for several days to allow particles to settle out. Settled particles are added to the sludge from the primary treatment. The remaining water is disinfected, using chlorine, ozone, or ultraviolet light to kill any remaining pathogens. The disinfected water is then released into a nearby river or lake, where it is once again part of the global water cycle.

The combined sludge from the primary and secondary treatments must be taken away from the sewage treatment plant. To reduce the volume of material prior to transport and help remove many of the pathogens, sludge is typically exposed to bacteria that can digest it. After digestion, most of the water is removed from the sludge, which reduces its volume and weight. This final form of sludge can be placed into a landfill, burned, or converted into fertilizer pellets for agricultural fields, lawns, and gardens.

Sewage treatment plants are very effective at breaking down the organic matter into carbon dioxide and inorganic nutrients. Unfortunately, these inorganic nutrients can still have undesirable effects on the waterways into which they are released. Although nitrogen and phosphorus are important nutrients for increasing primary productivity, when high concentrations of these nutrients are released into bodies of water from sewage treatment plants, they fertilize the water, which can lead to unnaturally large increases in the abundance of algae and aquatic plants. In response to this problem, large sewage treatment plants are now developing tertiary treatments that remove nitrogen and phosphorus from the wastewater. The ultimate goal is to release wastewater that is similar in quality to the water body that is receiving it.

Legal Sewage Dumping

Sewage treatment plants are critical to human health because they remove a great deal of harmful organic matter and associated pathogens that cause human illness. It might surprise you to know that even in the most developed countries, raw sewage can sometimes be directly pumped into rivers and lakes. Sewage treatment plants are typically built to handle wastewater from local households and industries. However, many older sewage treatment plants also receive water from stormwater drainage systems. During periods of heavy rain, the combined volume of storm water and wastewater overwhelms the capacity of the plants. When this happens, the treatment plants are allowed to bypass their normal treatment protocol and pump vast amounts of water directly into an adjacent body of water.

How big a problem is this? According to the U.S. Environmental Protection Agency in 2017, overflows of raw sewage occur 23,000 to 75,000 times per year in the United States. This problem is concentrated in the older cities of the Northeast and Midwest, although it also occurs in cities such as Atlanta and San Francisco. In Indianapolis, for example, more than 3.8 billion liters (1 billion gallons) of raw sewage have been dumped into surrounding water bodies each year during periods of high rain. Around the country, 11 to 38 billion liters (3–10 billion gallons) are released and such incidents result in the contamination of drinking water, beaches, fish, and shellfish, and can lead to human illness. Between 1.8 million and 3.5 million illnesses are associated with swimming in sewage-contaminated water in the United States each year, and 500,000 illnesses are linked to drinking sewage-contaminated water. Illnesses caused by eating contaminated shellfish are estimated to cost $2.5 million to $22 million annually. The solution to this problem is straightforward though expensive. Municipalities facing sewage overflow need to modernize their sewage treatment systems to prevent the influx of storm water from overwhelming their capacity to treat human wastewater. Unfortunately, the cost of such work can be considerable. In the case of Indianapolis, for example, a 28-mile long tunnel is being constructed to hold the excess storm water and sewage during rain storms until it can later be treated at nearby sewage treatment plants; this construction project is part of a $2 billion upgrade.

Animal Feed Lots and Manure Lagoons

Although the impact of human waste on the environment and human health is generally understood, people are less familiar with the problem of animal waste. On a small scale, animal manure can contaminate waters when farm animals are allowed access to streams for food and water because, inevitably, they will defecate in the stream. As you may recall from Chapter 11, large-scale manure issues arise with concentrated animal feeding operations that raise thousands of cattle, hogs, and poultry. The manure from these operations contains digested food and often also contains a variety of hormones and antibiotics that farmers use to improve the growth and health of their animals.

Farms that raise thousands of animals in a single location often use a *manure lagoon* to dispose of the tremendous amount of manure produced (**FIGURE 41.7**). A **manure lagoon** is a large, human-made pond lined with rubber to prevent the manure from leaking into

Manure lagoon Human-made pond lined with rubber built to handle large quantities of manure produced by livestock.

DO THE MATH Building a Manure Lagoon

Preparing for the AP® Exam

Concentrated animal feeding operations typically use manure lagoons to hold the manure produced by the cattle that are being held. If an individual animal produces 50 L of manure each day and the average concentrated animal feeding operation holds 800 cattle on any given day, how much manure is produced each day?

$$\text{Daily manure production} = 50\ \text{L/animal} \times 800\ \text{animals} = 40{,}000\ \text{L}$$

YOUR TURN

1. If a manure lagoon needs to hold 30 days' worth of manure production for 900 cattle, what is the minimum capacity of the lagoon a farmer would need?
2. After the manure has broken down, the manure must be spread onto farm fields. A modern manure spreader can hold 40,000 L of liquid manure. How many trips will it take for the manure spreader to remove the 30 days' worth of manure that is held in the manure lagoon?

FIGURE 41.7 Feedlot and manure lagoon. Large-scale agricultural operations, such as this one in Georgia, produce great amounts of waste that can be held in lagoons until pumped out and removed. *(Jeff Vanuga/USDA, Natural Resources Conservation Service)*

the groundwater. After bacteria has broken down the manure—the same process that occurs in sewage treatment plants—the manure can be spread onto farm fields to serve as a fertilizer. One major risk of manure lagoons is the possibility of developing a leak in the liner that would allow the waste to seep into and contaminate the underlying groundwater. Another risk is a possible overflow into adjacent water bodies. Like human wastewater, overflow of animal waste into rivers can lead to disease outbreaks in humans and wildlife. Finally, the application of manure as fertilizer can create runoff that moves into nearby water bodies. "Do the Math: Building a Manure Lagoon" shows you how to calculate the appropriate size for a manure lagoon.

MODULE 41 AP® Review

Preparing for the AP® Exam

In this module, we learned that wastewater from livestock and humans can cause harmful effects on the water bodies that receive it. Some of the most important effects are the high oxygen demand associated with the decomposition of the organic matter that can cause dead zones, the production of inorganic nutrients such as nitrogen and phosphorus that can cause algal blooms, and the introduction of disease-causing organisms. Fortunately, humans have invented several technologies to handle wastewater including septic tanks, sewage treatment plants, and manure lagoons. These technologies rely in part on decomposition of organic matter and help reduce the presence of pathogens. In some cases, they also reduce the amount of inorganic nutrients before the wastewater is dumped into a natural water body. In the next module, we will examine how a number of heavy metals and other chemicals affect aquatic ecosystems.

AP® Practice Questions

Choose the best answer for the following.

1. Which is a point source of pollution?
   (a) farm runoff
   (b) a sewage treatment plant
   (c) storm runoff
   (d) a suburban community

2. Eutrophication
   (a) is the result of excess carbon.
   (b) causes low BOD.
   (c) can cause dead zones.
   (d) is rarely caused by human activities.

3. Fecal coliform bacteria
   (a) is used as an indicator of water quality.
   (b) is the cause of many waterborne diseases.
   (c) usually causes diarrhea.
   (d) is rarely found in septage.

4. Leach fields are
   (a) part of tertiary water treatment.
   (b) used to filter septage.
   (c) a result of sewage dumping.
   (d) often used with manure lagoons.

5. A manure lagoon is being built for a dairy with 700 cows, each of which produces 40 L of manure each day. How large must the lagoon be to hold 30 days' worth of manure?
   (a) 28,000 L
   (b) 120,000 L
   (c) 560,000 L
   (d) 840,000 L

MODULE 42

Heavy Metals and Other Chemicals

We have seen that some compounds such as nitrogen and phosphorus cause environmental problems by overfertilizing the water. Other inorganic compounds, including heavy metals (lead, arsenic, and mercury), acids, and synthetic compounds (pesticides, pharmaceuticals, and hormones), can directly harm humans and other organisms. In this module, we will examine how each of these chemicals enters water bodies and the effects that they can have on ecosystems and humans.

Learning Goals

After reading this module you should be able to

- explain the sources of heavy metals and their effect on organisms.
- discuss the sources and effects of acid deposition and acid mine drainage.
- explain how synthetic organic compounds can affect aquatic organisms.

Heavy metals are highly toxic to organisms

Heavy metals are a group of chemicals that can pose serious health threats to humans and other organisms. In this section, we will focus on three heavy metals that are of particular concern: lead, arsenic, and mercury.

Lead

Lead is rarely found in natural sources of drinking water, but it contaminates water that passes through lead-lined pipes and other materials that contain lead such as brass fittings and solder used to fasten pipes together. Fetuses and infants are the most sensitive to lead, and exposure can damage the brain, nervous system, and kidneys. A series of federal guidelines for building construction implemented during the past 3 decades now requires the installation of lead-free pipes, pipe fittings, and pipe solders. These changes and the increased use of water filtration systems in many homes are gradually reducing the problem of lead in drinking water, although it remains a concern in older houses and apartments.

A high-profile example of the problem of lead water pipes happened in Flint, Michigan, in 2014. For many years, the city had been buying its water from Detroit, which supplied water that contained an added chemical that coated pipes and prevented lead from leaching into household water. However Flint was struggling financially and determined it would save $5 million over two years if it did not buy the Detroit water but obtained water from the local Flint River. But Flint did not add the chemical to coat and protect the pipes. The Flint River has higher chloride concentrations, which can be corrosive to lead pipes. As a result, lead pipes in the Flint water system became quickly corroded, the water turned brown, and lead began to leach from the pipes into household water supplies (**FIGURE 42.1**).

The EPA guidelines state that there are health concerns for lead in water at 5 parts per billion (ppb). Water tests in 2015 revealed that concentrations of lead in Flint's water supply were much higher—the highest sample concentration measuring 13,000 ppb. Late in 2015, the Flint water system was reconnected to the Detroit water system, but the corroded pipes will likely continue to leach lead.

Exposure to lead attacks the nervous system and causes children to experience reduced IQ, attention problems, and even kidney damage. It is estimated that 8,000 children were exposed to high lead concentrations in the city of Flint. In an attempt to save $5 million in water costs, the city incurred an estimated at $400 million in current and future costs associated with lead poisoning. This is in addition to the $1.5 billion it will cost to replace the old, now corroded pipes.

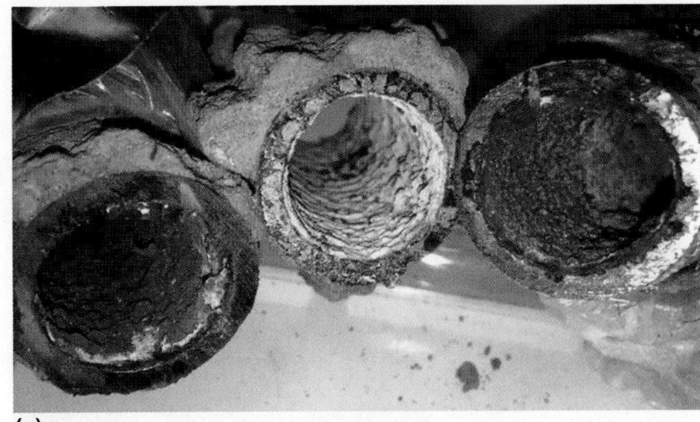

(a)

(b)

FIGURE 42.1 Lead contaminated water. (a) When the city of Flint, Michigan, switched their water source from Detroit to the Flint River and failed to add a chemical that prevents pipe corrosion, the old lead pipes in the water distribution system began to leak lead and poison residents of the city. (b) As a result of this corrosion, tap water in many homes and businesses often was contaminated by the corroding pipes. *(Courtesy Virginia Tech College of Engineering)*

Arsenic

Arsenic is a compound that occurs naturally in Earth's crust and can dissolve into groundwater. As a result, naturally occurring arsenic in rocks can lead to high concentrations of arsenic in groundwater and drinking water. Human activity also contributes to higher arsenic concentrations in groundwater. For example, mining breaks up rocks deep underground, and industrial uses of arsenic for items such as wood preservatives can add to the amount of arsenic found in drinking water. Fortunately, arsenic can be removed from water via fine membrane filtration, distillation, and reverse osmosis (see Figure 27.7 on page 318).

FIGURE 42.2 on page 512 indicates where high concentrations of arsenic have been found in well water throughout the United States. As you can see, concentrations are highest in the Midwest and West. Arsenic

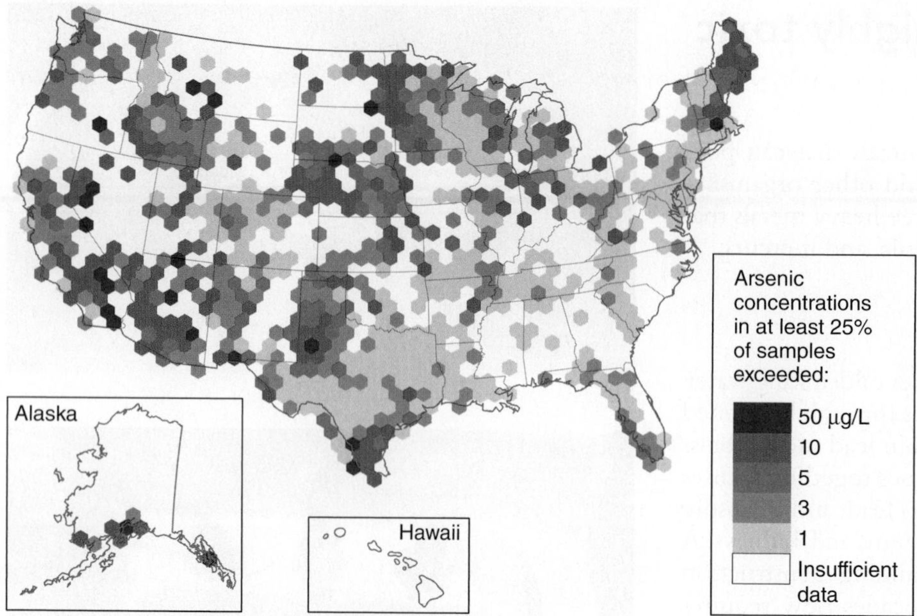

FIGURE 42.2 Arsenic in U.S. well water. The highest concentrations of arsenic are generally found in the upper Midwest and the West. *(Data from http://water.usgs.gov/nawqa/trace/pubs/geo_v46n11/fig3.html)*

in drinking water is associated with cancers of the skin, lungs, kidneys, and bladder. These illnesses can take 10 years or more after exposure to develop. Even very low concentrations of arsenic, such as those found in many wells around the world and measured in ppb, can cause severe health problems. In fact, though cancers can develop with less than 50 ppb of arsenic in drinking water, 50 ppb was set as the upper "safe" limit for U.S. drinking water from 1942 to 1999. In 1999, the EPA lowered the upper limit to 10 ppb, a compromise reached after much debate between environmental groups pushing for 5 ppb and water, mining, and wood preservative industries arguing that even the 10 ppb limit would be too expensive to implement. In 2001, the EPA delayed the implementation of the 10 ppb standard until more data could be evaluated. Later that year, the U.S. National Academy of Sciences recommended a standard of 5 ppb and the EPA decided to implement a standard of 10 ppb, which is still the standard today. As a result, after substantial investment required to improve many water treatment facilities, people in the United States will now have lower amounts of cancer-causing arsenic in their drinking water.

The problem of arsenic in drinking water is a worldwide issue. During the 1980s and 1990s, for example, water engineers in regions of Bangladesh and eastern India drilled millions of very deep wells in an attempt to find sources of groundwater that were not contaminated by local pollution. While this had the desired effect of obtaining less-polluted water, officials were not aware that this deeper groundwater was contaminated by naturally occurring arsenic from the rocks deep underground. Currently, 140 million people in this region drink arsenic-contaminated water and thousands of individuals have been diagnosed with arsenic poisoning. Current efforts to solve this problem include plans to collect uncontaminated rainwater as a source of drinking water and research to develop inexpensive filters that can remove arsenic from the well water. There is also the challenge of homeowners knowing whether they should have their water tested and, more importantly, having facilities available to do this testing.

Mercury

Mercury is another naturally occurring heavy metal found in increased concentrations in water as a result of human activities. **FIGURE 42.3** shows mercury releases from different regions of the world, as estimated by the

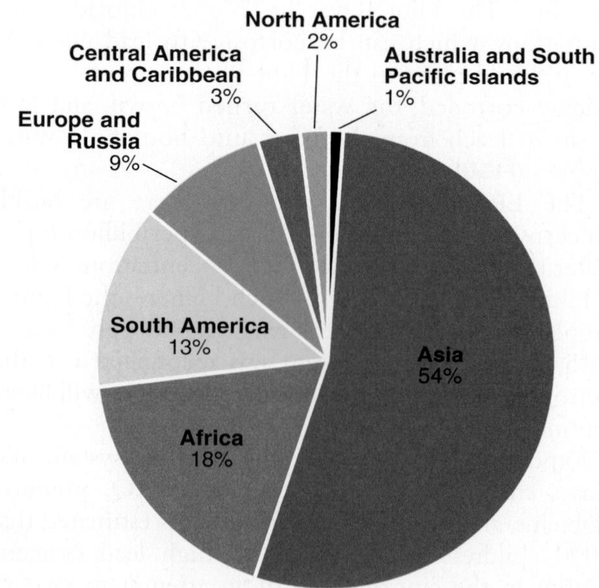

FIGURE 42.3 World mercury emissions. Mercury emissions from human activities vary greatly among regions of the world. *(Data from the 2018 Global Mercury Assessment.)*

2018 Global Mercury Assessment. Among regions of the world, 2 percent of human-produced mercury comes from North America, where mercury emissions have continued to decline, while more than half comes from Asia.

Approximately two-thirds of all mercury produced by human activities comes from the burning of fossil fuels, especially coal. Other important sources of mercury include the incineration of garbage, hazardous waste, and medical and dental supplies. One of the less well-known sources of mercury comes from the raw materials that go into the manufacturing of cement for construction. The limestone used to make cement can contain mercury, and this mercury is released during a heating process that is needed when making cement. Moreover, the source of heat for cement manufacturing is often coal that also releases mercury when burned.

Petroleum exploration is a source of both mercury and lead pollution. Each petroleum well produces roughly 681,374 L (180,000 gallons) of contaminated wastewater and mud over its lifetime. This water is usually dumped at the drilling site and, depending on the soils and topography, can either run into nearby waterways or infiltrate the soil and contaminate the underlying groundwater.

The mercury emitted by these activities eventually finds its way into water. Inorganic mercury (Hg) is not particularly harmful, but its release into the environment can be hazardous because of a chemical transformation it undergoes. In wetlands and lakes, bacteria convert inorganic mercury into methylmercury, which is highly toxic to humans. Methylmercury damages the central nervous system, particularly in young children and in the developing embryos of pregnant women. The result impairs coordination and the senses of touch, taste, and sight.

Human exposure to methylmercury occurs mostly through eating fish and shellfish. Methylmercury can move up the food chain in aquatic ecosystems, which results in the top consumers containing the highest concentrations of mercury in their bodies. Given that oceans are contaminated with mercury and that tuna are top predators, it is not surprising that these fish contain high concentrations of mercury. In 2008, a reporter for the *New York Times* purchased tuna sushi from 20 locations in New York City and, after analysis by a private laboratory, found that, at most restaurants, a diet of six pieces of sushi per week would exceed the EPA standard for human consumption of mercury. The EPA reported that the concentration of mercury in the tuna of the North Pacific Ocean had increased by 30 percent from 1990 to 2009 and this trend has continued through 2017. If China's plans to build more electric-generating plants that burn coal go forward, the EPA estimates that these concentrations in tuna will increase by another 50 percent by 2050.

What can be done about mercury pollution? In 2013 the United States announced a new agreement with more than 140 other countries to reduce global mercury pollution. In a step toward this goal, the EPA has proposed that cement-manufacturing plants reduce mercury emissions by 92 percent. New EPA rules are also being written to reduce mercury emissions from other major sources, including coal-burning power plants. These changes have allowed a reduction in mercury emissions in North America from 6 percent of the global contribution in 2013 to only 2 percent in 2018.

Acid deposition and acid mine drainage affect terrestrial and aquatic ecosystems

About 40 years ago, people throughout the northeastern United States, northern Europe, China, and Russia began to notice that the forests, soils, lakes, and streams were becoming more and more acidic. As a consequence, some trees were killed and some bodies of water became too acidic to sustain fish. After much debate, it became clear that the source of the lower pH of the water was the presence of very tall smokestacks of industrial plants that were burning coal and releasing sulfur dioxide and nitrogen dioxide into the air. These tall smokestacks kept the emissions away from local residents but sent the chemicals into the atmosphere where they were converted into sulfuric acid and nitric acid that returned to Earth hundreds of kilometers away. Acids deposited on Earth as rain and snow or as gases and particles that attach into the surfaces of plants, soil, and water are known as **acid deposition**. As described in Chapter 8, wet-acid deposition, also known as acid precipitation or acid rain, occurs in the form of rain and snow, whereas dry-acid deposition occurs as gases and particles that attach to the surfaces of plants, soil, and water. Acid deposition reduces the pH of water bodies from 5.5 or 6 to below 5, which can be lethal to many aquatic organisms, leaving these water bodies devoid of many species.

AP® Exam Tip

You should be able to differentiate between acidification and salinization. ●

To combat the problem of acids being released into the atmosphere, many coal-burning facilities have installed coal scrubbers. Coal scrubbers pass the hot gases produced during combustion through a limestone slurry. The limestone reacts with the acidic gases and removes them before the hot gases leave the smokestack.

Acid deposition Acids deposited on Earth as rain and snow or as gases and particles that attach to the surfaces of plants, soil, and water.

We will look at the issues raised by acid deposition in greater detail in Chapter 15 on air pollution.

Low pH in water bodies also occurs when very acidic water comes from below ground. This problem begins with the development of underground mines that, once abandoned, flood with groundwater. The combination of water and air allows pyrite, a type of rock, to break down and produce iron and hydrogen ions. The increase in hydrogen ions produces acidic water with a low pH. Water in the mine containing these ions can find its way up to the surface in the form of springs that feed into streams. As we discussed in Chapter 8, a similar effect occurs during mountaintop mining operations in which the tops of mountains are removed and the soil is dumped into stream valleys. The streams that are fed by springs from these mines are infiltrated with water that can have a pH close to zero. Low-pH water can cause many other harmful metal ions to become soluble, including zinc, copper, aluminum, and manganese. Although some water bodies such as bogs are naturally acidic and contain species that are adapted to acidic conditions, the combination of very low pH and toxic metals can produce environments that are too harsh for most organisms to survive. Because much of the dissolved iron precipitates out of solution as the low-pH water of the mine mixes with the high-pH water of the stream, these streams often have a striking red or yellow color (**FIGURE 42.4**).

Unfortunately, many mining companies responsible for making streams uninhabitable for fish and other organisms are no longer in business. As a result, they are often not held accountable for the environmental damage they caused. This presents a major challenge for state and local governments that bear the cost of dealing with the problem of acidic water from mines and the toxic metals that the companies produced. Researchers are currently investigating a number of strategies to counteract the acidity of these streams. One approach has been to pass stream water through a limestone treatment facility that raises the pH of the water and removes toxic metals to levels tolerable to stream organisms.

FIGURE 42.4 Acid mine drainage. The low pH of water emerging from abandoned mines mixes with stream water to lower the pH of streams, causing iron to precipitate out of solution and form a rusty red oxidized iron. This problem occurs around the world, including this stream in Colline Metallifere, Italy. *(Elbardamu/Alamy)*

Synthetic organic compounds are human-produced chemicals

Synthetic, or human-made, compounds can enter the water supply either from industrial point sources where they are manufactured or from nonpoint sources when they are applied over very large areas. These organic (carbon-containing) compounds include pesticides, pharmaceuticals, military compounds, and industrial compounds. Synthetic organic compounds have a variety of effects on organisms. They can be toxic, cause genetic defects, and, in the case of compounds that resemble animal hormones, interfere with growth and sexual development.

Pesticides and Inert Ingredients

Pesticides serve an important role in helping to control pest organisms that pose a threat to crop production and human health (**FIGURE 42.5**). Although natural pesticides such as arsenic have been used for centuries, the first generation of synthetic pesticides was developed during World War II. As we discussed in Chapter 11, these chemicals proved to be very effective in killing a variety of undesired plants (herbicides), fungi (fungicides), and insects (insecticides). In the decades that have followed, however, environmental scientists have identified a number of concerns about the unintended effects of pesticides.

Most pesticides do not target particular species of organisms, but generally kill a wide variety of related organisms. For example, an insecticide that is sprayed to kill mosquitoes is typically lethal to many other species of invertebrates, including insects that might be desirable as predators of the mosquito. Some pesticides are lethal to unrelated species. For example, researchers have recently discovered that the insecticide endosulfan, a chemical designed to kill insects, is highly lethal to amphibians even at very low concentrations. Even a pesticide that is not directly lethal to a species can indirectly affect organisms by causing a chain reaction through the community.

Synthetic pesticides are generally designed specifically to target particular aspects of a pest's physiology. However, they can also alter physiological functions of other species. For example, most insecticides target the nervous system of a particular pest, yet they can have unintended impacts on other pests as well as on many nonpest species. The insecticide DDT (dichlorodiphenyltrichloroethane) is a good example. While DDT was designed to alter nerve transmissions in insects, the

FIGURE 42.5 Applying pesticides. Pesticides provide benefits to humans, but they also can have unexpected effects on humans and other nonpest organisms that are not fully understood and have not been adequately investigated. This airplane is spraying pesticides over crops in New South Wales, Australia. *(incposterco/Getty Images)*

chemical can move up an aquatic food chain all the way to eagles that consume fish. Eagles that consumed DDT-contaminated fish produced eggs with thinner shells that would prematurely break during incubation. Despite the fact that DDT was designed to kill mosquitoes, its unintended impact was the thinning of bird eggshells. After the United States banned the spraying of DDT in 1972, the bald eagle and other birds of prey such as the peregrine falcon increased in numbers. However, DDT is still manufactured in developed nations and sprayed in developing countries as a preferred way to control the mosquitoes that carry the deadly malaria parasite.

Another concern about pesticides is the effect of inert ingredients added to commercial formulations. Inert ingredients are additives that make a pesticide more effective—for example they allow it to dissolve in water for spraying or to penetrate inside a pest species. Although the term *inert* may suggest that these chemicals are harmless, this is often not true. The popular herbicide Roundup, for example, is composed of a chemical that is highly effective at killing plants but has difficulty getting past the waxy outer layer of leaves without the help of an added inert ingredient. Since inert ingredients are legally classified as trade secrets and most are not required to be tested for safety, their effects are not always known before a product comes to market. In the case of Roundup, recent research has discovered that the herbicide is highly toxic to amphibians, not because of the plant-killing chemical but because of the inert ingredients. It appears that the same properties that allow the penetration of leaves also allow the penetration of tadpole gill cells. The gills burst and the tadpoles suffocate. Unintended consequences such as these have roused interest in both Europe and North America to require that inert ingredients be tested for potentially harmful effects.

Pharmaceuticals and Hormones

While most people know that pesticides are commonly found in the environment, they often are surprised to learn that pharmaceutical drugs are also common. For example, the U.S. Geological Survey tested 139 streams across the United States for a variety of chemical contaminants. **FIGURE 42.6** on page 516 shows data for the frequency of detection. Among the different types of chemicals that were present at detectable levels, approximately 50 percent of all streams contained antibiotics and reproductive hormones, 80 percent contained nonprescription drugs, and 90 percent contained steroids. In a follow-up study in the American Midwest, the U.S. Geological Survey sampled 100 streams and reported in 2017 that the average stream contained 52 pesticides. In most cases the concentrations of these chemicals are quite low and currently are not thought to pose a risk to environmental or human health. Some chemicals such as hormones, however, operate at very low concentrations inside the tissues of organisms and

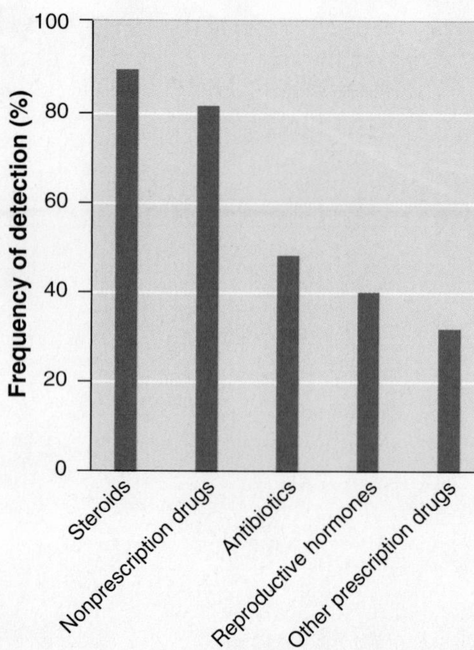

FIGURE 42.6 Contaminants in streams. Streams contain a wide variety of chemicals, including pharmaceutical drugs and hormones. These come from a combination of wastewater inputs, agriculture, forestry, and industry. (Data from D.W. Kolpin et.al. 2002. Pharmaceuticals, hormones, and other organic wastewater contaminants in U.S. streams, 1999–2000: A national reconnaissance. *Environmental Science and Technology* 36: 1202–1211.)

we have a poor understanding of their effects. As noted in our discussion of the Chesapeake Bay, it is now clear that low concentrations of pharmaceutical drugs that mimic estrogen are connected to male fish growing eggs in their testes. The extent of hormone effects on humans and wildlife around the world is currently unknown but environmental scientists are showing increased attention to these effects. We will talk more about the ways that chemicals can alter the endocrine systems of animals in Chapter 17.

Military Compounds

Perchlorates, a group of harmful chemicals used for rocket fuel, sometimes contaminate the soil in regions of the world where military rockets are manufactured, tested, or dismantled. Perchlorates come in many forms. The U.S. space shuttle, for example, used a booster rocket that contained 70 percent ammonium perchlorate.

Perchlorates A group of harmful chemicals used for rocket fuel.

Polychlorinated biphenyls (PCBs) A group of industrial compounds used to manufacture plastics and insulate electrical transformers, and responsible for many environmental problems.

FIGURE 42.7 A river on fire. In 1952, the polluted Cuyahoga River caught fire after a spark ignited the film of industrial pollution that was floating on the surface of the water. *(Anonymous/AP Images)*

Perchlorates easily leach from contaminated soil into the groundwater, where they can persist for many years. Human exposure to perchlorates comes primarily through consumption of contaminated food and water. In the human body, perchlorates can affect the thyroid gland and reduce the production of hormones necessary for proper functioning of the human body.

Industrial Compounds

Industrial compounds are chemicals used in manufacturing. Unfortunately, it used to be common for manufacturers in the United States to dispose of industrial compounds directly into bodies of water. One of the most widely publicized consequences of this practice occurred in the Cuyahoga River of Ohio. For more than 100 years, industries along the river dumped industrial wastes that formed a slick of pollution along the surface, killing virtually all animal life. The problem became so bad that the river actually caught fire and burned several times over the decades (**FIGURE 42.7**). The fire on the river in 1969 garnered national attention and led to a movement to clean up America's waterways. Today, the Cuyahoga River and most other rivers in the United States are much cleaner because of legislation that substantially reduced the amount of industrial and other waste that can be legally dumped into waterways.

Polychlorinated biphenyls (PCBs) are a group of industrial compounds that were once used to manufacture plastics and insulate electrical transformers. PCBs have caused many environmental problems. Ingested PCBs are lethal and carcinogenic, or cancer-causing. Although the manufacture of PCBs ended in 1979 and they are no longer used in the United States, because of their long-term persistence they are still present in the environment. One particularly high-profile case involves two General Electric manufacturing plants in New York State that dumped 590,000 kg (1.3 million pounds) of PCBs into the Hudson River from 1947 to 1977. In 2002, the EPA ruled that General Electric must pay for the dredging and removal of approximately 2.03 million cubic meters (2.65 million cubic yards) of PCB-contaminated sediment from a 64-km (40-mile) stretch of the upper Hudson River in New York State. General Electric argued that the dredging should not occur because it would stir up the sediments and re-suspend the PCBs in the river water, causing further problems. The courts and EPA scientists disagreed and in 2009 the dredging of the PCBs finally began. The dredging was completed in 2015 and today researchers are continuing to monitor PCB levels in the water, soil, and fish. It is expected that the Hudson River will still require 15 to 30 years before PCB levels are low enough in fish to allow consumption by humans. As this case demonstrates, though it can take decades of scientific research and debate, in the end we can partially reverse the contamination of our water bodies.

While PCBs have long been a concern, there is a growing uneasiness over compounds known as PBDEs (polybrominated diphenyl ethers). PBDEs are most commonly known as flame retardants added to a wide variety of items that include construction materials, furniture, electrical components, and clothing. They make buildings and their contents considerably less flammable than they would be otherwise. Since the 1990s, however, scientists have been detecting PBDEs in some unexpected places, including fish, aquatic birds, and human breast milk. Exposure to some types of PBDEs can lead to brain damage, especially in children. As a result, the European Union and several states, including Washington and California, have banned the manufacture of several types of PBDEs.

MODULE 42

Preparing for the AP® Exam

AP® Review

In this module, we learned that a wide variety of chemicals in the water can have harmful effects on humans and other organisms. These chemicals include heavy metals, acids, and synthetic organic compounds. Through legislation and cleanup efforts, some of these compounds are now found less frequently in the environment. However, some chemicals continue to make their way into water bodies and others persist over time. In the next module, we will discuss the effects of oil pollution in bodies of water.

AP® Practice Questions

Choose the best answer for the following.

1. Arsenic is
   (a) found naturally in groundwater.
   (b) often discovered due to its rapid health effects.
   (c) inexpensive to remove from drinking water.
   (d) primarily found in water in North America.

2. Which is NOT true of acid deposition?
   (a) It is primarily due to the burning of coal.
   (b) It is treated using limestone.
   (c) It can occur as a result of mining.
   (d) It causes increased solubility of many ions.

3. What caused the eggshells of some birds to become thin and break?
   (a) heavy metal concentrations in prey species
   (b) an inert ingredient in herbicides
   (c) an insecticide ingested by prey species
   (d) a pesticide used to control pest birds such as crows

4. Mercury
   (a) is harmless once converted into methylmercury.
   (b) exposure often occurs through shellfish.
   (c) is most concentrated in herbivores.
   (d) can be safely trapped during the production of concrete.

5. Which hazardous material is known to cause cancer?
   (a) lead
   (b) mercury
   (c) perchlorates
   (d) PCBs

MODULE 43

Oil Pollution

In Chapter 12 we noted that the pollution of Earth's oceans and shorelines from crude oil and other petroleum products is an ongoing problem. Petroleum products are highly toxic to many marine organisms, including birds, mammals, and fish, as well as to algae and microorganisms that form the base of the aquatic food chain. Oil is a persistent substance that can spread below and across the surface of the water for hundreds of kilometers and leave shorelines with a thick, viscous covering that is extremely difficult to remove. In this section, we will examine the many sources of oil pollution and then talk about some of the ways currently used to clean up oil and to reduce its harmful effects.

Learning Goals

After reading this module you should be able to

- identify the major sources of oil pollution.
- explain some of the current methods to remediate oil pollution.

There are several sources of oil pollution

There are many different sources of oil pollution in water. Oil and other petroleum products can enter the oceans as spills from oil tankers. One of the best-known spills involved the tanker *Exxon Valdez* that ran aground off the coast of Alaska in 1989 (**FIGURE 43.1**). As we discussed in Chapter 12, the ship spilled 42 million liters (11 million gallons) of crude oil that spread across the surface of several kilometers of ocean and coastline. The spill killed 250,000 seabirds, 2,800 sea otters, 300 harbor seals, and 22 killer whales. Cleanup efforts went on for 2 decades.

FIGURE 43.1 The *Exxon Valdez* oil spill. In 1989, the oil tanker ran aground and spilled millions of liters of crude oil onto the shores of Alaska, where the oil killed thousands of animals and harmed many others such as this red-necked grebe (*Podiceps grisegena*) on Knight's Island, located about 35 miles from the spill. *(AP Images)*

In 2009, 20 years after the *Exxon Valdez* spill, scientists evaluated the state of the contaminated Alaskan ecosystem. They concluded that the harmed populations of many species have rebounded, including bald eagles and salmon. However, several species have not yet rebounded, including killer whales (*Orcinus orca*) and sea otters (*Enhydra lutris*). Nor has the oil been completely removed from the environment. Pits dug into the shoreline suggest that approximately 55,000 L (14,500 gallons) of oil remain. It is estimated that this oil will take more than 100 years to break down and the long-lasting effects will only become apparent over the coming decades.

For its part, Exxon has paid $1 billion for the cleanup and $500 million in damages. The company also changed the ship's name, although the ship has been banned from carrying oil in North America. The *Valdez* accident sparked new rules for oil tankers in North America. The *Exxon Valdez* had a single-hull design, but tankers must now have a double-hull design with two steel walls to contain leaking oil.

Offshore drilling is another source of oil pollution. There are approximately 5,000 offshore oil platforms in North America and another 3,000 worldwide. Drilling platforms often experience leaks. The best estimate for the amount of petroleum leaking into North American

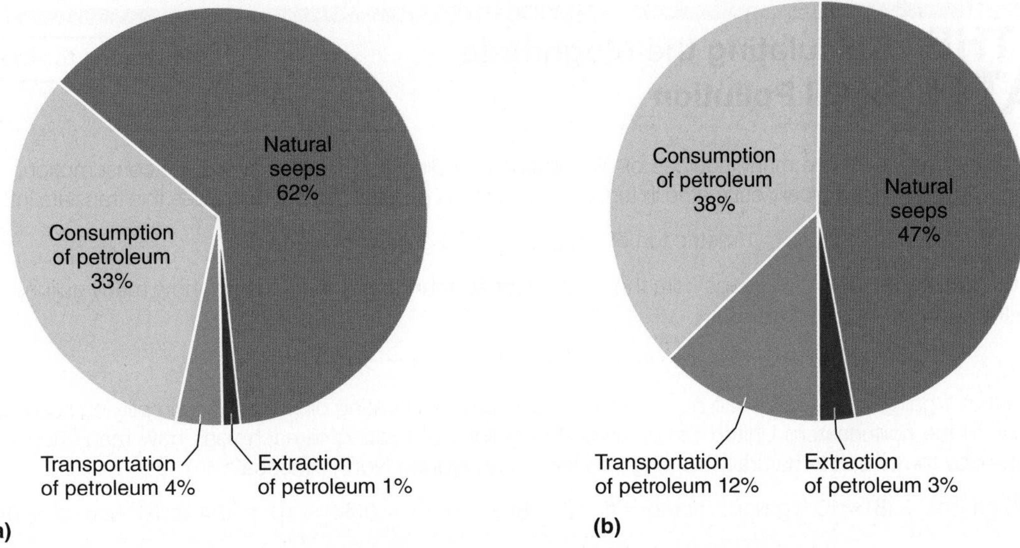

FIGURE 43.2 Sources of oil in the ocean. Oil contamination in the ocean, both (a) in North America and (b) worldwide, comes from a variety of sources including natural seeps, extraction of oil from underneath the ocean, transport of oil by tanker or pipeline, and consumption of petroleum-based products. *(Data from http://oceanworld.tamu.edu/resources/oceanography-book/contents.htm)*

waters is 146,000 kg (322,000 pounds) per year. In other parts of the world, antipollution regulations are often less stringent. Estimates of the amount of petroleum leaking into the ocean annually from foreign oil platforms range from 0.3 million to 1.4 million kilograms (0.6 million to 3.1 million pounds).

One of the most famous oil leaks from an offshore platform occurred in 2010 on a BP operation in the Gulf of Mexico. In this case, an explosion on the *Deepwater Horizon* platform caused a pipe to break on the ocean floor nearly 1.6 km (1 mile) below the surface of the ocean. From the time of the explosion in April until the well was sealed in August 2010, the broken pipe released an estimated 780 million liters (206 million gallons) of crude oil into the Gulf of Mexico (see Chapter 12, page 414). This spill contaminated beaches, wildlife, and the estuaries that serve as habitats for the reproduction of commercially important fish and shellfish. The magnitude of the oil spill was nearly 20 times larger than that of the *Exxon Valdez*. However, because much of the oil spilled into the ocean, scientists may not be able to assess the full impact of the oil spill for several decades. The accident has the potential to become one of the largest environmental disasters in history.

In addition to oil spills, oil pollution in the ocean occurs naturally. In fact, the U.S. National Academy of Sciences recently estimated that natural releases of oil from seeps in the bottom of the ocean account for 60 percent of all oil in the waters surrounding North America and 45 percent of all oil in water worldwide. **FIGURE 43.2** shows the proportion of different sources of oil in water for both North American and worldwide marine waters. In the waters controlled by the United States, the ocean seeps more than 270,000 L (70,000 gallons) of oil every day. This means that when we assess the environmental impact of oil in our oceans, we must consider the combination of both natural and anthropogenic releases of oil. "Do the Math: Calculating the Magnitude of Oil Pollution" on page 520 provides an opportunity to work with converting units of measurement to determine the amount of oil seepage.

There are ways to remediate oil pollution

Since the 1989 *Exxon Valdez* oil spill, researchers have been investigating how best to remediate oil spills. Contaminated mammals and waterfowl must be cleaned by hand. Bird feathers that are covered with oil, for example, become heavy and lose their ability to insulate. The best approach to cleaning up the spilled oil, however, is not always clear.

Oil spilled in the ocean can either float on the surface or remain far below in the form of underwater plumes. For oil floating on the surface of the open ocean, a common approach is to contain the oil within an area and then suck it off the surface of the water. Containment occurs by laying out oil containment booms that consist of plastic barriers floating on the surface of the water and extending down into the water for several meters. These plastic walls keep the floating oil from spreading further. Once the oil is contained, boats equipped with giant vacuums suck up as much

DO THE MATH Calculating the Magnitude of Oil Pollution

Preparing for the AP® Exam

As we have discussed, there are many sources of oil pollution including from natural seeps, oil consumption, extraction, and transportation. Oil pollution is typically estimated in terms of metric tons of crude oil, but how does this translate into gallons of oil?

$$1 \text{ metric ton of crude oil} = 320 \text{ gallons of crude oil}$$

Given that 160,000 metric tons of oil seeps from the ocean floor surrounding North America, how many gallons of crude oil are seeping into the ocean in this region?

$$160{,}000 \text{ metric tons} \times 320 \text{ gallons/ton} = 5.12 \times 10^7 \text{ gallons}$$

If we assume that 1 gallon of crude oil can be used to make 1 gallon of heating oil (which is a simplifying assumption), and the average house in the northeastern United States uses 800 gallons of heating oil each year, how many northeastern homes could be heated by the oil that is naturally seeping into the ocean around North America?

$$(5.12 \times 10^7) \text{ gallons} \div (8 \times 10^2) \text{ gallons/home} = (5.12 \div 8) \times (10^{7-2}) = 0.64 \times 10^5 = 6.4 \times 10^4 \text{ homes} = 64{,}000 \text{ homes}$$

YOUR TURN

1. The amount of pollution from oil extraction activities is 3,000 metric tons. How many gallons does this represent?
2. How many homes in the northeastern United States could be heated by this amount of oil?

oil as possible (**FIGURE 43.3**). In shallow areas and along the coastline, absorbent materials are used to suck up the spilled oil.

A second approach to treating oil floating on the surface is to apply chemicals that help break up and disperse the oil before it hits the shoreline and causes damage to the coastal ecosystems. Although the dispersants can be effective, they can also be toxic to marine life. Current research is examining ways to make chemical dispersants more environmentally friendly.

A third approach is to burn the oil slicks. This was done during the *Deepwater Horizon* oil spill to rapidly remove floating crude oil. However, the burning causes a great deal of air pollution and leaves behind a thick by-product that can be harmful to wildlife.

FIGURE 43.3 Oil-spill containment. Floating plastic walls can contain oil spills while the oil is sucked off the surface of the water. This worker is using a vacuum to clean up the Kalamazoo River in Battle Creek, Michigan. *(Jim West/Alamy Stock Photo)*

A fourth approach to cleaning up oil uses genetically engineered bacteria. Several years ago scientists discovered a naturally occurring bacterium that obtained its energy by consuming oil emerging from natural seeps. These bacteria were typically rare in the ocean, but were very abundant in areas where oil spills or seeps occurred. Scientists are currently trying to determine the genes that confer the bacteria's ability to consume oil and hope to insert copies of these genes into genetically modified bacteria to consume oil spills even faster.

Research on oil spill cleanups continues today. In 2016, researchers discovered that specially designed sponges could absorb oil while repelling seawater. These sponges can be used repeatedly and absorb 50 times their own weight. In 2017, researchers discovered a new method to burn oil slicks by heating the oil to allow more complete combustion with cleaner emissions from the fires.

In contrast to oil floating on the surface of the water, oil in underwater plumes persists as a mixture of water and oil, similar to the mixture of vinegar and oil in a salad dressing. In the case of the BP platform explosion in the Gulf of Mexico, scientists reported observing an oil plume moving approximately 1,000 m (3,000 feet) below the surface of the ocean. The plume was approximately 24 to 32 km (15–20 miles) long, 8 km (5 miles) wide, and hundreds of meters thick. There is currently no agreed-upon method of removing underwater plumes from the water.

When spilled oil comes to shore, the best solution in not always clear. For example, there is some debate over the best way to treat rocky coastlines after an oil spill. Scientists have been monitoring parts of Prince William Sound that were treated in different ways after the *Exxon Valdez* spill. Workers cleaned some areas with high-pressure hot water to remove the oil. Unfortunately, the hot water sprayers not only removed the oil, but also removed most of the marine life and, in some cases, the fine-grained sediments containing nutrients. Without the fine-grained sediment, many organisms were unable to recolonize the coast.

Other parts of the coastline received no human intervention. Over the years since the spill, the repeated action of waves and tides slowly removed much of the oil. However, the remaining oil existing in crevices of the rocky shoreline continues to have a negative effect on organisms that live among the rocks. Thus, leaving the oil on beaches also poses problems. At present, there is no clear consensus on the best way to respond to oil spills on coastlines.

MODULE 43 AP® Review

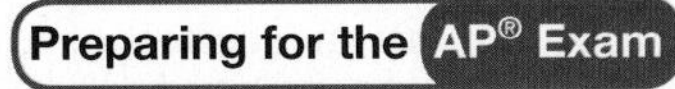

In this module, we learned that oil pollution in water comes from both natural sources and human activity. Oil pollution can be devastating to plants and animals. We have developed several ways to help clean up oil spills. In the next module, we will examine nonchemical forms of pollution.

AP® Practice Questions

Choose the best answer for the following.

1. The *Exxon Valdez* oil spill
   (a) was cleaned up within 2 years of the spill.
   (b) led to new regulations for oil tankers.
   (c) was larger than the *Deepwater Horizon* spill.
   (d) has not significantly changed the ecosystem 20 years later.

2. Approximately what percent of the oil in marine waters worldwide is due to natural causes?
   (a) 20 percent
   (b) 35 percent
   (c) 45 percent
   (d) 60 percent

3. Which is NOT used in the cleanup of oil spills?
   (a) clumping agents applied to underwater plumes
   (b) chemical dispersants
   (c) large vacuums
   (d) absorbent materials along the coast

MODULE 44

Nonchemical Water Pollution

Learning Goals

After reading this module you should be able to

- identify the major sources of solid waste pollution.
- explain the harmful effects of sediment pollution.
- discuss the sources and consequences of thermal pollution.
- understand the causes of noise pollution.

When we think about water pollution, we most commonly envision scenes of dirty water contaminated with toxic chemicals. Though such scenarios certainly receive a great deal of public attention, there are other, less familiar types of water pollution. In this module, we will examine four types of nonchemical water pollution: solid waste, sediment, heat, and noise.

Solid waste pollution includes garbage and sludge

Solid waste consists of discarded materials that do not pose a toxic hazard to humans and other organisms. Much solid waste is what we call garbage and the sludge produced by sewage treatment plants. In the United States, solid waste is generally disposed of in landfills, but in some cases it is dumped into bodies of water and can later wash up on coastal beaches. In 1997, scientists discovered a large area of solid waste, composed mostly of discarded plastics, floating in the North Pacific. This area, named the Great Pacific Garbage Patch, appears to collect much of the solid waste that is dumped into waters and concentrates it in the middle of the rotating currents in an area the size of Texas. Other ocean currents appear to also have the ability to concentrate garbage. Given the vastness of these areas, no one is certain how much solid waste is floating but current estimates are in the range of hundreds of millions of kilograms.

Garbage on beaches and in the ocean is not only unsightly, it is dangerous to both marine organisms and people. Plastic rings from beverage six-packs, for instance, can strangle many animals. Medical waste poses a threat to people on the beach, particularly children. During the 1970s and 1980s, the public turned its attention to garbage dumping off the coasts of the United States by both municipalities and cruise ships. The outcry grew when garbage was found to include medical waste such as used hypodermic needles. As a result, the practice of dumping garbage in the ocean was curtailed in the early 1980s. The problem remains in many developing countries where there is often no political mechanism to prevent dumping or it is economically difficult to manage proper garbage disposal (**FIGURE 44.1**).

Another major source of solid waste pollution is the coal ash and coal tailings that remain behind when coal is burned. As we saw in Chapter 12, such waste contains

FIGURE 44.1 A river of garbage. Environmental regulations have greatly reduced the amount of solid waste that is dumped in U.S. waters, but other parts of the world such as the Citarum River in Indonesia still face a major environmental challenge. *(REUTERS/Dadang Tri)*

a number of harmful chemicals including mercury, arsenic, and lead. The solid waste products from burning coal are typically dumped into landfills, ponds, or abandoned mines where it can contaminate groundwater. In the United States, the Environmental Protection Agency considers the waste from burning coal and other fossil fuels to be "special waste" that is exempt from federal regulations for the disposal of hazardous waste. In addition, most states have either no regulations or weak regulations governing the disposal of coal ash and coal tailings.

Sediment pollution consists of soil particles that are carried downstream

As noted in our discussion of the Chesapeake Bay, sediments are particles of sand, silt, and clay carried by moving water in streams and rivers that eventually settle out in another location where water movement is slowed, such as where streams empty into lakes and rivers empty into oceans forming deltas. The transport of sediments by streams and rivers is a natural phenomenon, but sediment pollution is the result of human activities that can substantially increase the amount of sediment entering natural waterways (**FIGURE 44.2**).

Numerous human activities lead to increased sedimentation. Construction of buildings, for example, requires digging up the soil, and the bare ground can lose some of its soil to erosion. As we saw in Chapters 8 and 10, plowed agricultural fields are susceptible to erosion from rain and wind. Sediments can also enter streams and rivers when natural vegetation is removed from the edge of a water body and replaced with either crops or domesticated animals that continually disturb the soil when they come to drink. According to recent estimates, 30 percent of all sediments in our waterways comes from natural sources while 70 percent comes from human activities.

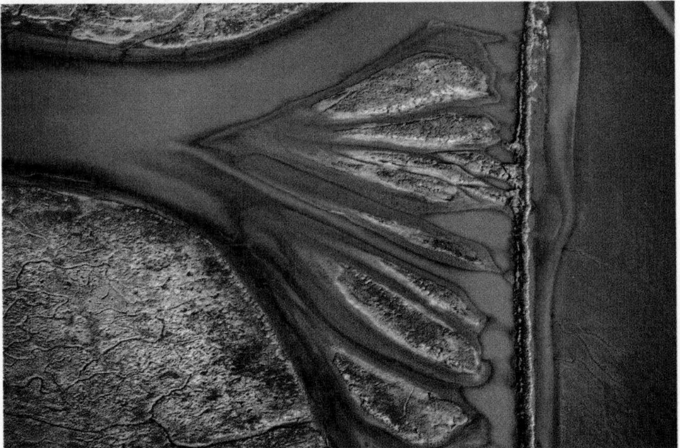

FIGURE 44.2 Sediments carried by rivers. Some rivers, such as the Fraser River Delta as it enters the Pacific Ocean, carry a large amount of sediment that gets emptied into lakes and oceans to form deltas. *(National Geographic Creative/Alamy)*

AP® Exam Tip

Be sure that you can describe how human activity increases sediment load in rivers. ●

What is the effect of increased sediment caused by human activities? Perhaps the most noticeable is that waterways become brown and cloudy due to the suspension of soil particles in the water. Increased sediment in the water column reduces the infiltration of sunlight, which can reduce the productivity of aquatic plants and algae. Because sediments can also clog gills, they sometimes hinder the ability of fish and other aquatic organisms to obtain oxygen. In locations where water moves slowly, the sediments settle out and accumulate on the bottom of the water body. This accumulation can clog the gills of bottom dwellers such as oysters or clams. Since the sediments can contain nutrients, this pollution may also contribute to increased nutrients coming into the ecosystem. In total, the U.S. Environmental Protection Agency estimates that sediment pollution costs $16 billion annually in environmental damage.

Thermal pollution causes substantial changes in water temperatures

A third type of nonchemical water pollution, **thermal pollution**, occurs when human activities cause a substantial change in the temperature of water. Thermal pollution is most common when an industry removes cold water from a natural supply, uses it to absorb heat that is generated in a manufacturing process, and returns the heated water back to the natural supply. For example, we saw in Chapter 12 that electric power plants use nearly half of all water extracted; they remove cold water from rivers, lakes, or oceans, cool the steam converted from water back into water, and then return the water to nature at temperatures that are 10°C to 15°C (18°F − 27°F) warmer. A variety of other industries make use of water for cooling, including steel mills and paper mills that need to cool their machines.

AP® Exam Tip

Remember that thermal pollution is caused not by global warming but by industrial activities that return heated water back to the natural water body. ●

Thermal pollution Nonchemical water pollution that occurs when human activities cause a substantial change in the temperature of water.

FIGURE 44.3 Thermal pollution. Nuclear reactors, such as the Three Mile Island nuclear plant in Middletown, Pennsylvania, use water to generate steam. To cool this steam, they either use cooling towers or empty the water into holding ponds. In both cases, the water must be cooled before it is returned to the natural source of water such as a lake or river. *(Bloomberg/Getty Images)*

Species in a given community are generally adapted to a particular natural range of temperatures. Therefore a dramatic change in temperature can kill many species, a phenomenon called **thermal shock**. High temperatures also cause organisms to increase their respiration rate. But because warmer water does not contain as much dissolved oxygen as cold water, an increase in respiration and a decrease in oxygen will cause many animals to suffocate. In recent years, steps have been taken to help reduce thermal pollution, including pumping the heated water into outdoor holding ponds where it can cool further before being pumped back into natural water bodies.

In the United States, the EPA regulates how much heated water can be returned to natural water bodies. Compliance can become a real challenge in the summer when high demand for electricity causes an increased demand for cooling water even though the water in rivers and lakes will then probably be at its lowest volume and at its highest temperature. One common solution to this problem has been the construction of cooling towers that release the excess heat into the atmosphere instead of into the water. A cooling tower relies on the cooling power of evaporation to reduce the temperature of the water, much as we depend on our own sweat and a breeze to cool ourselves on a hot day. Some industries have built closed systems in which they cool the hot water in a cooling tower and then recycle the water to be heated again. In this way, the industries do not extract water from natural water bodies, nor do they release any heated water back into nature (**FIGURE 44.3**).

Thermal shock A dramatic change in water temperature that can kill organisms.

Noise pollution may interfere with animal communication

It may strike you as odd to think of noise as a type of water pollution. Indeed, noise pollution has received the least amount of attention from environmental scientists and, as a result, we know the least about it. When we think of noise pollution, what often comes to mind is the sound of city traffic. However, noise pollution also occurs in the water. For example, sounds emitted by ships and submarines that interfere with animal communication are the major concern. Especially loud sonar could negatively affect species such as whales that rely on low-frequency, long-distance communication (**FIGURE 44.4**). Several instances of beached whales in the Bahamas, the Canary Islands, and the Gulf of California have been connected to the use of military sonar and the use of loud, underwater air guns by energy companies searching for oil deposits under the oceans.

In 2003 a federal judge rejected the U.S. Navy's request to install a network of long-range sonar systems across the ocean floor to detect incoming submarines because of suspected negative impacts on endangered whales and other species of marine animals. In 2008, however, the U.S. Supreme Court ruled that the president of the United States could exempt the Navy from environmental laws that were related to potential sonar effects on ocean life. To better understand noise pollution in the ocean, the U.S. National Oceanic and Atmospheric Administration announced in 2012 that had it completed the first step in mapping the noises across different regions of the ocean. By 2014, NOAA concluded that the Navy's activities would have a negligible effect on fish and whales in the ocean. At the same time, an increased awareness of noise pollution in the ocean has inspired some ship builders to design ships equipped with quieter propellers.

FIGURE 44.4 Noise pollution. Noise from navy sonar and oil exploration may interfere with the normal behavior of whales. *(Lazareva/Getty Images)*

MODULE 44 AP® Review

Preparing for the AP® Exam

In this module, we learned that there are several nonchemical forms of water pollution. Solid waste pollution comes from garbage, sludge, and the ash and tailings produced by burning coal. Sediment pollution consists of soil particles that erode from the land and are carried downstream in streams and rivers where they can reduce light penetration, clog the gills of aquatic organisms, and add nutrients to aquatic ecosystems. Thermal pollution typically happens when industrial processes take in cool water from water bodies, use the water to cool their equipment, and then return much warmer water to the water body. Finally, noise pollution from ships, sonar, and air guns exploring for energy deposits has the potential to interfere with the communication of aquatic animals in the ocean, including whales. In the next module, we will examine how water pollution laws have been designed to control some of the major sources of pollution.

AP® Practice Questions

Choose the best answer for the following.

1. Solid waste
   (a) sinks to the ocean floor.
   (b) dumping in the ocean is prohibited worldwide.
   (c) can include coal-burning byproducts.
   (d) is rarely toxic to humans.

2. Sediments in water
   (a) decreases the solubility of oxygen.
   (b) clogs the gills of some aquatic animals.
   (c) results in decreased nutrient availability.
   (d) is primarily due to industrialization.

3. Thermal pollution
   (a) is primarily a problem in the winter.
   (b) is rarely lethal.
   (c) is not regulated in the United States.
   (d) has been reduced by the use of cooling towers.

4. The use of sonar
   (a) can reduce the productivity of some algae.
   (b) has little effect on aquatic ecosystems.
   (c) has a positive effect on some fish species.
   (d) disrupts communication among whales.

MODULE 45

Water Pollution Laws

A country's water quality improves when its citizens demand it and the country is affluent enough to afford measures to clean up pollution and to take steps to prevent it in the future. In this module we will look at U.S. laws that protect water from pollution and ensure safe drinking water. We will also examine how laws in developing nations are changing to address water pollution.

Learning Goals

After reading this module you should be able to

- explain how the Clean Water Act protects against water pollution.
- discuss the goals of the Safe Drinking Water Act.
- understand how water pollution legislation is changing in developing countries.

The Clean Water Act protects water bodies

As recently as the 1960s, water quality was very poor in much of the United States, but a growing awareness of the problem encouraged a series of laws to fight water pollution. The Federal Water Pollution Control Act of 1948 was the first major piece of legislation affecting water quality. In 1972, the act was expanded into the **Clean Water Act,** which supports the "protection and propagation of fish, shellfish, and wildlife and recreation in and on the water" by maintaining and, when necessary, restoring the chemical, physical, and biological properties of surface waters. Note that this objective does not include the protection of groundwater.

The Clean Water Act originally focused mostly on the chemical properties of surface waters. More recently, there has been an increased focus on ensuring that the biological properties of the waters also receive attention, including the abundance and diversity of various species.

Clean Water Act Legislation that supports the "protection and propagation of fish, shellfish, and wildlife and recreation in and on the water" by maintaining and, when necessary, restoring the chemical, physical, and biological properties of surface waters.

Safe Drinking Water Act Legislation that sets the national standards for safe drinking water.

Maximum contaminant level (MCL) The standard for safe drinking water established by the EPA under the Safe Drinking Water Act.

Most importantly, the Clean Water Act issued water quality standards that defined acceptable limits of various pollutants in U.S. waterways. To help enforce these limits, the act allowed the EPA and state governments to issue permits to control how much pollution industries can discharge into the water. Over time, more and more categories of pollutants have been brought under the jurisdiction of the Clean Water Act, including animal feedlots and storm runoff from municipal sewer systems.

The Safe Drinking Water Act protects sources of drinking water

In addition to the Clean Water Act, other legislation has been passed to regulate water pollution, including the **Safe Drinking Water Act** (1974, 1986, 1996), which sets the national standards for safe drinking water. Under the Safe Drinking Water Act, the EPA is responsible for establishing **maximum contaminant levels (MCL)** for 77 different elements or substances in both surface water and groundwater. This list includes some well-known microorganisms, disinfectants, organic chemicals, and inorganic chemicals (**TABLE 45.1**). These maximum concentrations consider both the concentration of each compound that can cause harm as well as the feasibility and cost of reducing the compound to such a concentration.

MCLs are somewhat subjective and are subject to political and economic pressures. For example, despite

TABLE 45.1	The maximum contaminant levels (MCL) for a variety of contaminants in drinking water as determined by the U.S. Environmental Protection Agency, in parts per billion (ppb)	
Contaminant category	**Contaminant**	**Maximum contaminant level (ppb)**
Microorganism	*Giardia*	0
Microorganism	Fecal coliform	0
Inorganic chemical	Arsenic	10
Inorganic chemical	Mercury	2
Organic chemical	Benzene	5
Organic chemical	Atrazine	3

Data from: U.S. Environmental Protection Agency, http://www.epa.gov/safewater/contaminants/index.html.

the evidence that 50 ppb of arsenic caused harm in humans, the MCL for arsenic was kept at 50 ppb for many years because of concerns that many communities could not afford to reduce levels to 10 ppb. As noted earlier in this chapter, the MCL for arsenic was finally reduced to 10 ppb in 2001.

AP® Exam Tip

You should be able to differentiate between the treatments for sewage wastewater and drinking water, as well as the laws that apply to each. ●

What has been the impact of these water pollution laws? In general, they have been very successful. The EPA defines bodies of water in terms of their designated uses, including aesthetics, recreation, protection of fish, and as a source of safe drinking water. The EPA then determines if a particular waterway fully supports all of the designated uses. According to the EPA's most recent report in 2017, 47 percent of wetlands, 44 percent of all rivers and streams, 30 percent of lakes and ponds, and 22 percent of bays and estuaries in the United States now fully support their designated uses. This is a large improvement from decades past but, as **TABLE 45.2** shows, we still have a lot of work to do to improve the remaining waterways. Today, the water in municipal water systems in the United States is generally safe. Water regulations have greatly reduced contamination of waters and nearly eliminated major point sources of water pollution. But nonpoint sources such as oil from parking lots and nutrients and pesticides from suburban lawns are not covered under existing regulations. In addition, the U.S. government has exempted fracking for natural gas (see Chapter 1) from the Safe Drinking Water Act, despite the fact that fracking injects a suite of harmful chemicals deep into the ground.

TABLE 45.2	The current leading causes and sources of impaired waterways in the United States	
	Causes of impairment	**Sources of impairment**
Wetlands	Organic enrichment, mercury, arsenic, selenium	Agriculture, atmospheric deposition, petroleum and natural gas production
Streams and rivers	Bacterial pathogens, sediments, excess nutrients	Agriculture, atmospheric deposition, water diversions, dam construction
Lakes, ponds, and reservoirs	Mercury, PCBs, nutrients	Atmospheric deposition, agriculture
Bays and estuaries	Bacterial pathogens, oxygen depletion, mercury	Atmospheric deposition, municipal discharges including sewage

Source: Data from U.S. Environmental Protection Agency. 2017. *National Water Quality Inventory: Report to Congress.*

Water pollution legislation is becoming more common in the developing world

If we look at water pollution legislation around the world, there is a clear difference between developed and developing countries. Developed countries, including those in North America and Europe, experienced tremendous industrialization many decades ago and widely polluted their air and water at that time. More recently they have addressed the problems of pollution by cleaning up polluted areas and by passing legislation to prevent pollution in the future. Developing countries are still in the process of industrializing. They are less able to afford water-quality improvements such as wastewater treatment plants or the costs associated with restrictive legislation. Moreover, political instability and corruption often make enforcement of legislation difficult. In some cases, contaminating industries move from developed countries to developing countries. Although the developing countries suffer from the additional pollution, they benefit economically from the additional jobs and spending that the new industries bring with them.

Water pollution problems are prevalent in many of the developing nations of Africa, Asia, Latin America, and eastern Europe. China and India, for example, have undergone rapid industrialization and have many areas of dense human population—a recipe for major water pollution problems. However, as a nation becomes more affluent, people often show more interest in the environment and resources available to address environmental issues. In Brazil, for example, industrialization began to take off in the 1950s. By the 1990s, the Tietê River, which passes through the large city of São Paulo, was badly polluted. More than a million Brazilians signed a petition in 1992 requesting that the government regulate the industrial and municipal pollution being dumped into the river. Today the Tietê River is much cleaner (**FIGURE 45.1**).

FIGURE 45.1 The Tietê River in Brazil. The Tietê River, which passes through the large city of São Paolo, was badly polluted in the 1950s but is much cleaner today. *(Pulsar Imagens/Alamy)*

MODULE 45 AP® Review

Preparing for the AP® Exam

In this module, we looked at the two major water pollution laws in the United States. The Clean Water Act is designed to protect surface water from pollution, but it does not protect groundwater. In contrast, the Safe Water Drinking Act sets maximum levels for microorganisms, disinfectants, organic chemicals, and inorganic chemicals in drinking water. Whereas developed countries have gone through periods of large-scale water pollution as they industrialized, they have since passed many laws to improve water quality. Developing countries are currently experiencing industrialization and widespread water pollution. As these countries become more affluent, it is expected that they will also turn their attention to improving water quality.

AP® Practice Questions

Choose the best answer for the following.

1. The first legislation on water quality was
   (a) the Federal Water Pollution Control Act.
   (b) the Clean Water Act.
   (c) the Safe Drinking Water Act.
   (d) the Resource Conservation and Recovery Act.
2. Maximum contaminant levels for groundwater were set in
   (a) the Clean Water Act.
   (b) the Safe Drinking Water Act.
   (c) the Resource Conservation and Recovery Act.
   (d) the Water Quality Act.
3. Which water quality issue is not covered in existing legislation?
   (a) organic chemicals
   (b) offshore drilling
   (c) nonpoint sources
   (d) groundwater
4. Which does NOT contribute to poor water quality in developing countries?
   (a) high population density
   (b) rapid industrialization
   (c) high unemployment
   (d) relocation of industry

Working Toward Sustainability

Purifying Water for Pennies

As we discussed in this chapter, having access to clean drinking water is a worldwide problem. Nearly 1 billion people drink unsafe water that contains soil sediments, pesticides, heavy metals, and disease-causing organisms. The problem of unsafe drinking water is largely a problem of developing countries that lack the financial resources to build proper sewage and water treatment facilities. Any solution to this challenge would need to be both low cost and effective.

Nearly a decade ago, the Procter & Gamble Company teamed up with the Centers for Disease Control and Prevention to come up with a solution. Together they developed a powder with two components that purify water. The first component is known as a flocculant; it attaches to soil sediments, heavy metals, and pesticides that are in the water and forces them to settle out of the water. The second component is a chlorine compound that kills 99.99 percent of all harmful bacteria and viruses. The first step is to mix a small amount of the powder into 10 L of water and allow the flocculant to work for 5 minutes. Once the flocculant settles to the bottom, it is filtered out through a piece of cotton fabric. After an additional 20 minutes, the chlorine has killed bacteria and viruses. In short, for a small amount of effort, a person can create 10 L of safe drinking water in just 30 minutes.

Although Procter & Gamble typically develops products designed to make a profit, this project was different. Purifying powder is needed in developing countries with populations that cannot afford to pay much even for something as essential as clean drinking water. Procter & Gamble decided to create a nonprofit organization, called the Children's Safe Drinking Water Program, that would supply the powder to people in developing countries at no profit. They created small powder packets that could clean 10 L of contaminated water but were not much bigger than a fast-food ketchup packet. The powder is inexpensive for Procter & Gamble to manufacture and the company is able to sell the packets at 3.5 cents each. Partnering with more than 150 governments and humanitarian groups including UNICEF, the company distributes the packets throughout the world. In 2017, Procter & Gamble announced that they had distributed enough packets to filter 13 billion liters of water and estimated that this saved tens of thousands of lives.

As the water purification program became known around the world, Procter & Gamble realized that the product they created with no expectation for profit could be sold at a profit to wealthier people who need to purify water that they collect from streams and lakes when they go hiking and camping. In 2007, the company announced that it would begin selling the packets to consumers under the brand name PUR, for $2.50 each. This price yields a considerable profit, part of which supports the goals of the nonprofit program.

The Children's Safe Drinking Water Program is an excellent example of new technologies that can be developed to be effective, inexpensive solutions to major environmental problems. It also demonstrates that such solutions can ultimately be profitable while providing a positive corporate image.

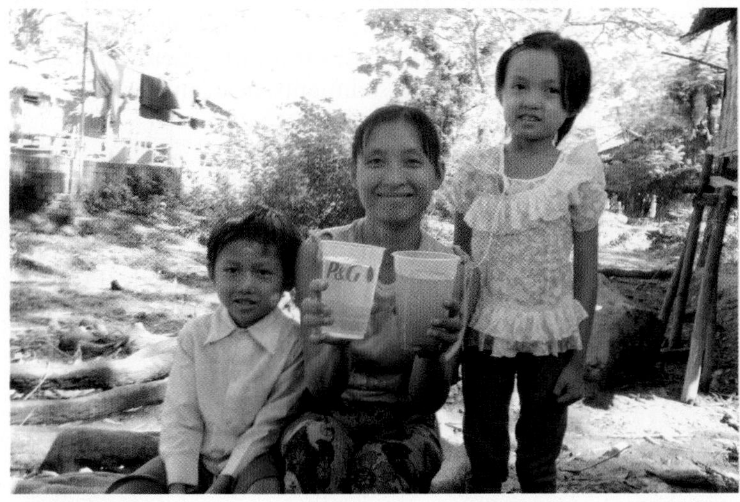

Making safe drinking water. Nearly 1 billion people do not have access to safe drinking water. Procter & Gamble partnered with many other organizations to distribute its water purifying powder technology to people in developing countries. The powder technology kills bacteria and viruses and removes parasites and solid materials. In this photo, a family poses with a cup of typical dirty water in one hand and the same water after adding the power in the other hand. *(Courtesy P&G)*

Critical Thinking Questions

1. Why is it important for a water-purifying system to contain both chlorine and a flocculant?
2. How might a company benefit when it produces a product that purifies water at no profit?

References

Deutsch, C. H. 2007. Procter & Gamble to benefit in all but name from water purifier. *New York Times*, July 23.

P&G Children's Safe Drinking Water. http://www.csdw.org/csdw/index.shtml.

Water-purification plant the size of a fast-food ketchup packet saves lives. 2013. *ScienceDaily*, September 9. http://www.sciencedaily.com/releases/2013/09/130909092343.htm

Water purifying packet. The contents of each small packet can remove soil and kill harmful organisms in 10 L of water. *(TOM UHLMAN/AP Images)*

Chapter 14 Review

In this chapter, we learned about the many sources of water pollution and their effects on humans and the environment. One prominent source of water pollution is the wastewater produced by humans and livestock. Wastewater can bring excessive nutrients to water bodies and be a source of disease-carrying organisms. Fortunately, a number of technologies exist to treat wastewater including septic tanks, sewage treatment facilities, and manure lagoons. Water pollution can also occur when heavy metals, pesticides, pharmaceuticals, and acids find their way into groundwater or surface waters. As we continue to extract energy from the ground, we face the risk of pollution from the pipelines and ships that carry petroleum products. Fortunately, new technologies present options for cleaning up some of the petroleum products that have leaked or spilled. We also face pollution risks from nonchemical sources including solid waste, sediments, heat, and noise.

Key Terms

Water pollution
Wastewater
Point source
Nonpoint source
Biochemical oxygen demand (BOD)
Dead zone
Eutrophication
Cultural eutrophication
Indicator species
Fecal coliform bacteria
Septic system
Septic tank
Sludge
Septage
Leach field
Manure lagoon
Acid deposition
Perchlorates
Polychlorinated biphenyls (PCBs)
Thermal pollution
Thermal shock
Clean Water Act
Safe Drinking Water Act
Maximum contaminant level (MCL)

Learning Goals Revisited

Module 41 Wastewater from Humans and Livestock

Discuss the three major problems caused by wastewater pollution.

Because wastewater contains organic matter, decomposition by microbes can cause declines in the amount of oxygen available in water bodies and produce dead zones. Dead zones also occur where wastewater adds enough nutrients to a water body to cause eutrophication. Finally, wastewater can be a source of many harmful pathogens including viruses, bacteria, and protists.

Explain the modern technologies used to treat wastewater.

At low residential densities, septic systems can be used to treat sewage. At higher residential densities such as in cities, sewage treatment plants treat the collected sewage from thousands of people. For livestock operations that raise large numbers of animals, manure lagoons hold manure until it is broken down by bacteria and can be applied to agricultural fields.

Module 42 Heavy Metals and Other Chemicals

Explain the sources of heavy metals and their effect on organisms.

Among the heavy metals of most concern are lead, mercury, and arsenic. Lead can harm the nervous system and kidneys of fetuses; the major source of lead is old water pipes. Arsenic can cause cancers of the skin, lungs, kidneys, and bladder; it exists naturally in rocks that release arsenic into groundwater that people drink. Mercury is not particularly toxic, but bacteria in water convert it into methylmercury, which can damage the nervous systems of animals, including humans. Major sources of mercury include coal burning, garbage incineration, medical supplies, and cement manufacturing.

Discuss the sources and effects of acid deposition and acid mine drainage.

The major sources of acid deposition are industrial plants that burn coal and release sulfur dioxide and nitrogen dioxide into the atmosphere where they are converted into sulfuric acid and nitric acid. The major source of acid mine drainage is abandoned mines that contain pyrite rocks. When the mines become flooded with groundwater, rocks release hydrogen ions and any water that leaves the mine and goes into a stream can have a pH close to zero. Both sources of acid pollution cause a decline in aquatic pH to a point that many organisms cannot survive.

Explain how synthetic organic compounds can affect aquatic organisms.

Pesticides are designed to kill particular organisms, but their presence in water bodies can have lethal effects on a wide range of species. Pharmaceuticals including hormones can alter the physiology of organisms at very low concentrations and have a range of effects, including causing males to grow eggs inside their testes. Military compounds such as perchlorates used for rocket fuel and industrial compounds such as PCBs and PBDEs can also make their way into groundwater and water bodies.

Module 43 Oil Pollution

Identify the major sources of oil pollution.

The major sources of oil pollution are natural leaks from the ocean floor, oil spills from tankers, and drilling for undersea oil using offshore platforms.

Explain some of the current methods to remediate oil pollution.

When an oil spill occurs, several steps can be taken to remediate the problem. Contaminated wildlife are often washed to remove oil. Floating oil slicks can be surrounded by containment booms and then vacuumed off the surface of the water. Oil slicks can also be treated with chemicals that break up the oil before it can come to shore. Genetically modified bacteria are also being developed to consume oil in the water. There is currently no agreement on how to remove underwater plumes of oil.

Module 44 Nonchemical Water Pollution

Identify the major sources of solid waste pollution.

The major sources of solid waste are discarded materials from households and industries, sludge from wastewater treatment plants, and coal ash and coal tailings that are produced when coal is burned.

Explain the harmful effects of sediment pollution.

Sediments that make their way into water bodies from natural sources, agricultural fields, and construction sites can reduce the ability of light to transmit into the water, which makes it difficult for plants and algae to grow. Sediments also can clog the gills of aquatic organisms.

Discuss the sources and consequences of thermal pollution.

Thermal pollution occurs when industries pump in cool water from a local water body, use the water for cooling purposes, and then return the much warmer water back to the water body. Many organisms cannot survive the shock of having the temperature of the environment substantially increase.

Understand the causes of noise pollution.

Noise pollution is primarily an issue of concern in oceans where noises come from the propellers of large ships, long-range sonar, and loud air guns that are used to explore for oil located below the ocean floor.

Module 45 Water Pollution Laws

Explain how the Clean Water Act protects against water pollution.

The Clean Water Act has a goal of maintaining and, when necessary, restoring the chemical, physical, and biological properties of surface waters. It does this by setting water quality standards for various pollutants.

Discuss the goals of the Safe Drinking Water Act.

The Safe Drinking Water Act sets maximum contaminant levels that can be found in sources of drinking water including groundwater. The contaminants include microorganisms, disinfectants, organic chemicals, and inorganic chemicals.

Understand how water pollution legislation is changing in developing countries.

Developed countries have gone through a period of industrialization and water pollution followed by steps to improve their water quality. Most developing countries are in the stage of industrializing and, as a result, they commonly have problems of polluted surface and groundwater. As developing countries rise in affluence, they begin to have the desire and the financial ability to improve the quality of their water.

Practice Math and Graphing

Preparing for the AP® Exam

Answer the following questions. Be sure to show all your work.

1. Practice Math

The nitrogen pollution coming into the Chesapeake Bay is often separated into two geographic sources. The Eastern Shore is a small area of land to the east of the bay and represents 7 percent of the entire watershed. The remainder of the watershed extends from Virginia to central New York and represents 93 percent of the watershed. Researchers have quantified the amount of nitrogen coming in from each region due to crops and due to manure applications.

Given that 1 kg = 2.2 pounds and 1 ha = 2.5 acres, then 1 kg/ha = 2.2 pounds/2.5 acres = 0.9 pounds/acre.

The following table provides nitrogen values in units of pounds per acre. Convert these values into units of kilograms per hectare.

Region	Source	Nitrogen (pounds/ acre)	Nitrogen (kg/ha)
Eastern Shore	Crop production	6.2	
Eastern Shore	Manure applications	1.5	
Remainder of watershed	Crop production	3	
Remainder of watershed	Manure applications	0.6	

Data from Ator, S.W., et al. 2011. *Sources, Fate, and Transport of Nitrogen and Phosphorus in the Chesapeake Bay Watershed:An Empirical Model*. U.S. Geological Survey Scientific Investigations Report 2011–5167.

2. Practice Graphing

(a) Use your answers from "Practice Math" to convert the four sources of nitrogen into percentages.

(b) Create a pie chart that illustrates the percent contribution of the four nitrogen sources.

(c) If you were to prioritize your efforts to control nitrogen inputs into the Chesapeake Bay, where would you concentrate your efforts?

Section 1: Multiple-Choice Questions

Choose the best answer for questions 1–17.

1. Which statement about nonpoint source (NPS) pollution is NOT correct?
   (a) NPS results from rain or snowmelt moving over or permeating through the ground.
   (b) NPS is more difficult to control, measure, and regulate than point source pollution.
   (c) NPS includes sediment from improperly managed construction sites as a pollutant.
   (d) NPS is water pollution that originates from a distinct source such as a pipe or tank.

2. Human wastewater results in which water-pollution problem?
   I. Decomposition of organic matter reduces dissolved oxygen levels.
   II. Decomposition of organic matter releases great quantities of nutrients.
   III. Pathogenic organisms are carried to surface waters.
   (a) I only
   (b) II only
   (c) I and III
   (d) I, II, and III

3. Which indicates that a body of water is contaminated by human wastewater?
   (a) low BOD and a fecal coliform bacteria count of zero
   (b) high levels of nutrients, such as nitrogen and phosphorus, and high BOD
   (c) low BOD and low levels of nutrients, such as nitrogen and phosphorus
   (d) low levels of nutrients, such as nitrogen and phosphorus, and a fecal coliform bacteria count of zero

4. Both septic systems and sewage treatment plants utilize bacteria to break down organic matter. Where in each system does this process occur?
   (a) septic tank and leach field; primary treatment and secondary treatment
   (b) septic tank only; primary treatment and chlorination
   (c) leach field only; secondary treatment only
   (d) septic tank and leach field; secondary treatment only

5. Under which circumstance is a sewage treatment plant legally permitted to bypass normal treatment protocol and discharge large amounts of sewage directly into a lake or river?
   (a) when the population of the surrounding community surpasses the plant's capacity
   (b) when combined volumes of storm water and wastewater exceed the capacity of an older plant
   (c) when a permit to modernize the plant is denied by the Environmental Protection Agency
   (d) when an extended period of drought restricts water flow in a lake or river

6. Tertiary treatment of wastewater
   (a) removes pathogens.
   (b) reduces sediment.
   (c) reduces eutrophication.
   (d) reduces the amount of sludge.

7. Which inorganic substance is naturally occurring in rocks, soluble in groundwater, and toxic at low concentrations?
   (a) mercury
   (b) lead
   (c) PCBs
   (d) arsenic

8. Acid mine drainage occurs when acidic water formed belowground makes its way to the surface; the acidic water is formed in flooded abandoned mines where the underground water
   (a) reacts with a type of rock, pyrite, which releases iron and hydrogen ions.
   (b) reacts with sulfur dioxide and nitrogen dioxide to form sulfuric and nitric acids.
   (c) flushes out the chemicals used in the mining process.
   (d) permeates a limestone layer that lowers the pH.

9. Which is NOT a problem that results from the use of pesticides?
   (a) Most pesticides are not target-specific and kill other related and nonrelated species.
   (b) Pesticide runoff enters surface waters and increases the solubility of heavy metals.
   (c) Pesticides target specific physiological functions, but also disrupt other functions.
   (d) Most inert ingredients are not tested for safety and may pose unacceptable risks.

10. In the northeastern United States, during what time(s) of the year are lakes located near agricultural fields in temperate ecosystems most likely to experience the highest biochemical oxygen demand?
    (a) throughout the entire growing season
    (b) primarily during the spring
    (c) primarily during the summer
    (d) during the fall

11. You want to conduct a natural experiment to test whether leakage from a manure lagoon has an adverse effect on a nearby stream. What is an appropriate null hypothesis for your experiment?
    (a) There will be a higher abundance of an indicator species in the stream before the lagoon began leaking than after it began leaking.
    (b) There will be a lower abundance of an indicator species in the stream before the lagoon began leaking than after it began leaking.
    (c) There will be no difference in the abundance of phytoplankton in the stream before the lagoon began leaking and after it began leaking.
    (d) There will be no difference in the abundance of an indicator species in the stream before the lagoon began leaking and after it began leaking.

12. Which mineral is ultimately responsible for the yellow color of streams that drain from abandoned mines?
    (a) pyrite rock
    (b) aluminum
    (c) coal
    (d) sulfur

13. Which is NOT an example of a consequence of biomagnification?
    (a) incidences of human cancer associated with spraying crops with pesticides
    (b) suffocation of fish as a result of inert pesticide ingredients
    (c) thinning of bird egg shells
    (d) increasing concentrations of iron at higher levels of a food chain

14. A group of researchers monitored the health of all streams in a single watershed. Several streams in one region of the watershed had a high silt content, large boulders, and a low diversity of insects and fish. What would be the most direct inference from this description?
    (a) Most of the sediment load in the stream is probably from a natural source.
    (b) The surrounding area probably has a high amount of impervious surface.
    (c) There is probably a high biochemical oxygen demand in the benthos of the stream.
    (d) The streams have very low resilience.

15. What made the cleanup of Chesapeake Bay difficult?
    I. The companies that polluted it were no longer in business.
    II. The bay's watershed is large and encompasses several states.
    III. Tides and currents impeded the cleanup work.
    (a) I only
    (b) II only
    (c) III only
    (d) I and II

16. A forest surrounding a stream is cleared to create pasture for cattle grazing. Which is likely to occur in the stream?
    (a) lower productivity due to increased erosion and sedimentation
    (b) thermal shock as water temperature rises due to reduced vegetative cover
    (c) eutrophication caused by increased availability of sunlight
    (d) dead zones caused by increasing heavy metal concentrations

Question 17 uses the following graph.

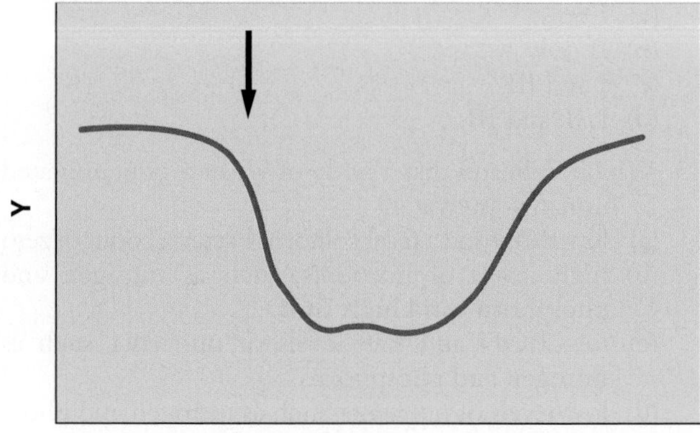

17. The graph plots variable Y as it changes over the course of a stream. The arrow indicates an area where wastewater from a sewage treatment plant is released into the stream. Which does Y most likely represent?
    (a) BOD
    (b) pH
    (c) dissolved oxygen
    (d) fecal coliform bacteria

Section 2: Free-Response Questions

Write your answer to each part clearly. Support your answers with relevant information and examples. Where calculations are required, show your work.

1. Answer questions a–d using the graph below, which contains data collected by the Maryland Department of Natural Resources and the Chesapeake Bay Program.
   (a) Calculate the differences in crab population from 1990–1997, 1998–2009, and 2010–2017. Predict the average blue crab population for 2010–2020 and explain your answer. (4 points)
   (b) Identify and explain three possible factors related to water pollution that could have contributed to the decline in the total blue crab population in the Chesapeake Bay. (3 points)
   (c) Select ONE factor stated in (b) and describe how that source of water pollution could be managed and controlled. (2 points)
   (d) What federal legislation would apply to the Chesapeake Bay and the blue crabs? (1 point)

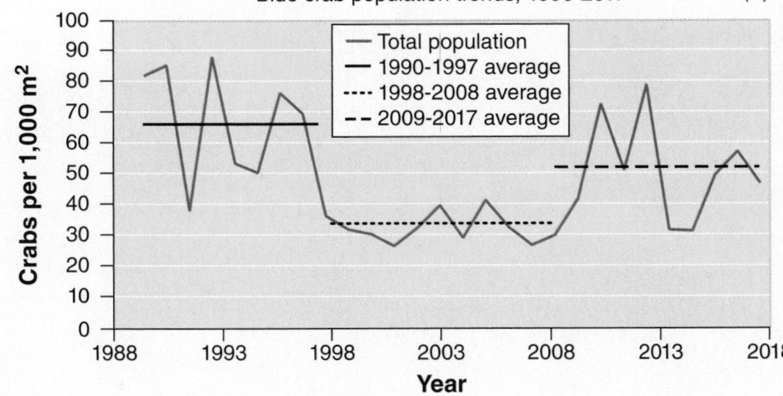

2. A researcher collects the following data from two polluted streams. One stream is affected by acid mine drainage from an abandoned copper mine. The other stream is affected by cultural eutrophication due to runoff from several nearby farms.

	Stream A	Stream B
pH	7.5	2.0
BOD	230 mg	15 mg
Aluminum	0.001 mg/L	1 mg/L

   (a) Which stream is affected by acid mine drainage, and which by eutrophication? Explain your answer for each using at least TWO pieces of evidence. (3 points)
   (b) If you were shown photos of these streams, how might you be able to tell that they were affected by eutrophication and acid mine drainage? (2 points)
   (c) Both streams are affected by nonpoint source pollution. Give ONE example of point source and ONE example of nonpoint source water pollution. Why is nonpoint source pollution harder to control and regulate than point source pollution? (2 points)
   (d) Researchers have monitored the population of northern leopard frogs in and around the stream affected by agricultural runoff since 1980. In 1994, the farms surrounding the stream changed their weed management practices.
      (i) Plot the data below. (2 points)
      (ii) Describe a potential reason behind the decline in frog population beginning in 1995. (1 point)

Northern leopard frog population (frogs/square mile)	Year
2000	1980
2050	1985
2030	1990
1700	1995
1500	2000
900	2005
600	2010

In this recent aerial view of Chattanooga, Tennessee, Lookout Mountain is clearly visible in the background. Fifty years ago, this view was often obscured by haze. *(© Rock Creek Aviation)*

Atmospheric Pollution and Stratospheric Ozone Depletion

CHAPTER 15

CASE STUDY

Cleaning Up in Chattanooga

The fall of 2016 was not kind to Chattanooga, Tennessee. Nearby wildfires covered downtown Chattanooga with a smoky haze resulting in a number of poor air quality days in a city that is generally known for very good air quality. A few years earlier, dust storms in the west had also reduced the generally good air quality of the city. Protecting air quality in 2020 and beyond in the United States is something of a challenge. Regulators and government officials can impose standards, but a vast array of natural and human-caused pollutants can conspire to cause local air pollution problems for any number of reasons. Local geography, direction of prevailing winds, the activities of other cities and sometimes other countries, and many other factors can affect air quality. Unfortunately, most of these factors are beyond the control of one city. The story of Chattanooga is representative of many cities in the United States and around the world.

Chattanooga sits along the Tennessee River in a natural basin formed by the Appalachian Mountains, one of which—Lookout Mountain—rises 600 m (1,970 feet) over the city. After the Civil War, foundries, textile mills, and other industrial plants were quickly built and Chattanooga became one of the leading manufacturing centers in the nation.

> By 1957, Chattanooga had the third-worst particulate pollution in the country and rates of respiratory diseases were well above the national average.

The economic boom in Chattanooga had an environmental cost. Like Los Angeles and many other highly polluted cities, Chattanooga is located in a bowl formed by surrounding mountains. This geography traps pollutants that hover above the city. By 1957, Chattanooga had the third worst particulate pollution in the country and rates of respiratory diseases were well above the national average. Over the next decade conditions worsened and by the 1960s, people were often unable to see Lookout Mountain even from a distance of a quarter mile. In 1969, a U.S. survey of the nation's air quality confirmed what many Chattanooga residents suspected: Their city topped the list of the worst cities in the United States for particulate air pollution.

Obviously the poor quality of the air needed to be addressed. In 1969, Chattanooga and the county in which it is located, created its own air pollution legislation by enacting the Air Pollution Control Ordinance. It controlled the emissions of sulfur oxides, allowed open burning by permit only, placed regulations on odors and dust, outlawed visible automobile emissions, capped the sulfur content of fuel at 4 percent, and limited visible emissions from industry. At the same time, the city and county governments implemented new pollution monitoring techniques to ensure that the ordinance was being followed.

The city and county governments, along with private industry, spent approximately $40 million on the cleanup effort. Actions to improve air quality did not hinder business, as some people feared, but created new industrial opportunities related to the cleanup effort, such as the establishment of a local manufacturer of smokestack scrubbers. As a result of all these measures, in 1972—just 3 years after passage of the city ordinance and 2 years after the passage of the Federal Clean Air Act—Chattanooga achieved compliance with Clean Air Act air-quality standards.

The people of Chattanooga and the local governments recognized that keeping their air clean and maintaining economic sustainability would be an ongoing effort. To maintain their newly improved air quality, the city government and local businesses began several programs. One such program was a comprehensive recycling program, chosen as an alternative to a waste incinerator that would have added more particulates to the air. Public and private sectors successfully partnered to achieve both environmental and economic sustainability in creating the largest municipal fleet of electric buses in the United States, manufactured by another local business.

Unfortunately, while Chattanooga's efforts dramatically reduced the levels of particulate pollutants, the concentration of ozone, mostly from automotive pollutant precursors within and beyond the city limits, continued to climb. Ozone concentrations exceeded the 1997 standard of 0.08 parts per million by volume set by the Environmental Protection Agency. Chattanooga has responded to the new air pollution problem in the same way it faced the particulate pollutant problem of the 1960s—through a combined effort of government, the public, and local industries. The city and county governments formed an Early Action Compact with the EPA, agreeing to improve ozone concentrations ahead of EPA requirements in return for not being labeled a "nonattainment area," a designation that can result in the loss of federal highway funds and create a negative image that makes industrial recruitment and economic development more difficult. Like the 1969 Air Pollution Control Ordinance, the new Early Action Compact calls for a concerted effort by private and public sectors, and includes educating people on how they can take actions to limit ozone production on high-ozone days.

Chattanooga attained the 0.08 parts per million standard in 2007, 2 years ahead of schedule. Then, national legislation again lowered the ozone standard, this time to 0.075 parts per million. By 2011, before the dust storms from the west and the local forest fires, Chattanooga had met the new, lower ozone standard. In 2017, the American Lung Association rated Chattanooga one of the cleanest cities in the United States. However, residents know that to achieve their goals of an economically vibrant city with clean air, they must continue to encourage cooperation among government, people, and business. They also need fewer dust storms from the west and fewer nearby forest fires, which are, unfortunately, not in their control.

Sources: Chattanooga Area Chamber of Commerce, *Summary of the Chattanooga Area Chamber of Commerce's Position on Strengthening the National Ambient Air Quality Standard for Ozone,* 2010, www.chattanoogachamber.com; S. Johnson, Report: Chattanooga air quality among best in U.S., *Chattanooga Times Free Press*, April 19, 2017.

Because air is a common resource that covers the planet, air pollution crosses many system boundaries. Air pollution occurs over terrestrial, aquatic, and marine natural systems and it also occurs in human-made indoor systems. To understand air pollution and its effects, we need to examine the wide variety of air pollutants, where they come from, and what happens to them after they are released into the atmosphere. In this chapter, we will identify the major air pollutants found around the globe and we will discuss the specific air pollution situations that occur with photochemical smog and acid deposition. We will review several air pollution control measures and examine stratospheric ozone depletion. We conclude the chapter with a discussion of indoor air pollution.

MODULE 46

Major Air Pollutants and Their Sources

Learning Goals

After reading this module, you should be able to

- identify and describe the major air pollutants.
- describe the sources of air pollution.

Air pollution is defined as the introduction of chemicals, particulate matter, or microorganisms into the atmosphere at concentrations high enough to harm plants, animals, and materials such as buildings, or to alter ecosystems. Generally, the term *air pollution* refers to pollution in the troposphere, the first 16 km (10 miles) of the atmosphere above the surface of Earth. Tropospheric pollution is also sometimes called ground-level pollution. Air pollution can occur naturally, from sources such as volcanoes and fires, or it can be anthropogenic, from sources such as automobiles and factories.

In this module, we will examine the major air pollutants that occur in the troposphere and where they come from.

Air pollution is a global system

Since one of the major repositories for air pollutants is the atmosphere, which envelops the entire globe, we must think of the air pollution system as a global system. In fact, evidence appears to link air pollution across long distances. For example, in recent years, air pollution in Asia has been responsible for acidic rainfall on the West Coast of the United States (**FIGURE 46.1**).

The air pollution system has many inputs, which are the sources of pollution. It also has many outputs, which are components of the atmosphere and biosphere that remove air pollutants. It is difficult to conceptualize this system because the inputs do not originate from just one location. Air pollution inputs can come from automobiles on the ground, airplanes in the sky, or vegetation (tree leaves) 100 feet in the air. Similarly, air pollution can be removed or altered by vegetation, soil, and components of the atmosphere such as clouds, particles, or gases.

As a starting point for understanding the global air pollution system, we will identify the major pollutants and determine where they come from.

FIGURE 46.1 Particulate pollution and visibility. Particulates and sulfate aerosols are most responsible for causing pollution and reducing visibility in cities, such as this location in China. Pollution can also be transported long distances and cause problems far from the source. *(Natalie Behring/Panos Pictures)*

Air pollution The introduction of chemicals, particulate matter, or microorganisms into the atmosphere at concentrations high enough to harm plants, animals, and materials such as buildings, or to alter ecosystems.

Classifying Pollutants

Even though air pollution has been with us for millennia, both the specific definition of pollution and the classification of a substance as a pollutant have evolved. The atmosphere is a public resource—in effect, a global commons—and consequently the science of air pollution is closely intertwined with political and social perspectives. In formulating the U.S. Clean Air Act of 1970 and subsequent amendments, legislators used information from environmental scientists and human health scientists on the most important air pollutants to monitor and control. The original act identified six pollutants that significantly threaten human well-being, ecosystems, and structures: sulfur dioxide, nitrogen oxides, carbon monoxide, particulate matter, tropospheric ozone, and lead. These were called criteria air pollutants because under the Clean Air Act, the EPA must specify allowable concentrations of each pollutant.

Although carbon dioxide was not included among the major air pollutants identified in the 1970s, today it is widely accepted that carbon dioxide is altering ecosystems in a substantial way. In 2007, the U.S. Supreme Court ruled that carbon dioxide should be considered an air pollutant under the Clean Air Act, and in 2012, a federal appeals court agreed that the EPA is required to impose limits on harmful greenhouse gas emissions, including carbon dioxide. In addition, volatile organic compounds and mercury, though not officially listed in the Clean Air Act, are commonly measured air pollutants that have the potential to be harmful.

The sources and effects of the major air pollutants, including the six criteria air pollutants, are summarized in **TABLE 46.1**. Let's take a closer look at each of these major air pollutants.

Sulfur Dioxide

Sulfur dioxide (SO_2) is a corrosive gas that comes primarily from combustion of fuels such as coal and oil. It is a respiratory irritant and can adversely affect plant tissue as well. Because all plants and animals contain sulfur in varying amounts, the fossil fuels derived from their remains contain sulfur. When these fuels are combusted, the sulfur combines with oxygen to form sulfur dioxide. Sulfur dioxide is also released in large quantities during volcanic eruptions and can be released, though in much smaller quantities, during forest fires.

Nitrogen Oxides

Nitrogen oxides are generically designated NO_X, with the X indicating that there may be either one or two oxygen atoms per nitrogen atom: nitrogen oxide (NO), a colorless, odorless gas; and nitrogen dioxide (NO_2), a pungent, reddish-brown gas, respectively. When we use the term *nitrogen oxides* in our discussion, we will be referring to either nitrogen oxide or nitrogen dioxide since they easily transform from one to the other in the atmosphere. The atmosphere is 78 percent nitrogen gas (N_2), and all combustion in the atmosphere leads to the formation of some nitrogen oxides. Motor vehicles and stationary fossil fuel combustion are the primary anthropogenic sources of nitrogen oxides. Natural sources include forest fires, lightning, and microbial activity in soils. Atmospheric nitrogen oxides play a role in forming ozone and other components of smog. We will take a closer look at this process later in the chapter.

Particulate matter (PM) Solid or liquid particles suspended in air. *Also known as* **Particulates; Particles.**

Carbon Oxides

Carbon monoxide (CO) is a colorless, odorless gas that is formed during *incomplete* combustion of most matter, and therefore is a common pollutant in vehicle exhaust and most other combustion processes. Carbon monoxide can be a significant component of air pollution in urban areas. It also can be a dangerous indoor air pollutant when exhaust systems on natural gas heaters malfunction. Carbon monoxide is a particular problem in developing countries, where people may cook with manure, charcoal, or kerosene within poorly ventilated structures.

Carbon dioxide (CO_2) is a colorless, odorless gas that is formed during the *complete* combustion of most matter, including fossil fuels and biomass. It is absorbed by plants during photosynthesis and is released during respiration. In general, the complete combustion of matter that produces carbon dioxide is more desirable than the incomplete combustion that produces carbon monoxide and other pollutants. However, burning fossil fuels has contributed additional carbon dioxide to the atmosphere and led to its becoming a major pollutant. It recently exceeded a concentration of 410 parts per million in the atmosphere and has been steadily increasing each year. This topic will be covered in more detail in Chapter 19 where we discuss the issue of climate change.

Particulate Matter

Particulate matter (PM), also called **particulates** or **particles**, is solid or liquid particles suspended in air. **FIGURE 46.2** on page 542 shows the sources of particulate matter and its effects. Particulate matter comes from the combustion of wood, animal manure and other biofuels, coal, oil, and gasoline. It is most commonly known as a class of pollutants released from the combustion of fuels such as coal and oil. Diesel-powered vehicles give off more particulate matter, in the form of black smoke, than do gasoline-powered vehicles. Particulate matter can also come from road dust and rock-crushing operations. Volcanoes, forest fires, and dust storms are important natural sources of particulate matter.

TABLE 46.1 Major air pollutants

Compound	Symbol	Human-derived sources	Effects/Impacts
Criteria air pollutants			
Sulfur dioxide	SO_2	• Combustion of fuels that contain sulfur, including coal, oil, gasoline	• Respiratory irritant, can exacerbate asthma and other respiratory ailments • SO_2 gas can harm stomata and other plant tissue • Converts to sulfuric acid in atmosphere, which is harmful to aquatic life and some vegetation
Nitrogen oxides	NO_X	• All combustion in the atmosphere including fossil fuel combustion, wood, and other biomass burning	• Respiratory irritant, increases susceptibility to respiratory infection • An ozone precursor, leads to formation of photochemical smog • Converts to nitric acid in atmosphere, which is harmful to aquatic life and some vegetation • Contributes to over-fertilization of terrestrial and aquatic systems
Carbon monoxide	CO	• Incomplete combustion of any kind • Malfunctioning exhaust systems and poorly ventilated cooking fires	• Bonds to hemoglobin, thereby interfering with oxygen transport in the bloodstream • Causes headaches at low concentrations • Can cause death with prolonged exposure at high concentrations
Particulate matter	PM_{10} (smaller than 10 micrometers) $PM_{2.5}$ (2.5 micrometers and smaller)	• Combustion of coal, oil, and diesel, and of biofuels such as manure and wood • Agriculture, road construction, and other activities that mobilize soil, soot, and dust	• Can exacerbate respiratory and cardiovascular disease and reduce lung function • May lead to premature death • Reduces visibility and contributes to haze and smog
Lead	Pb	• Gasoline additive, oil and gasoline, coal, old paint	• Impairs central nervous system • At low concentrations, can have measurable effects on learning and ability to concentrate
Ozone	O_3	• A secondary pollutant formed by the combination of sunlight, water, oxygen, VOCs, and NO_X	• Reduces lung function and exacerbates respiratory symptoms • Degrades plant surfaces • Damages materials such as rubber and plastic
Other air pollutants			
Volatile organic compounds	VOC	• Evaporation of fuels, solvents, paints • Improper combustion of fuels such as gasoline	• A precursor to ozone formation
Mercury	Hg	• Coal, oil, gold mining	• Impairs central nervous system • Bioaccumulates in the food chain
Carbon dioxide	CO_2	• Combustion of fossil fuels and clearing of land	• Affects climate and alters ecosystems by increasing greenhouse gas concentrations

Particulate matter ranges in size from 0.01 micrometer (μm) to 100 μm (1 micrometer = 0.000001 m). For comparison, a human hair has a diameter of roughly 50 to 100 μm. Particulate matter larger than 10 μm is usually filtered out by the nose and throat; particulate matter of this size is not regulated by the EPA. Particles smaller than 10 μm are called Particulate Matter-10, written as PM_{10}, and are of concern to air pollution scientists because they are not filtered out by the nose and throat and can be deposited deep within the respiratory tract. Particles of 2.5 μm and smaller, called $PM_{2.5}$, are an even greater health concern because they can travel further within the respiratory tract and they tend to be composed of more toxic substances than particles in larger size ranges.

Particulate matter also scatters and absorbs sunlight. If the atmospheric concentration of particulate matter is high enough, as it would be immediately following a large forest fire or a volcanic eruption, incoming solar radiation in the region will be reduced enough to affect photosynthesis. This happened in 1816, a year after a large volcanic eruption in Java released more than

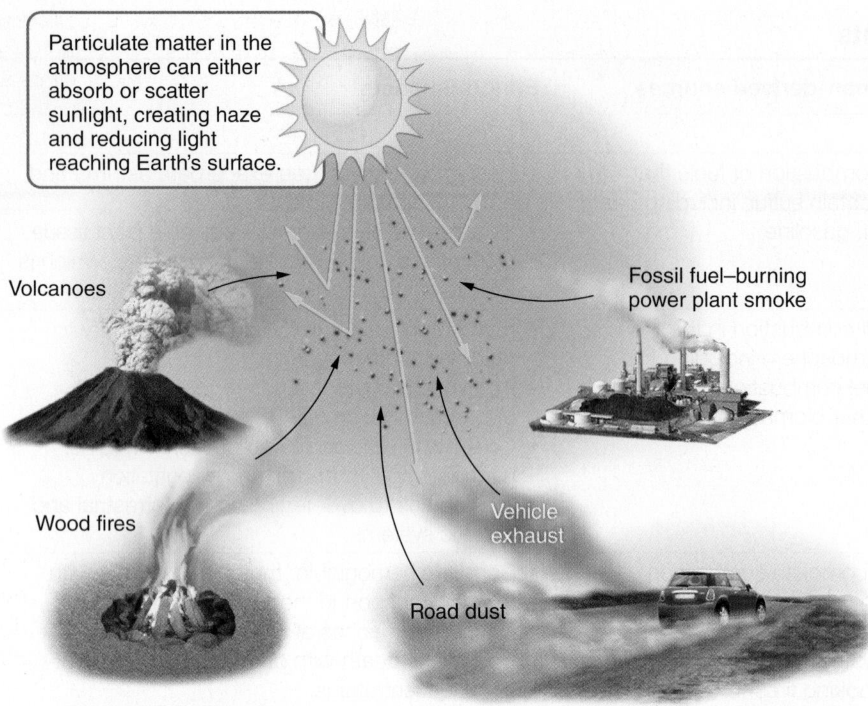
Particulate matter in the atmosphere can either absorb or scatter sunlight, creating haze and reducing light reaching Earth's surface.
Volcanoes
Fossil fuel–burning power plant smoke
Wood fires
Vehicle exhaust
Road dust

FIGURE 46.2 The sources of particulate matter and its effect. Particulate matter can be natural or anthropogenic. Particulate matter in the atmosphere ranges considerably in size and can absorb or scatter light, which creates a haze and reduces the light that reaches the surface of Earth.

150 million metric tons of particles that slowly spread around the globe. That year is commonly referred to as the year without a summer.

Reduced visibility, also known as **haze**, occurs primarily when particulate matter from air pollution scatters light. But as we will see in the next section, ozone and photochemical oxidants can also play an important indirect role in the formation of haze.

Photochemical Oxidants

Oxides are reactive compounds that remove electrons from other substances. **Photochemical oxidants** are a class of air pollutants formed as a result of sunlight (*photo*) acting on chemical compounds such as nitrogen oxides and sulfur dioxide. There are many photochemical oxidants, and they are generally harmful to plant tissue, human respiratory tissue, and construction materials. However, environmental scientists frequently focus on *ozone*. **Ozone (O_3)** is a secondary pollutant made up of three oxygen atoms bound together. It is harmful to both plants and animals and impairs respiratory function.

In the presence of sulfur and nitrogen oxides, photochemical oxidants can enhance the formation of certain particulate matter, which contributes to scattering light. The resulting mixture is called **smog**, a mixture of oxidants and particulate matter. The word is derived by combining *smoke* and *fog*. Smog is partly responsible for the hazy view and reduced sunlight observed in many cities. Smog can be divided into two categories: *photochemical smog* and *sulfurous smog*. **Photochemical smog** is dominated by oxidants such as ozone and is sometimes called **Los Angeles–type smog** or **brown smog**. **Sulfurous smog** is dominated by sulfur dioxide and sulfate compounds and is also called **London-type smog**, **gray smog**, or **industrial smog.**

Atmospheric brown cloud is a relatively new descriptive term that has been given to the combination of particulate matter and ozone. Derived primarily from combustion of fossil fuel and burning biomass, atmospheric brown clouds have been observed in cities and throughout entire regions, especially in Asia. The brownish tint that characterizes these clouds of pollution is typically caused by the presence of black or brown light-absorbing carbon particles and/or nitrogen dioxide.

In addition to human health problems, particulate matter and photochemical oxidants also cause economic harm, since poor visibility in popular vacation destinations can reduce tourism revenues for recreation areas, such as lower incomes for hotels and restaurants in these areas. We will examine photochemical smog in greater detail in the next module.

Lead and Other Metals

Lead (Pb) is a trace metal that occurs naturally in rocks and soils. It is present in small concentrations in fuels including oil and coal. Lead compounds were added to gasoline for many years to improve vehicle performance. During that time, lead compounds released into the air traveled with the prevailing winds and were deposited on the ground by rain or snow. They became pervasive around the globe, including in polar regions far from combustion sources.

Haze Reduced visibility.

Photochemical oxidant A class of air pollutants formed as a result of sunlight acting on compounds such as nitrogen oxides.

Ozone (O_3) A secondary pollutant made up of three oxygen atoms bound together.

Smog A type of air pollution that is a mixture of oxidants and particulate matter.

Photochemical smog Smog that is dominated by oxidants such as ozone. *Also known as* **Los Angeles–type Smog; Brown smog.**

Sulfurous smog Smog dominated by sulfur dioxide and sulfate compounds. *Also known as* **London-type smog; Gray smog; Industrial smog.**

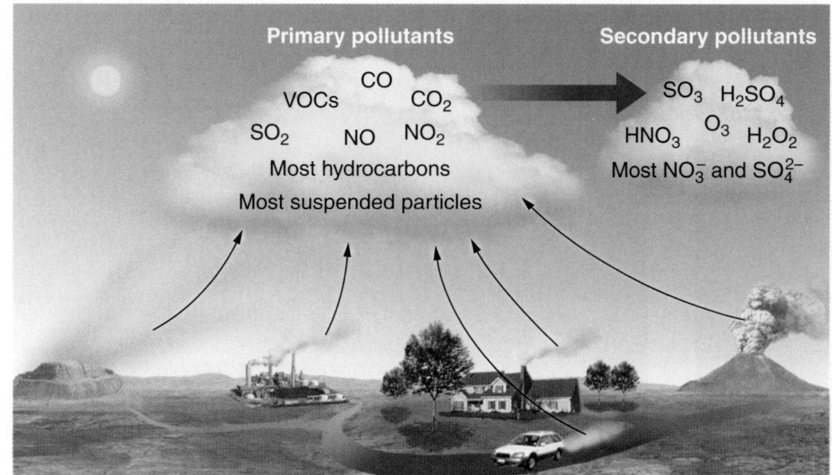

FIGURE 46.3 Primary and secondary air pollutants. The transformation from primary to secondary pollutant requires a number of factors including sunlight, water (clouds), and the appropriate temperature.

Lead was phased out as a gasoline additive in the United States between 1975 and 1996, and since then its concentration in the air has dropped considerably. A campaign to phase out lead use in gasoline globally is still underway. Another persistent source of lead is lead-based paint in older buildings. When the paint peels off, the resulting dust or chips can be toxic to the central nervous system and can affect learning and intelligence, particularly for young children who, attracted by the sweet taste, may ingest the dust or chips.

Mercury (Hg), another trace metal, is also found in coal and oil and, like lead, is toxic to the central nervous system of humans and other organisms. The EPA regulates mercury through its hazardous air pollutants program. As a result of the release of mercury into the air, primarily from the combustion of fossil fuels, especially coal, the concentrations of mercury in both air and water have increased dramatically in recent years. As we will see in Chapter 17, mercury concentrations in some fish have also increased. People who eat these fish increase their own mercury concentrations—an example of the interconnectedness of air pollution, air, water, aquatic health, and human health. Over the past 30 years, mercury emissions in the United States from waste incinerators have been reduced substantially. Because coal-fired electricity generation plants remain the largest uncontrolled source of mercury, emissions standards for coal plants will likely be the focus of future regulations.

Volatile Organic Compounds

Organic compounds that evaporate at typical atmospheric temperatures are called **volatile organic compounds (VOCs)**. Many VOCs are hydrocarbons—compounds that contain carbon-hydrogen bonds, such as gasoline, lighter fluid, dry-cleaning fluid, oil-based paints, and perfumes. Compounds that give off a strong aroma are often VOCs since the chemicals are easily released into the air. VOCs play an important role in the formation of photochemical oxidants such as ozone. VOCs are not necessarily hazardous; many, such as VOCs given off by conifer trees, cause no direct harm. VOCs are not currently considered a criteria air pollutant, but because they can lead to the formation of photochemical oxidants, they have the potential to be harmful and are therefore of concern to air pollution scientists.

Primary and Secondary Pollutants

When trying to understand pollution, and when attempting to reduce pollution emissions, it is important to understand if a particular pollutant is coming directly from an emission source such as a smokestack or tailpipe, or if it has undergone transformations after emission. Depending on the source, pollutants in the air can be categorized as *primary* or *secondary.*

AP® Exam Tip

Make sure that you know the difference between primary and secondary pollutants, and can describe how each type of pollutant is formed. ●

Primary Pollutants

Primary pollutants are polluting compounds that come directly out of a smokestack, exhaust pipe, or natural emission source. As you can see in **FIGURE 46.3**, they include CO, CO_2, SO_2, NO_X, and most suspended particulate matter. Many VOCs are also primary pollutants. For example, as gasoline is burned in a car, it volatilizes from a liquid to a vapor, some of which is emitted from the exhaust pipe in an uncombusted form. The effect is more pronounced if the car is not operating efficiently. The resulting VOC becomes a primary air pollutant.

Volatile organic compound (VOC) An organic compound that evaporates at typical atmospheric temperatures.

Primary pollutant A polluting compound that comes directly out of a smokestack, exhaust pipe, or natural emission source.

(a)

(b)

FIGURE 46.4 Natural sources of air pollution. There are many natural sources of air pollution, including volcanoes, lightning strikes, forest fires, and plants. (a) A forest fire in Ojai, California, produces air pollution. (b) The Great Smoky Mountains were named for the natural air pollutants that reduce visibility and give the landscape a smoky appearance. *(a: Bill Boch/Getty Images; b: Bill Lea/Dembinsky Photo Associates/Alamy)*

Secondary Pollutants

Secondary pollutants are primary pollutants that have undergone transformation in the presence of sunlight, water, oxygen, or other compounds. Because solar radiation provides energy for many of these transformations, and because water is usually involved, the conversion to secondary pollutants occurs more rapidly during the day and in wet environments.

Ozone is an example of a secondary pollutant. Ozone is formed in the atmosphere as a result of the emission of the primary air pollutants NO_X and VOCs in the presence of sunlight. The main components of acid deposition—sulfate (SO_4^{2-}) and nitrate (NO_3^-)—are also secondary pollutants. Both of these secondary pollutants will be discussed more fully below.

When trying to control secondary pollutants, it is necessary to consider the primary pollutants that create them, as well as factors that may lead to the formation, breakdown, or reduction in the secondary pollutants themselves. For example, when municipalities such as Chattanooga try to reduce ozone concentrations in the air, as described at the beginning of this chapter, they focus on reducing the compounds that lead to ozone formation—NO_X and VOCs—rather than on the ozone itself.

AP® Exam Tip

You should be able to describe both the environmental and human health effects of secondary pollutants. ●

Secondary pollutant A primary pollutant that has undergone transformation in the presence of sunlight, water, oxygen, or other compounds.

Air pollution comes from both natural and human sources

Environmental scientists and concerned citizens direct much of their attention toward air pollution that comes from human activity. But human activity is not the only source of air pollution because processes in nature also cause air pollution. In this section we will examine both natural and human sources of air pollution.

Natural Emissions

Volcanoes, lightning, forest fires, and plants both living and dead all release compounds that can be classified as pollutants. Volcanoes release sulfur dioxide, particulate matter, carbon monoxide, and nitrogen oxides. Lightning strikes create nitrogen oxides from atmospheric nitrogen. Forest fires release particulate matter, nitrogen oxides, and carbon monoxide (**FIGURE 46.4a**). Living plants release a variety of VOCs, including ethylene and terpenes. The fragrant smell from conifer trees such as pine and fir and the smell from citrus fruits are mostly from terpenes; though we enjoy their fragrance, they can be precursors to photochemical smog. Long before anthropogenic pollution was common, the natural VOCs from plants gave rise to smog and photochemical oxidant pollution—hence the names of the forested mountain ranges in the southeastern United States, the Blue Ridge and the Smoky Mountains (Figure 46.4b). Large nonindustrial areas such as agricultural fields can give rise to particulate matter when they are plowed, as happened in the Dust Bowl of the 1930s. *The Encyclopedia of Earth,* using data in part from the Intergovernmental Panel on

Climate Change, estimates that across the globe, sulfur dioxide emissions are 30 percent natural, nitrogen oxide emissions are 44 percent natural, and volatile organic compound emissions are 89 percent natural. However, in certain locations, such as North America, the anthropogenic contribution is much greater, perhaps as much as 95 percent for nitrogen oxides and sulfur dioxide.

The effects of these various compounds, especially when major natural events occur, depend in part on natural conditions such as wind direction. In May 1980, the prevailing westerly winds carried volcanic emissions from Mount St. Helens in Washington State and distributed particulate matter and sulfur oxides from the volcano across the United States. The rainfall pH was noticeably lower in the eastern United States that summer.

Anthropogenic Emissions

In contrast to natural emissions, emissions from human activity are monitored, regulated, and in many cases controlled. The EPA reports periodically on the emission sources of the criteria air pollutants for the entire United States, listing pollution sources in a variety of categories, such as on-road vehicles, power plants, industrial processes, and waste disposal. **FIGURE 46.5** shows some of the most recent data. Mobile sources, primarily vehicles, also sometimes referred to as the general category of *transportation,* are the largest sources of carbon monoxide and nitrogen oxides. Electricity generation, 30 percent of which, as we know from Chapter 12, is fueled by coal, along with other stationary source fuel combustion, is the major source of anthropogenic sulfur dioxide. Particulate matter comes from a variety of sources including natural and human-made fires, agriculture, road dust, and the generation of electricity.

The Clean Air Act and its various amendments require that the EPA establish standards to control pollutants that are harmful to "human health and welfare." The term *human health* means the health of the human population and includes the elderly, children, and sensitive populations such as those with asthma. The term *welfare* refers to visibility, the status of crops, natural vegetation, animals, ecosystems, and buildings. Through the National Ambient Air Quality Standards (NAAQS), the EPA periodically specifies concentration limits for each air pollutant. For each pollutant the NAAQS note a concentration that should not be exceeded over a specified time period. Our discussion of air pollution in Chattanooga at the beginning of this chapter partially described the standard for ozone: For each locality in the United States, the average ozone concentration for any 8-hour period should not exceed 0.075 parts of ozone per million parts of air by volume more than 4 days per year, averaged over a 3-year period. If a locality violates the ozone air-quality standard and does not make an attempt to improve air quality, it is subject to penalties.

Each year, the EPA issues a report that shows the national level of the six criteria air pollutants relative to

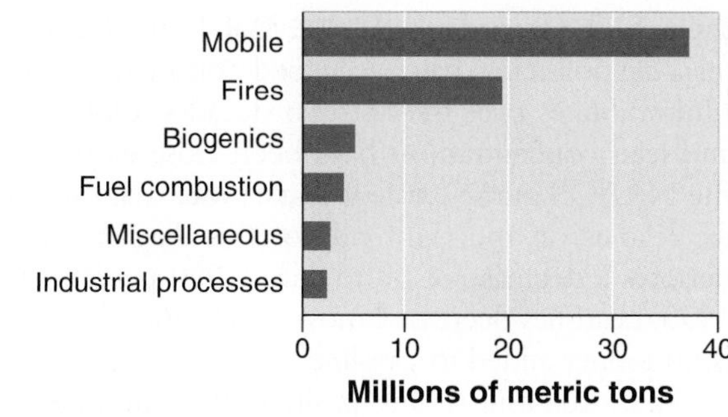

(a) Carbon monoxide

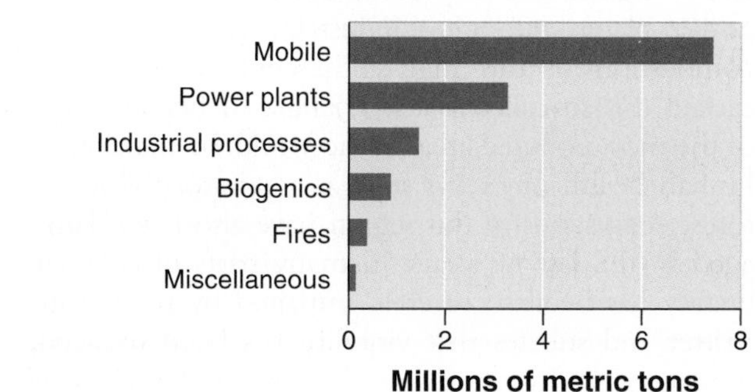

(b) Nitrogen oxides

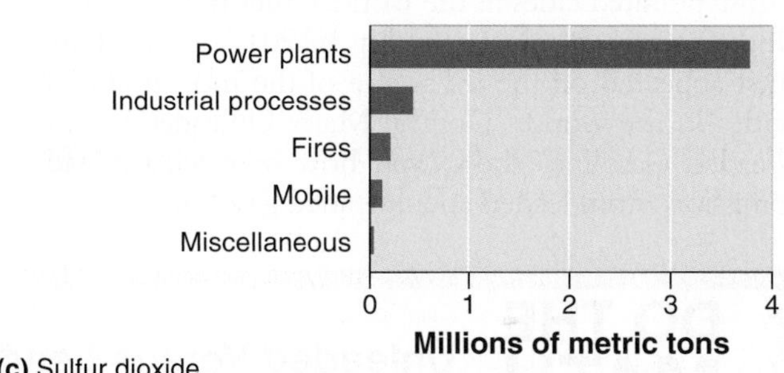

(c) Sulfur dioxide

(d) Particulate matter ($PM_{2.5}$)

FIGURE 46.5 Emission sources of criteria air pollutants for the United States. Recent EPA data show that vehicles are the largest source of (a) carbon monoxide and (b) nitrogen oxides. The major source of (c) anthropogenic sulfur dioxide is the power plants that generate electricity with coal. Among the sources of (d) particulate matter are natural and human-made fires, agriculture and road dust. *(Data from http//www.epa.gov/air/emisions/index.htm.)*

the published standards. **FIGURE 46.6** shows that all criteria air pollutants have decreased considerably in the United States over the last two decades. Only ozone and lead concentrations have been close to or above the NAAQS in the last few years. Note that the *y* axis for lead is on the right side of this diagram, and it denotes a decrease of more than 1,000 percent since 1995. Lead has decreased most significantly because it is no longer added to gasoline.

The situation is less positive in other parts of the world. Large areas in Germany, Poland, and the Czech Republic contain a great deal of "brown" coal or lignite that provides fuel for nearby coal-fired power plants and other industries. Emissions from combustion of this high-sulfur-content coal once caused this so-called Black Triangle to become one of the most polluted areas in the world. In addition to human health problems such as respiratory illnesses, forest ecosystems in this region have also been damaged in the last 40 years. In many parts of Asia, air quality has been so severely impaired by particulate matter and sulfates that visibility has been reduced, in some cases by more than 20 percent. A variety of nongovernmental and environmental organizations have prepared lists indicating the top 10 or top 20 most-polluted cities in the world. Cities in China and India usually dominate the list. In 2017, Zabol, Iran, first appeared on the list as one of the most polluted cities in the world. "Do The Math: Unleaded Versus Leaded Gasoline" shows you how to calculate lead emissions from leaded and unleaded gasoline.

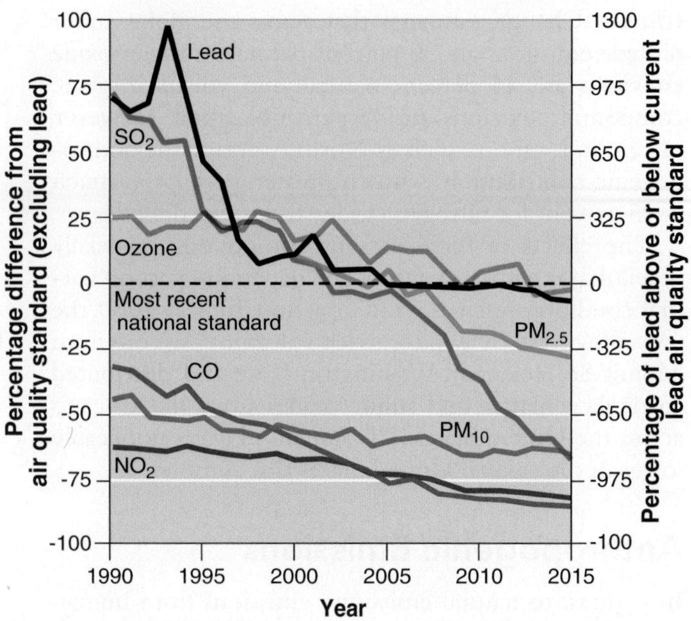

FIGURE 46.6 Criteria and other air pollutant trends. Trends in the criteria air pollutants in the United States between 1990 and 2015 are shown. All criteria air pollutants have decreased during this time period. The decrease for lead is the greatest. (Note that the *y* axis on right side showing lead is a different scale.) *(Data from New England Journal of Medicine http://www.nejm.org/doi/full/10.1056/NEJMms1615242.)*

DO THE MATH Unleaded Versus Leaded Gasoline

Preparing for the AP® Exam

In 1970, when gasoline in the United States contained the additive tetra-ethyl lead, there was roughly 0.5 g of lead added to each liter of gasoline. How much lead was emitted to the air per year by one car if that car burned 2,000 liters of gasoline per year?

$$0.5 \text{ g lead/liter} \times 2{,}000 \text{ liters/year} = 1{,}000 \text{ g of lead released in a year from one car in 1970}$$

YOUR TURN Although lead is not added to gasoline today, gasoline typically contains about 0.01 g of lead per liter from lead that is in the petroleum before the gasoline is refined. How much lead is released per year in a car that burns 2,000 liters of unleaded gasoline per year?

MODULE 46 AP® Review

Preparing for the AP® Exam

In this module, we have seen that that air pollution occurs in a global system across the troposphere. The major air pollutants are sulfur dioxide, which comes from sulfur contained in coal and oil; nitrogen oxides, which come from combustion; and carbon monoxide, which comes from incomplete combustion. Combustion of coal and oil is the most common source of particulate matter, but it can also be released during the combustion of biofuels such as wood and animal manure. Photochemical oxidants are a class of pollutants that form in the presence of sunlight, nitrogen, and sulfur oxides. There are both natural and human-generated sources of air pollution. Presently, the most polluted cities in the world are found in China and India. These cities suffer from smog, which is the focus of the next module.

AP® Practice Questions

Choose the best answer for the following.

1. Which is NOT a criteria air pollutant?
   (a) sulfur dioxide
   (b) lead
   (c) carbon dioxide
   (d) particulate matter
2. A secondary pollutant
   (a) forms in the stratosphere.
   (b) is transformed by sunlight or water.
   (c) cannot be directly tracked.
   (d) does not directly harm humans.
3. Which is a source of sulfur dioxide found in nature?
   (a) forest fires
   (b) lightning strikes
   (c) plant emissions
   (d) volcanoes
4. Which size of particulate matter causes the greatest health concern?
   (a) $PM_{2.5}$
   (b) PM_{10}
   (c) PM_{100}
   (d) PM_{30}
5. Carbon monoxide
   (a) increases lung cancer rates.
   (b) leads to the formation of photochemical smog.
   (c) is most problematic in rural areas.
   (d) is produced by incomplete combustion.

MODULE 47

Photochemical Smog and Acid Rain

The air quality in the United States has greatly improved in recent decades. However, some cities and regions of the United States continue to experience intermittent air pollution, often related to smog formation. In this module, we will examine photochemical smog, which is a problem in the United States and elsewhere in the world, and acid rain, which is no longer a problem in the United States but has become a problem in Asia.

Learning Goals

After reading this module, you should be able to

- explain how photochemical smog forms and why it is still a problem in the United States.
- describe how acid deposition forms and why it has improved in the United States and become worse elsewhere.

Photochemical smog remains an environmental problem in the United States

A recent news article stated that "More than 4 in 10 Americans live in counties where the air is unhealthy to breathe." You might think this was from a newspaper in the 1970s, before the Clean Air Act was fully in effect. But in *State of the Air 2017*, the American Lung Association reported that over 125 million people within the United States (out of 325 million total) were exposed to air that did not comply with the maximum allowable ozone concentration of 0.075 parts of ozone per million parts of air over an 8-hour period. Although sulfur, nitrogen, and carbon monoxide pollution have been reduced well below the specified standards since the Clean Air Act was implemented, controlling photochemical smog and ozone present especially difficult challenges. The reason lies in the chemistry of smog formation and the behavior of the atmosphere during changing weather conditions. These factors make smog formation very complex and difficult to predict and difficult to reduce.

The Chemistry of Ozone and Photochemical Smog Formation

As we mentioned earlier, the term *smog* was originally used to describe the combination of smoke, fog, and sometimes sulfur dioxide that used to occur in cities that burned large quantities of coal. Today, Los Angeles–type brown photochemical smog is still a problem in many U.S. cities. The formation of this photochemical smog is complex and still not well understood. A number of pollutants are involved and they undergo a series of complex transformations in the atmosphere that involve sunlight, water, and the presence of VOCs.

FIGURE 47.1 shows a portion of the chemical process that creates photochemical smog. The first part of the process, shown in Figure 47.1a, takes place during the day, in the presence of sunlight. When an abundance of nitrogen oxides are present in the atmosphere, with very few VOCs present, nitrogen dioxide (NO_2) splits to form nitrogen oxide (NO) and a free oxygen atom (O). In the presence of energy inputs from sunlight, this free oxygen atom combines with diatomic oxygen (O_2) to form ozone (O_3). With abundant nitrogen dioxide and abundant sunlight, ozone accumulates in the atmosphere.

Figure 47.1b shows that a few hours later, when sunlight intensity decreases and nitrogen oxide is still present in the atmosphere, the ozone combines with nitrogen oxide (NO), and re-forms into $O_2 + NO_2$. This is referred to as ozone destruction and it is a natural process that happens in the latter part of the day and evening.

Thermal inversion A situation in which a relatively warm layer of air at mid-altitude covers a layer of cold, dense air below.

Volatile organic compounds come from human activity such as spilling of gasoline on pavement and from natural sources such as forests. When volatile organic compounds are absent or in small supply, the cycle of ozone formation and destruction generally takes place on a daily basis and relatively small amounts of photochemical smog form.

As shown in Figure 47.1c, a different scenario occurs when VOCs are present in larger quantities. The first part is the same: Sunlight causes nitrogen dioxide to break apart into nitrogen oxide and a free oxygen atom. The free oxygen atom combines with diatomic oxygen to form ozone. However, because VOCs have combined with nitrogen oxide in a strong bond, nitrogen oxide is no longer available to combine with ozone. Since the nitrogen oxide is not available to break down ozone by recombining with it, a larger amount of ozone accumulates. This explains, in part, the daytime accumulation of ozone in urban areas with an abundance of both VOCs and nitrogen dioxide.

Although smog is associated with urban areas, it is not limited to such areas. Trees and shrubs in rural areas produce VOCs that can contribute to the formation of photochemical smog, as do forest fires that begin naturally.

Atmospheric temperature influences the formation of smog in several important ways. Emissions of VOCs from vegetation such as trees, as well as from evaporation of volatile liquids like gasoline, increase as the temperature increases. NO_X emissions from fuel combustion by electric utilities are also greater with air-conditioning demands for electricity increasing on the hottest days. Moreover, many of the chemical reactions that form ozone and other photochemical oxidants proceed more rapidly at higher temperatures. These and other factors increase smog concentrations when temperatures are higher.

AP® Exam Tip

Previous AP® Environmental Science Exams have asked students to describe the natural formation and destruction of tropospheric ozone discussed in this module. Be sure that you understand this process and can write the basic chemical equations for natural ozone accumulation and formation, as outlined in Figure 47.1. ●

Thermal Inversions

Temperature also influences air pollution conditions in more complex ways. Normally, temperature decreases as altitude increases. As shown in **FIGURE 47.2a** on page 550, the warmest air is closest to Earth. This warm air, which is less dense than the colder air above it, can easily rise, dispersing pollutants into the upper atmosphere. This allows pollutants from the surface to be reduced or diluted by all of the atmosphere above. However, during a **thermal inversion**—shown in Figure 47.2b—a relatively warm layer of air at mid-altitude covers a layer of cold, dense

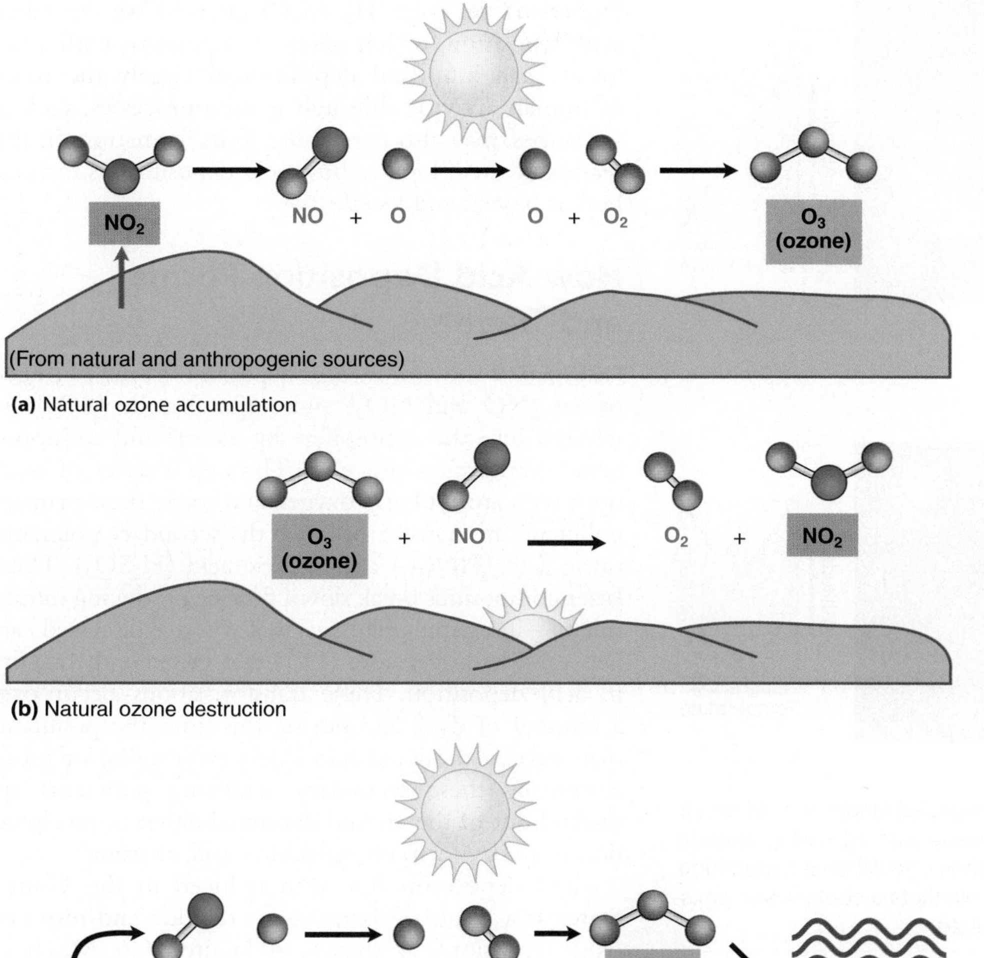

FIGURE 47.1 Tropospheric ozone and photochemical smog formation. (a) In the absence of VOCs, ozone will form during the daylight hours. (b) After sunset, the ozone will break down. (c) In the presence of VOCs, ozone will form during the daylight hours. The VOCs combine with nitrogen oxides to form photochemical oxidants, which reduce the amount of ozone that will break down later and contribute to prolonged periods of photochemical smog.

air below it. The warm layer of air trapped between the two cooler layers is known as an **inversion layer**. Because the air closest to the surface of Earth is denser than the air above it, the cool air and the pollutants within it do not rise. Thus, the inversion layer traps emissions that then accumulate beneath it, and these trapped emissions can cause a severe pollution event. Thermal inversions that create pollution events are particularly common in some cities, where high concentrations of vehicle exhaust and industrial emissions are easily trapped by the inversion layer.

Thermal inversions can also lead to other forms of pollution. A striking example occurred in spring 1998 in the northern Chinese city of Tianjin. A cold spell that occurred after the city had shut off its district heating system for the season led many households to use individual coal-burning stoves for heat. A temperature inversion trapped the carbon monoxide and particulate matter from the coal used in these stoves and caused over 1,000 people to suffer carbon monoxide poisoning, or respiratory ailments from the polluted air. Eleven people died.

Inversion layer The layer of warm air that traps emissions in a thermal inversion.

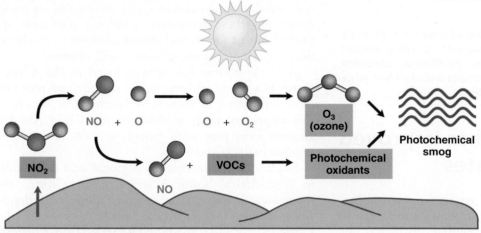

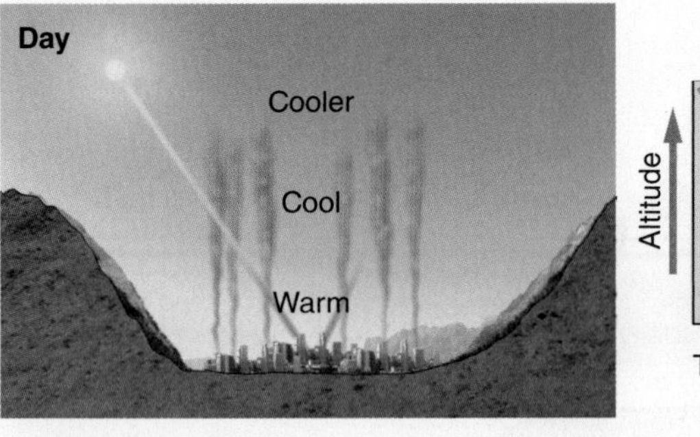

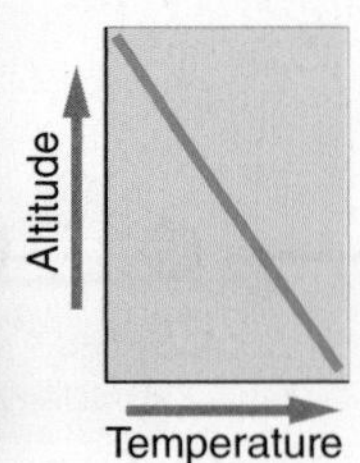

(a) Normal conditions

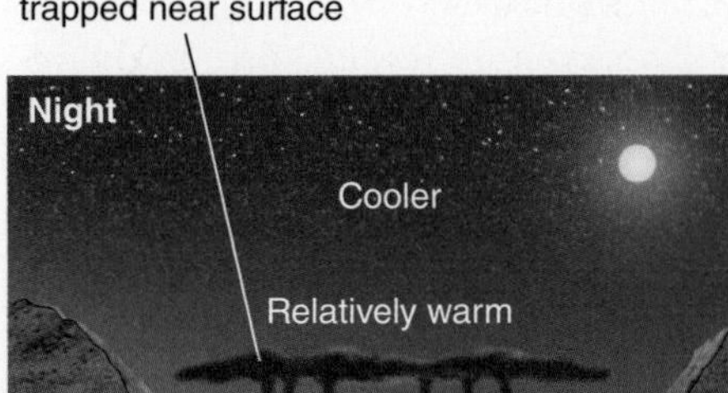

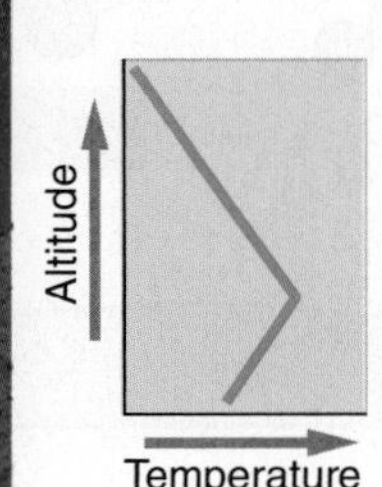

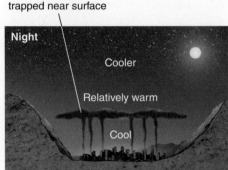

(b) Thermal inversion

FIGURE 47.2 A thermal inversion. (a) Under normal conditions, where temperatures decrease with increasing altitude, emissions rise into the atmosphere. (b) When a midaltitude, relatively warm inversion layer blankets a cooler layer, emissions are trapped and accumulate.

Acid deposition has improved in the United States

All rain is naturally somewhat acidic; the reaction between water and atmospheric carbon dioxide lowers the pH of precipitation from neutral 7.0 to 5.6 (see Figure 4.7 on page 41). In Chapter 14 we described acid deposition, which refers to deposition with a pH lower than 5.6. Acid deposition is largely the result of human activity, although natural processes, such as volcanoes, may also contribute to its formation. In this section we will look at how acid deposition is formed, how it travels, and its effects.

How Acid Deposition Forms and Travels

FIGURE 47.3 shows how acid deposition forms. Nitrogen oxides (NO and NO_2) and sulfur dioxide (SO_2) are released into the atmosphere by natural and anthropogenic combustion processes. Through a series of reactions with atmospheric oxygen and water, these primary pollutants are transformed into the secondary pollutants nitric acid (HNO_3) and sulfuric acid (H_2SO_4). These latter compounds break down further, producing nitrate, sulfate—inorganic pollutants that we have discussed earlier—and hydrogen ions (H^+) that generate the acidity in acid deposition. These transformations occur over a number of days, and during this time, the pollutants may travel a thousand kilometers (600 miles) or more. Eventually, these secondary acidifying pollutants are washed out of the air and deposited either as precipitation or in dry form on vegetation, soil, or water.

Acid deposition has been reduced in the United States as a result of lower sulfur dioxide and nitrogen oxide emissions, as shown in Figure 46.6. Much of this improvement is a result of the Clean Air Act Amendments that were passed in 1990 and implemented in 1990 and 1995.

Studies have documented regional acid deposition in West Africa, South America, Japan, China, and many areas in eastern and central Europe. Acid deposition crosses international borders between the United States and Canada and is carried from England,

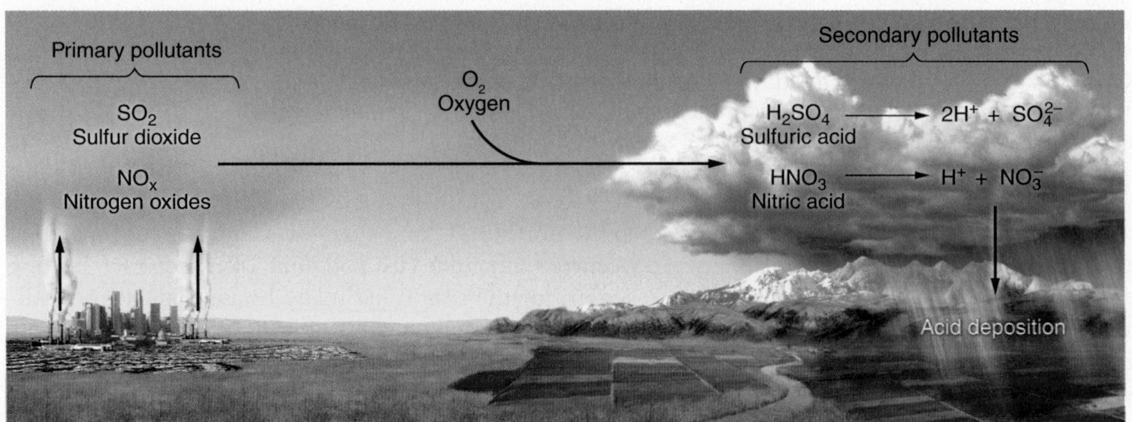

FIGURE 47.3 Formation of acid deposition. The primary pollutants sulfur dioxide and nitrogen oxides are precursors to acid deposition. After transformation to the secondary pollutants—sulfuric and nitric acid—dissociation occurs in the presence of water. The resulting ions—hydrogen, sulfate, and nitrate—cause the adverse ecosystem effects of acid deposition.

Germany, and the Netherlands to Scandinavia. Because of this mobility, the precursors to acid deposition emitted in one region may have a significant impact on another region or another country. For example, over the years, there have been legislative and legal attempts to restrict emissions from coal-burning power plants in the midwestern United States that fall as acid deposition in Canada. Many of the regions that once received high amounts of acid deposition are receiving less deposition today. However, recently observed acid deposition documented along the West Coast of the United States is believed to be the result of coal combustion in China; sulfur and nitrogen oxides are released in China and elsewhere in Asia and are carried by the prevailing westerlies from one continent to another, across the Pacific Ocean. When they reach the western United States, they are deposited on the ground in rain and snow and intercepted by vegetation in California, Oregon, and Washington State.

Effects of Acid Deposition

As we saw in Chapter 14, acid deposition in the United States increased substantially from the 1940s through the 1990s due to human activity. It had a variety of effects on materials, on agricultural lands, and on both aquatic and terrestrial natural habitats. Newspaper headlines in the United States and Europe in the 1980s contained frequent reports about adverse effects of acid deposition on forests, lakes, and streams.

Effects of acid deposition may be direct, such as a decrease in the pH of lake water, or indirect. It is often difficult to determine whether an effect is direct or indirect, making remediation challenging. The greatest effects of acid deposition have been on aquatic ecosystems. Lower pH of lakes and streams in areas of northeastern North America, Scandinavia, and the United Kingdom has caused decreased species diversity of aquatic organisms. As we saw in Chapter 6, many species are able to survive and reproduce only within a narrow range of environmental conditions. Many amphibians, for instance, will survive when the pH of a lake is 6.5, but when the lake acidifies to pH 6.0 or 5.5, the same organism will begin to have developmental or reproductive problems. In water below pH 5.0, most salamander species cannot survive.

Lower pH can also lead to mobilization of metals, an indirect effect. When this happens, metals bound in organic or inorganic compounds in soils and sediments are released into surface water. Because metals such as aluminum and mercury can impair the physiological functioning of aquatic organisms, exposure can lead to species loss. Decreased pH can also affect the food sources of aquatic organisms, creating indirect effects at several trophic levels. On land, at least one species of tree, the red spruce (*Picea rubens*), at high elevations of the northeastern United States was shown to have been harmed by acid deposition. It is likely that these trees have been harmed by both the acidity of the deposition as well as by the nitrate and sulfate ions.

People are not harmed by direct contact with precipitation at the acidities commonly experienced in the United States or elsewhere in the world because human skin is a sufficiently robust barrier. Human health is more affected by the precursors to acid deposition such as sulfur dioxide and nitrogen oxides.

Acid deposition can, however, harm human-built structures such as statues, monuments, and buildings. For example, buildings from ancient Greece such as those on and near the Acropolis in Athens, many of which have stood for approximately 2,000 years, have been seriously eroded over the last half century by acid deposition (**FIGURE 47.4**). The damage happens because acid deposition reacts with building materials. When the hydrogen ion in acid deposition interacts with limestone or marble, the calcium carbonate

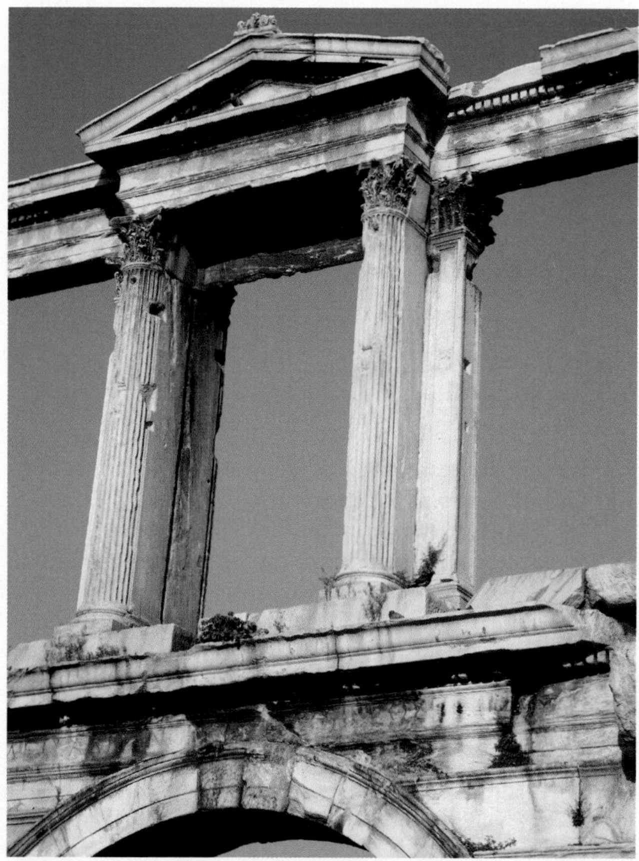

FIGURE 47.4 Material damage from acid deposition. Hadrian's Arch, near the Acropolis in Athens, Greece, has been damaged by acid deposition. It is made of marble, which contains calcium carbonate, and is susceptible to deterioration from acid deposition and acids in the air. *(Sites & Photos/HIP/The Image Works)*

reacts with H^+ and gives off Ca^{2+}. In the process, the calcium carbonate material is partially dissolved. The more acidic the precipitation, the more hydrogen ions there are to interact with the calcium carbonate. In the case of the Acropolis and some other stone structures, other components of acidic deposition, including gaseous sulfur dioxide (SO_2) or sulfuric acid vapor, have contributed to the deterioration. Acid deposition also erodes many exposed painted surfaces, including automobile finishes.

MODULE

47 AP® Review

Preparing for the AP® Exam

In this module, we have seen that that photochemical smog accumulates in the troposphere in the presence of sunlight when nitrogen oxides and volatile organic compounds are present. Despite many improvements in air quality in the United States, photochemical smog is still a problem in a number of urban and some relatively rural areas. Air pollution is intensified by thermal inversions, when pollutants in a layer of cold air are trapped beneath a layer of warm air. Acid deposition forms when the oxides of sulfur and nitrogen undergo transformations in the atmosphere in the presence of water. It travels hundreds of kilometers and is deposited in precipitation. Acid deposition can adversely affect material structures, vegetation, soils, and aquatic systems. In the next module, we shall look at how acid deposition has been reduced significantly in the United States in recent decades.

AP® Practice Questions

Choose the best answer for the following.

1. High levels of photochemical smog are due to
   I. nitrogen dioxide.
   II. sulfur dioxide.
   III. VOCs.
   (a) I only
   (b) I and II
   (c) II and III
   (d) I and III

2. Recent increases in acid deposition in the western United States are due to
   (a) increased emissions in the United States.
   (b) decreased precipitation due to climate change.
   (c) increased emissions in China.
   (d) increased precipitation due to climate change.

3. Thermal inversions
   (a) increase the rate of smog formation.
   (b) trap high concentrations of pollution at ground level.
   (c) result in increased levels of acid deposition.
   (d) are caused by high levels of precipitation.

4. Acid deposition forms as a result of
   (a) nitrogen oxides and photochemical smog.
   (b) carbon monoxide and VOCs.
   (c) sulfur dioxide and VOCs.
   (d) nitrogen oxides and sulfur dioxide.

5. The effects of acid deposition include
   (a) increased cancer rates in aquatic organisms.
   (b) decreased aquatic biodiversity.
   (c) decreased mobilization of metals.
   (d) increased pH of lake water.

MODULE 48

Pollution Control Measures

In our discussion of energy and energy choices in Chapters 12 and 13, we observed that sustainability is best achieved by considering conservation and efficiency first. Similarly, we will require less energy and resources to clean up pollution if we seek ways to avoid creating it in the first place. Preventing pollution is usually much less expensive and energy intensive then controlling it. Unfortunately, pollution prevention is not always possible.

Learning Goals

After reading this module, you should be able to

- explain strategies and techniques for controlling sulfur dioxide, nitrogen oxides, and particulate matter.
- describe innovative pollution control measures.

Pollution control includes prevention, technology, and innovation

As with other types of pollution, the best way to decrease air pollution emissions is to avoid them in the first place. This can be achieved through the use of fuels that contain fewer impurities. Coal and oil, for example, occur naturally with different sulfur concentrations and are available for purchase at a variety of sulfur concentrations. In addition, during refining and processing, the concentration of sulfur can be reduced in both fuels.

Use of a low-sulfur coal or oil is certainly one of the best means of controlling air pollution, although typically low-sulfur coal or oil is more expensive to purchase than coal or oil containing higher sulfur concentrations. As we discussed in Chapters 12 and 13, other ways to reduce air pollution include increased efficiency and conservation: Use less fuel and you will produce less air pollution. While these measures reduce emissions by a certain amount, wherever fuel is combusted, pollution will be emitted. So ultimately, most attempts to reduce air pollution will depend on the control of pollutants after combustion. There are a number of approaches to controlling emissions, as we will discuss in the next section.

Control of Sulfur and Nitrogen Oxide Emissions

Sulfur and nitrogen oxides are common air pollutants in the United States and they cause a variety of environmental problems, including acid deposition that we described in the previous module. A substantial number of air pollution control measures have been directed toward sulfur and nitrogen oxides. Sulfur dioxide emissions from coal exhaust can be reduced by a process known as fluidized bed combustion. In this process, granulated coal is burned in close proximity to calcium carbonate. The heated calcium carbonate absorbs sulfur dioxide and produces calcium sulfate, which can be used in the production of gypsum wallboard, also known as sheetrock, for houses. Some of the sulfur oxide that does escape the combustion process can be captured by other methods after combustion.

The atmosphere of Earth is 78 percent nitrogen gas and, as a result, nitrogen oxides are produced in virtually all combustion processes. Hotter burning conditions and the presence of oxygen allow proportionally more nitrogen oxide to be generated per unit of fuel burned. In order to reduce nitrogen oxide emissions, burn temperatures must be reduced and the amount of oxygen must be controlled—a procedure that is sometimes utilized in factories and power plants by means of certain air pollution control technologies. However, lowering temperatures and oxygen supply can result in less-complete combustion, which reduces the efficiency of the process and increases the amount of particulates and carbon monoxide. Finding the exact mix of air, temperature, oxygen, and other factors is a significant challenge.

Nitrogen oxide emissions from automobiles have also been reduced significantly in the United States over the last 35 years. Beginning in 1975, all new automobiles sold in the United States were required to include a catalytic converter, which reduces nitrogen oxide and carbon monoxide emissions. In order to operate properly, the precious metals in the catalytic converter (mostly platinum and palladium) cannot be exposed to lead. Therefore gasoline could no longer contain lead. As we saw in Figure 46.6, the change in gasoline formulation caused a significant reduction in emissions of lead from automobiles and in lead concentrations in the atmosphere. At the same time, improvements in the combustion technology of power plants and factories also reduced emissions of nitrogen oxides.

Control of Particulate Matter

The removal of particulate matter is the most common means of pollution control. Sometimes the process of removing particulate matter also removes sulfur. There are a variety of methods used to remove particulate matter. The simplest is gravitational settling, which relies on gravity as the exhaust travels through the smokestack. The particles simply settle out to the bottom. The ash residue that accumulates must be disposed of in a landfill. Depending on the fuel that was burned, the ash may contain sufficiently high concentrations of metals that require special disposal. This subject will be covered in more detail in Chapter 16 on solid waste.

A variety of different pollution control devices remove particulate matter and sometimes other compounds after combustion. Each has its advantages and disadvantages and all of them use energy—most commonly electricity—which generates additional pollution. Fabric filters are a type of filtration device that allow gases to pass through them but remove particulate matter. Often called baghouse filters, certain fabric filters can remove almost 100 percent of the particulate matter emissions. Electrostatic precipitators also remove particulate matter, by using an electrical charge to make particles coalesce so they can be removed. Polluted air enters the precipitator and the electrically charged particles within are attracted to negative or positive charges on the sides of the precipitator. The particles collect and relatively clean gas exits the precipitator. A scrubber, shown in **FIGURE 48.1**, uses a combination of water and air that actually separates and removes particles. Particles are removed in the scrubber in a liquid or sludge form and clean gas exits. Borrowing from the concept utilized in the electrostatic precipitator, particles are sometimes ionized before

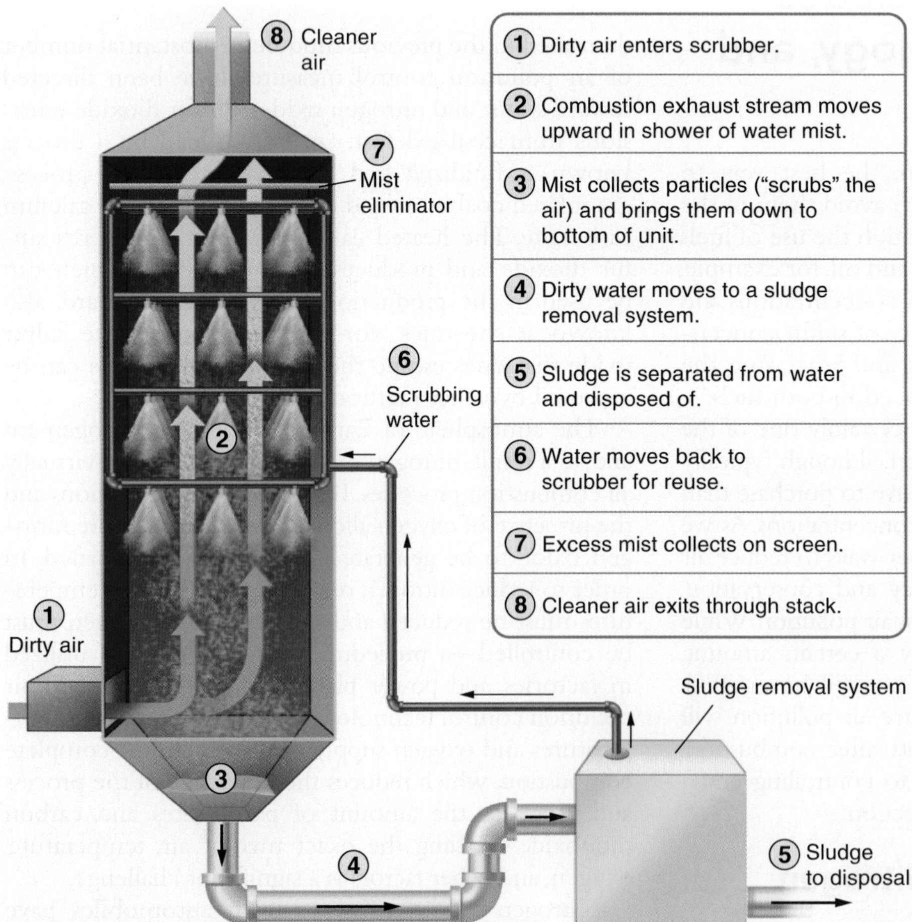

FIGURE 48.1 The scrubber. In this air pollution control device, particles are "scrubbed" from the exhaust stream by water droplets. A water-particle "sludge" is collected and processed for disposal.

entering the scrubber to increase its efficiency. Scrubbers also remove sulfur dioxide. All three types of pollution control devices, because they use additional energy and increase resistance to air flow in the factory or power plant, require the use of more fuel and result in increased carbon dioxide emissions.

Devices such as the electrostatic precipitator and the scrubber have helped reduce pollution significantly before it is released into the atmosphere. It is much harder—if not impossible—to remove pollutants from the environment after they have been dispersed over a wide area.

Smog Reduction

We have seen that many cities in the United States and around the world continue to have smog problems. Because the main component of photochemical smog—ozone—is a secondary pollutant, control efforts must be directed toward reducing the precursors, or primary pollutants. Historically, most local smog reduction measures have been directed primarily at reducing emissions of VOCs in urban areas. As noted earlier, with fewer VOCs in the air there are fewer compounds to interact with nitrogen oxides, and thus more nitrogen oxide will be available to recombine with ozone. More recently, regional efforts to control ozone have focused on reducing nitrogen oxide emissions, which appears to be a more effective method of controlling smog in areas away from urban centers.

Around the world, people are implementing innovative pollution control measures

A number of cities around the world, including those in China, Mexico, and England, have taken innovative and often controversial measures to reduce smog levels. Municipalities have passed measures, for example, to reduce the amount of gasoline spilled at gasoline stations, restrict the evaporation of dry-cleaning fluids, or restrict the use of lighter fluid (a VOC) for starting charcoal barbecues. Both urban and suburban areas have taken additional actions such as calling for a reduction in the use of wood-burning stoves or fireplaces that would reduce emissions of not only nitrogen oxide but also particulate matter, VOCs, and carbon monoxide. A number of California municipalities even discussed reducing the number of bakeries within certain areas, as the emissions from rising bread contain VOCs. This proposal was not very popular, as you can imagine, but emissions from bakeries along with many other businesses are sometimes regulated by local air-quality ordinances.

Since cars are responsible for large emissions of nitrogen oxides and VOCs in urban areas, and these two compounds are the major contributors to smog formation, some municipalities have tried to achieve lower smog concentrations by restricting automobile use. A number of cities, including Mexico City, have instituted plans permitting automobiles to be driven only every other day—for example, those with license plates ending in odd numbers may be used on one day and those with even-numbered license plates on alternate days. In Mexico City, as of 2018, the plan had not had much of an impact. In China, during the 2008 Beijing Olympics, the government successfully expanded public transportation networks, imposed motor vehicle restrictions, and temporarily shut down a number of industries as a way to reduce photochemical smog and improve visibility. For a short period, the air pollution control measures were successful. However, long-term improvements in air quality in China have been harder to achieve (**FIGURE 48.2**).

In 1990 and again in 1995, scientists, policy makers, and academics collaborated on amendments to the Clean Air Act that would allow the free market to determine the least expensive ways to reduce emissions of sulfur dioxide. The free-market program was implemented in two phases between 1995 and 2000, and approximately 3,000 power plants are now covered under the Acid Rain Program of the act. So far, each phase has led to significant reductions in sulfur emissions.

(a)

(b)

FIGURE 48.2 Reducing photochemical smog. The photos show the same locations in Beijing, China, (a) during the Beijing Olympics in 2008 and (b) after the pollution restrictions were removed. *(PETER PARKS/Getty Images)*

One of the most innovative aspects of the Clean Air Act amendments was the provision for the buying and selling of allowances that authorized the owner to release a certain quantity of sulfur. Each allowance authorizes a power plant or industrial source to emit one ton of SO_2 during a given year. Sulfur allowances are awarded annually to existing sulfur emitters proportional to the amounts of sulfur they were emitting before 1990, and the emitters are not allowed to emit more sulfur than the amount for which they have permits. At the end of a given year, the emitter must possess a number of allowances at least equal to its annual emissions. In other words, a facility that emits 1,000 tons of SO_2 must possess at least 1,000 allowances that are usable in that year. Facilities that emit quantities of SO_2 above their allowances must pay a financial penalty.

Sulfur allowances can be bought and sold on the open market by anyone. If emitters wanted to exceed their allowance level—say, because they intended to increase their industrial output—they would be required to purchase more allowances from another source. If, on the other hand, a company decreased its sulfur emissions more than it needed to in order to comply with its allowance amount, it could sell any unused sulfur emission allowances. Over time, the number of allowances distributed each year has been gradually reduced: The total SO_2 emissions from all sources in the United States have declined from 23.5 million metric tons (26 million U.S. tons) in 1982 to 10.3 million metric tons (11.4 million U.S. tons) in 2008. "Do the Math: Calculating Annual Sulfur Reductions" shows you how to calculate these decreases as percentages. The overall economic cost for achieving these reductions has been about one-quarter of the original cost estimate. Global change researchers have used the sulfur allowance example as a model for the more recent experiments with buying and selling carbon dioxide allowances.

DO THE MATH Calculating Annual Sulfur Reductions

Preparing for the AP® Exam

The text noted that the total SO_2 emissions from all sources in the United States declined from 23.5 million metric tons (26 million U.S. tons) in 1982 to 10.3 million metric tons (11.4 million U.S. tons) in 2008. Calculate the total percentage reduction and the annual percentage reduction of SO_2 emissions.

$$23.5 \text{ million metric tons} - 10.3 \text{ million metric tons} = 13.2 \text{ million metric tons total reduction}$$

Divide the reduction by the original amount and multiply by 100 to obtain a percent reduction:

$$13.2 \text{ million} \div 23.5 \text{ million metric tons} \times 100\% = 56\%$$

The total reduction was 56 percent.

To calculate the reduction per year, divide 56 percent by the number of years from beginning to end:

$$2008 - 1982 = 26 \text{ years}$$

$$56\% \div 26 \text{ years} = 2.2\%/\text{year}$$

YOUR TURN In the United States, nitrogen oxide emissions decreased from 22.6 million metric tons (25 million U.S. tons) in 1990 to 17.2 million metric tons (19 million U.S. tons) in 2005. Calculate the total percentage reduction and the annual percentage reduction for nitrogen oxides.

MODULE 48 AP® Review

Preparing for the AP® Exam

In this module, we have seen that the best approach for decreasing air pollution emissions is by avoiding them in the first place. This can be done by choosing fuels that contain fewer impurities or by increasing the efficiency of operations. After these options have been considered and implemented, a number of pollution control technologies are available to prevent pollutants from entering the air. The baghouse filter physically removes particles and the electrostatic precipitator uses an electrical charge to attract and remove pollutant particles. The scrubber literally scrubs the exhaust stream with water droplets, creating a sludge that can be collected and removed. Innovative techniques for reducing pollution include putting a price on the right to emit certain pollutants and allowing pollution permits bought and sold on the free market to bring down emissions in the most cost-efficient way possible. In the next module we will look at how pollutants caused a depletion of stratospheric ozone and the international agreement that led to a successful resolution of the problem.

AP® Practice Questions

Choose the best answer for the following.

1. Which method is used to reduce nitrogen oxide emissions?
   (a) installation of catalytic converters
   (b) increased temperature of combustion
   (c) the addition of oxygen to combustion processes
   (d) the use of fluidized bed combustion

2. Which is NOT used to prevent the emission of particulate matter?
   (a) gravitational settling
   (b) fabric filters
   (c) electrostatic precipitators
   (d) catalytic converters

3. Which pollution control method was proposed by an amendment to the Clean Air Act?
   (a) allowing tolls to limit the use of automobiles
   (b) a market for sulfur emissions
   (c) the regulation of radon emissions
   (d) the use of catalytic converters

4. If carbon monoxide emissions decreased from 145 million tons annually to 80 million tons annually, by what percentage have emissions been reduced?
   (a) 41 percent
   (b) 45 percent
   (c) 47 percent
   (d) 52 percent

MODULE 49

Stratospheric Ozone Depletion

We have seen that tropospheric or ground-level pollution contributes to a number of problems in the natural world, exacerbates asthma and breathing difficulties in humans, and contributes to the incidence of cancer. Now we turn to the effects of certain pollutants in the stratosphere that have a substantial impact on the health of humans and ecosystems. In the troposphere, ozone is an oxidant that can harm respiratory systems in animals and damage a number of structures in plants. However, in the stratosphere ozone forms a necessary, protective shield against radiation from the Sun; it absorbs ultraviolet light and prevents harmful ultraviolet radiation from reaching Earth.

Learning Goals

After reading this module, you should be able to

- explain the benefits of stratospheric ozone and how it forms.
- describe the depletion of stratospheric ozone.
- explain efforts to reduce ozone depletion.

Stratospheric ozone is beneficial to life on Earth

The Sun radiates energy at many different wavelengths, including the ultraviolet range (see Figure 5.1 on page 48). The ultraviolet wavelengths are further classified into three groups: UV-A, or low-energy ultraviolet radiation, and the shorter, higher-energy UV-B and UV-C wavelengths. UV radiation of all types can damage the tissues and DNA of living organisms. Exposure to UV-B radiation increases the risk of skin cancer and cataracts, and suppresses the immune system in humans. Exposure to UV-B is also harmful to the cells of plants, and it reduces their ability to convert sunlight into usable energy. UV-B exposure can therefore harm entire biological communities. For example, losses of phytoplankton—the microscopic algae that form the base of many marine food chains—will cause the depletion of fisheries.

As we saw in Chapter 4, a layer of ozone in the stratosphere (see Figure 9.1 on page 110) absorbs ultraviolet radiation, filtering out harmful UV rays from the Sun. It is easy to confuse stratospheric ozone with tropospheric, or ground-level, ozone that we discussed earlier in this chapter because it *is* the same gas, O_3. However, stratospheric ozone occurs higher in the atmosphere where its ability to absorb ultraviolet radiation and thereby shield the surface below makes stratospheric ozone critically important to life on Earth. In this section we will examine the formation of stratospheric ozone and look at how it breaks down.

AP® Exam Tip

The AP® Environmental Science Exam regularly asks questions about the difference between stratospheric ozone and ground level ozone. Make sure that you know the difference between the two. ●

Formation of Stratospheric Ozone

When solar radiation strikes O_2 in the stratosphere, 16 to 50 km (10–31 miles) above Earth's surface, a series of chemical reactions begins that produces a new molecule: ozone (O_3).

In the first step, UV-C radiation breaks the molecular bond holding an oxygen molecule together:

$$O_2 + \text{UV-C} \rightarrow O + O$$

This happens to only a few oxygen molecules at any given time. The vast majority of the oxygen in the atmosphere remains in the form O_2.

In the second step, a free oxygen atom (O) produced in the first reaction encounters an oxygen molecule, and they form ozone:

$$O + O_2 \rightarrow O_3$$

Both UV-B and UV-C radiation can break a bond in this new ozone molecule, forming molecular oxygen and a free oxygen atom once again:

$$O_3 + \text{UV-B or UV-C} \rightarrow O_2 + O$$

Thus formation of ozone in the presence of sunlight and its subsequent breakdown is a cycle that can occur indefinitely as long as there is UV energy entering the atmosphere. Under normal conditions, the amount of ozone in the stratosphere remains at steady state.

Breakdown of Stratospheric Ozone

We rely on refrigeration to keep our foods safe and edible, and on air conditioning to keep us comfortable in hot weather. For many years, the same chemicals that made refrigeration and air conditioning possible were also used in a host of other consumer items, including aerosol spray cans and products such as Styrofoam. These chemicals, called chlorofluorocarbons, or CFCs, were considered essential to modern life, and producing them was a multibillion-dollar industry. CFCs were considered "safe" because they are both nontoxic and nonflammable. But it turned out that these chemicals had adverse effects in one part of the upper atmosphere, the stratosphere, by promoting the breakdown of ozone.

CFCs introduce chlorine (Cl) into the stratosphere. When chlorine is present, it can attach to an oxygen atom in an ozone molecule, thereby breaking the bond between that atom and the molecule and forming chlorine monoxide (ClO) and O_2:

$$O_3 + Cl \rightarrow ClO + O_2$$

Subsequently, the chlorine monoxide molecule reacts with a free oxygen atom, which pulls the oxygen from the ClO to produce free chlorine again:

$$ClO + O \rightarrow Cl + O_2$$

When we consider these reactions together, we see that chlorine starts out and ends up as a free Cl atom. In contrast, an ozone molecule and a free oxygen atom are converted into two oxygen molecules. A substance that aids a reaction but does not get used up itself is called a catalyst. A single chlorine atom can catalyze the breakdown of as many as 100,000 ozone molecules until finally one chlorine atom finds another and the process is stopped. In the process, the ozone molecules are no longer available to absorb incoming UV-B radiation. As a result, the UV-B radiation can reach Earth's surface and cause harm to biological organisms.

AP® Exam Tip

Make sure you understand the source of ozone depletion and can outline the chemical reactions involved in the formation and destruction of stratospheric ozone. ●

Humans have contributed to significant destruction of the ozone layer

Ozone formation and ozone destruction have occurred for many years. But the use of CFCs as refrigerants starting in the 1920s and their increased use since that time led to the more rapid destruction of stratospheric ozone. After decades of CFC use, its effects became apparent.

Depletion of the Ozone Layer

In the mid-1980s, atmospheric researchers noticed that stratospheric ozone in Antarctica had been decreasing each year, beginning in about 1979. Since the late 1970s, global ozone concentrations had decreased by more than 10 percent. Depletion was greatest at the poles, but occurred worldwide. One set of observations from New Zealand showed an erratic but gradually decreasing trend of ozone concentrations from 1980 through 2016. The graph in **FIGURE 49.1** shows these results.

Researchers also determined that, in the Antarctic, ozone depletion was seasonal: Each year the depletion occurred from roughly August through November (late winter through early spring in the Southern Hemisphere). The depletion caused an area of severely reduced ozone concentrations over most of Antarctica, creating what has come to be called the "ozone hole." A depletion of ozone also occurs over the Arctic in January through April, but it is not as severe, varies more from year to year, and does not cause a hole as in the Antarctic.

The cause of the formation of the ozone hole, which has received a great deal of media attention and has been studied intensively, is complex. It appears that extremely cold weather conditions during the polar winter cause a buildup of ice crystals mixed with nitrogen oxide. This in turn provides the perfect surface for the formation of the stable molecule Cl_2, which accumulates as atmospheric chlorine interacts with the ice crystals. When the Sun reappears in the spring, UV radiation breaks down this molecule into Cl again, which in turn catalyzes the destruction of ozone as described above. Because almost no ozone forms in the dark of the polar winter, a large amount of thinning occurs. Only after the temperatures warm up and the chlorine gets diluted by air coming from outside the polar region does the thinning diminish. In contrast, the overall global trend of decreasing stratospheric ozone concentration is not related to temperature but is caused by the breakdown reactions described earlier that result from increased concentrations of chlorine in the atmosphere.

Decreased stratospheric ozone has led to a rise in the amount of UV-B radiation that reaches the surface of Earth. A United Nations study showed that in mid-latitudes in North America, UV radiation at the surface of Earth increased about 4 percent between 1979 and 1992. Since that time through the present, UV radiation in North America has been approximately level. For plants, both on land and in water, increased exposure to UV-B radiation can be harmful to cells and can reduce photosynthetic activity, which could have an adverse impact on ecosystem productivity, among other things. In humans, particularly those with lighter skin, increasing exposure to UV-B radiation is correlated with increased risks of skin cancer, cataracts, and other eye problems, and with a suppressed immune system. Significant increases in skin cancers have already been recorded, especially in countries near the Antarctic ozone hole such as Chile and Australia.

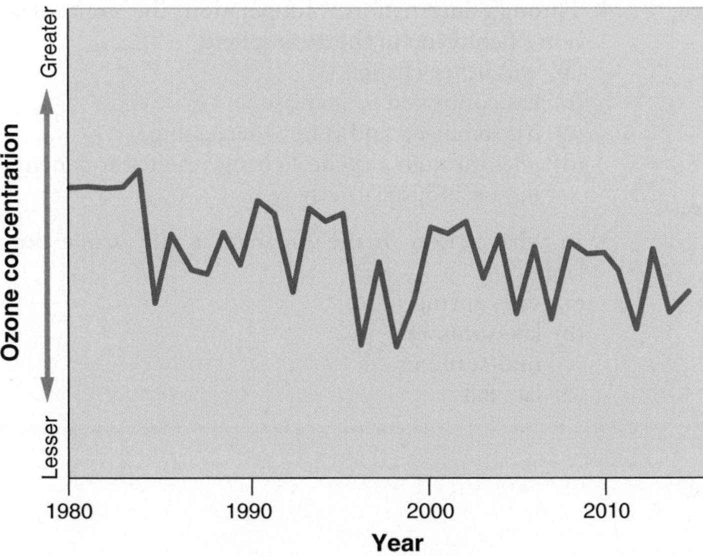

FIGURE 49.1 Stratospheric ozone concentration. These data for one area of New Zealand show a generally decreasing trend from 1970 to 2016. *(Data from https://data.mfe.govt.nz/table/89463-annual-ozone-concentrations-19792016/data/.)*

Efforts to reduce ozone depletion have been mostly effective

In response to the decrease in stratospheric ozone, 24 nations in 1987 signed the Montreal Protocol on Substances That Deplete the Ozone Layer. This was a commitment to reduce CFC production by 50 percent by the year 2000. It was the most far-reaching environmental treaty to date, in which global CFC exporters like the

United States appeared in some ways to prioritize the protection of the global biosphere over their short-term economic self-interest. More than 180 countries eventually signed a series of increasingly stringent amendments that required the elimination of CFC production and use in the developed world by 1996. In total, the protocol addressed 96 ozone-depleting compounds.

Because of these efforts, the concentration of chlorine in the stratosphere peaked at about 4 ppb and is now decreasing. The chlorine concentration reduction process is slow because CFCs are not easily removed from the stratosphere and in some recent years ozone depletion has continued to reach record levels. With the stabilization of chlorine concentrations, stratospheric ozone depletion should decrease in subsequent decades. However, a few reports in 2018 have suggested that perhaps ozone recovery in the lower stratosphere is not preceding as previously reported. It may take a number of additional years of measurements to obtain resolution of this issue. If stratospheric ozone concentrations do recover, it is thought that the number of additional skin cancers should eventually decrease as well, although this effect may take decades to observe due to the long time it takes for these cancers to develop.

MODULE 49 AP® Review

Preparing for the AP® Exam

In this module, we have seen that there is a natural process of ozone formation and ozone destruction in the stratosphere. As a result, with relatively constant ozone concentrations, harmful ultraviolet radiation is absorbed in the upper atmosphere and does not reach ground level where it could be harmful to plants and animals, including humans. However, the introduction of human-synthesized chlorofluorocarbons led to increased amounts of chlorine in the stratosphere, leading to destruction of stratospheric ozone. This caused an increase in UV-B radiation in certain locations on Earth. Since nations signed the Montreal Protocol on Substances That Deplete the Ozone Layer, stratospheric ozone depletion has begun to slow. In the next module we will turn from outdoor air pollution to indoor air pollution that occurs in houses and other living environments.

AP® Practice Questions

Choose the best answer for the following.

1. The formation of ozone begins when an O_2 molecule is split by
   (a) UV-A radiation.
   (b) UV-B radiation.
   (c) UV-C radiation.
   (d) UV-A or UV-B radiation.
2. How many ozone atoms can a single chlorine atom break down?
   (a) 5,000
   (b) 10,000
   (c) 50,000
   (d) 100,000
3. Which is NOT a result of the reduction in stratospheric ozone?
   (a) reduced photosynthetic activity
   (b) increased skin cancer
   (c) increased eye problems
   (d) increased birth defects
4. Through international cooperation, the concentration of chlorine in the atmosphere
   (a) will never change.
   (b) has continued to increase.
   (c) has stabilized and is now decreasing.
   (d) goes through a cyclical change each year depending on industrial activity.
5. In what season in the Antarctic is the ozone hole largest?
   (a) early spring
   (b) late summer
   (c) mid-summer
   (d) late fall

MODULE 50

Indoor Air Pollution

When we think of air pollution, we usually don't associate it with air inside our buildings, but indoor air pollution actually causes more deaths each year than does outdoor air pollution. Most of these deaths occur in the developing world. The amount of time one spends indoors depends on culture, climate, and economic situation. The quality of indoor air is highly variable and when polluting activities take place indoors, exposure to pollutants in a confined space can be a significant health risk.

Learning Goals

After reading this module, you should be able to

- explain how indoor air pollution differs in developing and developed countries.
- describe the major indoor air pollutants and the risks associated with them.

Indoor air pollution is a significant hazard in developing and developed countries

Although it generally receives less attention than outdoor air pollution, indoor air pollution is a hazard all over the world. The reasons for indoor air pollution and its characteristics differ between the developing world and the developed world.

Developing Countries

In Chapter 13 we saw that around the world, between 2 billion and 3 billion people use wood, animal manure, or coal indoors for heat and cooking. Biomass and coal are usually burned in open-pit fires that lack the proper mix of fuel and air to allow complete combustion. Usually, there is no exhaust system and little or no ventilation available in the home, which makes indoor air pollution from carbon monoxide and particulates a particular hazard in developing countries (**FIGURE 50.1**). Exposure to indoor air pollution from cooking and heating increases the risk of acute respiratory infections, pneumonia, bronchitis, and even cancer. The World Health Organization estimates that indoor air pollution is responsible for more than 4 million deaths annually worldwide. More than 90 percent of deaths attributable to indoor air pollution occur in developing countries.

Developed Countries

There are a number of factors that have caused the quality of air in homes in developed countries to take on greater importance in recent decades. First of all, people in much of the developed world have begun to spend more and more time indoors. Although improved insulation and tightly sealed building envelopes reduce energy consumption, these tightly sealed buildings also

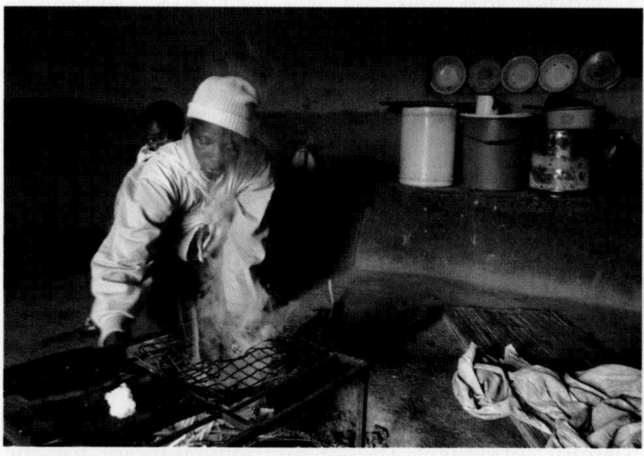

FIGURE 50.1 Indoor air pollution in the developing world. This photo shows a woman and her child in their home in Zimbabwe. *(Crispin Hughes/ Panos Pictures)*

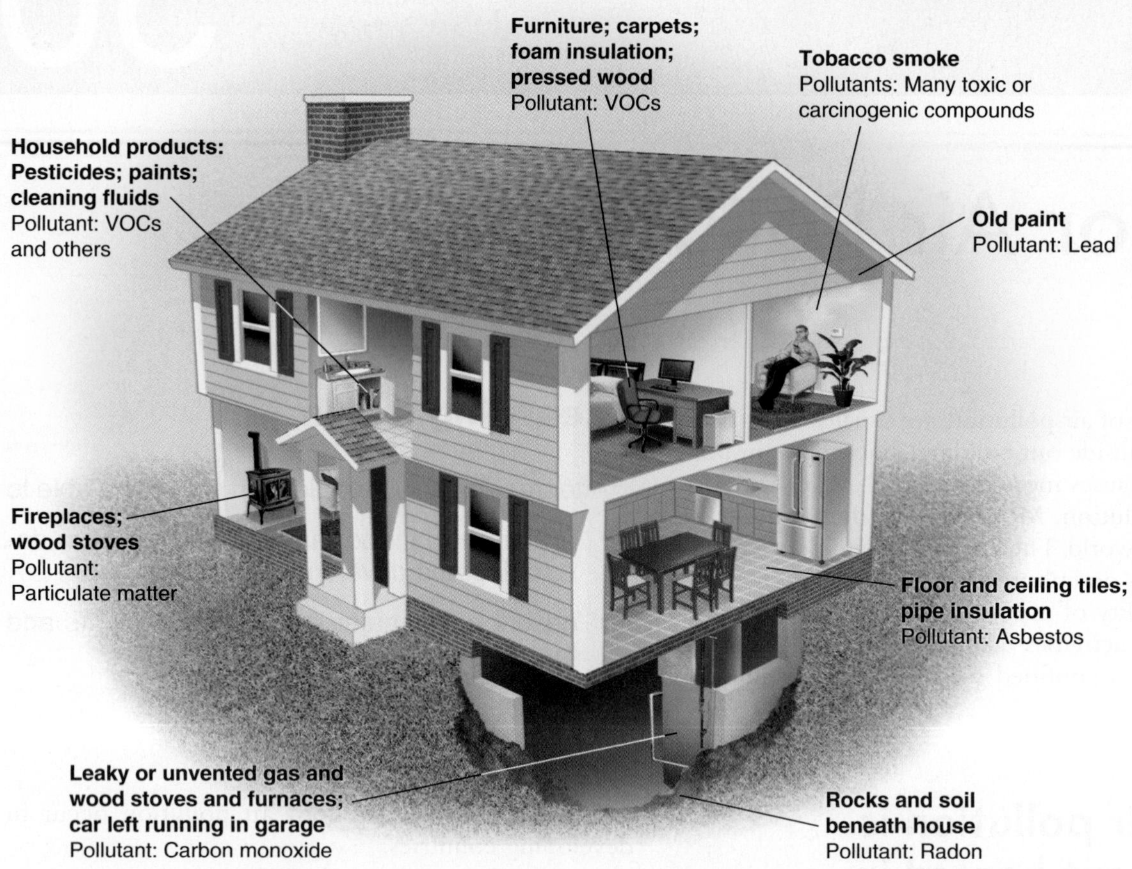

FIGURE 50.2 Some sources of indoor air pollution in the developed world. A typical home in the United States may contain a variety of chemical compounds that could, under certain circumstances, be considered indoor air pollutants.

keep existing air in contact with the inhabitants of homes, schools, and offices for greater amounts of time. Finally, an increasing number of materials in the home and office are made from plastics and other petroleum-based materials that can give off chemical vapors. As **FIGURE 50.2** shows, all of these factors combine to allow many possible sources of indoor air pollution to impact occupants.

Most indoor air pollutants differ from outdoor air pollutants

Because a house is a closed system with an abundance of manufactured materials, there is ample opportunity for indoor air pollutants to accumulate and for the occupants of that house to come into contact with harmful substances. Indoor air pollutants are for the most part different from outdoor pollutants, although, as we will see, carbon monoxide is one pollutant that causes problems both indoors and outdoors.

Carbon Monoxide

We have already described carbon monoxide as an outdoor air pollutant, but it can be even more dangerous as an indoor air pollutant. It occurs as a result of malfunctioning exhaust systems on household furnace heating systems, most commonly natural gas heaters. When the exhaust system malfunctions, exhaust air escapes into the living space of the house. Because natural gas burns relatively cleanly with little odor, a malfunctioning natural gas burner can cause the colorless, odorless carbon monoxide to accumulate in a house without the occupants noticing. This is especially problematic if the occupants are asleep. In the body, carbon monoxide binds with hemoglobin more efficiently than oxygen, thereby interfering with oxygen transport in the blood. Extended exposure to high concentrations of carbon monoxide in air can lead to oxygen deprivation in the brain and, ultimately, death. That is why carbon monoxide detectors are so important in homes and apartments and why they should be maintained and tested periodically.

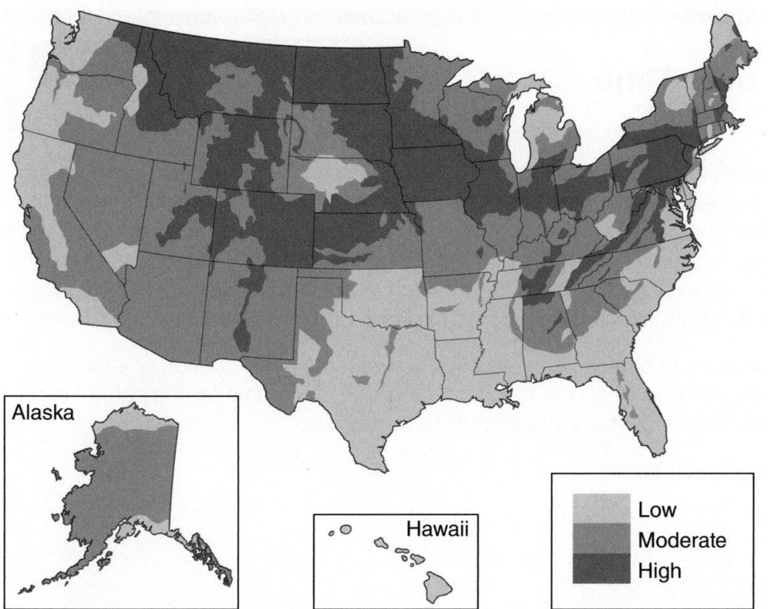

FIGURE 50.3 Potential radon exposure in the United States. Depending on the underlying bedrock and soils, the potential for exposure to radon exists in houses in certain parts of the United States. *(Data from U.S. Geological Survey and http://www.epa.gove/radon/zonemap.html.)*

Asbestos

Asbestos is a long, thin, fibrous silicate mineral with insulating properties. For many years it was used as an insulator on steam and hot water pipes and in shingles for the siding of buildings. The greatest health risks from asbestos have been respiratory diseases such as asbestosis and lung cancer found at very high rates among those who have mined asbestos. In manufactured form, asbestos is relatively stable and not dangerous until it is disturbed. When insulating materials become old or are damaged or disrupted, however, the fine fibers can become airborne and can enter the respiratory tract. In the United States, asbestos is no longer used as an insulating material, but it can still be found in older buildings, including schools. Removal of asbestos insulation must be done under tightly controlled conditions so that the fibers, typically less than 10 microns in diameter, cannot enter the air inside the building. Some studies have shown that when asbestos removal is complete, the concentration of asbestos in the air of the remediated building can be greater in the year after removal than during the year before removal. For this reason, it is absolutely necessary that asbestos removal be carefully done by qualified asbestos abatement personnel.

Radon

Radon-222, a radioactive gas that occurs naturally from the decay of uranium, exists in granitic and some other rocks and soils in many parts of the world. The map in **FIGURE 50.3** shows areas for potential radon exposure in the United States. Humans can receive significant exposure to radon if it seeps into a home through cracks in the foundation, or from underlying rock, soil, or groundwater. Radon-222 decays within 4 days to a radioactive daughter product, polonium-210. Either the radon or the polonium can attach to dust and other particles in the air and then be inhaled by the inhabitants of the home.

The EPA, the federal agency most responsible for identifying, measuring, and addressing environmental risks, estimates that about 21,000 people die each year from radon-induced lung cancer. This is 14 percent of yearly lung cancer deaths, and makes radon the second leading cause of lung cancer, after smoking. The EPA suggests that people test their homes for airborne radon. If radon air concentrations are determined to be high, it is important to increase ventilation in the home and ventilate the underground space below the house. Other relatively inexpensive actions, such as sealing cracks in the basement, can be beneficial if radon is coming from underlying soil and bedrock. "Do The Math: Averaging Radon over Time" on page 564 asks you to calculate radon concentrations in a home based on the readings of two radon collection devices.

VOCs in Home Products

Many volatile organic compounds are used in building materials, furniture, and other home products such as glues and paints. One of the most toxic of these compounds is formaldehyde, which is used widely to manufacture a variety of building products such as particle board and carpeting glue. Formaldehyde is common in new homes and new products made from pressed wood, such as certain bookcases. Bookcases made from actual wood and most hardwood flooring do not contain formaldehyde. The pungent smell that you may have noticed in a new home or one with new carpeting comes from formaldehyde, which is volatile and emits gases over time.

Asbestos A long, thin, fibrous silicate mineral with insulating properties, which can cause cancer when inhaled.

DO THE MATH Averaging Radon over Time

Preparing for the AP® Exam

Since radon in homes is believed to cause approximately 21,000 deaths in the United States each year, homeowners are encouraged to measure radon concentrations in their homes at least once. There are short-term (for example, 4 days) and long-term (up to 1 year) radon collection devices available. Longer measurements times are likely to produce more accurate readings. After the designated collection time period, the device is mailed to a laboratory and the air concentration of radon is estimated from the collection device. The EPA has established an action level for home radon: If concentrations of 4.0 pCi/L (picoCuries per liter) or greater are found, the homeowner should install abatement equipment to lower radon concentrations in the home.

For one brand of short-term collection device, the homeowner should place two devices within 1 meter (3 feet) of each other and leave them undisturbed for 4 days. These devices are monitoring the same space at the same time for the same length of time. The concentration of radon in the home is determined by the average of the two devices.

As an example, imagine two devices with the following readings:

Device 1 reading: 5 pCi/L

Device 2 reading: 2 pCi/L

What is the concentration of radon in the home?

The average of 5 pCi/L and 2 pCi/L is

$$(5 \text{ pCi/L} + 2 \text{ pCi/L} = 7 \text{ pCi/L}) \div 2 = 3.5 \text{ pCi/L}$$

3.5 pCi/L is less than 4.0 pCi/L so radon abatement is not recommended by the EPA.

YOUR TURN A homeowner uses two different short-term radon collection devices to measure the same space. One device measured the air for 4 days and the result was 1 pCi/L. The other device measured the air for 8 days and the result was 6 pCi/L. Calculate an estimate of the radon concentrations in the home, taking into consideration the different measurement periods.

A high enough concentration in a confined space can cause a burning sensation in the eyes and throat, and breathing difficulties and asthma in some people. There is evidence that people develop a sensitivity to formaldehyde over time; though they may not be very sensitive at first, with continued exposure they can experience irritation from increasingly smaller exposures. Formaldehyde has been shown to cause cancer in laboratory animals and has recently been suspected of being a human carcinogen.

Many other consumer products such as detergents, dry-cleaning fluids, deodorizers, and solvents may contain VOCs and can be harmful if inhaled. Plastics, fabrics, paints, construction materials, and synthetic carpets may also release VOCs over time. In response to concerns over manufactured products in the home, some homeowners choose to utilize wood flooring rather than carpeting, or natural fiber carpeting rather than synthetics. While such actions will not ensure zero exposure to VOCs, it will decrease the likelihood of exposure.

Sick Building Syndrome

In newer buildings in developed countries in the temperate zone, more and more attention is being given to insulation and the prevention of air leaks in order to reduce the amount of heating or cooling necessary for a comfortable existence. This efficiency improvement reduces energy use but may have the unintended side effect of allowing the buildup of toxic compounds and pollutants in an airtight space. In fact, such a phenomenon has been observed often enough in new or renovated buildings to be given a name: **sick building syndrome**, which describes a buildup of toxic pollutants in airtight spaces such as newer buildings. Because new buildings contain many products made with synthetic materials and glues that may not have fully dried out, a significant amount of off-gassing occurs, which usually means that the indoor levels of VOCs, hydrocarbons, and other potentially toxic materials are quite high. Sick building syndrome has been observed particularly in office buildings, where large numbers of workers have reported a variety of maladies such as headaches, nausea, throat or eye irritations, and fatigue.

The EPA has identified four specific reasons for sick building syndrome: inadequate or faulty ventilation; chemical contamination from indoor sources such as glues, carpeting, furniture, cleaning agents, and copy machines; chemical contamination in the building from outdoor sources such as vehicle exhaust transferred through building air intakes; and biological contamination from inside or outside, such as molds and pollen.

Sick building syndrome A buildup of toxic pollutants in an airtight space, seen in newer buildings.

MODULE **50** AP® Review

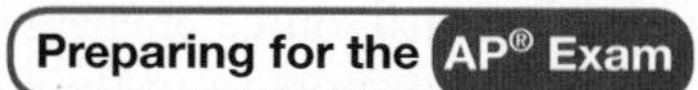

In this module, we have seen that indoor air pollution is a problem around the world, but that its manifestations in the developing and the developed worlds differ. Carbon monoxide from combustion for cooking is the biggest problem in the developing world. In the developed world, houses sealed up for greater energy efficiency are often the causes of exposure to radon, carbon monoxide, and volatile organic compounds for people who spend substantial amounts of time indoors.

AP® Practice Questions

Choose the best answer for the following.

1. What percentage of worldwide deaths due to indoor air pollution occurs in developing nations?
   (a) 50 percent
   (b) 60 percent
   (c) 70 percent
   (d) 90 percent

2. Which outdoor air pollutant is also a significant indoor air pollutant?
   (a) sulfur dioxide
   (b) nitrogen oxides
   (c) carbon monoxide
   (d) lead

3. The primary source of radon is
   (a) electronics.
   (b) indoor fires.
   (c) household chemical fumes.
   (d) rocks and soils.

4. Sick building syndrome
   (a) occurs most often in old buildings.
   (b) is a primary cause of lung cancer.
   (c) is a result of off-gassing.
   (d) results from too much ventilation.

5. Asbestos
   (a) causes respiratory ailments and cancer.
   (b) can be easily removed and treated.
   (c) can be a problem in new construction.
   (d) causes skin irritation, nausea, and fatigue.

Working Toward Sustainability

A New Cook Stove Design

In China, India, and sub-Saharan Africa, people in 80 to 90 percent of households cook food using wood, animal manure, and crop residues as their fuel. Since women do most of the cooking, and young children are with the women of the household for much of the time, it is the women and young children who receive the greatest exposure to carbon monoxide and particulate matter. When biomass is used for cooking, concentrations of particulate matter in the home can be 200 times higher than the exposure limits recommended by the U.S. EPA. A wide range of diseases has been associated with exposure to smoke from cooking. Earlier in this chapter, we described that indoor air pollution is responsible for 4 million deaths annually around the world, and indoor cooking is a major source of indoor air pollution.

There are hundreds of projects underway around the world to enable women to use more efficient cooking stoves, ventilate cooking areas, cook outside whenever possible, and change customs and practices that will reduce their exposure to indoor air pollution. The use of an efficient cook stove will have the added benefit of consuming less fuel. This improves air quality and reduces the amount of fuel needed, which reduces the amount of nearby vegetation that must be collected and also reduces the amount of time that a woman must spend searching for fuel.

Increasing the efficiency of the combustion process requires the proper mix of fuel and oxygen. One effective method of ensuring a cleaner burn is the use of a small fan to facilitate greater oxygen delivery. However, because most homes in developing countries with significant indoor air pollution problems do not have access to electricity, some sort of internal source of energy for the fan is needed.

Two innovators from the United States developed a cook stove for backpackers and other outdoor enthusiasts who needed to cook a hot meal with little impact on the environment. The Biolite CampStove, which works by burning solid biomass and weighs only 935 grams (2.06 pounds), needs no gasoline or batteries—a desirable feature for people carrying all their belongings on their backs. The innovators managed to generate the electricity by adding a small semiconductor that generates electricity from the heat of the stove.

The inventors of the CampStove soon realized that the same technology could make an important contribution in the developing world. For home use in the developing world, they created the Biolite HomeStove, which weighs 8 kg (17.64 lbs). Like the CampStove, it physically separates the solid fuel from the gases that form when the fuel is burned, which allows the stove to burn the gases. In addition, a small electric fan, located inside the stove, harnesses energy from the heat of the fire and moves air through the stove at a rate that ensures complete combustion. The result is a more efficient burn, less fuel use, and less release of carbon monoxide and particulate matter. The stove generates enough electricity to power the fan and has additional capacity to power a small USB port suitable for charging a cell phone or flashlight.

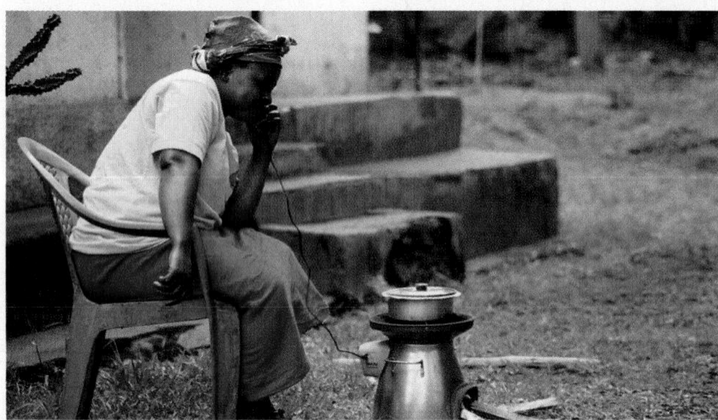

BioLite HomeStove. This small stove and others like it have the potential to reduce the amount of firewood needed to cook a meal, as well as lower the amount of indoor air pollution emitted. This woman in a village in Kenya is also using the stove to charge her cell phone. *(Courtesy BioLite Inc.)*

The Biolite CampStove stove is readily available on consumer goods websites; the Biolite HomeStove is sold in Kenya and Uganda. Biolite stoves won awards in 2009, 2012, and 2014 for low emissions, innovation, and promoting social good. There are many other small stoves under development and in use around the world that increase combustion and reduce carbon monoxide and particulate emissions. The challenge is to make them affordable and available so that large numbers of people in the developing world will use them. Other promising ways to reduce fuel use and improve indoor air quality include the solar cooker shown in Figure 39.2 on page 470.

Critical Thinking Questions

1. Why are women and children exposed more often to indoor air pollution than men in developing countries?
2. How can technology offer solutions to cooking over open fires?

References

Bilger, B. 2009. Annals of Invention, Hearth Surgery, The New Yorker, December 21, p. 84; https://www.newyorker.com/magazine/2009/12/21/hearth-surgery

https://www.bioliteenergy.com/, homepage of BioLite stove.

CHAPTER 15 Review

In this chapter, we examined the major air pollutants and their natural and anthropogenic sources. We found that photochemical smog and acidic deposition are two air pollution problems that have had different outcomes, at least for now. Smog is still a problem in many locations around the world while acidic deposition has become less of a problem in North America and Europe. There are a variety of measures for controlling air pollution including pollution prevention and devices that remove pollutants from smokestacks before it is released into the atmosphere. Stratospheric ozone depletion has occurred because of the release of chlorofluorocarbons (CFCs) from refrigeration and air-conditioning units. Due to an international agreement, the Montreal Protocol on Substances That Deplete the Ozone Layer, there was a significant reduction in the use of CFCs and stratospheric ozone depletion has been reduced. Indoor air pollution is a problem that occurs around the world, although with causes and pollutants that differ between developing and developed countries.

Key Terms

Air pollution
Particulate matter (PM)
Particulates
Particles
Haze
Photochemical oxidant
Ozone (O_3)
Smog
Photochemical smog
Los Angeles–type smog
Brown smog
Sulfurous smog
London-type smog
Gray smog
Industrial smog
Volatile organic compound (VOC)
Primary pollutant
Secondary pollutant
Thermal inversion
Inversion layer
Asbestos
Sick building syndrome

Learning Goals Revisited

Module 46 Major Air Pollutants and Their Sources

Identify and describe the major air pollutants.

Sulfur oxides, nitrogen oxides, carbon monoxide, carbon dioxide, particulates, and ozone are some of the major ground-level air pollutants. Carbon dioxide was not originally considered one of the major pollutants but, because it alters ecosystems, it is now considered an air pollutant.

Describe the sources of air pollution.

Air pollution comes from both natural and human sources. Human activities that release these pollutants or their precursors include transportation, generation of electricity, space heating, and industrial processes.

Module 47 Photochemical Smog and Acid Rain

Explain how photochemical smog forms and why it is still a problem in the United States.

Smog forms when sunlight, nitrogen oxides, and volatile organic compounds are present. The secondary pollutant ozone is a major component of photochemical smog. Sulfur is the dominant ingredient in sulfurous smog. Smog impairs respiratory function in human beings. Because of an abundance of both nitrogen oxides and volatile organic compounds, smog occurs during daylight hours in many parts of the world, including the United States.

Describe how acid deposition forms and why it has improved in the United States and become worse elsewhere.

Acidic deposition, which is composed of hydrogen, sulfate, and nitrate ions, forms from both sulfur dioxide and nitrogen oxides. Acid deposition is harmful to aquatic organisms and can reduce forest productivity in sensitive ecosystems. Due to reductions in sulfur emissions in the United States, acidic deposition is much less of a problem than it used to be. Asia is a location today where acidic deposition is an environmental problem.

Module 48 Pollution Control Measures

Explain strategies and techniques for controlling sulfur dioxide, nitrogen oxides, and particulate matter.

Air pollution is best controlled by increasing the efficiency of processes that cause pollution, thereby reducing emissions, or by removing pollutants from fuel before combustion. After combustion occurs, filters and scrubbers remove pollutants from the exhaust stream before they can be released into the environment. The use of filters and scrubbers is preferable to trying to remove pollutants after they have been distributed throughout the environment.

Describe innovative pollution control measures.

Some innovative approaches to pollution control include targeting specific sources of certain kinds of pollution. Allowing for the buying and selling of permits to emit sulfur has been an effective means of reducing sulfur dioxide emissions and subsequent acidic deposition.

Module 49 Stratospheric Ozone Depletion

Explain the benefits of stratospheric ozone and how it forms.

Although ozone is the same gas that is a component of photochemical smog in the troposphere, in the stratosphere it is an important gas that absorbs harmful ultraviolet radiation. Its presence allows for plant and animal life to exist on the surface of Earth.

Describe the depletion of stratospheric ozone.

Chlorine-containing compounds such as CFCs that were part of refrigeration and air-conditioning systems led to a reduction in stratospheric ozone. When the chlorine in a CFC attaches to one of the oxygen atoms in an ozone molecule, it breaks apart the ozone molecule, which causes a reduction in ozone concentration.

Explain efforts to reduce ozone depletion.

International efforts to reduce CFC emissions have helped stratospheric ozone concentrations to recover.

The Montreal Protocol on Substances that Deplete the Ozone Layer is regarded as one of the most successful international agreements of modern times.

Module 50 Indoor Air Pollution

Explain how indoor air pollution differs in developing and developed countries.

Indoor air pollution is a different kind of problem in developing and in developed countries. Cooking over open fires with inadequate ventilation exposes occupants to carbon monoxide in the developing world. Relatively well-insulated living spaces in the developed world expose occupants to higher concentrations of carbon monoxide, radon, and certain volatile organic compounds.

Describe the major indoor air pollutants and the risks associated with them.

Carbon monoxide and particulates are the most harmful aspects of indoor air pollution and account for 4 million deaths annually worldwide. Ninety percent of those deaths occur in the developing world. Cooking over open fires in the developing world exposes women and children to particulate matter and carbon monoxide pollution.

Practice Math and Graphing

Preparing for the AP® Exam

1. Practice Math

As Module 50 discussed, radon in homes is believed to cause approximately 21,000 deaths in the United States each year. The EPA has established an action level for home radon: If concentration of 4.0 pCi/L (picoCuries per liter) or greater are found, the homeowner should install abatement equipment to lower radon concentrations.

(a) Use the information from five homes presented below to determine the average radon concentration and a maximum concentration found in each home. Complete the table with your calculations.

(b) For each row of the table, determine if the house should have radon abatement based on

i. if the average is the most important indicator.

ii. if the maximum is the most important indicator.

Home radon concentration	First reading pCi/L	Second reading pCi/L	Third reading pCi/L	Average	Max	Abatement yes or no if average most important	Abatement yes or no if maximum most important
1	3.2	3.5	3.9				
2	3.8	3.5	4.9				
3	4.5	6.5	12.7				
4	5.5	3.3	1.9				
5	9.6	4.4	8.3				

2. Practice Graphing

The table lists $PM_{2.5}$ air concentrations in Beijing, China, from 2008 through 2014. Plot the data on a graph with year on the *x* axis and $PM_{2.5}$ concentrations on the *y* axis.

Year	$PM_{2.5}$ air concentration $\mu g/m^3$
2008	82
2009	101
2010	103
2011	99
2012	91
2013	100
2014	133

Section 1: Multiple-Choice Questions

Choose the best answer for questions 1–19.

Questions 1–4 refer to the selections (a)–(d) below. Match the lettered item with the numbered descriptor.

(a) CO
(b) NO_2
(c) SO_2
(d) PM_{10}

1. a pungent reddish-brown gas often associated with photochemical smog

2. a corrosive gas from burning coal often associated with industrial smog

3. a dangerous indoor air pollutant

4. emitted from both diesel and burning wood

5. All of the following are examples of primary air pollutants except
(a) sulfur dioxide.
(b) carbon dioxide.
(c) tropospheric ozone.
(d) nitrogen oxide.

6. The greatest emission of sulfur dioxide comes from
(a) on-road vehicles.
(b) biofuels.
(c) industrial processes.
(d) power plants.

7. The largest amount of nitrogen oxide emissions comes from
(a) vehicles.
(b) fossil fuel combustion.
(c) industrial processes.
(d) electricity generation.

8. The accumulation of tropospheric ozone during the middle of the day depends mainly upon the atmospheric concentration of nitrogen oxides and
(a) carbon dioxide.
(b) volatile organic compounds.
(c) chlorofluorocarbons.
(d) sulfates and nitrates.

9. Air pollution generally refers to pollution in the
(a) stratosphere.
(b) troposphere.
(c) lithosphere.
(d) mesosphere.

10. The effects of acid deposition include all of the following EXCEPT
(a) mobilization of metal ions from the soil into surface water.
(b) increased numbers of salamanders in ponds and streams.
(c) reduced food sources for aquatic organisms.
(d) erosion of marble buildings and statues.

11. The World Health Organization estimates that 4 million deaths from indoor air pollution are mostly due to:
(a) developed world inner-city poor air quality.
(b) developing world cooking exposure.
(c) workers exposed in office buildings.
(d) developing world workers in the automotive industry.

12. Two major factors involved in the conversion of primary pollutants into secondary pollutants are
(a) sunlight and water.
(b) sulfates and sunlight.
(c) water and volatile organics.
(d) nitrogen oxides and sulfates.

13. The pollutant least likely to be emitted from a smokestack would be
(a) carbon monoxide.
(b) carbon dioxide.
(c) ozone.
(d) sulfur dioxide.

14. All of the following are sources of particulate matter EXCEPT
(a) combustion of biofuels.
(b) incomplete natural gas combustion.
(c) road construction.
(d) volcanoes.

15. The EPA identifies all of the following as reasons for sick building syndrome EXCEPT
(a) faulty ventilation systems.
(b) emissions from carpets and furniture.
(c) contamination from molds and pollen.
(d) high levels of radon in the basement.

16. Which statement regarding the decreased concentrations of stratospheric ozone is correct?
(a) Increased photosynthetic activity has been measured in phytoplankton around Antarctica.
(b) Significant increases in skin cancers have already occurred.
(c) Although the Montreal Protocol led to a reduction in the use of CFCs, it will have little effect on stratospheric ozone concentrations in the long term.
(d) There is no correlation between the incidence of suppressed immune systems and the lower concentrations of stratospheric ozone.

17. Which of these acts is a catalyst and repeatedly breaks down ozone molecules?
    (a) bromine
    (b) chlorine
    (c) fluorine
    (d) serpentine
18. Pollutants can be categorized as primary or secondary. Which pairing is NOT correct?
    (a) secondary: ozone
    (b) primary: sulfate
    (c) secondary: nitrate
    (d) primary: carbon monoxide
19. A thermal inversion
    (a) occurs when cool and warm air are intermixed throughout the year.
    (b) occurs when air pollution from China reaches the West Coast of the United States.
    (c) occurs when cool air is present through the troposphere.
    (d) occurs when a warm air layer overlies a cooler layer.

Section 2: Practice Free-Response Questions

Write your answer to each part clearly. Support your answers with relevant information and examples. Where calculations are required, show your work.

1. The table below shows the ambient air data collected for Pittsburgh, Pennsylvania. Examine the data and answer the following questions.
    (a) Based on the National Ambient Air Quality Standards (NAAQS), average ozone concentrations are not to exceed 0.075 ppm (75 ppb) in any 8-hour period. In 2008, was Pittsburgh in compliance with this standard? Discuss how this NAAQS may not truly reflect the overall air quality. (2 points)
    (b) Ozone is classified as a secondary pollutant. Identify the primary pollutants necessary for its formation and describe how tropospheric ozone is formed. (2 points)
    (c) Identify two relationships between the data presented. Apply these relationships to your answer in (b) to explain the pattern of ozone concentrations in Pittsburgh. (4 points)
    (d) Explain how the same ozone that is harmful in the troposphere is beneficial in the stratosphere. (2 points)

2008 Monthly ambient air monitoring report, Pittsburgh

Month	Monthly maximum ozone levels (ppb)	Monthly average ozone levels (ppb)	Monthly average solar radiation (watts/m^3)
January	37	14	65
February	49	15	63
March	56	23	86
April	76	31	81
May	75	27	152
June	77	32	208
July	95	31	215
August	92	27	204
September	89	20	153
October	48	14	109
November	57	12	64
December	30	14	45

Source: http://www.ahs.dep.pa.gov/aq_apps/aadata/

2. Read the following story and answer the questions below.

Sick Building Syndrome Can be Remedied by Ultraviolet Light

Sick building syndrome (SBS) can occur when bacteria builds up in the ventilation systems of office buildings. In some cases, office workers experience a variety of symptoms including respiratory distress, headaches, sore throats, congestion, and eye irritation.

Researchers tested UVGI (ultraviolet germicidal irradiation) in the ventilation systems of three buildings where office workers suffered from SBS symptoms. UVGI kills the bacteria and molds in ventilation systems. The researchers compared symptoms of the office workers when the UVGI system was on and when it was off.

The UVGI was associated with a 20 percent reduction in all symptoms. Respiratory problems dropped by 40 percent and nasal congestion dropped by 30 percent. Muscular complaints were reduced by 50 percent. Installing the systems is relatively inexpensive and can increase worker health and productivity.

(a) According to the story, what is the cause of sick building syndrome and what are the advantages of the solution they tested? (3 points)
(b) What other sources of indoor pollution contribute to sick building syndrome? Will ultraviolet light be effective against these sources? Suggest one additional control measure that would be effective against SBS. (4 points)
(c) Will the solution mentioned in the story be effective against indoor pollution in developing countries? Discuss why or why not. (3 points)

3. Temperature can affect air pollution in a variety of ways.
(a) Describe how a thermal inversion occurs and how it influences air pollution. (3 points)
(b) Describe how smog is created and name TWO ways that atmospheric temperature influences its formation. (4 points)
(c) Describe how cold weather conditions contribute to the formation of the "ozone hole" over Antarctica. (3 points)

This earthmover is preparing solid waste for compaction. *(Stephen Wilkes/Getty Images)*

Waste Generation, Terrestrial Pollution, and Waste Disposal

CHAPTER 16

CASE STUDY

Paper or Plastic?

Polystyrene is a plastic polymer that has high insulation value. More commonly known by its trade name, Styrofoam, it is particularly useful for food packaging because it minimizes temperature changes in both food and beverages. Polystyrene is lighter, insulates better, and is less expensive than the alternatives. However a number of years ago, polystyrene was deemed harmful to the environment because, like all plastics, it is made from petroleum and because it does not decompose in landfills. In response to public sentiment, most food businesses greatly reduced or eliminated their use of polystyrene. All over the country, schools, businesses, and public institutions began purging their cafeterias of polystyrene cups and most have replaced them with disposable paper cups. This process was still going on in 2018 when a major doughnut, coffee, and baked goods restaurant in the United States announced that it would phase out polystyrene by 2020.

But is the elimination of polystyrene actually an environmental victory? It is hard to quantify the exact environmental benefits and costs of using a paper cup versus using a polystyrene cup. For example, because a paper cup does not insulate as well as a Styrofoam cup, paper cups filled with hot drinks are usually too hot to hold and vendors often wrap them in a cardboard band that becomes additional waste. To fully quantify the environmental costs and benefits of each type of cup, one must create a list of inputs and outputs related to their manufacture, use, and disposal. When we make a list of all the materials and all the energy required to produce and then dispose of each type of cup, we find that it is not easy to determine which choice is better for the environment.

> Today, there is still no definitive answer as to whether the paper cup or the polystyrene cup causes less harm to the environment.

One study found that making a paper cup requires approximately 2 grams of petroleum along with 33 grams of wood and bark, which are renewable materials. A polystyrene cup requires 3 grams of petroleum, a nonrenewable material, but no wood or bark. Manufacturing the paper cup requires about twice as much energy, and much more water. A paper cup of the exact same size as a polystyrene cup is substantially heavier, which means it requires more energy to transport a paper cup to the location where it will be used. Air emissions are different in the manufacturing of each type of cup and it is difficult to say which emissions are more harmful to the environment. Since more energy is needed to make and transport a paper cup, it is reasonable to assume that using it generates more air pollution. A paper cup is normally used once or at most a few times while the polystyrene cup can, at least in theory, be reused many times. However, both types of cups are usually thrown away after one use. There has been concern among some scientists—but no consensus—that a polystyrene cup might leach chemicals from the plastic into the coffee; if this is true, using a paper cup could pose less risk to human health. However, without proper disposal, the bleach used to make paper cups in a paper mill, along with small amounts of the associated by-product, dioxin, can cause harm to aquatic life when the water is discharged into rivers and streams. Incineration of both types of cup could yield a small amount of energy. In a landfill, the paper cup will degrade and eventually produce methane gas, while the polystyrene cup, because it is made of an inert material, will remain there for a very long time.

Weighing these and other factors, one study concluded that a polystyrene cup is more desirable than a paper cup for one-time use. However, critics of that study felt that the author did not account for all relevant factors, including *toxicity*. **Toxicity** is harm, illness, or death caused by chemical means through ingestion, inhalation, or absorption. The production of polystyrene exposes workers to toxic emissions. Critics also felt the study did not consider the impact of both paper and polystyrene cups on global carbon dioxide emissions, or the possibility of making the cup from materials other than petroleum or paper. Today, there is still no definitive answer as to whether the paper cup or the polystyrene cup causes less harm to the environment.

There is widespread agreement that paper and Styrofoam are not the only options. Reusable mugs are a possibility but they would require consideration of a host of different issues such as greater inputs for manufacturing, and energy consumption, as well as the water and other resources needed to clean them after each use.

These types of studies illustrate that analyzing the environmental effects of the products we use is complex since it involves the synthesis of many aspects of environmental studies. Not only does it include science, ethics, and social judgments, it also necessitates a systems-based understanding of waste generation, waste reduction, and waste disposal.

Sources: M. B. Hocking, Paper versus polystyrene: A complex choice, *Science* 251 (1991): 504–505, DOI: 10.1126/science.251.4993.504; E. van der Harst, J. Potting, and C. Kroeze, Comparison of different methods to include recycling in LCAs of aluminum cans and disposable polystyrene cups, *Waste Management* 48 (2016): 565–583, 10.1016/j.wasman.2015.09.027.

As life in many countries has become increasingly dependent on disposable items, the generation of solid waste has become more of a problem for both the natural and human environments. In this chapter, we examine solid waste, something that only humans generate. We examine methods of reducing waste and ways of recycling. Then we describe the principal methods of getting rid of waste: landfills and incineration. We also consider the most toxic forms of waste: hazardous waste. We conclude the chapter with a discussion of innovative ways to think about solid waste from a systems perspective.

Toxicity Harm, illness, or death caused by chemical means through ingestion, inhalation, or absorption.

MODULE 51

Only Humans Generate Waste

Throughout this book, we have examined systems in terms of inputs, outputs, and internal changes. We will begin our discussion of solid waste in the same way—by considering the inputs and outputs of materials that end up becoming solid waste. We will define solid waste and examine the contents of the waste stream and waste generation trends. We will conclude the module with an examination of electronic waste and the specific challenges that it raises.

Learning Goals

After reading this module, you should be able to

- explain why we generate waste and describe recent waste disposal trends.
- describe the content of the solid waste stream in the United States.

Humans generate waste that other organisms cannot use

Humans are the only organism that produces waste others cannot use. To explore this further, we need to learn why materials generated by humans become waste and what that waste contains. Although this seems like a simple question, it touches upon recent U.S. history, human behavior, and many other topics, including some that go beyond environmental science. We need to establish that waste can be viewed as a system, just like other materials. And it is necessary to describe how we got here—what features of U.S. society allowed us to generate the quantities of waste that we do.

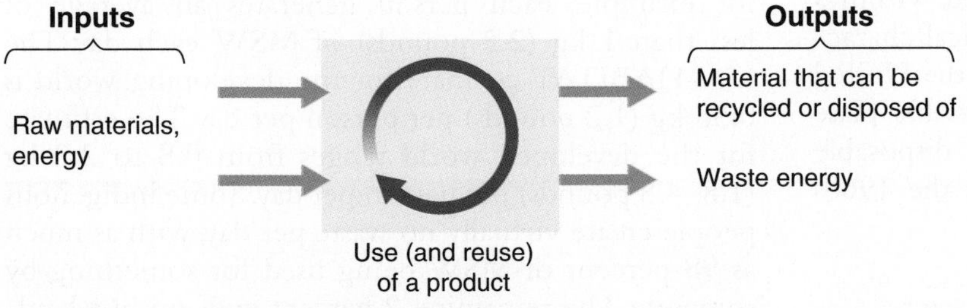

FIGURE 51.1 The solid waste system. Waste is a component of a human-dominated system in which products are manufactured, used, and eventually disposed of (arrows are not proportional). At least some of the waste of this system may become the input of another system.

Waste as a System

In an ecological system, plant materials, nutrients, water, and energy are the inputs. In a human system, inputs are very similar but contain materials manufactured by humans as well as natural materials. Within this system, humans use these inputs and materials to produce goods. And, as in any system, outputs are generated. We call these outputs **waste**, which is defined as material outputs that are not useful or are not consumed. Energy waste is also an output. **FIGURE 51.1** shows a diagram of the relationship between inputs and outputs in a human system.

If waste is the nonuseful output of a system, how do we determine what is useful? The detritivores we described in Chapter 3 recycle waste from animals and plants; they use the energy and nutrients they obtain and turn the remainder into compost or humus that nourishes other organisms. Dung beetles, for example, live on the energy and nutrients contained within the dung of elephants and other animals; in the natural world, this is not waste, it is food (**FIGURE 51.2** on page 576). Even humans make use of animal waste—for fertilizer, heat, and cooking fuel. In most situations, the waste of one organism becomes a source of energy for another.

Waste Material outputs from a system that are not useful or consumed.

FIGURE 51.2 A dung beetle. This dung beetle is using elephant waste as a resource. The waste of most organisms in the natural world ends up being a resource for other organisms. *(Michael Potter11/Shutterstock)*

The Throw-Away Society

Until a society becomes relatively wealthy, it generates little waste. Every object that no longer has value for its original purpose becomes useful for another purpose. In 1900 in the United States, virtually all metal, wood, and glass materials were recycled, although no one called it recycling back then. Those who collected recyclables were called junk dealers, or scrap metal dealers. For example, if a wooden bookcase broke and it could not be repaired, the pieces could be used to make a step stool. When the step stool broke, the wood was burned in a wood stove to heat the house. After World War II and with the rapid population growth that occurred in the United States, consumption patterns changed. The increasing industrialization and wealth of the United States, as well as cultural changes, made it possible for people to purchase household conveniences that could be used and then thrown away. Families were large, and people were urged to buy "labor-saving" household appliances and to dispose of them as soon as a new model was available. **Planned obsolescence** is the process of designing a product so that it will need to be replaced within a few years. Planned obsolesce became a typical characteristic of many products manufactured in the United States, from toasters to automobiles. Disposable plate "TV dinners," throw-away napkins, and disposable plates and forks also became common. In the 1960s disposable diapers became widely available and eventually replaced reusable cloth diapers. The components of household materials also changed. Objects were more likely to contain mixtures of different materials, which makes them harder to use for another purpose and difficult to recycle. The United States became the leader of what came to be known as the "throw-away society."

Planned obsolescence The process of designing a product so that it will need to be replaced within a few years.

Municipal solid waste (MSW) Refuse collected by municipalities from households, small businesses, and institutions.

Refuse collected by municipalities from households, small businesses, and institutions such as schools, prisons, municipal buildings, and hospitals is known as **municipal solid waste (MSW)**. The Environmental Protection Agency (EPA) estimates that approximately 60 percent of MSW comes from residences and 40 percent from commercial and institutional facilities. Other kinds of waste generated in the United States in addition to MSW include agricultural waste, mining waste, and industrial waste. Waste other than MSW is typically deposited and processed on-site rather than transferred to a different location for disposal. Although some of these other categories generate a much greater percentage of yearly total solid waste, this chapter focuses on MSW.

FIGURE 51.3 shows the trend toward greater generation of MSW both overall and on a per capita basis from 1960 to 2014. In the first 47 years of this period, the total amount of MSW generated in the United States increased from 80 million metric tons (88 million U.S. tons) to 227 million metric tons (250 million U.S. tons) per year. In the last several years for which there are data, the total amount of MSW has leveled off or in some years decreased by a small amount. The increase for all but the last several years can be explained in part because of population growth and in part because individuals have been generating increasing amounts of MSW. In 2014, average waste generation was 2.0 kg (4.4 pounds) of MSW per person per day. Waste generation varies by season of the year, socioeconomic status of the individual generating the waste, and even geographic location within the country.

Waste generation in much of the rest of the world stands in contrast to the United States. In Japan, for example, each person generates an average of less than 1 kg (2.2 pounds) of MSW each day. The UN-HABITAT estimate for the developing world is 0.55 kg (1.2 pounds) per person per day. The estimate for the developed world ranges from 0.8 to 2.2 kg (1.8–4.8 pounds) per person per day. Some indigenous people create virtually no waste per day, with as much as 98 percent of MSW being used for something by someone. The remaining 2 percent ends up in a landfill or waste pile. Even there, impoverished people scavenge and reuse some of the discarded material (**FIGURE 51.4**).

Developing countries have become responsible for a greater portion of global MSW, in part because of their

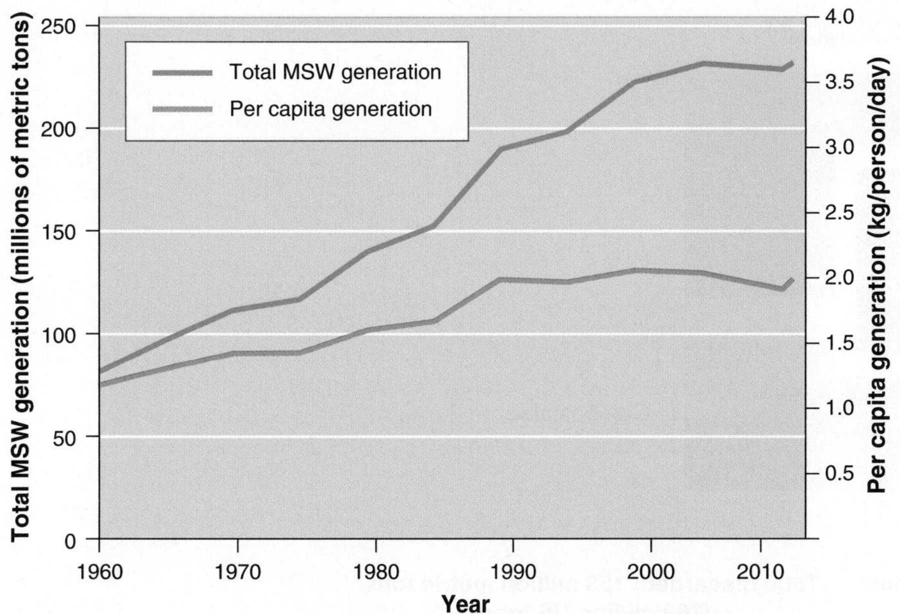

FIGURE 51.3 Municipal solid waste generation in the United States, 1960–2014. Total MSW generation and per capita MSW generation had been increasing from 1960 through 2008. They have recently started to level off. *(Data from Advancing Sustainable Materials Management: 2014 Fact Sheet, U.S. EPA, 2016.)*

growing populations. In addition, as developing countries produce more of the goods used in the developed world, they generate more waste in the production process for these goods. For example, computers sold to consumers in the United States are assembled in such places as Taiwan, Singapore, and China, and the waste products generated are disposed of at the manufacturing location.

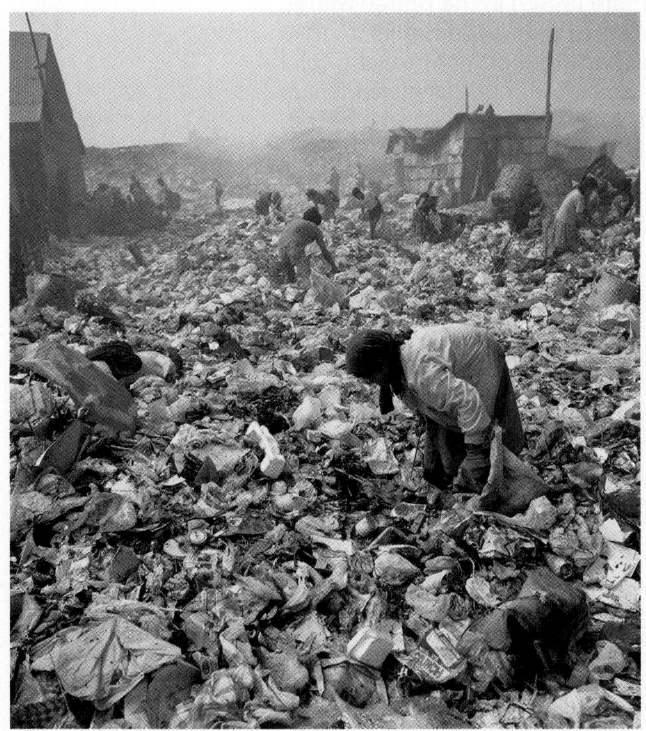

FIGURE 51.4 A large dump in Manila, Philippines. Throughout the world, impoverished people scavenge dumps. *(Stockbyte/Getty Images)*

The solid waste stream contains materials from many sources

We have seen that MSW is made up of the things we use and then throw away. The goods that we use are generally a combination of organic items, fibers, metals, and plastics made from petroleum. A certain amount of waste is generated during any manufacturing process. Waste is also generated from the packaging and transporting of goods.

Depending on the particular materials, products, and goods that consumers use, such items can remain in the consumer-use system for a long time. For example, a ceramic plate or drinking mug might last for 5 to 10 years. In most cases, a disposable paper cup leaves the system within minutes or hours after it is used. Ultimately, all products wear out, lose their value, or are discarded. At this point they enter the **waste stream**—the flow of solid waste that is recycled, incinerated, placed in a solid waste landfill, or disposed of in another way. In this section we will look at the composition of MSW and then explore e-waste in more detail.

Composition of Municipal Solid Waste

FIGURE 51.5a on page 578 shows the data for MSW composition in the United States in 2014 by category before any recycling has occurred. The category "paper," which

Waste stream The flow of solid waste that is recycled, incinerated, placed in a solid waste landfill, or disposed of in another way.

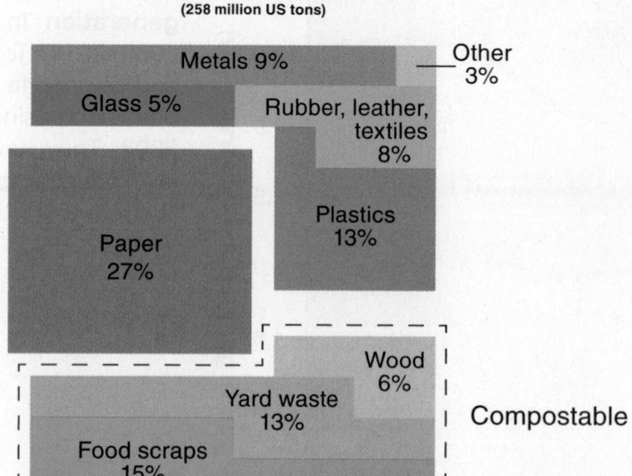

(a) Original waste stream

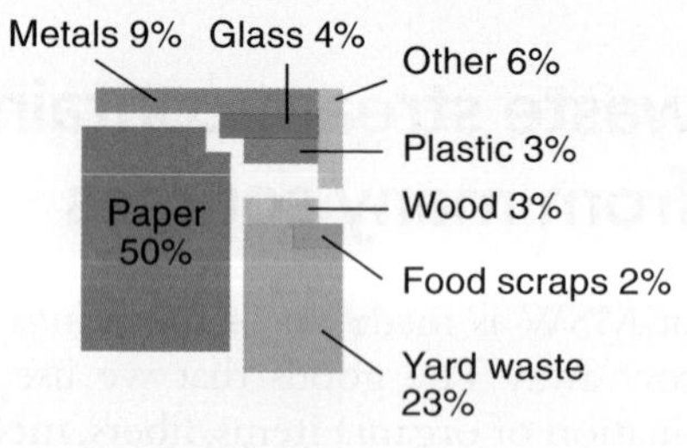

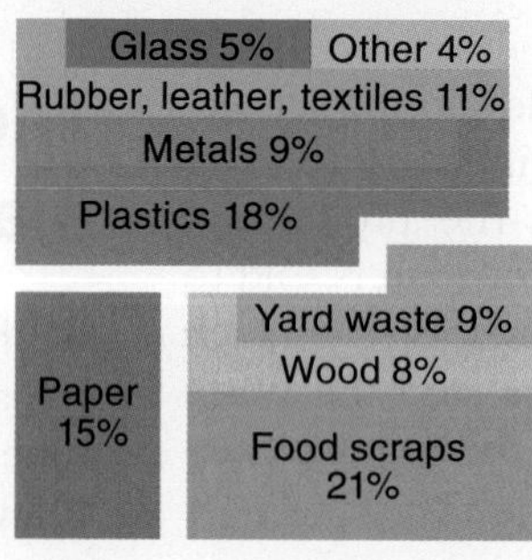

(b) After recovery and disposal

FIGURE 51.5 Composition and sources of municipal solid waste (MSW) in the United States during the last decade. (a) The composition, by weight, of MSW in the United States in 2014 before recycling. Paper, food, and yard waste make up more than half of the MSW by weight. Total does not add up to 100 percent because of rounding. (b) The breakdown of the material that is recovered and the material that is discarded. Paper makes up half of the material that is recovered. Food and yard waste make up almost one-third of material that is discarded. *(Source: After Source: Advancing Sustainable Materials Management: 2014 Fact Sheet. U.S. EPA (2016).)*

includes newsprint, office paper, cardboard, and boxboard such as cereal and food boxes, made up 27 percent of the 234 million metric tons (258 million U.S. tons) of waste generated before recycling. The fraction of paper in the solid waste stream has been decreasing; less than a decade ago it was 40 percent of MSW. Organic materials other than paper products make up another large category, with yard waste and food scraps together making up 28 percent of MSW. Wood, which includes construction debris, accounts for another 6 percent. So, not including paper products, which are more easily recycled, roughly 34 percent of current MSW could be composted, although some wood construction debris is difficult to compost because of its size and thickness. The combination of all plastics makes up approximately 13 percent of MSW.

As Figure 51.5b shows, 35 percent of the material that could potentially end up in a landfill or an incinerator is recovered. Roughly half of the recovered material is paper. Yard waste accounts for another large portion of recovered material, 23 percent.

After roughly one-third of our MSW is recycled, the resulting 153 million tons (169 million U.S. tons) that do end up in the landfill or incinerator have a different composition. As Figure 51.5b shows, paper represents a much smaller part of the diagram than in Figure 51.5a, largely because paper is so easily and frequently recycled. Food waste becomes a large part of the diagram because there are fewer composting programs available in the United States, although there has been an increase in municipal composting programs in recent years. Plastic is another component that increases, from 13 percent before recycling to 18 percent after materials have been recycled, in part because of the difficulty of recycling certain plastics.

Long-term viability is another way to consider MSW: Durable goods will last for years, nondurable goods are disposable, and compostable goods are those largely made up of organic material that can decompose under proper conditions. Containers and packaging make up 30 percent of MSW and are typically intended for one use. Food and yard waste account for 28 percent, and nondurable goods such as newspaper, white paper, printed products like telephone books, clothing, and plastic items like utensils and cups are 21 percent of the solid waste stream. Durable goods such as appliances, tires, and other manufactured products make up 20 percent of the waste stream. In addition to considering waste by weight, there is sometimes merit in considering waste by volume, especially when considering how much can be transported per truckload and how much will fit in a particular landfill.

FIGURE 51.6 Electronic waste recycling in New Delhi, India. Much of the recycling is done without protective gear and respirators that would typically be used in the United States. *(Bloomberg/Getty Images)*

E-Waste

Electronic waste, or e-waste, is one component of MSW that is small by weight but very important and rapidly increasing. Consumer electronics, including televisions, computers, portable music players, and cell phones, account for roughly 2 percent of the waste stream. This may not sound like a large amount but the environmental effect of these discarded objects is far greater than their weight. The older-style cathode-ray-tube (CRT) television or computer monitor contains 1 to 2 kg (2.2–4.4 pounds) of the heavy metal lead as well as other toxic metals such as mercury and cadmium. These metals may eventually leach out of the bottom of the landfill into groundwater or surface water. The toxic metals and other components can be extracted, but at present there is little formalized infrastructure or incentive to recycle them. However, many communities have begun voluntary programs to divert e-waste from landfills.

In the United States, most electronic devices are not designed to be easily dismantled after they are discarded. It generally costs more to recycle a computer than to put it in a landfill. The EPA estimates that approximately 30 percent of televisions and computer products discarded in 2012 were sent to recycling facilities. Unfortunately, much e-waste from the United States is exported to Asia, where adults as well as some children separate valuable metals from other materials using fire and acids in open spaces with no protective clothing and no respiratory gear (**FIGURE 51.6**). So even when consumers in the United States do send electronic products to be recycled, there is a good chance that the recycling will not be done properly.

MODULE 51 AP® Review

Preparing for the AP® Exam

In this module, we have seen that creating waste that is unusable by other organisms is a uniquely human characteristic. The refuse we dispose of is called municipal solid waste. The generation of MSW, both total and on a per capita basis, has increased steadily in the last 45 years although in recent years it has leveled off. Paper, food waste, and yard waste make up almost 60 percent of MSW. A fraction of each is reused, recycled, or composted. Of the approximately 235 million metric tons of MSW that could be disposed of, a little more than one-third ends up being diverted from the landfill for reuse, recycling, or composting. E-waste is a relatively small component of MSW but because of the toxicity of the metals it often contains, its disposal is a significant problem. In the next module, we will examine diversion from the landfill more closely.

AP® Practice Questions

Choose the best answer for the following.

1. Which played an important role in the development of the "throw-away" society?
   (a) the increased use of glass and metals
   (b) objects made of many materials
   (c) attitude changes after World War I
   (d) the shift in manufacturing to developing nations

2. On average, how much municipal solid waste is generated per person each day in the United States?
   (a) 0.5 kg
   (b) 1 kg
   (c) 2 kg
   (d) 4 kg

3. The material that makes up the highest proportion of MSW is
   (a) plastic.
   (b) rubber, leather, and textiles.
   (c) paper and paperboard.
   (d) food.

4. Electronic waste
   (a) accounts for over 10 percent of the waste stream.
   (b) is almost always recycled.
   (c) contains few toxic components.
   (d) is more expensive to recycle than to put in a landfill.

5. Approximately how much MSW is recovered before it enters a landfill or incinerator?
   (a) 15 percent
   (b) 20 percent
   (c) 35 percent
   (d) 45 percent

The Three Rs and Composting

Almost every schoolchild in the United States has heard the phrase "reduce, reuse, and recycle." In recent years, composting has been added to the list of actions one should take before adding material to the waste stream. In this module we will examine the *three Rs* and composting.

Learning Goals

After reading this module, you should be able to

- describe the three Rs.
- understand the process and benefits of composting.

The three Rs divert materials from the waste stream

Starting in the 1990s, people in the United States began to promote the idea of diverting materials from the waste stream with a popular phrase "**Reduce, Reuse, Recycle**," also known as **the three Rs**. The phrase incorporates a practical approach to the subject of solid waste management, with the techniques presented from the most environmentally beneficial to the least (**FIGURE 52.1**).

Reduce

"Reduce" is the first choice among the three Rs because reducing inputs is the optimal way to achieve a reduction in solid waste generation. This strategy is also known as waste minimization and waste prevention. If the input of materials to a system is reduced, the outputs will also be reduced; in terms of waste, this means that when less material is used, there will be less

FIGURE 52.1 Reduce, Reuse, Recycle. This popular slogan emphasizes the actions to take in the proper order, from the most environmentally effective to the least. *(Christopher Steer/iStockphoto.com)*

material to discard. One approach, known as **source reduction**, seeks to cut waste by reducing the use of potential waste materials in the early stages of design and manufacture. In many cases, source reduction also increases energy efficiency because it means that manufacturing produces less waste to begin with and can minimize disposal processes. Since fewer resources are being expended, source reduction also provides economic benefits.

Source reduction can be implemented both on individual and on corporate or institutional levels. For example, if an instructor has two pages of handout material for a class, she could reduce her paper use by 50 percent if she provides her students with double-sided photocopies. A copy machine that is designed so that it can automatically make copies on both sides of the page might have required more materials and energy during manufacturing than a copy machine that prints on only one side. However by making potentially half as many copies as a single-side-only machine during its lifetime, the overall energy used to provide photocopies to a school will probably be less. Further source reduction could be achieved if the instructor did not hand out any sheets of paper at all but sent copies to the class electronically, and the students refrained from printing the documents.

Source reduction in manufacturing can happen in several ways. If the company creates new packaging that provides the same amount of protection to the product with less material, successful source reduction has occurred. Consider the incremental source reduction that occurred with purchasing music. Music compact discs were packaged in large plastic sleeves that were three times the size of the CD. Today, when you can find a CD for sale, they are wrapped with a small amount of plastic material that just covers the CD case. However, most people no longer purchase CDs but prefer to download or listen to music from the Web. Less wrapping in CD packaging is an example of source reduction on the corporate level. Not purchasing CDs at all is an example of source reduction on the individual level.

Source reduction can also be achieved by material substitution. In an office where workers drink water and coffee from paper cups, providing every worker with a reusable mug will reduce MSW. In some categorization schemes, this could be considered reuse rather than source reduction. Nevertheless, cleaning the mugs will require water, energy to heat the water, soap, and processing of wastewater. The break-even point, beyond which there are gains achieved by using a ceramic mug (though this depends on a variety of factors), might be approximately 50 uses. The break-even point is shorter for a reusable plastic mug, in part because less energy is required to manufacture the plastic mug and, because it is lighter, less energy is used to transport it from the manufacturing facility to the point of use. Source reduction may also involve substituting less toxic materials or products in situations where manufacturing utilizes or generates toxic substances. For example, switching from an oil-based paint that contains toxic petroleum derivatives to a relatively nontoxic latex paint is a form of source reduction.

All of these examples achieve a reduction in material use and, ultimately, material waste, without an additional expenditure of materials or energy. For that reason, reduction is the first "R"—and it is the most environmentally beneficial.

Reuse

Reuse of a product or material that would otherwise be discarded, rather than disposal, allows a material to cycle within a system longer before it becomes an output. In other words, its mean residence time in the system is greater. Optimally, no additional energy or resources are needed for the object to be reused. For example, a mailing envelope can be reused by covering the first address with a label and writing the new address over it. Here we are increasing the residence time of the envelope in the system and reducing the waste disposal rate. Or we could reuse a disposable polystyrene cup more than once, though reuse might involve cleaning the cup, which would add some energy cost and generate some wastewater. Sometimes reuse may involve repairing an existing object, which costs time, labor, energy, and materials.

Energy may also be required to prepare or transport an object for reuse by someone other than the original user.

Reduce, reuse, recycle A popular phrase promoting the idea of diverting materials from the waste stream. *Also known as* **the three Rs**.

Source reduction An approach to waste management that seeks to cut waste by reducing the use of potential waste materials in the early stages of design and manufacture.

Reuse Using a product or material that was intended to be discarded.

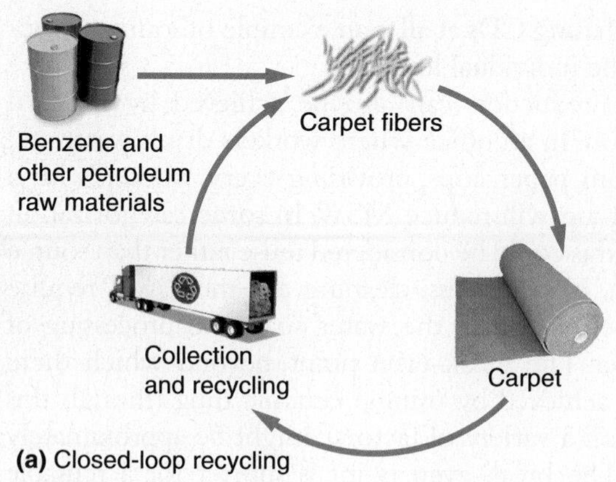

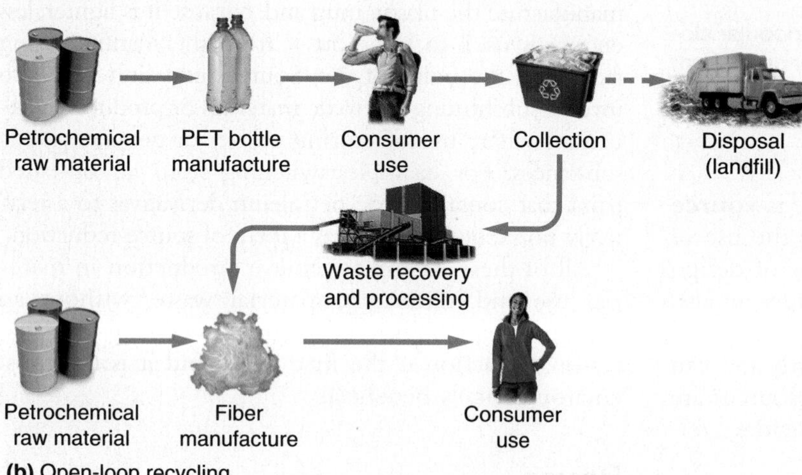

FIGURE 52.2 Closed- and open-loop recycling. (a) In closed-loop recycling, a discarded carpet can be recycled into a new carpet, although some additional energy and raw material are needed. (b) In open-loop recycling, a material such as a beverage container is used once and then recycled into something else, such as a fleece jacket.

For example, certain companies reuse beverage containers by shipping them to the bottling factory where they are washed, sterilized, and refilled. Although energy is involved in the transport and preparation of the containers, it is still less than the energy that would be required for recycling or disposal.

AP® Exam Tip

You should be able to describe ways to divert waste form the waste stream and be ready to discuss which action is the most environmentally beneficial and why. ●

We have noted that reuse is still common in many countries and that it was common practice in the United States before we became a "throw-away society." However, it is still practiced in many ways that we might not think of as reuse. For example, people often reuse newspapers for animal bedding or art projects. Many businesses and universities have surplus equipment agents who help find a home for items no longer needed. Flea markets, swap meets, and even popular websites such as eBay, craigslist, and The Freecycle Network are all agents of reuse. Reuse sometimes involves the expenditure of energy and generates waste. For example, in order to transport, wash, and sterilize the beverage bottles in the example previously cited, energy is expended and wastewater is generated. So reuse does have environmental costs that typically exceed reducing the use of something, but it is preferable to using new material.

Recycling The process by which materials destined to become municipal solid waste (MSW) are collected and converted into raw materials that are then used to produce new objects.

Closed-loop recycling Recycling a product into the same product.

Recycle

The third "R" is **recycling**, the process by which materials destined to become MSW are collected and converted into raw materials that are then used to produce new objects. We divide recycling into two categories: *closed-loop* and *open-loop.* **FIGURE 52.2** shows the process for each. **Closed-loop recycling**, shown in Figure 52.2a, is the recycling of a product into the same product. Aluminum cans are a familiar example; they are collected, brought to an aluminum plant, melted down, and made into new aluminum cans. This process is called a closed loop because in theory it is possible to keep making aluminum cans from only old aluminum cans almost indefinitely; the process is thus similar to a closed system. In **open-loop recycling**, shown in Figure 52.2b, one product, such as plastic soda bottles, is recycled into another product, such as polar fleece jackets. Although recycling plastic bottles into other materials avoids sending the plastic bottles to a landfill, it does not reduce demand for the raw material—in this case petroleum—to make plastic for new bottles.

Recycling is not new in the United States, but over the past 25 years it has been embraced enthusiastically by both individuals and municipalities in the belief that it measurably improves environmental quality. The graph in **FIGURE 52.3** shows both the increase in the weight of MSW in the United States from 1960 to 2014 and the increase in the percent of waste that was recycled over the same period of time. Recycling rates have increased in the United States since 1975, and today we recycle roughly one-third of MSW. To get an idea of how recycling rates are calculated, see "Do the Math: Calculating Recycling Rates" on page 584. In Germany, recycling rates are 56 percent, and some colleges and universities in the United States report recycling rates of 60 percent for their campuses.

Extracting resources from Earth requires energy, time, and usually a considerable financial investment. As we have seen, these processes generate pollution. Therefore, on many levels, it makes sense for manufacturers to utilize resources that have already been extracted. In the past decade many communities have adopted zero-sort

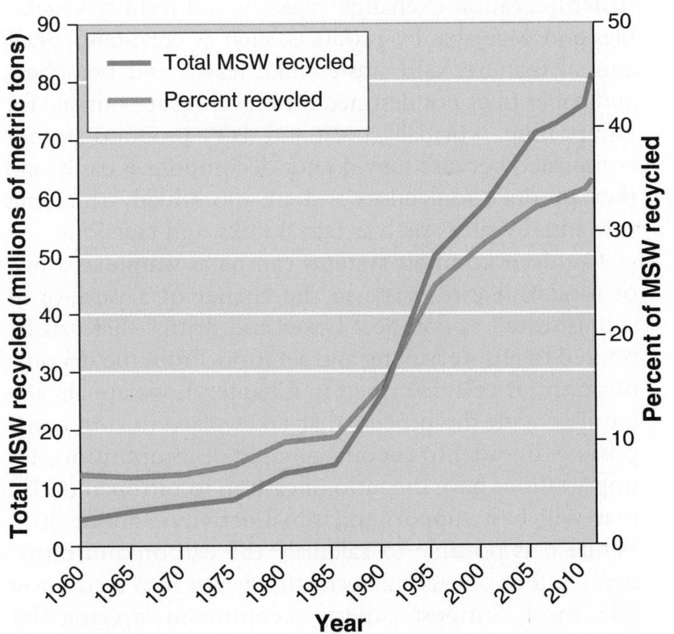

FIGURE 52.3 Total weight of municipal solid waste recycled and percent of MSW recycled in the United States over time. Both the total weight of MSW that is recycled and the percentage of MSW that is recycled have increased over time. *(Data from Advancing Sustainable Materials Management: 2014 Fact Sheet, U.S. EPA, 2016.)*

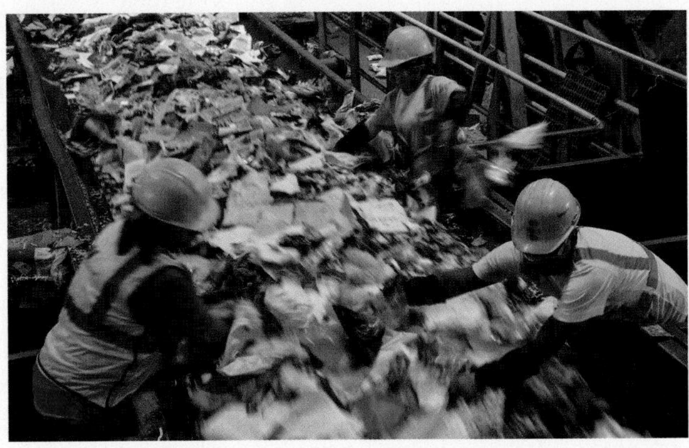

FIGURE 52.4 A mixed single-stream solid waste recycling facility in Elkridge, Maryland. With single-stream recycling, also called no-sort or zero-sort recycling, consumers no longer have to worry about separating materials. This facility takes in and sends out a thousand tons of material daily. *(The Washington Post/Getty Images)*

recycling programs. These programs allow residents to mix all types of recyclables in one container that they deposit on the curb outside the home or bring to a transfer station. This saves time for residents who were once required to sort materials. At the sorting facility, workers sort the materials destined for recycling into whatever categories are in greatest demand at a given time and offer the greatest economic return (**FIGURE 52.4**). The markets for glass and paper are highly volatile. While there is always a demand for metals such as aluminum and copper, demand for recycled newspapers fluctuates. When newspaper is in low demand, the single-stream sorting facility might pull out newsprint to sell or give to local stables for use as horse bedding. At other times, when demand for paper is higher, the newsprint might be kept with other paper to be recycled into new paper products. Lately, the market for recyclable materials has experienced even greater uncertainty because China has stopped accepting recyclables from other countries, citing too many non-recyclable and contaminated products mixed in with materials designated for recycling.

Nevertheless, because recycling requires time, processing, cleaning, transporting, and possible modification before the waste is usable as raw material, it does require more energy than reducing or reusing materials. Such costs caused New York City to make a controversial decision about 20 years ago to suspend glass and plastic recycling. This was a major shift in policy for the city, which had been encouraging as much recycling as possible, including collection of mixed recyclables—glass, plastic, newspapers, magazines, and boxboard—at the same time as waste materials. After collection, the recyclables were sent to a facility where they were sorted.

Open-loop recycling Recycling one product into a different product.

DO THE MATH Calculating Recycling Rates

Preparing for the AP® Exam

Imagine that it is the year 2030. People in the United States are able to recycle 40 million metric tons of material that was destined to be MSW. The total amount of material destined to be MSW before recycling is 80 million metric tons. What would be the percent recycling rate in 2030?

Divide the amount recycled by the total amount that would have gone to the landfill without recycling:

$$40 \text{ million metric tons} \div 80 \text{ million metric tons} = 0.5$$

To convert to percentage, multiply by 100%:

$$0.5 \times 100\% = 50\%$$

YOUR TURN Using Figure 52.3, determine the total MSW generated in 1960.

According to city officials, the entire process of sorting glass and plastic from other recyclables and selling the material was not cost effective. After 2 years, the recycling of materials was reinstated in New York City. Today, the goal in most recycling programs is to maximize diversion from the landfill, even if that means collecting materials that have little economic value. However, many communities periodically have difficulty finding buyers for glass and plastic since the price paid for these materials fluctuates widely.

The New York City case is just one example of why recycling is the last choice among the three Rs. This doesn't necessarily suggest that we should abandon recycling programs. Not only does it work well for materials such as paper and aluminum, but it also encourages people to be more aware of the consequences of their consumption patterns. Nevertheless, in terms of the environment, source reduction and reuse are preferable. The environmental implications of recycling are considered further in "Science Applied 7: Is Recycling Always Good for the Environment?" that follows Chapter 17 on page 644.

Composting is becoming more popular

While diversion from the landfill is usually referred to as the three Rs, there is one more diversion pathway that is equally important, if not more so. Organic materials such as food and yard waste that end up in landfills cause two problems. Like any material, they take up space, but unlike glass and plastic that are chemically inert, organic materials are also unstable. As we will see later in this chapter, the absence of oxygen in landfills causes organic material to decompose anaerobically, which produces methane gas, a much more potent greenhouse gas than carbon dioxide.

An alternate way to treat organic waste is through *composting.* **Composting** creates organic matter (humus) that has decomposed under controlled conditions to produce an organic-rich material that enhances soil structure, cation exchange capacity, and fertility. Vegetables and vegetable by-products, such as cornstalks, grass, animal manure, yard wastes (like leaves and branches), and paper fiber not destined for recycling are suitable for composting. Normally, meat and dairy products are not composted because they do not decompose as easily, and they produce foul odors and are more likely to attract unwanted visitors such as rats, skunks, and raccoons.

Outdoor compost systems can be as simple as a pile of food and yard waste in the corner of a yard, or as sophisticated as compost boxes and drums that can be rotated to ensure mixing and aeration. From the decomposition process described in Chapter 3, we are already familiar with the process that takes place during composting. In order to encourage rapid decomposition, it is important to have the ratio of carbon to nitrogen (C:N) that will best support microbial activity—about 30:1. While it is possible to calculate the carbon and nitrogen content of each material that is put into a compost pile, most compost experts recommend layering dry material such as leaves or dried cut grass—normally brown material—with wet material such as kitchen vegetables—normally green material. This will provide the correct carbon to nitrogen ratio for optimal composting. Frequent turning or agitation is usually necessary to ensure that decomposition processes are aerobic

Composting Creation of organic matter (humus) by decomposition under controlled conditions to produce an organic-rich material that enhances soil structure, cation exchange capacity, and fertility.

FIGURE 52.5 Composting. Good compost has a pleasant smell and will enhance soil quality by adding nutrients to the soil and by improving moisture and nutrient retention. In a compost pile that is turned frequently, compost can be ready to use in a few months. *(Marina Lohrbach/Shutterstock)*

and to maintain appropriate moisture levels—otherwise the compost pile will produce methane and associated gases and emit a foul odor. If the pile becomes particularly dry, water needs to be added. Although many people assume that a compost pile must smell bad, with proper aeration and the appropriate amount of moisture, the only odor will be that of fresh compost in 2 to 3 months' time (**FIGURE 52.5**).

Large-scale composting facilities currently operate in many municipalities in the United States. Some facilities are indoors, but most employ the same basic process we have described, though on a much larger scale. **FIGURE 52.6** shows the process. Organic material is piled up in long, narrow rows of compost. The material is turned frequently, exposing it to a combination of air and water that will speed natural aerobic decomposition. As with household composting, the organic material must include the correct combination of green (fresh) and brown (dried) organic material so that the ratios of carbon and nitrogen are optimal for bacteria. Various techniques are used to turn the organic material over periodically, including the use of rotating blades that move through the piles of organic material or a front loader that turns over the piles. The respiration activity of the microbes generates enough heat to kill any pathogenic bacteria that may be contained in food scraps, which is typically a concern only in large municipal composting systems. If the pile becomes too hot, it should be turned more frequently. If the pile doesn't become hot enough, operators should check to make sure the C:N ratio is optimal, or they should slow the turnover rate. Within a matter of weeks, the organic waste becomes compost. Large-scale municipal composting systems with relatively little mechanization and labor may take up to a year to create a finished compost.

It is not necessary to have an outdoor space to compost household waste; composting is possible even in a

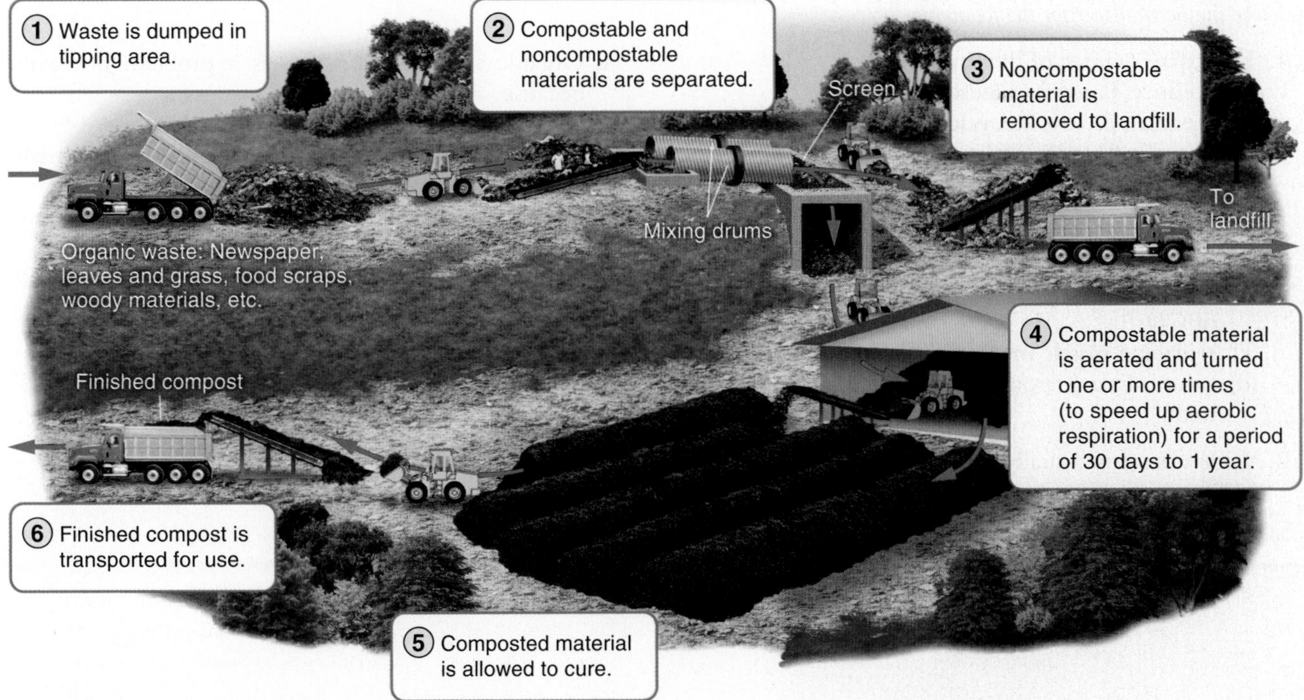

FIGURE 52.6 A municipal composting facility. A typical facility collects almost 100,000 metric tons of food scraps and paper per year and turns it into usable compost. Most facilities have some kind of mechanized system to allow mixing and aeration of the organic material, which speeds conversion to compost.

city apartment or a dorm room. It is even possible to set up a composting system in a kitchen or basement. The very popular book *Worms Eat My Garbage: How to Set Up and Maintain a Worm Composting System* by Mary Appelhof has encouraged thousands of individuals across the country to compost kitchen waste using red wiggler worms. A small household recycling bin is large enough to serve as a worm box. As with an outdoor compost pile, a properly maintained worm box does not give off bad odors.

The composting process does take time and space. Source separation can be an inconvenience or, in some situations, not possible. Also, in certain environments, storing materials before they are added to the compost pile can attract flies or vermin. Finally, the compost pile itself can attract unwanted animals such as rats, skunks, raccoons, and even bears. But because compost is high in organic matter, which has a high cation exchange capacity and contains nutrients, it enhances soil quality when added to agricultural fields, gardens, and lawns.

MODULE 52 AP® Review

Preparing for the AP® Exam

In this module, we have seen that reduce, reuse, recycle, and compost are appropriate actions to take before sending a solid waste material into the waste stream. It is important to observe the "three Rs" in the order stated here, because each action uses less energy than the next action. Composting is another pathway for diverting material from the waste stream. The diversion of compostable materials from the landfill reduces methane production and produces a desirable end product, decomposed organic matter. In the next module, we will see what happens to material that is not diverted from the solid waste stream.

AP® Practice Questions

Choose the best answer for the following.

1. The correct order of the three Rs is
   (a) Reduce, Recycle, Reuse.
   (b) Reduce, Reuse, Recycle.
   (c) Recycle, Reuse, Reduce.
   (d) Recycle, Reduce, Reuse.
2. Which is NOT a form of source reduction?
   (a) printing pages double-sided instead of single-sided
   (b) purchasing digital versions of music instead of CDs
   (c) replacing plastic mugs with disposable paper cups
   (d) substituting a nontoxic material for something toxic
3. Which material usually uses closed-loop recycling?
   (a) aluminum
   (b) glass
   (c) paper
   (d) cardboard
4. Organic matter in landfills is a problem primarily because
   (a) it contains bacteria that spreads disease.
   (b) as it breaks down it can dissolve the containment.
   (c) it creates excessive heat.
   (d) it produces methane gas.
5. For composting to work effectively, the compost
   (a) must be kept very wet.
   (b) must be mixed.
   (c) must be anaerobic.
   (d) must be protected from high temperatures.

MODULE 53

Landfills and Incineration

In the 1930s, public opposition to open dumps in the United States began to increase. Municipalities started to dispose of MSW in holes created when earth material was removed to be used for construction purposes. When people filled those holes with waste, the sites became known as landfills. Though open dumps are now rare in developed countries, they still exist in the developing world, where they pose a considerable health hazard.

In this module, we will examine the basic design and implementation of landfills and incinerators in the United States.

Learning Goals

After reading this module, you should be able to

- describe the goals and function of a solid waste landfill.
- explain the design and purpose of a solid waste incinerator.

Landfills are the primary destination for MSW

As **FIGURE 53.1** shows, in the United States a third of our waste is recovered through reuse and recycling, while more than half is discarded. The remainder is converted into energy through incineration. In the United States, MSW that is not diverted from the waste stream ends up becoming either buried in landfills or incinerated.

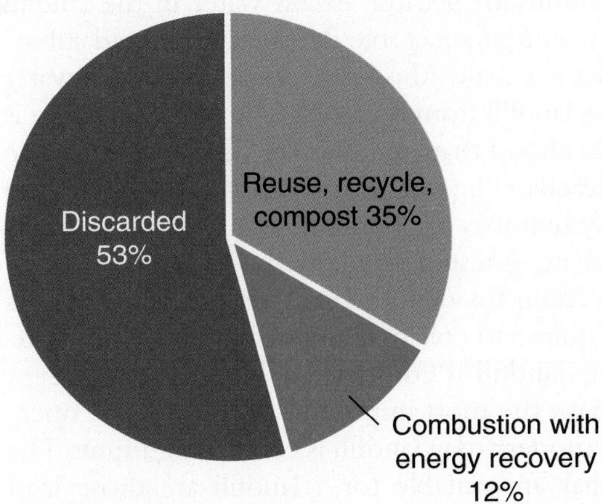

FIGURE 53.1 The fate of municipal solid waste in the United States. The majority of MSW is disposed of in landfills. *(Data from Advancing Sustainable Materials Management: 2014 Fact Sheet, U.S. EPA, 2016.)*

Initially, there were few concerns about what material went into a landfill. Those responsible for collecting and disposing of solid waste in landfills did not recognize the many problems associated with landfills, such as components of the MSW generating harmful runoff and *leachate*. **Leachate** is a liquid that contains elevated levels of pollutants as a result of having passed through municipal solid waste (MSW) or contaminated soil. Nor did they recognize the harm a landfill could cause when it was located near sensitive features of the landscape such as aquifers, rivers, streams, drinking-water supplies, and human habitation.

Today, some environmental scientists believe we should not use landfills at all; in later sections of this chapter we will discuss alternative means of waste disposal. Although landfills are still a component of solid waste management in the United States, we can make them much less harmful than they have been in the past. In this section we will examine landfill basics, how a landfill is sited, and the problems of using landfills.

Landfill Basics

Many people believe that a great deal of biological and chemical activity occurs in landfills. They imagine that organic matter breaks down and that landfills shrink over time as the material in them is converted into carbon dioxide that is released to the atmosphere. Professor William Rathje (1945–2012), then at the University of

Leachate Liquid that contains elevated levels of pollutants as a result of having passed through municipal solid waste (MSW) or contaminated soil.

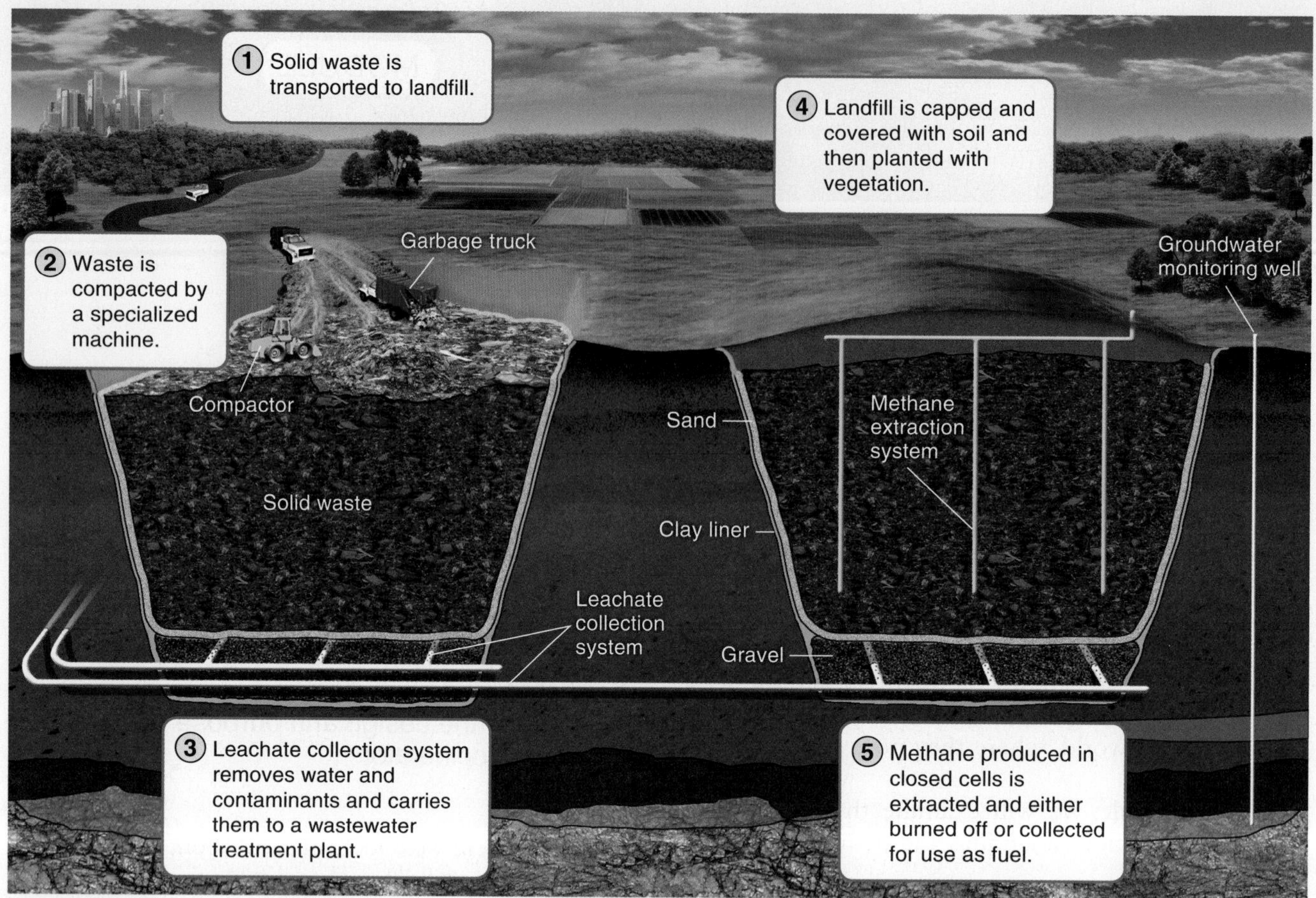
1 Solid waste is transported to landfill.
4 Landfill is capped and covered with soil and then planted with vegetation.
2 Waste is compacted by a specialized machine.
Garbage truck
Groundwater monitoring well
Compactor
Sand
Methane extraction system
Solid waste
Clay liner
Leachate collection system
Gravel
3 Leachate collection system removes water and contaminants and carries them to a wastewater treatment plant.
5 Methane produced in closed cells is extracted and either burned off or collected for use as fuel.

FIGURE 53.2 A modern sanitary landfill. A landfill constructed today has many features to keep components of the solid waste from entering the soil, water table, or nearby streams. Some of the most important environmental features are the clay liner, the leachate collection system, the cap—which prevents additional water from entering the landfill—and, if present, the methane extraction system.

Arizona, challenged these ideas by using archaeological tools to examine modern-day garbage in landfills. Rathje and his colleagues found newspapers with headlines still legible 40 years after being deposited, demonstrating that little decomposition had taken place. Today it is widely accepted among environmental scientists that decomposition takes place only in those areas of a landfill where the correct mixtures of air, moisture, and organic material are present. Because most areas do not contain this necessary mixture, a landfill will probably remain the same size it was when capped.

Given the lack of decomposition that occurs in a landfill, design and construction must proceed with care. In the United States today, repositories for MSW, known as **sanitary landfills**, are engineered ground facilities designed to hold MSW with as little contamination of the surrounding environment as possible. These facilities, like the one illustrated in **FIGURE 53.2**, generally utilize a variety of technologies that safeguard against the problems of traditional dumps.

Sanitary landfills are constructed with a clay or plastic lining at the bottom. Clay is often used because it impedes water flow and retains positively charged ions, such as metals. A system of pipes is constructed below the landfill to collect leachate, which is sometimes recycled back into the landfill. Finally, a cover of soil and clay, called a cap, is installed when the landfill reaches capacity.

Sanitary landfill An engineered ground facility designed to hold municipal solid waste (MSW) with as little contamination of the surrounding environment as possible.

The input of rainfall and other water sources are kept to a minimum because excess water in the landfill increases the rate of anaerobic decomposition and subsequent methane release. Also, with a large amount of water entering the landfill from both MSW and rainfall, there is a greater likelihood that some of that water will leave the landfill as leachate. Leachate that is not captured by the collection system may leach into nearby soils and groundwater. Leachate is tested regularly for its toxicity and if it exceeds certain toxicity standards, the landfill operators could be required to collect it and treat it as a toxic waste.

Once the landfill is constructed, it is ready to accept MSW. Perhaps the most important component of operating a safe modern-day landfill is controlling inputs. The materials that are suitable for a landfill are those least likely to cause environmental damage through leachate or by generating methane. Composite materials made of plastic and paper, such as juice boxes for children, are good candidates for a landfill because they are difficult to recycle, while aluminum and other metals such as copper may contribute to leaching. In addition, metals

FIGURE 53.3 Reclamation of a landfill. The Fresh Kills landfill in Staten Island, New York, was open from 1947 to 2001. It is currently in the process of being converted into a 890-ha (2,200-acre) park and is scheduled to open in the early 2020s. The mounds of landfill have been capped and planted with vegetation. *(Richard Levine/Alamy)*

are valuable as recyclables. Therefore, aluminum and copper should never go into a landfill. Glass and plastics are both chemically inert, making them suitable for a landfill when reuse or recycling is not possible. Toxic materials (such as household cleaners, oil-based paints, and automotive additives like motor oil and antifreeze), consumer electronics, appliances, batteries, and anything that contains substantial quantities of metals should not be deposited in landfills. All organic materials, such as food and garden scraps and yard waste, are potential sources of methane and should not be placed in landfills.

AP® Exam Tip

Many past AP® Environmental Science Exams have asked students to describe and explain landfill design and function. Be sure you know how a landfill works. ●

The MSW added to a landfill is periodically compacted into compartments or "cells," which reduce the volume of solid waste, thereby increasing the capacity of the landfill. The cells are covered with soil, which minimizes the amount of water that enters them and so reduces odor and anaerobic decomposition. When a landfill is full, it must be closed off from the surrounding environment so that the input and output of water are reduced or eliminated. Once a landfill is closed and capped, the MSW within it is more or less sealed off. Some water may enter from the outside environment, but this is minimal if the landfill is well-designed and properly sealed. The design and topography of the landfill cap, which is a combination of soil, clay, and sometimes plastic, encourage water to flow off to the sides, rather than into the landfill. Closed landfills can be reclaimed, meaning that some sort of herbaceous, shallow-rooted vegetation can be planted on the topsoil layer, both for aesthetic reasons and as a way to reduce soil erosion. Construction on the landfill is normally restricted for many years, although parks, playgrounds, and even golf courses have been built on reclaimed landfills (**FIGURE 53.3**).

A municipality or private enterprise constructs a landfill at a tremendous cost. These costs are recovered by charging a fee for waste delivered to the landfill, called a **tipping fee** because each truckload is put on a scale and, after the MSW is weighed, it is tipped into the landfill. Tipping fees at solid waste landfills average $50 per ton in the United States, although in certain regions, such as the Northeast, fees can be twice that much. These fees create an economic incentive to reduce the amount of waste that goes to the landfill. Many localities accept recyclables at no cost but charge for disposal of material destined for a landfill. This practice encourages individuals to separate recyclables from MSW. Some localities mandate that recyclable material be removed from the waste stream and disposed of separately. However, if tipping fees become too high, and regulations too stringent, a locality may inadvertently encourage illegal dumping of waste materials in locations other than the landfill and recycling center.

Choosing a Site for a Sanitary Landfill

Many considerations go into **siting** a landfill, which means designating its location. A landfill should be located in a loam or clay loam soil lined with clay or plastic to reduce migration of contaminants. It should be located

Tipping fee A fee charged for disposing of material in a landfill or incinerator.

Siting The designation of a landfill location, typically through a regulatory process involving studies, written reports, and public hearings.

away from rivers, streams, and other bodies of water and drinking-water supplies. A landfill should also be sufficiently far from population centers so that trucks transporting the waste and animal scavengers such as seagulls and rats present minimal risks to people. However, the energy needed to transport MSW must also be considered in siting; as distance from a population center increases, so does the amount of energy required to move MSW to the landfill. Regional landfills, though, are becoming more common because sending all waste to a single location often offers the greatest economic advantage.

A landfill siting is always highly controversial and sometimes politically charged. Because landfills are unsightly and smell bad they are not considered desirable neighbors. Landfill siting has been the source of considerable environmental injustice. People with financial resources or political influence often adopt what has been popularly called a "not-in-my-backyard," or NIMBY, attitude about landfill sites. Because of this, a site may be chosen not because it meets the safety criteria better than other options but because its neighbors lack the resources to mount an effective opposition.

In Fort Wayne, Indiana, for example, the Adams Center Landfill was located in a densely populated, low-income, and predominantly minority neighborhood. A University of Michigan environmental justice study quotes Darrell Leap, a hydrogeologist and professor at Purdue University, as saying that on a scale of 1 to 10, with 10 being a geologically ideal site, the Adams Center Landfill would rate a "3, possibly 4" because the site held a substantial risk of water contamination. When the communities surrounding the Adams Center Landfill learned of the report and this danger, they protested the renewal of the federal permit and fought expansion of the landfill at both the state and local levels. Ultimately, the Indiana Department of Environmental Management closed the landfill.

Environmental Consequences of Landfills

Though sanitary landfills, as we have seen, are an improvement over open dumps, they present many problems. Locating landfills near populations that do not have the resources to object is a global problem. No matter how careful the design and engineering, there is always the possibility that leachate from a landfill will contaminate underlying and adjacent waterways. The EPA estimates that some leaching has occurred at all landfills in the United States. Even after a landfill is closed, the potential to harm adjacent waterways remains. The amount of leaching, the substances that have leached out, and how far they will travel are impossible to know in advance. To get an idea of how much leachate is generated from a landfill and how much might be collected, see "Do the Math: How Much Leachate Might Be Collected?"

The risk to humans and ecosystems from leachate is uncertain. Public perception is that landfill contaminants pose a great threat to human health, though the EPA has ranked this risk as fairly low compared with other risks such as global climate change and air pollution. But methane and other organic gases generated from decomposing organic material in landfills do release greenhouse gases, which contributes to climate change.

When solid waste is first placed in a landfill, some aerobic decomposition may take place, but shortly after the waste is compacted into cells and covered with soil, most of the oxygen is used up. At this stage, anaerobic decomposition begins, a process that generates methane and carbon dioxide—both greenhouse gases—as well as other gaseous compounds. The methane also creates an explosion hazard. For this reason, landfills are vented so that methane does not accumulate in highly explosive quantities. In recent years, more and more landfill operators

DO THE MATH How Much Leachate Might Be Collected?

Preparing for the AP® Exam

Annual precipitation at a landfill in the town of Fremont is 100 mm per year, and 50 percent of this water runs off the landfill without infiltrating the surface. The landfill has a surface area of 5,000 m^2. Underneath the landfill, the town installed a leachate collection system that is 80 percent effective. Any leachate not collected by the system enters the surrounding soil and groundwater. This leachate contains cadmium and other toxic metals.

Calculate the volume of water in cubic meters (m^3) that infiltrates the landfill per year.

$$100 \text{ mm/year} = 0.1 \text{ m/year}$$

$$0.1 \text{ m/year} \times 5{,}000 \text{ m}^2 \times 50\% = 250 \text{ m}^3$$

So the volume of leachate in m^3 that is treated per year is: $250 \text{ m}^3 \times 80\% = 200 \text{ m}^3/\text{year}$

YOUR TURN In a neighboring landfill, 70 percent of annual precipitation runs off without infiltrating the surface and the leachate collection system is 90 percent effective. Calculate in cubic meters (m^3) the volume of leachate that is treated each year.

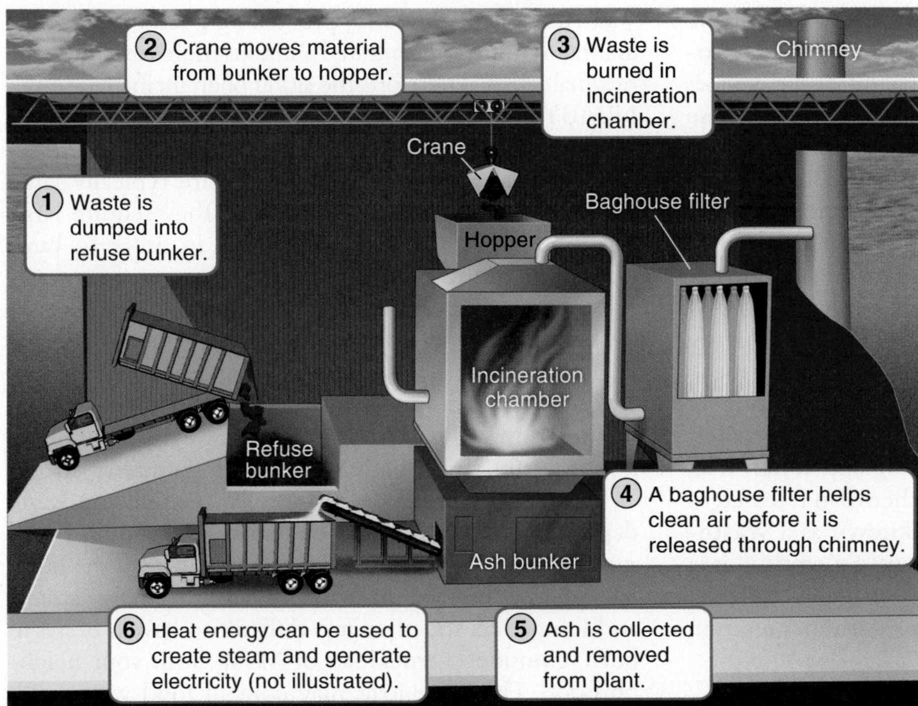

2 Crane moves material from bunker to hopper.
3 Waste is burned in incineration chamber.
Chimney
Crane
1 Waste is dumped into refuse bunker.
Baghouse filter
Hopper
Incineration chamber
Refuse bunker
4 A baghouse filter helps clean air before it is released through chimney.
Ash bunker
6 Heat energy can be used to create steam and generate electricity (not illustrated).
5 Ash is collected and removed from plant.

FIGURE 53.4 A municipal mass-burn waste-to-energy incinerator. In this plant, MSW is combusted and the exhaust is filtered. Remaining ash is disposed of in a landfill. The resulting heat energy is used to make steam, which turns a generator that generates electricity in the same manner as was illustrated in Figure 34.7 on page 424.

are collecting the methane the landfill produces and using it to generate heat or electricity. An even more desirable environmental choice would be keeping organic material out of landfills entirely by using it to make compost.

Incineration is another way to treat waste materials

Given all the problems of landfills, people have turned to a number of other means of solid waste disposal, including *incineration*. **Incineration** is the process of burning waste materials to reduce their volume and mass and, sometimes, to generate electricity or heat. More than three-quarters of the material that constitutes municipal solid waste is easily combustible. Because paper, plastic, and food and yard waste are composed largely of carbon, hydrogen, and oxygen, they are excellent candidates for incineration. An efficient incinerator operating under ideal conditions may reduce the volume of solid waste by up to 90 percent and the weight of the waste by approximately 75 percent, although the reductions vary greatly depending on the incinerator and the composition of the MSW.

Incineration Basics

FIGURE 53.4 shows a mass-burn municipal solid waste incinerator. Typically at such an incinerator, MSW is sorted and certain recyclables are diverted to recycling centers. The remaining material is dumped from a refuse truck onto a platform where certain materials such as metals are identified and removed. A moving grate or other delivery system transfers the waste to a furnace. Combustion rapidly converts much of the waste into carbon dioxide and water, which are released into the atmosphere along with heat.

Particulates, more commonly known as *ash* in the solid waste industry, are an end product of combustion. **Ash** is the residual nonorganic material that does not combust during incineration. Residue collected underneath the furnace is known as **bottom ash** and residue collected beyond the furnace is called **fly ash**.

Because incineration often does not operate under ideal conditions, ash typically fills roughly one-quarter the volume of the precombustion material. Disposal of this ash is determined by its concentration of toxic metals. The ash is tested for toxicity by leaching it with a weak acid. If the leachate is relatively low in concentration of contaminants such as lead and cadmium, the ash can be disposed of in a conventional solid waste landfill. Ash deemed safe can also be used for other purposes such as fill in road construction or as an ingredient in cement blocks and cement flooring. If deemed toxic, the ash goes to a special ash landfill designed specifically for toxic substances.

Metals and other toxins in the MSW may be released to the atmosphere or may remain in the ash, depending on the pollutant, the specific incineration process, and

Incineration The process of burning waste materials to reduce volume and mass, sometimes to generate electricity or heat.

Ash The residual nonorganic material that does not combust during incineration.

Bottom ash Residue collected at the bottom of the combustion chamber in a furnace.

Fly ash The residue collected from the chimney or exhaust pipe of a furnace.

the type of technology used. Exhaust gases from the combustion process, such as sulfur dioxide and nitrogen oxides, move through collectors and other devices that reduce their emission to the atmosphere. These collectors are similar in design to those described in Chapter 15 on air pollution. Acidic gases such as hydrogen chloride (HCl), which results from the incineration of certain materials including plastic, are recovered in a scrubber, neutralized, and sometimes treated further before disposal in a regular landfill or ash landfill.

Incineration also releases a great deal of heat energy, which is often used in a boiler immediately adjacent to the furnace either to heat the incinerator building or to generate electricity, using a process similar to that of a coal, natural gas, or nuclear power plant. When heat generated by incineration is used rather than released to the surrounding environment it is known as a **waste-to-energy** system. Although energy generation is a positive benefit of incineration, as we shall see, there are a number of environmental problems with incineration as a method of waste disposal.

Environmental Consequences of Incineration

Though incinerators address some of the problems of landfills, they also have shortcomings. To cover the costs of construction and operation, incineration facilities also charge tipping fees. Tipping fees are generally higher at incinerators than at landfills; national averages are around $70 per U.S. ton. We have seen that an incinerator may release air pollutants such as organic compounds from the incomplete combustion of plastics and metals contained in the solid waste that was burned. Some environmental scientists believe that incinerators are a poor solution to solid waste disposal because they produce air pollution as well as ash that is more concentrated and thus more toxic than the original MSW. Therefore, the siting of an incinerator raises NIMBY and environmental justice issues similar to those of landfill siting.

Waste-to-energy A system in which heat generated by incineration is used as an energy source rather than released into the surrounding environment.

In addition, because incinerators are typically large and expensive to build and operate, they require large quantities of MSW on a daily basis in order to burn efficiently and to be profitable. As a means of supporting these costs, communities that use incinerators may be less likely to encourage recycling. One solution is to use rate structures and other programs to encourage MSW reduction and diversion, with the goal of using incineration only as a last resort. However, in order to be successful, this usually must be a community effort.

Incinerators may not completely burn all the waste deposited in them. Plant operators can monitor and modify the oxygen content and temperature of the burn, but because the contents of MSW are extremely variable and lumped all together, it is difficult to have a uniform burn. Consider a truckload of MSW from your neighborhood. The same load may contain food waste with high moisture content and, right next to that, packaging and other dry, easily burnable material. It is difficult for any incinerator—even a state-of-the-art modern facility—to burn all of these materials uniformly.

Inevitably, MSW contains some toxic material. The concentration of toxics in MSW is generally quite low relative to all the paper, plastic, glass, and organics in the waste. However, rather than being dissipated to the atmosphere, most metals remain in the bottom ash or are captured in the fly ash. As we have already mentioned, incinerator ash that is deemed toxic must be disposed of in a special landfill for toxic materials.

As we have seen, there is no ideal choice for waste disposal. Sometimes the decision between whether to construct a landfill or build an incinerator is in part a decision about the kind of pollution a community prefers. The best choice is the production of less material for either the landfill or the incinerator.

MODULE 53 AP® Review

Preparing for the AP® Exam

In this module, we have seen that landfills and incinerators are the two ways to dispose of material not diverted from the waste stream. Landfills are designed to keep MSW dry and isolated so that the material in the landfill does not contaminate the surrounding environment. Leachate can contaminate nearby land and waterways if it escapes. Anaerobic decomposition in landfills can lead to the formation of methane, a potent greenhouse gas. Incinerators can generate a variety of air pollutants. Any metals contained within MSW that is incinerated may be released into the air. Bottom ash is particulate matter that accumulates underneath the combustion furnace. Sometimes, ash from incinerators can be considered toxic and must be deposited in a special ash landfill. Hazardous waste is the subject of our next module.

AP® Practice Questions

Choose the best answer for the following.

1. Which material, when placed in a landfill, is most likely to cause problems as a result of leaching?
   (a) paper
   (b) food matter
   (c) glass
   (d) aluminum

2. A 500 m^2 landfill experiences 150 mm of rain each year and 60 percent of the rain is runoff. If the landfill has a 90 percent effective leachate collection system, how much leachate escapes each year?
   (a) 3 m^3
   (b) 5 m^3
   (c) 16 m^3
   (d) 27 m^3

3. Incineration of waste is primarily used
   (a) to generate heat or electricity.
   (b) to reduce waste volume and mass.
   (c) to eliminate heavy metals.
   (d) when there is no other option.

4. NIMBY describes
   (a) the costs associated with the use of incinerators.
   (b) factors contributing to leaching from landfills.
   (c) attempts to develop better landfills.
   (d) an attitude about the placement of landfills.

5. Which is NOT a detriment of waste incineration compared with landfills?
   (a) increased toxicity of waste
   (b) increased cost to dispose of waste
   (c) increased space taken up by solid waste
   (d) increased air pollution

MODULE 54

Hazardous Waste

When solid waste material is deemed toxic or otherwise harmful to people or natural ecosystems, it should not be placed in a MSW landfill or incinerated. In these cases, special means of handling and disposal are required. In this module we will discuss the proper treatment of hazardous waste and some of the legislation concerning it.

Learning Goals

After reading this module, you should be able to

- define hazardous waste and discuss the issues involved in handling it.
- describe regulations and legislation regarding hazardous waste.

Hazardous waste requires proper handling and disposal

Hazardous waste is liquid, solid, gaseous, or sludge waste material that is harmful to humans, ecosystems, or materials. The U.S. EPA identifies four characteristics

Hazardous waste Liquid, solid, gaseous, or sludge waste material that is harmful to humans, ecosystems, or materials.

of hazardous waste: ignitability—it can catch on fire; corrosivity—it can cause materials to corrode or degrade; reactivity—it is not stable under normal conditions; toxicity—its chemical components are harmful or fatal when ingested or absorbed through the skin. Although the terms toxic and hazardous are often used interchangeably, toxic waste is a category of hazardous waste. While many substances are hazardous and can cause harm, substances that are toxic cause harm through chemical means when ingested or absorbed through the skin.

AP® Exam Tip

You should be able to differentiate between toxic waste and hazardous waste. Free-response questions on the AP® Environmental Science Exam often require specificity in the answers to earn points. ●

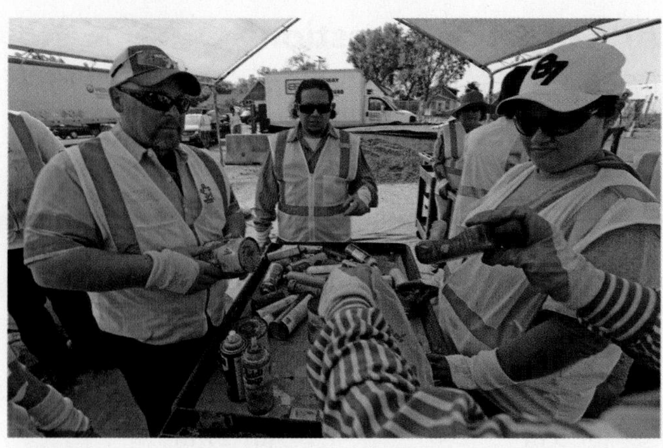

FIGURE 54.1 A typical household hazardous waste collection site in Minneapolis, Minnesota. Residents are encouraged to keep hazardous household waste separate from their regular household waste. Collections are held periodically. *(ZUMA Press Inc/Alamy)*

According to the EPA, in 2015 over 26,000 individual hazardous waste generators in the United States produced about 34 million metric tons (38 million U.S. tons) of hazardous waste each year. Only about 4 percent of that waste was recycled. The majority of hazardous waste is the by-product of industrial processes such as textile production, cleaning machinery, and manufacturing computer equipment, but it is also generated by small businesses such as dry cleaners, automobile service stations, and small farms. Even households generate hazardous waste—total waste estimates range from 0.5 million metric tons to 1.5 million metric tons (0.6 million U.S. tons to 1.7 million U.S. tons) per year in the United States, including materials such as oven cleaners, batteries, and lawn fertilizers. All of these materials have a much greater likelihood of causing harm to humans and ecosystems than other materials and should not be disposed of in regular landfills.

Most municipalities do not have regular collection sites for hazardous waste or household hazardous waste. Rather, homeowners and small businesses are asked to keep their hazardous waste in a safe location until periodic collections are held (**FIGURE 54.1**). Every aspect of the treatment and disposal of hazardous waste is more expensive and more difficult than the disposal of ordinary MSW. Households use numerous substances, such as oil-based paints, motor oil, or chemical cleaners, that are easy to purchase but become regulated hazardous waste as soon as the municipality collects them. Hazardous waste must be treated before disposal. Treatment, according to the EPA definition, means making it less environmentally harmful. To accomplish this, the waste must usually be altered through a series of chemical procedures.

Collection sites designated as hazardous waste collection facilities must be staffed with specially trained personnel. Sometimes the materials brought to a collection facility are unlabeled and unknown and must be treated with extreme caution. Ultimately, the wastes may be sorted into a number of categories, such as fuels, solvents, and lubricants. Some items, such as paint, can be reused, while others are sent to a special facility for treatment.

To an even greater degree than with other waste, there are no good options for disposing of hazardous waste. Hazardous waste landfills are more expensive to construct and require much more monitoring than a conventional MSW landfill. They must be monitored for at least 30 years after they are closed to additional inputs. The best recommendation for disposing of hazardous waste is to avoid creating the waste in the first place. In the case of household hazardous waste, many community groups and municipalities encourage consumers to substitute products that are less toxic or to use as little of the toxic substances as possible. For example, a can of household oven cleaner can be purchased at a grocery store, hardware store, or home center for roughly $5. Once the consumer brings it home, it is potentially household hazardous waste, and if it were to be disposed of before being completely used, the disposal cost might be $25 to $50, which is many times the purchase price. A consumer who brings a product to a household hazardous waste collection site typically does not pay for product disposal directly; the municipality or some other entity bears the cost. Therefore, product substitution or product avoidance is often the best solution to the household hazardous waste problem.

Legislation oversees and regulates the treatment of hazardous waste

The dangers that hazardous waste present have brought disposal to the attention of the general public. At times in the past few decades, hazardous waste sites have become commonly known household names. Because lawmakers and the general public have mandated that hazardous waste be regulated, there are a number of laws and acts that specifically cover hazardous waste. We will discuss two of the most significant laws in this section.

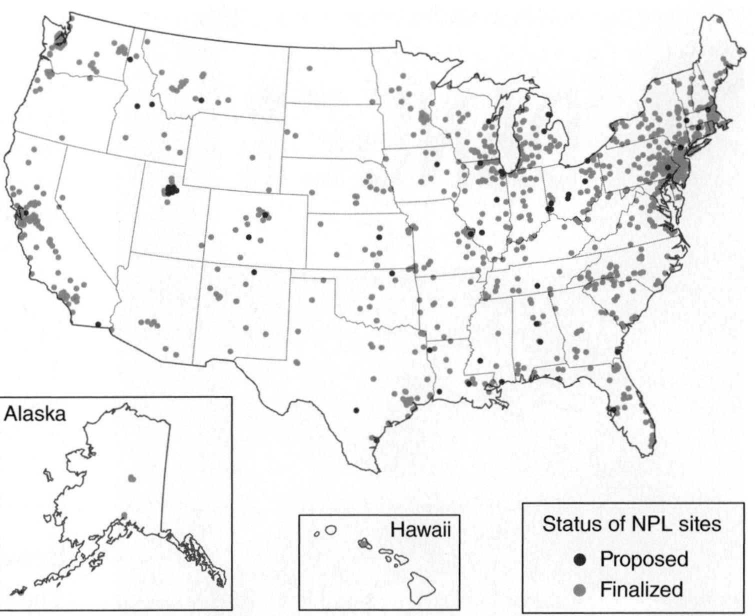

FIGURE 54.2 Distribution of NPL (Superfund) sites in the United States. Under Superfund, the EPA maintains the National Priorities List (NPL) of contaminated sites that are eligible for cleanup funds. (U.S. EPA (2018).)

Regulation and Oversight of Handling Hazardous Waste

Regulation and handling of hazardous waste in the United States falls under two pieces of federal legislation, the U.S. Resource Conservation and Recovery Act (RCRA) and the Comprehensive Environmental Response, Compensation, and Liability Act (CERCLA).

In 1976, RCRA expanded previous solid waste laws. Its main goal was to protect human health and the natural environment by reducing or eliminating the generation of hazardous waste. Under RCRA's provision for "cradle-to-grave" tracking, the EPA maintains lists of hazardous wastes and works with businesses and state and local authorities both to minimize hazardous waste generation and to make sure that the waste is tracked until proper disposal. In 1984, RCRA was modified with the federal Hazardous and Solid Waste Amendments (HSWA), which encouraged waste minimization and phased out the disposal of hazardous wastes on land. The amendments also increased law enforcement authority in order to punish violators.

CERCLA, usually referred to as the **Superfund Act**, is a 1980 U.S. federal act that imposes a tax on the chemical and petroleum industries, funds the cleanup of abandoned and nonoperating hazardous waste sites, and authorizes the federal government to respond directly to the release or threatened release of substances that may pose a threat to human health or the environment. The Superfund Act is well known because of a number of sensational cases that have fallen under its jurisdiction. Originally passed in 1980 and amended in 1986, this legislation has several parts. First, it imposes a tax on the chemical and petroleum industries. The revenue from this tax is used to fund the cleanup of abandoned and nonoperating hazardous waste sites where a responsible party cannot be established. The name *Superfund* came from this provision. CERCLA also authorizes the federal government to respond directly to the release or threatened release of substances that may pose a threat to human health or the environment.

Under Superfund, the EPA maintains the National Priorities List (NPL) of contaminated sites that are eligible for cleanup funds. The map in **FIGURE 54.2** shows the location of the NPL sites. As of 2018, there were 1,342 Superfund sites—at least 1 in every state. New Jersey, with 114 Superfund sites, has the most. California has the next highest number with 98 Superfund sites, followed by Pennsylvania with 95. New York has 86 sites. For a long time, very little Superfund money was disbursed and few sites underwent remediation. By 2018, almost 400 sites had been appropriately remediated and removed from the Superfund list.

Perhaps the best-known Superfund site is Love Canal, a neighborhood within Niagara Falls, New York (**FIGURE 54.3** on page 596). Originally a hazardous waste landfill, Love Canal was covered with fill and topsoil and used as a site for a school and a housing development. In 1978 and 1980, known cancer-causing wastes were found in the basements of homes in the area. These substances included the solvents benzene, dioxin, (which is a by-product of chemical manufacturing), and

AP® Exam Tip

Make sure you know the provisions of the Superfund Act (CERCLA) and the U.S. Resource Conservation and Recovery Act (RCRA). These laws are frequently tested in multiple-choice questions on the AP® Environmental Science Exam. ●

Superfund Act The common name for the Comprehensive Environmental Response, Compensation, and Liability Act (CERCLA); a 1980 U.S. federal act that imposes a tax on the chemical and petroleum industries, funds the cleanup of abandoned and nonoperating hazardous waste sites, and authorizes the federal government to respond directly to the release or threatened release of substances that may pose a threat to human health or the environment.

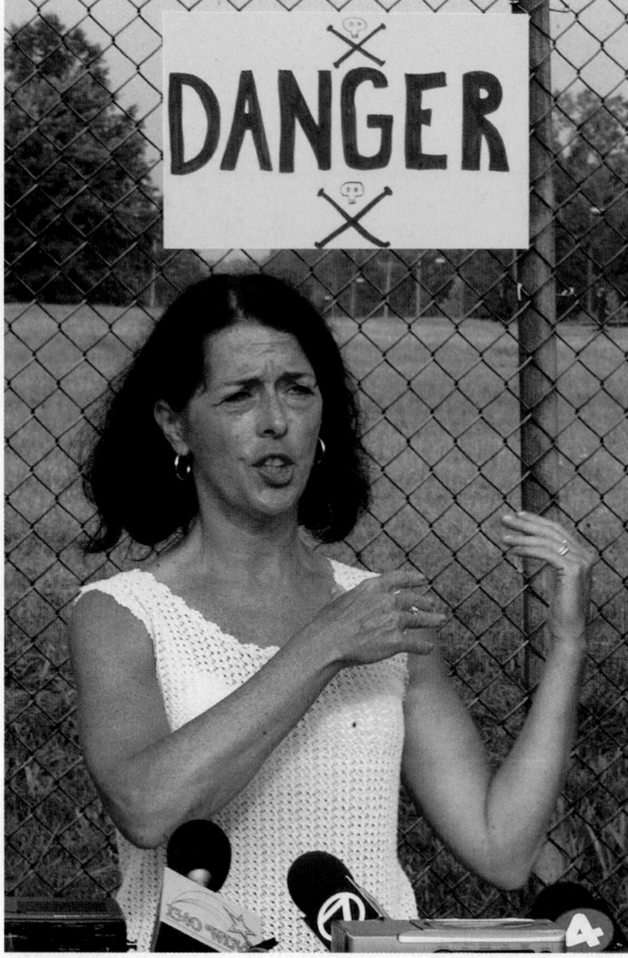

FIGURE 54.3 Lois Gibbs and Love Canal, New York. Love Canal became a symbol of hazardous chemical pollution in the United States in the 1970s. Lois Gibbs became a spokesperson for the neighborhood of Love Canal and is credited with bringing national attention to her community. *(Harry Scull Jr./Getty Images)*

a degreasing agent known as trichloroethylene. When it became clear that a disproportionately large number of illnesses, possibly connected to the chemical waste, had been diagnosed in the local population, the situation attracted national attention. Much of the attention that Love Canal received was due to the mother of a then 5-year-old boy in the school near Love Canal. Lois Gibbs, who at the time had no previous experience in community activism, is credited with drawing attention to the problems created by Love Canal and other locations. The contamination was so bad that in 1983 Love Canal was listed as a Superfund site and the inhabitants of the area were evacuated. In 1994, the EPA removed Love Canal from the National Priorities List because the physical cleanup had been completed and the site was no longer deemed a threat to human health. However, the site remains a symbol of the dangers presented by improper disposal of hazardous waste in a residential neighborhood. It also shows that a private citizen with no previous experience in community activism can greatly improve the safety of a neighborhood.

Brownfields Contaminated industrial or commercial sites that may require environmental cleanup before they can be redeveloped or expanded.

Since its inception, many have observed that CERCLA has not had enough funding to clean up the numerous hazardous waste sites around the country. The listing of NPL sites in a state does not necessarily include all contaminated sites within that state. For example, the Department of Environmental Protection in New Jersey, a state that has 114 listed Superfund sites, believes that there are more than 9,000 sites where soil or groundwater is contaminated with hazardous chemicals.

Brownfields

The Superfund designation is reserved strictly for those locations with the highest risk to public health, and Superfund sites are managed solely by the federal government. In 1995, the EPA created the *Brownfields* Program. **Brownfields**, like Superfund sites, are contaminated industrial or commercial sites that may require environmental cleanup before they can be redeveloped or expanded. The Brownfields Program assists state and local governments in cleaning up contaminated industrial and commercial land that did not achieve conditions necessary to be in the Superfund category. Old factories, industrial areas and waterfronts, dry cleaners, gas stations, landfills, and rail yards are common examples of brownfield sites. Brownfields legislation has prompted the revitalization of several sites throughout the country. One notable instance is Seattle's Gas Works Park. In 1962, the city of Seattle purchased the land, which had housed a coal gasification plant, to rehabilitate the site into a park. After undergoing chemical abatement and environmental cleanup, the park has become a distinctive landmark for the city and is the site of many public events throughout the year.

The Brownfields Program has been criticized as an inadequate solution to the estimated 450,000 contaminated locations throughout the country. Since the cleanup is managed entirely by state and local governments, brownfields management can vary widely from region to region. Furthermore, the brownfields legislation lacks legal liability controls to compel polluters to rehabilitate their properties. Without legal recourse, many brownfield sites remain unused and contaminated, posing a continued risk to public health.

International Consequences

Because of the difficulties involved in disposing of hazardous waste, municipalities and industries sometimes try to send the waste to countries with less stringent regulations. There are numerous reports of garbage and ash barges that travel the oceans looking for a developing country willing to accept hazardous waste from the United States in exchange for a cash payment. Perhaps the most famous is the story of the cargo vessel *Khian Sea,* which left Philadelphia in 1986 with almost 13,000 metric tons of hazardous ash from an incinerator. It traveled to a number of countries in the Caribbean in search of a dumping place. Some of the ash was dumped in Haiti, and some was dumped in the ocean. In 1996, the United States ordered that the ash dumped in Haiti be retrieved and returned to the United States. After being held at a dock in Florida, the ash was deemed nonhazardous by the EPA and other agencies and in 2002, the ash was placed in a landfill in Franklin County, Pennsylvania, not very far from its source of origin.

In 2003, a notable instance of a waste transfer in the other direction occurred. A Pennsylvania company that specializes in recovering mercury from a wide variety of products agreed to accept 270 metric tons of mercury waste generated by a company in the state of Tamil Nadu, India, during the manufacture of thermometers. India had no facilities for recycling mercury waste, and most of the thermometers manufactured at the plant were shipped to the United States and Europe, so the transfer of material seemed appropriate. The mercury was shipped from India to the United States, where the mercury was concentrated, purified, and then sold to industrial users of the metal. Today, there are many fewer stories of hazardous waste being shipped from country to country and being "dumped" on countries not equipped to safely handle hazardous waste. Since 2018, all U.S. exporters of hazardous waste must file appropriate information in an EPA automated export system, which allows monitoring and presumably has reduced infractions.

MODULE 54 AP® Review

Preparing for the AP® Exam

Hazardous wastes are a special category of waste material that is harmful to humans or ecosystems. Much of this waste is composed of byproducts from industrial processes. Hazardous wastes must be handled and treated separately from the MSW stream. There are a number of legislative acts that specifically address hazardous waste. There are many fewer stories today compared to decades ago of hazardous waste being disposed of in other countries, which sometimes resulted in the unsafe handling and treatment of hazardous waste.

AP® Practice Questions

Choose the best answer for the following.

1. Hazardous waste
   (a) costs much less to recycle than regular waste.
   (b) is primarily generated by individual households.
   (c) should be disposed of in landfills intermixed with MSW.
   (d) can include certain household items such as paints and oil.

2. Which legislation calls for listing hazardous waste to use in cradle-to-grave tracking?
   (a) Resource Conservation and Recovery Act
   (b) Comprehensive Environmental Response, Compensation, and Liability Act
   (c) Hazardous and Solid Waste Amendments
   (d) Superfund Act

3. The Brownfields Program
   (a) includes hazardous locations with the highest risk to public health.
   (b) attempts to prevent pollution in sites near agricultural lands.
   (c) is managed primarily by state and local governments.
   (d) has made substantial progress in cleaning the 450,000 identified sites.

4. Why might hazardous waste disposal in the United States be an international issue?
   (a) Lower disposal costs elsewhere encourages U.S. industries to export waste.
   (b) Air pollution from waste treatment often crosses international boundaries.
   (c) The United States continues to dump its hazardous waste into oceans.
   (d) The United States produces the majority of the world's hazardous waste.

MODULE 55

New Ways to Think About Solid Waste

Learning Goals

After reading this module, you should be able to

- explain the purpose of life-cycle analysis.
- describe alternative ways to handle waste and waste generation.

Throughout this chapter we have described a variety of ways of managing solid waste, from creating less of it to burying it in a landfill to burning it. Each method has both benefits and drawbacks. Because there is no obvious best method and because waste is a pervasive fact of contemporary life, the problem of waste disposal seems overwhelming. How can an individual, a small business, an institution, or a municipality best address solid waste? The answer is highly specific to each case and varies by region. Every method of waste disposal will have adverse environmental effects; the challenge is to find the least detrimental option. How do we decide which choices are best? Life-cycle analysis and a holistic approach are two useful strategies to help determine what we should do with our solid waste.

Life-cycle analysis considers materials used and released throughout the lifetime of a product

Recall the beginning of the chapter where we attempted to make an objective assessment of solid waste disposal options by comparing paper and polystyrene cups. This process, known as **life-cycle analysis**, and also known as **cradle-to-grave analysis**, is an important systems tool that examines the materials used and released throughout the lifetime of a product—from the procurement of raw materials through their manufacture, use, and disposal. As we saw when we compared a paper cup with a polystyrene cup, the full inventory of a life-cycle analysis sometimes yields surprising results.

Life-cycle analysis A systems tool that examines the materials used and released throughout the lifetime of a product—from the procurement of raw materials through their manufacture, use, and disposal. *Also known as* **cradle-to-grave analysis.**

In theory, conducting a life-cycle analysis should help a community determine whether incineration is more or less desirable than using a landfill. However, such an analysis has limitations. For example, it can be difficult to determine the overall environmental impact of a specific material. It is not possible to know whether the particulates and nitrogen oxides released from incinerating food waste are better or worse for the environment than the amount of methane that might be released if the same food waste were placed in a landfill. So in the case of waste that contains food matter, it is not possible to directly compare the full environmental impact of disposal in a landfill versus incineration. Similarly, we saw that it is a challenge to compare the sulfur dioxide released when making a paper cup with the volatile organic compounds released when making a plastic cup. Although life-cycle analysis may not be able to determine absolute environmental impact, it can be very helpful in assessing other considerations, especially those related to economics and energy use, which are usually relatively easy to quantify.

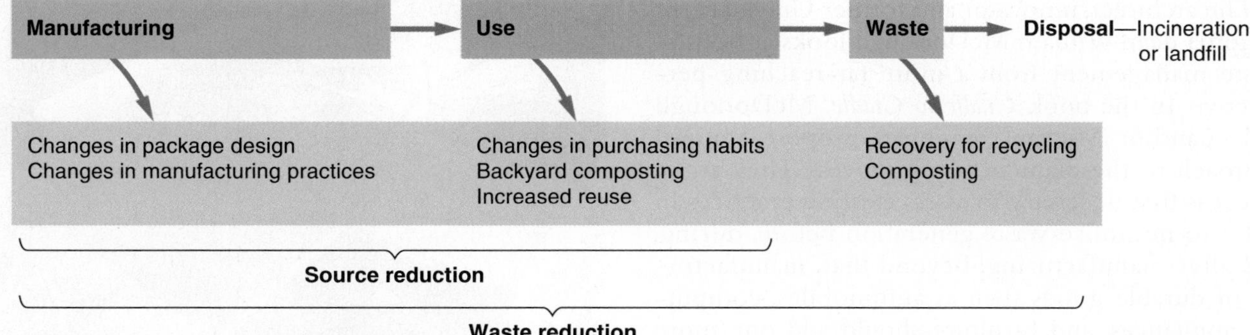

FIGURE 55.1 A holistic approach to waste management. Depending on the kind of waste and the geographic location, reducing waste can take much less time and money than disposing of it. Horizontal arrows indicate the waste stream from manufacture to disposal and curved arrows indicate ways in which waste can either be reduced or removed from the stream, thereby reducing the amount of waste incinerated or placed in landfills.

In terms of economics, the municipality might compare the costs of different disposal methods. For example, a glass manufacturing plant might pay $5 per ton for green glass that it will recycle into new glass. Economically, a municipality might do better if it received $5 for a ton of glass from a bottle manufacturing plant than if it paid a $50 per ton tipping fee so it could throw the material into a landfill. But the municipality must also consider the lower cost of transporting the glass to a relatively close landfill rather than to a distant glass plant.

In some parts of the country, the cost of waste disposal is covered by local taxes; in other locations, municipalities, businesses, or households may have to pay directly for disposal of their solid waste. Whether direct or indirect, there is always a cost to waste disposal. Normally, disposal of recyclables costs less than material destined for the landfill because the landfill always involves a tipping fee while recyclables either incur a lower tipping fee or generate revenue. However, as we have seen, costs change depending on many factors, including market conditions. For example, in a particular year, there may not be a market for recycled newspapers in the United States, but in the following year a large increase in overseas demand for newsprint may cause the price to go up in the United States. It is therefore essential for municipalities to have many choices and to be able to modify these choices as market environments change.

From the perspective of energy use, a life-cycle analysis should also consider the energy content of gasoline or diesel fuel used and the pollution generated in trucking material to each destination, as well as the monetary, energy, and pollution savings achieved if the new glass is made from old glass rather than from raw materials (sand, potash, lime). Reconciling all these competing factors is very challenging and the ultimate decisions based on such analyses are often debatable.

Integrated waste management is a more holistic approach

A more holistic method seeks to develop as many options as possible, emphasizing reduced environmental harm and cost. **Integrated waste management** employs several waste reduction, management, and disposal strategies in order to reduce the environmental impact of MSW. Such options include a major emphasis on source reduction and include any combination of recycling, composting, use of landfills, incineration, and whatever additional methods are appropriate to the particular situation. **FIGURE 55.1** shows how a nation or a community could consider a series of steps, starting with source reduction during manufacturing and procurement of items. After that, behavior related to use and disposal can be considered and possibly altered in order to obtain the desired outcome: less generation of MSW. According to this approach, no community should be forced into any one method of waste disposal. If a region makes a large investment in an incinerator, for example, there is a risk that it would then need to attract large quantities of waste to pay for that incinerator, thereby reducing the incentive to recycle or use a landfill. Landfill space may be abundant or scarce, depending on the location of the community. If the municipality is free to consider all options, it can make the choice or choices that are efficient, cost effective, and least harmful to the environment.

Integrated waste management An approach to waste disposal that employs several waste reduction, management, and disposal strategies in order to reduce the environmental impact of MSW.

The architect, innovator and former University of Virginia dean William McDonough looks at holistic waste management from a more far-reaching perspective. In the book *Cradle to Cradle,* McDonough and coauthor Michael Braungart propose a novel approach to the manufacturing process. They argue that it is first necessary to assess existing practices in order to minimize waste generation before, during, and after manufacturing. Beyond that, manufacturers of durable goods such as automobiles, computers, appliances, and furniture should add one more component to the design of the product: Be certain that it is designed so that when they are no longer useful, components can be recycled with as little of the material as possible becoming part of the waste stream.

Some industries have developed new approaches to waste. Most automobile manufacturers, for example, maintain that they design and build their cars so that they can be easily taken apart and materials of different composition easily separated to allow recycling. The EPA reported in 2016 that 75 percent by weight of automobiles were recycled and that most, but not all, of the recycled materials were metals. Certain carpet manufacturers design their carpets so that when worn out they can be easily recycled into new carpeting (**FIGURE 55.2**). This is typically done by making a base that is extremely durable with a top portion of the carpet that can be changed when the color fades, is worn out, or is no longer desired. Finally, McDonough and Braungart point out that many organisms in the natural world, such as the turtle, produce very hard, impact-resistant materials, such as a shell, without producing any toxic waste. They suggest that humans should examine how a turtle creates such a hard shell without the production of toxic wastes. Humans can use this example as a goal for other kinds of production where no toxic wastes would be produced. More recently, McDonough and Braungart have introduced the term *upcycle* to describe the conversion of a waste material to something of higher quality and greater value than the original product.

FIGURE 55.2 A recyclable carpet. FLOR carpet tiles are designed to be easily replaced and easily recycled when the carpet wears out. *(Glenn Asakawa/Getty Images)*

Although we can still greatly improve our handling of solid waste, there is some evidence that the nation as a whole has taken source reduction seriously; per capita waste generation has been approximately level since 1990. It appears that recycling has been taken seriously, too, since recycling rates have increased since 1985. However, considering both the amount of waste generated in this country and the recycling rates found in other countries, we have a long way to go.

MODULE 55 AP® Review

Preparing for the AP® Exam

In this module, we have seen that there are new and innovative ways to think about and handle MSW. Life-cycle analysis allows the determination of all of the materials and energy that go into the manufacture, use, and disposal of a product, with the goal of minimizing environmental impact. Life-cycle analysis is sometimes called cradle-to-grave analysis because it tracks an object from the time it is manufactured or born (cradle) until it is disposed of (grave). Sometimes life-cycle analysis can inform decisions about existing manufacturing practices. In some cases, the manufacturing process can be modified so that less waste is generated and, in addition, after a useful lifetime, the manufactured product can be remanufactured into something else.

AP® Practice Questions

Choose the best answer for the following.

1. Life-cycle analysis
   (a) examines the total environmental impact an object will make when it is discarded.
   (b) examines the materials and energy associated with an object from extraction of materials to disposal.
   (c) is used to analyze the least environmentally harmful way to dispose of an object.
   (d) considers only the environmental costs associated with an object from extraction of materials to disposal.

2. Integrated waste management
   (a) suggests that communities should have multiple options for waste disposal.
   (b) focuses on recycling waste.
   (c) is a method of educating individuals about the best method of waste disposal.
   (d) often ignores the benefits of composting.

3. Which is NOT part of the cradle to cradle concept proposed by William McDonough?
   (a) minimize waste generation during manufacturing
   (b) evaluate existing practices before making modifications
   (c) mimic the natural world in the production of hard materials
   (d) integrate all forms of waste management

Working Toward Sustainability

Recycling E-Waste in Africa

Electronic waste is a small part of the waste stream but it contains a large fraction of waste that ends up in landfills, dumps, and other locations. Toxic metals such as lead, mercury, and cadmium as well as carcinogenic organic compounds are common in electronic waste such as cell phones and laptops. There is an estimated 50 million metric tons (55 million U.S. tons) of e-waste generated worldwide and recycling rates vary around the world. Switzerland and Norway are two countries that generate large quantities of e-waste per year, but their e-waste recycling rates are in the 60 to 70 percent range. In the United States, the recycling rate for e-waste is 22 percent. Some countries recycle almost none of their electronic waste. Globally, the fate of 76 percent of e-waste is officially classified as "unknown." This suggests that it is probably dumped with MSW or recycled under unregulated and unsafe conditions. For e-waste that is collected separately, fully assembled electronic devices are usually exported for disassembly and recycling elsewhere. Not only does this practice export the pollution burden, it also uses unnecessary energy and space to export the entire device rather than just the components that need specialized e-waste recycling.

A United Nations University report from 2017 estimated that none of the e-waste that is generated in or transported to Africa is recycled. This statistic does not represent the actual amount of e-waste recycling that is taking place. There are encouraging stories from around the African continent that demonstrate an increase in e-waste recycling. Sometimes it occurs as a result of a private venture and sometimes as a government sponsored endeavor. South Africa (population 57 million) and the east African nation of Rwanda (population 12 million) provide an example of each.

A private company in South Africa called New Earth Waste Solutions (NEWS) has established 5 e-waste recycling facilities across South Africa with four goals: (1) increase the e-waste recycling rate in South Africa while at the same time; (2) comply with environmental and human health regulations; (3) extract valuable metals from the e-waste; and (4) provide jobs for underemployed local residents. NEWS disassembles e-waste, which is done by relatively unskilled laborers who might not otherwise have a reliable and well-paying job. The disassembly takes place in a controlled environment that is safe for the workers and it prevents the loss to the environment of metals and organic compounds that might otherwise contaminate soils and waterways. NEWS was founded by a diverse directorate including South African musician and social activist Johnny Clegg, who has strong ties to the Zulu community and writes about the environment in a number of his songs. Clegg and other members of the directorate have worked hard to create jobs in Zulu communities where unemployment is very high. In 2017, there were between 10 and 20 e-waste factories operating around South Africa. Perhaps as much as 100,000 metric tons of e-waste were recycled out of an estimated total of 330,000 metric tons (363,000 U.S. tons) of e-waste generated, which is a 30 percent recycling rate. One study estimated that there were an estimated 30 jobs created for every 1,000 metric tons (1,100 U.S. tons) of e-waste diverted.

In Rwanda, a national e-waste management strategy was established with the goal of increasing the collection of e-waste and other recyclables and establishing recycling industries, again with a goal of increasing employment. Rwanda is one of only a few African countries to have an explicitly stated national e-waste policy. A project was initiated in 2014 and by late 2017, an e-waste recycling facility was constructed by the government and began operating. Although numbers are preliminary, it appears that Rwanda may be collecting and recycling about 240 metric tons (265 U.S. tons) of e-waste per year. Based on estimates that e-waste is generated at a rate of 10,000 metric tons (11,000 U.S. tons) per year, this is only a 2.4 percent recycling rate. But the program has just started, and the expectation is that the factory should be able to recycle 7,000 metric tons (7,700 U.S. tons) of e-waste per year (which would be a 70 percent e-waste recycling rate) so percentage recovery is expected to increase.

An e-waste collection site in Rwanda. This worker is safely vacuuming glass and metals from the picture tube of an old television at the Rwandan National E-Waste Facility in Eastern Province, Rwanda. *(Rwanda Green Fund)*

Although both relatively recent, these examples suggest that e-waste recycling is happening in southern and eastern Africa with some encouraging results.

Critical Thinking Questions

1. What are some of the barriers to e-waste recycling?
2. Are you aware of any e-waste recycling activities in your community? Are they well publicized? Can you think of ways to enhance them?

References

Fayiga, A.O., M.O. Ipinmoroti, and T. Chirenje. 2018. Environmental pollution in Africa. *Environment, Development and Sustainability* 20:41–73.

Baldé, C.P., V. Forti, V. Gray, R. Kuehr, and P. Stegmann. 2017. *The Global E-Waste Monitor, 2017*. United Nations University Press.

Chapter 16 Review

In this chapter, we examined municipal solid waste. We observed that waste is unique to human beings. Total MSW and per capita waste generation in the United States has leveled off in recent years. The diversion of material from the solid waste stream can occur from reduction, reusing, recycling, and composting. The total amount and per capita recycling had been increasing for a number of years but has leveled off since roughly 2008. Waste disposal relies on landfills or incineration, both of which have benefits and adverse consequences. Hazardous waste is a special category of waste that must be handled and disposed of with particular care. There is national legislation that addresses hazardous waste. Life-cycle analysis allows a holistic approach to studying the entire waste stream from the creation of materials through their use and ultimate disposal.

Key Terms

Waste
Planned obsolescence
Municipal solid waste (MSW)
Waste stream
Reduce, reuse, recycle
The three Rs
Source reduction
Reuse
Recycling
Closed-loop recycling
Open-loop recycling
Composting
Leachate
Sanitary landfill
Tipping fee
Siting
Incineration
Ash
Bottom ash
Fly ash
Waste-to-energy
Hazardous waste
Superfund Act
Brownfields
Life-cycle analysis
Cradle-to-grave analysis
Integrated waste management

Learning Goals Revisited

Module 51 Only Humans Generate Waste

Explain why we generate waste and describe recent waste disposal trends.

From an ecological and systems perspective, waste is composed of the nonuseful products of a system. Much of the solid waste problem in the United States stems from the attitudes of the "throw-away society" adopted after World War II. While the United States may be a major generator of solid waste, the lifestyle and goods disseminated around the world have made solid waste a global problem.

Describe the content of the solid waste stream in the United States.

Solid waste in the United States is composed of primarily paper, food, and yard waste. The total amount and per capita waste generation was steadily increasing in the United States until the last few years, during which there has been a slight decline.

Module 52 The Three Rs and Composting

Describe the three Rs.

The three Rs—Reduce, Reuse, and Recycle—divert materials from the waste stream. Reduce refers to activities that encourage a reduction in the use and disposal of materials. Reuse refers to using items multiple times whenever possible. Recycling refers to returning an object to a manufacturing plant where it is turned into the same or another object made from the same material. They are listed in order of least environmental impact to greatest environmental impact.

Understand the process and benefits of composting.

Composting is the diversion of organic material such as food and yard waste from the waste stream and allowing it to decompose into organic soil (humus). Composting, source reduction, and reuse generally have lower energy and financial costs than recycling, but all are important ways to minimize solid waste production.

Module 53 Landfills and Incineration

Describe the goals and function of a solid waste landfill.

Currently, most solid waste in the United States is buried in landfills. Contemporary landfills entomb the garbage and keep water and air from entering and leachate from escaping. The potential for toxic leachate to contaminate surrounding waterways is one major concern. The other concern is the generation of methane gas. Siting of landfills often raises issues of environmental justice.

Explain the design and purpose of a solid waste incinerator.

Incineration is an alternative to landfills. Its main benefit is that it reduces the waste material to roughly one-quarter of its original volume. In addition, waste-to-energy incineration often uses the excess heat produced to generate electricity. However, incineration generates air pollution and ash, which can sometimes contain a high concentration of toxic substances and require disposal in a special ash landfill.

Module 54 Hazardous Waste

Define hazardous waste and discuss the issues involved in handling it.

Hazardous waste is a special category of material that is especially toxic to humans and the environment. It includes industrial by-products and some household items such as batteries and oil-based paints, all requiring special means of disposal.

Describe regulations and legislation regarding hazardous waste.

Though a variety of regulations and legislation have been implemented to address issues of hazardous waste, many problems remain. CERCLA, also called the Superfund Act, is probably the most well-known regulation concerning hazardous waste because it provides for the cleanup of hazardous waste sites.

Module 55 New Ways to Think About Solid Waste

Explain the purpose of life-cycle analysis.

Life-cycle analysis tracks material "from cradle to grave." Using life-cycle analysis and integrated waste management—which draws on all the available treatment methods—we can make optimal decisions regarding our solid waste.

Describe alternative ways to handle waste and waste generation.

The best solution is to design products with a strategy for their ultimate reuse or their dismantling and recycling. This approach has become more common in recent years.

Practice Math and Graphing

Preparing for the AP® Exam

1. Practice Math

The table shows the amounts of waste a factory generated in the first 5 years of operation. If by year 5 the factory was able to implement a source reduction of 25 percent, and assuming waste generation remained proportionately the same, how much waste would the factory generate in year 5?

	Waste generated (tons)
Year 1	1,472
Year 2	1,566
Year 3	1,900
Year 4	1,950
Year 5	2,000

2. Practice Graphing

Using the table from "Practice Math," plot the amount of waste generation from the factory in years 1 through 4 with a vertical bar graph. Plot year 5 using your calculation for a 25 percent waste reduction. Show the year on the *x* axis and the amount of waste in tons on the *y* axis.

Chapter 16 AP® Environmental Science Practice Exam

Preparing for the AP® Exam

Section 1: Multiple-Choice Questions

Choose the best answer for questions 1–20.

1. Which is NOT a reason to keep household batteries out of landfills?
   (a) They can leach toxic metals.
   (b) Their decomposition can contribute to greenhouse gas emissions.
   (c) They can be recycled, which would reduce the need for new raw materials.
   (d) They can be recycled, which would reduce the need for additional energy.

2. All of the following are desired outcomes of MSW incineration EXCEPT
   (a) extracting energy.
   (b) reducing volume.
   (c) prolonging the life of landfills.
   (d) increasing air pollution.

3. In the last 15 years, MSW per capita in the United States has
   (a) decreased drastically.
   (b) decreased, then increased drastically.
   (c) increased drastically.
   (d) stayed the same.

4. The EPA estimates that approximately ______ percent of municipal solid waste comes from residences and ______ percent comes from commercial and institutional facilities.
   (a) 30; 70
   (b) 40; 60
   (c) 50; 50
   (d) 60; 40

5. Which material constitutes the largest component of municipal solid waste?
   (a) metals
   (b) yard waste
   (c) food scraps
   (d) paper

6. Increasing tipping fees can cause
   (a) decreased rates of recycling.
   (b) increases in illegal dumping.
   (c) incentives for proper waste disposal.
   (d) reduced use of hazardous material.

7. From an environmental waste perspective, which of the following is the most desirable?
   (a) reduce
   (b) reuse
   (c) recycle
   (d) compost

8. In the United States, how much of generated waste ends up being recycled?
   (a) approximately one quarter
   (b) approximately one third
   (c) roughly half
   (d) more than three-quarters

9. Which legislation imposes a tax on the chemical and petroleum industries to generate funds to pay for the cleanup of hazardous substances?
   (a) RCRA
   (b) Cradle-to-Grave Act
   (c) HSWA
   (d) CERCLA

10. Of the following, which contributes most to the production of methane?
    (a) packaging
    (b) e-waste
    (c) plastics
    (d) yard waste

For questions 11–13, refer to the following lettered choices and choose the compound that is most associated with each numbered statement.

(a) benzene
(b) calcium carbonate
(c) methane
(d) hydrochloric acid

11. It may be present in the emissions from waste incinerators.

12. It contaminated the land and water near the housing development Love Canal in New York.

13. It is potentially a source of energy.

14. Roughly half of the material in the solid waste stream that is recovered before ending up in a landfill or incinerator is
    (a) yard waste.
    (b) paper products.
    (c) plastic products.
    (d) glass products.

15. Which statement regarding composting is NOT true?
    (a) Composting can be a method for diverting material from the landfill.
    (b) In order for compost to be made in a reasonable amount of time, the material must be mixed frequently.
    (c) Composting results in the release of carbon dioxide.
    (d) Efficient composting requires an abundance of anaerobic bacteria.

16. Which statement best describes the difference between the Superfund and Brownfield programs?
    (a) Superfund sites are currently hazardous to human health whereas Brownfields are potentially hazardous to humans in the future.
    (b) The Superfund program only funds the cleanup of wastes deemed hazardous under the Resource Conservation and Recovery Act, whereas the Brownfield program funds the cleanup of all hazardous wastes.
    (c) Superfund sites are generally reclaimed for human uses (e.g., recreation) whereas Brownfields are cleaned but never reclaimed.
    (d) Superfund site designation is reserved for sites that are the highest risk to public health whereas Brownfields are characterized as less dangerous.

17. Waste products from petroleum operations are harmful when ingested and can catch on fire. The EPA would therefore designate them as
    I. ignitable.
    II. toxic.
    III. hazardous.
    (a) II only
    (b) III only
    (c) I and II
    (d) I, II and III

18. Which of the following activities would CERCLA, or the Superfund Act, help fund?
    (a) the creation of a new regional landfill
    (b) the creation of an e-waste recycling site
    (c) the cleanup of a Brownfields site
    (d) the cleanup of an abandoned waste disposal plant

19. In 2014 in the United States, about 35 percent of MSW was recycled and composted, which was equivalent to 89 million tons. In addition, 33 million tons were combusted in a waste-to-energy system. How much total MSW did the United States generate in 2014?
    (a) 225 million tons
    (b) 254 million tons
    (c) 291 million tons
    (d) 315 million tons

20. Which is NOT likely to provide an incentive for a municipality to initiate a recycling program?
    (a) tipping fees that exceed the value of recycled goods
    (b) a stable demand, but low supply of recycled cardboard
    (c) construction of a landfill with a low tipping fee
    (d) initiation of an integrated waste management program

Section 2: Free-Response Questions

Write your answer to each part clearly. Support your answers with relevant information and examples. Where calculations are required, show your work.

1. The total amount of municipal solid waste (MSW) generated in the United States increased from 80 million metric tons (88 million U.S. tons) in 1960 to 232 million metric tons (255 million U.S. tons) in 2007.
    (a) Describe reasons for this increase and explain how the United States became the leader of the "throw-away society." (2 points)
    (b) Explain why reducing is more favorable than reusing, which in turn is more favorable than recycling. (4 points)
    (c) Describe the process of composting and compare a home composting system with that of a large-scale municipal facility. (4 points)

2. The City of Philadelphia recently replaced 1 out of every 10 trash bins with solar-powered trash compactors. The compactor is an enclosed unit with a door that opens for trash disposal. The compactor automatically detects when the bin is full and uses a solar-powered mechanical crusher to compact the contents. When the compactor needs to be emptied, it sends an electronic signal. Use of solar-powered compactors has increased the capacity of public trash bins and has reduced the number of trash collection visits to each bin from 17 times per week to 5 times per week.
   (a) Describe four positive externalities of installing solar-powered trash compactors. (2 points)
   (b) Describe six cradle-to-grave components of solar-powered trash compactors. (2 points)
   (c) Suggest one way that the installation of solar-powered trash compactors can reverse the effects of urban blight. (2 points)
   (d) The price of a regular trash bin is \$300, and it has a lifespan of 20 years. The price of a solar-powered trash compactor is \$4,000, and it has a lifespan of 10 years; it also requires approximately \$150 in maintenance costs each year. On average, a trash collection visit costs \$5 in fuel and \$20 in employee salary. Based on this information, are solar-powered trash compactors economically beneficial? (2 points)
   (e) Describe two ways that you might determine if solar-powered trash compactors are environmentally beneficial. (2 points)

3. In the United States, MSW that is not incinerated or recovered through reuse and recycling is buried in landfills.
   (a) Name THREE design features of sanitary landfills. (3 points)
   (b) Why does a landfill not generally change size after it is capped? (2 points)
   (c) An important part of operating a safe landfill is controlling inputs. Name ONE material that is a good candidate for a landfill, and ONE that is not, and explain why for each. (3 points)
   (d) One way to decrease the amount of material going into a landfill is source reduction. Describe source reduction and give ONE example. (2 points)

Outbreaks of the *E. coli* bacteria in food can make people sick and cause some people to die. As a result, food contaminated with *E. coli* must be discarded. *(Denis Doyle/Getty Images)*

Human Health and Environmental Risks

CHAPTER 17

MODULE 56 Human Diseases

MODULE 57 Toxicology and Chemical Risks

MODULE 58 Risk Analysis

CASE STUDY

Deadly *E. coli* outbreaks

In 2015, the Chipotle Mexican Grill restaurant chain temporarily closed 43 of its restaurants in Oregon and Washington State after dozens of customers became ill. The victims experienced nausea, stomach cramps, vomiting, and diarrhea. The Center for Disease Control and Prevention (CDC) stepped in and determined that the victims were suffering from infections from the *E. coli* bacteria.

Unfortunately, outbreaks of *E. coli* infections seem to happen every year in different parts of the country. The first major outbreak of *E. coli* infections occurred in 1992–1993 at Jack in the Box restaurants in California, Washington, Oregon, and Idaho; more than 700 people became infected and three died from eating hamburgers contaminated with *E. coli*. A decade later, the ConAgra Beef Company detected *E. coli* contamination in its meat and had to recall 8.7 million kilograms (19 million pounds) of beef that had been distributed to 21 states. Several years later, Americans learned of another major *E. coli* source in spinach grown in California and distributed around the country. In 2017, the CDC reported an outbreak of *E. coli* infections coming from two I. M. Healthy products: soy nut butter and granola. As a result, the products were recalled by the company.

How does *E. coli* cause people to become sick? You might recall from Chapter 14 that *E. coli* is a species of fecal coliform bacteria that we use to detect sewage inputs to water bodies. However, we also mentioned that this bacteria is typically not harmful to people. In fact, most strains of *E. coli* live in our digestive system and help us to digest food. However, some strains of *E. coli* produce a chemical known as shiga toxin, which causes people to become sick. While most people can recover from the infection by *E. coli* in 5 to 7 days, a small percentage of victims experience life-threatening kidney failure.

> Many human disease outbreaks are caused by pathogens that are transmitted from animals to humans, often due to human activities that alter the environment.

How do foods continue to get contaminated by *E. coli* and how can we help prevent future infections? Researchers believe that the bacteria are spread to food most commonly through improper hygiene. Because *E. coli* is an intestinal bacteria, it is critical for food handlers to thoroughly wash their hands after using the bathroom. In the case of the spinach outbreak, the CDC traced the source of the *E. coli* outbreak to cattle waste from a cattle pasture near the spinach farm. Investigators suspect that wild boars traveled through the cattle pasture, inadvertently stepped in contaminated cow pies, and then broke through a fence and into the spinach field where they spread the *E. coli*. To remedy this problem, the farmers had to improve their fencing to prevent wildlife from moving between the cattle pastures and spinach fields. On the consumer end, potential bacterial contamination in meat can be addressed by cooking the meat to a temperature high enough to kill the bacteria, 60°C to 71°C (140°F – 160°F). Unfortunately, foods such as spinach are often eaten raw which can make it harder to avoid consuming harmful bacteria.

The story of *E. coli* infections highlights the fact that human health can be affected by both infections and toxins. It also underscores that many human disease outbreaks are caused by pathogens that are transmitted from animals to humans, often due to human activities that alter the environment and allow such transmission to be more likely. By understanding the causes and contributing factors of infections and toxins, and the risks that they pose to us, we can do a better job of protecting human health.

Sources: E. coli outbreaks fast facts. *CNN*, July 19, 2017. http://www.cnn.com/2013/06/28/health/e-coli-outbreaks-fast-facts/index.html; FDA shuts down soy nut butter maker linked to *E. coli* outbreak. *Food Safety News*, March 31, 2017. http://www.foodsafetynews.com/2017/03/fda-shuts-down-soy-nut-butter-maker-linked-to-e-coli-outbreak/#.WkuW2ktG3XE; Shiga toxin-producing *E. coli* and food safety. Center for Disease Control and Prevention. https://www.cdc.gov/features/ecoliinfection/index.html.

The number of health risks that we face in our lives can feel overwhelming. Sometimes it seems that we hear new warnings every day. How do we evaluate and manage these risks? The first step in understanding health risks is to consider the three major categories of risk that can be detrimental to human health: physical, biological, and chemical. Physical risks include environmental factors, such as natural disasters, that can cause injury and loss of life. Physical risks also include less dramatic factors such as excessive exposure to ultraviolet radiation from the Sun, which causes sunburn and skin cancer, and exposure to radioactive substances such as radon, which we discussed in Chapter 15. Biological risks, which cause the most human deaths, are those risks associated with disease. Chemical risks are associated with exposure to chemicals ranging from those that occur naturally, such as arsenic, to those that are manufactured, such as synthetic chemicals and pesticides. In this chapter, we will focus on biological and chemical risks, and we will determine which of those risks are common and the current state of our understanding about each of them. We will look at how to assess and manage these risks. As we will see, although many health risks exist in both the developed and developing worlds, we can do a great deal to manage these risks and improve our lives.

MODULE 56

Human Disease

Humans face a wide range of potential diseases. In this module, we will explore different types of diseases and discuss the risk factors that make some people more likely to contract diseases. We will then discuss a number of diseases that have been important in human history as well as the diseases that have become prevalent in recent decades.

Learning Goals

After reading this module you should be able to

- identify the different types of human diseases.
- understand risk factors for human chronic diseases.
- discuss historically important human diseases.
- identify major emergent infectious diseases.
- discuss future challenges for improving human health.

There are different types of human diseases

FIGURE 56.1 on page 612 shows causes of human deaths worldwide. Approximately three-quarters of these causes are *diseases*. A **disease** is any impaired function of the body with a characteristic set of symptoms. **Infectious diseases** are diseases caused by infectious agents, known as pathogens. Examples include pneumonia and sexually transmitted diseases. Infectious diseases cause about one-quarter of worldwide deaths. Noninfectious diseases are not caused by pathogens; these include most cardiovascular diseases, respiratory and digestive diseases, and most cancers.

The pathogens that cause most infectious diseases include viruses, bacteria, fungi, protists, and a group of parasitic worms called helminths. However, only four types of infectious disease account for nearly 80 percent of all deaths attributed to infectious diseases: respiratory infections, HIV/AIDS, tuberculosis, and diarrheal diseases. Two other types of infectious diseases that cause 3 to 5 percent of infectious deaths globally are malaria, measles, and tetanus. We will discuss many of these infectious diseases later in the chapter.

All diseases can also be categorized as either *acute* or *chronic*. **Acute diseases** rapidly impair the functioning of a person's body. In some cases, such as a disease called Ebola hemorrhagic fever that we will discuss later in this chapter, death can come in a matter of days or weeks. In contrast, **chronic diseases** slowly impair the functioning of a person's body. Heart disease and most cancers, for example, are chronic diseases that develop over several decades.

Numerous risk factors exist for chronic disease in humans

Numerous factors cause people to be at a greater risk for chronic diseases such as cancer, cardiovascular diseases, diabetes, and chronic infectious diseases. The World Health Organization (WHO) has found that these risk factors differ substantially between low- and high-income countries. **FIGURE 56.2** on page 612 shows the WHO data for a variety of risks.

Disease Any impaired function of the body with a characteristic set of symptoms.

Infectious disease A disease caused by a pathogen.

Acute disease A disease that rapidly impairs the functioning of an organism.

Chronic disease A disease that slowly impairs the functioning of an organism.

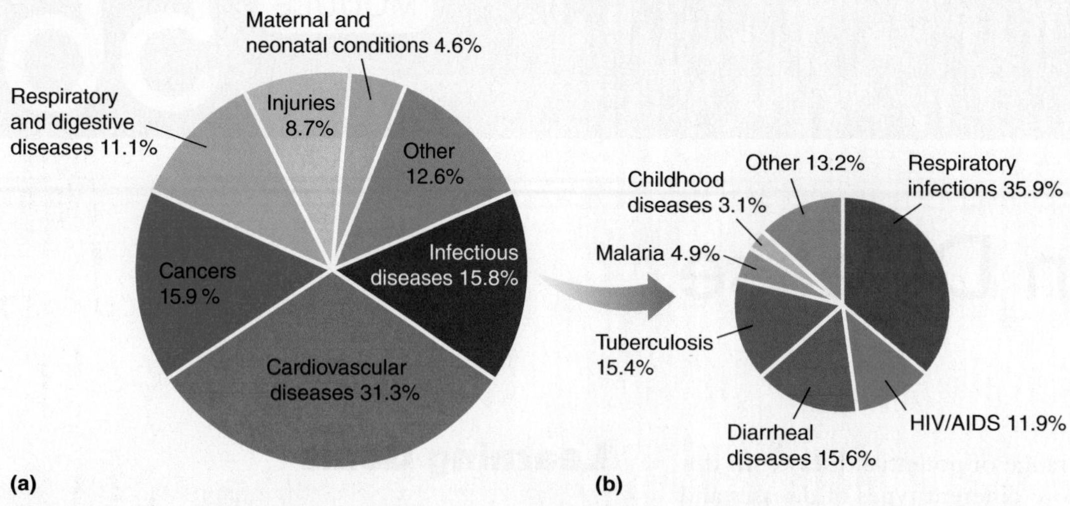

FIGURE 56.1 Leading causes of death in the world. (a) More than three-quarters of all world deaths are caused by diseases, including respiratory and digestive diseases, various cancers, cardiovascular diseases, and infectious diseases. (b) Among the world's deaths caused by infectious diseases, 94 percent are caused by only six types of diseases. *(Data from Global Health Estimates 2015: Deaths by Cause, Age, Sex, by Country and by Region, 2000-2015. Geneva, World Health Organization, 2016.)*

In low-income countries, the top risk factors leading to chronic disease are associated with poverty and include underweight children, unsafe drinking water, poor sanitation, and malnutrition. In Chapter 11 we looked at some of the factors that contribute to malnutrition, including political upheaval, poverty, and high food prices. As an example of poverty leading to chronic disease, nearly half of the children under the age of 5 who die from pneumonia succumb to the disease because they suffer from poor nutrition. Similarly, nearly three-quarters of children who die from diarrhea are also malnourished. Children with good

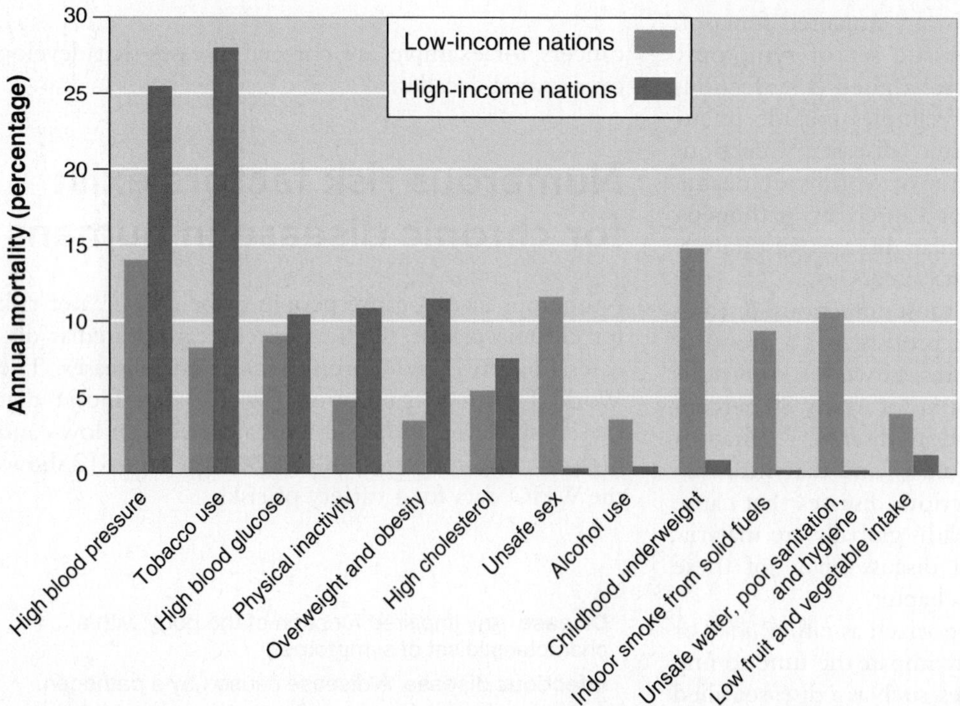

FIGURE 56.2 Leading health risks in the world. If we consider all deaths that occur and separate them into different causes, we can examine which categories cause the highest percentage of all deaths. The leading health risks for low-income countries include issues related to low nutrition and poor sanitation. The leading risks for high-income countries include issues related to tobacco use, inactivity, obesity, and urban air pollution. *(Data from World Health Organization, 2009.)*

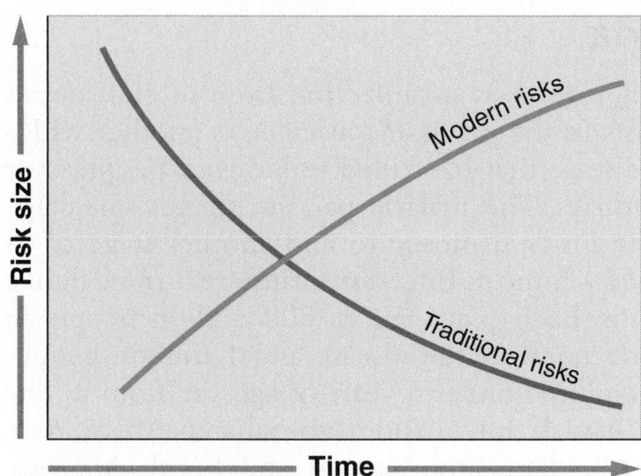

FIGURE 56.3 The transition of risk. As a nation becomes more developed over time and attains higher income levels, the risks of inadequate nutrition and sanitation decline while the risks of tobacco, obesity, and poor urban air quality rise.

nutrition are better able to fight infectious diseases and to survive.

Affluence changes the major health risk factors for chronic disease. Because people in high-income countries can afford better nutrition and proper sanitation, fewer die at a young age from diseases such as pneumonia and diarrhea. Risk factors for chronic disease in high-income countries include increased availability of tobacco, and a combination of less-active lifestyles, poor nutrition, and overeating that leads to high blood pressure and obesity.

The change in risk factors between low- and high-income countries occurs over time as a given country becomes more affluent. The graph in **FIGURE 56.3** illustrates how this transition in economic development affects health risk. People living in low-income countries face the challenge of fulfilling basic needs such as food and proper sanitation. As a result, their greatest causes of mortality are things such as children being underweight and unsafe water. As a country begins to accumulate wealth, the health risks to its citizens will change in a predictable fashion. In high-income countries, people are more able to have sufficient food and high-quality sanitation. However, their major causes of mortality are related to smoking, high blood pressure, obesity, and inactivity.

Some infectious diseases have been historically important

Diseases can have both genetic and environmental causes and these two factors often interact, especially for diseases caused by pathogens such as fungi, bacteria, and viruses. These pathogens have evolved a wide variety of pathways for infecting humans, including transmission through food, water, other humans, and other animals. **FIGURE 56.4** illustrates some of these relationships.

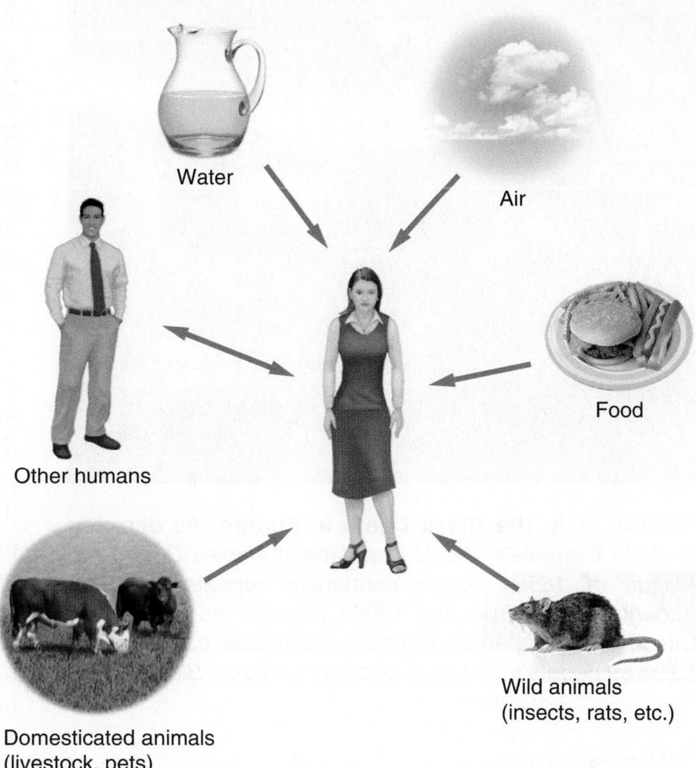

FIGURE 56.4 Pathways of transmitting pathogens. Pathogens have evolved a wide variety of ways to infect humans.

Historically, disease-causing pathogens have taken a large toll on human health and mortality. When a pathogen causes a rapid increase in disease, we call it an **epidemic**. When an epidemic occurs over a large geographic region such as an entire continent, we call it a **pandemic**. Among the diversity of human diseases that have caused epidemics and pandemics, we will consider both those that have been historically important and those that have emerged recently. In Chapters 7 and 14 we looked at several diseases historically associated with poor sanitation and unsafe drinking water including cholera, hepatitis, and diarrheal diseases. Additional historic diseases that are passed between hosts include *plague, malaria,* and *tuberculosis.*

Plague

Plague is caused by an infection from a bacterium (*Yersinia pestis*) that is carried by fleas. Fleas attach to

Epidemic A situation in which a pathogen causes a rapid increase in disease.

Pandemic An epidemic that occurs over a large geographic region.

Plague An infectious disease caused by a bacterium (*Yersinia pestis*) that is carried by fleas.

FIGURE 56.5 The Black Death in Europe. As depicted in Carlo Coppola's *The Marketplace in Naples During the Plague of 1656,* plague pandemics repeatedly swept through Europe from the 1300s through the 1800s and killed millions of people. Because the disease caused black sores on people's bodies, it also had the name *Black Death*. *(Roger-Viollet/The Image Works)*

rodents such as mice and rats, which gives the fleas tremendous mobility. One of the most well-known diseases of human history, plague also carries several historical names including *bubonic plague* and *Black Death*. When humans live in close contact with mice and rats, the bacterium can be transmitted either by flea bites or by handling the rodents. Individuals who become infected often experience swollen glands, black spots on their skin, and extreme pain. Plague is estimated to have killed hundreds of millions of people throughout history, including nearly one-fourth of the European population in the 1300s (**FIGURE 56.5**). The last major pandemic of plague occurred in Asia in the early 1900s. Today there are still occasional small outbreaks of plague around the world. For example, on the island of Madagascar off the eastern coast of Africa, there was an outbreak of plague in 2017 that infected more than 2,000 people and killed more than 200 of them. The plague has also continued to infect small numbers of people each year in Arizona, California, Colorado, and New Mexico; these infections occur because a small number of rodents in the American Southwest continue to carry the bacteria. Fortunately, modern antibiotics are highly effective at killing the bacterium and preventing human death.

Malaria An infectious disease caused by one of several species of protists in the genus *Plasmodium*.

Tuberculosis A highly contagious disease caused by the bacterium *Mycobacterium tuberculosis* that primarily infects the lungs.

Malaria

Malaria, caused by an infection from several species of protists in the genus *Plasmodium,* is another widespread disease that has killed millions of people over the centuries. The malaria parasite spends one stage of its life inside a mosquito and another stage of its life inside a human. Infections cause recurrent flulike symptoms. Each year, 350 to 500 million people in the world contract the disease and 1 million people, mostly children under 5 years of age, die from it. The regions hardest hit include sub-Saharan Africa, Asia, the Middle East, and Central and South America. Since 1951, the malaria parasite has been eliminated from the United States by mosquito eradication programs. Although more than 1,000 cases of malaria are diagnosed in the United States each year, these are people who have returned from regions of the world where the malaria parasite lives.

The traditional approach to combating malaria was widespread spraying of insecticides such as DDT to eradicate the mosquitoes. Eradication efforts have proven to be ineffective in many parts of the world. Moreover, as we will see later in this chapter, the widespread use of many insecticides can create new problems. At the end of this chapter, "Working Toward Sustainability: The Global Fight Against Malaria" on page 637 examines the latest approaches toward combating malaria.

Tuberculosis

Tuberculosis is a highly contagious disease caused by a bacterium (*Mycobacterium tuberculosis*) that primarily infects the lungs. Tuberculosis is spread when a person coughs and expels the bacteria into the air. The bacteria can persist in the air for several hours and infect a person who inhales them. Symptoms of an infection include feeling weak, sweating at night, and coughing up blood. As is the case with many pathogens, a person can be infected but not develop the tuberculosis disease. Indeed, it is estimated that one-third of the world's population is infected with tuberculosis. Each year 9 million people develop the disease and 2 million die.

A year-long course of antibiotics can treat most tuberculosis infections. In countries such as the United States, where the appropriate antibiotics are readily available, there has been a dramatic drop in both the number of new cases and the number of deaths from tuberculosis since the mid-1950s. As shown in **FIGURE 56.6**, the decline in tuberculosis elsewhere in the world has been much slower. Today, tuberculosis remains the leading cause of death by disease in the developing world. In these countries, the medicines are not as available or affordable and those who receive the medicine sometimes do not take the full course

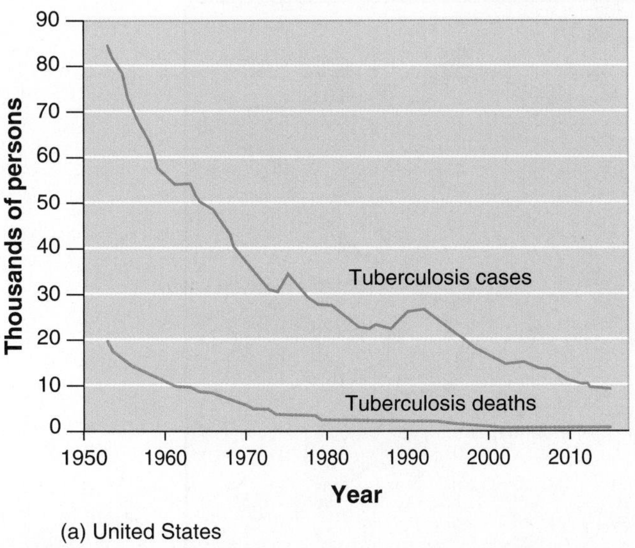

(a) United States

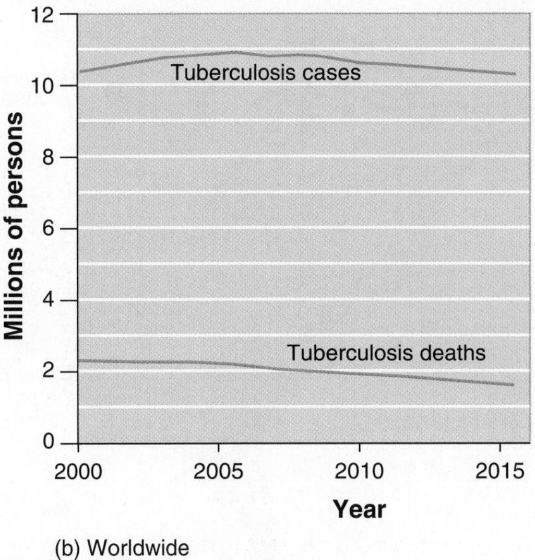

(b) Worldwide

FIGURE 56.6 Tuberculosis cases and deaths. (a) Due to effective and available medicines, tuberculosis has gone from being one of the most deadly diseases in the United States to a disease that rarely kills. (b) Worldwide, tuberculosis has continued to infect and kill millions of people, especially in low- and middle-income countries. *(Data from https://www.cdc.gov/tb/statistics/reports/2016/pdfs/H-2016-Surveillance-Report-table1.pdf, Center for Disease Control. 2016. Reported Tuberculosis in the United States, World Health Organization. 2017. Global Tuberculosis Report.)*

at the prescribed dose. When a patient stops taking antibiotics before the last bacteria have been killed, there are two consequences. First, the pathogen can quickly rebuild its population inside the person's body. Second, because the last few bacteria are generally the most drug-resistant, stopping the antibiotics before the bacteria are eradicated selects for drug-resistant strains. Drug-resistant strains of tuberculosis are becoming a major concern, particularly in parts of Africa and in Russia, where up to 20 percent of the people infected with tuberculosis carry a drug-resistant strain. Such strains are much harder to kill and therefore require newer antibiotics that can cost 100 times more than the traditional drugs.

Emergent infectious diseases pose new risks to humans

In recent decades, we have witnessed the appearance of many **emergent infectious diseases**, which are defined as infectious diseases that were previously not described or have not been common for at least the prior 20 years. **FIGURE 56.7** on page 616 locates some of the best known emergent infectious diseases. Since the 1970s, the world has observed an average of one emergent disease each year. Many of these new diseases have come from pathogens that normally infect animal hosts but then unexpectedly jump to human hosts. This typically occurs because the diseases can mutate rapidly and eventually produce a genotype that can infect humans. Some of the most high-profile diseases that have jumped from animals to humans include HIV/AIDS, Ebola, mad cow disease, bird flu, SARS, and West Nile virus. These emerging diseases are of increasing concern because of the increased movement of people and cargo throughout the world during the past century. In fact, currently diseases can spread to nearly any place on Earth within 24 hours.

HIV/AIDS

In the late 1970s, rare types of pneumonia and cancer began appearing in individuals with weak immune systems. The condition responsible for the weakened immune systems was the disease **Acquired Immune Deficiency Syndrome (AIDS)**, which is caused by a virus known as **Human Immunodeficiency Virus (HIV)**. This virus spreads through sexual contact and blood transfusions, from mothers to their fetuses, and among drug users who share unsanitized needles.

The origin of this new virus remained a mystery until 2006 when researchers found a genetically similar virus in a wild population of chimpanzees living in the African

Emergent infectious disease An infectious disease that has not been previously described or has not been common for at least 20 years.

Acquired Immune Deficiency Syndrome (AIDS) An infectious disease caused by the human immunodeficiency virus (HIV).

Human Immunodeficiency Virus (HIV) A type of virus that causes Acquired Immune Deficiency Syndrome (AIDS).

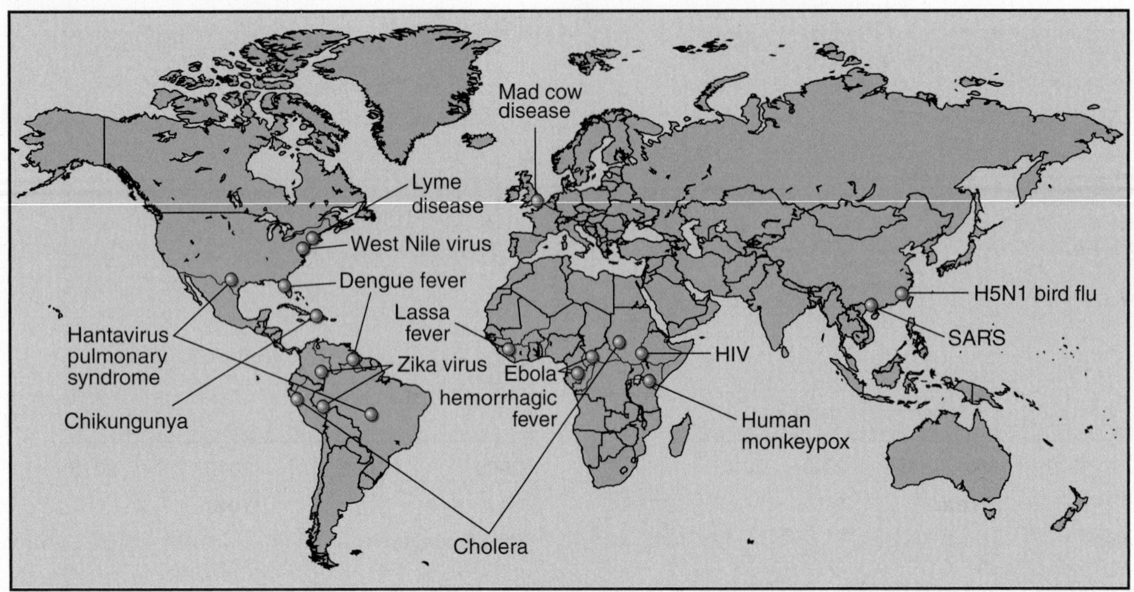

FIGURE 56.7 The emergence of new diseases. Since the 1970s, new diseases, or diseases that have been rare for more than 20 years, have been appearing throughout the world at a rate of approximately one per year. *(Data from https://www.niaid.nih.gov/sites/default/files/public%3A//images/news/main%20map.jpg.)*

nation of Cameroon (**FIGURE 56.8**). The researchers hypothesized that local hunters were exposed to the virus when butchering or eating the chimps (a common practice in this part of the world). With this exposure, the virus was able to infect a new host, humans. According to the World Health Organization, more than 70 million people in the world have been infected with HIV and about 35 million people have died from HIV-related illnesses, including 1 million people in 2016.

Fortunately, new antiviral drugs are able to maintain low HIV populations inside the human body and thereby substantially extend life for those who are infected with the virus. From the lessons learned in combating other diseases such as tuberculosis, combinations of antiviral drugs are being used to reduce the risk that the virus will evolve resistance to any single drug. Unfortunately, many of these drugs are expensive and most people living in low-income countries cannot afford them, although this is changing with wider availability and improved distribution of these drugs to the poor.

Ebola Hemorrhagic Diseases

In 1976, researchers first discovered **Ebola hemorrhagic fever**, an infectious disease with high death rates, caused by several species of Ebola viruses. First discovered in the Democratic Republic of Congo near the Ebola River, the virus has infected several hundred humans and a variety of other primates from several countries in central Africa. Infections have been sporadic since Ebola was first discovered, but there was a large outbreak in 2014 that infected thousands of people. The Ebola virus is of

Ebola hemorrhagic fever An infectious disease with high death rates, caused by several species of Ebola viruses.

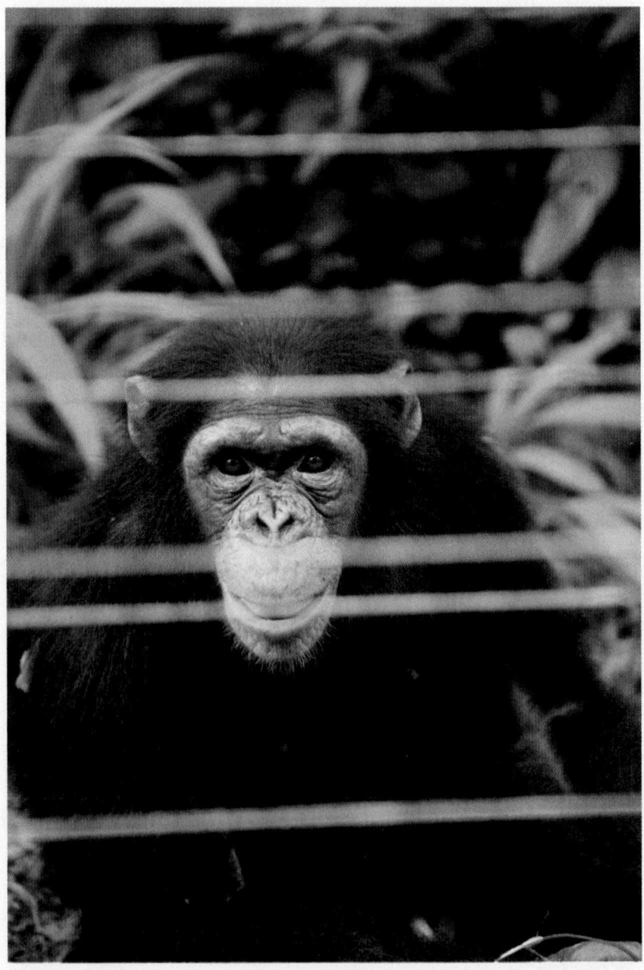

FIGURE 56.8 The source of HIV. In 2006, researchers found that chimpanzees in Cameroon carried a virus that was genetically very similar to HIV. Thus, these chimps are the most likely source of this emerging human disease. *(Tim E White/Alamy)*

particular concern because it kills a large percentage of those infected. Infected individuals have suffered a 50 to 90 percent death rate from different outbreaks of the disease. Those infected quickly begin to experience fever, vomiting, and sometimes internal and external bleeding (**FIGURE 56.9**). Death occurs within 2 weeks, and currently only experimental drugs are available to fight the virus. Unlike the progress that has been made with identifying the origin of HIV, the natural source of the Ebola virus has been difficult to determine. Because the virus also kills other primates at high rates, leaving no primate hosts for the virus, primates are not a likely long-term source of the virus. In 2013, however, researchers discovered that fruit bats carried the Ebola virus and were likely the reservoir species that spread the virus to primates. In 2017, a new experimental Ebola vaccine showed excellent protection against the pathogen and it was approved for use in the Democratic Republic of Congo.

Mad Cow Disease

In the 1980s, scientists first described a neurological disease, later known as **mad cow disease**, in which *prions* mutate into deadly pathogens and slowly damage a cow's nervous system. The cow loses coordination of its body (a condition compared with a person going mad), and then dies (**FIGURE 56.10**). Scientists now know that small, beneficial proteins in brains of cattle, called **prions**, occasionally mutate into deadly proteins that act as pathogens and subsequently cause mad cow disease. Prions are not well understood and represent a new category of pathogen.

In 1996, scientists in Great Britain announced that mad cow disease, also known as bovine spongiform encephalopathy (BSE), could be transmitted to humans who ate meat from infected cattle. Unlike harmful bacteria that can be killed with proper cooking, prions are difficult to destroy by cooking. Infected humans developed variant Creutzfeldt-Jakob Disease (CJD) and suffered a fate similar to the infected cattle.

FIGURE 56.9 Ebola hemorrhagic fever. The Ebola virus is highly lethal to humans and there are only experimental drugs for treatment. When treating a person infected with the virus, such as this patient who escaped hospital quarantine from Elwa hospital in Monrovia, Liberia, researchers and medical workers have to exercise extreme caution to avoid getting infected. *(REUTERS/Reuters TV)*

FIGURE 56.10 Mad cow disease. Cows that have been fed the remains of dead cows and sheep can become infected with harmful prions. These prions damage the nervous system and cause the cows to develop glazed eyes, body tremors, and a loss of coordination, eventually leading to death. Humans who consume beef from infected cows can become infected and suffer a similar fate. *(C. E. V./Science Source)*

Mutant prions cannot be transmitted among cattle that only live together. Transmission requires an uninfected cow to consume the nervous system of an infected cow. As a result, when cattle feed on grass together in a pasture, a rare mutation in a prion would be restricted to a single cow and not spread to other cattle. In the 1980s, however, the diets of European cattle commonly included the ground-up remains of dead cattle as a source of additional protein. When these remains happened to contain a mutant prion, the prions spread rapidly through the entire cattle population and, in turn, infected humans who ate the beef. In Britain, as of 2015, a total of 180,000 cattle have become infected and 177 people have died. It is estimated that several thousand people are currently infected, but the prions can exist in the human body for many years before they begin to cause symptoms of the disease. The European Union temporarily banned British beef imports in 1996 and the British government destroyed tens of thousands of cattle. Since that time, the disease has been found in several other countries including Canada and the United States, but only in a few cattle and no humans. Today, new rules exist that forbid the feeding of animal remains to cattle. As a result, the current risk of mad cow disease to humans has been greatly reduced and only five cases were detected in the United States from 2003 to 2017.

Swine Flu and Bird Flu

Humans commonly contract many types of flu viruses. As we saw in Chapter 5, the Spanish flu of 1918 killed up to

Mad cow disease A disease in which prions mutate into deadly pathogens and slowly damage a cow's nervous system.

Prion A small, beneficial protein that occasionally mutates into a pathogen.

100 million people. Spanish flu was a type of influenza, known as **swine flu**, caused by the H1N1 virus. This virus is similar to a flu virus that humans normally contract, but H1N1 normally infects only pigs. Occasionally, however, the flu jumps from pigs to humans. Another pandemic of swine flu occurred around the world in 2009 and 2010 and it caused more than 18,000 deaths. The most recent outbreaks have occurred in Venezuela in 2013 and in India in 2017. In India, more than 20,000 people became ill and at least 1,000 of them died. Fortunately, there are vaccines available to help prevent swine flu infections and there are drugs available to combat the swine flu virus if a person becomes infected.

In 2006, reports emerged from Asia that a related virus, known as H5N1, or **bird flu**, had jumped from birds to people, primarily to people who were in close contact with birds (**FIGURE 56.11**). Infections are rarely deadly to wild birds but can frequently cause domesticated birds such as ducks, chickens, and turkeys to become very sick and die. Humans often contract a variety of flu viruses. Because humans have no evolutionary history with the H5N1 virus, they have few defenses against it. As of 2016, more than 800 people had become infected by H5N1 and over half of them died. Governments responded to this risk by destroying large numbers of infected birds. Currently the H5N1 virus is not easily passed among people, but if a future mutation makes transmission easier, scientists estimate that H5N1 has the potential to kill 150 million people.

FIGURE 56.11 Bird flu. A farm worker feeds a large number of chickens on a farm in the Republic of Niger in western Africa. The virus that causes bird flu normally infects only birds. In 2006, however, the virus began jumping to human hosts where people and birds were in close contact. *(PIUS UTOMI EKPEI/Getty Images)*

SARS and MERS

In 2003, an unusual form of pneumonia was spreading through human populations in Southeast Asia that was eventually named **severe acute respiratory syndrome (SARS)**. While some of the respiratory symptoms are similar to bird flu and swine flu, SARS is a disease caused by a different type of virus known as a coronavirus. The virus can spread from one person to another and, during this outbreak, there were more than 8,000 people infected and nearly 10 percent of them died. Another coronavirus, *Middle Eastern Respiratory Syndrome (MERS)*, appeared on the Arabian Peninsula in 2012. Researchers are investigating the possibility that this disease originated from an animal source. To date, approximately 400 people have died from MERS.

West Nile Virus

The **West Nile virus** lives in hundreds of species of birds and is transmitted among birds by mosquitoes. Although the virus can be highly lethal to some species of birds, including blue jays (*Cyanocitta cristata*), American crows (*Corvus brachyrhynchos*), and American robins (*Turdus migratorius*), most species of birds survive the infection. During the latter half of the twentieth century there were increasing reports that the virus could sometimes infect horses and humans who had been bitten by mosquitoes. The first human case was identified in 1937 in the West Nile region of Uganda, thus giving the virus its name. In humans, the virus causes an inflammation of the brain leading to illness and sometimes death. In 1999, the virus appeared in New York and quickly spread throughout much of the United States. **FIGURE 56.12** shows the history of the West Nile virus in the United States. The highest numbers of infections and deaths from the virus occurred in 2002 and 2003. Increased efforts to combat mosquito populations and protect against mosquito bites are causing a decline in the disease, although there was a brief spike in West Nile Virus cases in 2012.

Swine flu A type of flu caused by the H1N1 virus.

Bird flu A type of flu caused by the H5N1 virus.

Severe acute respiratory syndrome (SARS) A type of flu caused by a coronavirus.

West Nile virus A virus that lives in hundreds of species of birds and is transmitted among birds by mosquitoes.

Lyme disease A disease caused by a bacterium (*Borrelia burgdorferi*) that is transmitted by ticks.

Lyme disease

Lyme disease is a disease caused by a bacterium (*Borrelia burgdorferi*) that is transmitted by ticks. The primary vector is the black-legged tick (*Ixodes scapularis*), which is also known as the deer tick (**FIGURE 56.13**). After hatching from eggs on the forest floor, the tick spends its first stage of life attached to birds and small rodents such as mice and chipmunks. If a bird or rodent is infected with the bacterium, it can be transferred to the tick feeding on the host's blood. After dropping off these hosts and spending the winter in the leaves of the forest, they attach themselves to larger mammals, including deer and people. It is when they attach their mouthparts to people that they transfer the disease-causing bacteria.

The CDC currently estimates that there are between 20,000 and 30,000 cases of Lyme disease in the United States annually, with most infections happening in the

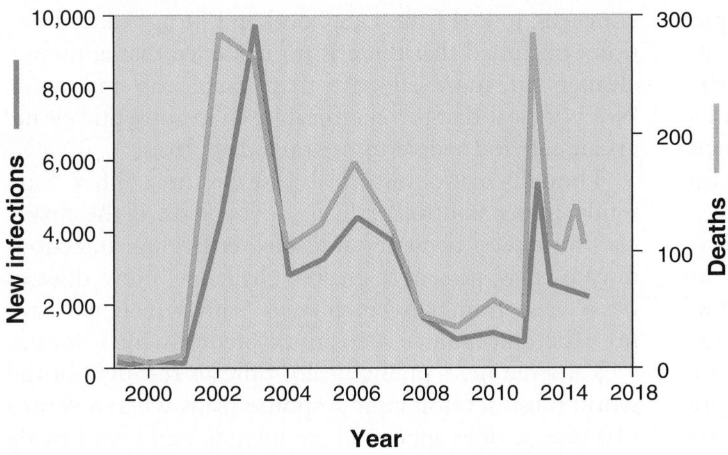

FIGURE 56.12 West Nile virus in the United States. Following the first appearance of West Nile virus in the United States in 1999, the number of human infections and deaths rapidly increased. Efforts to control populations of mosquitoes that carry the virus are helping to reduce the prevalence of the disease. *(Data from https://www.cdc.gov/westnile/statsmaps/cumMapsData.html.)*

northeastern United States. Infected people often experience a red bullseye at the site of tick attachment. This is followed by symptoms that resemble the flu, arthritis, and various neurological disorders. Most victims can be cured using modern antibiotics, although some people have persistent health problems years after becoming infected.

Lyme disease was first discovered in school children living in Lyme, Connecticut, in the 1970s, which is how the disease got its name. However, researchers have recently discovered the bacteria in a frozen mummy found in the European Alps, suggesting that the disease may have been infecting humans for more than 5,000 years.

FIGURE 56.13 Deer tick. When deer ticks attach to birds and rodents infected by the Lyme bacteria, the pathogen can be transferred to the tick. After the tick drops off the bird or rodent and attaches to a person, the pathogen can be transferred to the person. *(Jiri Prochazka/Shutterstock.com)*

Zika virus disease

The **Zika virus disease** is caused by a pathogen that causes babies to be born with unusually small heads and damaged brains. The virus is carried by mosquitoes and transmitted to humans when bitten by the mosquito or through sexual contact with an infected person. Most infected people only experience mild effects for a few days or weeks after being infected, such as rashes, fevers, and headaches. However, the major risk is that the virus can be passed between pregnant mothers and their fetuses. When this happens, the babies become infected and the brain damage begins.

Zika is a very recent emerging infectious disease. Although it was first discovered in a monkey in Uganda in 1947, it was only found in 14 people around the world prior to 2007. In that year, more than 10,000 people became infected on Yap Island in Southeast Asia. In 2013, more than 28,000 people became infected in Tahiti. Today, it is present in Central America, South America, Central Africa, and Southeast Asia. In 2015, the virus appeared in Brazil and more than 1 million people became rapidly infected. There is currently no known treatment for the Zika virus and attempts to control its spread are focused on reducing mosquito populations.

Human health faces a number of future challenges

While humans face a large number of health risks, we have an excellent understanding of important risk factors and the ways to combat many historical and emerging infectious diseases. Combating diseases in low-income countries requires improvements in nutrition, wider availability of clean drinking water, and proper sanitation. In high-income countries, we need to promote healthier lifestyle choices such as increased physical activity, a balanced diet, and limiting excess food consumption and tobacco use. In all countries, continued education is needed to reduce the spread of diseases such as HIV and tuberculosis.

As we combat many diseases, an issue of growing concern is the ability of pathogens to evolve resistance. As we noted in the case of tuberculosis, patients often feel much better long before they complete the full year of prescribed medicine. Because they feel better or because they cannot afford a full year of drugs, they stop the drug treatment early. This allows a small fraction of highly resistant bacteria to survive in the body,

Zika virus disease A disease caused by a pathogen that causes fetuses to be born with unusually small heads and damaged brains.

reproduce, and then potentially spread to other people. When a pathogen evolves resistance to one drug, physicians often prescribe a different drug to combat the pathogen. Over time, however, some pathogens such as tuberculosis have evolved multiple drug resistance. Without new drugs, little can be done for patients with pathogen strains that possess multiple drug resistance.

A similar issue occurs with the increased use of antiseptic cleaners that are designed to kill microbes such as bacteria. Antiseptic cleaners include a large number of antibacterial soaps; these products typically kill a large proportion of harmful pathogens, but not all of them. As a result of our efforts to wipe out pathogens, we are inadvertently selecting for pathogens that possess a stronger resistance to our efforts. Moreover, these products also move through wastewater and into streams, rivers, and lakes. Ironically, it is unclear that antiseptic products kill microbes any better than plain soap. Because of these concerns, in 2013 the U.S. Food and Drug Administration concluded that there is no evidence that antiseptic cleaners are more effective than plain soap and water. Nor is it clear that the chemicals used in antiseptic cleaners are safe for people to use on a daily basis.

Though many historical diseases are either currently under control or likely to be soon if the financial resources become available, emerging infectious diseases may present a greater challenge. New diseases often arise from new pathogens with which we have no experience. Since we cannot predict which diseases will emerge next, public health officials throughout the world must develop rapid response plans when a particular disease does appear. These include rapid worldwide notification of newly identified diseases and strategies to isolate infected persons, which will slow the spread of the disease and provide time for researchers to develop appropriate tactics to combat the threat.

MODULE 56 AP® Review

Preparing for the AP® Exam

In this module, we examined biological risks. We learned that human health risks include physical, chemical, and biological risks. Biological risks are those associated with diseases that are either noninfectious, such as heart disease, or infectious, such as diseases caused by pathogens. Diseases can also be categorized as either acute or chronic and the risk factors for chronic diseases differ between low- and high-income countries. Some of the infectious diseases that have a long history of harming humans include plague, malaria, and tuberculosis. Other diseases have emerged more recently and include HIV/AIDS, mad cow disease, swine flu, bird flu, SARS, West Nile virus, Zika virus disease, and Lyme disease. Knowing the disease risk factors and the ways to combat these diseases will help to reduce their effect on humans, providing that we remember issues related to the evolution of drug resistance. In the next module, we will look to chemical risks.

AP® Practice Questions

Choose the best answer for the following.

1. An infectious disease is always
   (a) caused by a virus.
   (b) transmitted between humans and animals.
   (c) treatable with antibiotics.
   (d) caused by a pathogen.

2. A disease that rapidly impairs a body's function is
   (a) infectious. (c) chronic.
   (b) acute. (d) pathogenic.

3. Tuberculosis
   (a) is transmitted by mosquitoes.
   (b) has been almost entirely eliminated in the world.
   (c) is caused by a virus.
   (d) has strains that have developed resistance to antibiotics.

4. Prions are pathogens that are responsible for
   (a) Ebola hemorrhagic fever.
   (b) mad cow disease.
   (c) bird flu.
   (d) SARS.

5. Which is NOT a step to be taken in the future for combating diseases?
   (a) improving nutrition
   (b) the use of many antiseptic cleaners
   (c) increased education
   (d) proper sanitation

MODULE 57

Toxicology and Chemical Risks

The complexity of the biological risks humans face is matched by the complexity of chemical risks. Our modern society has developed an incredible array of chemicals to improve human health and food production, including pharmaceuticals, insecticides, herbicides, and fungicides. We have also seen that chemical by-products from manufacturing and the generation of energy can be harmful to humans and the environment. Even beneficial chemicals, when released into the environment, can harm humans and other organisms. Many pharmaceuticals, for example, have unexpected consequences when released into the environment. In this module we will look at the types of chemicals that can have harmful effects. We will see how scientists study these chemicals, and what effect the chemicals have on humans.

Learning Goals

After reading this module you should be able to

- identify the major types of harmful chemicals.
- explain how scientists determine the concentrations of chemicals that harm organisms.

Many types of chemicals can harm organisms

Chemicals can have many different effects on organisms, and some of the most harmful are common in our environment; **TABLE 57.1** on page 622 lists those of current concern. They can be grouped into five categories: *neurotoxins, carcinogens, teratogens, allergens,* and *endocrine disruptors.*

Neurotoxins

Neurotoxins are chemicals that disrupt the nervous systems of animals. Many insecticides, for example, are neurotoxins that interfere with an insect's ability to control its nerve transmissions. Insects and other invertebrates are highly sensitive to neurotoxin insecticides. These animals can become completely paralyzed, cannot obtain oxygen, and quickly die. Other important neurotoxins include lead and mercury. As we discussed in Chapters 14 and 15, lead and mercury are very harmful heavy metals that can damage the human kidneys, brain, and nervous system. Since the federal government required a gradual elimination of lead in gasoline and paint in the 1970s, lead exposure in the United States has declined by about 90 percent. However, lead contamination in children remains a serious problem in low-income neighborhoods due to the presence of old lead paint in buildings. Mercury also remains a major problem.

Carcinogens

Carcinogens are chemicals that cause cancer. Carcinogens cause cell damage and lead to uncontrolled growth of these cells either by interfering with the normal metabolic processes of the cell or by damaging the genetic material of the cell. Carcinogens that cause damage to the genetic material of a cell are called **mutagens** (although not all mutagens are carcinogens). Some of the most well-known carcinogens include asbestos, radon, formaldehyde, and the chemicals found in tobacco.

Teratogens

Teratogens are chemicals that interfere with the normal development of embryos or fetuses. One of the most infamous teratogens was the drug thalidomide, prescribed to pregnant women during the late 1950s and early 1960s to combat morning sickness. Sadly, tens of thousands of these mothers around the world gave

Neurotoxin A chemical that disrupts the nervous systems of animals.

Carcinogen A chemical that causes cancer.

Mutagen A type of carcinogen that causes damage to the genetic material of a cell.

Teratogen A chemical that interferes with the normal development of embryos or fetuses.

TABLE 57.1 Some chemicals of major concern

Chemical	Sources	Type	Effects
Lead	Paint, gasoline	Neurotoxin	Impaired learning, nervous system disorders, death
Mercury	Coal burning, fish consumption	Neurotoxin	Damaged brain, kidneys, liver, and immune system
Arsenic	Mining, groundwater	Carcinogen	Cancer
Asbestos	Building materials	Carcinogen	Impaired breathing, lung cancer
Polychlorinated biphenyls (PCBs)	Industry	Carcinogen	Cancer, impaired learning, liver damage
Radon	Soil, water	Carcinogen	Lung cancer
Vinyl chloride	Industry, water from vinyl chloride pipes	Carcinogen	Cancer
Alcohol	Alcoholic beverages	Teratogen	Reduced fetal growth, brain and nervous system damage
Atrazine	Herbicide	Endocrine disruptor	Feminization of males, low sperm counts
DDT	Insecticide	Endocrine disruptor	Feminization of males, thin eggshells of birds
Phthalates	Plastics, cosmetics	Endocrine disruptor	Feminization of males

birth to children with defects before the drug was taken off the market in 1961 (**FIGURE 57.1**). One of the most common modern teratogens is alcohol. Excessive alcohol consumption reduces the growth of the fetus and damages the brain and nervous system of the fetus, a condition known as fetal alcohol syndrome. This is why physicians recommend that women not consume alcoholic beverages while they are pregnant.

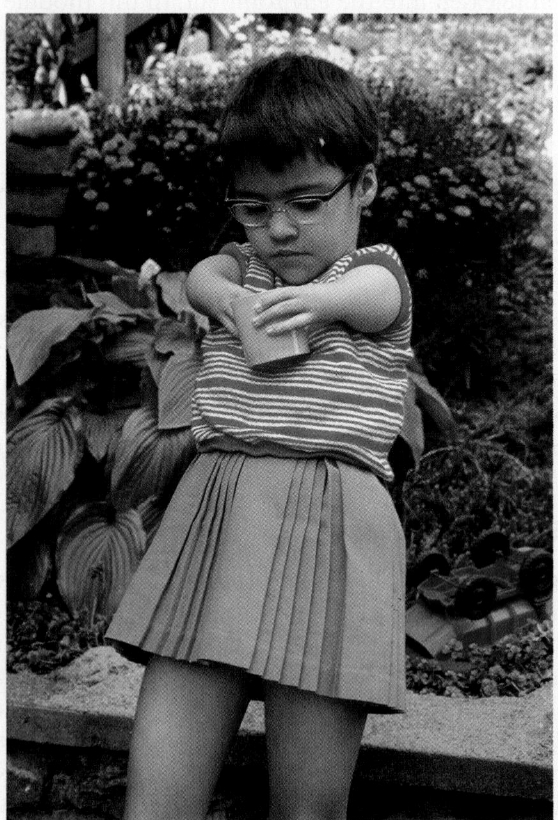

FIGURE 57.1 The effects of thalidomide. Thalidomide was widely prescribed to pregnant women in the late 1950s to alleviate the symptoms of morning sickness, but it had the unanticipated effect of causing birth defects in tens of thousands of newborn children. *(Leonard McCombe/Getty Images)*

AP® Exam Tip

You should be able to identify specific examples of chemical toxins, their source, and their human health impacts. ●

Allergens

Allergens are chemicals that cause allergic reactions. Although allergens are not pathogens, allergens are capable of causing an abnormally strong response from the immune system. In some cases, this response can cause breathing difficulties and even death. Typically, a given allergen only causes allergic reactions in a small fraction of people. Some common chemicals that cause allergic reactions include the chemicals naturally found in peanuts and milk and several drugs including penicillin and codeine.

Allergen A chemical that causes allergic reactions.

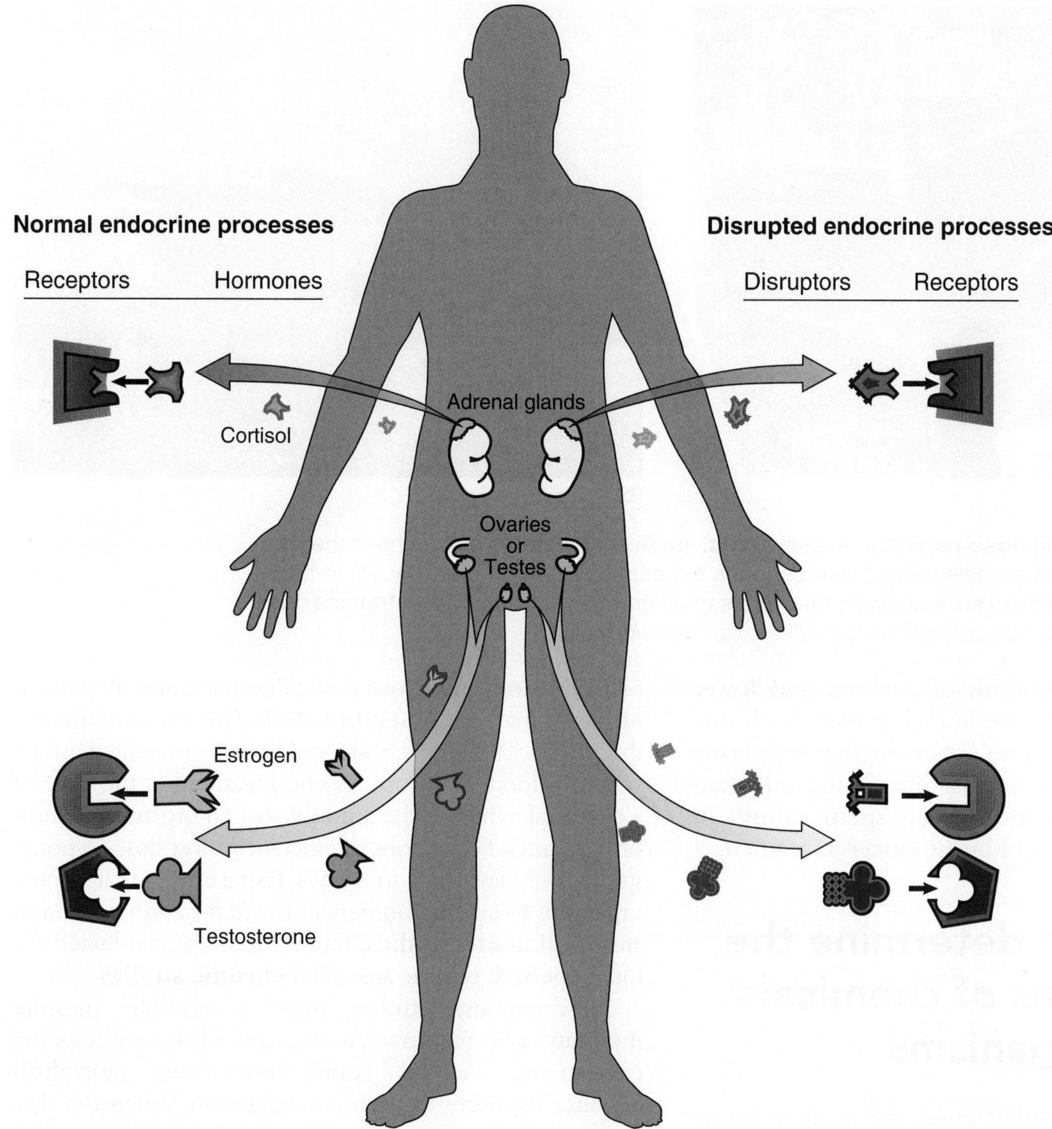

FIGURE 57.2 Endocrine disruption. In normal endocrine processes, hormones bind with receptors on cells to regulate the functioning of the body including growth, metabolism, and the development of reproductive organs. Hormone-disrupting chemicals mimic the hormones in the body and also bind to receptive cells and cause the cell to respond in ways that are not beneficial to the organism.

Endocrine Disruptors

Endocrine disruptors are chemicals that interfere with the normal functioning of hormones in an animal's body. Hormones are normally manufactured in the endocrine system and released into the bloodstream in very low concentrations. As the hormones move through the body, they bind to specific cells. Binding stimulates the cell to respond in a way that regulates the functioning of the body including growth, metabolism, and the development of reproductive organs. As **FIGURE 57.2** shows, an endocrine disruptor can bind to receptive cells and cause the cell to respond in ways that are not beneficial to the organism.

One high-profile example of endocrine disruptors in our environment is the group of reproductive hormones that can be found in wastewater. As we discussed in Chapter 14, wastewater may contain hormones from a variety of sources including animal-rearing facilities, human birth control pills, and pesticides that mimic animal hormones. In waterways exposed to hormones through wastewater, scientists are increasingly finding that male fish, reptiles, and amphibians are becoming feminized; males possess testes that have low sperm counts and, in some cases, testes that produce both eggs and sperm. Males normally convert the female hormone estrogen into the male chemical testosterone. Reproductive hormones in wastewater can interfere with the production of testosterone, which causes males

Endocrine disruptor A chemical that interferes with the normal functioning of hormones in an animal's body.

(a)

(b)

FIGURE 57.3 Conducting dose-response experiments. (a) Researchers determine how chemicals affect the mortality of animals using dose-response experiments in the laboratory. (b) In the experiment shown, researchers are examining the effects of different insecticide concentrations on the survival of tadpoles. *(a: Photo courtesy of Rick Relyea; b:Courtesy of Jason Hoverman)*

to have higher concentrations of estrogen and lower concentrations of testosterone in their bodies. Such discoveries raise serious concerns about whether endocrine disruptors might affect the normal functioning of human hormones. These effects include low sperm counts in men and an increased risk of breast cancer in women.

Scientists can determine the concentrations of chemicals that harm organisms

To assess the risk a chemical poses, we need to know the concentrations that cause harm. Scientists have three techniques to determine harmful concentrations: dose-response studies, prospective studies, and retrospective studies.

Dose-Response Studies

Dose-response studies expose animals or plants to different amounts of a chemical and then look for a variety of possible responses including mortality or changes in behavior or reproduction. For example, dose-response studies of aquatic animals such as tadpoles are used to determine the concentrations of various pesticides that cause 50 percent of the animals to die (**FIGURE 57.3**). The concentration of the chemicals being considered can be measured in air, water, or food. They can also be measured as the dose of a chemical, which is the amount that an organism absorbs or consumes. For reasons of efficiency, most dose-response studies only last for 1 to 4 days. Experiments that expose organisms to an environmental hazard for a short duration are called **acute studies**. Studies that are conducted for longer periods of time are called **chronic studies**.

Dose-response study A study that exposes organisms to different amounts of a chemical and then observes a variety of possible responses, including mortality or changes in behavior or reproduction.

Acute study An experiment that exposes organisms to an environmental hazard for a short duration.

Chronic study An experiment that exposes organisms to an environmental hazard for a long duration.

LD50 The lethal dose of a chemical that kills 50 percent of the individuals in a dose-response study.

Dose-response studies most commonly measure mortality as a response. At the end of a dose-response experiment, scientists count how many individuals die after exposure to each concentration. When the data are graphed, they generally follow an S-shaped curve, like the one in **FIGURE 57.4**. If you examine the purple curve, you will see that at the lowest dose no individuals die. At slightly higher doses, a few individuals die. The dose at which an effect can be detected is called the threshold. These individuals generally are in poorer health or genetically are not very tolerant of the chemical. As the dose is further increased, many more individuals begin to die. At the highest concentrations all individuals die.

To compare the harmful effects of different chemicals scientists measure the **LD50**, which is an abbreviation for the lethal dose that kills 50 percent of the individuals in a dose-response study. The LD50 value helps assess the relative toxicity of a chemical to a particular species. For example, scientists can compare the LD50 value of a new chemical with the LD50 value of thousands of previously tested chemicals to determine whether the new chemical is more or less lethal to a given organism than other chemicals.

Although the vast majority of toxicology studies are only conducted for a few days, chronic studies will often last from the time an organism is very young to when it is old enough to reproduce. For some species such as fish, chronic

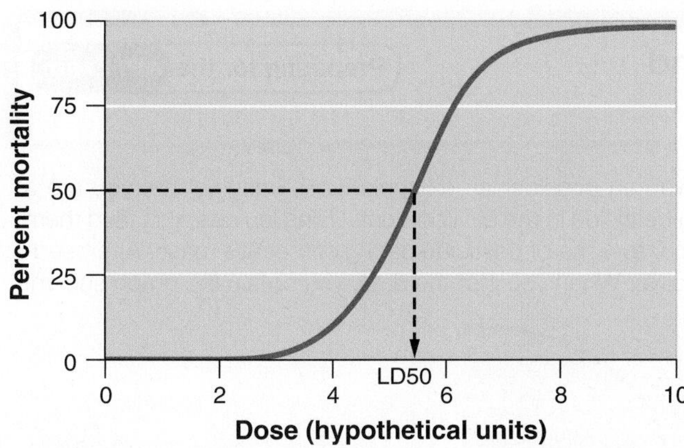

FIGURE 57.4 LD50 studies. To determine the dose of a chemical that causes a 50 percent death rate, scientists expose animals to different doses of a chemical and determine what proportion of the animals die at each dose. Such an experiment typically produces an S-shaped curve.

experiments can take several months. The goal of chronic studies is to examine the long-term effects of chemicals, including how they affect survival and reproduction.

Not all dose-response experiments measure death as a response to chemicals. In many cases, scientists are interested in other harmful effects, including chemicals acting as a teratogen, carcinogen, or neurotoxin. When exposure to a chemical does not kill an organism but impairs its behavior, physiology, or reproduction, we say the chemical has **sublethal effects**. In these cases, the experiments are conducted to determine the **ED50**, which is the effective dose that causes 50 percent of the individuals in a dose-response study to display a harmful, but nonlethal, effect. In addition to quantifying LD50 and ED50 values, researchers commonly determine the **No-observed-effect level (NOEL)**, which is the highest concentration of a chemical that causes no lethal or sublethal effects. The NOEL indicates how much of a chemical an organism can experience with no harmful effect.

AP® Exam Tip

You should be able to read and interpret an LD50 graph for the AP® Environmental Science Exam. ●

Testing Standards

In the United States, chemicals that affect humans and other species are regulated by the Environmental Protection Agency (EPA). The Toxic Substances Control Act of 1976 gives the EPA the authority to regulate many chemicals, but does not include food, cosmetics, and pesticides. Pesticides are regulated under a separate law—the Federal Insecticide, Fungicide, and Rodenticide Act of 1996. Under this act, a manufacturer must demonstrate that a pesticide "will not generally cause unreasonable adverse effects on the environment."

Because no chemical can be tested on every one of the approximately 10 million species of organisms on Earth, scientists have devised a system of testing a few species—a bird, mammal, fish, and invertebrate—that are thought to be among the most sensitive in the world. The particular species tested from each of the four animal groups can vary, depending on which species is thought to be the most sensitive to a particular chemical. The reasoning for this is that regulations devised to protect the most sensitive species in a group will automatically protect all other species in that group. Since conducting LD50 studies on humans would be unethical, results from studies conducted on mice and rats are extrapolated to humans. For nonhuman animals, test results from mice and rats are used to represent all mammals, birds such as pigeons and quail are used to represent all birds, fish such as trout are used to represent all fish, and invertebrates such as water fleas are used to represent all invertebrates.

You might have noticed that the groups of tested animals do not include amphibians or reptiles. Unfortunately, the standards for testing chemicals were set up before there was much interest in protecting amphibians and reptiles. Currently, test results from fish are used to represent aquatic amphibians and reptiles, whereas test results from birds are used to represent terrestrial amphibians and reptiles. Because amphibians and reptiles are now experiencing population declines throughout the world, there is increased interest in requiring tests on species from these two groups as well.

Using the LD50 and ED50 values from dose-response experiments, regulatory agencies such as the EPA can determine the concentrations in the environment that should cause no harm. For most animals, a safe concentration is obtained by taking the LD50 value and dividing it by 10. The logic is that if the LD50 value causes 50 percent of the animals to die, then 10 percent of the LD50 value should cause few or no individuals to die.

The regulatory agencies, however, are much more conservative in setting concentrations for humans. Scientists determine the LD50 or ED50 values for rats or mice and then divide by 10 to determine a safe concentration for rats and mice. This value is divided by 10 again to reflect that rats and mice may be less sensitive to a chemical than humans. Finally, this value is often divided by 10 again to ensure an extra level of caution. In short, the LD50 and ED50 values obtained from rats and mice are divided by 1,000 to set the safe values for humans. "Do the Math: Estimating LD50 Values and Safe Exposures" on page 626 shows you how to make this calculation.

Sublethal effect The effect of an environmental hazard that is not lethal, but which may impair an organism's behavior, physiology, or reproduction.

ED50 The effective dose of a chemical that causes 50 percent of the individuals in a dose-response study to display a harmful, but nonlethal, effect.

No-observed-effect level (NOEL) The highest concentration of a chemical that causes no lethal or sublethal effects.

DO THE MATH Estimating LD50 Values and Safe Exposures

Preparing for the AP® Exam

Using our knowledge of how scientists conduct LD50 studies, we can consider an example. Let's imagine that you are a scientist charged with determining the safe levels for mammals of a pesticide in the environment. Using lab rats, you feed them a diet that contains different amounts of the pesticide, ranging from 0 to 4 mg of pesticide per kg of the rat's mass. After feeding them these diets for 4 days, you count how many rats are still alive. When you plot the data, you obtain the graph shown.

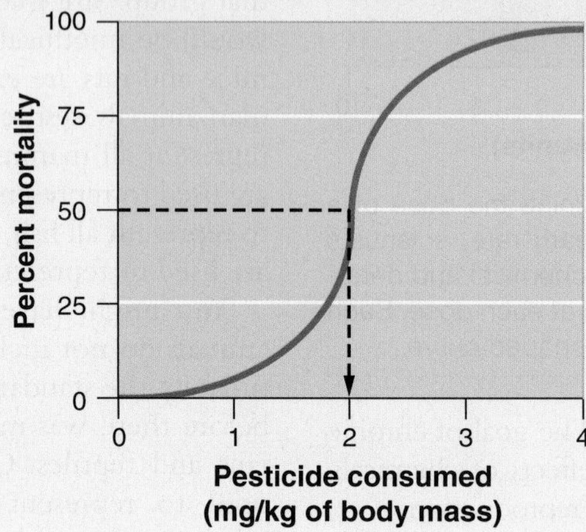

What is the LD50 value for lab rats? To determine this, we can draw a horizontal line at the point of 50 percent mortality on the *y* axis. Where this line intersects the purple line, we can draw another line straight down to the *x* axis. This second line crosses the *x* axis at 2 mg/kg of mass.

Based on this LD50 study, what amount of pesticide would be considered safe for mammals to ingest? Recall that we can calculate this number by taking the LD50 value and dividing it by 10. Thus the safe amount of pesticide for a rat is:

$$\frac{2 \text{ mg/kg of mass}}{10} = 0.2 \text{ mg/kg of mass}$$

YOUR TURN Using the same LD50 study, what amount of pesticide would be considered safe for a human to ingest?

Retrospective versus Prospective Studies

Estimating the effects of chemicals on humans is a major challenge. We have seen that one approach is to conduct dose-response experiments on rats and mice and extrapolate the results to humans. An alternative approach is to examine large populations of humans or animals that are exposed to chemicals in their everyday lives and then determine whether these exposures are associated with any health problems. Such investigations fall within the study of epidemiology, a field of science that strives to understand the causes of illness and disease in human and wildlife populations. There are two ways of conducting this type of research: *retrospective studies* and *prospective studies.*

Retrospective studies monitor people who have been exposed to an environmental hazard, such as a harmful chemical, at some time in the past. In such studies, scientists identify a group of people who have been exposed to a potentially harmful chemical and a second group of people who have not been exposed to the chemical. Both groups are then monitored for many years to see if the exposed group experiences more health problems than the unexposed group. In 1984, for example, there was an accidental release of methyl isocyanate gas from a Union Carbide pesticide factory in Bhopal, India (**FIGURE 57.5a**). More than 36,000 kg (80,000 pounds) of hazardous gas spread through the city of 500,000 inhabitants. An estimated 2,000 people died that night and another 15,000 died later from effects related to the exposure. For more than 2 decades scientists have been monitoring many citizens of Bhopal to determine if survivors of the accident have developed any additional health problems. The retrospective studies have found that approximately 100,000 people are still suffering illnesses from the accidental exposure to the gas. The survivors have higher rates of genetic abnormalities, infant mortality, kidney failure, and

Retrospective study A study that monitors people who have been exposed to an environmental hazard, such as a harmful chemical, at some time in the past.

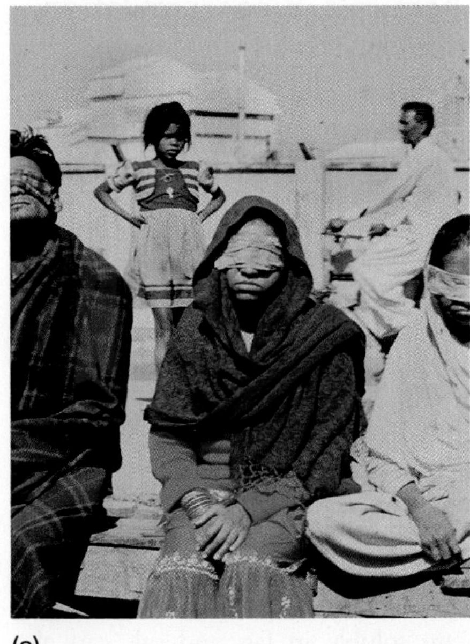

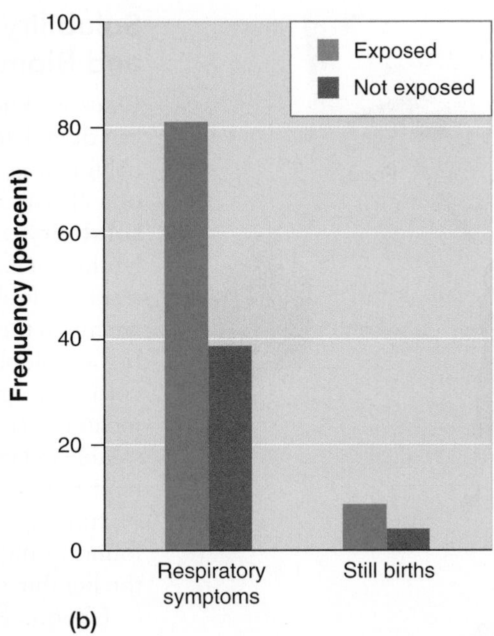

(a) (b)

FIGURE 57.5 The chemical disaster in Bhopal, India. (a) In 1984, a massive release of methyl isocyanate gas killed and injured thousands of people. (b) Retrospective studies that followed the survivors of the accident have identified a large number of longer-term health effects from the accident. *(Data from P. Cullinan, S.D. Acquilla, and V.R. Dhara, Long-term morbidity in survivors of the 1984 Bhopal gas leak, National Medical Journal of India, Jan.–Feb. (1996) 9(1): 5–10. Photo by STR/Getty Images)*

learning disabilities. As shown in Figure 57.6b, they also have higher rates of respiratory problems and stillbirths.

In contrast to retrospective studies, **prospective studies** monitor people who might become exposed to an environmental hazard, such as a harmful chemical, at some time in the future. In this case, scientists might select a group of 1,000 participants and ask them to keep track of the food they eat, the tobacco they use, and the alcohol they drink over a period of several decades. As time passes, the researchers can determine if the habits of the participants are associated with any future health problems. Prospective studies can be quite challenging because a participant's habits, such as tobacco use, can also be associated with many other risk factors, such as socioeconomic status. Of particular concern is when multiple risks cause **synergistic interactions**, in which two risks together cause more harm than expected based on the separate effects of each risk alone. For example, the health impact of a carcinogen such as asbestos can be much higher if an individual also smokes tobacco.

Studies of lead in children are often prospective. In one study, researchers at Harvard University looked at the effects of lead on children's intelligence by following 276 children in Rochester, New York, from 6 months to 5 years of age. IQ tests are reliable at the age of 5. In addition to lead exposure, the researchers also accounted for other factors that might affect childhood IQ including the mother's IQ, exposure to tobacco, and the intellectual environment of their homes. After controlling for these other factors, the researchers found that among children who had been exposed to lead in the environment—primarily from breathing lead dust and consuming lead paint chips—those with higher lead exposures scored lower on subsequent IQ tests. Such prospective studies can help regulators determine acceptable levels of chemical exposure.

Factors That Determine the Concentrations of Chemicals That Organisms Experience

Knowing the concentrations of chemicals that can harm humans or other animals is important, but it is only useful when combined with information about the concentrations that an individual might actually experience in the environment. If a chemical is quite harmful at some moderate concentration but individuals only experience lower concentrations of that chemical, we might not be particularly concerned. Therefore, to identify and understand the effects of chemical concentrations that organisms experience, we need to know something about how the chemicals behave in the environment.

Prospective study A study that monitors people who might become exposed to an environmental hazard, such as a harmful chemical, at some time in the future.

Synergistic interaction A situation in which two risks together cause more harm than expected based on the separate effects of each risk alone.

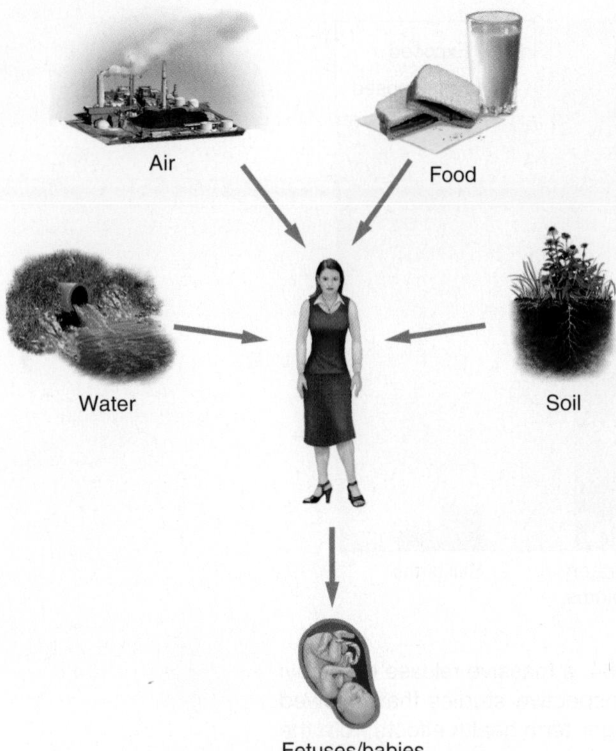

FIGURE 57.6 Routes of exposure. Despite a multitude of potential routes of exposure to chemicals, most chemicals have a limited number of major routes.

Routes of Exposure

The ways in which an individual might come into contact with an environmental hazard, such as a chemical, are known as **routes of exposure**. As **FIGURE 57.6** illustrates, the full range of possibilities is complex because it includes potential exposures from the air, from water used for drinking, bathing, or swimming, from food, and from the environments of places where people live, work, or visit. For any particular chemical, however, the major routes of exposure are usually limited to just a few of the many possible routes. For example, bisphenol A is a chemical used in manufacturing hard plastic items such as toys, food containers, and baby bottles. Recent research has raised concerns that bisphenol A may be responsible for early puberty and increased rates of cancer. While these effects are being debated and investigated, it is clear that a child's routes of exposure to bisphenol A are limited to toys, food containers, and baby bottles.

Route of exposure The way in which an individual might come into contact with an environmental hazard, such as a chemical.

Solubility How well a chemical dissolves in a liquid.

Bioaccumulation An increased concentration of a chemical within an organism over time.

Biomagnification The increase in chemical concentration in animal tissues as the chemical moves up the food chain.

Solubility of Chemicals, Bioaccumulation, and Biomagnification

Once we know the potential routes of exposure, scientists can then determine the chemical's *solubility* and its potential for *bioaccumulation* and *biomagnification*. The movement of a chemical in the environment depends in part on its **solubility**, which is how well a chemical can dissolve in a liquid. For example, some chemicals such as herbicides are readily soluble in water whereas others such as insecticides are much more soluble in fats and oils. When a chemical is highly soluble in water, it can be washed off surfaces, percolate into groundwater, and run off into surface waters including rivers and lakes. In contrast, chemicals that are soluble in fats and oils are not very soluble in water so they tend not to be found percolating into the groundwater or running off into surface waters. Instead, they can be found in higher concentrations bound to soils, including the benthic soils that underlie bodies of water.

Chemicals that are soluble in fats and oils can also become stored in the fatty tissues of animals. For example, in Chapter 11 we mentioned that DDT accumulates in the fatty tissues of aquatic birds such as pelicans and those, such as eagles, that feed on aquatic animals. This process, known as **bioaccumulation**, is the increased concentration of a chemical within an organism over time. The process of bioaccumulation begins when an individual is exposed to small amounts of a chemical from the environment and incorporates the chemical into its tissues, typically its fat tissues. Fish, for example, are exposed to low concentrations of methyl mercury when they drink water, pass water over their gills to breathe, and consume food that contains mercury. A fish stores mercury in its fat tissues and, over time, the mercury accumulates. The rate of accumulation for any animal will depend on the concentration of the chemical in the environment, the rate at which the animal takes up each source of the chemical, the rate at which the chemical breaks down inside the animal, and the rate at which it is excreted by the animal.

Biomagnification is the increase in chemical concentration in animal tissues as the chemical moves up the food chain. In this way, the original concentration in the environment is magnified to occur at a much higher concentration in the top predator of the community. The classic example of biomagnification is the case of DDT, an insecticide that has been widely used to kill insect pests in agriculture and to kill the mosquitoes that carry malaria and other diseases. DDT is not soluble in water, so when sprayed over water it quickly binds to particulates in the water and the underlying soil or is quickly taken up by the tiny zooplankton that act as primary consumers on algae. As we see in **FIGURE 57.7**, the very low concentration of DDT in the water bioaccumulates in the bodies of the zooplankton where it becomes approximately 1,000 times more concentrated. Small fish eat the zooplankton for many weeks or months and the DDT is further concentrated approximately sixfold. Large fish spend their lives eating

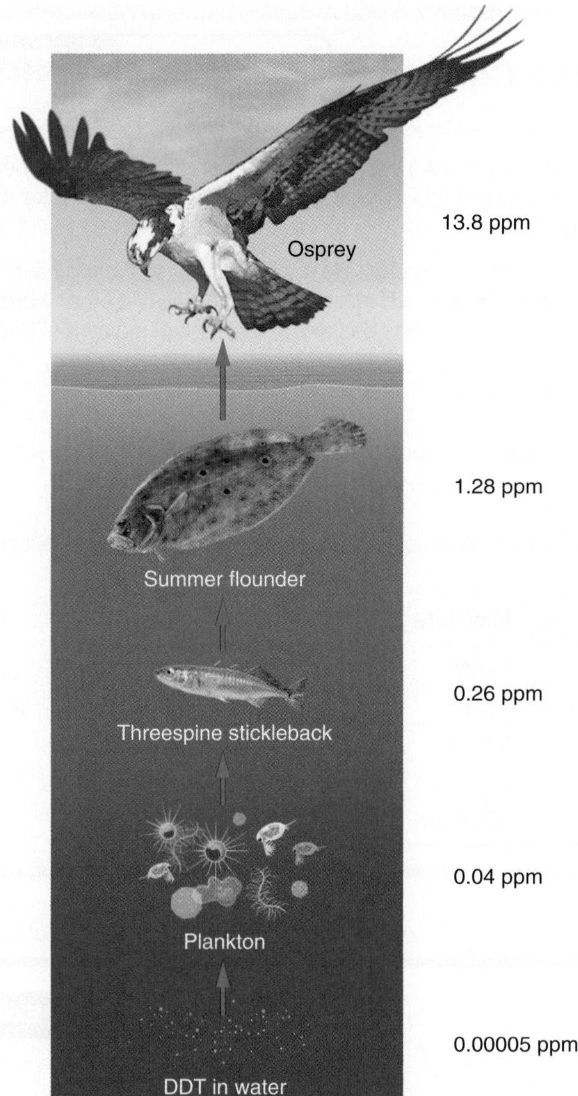

FIGURE 57.7 The biomagnification of DDT. The initial exposure is primarily in a low trophic group such as the plankton in a lake. Consumption causes the upward movement of the chemical where it is accumulated in the bodies at each trophic level. The combination of bioaccumulation at each trophic level and upward movement by consumption allows the concentration to magnify to the point where it can be substantially more concentrated in the top predator than it was in the water. *(Data from G. M. Woodwell, C. F. Wurster, Jr., and Peter A. Isaacson, DDT Residues in an East Coast Estuary: A Case of Biological Concentration of a Persistent Insecticide, Science, New Series, 156 (3776) (May 12, 1967): 821–824, http://www.jstor.org/stable/1722018.)*

the contaminated smaller fish and the DDT in the large fish is further concentrated approximately fivefold. Finally, fish-eating birds such as pelicans and eagles spend years eating the large fish and further magnify the DDT in their own bodies. Because of biomagnification along the food chain, the concentration of DDT in the birds is nearly 276,000 times higher than the concentration of DDT in the water. The concentrated DDT in the fish-eating birds causes them to produce thin-shelled eggs that often break when the parent birds try to incubate the eggs. This was a primary cause in the decline of these birds in the 1960s. Since DDT was banned in the United States in 1972, the populations of fish-eating birds have dramatically increased.

TABLE 57.2 The persistence of various chemicals in the environment

Chemical	Source	Half-Life
Malathion	Insecticide	1 day
Radon	Rocks and soil	4 days in air
Vinyl chloride	Industry, water from vinyl chloride pipes	4.5 days in air
Phthalates	Plastics, cosmetics	2.5 days in water
Roundup	Herbicide	7 to 70 days in water
Atrazine	Herbicide	224 days in wetland soils
Polychlorinated biphenyls (PCBs)	Industry	8 to 15 years in water
DDT	Insecticide	30 years in soil

Persistence

The **persistence** of a chemical refers to how long the chemical remains in the environment. Persistence depends on a number of factors including temperature, pH, whether the chemical is in water or soil, and whether it can be degraded by sunlight or broken down by microbes. Scientists often measure persistence by observing the time needed for a chemical to degrade to half its original concentration, known as the half-life of the chemical, which we discussed in Chapter 12. **TABLE 57.2** lists the persistence of various chemicals in the environment measured according to half-life. DDT, for example, has a half-life in soil of up to 30 years. Thus, even after DDT is no longer sprayed in an area, half of the chemical that was absorbed in the soil would still be present after 30 years, and one-fourth would be present after 60 years. Such synthetic, carbon-based molecules that break down very slowly in the environment are known as *persistent organic pollutants (POPs)*. Chemicals that cause harmful effects on humans and other organisms may become even larger risks when they persist for many years. For this reason, many modern chemicals are designed to break down much more rapidly so that any unintended effects will be short-lived. "Do the Math: Estimating Half-Lives of Toxic Chemicals" shows you how half lives of chemicals are calculated.

Persistence The length of time a chemical remains in the environment.

DO THE MATH Estimating Half-Lives of Toxic Chemicals

Preparing for the AP® Exam

As we have discussed, it is important that we know something about the persistence of toxic chemicals in the environment to evaluate how much a person, plant, or animal might experience. Using the table below, quantify how long it will take the insecticide malathion to break down to one-eighth of its initial concentration.

We know that the amount of any chemical is reduced by half after the duration of a half-life (by definition), so we will have ½ as much malathion after one day, ¼ as much malathion after 2 days, and 1/8 as much malathion after 3 days. In other words, we will have 1/8 as much of any chemical after the time required for three half-lives has passed. In the case of malathion, three half-lives requires 3 days.

$$3 \text{ half-lives} \times 1 \text{ day/half-life} = 3 \text{ days}$$

As another example, how much time will it take until we have 1/8 as much radon present?

$$3 \text{ half-lives} \times 4 \text{ days/half-life} = 12 \text{ days}$$

YOUR TURN Based on the half-lives provided in the table, how long will it take to have only 1/8 as much Roundup, atrazine, and PCBs?

Chemical	Source	Half-Life
Malathion	Insecticide	1 day
Radon	Rocks and soil	4 days
Roundup	Herbicide	40 days
Atrazine	Herbicide	224 days
PCBs	Industry	10 years

MODULE 57 AP® Review

Preparing for the AP® Exam

In this module, we learned there are many types of chemicals that can potentially cause harmful effects in humans. Neurotoxins disrupt the nervous systems of animals, carcinogens cause cancer, teratogens cause abnormal development in embryos and fetuses, allergens cause abnormally high immune responses, and endocrine disruptors interfere with the normal functioning of hormones. For each of these types of chemicals, scientists can determine the concentrations of chemicals that will harm organisms using short-term LD50 to assess lethal effects and ED50 studies to assess sublethal effects. Scientists can also use chronic studies that examine long-term effects of chemical exposure including both retrospective and prospective studies. Once we understand how different concentrations of a chemical can affect an organism, we can determine the concentrations that an organism could experience by examining solubility, bioaccumulation, biomagnification, and the persistence of chemicals in the environment. In the next module, we will examine how scientists analyze the risk that chemicals or any other environmental hazard poses to humans and other species.

AP® Practice Questions

Choose the best answer for the following.

1. Atrazine and DDT are examples of
   (a) neurotoxins.
   (b) carcinogens.
   (c) allergens.
   (d) endocrine disruptors.

2. Teratogens
   (a) interfere with embryo and fetus development.
   (b) disrupt the circulatory system.
   (c) alter the function of hormones.
   (d) suppress the immune system.

3. If 1 mg/kg of mass of a pesticide is the LD50 for rats in an experiment, what would be considered the safe exposure for humans?
   (a) 1 mg/kg
   (b) 0.1 mg/kg
   (c) 0.01 mg/kg
   (d) 0.001 mg/kg
4. Which is NOT a cause of high concentrations of DDT in fish-eating birds?
   (a) bioaccumulation
   (b) biomagnification
   (c) synergistic interactions
   (d) solubility
5. A prospective study
   (a) determines synergistic interactions of toxins.
   (b) measures the effect of a particular event after it has occurred.
   (c) monitors individuals who might be exposed to harmful chemicals in the future.
   (d) determines the number of individuals who might be affected by a particular chemical.

Risk Analysis

Learning Goals

After reading this module you should be able to

- explain the processes of qualitative versus quantitative risk assessment.
- understand how to determine the amount of risk that can be tolerated.
- discuss how risk management balances potential harm against other factors.
- contrast the innocent-until-proven-guilty principle and the precautionary principle.

Most people face some kind of environmental hazard every day. The hazards we face may be voluntary, as when we make a decision to smoke tobacco, or they may be involuntary, as when we are exposed to air pollution. When assessing the risk of different environmental hazards, regulatory agencies, environmental scientists, and policy makers usually follow three steps for risk analysis: risk assessment, risk acceptance, and risk management. In this module, we will examine each of the three steps.

Risk assessment estimates potential harm

As illustrated in **FIGURE 58.1** on page 632, risk assessment is the first of the three steps involved in risk analysis. Risk analysis seeks to identify a potential hazard and determine the magnitude of the potential harm. There are two types of risk assessment—qualitative and quantitative. Each of us has some idea of the risk associated with different *environmental hazards*. For our purposes, an **environmental hazard** is anything in our environment that can potentially cause harm. Environmental hazards include substances such as pollutants or other chemical contaminants, human activities such as driving cars or flying in airplanes, or natural catastrophes such as volcanoes and earthquakes.

We generally make qualitative judgments in which we might categorize our decisions as having low, medium, or high risks. When we choose to slow down on a wet highway or to buy a more expensive car because we feel

Environmental hazard Anything in the environment that can potentially cause harm.

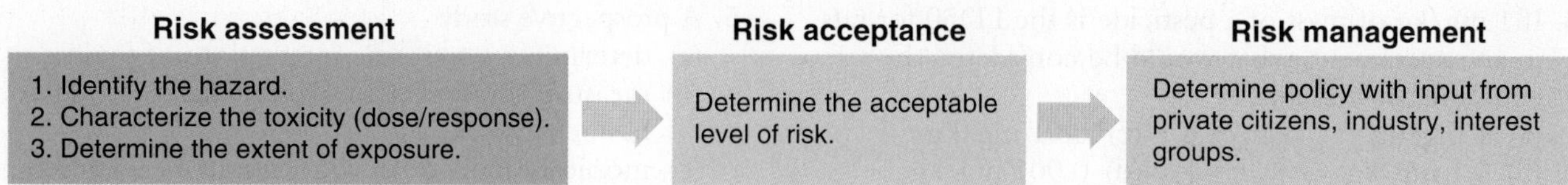

Risk assessment
1. Identify the hazard.
2. Characterize the toxicity (dose/response).
3. Determine the extent of exposure.
Risk acceptance
Determine the acceptable level of risk.
Risk management
Determine policy with input from private citizens, industry, interest groups.

FIGURE 58.1 The process of risk analysis. Risk analysis involves risk assessment, risk acceptance, and risk management.

it is safer, we are making qualitative judgments of the relative risks of various decisions. We are making judgments that are based on our perceptions but that are not based on actual data. It would be unusual for us to consider the actual probability—that is, the statistical likelihood—of an event occurring and the probability of that event causing us harm. Because our personal risk assessments are not quantitative, they often do not match the actual risk. For example, some people find air travel very stressful because they are afraid the plane might crash. These same people often prefer riding in a car, which they perceive to be much safer. Or, a person may be very cautious while walking in an area with heavy traffic but never consider the health dangers of smoking or a lack of exercise. To manage our risk effectively, we need to ask how closely our perceptions of risk match the reality of actual risk.

In the United States, the probability of death from various hazards can be calculated from data kept by the government. By looking at the total number of people who die in a year and their causes of death, researchers are able to determine the probability that an individual will die from a particular cause. **FIGURE 58.2** provides current data on causes of death in the United States. Because these risk estimates are based on real data, they are quantitative rather than qualitative. If we examine this figure, we see that the probability of dying in an automobile is far greater than the probability of dying in an airplane. Similarly, the probability of dying from heart disease is monumentally

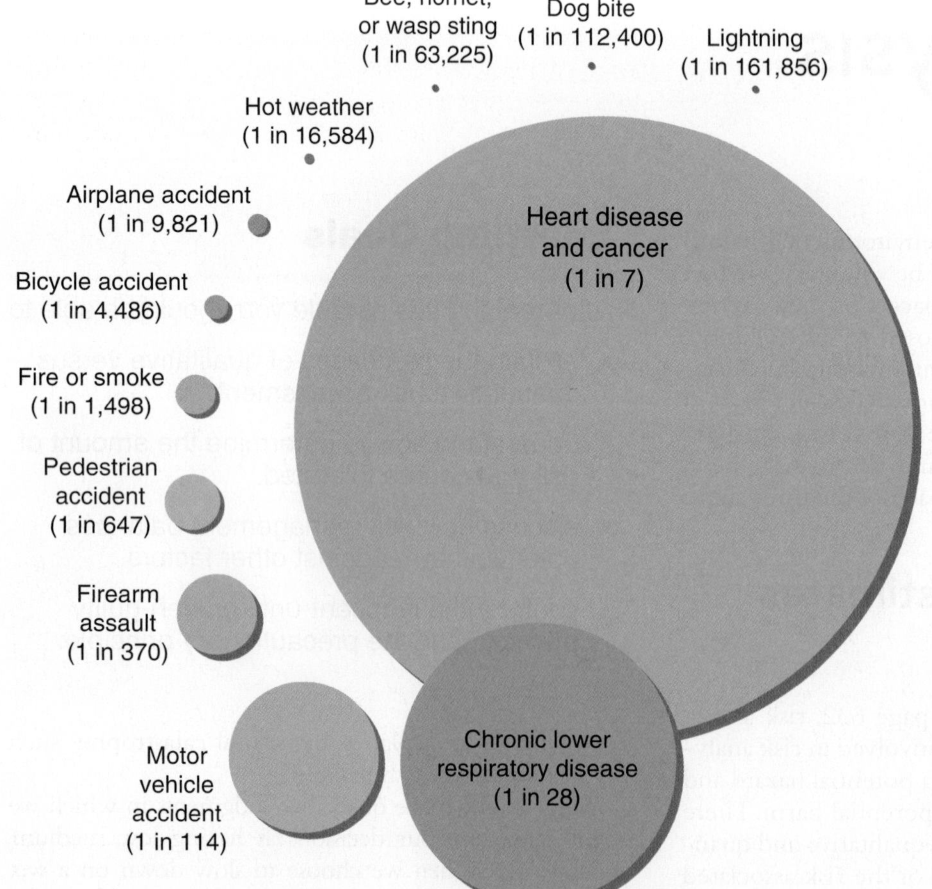

FIGURE 58.2 The probabilities of death in the United States. Some causes of death that people perceive as having a high probability of occurring, such as dying in an airplane crash, actually have a low probability of occurring. In contrast, some causes of death that people rate as having a low probability of occurring, such as dying from heart disease, actually have a very high probability of occurring. *(Data from http://www.nsc.org/learn/safety-knowledge/Pages/injury-facts-chart.aspx.)*

greater than the risk of dying in a pedestrian accident. These numbers underscore the fact that our perceptions of risk can often be very different from the actual risk. Because a catastrophic event, such as a nuclear plant meltdown or a plane crash, can do a great deal of harm and receives great media attention, people believe that it is very risky to use nuclear reactors or to fly in airplanes. However, as these events rarely occur, the risk of harm is low. In contrast, we tend to downplay the risk of activities that provide us with cultural, political, or economic advantages such as drinking alcohol or working in a coal mine.

Quantitative Risk Assessment

The most common approach to conducting a quantitative risk assessment can be expressed with a simple equation:

$$\text{Risk} = \text{probability of being exposed to a hazard} \times \text{probability of being harmed if exposed}$$

Using this equation, we could ask whether it is riskier in a year to fly on commercial airlines for 1,609 km (1,000 miles) per year or to eat 40 tablespoons of peanut butter, which contains tiny amounts of a carcinogenic chemical produced naturally by a fungus that sometimes occurs in peanuts that are used to make peanut butter. The risk of dying in a plane crash depends on the probability of experiencing a plane crash, which is very low, multiplied by the probability of dying if the plane does crash, which approaches 100 percent. The risk of dying of cancer from consuming peanut butter depends on your probability of eating peanut butter, which may be near 100 percent, multiplied by the probability that consuming peanut butter will cause you to develop lethal cancer, which is very low. It turns out that both behaviors produce a 1 in 1 million chance of dying. This example demonstrates a fundamental rule of risk assessment: The risk of a rare event that has a high likelihood of causing harm can be equal to the risk of a common event that has a low likelihood of causing harm.

Quantitative risk assessments bring together tremendous amounts of data. The estimates of harm can come from acute and chronic dose-response experiments, retrospective studies, and prospective studies. The estimates of which concentrations of a chemical an organism will experience in the environment incorporates the concentrations found in nature, routes of exposure, solubility, persistence, and the potential for the chemical to bioaccumulate or biomagnify. Together, these two groups of data can be used to estimate the probability of harm.

A Case Study in Risk Assessment

As we saw in our discussion of water pollution in Chapter 14, from the 1940s to the 1970s some companies manufacturing electrical components dumped PCBs (polychlorinated biphenyls) into rivers. Beginning in the 1960s, there was increasing evidence that PCBs might have harmful health effects on organisms that came into contact with them, including liver damage in animals and impaired learning in human infants.

Once the EPA identified PCBs as a potential hazard, it began a risk assessment. The agency brought together a range of data. First, scientists had to determine which concentrations of PCBs might cause cancer. To accomplish this objective, they examined dose-response studies on laboratory rats exposed to different concentrations of PCBs. They also examined retrospective studies of cancer cases in workers employed by industries that used PCBs. Next, they had to determine what concentrations people might experience. To accomplish this, scientists examined data on current concentrations in the air, soil, and water and considered the half-life of the chemical. Because PCBs were found throughout the environment and because they are very persistent, the probability of coming into contact with PCBs was considered relatively high. They also considered the potential routes of exposure: eating contaminated fish, drinking contaminated water, and breathing contaminated air.

The final result of the risk assessment on PCBs showed that the risk from eating contaminated fish is higher than the risk from drinking contaminated water and much higher than breathing contaminated air. As a result, signs were posted on the Hudson River and at Silver Lake in Massachusetts instructing anglers not to consume any fish they caught (**FIGURE 58.3**). With limited fish consumption, the EPA concluded that the absolute risk of an individual developing cancer from PCB exposure was low. However, the risk was high enough to cause the EPA to recommend a dredging

FIGURE 58.3 The outcome of a risk assessment of PCBs. Based on a risk assessment of humans consuming fish, the EPA determined that the fish living in the Hudson River in New York State and in Silver Lake in Massachusetts had unacceptably high concentrations of PCBs due to illegal dumping of PCBs by General Electric. As a result, anglers were not allowed to keep and consume the fish that they caught from those water bodies. *(Suzanne DeChillo/New York Times/Redux)*

of the Hudson River to remove a large fraction of the PCBs that had settled at the bottom of the river. As we discussed in Chapter 14, this decision led to a long court battle between the EPA and General Electric, the company that had been dumping PCBs into the Hudson River. The dredging was finally initiated in 2009 and completed in 2015. While the dredging has removed a large amount of the contaminated river mud, scientists estimate that it may be another 15 to 30 years before people can safely consume any fish from the river.

Risk acceptance determines how much risk can be tolerated

Once the risk assessment is completed, the second step in risk analysis is to determine risk acceptance—the level of risk that can be tolerated. Risk acceptance may be the most difficult of the three steps in the risk-analysis process. No amount of information on the extent of the risk will overcome the conflict between those who are willing to live with some amount of risk and those who are not. Even among those people who are willing to accept some risk, the precise amount of acceptable risk is open to heated disagreement. For example, according to the EPA, a risk of 1 in 1 million is acceptable for most environmental hazards. Some people believe this is too high. Others feel that a risk such as a 1 in 1 million chance of death from radiation leaks is a small price to pay for the electricity generated by nuclear energy. While personal preferences will always complicate the determination of risk acceptance, environmental scientists, economists, and others can help us weigh the options as objectively as possible by providing accurate estimates of the costs and benefits of activities that affect us and the environment.

Risk management balances potential harm against other factors

Risk management, the third step of the risk-analysis process, seeks to balance possible harm against other considerations. Risk management integrates the scientific data on risk assessment and the analysis of acceptable levels of risk with a number of additional factors including economic, social, ethical, and political issues. Whereas risk assessment is the job of environmental scientists, risk management is a regulatory activity that is typically carried out by local, national, or international government agencies.

The regulation of arsenic in drinking water provides an excellent example of the difference between risk assessment and risk management. As we saw in Chapter 14, despite the fact that scientists knew that 50 ppb of arsenic could cause cancer in people, from 1942 to 1999 the federal government set the acceptable concentration of arsenic at 50 ppb. In 1999, the EPA announced it was lowering the maximum concentration of arsenic in drinking water to 10 ppb, which matched the standards set by the European Union and the World Health Organization. This regulation threatened to place a large financial burden on mining companies that produced arsenic as a by-product of mining, and an economic burden on several municipalities in western states with naturally high concentrations of arsenic in their drinking water. Both groups lobbied hard against the lower arsenic limits. In 2001, weeks before the new lower limits were to go into effect, the EPA announced that it would return to the 50 ppb. The agency argued that further risk assessments needed to be conducted and any risk assessment had to be balanced by economic interests. Later in 2001, the National Academy of Sciences concluded that the acceptable amount of arsenic was a mere 5 ppb, which was lower than some previous estimates. This new risk assessment played a key role in striking a balance between the scientific data and economic interests and the EPA revised its ruling, ultimately setting the safe arsenic concentration at 10 ppb.

Worldwide standards of risk can be guided by two different philosophies

There are currently about 80,000 regulated chemicals in the world but they are not regulated the same way around the globe. A key factor determining the type of chemical regulation is whether the regulations are guided by the *innocent-until-proven-guilty principle* or the *precautionary principle,* both illustrated in **FIGURE 58.4**. The **innocent-until-proven-guilty principle** is based on the belief that a potential hazard should not be considered a real hazard until the scientific data definitively demonstrates that it actually causes harm. This strategy allows beneficial chemicals to be sold sooner. The downside is that harmful chemicals can affect humans or wildlife for decades before sufficient scientific evidence accumulates to confirm that they are harmful.

In contrast, the **precautionary principle** is based on the belief that when a hazard is plausible but not yet certain, we should take actions to reduce or remove the hazard. The plausibility of the risk cannot be speculation;

Innocent-until-proven-guilty principle A principle based on the belief that a potential hazard should not be considered an actual hazard until the scientific data definitively demonstrate that it actually causes harm.

Precautionary principle A principle based on the belief that action should be taken against a plausible environmental hazard.

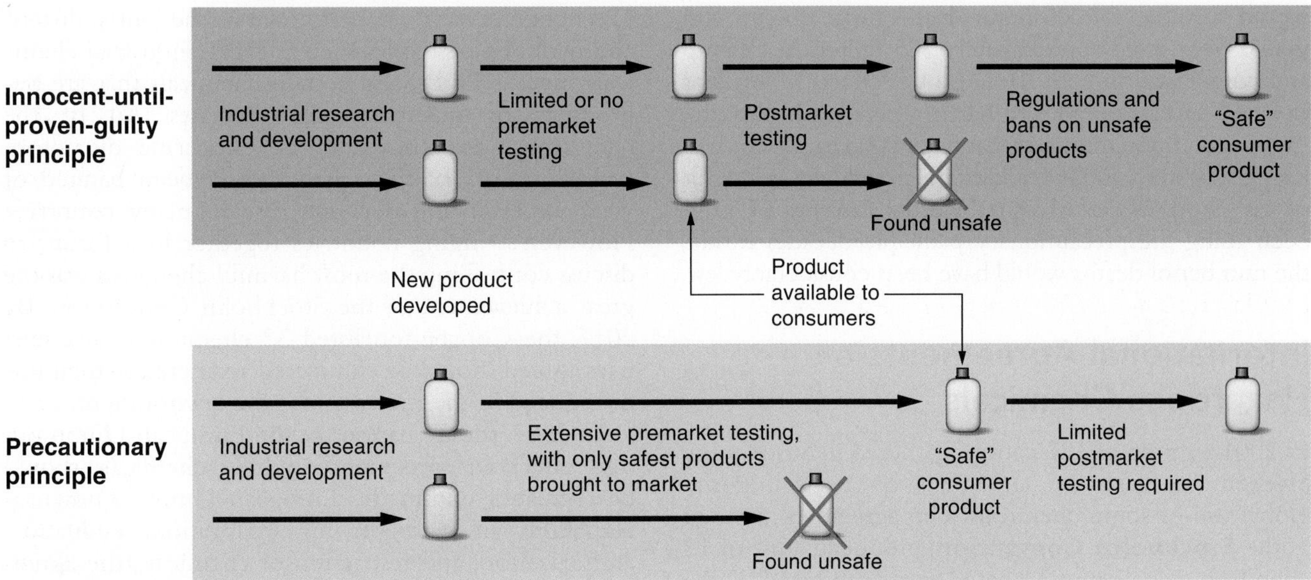

FIGURE 58.4 The two different approaches to managing risk. The innocent-until-proven-guilty principle requires that researchers prove harm before the chemical is restricted or banned. The precautionary principle requires that when there is scientific evidence that demonstrates a plausible risk, the chemical must then be further tested to demonstrate it is safe before it can continue to be used.

it must have a scientific basis. In addition, the intervention should be in proportion to the potential harm that might be caused by the hazard. This approach allows fewer harmful chemicals to enter the environment. However, if the initial assessment indicates a plausible risk and the chemical ultimately proves harmless but beneficial, its introduction can be delayed for many years. Moreover, the slower pace of approval can reduce the financial motivation of manufacturers to invest in research for new chemicals. In short, there is a tradeoff between greater safety with slower introduction of beneficial chemicals versus greater potential risk with a greater rate of discovery of helpful chemicals. Use of the precautionary principle has been growing throughout many parts of the world and was instituted by the European Union in 2000. The United States, however, continues to use the innocent-until-proven-guilty principle.

The benefit of the precautionary principle can be illustrated using the case of asbestos. Asbestos is a white, fibrous mineral that is very resistant to burning. This made asbestos a popular building material throughout much of the twentieth century. It is now widely accepted that dust from asbestos can cause a number of deadly diseases including asbestosis (a painful inflammation of the lungs) and several types of cancer. When asbestos was first mined in 1879, there was no evidence that it harmed humans. The first report of deaths in humans was in 1906 and the first experiment showing harmful effects in rats was conducted in 1911. In 1930, it was reported that 66 percent of workers in an asbestos factory suffered from asbestosis. In 1955, researchers found that asbestos workers had a higher risk of lung cancer than other groups. In 1965, a study linked a rare form of cancer with workers who were exposed to asbestos dust. Despite all of the growing scientific evidence that asbestos was harming human health, little was done to reduce the exposure of workers and the public. Indeed, it was not until 1998 that the European Union banned asbestos. Today, workers go to great lengths not to be exposed to asbestos dust such as by wearing protective suits and using respirators (**FIGURE 58.5**). A study in the Netherlands estimated that had asbestos been banned in 1965 when the harm to health became clear, the country

FIGURE 58.5 The risks of asbestos dust. Despite nearly a century of studies on the risks of asbestos dust to human health, only recently have workers been required to go to great lengths to prevent exposure. Today, they dress in chemical suits and respirators when removing asbestos from a building. Applying the precautionary principle would have required protection of workers many decades earlier and saved hundreds of thousands of lives. *(Phanie/Alamy)*

would have had 34,000 fewer deaths from asbestos and would have saved approximately $25 billion in cleanup and compensation costs. Because the effects of asbestos can take several decades to harm a person's health, the European Union estimates that from 2005 to 2040 it will have 250,000 to 400,000 additional people die as a result of past exposures to asbestos. Had the European Union been using the precautionary principle decades earlier, the number of deaths would have been considerably less.

International Agreements on Hazardous Chemicals

In 2001 a group of 127 nations gathered in Stockholm, Sweden, to reach an agreement on restricting the global use of some chemicals. The agreement, known as the **Stockholm Convention**, produced a list of 12 chemicals to be banned, phased out, or reduced. These 12 chemicals came to be known as the "dirty dozen" and included pesticides such as DDT, industrial chemicals such as PCBs, and certain chemicals that are by-products of manufacturing processes. All of the chemicals were known to be endocrine disruptors, and a number of them had already been banned or were experiencing declining use in many countries. However, bringing countries together in a forum to discuss controlling the most harmful chemicals was the great achievement of the Stockholm Convention. By 2017, the Convention listed 32 chemicals that countries agreed should be eliminated, restricted in their use, or reduced in their release into the environment.

In 2007, the 27 nations of the European Union put into effect an agreement on how chemicals should be regulated within the European Union. Known as **REACH**, an acronym for registration, evaluation, authorization, and restriction of chemicals, the agreement embraces the precautionary principle by putting more responsibility on chemical manufacturers to confirm that chemicals used in the environment pose no risk to people or the environment. This regulation was enacted because many chemicals used for decades in the European Union had not been subjected to rigorous risk analyses. The new regulations were phased in through 2018 to permit sufficient time for chemical manufacturers to complete the required testing.

Stockholm Convention A 2001 agreement among 127 nations concerning 12 chemicals to be banned, phased out, or reduced.

REACH A 2007 agreement among the nations of the European Union about regulation of chemicals; the acronym stands for registration, evaluation, authorization, and restriction of chemicals.

MODULE 58 AP® Review

Preparing for the AP® Exam

In this module, we learned that we can examine the environmental hazards faced by humans by using the process of risk analysis. Risk analysis begins with a risk assessment, which can be either qualitative or quantitative. Quantitative risk assessments are typically preferred because a person's perception of risk can be quite different from the actual risk. Once we assess the level of risk posed by a hazard, we need to determine how much risk humans are willing to accept. With this information on risk assessment and risk tolerance from environmental scientists, government regulators balance the risk of a particular environmental hazard against numerous other factors including economic, social, ethical, and political issues. Balancing this risk can be done using the innocent-until-proven-guilty principle or the precautionary principle, with each principle having different costs and benefits. In recent years, there has been worldwide agreement on a number of hazardous chemicals that will be banned, phased out, or reduced.

AP® Practice Questions

Choose the best answer for the following.

1. Which is NOT an environmental hazard?
   (a) air pollutants
   (b) driving a car
   (c) smoking tobacco
   (d) cancer

2. What does the EPA consider the limit of acceptable risk for environmental hazards?
   (a) 1 in 10,000
   (b) 1 in 100,000
   (c) 1 in 1,000,000
   (d) 1 in 10,000,000

3. The Stockholm Convention
   (a) was an international agreement to ban a number of endocrine disruptors.
   (b) led to the REACH agreement on chemical evaluation.
   (c) was an international agreement on asbestos and other hazardous materials.
   (d) was an agreement between the United States and the European Union to ban many carcinogens.
4. Which risk has the highest probability of death in the United States?
   (a) motor vehicle accident
   (b) drowning
   (c) fire
   (d) firearm assault
5. The precautionary principle
   (a) decreases financial incentives for chemical development.
   (b) was used in considering the use of asbestos.
   (c) is primarily used in the United States.
   (d) increases the risk of harmful chemicals being used.

Working Toward Sustainability

The Global Fight Against Malaria

Bill Gates is best known as the founder of Microsoft, the computer software company, but he is also an active philanthropist. In 2007 he stood up in front of a large group of scientists in Seattle, Washington, and declared that the world needed to eradicate malaria. In challenging the scientists of the world, he asked, "Why would anyone want to follow a long line of failures by becoming the umpteenth person to declare the goal of eradicating malaria?"

Bill Gates knew the history of malaria. People have died from this disease for thousands of years. In modern times, 350 million to 500 million people are infected each year and 1 million of them die. Most malaria cases are in Africa and most of those who die are children. Several eradication efforts have been attempted over the past 6 decades, mostly focused on eliminating the mosquitoes that carry the malaria pathogen. In the United States, eradication was achieved in 1951 through widespread spraying of the insecticide DDT as well as numerous other public health measures. The spraying program became controversial in the 1960s and 1970s because DDT was found to be widely distributed around the globe and it was linked to the thinning of egg shells in large birds of prey due to bioaccumulation and biomagnification. DDT is still sprayed in many parts of the world to assist in the eradication of malaria, but malaria persists.

Malaria is difficult to combat for a number of reasons. First, mosquito populations that are reduced by spraying insecticides can rebound quickly. In Sri Lanka, for example, consistent spraying to kill mosquitoes reduced the number of malaria cases from 1 million to a mere 18. Because of this great success, the spraying program was stopped, but within a few years, malaria cases rapidly increased to a half million. In short, the spraying program was ended before the job was done. Moreover, if one country is spraying to kill mosquitoes and neighboring countries are not, mosquitoes will continue to enter from the neighboring countries. Mosquitoes can also rapidly evolve resistance to insecticides such as DDT. In addition, the malaria pathogen can rapidly evolve resistance to antimalarial drugs.

Combating malaria. By distributing medicine and nets impregnated with insecticide to households in Africa, childhood death from malaria has declined by as much as 60 percent. *(Paula Bronstein/Getty Images)*

Finally, eradicating malaria is expensive. Typically, countries use multiple strategies including insecticide spraying, antimalarial drugs, and the distribution of mosquito tents in which people can sleep and avoid being bitten during the night. Collectively, these strategies can carry a price tag that many low-income countries cannot afford. Additionally, other social and economic priorities of these countries, as well as social disruption, have precluded or curtailed malaria control programs.

So why did Bill Gates think there was now a possibility of eradicating malaria? Earlier that year, scientists had reported that a new drug, combined with a new style of mosquito net, produced large reductions in malaria cases—for example, as much as 97 percent in Uganda. The new nets, which were impregnated with more modern insecticides, could last 3 to 5 years. This was a big improvement from the earlier nets, which only lasted no more than 3 months. In addition to having a new drug and longer-lasting nets, the key to the success of the Ugandan program was to pay for and distribute the drug and nets to everyone who needed them. Employing this strategy around the world is an expensive endeavor, and that's where Bill Gates comes in.

The Bill and Melinda Gates Foundation funds projects that have been historically underfunded. Equally important, by throwing its prominent name and financial resources behind a cause like malaria eradication, the foundation can rally significant financial support from other foundations and from developed countries. Western governments joined the movement and increased malaria funding from $50 million to $1.1 billion. This gave new hope to the declared goal of eradicating malaria from the globe within 50 years.

Many challenges remain. One of the largest is simply organizing distribution systems to hand out the drugs and millions of mosquito tents. In some regions, there are no roads into the villages and the items must be brought there by foot or by boat. There is also the challenge to continue research into new strategies against the pathogen and the mosquito. Beginning in 2001, a new antimalaria drug proved very effective against the pathogen and has been quite inexpensive to manufacture. The manufacturer of this drug agreed to sell it at less than the cost of manufacturing it, making the drug a very attractive option for low-income countries. As of 2017, the company had provided more than 800 million treatments. However, mosquitoes have started to show signs of evolving increased tolerance to the drug, so the company is already working on new drugs to combat the malaria pathogen.

Another possibility is the development of a vaccine that would provide immunity to malaria infections. For nearly 30 years, the pharmaceutical company GlaxoSmithKline has worked to develop a malaria vaccine, but the company was reluctant to fully fund a study of the effectiveness of the vaccine in African children. The Gates Foundation provided $200 million to help fund the study and the results have been very encouraging. In 2013, it was announced that children vaccinated between 5 and 17 months of age experienced 46 percent fewer cases of malaria than similar children who were not vaccinated. The researchers conducting the study will continue to follow these children as part of a prospective study to determine if the vaccination continues to protect the children throughout their lives. In 2018, the World Health Organization will be conducting larger vaccine trials in the African countries of Malawi, Ghana, and Kenya.

Today there is evidence that Bill Gates's dream of eradicating malaria is gaining ground. In 2016, researchers reported that the rate of malaria cases in sub-Saharan Africa declined by 57 percent from 2000 to 2015. Most experts agree that malaria cases could be reduced by at least 85 percent in most African countries. The reduction in illness and death would also be highly beneficial to the economies of these low-income countries by reducing health costs and creating a healthier, and therefore more productive, workforce. The success of the global fight against malaria critically depends on sustained financial support from foundations and governments, continued discovery of new drugs and vaccines, and the recognition that we cannot stop fighting malaria until the job is done.

Critical Thinking Questions

1. If mosquitoes vary in their resistance to insecticides, what might you predict about the long-term success of trying to eliminate all of the mosquitoes with insecticide as a way to eliminate malaria?
2. How might a reduction in illness and death from malaria affect the economies of low-income countries in terms of future health costs and the health and productivity of their workforce?

Sources

Malaria: Kenya, Ghana and Malawi get first vaccine. 2017. BBC News, April 24. http://www.bbc.com/news/health-39666132;

McNeil, D., Jr. 2008. Eradicate malaria? Doubters fuel debate. *New York Times,* March 4. http://www.nytimes.com/2008/03/04/health/04mala.html;

Novartis Speeds New Anti-Malarial as Older Drug Loses Potency. 2017. Bloomberg, August 21. https://www.bloomberg.com/news/articles/2017-08-21/novartis-speeds-new-anti-malarial-as-older-drug-loses-potency.

Chapter 17 Review

In this chapter, we learned about human diseases and chemicals that can affect human health and how we analyze the risk of environmental hazards. Human diseases can be categorized as either acute or chronic and can be infectious or not. We reviewed many of the historically important infectious diseases and then discussed the modern problem of emerging infectious diseases. In addition to these biological risks, we also need to consider chemical risks to humans and other species. Chemical risks are assessed by experiments that determine the LD50 or ED50 for various species and by following a large sample of individuals using prospective and retrospective studies. Such risk assessments can be combined with data on risk tolerance to help in risk management, which weighs the assessed risk against social, economic, and political considerations. In conducting risk management, regulators in some regions of the world use the precautionary principle while regulators in other countries, including the United States, use the guilty-until-proven-innocent principle.

Key Terms

Disease
Infectious disease
Acute disease
Chronic disease
Epidemic
Pandemic
Plague
Malaria
Tuberculosis
Emergent infectious disease
Acquired Immune Deficiency Syndrome (AIDS)
Human Immunodeficiency Virus (HIV)
Ebola hemorrhagic fever
Mad cow disease
Prion
Swine flu
Bird flu
Severe acute respiratory syndrome (SARS)
West Nile virus
Lyme disease
Zika virus disease
Neurotoxin
Carcinogen
Mutagen
Teratogen
Allergen
Endocrine disruptor
Dose-response study
Acute study
Chronic study
LD50
Sublethal effect
ED50
No-observed-effect level (NOEL)
Retrospective study
Prospective study
Synergistic interaction
Route of exposure
Solubility
Bioaccumulation
Biomagnification
Persistence
Environmental hazard
Innocent-until-proven-guilty principle
Precautionary principle
Stockholm Convention
REACH

Learning Goals Revisited

Module 56 Human Diseases

Identify the different types of human diseases.

Human diseases can be categorized as either infectious or noninfectious. Infectious diseases are caused by pathogens such as viruses, bacteria, fungi, protists, and helminths. Human diseases can also be categorized as either acute, which means they rapidly impair a body's functions, or chronic, which means they slowly impair a body's functions.

Understand the risk factors for human chronic diseases.

Risk factors for human health differ between low- and high-income countries. In low-income countries, the top risk factors include unsafe drinking water, poor sanitation, and malnutrition. In high-income countries, the top risk factors include tobacco use, less active lifestyles, poor nutrition, and overeating that leads to high blood pressure and obesity.

Discuss the historically important human diseases.

Among the historically important infectious diseases, plague is caused by a bacterium that is carried by fleas, malaria is caused by several different species of protists, and tuberculosis is caused by a bacterium that primarily infects the lungs.

Identify the major emergent infectious diseases.

Among the emerging infectious diseases, HIV/AIDS is caused by a virus that weakens the immune system, Ebola hemorrhagic fever is caused by a highly lethal virus, and mad cow disease is caused by a prion that damages the nervous system. The viruses that cause bird flu and swine flu are easily spread and sometimes lethal, SARS is a type of pneumonia caused by a virus, and West Nile virus normally infects birds but can be transmitted to humans by mosquitoes. In addition, Lyme disease is caused by a bacterium carried by ticks and Zika is a virus carried by mosquitoes.

Discuss the future challenges for improving human health.

The future challenges for improving human health include improving nutrition and sanitation in low-income regions of the world and promoting healthier lifestyles in high-income regions of the world. We also need to educate people about the importance of taking the full duration of medicines that combat pathogens to prevent the evolution of drug-resistant strains of pathogens. Finally, we need to continue to develop rapid response to emerging infectious diseases to reduce the probability of them spreading worldwide.

Module 57 Toxicology and Chemical Risks

Identify the major types of harmful chemicals.

The major types of harmful chemicals are neurotoxins, carcinogens, teratogens, allergens, and endocrine disruptors. Neurotoxins disrupt the nervous systems of animals and they include insecticides, lead, and mercury. Carcinogens are cancer-causing chemicals that include asbestos, formaldehyde, radon, and chemicals from tobacco. Teratogens interfere with the normal development of embryos or fetuses and include thalidomide and alcohol. Allergens cause abnormally strong immune responses and a given allergen typically only affects a small fraction of people. Endocrine disruptors such as hormones from animal rearing facilities interfere with the normal functioning of hormones in organisms.

Explain how scientists determine the concentrations of chemicals that harm organisms.

Scientists can conduct LD50 experiments to determine lethal effects of chemicals and ED50 experiments to determine sublethal effects of chemicals. They can also follow large groups of individuals backward in time using retrospective studies or forward in time using prospective studies. Once we know the chemical concentrations that can cause harm, we also need to determine the routes of exposure by which an individual may come in contact with the chemical as well as the chemical's solubility and its potential to bioaccumulate and biomagnify.

Module 58 Risk Analysis

Explain the processes of qualitative versus quantitative risk assessment.

For a given environmental hazard, we can qualitatively categorize risks as relatively low, medium, or high. However, the actual risk of a given hazard may be quite different from our qualitative assessments. Quantitative assessments use actual data to determine the actual probability of various risks, either based on government death statistics or by calculating the probability of being exposed to a hazard multiplied by the probability of being harmed if exposed.

Understand how to determine the amount of risk that can be tolerated.

Individuals differ in how much risk they are willing to tolerate. For most environmental hazards, we often set risk tolerance at 1 in 1 million.

Discuss how risk management balances potential harm against other factors.

Understanding the level of risk is important, but we must also assess the effects of trying to reduce the risk. Such effects include economic, political, and social considerations that collectively can come to some compromise that balances all of these factors. Whereas risk assessment is conducted by environmental scientists, risk management is typically conducted by local, national, or international government agencies.

Contrast the innocent-until-proven-guilty principle and the precautionary principle.

According to the innocent-until-proven-guilty principle, a potential hazard should not be considered harmful until it can definitively be demonstrated to cause harm. According to the precautionary principle, when a hazard is plausible but not yet certain, we should reduce or remove the hazard.

Practice Math and Graphing

Preparing for the AP® Exam

Answer the following questions. Be sure to show all your work.

1. Practice Math

To determine the LD50 of a new pesticide, a scientist exposes shrimp to a range of concentrations. The scientist places the shrimp in tubs of water, with 12 shrimp in each tub. The table shows how many shrimp are dead after 24 hours of being exposed to the various concentrations. Based on these data, calculate the percentages of dead shrimp in each concentration.

Concentration (mg/L)	Number dead	Percent dead
1	0	
2	1	
3	3	
4	6	
5	9	
6	11	
7	12	

2. Practice Graphing

(a) Using your percentage data in the above table, create a graph that shows the relationship between chemical concentration and the percentage of dead shrimp. Then draw a vertical and horizontal line to indicate the LD50 value of the chemical.

(b) Based on this graph, what is the no-observed-effect level for shrimp exposed to this chemical?

Chapter 17 AP® Environmental Science Practice Exam

Preparing for the AP® Exam

Section 1: Multiple-Choice Questions

Choose the best answer for questions 1–18.

1. Most people die from
   (a) infectious diseases.
   (b) respiratory and digestive diseases.
   (c) cancers.
   (d) cardiovascular diseases.

2. Which statement about the relationship between health risks and income is CORRECT?
   (a) A major risk in high-income countries is poor sanitation.
   (b) A major risk in low-income countries is obesity.
   (c) A major risk in low-income countries is a lack of food.
   (d) The major risks in high- and low-income countries are similar.

3. Which statement about historical infectious diseases is NOT true?
   (a) Plague is a disease that is carried by fleas attached to rodents.
   (b) Malaria is a disease that is carried by rodents.
   (c) Tuberculosis is a disease that is transmitted through the air.
   (d) Historically important infectious diseases still pose a health risk.

4. Which statement about emerging infectious diseases is NOT true?
   (a) HIV is a virus that most likely came from chimps.
   (b) Ebola hemorrhagic fever causes a high rate of death.
   (c) Mad cow disease is spread when cows are fed grain in large feedlots.
   (d) Bird flu is a virus that jumps from birds to people.

5. Which is NOT an example of an infectious disease?
   (a) AIDS
   (b) pneumonia
   (c) malaria
   (d) leukemia

6. Which toxins cause birth defects?
   (a) neurotoxins
   (b) carcinogens
   (c) teratogens
   (d) endocrine disruptors

7. Which statement about dose-response studies is NOT true?
   (a) Dose-response studies test chemicals across a range of concentrations.
   (b) Dose-response studies only test for lethal effects.
   (c) Dose-response studies can last for days or months.
   (d) LD50 values are divided by 10 to determine safe concentrations for wildlife.

8. Which statement about retrospective and prospective toxicity studies is TRUE?
   (a) Prospective studies are only conducted on wild animals.
   (b) Retrospective studies monitor health effects from future chemical exposures.
   (c) Prospective studies monitor health effects from future chemical exposures.
   (d) Prospective studies monitor health effects from past chemical exposures.

9. The concentration of chemical exposure depends on
   I. the persistence of the chemical.
   II. the solubility of the chemical.
   III. The LD50 of the chemical.
   (a) I only
   (b) II only
   (c) I and II
   (d) II and III

10. Which statement is NOT correct?
    (a) Risk assessment quantifies the potential harm that a chemical poses.
    (b) Risk assessment does not include social, political, and economic considerations.
    (c) Risk acceptance determines the amount of tolerated risk.
    (d) Risk management includes social, political, and economic considerations.

11. What is NOT true about the two approaches to regulating chemicals?
    (a) The innocent-until-proven-guilty principle assumes chemicals are safe unless harm can be demonstrated.
    (b) The precautionary principle is used in the United States.
    (c) The innocent-until-proven-guilty principle allows rapid approval of chemicals by regulatory agencies but increases the risk that harmful chemicals will be approved.
    (d) The precautionary principle can cause delays in the use of beneficial chemicals but reduces the risk of harmful chemicals being approved.

12. The majority of emergent infectious diseases arise
    (a) when people generate resistant bacteria by not completing antibiotic dosages.
    (b) when a disease unexpectedly jumps from an animal to human host.
    (c) from prion mutations.
    (d) from sexual transmission.

13. Which could be a set of treatment and control groups for a retrospective study on the effects of endocrine disruptors on humans?
    (a) a group of construction workers that have been exposed to asbestos and a group of construction workers that have been exposed to PCBs
    (b) a group of children with mothers exposed to PCBs during pregnancy and a group of children who were exposed to DDT after they were born
    (c) residents of a town that was downwind of a commercial farm that sprayed pesticides and residents of a town that was downwind of an organic farm
    (d) humans raised when DDT was used and humans raised after DDT was no longer used

14. Which is an application of the innocent-until-proven-guilty principle?
    (a) policies enacted as a result of the Stockholm Convention
    (b) allowing the use of a pesticide that has a low risk to human health
    (c) banning the use of certain nanoparticles until further research has assessed their risk to human health
    (d) laws that permit the sale of drugs for human use only if the drug has been tested for toxicity

15. After spraying pesticide on the farm's crops, a farmer notices that the population of amphibians in a nearby wetland has begun to decrease. The concentration of pesticide in the water is well below the LD50 value for amphibians. Based on this information, which is the most likely explanation for the decrease in amphibian population?
   (a) There was a lack of nearby source populations to recolonize the wetland.
   (b) There were synergistic interactions between the pesticide and environmental factors.
   (c) An emerging infectious disease was killing susceptible individuals.
   (d) Wetland conditions allowed for a longer persistence time of the pesticide than laboratory studies revealed.

16. Which is NOT one of the four infectious diseases that account for almost 80 percent of all infectious deaths?
   (a) malaria
   (b) diarrheal diseases
   (c) HIV/AIDS
   (d) respiratory infections

17. Scientists test the response of mice to a chemical, producing the following graph.

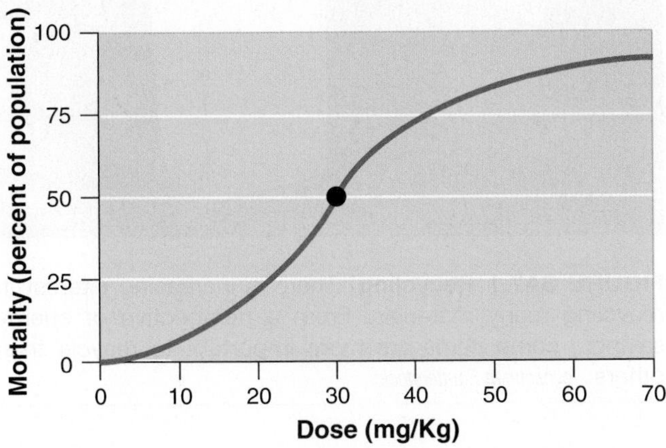

   Calculate the concentration of the chemical that would be considered safe for humans.
   (a) 0.03 mg/kg
   (b) 0.7 mg/kg
   (c) 3 mg/kg
   (d) 30 mg/kg

18. Which host/disease pairing is CORRECT?
   (a) Lyme disease; rodents
   (b) tuberculosis; birds
   (c) HIV/AIDS; fruit bats
   (d) Ebola; deer

Section 2: Free-Response Questions

Write your answer to each part clearly. Support your answers with relevant information and examples. Where calculations are required, show your work.

1. You are an employee of the Environmental Protection Agency. You are given the task of conducting risk management for spraying insecticides to kill the mosquitoes that carry West Nile virus.
   (a) How might you determine the proper concentration needed to kill mosquitoes? (2 points)
   (b) How might you determine whether the concentration used to kill mosquitoes might also kill other species of insects? (2 points)
   (c) If you knew the LD50 value of the insecticide for humans, what concentration would be considered the safe upper limit for humans? (2 points)
   (d) Given the information you have accumulated as part of your risk assessment, describe the factors that might be important in the risk management of spraying insecticides to kill the mosquitoes that carry West Nile virus. (4 points)

2. Given the differences in health risks that exist between low- and high-income countries, consider the following issues.
   (a) What strategies might you use to reduce the health risks of low-income countries? (3 points)
   (b) What strategies might you use to reduce the health risks of high-income countries? (3 points)
   (c) Suppose a low-income country discovers oil and is projected to become a high-income country within a decade. What changes in the country's health care system might you suggest? (4 points)

3. A scientist is assessing how atrazine, a commonly used herbicide, affects humans and other organisms, as well as the environment.
   (a) Design one retrospective study AND one prospective study to assess the effects of atrazine. (4 points)
   (b) What are the challenges in creating a control for a prospective study on how atrazine affects human health? (2 points)
   (c) Describe how atrazine's solubility could influence its movement in the environment. (2 points)
   (d) Describe the differences between bioaccumulation and biomagnification. (2 points)

pryzmat/Shutterstock

science applied 7

Is Recycling Always Good for the Environment?

One of the three ways to reduce solid waste is to recycle. As we discussed in Chapter 16, when we recycle items such as paper, plastic, bottles, and cans, less material ends up in landfills and fewer natural resources need to be extracted to produce these items in the future. The EPA estimates that Americans recycle 81 million metric tons (89 million U.S. tons) of trash. This represents about 35 percent of all the trash that we generate, with approximately 159 kg (350 pounds) of recycling per household per year. As impressive as that sounds, it is estimated that the average household generates more than twice that amount of recyclables, so there is more work to be done.

At first glance, recycling appears to make a lot of sense both economically and environmentally. Indeed, many state and local governments have encouraged or required recycling programs and the public generally associates recycling with being good for the environment (**FIGURE SA7.1**). But what do the data tell us? When we decide to recycle, what are the measurable benefits for the environment? How do these benefits compare with benefits from other decisions we make, such as the type of car we drive? The answers to these questions may surprise you.

FIGURE SA7.1 Recycling. There is increasing interest in recycling many materials. From a perspective of energy savings, some items are more important to recycle than others. *(sunsetman/Shutterstock)*

How do we begin to assess the benefits of recycling?

To determine the overall effect of recycling any type of waste, we need to consider the full range of costs and benefits of recycling and then compare these with the costs and benefits of manufacturing the same item from raw materials. For example, to assess the benefits of recycling paper, we need to compare the cost of recycling old paper into new paper products versus the cost of manufacturing new paper products from trees.

As we saw in Chapter 16, the best way to answer these questions is to complete a life-cycle analysis. Let's look at two examples: aluminum cans and plastic containers. To compare the environmental and economic costs of transporting and manufacturing these items from recycled materials versus raw materials, we begin at the manufacturing facility.

The recycling of aluminum, primarily from aluminum cans, is widespread in the United States. According to the Aluminum Association, more than 57 billion cans were recycled in the United States in 2016, which represents nearly 64 percent of all aluminum cans that were manufactured in that year (**FIGURE SA7.2**). To manufacture aluminum cans from raw materials, aluminum ore or bauxite must be mined and processed into pure aluminum. Not only does mining have environmental

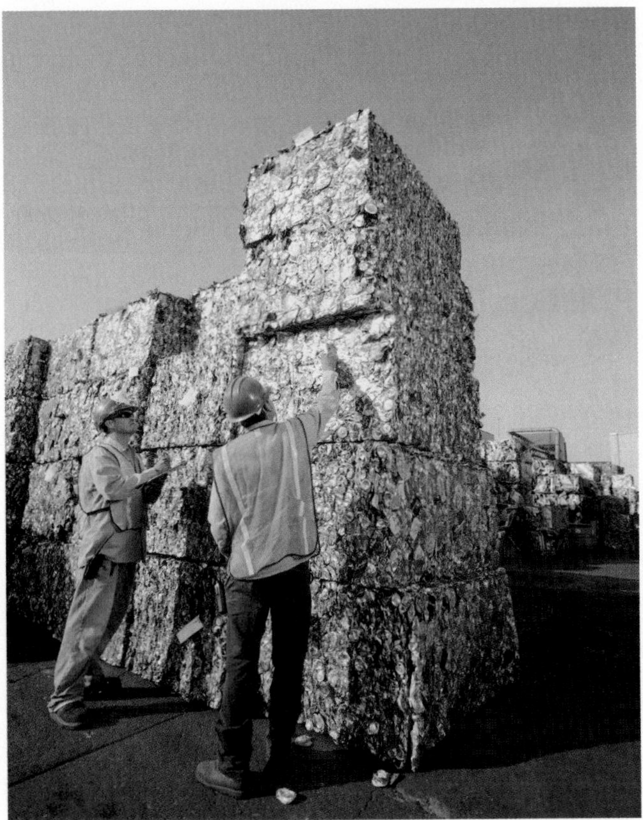

FIGURE SA7.2 Recycling aluminum cans. Converting old aluminum cans into new aluminum cans requires only 6 percent of the energy used to convert aluminum ore from a mine into new aluminum cans. *(Erik Isakson/AGE Fotostock)*

impacts as discussed in Chapter 8, but this processing of aluminum from ore also takes a substantial amount of energy. In contrast, manufacturing aluminum cans from recycled cans requires only 5 percent of this energy. In short, making new cans from recycled cans saves a large amount of energy and therefore saves manufacturers a lot of money. In fact, according to the EPA, recycling 0.9 metric ton (1 U.S. ton) of aluminum cans saves about 32 barrels of oil. When this energy comes from burning fossil fuels, it also means that manufacturing recycled cans reduces the amount of carbon dioxide and other pollutants in the atmosphere. Moreover, aluminum can be recycled over and over again without any loss of quality. In fact, 75 percent of all aluminum ever produced is still in circulation today thanks to recycling efforts. However, Americans still throw nearly $1 billion worth of aluminum cans into landfills each year, so there is room for improvement.

The recycling of plastic containers is also widely practiced. According to the American Chemistry Council and the Association of Postconsumer Plastic Recyclers, the recycling of plastic continues to grow with 1.3 million kilograms (2.9 million pounds) of bottles recycled in 2016, representing 30 percent of all plastic bottles that are manufactured. The cost of energy required to make new plastic bottles from raw material—in this case petroleum—is substantially less than the cost of recycling plastic bottles. As a result, manufacturing plastic bottles from recycled plastic bottles results in much smaller energy savings; washing and reprocessing plastic bottles requires nearly 50 percent of the energy required in manufacturing plastic bottles from oil. In addition, recycled plastic typically degrades in quality, so plastic from recycled bottles is typically used for products such as carpets and insulation for jackets and sleeping bags rather than new plastic bottles. This means that the economic and environmental benefits of using recycled plastic are much smaller than the economic benefits of using recycled aluminum.

What other costs of recycling do we need to consider?

Regardless of the type of material that is being recycled, we have to remember that there are several additional costs of recycling beyond the cost of energy used in manufacturing. To understand these costs, let's start at your house. If you rinse out your cans and bottles before recycling, energy is needed to get the water to your sink, particularly if you use hot water. If you use hot water to rinse out the peanut butter from a plastic peanut butter jar, for example, you are likely using more energy to clean the jar than is saved when you recycle the jar.

After they are cleaned, the materials to be recycled must be transported to a central recycling facility. Depending on location, the homeowner must either set out recycled items on the curb for pickup by a collection truck (**FIGURE SA7.3**) or bring them to a central

FIGURE SA7.3 Transportation costs. Although early recycling programs had garbage trucks make separate trips to pick up trash and recycling materials, modern trucks have separate compartments that allow both garbage and recycled items to be picked up at the curb in a single trip. A single trip saves time and money, and reduces consumption of fossil fuels as well as the production of air pollutants. *(©Rob Crandall/The Image Works)*

facility. Both scenarios require burning fossil fuels for transportation. Additional fossil fuels must be consumed to transport the recycled items from the collection facility to the manufacturing facility. Although transportation costs will vary among different towns and cities, in terms of energy consumed and pollutants produced, they reduce the benefits of recycling. However, we can easily compare the cost of transporting recycled materials to manufacturers against the cost of transporting raw materials from their source, such as an aluminum mine. In addition, when homeowners pay for transporting the items to the collection center through taxes or trash collection fees, they will avoid the costs of putting the waste in a landfill.

What other benefits of recycling do we need to consider?

The primary argument for recycling is that it reduces the need for raw materials and keeps solid waste out of landfills. During the 1990s, there was a growing concern that the United States was running out of landfill space and that recycling was critical to extending the life of existing landfills. While it is true that many landfills are nearing capacity, particularly in the northeastern United States, there is still a large amount of land throughout the country that could serve as landfill space if people in those areas agreed to the construction of new landfills.

Reducing the amount of solid waste going into landfills allows existing landfills to operate longer. This, in turn, reduces the costs of closing and monitoring existing landfills. It also reduces the costs of building more landfills in the future as well as the costs of trucking the waste to new landfills likely to be farther away. Increased trucking raises both the economic cost and the environmental impact.

When we consider how we can improve the environment it is often helpful to gather the scientific data to make objective comparisons rather than simply make decisions based on perceptions. In the case of recycling, the analysis of the data makes it clear that recycling certain materials will have much greater environmental and economic benefit than recycling other materials. This helps us understand why manufacturers might be much more inclined to promote the recycling of certain items such as aluminum cans. Identifying the full range of costs and benefits also helps us identify the complexity of the question. The energy costs of recycling, for example, are wide-ranging and include homeowner costs, transportation costs, manufacturing costs, and landfill costs. By identifying all of the costs and benefits, we can strive to design more efficient recycling programs.

Questions

1. Why is it much more beneficial to recycle aluminum than plastic?
2. Why is it important to take a life-cycle approach when considering the benefits of recycling?

Preparing for the AP® Exam

Practice Free-Response Question

Write your answer to each part clearly. Support your answers with relevant information and examples. Where calculations are required, show your work.

Suppose you are planning a party and want to determine the most environmentally friendly way to serve drinks to your friends. You can choose either one-use, recyclable plastic cups or glass cups that can be kept and reused.

(a) List four factors that are likely to increase or decrease the environmental benefits of using recyclable plastic cups. (2 points)

(b) Suggest three reasons why using plastic cups at a party may be less environmentally costly than using glass cups. (3 points)

(c) To reduce the amount of waste that ends up in landfills, engineers have developed plastic cups made with biodegradable plastic that are designed to be composted, where oxygen can help break the plastics down into very small pieces that can then be further degraded by microbes. However, many of them end up in landfills because people do not have compost piles or do not understand that they can be composted. Despite their biodegradability, these cups persist for many years in landfills. Why might biodegradable plastic persist in a landfill for a long time? (2 points)

(d) Suggest one method to reduce, one method to reuse, and one method to recycle waste at your party. (3 points)

References

The Aluminum Association. http://www.aluminum.org/industries/production/recycling.

The Association of Plastic Recyclers. 2017. *2016 United States National Postconsumer Plastic Bottle Recycling Report*. https://plastics.americanchemistry.com/2016-US-National-Postconsumer-Plastic-Bottle-Recycling-Report.pdf.

United States EPA. 2016. *Advancing Sustainable Materials Management: 2014 Fact Sheet*. https://www.epa.gov/sites/production/files/2016-11/documents/2014_smmfactsheet_508.pdf.

Section 1: Multiple-Choice Questions

Choose the best answer for questions 1–25.

1. The owner of a rural home poured several liters of concentrated bleach into a sink drain connected to a buried septic system. Which is the owner most likely to notice after a few weeks?
   (a) Settled sludge will be more sterile and dispersed into the leach field.
   (b) Sludge has started to accumulate faster in the septic tank.
   (c) There is a lower abundance of pathogens in groundwater.
   (d) The septage released from the septic tank is cleaner and safer for the environment.

2. Which source of mercury pollution is likely to harm humans?
   I. mercury-based compounds that are absorbed by marine plankton
   II. inorganic or synthetic mercury-based compounds
   III. wetlands with mercury concentrations below the legal limit set by the EPA
   (a) I only
   (b) II only
   (c) I and II
   (d) I, II, and III

3. To remediate an oil spill in the middle of the ocean, the use of dispersants
   (a) is less effective than using hot water sprayers.
   (b) can prevent oil from reaching the shorelines.
   (c) removes the oil from the ocean.
   (d) is an effective way to break up oil plumes under the ocean's surface.

4. Although thermal and sediment pollution have different sources, they both can
   (a) lead to respiratory problems in aquatic animals.
   (b) contaminate waterways with heavy metals.
   (c) lower the temperature of water.
   (d) harm fish in the open ocean.

5. Which is NOT regulated by EPA Clean Water Act or Safe Drinking Water Act?
   (a) levels of arsenic in drinking water
   (b) septage disposal from sewage treatment plants
   (c) nonpoint sources of oil pollution
   (d) inorganic chemicals in groundwater

6. Which group consists entirely of secondary pollutants?
   (a) carbon dioxide, methane, lead
   (b) nitrogen dioxide, sulfur dioxide, ozone
   (c) VOC, mercury, carbon dioxide
   (d) nitrate, ozone, sulfate

7. ______ is NOT likely to lead to the formation of chemical smog.
   (a) Combustion of gasoline
   (b) Industrial release of CO_2
   (c) A thermal inversion
   (d) Intense sunlight

8. Sulfur dioxide emissions can be reduced by
   (a) selling sulfur emission credits.
   (b) increasing the temperature at which coal is combusted.
   (c) installing catalytic converters in cars.
   (d) using electrostatic precipitators on power plants.

Question 9 refers to the following equations:

$$i + O_2 \rightarrow O_3$$

$$O_3 + ii \rightarrow O_2 + O$$

9. Roman numerals i and ii refer to which compounds?
   (a) O, chlorofluorocarbon
   (b) O_2, 2O
   (c) UV-C, O_2,
   (d) O, UV-C

10. Which is likely to increase the risk of sick building syndrome?
    (a) replacing an old ventilation system with a modern one
    (b) replacing old hardwood floors with new carpeting
    (c) installing carbon monoxide detectors
    (d) using low VOC (volatile organic compounds) paint

11. Which factor is most likely to reduce the total amount of material that ends up as municipal solid waste in a landfill?
    (a) utilizing packing materials that are recyclable
    (b) increasing production of one-time-use materials
    (c) lowering the cost of disposable products
    (d) manufacturing products that are less durable

12. Open-loop recycling refers to
    (a) the recycling of a product into a product that will enter a different waste stream.
    (b) the recycling of a product into compost.
    (c) a process where only part of the waste is ultimately recycled.
    (d) a process where one product is recycled into a different product.

13. Which factor is least likely to contribute to a decision on where to locate a landfill?
    (a) proximity of landfill to residential neighborhoods
    (b) location of underlying aquifers
    (c) underlying soil texture
    (d) net primary productivity of habitat

Question 14 refers to the following table on the costs associated with landfills and incinerators

Landfill		Incinerator	
Transporting material to landfill	$10/ton	Transporting material to incinerator	$10/ton
Tipping fee to dump waste in landfill	$60/ton	Tipping fee to incinerate waste in incinerator and remove ash	$90/ton

14. Assume an incinerator can produce 500 kWh of energy by incinerating 1 ton of waste, and can sell electricity for $0.06/kwh. What is the overall dollar cost of incineration relative to a landfill?
(a) Incinerator is $10 more per ton.
(b) Incinerator is $20 more per ton.
(c) They are the same.
(d) Incinerator is $10 less per ton.

15. The Comprehensive Environmental Response, Compensation, and Liability Act is referred to as "Superfund" because it
(a) imposes taxes collected from chemical and petroleum industries to fund the cleanup of abandoned hazardous waste sites.
(b) provides a fund to compensate workers exposed to hazardous situations during national emergencies such as hurricanes and floods.
(c) provides a fund to help other countries design environmental responses to emergencies involving the accidental release of hazardous waste.
(d) distributes tax revenues from various government agencies to chemical and petroleum industries.

16. Which is an example of an integrated waste management strategy?
(a) Integrating the waste management goals of several small townships into a regional waste management plan.
(b) Training waste management workers on how to sort recyclables from nonrecyclables.
(c) Simultaneously developing several waste management strategies to reduce the amount of waste that must be incinerated or placed in landfills.
(d) Collecting state taxes to fund the development of more landfills.

17. HIV, H1N1, and mad cow disease are all similar in that
(a) they can be sexually transmitted.
(b) they have been largely eradicated from developed nations.
(c) they originated from an animal other than humans.
(d) humans infected with these diseases cannot be cured.

Questions 18 and 19 refer to the following graph:

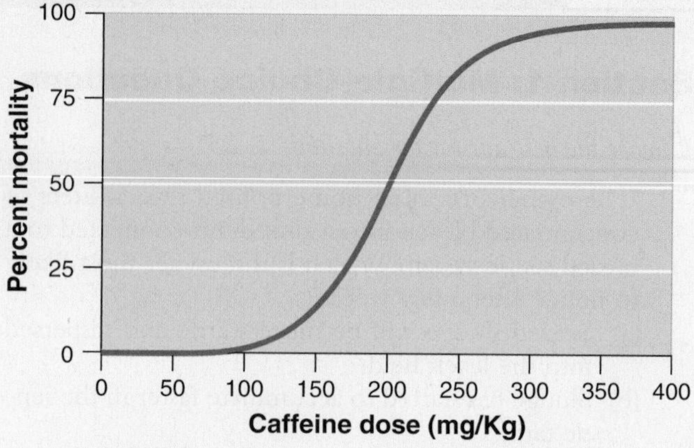

18. Given the mortality curve in the figure for a rat, what would be the safe dosage of caffeine for a rat?
(a) 10 mg
(b) 20 mg
(c) 40 mg
(d) 100 mg

19. What would be the safe dosage of caffeine for a human?
(a) 0.1
(b) 0.2
(c) 0.3
(d) 0.5

20. Which factor is the most difficult to predict when quantifying the risk of human exposure to a given contaminant?
(a) route of contaminant exposure
(b) synergistic interactions with other contaminants
(c) biomagnification of a contaminant
(d) persistence of a contaminant in water

21. Under the Safe Drinking Water Act, the EPA is responsible for
(a) regulating pollution caused by fracking for natural gas.
(b) setting maximum contaminant levels.
(c) mitigating the effects of cultural eutrophication.
(d) eradicating fecal coliform from all surface waters.

22. What is usually the best solution to household production of hazardous waste?
(a) hazardous waste landfills
(b) product substitution
(c) source separation
(d) reuse and recycling

23. Which is a chronic disease?
I. Ebola
II. tuberculosis
III. HIV/AIDS
(a) I only
(b) II only
(c) I and II
(d) II and III

24. Which contributes to the creation of drug-resistant strains of bacteria?
    (a) using newer, untested antibiotics
    (b) patients stopping antibiotics too early
    (c) diseases mutating and jumping from animals to humans
    (d) decreased numbers of people taking vaccines

25. Which statement about ozone is FALSE?
    (a) It is the main component of photochemical smog.
    (b) It can harm organisms in the stratosphere.
    (c) It is formed from nitrogen dioxide and sunlight.
    (d) It protects organisms from harmful UV-B radiation.

Section 2: Free-Response Questions

Write your answer to each part clearly. Support your answers with relevant information and examples. Where calculations are required, show your work.

1. Concentrated Animal Feeding Operations (CAFOs) house thousands of cows that are concentrated in close living quarters. Manure generated at CAFOs is often washed into manure lagoons where it decomposes anaerobically until the manure is sprayed over fields as fertilizer. This practice has several problematic consequences. First, the anaerobic decomposition of manure releases methane and sulfur compounds including sulfur dioxide. Second, manure from lagoons can potentially leak into surrounding soils, streams, and lakes. Third, to make room for additional manure in the lagoons, more manure is often sprayed on agricultural fields than is needed for plant growth.
    (a) Explain TWO reasons why the release of sulfur dioxide poses an environmental and health risk. (2 points)
    (b) Name TWO reasons why the leakage of manure from lagoons or the overspraying of manure on agricultural fields poses environmental risks for nearby soils, streams, and lakes. (2 points)
    (c) Suggest a retrospective study that could be conducted to determine the effect of manure lagoons on nearby residents. (2 points)
    (d) List TWO possible ways to determine if there is leakage in the soil surrounding a manure lagoon. (2 points)
    (e) Which TWO legislative acts require remediation of groundwater and stream water in cases where a manure lagoon leak occurs? (2 points)

2. During the first half of the twentieth century, residents in Los Angeles were allowed to have "backyard incinerators" in which they could burn their trash. Although these incinerators reduced the amount of trash going to a landfill, they also created a substantial amount of air pollution that led to severe smog. Initially, Los Angeles attempted to reduce smog by limiting the use of backyard incinerators to the hours between 4 AM and 7 AM. Backyard incinerators were ultimately banned in 1957.
    (a) Describe the chemical process that generates smog, and how backyard incinerators are likely to contribute to that process. (2 points)
    (b) How would limiting the incineration of trash to the hours between 4 AM and 7 AM potentially reduce the production of smog? (2 points)
    (c) Define a thermal inversion and describe how it affects smog. (2 points)
    (d) Much of the ash from backyard incinerators was either used in gardens or disposed of in landfills. What is the environmental risk of this practice? (2 points)
    (e) How do modern incinerators control the release of both bottom and fly ash? (2 points)

3. Humans have many ways of handling wastewater. Homes in rural areas often have their own septic system, while more urban areas require sewage treatment plants.
    (a) Name the two parts of a septic system, and describe the function of each. What is one advantage of a septic system? (5 points)
    (b) An old sewage treatment plant services a town of 10,000 people. Each person produces an average of 150 liters of wastewater per day. If the plant is required to hold 20 days' worth of wastewater for the town, what must be its minimum capacity? (2 points).
    (c) Assume the sewage treatment plant meets only the minimum required size. If the town experiences a large storm event that injects 40 million additional liters of water into the plant, how might this impact the plant's functioning and what immediate steps might be needed to address the extra water? What would the capacity of the plant have to be in order to deal with the additional water? (2 points)
    (d) A town council is concerned about how a recently built sewage treatment plant's release of untreated sewage during a large storm event might be harming a nearby reservoir that supplies much of the town's drinking water. What could the council test to make sure that water from the reservoir is safe to drink? (1 point)

This serpent sea star and sea fan are located in the Northeast Canyons and Seamounts Marine National Monument.
(NORTHEAST U.S. CANYONS 2013 SCIENCE TEAM/National Geographic Creative)

Conservation of Biodiversity

CHAPTER 18

CASE STUDY

Modern Conservation Legacies

The biodiversity of the world is currently declining at such a rapid rate that many scientists have declared that we are in the midst of a sixth mass extinction. There are many causes of this decline, but all are related to human activities ranging from habitat destruction to overharvesting plant and animal populations. In response to this crisis, there is growing interest in conserving biodiversity by setting aside areas that are protected from many human activities.

The conservation of biodiversity has a long history. The United States, for example, has been protecting habitats as national parks, national monuments, national forests, and wilderness areas for more than a century. Yellowstone National Park was the first national park in the United States, designated in 1872 by President Ulysses Grant. During the presidency of Theodore Roosevelt (1901–1909), nearly 93 million hectares (230 million acres) received federal protection. This included the creation of more than a hundred national forests, although much of this land was set aside to ensure a future supply of trees for lumber and therefore lacked complete protection.

In contrast to the long history of protecting terrestrial habitats, efforts to protect marine habitats are relatively recent. One of the most expansive efforts in the United States was made during the administration of George W. Bush. From 2006 to 2009, President Bush designated a total of 95 million hectares (215 million acres) of marine habitats as protected around the northwestern Hawaiian Islands and other U.S. Pacific islands. In the northwestern Hawaiian Islands, 36 million hectares (90 million acres) of these marine habitats were set aside as the Papaha–naumokua–kea Marine National Monument. This protected region is immense, covering an area about the size of California. In 2016, President Obama nearly quadrupled the areas of this monument by setting aside an additional 115 million hectares (283 million acres).

> Many scientists have declared that we are in the midst of a sixth mass extinction.

The marine ecosystem that surrounds the Hawaiian Islands contains a great deal of biodiversity—more than 7,000 marine species, approximately one-fourth of which are found nowhere else in the world. Unfortunately, in recent decades human activities have caused a decline in this diversity. The human causes of declining diversity are wide ranging. Although Hawaii has only 1.4 million residents, 9 million tourists visit each year. Individual anglers and commercial fishing operations have exploited marine life, including coral and fish. In addition to this exploitation, there are hundreds of metric tons of old fishing equipment lying at the bottom of the ocean that sometimes wash up on shore, entangling wildlife in old fishing lines. Invasive species of algae also dominate some areas.

The Papaha–naumokua–kea monument presents an opportunity for improving the Hawaiian marine environment. As a national monument, the area is protected from fishing, harvesting of coral, and the extraction of fossil fuels. As of 2017, 532 metric tons (586 tons) of marine debris have been removed from the shorelines and coral reefs, and efforts are under way to clean out much of the invasive algae. It is expected that the biodiversity of the area will quickly respond to these efforts. As the populations of organisms increase in the protected areas, individuals will disperse and add to the populations in the larger surrounding area. In this way, the protected area can serve as a constant supply of individuals to help neighboring areas maintain their diversity of species.

In the United States and the rest of the world, conserving the biodiversity of marine areas by creating marine reserves is a relatively new activity for governments, but the idea is gaining ground. In the Galápagos Islands, where Charles Darwin studied the evolution of finches, the nation of Ecuador recently designated a marine reserve that extends 64 km (40 miles) into the ocean from the islands and allows only limited fishing. Marine reserves have also been designated by Russia, the United Kingdom, Australia, Canada, and Belize.

Efforts to protect critical wildlife habitats continue today. For example, in 2015 the Obama administration set aside the first marine monument in the Atlantic Ocean. Known as the Northeast Canyons and Seamounts Marine National Monument, it is located on the edge of the continental shelf east of New York City and covers 1.3 million hectares (3.1 million acres). This area is home to several species of whales and a diversity of unique deep-sea species including corals. While recreational fishing will still be allowed, commercial fishing is being phased out over a 7-year period.

As more countries develop marine reserves, we have to make sure these areas are large enough to allow long-term protection of local species and we must consider how each new reserve is positioned relative to other reserves so that individuals are able to move among them. Furthermore, countries must decide what human activities will be allowed in each reserve, perhaps protecting a core area and allowing tourism, fishing, or extraction of fossil fuels to occur in more distant areas of the reserve. This is a current question for many protected areas in the United States, where some residents fear federal government overreach in setting aside areas to reduce or ban exploitation of the resources.

Sources: J. H. Davis, Obama creates Atlantic Ocean's first marine monument, *New York Times*, September 15, 2016. https://www.nytimes.com/2016/09/16/us/politics/obama-to-create-atlantic-oceans-first-marine-monument.html; L. Parker, See the wild places that may lose protections as national monuments, *National Geographic*, September 18, 2017. https://news.nationalgeographic.com/2017/09/national-monuments-shrink-trump/; P. Thomas, President Bush to add marine reserves; not all are applauding, *Los Angeles Times*, January 6, 2009. http://latimesblogs.latimes.com/outposts/2009/01/news-flash-pres.html

Preserving habitats is one important way to protect against declines in the world's biodiversity. In this chapter, we will examine declines in biodiversity at multiple levels including declines in the genetic diversity of wild plants and animals, declines in the genetic diversity of domesticated plants and animals, and declines in the species of large taxonomic groups. We will also investigate the major causes of these declines, which include habitat loss, overharvesting, and the introduction of species from other regions of the world. To help curb the loss of biodiversity, we have a number of laws and international agreements that are in effect. Approaching these efforts with an understanding of the concepts of metapopulations, island biogeography, and biosphere reserves can help us succeed in protecting large ecosystems.

MODULE 59

The Sixth Mass Extinction

Learning Goals

After reading this module you should be able to

- explain the global decline in the genetic diversity of wild species.
- discuss the global decline in the genetic diversity of domesticated species.
- identify the patterns of global decline in species diversity.
- explain the values of ecosystems and the global declines in ecosystem function.

In Chapter 5, we noted that the world has experienced five major extinctions during the past 500 million years. Many scientists have suggested that we may currently be in the midst of a sixth mass extinction event. In the most recent assessment made in 2014, one group of scientists estimated that the world is currently experiencing approximately 1,000 species extinctions per year. Other scientists have estimated much lower rates of extinction. The uncertainty around these highly variable estimates depends on the assumption made about how many species (named and unnamed) live on Earth and whether the rates of extinction observed in well-documented groups of plants or animals can be applied to all plants and animals. Regardless of the exact estimate of extinction rate, this current sixth mass extinction is unique because it is happening over a relatively short time and it is the first mass extinction to occur since humans have been present on Earth.

In this module, we will examine the declines in biodiversity of Earth at various levels of complexity including genetic diversity, species diversity, and ecosystem function. In each case, we will examine the roles that humans have played in the decline of biodiversity.

We are experiencing global declines in the genetic diversity of wild species

At the lowest level of complexity, environmental scientists are concerned about conserving genetic diversity. Populations with low genetic diversity are not well suited to surviving environmental change and they are prone to inbreeding depression, as we discussed in Chapter 6. Inbreeding depression by parents that each carry a harmful recessive mutation causes some of their offspring to receive two copies of the harmful mutation and, as a result, causes the offspring to have a poor chance of survival and reproduction. High genetic diversity ensures that a wider range of genotypes is present, which reduces the probability that an offspring will receive the same harmful mutation from both parents. In addition, high genetic diversity improves the probability of surviving future change in the environment. High genetic diversity produces a wide range of phenotypes which increases the chances that at least some individuals can survive and reproduce under changing environmental conditions.

AP® Exam Tip

You should be able to identify and discuss the ecological benefits of conservation as well as the economic pressures involved in setting aside land and resources for protection. ●

Some declines in genetic diversity have nonhuman causes. Cheetahs, for example, possess very low genetic diversity. Researchers have determined that this condition is the result of a population bottleneck that occurred approximately 10,000 years ago (see Figure 15.10 on page 166). Other declines in genetic diversity have human causes. For example, we discussed in Chapter 5 that the Florida panther once roamed throughout the southeastern United States

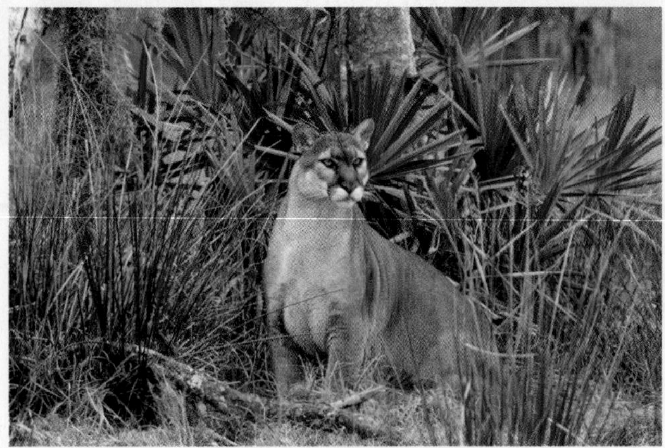

FIGURE 59.1 Declines in genetic diversity. The Florida panther was reduced to such a small population that it suffered severe effects of inbreeding. In recent years the introduction of new genotypes from a Texas population has allowed the Florida panther to rebound. *(©Tom & Pat Leeson/AGE Fotostock)*

(**FIGURE 59.1**). Because of hunting and habitat destruction, the population of the Florida panther shrank to only a small group in south Florida and this led to inbreeding. This inbreeding caused a number of harmful defects that caused the population to decline even further. After scientists released 8 panthers from Texas into Florida to add genetic diversity, the Florida panther population increased from 20 to nearly 100 individuals.

We are also experiencing global declines in the genetic diversity of domesticated species

Although declining genetic variation of plants and animals in the wild is of great concern to scientists, there are also major concerns about declining genetic variation in the domesticated species of crops and livestock on which humans depend. The United Nations notes that while there are 38 species of domesticated animals, the majority of livestock species comes from seven species of mammals (donkeys, buffalo, cattle, goats, horses, pigs, and sheep) and four species of birds (chickens, ducks, geese, and turkeys). In total, humans have bred the 38 species of domesticated animals into approximately 8,800 different varieties. Across the world, these species have been bred by humans for a variety of characteristics including adaptations that allow them to survive local climates. For example, humans have bred for a tremendous diversity of traits in cattle, as illustrated in **FIGURE 59.2**. This wide variety of adaptations, which is produced by a great deal of genetic variation, could be used for adapting to changing environmental conditions in the future or resisting new diseases. Unfortunately, livestock producers have concentrated their efforts on the breeds that are most productive in terms of meat

FIGURE 59.2 The genetic diversity of livestock. Over thousands of years, humans have selected for numerous breeds of domesticated animals to thrive in local climatic conditions and to resist diseases common in their local environments. Modern breeding, which focuses on productivity, has caused the decline or extinction of many of these animal breeds.

or milk production, so much of this genetic variation is being lost. The United Nations assessed the state of domesticated breeds of animal on 2015 and concluded that 17 percent of the breeds are at risk of extinction. However, this percentage may be much higher because there are insufficient data of 58 percent of the breeds to assess whether any of them are also declining. You can view the global data for cows, sheep, goats, pigs, and chickens in **FIGURE 59.3**.

A similar story exists for crop plants. A century ago, most of the crops that humans consumed were composed of hundreds or thousands of unique genetic varieties. Each variety grew well under specific environmental conditions and was usually resistant to local pests. In addition, each variety often had its own unique flavor. As we saw in Chapter 11, the green revolution in agriculture focused on techniques that increased productivity. Farmers planted fewer varieties, concentrating on those with higher yields. Fertilizers and irrigation helped humans control many of the abiotic conditions, allowing fewer but higher-yielding varieties to be grown across large regions of the world. For example, at the turn of the twentieth century, farmers grew approximately 8,000 varieties of apples. Today, that number has been reduced to about 100, and considerably fewer are available in your local grocery store.

Planting only a few varieties leaves us open to crop loss if the abiotic or biotic environment changes. For example, in the 1970s, a fungus spread through cornfields of the southern United States and killed half the crop. Although the fungus was uncommon, the high-yielding variety of corn that most farmers planted turned out to be susceptible to it. Following this crisis, scientists modified this high-yielding corn by adding a gene from a variety that is resistant to the fungus. Had the resistant variety not been preserved, this gene would not have been available.

The nations of the world have recognized the problem of declining seed diversity and have responded by storing seed varieties in specially designed warehouses to preserve genetic diversity. In fact, there are currently more than 1,700 such storage facilities around the world. However, many of these facilities are at risk from war and natural disasters. In the past decade, nations and philanthropists have funded an international storage facility known as the Svalbard Global Seed Vault (**FIGURE 59.4**). This facility consists of a tunnel built into the side of a frozen mountain on an island above the Arctic Circle of northern Norway. It was designed to resist a wide range of possible calamities, including natural disasters and global warming. Should the environment change in future years, either in terms of abiotic conditions or because of emergent diseases,

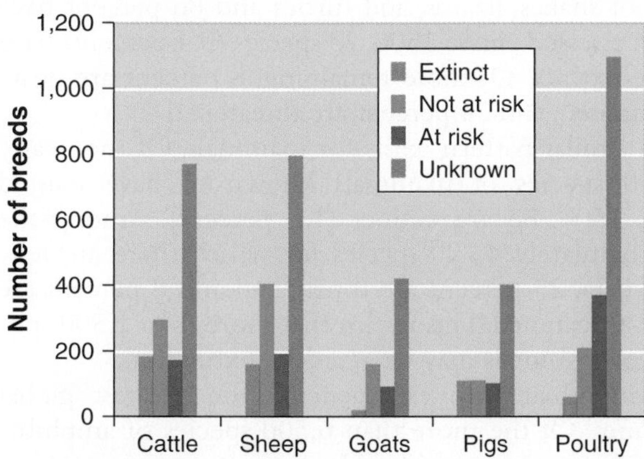

FIGURE 59.3 The global decline in genetic diversity of domesticated animals. Some of the most common species of livestock include cows, sheep, goats, pigs, and chickens. Across all domesticated breeds, more than 400 have gone extinct and 17 percent are at risk of extinction. The status of another 58 percent is unknown due to insufficient data on their population sizes. (Data from Food and Agriculture Commission of the United Nations (2015). *The Second Report on the State of the World's Animal Genetic Resources for Food and Agriculture.*)

FIGURE 59.4 A global seed bank. The Svalbard Global Seed Vault in northern Norway is an international storage area for many varieties of crop seeds from throughout the world. *(Paul Fearn/Alamy)*

the seed bank will be available to help scientists address the challenge. The Svalbard facility opened in 2008 with a capacity of 4.5 million seed varieties with each variety represented by more than 500 seeds. As of 2018, more than 890,000 seed samples had been sent to Svalbard for long-term storage.

Species diversity has declined around the world

Extinction occurs when the last member of a species dies. These major extinction events are characterized as a loss of at least 75 percent of all species within a period of 2 million years. Scientists estimate that as a result of these multiple mass extinctions and many minor extinctions, nearly 99 percent of the 4 billion species that have existed on Earth have gone extinct. However, because each of these mass extinction events has been followed by high rates of speciation that produced new species over millions of years, we still have millions of species on Earth.

To understand the current loss of species around the world, we can look at how particular groups of species are declining. When considering the status of a species, we use one of five categories defined by the International Union for Conservation of Nature (IUCN). Data-deficient species have no reliable data to assess their status; they may be increasing, decreasing, or stable. Species for which we have reliable data are placed in one of four categories. Extinct species are those that were known to exist as recently as the year 1500 but no longer exist today. The IUCN defines **threatened species** as those that have a high risk of extinction in the future and **near-threatened species** are very likely to become threatened in the future. **Least-concern species** are widespread and abundant. These categories provide a mechanism for comparing the status of different groups of species.

Evaluating the status of different plant and animal groups presents several challenges. Many species fall under the category of data-deficient. At the same time, we are still discovering many new species, particularly in remote areas of the world. Since the number of species known to science constantly increases, it is not possible to evaluate every species and our estimates of what fraction of species are declining will constantly change. Finally, the work is expensive. Making an assessment for even one group of species, such as birds or mammals, requires thousands of scientists and millions of dollars.

Threatened species According to the International Union for Conservation of Nature (IUCN), species that have a high risk of extinction in the future.

Near-threatened species Species that are very likely to become threatened in the future.

Least-concern species Species that are widespread and abundant.

Of the estimated 10 million species that currently live on Earth, ranging from bacteria to whales, only about 50,000 have been assessed to determine whether their populations are increasing, stable, or declining. Across all groups of organisms that have been assessed, nearly one-third are threatened with extinction. Some of the best data are for conifers, birds, reptiles, mammals, amphibians, and fish. **FIGURE 59.5** shows the data for those species that are not yet extinct.

Conifers are one of the best documented plant groups, mostly because there are only 606 species on Earth. This includes species of pines, spruces, hemlocks, firs, cedars, and redwoods. Ninety-five percent of conifer species have sufficient data for assessment. Since the year 1500, no conifer has gone extinct. Of those assessed, 50 percent are categorized as of least-concern, 16 percent are near-threatened, and 34 percent are threatened.

Over the past 500 years, nearly 11,000 bird species have existed and 156 (1.4 percent) have gone extinct. Of those remaining, 77 percent are of least concern and 23 percent are threatened or near-threatened. Among the 800 species of birds living in the United States, nearly one-third are experiencing declining populations. These include 40 percent of bird species that live in grasslands and 30 percent of bird species that live in arid regions. Multiple threats, including reduced habitat and rising sea levels, have caused a growing concern for all species of birds that live on coastlines or on islands.

In terms of percent declines, reptiles are faring worse than birds. There are approximately 5,000 species of snakes, lizards, and turtles and 86 percent have been assessed. Since 1500, 28 species (0.4 percent) have gone extinct. Of those remaining, 8 percent are near-threatened, and 24 percent are threatened.

A similar pattern exists for mammals. Of the nearly 5,560 species of mammals known to have existed after 1500, 83 are extinct (1.5 percent). Among the approximately 4,700 species for which there are reliable data, 25 percent are threatened and 8 percent are near-threatened. This means that more than 1,500 species of mammals may be at risk of extinction.

Amphibians are experiencing the greatest global declines. Of the more than 6,500 species of amphibians, 35 species are extinct. However, a recent assessment of amphibian populations suggests that the number of extinctions may accelerate in the coming decades. Among the approximately 5,000 species for which reliable data exist, 50 percent are either threatened or near-threatened. This means that nearly 2,500 species of amphibians are declining around the world.

Finally the fish species of the world are also experiencing declines. Of the nearly 16,000 species of fish, nearly 80 percent have adequate data to assess their

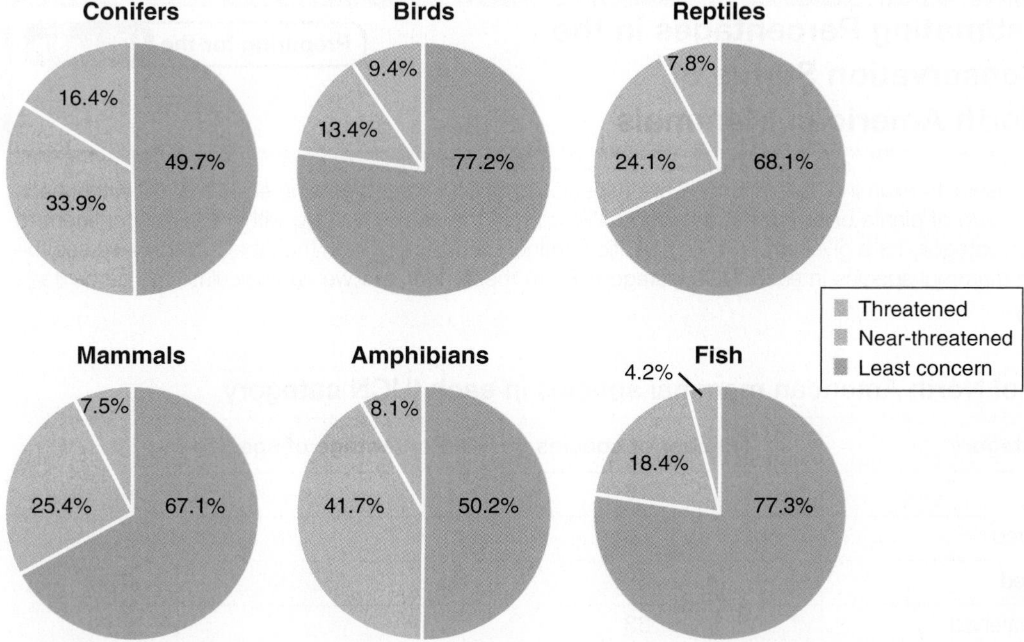

FIGURE 59.5 The decline of species. Based on those species for which scientists have reliable data, 50.3 percent of conifers, 22.8 percent of birds, 31.9 percent of reptiles, 32.9 percent of mammals, 48.8 percent of amphibians, and 22.6 percent of fish are currently classified as threatened or near-threatened with extinction. *(Data from International Union for Conservation of Nature (2017), https://goo.gl/ZJFSbl, https://goo.gl/SVYrfB)*

status. Seventy species of fish (1.1 percent) are extinct in nature. Of those remaining, 18 percent are threatened and 4 percent are near-threatened.

Many other groups of organisms are also experiencing large declines, but complete assessments have not yet been conducted because of the time and money required for each assessment. However, from the groups that have been assessed, it is clear that all of the groups have experienced a substantial number of extinctions and that a large percentage of species are declining. These results suggest that when the assessments are complete for the remaining major groups of plants and animals, the news will most likely not be good. "Do the Math: Estimating Percentages in the Conservation Status of North American Mammals" on page 657 will help you understand how we obtain these percentages from the number of species in each conservation category.

Ecosystem values and the global declines in ecosystem function

Given that we rely on a relatively small number of the millions of species on Earth for our essential needs, why should we care about the millions of other species that live in various ecosystems? To understand the value of ecosystems, we can consider both *intrinsic values* and *instrumental values*.

Many people believe that ecosystems have **intrinsic value**—that is, that ecosystems are valuable independent of any benefit to humans. These beliefs may grow out of religious or philosophical convictions. People who believe that ecosystems are inherently valuable may argue that we have a moral obligation to preserve them. They may equate the obligation of protecting ecosystems with our responsibility toward people or animals that might need our help to survive. People who argue that ecosystems are valuable independent of any benefit to humans generally believe that environmental policy and the protection of ecosystems should be driven by this intrinsic value.

An ecosystem may also have **instrumental value**, meaning that it has worth as an instrument or tool that can be used to accomplish a goal. Instrumental values, which include the value of items such as crops, lumber, and pharmaceutical drugs, can be thought

Intrinsic value Value independent of any benefit to humans.

Instrumental value Worth as an instrument or a tool that can be used to accomplish a goal.

DO THE MATH Estimating Percentages in the Conservation Status of North American Mammals

Preparing for the AP® Exam

As we have seen, it can be useful to examine both the number of species and the percentage of species that fall within each IUCN category for a given group of plants or animals in the world. We can do the same exercise within a given continent to see the percentages in each category for a given group on a specific continent. For example, in the table provided we see the number of North American mammal species in each IUCN category. From these numbers we can calculate the percentage of species in each category.

Number of North American mammal species in each IUCN category

IUCN category	Number of species	Percentage of species
Extinct	6	
Endangered	86	
Threatened	36	
Near-threatened	33	
Least-concern	546	
Data deficient	37	
Total		

Data from: *Smithsonian Museum of Natural History.* https://naturalhistory.si.edu/mna/image_menu.cfm

To calculate the percentages, we first have to determine the total number of mammal species:

$$\text{Total} = 6 + 86 + 36 + 33 + 546 + 37 = 744$$

To determine the percentage of species that are extinct, we divide the number of species in that category, by the total number of mammal species, and then multiply the decimal fraction by 100 to make it into a percentage:

$$\text{Percent extinct} = \text{Number of extinct mammal species} \div \text{Total number of mammal species} \times 100$$

$$= 6 \div 744 = 0.008$$

$$0.008 \times 100\% = 0.8\%$$

YOUR TURN Complete the table by calculating the percentages of mammal species in the remaining IUCN categories. Please round off to the nearest tenth of a percentage.

of in terms of how much economic benefit a species bestows. As noted in Chapter 1, we often refer to these instrumental values as ecosystem services. When calculating the instrumental value of various ecosystem services, we can consider five categories: *provisions, regulating services, support systems, resilience,* and *cultural services*.

Provision A good that humans can use directly.

Provisions

Goods produced by ecosystems that humans can use directly are called **provisions**. Examples include lumber, food crops, medicinal plants, natural rubber, and furs. Of the top 150 prescription drugs sold in the United States, about 70 percent come from natural sources. For example, Taxol, a potent anticancer drug, was originally discovered by a botanist in the bark of the Pacific yew (*Taxus brevifolia*), a rare tree that grows in forests of the Pacific Northwest (**FIGURE 59.6**). Once approved by the FDA,

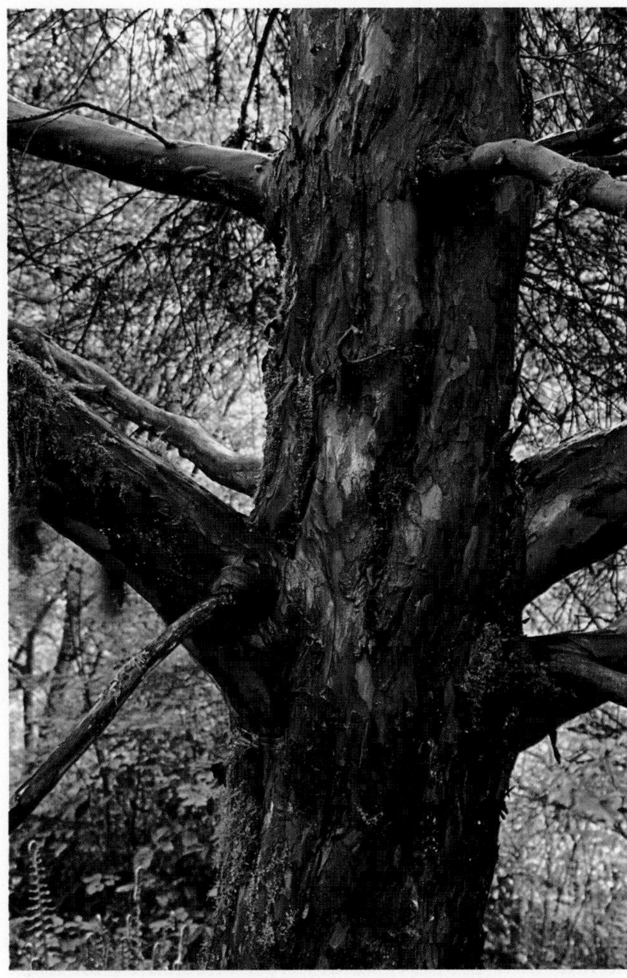

FIGURE 59.6 Provisions. Scientists discovered that the bark of the Pacific yew contains a chemical that has anticancer properties. *(Ray Pfortner/Getty Images)*

the synthetic version of this single drug has had annual sales of over $1.5 billion and to date has helped more than a million cancer patients. There is no way to estimate the potential value of natural pharmaceuticals that have yet to be discovered, but currently more than 800 natural chemicals have been identified as having potential uses to improve human health. Therefore, our best strategy may be to preserve as much biodiversity as we can to improve our chances of finding the next critical drug.

Regulating Services

Natural ecosystems help to regulate environmental conditions. For example, humans currently add about 8 gigatons of carbon to the atmosphere annually (1 gigaton = 1 trillion kilograms), but only about 4 gigatons of carbon remain there. The rest is removed by natural ecosystems, such as tropical rainforests and oceans, which provide us with more time to address climate

FIGURE 59.7 Regulating services. Tropical rainforests play a major role in regulating the amount of carbon in the atmosphere. *(JOHN PONTIER/Earth Scenes/Animals Animals)*

change than we would otherwise have (**FIGURE 59.7**). As we have already seen, ecosystems also are important in regulating nutrient and hydrologic cycles.

Support Systems

Natural ecosystems provide numerous support services that would be extremely costly for humans to generate. One example is pollination of food crops (**FIGURE 59.8**). The American Institute of Biological Sciences estimates that crop pollination in the United States by native species of bees and other insects, hummingbirds, and bats is worth roughly $3 billion in added food production. In addition to providing habitat for animals that pollinate crops, ecosystems offer natural pest control services because they provide habitat for predators that prey on

FIGURE 59.8 Support systems. Pollinators such as this honeybee on a cherry tree play an essential role in ensuring the pollination of food crops. *(Steffan & Alexandria Sailer/Ardea/Earth Scenes/Animals Animals)*

agricultural pests. Although organic farmers, who rarely use synthetic pesticides, gain the most from these pest controls, conventional agriculture benefits as well.

Healthy ecosystems also filter harmful pathogens and chemicals from water, leaving humans with water that requires relatively little treatment prior to drinking. Without these water-filtering services, humans would have to build many new water treatment facilities that use expensive filtration technologies. New York City, for example, draws its water from naturally clean reservoirs in the Catskill Mountains. But residential development and tourism in the area have threatened to increase contamination of the reservoirs with silt and chemicals. Building a filtration plant adequate to address these problems would cost \$6 billion to \$8 billion. For this reason, New York City and the U.S. Environmental Protection Agency have been working to protect sensitive regions of the Catskills.

Resilience

We have already seen that resilience ensures an ecosystem will continue to exist in its current state, which means it can continue to provide benefits to humans. Resilience depends greatly on species diversity. For example, several different species may perform similar functions in an ecosystem but differ in their susceptibility to disturbance. If a pollutant kills one plant species that contains nitrogen-fixing bacteria, but does not kill all plant species that contain nitrogen-fixing bacteria, the ecosystem can still continue to fix nitrogen (**FIGURE 59.9**).

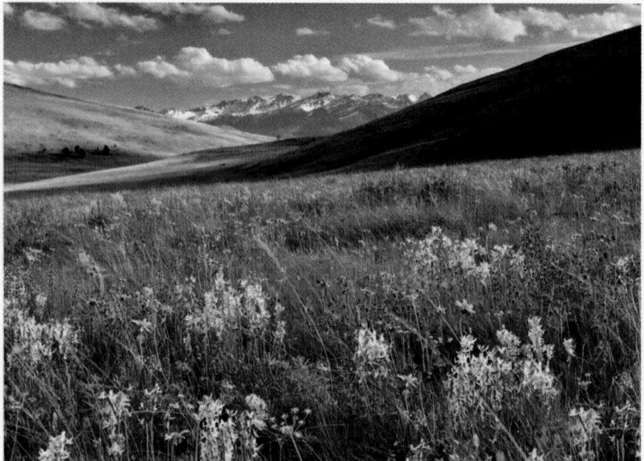

FIGURE 59.9 Species diversity as a component of resilience. The Zumwalt Prairie ecosystem in Oregon contains a high diversity of grasses and wildflowers, including many species of nitrogen-fixing wildflowers. If one nitrogen-fixing species is eliminated, the lost function can be compensated for by another nitrogen-fixing species. *(Dennis Frates / Alamy)*

Cultural Services

Ecosystems provide cultural or aesthetic benefits to many people. The awe-inspiring beauty of nature has instrumental value because it provides an aesthetic benefit for which people are willing to pay (**FIGURE 59.10**). Similarly, scientific funding agencies may award grants to scientists for research that explores biodiversity with no promise of any economic gain. Nevertheless, the research itself has instrumental value because the scientists and others benefit from these activities by gaining knowledge. While intellectual gain and aesthetic satisfaction may be difficult to quantify, they can be considered cultural services that have instrumental value.

The Monetary Value of Ecosystem Services

Most economists believe that the instrumental values of an ecosystem can be assigned monetary values, and they are beginning to incorporate these values into their calculations of the economic costs and benefits of various human activities. However, assigning a dollar value is easier for some categories of ecosystem services than for others. In 2014, a team of scientists and economists attempted to estimate the total value of ecosystem services to the human economy. They considered replacement value—the cost to replace the services provided by natural ecosystems. They also looked at other factors, such as how property values were affected by location relative to these services—for example, oceanfront housing. Finally, they considered how much time or money people were willing to spend to use these services—for example, whether they were willing to pay a fee to visit a national park. Using this method, researchers estimated that ecosystem services were worth over \$125 trillion per year, which is about twice the entire global economy.

The Decline of Ecosystem Services

Because species help determine the services that ecosystems can provide, we would expect declines in species diversity to be associated with declines in ecosystem function. In the Millennium Ecosystem Assessment conducted by the United Nations, scientists from around the world examined the current state of 24 ecosystem functions, including food production, pollination, water purification, and the cycling of nutrients such as nitrogen and phosphorus. Of these 24 different ecosystem functions, 15 were found to be declining or used at a rate that cannot be sustained. If we want to improve ecosystem functions, we need to improve the fate of the species and ecosystems that provide these services.

FIGURE 59.10 Cultural services. Many natural areas, such as this scene from the Grand Tetons National Park, provide aesthetic beauty valued by humans. *(Buddy Mays/Getty Images)*

MODULE 59 AP® Review

Preparing for the AP® Exam

In this module, we have seen that declines in biodiversity are happening at a rate that may indicate the start of a sixth mass extinction. At the genetic level, a loss of diversity is a concern because it places species under the threat of inbreeding depression that can cause population declines. We have observed declines in the genetic diversity of both wild species and domesticated species such as crops and livestock. At the species level, we have learned that major groups of organisms, including mammals, fish, and amphibians, have all experienced extinctions and contain large groups that are threatened or near-threatened. This global decline in species affects the functioning of ecosystems and reduces the intrinsic and instrumental values that these ecosystems provide. In the next module, we will examine the many causes of these declines in biodiversity.

AP® Practice Questions

Choose the best answer for the following.

1. In a major extinction event, what is the minimum percentage of species that goes extinct?
   (a) 40 percent
   (b) 50 percent
   (c) 75 percent
   (d) 90 percent

2. What factor has played the largest role in decreased diversity of domesticated species?
   (a) a focus on increased yields
   (b) the use of genetic engineering
   (c) the adaptation to specific growing environments
   (d) increased seed storage efforts

3. Which group of organisms has had the greatest number of extinctions since 1500?
   (a) birds
   (b) amphibians
   (c) reptiles
   (d) fish

4. The intrinsic value of an ecosystem
   (a) is the total monetary worth of its features.
   (b) is based on the goods the ecosystem produces.
   (c) considers the potential benefits that could be discovered in the ecosystem.
   (d) is the value it has independent of humans.

5. Which is NOT a category of instrumental value?
   (a) regulating services
   (b) resilience
   (c) provisions
   (d) diversity

MODULE 60

Causes of Declining Biodiversity

We have seen that declines in biodiversity are happening around the globe. In "Science Applied 2: How Should We Prioritize the Protection of Species Diversity?" on page 185, we discussed biodiversity hotspots, which are areas around the world rich in biodiversity. Many organisms in these biodiversity hotspots face threats of extinctions due to human activities. However, threats to biodiversity exist throughout the world. In this module, we build on the basics of population and community ecology from Chapter 5 to understand how a number of factors can affect biodiversity.

Learning Goals

After reading this module you should be able to

- discuss how habitat loss can lead to declines in species diversity.
- explain how the movement of exotic species affects biodiversity.
- describe how overharvesting causes declines in populations and species.
- understand how pollution reduces populations and biodiversity.
- identify how climate change affects species diversity.

Habitat loss is the major cause of declining species diversity

For most species, the greatest cause of decline and extinction is habitat loss. In modern times, the primary cause of habitat loss is human development that removes natural habitats and replaces them with homes, industries, agricultural fields, shopping malls, and roads. Many species can only thrive in a particular habitat within a narrow range of abiotic and biotic conditions. Species requiring such specialized habitats are particularly prone to population declines, especially when their favored habitat is limited, which restricts their distribution to a specific geographic area suitable only for a small population.

Altering distinctive characteristics of a habitat, such as removing trees or damming streams, has an effect on the organisms that live in that habitat. For example, for thousands of years the northern spotted owl (*Strix occidentalis caurina*) lived in old-growth forests—those dominated by trees that are hundreds of years old—in the northwestern United States and southwestern Canada. This habitat provided the ideal sites for nesting, roosting, and catching small mammals to eat. The removal of the old trees, for lumber and housing developments, has transformed much of the former old-growth forest into a different habitat (**FIGURE 60.1**). This habitat alteration reduces the number of northern spotted owls because they have fewer trees in which to nest and less forest in which to find food.

The map in **FIGURE 60.2** on page 664 shows the changing face of forest habitats from 2001 to 2015. As we saw in Chapter 6, much of the forest in the United States during the 1700s and 1800s was logged for lumber and cleared for agriculture. In recent decades, forested land has been increasing, although humans have often planted the new forests, which have a lower diversity of species than the original forests. At the same time, developing countries in South America, sub-Saharan Africa, and Southeast Asia are clearing their forests much as the United States and Europe did in years past. As a result, large declines in forest cover are occurring in developing countries that were once forested. It is currently unclear whether these countries will follow the pattern of Europe and North America and eventually allow their forested areas to increase.

Although deforestation receives a lot of attention, many other habitats are also being lost. According to the most recent Millennium Ecosystem Assessment, approximately 70 percent of the woodland/shrubland ecosystem that borders the Mediterranean Sea has been lost. Similarly, across the globe we have lost nearly 50 percent of grassland habitats and 30 percent of desert habitats. Wetlands exhibit a mixed picture. The amount of wetland habitat is less than half of what existed in the United States during the 1600s, with some states like California having lost 90 percent of its original wetlands. A 2013 study found that the combination of hurricanes, rising sea levels, and human development has caused the loss of more than 145,000 hectares (360,000 acres) of freshwater and saltwater wetlands in just 4 years.

(a)

(b)

FIGURE 60.1 Habitat loss. Habitat loss caused by humans is the largest threat to biodiversity. (a) Old-growth forests, such as this one in the Mount Rainier National Park in Washington State, serve as habitat for spotted owls. (b) After clear-cutting an old-growth forest for timber, such as this site in the Olympic National Forest in Washington State, the forest provides a very different habitat. It may take hundreds of years before the habitat again becomes suitable for the spotted owl. *(a: Stephen Matera / DanitaDelimont.com; b: Kent Foster/Science Source)*

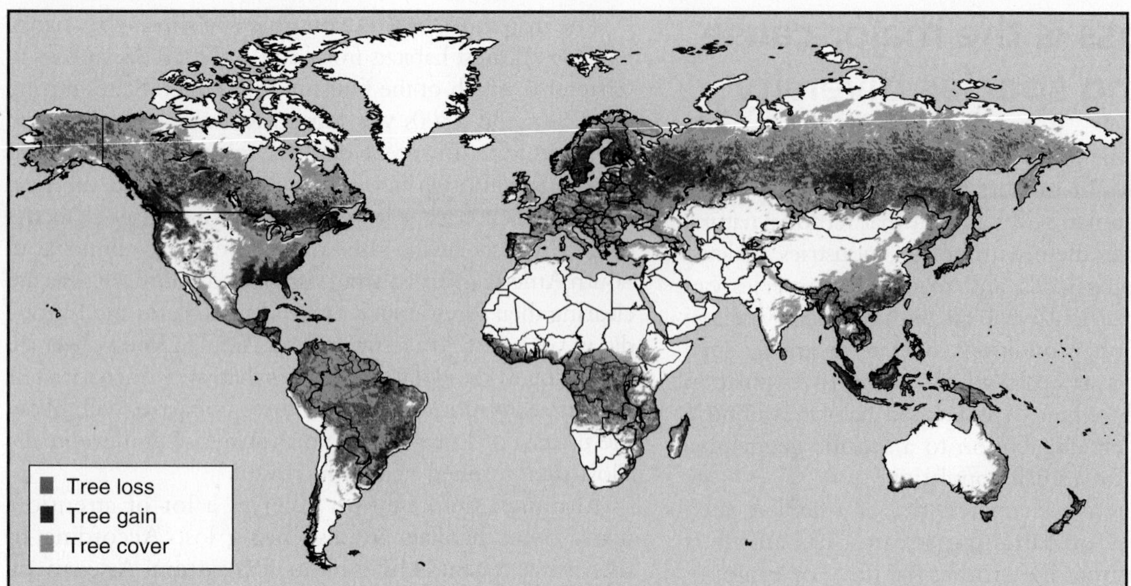

FIGURE 60.2 Changing forests. Some regions of the world experienced large declines in the amount of forested land from 2001 to 2015 while other regions have shown little change or have seen increases in forest cover. *(Data from Global Forest Watch, http://www.globalforestwatch.org/map. Source: Hansen/UMD/Google/USGS/NASA, accessed through Global Forest Watch.)*

In marine systems, there has been a sharp decline in the amount of living coral in the Caribbean Sea, from a high of 50 percent live coral in the 1970s to a mere 8 percent by 2012. Living coral provide habitat for thousands of other species, which makes them particularly vital to the persistence of marine habitats. The decline in coral is the result of human impacts including the warming of oceans (associated with global warming) and increased pollution, which lead to coral bleaching, and the removal of coral by collectors. In 2018, scientists reported that the occurrence of coral bleaching at any given site is about five times more frequent today than it was in the 1980s; rather than occurring every 25 to 30 years, the average bleaching event now occurs once every 6 years. This loss of coral habitat is occurring at a rapid rate throughout the world.

A species may decline in abundance or become extinct even without complete habitat destruction since a reduction in the size of critical habitat also can lead to extinctions through a variety of processes. As we saw in the case of the Florida panther, a smaller habitat supports a smaller population, reducing genetic diversity. Less habitat also reduces the variety of physical and climatic options available to individuals during periods of extreme conditions. The presence of cooler, high-altitude areas in a habitat, for example, allows animals a place to move during periods of hot weather. Also, loss of habitat can restrict the movement of migratory or highly mobile species. While many species can thrive in small habitats, other species, such as mountain lions, wolves, and tigers, require large tracts of relatively uninhabited, undisturbed land.

Smaller habitats can also cause increased interactions with other harmful species. For example, many songbirds in North America live in forests. When these birds make their nests near the edge of the forests—where the forest meets a field—they often have to contend with the brown-headed cowbird (*Molothrus ater*). The cowbird is a nest parasite—it does not build its own nest, but lays its eggs in the nests of several other species of birds (**FIGURE 60.3**). In this way the cowbird tricks forest birds into raising its offspring, which takes food away from the forest birds' own offspring. In some cases, the host bird will simply abandon the nest. As forests are broken up into smaller fragments, the proportion of forest near the edge increases and, therefore, the number of bird nests that are susceptible to brown-headed cowbirds increases. Over time, increased fragmentation has allowed brown-headed cowbirds to cause declines in many species of North American songbirds.

Native species Species that live in their historical range, typically where they have lived for thousands or millions of years.

Exotic species A species living outside its historical range. *Also known as* **alien species**.

Exotic species are moving around the world

Native species are species that live in their historical range, typically where they have lived for thousands or millions of years. In contrast, **exotic species**, also known as **alien species**, are species that live outside their historical range. For example, honeybees (*Apis mellifera*) were introduced to North America in the 1600s to provide a source of honey for European colonists.

(a)

(b)

FIGURE 60.3 Habitat fragmentation. (a) Increased fragmentation of forests has caused forest songbirds to come into increasing contact with the brown-headed cowbird. (b) The cowbird does not make its own nest. Instead, it lays its brown, spotted eggs in the nests of other species, such as this nest containing four blue eggs of the chipping sparrow (*Spizella passerina*). *(a: Jim Zipp/Ardea/Animals Animals; b: Paul J. Fusco/Science Source)*

Red foxes, now abundant in Australia, were introduced there in the 1800s for the purpose of fox hunts, which were popular in Europe at the time.

During the past several centuries, humans have frequently moved animals, plants, and pathogens around the world. Some species are also moved accidentally. For example, rats that stowed away in shipping containers ended up on distant oceanic islands. Because these islands never had rats or other ground predators, there had never been any natural selection against nesting on the ground, and numerous island bird species had evolved to nest on the ground. When the rats arrived in places such as Hawaii, they found the eggs and hatchlings from ground nests an easy source of food, resulting in a high rate of extinction in ground-nesting birds. Similar accidental movements have occurred for many pathogens, including exotic fungi that were introduced to North America nearly a century ago and have since killed nearly all American elm (*Ulmus americana*) and American chestnut (*Castanea dentata*) trees in eastern North America. Similarly, an exotic protist that causes avian malaria has driven many species of Hawaiian birds to extinction. Other movements of exotic species are intentional, such as exotic plants that are sold in greenhouses for houseplants and outdoor landscape plants, or exotic animals that are sold as pets or to game ranches that raise exotic species of large mammals for hunting.

In most cases, exotic species fail to establish successful populations when they are introduced to a new region. For the small percentage of introductions that are successful, exotic species can live in their new surroundings and have no negative effect on the native species. In other cases, however, the exotic species rapidly increase in population size and cause harmful effects on native species. When exotic species spread rapidly across large areas and cause harm, we call them **invasive species**. Rapid spread of invasive species is possible because invasive species, which have natural enemies in their native regions that act to control their population, often have no natural enemies in the regions where they are introduced. Two of the best-known examples of invasive exotic species in North America are the kudzu vine (*Pueraria lobata*) and the zebra mussel (*Dreissena polymorpha*).

AP® Exam Tip

Make sure you can explain why an exotic species might be more successful than an indigenous species. Also know some specific reasons why an exotic species might not be able to out-compete a native species. ●

The kudzu vine is native to Japan and southeast China but was introduced to the United States in 1876. Throughout the early 1900s, farmers in the southeastern states were encouraged to plant kudzu to help reduce erosion in their fields. By the 1950s, it became apparent that the southeastern climate was ideal for kudzu, with growth rates of the vine approaching 0.3 m (1 foot) *per day*. Because herbivores in the region do not eat kudzu, the species has no enemies and can spread rapidly. The vine grows up over most wildflowers and trees and shades them from the sunlight,

Invasive species A species that spreads rapidly across large areas and causes harm.

causing the plants to die. Indeed, the vine grows over just about anything that does not move (**FIGURE 60.4**). Kudzu currently covers approximately 3 million hectares (7.4 million acres) in the United States.

The zebra mussel is native to the Black Sea and the Caspian Sea in eastern Europe and western Asia. Over the years, large cargo ships that traveled in these seas unloaded their cargo in the ports of the Black Sea and Caspian Sea and then pumped seawater into the holding tanks to ensure that the ship sat low enough in the water to remain stable. This water that is pumped into the ship is called ballast water. When the ships arrived in the St. Lawrence River and the Great Lakes, they loaded on new cargo and no longer needed the weight of the ballast water, which they pumped out of the ship into local waters. One consequence of transporting ballast water from Asia to North America is that many aquatic species from Asia, including zebra mussels, have been introduced into the aquatic ecosystems of North America. Because the St. Lawrence River and the Great Lakes provided an ideal ecosystem for the zebra mussel, and because a single zebra mussel can produce up to 30,000 eggs, the mussel spread rapidly through the Great Lakes ecosystem. On the positive side, because the mussels feed by filtering the water, they remove large amounts of algae and some contaminants, which helps to counteract the cultural eutrophication that has occurred in the Great Lakes ecosystem. On the negative side, the zebra mussels physically crowd out many native mussel species and the zebra mussels can consume so much algae that they negatively affect native species that also need to consume the algae. Moreover, the invasive mussels can achieve such high densities that they can clog intake pipes and impede the flow of water on which industries and communities rely.

A new threat to the Great Lakes is the silver carp (*Hypophthalmichthys molitrix*), a fish that is native to Asia but has been transported around the world in an effort to consume excess algae that accumulates in aquaculture operations and the holding ponds of sewage treatment plants. After being brought to the United States, some of the fish escaped and rapidly spread through many of the major river systems, including the Mississippi River. Over the years, the carp population has expanded northward, and by 2010 it approached a canal where the Mississippi River connects to Lake Michigan. Although researchers detected the DNA of the carp in water samples from the Great Lakes from 2009 to 2011, substantial netting efforts in the spring of 2013 failed to find any of the fish in the lakes. As of 2018, it appeared that the carp had not yet found its way into the Great Lakes. There are two major concerns about this invading fish. First, scientists worry that it will outcompete native species of fish that also consume algae. Second, the silver carp has an unusual behavior; it jumps out of the water when startled by passing boats (**FIGURE 60.5**). Given that the carp can grow to 18 kg (40 pounds) and jump up to 3 m (10 feet) into the air, this poses a major safety issue to boaters.

Around the world, invasive exotic species pose a serious threat to biodiversity by acting as predators, pathogens, or superior competitors to native species. Some of the most complete data exist in Europe. During the past 100 years, Europe has experienced a steady increase of nearly 2,000 exotic species in terrestrial ecosystems. Additional species have been introduced into freshwater and marine ecosystems. A number of efforts are currently being used to reduce the introduction of invasive exotic species, including the inspection of goods coming into a country and the prohibition of wooden packing crates made of untreated wood that could contain insect pests.

FIGURE 60.4 The spread of the exotic kudzu vine. The fast-growing kudzu vine is native to Asia but was introduced to the United States to control erosion. It has since spread rapidly, growing over the top of nearly anything that does not move, like this barn in Rose Hill, North Carolina. *(Gilbert S Grant/Getty Images)*

Overharvesting causes declines in populations and species

Hunting, fishing, and other forms of harvesting are the most direct human influences on wild populations of plants and animals. Most species can be harvested to

FIGURE 60.5 Silver carp invading the Mississippi River. The silver carp has been introduced into the Mississippi River from Asia and is quickly heading toward an invasion of the Great Lakes. *(Chris Olds/U.S. Fish & Wildlife Service)*

some degree, but a species is overharvested when individuals are removed at a rate faster than the population can replace them. In the extreme, overharvesting of a species can cause extinction. In the seventeenth century, for example, ships sailing from Europe stopped for food and water at Mauritius, an uninhabited island in the Indian Ocean. On Mauritius, the sailors would hunt the dodo (*Raphus cucullatus*), a large flightless bird that had no innate fear of humans because it had never seen humans during its evolutionary history (**FIGURE 60.6**). The dodo, unable to protect itself from human hunters and the rats (introduced by humans) that consumed dodo eggs and hatchlings, became extinct in just 80 years. This same scenario appears to have taken place with many other large animal species as well. These animals include the giant ground sloths, mammoths, American camels of North and South America, and the 3.7 m (12 feet) tall moa birds of New Zealand. Each species became extinct soon after humans arrived, suggesting that the animals' demise may have been due to overharvesting.

Overharvesting has also occurred in the more recent past. In the 1800s and early 1900s, for example, market hunters slaughtered wild animals to sell their parts on such a scale that many species, including the American bison, declined dramatically. Bison were once abundant on the western plains, with estimates ranging from 60 to 75 million individuals. By the late 1800s fewer than 1,000 were left. This means that 99.999 percent of all bison were killed. Following enactment of legal protections, the bison population today has increased to more than 500,000, including both wild bison and bison raised commercially for meat.

FIGURE 60.6 Overharvesting. The dodo was a large flightless bird that served as an easy source of meat for sailors and settlers on the island of Mauritius. Because it evolved on an island with no large predators or humans, the dodo had no instinct to fear humans. *(Stock Montage/Getty Images)*

Not all species harvested by market hunters fared as well as the American bison. The passenger pigeon was once one of the most abundant species of birds in North America. Population estimates range from 3 to 5 billion birds in the nineteenth century. In fact, during annual migrations, people observed continuous flocks of pigeons flying overhead for 3 days straight in densities that blocked out most of the sun. Breeding flocks could cover 40,000 ha (100,000 acres) with 100 nests built into each tree. With such high densities, market hunters could shoot or net the birds in very large numbers and fill train cars with harvested pigeons to be sold in eastern cities. This overharvesting, combined with the effects of forest clearing for agriculture, caused the passenger pigeon to decline quickly. The last passenger pigeon died in 1914 at the Cincinnati Zoo.

During the past century, regulations have been passed to prevent the overharvesting of plants and animals. In the United States, for example, state and federal regulations restrict hunting and fishing of game animals to particular times of the year and limit the number of animals that can be harvested. Similar agreements have been reached among countries through international treaties. In general, these regulations have proven very successful in preventing species declines caused by overharvesting. In some regions of the world, however, harvest regulations are not enforced and illegal poaching, especially of large, rare animals that include tigers, rhinoceroses, and apes, continues to threaten species with extinction. Harvesting of rare plants, birds, and coral reef dwellers for private collections has also jeopardized these species.

AP® Exam Tip

You should be able to identify both anthropogenic and nonanthropogenic causes of extinction. ●

Plant and Animal Trade

For some species, the legal and illegal trade in plants and animals represents a serious threat to their ability to persist in nature. One of the earliest laws in the United States to control the trade of wildlife was the **Lacey Act**. First passed in 1900, the act originally prohibited the transport of illegally harvested game animals, primarily birds and mammals, across state lines. Over the years, a number of amendments have been added so that the Lacey Act today forbids the interstate shipping of all illegally harvested plants and animals.

Lacey Act A U.S. act that prohibits interstate shipping of all illegally harvested plants and animals.

At the international level, the United Nations **Convention on International Trade in Endangered Species of Wild Fauna and Flora**, also known as CITES, was developed in 1973 to control the international trade of threatened plants and animals. Today, CITES is an international agreement among 182 countries throughout the world. The IUCN maintains a list of threatened species known as the **Red List**. Each member country assigns a specific agency to monitor and regulate the import and export of animals on the list. For example, in the United States, the U.S. Fish and Wildlife Service conducts this oversight.

Despite such international agreements, much illegal plant and animal trade still occurs throughout the world. In 2012, a report by the Congressional Research Service estimated that illegal trade in wildlife was worth $5 billion to $20 billion annually. In 2017, it was estimated that 50,000 illegal shipments of wildlife and wildlife parts had been sent to United States ports over a 10-year period. In some cases, animals are sold for fur or for body parts that are thought to have medicinal value. For example, 1 kg (2.2 pounds) of rhino horn can sell for $50,000. In other cases, rare animals are in demand as pets. For example, in 2001 a population of the Philippine forest turtle (*Siebenrockiella leytensis*), once thought to be extinct, was discovered on a single island in the Philippines (**FIGURE 60.7**). This animal, one of the most endangered species in the world, cannot be traded legally, but demand for it as a pet has caused it to be sold illegally and the last remaining population has declined sharply in only a few years. A single turtle sells for $50 to $75 in the Philippines and up to $2,500 in the United States and Europe. Today, the turtle is critically endangered. Similar cases of illegal trade occur in rare species of trees for lumber such as big-leaf mahogany (*Swietenia macrophylla*), rare species of plants for medicine such as goldenseal (*Hydrastis canadensis*), and many rare species of orchids for their beautiful flowers.

Sometimes even when trade in a particular species is legal, it can pose a potential long-term threat to species persistence. In the southwestern United States, for example, there is a growing movement to reduce water use by replacing grass lawns with desert landscapes. One of the unintended consequences is the increased demand for cacti and other desert plants that are collected from the wild. Sales are currently estimated to be $1 million annually. Given the slow growth of desert plants, this increased demand is causing heightened concern for these plant populations in the wild.

Convention on International Trade in Endangered Species of Wild Fauna and Flora (CITES) A 1973 treaty formed to control the international trade of threatened plants and animals.

Red List A list of worldwide threatened species.

FIGURE 60.7 Species declines due to the pet trade. The only remaining population of this turtle lives on a single island in the Philippines. Although protected by law, illegal trade has caused a rapid decline of this species in the wild. *(NHPA/Photoshot)*

Pollution can have harmful effects on species

In Chapters 14 and 15, we saw how water and air pollution harm ecosystems. Threats to biodiversity come from toxic contaminants such as pesticides, heavy metals, acids, and oil spills. Other contaminants, such as endocrine disrupters, can have nonlethal effects that prevent or inhibit reproduction. Pollution sources that cause declines in biodiversity also include the release of nutrients that cause algal blooms and dead zones as well as thermal pollution that can make water bodies too warm for species to survive.

In 2010, for example, an oil platform in the Gulf of Mexico owned by BP and named the *Deepwater Horizon* exploded, causing a massive release of oil that lasted for several months. As we discussed in Chapters 12 and 14, the release of oil caused a tremendous amount of death across a wide range of animal species including sea turtles, pelicans, fish, and shellfish. In response to the massive oil spill, BP released hundreds of thousands of liters of oil dispersant, a chemical designed to break up large areas of oil into tiny droplets that can be consumed by specialized species of bacteria. However, the dispersant is also toxic to many species of animals. It is estimated that the oil spill and subsequent dispersant killed tens of thousands of sea turtles; in the northern Gulf it is estimated to have killed 12 percent of brown pelicans (*Pelecanus occidentalis*) and 32 percent of laughing gulls (*Leucophaeus atricilla*). The total impact of the spilled oil and applied dispersants on the wildlife of the Gulf of Mexico may not be known for many years.

Climate change has the potential to affect species diversity

We have mentioned climate change in previous chapters and will discuss it in detail in Chapter 19. As a threat to biodiversity, the primary concern about climate change is its effect on patterns of temperature and precipitation in different regions of the world. In some regions, a species may be able to respond to warming temperatures and changes in precipitation by migrating to a place where the climate is well suited to the species niche. In other cases, this is not possible. For example, in southwestern Australia, a small woodland/shrubland peninsula exists on the edge of the continent with a much larger area of subtropical desert farther inland (see Figure 12.3 on page 127). Scientists expect conditions on the peninsula to become drier during the next 70 years. If this occurs, many species of plants in this small ecosystem will not have a nearby hospitable environment to which they can migrate, since the surrounding desert ecosystem is already too dry for them. An examination of 100 species of plants in the area (all from the genus *Banksia*) has led scientists to project that 66 percent of the species will decline in abundance and up to 25 percent will become extinct. As we will see in Chapter 19, many species in the world are expected to be affected by climate change.

MODULE 60 AP® Review

Preparing for the AP® Exam

In this module, we learned that the biodiversity of our planet is declining for a number of reasons. The primary causes of this decline are the loss of habitats and the fragmentation of habitats that species need to survive and reproduce. Exotic species are being moved around the world with the increased global movement of people and materials. Many populations of these exotic species remain small and cause no discernible harm, but some become invasive species that spread quickly and have harmful effects on native species. Overharvesting of species can cause larger declines in population sizes and, in some cases, extinction. Current regulations within and among countries are designed to limit harvesting to sustainable levels, although these regulations are not always successfully enforced. Toxic compounds including pesticides, heavy metals, and spilled oil can also be detrimental to species either by direct lethal effects or by altering communities and ecosystems. Finally, climate change has the potential to alter populations and the long-term persistence of species, but more time is needed to determine if these predictions will come true. In the next module, we will examine past and current efforts to conserve biodiversity.

AP® Practice Questions

Choose the best answer for the following.

1. The most significant cause of species decline and extinction throughout the world is
   (a) habitat loss.
   (b) overharvesting.
   (c) climate change.
   (d) invasive species.

2. Invasive species are
   (a) usually not a threat to biodiversity.
   (b) rare in island habitats.
   (c) successful due to a lack of natural enemies.
   (d) often unable to compete effectively in the new environment.

3. Passenger pigeons were driven extinct primarily by
   (a) habitat loss.
   (b) overharvesting.
   (c) pollution.
   (d) invasive species.

4. The Lacey Act
   (a) provides protected habitats for a number of threatened species.
   (b) forbids the interstate shipping of illegally harvested plants and animals.
   (c) provides harvesting quotas and prevents overharvesting.
   (d) prevents the spread of invasive species to the United States.

5. The primary impact of climate change on species diversity is expected to be
   (a) an increased variability in weather.
   (b) decreased precipitation worldwide.
   (c) changes in available habitat because of changing temperatures.
   (d) the increased ability of species to disperse.

MODULE 61

The Conservation of Biodiversity

It is important that we consider how to protect and increase biodiversity because of the large number of factors that can reduce biodiversity. There are two general approaches to conserving biodiversity: the single-species approach and the ecosystem approach. In this module, we explore each of these approaches.

Learning Goals

After reading this module you should be able to

- identify legislation that focuses on protecting single species.
- discuss conservation efforts that focus on protecting entire ecosystems.

Conservation legislation often focuses on single species

The single-species approach to conserving biodiversity focuses our efforts on one species at a time. When a species declines significantly, the natural response is to encourage a population rebound by improving the conditions in which that species exists. This might be accomplished by providing additional habitat, reducing the harvest, or reducing the presence of a contaminant that is impairing survival or reproduction. When the population of a species has declined to extremely low numbers, sometimes the remaining few individuals will be captured and brought into captivity. Captive animals are bred with the intention of returning the species to the wild. A well-known example of captive breeding occurred with the California condor. As we discussed in Chapter 6, the condor had declined to a mere 22 birds in 1987. Thanks to captive breeding and several improvements in the condor's habitat, the population in 2013 was more than 400 birds. Programs such as these are a major function of zoos and aquariums around the world. During the past 50 years, legislation has been passed to help many more species recover.

AP® Exam Tip

Be able to identify ecological, economic, and social benefits of biodiversity. ●

Marine Mammal Protection Act A 1972 U.S. act to protect declining populations of marine mammals.

The Marine Mammal Protection Act

In the United States, the single-species approach to conservation formed the foundation of the *Marine Mammal Protection Act* and the *Endangered Species Act*. The **Marine Mammal Protection Act** prohibits the killing of all marine mammals in the United States and prohibits the import or export of any marine mammal body parts. Only the U.S. Fish and Wildlife Service and the National Marine Fisheries Service are allowed to approve any exceptions to the act. The act was passed in 1972 in response to declining populations of many marine mammals, including polar bears, sea otters, manatees, and California sea lions (*Zalophus californianus*) (**FIGURE 61.1**).

FIGURE 61.1 Protected marine mammals. The Marine Mammal Protection Act protects marine mammals in the United States such as the sea otter from being killed. *(Hal Beral / V&W / The Image Works)*

The Endangered Species Act

In Chapter 10, we noted that the Endangered Species Act is a 1973 law designed to protect species from extinction. This act authorizes the U.S. Fish and Wildlife Service to determine which species can be listed as threatened species or *endangered species* and prohibits the harming of such species, including prohibitions on the trade of listed species, their fur, or their body parts. Earlier in this chapter we discussed the international definitions of threatened and near-threatened as used by the IUCN (see page 656): Threatened species have a high risk of extinction in the future and near-threatened species are very likely to become threatened in the future. In the United States, an **endangered species** is defined as a species that is in danger of extinction within the foreseeable future throughout all or a significant portion of its range whereas a **threatened species** is defined as any species that is likely to become an endangered species within the foreseeable future throughout all or a significant portion of its range. As you can see, the U.S. definition of endangered is similar to the international definition for threatened and the U.S. definition of threatened is similar to the international definition of near-threatened.

The Endangered Species Act was first passed in 1973 and has been amended several times since then. From an international perspective, the act also implements the international CITES agreement that we discussed in the previous module. To assist in the conservation of threatened and endangered species, the act authorizes the government to purchase habitat that is critical to the conservation of these species and to develop recovery plans to increase the population of threatened and endangered species. This is often one of the most important steps in allowing endangered species to persist. For example, grizzly bears in the Greater Yellowstone Ecosystem had declined to 136 individuals in 1975. After being listed as endangered, steps were taken to protect the bear's favored habitat and today it numbers nearly 700 individuals. (**FIGURE 61.2**).

FIGURE 61.2 Bringing back endangered species. Habitat protection and reduced contaminants in the environment have allowed grizzly populations to increase to the point where they could be taken off the Endangered Species List. These grizzlies live in Yellowstone National Park. *(James Hager/Getty Images)*

As of 2018, the species that have been listed as threatened or endangered in the United States include 272 invertebrate animals, 442 vertebrate animals, and 947 plants. Although the listing process can take several years, once listed many threatened and endangered species have experienced stable or increasing populations. Indeed, some species have experienced sufficient increases in numbers to be removed from the endangered species list; these include the bald eagle, peregrine falcon, American alligator, and the eastern Pacific population of the gray whale (*Eschrichtius robustus*). Other species are currently increasing in number and may be taken off the list in the future. The gray wolf, for example, was reintroduced into Yellowstone National Park to help improve the species' abundance in the United States and it is now no longer endangered.

The Endangered Species Act has sparked a great deal of controversy in recent years because it can restrict certain human activities in areas where listed species live, including how landowners use their land. For example, some construction projects have been prevented or altered to accommodate threatened or endangered species. Organizations whose activities are restricted by the Endangered Species Act often try to pit the protection of listed species against the jobs of people in the region. In the 1990s, for example, logging companies wanted to continue logging the old-growth forest of the Pacific Northwest. As we discussed earlier in this chapter, these forests are home to the threatened northern spotted owl and many other species that depend on old-growth forest. Automation had caused a large decline in the number of logging jobs over the preceding several decades, and many loggers perceived the Endangered Species Act as a further threat to their livelihood. They denounced the act because they said it placed more value on the spotted owl than it did on the humans who depended on logging. In the end, a compromise allowed continued logging on some of the old-growth forest while the rest became protected habitat.

During the past decade, several politicians and their constituents have attempted to weaken the Endangered Species Act. However, strong support from the public and scientists has allowed it to retain much of its original power to protect threatened and endangered species. The biggest current challenge is a lack of sufficient funds and personnel required to implement the law.

Endangered species A species that is in danger of extinction within the foreseeable future throughout all or a significant portion of its range.

Threatened species According to U.S. legislation, any species that is likely to become an endangered species within the foreseeable future throughout all or a significant portion of its range.

The Convention on Biological Diversity

Protecting biodiversity is an international concern. In 1992, world nations came together and created the **Convention on Biological Diversity**, which is an international treaty to help protect biodiversity. The treaty had three objectives: conserve biodiversity, sustainably use biodiversity, and equitably share the benefits that emerge from the commercial use of genetic resources such as pharmaceutical drugs.

In 2002, the convention developed a strategic plan to achieve a substantial reduction in the worldwide rate of biodiversity loss by 2010. The nations that signed this agreement recognized both the instrumental and intrinsic values of biodiversity. In 2014, the convention evaluated the current trends in biodiversity around the world and concluded that the goal had not been met. They identified the following trends:

- Habitat loss has been slowed in some regions of the world, but remains high in other regions.
- Many fish species continue to be overharvested.
- On average, species at risk of extinction have moved closer to extinction.
- There is high variation around the world in the preservation of ecosystem services.
- Natural habitats are becoming smaller and more fragmented.

Collectively, the message emerging from the convention is not very positive. From the perspectives of genetic diversity, species diversity, and ecosystem services, all of the trends continue to move in the wrong direction.

AP® Exam Tip

Make sure you can discuss legislation that focuses on the protection of biodiversity, including the Marine Mammal Protection Act, the Endangered Species Act, and the Convention on Biological Diversity. ●

Some conservation efforts focus on protecting entire ecosystems

Awareness of a potential sixth mass extinction in which humans have played a major role has brought a growing interest in the ecosystem approach to conserving biodiversity. This approach recognizes the benefit of preserving particular regions of the world, such as biodiversity hotspots. Protecting entire ecosystems has been one of the major motivating factors in setting aside national parks and marine reserves. In some cases, these areas were originally protected for their aesthetic beauty, but today they are also valued for their communities of organisms. The amount of protected land has increased dramatically throughout the world since 1960. As an example of this increase, **FIGURE 61.3** shows changes in the amount of protected land worldwide since 1900.

When protecting ecosystems to conserve biodiversity, a number of factors must be taken into consideration including the size and shape of the protected area. We must also consider the amount of connectedness to other protected areas and how best to incorporate conservation while recognizing the need for sustainable habitat use for human needs.

Convention on Biological Diversity An international treaty to help protect biodiversity.

The Size, Shape, and Connectedness of Protected Areas

A number of questions arise when we consider protecting areas of land or water. For example, how large should the designated area be? Should we protect a single large area or several smaller areas? Does it matter whether protected areas are isolated or if they are near other protected areas? To help us answer these questions, we can return to our discussion of the theory of island biogeography from Chapter 6.

As you may recall, the theory of island biogeography looks at how the size of islands and the distance between islands and the mainland affect the number of species that are present on different islands. Larger islands generally contain more species because they support larger populations of each species, which makes them less susceptible to extinction. Larger islands also contain more species because they typically contain more habitats and, therefore, provide a wider range of niches for different species to occupy. The distance between an island and the mainland, or between one island and another, is another crucial factor, since more species are capable of dispersing to close islands than to islands farther away.

Although the theory of island biogeography was originally applied to oceanic islands, it has since been applied to islands of protected areas in the midst of less hospitable environments. For example, we can think of all the state and national parks, natural areas, and wilderness areas as islands surrounded by environments subject to high levels of human activity, including agricultural fields, logged forests,

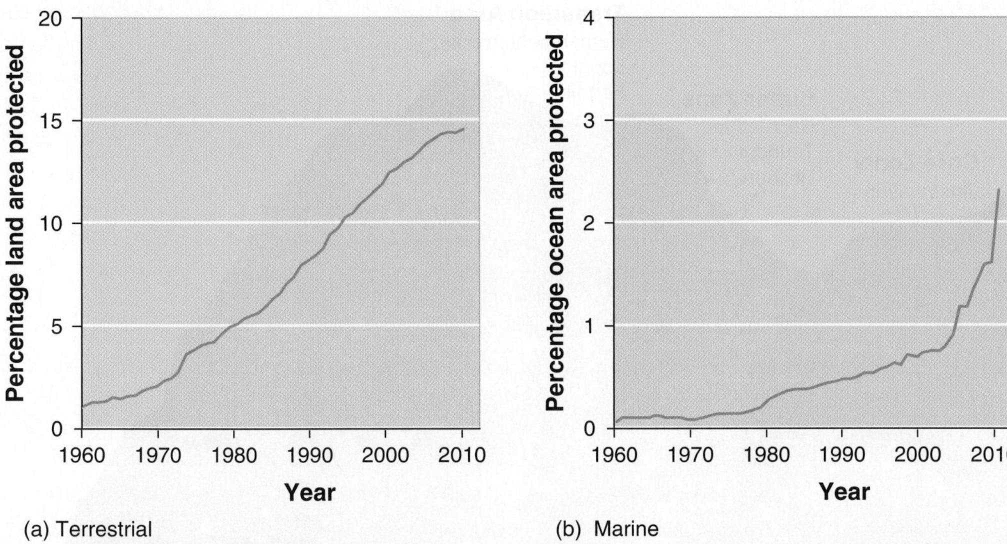

FIGURE 61.3 Changes in protected land. Since the 1960s, there has been a large increase in the amount of (a) terrestrial and (b) marine habitat that is under various types of protection throughout the world. *(Data from Secretariat of the Convention on Biological Diversity,* Global Biodiversity Outlook 4 *(2014), https://www.cbd.Int/gbo/gbo4/publication/gbo4-en-hr.pdf, https//www.cbd.int/gbo2/.)*

housing developments, and cities (**FIGURE 61.4**). These areas provide habitats for species and places to stop and rest for migrating species. Applying the theory of island biogeography from this perspective gives us some idea of the best ways to design and manage protected areas. For example, when protected areas are far apart, it is less likely that species can travel among them. This means that when a species has been lost from one ecosystem, it will be harder for individuals of that species from other ecosystems to recolonize it. So when we create smaller areas, they should be close enough for species to move among them easily.

Decisions regarding the design of protected areas can also be informed by the concept of metapopulations. As we learned in Chapter 6, a metapopulation is a collection of smaller populations connected by occasional dispersal of individuals along habitat corridors. Each population fluctuates somewhat independently of the other populations and a population that declines or goes extinct, due to a disease for example, can be rescued by dispersers from a neighboring population. So if we set aside multiple protected areas, and recognize the need for connecting habitat corridors, a species is more likely to be protected from extinction by a decimating event such as a disease or natural disaster that could eliminate all individuals in a single protected area.

FIGURE 61.4 Habitat islands. Central Park in New York City is an extreme example of an island of hospitable habitat surrounded by an urban environment that is not hospitable to most species. *(Michael S. Yamashita/National Geographic Stock)*

The concepts of island biogeography and metapopulations raise an interesting dilemma for conservation efforts. If we have limited resources to protect the biodiversity of a region, should we protect a single large area or several small areas? A single large area would support larger populations, but a species is more likely to survive a disease or natural disaster if it occupies several different areas. The debate over the best approach is known as SLOSS, which is an acronym for "single large or several small." While both approaches have their merits, in reality, human development and other factors often mean that only one of the two strategies is available. For example, due to human development of a region, there may simply not be a single large area available to protect, so the only available strategy is to protect several small areas. A final consideration regarding the size and shape of protected areas is the amount of *edge habitat* that an area contains. **Edge habitat** occurs where

Edge habitat Habitat that occurs where two different communities come together, typically forming an abrupt transition, such as where a grassy field meets a forest.

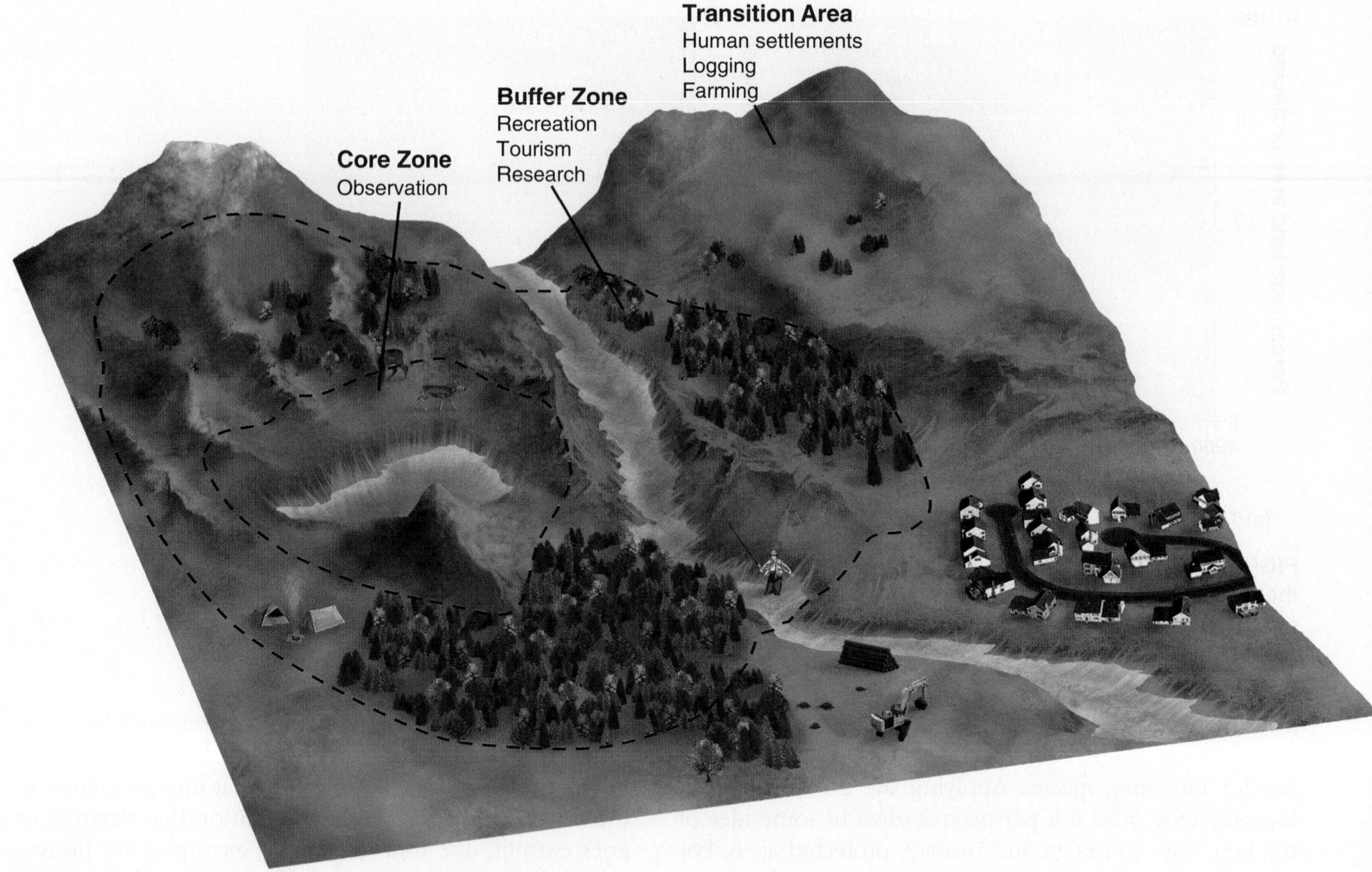

FIGURE 61.5 Biosphere reserve design. Biosphere reserves ideally consist of core areas that have minimal human impact and outer zones that have increasing levels of human impacts.

two different communities come together, typically forming an abrupt transition, such as where a grassy field meets a forest. While some species will live in either field or forest, others, like the brown-headed cowbird, specialize in living at the forest edge. So another challenge of protecting many small areas is the comparatively larger amount of edge habitat. When we protect several small forests, for example, the proliferation of species such as the cowbird in this larger amount of edge habitat can have a detrimental effect on songbirds that typically live farther inside a forest.

Biosphere Reserves

In Chapter 10 we saw that managing national parks and other protected areas so they serve multiple users can be a challenge. While we want to make places of great natural beauty available to everyone, when large numbers of people use an area for recreation, at least some degradation is very likely. To address this problem, the United Nations Educational, Scientific and Cultural Organization (UNESCO) developed the innovative concept of *biosphere reserves*. **Biosphere reserves** are protected areas consisting of zones that vary in the amount of permissible human impact. These reserves protect biodiversity without excluding all human activity. **FIGURE 61.5** shows the different zones in a hypothetical biosphere reserve. The central core is an area that receives minimal human impact and is therefore the best location for preserving biodiversity. A buffer zone encircles the core area. Here, modest amounts of human activity are permitted, including tourism, environmental education, and scientific research facilities. Farther out is a transition area containing sustainable logging, sustainable agriculture, and residences for the local human population.

Designing reserves with these three zones represents an ideal scenario. In reality, biosphere reserves can take many forms depending on their location,

Biosphere reserve Protected area consisting of zones that vary in the amount of permissible human impact.

though all attempt to maintain low-impact core areas. As of 2018 there are 669 biosphere reserves worldwide, with a total of 120 nations participating. The number of biosphere reserves in the United States had grown to 47 by 2016, but 17 of these were withdrawn in 2017.

One well-known biosphere reserve is Big Bend National Park in Texas. The park itself serves as the core area and receives relatively little human impact, although hikers are permitted to walk through the beautiful desert landscapes and tree-covered mountain peaks. The park contains several dozen threatened and endangered plant and animal species. It also contains more than 1,000 species of plants and 400 species of birds, many of which pass through the Big Bend region during their annual spring and fall migrations. Outside the boundaries of the park is a region of increased human impact including tourist facilities, human settlements, and agriculture (**FIGURE 61.6**).

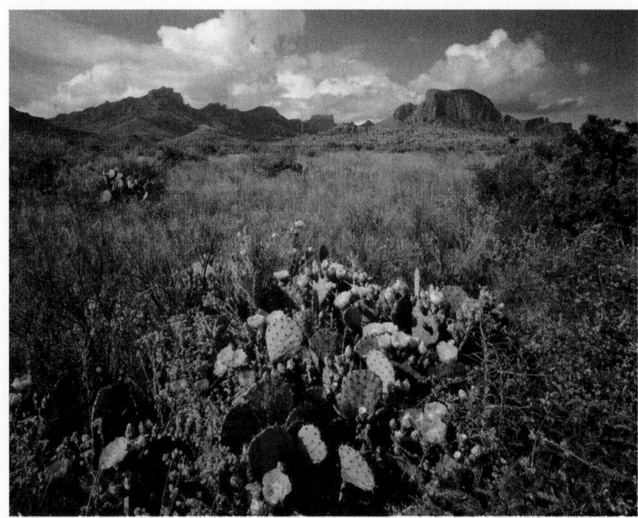

FIGURE 61.6 Biosphere reserve. Big Bend National Park, located in southwest Texas, serves as a low-impact core area of the Big Bend biosphere reserve. *(Tom Till/Alamy)*

MODULE 61 AP® Review

Preparing for the AP® Exam

In this module, we learned that efforts to conserve biodiversity focus on either single species or entire ecosystems. The single-species approach is often the approach taken by conservation legislation. In the United States, single-species legislation includes the Marine Mammal Protection Act and the Endangered Species Act. Internationally, there are treaties that protect species, including CITES and the Convention on Biological Diversity. A number of conservation efforts have focused on protecting entire ecosystems by considering the concepts of island biogeography and metapopulations. To reach a compromise between complete protection, on the one hand, and human use of habitats, on the other, scientists have developed the concept of biosphere reserves in which core areas receive greater protection while outer areas are allowed to have sustainable impacts.

AP® Practice Questions

Choose the best answer for the following.

1. The Marine Mammal Protection Act
   (a) allows states to make exceptions regarding the killing of marine mammals.
   (b) prohibits the killing of all marine mammals.
   (c) allows the import of marine mammal body parts.
   (d) protects sharks as well as marine mammals.

2. Which is NOT true of the Endangered Species Act?
   (a) It is an example of the single-species approach to conservation.
   (b) It prohibits the hunting or harvesting of some listed species.
   (c) It includes the development of recovery plans for listed species.
   (d) It uses a different definition for *threatened* than the IUCN.

3. Problems with protecting many small habitats include
   I. increased proportions of edge habitats.
   II. increased dispersal between populations.
   III. the need for corridors between some protected species.
   (a) I only
   (b) I and III
   (c) II only
   (d) II and III

4. Which is a trend identified by the Convention on Biological Diversity?
   (a) Very few plant species are at risk of extinction.
   (b) The genetic diversity of crops is increasing.
   (c) Marine species are affected most by recent biodiversity losses.
   (d) Natural habitats are becoming smaller and more fragmented.

5. According to the theory of island biogeography
   (a) when conservation areas are close to each other, more species will persist.
   (b) species on islands far from the mainland are at the least risk of extinction.
   (c) multiple small conservation areas will protect species better than one large area of the same size.
   (d) edge habitat is important to protect for increased diversity.

Working Toward Sustainability

Swapping Debt for Nature

Preserving biodiversity is expensive. A case in point is the money required to set aside terrestrial or aquatic areas for protection. As an example, if the land is privately owned, it must be purchased. Indirect costs can also be high. Not using the land, water, or other natural resources—such as wood materials, metals, and fossil fuels—results in lost income. Finally, the costs of maintaining the protected area can be prohibitive, ranging from monitoring the biodiversity to hiring guards to prevent illegal activities such as poaching. Given the fact that preserving biodiversity is expensive, how can the developing nations of the world, which contain so much biodiversity but have such little wealth, afford it?

In 1984, Thomas Lovejoy from the World Wildlife Fund came up with an idea that would help protect large areas of land but at the same time improve the economic conditions of developing countries. Lovejoy observed that developing nations possessed much biodiversity but were often deep in debt to wealthier, developed countries. Developing countries borrowed large amounts for the purpose of improving economic conditions and political stability. While the developing countries were slowly repaying their loans with interest, some had fallen so far behind on these payments that it seemed unlikely the loans would ever be repaid in full. These debtor countries had little money left over for investment in an improved environment after they had paid their loans to developed countries. Lovejoy considered the possibility that the wealthy countries might be willing to let debtor nations swap their debt in exchange for investing in the conservation of the biodiversity of the debtor nations.

The "debt-for-nature" swap has been used several times in Central and South America. In these swaps, the United States government and prominent environmental organizations provide cash to pay down a portion of a country's debt to the United States. The debt is then transferred to environmental organizations within that country with the debtor government making payments to the environmental organizations rather than to the United States. This does not mean that the country is out of debt, just that it now sends its loan payments to the environmental organizations for the purpose of protecting the country's biodiversity. In short, the indebted country switches from sending its money out of the country to investing in its own environmental conservation.

One of the largest debt-for-nature swaps recently happened in the Central American country of Guatemala. The United States government paired with two conservation organizations to provide $17 million to Guatemala. Over a period of 15 years, this amount, with interest, would have grown to more than $24 million, or about 20 percent of Guatemala's debt to the United States. In exchange, Guatemala agreed to pay $24 million over 15 years to improve conservation efforts in four areas of the country, including the purchase of land, the prevention of illegal logging, and future grants to conservation organizations helping to document and preserve the local biodiversity. The four areas include two ecosystems—mangrove forests and tropical forests. Each forms a core area within a biosphere reserve that contains a large number of rare and endangered species including the jaguar (*Panthera onca*). More than twice the size of Yellowstone National Park in the United States, this reserve offers important protection to biodiversity while also preserving historic Mayan temples that are part of Guatemala's cultural heritage and allowing sustainable use of some of the forest by local people.

Since the program began in 1998, the United States has used the debt-for-nature swap to protect tropical forests in 15 countries from Central America to the Philippines. To take part in the swap program, the countries are required to have a democratically elected government, a plan for improving their economies, and an agreement to cooperate with the United States on issues related to combating drug trafficking and terrorism. The results of these agreements have been encouraging. In Belize, for example, a debt-for-nature swap allowed 9,300 ha (23,000 acres) to be protected and an additional 109,000 ha (270,000 acres) to be managed for conservation. In Peru, a $10.6 million debt-for-nature swap led to the protection of more than 11 million ha (27 million acres) of tropical forest.

Costa Rica has made tremendous use of the debt-for-nature program with a total debt forgiveness of $26 million since 2007. In 2017, it announced its seventh such arrangement, which involved debt forgiveness of $1 million. As part of its announcement, the Costa Rican government invited conservation groups in the country to submit proposals that were focused on conservation topics, including monitoring the abundance of wildlife species, conducting research on endangered plant species, and recovering degraded areas of tropical rain forest. Although these arrangements are only currently being applied to tropical forests, there is no inherent reason that this unique, modern-day conservation strategy would not also work in many other developing countries around the world.

Swapping debt for nature in Costa Rica. Costa Rica has exceptional biodiversity, including the tropical forest surrounding the Rio Celeste waterfall. For more than a decade, Costa Rica has been forgiven more than $20 million in debt in exchange for conserving its biodiversity. *(Gonzalo Azumendi/Getty Images)*

Critical Thinking Questions

1. In debt-for-nature swaps, why might the United States require that developing countries receiving such assistance have a plan for improving their economies?
2. How might the debt-for-nature program promote the goals of the Convention on Biological Diversity?

References

Anders, W. 2017. U.S.–Costa Rica debt-for-nature swap will provide $1M for forest conservation. *Costa Rica Star*, September 13. https://news.co.cr/us-costa-rica-debt-for-nature-swap-will-provide-1m-for-forest-conservation/65624/

How debt-for-nature swaps protect tropical forests. The Nature Conservancy. http://www.nature.org/ourinitiatives/regions/centralamerica/guatemala/guatemala-debt-for-nature-swap-is-a-win-for-tropical-forest-conservation.xml

Lacey, M. 2006. U.S. to cut Guatemala's debt for not cutting trees. *New York Times*, October 2. http://www.nytimes.com/2006/10/02/world/americas/02conserve.html

U.S.–Brazil debt for nature swap to protect forests. *BBC News*, August 12, 2010. http://www.bbc.co.uk/news/world-latin-america-10958695

In this chapter, we examined the state of the world's biodiversity. We learned that the genetic diversity of many wild and domesticated populations has declined substantially over the past century. In addition, the species diversity of most major taxonomic groups has also declined, with large proportions of birds, mammals, and amphibians being threatened or near-threatened. These declines in species diversity also lead to declines in the intrinsic and instrumental values of ecosystems. The causes underlying these declines in diversity are wide-ranging and may include any combination of habitat loss, intrusion of exotic species, overharvesting, pollution, and climate change. While legislation to reverse these declines has focused on single species, conservation efforts have applied the concepts of metapopulations, island biogeography, and biosphere reserves to protect large areas of habitat and thereby protect large ecosystems.

Key Terms

Threatened species (IUCN)
Near-threatened species
Least-concern species
Intrinsic value
Instrumental value
Provision
Native species
Exotic species
Alien species
Invasive species
Lacey Act
Convention on International Trade in Endangered Species of Wild Fauna and Flora (CITES)
Red List
Marine Mammal Protection Act
Endangered species
Threatened species (U.S.)
Convention on Biological Diversity
Edge habitat
Biosphere reserve

Learning Goals Revisited

Module 59 The Sixth Mass Extinction

Explain the global decline in the genetic diversity of wild species.

Declines in the abundance of individuals in a population can lead to reductions in genetic diversity that cause inbreeding depression. Inbreeding depression can cause offspring to inherit two copies of a harmful mutation and experience reduced survival and reproduction.

Discuss the global decline in the genetic diversity of domesticated species.

Humans have bred a wide variety of domesticated plants and animals, but in recent decades farmers have focused on the most productive varieties and many of the other varieties have disappeared over time. Such reductions in genetic diversity limit the options available to respond to new diseases or changing environmental conditions.

Identify the patterns of global decline in species diversity.

Of the estimated 10 million species on Earth, only about 50,000 have been assessed to determine whether their populations are increasing, stable, or declining. In examining those groups with the most complete data, scientists have found that every group has a substantial percentage of species that are threatened or near-threatened.

Explain the values of ecosystems and the global declines in ecosystem function.

Ecosystems can have intrinsic values, which are independent of any benefit to humans, or they can have instrumental values, which provide a benefit to humans and can be assigned a monetary value. Instrumental values include provisions regulating services, support systems, resilience, and cultural services. Recent assessments of ecosystem function have found that more than half of those assessed are either declining or used at a rate that cannot be sustained.

Module 60 Causes of Declining Biodiversity

Discuss how habitat loss can lead to declines in species diversity.

The loss of habitat means that fewer individuals can be sustained in the habitat that remains. Smaller populations can then suffer from inbreeding depression. A reduction in habitat can also prevent the normal migration of species to important seasonal habitats and cause increased interactions with other species that have negative effects.

Explain how the movement of exotic species affects biodiversity.

Exotic species are those that are moved to new parts of the world where they are not native. Some of these species spread rapidly in their new locations and cause the demise of native species either as competitors, predators, herbivores, or pathogens.

Describe how overharvesting causes declines in populations and species.

Overharvesting plants and animals at rates that exceed the production of new individuals can cause population declines and even extinctions. Many of these extinctions have occurred due to unregulated harvesting in the past. However, in most parts of the world, governments have imposed harvest regulations to ensure that harvests occur in a sustainable manner.

Understand how pollution reduces populations and biodiversity.

Some pollutants can have direct lethal effects on species. Many other pollutants, however, can have sublethal effects that prevent or inhibit reproduction or alter ecosystems in ways that indirectly harm species.

Identify how climate change affects species diversity.

Climate change has the potential to alter the distribution of environmental conditions around the world. When conditions change and species are unable to move to more hospitable conditions, scientists predict that these species will either decline in abundance or go extinct.

Module 61 The Conservation of Biodiversity

Identify legislation that focuses on protecting single species.

The primary pieces of legislation in the United States to protect species are the Marine Mammal Protection Act and the Endangered Species Act. Internationally, nations created the Convention on Biological Diversity in order to conserve biodiversity, to use biodiversity sustainably, and to share equitably the benefits that emerge from the commercial use of biodiversity.

Discuss conservation efforts that focus on protecting entire ecosystems.

There has been a continual increase in the amount of aquatic and terrestrial habitats that have been protected around the world. When preserving such habitats, scientists consider the size, shape, and connectedness of these habitats as well as the presence of edge habitats. They have also incorporated the need to balance human use and habitat protection by designing biosphere reserves.

Practice Math and Graphing

Preparing for the AP® Exam

1. Practice Math

In the United States, there are endangered species in all of the major vertebrate groups. Given the table below, calculate the percentage of endangered species in each of the five groups of vertebrates.

Vertebrate group	Number of endangered species	Percent of endangered species
Mammals	95	
Birds	102	
Reptiles	45	
Amphibians	36	
Fishes	164	
Total		

2. Practice Graphing

(a) Using the percentage data you calculated in "Practice Math," create a pie graph that shows the percentages of endangered species among the five groups of vertebrates in the United States.

(b) How might the total number of species in each vertebrate group affect the percentage of endangered species among the five groups?

Section 1: Multiple-Choice Questions

Choose the best answer for questions 1–17.

1. Which is a cause of declining global biodiversity?
   I. pollution
   II. habitat loss
   III. overharvesting
   (a) I
   (b) I and III
   (c) II and III
   (d) I, II, and III

2. Which statement about global biodiversity is TRUE?
   (a) Species diversity is decreasing but genetic diversity is increasing.
   (b) Species diversity is decreasing and genetic diversity is decreasing.
   (c) Species diversity is increasing but genetic diversity is decreasing.
   (d) Declines in genetic diversity are occurring in wild plants but not in crop plants.

3. Which group of animals is declining in species diversity around the world?
   I. fish and amphibians
   II. birds and reptiles
   III. mammals
   (a) I only
   (b) I and III
   (c) II and III
   (d) I, II, and III

4. Which species was historically overharvested?
   (a) brown-headed cowbird
   (b) honeybee
   (c) dodo bird
   (d) zebra mussel

5. Which statement is NOT correct regarding the genetic diversity of livestock?
   (a) The use of only the most productive breeds improves genetic diversity.
   (b) Livestock come from very few species.
   (c) The genetic diversity of livestock has declined during the past century.
   (d) Different breeds are adapted to different climatic conditions.

6. Which statement is NOT correct about invasive exotic species?
   (a) They often have no major predators or herbivores.
   (b) Most introduced species become established in new regions.
   (c) A well-known invasive exotic plant is the kudzu vine.
   (d) A well-known invasive exotic animal is the zebra mussel.

7. Which is an example of the single-species approach to conservation?
   I. the Endangered Species Act
   II. the Marine Mammal Protection Act
   III. the Biosphere Reserve
   (a) I only
   (b) I and II
   (c) I and III
   (d) I, II, and III

8. Which is NOT an example of how the Endangered Species Act can affect human activities?
   (a) It has been used to prevent new construction.
   (b) It has prevented logging in particular areas.
   (c) It prevents the killing of listed species.
   (d) It prevents human use of biosphere reserves.

9. Which does NOT illustrate the principles of island biogeography?
   (a) A larger protected area should contain more species.
   (b) Protected areas that are closer together should contain more species.
   (c) National parks can be thought of as islands of biodiversity.
   (d) A larger protected area will have fewer habitats.

10. In a biosphere reserve
    (a) sustainable agriculture and tourism are permitted in different zones.
    (b) human activities are allowed throughout the reserve.
    (c) human activities are restricted to the central core of the reserve.
    (d) sustainable agriculture is permitted, but tourism is not.

11. The Svalbard Global Seed Vault was created to address the problem of
    (a) climate change.
    (b) emerging diseases.
    (c) declining genetic diversity of crop plants.
    (d) disease resistance.

12. The international agreement known as CITES was created to
    (a) monitor populations of endangered species on the Red List.
    (b) determine if a species is endangered or not.
    (c) control the international trade of threatened plants and animals.
    (d) reduce the spread of invasive species.

13. In the "debt-for-nature" swap, the debt of a country is transferred from the ________ to the ________.
   (a) UN; Nature Conservancy
   (b) World Bank; local environmental organizations
   (c) U.S. government; local environmental organizations
   (d) IUCN; Nature Conservancy

14. Which is NOT associated with smaller habitat sizes?
   (a) a decrease in genetic diversity
   (b) a decrease in edge habitat
   (c) loss of habitat diversity
   (d) faster extinction rates

15. Which agreement regulates the shipping of endangered or threatened species within the United States?
   (a) the Convention on International Trade in Endangered Species
   (b) the Lacey Act
   (c) the Endangered Species Act
   (d) National Environmental Policy Act

16. The Galápagos Islands, located off the coast of Ecuador, contain a rich diversity of endemic plants and animals. Several of the islands are entirely off-limits to tourists, whereas other islands allow tourists on the outer edges and on designated trails. Which best describes the Galápagos Islands?
   (a) a wildlife refuge
   (b) a biosphere reserve
   (c) multiple-use lands
   (d) a managed resource protection area

17. Which is true of invasive species?
   (a) Invasive species are defined as species that come from other countries.
   (b) By definition, all exotic species are invasive.
   (c) All invasive species spread rapidly.
   (d) Invasive species typically have few ecological interactions.

Section 2: Free-Response Questions

Write your answer to each part clearly. Support your answers with relevant information and examples. Where calculations are required, show your work.

1. The conservation of biodiversity is an international problem.
   (a) Name and describe ONE U.S. law that is intended to prevent the extinction of species. (4 points)
   (b) Name and describe ONE international treaty that is intended to prevent the extinction of species. (4 points)
   (c) Explain the benefits of taking an ecosystem approach, as opposed to a single-species approach, to conserving biodiversity. (2 points)

2. Tropical rainforests are home to a tremendous diversity of species. You have been asked to develop a plan to protect this diversity.
   (a) Describe the advantages and disadvantages of protecting a single large area versus several small areas. (2 points)
   (b) How might increasing the amount of edge habitat affect species that typically live deep in the forest? (3 points)
   (c) Discuss the merits of preserving individual species that are threatened and endangered versus preserving the function of the ecosystem. (3 points)
   (d) Describe three characteristics of organisms that would make them particularly vulnerable to extinction. (2 points)

3. An environmental organization is concerned about protecting a little-studied, endangered species of frog that has recently been listed as endangered by the IUCN. It is considering whether to create one or several forested reserves throughout a province. The frog has recently lost much of its forest habitat to agriculture, and currently exists in one small forested area surrounded by agricultural fields.
   (a) Explain why the frog species has only recently been listed as endangered by the IUCN. (1 point)
   (b) The frog population likely has low genetic diversity due to its small population. Describe TWO potential risks of low genetic diversity. (2 points)
   (c) As part of its appeal to the government, the environmental organization argues that local citizens will gain many benefits from the reserves, due to the ecosystem services provided by the forests. Name and describe THREE of the categories of ecosystem services that the organization could use to convince the government to support its plan. (3 points)
   (d) The environmental organization has to decide between creating one large reserve for the frog versus several smaller reserves. Describe the costs and benefits of each structure. (2 points)
   (e) Suppose the primary reason for the decline in the frog species is an invasive predator that tends to live in edge habitat. How might this information affect the environmental organization's plan for the number and size of the reserves? (2 points)

Global warming in the Great Barrier Reef of Australia has caused a dramatic increase in the proportion of female green sea turtles. *(Dave Fleetham/Getty Images)*

Global Change

CHAPTER 19

CASE STUDY

Sea Turtle Responses to a Warming World

When we think about global warming, we typically think about small increases in the average temperature of Earth over decades, causing glaciers to melt and sea levels to rise. Given that the temperature increases are small, we might expect plants and animals to be minimally affected. It was therefore shocking when researchers discovered that warming temperatures were causing green sea turtles (*Chelonia mydas*) to change their sex from male to female.

In 2018, researchers reported that they had examined the sex of green sea turtles that were feeding around the Great Barrier Reef of Australia. They used genetic markers that told them whether the turtles had originally hatched from cooler southern beaches or warmer northern beaches around the reef. In this species, turtles from the northern and southern populations can be found feeding together, but when the turtles are ready to breed, they go back to the place where they hatched. While most animal populations are comprised of about 50 percent females and 50 percent males, the green sea turtles from southern beaches had a modest bias, with 65 to 69 percent females. However, those from the northern beaches had an extreme bias; juvenile turtles were 99.1 percent female, subadult turtles were 99.8 percent female, and adults were 86.8 percent female. Given that these turtles take 25 years to reach adulthood, these data suggest that the cause of the most extreme bias, found in the juveniles and subadults, has occurred in the past 25 years.

Researchers discovered that warming temperatures were causing green sea turtles to change their sex from male to female.

How did these turtles come to produce so many females? It turns out that in some species of reptiles and fish, an individual's sex is not determined by the genes, such as the X and Y chromosomes that determine sex in humans. Instead, sex is determined by temperature. In the case of green sea turtles, if the hatchlings develop in sandy nests on beaches with warmer temperatures, they become female. If they experience cooler nest temperatures, they develop into males. When the researchers examined weather records for the southern and northern nesting regions of Australia, they found that the threshold temperature for creating females was consistently exceeded in the northern populations starting in 1990. Such biased sex ratios are of great concern to researchers; as the 99 percent female juveniles and subadults become breeding adults over the next 2 decades, they will likely have a difficult time finding males with which they can breed in the northern nesting sites.

The problem of warming global temperatures altering sex ratios is not limited to the green sea turtles. In New Zealand, researchers have been examining the sex ratios of a lizard species known as the tuatara (*Sphenodon punctatus*). The sex ratios of tuataras are also temperature dependent, but in this species warmer nest temperature cause more male offspring to be produced. During surveys of the tuataras from 1988 to 1998, the researchers found that the populations were modestly biased with 62 percent males. Subsequent surveys from 2005 to 2012 found that the bias had increased to 70 percent male.

Such a biased sex ratio not only reduces the number of females that can lay eggs in the future; it also causes the males to compete more with each other and to harass females more as they try to court them. During a 24-year period, researchers found that increased competition for females caused a decline in body condition among both males and females. Based on these results, the researchers examined worst-case scenarios for global warming in New Zealand, which is a 3.3°C to 4°C temperature increase. Under this scenario, they predict that the tuatara hatchlings will become 100 percent male and the population will go extinct. As in the case of the green sea turtles, an increase in global temperatures of just a few degrees can have major effects on the species experiencing these temperatures.

Sources: K.L. Grayson et al., Sex ratio bias and extinction risk in an isolated population of tuatara (*Sphenodon punctatus*), PLOS ONE 9: e94214 (2014); M.P. Jensen et al., Environmental warming and feminization of one of the largest sea turtle populations in the world, *Current Biology* 28 (2018):154–159.

The temperature-induced changes in reptile sex ratios is just one example of many changes taking place on Earth over the past few decades. In this chapter, we will examine how humans have altered the world's climate and explore the underlying causes of these climate changes. We will also investigate the observed consequences of these changes for humans and other species, and consider predictions of future consequences. This chapter ties together many of the themes we have developed throughout the book: the interconnectedness of the systems on Earth, the environmental indicators that enable us to measure and evaluate the environmental status of Earth, and the interaction of environmental science and policy.

MODULE 62

Global Climate Change and the Greenhouse Effect

In this module, we will consider the distinctions among global change, global climate change, and global warming. We will then explore the processes that underlie changes in global climates and, more specifically, global temperatures.

Learning Goals

After reading this module, you should be able to

- distinguish among global change, global climate change, and global warming.
- explain the process underlying the greenhouse effect.
- identify the natural and anthropogenic sources of greenhouse gases.

Global change includes global climate change and global warming

Throughout this book, we have highlighted a wide variety of ways in which the world has changed as a result of a rapidly growing human population. Human activity has placed increasing demands on natural resources such as water, trees, minerals, and fossil fuels. We have also emitted growing amounts of carbon dioxide, nitrogen compounds, and sulfur compounds into the atmosphere. Our agricultural methods depend on chemicals, including fertilizers and pesticides. Finally, a growing population faces challenges of waste disposal, sanitation, and the spread of human diseases.

Change that occurs in the chemical, biological, and physical properties of the planet is referred to as **global change**. As you can see in **FIGURE 62.1** on page 686, some types of global change are natural and have been occurring for millions of years. Global temperatures, for example, have fluctuated over millions of years. During periods of cold temperatures, Earth has experienced ice ages. In modern times, however, the rates of change have often been much higher than those that occurred historically. Many of these changes are the result of human activities, and they can have significant, sometimes cascading, effects. For example, as we saw in Chapter 17, emissions from coal-fired power plants and waste incinerators have increased the amount of mercury in the air and water, with concentrations roughly triple those of preindustrial levels. This mercury bioaccumulates in fish caught thousands of kilometers away from the sources of pollution. Because mercury has harmful effects on the nervous system of children, women who might become pregnant and children are advised to avoid eating top predator fish such as swordfish and tuna. Far-reaching effects on this scale were unimaginable just 50 years ago.

One type of global change of particular concern to scientists is **global climate change**, which refers to changes in the average weather that occurs in an area over a period of years or decades. Changes in climate can be categorized as either natural or anthropogenic. For example, you might recall from Chapter 4 that El Niño events, which occur every 3 to 7 years, alter global patterns of temperature and precipitation (see Figure 11.3 on page 124). Anthropogenic activities such as fossil fuel combustion and deforestation also have

Global change Change that occurs in the chemical, biological, and physical properties of the planet.

Global climate change Changes in the average weather that occurs in an area over a period of years or decades.

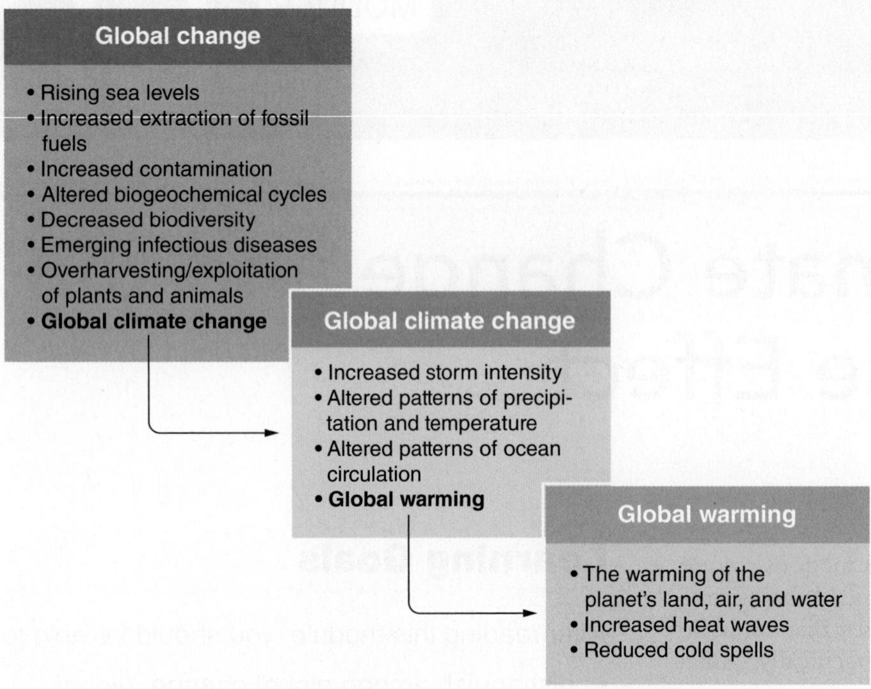

Global change
• Rising sea levels
• Increased extraction of fossil fuels
• Increased contamination
• Altered biogeochemical cycles
• Decreased biodiversity
• Emerging infectious diseases
• Overharvesting/exploitation of plants and animals
• Global climate change
Global climate change
• Increased storm intensity
• Altered patterns of precipitation and temperature
• Altered patterns of ocean circulation
• Global warming
Global warming
• The warming of the planet's land, air, and water
• Increased heat waves
• Reduced cold spells

FIGURE 62.1 Global change. Global change includes a wide variety of factors that are changing over time. Global climate change refers to those factors that affect the average weather in an area of Earth. Global warming refers to changes in temperature in an area.

major effects on global climates. **Global warming** refers to a specific aspect of climate change: the warming of the oceans, land masses, and atmosphere of Earth.

Solar radiation and greenhouse gases make our planet warm

The physical and biogeochemical systems that regulate temperature at the surface of Earth—the concentrations of gases, distribution of clouds, atmospheric currents, and ocean currents—are essential to life on our planet. It is therefore critical that we understand how the planet is warmed by the Sun and how the *greenhouse effect* contributes to the warming of Earth.

The Sun–Earth Heating System

The ultimate source of almost all energy on Earth is the Sun. In the most basic sense, the Sun emits solar radiation that strikes Earth. As the planet warms, it emits radiation back toward the atmosphere. However, the types of energy radiated from the Sun and Earth are different (see Figure 5.1 on page 48). Because the Sun is very hot, most of its radiated energy is in the form of high-energy visible radiation and ultraviolet radiation—also known as visible light and ultraviolet light. When this radiation strikes Earth, the planet warms and radiates energy. Earth is not nearly as hot as the Sun, so it emits most of its energy as infrared radiation—also known as infrared light. We cannot see infrared radiation, but we can feel it being emitted from warm surfaces like the heat that radiates from an asphalt road on a hot day.

Differences in the types of radiation emitted by the Sun and Earth, in combination with processes that occur in the atmosphere, cause the planet to warm. Using **FIGURE 62.2**, we can walk through each step of this process. As radiation from the Sun travels toward Earth, about one-third of the radiation is reflected back into space. Although some ultraviolet radiation is absorbed by the ozone layer in the stratosphere, the remaining ultraviolet radiation, as well as visible light, easily passes through the atmosphere. Once it has passed through the atmosphere, this solar radiation strikes clouds and the surface of Earth. Some of this radiation is reflected from the surface of the planet back into space. The remaining radiation is absorbed by clouds and the surface of Earth, which become warmer and begin to

Global warming The warming of the oceans, land masses, and atmosphere of Earth.

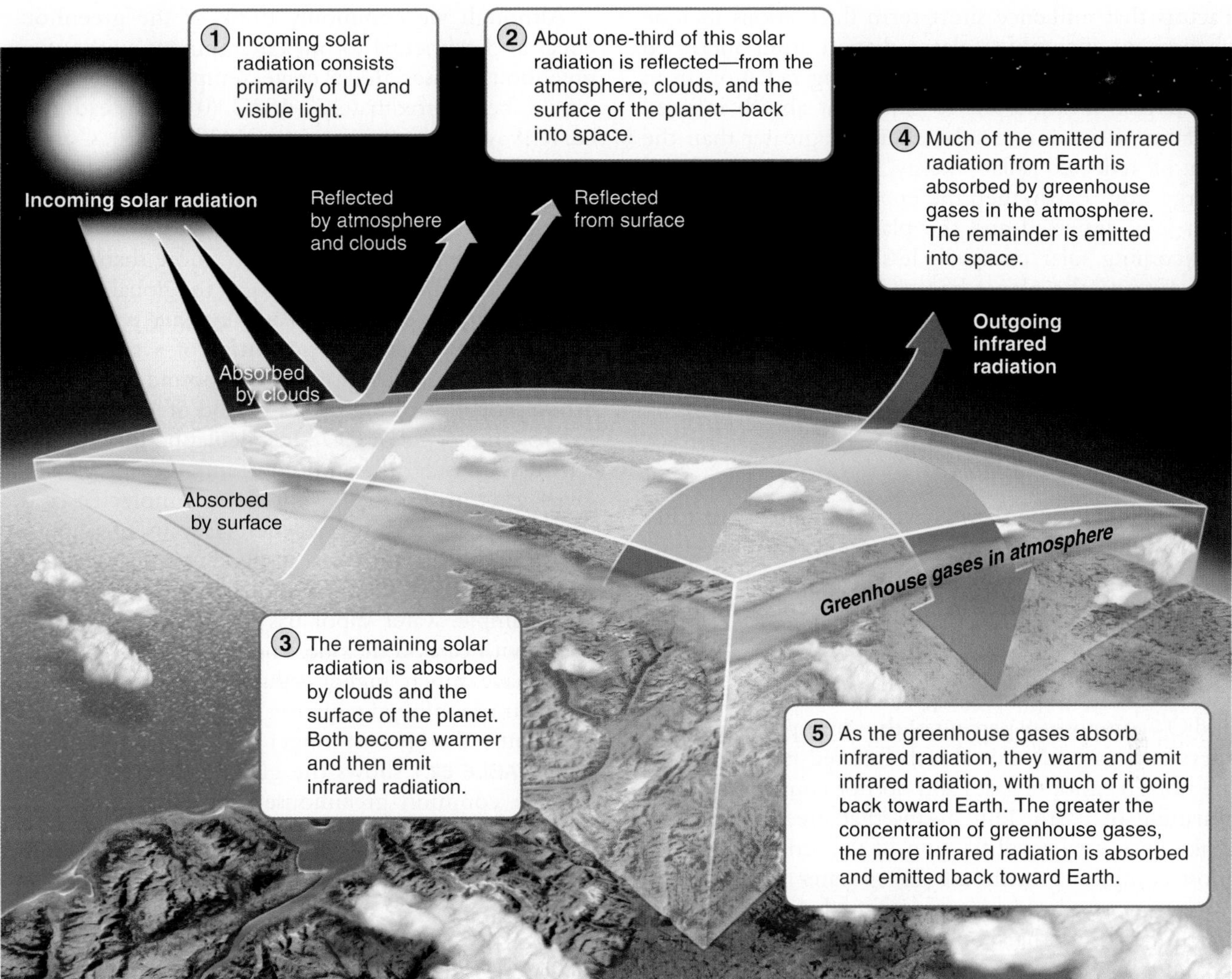

1 Incoming solar radiation consists primarily of UV and visible light.
2 About one-third of this solar radiation is reflected—from the atmosphere, clouds, and the surface of the planet—back into space.
4 Much of the emitted infrared radiation from Earth is absorbed by greenhouse gases in the atmosphere. The remainder is emitted into space.
Incoming solar radiation
Reflected by atmosphere and clouds
Reflected from surface
Outgoing infrared radiation
Absorbed by clouds
Absorbed by surface
Greenhouse gases in atmosphere
3 The remaining solar radiation is absorbed by clouds and the surface of the planet. Both become warmer and then emit infrared radiation.
5 As the greenhouse gases absorb infrared radiation, they warm and emit infrared radiation, with much of it going back toward Earth. The greater the concentration of greenhouse gases, the more infrared radiation is absorbed and emitted back toward Earth.

FIGURE 62.2 The greenhouse effect. When the high-energy radiation from the Sun strikes the atmosphere, about one-third is reflected from the atmosphere, clouds, and the surface of the planet. Much of the high-energy ultraviolet radiation is absorbed by the ozone layer where it is converted to low-energy infrared radiation. Some of the ultraviolet radiation and much of the visible light strikes the land and water of Earth where it is also converted into low-energy infrared radiation. The infrared radiation radiates back toward the atmosphere where it is absorbed by greenhouse gases that radiate much of it back toward the surface of Earth. Collectively, these processes cause warming of the planet.

emit lower-energy infrared radiation back toward the atmosphere. Unlike ultraviolet and visible radiation, infrared radiation does not easily pass through the atmosphere. It is absorbed by gases, which causes these gases to become warm. The warmed gases emit infrared radiation out into space and back toward the surface of Earth. The infrared radiation that is emitted toward Earth causes Earth's surface to become even warmer. This absorption of infrared radiation by atmospheric gases and reradiation of the energy back toward Earth is the **greenhouse effect**.

The greenhouse effect gets its name from the idea that solar radiation causes a gardener's greenhouse to become very warm. However, the process by which actual greenhouses are warmed by the Sun involves glass windows holding in heat whereas the process by which Earth is warmed involves greenhouse gases radiating infrared energy back toward the surface of the planet.

In the Sun–Earth heating system, the net flux of energy is zero; the inputs of energy to Earth equal the outputs from Earth. Over the long term—thousands or millions of years—the system has been in a steady state. However, in the shorter term—over years or decades—inputs can be slightly higher or lower than outputs.

Greenhouse effect Absorption of infrared radiation by atmospheric gases and reradiation of the energy back toward Earth.

Factors that influence short-term fluctuations include changes in incoming solar radiation from increased solar activity and changes in outgoing radiation from an increase in atmospheric gases that absorb infrared radiation. If incoming solar energy is greater than the sum of reflected solar energy and radiated infrared energy from Earth, then the energy accumulates faster than it is dispersed and the planet becomes warmer. If incoming solar energy is less than the sum of the two outputs, the planet becomes cooler. Such natural changes in inputs and outputs cause natural changes in the temperature of Earth over time.

The Gases That Cause the Greenhouse Effect

Throughout this book we have seen that certain gases in the atmosphere can absorb infrared radiation emitted by the surface of the planet and radiate much of it back toward the surface. As we have seen, gases in the atmosphere that absorb infrared radiation are known as greenhouse gases.

The two most common gases in the atmosphere, N_2 and O_2, compose 99 percent of the atmosphere. Because these two gases do not absorb infrared radiation, they are not greenhouse gases and do not contribute to the warming of Earth. This means that greenhouse gases make up a very small fraction of the atmosphere. The most common greenhouse gas is water vapor (H_2O). Water vapor absorbs more infrared radiation from Earth than any other compound, although a molecule of water vapor does not persist nearly as long as other greenhouse gases. Other important greenhouse gases include carbon dioxide (CO_2), methane (CH_4), nitrous oxide (N_2O), and ozone (O_3). All of these gases have been a part of the atmosphere for millions of years, and have kept Earth warm enough to be habitable. In the case of ozone, we have seen that its effects on Earth are diverse. Ozone in the stratosphere is beneficial because it filters out harmful ultraviolet radiation. In contrast, ozone in the lower troposphere acts as a greenhouse gas and can cause increased warming of Earth. It also is an air pollutant in the lower troposphere because it can cause damage to plants and human respiratory systems. There is one other type of greenhouse gas, chlorofluorocarbons (CFCs), which does not exist naturally. It occurs in the atmosphere exclusively due to production of CFCs by humans and, as we discussed in Chapter 15, these CFCs have contributed to a hole in the ozone layer over Antarctica.

Greenhouse warming potential An estimate of how much a molecule of any compound can contribute to global warming over a period of 100 years relative to a molecule of CO_2.

Although we commonly think of the greenhouse effect as detrimental to our environment, without any greenhouse gases the average temperature on Earth would be approximately −18°C (0°F) instead of its current average temperature of 14°C (57°F). Concern about the danger of greenhouse gases is based on our understanding that an increase in the concentration of these gases—as has occurred due to human activities—can cause the planet to warm even more than usual.

The contribution of each gas to global warming depends in part on its *greenhouse warming potential*. The **greenhouse warming potential** of a gas estimates how much a molecule of any compound can contribute to global warming over a period of 100 years relative to a molecule of CO_2. In calculating this potential, scientists consider the amount of infrared energy that a given gas can absorb and how long a molecule of the gas can persist in the atmosphere. Because greenhouse gases can differ a great deal in these two factors, greenhouse warming potentials span a wide range of values. For example, water vapor has a lower potential compared with carbon dioxide. The remaining greenhouse gases have much higher values, either because they absorb more infrared radiation than a molecule of CO_2 or because they persist longer in the atmosphere than CO_2. **TABLE 62.1** shows the global warming potential for five common greenhouse gases. Compared with CO_2, the greenhouse warming potential is 25 times higher for methane (CH_4), nearly 300 times higher for nitrous oxide (N_2O), and up to 13,000 times higher for CFCs.

The effect of each greenhouse gas depends on both its warming potential and its concentration in the atmosphere. Although carbon dioxide has a relatively low warming potential, it is much more abundant than most other greenhouse gases, except for water vapor, which can have a concentration similar to carbon dioxide. While human activity appears to have little effect on the amount of water vapor in the atmosphere, it has caused substantial increases in the amount of the other greenhouse gases. Among these, carbon dioxide remains the greatest contributor to the greenhouse effect because its concentration is so much higher than any of the others. As a result, scientists and policy makers focus their efforts on ways to reduce carbon dioxide in the atmosphere.

Given what we now know about how greenhouse gases work, the concentrations of each gas, and how much infrared energy each gas absorbs, we can understand how changes in the concentrations of greenhouse gases can contribute to global warming. Increasing the concentration of any historically present greenhouse gas should cause more infrared radiation to be absorbed in the atmosphere, which will then radiate more energy back toward the surface of the planet and cause the planet to warm. Likewise, producing new greenhouse gases that can make their way into the atmosphere,

TABLE 62.1 The major greenhouse gases (parts per million)

Greenhouse gas	Concentration in 2017	Global warming potential (over 100 years)	Duration in the atmosphere
Water vapor	Variable with temperature	<1	9 days
Carbon dioxide	407 ppm	1	Highly variable (ranging from years to hundreds of years)
Methane	1.85 ppm	25	12 years
Nitrous oxide	.33 ppm	300	114 years
Chlorofluorocarbons	0.0007 ppm	1,600 to 13,000	55 to >500 years

Data from: The National Oceanic and Atmospheric Administration, www.esrl.noaa.gov/gmd/aggi, and The United Nations Framework Convention on Climate Change.

such as CFCs, should also cause increased absorption of infrared radiation in the atmosphere and further cause the planet to warm.

AP® Exam Tip

You should be able to identify the various greenhouse gases and discuss how humans have influenced increased concentrations of these gases. ●

Sources of greenhouse gases are both natural and anthropogenic

As we have seen, greenhouse gases include a variety of compounds such as water vapor, carbon dioxide, methane, nitrous oxide, and CFCs. These gases have natural and anthropogenic sources. After reviewing the different sources of greenhouse gases, we will discuss the relative ranks of the different anthropogenic sources.

Natural Sources of Greenhouse Gases

Natural sources of greenhouse gases include volcanic eruptions, decomposition, digestion, denitrification, evaporation, and evapotranspiration.

Volcanic Eruptions

Over the scale of geologic time, volcanic eruptions can add a significant amount of carbon dioxide to the atmosphere. Other gases and the large quantities of ash released during volcanic eruptions can also have important, short-term climatic effects. A volcanic eruption emits a large quantity of ash into the atmosphere. The ash reflects incoming solar radiation back out into space, which has a cooling effect on Earth. In 1991, for example, Mount Pinatubo in the Philippines erupted and spewed millions of tons of ash into the atmosphere, as far as 20 km (12 miles) high (**FIGURE 62.3**). The large amount of ash in the atmosphere reduced the amount of radiation striking Earth, which caused a 0.5°C (0.9°F) decline in the temperature on the planet's surface. Because the ash and small particles eventually settle out of the atmosphere, such effects usually last only a few years.

Decomposition and Digestion

When decomposition occurs under high-oxygen conditions, the dead organic matter is ultimately converted into carbon dioxide. As we saw in our discussion of landfills in Chapter 16, methane is created when there is not enough oxygen available to produce carbon dioxide. This is a common occurrence at the bottom of wetlands where plants and animals decompose and

FIGURE 62.3 Ash from volcanic eruptions. Volcanic eruptions, such as this eruption of Mount Pinatubo in the Philippines in 1991, send millions of tons of ash into the atmosphere where it can absorb incoming solar radiation, reradiate it back to space, and cause Earth to cool. *(Exactostock/SuperStock)*

oxygen is in low supply. Wetlands are the largest natural source of methane.

A similar situation occurs when certain animals digest plant matter. Animals that consume significant quantities of wood or grass, including termites and grazing antelopes, require gut bacteria to digest the plant material. Because the digestion occurs in the animal's gut, the bacteria do not have access to oxygen and methane is produced as a by-product. A single termite colony can contain more than a million termites (**FIGURE 62.4**). Termites are abundant throughout the world—especially in the tropics—and represent the second largest natural source of methane.

Denitrification

As we learned in Chapter 3, nitrous oxide (N_2O) is a natural component of the nitrogen cycle that is produced through the process of denitrification. Denitrification occurs in the low-oxygen environments of wet soils and at the bottoms of wetlands, lakes, and oceans. (Figure 7.3 on page 87 shows the nitrogen cycle.) In these environments, nitrate is converted to nitrous oxide gas, which then enters the atmosphere as a powerful greenhouse gas.

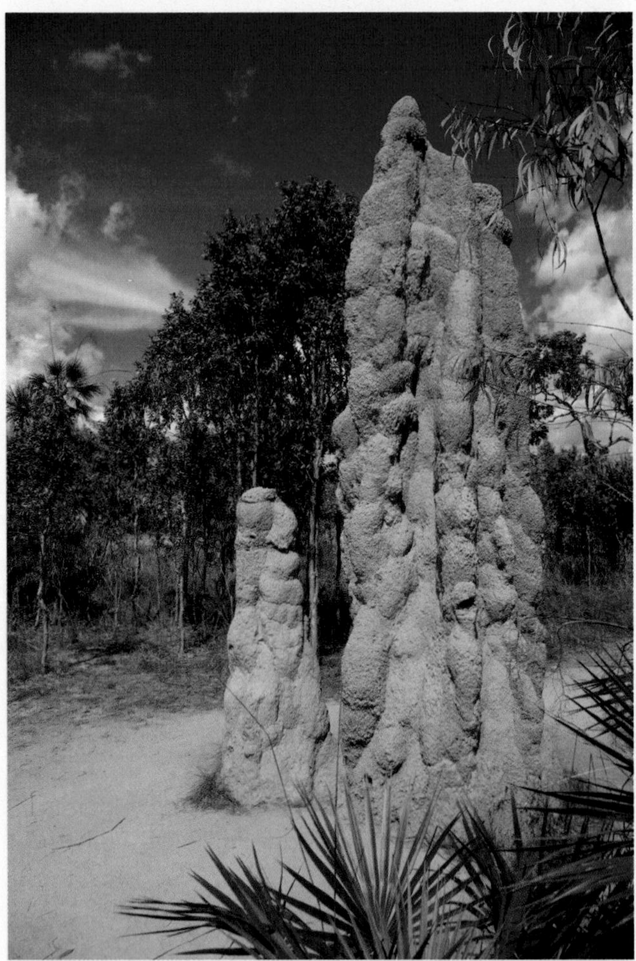

FIGURE 62.4 Termites and methane. The bacteria that live in the anaerobic gut environment of herbivores such as termites produce methane as a by-product of their digestive activities. Because termite colonies, such as this one in Australia, can achieve population sizes of more than one million, collectively they can produce large amounts of methane. *(Anton Harder/Shutterstock)*

Evaporation and Evapotranspiration

As we stated earlier, water vapor is the most abundant greenhouse gas in the atmosphere and the greatest natural contributor to global warming. In Chapter 3 we examined the role of water vapor in the hydrologic cycle. (Figure 7.1 on page 83 shows the hydrologic cycle.) Water vapor is produced when liquid water from land and water bodies evaporates and by the evapotranspiration process of plants. Because the amount of evaporation into water vapor varies with climate, the amount of water vapor in the atmosphere can vary regionally.

Anthropogenic Sources of Greenhouse Gases

As shown in **FIGURE 62.5**, there are many anthropogenic sources of greenhouse gases. The most significant of these are the burning of fossil fuels, agricultural practices, deforestation, landfills, and industrial production of new greenhouse chemicals.

Burning Fossil Fuels

Tens to hundreds of millions of years ago, organisms were sometimes buried without first decomposing into carbon dioxide. In Figure 7.2 on page 85 we outlined the process by which the carbon contained in these organisms, called fossil carbon, is slowly converted to fossil fuels deep underground. When humans burn these fossil fuels, we produce CO_2 that goes into the atmosphere. Because of the long time required to convert carbon into fossil fuels, the rate of putting carbon into the atmosphere by burning fossil fuels is much greater than the rate at which producers take CO_2 out of the air and both the producers and consumers contribute to the pool of buried fossil carbon.

Because fossil fuels differ in how they store energy, each type of fossil fuel produces different amounts of carbon dioxide. For a given amount of energy, burning coal produces the most CO_2. In comparison, burning oil produces 85 percent as much CO_2 as coal, and burning natural gas produces 56 percent as much. As we saw in Chapter 12, from the perspective of CO_2 emissions, natural gas is considered better for the environment than coal. The production of fossil fuels, such as the mining of coal, and the combustion

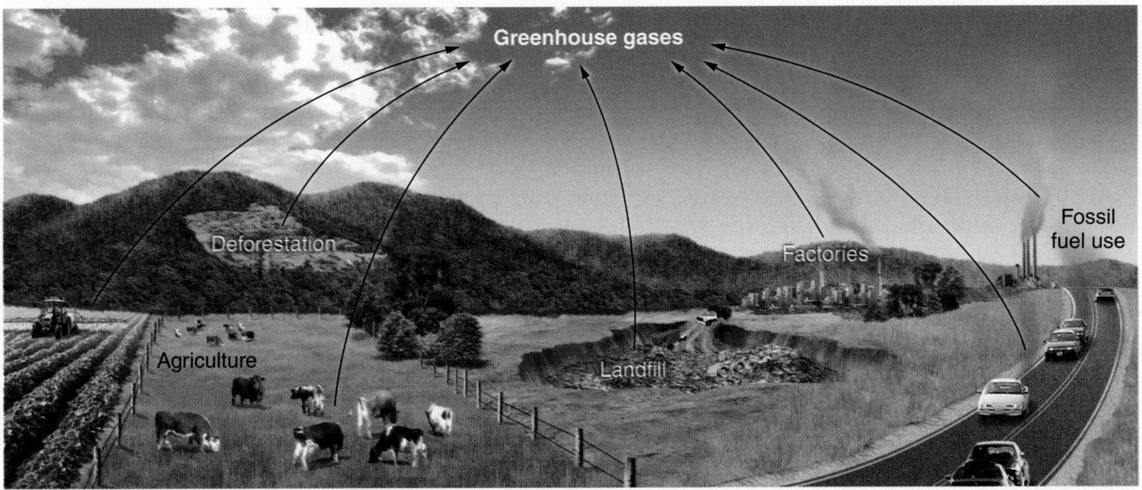

FIGURE 62.5 Anthropogenic sources of greenhouse gases. Human activities are a major contributor of greenhouse gases, including CO_2, methane, and nitrous oxide. These activities include the use of fossil fuels, agricultural practices, the creation of landfills, and the industrial production of new greenhouse gases.

of fossil fuels can also release methane and, in some cases, nitrous oxide.

Particulate matter may also play an important role in global warming. Although particulate matter, also known as black soot, may reflect solar radiation under some conditions, recent findings suggest that it may be responsible for up to one-quarter of observed global warming during the past century. Particulates that fall on ice and snow in the higher latitudes absorb more energy of the Sun by lowering the albedo. As the snow and ice begin to melt, the particulates become more concentrated on the surface. The increased concentration raises the amount of solar radiation absorbed, which increases melting. This positive feedback system might help explain warming that occurred early in the last century, when atmospheric concentrations of greenhouse gases had not yet increased much but soot from the burning of coal was widespread.

Agricultural Practices

Agricultural practices can produce a variety of greenhouse gases. Agricultural fields that are overirrigated, or those that are deliberately flooded for cultivating crops such as rice, create low oxygen environments similar to wetlands and therefore can produce methane and nitrous oxide. Synthetic fertilizers, manures, and crops that naturally fix atmospheric nitrogen—for example, alfalfa—can create an excess of nitrates in the soil that are converted to nitrous oxide by the process of denitrification.

Raising livestock can also produce large quantities of methane. Many livestock such as cattle and sheep consume large quantities of plant matter and rely on gut bacteria to digest this cellulose. As we saw in the case of termites, gut bacteria live in a low-oxygen environment and digestion in this environment produces methane as a by-product. Manure from livestock operations will decompose to CO_2 under high-oxygen conditions, but in low-oxygen conditions, for example in manure lagoons that are not aerated, it will decompose to methane.

Deforestation

Each day, living trees remove CO_2 from the atmosphere during photosynthesis, and decomposing trees add CO_2 to the atmosphere. This part of the carbon cycle does not change the net atmospheric carbon because the inputs and outputs are approximately equal. However, when forests are destroyed by burning or decomposition and not replaced, as can happen during deforestation, the destruction of vegetation will contribute to a net increase in atmospheric CO_2. This is because the mass of carbon that made up the trees is added to the atmosphere by combustion or decomposition. The shifting agriculture described in Chapter 11, which involves clearing forests and burning the vegetation to make room for crops, is a major source of both particulates and a number of greenhouse gases, including carbon dioxide, methane, and nitrous oxide.

Landfills

As we saw in Chapter 16, landfills receive a great deal of household waste that slowly decomposes under layers of soil. When the landfills are not aerated properly, they create a low-oxygen environment, like wetlands, in which decomposition causes the production of methane as a by-product.

Industrial Production of New Greenhouse Chemicals

The creation of new industrial chemicals often has unintended effects on the atmosphere. In Chapter 15 we looked at CFCs, the family of chemicals that serves as refrigerants used in air conditioners, freezers, and refrigerators. CFCs were used in the past until scientists discovered that they were damaging the protective ozone layer. As we discussed, the nations of the world joined together to sign the Montreal Protocol on Substances that Deplete the Ozone Layer, which phased out the production and use of CFCs by 1996. Unfortunately, many of the alternative refrigerants that are less harmful to the ozone layer, including a group of gases known as hydrochlorofluorocarbons (HCFCs), still have very high greenhouse warming potentials. As a result, developed countries will phase out the use of HCFCs by 2030.

Ranking the Anthropogenic Sources of Greenhouse Gases

We have seen that there are multiple anthropogenic sources of greenhouse gases. What is the relative contribution of each source? **FIGURE 62.6** shows the major anthropogenic sources of greenhouse gases in the United States. Figure 62.6a shows that the three major contributors of methane in the atmosphere are the digestive processes of livestock, landfills, and the production of natural gas and petroleum products. The major contributor of nitrous oxide, shown in Figure 62.6b, is agricultural soil because it receives nitrogen from synthetic fertilizers, combustion, and industrial production of fertilizers and other products. Finally, looking at the numbers for carbon dioxide in Figure 62.6c, we see that approximately 93 percent of all CO_2 emissions come from industrial processes and the burning of fossil fuels for transportation, residents, businesses, and generating electricity.

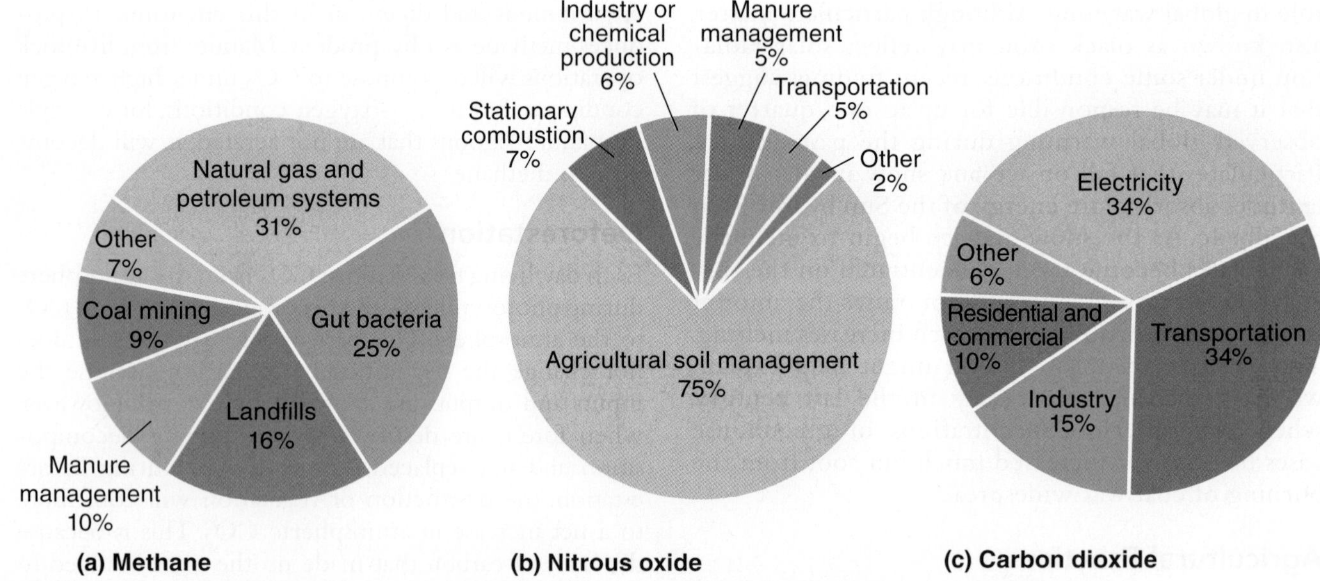

FIGURE 62.6 Anthropogenic sources of greenhouse gases in the United States. (a) The largest contributions of methane in the atmosphere arise from gut bacteria that help many livestock species digest plant matter, landfills that experience decomposition in low-oxygen environments, and the production, storage, and transport of natural gas and petroleum products from which methane escapes. (b) The largest contributions of nitrous oxide in the atmosphere arise from the agricultural soils that obtain nitrogen from applied fertilizers, combustion, and industrial production of fertilizers and other products. (c) Nearly all anthropogenic CO_2 emissions come from the burning of fossil fuels. *(Data from https://www.epa.gov/ghgemissions/overview-greenhouse-gases - carbon-dioxide; https://www.epa.gov/ghgemissions/overview-greenhouse-gases - methane; https://www.epa.gov/ghgemissions/overview-greenhouse-gases - nitrous-oxide.)*

MODULE 62 AP® Review

Preparing for the AP® Exam

In this module, we considered global change, global climate change, and global warming. In a natural process known as the greenhouse effect, visible light and ultraviolet light from the Sun strike our planet and this energy is converted to infrared radiation that is emitted back to the atmosphere. A tiny percentage of gases in the atmosphere, known as greenhouse gases, absorb this infrared radiation and emit a portion of it back to Earth and this causes the planet to warm even more. Although most greenhouse gases have natural sources, human activities have increased the concentration of these gases in the atmosphere and produced new chemicals that are potent greenhouse gases. In the next module, we will examine the evidence that this increase in greenhouse gases has caused our planet to become warmer during the past two centuries.

AP® Practice Questions

Choose the best answer for the following.

1. The greenhouse gas with the highest greenhouse warming potential is
   (a) carbon dioxide.
   (b) methane.
   (c) chlorofluorocarbon.
   (d) nitrous oxide.

2. Methane is naturally produced by
   (a) decomposition.
   (b) volcanic eruptions.
   (c) denitrification.
   (d) forest fires.

3. The greenhouse effect is due to
   (a) the absorption and reradiation of infrared radiation by the atmosphere.
   (b) the reflection of ultraviolet radiation by the atmosphere.
   (c) the absorption of ultraviolet radiation by the atmosphere.
   (d) the reflection of infrared radiation from Earth's surface.

4. Most nitrous oxide emissions are from
   (a) fossil fuel-combustion.
   (b) agricultural practices.
   (c) refrigerants.
   (d) industrial processes.

5. Particulate matter can increase global warming by
   (a) reacting with chlorofluorocarbons.
   (b) reducing the surface absorption of ultraviolet radiation.
   (c) reflecting radiation.
   (d) lowering surface albedo.

MODULE 63

The Evidence for Global Warming

Now that we have reviewed greenhouse gases and their role in global warming, we can explore the evidence that an increased concentration of greenhouse gases is causing Earth to become warmer. If Earth is becoming warmer, we can begin to evaluate whether or not this warming is caused by human activities that have released greenhouse gases. One way to make these assessments is to determine gas concentrations and temperatures from the past and compare them to gas concentrations and temperatures in the present day. We can also use information about changes in gas concentrations and temperatures to predict future climate conditions. In this module, we will examine how greenhouse gas concentrations have changed over time and how these changes are linked to global warming.

Learning Goals

After reading this module, you should be able to

- explain how CO_2 concentrations have changed over the past 7 decades and how emissions compare among the nations of the world.
- explain how temperatures have increased since records began in 1880.
- discuss how we estimate temperatures and levels of greenhouse gases over the past 500,000 years and into the future.
- explain the role of feedbacks on the impacts of climate change.

CO_2 concentrations have been increasing for the past 7 decades

In 1988 the United Nations and the World Meteorological Organization created the Intergovernmental Panel on Climate Change (IPCC), a group of more than 3,000 scientists from around the world working together to assess climate change. Their mission is to understand the details of the global warming system, the effects of climate change on biodiversity and energy fluxes in ecosystems, and the economic and social effects of climate change. The IPCC enables scientists to assess and communicate the state of our knowledge and to suggest research directions that would improve our understanding in the future. This effort has produced an excellent understanding of how greenhouse gases and temperatures are linked.

Measuring CO_2 Concentrations in the Atmosphere

Through the work of the IPCC, we now understand that CO_2 is an important greenhouse gas that can contribute to global warming, but we didn't always realize this. In the first half of the twentieth century, most scientists believed that if any excess CO_2 were being produced, it would be absorbed by the oceans and vegetation. In addition, because the concentration of atmospheric CO_2 was very low compared to gases such as oxygen and nitrogen, it was difficult to measure accurately.

Charles David Keeling was the first to overcome the technical difficulties in measuring CO_2. When Keeling set out to measure the precise level of CO_2 in the atmosphere, most atmospheric scientists believed that two measurements several years apart would be sufficient to answer the question of whether human activities were causing increased concentrations of CO_2 in the atmosphere. Keeling did not agree and in 1958 he began collecting data throughout the year at the Mauna Loa Observatory in Hawaii. After just 1 year of work, Keeling found that CO_2 levels varied seasonally and that the concentration of CO_2 increased from year to year. His results prompted him to take measurements for several more years, and he and his students have continued this work into the twenty-first century. The results, shown in **FIGURE 63.1**, confirm Keeling's early findings; although CO_2 concentrations vary between seasons, there is a clear trend of rising CO_2 concentrations across the years. This increase over time is correlated to increased human emissions of carbon from the combustion of fossil fuels and net destruction of vegetation. "Do the Math: Projecting Future Increases in CO_2" gives you an opportunity to estimate the increase in CO_2 concentrations by the end of the century.

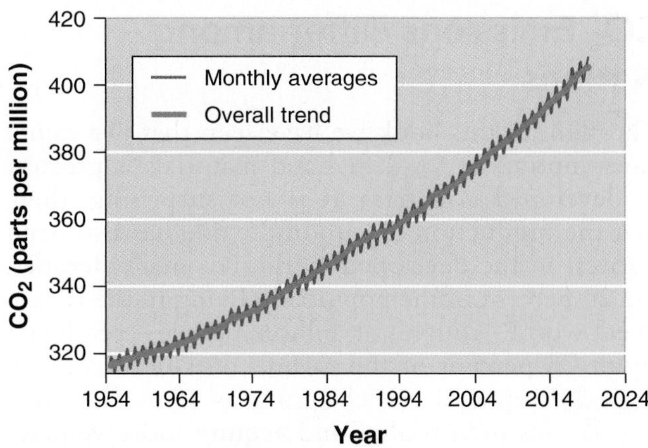

FIGURE 63.1 Changes in atmospheric CO_2 over time. Carbon dioxide levels have risen steadily since measurement began in 1958. *(Data from https://www.esrl.noaa.gov/gmd/ccgg/trends/full.html).*

AP® Exam Tip

Graphs that show changes over time, like Figure 63.1, frequently appear on the AP® Environmental Science Exam. You should be able to describe the general trends shown in such graphs and explain fluctuations. ●

What causes the seasonal variation? Each spring, as deciduous trees, grasslands, and farmlands in the Northern Hemisphere turn green, they increase their absorption rates of CO_2 to carry out photosynthesis. At the same time, bodies of water begin to warm and the algae and plants also begin to photosynthesize. In doing so, these producers take up some of the CO_2 in the atmosphere. Conversely, in the fall, as leaves drop, crops are harvested, and bodies of water cool, the uptake of atmospheric CO_2 by algae and plants declines and the amount of CO_2 in the atmosphere increases.

AP® Exam Tip

Do the Math, "Projecting Future Increases in CO_2," contains two problems that have appeared on past exams. Make sure you understand how to calculate the average annual increase in CO_2 and how to predict the concentration of CO_2. ●

DO THE MATH Projecting Future Increases in CO_2

Preparing for the AP® Exam

Because Charles David Keeling and his colleagues began measuring CO_2 in 1958, we have an excellent record of how CO_2 concentrations have changed in the atmosphere over time. From 1959 to 2017, the concentration of CO_2 in the atmosphere increased from 316 to 407 ppm (parts per million).

Based on these two points in time, what has been the average annual increase of CO_2 in the atmosphere?

$$\text{Time} = 2017 - 1959 = 58 \text{ years}$$

$$\text{Increase in } CO_2 = 407 \text{ ppm} - 316 \text{ ppm} = 91 \text{ ppm}$$

$$\text{Average annual increase in } CO_2 = 91 \text{ ppm} \div 58 \text{ years} = 1.57 \text{ ppm/year}$$

YOUR TURN

1. If the annual rate of CO_2 increase is 1.57 ppm, what concentration of CO_2 do you predict for the year 2100? Round to the nearest whole number.
2. From 2007 to 2017, the rate of increase grew to 2.3 ppm per year. Based on this faster rate, what concentration of CO_2 do you predict for the year 2100? Round to the nearest whole number.

CO_2 Emissions Differ among Nations

Throughout this book we have seen that per capita consumption of fossil fuel and materials is greatest in developed countries. It is not surprising, then, that the production of carbon dioxide has also been greatest in the developed world. For many decades, the 20 percent of the population living in the developed world—roughly 1 billion people—produced about 75 percent of the carbon dioxide. However, these percentages are changing as some developing nations industrialize and acquire more vehicles that burn fossil fuels. In 2009, developing countries surpassed developed countries in the production of CO_2.

Development has been especially rapid in China and India, which together contain one-third of the world's population. From 2000 to 2009, China more than doubled its emissions of carbon dioxide as the country built many new coal-powered electrical plants that increased its ability to burn coal. **FIGURE 63.2a** shows that as of 2015 China was the leading emitter of CO_2. China emits more than 9,000 million metric tons of CO_2, representing 28 percent of all global CO_2 emissions. The United States was in second place, emitting nearly 5,000 million metric tons. This represents 15 percent of all global CO_2 emissions, yet the United States contains only 5 percent of the world's population. If we consider the amount of per capita CO_2 emissions, shown in Figure 63.2b, we obtain a very different picture of which countries produce the most CO_2. Of the top 10 emitting countries, the United States, Saudi Arabia, and Canada are the leading per capita emitters of CO_2. Despite the fact that China and India rank among the top producers of CO_2, their per capita production ranks them ninth and tenth, respectively. This reflects the fact that these two countries both have very large populations.

Global temperatures have steadily increased since records began in 1880

Before we can determine if global temperature increases are a recent phenomenon and if these increases are unusual, we must establish how the temperatures of Earth have changed in the past. Since about 1880, there have been enough direct measurements of land and ocean temperatures that NASA Goddard Institute for Space Studies has been able to generate a graph of global temperature change over time. This graph, updated monthly, is shown in **FIGURE 63.3**. Comprising thousands of measurements from around the world, the graph shows global temperatures have increased 1.1°C (2.0°F) from 1880 through 2017. In fact, of the 18 warmest years since 1880, 17 of them have occurred between 2000 and 2017. The one remaining warm year was in 1998.

While an increase in average global temperature of 1.1°C (2.0°F) may not sound very substantial, it is not evenly distributed around the globe. As the map in **FIGURE 63.4** shows, some regions, including parts of Antarctica, have experienced cooler temperatures.

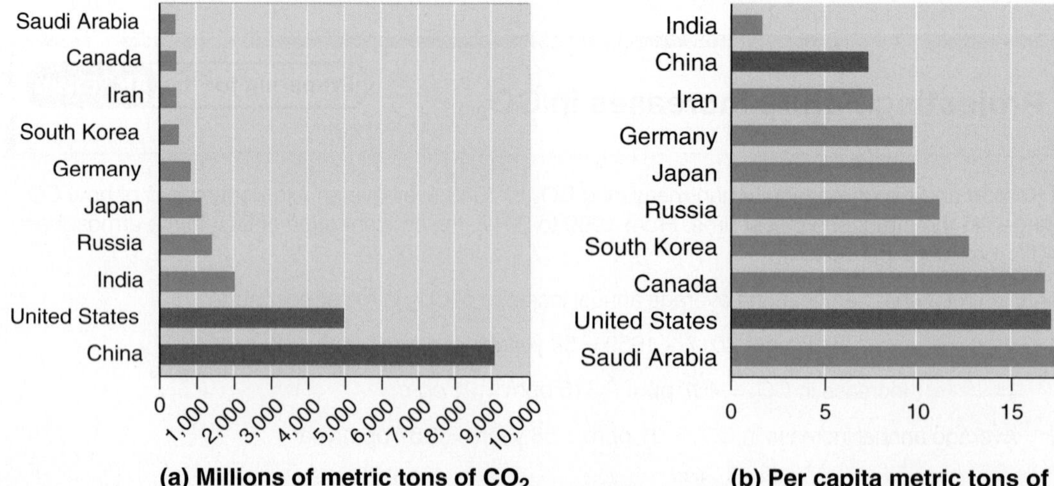

FIGURE 63.2 CO_2 emissions by country in 2015. (a) When we consider the total amount of CO_2 produced by a country, we see that the largest contributors are the developed and rapidly developing countries of the world. (b) Some major CO_2 emitters, such as China and India, have relatively low per capita CO_2 emissions. *(Data from International Energy Agency. 2017. CO_2 emissions from fuel combustion. Highlights. 162 pp.)*

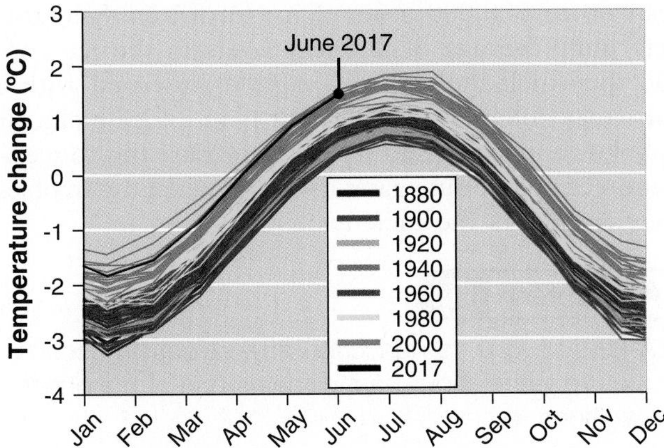

FIGURE 63.3 Changes in mean global temperatures over time. Although annual mean temperatures can vary from year to year, temperatures have exhibited a slow increase from 1880 to 2017. This pattern becomes much clearer when scientists compute the average temperature each year for the past 5 years. Note that the zero value is set at the mean temp from 1980–2015. *(Data from https://data.giss.nasa.gov/gistemp/news/20170714/cycle_201706_1600.tif.)*

Some regions, including areas of the oceans, have experienced no change in temperature. Finally, some regions, such as those in the extreme northern latitudes, have experienced increases of 1°C to 4°C (1.8°F to 7.2°F). The substantial increases in temperatures in the northern latitudes have caused, among other problems, nearly 45 percent of the northern ice cap to melt.

The data collected by the NASA clearly demonstrate that the globe has been slowly warming during the past 120 years. However, it is possible that such changes in temperature are simply a natural phenomenon. If we want to know whether these changes are typical, we must examine a much longer span of time.

Scientists can estimate global temperatures and greenhouse gas concentrations for over 500,000 years

Since no one was measuring temperatures thousands of years ago, we must use indirect measurements. Common indirect measurements include changes in the species composition of organisms that have been preserved over millions of years and chemical analyses of air bubbles formed in ice long ago.

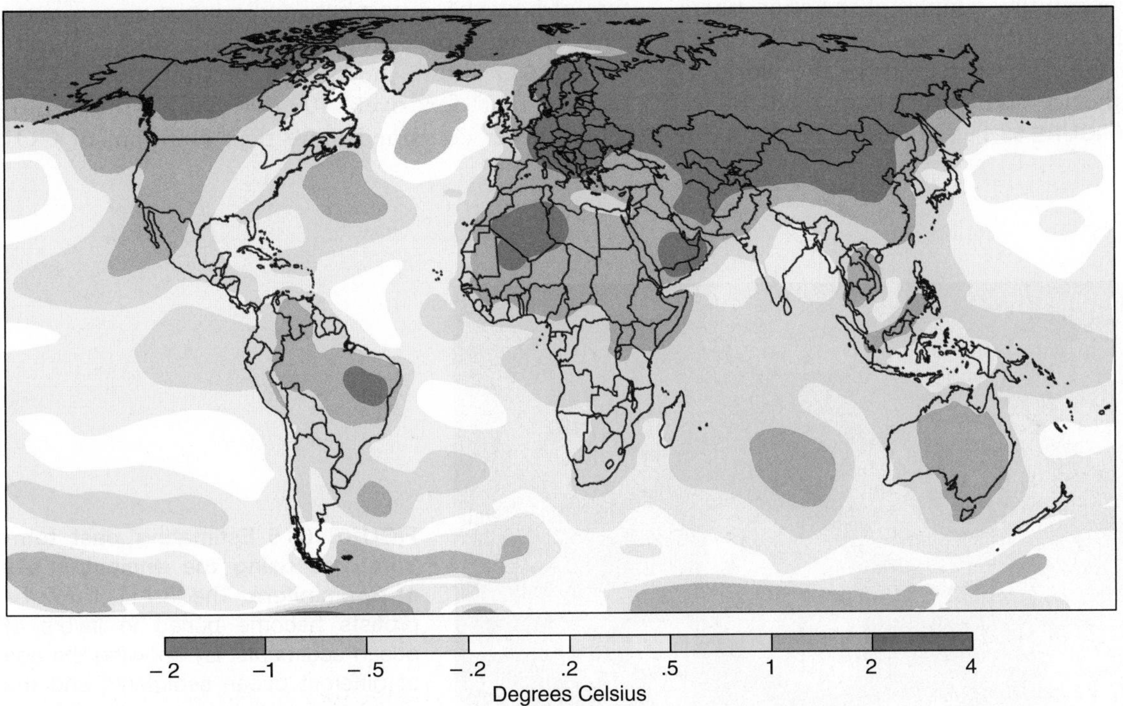

FIGURE 63.4 Changes in mean annual temperature in different regions of the world. In 2010, some regions became cooler, some regions had no temperature change, and the northern latitudes became substantially warmer than the long-term average temperature. The surface temperatures plotted on the map represent differences relative to the average temperature from 1951 to 1980. *(Data from https://www.nasa.gov/press-release/long-term-warming-trend-continued-in-2017-nasa-noaa.)*

Changing Species Compositions

One commonly used biological measurement is the change in species composition of a group of small protists, called foraminifera. Foraminifera are tiny, marine organisms with hard shells that resist decay after death (**FIGURE 63.5**). In some regions of the ocean floor, the tiny shells have been building up in sediments for millions of years. The youngest sediment layers are near the top of the ocean floor whereas the oldest sediment layers are much deeper. Fortunately, different species of foraminifera prefer different water temperatures. As a result, when scientists identify the predominant species of foraminifera in a layer of sediment, they can infer the likely temperature of the ocean at the time the layer of sediment was deposited. By examining thousands of sediment layer samples, we can gain insights into temperature changes over millions of years.

Air Bubbles in Ancient Ice

Scientists can determine changes in greenhouse gas concentrations and temperatures over long periods of time by examining ancient ice. In cold areas such as Antarctica and at the top of the Himalayas, the snowfall each year eventually compresses to become ice. Similar to marine sediments, the youngest ice is near the surface and the oldest ice is much deeper. During the process of compression, the ice captures small air bubbles. These bubbles contain tiny samples of the atmosphere that existed at the time the ice was formed. Scientists have traveled to these frozen regions of the world to drill deep into the ice and extract long tubes of ice called ice cores (**FIGURE 63.6**). Samples of ice cores can span up to 500,000 years of ice formation. Scientists determine the age of different layers in the ice core and then melt the ice from a piece associated with a particular time period. When the piece of ice melts, air bubbles are released and scientists measure the concentration of greenhouse gases in the air when the bubbles were trapped in the ancient ice.

AP® Exam Tip

Make sure you can describe how scientists use ice cores to learn about past climates and atmospheric conditions. ●

Oxygen atoms in melted ice cores can also be used to determine temperatures from the distant past. Oxygen atoms occur in two forms, or isotopes: light oxygen, also known as oxygen-16 (^{16}O), contains 8 protons plus 8 neutrons. In contrast, heavy oxygen, also known as oxygen-18 (^{18}O), contains 8 protons plus 10 neutrons. Ice formed during a period of warmer temperatures contains a higher percentage of heavy oxygen than ice formed during colder temperatures. By examining changes in the percentage of heavy oxygen atoms from different layers of the ice core, we can indirectly estimate temperatures from hundreds of thousands of years ago.

Combining data from different biological and physical measurements, researchers have created a picture of how the atmosphere and temperature of Earth have changed over hundreds of thousands of years. **FIGURE 63.7** shows the pattern of atmospheric CO_2. Notice that for over 400,000 years, the atmosphere never contained more than 300 ppm of CO_2.

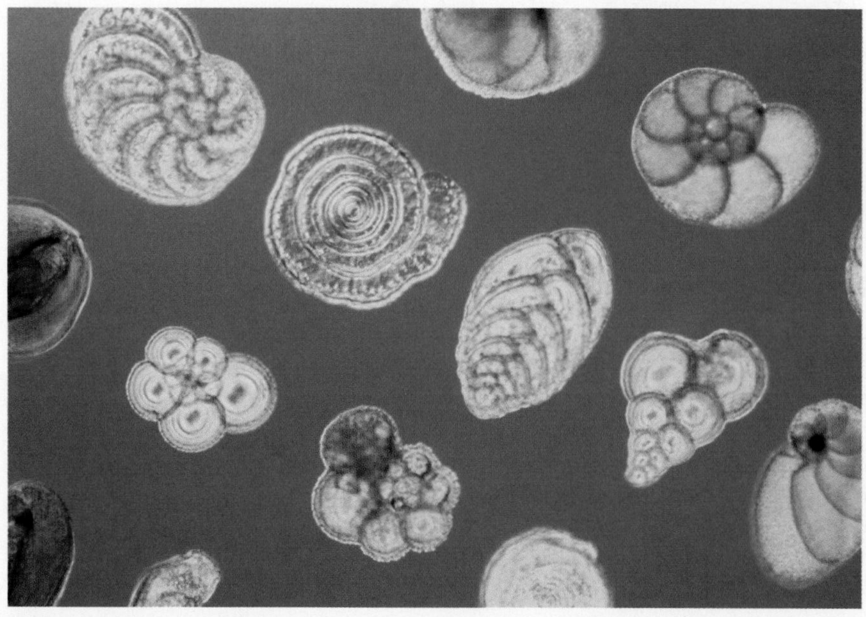

FIGURE 63.5 Estimating past temperatures using the ancient shells of foraminifera. The tiny shells of the protists become buried in layers of ocean sediments. By knowing the age of different ocean sediments and the preferred temperature of different species of foraminifera, scientists can indirectly estimate ocean temperature changes over time based on which species are found in each sediment layer. *(Astrid & Hanns-Frieder Michler/Science Source)*

(a)

(b)

FIGURE 63.6 Estimating past greenhouse gas concentrations and past temperatures using ice cores. (a) Ice cores are extracted from very cold regions of the world such as this glacier on Mount Sajama in Bolivia. (b) Ice cores have tiny trapped air bubbles of ancient air that can provide indirect estimates of greenhouse gas concentrations and global temperatures. *(a: George Steinmetz/Getty Images; b: Marc Steinmetz/Visum/The Image Works)*

In contrast, from 1958 to 2017 the concentration of CO_2 in the atmosphere has rapidly climbed from 310 to 407 ppm. This means that the rise of CO_2 in the atmosphere during the past 50 years is unprecedented.

During the past 800,000 years, CO_2 is not the only greenhouse gas whose concentration has increased. As you can see in **FIGURE 63.8** on page 700, methane and nitrous oxide show a pattern of increase that is similar to the pattern we saw for CO_2. For all three gases, concentrations have varied over the previous 800,000 years. After 1800, however, concentrations of the three gases all rose dramatically. Given what we now know about the anthropogenic sources of greenhouse gases, this increase in greenhouse gases occurred because this time period marks the start of the Industrial Revolution when humans began burning large amounts of fossil fuel and producing a variety of greenhouse gases.

FIGURE 63.9 on page 701 charts historic temperatures and CO_2 concentrations. Looking at the blue line, we see that temperatures have changed dramatically over the past 400,000 years. Most of these rapid shifts occurred during the onset of an ice age or during the transition from an ice age to a period of warm temperatures after an ice age. Because these changes occurred before humans could have had an appreciable effect on global systems, scientists suspect the changes were caused by small, regular shifts in the orbit of Earth. The path of the orbit, the amount of tilt on Earth's axis, and the position relative to the Sun all change regularly over hundreds of thousands of years. These changes alter the

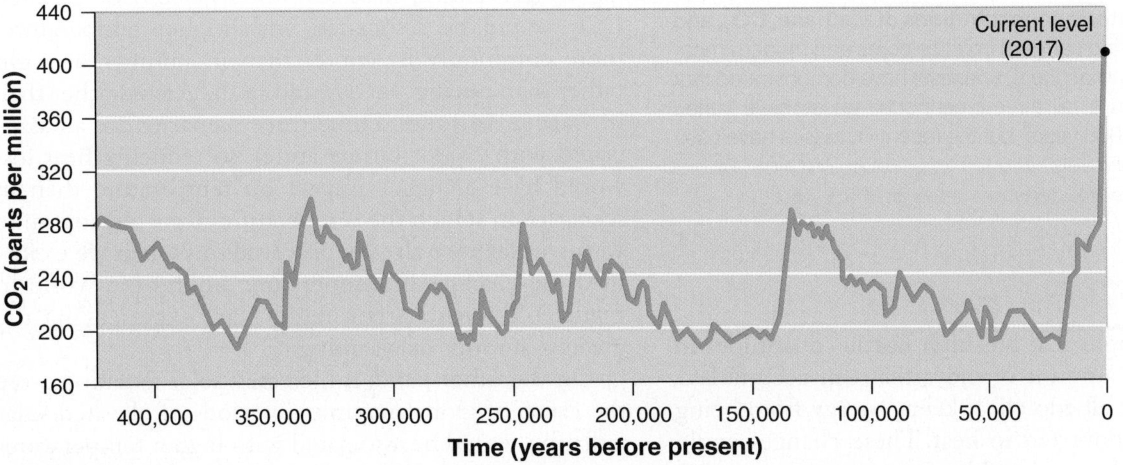

FIGURE 63.7 Historic CO_2 concentrations. Using a variety of indirect indicators including air bubbles trapped in ancient ice cores, scientists have found that for more than 400,000 years CO_2 concentrations never exceeded 300 ppm. After 1950, CO_2 concentrations have sharply increased to their current level of more than 400 ppm. *(Data from https://data.giss.nasa.gov/gistemp/graphs_v3/.)*

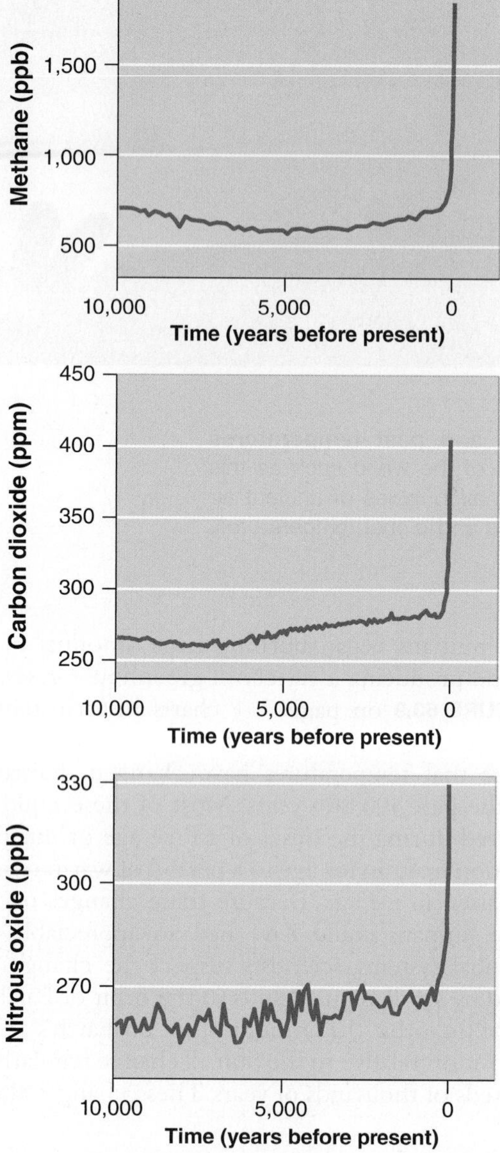

FIGURE 63.8 Historic concentrations of methane, CO_2, and nitrous oxide. Using samples from ice cores and modern measurements of the atmosphere, scientists have demonstrated that the concentrations of all three greenhouse gases have varied over the past 800,000 years, but the recent increases have risen to unprecedented levels. *(Data from https://www.epa.gov/climate-indicators/climate-change-indicators-atmospheric-concentrations-greenhouse-gases.)*

amount of sunlight that hits high northern latitudes in the winter, the amount of snow that can accumulate, and the way the albedo effect keeps energy from being absorbed and converted to heat. These changes could give rise to fairly regular shifts in temperature over a long period of time.

The more important insight from Figure 63.9 is the close correspondence between historic temperatures and CO_2 concentrations. But the graph does not tell us the nature of this relationship. Did periods of increased CO_2 cause increased temperature; did periods of increased temperature cause increased production of CO_2; or is another factor at work? Scientists believe that the relationship between fluctuating levels of CO_2 and the temperature is complex and that both factors play a role. As we know, the increase of CO_2 in the atmosphere causes a greater capacity for warming through the greenhouse effect. However, when Earth experiences higher temperatures, the oceans warm and cannot contain as much CO_2 gas and, as a result, they release CO_2 into the atmosphere. What ultimately matters is the net movement of CO_2 between the atmosphere and the oceans and how these different feedback loops work together to affect global temperatures.

Greenhouse Gases versus Increased Solar Radiation

We have seen that the surface temperature of Earth has increased roughly 1.1°C (2.0°F) over the past 137 years. But larger changes in temperature have occurred over the past 400,000 years without human influence. How can we tell if the recent changes are anthropogenic? One explanation for warming temperatures during the past century is an increase in solar radiation. Another possibility is that warming is caused by increased CO_2 in addition to warming caused by natural fluctuations in solar radiation. In other words, both factors might be important. Unfortunately, simply looking at temperature and CO_2 data averages around the globe will not allow us to determine if either of these two possibilities is correct.

One way to approach the problem is to look for more detailed patterns in temperature changes. For example, if increased CO_2 concentrations caused global warming by preventing heat loss, then periods of elevated CO_2 would be associated with higher temperatures more commonly in winter than in summer, at night rather than during the day, and in the Arctic rather than in warmer latitudes. These three scenarios are all associated with colder temperatures, so reducing heat loss would have a bigger impact on temperature than in scenarios in which the temperatures were already quite warm. In fact, we already observed this when we examined the changes in temperature around the world in Figure 63.4—the Arctic regions are experiencing the greatest amount of warming.

On the other hand, if increased solar radiation were the cause of global warming, periods of elevated solar radiation would be associated with higher temperatures more commonly when the Sun is shining more—namely in the summer, during the day, and at low latitudes. These times and locations on Earth receive the greatest amount of sunlight, so an increased intensity

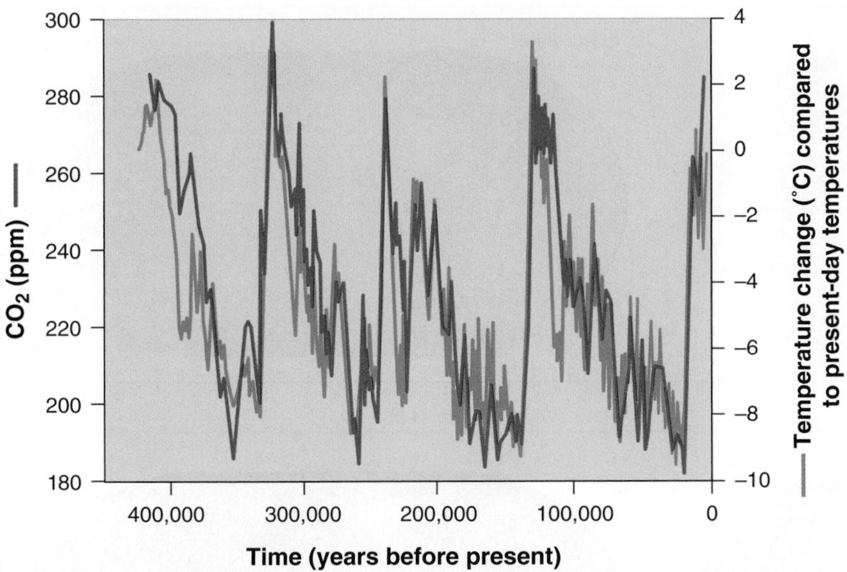

FIGURE 63.9 Historic temperature and CO_2 concentrations. Ice cores used to estimate historic temperatures and CO_2 concentrations indicate that the two factors vary together. *(Data from http://www.ncdc.noaa.gov/paleo/globalwarming/temperature-change.html.)*

of solar radiation would cause these times and places on Earth to warm more than other times and places. When scientists make these types of detailed comparisons, they find that the patterns in temperature change are strongly consistent with increased greenhouse gases such as CO_2 and not consistent with increased solar radiation. This body of evidence led the IPCC to conclude that most of the observed increase in global average temperatures since the mid-twentieth century has been the result of increased concentrations of anthropogenic greenhouse gas.

Climate Models and the Prediction of Future Global Temperatures

Just as indirect indicators can help us get a picture of what the temperature has been in the distant past, computer models can help us predict future climate conditions. Researchers can determine how well a model approximates real-world processes by applying it to a time in the past for which we have accurate data on conditions such as air and ocean temperatures, CO_2 concentration, extent of vegetation, and sea ice coverage at the poles. Modern models reproduce recent temperature fluctuations well over large spatial scales. From this work modelers are fairly confident that climate models capture the most significant features of today's climate.

Although climate models cannot forecast future climates with total accuracy, as the models improve scientists have been able to place more confidence in their predictions of temperature change, although they have had more difficulty predicting changes in precipitation. Because assumptions vary among different climate models, when multiple models predict similar changes, we can have increased confidence that the predictions are robust. **FIGURE 63.10** on page 702 shows recent predictions based on the data from climate models. Scientists generally agree that average global temperatures will rise by 1.8°C to 4°C (3.2°F–7.2°F) by the year 2100, depending on whether CO_2 emissions experience slow, moderate, or high growth over time. As you can see in the figure, northern latitudes will experience temperatures well above these averages, which is a pattern similar to what is already happening (see Figure 63.4).

Feedbacks can increase or decrease the impact of climate change

The global greenhouse system is made up of several interconnected subsystems with many potential positive and negative feedbacks. As we saw in Chapter 2 with population systems, positive feedbacks amplify changes. Because of this, positive feedback often leads to an unstable situation in which small fluctuations in inputs lead to large observed effects. On the other hand, negative feedbacks dampen changes. When we think about how anthropogenic greenhouse gases will affect Earth, we must ask whether positive or negative feedbacks will predominate. We do not currently have enough evidence to settle this question conclusively, but we can

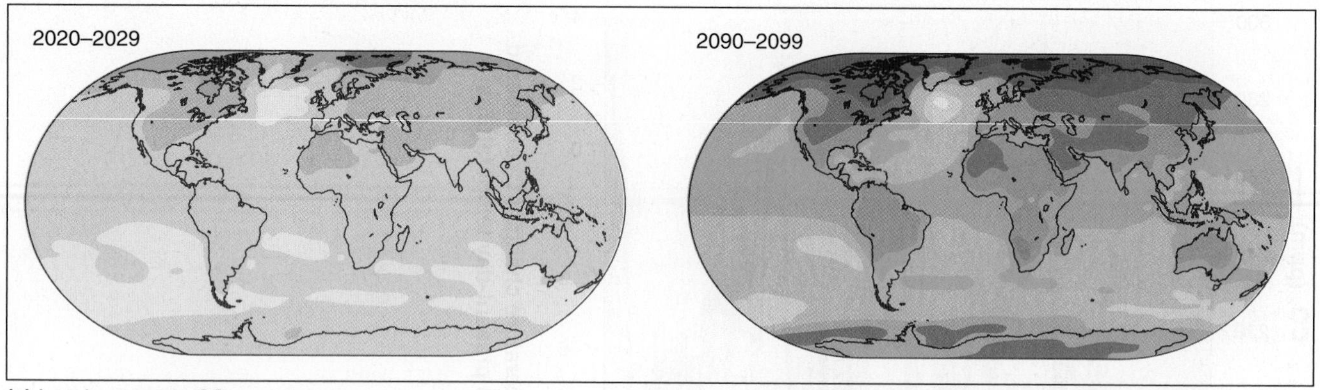

(a) Low increase in CO_2

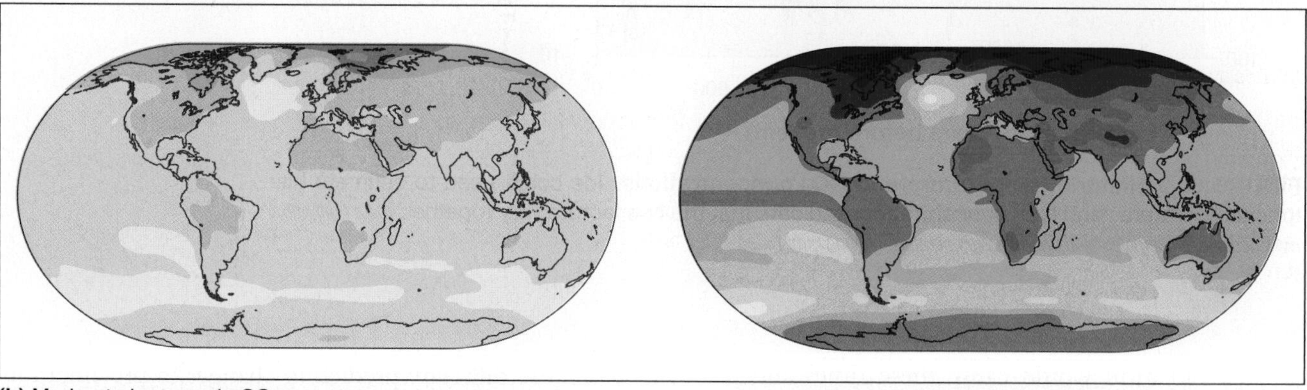

(b) Moderate increase in CO_2

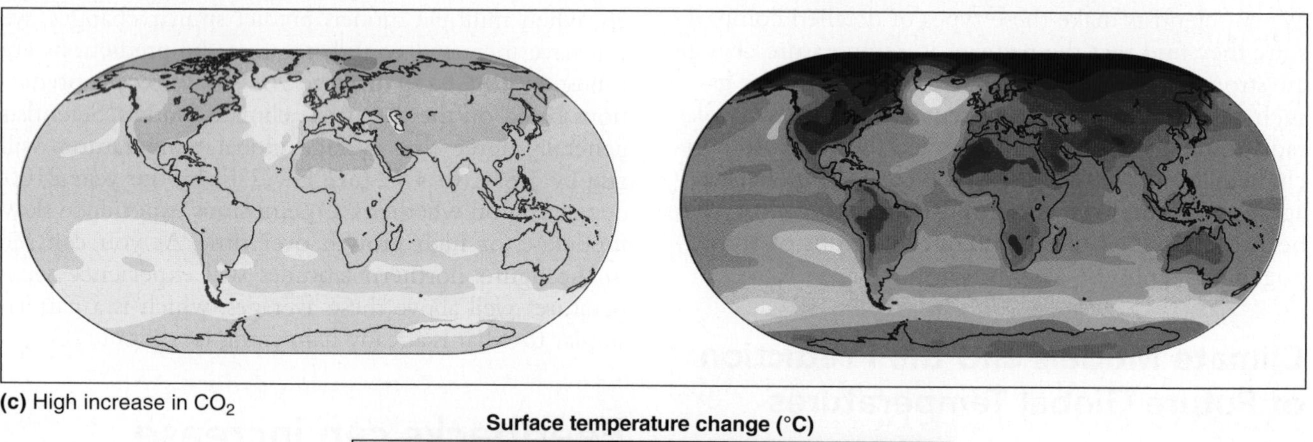

(c) High increase in CO_2

Surface temperature change (°C)

0 1 2 3 4 5 6 7 8

FIGURE 63.10 Predicted increase in global temperatures by 2100. The predictions depend on whether we expect (a) low, (b) moderate, or (c) high increases in how much CO_2 the world emits during the current century. These changes in temperature are relative to the mean temperatures from 1961 to 1990. *(Data from Moberg, A., et al. 2005. Highly variable Northern Hemisphere temperatures reconstructed from low- and high-resolution proxy data. Nature 433:613–617.)*

examine some of the feedback cycles and the way they influence temperatures on Earth.

Positive Feedbacks

There are many ways that a rise in temperatures could create a positive feedback. For example, global soils contain more than twice as much carbon as the amount currently in the atmosphere. As shown in **FIGURE 63.11a**, higher temperatures are expected to increase the biological activity of decomposers in these soils. Because this decomposition leads to the release of additional CO_2 from the soil into the atmosphere, the temperature change will be amplified even more.

A similar, but more troubling, scenario is expected in tundra biomes containing permafrost. As atmospheric

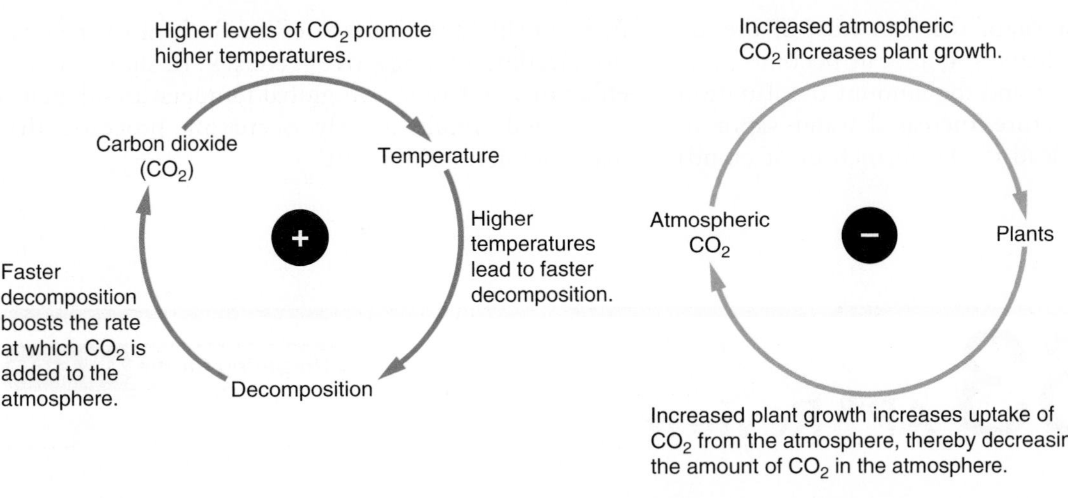
Higher levels of CO_2 promote higher temperatures.
Carbon dioxide (CO_2)
Temperature
+
Higher temperatures lead to faster decomposition.
Faster decomposition boosts the rate at which CO_2 is added to the atmosphere.
Decomposition
Increased atmospheric CO_2 increases plant growth.
Atmospheric CO_2
−
Plants
Increased plant growth increases uptake of CO_2 from the atmosphere, thereby decreasing the amount of CO_2 in the atmosphere.
(a) Positive feedback system
(b) Negative feedback system

FIGURE 63.11 Global change feedback systems. (a) Temperature and CO_2 represent a positive feedback system. When the concentration of CO_2 increases in the atmosphere, it can cause global temperatures to increase. This in turn can cause more rapid decomposition, thereby releasing even more CO_2 into the atmosphere. (b) Carbon dioxide and producers represent a negative feedback system. Increased CO_2 in the atmosphere from anthropogenic sources can be partially removed by increased photosynthesis by producers.

concentrations of CO_2 from anthropogenic sources increase, the Arctic regions become substantially warmer and the frozen tundra begins to thaw. As it thaws, the tundra develops areas of standing water with little oxygen available under the water as the thick organic layers of the tundra begin to decompose. As a result, the organic material experiences anaerobic decomposition that produces methane, a stronger greenhouse gas than CO_2, which should lead to even more global warming.

Negative Feedbacks

One of the most important negative feedbacks occurs as plants respond to increases in atmospheric carbon. Figure 63.11b shows this cycle. Because carbon dioxide is required for photosynthesis, an increase in CO_2 can stimulate plant growth. The growth of more plants will cause more CO_2 to be removed from the atmosphere. This negative feedback, which causes carbon dioxide and temperature increases to be smaller than they otherwise would have been, appears to be one of the reasons why only about half of the CO_2 emitted into the atmosphere by human activities has remained in the atmosphere.

A second negative feedback exists in the oceans. As CO_2 concentrations increase in the atmosphere, more CO_2 is absorbed by the oceans. Although this is beneficial because it reduces CO_2 in the atmosphere, it causes harmful effects to the oceans. When CO_2 dissolves in water, much of it combines with water molecules to form carbonic acid (H_2CO_3). Since this is an equilibrium reaction, an increase in ocean CO_2 causes more CO_2 to be converted to carbonic acid, which lowers the pH of the water in a process known as ocean acidification, described in Chapter 2. Ocean acidification is of particular concern for the wide variety of species that build shells and skeletons made of calcium carbonate, including corals, mollusks, and crustaceans. As pH decreases, the calcium carbonate in these organisms can begin to dissolve and the ocean's saturation point for calcium carbonate declines, which makes it harder for organisms to acquire the material they need to build their shells and skeletons.

The Limitations of Feedbacks

Most of the feedbacks we have discussed are limited by features of the systems in which they take place. For example, the soil-carbon feedback is limited by the amount of carbon in soils. While warming soils could add large amounts of CO_2 to the atmosphere for a time, eventually soil stocks will become so low that biological activity falls back to earlier rates. The enhanced CO_2 uptake by plants is also limited: Studies indicate that only some plants benefit from CO_2 fertilization, and often the growth is enhanced only until another factor becomes limiting, such as water or nutrients.

The magnitude and direction of many feedbacks are complex. For example, water vapor has both positive and negative feedbacks and there are limits to each. As temperatures increase, water can evaporate into the atmosphere more easily. Because water vapor is a greenhouse gas, this increased evaporation will lead to further warming. There is a limit, however, to the amount of water vapor that can exist in the air. As we discussed in Chapter 4, air can become saturated with water vapor and the amount of saturation changes with temperature. Increased water vapor in the atmosphere can lead to the formation of clouds that can shield the surface of Earth from solar radiation, leading to a negative feedback. In short, the net effect of water vapor on global temperatures depends on several simultaneously occurring processes that make predictions difficult.

MODULE 63 AP® Review

Preparing for the AP® Exam

In this module, we learned that the concentrations of greenhouse gases have been steadily increasing in the past century and that the production of these gases differs among nations. At the same time, global temperatures have also increased since when direct measurements were first made. We also saw how scientists can estimate temperatures and greenhouse gas concentrations for over 500,000 years using changes in species composition of foraminifera and examining the concentrations of gases in bubbles of air that are frozen in ancient ice. Together, these data suggest a close relationship between changes in CO_2 and changes in global temperatures for more than 400,000 years. Using climate models, we can predict future changes in global temperatures under different scenarios of small, medium, or high increases in CO_2 concentrations. In the next module, we will examine the consequences of current and future changes in global temperatures.

AP® Practice Questions

Choose the best answer for the following.

1. What is the evidence that solar radiation is less important to global warming than the increase in greenhouse gases?
   (a) Temperatures have increased more in the summer than the winter.
   (b) Temperatures have increased more in the winter than the summer.
   (c) There is a historic correlation of CO_2 and temperature.
   (d) There is a lack of temperature change in some areas.

2. If the annual rate of CO_2 increase is 2.3 ppm and the concentration in 2017 is 407 ppm, what concentration would you expect in 2047?
   (a) 420 ppm
   (b) 476 ppm
   (c) 505 ppm
   (d) 525 ppm

3. How much has the average global temperature increased from 1880 to 2017?
   (a) 0.5°C
   (b) 1.1°C
   (c) 2.0°C
   (d) 3.1°C

4. Which data are used to estimate historic temperatures and carbon dioxide concentrations?
   I. marine organism fossils
   II. air trapped in ice
   III. glacier depth
   (a) I only
   (b) I and II
   (c) I and III
   (d) I, II, and III

5. Which is an example of a negative feedback?
   (a) Higher air temperatures cause increased ocean evaporation.
   (b) Lower concentrations of CO_2 cause increased absorption of oceanic CO_2.
   (c) Low albedo causes decreased reflection of sunlight.
   (d) Higher CO_2 concentrations cause increased photosynthesis.

MODULE 64

Consequences of Global Climate Change

A wide range of environmental indicators demonstrate that global warming is affecting global processes and contributing to overall global change. In many cases, we have clear evidence of how global warming is having an effect. In other cases, we can use climate models to make predictions about future changes. As with all predictions of the future, there is a fair degree of uncertainty regarding the future effects of global warming. In this module, we will discuss how global warming is expected to affect the environment and organisms living on Earth.

Learning Goals

After reading this module, you should be able to

- discuss how global climate change has affected the environment.
- explain how global climate change has affected organisms.
- identify the future changes predicted to occur with global climate change.
- explain the global climate change goals of the Kyoto Protocol.

Global climate change is already affecting the environment

Warming temperatures are expected to have a wide range of impacts on the environment. Many of these effects are already happening, including melting of polar ice caps, glaciers, and permafrost and rising sea levels. Other effects are predicted to occur in the future, including an increased frequency of heat waves, fewer and less-intense cold spells, altered precipitation patterns and storm intensity, and shifting ocean currents.

Polar Ice

As we have seen, the Arctic has already warmed by 1°C to 4°C (1.8°F–7.2°F). **FIGURE 64.1** illustrates the extent of the reduction in the size of the ice cap that surrounds the North Pole. These data are collected in September of each year, which is about the time that the sea ice has reached its minimum extent. As you can see, the extent of sea ice

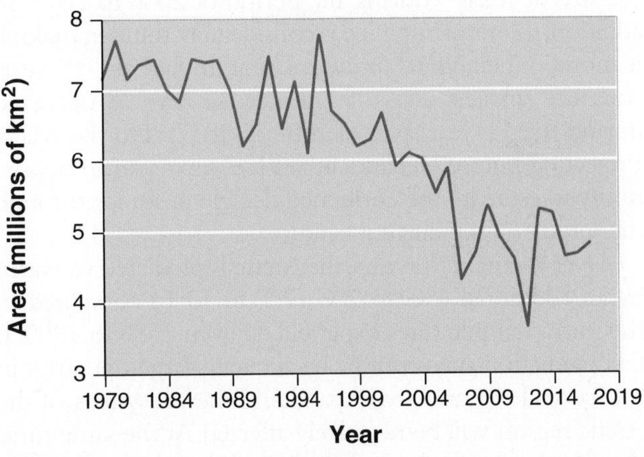

FIGURE 64.1 The melting polar ice cap. Because northern latitudes have experienced the greatest amount of global warming, the extent of the ice cap near the North Pole has been declining over the past 3 decades. The polar ice cap reaches its minimum late in the summer of each year, so we can look for a trend by examining the extent of ice cover each year during September. From 1979 to 2017, the polar ice has declined an average of 13 percent per decade. *(Data from http://nsidc.org/arcticseaicenews/2017/10/.)*

DO THE MATH Projecting Future Declines in Sea Ice

Preparing for the AP® Exam

As we have just learned, the polar sea ice has been declining at a rate of 13 percent per decade since 1979, which means that 87 percent remains after each decade. Given that the extent of sea ice was approximately 8 million km^2 in 1979, how much sea ice should there be in 1989, 1999, and 2009? To do this calculation, we have to calculate the amount of sea ice decline in each of the past 3 decades (rounding to the nearest hundredth):

$$\begin{aligned}\text{Sea ice in 1989} &= \text{Sea ice in 1979} \times 87\% \\ &= 8 \text{ million km}^2 \times 0.87 \\ &= 6.96 \text{ million km}^2\end{aligned}$$

$$\begin{aligned}\text{Sea ice in 1999} &= \text{Sea ice in 1999} \times 87\% \\ &= 6.96 \text{ million km}^2 \times 0.87 \\ &= 6.06 \text{ million km}^2\end{aligned}$$

$$\begin{aligned}\text{Sea ice in 2009} &= \text{Sea ice in 1999} \times 87\% \\ &= 6.06 \text{ million km}^2 \times 0.87 \\ &= 5.27 \text{ million km}^2\end{aligned}$$

YOUR TURN

1. If the annual rate of sea ice melting were to continue declining at this same rate, how much ice should be left by 2049?
2. What percentage of the original 1979 sea ice will remain in 2049?

fluctuates, but there is an overall trend of a 13 percent decline per decade from 1979 to 2017. Compared with the average amount of ice present from 1979 to 2000, scientists found that there was 45 percent less ice—a total area twice the size of Texas—during the period of 2006 to 2012. In addition, the remaining ice is considerably thinner, making it more vulnerable to future melting. In fact, the 10 years with the smallest areas of summer ice have all occurred during the 11-year span of 2007 to 2017. "Do the Math: Projecting Future Declines in Sea Ice" gives you an opportunity to estimate the continued decline in sea ice through the middle of the current century.

Over the next 70 years, the Arctic is predicted to warm by an additional 4°C to 7°C (7°F to 13°F) compared to the mean temperatures experienced from 1980 to 1990. If this prediction turns out to be accurate, large openings in sea ice will continue to expand and the ecosystem of the Arctic region will be negatively affected. At the same time, though, there may also be benefits to humans. For example, the opening in the polar ice cap could create new shipping lanes that would reduce by thousands of kilometers the distance some ships have to travel. Also, it is estimated that nearly one-fourth of all undiscovered oil and natural gas lies under the polar ice cap and a melted polar ice cap might make these fossil fuels more easily obtainable. However, the combustion of these fossil fuels would further facilitate global warming, representing another example of a positive feedback.

In addition to the polar ice cap in the Arctic, Greenland and Antarctica have also experienced melting. As you can see in **FIGURE 64.2**, sea ice mass has been measured in Antarctica and Greenland from 2000 to 2017. During this time, Antarctica has lost nearly 2,000 gigatons (about 4,400 trillion pounds) of ice while Greenland lost nearly 3,800 gigatons (8,378 trillion pounds) of ice. The polar ice cap in Antarctica is particularly interesting. It has shown a small increase in total area over the past 3 decades. However, the mass of the ice cap includes both area and thickness. Because the ice cap is losing thickness due to melting, its overall mass has been reduced. Current evidence shows that the melting rate of these ice-covered regions is continuing to increase. As we will see, such large amounts of melted ice have caused sea levels to rise.

Glaciers

As we discussed in Chapter 9, global warming has caused the melting of many glaciers around the world. Glacier National Park in northwest Montana, for example, had 150 individual glaciers in 1850 but has only 25 glaciers today. It is estimated that by 2030 Glacier National Park will no longer have any glaciers. The loss of glaciers is not simply a loss of an aesthetic natural wonder. In many parts of the world the melting of glaciers starting each spring provides a critical source of water for many communities. Historically, these glaciers partially melted during the spring and summer and then grew back to their full size during the winter. However, as summers become warmer, glaciers are melting faster

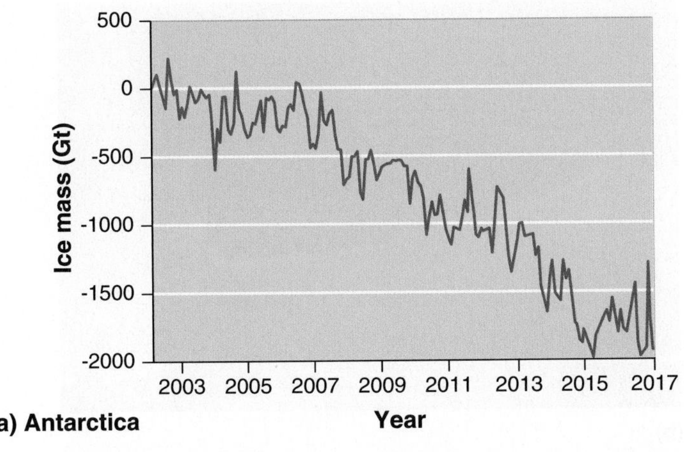

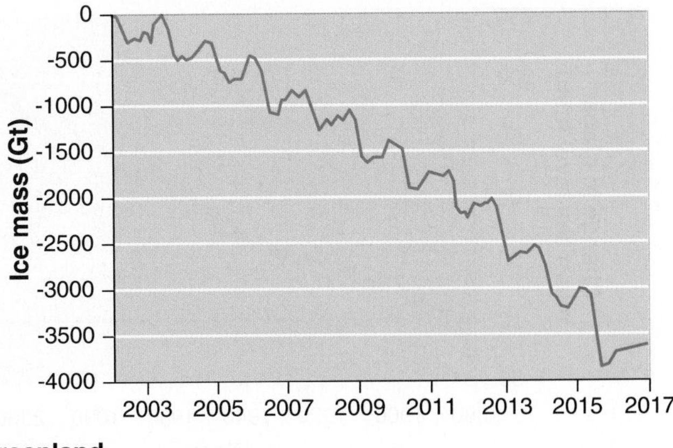

FIGURE 64.2 Declining ice in Antarctica and Greenland. Measurements of lost ice mass from 2002 to 2017 have detected declines in both (a) Antarctica and (b) Greenland. In both graphs, the y axes zero is set at 2002. *(Data from https://climate.nasa.gov/vital-signs/land-ice/.)*

than they can grow back in the winter, which leaves some people without a reliable water supply.

Permafrost

As warmer temperatures cause ice caps and glaciers to melt, it is perhaps not surprising that areas of permafrost are also melting. You may recall from the discussion on biomes in Chapter 4 that permafrost is permanently frozen ground that exists in the cold regions of high altitudes and high latitudes, which include the tundra and boreal forest biomes. About 20 percent of land on Earth contains permafrost; in some places it can be as much as 1,600 m (1 mile) thick. Melting of the permafrost causes overlying lakes to become smaller as the lake water drains deeper down into the ground. Melting can also cause substantial problems with human-built structures that are anchored into the permafrost, including houses and oil pipelines. As the frozen ground melts, it can subside and slide away.

Melting permafrost also means that the massive amounts of organic matter contained in the tundra will begin to decompose. Because this decomposition would be occurring in wet soils under low-oxygen conditions, it would release substantial amounts of methane, increasing the concentration of this potent greenhouse gas. This chain of events could produce a positive feedback in which the warming of Earth melts the permafrost, releasing more methane that causes further global warming.

Sea Levels

The rise in global temperatures affects sea levels in two ways. First, the water from melting glaciers and ice sheets on land adds to the total volume of ocean water. Second, as the water of the oceans becomes warmer, it expands. **FIGURE 64.3** shows that, as a result of both these effects, sea levels have risen 240 mm (9.5 inches) since 1870. Scientists predict that by the end of the twenty-first century, sea levels could rise an additional 130 to 540 mm (5 to 21 inches). This could endanger coastal cities and low-lying island nations by making them more vulnerable to flooding, especially during storms, with more saltwater intrusion into aquifers and increased soil erosion. Currently, 100 million people live within 1 m (3 feet) of sea level. The actual impact on these areas of the world will depend on the steps taken to mitigate these effects. For example, as we saw in Chapter 9, some countries may be able to build up their shorelines with dikes to prevent inundation from rising sea levels. Countries possessing less wealth are not expected to be able to respond as effectively to coastal flooding.

Global climate is already affecting organisms

The warming of the planet not only is affecting polar ice caps and sea levels but also is affecting living organisms. These effects range from temperature-induced changes in the timing of plant flowering and animal behavior to the ability of plants and animals to disperse to more hospitable habitats.

During the last decade, the IPCC reviewed approximately 2,500 scientific papers that reported the effects of warmer temperatures on plants and animals. The panel concluded that over the preceding 40 years, the growing season for plants had lengthened by 4 to 16 days in the Northern Hemisphere, with the greatest increases occurring in higher latitudes. Indeed, scientists are finding that in the Northern Hemisphere many species of plants now flower earlier, birds arrive at their breeding grounds earlier, and insects emerge earlier. At the same time, the ranges occupied by different species of plants, birds, insects, and fish have shifted toward both poles.

Rapid temperature changes have the potential to cause harm if organisms do not have the option of moving to more hospitable climates and do not have sufficient time to

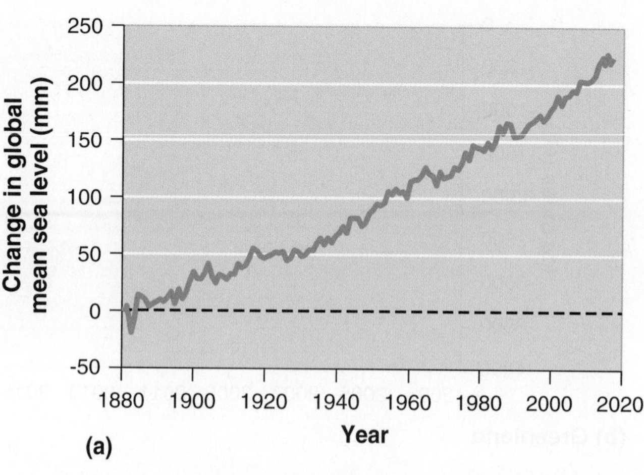

FIGURE 64.3 Rising sea levels. (a) Since 1870, sea levels have risen by 220 mm (9 inches). Future sea level increases are predicted to be 180 to 590 mm (7 to 23 inches) above 1999 levels by the end of this century. Note that sea level is graphed by setting the average sea level from 1993 to 2008 equal to zero. (b) Nearly 100 million people live within 1 m (3 feet) of sea level, such as on this island in the Maldives in the Indian Ocean. *(Data from https://www.climate.gov/news-features/understanding-climate/climate-change-global-sea-level) (b: VVO/Shutterstock.)*

evolve adaptations. Historically, organisms have migrated in response to climatic changes. This ability to migrate is one reason that temperature shifts have not been catastrophic over the past few million years. Today, however, fragmentation of certain habitats by roads, farms, and cities has made movement much more difficult. In fact, this fragmentation may be the primary factor that allows a warming climate to cause the extinction of species.

The pied flycatcher (*Ficedula hypoleuca*) is a bird that provides an interesting example. In the Netherlands, the pied flycatcher has evolved to synchronize the time that its chicks hatch with the time of peak abundance of caterpillars, a major source of food for the newly hatched birds (**FIGURE 64.4**). In 1980, the date of hatching preceded the peak in caterpillar abundance by a few days, so there was plenty of food for the new chicks. Twenty years later, warmer spring temperatures have caused trees to produce leaves earlier in the spring. Because the caterpillars feed on tree leaves, the peak abundance of caterpillars now occurs about 2 weeks earlier than it did in 1980. However, the hatching date of the pied flycatcher has not been affected by temperature change. Thus, by the time the chicks hatch, the caterpillars are no longer abundant and the hatchlings lack a major source of food, causing flycatcher populations in these areas to decline by 90 percent. There are other similar examples of the effects of rapid temperature change throughout the natural world.

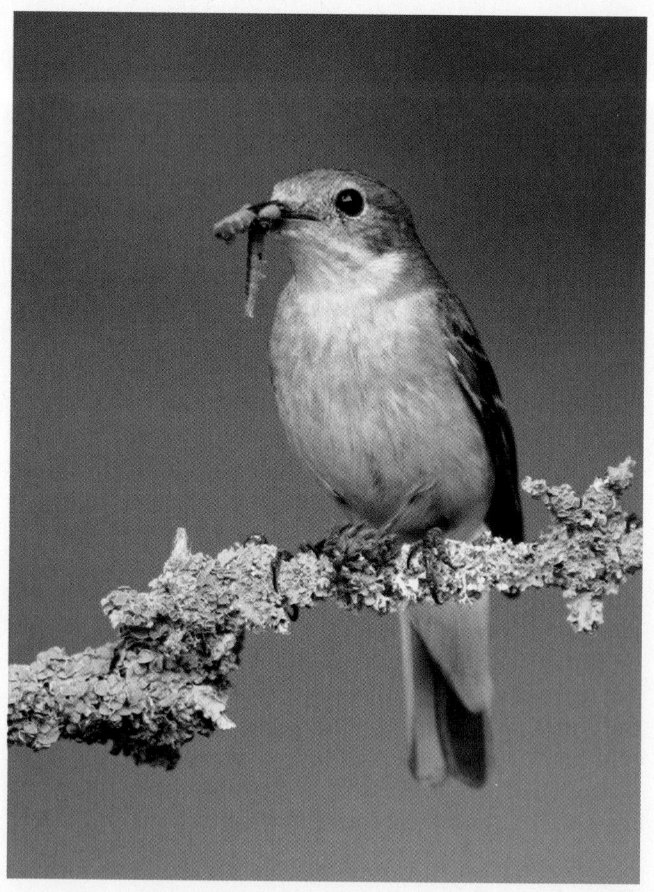

FIGURE 64.4 Effects of global warming on the pied flycatcher. Due to global warming in the Netherlands, the bird's main food source, a caterpillar, now becomes abundant 2 weeks earlier. However, the time when eggs hatch for the flycatcher has not changed. As a result, the birds hatch after the caterpillar population has begun to decline. *(Kats Edwin/AGE Fotostock)*

Corals are one group of organisms that are particularly sensitive to global warming because their range of temperature tolerance is quite small. With an increase in peak summer ocean temperatures of just 1°C, corals can experience "bleaching." Coral bleaching occurs when stressed corals eject their mutualistic algae, which provide corals with energy. The loss of algae causes the coral to turn white. The underlying causes of coral bleaching appear to be a combination of warming oceans, pollution, and sedimentation. While bleaching can be temporary, if it lasts for more than a short time, the corals die. While new corals should colonize regions at higher latitudes, it will take centuries before major new reef systems can be built. More coral bleaching is expected from global warming even if climate changes are kept relatively small.

As we noted earlier, the Arctic region is experiencing some of the most extreme effects of global warming. Polar bears live in the Arctic and they play a key role in the ecosystem by hunting for seals on the polar ice cap. The bears hunt for the holes in the ice and the bears pounce on any seals that come up for air. In many cases, the bears will consume only seal blubber, which is a concentrated source of the energy critical for an organism living in such a cold environment. The portion of the seal carcass that remains is a significant food source for other animals, including the Arctic fox (*Vulpes lagopus*). As the polar ice cap retreats far away from the land each summer, the polar bears can no longer reach the ice to hunt for seals. The problem for the bears is that today the sea ice melts 3 weeks earlier than it did 30 years ago. Because of the shorter time they are able to hunt for seals, male polar bears near Hudson Bay in Canada currently weigh 67 kg (150 pounds) less than they weighed 30 years ago. With less seal predation, there will be fewer seal carcasses for other animals like the Arctic fox to consume, so there is a cascade of effects when polar bears are affected by global warming.

In 2008, the United States classified polar bears as a threatened species because the decline in their ice habitat is expected to cause a decline in their population. In 2009, the five nations with polar bear populations (Canada, Greenland, Norway, Russia, and the United States) agreed that the polar bear should be classified as threatened throughout its entire global range. While acknowledging that pollution and hunting contributed to the bears' bleak future, the nations agreed that the effect of global warming on the ice cap poses the greatest threat to polar bears. In 2013, the United States and Russia went a step further by announcing an agreement to push for a CITES rule to ban the international trade in all products from polar bears including furs and teeth. While the plight of the polar bear has drawn attention to the effects of global warming, it is only one of many indicators that our world is rapidly changing because of human activity.

Global climate change is predicted to have additional effects in the future

Much of the controversy about global warming is not related to how the planet is already being affected, but rather what is predicted to happen in the future. Predicted future changes have some amount of uncertainty because they are based on computer models of complicated systems, such as the world's climates. It will take several decades to determine whether these predictions come true. In this section, we examine a number of these predicted effects including the frequency of heat waves, cold spells, precipitation patterns, storm intensity, and changes in ocean currents.

Heat Waves

As temperatures increase, long periods of hot weather—known as heat waves—are likely to become more frequent. Heat waves cause an increased energy demand for cooling the homes and offices where people live and work. For people who lack air conditioning in their homes, especially the elderly, heat waves increase the risk of death. Heat waves also cause heat and drought damage to crops, prompting the need for greater amounts of irrigation. The increased energy required for irrigation would raise the cost of food production.

Cold Spells

With global temperatures rising, minimum temperatures are expected to increase over most land areas, with fewer extremely cold days and fewer days below freezing. Such conditions would have two major positive effects for humans: fewer deaths due to cold temperatures and a decrease in the risk of crop damage from freezing temperatures. It may also make new areas available for agriculture that are currently too cold to grow crops. In addition, warmer temperatures would decrease the energy needed to heat buildings in the winter. However, a decrease in freezing temperatures that normally would cause the death of some pest species might allow these pest species to expand their range. The hemlock wooly adelgid (*Adelges tsugae*), for example, is an invasive insect from Asia that is causing the death of hemlock trees in North America by feeding on sap. Researchers have found that the range of the species is limited by cold temperatures. Warmer conditions in future decades are expected to allow this pest to expand its range and kill hemlock trees over a much larger area.

Precipitation Patterns

Because warmer temperatures should drive increased evaporation from the surface of Earth as part of the hydrologic cycle (see Figure 7.1 on page 83), global warming is projected to alter precipitation patterns. As mentioned earlier, climate models do not make consistent predictions about precipitation. Current models predict that some areas will experience increased rainfall, but the models differ in predicting which regions of the world would be affected. Regions receiving increased precipitation would benefit from an increased recharge to aquifers and higher crop yields, but they could also experience more flooding, landslides, and soil erosion. In contrast, other regions of the world are predicted to receive less precipitation, making it more difficult to grow crops and requiring greater efforts to supply water.

Storm Intensity

Although it is impossible to link any single weather event to climate change because of the multiple factors that are always involved, ocean warming may be increasing the intensity of Atlantic storms. For example, in 2005, hurricanes Katrina and Rita devastated coastal areas in Texas, Louisiana, and Mississippi. These hurricanes appear to have become as powerful as they did because waters in the Gulf of Mexico were unusually warm. Scientists at the National Center for Atmospheric Research concluded that climate change was responsible for at least half of this warming. As temperatures increase, such conditions should become more frequent, and hurricanes are likely to become more common farther north. The devastation in New Orleans did not come as a surprise; for many years scientists had warned that a strong hurricane could flood the city because of its position below sea level. Unfortunately, scientists caution that New Orleans is not the only American city that could be devastated by a powerful storm. Other cities at risk include New York, Miami, and Tampa. In 2017, the hurricane season was the most expensive in history for the United States, including hurricanes Harvey, Irma, and Maria that caused devastation in Texas, Florida, and Puerto Rico. As in 2005, these strong hurricanes were caused by unusually warm ocean waters.

Ocean Currents

Global ocean currents may shift as a result of more fresh water being released from melting ice. In Chapter 4 we saw that ocean currents have major effects on the climate of nearby continents. If the currents change, the distribution of heat on the planet could be disrupted. Scientists are particularly concerned about the thermohaline circulation, which, as we saw in Chapter 4, is a deep ocean circulation driven by water that comes out of the Gulf of Mexico and moves up to Greenland where it becomes colder and saltier and sinks to the ocean floor. This sinking water mixes with the deep waters of the ocean basin, resurfaces near the equator, and eventually makes its way back to the Gulf of Mexico (see Figure 11.2 on page 123). This circulating water moves the warm water from the Gulf of Mexico up toward Europe and moves cold water from the North Atlantic down to the equator. However, increased melting from Greenland and the northern polar ice cap could dilute the salty ocean water sufficiently to stop the water from sinking near Greenland and thereby shut off the thermohaline circulation. If this occurs, much of Europe would experience significantly colder temperatures.

Effects on Humans

Global warming and climate change could also affect many aspects of our lives. For example, some people may have to relocate from such vulnerable areas as coastal communities and some ocean islands. Poorer communities close to or along coastlines might not have the resources to rebuild on higher ground. If these communities do not obtain financial assistance, they will face severe consequences from flooding and saltwater intrusion. On the other hand, certain areas that have not been suitable for human habitation might become more hospitable if they become warmer, although other factors, such as water availability, might still limit their habitability.

Climate change has the potential to affect human health. Continued warming of the planet could affect the geographic range of temperature-limited disease vectors. For example, the mosquitoes that carry West Nile virus and malaria could spread beyond their current geographic range and bring health threats to regions that were once relatively untouched. As the climate changes, heat waves could cause more deaths to the very young, the very old, and those without access to air conditioning. Infectious diseases and bacterial and fungal illnesses might extend over a wider range than they do at present.

Climate changes will also have economic consequences. In northern locations, for example, warmer temperatures and shorter winters would drastically alter the character of northern communities that depend on snow for tourism. In the Alps, for example, many ski resorts are already adjusting to reduced snow on the mountains by catering to new groups of tourists who are more interested mountain biking than skiing (**FIGURE 64.5**). In warmer regions, the damage to corals reefs would negatively affect tourism as well. The economic impact on these types of tourist attractions depends on the rate of climate change and the ability of the tourism industry to adjust.

Assessing Uncertainty

How much controversy really exists regarding climate change? This question has been portrayed in many ways by various interest groups. Advocates in the environmental community talk of a "scientific consensus" on the topic of global warming while opponents of government regulation often speak of the global warming "controversy." The fundamental basis of climate change—that greenhouse gas concentrations are increasing and that this will lead to global warming—is not in dispute among the vast majority of scientists.

FIGURE 64.5 Adjusting tourism to climate change. In the French Alps, global warming has meant that many ski resorts are receiving less snow in the winter. In response, some ski resorts are altering their mountaintop facilities to cater to other types of tourists, including those who come in the summer because they want to leave hot cities and recreate in the cooler mountains. *(Patrice SCHREYER/RAPSODIA/Aurora Photos)*

Increases in greenhouse gases have been documented with real data and the ability of greenhouse gases to absorb and emit infrared radiation is a simple application of physics. Furthermore, the fact that the globe is warming is not in dispute. As we have seen, the data have clearly demonstrated increased global temperatures since direct measurements began in 1880; declines in the polar ice cap since measurements began in 1979; declines in Greenland and Antarctica ice sheets since measurements began in 2002; and rises in sea level since measurements began in 1880. As we discussed, what remains unclear is the likelihood that other changes in our climate have already started to happen.

In their 2007 and 2014 reports, the IPCC examined the existing data to quantify the likelihood that these additional types of climate change are already occurring and the likelihood that humans contributed to the changes. For example, the panel concluded that a decline in cold days and an increase in warm days has very likely occurred and that these changes very likely had a human contribution. They also concluded that heat waves, droughts, heavy precipitation, and hurricanes have likely increased since 1960, that these changes probably were influenced by human activities, and that these trends were likely to continue through the twenty-first century.

International agreements address climate change

Awareness of global change is a relatively recent phenomenon. In the past, most environmental issues could be dealt with at the national, state, or even local level. Global change is different because the scale of impact is so much larger and because the people and ecosystems affected can be extremely distant from the cause. In the case of climate change, many of the adverse effects are expected to be in the developing world, which has received disproportionately fewer benefits from the use of fossil fuels that led to the change in the first place. It would be impossible for one nation to pass legislation allowing it to avoid the impacts of climate change. To address the problem of global warming, the nations of the world must work together.

In 1997, representatives of the world's nations convened in Kyoto, Japan, to discuss how best to control the emissions contributing to global warming. At this meeting, they drew up the **Kyoto Protocol**, an international agreement which set a goal for global emissions of greenhouse gases from all industrialized countries to be reduced by 5.2 percent below their 1990 levels by 2012. Due to special circumstances and political pressures, countries agreed to different levels of emission restrictions, including a 7 percent reduction for the United States, an 8 percent reduction for the countries of the European Union, and a 0 percent reduction for Russia. Developing nations, including China and India, did not have emission limits imposed by the protocol. These developing nations argued that different restrictions on developed and developing countries are justified because developing countries are unfairly exposed to the consequences of global warming that in large part came from the developed nations. Indeed, the poorest countries in the world have only contributed to 1 percent of historic carbon emissions but are still affected by global warming. Thus, the approach was to have the countries that have historically emitted the

Kyoto Protocol An international agreement that sets a goal for global emissions of greenhouse gases from all industrialized countries to be reduced by 5.2 percent below their 1990 levels by 2012.

most CO_2 into the atmosphere pay most of the costs of reducing CO_2 emissions into the atmosphere.

The main argument for the Kyoto Protocol is grounded in the precautionary principle, which, as we saw in Chapter 17, states that in the face of scientific evidence that contains some uncertainty we should behave cautiously. In the case of climate change, this means that since there is sufficient evidence to suggest human activities are altering the global climate, we should take measures to stabilize greenhouse gas concentrations either by reducing emissions or by removing the gases from the atmosphere. The first option includes trying to increase fuel efficiency or switching from coal and oil to energy sources that emit less or no CO_2 such as natural gas, solar energy, wind-powered energy, or nuclear energy. The second option includes **carbon sequestration**—an approach that involves taking CO_2 out of the atmosphere. Methods of carbon sequestration might include storing carbon in agricultural soils or retiring agricultural land and allowing it to become pasture or forest, either of which would return atmospheric carbon to longer-term storage in the form of plant biomass and soil carbon. Researchers are also working on cost-effective ways of capturing CO_2 from the air, from coal-burning power plants, and from other emission sources. This captured CO_2 can then be compressed and pumped into abandoned oil wells or the deep ocean. Such technologies are still being developed, so their economic feasibility and potential environmental impacts are not yet known.

In developed countries, reductions in CO_2 emissions would require major changes to manufacturing, agriculture, or infrastructure at significant expense and economic impact. In 1997, before the Kyoto Protocol was finalized, the U.S. Senate voted unanimously (95–0) that the United States should not sign any international agreement that lacked restrictions on developing countries or any agreement that would harm the economy of the United States. Since it was never ratified by the U.S. Senate, the Kyoto Protocol is not legally binding on the United States.

In 2001, the Kyoto Protocol was modified to convince more developed nations to ratify it. At that time, the United States, under the administration of President George W. Bush (2001–2009), argued that there was too much uncertainty in global warming predictions to justify ratification of the protocol. The administration also argued that the costs of controlling carbon dioxide emissions would unfairly disadvantage businesses in the United States while businesses in China and India—two developing countries that were not significant emitters of greenhouse gases prior to 1990 but are significant emitters today—would have no reduction requirements.

Carbon sequestration An approach to stabilizing greenhouse gases by removing CO_2 from the atmosphere.

Ultimately, the United States argued that efforts to limit greenhouse gases should wait until there is more scientific evidence for global warming and the effects it produces before accepting the costs that the protocol would entail. Furthermore, the United States argued that all countries should be subject to emissions limits. Otherwise, polluting factories in developed countries could simply relocate to the developing countries, and little good would have been accomplished. Based on similar arguments, Canada, Russia, and Japan have since withdrawn adopting new targets for reduced CO_2 emissions beyond 2012. Proponents of reducing greenhouse gas emissions argue that profits gained from manufacturing new pollution-control technologies and savings in fuel costs through greater efficiency will offset any costs or decrease in short-term profits.

More recently, the U.S. government has taken stronger steps to regulate CO_2 emissions. In 2007, the U.S. Supreme Court ruled that the U.S. Environmental Protection Agency not only had the authority to regulate greenhouse gases as part of the Clean Air Act, but that it was required to do so. As a result, in 2009 the EPA announced it would begin regulating greenhouse gases for the first time. In 2010, the U.S. EPA began to look more closely at possible ways to regulate emissions of carbon dioxide. One proposal embraced by auto manufacturers has been to increase the fuel efficiency of vehicles. In 2012, the Obama Administration announced that the average fuel efficiency of cars and light trucks would increase from the current 12 km per liter (29 miles per gallon) to 16 km per liter (37 miles per gallon) by 2017 and 23 km per liter (55 miles per gallon) by 2025. This increase in efficiency is projected to cause a 50 percent reduction in CO_2 and other greenhouse gases from vehicles by 2025. The more fuel-efficient cars are expected to cost an extra \$1,000, but the reduced consumption of gasoline is expected to save the average driver \$3,000 over the lifetime of the vehicle. This would allow the United States to reduce greenhouse gases, invest in new automotive technology, reduce its consumption of fossil fuels, and save money. In 2017, the Trump Administration announced that it was going to review the fuel economy goals in response to auto manufacturers that were lobbying for lower fuel efficiency standards.

As of 2018, 192 countries have ratified the Kyoto Protocol, including most developed and developing countries, although more than 100 developing countries are exempt from any limits on CO_2 emissions including China and India. The United States is the only developed country that has not yet ratified the agreement.

Globally, total greenhouse gases increased by 19 percent between 1991 and 2014.Changes in CO_2 emissions in various countries have been mixed. For example, according to the World Bank, the total emissions of greenhouse gases from 1991 to 2014 have decreased by 24 percent for Germany, 15 percent for Russia, and 13 percent for the United States. Emissions in the United States have been declining sharply since 2007 for a variety of reasons including a slower economy, increased fuel efficiency, and an increase in the use of natural gas rather than coal for power generation. In contrast, 2014 greenhouse gas emissions in Canada have returned to their earlier 1991 levels, but emissions in Australia have increased by 80 percent.

In 2015, the nations of the world gathered once again, this time in Paris, to discuss what actions could be taken to combat global climate change. The **Paris Climate Agreement** (also known as the **Paris Climate Accord**) was a pledge by 195 countries to keep global warming less than 2°C above pre-industrial levels. To achieve this goal, each country individually decides how it will help in the global effort and reports on its progress periodically. The agreement goes into effect in 2020 and there is no enforcement of each country's targets by the United Nations. In 2017, the Trump administration announced that the United States would pull out of this agreement, citing President Trump's desire to promote the increased use of coal in the United States.

Paris Climate Agreement A pledge by 195 countries to keep global warming less than 2°C above pre-industrial levels. *Also known as the* **Paris Climate Accord**.

MODULE 64 AP® Review

Preparing for the AP® Exam

In this module, we learned that the warming of Earth due to human activities has had a number of effects. In regard to the environment, global warming has caused declines in the polar ice cap, declines in the ice masses of Greenland and Antarctica, melting of glaciers, thawing of permafrost, and increases in sea level. In regard to organisms, global warming has caused changes in the dates of flowering, bird migrations, insect emergence, and the length of the growing season. Scientists predict a number of additional effects with various levels of certainty, including more extreme temperatures, changes in global patterns of precipitation, more intense storms, and altered ocean currents. These predicted changes could affect many aspects of human life including where humans can live as well as their health.

AP® Practice Questions

Choose the best answer for the following.

1. Global warming might limit the availability of fresh water in many areas because of
   (a) melting glaciers.
   (b) thawing permafrost.
   (c) more frequent heat waves.
   (d) changes in storm intensity.

2. Effects of climate change on organisms include all of the following EXCEPT
   (a) increased growing seasons.
   (b) disruption of animal life cycles.
   (c) coral bleaching.
   (d) decreased species ranges in temperate areas.

3. What is NOT a potential negative effect of climate change on agriculture?
   (a) increased range of pests
   (b) increased risk of droughts
   (c) increased damage from severe weather events
   (d) increased rates of photosynthesis

4. How might climate change increase the range of pests?
   (a) higher intensity weather events
   (b) changing precipitation patterns
   (c) increasing number of heat waves
   (d) decreased duration of cold spells

5. In 1997 the United States did not ratify the Kyoto Protocol because
   (a) it required the same emission reductions from all signatories.
   (b) it did not limit the emissions of developing nations.
   (c) current technology was unable to meet the required goals.
   (d) the target emission reductions were seen as impossible to achieve.

Working Toward Sustainability

Cities, States, and Businesses Lead the Way to Reduce Greenhouse Gases

Although the United States Senate never ratified the original Kyoto Protocol, many state and local governments felt they had waited long enough for change at the federal level. In 2005, mayors from 141 cities and both major political parties gathered in San Francisco to organize their own efforts to reduce the causes and consequences of global warming. Their goal was to reduce greenhouse emissions in their own cities by the same 7 percent that the United States had agreed to in the Kyoto Protocol. As of 2018, a total of 1,060 out of 1,139 mayors of U.S. cities had signed the U.S. Conference of Mayors Climate Protection Agreement. Among the reasons the mayors cited for supporting this agreement were concerns in their communities over increasing droughts, reduced supplies of fresh water due to melting glaciers, and rising sea levels in coastal cities.

Similar actions are being taken at the state level. In the northeastern United States, for example, nine states have joined together collectively to form the Regional Greenhouse Gas Initiative to control regional production of greenhouse gases. A similar group emerged in western North America when seven western states and four Canadian provinces joined together in 2007 to form the Western Climate Initiative. For both groups, the goal was to regulate greenhouse emissions. By 2018, northeastern group continued to work together while the western group had a reduced membership that included only California and three Canadian provinces.

A number of large businesses are also joining in efforts to reduce greenhouse gases. General Electric, for example, announced in 2014 that it would reduce its emissions by 20 percent by 2020 (compared to its emissions in 2011). Two years later, the company announced that they had already reduced their emissions by 18 percent. They have invested $20 billion for research and development of technologies that can reduce greenhouse gases and this technology generated more than $270 billion in revenues, which confirmed that creating technology that would reduce greenhouse emissions was a profitable endeavor.

A growing number of companies including Google, Microsoft, ExxonMobil, and WalMart recognize the scientific evidence of human-caused global climate change and they are increasingly addressing the issue. For example, WalMart announced in 2017 that it was embarking on a new plan to reduce their greenhouse gas emissions. Calling the endeavor "Project Gigaton," they pledged to work with their suppliers to find ways to collectively reduce their emissions by 1 gigaton by 2030. Doing so would have the same effect as taking 211 million vehicles off the road for a year in the United States.

Innovative technologies to reduce greenhouse emissions. Companies such as General Electric have invested billions of dollars into technologies that are more energy efficient and produce less greenhouse gas. The Evolution® Series Tier 3 locomotive uses less fuel and has 40 percent lower emissions than previous locomotives. *(Courtesy, GE Transportation)*

From these stories, it is clear that progress on reducing greenhouse gases that cause global warming does not have to wait for national and international agreements to take effect. The public overwhelmingly understands that Earth is warming, states and cities are pushing forward with solutions that save money, and large corporations have come to understand that reducing emissions can reduce costs and improve profits over the long term. In short, curbing greenhouse gases and global warming is not only good for humans and the environment, it can be good for business and new jobs as well.

Critical Thinking Questions

1. What data might city mayors use to support their assertion that humans are causing global warming?
2. Why is it more effective for states and provinces to create regional partnerships to combat global warming rather than doing so alone?

References

Barringer, F. 2013. States group calls for 45% cut in amount of carbon emissions allowed. *New York Times*, February 7. http://www.nytimes.com/2013/02/08/business/energy-environment/states-group-calls-for-45-cut-in-amount-of-carbon-emissions-allowed.html.

Davenport, C. 2013. Large companies prepared to pay price on carbon. *New York Times*, December 5. http://www.nytimes.com/2013/12/05/business/energy-environment/large-companies-prepared-to-pay-price-on-carbon.html.

Frangoul, A. 2017. WalMart aims to slash one gigaton of emissions by 2030. *CNBC*. 20 April. https://www.cnbc.com/2017/04/20/walmarts-project-gigaton-aims-to-slash-emissions.html.

Chapter 19 Review

In this chapter, we learned the distinctions among global change, global climate change, and global warming. Global warming is an inherently natural process whereby a tiny percentage of gases in the atmosphere, known as greenhouse gases, absorb infrared radiation from Earth and emit some of this energy back to Earth. This warms the planet. Environmental scientists are concerned that human activities have caused a higher concentration of greenhouse gases in the atmosphere that is responsible for a gradual warming of the planet above that observed a century ago. The average amount of warming around the world has been relatively small at 0.8°C but some regions such as high northern latitudes have experienced increases of up to 4°C. This warming has caused a decline in polar ice, a decline in glaciers, an increased thawing of permafrost, and an increase in sea level. It has also affected the timing of plant flowering, bird migration, insect emergence, and the length of growing seasons. While these effects have already been observed, other climate changes are also predicted to occur including more extreme temperatures, more intense storms, changing patterns of precipitation, and altered ocean current. The Kyoto Protocol was designed to reduce the global emissions of greenhouse gases, but the goal set by nations around the world has not yet been achieved.

Key Terms

Global change
Global climate change
Global warming
Greenhouse effect
Greenhouse warming potential
Kyoto Protocol
Carbon sequestration
Paris Climate Agreement
Paris Climate Accord

Learning Goals Revisited

Module 62 Global Climate Change and the Greenhouse Effect

Distinguish among global change, global climate change, and global warming.

Global change refers to changes in the chemical, biological, and physical properties of the planet. One aspect of this is global climate change, which refers more specifically to the average weather that occurs in an area over a period of years or decades. One aspect of global climate change is global warming, which refers to the warming of the oceans, land masses, and atmosphere.

Explain the process underlying the greenhouse effect.

The greenhouse effect occurs when high-energy visible and ultraviolet light strike Earth and are emitted back from Earth as infrared radiation. Some of this infrared radiation is absorbed by greenhouse gases in the atmosphere, which make up a very small percentage of all atmospheric gases. The greenhouse gases subsequently emit infrared radiation, some of which head back to Earth and this causes the planet to become warmer.

Identify the natural and anthropogenic sources of greenhouse gases.

The natural and anthropogenic sources of greenhouse gases include water vapor, carbon dioxide, methane, and nitrous oxide. Chlorofluorocarbons are a type of greenhouse gas that only has an anthropogenic source.

Module 63 The Evidence for Global Warming

Explain how CO_2 concentrations have changed over the past 7 decades and how emissions run on compare among the nations of the world.

CO_2 concentrations have continually increased since atmospheric measurements first began in 1958. These concentrations have increased from 310 ppm in 1958 to nearly 400 ppm in 2013. The major producers of CO_2 include large developed countries, such as the United States and Russia, and rapidly growing developing countries, such as China and India. When we consider the per capita production of CO_2, we see that some of the largest producers of CO_2 have relatively low per capita production of CO_2.

Explain how temperatures have increased since records began in 1880.

Direct measurements of land and sea temperatures have taken place since 1880. Averaged across the globe, mean annual temperatures have increased by 0.8°C. However, some regions have become a bit cooler while other regions have become much warmer with increases up to 4°C.

Discuss how we estimate temperatures and greenhouse gases over the past 500,000 years and into the future.

To estimate changes in global temperatures over 500,000 years, we can use changes in the species composition of foraminifera that are found in ocean sediments. We can also examine the ratio of ^{16}O and ^{18}O atoms in the air bubbles that are preserved in ancient ice. These air bubbles can also be used to determine the concentrations of various greenhouse gases from different time periods in the past. We can predict future climate changes using climate models designed to understand processes that affect climate, such as air and ocean temperatures, CO_2 concentration, extent of vegetation, and sea ice coverage at the poles. The models use estimates about how changes in these factors will change in the future to predict how climates will change.

Explain the role of feedbacks on the impacts of climate change.

Feedbacks in the environment can be either positive or negative. Positive feedbacks, such as global warming that causes higher rates of soil decomposition, can amplify the effects of global warming. Negative feedbacks, such as plants responding to increased CO_2 concentrations, can reduce the effects of global warming.

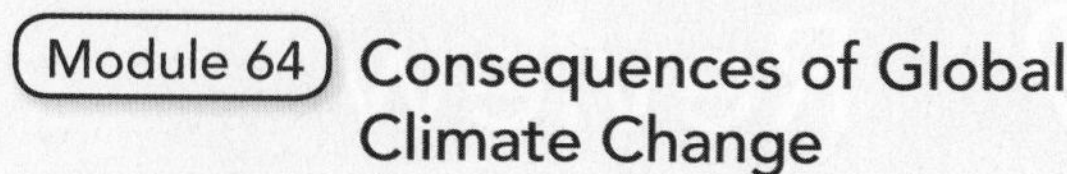

Module 64 Consequences of Global Climate Change

Discuss how global climate change has affected the environment.

Global warming has already caused a decline in the ice mass of the Arctic polar ice cap and in Antarctica and Greenland. It has also caused a decline in glaciers around the world, an increase in the thawing of permafrost, and an increase in sea level.

Explain how global climate change has affected organisms.

Warmer global temperatures have caused longer growing seasons. Warmer temperatures have also caused many species of plants to flower earlier, and they have changed the times when animals breed and insects emerge.

Identify the future changes predicted to occur with global climate change.

Future global changes include longer periods of cold and warm temperatures, more intense storms, changes in global precipitation patterns, and the alteration of ocean currents. These changes may not only affect wild plants and animals but also may influence where humans can live and how humans are affected by diseases.

Explain the global climate change goals of the Kyoto Protocol.

The Kyoto Protocol is an international agreement to reduce the concentrations of greenhouse gases by 5.2 percent below 1990 levels by 2012. Different developed countries agreed to different emission limits and the protocol did not impose emission limits for developing countries. Today, some developed countries have accomplished substantial reductions in the emissions whereas others have not. Collectively, the global goal has not been met.

Practice Math and Graphing

Preparing for the AP® Exam

Answer the following questions. Be sure to show all your work.

1. Practice Math

Figure 63.2 shows the current amount of CO_2 emissions for the top 10 countries. In the table shown, you can see the CO_2 emissions in 2015, with a comparison to emissions for each country in 1990.

(a) Based on these data, calculate the percent change in CO_2 emissions from 1990 to 2015. Round to the nearest whole number.

(b) Which country has increased its emissions by the greatest percentage over the 25-year period and which country has decreased its emissions by the greatest percentage?

Country	CO_2 emissions in 2015 (billions of metric tons)	CO_2 emissions in 1990 (billions of metric tons)	Percent change
China	9.1	2.1	
United States	5.0	4.8	
India	2.1	0.5	
Russian	1.5	2.2	
Japan	1.1	1.0	
Germany	0.9	0.9	
South Korea	0.6	0.2	
Iran	0.6	0.2	
Canada	0.5	0.4	
Saudi Arabia	0.5	0.2	

Data from: International Energy Agency. 2017. CO_2 emissions from fuel combustion.

2. Practice Graphing

Using the percentage data that you calculated in "Practice Math", create a bar graph that shows how each country has altered its CO_2 emissions in 2015 as a percentage of its emissions in 1990.

Chapter 19 AP® Environmental Science Practice Exam

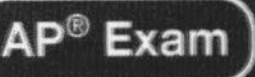

Section 1: Multiple-Choice Questions

Choose the best answer for questions 1–17.

1. Which activity causes a cooling of Earth?
   (a) volcanic eruptions
   (b) emissions of anthropogenic greenhouse gases
   (c) evaporation of water vapor
   (d) combustion of fossil fuels

2. In regard to the greenhouse effect, which statement is NOT true?
   (a) Ultraviolet and visible radiation are converted to infrared radiation at the surface of Earth.
   (b) Infrared radiation is absorbed by greenhouse gases.
   (c) Greenhouse gases were not historically present in the atmosphere.
   (d) Ultraviolet radiation is absorbed by ozone.

3. Which is NOT a greenhouse gas?
   (a) carbon dioxide
   (b) methane
   (c) nitrous oxide
   (d) nitrogen

4. Which factor is important when considering the effect of a greenhouse gas on global warming?
   I. the amount of infrared radiation the gas can absorb
   II. how long the gas remains in the atmosphere
   III. the concentration of the gas in the atmosphere

   (a) I
   (b) I and III
   (c) II and III
   (d) I, II, and III

5. Which greenhouse gas is NOT correctly paired with one of its sources?
   (a) nitrous oxide: landfills
   (b) water vapor: evaporation
   (c) nitrous oxide: automobiles
   (d) CO_2: deforestation

6. Carbon sequestration
   (a) is a process to remove CO_2 from the atmosphere.
   (b) is a method for preventing carbon emissions from landfills.
   (c) is a method for emissions reduction that focuses on improved efficiency.
   (d) is the release of carbon from soils due to warming.

7. Which statement about global warming is TRUE?
   (a) The planet is not warming.
   (b) The planet is warming, but humans have not played a role.
   (c) The planet has had many periods of warming and cooling in the past.
   (d) Greenhouse gases compose only a small fraction of the atmosphere, so they cannot be important in causing global warming.

8. Which sources of data have been used to assess changes in global CO_2 and temperature?
   I. air bubbles in ice cores from glaciers
   II. thermometers placed around the globe
   III. CO_2 sensors placed around the globe
   (a) I and II
   (b) I and III
   (c) II and III
   (d) I, II, and III

9. Which is a true statement about feedback loops that occur with climate change?
   (a) All feedback loops are positive.
   (b) All feedback loops are negative.
   (c) Increased soil decomposition under warmer temperatures represents a positive feedback loop.
   (d) Increased plant growth under higher CO_2 concentrations represents a positive feedback loop.

10. Which predicted consequence of global warming has not yet occurred?
    (a) melting ice caps
    (b) rising sea levels
    (c) shutting down the thermohaline circulation of the ocean
    (d) altered breeding times and flowering times of animals and plants

11. Which statement regarding the Kyoto Protocol is TRUE?
    (a) All nations agreed to reduce their emission of greenhouse gases.
    (b) All nations agreed to stop emitting greenhouse gases.
    (c) Developed nations agreed to different levels of emission reductions.
    (d) Developing nations agreed to reduce their emission of greenhouse gases.

12. Which is NOT true regarding greenhouse gases?
    (a) Greenhouse gases with high global warming potential persist in the atmosphere for tens to hundreds of years.
    (b) Greenhouse gases absorb infrared radiation reflected by Earth's surface.
    (c) Greenhouse gases with the highest warming potential include compounds derived from both natural and anthropogenic sources.
    (d) Some greenhouse gases filter out harmful ultraviolet radiation.

13. Which phenomenon is part of a negative feedback with regard to global warming?
    (a) increased absorption of atmospheric CO_2 by the ocean
    (b) methane production from peatland decomposition
    (c) increased evapotranspiration from plant growth
    (d) CO_2 release from forest fires during dry spells

14. Which statement is NOT true regarding U.S. carbon emission output and politics?
    (a) The United States was the only developed nation that did not sign the Kyoto protocol.
    (b) Total carbon emissions in the United States have increased over the past 2 decades.
    (c) A switch from coal to natural gas has led to a major decrease in carbon emissions.
    (d) The U.S. EPA is legally required to regulate greenhouse gases.

15. Which country is among the top three emitters of CO_2 per capita?
   I. Canada
   II. India
   III. United States
   (a) I only
   (b) II only
   (c) III only
   (d) I and III

16. Human activities affect the concentrations of all the following greenhouse gases EXCEPT
   (a) carbon dioxide.
   (b) methane.
   (c) water vapor.
   (d) nitrous oxide.

17. Which is NOT a natural source of greenhouse gas?
   (a) eutrophication
   (b) volcanic eruption
   (c) evapotranspiration
   (d) termites

Section 2: Free-Response Questions

Write your answer to each part clearly. Support your answers with relevant information and examples. Where calculations are required, show your work.

1. During a debate on climate change legislation in 2018, a U.S. congressman declared that human-induced global warming was a "hoax" and that "there is no scientific consensus."
   (a) If you were a member of Congress, what points might you raise in the debate to demonstrate that global warming is real? (4 points)
   (b) What points might you raise to demonstrate that global warming has been influenced by humans? (4 points)
   (c) What are some human health and economic effects that could occur because of global warming? (2 points)

2. Given what you have learned about global warming and global climate change:
   (a) what actions might you propose in the United States to reduce CO_2 emissions? (3 points)
   (b) what actions might you propose in the United States to reduce methane emissions? (3 points)
   (c) what actions might you propose in the United States to reduce nitrous oxide emissions? (3 points)
   (d) what evidence have scientists used to support the assertion that global warming is happening? (1 point)

3. Although there are natural sources of greenhouses gases, human activity is exponentially increasing the concentrations of many, resulting in the greenhouse effect and global climate change. You are discussing this with your aunt, who is curious about the evidence that human-induced climate change is occurring.
   (a) Your aunt observes that the past winter was particularly cold, and suggests that this indicates climate change is not occurring. Explain the difference between climate and weather, and why a single cold winter is not evidence that climate change is not happening. (3 points)
   (b) Explain how the greenhouse effect works. (3 points)
   (c) Describe TWO methods scientists have used to determine climate conditions in the past. (4 points)

Poverty and environmental degradation are evident along the border between Mexico and the United States, as seen here in Ciudad Juárez, Mexico. *(Julio Etchart/Alamy)*

Sustainability, Economics, and Equity

CHAPTER 20

MODULE 65 Sustainability and Economics

MODULE 66 Regulations and Equity

CASE STUDY

Assembly Plants, Free Trade, and Sustainable Systems

The United States–Mexico border has been a zone of tension between economics and the environment for quite some time. While people in Mexico call the border with the United States la línea, or "the line," the border, which stretches nearly 3,200 km (2,000 miles), is more a network of passageways than a division. Trade among people and cultures across this international boundary affects the economy and environment of both countries. The maquiladora, or assembly plant, industry strongly connects the Mexican and American economies with many economic, social, and environmental implications.

Established decades ago, the maquiladora industry allows international companies to import materials and equipment free of tariffs to Mexico and then to export the finished product to markets in other countries. In 1994, the United States, Canada, and Mexico passed the North American Free Trade Agreement (NAFTA), which was intended to increase trade among the three countries by further reducing tariffs and other taxes as well as regulations. After NAFTA and the subsequent reduction of tariffs, the use of maquiladoras increased significantly, with export of assembled products tripling between 1995 and 2000. Twenty years later, the city of Juarez, in the Mexican state of Chihuahua, south of El Paso, Texas, has 400 factories that employ over 300,000 people. Maquiladoras, which export 90 percent of their products to the United States, comprise 80 percent of the economy in the northern border region and a sizeable portion of Mexico's total GDP. While the additional employment has been welcome in this economically depressed area, there have been many negative consequences as well, including industrial pollution, unsafe working conditions, numerous workplace accidents, and discrimination. In addition, the maquiladoras raise questions of social justice because in some cases the profit is sent to other countries.

If we could give equal attention to economic profit, environmental integrity, and human welfare, could we ultimately create more sustainable development?

In terms of the environment, maquiladora operations often contaminate the border region with toxic industrial waste. Environmental regulations are lenient or nonexistent, and the majority of companies do not comply with mandates that maquiladora waste be shipped to the company's home country. Disposal of toxic chemicals and heavy metals into the local environment causes groundwater and surface water pollution and significant harm to human health. Many maquiladora employees are women of reproductive age, a population that is particularly vulnerable to toxic chemicals.

In addition to pollution from the manufacturing processes, an increase in the human population in maquiladora areas has added greatly to other environmental problems. Many municipalities in which maquiladoras are situated do not have sewage treatment facilities or trash collection capabilities. The solid waste pollutes water sources, and seasonal floods spread garbage throughout the areas.

Social abuses also occur in this system. Employers often test women for pregnancy before they are hired, and those who become pregnant may be illegally fired. Managers employ underage workers. Factory conditions are hazardous, and employees are often unaware of risks because of the lack of "right to know" laws and an absence of warning signs in Spanish. Companies exploit the poverty of the region by offering wages that barely support employee needs. An average maquiladora worker earns the equivalent of a few dollars per day, and these wages have remained stagnant for years even as living costs have risen.

Sometimes the profits from these factories do not enter the Mexican economy, but rather go to the home countries of the companies that run the plants. Many observers believe that northern Mexico pays the social and environmental costs for the maquiladora industry, while foreign corporations reap the benefits.

Free trade and globalization agreements like NAFTA are designed to enhance developing economies by facilitating international business. However, in northern Mexico, increased free trade has stimulated an industry that in some cases may sacrifice social well-being and environmental health. Nevertheless, because many people are employed in the maquiladora industry, money does enter the local economy and helps individuals. Environmental scientists interested in human social welfare and the well-being of the environment look at situations such as these and ask: If we could give equal attention to economic profit, environmental integrity, and human welfare, could we ultimately create more sustainable development?

Sources: J. Carrillo and R. Zarate, The evolution of maquiladora best practices: 1965–2008, *Journal of Business Ethics* (2009) 88: 335–348; J. C. Castillo and G. de Vries, The domestic content of Mexico's maquiladora exports: A long-run perspective, *Journal of International Trade and Economic Development* (2018) 27: 200–219.

Throughout this book, we have seen that economic development, social justice, and sustainable environmental practices are often in conflict. In recent years, environmental scientists have begun to address these relationships by using the tools from economics and other fields to help find ways in which we can achieve a sustainable, equitable, and prosperous existence for all inhabitants on Earth. However, it is difficult to expect people to be concerned about the welfare of the planet on which they live if they have not met their own basic needs of water, food, health, and housing. Today, it is common to observe that the quality of the environment and the quality of human life are linked. In this chapter, we will begin to explore some of these connections.

MODULE 65

Sustainability and Economics

Learning Goals

After reading this module you should be able to

- explain why efforts to achieve sustainability must consider both sound environmental science and economic analysis.
- describe how economic health depends on the availability of natural capital and basic human welfare.

Sustainability is a relatively new and evolving concept in contemporary environmental science. We have seen that something is sustainable when it meets the needs of the present generation without compromising the ability of future generations to meet their own needs. Although human needs can be defined in various ways, for our purposes, we identify the basic necessities as access to food; water; shelter; education; and a healthy, disease-free existence. In order for these five necessities to be available, there must be functioning environmental systems that provide us with breathable air; drinkable water; and productive land for growing food, fiber, and other raw materials—the ecosystem services that we have described in this book.

The quest to obtain resources and increase **well-being**—the status of being healthy, happy, and prosperous—has caused individuals and nations to exploit and degrade natural resources such as air, land, water, wildlife, minerals, and even entire ecosystems. To address questions of sustainability, we need to be able to understand where human well-being and the condition of environmental systems are in conflict. To do this, we will consider economic analysis, ecological economics and ecosystem services, and the role of regulatory agencies in bringing about environmental regulation and protection.

Achieving sustainability requires both sound environmental science and economic analysis

Researchers and policy makers have experimented with a variety of techniques to encourage consumers to change their behavior in ways that would benefit the environment. We explored some of these techniques in Chapter 10 where we discussed externalities and in Chapter 15 where we discussed the buying and selling of air pollution allowances as well as charging a fee or tax for the use of certain resources or for the emission of certain pollutants. **Economics** is the study of how humans allocate scarce resources in the production, distribution, and consumption of goods and services.

Throughout this text, we have already applied many concepts from the field of economics. When we looked at the problem of externalities and pollution, we were using economic theory. Life-cycle analysis is very similar to the cost-benefit analysis that economic policy makers use. In this section, we will look at some basic economic concepts and learn how they can be applied to environmental issues.

Supply, Demand, and the Market

In today's world, most economies are market economies. In the simplest sense, a market occurs wherever people engage in trade. In a market economy, the cost of a good is determined by supply and demand. When a good is in great demand and wanted by many people, producers are typically unable to provide an unlimited supply of that good. Price is the method that producers

Well-being The status of being healthy, happy, and prosperous.

Economics The study of how humans allocate scarce resources in the production, distribution, and consumption of goods and services.

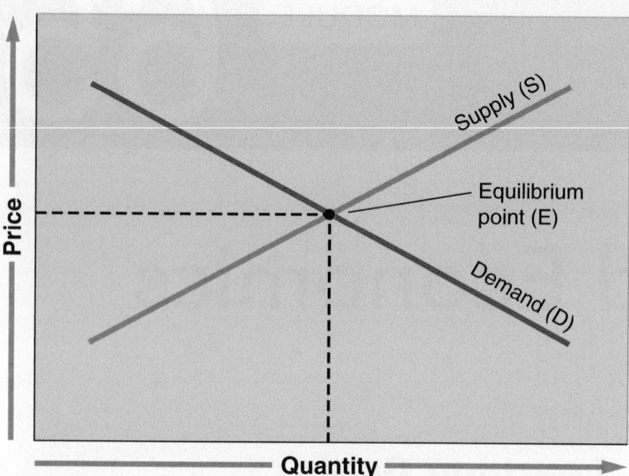

FIGURE 65.1 Supply and demand. A manufacturer will supply a certain number of units of an item based on the revenue that will be received. A consumer will demand a certain number of units of that item based on the price paid. The intersection of the supply and demand curves determines the market equilibrium point for that item.

and consumers use to communicate the value of an item and to allocate the scarce item.

The graph shown in **FIGURE 65.1** illustrates the relationship between supply, demand, and price. The supply curve (S) shows how many units that suppliers of a given product or service—for example, T-shirts—are willing to provide at a particular price. Factors that influence the supply of a good include input prices (the cost of the resources used to produce the item), technology, expectations about future prices, and the number of people selling the product. For example, if you are the only person selling T-shirts and many people want them, you will be willing to make the investment required to produce many T-shirts. However, if a new T-shirt seller comes along, because you will be concerned that you will not sell as many, you will decrease your production because you now must share the market with another supplier.

The demand curve (D) shows how much of a particular good consumers want to buy. Factors that influence demand include income, prices of related goods, tastes, expectations, and the number of people who want the good. For example, if your boss gives you a raise, you may feel like you can afford that T-shirt you have been wanting to buy.

Notice that the demand curve slopes downward. In other words, as the price of T-shirts rises, the demand for them declines. This illustrates the law of demand, which states that when the price of a good rises, the quantity demanded falls and when the price falls, the quantity demanded rises. Conversely, the supply curve slopes upward. This reflects the law of supply, which states that when the price of a good rises, the quantity supplied of that good will rise and when the price of a good falls, the quantity supplied will fall.

The laws of supply and demand make intuitive sense. After all, if you are selling T-shirts and you find that your profits have shrunk, you are more likely to use your resources to produce and sell something more popular, and more profitable. If you are a consumer of T-shirts, the less expensive they are, the more you are inclined to buy.

With these different interests, how do demand and supply ever meet? In a market system, without any restrictions such as taxes or other regulations, the price of a good will come to an equilibrium point (E) where the two curves on the graph intersect. Here the quantity demanded and the quantity supplied are exactly equal. At this price, suppliers find it worthwhile to supply exactly as many T-shirts as consumers are willing to buy.

Unfortunately, markets—composed of many buyers and sellers—do not always take all costs of production into account. We have already seen that this is the case in situations of land degradation where people, organizations, or even governments deplete or damage a natural resource because they do not bear any direct costs for doing so. As we saw in Chapter 10, the cost of using a resource that is not included in the purchase price is called an externality. When we pollute air or water without directly paying for it, that is also an externality. When we account for the costs of externalities created by manufacturing a good or offering a service, the price changes. This, in turn, affects demand and supply.

Let's look at the example of coal. The dollar cost of coal-generated electricity includes the cost of the coal, the cost of paying people to operate the power plant, and the cost of electricity distribution to customers. However, the cost to the environment of emitting sulfur dioxide, carbon dioxide, and other waste products, all of which are negative externalities, is largely missing from the price customers pay. We know that these negative externalities certainly add costs, both financially and in terms of the well-being of people living downwind from the power plant. For example, someone with a respiratory ailment could incur greater medical expenses because of increased sulfur dioxide and particulates in the air. There may be provisions requiring polluters to pay some of the costs related to these emissions, but often these payments are not sufficient to cover the total cost of the pollution. In addition, they often do not reach the affected individuals or groups.

If the dollar cost of a good included externalities such as the expenses incurred by emitting pollutants into the air, or the expenses related to removing the pollutants before they were emitted, then the cost for most items produced would be greater. This could only occur if a tax were imposed by a regulatory agency. When the cost of production rises due to this tax, the supply curve shifts to the left, from S to S_1 as shown

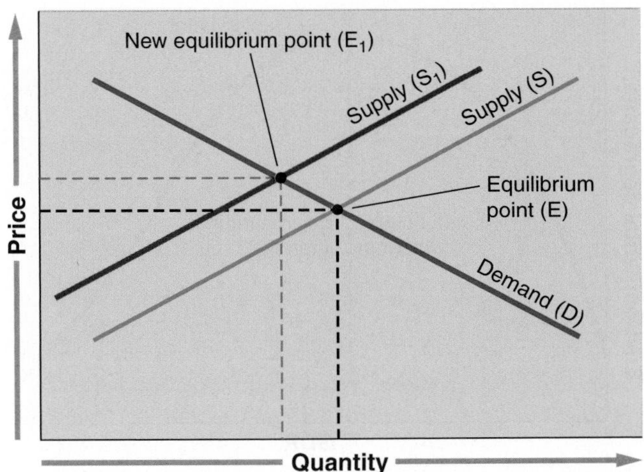

FIGURE 65.2 Supply and demand with externalities. When the cost of emitting pollutants is included in the price of a good, for any given quantity of items, the price increases. This causes the supply curve to shift to the left, from S to S_1. Since the law of demand states that when the price of a good goes up, demand falls, the amount demanded falls, and the market reaches a new equilibrium, E_1.

in **FIGURE 65.2**. The new market equilibrium (E_1) is at a higher cost and, as a result, fewer items are manufactured and purchased. In other words, including the externalities raises the price and lowers the demand. Therefore the price that includes externalities is more reflective of the true cost of the item.

Measuring Wealth and Productivity

There are a variety of methods for measuring wealth and productivity. While there is no consensus on which method is most accurate and each has shortcomings, researchers do agree that measuring wealth and productivity can be a useful way of examining the health of an economy. In this section, we will look at the most common and widely accepted measure of wealth and productivity and then look at alternatives.

GDP

Economists use different national economic measurements to gauge the economic wealth of a country in terms of its productivity and consumption. Most of them do not take externalities into account. The most common of these measurements is gross domestic product (GDP), which refers to the value of all products and services produced in a year in a given country. GDP includes four types of spending: consumer spending, investments, government spending, and exports minus imports. As a measure of well-being, GDP has been criticized for a number of reasons. Because costs for health care contribute to a higher GDP, a society that has a great deal of illness would have a higher GDP than an equivalent society without a great deal of illness. Such an inclusion does not appear to be an accurate reflection of the "wealth" or "well-being" of a society. And because externalities such as pollution and land degradation are not included in GDP, measurement of GDP does not reflect the true cost of production. The Do The Math Box on page 726 asks you to calculate the percentage increase in GDP in Finland over a 70-year period.

AP® Exam Tip

Be able to list the factors that contribute to a country's GDP and describe them. You should also be able to identify factors affecting the economy that are not included in this economic measure. ●

Some social scientists maintain that the best way to improve the global environment is to increase the GDP in the less developed world. In Chapter 7, we examined the relationship between rising income and falling birth rates; as GDP increases, population growth slows. This, in turn, should lead to a reduction in anthropogenic environmental degradation. Wealthier, developed countries are able to purchase goods and services that will lead to environmental improvements—for example, pollution control devices like catalytic converters—and to use their resources more efficiently. On the other hand, as we have seen, developed countries use many more resources than developing countries, which leads to more environmental degradation.

The GPI

We have seen that GDP is an incomplete measurement of the economic status of a country because it only considers production. Some researchers attempt to address this shortcoming by using another measurement that is known as the *genuine progress indicator*. The **genuine progress indicator (GPI)** is a measure of economic status that includes personal consumption, income distribution, levels of higher education, resource depletion, pollution, and the health of the population. The GPI calculations for the United States have not been updated recently but the trends are believed to be roughly similar to those found in Finland. As shown in **FIGURE 65.3** for the country of Finland, while GDP rose steadily from 1945 through 2010, GPI reached a maximum in the early 1990s and has decreased since then. A number of countries, including England, Germany, and Sweden, have also recalculated their GDP using the GPI. They too have found that their

Genuine progress indicator (GPI) A measure of economic status that includes personal consumption, income distribution, levels of higher education, resource depletion, pollution, and the health of the population.

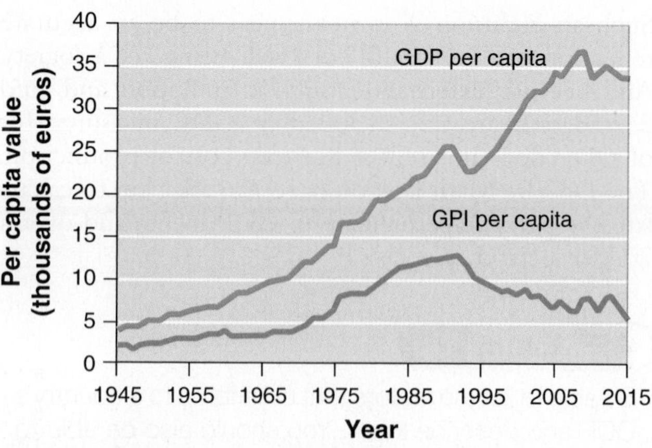

FIGURE 65.3 Genuine progress indicator versus gross domestic product, per capita, for Finland from 1945 to 2015. While gross domestic product measures the value of all products and services a country produces, the genuine progress indicator attempts to include the level of education, personal consumption, income distribution, resource depletion, pollution, and the health of the population. *(Data from UN Conference on Trade and Development. http://stats.unctad.org/Dgff2016/partnership/goal17/target_17_19.html)*

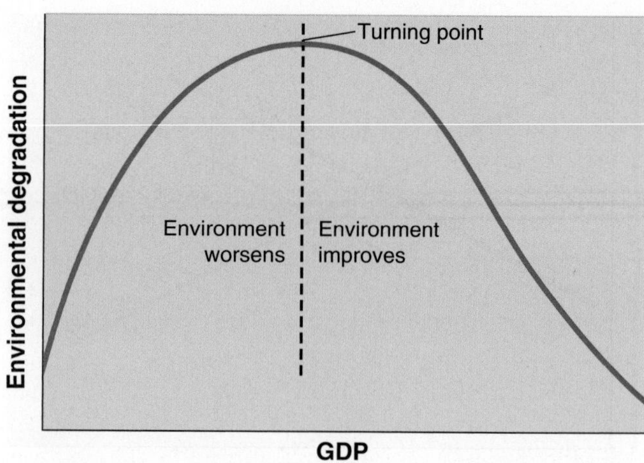

FIGURE 65.4 The Kuznets curve. This model suggests that as per capita income in a country increases, environmental degradation first increases and then decreases. In many respects, China is on the first part of this curve while the United States is on the second part of the curve.

overall wealth, when human and environmental welfare are included, has remained constant or declined over the last 3 decades. Although the parameters used to calculate GPI have been repeatedly modified by researchers, the general trends appear to be the same for all countries measured: While GDP continues to rise with time, GPI flattens out or decreases. "Do The Math: Calculating Percentage Increase in GDP" shows you how to calculate the percentage increase and change in GDP over time.

The Kuznets Curve

To address some of the shortcomings of GDP as a measurement of wealth, some environmental economists and scientists advocate using a model known as the Kuznets curve. The Kuznets curve, shown in **FIGURE 65.4**, suggests that as per capita income in a country increases, environmental degradation first increases and then decreases. The model is controversial because it is not easily applicable to all situations. For example, despite the increasing affluence of developed countries, carbon dioxide emissions and municipal solid waste (MSW) generation have both continued to increase. It is possible that these developed countries are not yet wealthy enough to deal with these problems effectively, but it is also possible that there are certain problems that cannot be solved simply with greater wealth. For example, as countries become wealthier, residents tend to use more fossil fuel for travel, consume more resources, and generate more waste.

DO THE MATH Calculating Percentage Increase in GDP

Preparing for the AP® Exam

As you can see in Figure 65.3, the GDP in Finland was €5,000 in 1945 and €35,000 in 2015. Calculate the percentage increase in GDP over that time period.

$$€35{,}000 - €5{,}000 = €30{,}000$$

€30,000 divided by the original value €5,000 =

$$€30{,}000 \div €5{,}000 = 6 \times 100\% = 600\%$$

There was a 600 percent increase in GDP in Finland between 1945 and 2015.

YOUR TURN Calculate the percentage change in GPI in Finland from 1945 to 2015.

Sometimes less developed countries experience technological leaps without going through each phase of technological development. These kinds of changes may influence the shape of the Kuznets curve or influence how well it characterizes a given situation. **Technology transfer** happens when less developed countries adopt technological innovations that were developed in wealthy countries. For example, in many less developed countries, a significant proportion of the population uses cell phones without ever having had access to a network of landline telephones. A situation in which less developed countries use new technology without first using the precursor technology is known as **leapfrogging.** Leapfrogging occurs whenever new technology develops in a way that makes the older technology unnecessary or obsolete. This allows the developing nations to take advantage of the expensive research, development, and experience of the more-developed nations.

Photovoltaic (PV) solar electricity is a particularly good example of leapfrogging. In industrialized nations, until recently, solar electricity has not been cost-competitive with gas- or coal-generated electricity. However, it has been very successful for a number of years in nations in Africa, Asia, and South America that lack the resources to build a reliable electrical distribution grid. Solar p.v. electricity is a small-scale energy source not dependent on outside connections to an electrical grid (see Chapter 13). In fact, it is possible that many less developed countries will continue to increase their use of solar PV and skip the step of building a nationwide electrical grid, much like what has happened with cell phones versus landlines for telephone service. Solar PV allows developing countries to produce and distribute their own electricity without investment in the massive infrastructure of an electrical distribution grid that would be needed in a developed country (**FIGURE 65.5**).

FIGURE 65.5 Solar panels in Africa. In areas where the electrical grid is not established and electrical submission lines are not present, the installation and use of photovoltaic solar cells may be less expensive and less environmentally disruptive than a traditional electrical infrastructure. *(Joerg Boethling/Alamy)*

Economic health depends on the availability of natural capital and basic human welfare

Capital, or the totality of our economic assets, is typically divided into three categories: natural, human, and manufactured. **Natural capital** refers to the resources of the planet, such as air, water, and minerals. **Human capital** refers to human knowledge and abilities. **Manufactured capital** refers to all goods and infrastructure that humans produce. While economists usually base their assessment of national wealth on productivity and consumption, environmental scientists point out that all economic systems require a foundation of natural capital. Without natural capital, humans would not be able to produce very much and would probably not survive.

Environmental and Ecological Economics

Some advocates of a purely free-market system believe that as long as market forces are left alone, human work and creativity will find solutions to problems of natural resource degradation and depletion. But as we have seen, externalities are not assessed appropriately if the cost of environmental degradation is not charged to the individuals responsible for that degradation. A **market failure** occurs when the economic system does not account for all costs. Among those economic thinkers who have sought ways to respond to market failures, many have become part of the discussion in the fields of *environmental economics* and *ecological economics*. **Environmental economics** is a subfield of economics that examines the costs and

Technology transfer The phenomenon of less developed countries adopting technological innovations developed in wealthy countries.

Leapfrogging The phenomenon of less developed countries using new technology without first using the precursor technology.

Natural capital The resources of the planet, such as air, water, and minerals.

Human capital Human knowledge and abilities.

Manufactured capital All goods and infrastructure that humans produce.

Market failure When the economic system does not account for all costs.

Environmental economics A subfield of economics that examines the costs and benefits of various policies and regulations that seek to regulate or limit air and water pollution and other causes of environmental degradation.

benefits of various policies and regulations that seek to regulate or limit air and water pollution and other causes of environmental degradation. **Ecological economics** is the study of economics as a component of ecological systems rather than as a distinctly separate field of study. Ecological economics is a method of understanding and managing the economy as a subsystem of both natural and human systems.

Environmental and ecological economists attempt to assign monetary value to intangible benefits and natural capital, a practice known as **valuation.** For example, they have developed methods for assessing the monetary value of a pristine nature preserve, a spotted owl, or a scenic view. One method is to calculate the revenue generated by people who pay for the benefit—for example, the amount tourists pay to visit a nature preserve would represent the dollar value of the preserve. Another method is to use surveys. They might ask a number of people how much they are willing to pay just to know that spotted owls exist, even if they are unlikely ever to see one. The most extensive assessments have attempted to determine the value of ecosystem services such as oxygen that plants produce or pollination that insects do. Although estimates vary, global ecosystem services might have a dollar value of approximately $125 trillion per year. The multi-organization Millennium Ecosystem Assessment categorized the variety of services that ecosystems provide for the benefit of humans. In many cases, it is possible to estimate the cost of a particular service if it were provided using technology rather than naturally. For example, as we discussed in Chapter 3, New York City could have constructed a massive water purification system at a known cost. Instead, it chose to protect watersheds in the Catskill Mountains, a region north of New York City that supplies the city with water, so the water would not need expensive purification. Accordingly, the ecosystem service of water purification has a known value that can be used to help calculate the total dollar value of ecosystem services.

Given all of the natural capital and ecological services distributed around the world, it is quite likely that human activities will generate multiple negative externalities. Economic tools can be used to incorporate the dollar cost of the externalities in the price of goods and services. We have seen examples of this with our discussions of charging for allowances to allow sulfur dioxide and carbon dioxide emissions. These economic tools can be used in many other ways as well. Typically, a tax or regulation calls for reducing externalities through a market-driven system. This system calls for the incorporation of negative environmental impacts of a commodity or service in its cost of production. For example, a car manufacturer would include in the cost of production for each car not only the cost of labor and natural resources, such as steel and water, but also the cost of the air pollution caused by the manufacturing process. Viewed this way, the cost of production of a car will immediately increase. Typically, the manufacturer would want to pass along at least part of this additional cost to the consumer by raising the price of the car. Calculating the full costs of a commodity or service by internalizing externalities will likely cause consumers to buy fewer items with high negative impacts because those impacts will be reflected in higher prices. The most obvious way for the costs of externalities to be included is by requiring the producer to pay them. This could be achieved through regulation, imposition of a tax, or some sort of public action mandating reparation for externalities or making it difficult for the company to produce its product in a way that pollutes. Much of the debate in environmental and ecological economics revolves around how best to impose the dollar cost on the producer.

Ecological economics The study of economics as a component of ecological systems.

Valuation The practice of assigning monetary value to intangible benefits and natural capital.

Sustainable Economic Systems

Critics of our current economic system maintain that it is based on maximizing the utilization of resources, energy, and human labor. This encourages the extraction of large amounts of natural resources and does not provide incentives that would reduce the amount of waste generated. A system analysis of the current economic situation, shown in **FIGURE 65.6a**, suggests that continuing with such a system is not sustainable. In the current system, large amounts of extracted resources and energy and relatively small amounts of ecosystem services are the inputs, and large amounts of waste are the outputs.

A sustainable economic system, depicted in Figure 65.6b, will rely more on ecosystem services and reuse of existing manufactured materials and less on resource extraction. In this system, there is greater reliance on ecosystem services and less reliance on resource extraction that requires energy. It also takes some of the waste stream and reuses it in the production and consumption cycle, as indicated by the arrow labeled "Waste stream recycling." Therefore, the cycle in Figure 65.6b would use more renewable energy, lessen negative externalities, and reuse more of the products that were destined for the waste stream. This model has led architects, environmental scientists, and engineers to a collaborative discussion of the optimal way to design, manufacture, use, and dispose of objects such as automobiles, houses, and consumer goods. Currently, a consumer purchases an object, such as an automobile or computer, and when that object has reached the end of its useful life, the consumer is responsible for its disposal. As we pointed out in Chapter 16, because the responsibility for the object rests with the consumer, there is no incentive for the manufacturer to make it easy to reuse or recycle the object. As we mentioned in Chapter 16, it is unfortunate that a can of a chemical oven cleaner purchased for $5, if unused, becomes hazardous waste when disposed of and may cost $25 for disposal. This type of situation has led to the cradle-to-grave and cradle-to-cradle analysis described in Chapter 16.

Because the cradle-to-cradle system includes human capital, resource, and energy inputs as well as a redirection of the waste stream, it gains even greater importance when we consider the entire economic system. **FIGURE 65.7** shows an alternative approach to the previous diagram. In this case, while there is some waste disposal, a good fraction of materials that are used up contribute to the raw materials for new items. The ultimate goal is to produce a good that at the end of its useful life—as it approaches its "grave"—can be easily reused to make a new product. That is, most or all of the parts of an old product will become the "cradle"—the beginning of life—for the same or other products. An automobile would be ideal for a cradle-to-cradle system because each individual car contains a ton or more of steel and other metals as well as rubber, plastic, and a host of other materials that can be reused or recycled.

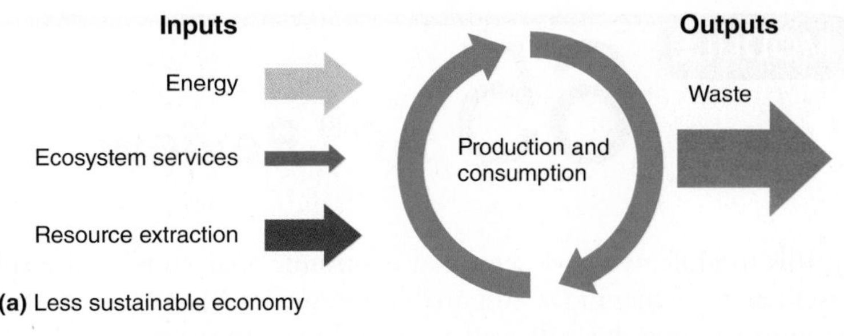

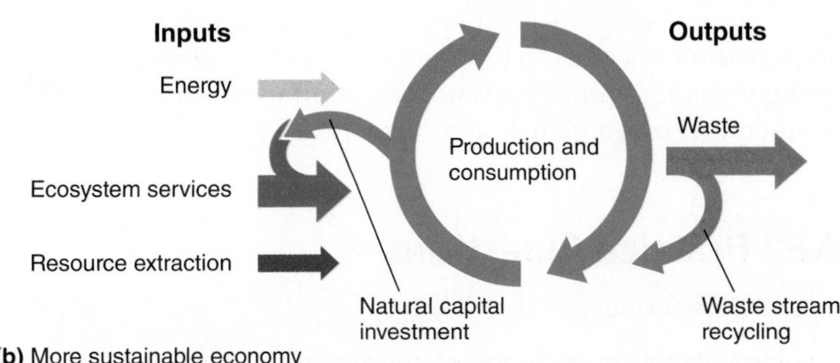

FIGURE 65.6 Systems diagrams of two economic systems. (a) A less sustainable system, like our current economy, is based on maximizing the utilization of resources and results in a fairly large waste stream. (b) A more sustainable system is based on greater use of ecosystem services, less resource extraction, and minimizing the waste stream.

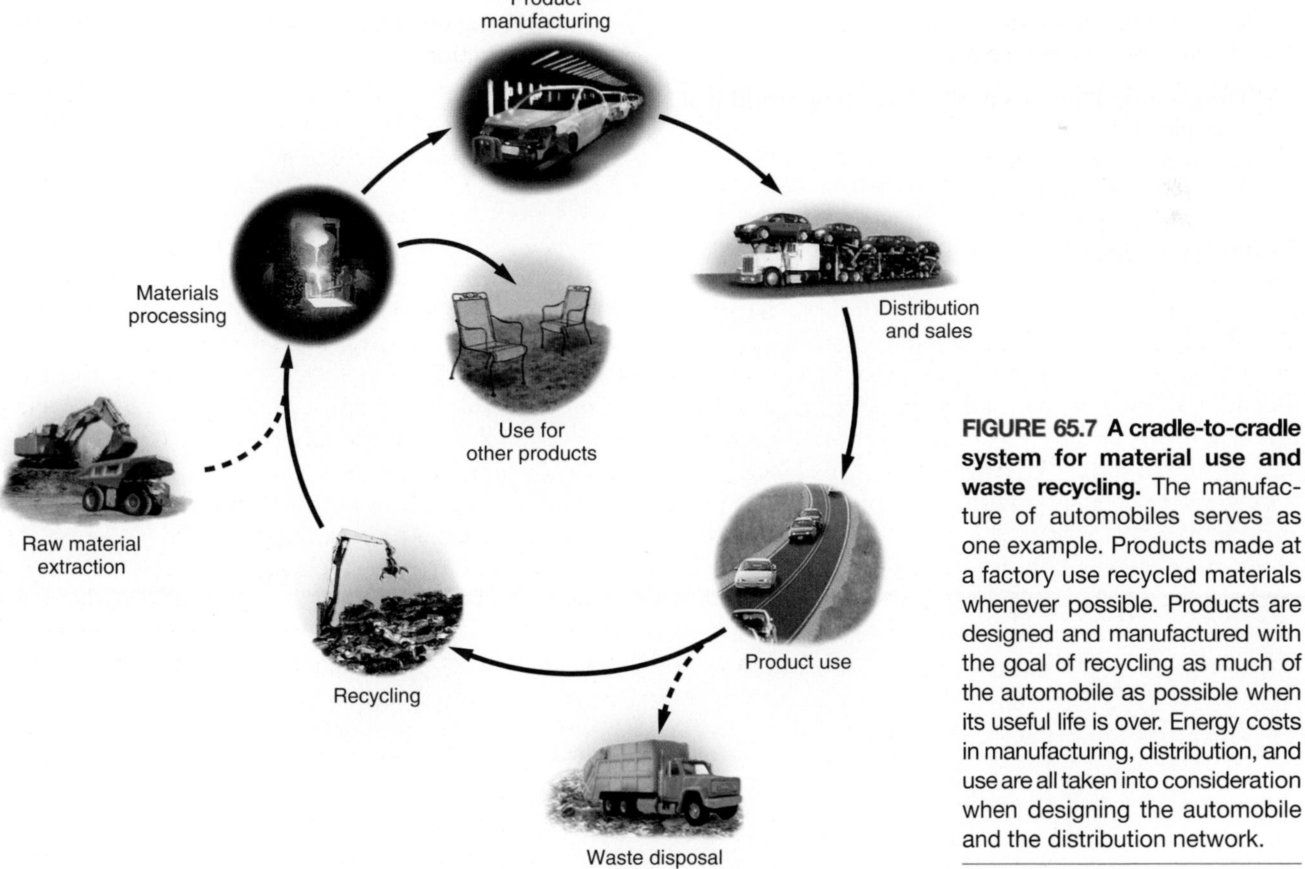

FIGURE 65.7 A cradle-to-cradle system for material use and waste recycling. The manufacture of automobiles serves as one example. Products made at a factory use recycled materials whenever possible. Products are designed and manufactured with the goal of recycling as much of the automobile as possible when its useful life is over. Energy costs in manufacturing, distribution, and use are all taken into consideration when designing the automobile and the distribution network.

MODULE

65 AP® Review

Preparing for the AP® Exam

In this module, we have seen that economic analysis is essential to achieving a sustainable existence. Market economics must be balanced with different measures of wealth and productivity, as well as other measures of the economic status of a country. Innovative techniques are sometimes needed to promote successful economic strategies along with successful protection of the environment. Environmental and ecological economics are subfields in economics that attempt to place value on benefits offered by the natural environment, some of which are hard to quantify. Cradle-to-grave and cradle-to-cradle systems analyses are tools used to merge sustainability with life-cycle analysis. In conjunction with applying economics to sustainability efforts, regulations are sometimes used as well. This is the focus of the first part of our next module.

AP® Practice Questions

Choose the best answer for the following.

1. How might the inclusion of an externality affect the supply and demand of a product?
   (a) It will increase price and decrease quantity demanded.
   (b) It will increase price and increase quantity demanded.
   (c) It will decrease price and increase quantity demanded.
   (d) It will decrease price and decrease quantity demanded.

2. Which is NOT included in the calculation of gross domestic product?
   (a) costs of health care
   (b) government spending
   (c) spending on durable goods
   (d) the costs of externalities

3. The use of cell phones in the developing world is an example of
   (a) an ecosystem service.
   (b) a positive externality.
   (c) natural capital.
   (d) leapfrogging.

4. Human capital includes
   (a) the goods that humans produce.
   (b) services that humans provide.
   (c) human knowledge and skills.
   (d) assets directly related to human survival.

5. Which is NOT an objective of a sustainable economic system?
   (a) giving priority to ecosystem health
   (b) using nonrenewable resources
   (c) placing value on ecosystems
   (d) relying on ecosystem services rather than resource extraction

MODULE 66

Regulations and Equity

We have examined the economic system and the roles of natural capital and human capital. This understanding provides us with the tools we need to evaluate different options for monitoring and managing human systems in a way that will result in the least amount of harm to the natural environment. Regulatory tools are also used to bring about the least environmental harm. Ultimately, the goal is freedom from exposure to environmental harm for all the people in the world. This is the final topic of the book: environmental equity.

Learning Goals

After reading this module you should be able to

- explain the role of agencies and regulations in efforts to protect our natural and human capital.
- describe the approaches to measuring and achieving sustainability.
- discuss the relationship among sustainability, poverty, personal action, and stewardship.

Agencies, laws, and regulations are designed to protect our natural and human capital

Many different techniques and approaches are used to influence how we treat the environment. Sometimes laws and regulations help achieve a certain outcome. Before examining some of the major laws and regulations in the United States and the world, we must familiarize ourselves with an important factor that shapes the way nations approach policy making—how people view the world.

Environmental Worldviews and Regulatory Approaches

We have seen that the approach a nation takes to the regulation of economic activity and the environment depends in part on that nation's stage of development. In addition, worldview and attitude toward risk also shape a nation's approach to economics and the environment.

Worldviews

An **environmental worldview** is a worldview that encompasses how one thinks the world works; how one views his or her role in the world; and what one believes to be proper environmental behavior. Three types of environmental worldviews dominate: human-centered, life-centered, and Earth-centered.

The **anthropocentric worldview** is a worldview that focuses on human welfare and well-being. In other words, nature has an instrumental value to provide for our needs. There are variations on this human-centered worldview. For example, those who favor a free-market approach to economics are optimistic about the results of unlimited competition and minimal government intervention. The planetary management school, while optimistic that we can solve resource depletion issues with technological innovations, believes that nature requires protection and that government intervention is at times necessary to provide this protection. **Stewardship,** a subset of the anthropocentric worldview, supports the careful and responsible management and care for Earth and its resources. The stewardship school of thought considers that while the natural world requires protection, it is also our ethical responsibility to be good managers of Earth.

Environmental worldview A worldview that encompasses how one thinks the world works; how one views one's role in the world; and what one believes to be proper environmental behavior.

Anthropocentric worldview A worldview that focuses on human welfare and well-being.

Stewardship The careful and responsible management and care for Earth and its resources.

The **biocentric worldview** is life-centered and holds that humans are just one of many species on Earth, all of which have equal intrinsic value. At the same time, this worldview considers that the ecosystems in which humans live have an instrumental value. There are various positions within the life-centered approach. While some consider that it is our obligation to protect a species, others consider that it is our obligation to protect every living creature.

The **ecocentric worldview** is Earth-centered. It places equal value on all living organisms and the ecosystems in which they live, and it demands that we consider nature free of any associations with our own existence. This worldview takes various forms. The environmental wisdom school, for example, believes that since resources on Earth are limited, we should adapt our needs to nature rather than adapt nature to our needs. The deep ecology school, meanwhile, insists that humans have no right to interfere with nature and its diversity. Our worldviews determine the decisions we make about our lives, our work, and the way we treat the planet.

Environmental worldviews can play a significant role in the policies a nation considers and how it implements them. For example, a nation or community that operates on an anthropocentric worldview might not concern itself with how economic activity affects the natural environment. A country with an ecocentric worldview might carefully regulate economic activity in order to protect ecosystems and the species within them. In practice, the policies and regulations of most nations represent a variety of worldviews depending on the particular nation and the specific resource or region of the biosphere that is being affected.

The Precautionary Principle

A nation's approach to environmental policy and regulation may also be influenced by whether or not it tends to follow the precautionary principle. In Chapter 17 we discussed the precautionary principle, which states that when the results of an action are uncertain—such as the effects caused by the introduction of a compound or chemical—it is better to choose an alternative known to be harmless. In many situations, scientific uncertainty complicates the estimation of the comparative risks of different actions. This is an important part of environmental decision making. In the United States, environmental law and policy has at times treated scientific uncertainty as a reason to discount or downplay scientific evidence of problems in the environment. Industrial and business groups have also used scientific uncertainty as a reason to avoid implementing expensive measures that would mitigate future environmental harms. Those who favor using the precautionary principle argue that if we wait for widespread scientific consensus about the adverse effects of a particular compound or action, we run the risk of creating an environmentally unsustainable and inequitable future.

Critics of the precautionary principle maintain that economic progress and human well-being will be hindered if we delay the use of a compound or implementation of an action until we verify that it is completely safe for the environment. There may also be an additional economic cost to waiting, or for choosing alternative means of achieving a goal.

For more than 2 decades, the International Union for Conservation of Nature, an organization composed of over 800 government and nongovernmental wildlife organizations, has repeatedly strengthened and reaffirmed the 1992 Convention on Biological Diversity by publishing guidelines that include applying the precautionary principle as a tool in reaching decisions about the sustainable use of plant and animal species. The guidelines emphasize using the "best science available" in deciding whether to list a species as endangered and whether to ban any activity that could jeopardize that species.

The 1987 Montreal Protocol on Substances That Deplete the Ozone Layer is an example of the precautionary principle applied to global change. When the protocol was adopted, there was still some scientific uncertainty about the evidence for the effect of CFCs on ozone depletion. You may recall from Chapter 14 that CFCs are chemicals that were used for refrigeration and other commercial applications. Despite the uncertainty about the effect of CFCs on the atmosphere, the protocol recommended eliminating their use. In this case, economic and political factors were balanced with the scientific findings to reach an agreement that CFCs should be phased out over a period of decades rather than immediately. Part of the success of the Montreal Protocol has been credited to the availability of an affordable and fairly effective replacement for CFCs. It has proven much more difficult to find affordable and effective replacements for the fossil fuels we currently use. Therefore, it is less likely that a similar scenario will unfold with respect to a reduction of greenhouse gases.

The precautionary principle is a relatively new and important part of environmental policy. It does not recommend or require any specific actions, but it does provide a reminder to environmental policy makers and managers that, in many cases, absolute scientific certainty may come too late when dealing with potentially serious environmental harms.

Biocentric worldview A worldview that holds that humans are just one of many species on Earth, all of which have equal intrinsic value.

Ecocentric worldview A worldview that places equal value on all living organisms and the ecosystems in which they live.

World Agencies

By considering the variety of worldviews presented earlier, and to the extent a particular country or agency subscribes to the precautionary principle, we can begin to understand more about the decision-making process that influences the various world agencies that have jurisdiction over global environmental issues. Global, national, or personal situations may prompt key beneficial decisions out of a sense of necessity and urgency. After World War II (1939–1945), leaders of the allied nations founded the **United Nations (UN),** a global institution dedicated to promoting dialogue among countries with the goal of maintaining world peace. When its charter was ratified in 1945, the UN had 51 member countries; by 2011, it had grown to 193, which is the number of member countries today. Since its establishment, the UN has created many internal agencies and institutions. Four of the many UN organizations relating to the environment are *the United Nations Environment Programme*, the *World Bank*, the *World Health Organization*, and the *United Nations Development Programme*.

The United Nations Environment Programme

The **United Nations Environment Programme (UNEP)** is a program of the United Nations responsible for gathering environmental information, conducting research, and assessing environmental problems. Headquartered in Nairobi, Kenya, UNEP is also the international agency responsible for negotiating certain environmental treaties. In particular, the Convention on Biological Diversity, the Convention on International Trade in Endangered Species (CITES), and the Montreal Protocol on Substances that Deplete the Ozone Layer are three important international treaties UNEP negotiated. The Global Environment Outlook (GEO) reports are prepared under the auspices of UNEP.

The World Bank

The **World Bank** is a global institution that provides technical and financial assistance to developing countries with the objectives of reducing poverty and promoting growth, especially in the poorest countries. The World Bank, with headquarters in Washington, D.C., cites four goals for economic development: (1) educating government officials and strengthening governments; (2) creating infrastructure; (3) developing financial systems, from microcredit to much larger systems; and (4) combating corruption. Critics of the World Bank maintain that there is too little consideration of environmental and ecological impacts when projects are evaluated and approved.

The World Health Organization

Headquartered in Geneva, Switzerland, the **World Health Organization (WHO)** is a global institution dedicated to the improvement of human health by monitoring and assessing health trends and providing medical advice to countries (**FIGURE 66.1**). It is the group within the UN responsible for human health, including combating the spread of infectious diseases, such as those that are exacerbated by global climate changes. This organization is also responsible for health issues in crises and emergencies created by storms and other natural disasters. The five key objectives of the WHO are: (1) promoting development, which should lead to improved health of individuals; (2) fostering health security to defend against outbreaks of emerging diseases; (3) strengthening health care systems; (4) coordinating and synthesizing health research, information, and evidence; and (5) enhancing partnerships with other organizations.

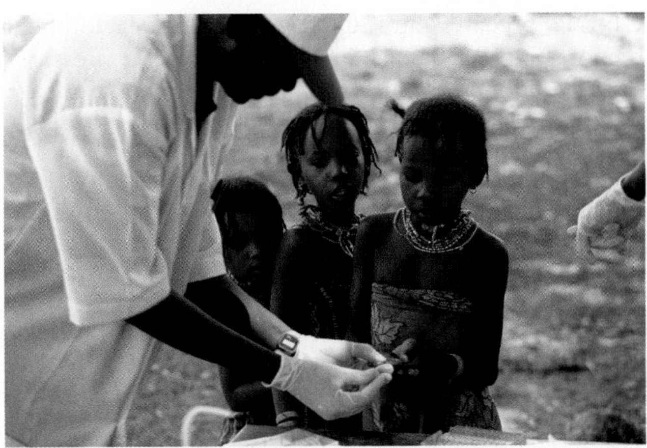

FIGURE 66.1 World Health Organization workers. A WHO worker in Chad draws blood from children for disease testing. *(Patrick Robert/Corbis via Getty Images)*

The United Nations Development Programme

The **United Nations Development Programme (UNDP)** is an international program that operates in 170 countries and territories around the world to advocate change that will help people obtain a better life through development. Headquartered in New York City, UNDP has a primary mission of addressing and facilitating issues of democratic governance, poverty reduction, crisis prevention and recovery, environment and energy issues, and prevention of the spread of HIV/AIDS.

United Nations (UN) A global institution dedicated to promoting dialogue among countries with the goal of maintaining world peace.

United Nations Environment Programme (UNEP) A program of the United Nations responsible for gathering environmental information, conducting research, and assessing environmental problems.

World Bank A global institution that provides technical and financial assistance to developing countries with the objectives of reducing poverty and promoting growth, especially in the poorest countries.

World Health Organization (WHO) A global institution dedicated to the improvement of human health by monitoring and assessing health trends and providing medical advice to countries.

United Nations Development Programme (UNDP) An international program that works in 170 countries around the world to advocate change that will help people obtain a better life through development.

UNDP prepares an annual Human Development Report (HDR) that is an extremely useful measurement tool for the status of the human population.

Other Agencies

There are also a great number of nongovernmental organizations (NGOs) that work on worldwide environmental issues. These include Greenpeace, the International Union for Conservation of Nature, World Wide Fund for Nature (formerly World Wildlife Fund), and Friends of the Earth International.

U.S. Agencies

In January 1969, offshore oil platforms 10 kilometers (6 miles) from Santa Barbara, California, began to leak oil. Roughly 11.4 million liters (3 million gallons) of oil spilled out over the next 11 days and the leak continued throughout the year. This was not the first oil spill during the 1960s, nor the largest, but its proximity to the southern California coast resulted in something new—vast media attention. Daily television news reports of dead seabirds, fish, and marine mammals, as well as large stretches of oil-soaked beaches, shocked the American public and government officials. The Santa Barbara oil spill caused a major shift in federal policy toward incorporating an awareness of how human society affects the environment.

The first Earth Day, April 22, 1970, was partially the result of public reaction to the Santa Barbara oil spill and to other environmental problems that surfaced during the 1960s, such as those documented by Rachel Carson in her book *Silent Spring* (**FIGURE 66.2**). Earth Day 1970 is the symbolic birthday of the modern expression of the view that the natural environment and human society are inextricably connected. It also signals the beginning of contemporary environmental policy. Before 1970, environmental policy focused primarily on biological and physical systems as economic resources for an industrial society. After 1970, sound environmental policy expanded to include the idea that economic benefits must be balanced by environmental science, environmental equity, and intergenerational equity—the interests of future generations in a healthy environment.

AP® Exam Tip

When a free-response question asks you to describe the benefits of a policy, first determine if the question is about ecological, societal, or economic benefits. If a question asks about ecological benefits, the answer should be directed toward factors in the environment such as plants or animals. If a question asks about social benefits, the answer should be directed toward human welfare and communities. If the question asks about economic benefits, the answer should be directed toward economic factors such as income, jobs, and economic policy. ●

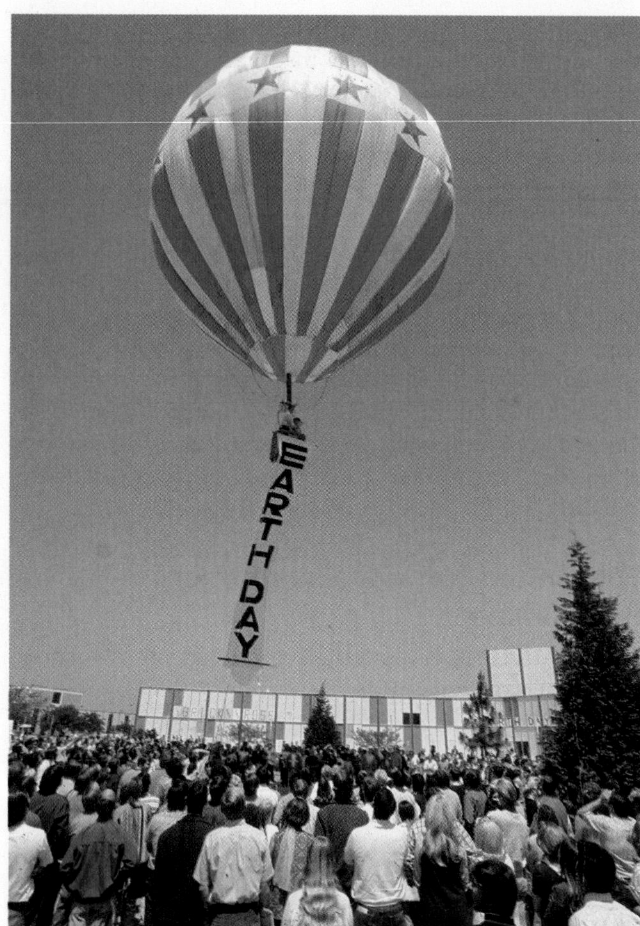

FIGURE 66.2 The first Earth Day, New York City, 1970. Large numbers of people gathered at many locations around the United States on April 22, 1970, to bring attention to the condition of Earth. *(Julian Wasser/The LIFE Images Collection/Getty Images)*

Since the early 1970s, several important U.S. agencies have been created to monitor human impact on the environment as well as to promote environmental and human health.

The Environmental Protection Agency

In 1970, President Richard Nixon signed the bill authorizing the creation of the **Environmental Protection Agency (EPA),** which oversees all governmental efforts related to the environment including science, research, assessment, and education. Headquartered in Washington, D.C., the EPA also writes and develops regulations and works with the Department of Justice, Department of State, and U.S. Native American governments to enforce those regulations.

The Occupational Safety and Health Administration

Also in 1970, President Nixon signed the act creating the **Occupational Safety and Health Administration (OSHA),** an agency of the U.S. Department of Labor that is responsible for the enforcement of health and safety regulations. Its main mission is to prevent injuries, illnesses, and deaths in the workplace. OSHA conducts inspections, workshops, and education efforts to achieve its goals. Limiting exposure to chemicals and pollutants in the workplace is one way that OSHA is involved in environmental protection.

The Department of Energy

In 1977, President Jimmy Carter signed an act creating the **Department of Energy (DOE),** which advances the energy and economic security of the United States. Among its top goals are scientific discovery, innovation, and environmental responsibility. Within the DOE, the Energy Information Agency gathers data on the use of energy in the United States and elsewhere.

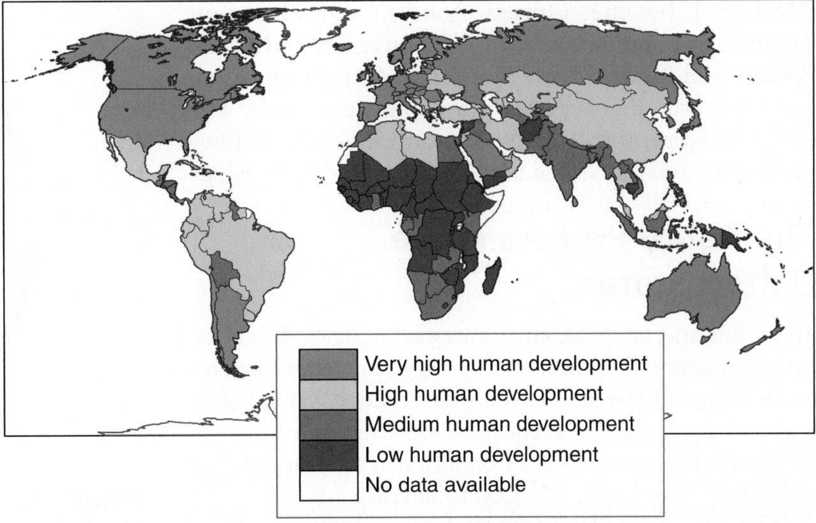

FIGURE 66.3 The human development index. The HDI is an index of well-being proposed by some as an alternative to GDP. Higher values indicate greater development. *(Data from UNDP 2016 Human Development Report, released on 21 March 2017. http://hdr.undp.org/en/countries)*

AP® Exam Tip

Be able to differentiate the role of the federal government, state government, and local government in environmental policy, and in economic issues related to environmental policy. ●

There are several approaches to measuring and achieving sustainability

Just as there are agencies, laws, and regulations designed to initiate and enhance sustainability, there are also a number of lenses through which to view the world, and a number of measurements or indexes used to evaluate sustainability. This section introduces some of these measurements and views. Eventually, some or all of these indices may become more directly involved in the measurement and assessment of sustainability.

Measuring Human Status

Despite the variety of economic indicators that are used around the world, there is still a call for a measurement that reports on the status of human beings with the specific goal of covering some of the noneconomic parameters such as levels of health and education. A variety of these are used and we describe two of them here.

The Human Development Index

The **human development index (HDI)** is a measurement index that combines three basic measures of human status: life expectancy; knowledge and education, as shown in adult literacy rate and educational attainment; and standard of living, as shown in per capita GDP and individual purchasing power. HDI was developed in 1990 by economists from Pakistan, England, and the United States, and it has been used since then by the UNDP in its annual HDR. As an index, HDI serves to rank countries in order of development and determine whether they are developed, developing, or underdeveloped. **FIGURE 66.3** shows the range of HDI categories and the distribution among countries. As you might expect, most of the developed countries have the highest HDI values.

The Global Multidimensional Poverty Index

The **global multidimensional poverty index (MPI)** is an internationally comparable measurement index

Environmental Protection Agency (EPA) The U.S. organization that oversees all governmental efforts related to the environment, including science, research, assessment, and education.

Occupational Safety and Health Administration (OSHA) An agency of the U.S. Department of Labor, responsible for the enforcement of health and safety regulations.

Department of Energy (DOE) The U.S. organization that advances the energy and economic security of the United States.

Human development index (HDI) A measurement index that combines three basic measures of human status: life expectancy; knowledge and education; and standard of living.

Global multidimensional poverty index (MPI) An internationally comparable measurement index that measures education, health, and living standards.

developed by the United Nations Development Program in conjunction with the University of Oxford Poverty and Human Development Initiative to measure acute poverty in over 100 developing countries. The index measures three things: education, health, and living standards. The Global MPI Index is updated yearly.

The Policy Process in the United States

To be fair and effective, environmental policies should be based on scientific indicators that suggest a certain behavior or action will be best for the environment. When policy makers believe there is adequate understanding of the science, and there is a course of preferred action for states or individuals, they begin a process to develop a policy.

The five basic steps in a policy cycle are problem identification, policy formulation, policy adoption, policy implementation, and policy evaluation. **FIGURE 66.4** depicts this process as a circular or reiterative process. As a policy is evaluated, the need for amendment might arise. When an amendment is initiated, it follows roughly the same steps. Many good environmental policies have had numerous amendments. For example, the Clean Air Act has been amended twice, and even the original Clean Air Act of 1970 was actually a modification of earlier clean air legislation.

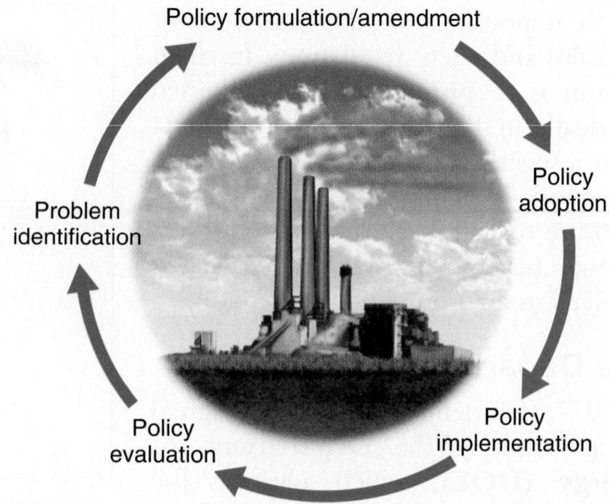

FIGURE 66.4 The environmental policy cycle. After an environmental problem is identified, environmental policy is formulated or modified. After a policy is adopted and implemented, it is evaluated and, if necessary, adjustments to the policy are made.

Legislative Approaches to Encourage Sustainability

United States governmental agencies have tried many ways to protect the environment, promote human safety and welfare and, in some cases, internalize externalities. The **command-and-control approach** is a strategy for pollution control that involves regulations and enforcement mechanisms. The **incentive-based approach** constructs financial and other incentives for lowering emissions based on profits and benefits. A combination of both approaches is likely to generate the maximum amount of desired changes.

Taxation is a major deterrent used to discourage companies from producing pollution and generating other negative impacts. A **green tax** is a tax placed on environmentally harmful activities or emissions in an attempt to internalize some of the externalities that may be involved in the life cycle of those activities or products. However, a tax alone may not be sufficient to achieve the desired results. Sometimes rebates or tax credits are given to individuals and businesses purchasing certain items such as energy-efficient appliances or building materials such as windows and doors.

In 1996, President Clinton's Council on Sustainable Development declared that "the essence of sustainable development is the recognition that the pursuit of one set of goals affects others and that we must pursue policies that integrate economic, environmental, and social goals." The **triple bottom line** is an approach to sustainability that considers three factors—economic, environmental, and social—when making decisions about business, the economy, and development. **FIGURE 66.5** shows that the intersection of these three factors is sustainability. There are many organizations and businesses that place one of these three factors at the top of a priority list. Some businesses strive for economic well-being—a sound financial bottom line—to the exclusion of human welfare or the environment. They may be regarded as successful within certain communities, but the triple bottom line concept emphasizes that to be a true success, there must be adequate treatment of both humans and environment. Paul Hawken, the author of *Natural Capitalism*, states the objective as, "Leave the world better than you found it, take no more than you need, try not to harm life or the environment, and make amends if you do."

Command-and-control approach A strategy for pollution control that involves regulations and enforcement mechanisms.

Incentive-based approach A strategy for pollution control that constructs financial and other incentives for lowering emissions based on profits and benefits.

Green tax A tax placed on environmentally harmful activities or emissions in an attempt to internalize some of the externalities that may be involved in the life cycle of those activities or products.

Triple bottom line An approach to sustainability that considers three factors—economic, environmental, and social—when making decisions about business, the economy, and development.

U.S. Policies for Promoting Sustainability

Of the many regulations that have been established in the last 50 years or so in the United States, there are at

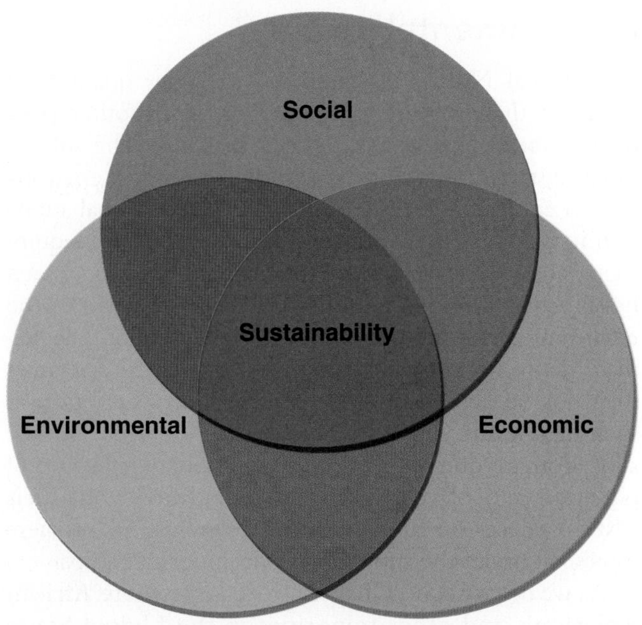

FIGURE 66.5 The triple bottom line. Sustainability is believed to be achievable at the intersection of the social, economic, and environmental factors that influence most development endeavors.

least seven important pieces of legislation that may help move the United States toward sustainability. All of these regulations have been discussed in other chapters and are summarized in **TABLE 66.1**.

Two major challenges of our time are reducing poverty and stewarding the environment

The classic environmental dichotomy is "jobs versus the environment." Those primarily concerned with human well-being ask how we can make demands for environmental improvements when there is so much poverty and injustice in the world. Those primarily concerned with the environment ask how we can focus exclusively on human suffering when an impoverished environment cannot support human health and well-being. The goal for most people is to make gains in both areas without sacrificing too much of either.

Poverty and Inequity

Approximately one-sixth of the human population—more than one billion people—lives in unsanitary conditions in informal settlements, slums, and shantytowns. Roughly one-tenth of the human population, 770 million people, earns less than U.S. $1.90 a day, which is the international poverty line. In the last 100 years, as developed countries have increased their GDPs and as many countries have modernized and developed their economies, the disparities between the rich and the poor have become greater. Poverty is simultaneously an issue of human rights,

TABLE 66.1 Major U.S. legislation for promoting sustainability

Act	Abbreviation	Year enacted	Purpose	Example of a success
National Environmental Policy Act	NEPA	1970	Enhance environment; monitor with a tool, the Environmental Impact Assessment	Protection of coral formation and sea turtles has occurred.
Occupational Safety and Health Act	OSHA	1970	Prevent occupational injury, illness, death from work-related exposure to physical and chemical harm	Worker training and knowledge of toxins has increased.
Endangered Species Act	ESA	1973	Protect animal and plant species from extinction	Bald eagle, peregrine falcon, and gray wolf populations have recovered.
Clean Air Act	CAA	1970	Promote clean air	Sulfur dioxide reductions from cap-and-trade have occurred.
Clean Water Act	CWA	1972	Promote clean water	Swimmable and fishable rivers across the United States have increased.
Resource Conservation and Recovery Act	RCRA	1976	Govern tracking and disposal of solid and hazardous waste	Numerous brownfields and contaminated lands have been cleaned up.
Comprehensive Environmental Response, Compensation, and Liability Act	CERCLA, also called Superfund	1980	Force and/or implement the cleanup of hazardous waste sites	Dozens of Superfund sites have been cleaned up around the United States.

economics, and the environment. Every human has a basic right to survival, well-being, and happiness—all directly threatened by poverty. Indebted individuals and nations are often unable to pay what they owe. In 2005, the eight major industrial countries of the world, known as the G8, canceled the debt of the 18 poorest countries. From an environmental standpoint, poverty increases overuse of the land, degradation of the water, and incidence of disease, among other problems.

In 2000, the United Nations offered an eight-point resolution listing what its member countries agreed were pressing issues that the world could no longer ignore. The member countries committed to reaching these Millennium Development Goals (MDGs), as outlined in the United Nations' Millennium Declaration, by 2015:

- Eradicate extreme poverty and hunger
- Achieve universal primary education
- Promote gender equality and empower women
- Reduce child mortality
- Improve maternal health
- Combat HIV/AIDS, malaria, and other diseases
- Ensure environmental sustainability
- Develop a global partnership for development

While some countries achieved some of their goals, many others did not. As with many environmental laws and their implementation, the distance from resolutions to results is immense, and not all developed countries have committed the resources that they had promised. In 2015, when it was apparent that many of the millennium goals would not be reached, the UN established a new agenda called Transforming Our World: The 2030 Agenda for Sustainable Development.

One proponent of the UN MDGs was Nobel Peace Prize Laureate Dr. Wangari Maathai (1940–2011) from Kenya (**FIGURE 66.6**). Dr. Maathai was the founder of the Green Belt Movement, a Kenyan and international environmental organization that empowers women by paying them to plant trees, some of which can be harvested for firewood after a few years. The Green Belt Movement is credited with replanting large expanses of land in East Africa, thereby reducing erosion and improving soil conditions and moisture retention. In addition, the trees that have been replanted, provided that they are not overharvested, offer a renewable source of fuel for cooking. The Green Belt Movement is considered a global sustainability success story promoting both individual human and environmental well-being. Dr. Maathai was also involved in environmental activism to achieve her goals, which sometimes caused her difficulties with certain governmental organizations.

Environmental Justice

The typical North American uses a larger quantity of resources than the average person in many other parts of the world. This situation is not equitable. The subject of fair distribution of Earth's resources known as environmental equity, has received increasing international attention in recent years. Beyond moral objections to inequity, there are concerns about sustainability. We have seen how increased resource use usually increases harm to the environment. As more and more people develop a legitimate desire for better living conditions, Earth's resources may not be able to support continued consumption at such high levels. Closely related to the equity of resource allocation are questions of the inequitable distribution of pollution and of environmental degradation with their adverse effects on humans and ecosystems. All of these topics fall under the subject of environmental equity.

As we discussed in Chapter 16 and elsewhere, African Americans and other minorities in the United States are more likely than Caucasians to live in an area with solid waste incinerators, chemical production plants, and other so-called "dirty" industries. In a number of now-famous studies in the 1980s and 1990s, investigators used the distribution of minority residents by postal zip code to relate race and class to the location of hazardous facilities and sites. In Atlanta, 83 percent of the African American population lived within the same zip code area as the 94 uncontrolled toxic waste sites, while 60 percent of the Caucasians lived in those areas. In Los Angeles, roughly 60 percent of Hispanics lived in the same areas as the toxic waste sites, while only 35 percent of the Caucasian population lived in those areas. One study in five southern states compared the size of specific landfill facilities with the percentage of minorities in the zip code area in which the landfill was located. The study concluded that the largest landfills

FIGURE 66.6 Wangari Maathai. Dr. Maathai was the founder of the Green Belt Movement in Kenya. *(Micheline Pelletier/Corbis via Getty Images)*

are located in areas that have the greatest percentage of minorities. An important issue that has not been entirely resolved, and that can vary from case to case, is whether the affected population or the hazardous facility came first to a given area. By knowing which came first, people and organizations attempting to remedy the situation will have a better idea of how to modify existing legislation and regulations to reduce the number of people who live in degraded environments.

More recently, it became clear that the subjects of disproportionate exposure to environmental hazards were not only African Americans, but all races in lower income brackets. Moreover, the problem was not limited to the United States. The concept was broadened and named environmental justice, which is both a social movement and an academic field of study. Those involved in environmental justice examine whether there is equal enforcement of environmental laws and elimination of disparities—intended or unintended—in the exposure to pollutants and other environmental harms affecting different ethnic and socioeconomic groups within a society. Delegates to the First National People of Color Environmental Leadership Summit in 1991 established 17 principles of environmental justice. Professor Robert Bullard of Texas Southern University has published books and papers in the academic field of environmental justice and has been involved in the social movement as well. He is probably best known for his 1990 book *Dumping in Dixie*, which demonstrated that minority and lower-socioeconomic groups were often the recipients of pollution from dumping of MSW and hazardous wastes. More recently, Professors Paul Mohai of the University of Michigan and Robin Saha of the University of Montana reassessed the unequal distribution of hazardous waste dumping in the United States and found that the situation is actually worse than previously reported. In particular, they believe they have resolved the issue of whether the hazardous waste facility or the lower-socioeconomic and minority population came first to an area. The authors maintain that the minority community in many cases was present first and that the hazardous waste facility, which came later, was specifically targeted to that community.

Individual and Community Action

There are a fair number of people who believe that whether or not governments and private agencies are able to achieve their goals, individuals can and must act to further their own goals of sustaining human existence on the planet. These individuals have begun to make attempts to live a sustainable existence without government incentives, taxes, or other measures. They have begun activities such as calculating their own ecological footprint, carbon footprint, energy footprint, and other metrics to determine how much of an impact they are making on Earth. From this starting point, they have begun to make changes in their consumption, behavior, and lifestyle to reduce that impact. Some people act on their own while others act through groups and organizations. They have adopted a philosophy represented by the saying, "If the people lead, the leaders will follow." Some individuals have joined together to organize communities centered around philosophies of sustainability.

Van Jones, a graduate of Yale Law School, was a community organizer in San Francisco working on civil rights and human justice issues when he decided to combine concerns about the environment and global climate change with the need for creating jobs in cities (**FIGURE 66.7**). He founded an organization called Green for All and published a book titled *The Green Collar Economy: How One Solution Can Fix Our Two Biggest Problems.* The two problems, as he addressed in his book, are global warming and urban poverty. He believes that creating green jobs, such as insulating buildings, constructing wind turbines and solar collectors, and building and operating mass transit systems, will improve the living conditions of some of the poorest people in the nation and reduce our impact on the environment. For Van Jones, this is a win-win solution—people will be employed and our emissions of global greenhouse gases will decrease.

Majora Carter exemplifies another model of how to participate in community activities. She was born and raised in the South Bronx section of New York City and is a Wesleyan University and New York University graduate. She founded a not-for-profit environmental justice organization before forming her own private sector firm. She advocates improving health and quality of life in communities by promoting economic development in a sustainable and environmentally sound way. She has attracted a great deal of attention by creating gardens and greenways in the South Bronx, while at the same time creating all types of employment opportunities, including green jobs.

FIGURE 66.7 A gathering of Green for All supporters in Oakland, California. Over the decades, individual and community action have influenced achievements in environmental quality and sustainability. *(©greenforall.org)*

MODULE

66 AP® Review

Preparing for the AP® Exam

In this module, we have seen that a series of worldviews can be used to approach environmental protection and regulation. A variety of world agencies such as the United Nations, the World Bank, and the World Health Organization have numerous environmental programs. In the United States, the Environmental Protection Agency and the Department of Energy are two of a number of environmental agencies. A variety of measures are used to assess sustainability and environmental well-being, including the human development index and the global multidimensional poverty index. The triple bottom line maintains that sustainability can be achieved at the intersection of the three factors—economic, environmental, and social—that influence most development endeavors. Reducing poverty and taking sound care of the environment are two challenges that are essential for sustainability. Achieving these goals involves addressing poverty and inequality, environmental justice, and individual and community action.

AP® Practice Questions

Choose the best answer for the following.

1. Which worldview considers ecosystems to have intrinsic value?
   I. anthropocentric
   II. biocentric
   III. ecocentric
   (a) I only
   (b) I and III
   (c) II only
   (d) II and III

2. Enacting legislation that restricts a chemical suspected of being harmful while there is still scientific uncertainty about that chemical is an example of
   (a) an anthropocentric worldview.
   (b) a biocentric worldview.
   (c) the precautionary principle.
   (d) technology leapfrogging.

3. In what way does the Occupational Safety and Health Administration (OSHA) contribute to environmental protection?
   (a) It develops regulations to limit emissions.
   (b) It limits human exposure to chemicals and pollutants.
   (c) It improves the quality of water.
   (d) It provides funds to clean contaminated sites.

4. The triple bottom line
   (a) is an approach to sustainability that considers economic, environmental, and social factors.
   (b) is a method for encouraging sustainability that includes incentives, regulations, and penalties.
   (c) consists of three measures of human status used in the Human Development Index.
   (d) is used by United Nations Development Programme to determine if a program is successful.

5. Which is an example of the command-and-control approach to encourage sustainability?
   (a) a cap-and-trade system for carbon emissions
   (b) a rebate for energy-efficient products
   (c) funding for solar energy projects
   (d) regulations that limit sulfur emissions and include a provision for fines

Working Toward Sustainability

Sustainable Housing, Sustainable Living

In most cities in the developing world, the lowest socioeconomic groups live in poor conditions in very low-quality housing, or they do not have housing at all. India, with its large population and relatively high growth rate, has more than its share of what might be called in the United States "low-income housing." But in India, this means large informal settlements on the outskirts of cities. Cardboard boxes and discarded industrial waste are often the dominant materials for housing. There is no running water or electricity and no system for disposing of MSW or human waste. Individuals do not have the ability to lock their front doors—in many cases, there are no front doors. Poverty, disease and crime are pervasive.

In Indore, India, an architect designed the Aranya Township to improve living conditions for poor citizens. Aranya means forest in Hindi, and the Aranya Township had many features of a forest. In an area 88 ha (217 acres) in size, just north of the Indore city center, 6,500 housing units were established. Each unit consisted of a core area with plumbing and a sewer drain, so that each unit had a bathroom and kitchen, and one additional room. There was space for each unit to expand, if and when the occupants had the financial resources to do so.

The housing units were built so that the courtyards between them had grassy areas and plantings and received direct sunlight during part of the day. But tall building corridors and plants shielded residents from the hottest sun of the day and allowed them to walk from one housing unit to another via courtyards and passageways. The housing development had many of the features that we described in Chapter 10 for cities in the United States. But this design was for residents in India, which had a much lower GDP. The housing units were affordable, durable, and were constructed with local, mostly low-environmental impact materials. The architect who designed this housing complex, Balkrishna Doshi, received the Pritzker Prize in 2018, which is sometimes called the Nobel Prize for architecture.

The Aranya Township housing project in Indore, India, is widely regarded as innovative for its application of architecture and sustainable design principles to housing for urban lower-income people in the developing world. Today, approximately 80,000 people live in the Aranya Township Housing project. Indore is also known for an exercise it conducted in the last decade related to global climate change. Indore developed a "City Resilience Strategy for Changing Climatic Scenarios." Resilience is the rate or time it takes for an ecosystem to return to its original state after disturbance. With respect to climate change, Indore city planners wanted to know how long it would take for the city to recover from a variety of climatic scenarios such as heavy rains and flooding. This required the city to assess its climate, physical environment, and social parameters for vulnerability to changes in precipitation, water supply, runoff from heavy rains, and many other factors. One of the scenarios the city considered was how runoff from city streets would be affected after intense precipitation caused the surrounding soils to become waterlogged. It became apparent that low lying areas in the city might be flooded, which would adversely affect lower-income housing (which is often located in the most vulnerable locations in a city). Sewage infrastructure may also be compromised. There were many other scenarios considered.

The city created vulnerability indices that allowed city planners to determine which conditions are most likely to occur and which parts of the city and which populations within the city were most likely to be affected. The city prioritized the strategies required to address each potential threat and launched projects to create awareness among citizens and, when possible, make modifications to city infrastructure that would allow the city to become more resilient. As with most cities, sufficient funds aren't available to implement all the changes identified from the study. But Indore has both examples of sustainable housing for low-income residents and a resiliency plan for responding to climate change. That's not a bad combination for a city in the second-largest (and soon to be the largest) country in the world.

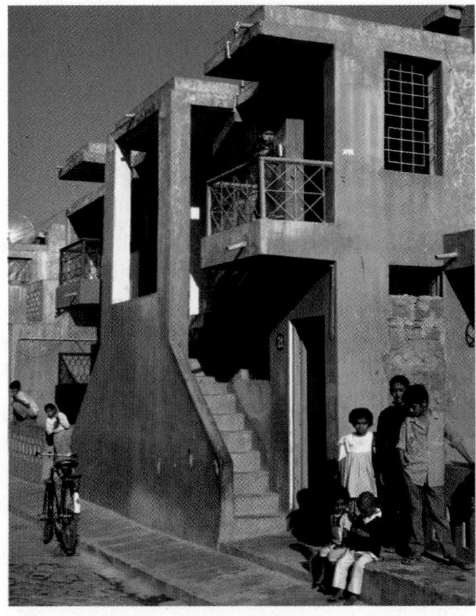

Aranya Community Housing. This housing project in India uses local, mostly low-environmental impact materials, and won an international architecture award. *(© Aga Khan Trust for Culture - Aga Khan Award for Architecture/Yatin Pandya)*

Critical Thinking Questions

1. What are some of the obstacles that prevent a city from establishing sustainable housing and living spaces?
2. What other steps are necessary to establish full sustainability in a city?

References

Aranya Low Cost Housing, March 7, 2018: http://www.architectmagazine.com/project-gallery/aranya-low-cost-housing_o; Indore City Resilience Strategy for Changing Climate Scenarios, May 2012: http://www.adaptationlearning.net/sites/default/files/resource-files/Indore_City%20Resilience%20Strategy.pdf

Chapter 20 Review

In this chapter, we identified sustainability as one of the ultimate goals of environmental science. It is often achieved through the use of sound economic and business practices as well as effective environmental regulations and laws. There are international agencies such as the United Nations Environment Programme and a number of U.S. agencies to enforce regulations and laws. The Environmental Protection Agency oversees all governmental efforts related to the environment. Different worldviews affect how people approach the goal of sustainability. Proper stewardship of the natural world and a reduction of poverty are challenges that must be met if people are to achieve sustainability.

Key Terms

Well-being
Economics
Genuine progress indicator (GPI)
Technology transfer
Leapfrogging
Natural capital
Human capital
Manufactured capital
Market failure
Environmental economics
Ecological economics
Valuation
Environmental worldview
Anthropocentric worldview
Stewardship
Biocentric worldview
Ecocentric worldview
United Nations (UN)
United Nations Environment Programme (UNEP)
World Bank
World Health Organization (WHO)
United Nations Development Programme (UNDP)
Environmental Protection Agency (EPA)
Occupational Safety and Health Administration (OSHA)
Department of Energy (DOE)
Human development index (HDI)
Global multidimensional poverty index (MPI)
Command-and-control approach
Incentive-based approach
Green tax
Triple bottom line

Learning Goals Revisited

Module 65 Sustainability and Economics

Explain why efforts to achieve sustainability must consider both sound environmental science and economic analysis.

Sustainable environmental systems must allow for maintaining air, water, land, and biosphere systems and must also maintain human well-being—the status of being healthy, happy, and prosperous. Sustainability will not be achieved if certain groups are exposed to a disproportionate share of dirty jobs or waste material in the home or workplace.

Describe how economic health depends on the availability of natural capital and basic human welfare.

Sustainable systems must include a consideration of externalities. Gross domestic product (GDP) is the value of all products and services produced in a year in a given country. Genuine progress indicator (GPI) includes measures of personal consumption, income distribution, levels of higher education, resource depletion, pollution, and health of the population. Economic assets, or capital, can come from the natural systems on Earth, from humans, or from the manufactured products made by humans. Valuing all three kinds of capital is essential to systems that are sustainable.

Module 66 Regulations and Equity

Explain the role of agencies and regulations in efforts to protect our natural and human capital.

Once a society believes it has enough scientific information to act with the intent of protecting or reducing harm to the environment, it must determine the rules and regulations it wishes to enact. A group of government agencies in the United States handles the areas that offer protection to the environment and humans. Policies are enacted through passage and modification of laws.

Describe the approaches to measuring and achieving sustainability.

The human development index combines life expectancy, knowledge and education, and standard of living as a measure of human status. The global multidimensional poverty index measures education, health, and living standards. A green tax can be used to internalize externalities or reduce environmental harm. The triple bottom line accounts for three factors—economic, environmental, and social—when making decisions about the environment and development. These ideas have led to a variety of policies in the United States for promoting sustainability.

Discuss the relationship among sustainability, poverty, personal action, and stewardship.

One-sixth of the world population has inadequate housing and inadequate income. People will need access to food, housing, clean water, and adequate medical care before they can be concerned about environmental sustainability. The UN Millennium Development Goals have established objectives for improving the status of people and the sustainability of the environment. The human-centered worldview maintains that humans have intrinsic value and nature provides for our needs. The life-centered worldview holds that humans are one of many species on Earth, all of which have value. The Earth-centered worldview places equal value on both all living organisms and ecosystems. Individual and community action can lead to sustainable actions occurring at a greater level worldwide.

Practice Math and Graphing

Preparing for the AP® Exam

Answer the following questions. Be sure to show all your work.

1. Practice Math

During the period from 1970 through 2000, the world gross domestic product increased by 67 percent. If the GDP was U.S. $6,000 per capita in 1970, what was the GDP per capita in 2000?

2. Practice Graphing

One measure of progress in reclaiming Superfund sites is a measure of how many sites are "ready for reuse," meaning that they have been sufficiently cleaned up. Using the data in the table, graph the number of sites ready for reuse each year between 2008 and 2016. Plot year on the x axis and number of sites on the y axis.

Year	Sites ready for reuse
2008	343
2009	409
2010	475
2011	540
2012	606
2013	662
2014	707
2015	752
2016	793

Chapter 20 AP® Environmental Science Practice Exam

Preparing for the AP® Exam

Section 1: Multiple-Choice Questions

Choose the best answer for questions 1–20.

Use the following graph to answer Question 1

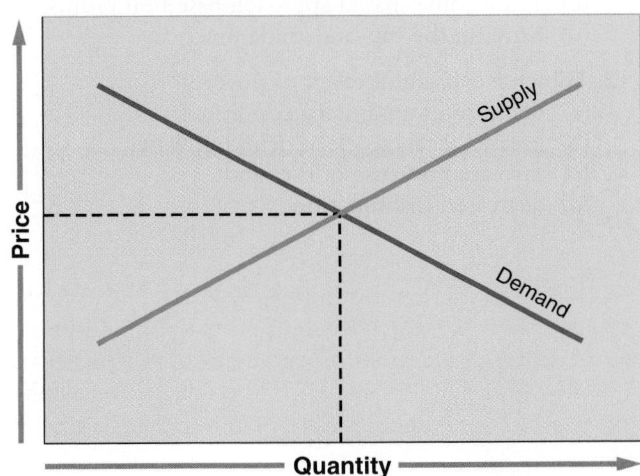

1. Which of the following can be inferred from the graph?
   I. A lower price results in a greater demand.
   II. A higher price results in a greater supply.
   III. Price changes as supply and demand fluctuate.
   (a) I only
   (b) II only
   (c) III only
   (d) I, II, and III

2. All of the following are examples of negative externalities EXCEPT
   (a) global climate change as a result of greenhouse gas emissions from burning coal, oil, and gasoline.
   (b) increased pollination rates of crop plants as a result of local beekeeping.
   (c) a pulp mill that produces paper and pollutes the surrounding water and air.
   (d) runoff of pesticides and fertilizers from a farm into a nearby river.

3. The genuine progress indicator (GPI) is more representative measure of the wealth and well-being of a country than gross domestic product (GDP) because GPI
   (a) measures productivity and consumption without taking externalities into account.
   (b) has risen in the United States while the GDP has remained fairly constant since 1970.
   (c) includes resource depletion, pollution, and health of the population in its calculation.
   (d) can be increased by higher health care costs and a greater incidence of illnesses.

4. Economic assets are the sum total of
   I. natural capital.
   II. human capital.
   III. manufactured capital.
   (a) II only
   (b) III only
   (c) I and III
   (d) I, II, and III

5. Valuation, according to environmental and ecological economics, would include all of the following EXCEPT
   (a) the revenue generated from tourists visiting a national park.
   (b) the cost of wastewater treatment provided by a natural wetland.
   (c) the benefits derived from medicinal plants found in tropical rainforests.
   (d) the profits realized from hiring more employees to increase production.

6. Full cost pricing by the internalization of externalities could result in which of the following?
   (a) higher prices and a reduction in the consumption of items with high negative impacts
   (b) lower prices and an increase in the consumption of items with high negative impacts
   (c) greater consumer demand for products with high negative impacts
   (d) lower production costs due to diminishing natural resources

7. Cradle-to-cradle and cradle-to-grave analyses of manufactured goods can best be described as the study of the
   (a) changes in the use of a product from one generation to the next.
   (b) life cycle of a product from its production to use to ultimate disposal.
   (c) use of resource extraction over the use of ecosystem services.
   (d) options for the disposal of solid waste generated by the product.

8. Recently the Los Angeles Unified School District adopted a new policy on the use of pesticides in schools. This policy assumes that the use of pesticides constitutes a risk to the health of children and the environment. Pesticides will be employed only after nonchemical methods have been explored. The pest control measure that is the least harmful will be implemented. This is an example of
   (a) the precautionary principle.
   (b) ecosystem services.
   (c) a market-driven approach.
   (d) sustainable use.

9. Which is a United Nations organization concerned with the environment?
   (a) World Resources Institute (WRI)
   (b) Occupational Safety and Health Administration (OSHA)
   (c) Department of Energy (DOE)
   (d) World Health Organization (WHO)

10. Which U.S. law contributes to sustainability by governing the tracking and disposal of solid and hazardous waste?
    (a) National Environmental Policy Act (NEPA)
    (b) Resource Conservation and Recovery Act (RCRA)
    (c) Clean Water Act (CWA)
    (d) Comprehensive Environmental Response, Compensation, and Liability Act (CERCLA)

11. Strategies to implement environmental laws and regulations include all of the following EXCEPT
    (a) green taxes on environmentally harmful activities or emissions.
    (b) buying and selling of pollution permits.
    (c) an incentive-based approach based on profits.
    (d) banning the cap-and-trade practice.

12. Which is a harmful effect of poverty?
    (a) decrease in unsanitary conditions
    (b) greater access to clean drinking water
    (c) increased overuse of the land
    (d) decreased malnutrition

13. The United Nations Millennium Declaration proposed to meet which goals?
    I. reduce environmental sustainability through economic development
    II. eliminate extreme poverty and hunger and reduce child mortality
    III. empower women and improve maternal health
    (a) I only
    (b) II only
    (c) III only
    (d) II and III

14. The following is a summary report for the distribution of environmental burdens for Fremont, California.

Population categories	Number of facilities emitting criteria air pollutants per square mile
African Americans	10
Caucasians	5
Low-income families	8
High-income families	4
Families below poverty threshold	9
Families above poverty threshold	4
Non-high school graduates	7
High school graduates	5

The information in this table reflects
(a) an environmental equity issue.
(b) an anthropocentric worldview.
(c) a biocentric worldview.
(d) an ecocentric worldview.

15. The idea that all people regardless of ethnic or socioeconomic status deserve equal environmental conditions is a central principle of
    (a) the triple bottom line.
    (b) the National Environmental Policy Act.
    (c) the United Nations Environment Programme.
    (d) environmental justice.

16. In the United States, which organization is most likely to address issues related to sick building syndrome in the work environment?
    (a) EPA
    (b) OSHA
    (c) DOE
    (d) WHO

17. Which is most likely to discourage environmental inequity?
    (a) NIMBY politics
    (b) higher carbon emissions in developed nations
    (c) technology transfer
    (d) racial segregation

18. A researcher is doing a study comparing human status, measured by education, health, and living standards, in a variety of countries. The best index for her study would likely be
    I. the human development index (HDI).
    II. the global multidimensional poverty index (MPI).
    III. the genuine progress indicator (GPI).
    (a) I only
    (b) II only
    (c) III only
    (d) I and II

19. In many developed countries, including England, Germany, and Sweden, ______ has risen with time, while ______ has remained constant or declined.
    (a) GDP; HDI
    (b) HDI; GDP
    (c) GPI; GDP
    (d) GDP; GPI

20. After an environmental science class, you are having a conversation with your friend Archana, who says she believes that all species are equally important, and it is our duty to protect them. She also thinks we should protect ecosystems because they provide a lot of services to humans. In her comments, Archana holds a(n) ______ worldview.
    (a) ecocentric
    (b) biocentric
    (c) anthropocentric
    (d) envirocentric

Section 2: Free-Response Questions

Write your answer to each part clearly. Support your answers with relevant information and examples. Where calculations are required, show your work.

1. Use the following information about gasoline consumption in the United States in 2008, when gasoline prices were among the highest in the last 20 years, to answer the questions below.
   - In 2008 the United States consumed approximately 138 billion gallons of gasoline.
   - The current federal tax on gasoline is 18.4 cents per gallon.
   - The national average cost of a gallon of regular unleaded gasoline in June 2008 was $4.00 per gallon.
   - Eighty percent of the federal gasoline tax is used to subsidize road construction.

   (a) Calculate the total amount of money spent in the United States on the purchase of gasoline in 2008 (when gasoline cost $4.00 per gallon). (2 points)

   (b)
   i. What percent of the cost per gallon is the gasoline tax?
   ii. How much revenue was generated by the gasoline tax in 2008?
   iii. How much was used in 2008 to subsidize road construction? (3 points)

   (c) Does the federal tax on gasoline qualify as a green tax? Explain your answer. (1 point)

   (d) Advocates of raising the gasoline tax suggest that the tax be increased to 80 cents per gallon. Identify two economic effects and two environmental effects of raising this tax. (4 points)

2. In 1997, the ecological economist Robert Costanza and his associates published a report titled *The Value of the World's Ecosystem Services and Natural Capital.* They estimated that if all the ecosystem services provided worldwide had to be paid for, the cost would average $33 trillion per year with a range from $16 trillion to $54 trillion. In that same year the global gross national product (GNP) was $18 trillion.

   (a) What is meant by ecosystem or ecological services? Give three specific examples and identify which United Nations organization might oversee these services. (4 points)

   (b) Define the term valuation. What would the worldwide consequences be if the world actually had to pay for ecosystem services and natural capital? (2 points)

   (c) Explain how this report could be used to develop a sustainable economic system. (2 points)

   (d) Which environmental worldview is most consistent with the concerns of environmental economics? Explain. (2 points)

3. Most people living in the town of Fremont use a particular type of combined pesticide/fertilizer on their lawns. At present, negative externalities, such as harm to nontarget species, are not included in the price of the good.

   (a) Draw a supply and demand graph for the pesticide/fertilizer in Fremont. Indicate the market equilibrium (E1). (2 points)

   (b) If negative externalities were taken into account in the pricing of the pesticide/fertilizer, how would this change the supply and demand graph? Show this graphically and explain your reasoning. (2 points)

   (c) Give TWO examples of potential negative externalities from using a combined pesticide/fertilizer. (2 points)

   (d) The town council is concerned about the potential negative impacts of the pesticide/fertilizer on the environment. At their meeting, they discuss some ways they might be able to reduce the amount used in Fremont.
   i. Describe the command-and-control approach and the incentive-based approach. (2 points)
   ii. Explain the approach you would recommend and provide ONE example of how the council could apply it in this particular situation. (2 points).

Juliane Golledge/Courtesy Zealandia

science applied 8

How can we bring back biodiversity?

Throughout this book we have discussed the decline of species in their native habitats and the introduction of non-native species, many of which rapidly spread and caused harm in their new surroundings. The island nation of New Zealand is an excellent example of how invasive species can harm native biodiversity. For millions of years, New Zealand had no land mammals, other than three species of bats. This means the island had no mammalian predators, so for millennia the birds of New Zealand evolved no defenses against mammalian predators, including some evolving to be flightless. In fact, many of these bird species are found nowhere else in the world. When humans arrived about 700 years ago, everything changed. They intentionally introduced Australian brushtail possums (*Trichosurus vulpecula*) and several species of weasels—for sources of fur—and accidentally introduced several species of rats. These introduced species have been preying on the birds of New Zealand for centuries, driving one-fourth of the bird species to extinction and many others to the brink of extinction (**FIGURE SA8.1**). In 2016, the New Zealand government announced a bold new initiative to reverse this decline in their unique biodiversity by removing every opossum, weasel, and rat from the country by 2050.

FIGURE SA8.1 Invasive mammals. Mammal predators, such as this rat (*Rattus rattus*), have decimated native populations in New Zealand for several centuries. *(Nga Manu Images NZ)*

Eliminating invasive species is a difficult task, especially for animals like rats that can rapidly increase their population size in just a few months. Past eradication efforts on small islands have successfully used traps and poisons to remove every last invasive predator, but those islands are much smaller than the two main islands of New Zealand. Moreover, the small islands have few people living on them, making it easier to find and eliminate invasive species.

Zealandia—the first step

One of the inspirations for the nationwide effort at eradicating the predatory mammals is a sanctuary known as Zealandia. Zealandia is a completely fenced 225-hectare (556-acre) area in which the invasive predators have been removed (**FIGURE SA8.2** on page 748). Endangered species of native birds, many of which have persisted only on neighboring predator-free islands, have been re-introduced to this mainland sanctuary and they now thrive. For example, the kaka parrot (*Nestor meridionalis*) is a rare native bird (**FIGURE SA8.3** on page 749); 14 individuals were introduced into Zealandia and this population has now expanded to several hundred. With an increased biodiversity of native birds flying and calling throughout the sanctuary, Zealandia now offers the public a sense of what New Zealand used to look and sound like before humans arrived.

With this inspiration, the New Zealand government unveiled its plan to expand the effort in what it is calling "Predator-Free 2050." The effort will involve the distribution of poisons from the air to cover large areas as well as new technologies using self-resetting traps that can kill large numbers of predators. Researchers are also investigating ways to design poisons that are specifically targeted to the invasive predators to avoid inadvertently poisoning native animals and people's pets. They anticipate having a targeted rat poison developed within a few years.

FIGURE SA8.2 Zealandia. The 225-hectare (556-acre) sanctuary is surrounded by a fence to keep out invasive predators and facilitate the conservation of New Zealand's declining bird species. The fence is located just inside the road surrounding the preserve. *(Courtesy Rob Suisted)*

Developing a targeted possum poison is thought to still be a decade away.

Opposition to the Predator-Free 2050 plan

Not everyone is on board with the movement to eliminate the non-native predators from New Zealand. Some people are concerned that controlling the predators sends a message to the public that the way to conserve nature is through widespread killing of thousands of animals. They argue that the predators did not choose to be introduced centuries ago; in fact, many of the predators were intentionally introduced by humans. Just because human preferences have changed, these opponents argue, this should not result in the non-native species being exterminated. Proponents counter this argument by pointing out that with or without the Predator-Free 2050 effort, thousands of animals will die. If the plan is enacted, the dying animals will be the non-native predators. If the plan were not enacted, the dying animals will be the native birds, including those that are already endangered.

Opponents of the plan have also argued that the people of New Zealand should simply learn to live with the exotic predators. After several centuries of being in the country, they argue that the non-native predators are now part of the modern ecosystem and that concept should be embraced. It is also noteworthy that other introduced species—including fallow deer (*Dama dama*), brown trout (*Salmo trutta*), and mallard ducks (*Anas platyrhynchos*)—are not part of the eradication program. These species are desired by hunters and anglers and do not have the same negative stigma as rats, possums, and weasels.

The Predator-Free 2050 plan also does not acknowledge two other common species of non-native mammals in New Zealand: domestic cats and dogs. While the plan does include the removal of feral cats, house cats are also a major cause of bird mortality. In the United States, for example, it is estimated that house cats, which are often allowed to roam outside of people's houses, kill 1.4 to 3.7 billion birds each year.

Looking forward

The key to the success of the Predator-Free 2050 plan will be to obtain widespread support from the public. This is particularly important because homes and businesses are potential places for small numbers of surviving individuals to mount a population rebound following the nationwide extermination effort. In short, the planning for Predator-Free 2050 has to consider social issues as much as it needs to consider biological issues. Fortunately for the proponents of the plan, there has been strong support from the residents. In fact, there are thousands of volunteer groups across the country that are trapping the animals to reduce their numbers.

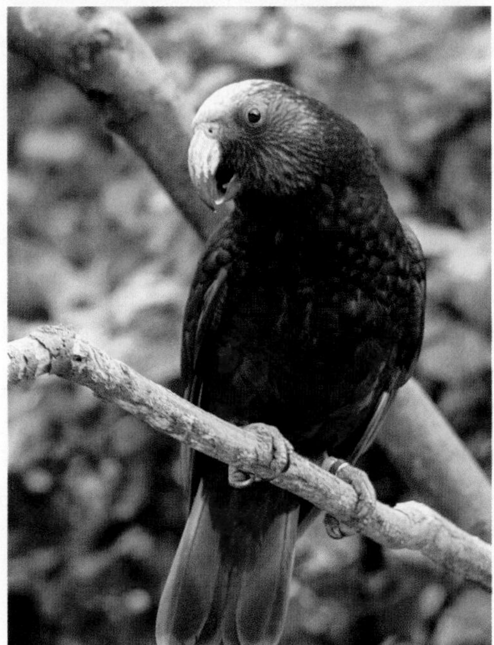

FIGURE SA8.3 Endangered New Zealand birds. Birds such as this kaka parrot are facing extinction due to predation by introduced predators. This bird was photographed inside the Zealandia sanctuary. *(Steve Attwood/Courtesy Zealandia)*

While the cost of the plan is expensive at U.S. $6 billion, it is estimated that the invasive predators currently cost the country $3 billion annually in crop losses. In addition, ecotourism is a major source of revenue for New Zealand, and ecotourism is likely to increase as New Zealand works to improve the populations of its many unique species of birds. Only time will tell whether the monumental effort to remove the invasive predators will be successful and whether the biodiversity of the unique birds will rebound.

References

Brown, K.V. 2018. The surprising way New Zealand could soon solve its predator problem. *Gizmodo*, January 22. https://gizmodo.com/the-surprising-way-new-zealand-could-soon-solve-its-pre-1822194179

Greshko, M. 2016. New Zealand announces plan to wipe out invasive predators. *National Geographic*, July 25. https://news.nationalgeographic.com/2016/07/new-zealand-invasives-islands-rats-kiwis-conservation/

Owens, B. 2017. Behind New Zealand's wild plan to purge all pests. *Nature*, January 11. https://www.nature.com/news/behind-new-zealand-s-wild-plan-to-purge-all-pests-1.21272

Questions

1. Given that declining biodiversity can be caused by both introduced predators and habitat loss, how might the government of New Zealand invest its limited amount of conservation funding?
2. How have proponents and opponents of the Predator-Free 2050 plan invoked arguments for intrinsic versus instrumental values of biodiversity?

Practice AP® Free-Response Question

Write your answer to each part clearly. Support your answers with relevant information and examples. Where calculations are required, show your work.

On the island of Macquarie—a small, uninhabited island between Australia and Antarctica—seal hunters introduced rats, feral cats, and European rabbits (*Oryctolagus cuniculus*) in the 1800s. The rats and cats wreaked havoc on the island's bird species while the rabbits consumed much of the island's vegetation. From 2011 to 2014, a major effort was undertaken that successfully removed every last rat, cat, and rabbit from the island.

(a) How might the fact that the island was uninhabited make the extermination of the invasive species more feasible? (2 points)

(b) What is the likely response of the island vegetation after the elimination of the invasive European rabbits? (2 points)

(c) In the eradication effort, cats were removed first. During a 15-year period of cat removal, the rabbit population dramatically increased from 10,000 to 100,000. What does this suggest about the relationship between the cats and rabbits? (2 points)

(d) Why might new technologies be required as successful eradication efforts on small islands such as Macquarie are now being applied to much larger islands such as those of New Zealand? (2 points)

(e) What is the potential risk of deploying poisons that are not designed to kill particular species of non-native species? (2 points)

Section 1: Multiple-Choice Questions

Choose the best answer for questions 1–25.

1. In a small town, local residents recently discussed a proposal to fill in a wetland near the planned site of a new apartment complex. A farmer objected to this proposal because the wetland provides a habitat for crop pollinators. A business owner objected because she was worried it would harm the town's image and reduce business. Which categories of instrumental values do these arguments consider?
   (a) The farmer is considering the intrinsic value of the wetland; the business owner is considering the instrumental value of the wetland.
   (b) Both are considering the intrinsic value of the wetland.
   (c) The farmer is considering the regulating services of the wetland; the business owner is considering the support services of the wetland.
   (d) The farmer is considering the support services of the wetland; the business owner is considering the cultural services of the wetland.

2. Which is NOT an international organization?
   (a) IUCN
   (b) USEPA
   (c) WWF
   (d) UNEP

3. Which phenomenon is NOT likely to be a potential consequence of habitat fragmentation?
   (a) a decrease in the proportion of edge habitat
   (b) an increase in the rate of inbreeding depression
   (c) loss of genetic diversity within populations
   (d) a decrease in the range of animal movement

4. In North America, honeybees are considered
   (a) a native species.
   (b) an exotic species.
   (c) an invasive species.
   (d) a threatened species.

5. In the United States, the definitions of *endangered* and *threatened* are
   (a) similar to IUCN definitions of threatened and near-threatened.
   (b) identical to the IUCN definitions of threatened and near-threatened.
   (c) similar to definitions set by UNEP.
   (d) designed to classify more species as endangered.

6. Which factor is least likely to be considered when applying the concept of SLOSS (single large or several small) to the conservation of a population of a particular species?
   (a) the genetic diversity of the population
   (b) the ability of individuals in the population to move between suitable habitats
   (c) the amount of edge habitat surrounding a suitable habitat
   (d) the IUCN risk categorization of the species

7. Overharvesting a species for sport, medicinal, or industrial purposes may alter the ______ associated with that species.
   I. intrinsic value
   II. instrumental value
   III. ecological interactions
   (a) I only
   (b) II only
   (c) III only
   (d) II and III

Question 8 refers to the following figure:

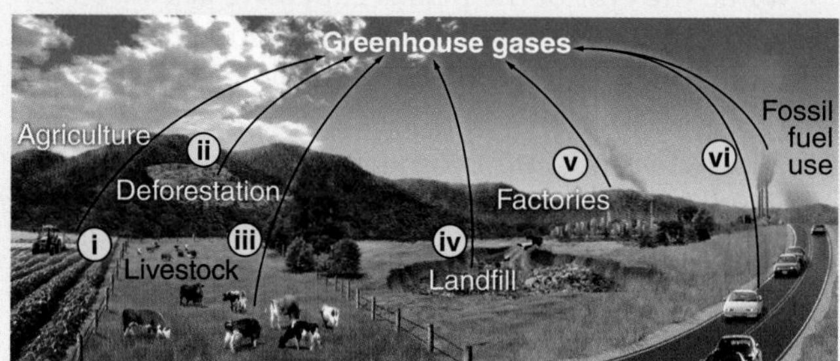

8. In the diagram of greenhouse gas sources, which refers to significant sources of methane (CH_4)?
   (a) i, ii, iv
   (b) i, ii, vi
   (c) ii and iii
   (d) iii and iv

9. Black soot contributes to global warming by
   (a) reflecting solar radiation.
   (b) lowering the albedo of snow.
   (c) releasing volatile organic compounds.
   (d) interacting with existing greenhouse gases.

10. From 1980 to 1990, the concentration of atmospheric CO_2 increased from 335 to 350 ppm. Based on these two time points, predict the concentration of atmospheric CO_2 in 2030 assuming that the rate of CO_2 increase remains the same.
   (a) 360 ppm
   (b) 395 ppm
   (c) 405 ppm
   (d) 410 ppm

11. Which factor is responsible for annual fluctuations in atmospheric CO_2 concentrations?
(a) primary production during the spring and summer
(b) the El Niño–Southern Oscillation
(c) deforestation
(d) combustion of fossil fuels during the summer

12. Scientists can analyze the ______ of foraminifera in ocean sediments to detect past changes in ______.
(a) species composition; CO_2 concentrations
(b) species composition; water temperatures
(c) quantity; global temperatures
(d) quantity; water temperatures

13. In a global warming scenario, which will most strongly contribute to a negative feedback loop?
(a) ocean acidification
(b) melting permafrost
(c) primary production
(d) evapotranspiration

14. Which is a potential consequence of the melting of glaciers and polar ice caps due to global warming?
I. Europe and North America experiencing warmer temperatures
II. loss of the thermohaline circulation
III. global rises in sea level
(a) I only
(b) II only
(c) I and II
(d) II and III

15. Ocean acidification is primarily caused by
(a) acid rain.
(b) the precipitation of base compounds.
(c) increased atmospheric CO_2.
(d) the melting of glaciers.

16. Which best describes the U.S. justification for not signing the Kyoto Protocol after it was modified in 2001?
(a) The U.S. government argued that the protocol imposed unreasonably stringent regulations on developing nations.
(b) It is impossible to reduce carbon emissions without causing substantial harm to the U.S. economy.
(c) The U.S. Senate voted unanimously that the United States should not sign any international agreement that lacked restrictions on developing countries.
(d) The agreement did not require sufficient reductions in the carbon emissions of China and India despite their large population sizes.

Question 17 refers to the following figure:

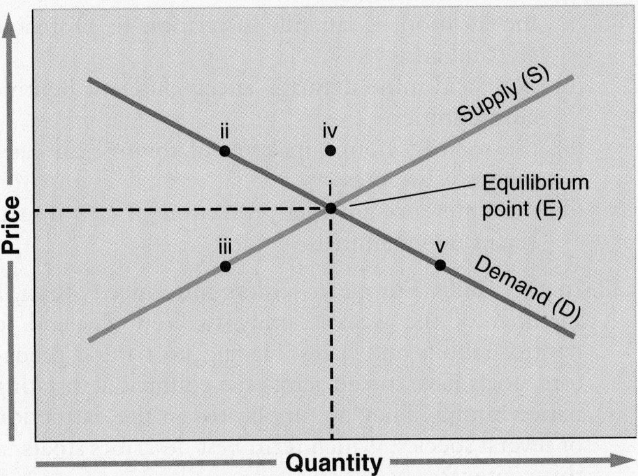

17. The graph depicts the supply and demand curves for production of energy-efficient light bulbs in the United States. Currently, the equilibrium point of the two curves is at *i*. Suppose that government subsidies are given to the light bulb manufacturers in order to decrease the overall cost of production and increase the quantity of bulbs produced. Where is the new equilibrium point most likely to lie?
(a) i
(b) ii
(c) iii
(d) v

18. Which challenges the conceptual basis of the Kuznets curve?
(a) technology transfer between developed nations
(b) leapfrogging by undeveloped nations
(c) increasing environmental degradation with increasing gross domestic product
(d) decreasing environmental degradation with increasing per capita income

19. According to the biocentric worldview
(a) we should adapt to nature rather than adapt nature to our needs.
(b) we can solve resource depletion with technological innovation but nature does require some protection.
(c) humans are one of many species on Earth, and each has equal intrinsic value.
(d) it is the ethical responsibility of humans to care for all species on the planet.

20. The United Nations Environment Programme
(a) seeks to improve human health by assessing health trends among countries.
(b) provides technical and financial assistance to developing countries.
(c) is responsible for gathering environmental information and conducting research.
(d) encourages the elimination of poverty through ecotourism.

21. Which research topic might be covered in the field of environmental justice?
(a) the location of landfills in relation to protected wetland areas
(b) how acid mine drainage affects different benthic communities
(c) the socioeconomic makeup of towns near hazardous waste sites
(d) the difference in legal protection of rare insects versus rare mammals

22. In the 1800s European settlers introduced stoats, a member of the weasel family, to New Zealand to control rabbits and hares. Having no natural predators, stoats have spread across the country, devastating native birdlife. They are implicated in the extinction of several species. Which term best describes stoats in this context?
(a) apex species
(b) ecosystem engineer
(c) invasive species
(d) endemic species

23. The triple bottom line approach to sustainability considers which factors when making decisions about business?
(a) social, political, environmental
(b) economic, social, environmental
(c) environmental, economic, social
(d) biological, economic, social

24. Which gas has the greatest global warming potential?
(a) water vapor
(b) methane
(c) carbon dioxide
(d) nitrous oxide

25. What caused the annual variation in carbon dioxide concentrations collected by Charles Keeling?
(a) the standard deviation of the equipment used to measure CO_2 concentration
(b) changes in emission standards in the top CO_2 producing countries
(c) seasonal variation in photosynthesis in the northern hemisphere
(d) fluctuation in infrared radiation absorption by clouds and the surface of Earth

Section 2: Free-Response Questions

Write your answer to each part clearly. Support your answers with relevant information and examples. Where calculations are required, show your work.

1. In 1997, tropical nations joined together in a massive plan to create a continuous terrestrial corridor that would conserve endemic species and allow dispersal of species between North and South America. Currently, the project is still underway and each nation is continually adding fragments of land to the corridor. Developers of this initiative have faced several problems. For example, many indigenous tribes exist throughout the corridor. Displacement from their land would mean the loss of the culture and heritage associated with those tribes. There are also challenges in purchasing and connecting fragmented land, as well as monitoring the success of the corridor.
(a) Identify one environmental worldview that advocates for allowing these tribes to persist on their land. Justify your answer by defining the worldview. (2 points)
(b) Describe how the concept of SLOSS can be used to overcome the challenges of connecting fragmented land. (4 points)
(c) Describe two ecosystem services that could be monitored to evaluate the health of protected land in the corridor. (2 points)
(d) Define the IUCN Red List and suggest how we can use the Red List to monitor the success of the corridor. (2 points)

2. Evidence indicates that atmospheric carbon dioxide has greatly increased over the past century and that this increase is associated with changes in average temperatures. Although these changes are likely to alter the current range of species, ecologists are concerned that these changes will also alter the timing and placement of ecological interactions.
(a) As temperatures warm during the spring, many species of butterfly migrate to northern latitudes to find food and breeding spots. As temperatures cool during the fall, they migrate south to warmer conditions.
i. Describe two ways in which migrating butterflies might suffer as a result of changes in global temperature. (2 points)
ii. Suggest one way that migrating butterflies might rapidly evolve over a few generations to cope with changes in global temperatures. (2 points)
(b) Describe how we can use ice cores from the Antarctic to determine if the abundance of carbon dioxide in the atmosphere has recently increased. (3 points)
(c) Describe THREE ways in which global warming might alter weather patterns in North America. (3 points)

3. Many people feel that climate change is one of most pressing and challenging environmental issues of our time.
   (a) One difficulty of addressing climate change is that we don't know exactly how the environment will respond to it. Describe ONE positive feedback loop and ONE negative feedback loop that may occur as a result of climate change. (4 points)
   (b) Over the years, nations around the world have come together to create agreements that address climate change. Compare and contrast the Paris Climate Agreement with the Kyoto Protocol in terms of goals and restrictions on developed and developing countries. How has the United States responded to these agreements? (4 points)
   (c) Describe the role of individual or community action in the context of climate change. (2 points)

Cumulative AP® Environmental Science Practice Exam 1

Section 1: Multiple-Choice Questions

Choose the best answer for questions 1–80.

Question 1 refers to the following table.

Net	Amount of organic matter
1	400 g
2	450 g
3	750 g
4	200 g
5	350 g

1. To estimate the amount of organic matter deposited on the soil within a forest covering 10,000 m^2, researchers set out five nets that can catch any material falling from above. They randomly place the nets throughout the forest. Each net has a surface area of 1 m^2. The results of their collection over 24 hours are listed in the table. Using this data, estimate how much organic matter was deposited throughout the entire forest during the time of collection.
   (a) 3,000 kg
   (b) 4,300 kg
   (c) 10,500 kg
   (d) 21,600 kg

2. Which is an example of a secondary pollutant?
   (a) volatile organic carbons
   (b) NO_X emissions from the burning of coal
   (c) photochemical oxidants of NO_X
   (d) carbon dioxide from car exhaust

3. Which example of water pollution describes a point source?
   (a) residents of a city in a developing country bathing in rivers
   (b) wastewater from an industrial chemical plant
   (c) acid rain
   (d) fertilizer runoff

4. Per capita energy consumption refers to
   (a) the total energy consumption of an entire country.
   (b) the total energy consumption per family.
   (c) the average energy consumption of a single person.
   (d) the average energy consumption of a state's capital.

Question 5 refers to the following figure.

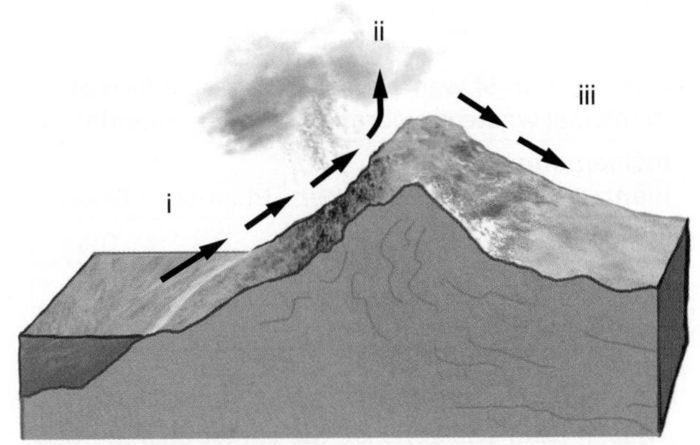

5. Which processes identify the labels associated with i, ii, and iii, respectively?
   (a) evaporation; adiabatic cooling; adiabatic heating
   (b) precipitation; evaporation; adiabatic heating
   (c) evapotranspiration; cloud formation; wind
   (d) evaporation; adiabatic heating; adiabatic cooling

6. One indicator of an improperly functioning septic system might be
   (a) scum that develops on the water's surface in a septic tank.
   (b) a large layer of sludge at the bottom of the septic tank.
   (c) large amounts of nutrients in septage.
   (d) detection of fecal coliform bacteria in a leach field.

7. Composting material efficiently requires
   (a) water, carbon dioxide, and oxygen.
   (b) oxygen, bacteria, and heat.
   (c) water, oxygen, and bacteria.
   (d) energy, oxygen, and bacteria.

Questions 8 and 9 refer to the following tables.

Energy density and percent ash remaining for different types of municipal solid waste		
Type of waste	**Average energy density**	**Average percent ash**
Food	5 kJ/g	5%
Rubber	23 kJ/g	10%
Plastics	32 kJ/g	10%
Glass	0.5 kJ/g	90%

Composition of waste received at five different municipal waste incineration plants (percentage)				
Incineration plant	**Food**	**Rubber**	**Plastics**	**Glass**
Plant A	25%	75%	0%	0%
Plant B	90%	0%	0%	10%
Plant C	0%	0%	80%	20%
Plant D	0%	15%	85%	0%

8. Which plant will generate the largest quantity of energy?
   (a) plant A
   (b) plant B
   (c) plant C
   (d) plant D

9. Which plant will provide the smallest quantity of energy per gram of ash generated?
   (a) plant A
   (b) plant B
   (c) plant C
   (d) plant D

10. Which represents a major way that humans have altered the nitrogen cycle?
    (a) production of fertilizer
    (b) burial of carbon
    (c) long-term mining of rocks
    (d) creation of algal blooms

11. Which would likely allow a home to retain warmth in colder seasons and stay cool for longer periods of time during warm seasons?
    (a) increasing the thermal mass of building materials
    (b) placing solar panels on the roof
    (c) using single-paned instead of double-paned windows
    (d) reducing the thickness of aboveground walls

12. You are asked to determine the ecological footprint required to build a new electric car. Which factor would you need to know to measure the footprint?
    I. the area of land exploited to obtain the raw materials to build the car
    II. the carbon footprint of nonelectric cars
    III. the efficiency of the electric car's engine
    (a) I only
    (b) II only
    (c) I and II
    (d) I and III

13. Suppose that you are buying a new washing machine. You have a choice to buy an Energy Star machine for $500 or a standard machine for $300. The Energy Star machine costs $1.00 per load of laundry, whereas the standard machine costs $1.50 per load of laundry. If you wash an average of three loads of laundry per week, how long will it take to recover the additional cost of the Energy Star unit?
    (a) 1.2 years
    (b) 2.6 years
    (c) 3.8 years
    (d) 6.4 years

14. In any given lake or pond, where is most plant life found?
    (a) intertidal zone
    (b) profundal zone
    (c) limnetic zone
    (d) littoral zone

15. In many parts of the world, park visitation is a "product" that consumers purchase by paying an entrance fee. Increasing the green taxes associated with national park entrance fees is likely to
    (a) result in undervaluation of national parks.
    (b) reduce demand for park visitation.
    (c) increase demand for parks while lowering the number of parks.
    (d) increase both supply and demand for park visitation.

16. Oxpeckers are a type of bird that likes to ride on the backs of large grazing animals. They eat the fleas, ticks, and lice that infest the hair of the animals. The relationship between the oxpecker and the animal on which it grazes is best described as
    (a) mutualistic.
    (b) commensal.
    (c) parasitic.
    (d) predatory.

17. Which describes a benefit of CAFOs?
    (a) CAFOs reduce the amount of animal waste produced.
    (b) CAFOs provide a high yield of animals per area of land.
    (c) CAFOs allow animals to roam freely around farmlands.
    (d) CAFOs use antibiotics to maintain the health of animals.

18. Suppose that a new type of lightbulb is 50 percent efficient at generating light. Energy production at the powerplant has a 30 percent efficiency, and the transmission of energy across powerlines is 80 percent efficient. What is the overall energy efficiency of generating light from this lightbulb?
    (a) 0.12 percent
    (b) 12 percent
    (c) 30 percent
    (d) 50 percent

19. What is a common feature of scrubbers, electrostatic precipitators, and bag house filters?
    (a) They generate solid waste that must be removed.
    (b) They increase the efficiency of electricity generation.
    (c) They remove pollutants using electric charges.
    (d) They remove gas pollutants.

20. Which is not an example of natural capital?
    (a) air
    (b) migratory mammals on the Serengeti
    (c) rainforests
    (d) the national highway system

21. How is electricity generated at hydroelectric, nuclear, and coal energy plants?
    (a) by using the kinetic energy of moving steam to rotate a turbine
    (b) by increasing the energy density of fuel resources
    (c) by collecting the energy emitted from radioactive material
    (d) by converting potential energy into kinetic energy of rotation

22. In the United States, citizens of minority backgrounds are more likely to be exposed to hazardous chemicals. What explains this phenomenon?
    I. environmental inequity
    II. poverty
    III. gender inequality
    (a) I only
    (b) II only
    (c) I and II
    (d) I and III

23. What piece of legislation in the United States defined acceptable limits of pollutants in waterways?
    (a) Clean Water Act
    (b) Water Reclamation Act
    (c) Safe Water Drinking Act
    (d) Water Pollution Act

24. A penguin chick is born with all white feathers. Researchers determine the coloring is caused by a genetic mutation. When the white-feathered penguin breeds with a black and white penguin, its offspring also have all white feathers. This phenomenon is best described as
    I. macroevolution.
    II. microevolution.
    III. a phenotypic change.
    (a) II only
    (b) III only
    (c) II and III
    (d) I, II, and III

25. Which is the cleanest form of nonrenewable energy?
    (a) solar
    (b) oil sands
    (c) natural gas
    (d) petroleum

26. For a lake in Michigan, researchers have determined that largemouth bass feed on smaller fish, which in turn feed on zooplankton. In 2016, there was 600,000 kg of zooplankton in the lake. In 2017, an accidental runoff of insecticide near the lake caused a 50 percent decline of the zooplankton population in the lake. Assuming 10 percent trophic efficiencies, what is the expected decline in largemouth bass?
    (a) 1,000 mg
    (b) 1,500 kg
    (c) 2,000 kg
    (d) 3,000 kg

27. In one year, researchers estimated that 5 billion tons of CO_2 was emitted from cars; 20 billion tons of CO_2 was emitted from industries; 20 billion tons of CO_2 was emitted from energy plants; and 4 billion tons of CO_2 was emitted from domestic usage. Assuming that these sources account for all major human-derived sources of CO_2 pollution, what percentage of CO_2 pollution can be attributed to domestic usage?
    (a) 2.3 percent
    (b) 3.0 percent
    (c) 8.2 percent
    (d) 22.5 percent

28. A forest contains two species of snakes. The hognose snake lives exclusively in the north part of the forest, and the black rat snake lives exclusively in the south part of the forest. Although the hognose can survive in the abiotic conditions of the southern forest and find plenty of prey in the south, the black rat snake is a better competitor for the southern forest prey. Which best explains why the hognose snake does not live in the southern forest?
   (a) The southern forest is not part of the hognose's realized niche.
   (b) The southern forest is not part of the hognose's fundamental niche.
   (c) The black rat snake cannot find enough resources in the northern forest.
   (d) The northern forest is not part of the black rat snake's realized niche.

29. Which is a major disadvantage of run-of-the-river power plants?
   (a) They rely on water impoundments.
   (b) Power generation is unreliable.
   (c) They lead to large amounts of silt deposition upstream.
   (d) They tend to be large and unsightly.

30. In urban areas situated in a region with moderate amounts of rainfall, a paved parking lot can contribute to
   (a) increased rates of flooding.
   (b) a reduction in total runoff.
   (c) increased periods of drought.
   (d) faster rates of aquifer recharge.

31. Water draining from an abandoned coal mine is likely to have
   (a) an excess of hydroxide ions and a low pH.
   (b) an excess of hydrogen ions and a low pH.
   (c) an excess of hydrogen ions and a high pH.
   (d) an excess of hydroxide ions and a high pH.

32. Which is most likely to result in the salinization of soil?
   (a) drip irrigation
   (b) mechanization
   (c) fertilizer application
   (d) flood irrigation

Question 33 refers to the following figure.

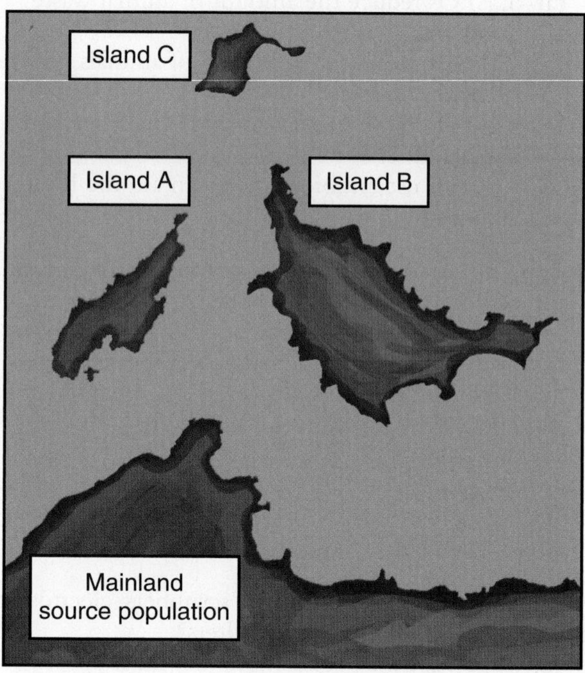

33. Consider how many species are likely to colonize islands A, B, and C and list the islands in order of increasing species richness.
   (a) island C, island B, island A
   (b) island C, island A, island B
   (c) island A, island B, island C
   (d) island A, island C, island B

34. Which could increase the energy return on energy investment of photovoltaic cells?
   I. increasing the amount of energy generated per cell
   II. increasing the lifetime of the cells
   III. decreasing the energy required to produce the cells
   (a) I only
   (b) III only
   (c) I and III
   (d) I, II, and III

35. Which is NOT one of the four major agencies that regulates public land?
   (a) Bureau of Land Management
   (b) National Park Service
   (c) National Forest Service
   (d) National Science Foundation

36. Which statement regarding electronic waste (e-waste) is TRUE?
   (a) Most e-waste in the United States is recycled.
   (b) Recycling e-waste involves handling many toxic metals and hazardous chemicals.
   (c) By federal regulation, e-waste must remain in the country of origin.
   (d) Recycling e-waste is better for the environment than reusing it.

37. Which is most likely to experience overshoots and die-offs?
(a) populations with a low carrying capacity
(b) *K*-selected species
(c) *r*-selected species
(d) populations with long generation times

Question 38 refers to the following table.

	MJ per passenger-kilometer	MJ per gallon of fuel	Cost per gallon of fuel
Bus (diesel)	1.7	150	$5.00
Bus (biodiesel)	1.7	120	$6.50
Passenger car	3.6	120	$4.00
Motorcycle	1.7	120	$4.00

38. Which mode of transportation is the least expensive for a single traveler?
(a) passenger car
(b) bus (biodiesel)
(c) bus (diesel)
(d) motorcycle

39. To examine long-term effects of a chemical on a study organism, it would be best to conduct a(n)
(a) dose-response study.
(b) acute study.
(c) extrapolative study.
(d) chronic study.

40. Which best describes the function of a keystone species?
(a) The species provides a vital service for humans.
(b) The species supports the growth of other species.
(c) The species has an effect on the ecosystem that is disproportionate to its abundance.
(d) The species engineers the ecosystems to alter the flow of rivers and streams.

41. Cogeneration is a method of energy production in which
(a) wasted heat during electricity production is used to warm buildings.
(b) two or more fuel types are used to generate electricity.
(c) both natural gas and coal are used to rotate turbines.
(d) a combination of renewable and nonrenewable sources is employed.

Question 42 refers to the following figure.

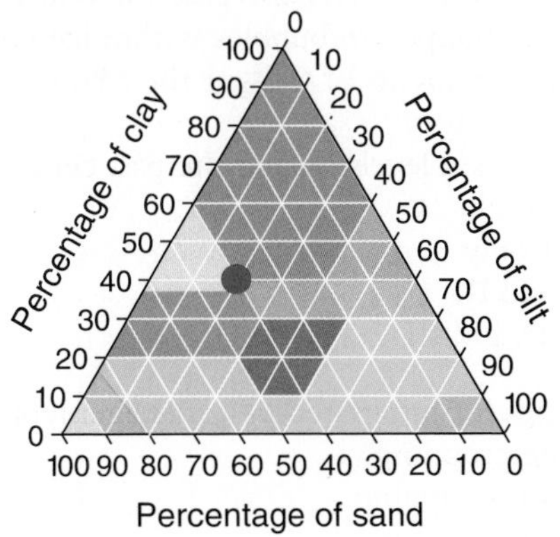

42. What is the composition of soil in the location indicated by the red dot?
(a) 60 percent sand, 60 percent silt, 40 percent clay
(b) 40 percent sand, 20 percent silt, 60 percent clay
(c) 40 percent sand, 20 percent silt, 40 percent clay
(d) 80 percent sand, 20 percent silt, 60 percent clay

43. Which is an example of an ecosystem provision?
(a) a tree that provides a home to nesting birds
(b) a tree that removes substantial amounts of CO_2 from the atmosphere
(c) a plant containing an anticancer compound
(d) nitrogen-fixing bacteria that provide nutrients to crops

44. Externalities are best defined as
(a) a cost of a good or service that is not assigned a monetary value.
(b) a tax imposed on a good to offset the high price of manufacturing.
(c) credit given to developing nations for atmospheric pollution.
(d) regulations placed on landowners.

45. *Ferroplasma acidiphilum* bacteria prefer to live in acidic discharge from coal mines, which reaches a pH level of about 2. Relative to bacteria living in neutral water, *Ferroplasma* bacteria are exposed to
(a) 100,000 times the number of hydrogen ions.
(b) 100,000 times the number of hydroxide ions.
(c) 10,000 times the number of hydrogen ions.
(d) 10,000 times the number of hydroxide ions.

46. If a nation has a replacement level fertility of 10, it might be an indication of
(a) low immigration rates.
(b) poor health care.
(c) increased energy use.
(d) high immigration rates.

47. Which provides evidence for long-term changes in atmospheric CO_2 over the past 400,000 years?
I. CO_2 trapped in bubbles within ice cores
II. measurements of CO_2 at the Manu Loa observatory
III. rising sea levels during the past century
(a) I only
(b) III only
(c) I and III
(d) I, II, and III

48. Strategies for integrated pest management include
I. intercropping.
II. monocropping.
III. crop rotation.
(a) I only
(b) II only
(c) I and II
(d) I and III

49. As a population grows smaller, which is likely to increase within that population?
I. inbreeding depression
II. competition
III. genetic diversity
(a) I only
(b) II only
(c) I and II
(d) I, II, and III

50. Most fresh water on Earth can be found in ice, glaciers, and
(a) the ocean.
(b) the ground.
(c) streams.
(d) the atmosphere.

51. What is one way that the realized niche of a species is likely to expand?
(a) A disturbance kills off a competing species within its fundamental niche.
(b) A massive catastrophe separates some individuals from the main population.
(c) A massive catastrophe causes a dramatic reduction in population size.
(d) There is an increase in the growth rate of the species.

52. A country in the fourth phase of demographic transition might experience population growth because:
(a) immigration is high.
(b) GDP is low.
(c) the death rate is low.
(d) infant mortality is high.

53. Sick building syndrome often occurs when
(a) contaminants accumulate in the air of a well-insulated building.
(b) a building falls into disrepair and contaminants are released.
(c) hazardous waste is improperly disposed.
(d) contaminants are released from asbestos insulation in old buildings.

54. Gardeners often add compost to the surface of gardens to increase productivity. Once applied, this layer of compost comprises the
(a) A soil horizon.
(b) B soil horizon.
(c) C soil horizon.
(d) E soil horizon.

55. Under the fishery management system of individual transferable quotas, a fisher who catches very few fish during a season can still earn a profit by
(a) selling quotas to other anglers.
(b) fishing for a longer period of time.
(c) selling quotas to the aquaculture industry.
(d) selling bycatch to other markets.

56. Which causes rapid changes to the structure of Earth's surface?
(a) glacial erosion
(b) root weathering
(c) freeze-thaw cycling
(d) earthquakes

Question 57 refers to the following figures.

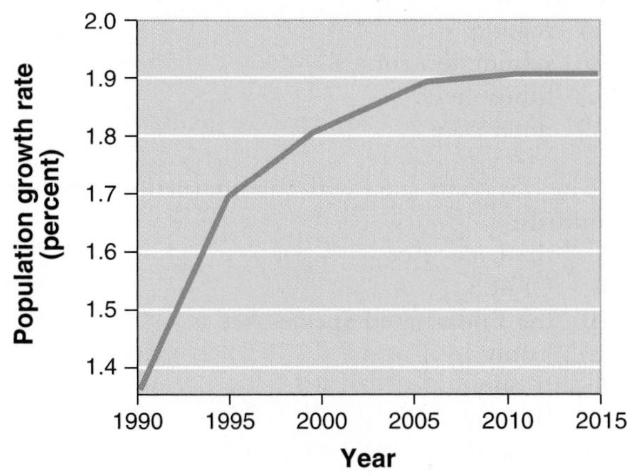

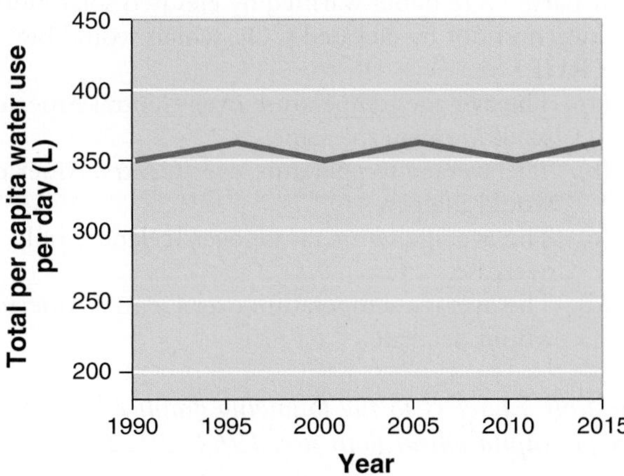

57. Which statement is accurate?
    (a) Total water use has decreased from 1990 to 2015.
    (b) Total water use has increased from 1990 to 2015.
    (c) There has been no change in total water use from 1990 to 2015.
    (d) Average per capita water use has doubled between 1990 and 2015.

58. For a given population, crude birth rate is 20 and crude death rate is 10. Immigration and emigration rates are 0. What is the doubling time?
    (a) 10 years
    (b) 27 years
    (c) 35 years
    (d) 70 years

59. In 1920, the United States had 7.21×10^8 acres of land classified as forested. On average, the country gained an additional 2.25×10^5 hectares of forest per year. Between 1920 and 2000, what percent change in forested land coverage occurred in the United States?
    (a) 2.5 percent
    (b) 6.2 percent
    (c) 9.0 percent
    (d) 10.4 percent

60. A clothes dryer uses 3,000 watts when the motor is running. A newer model uses 400 watts when the motor is running. For both models, a single drying cycle takes 1 hour to complete. Suppose that a family uses their clothes dryer four times per week and electricity costs $0.15 per kilowatt-hour. How much money will the family save in a single year if they replace their clothes dryer with a newer model?
    (a) $81
    (b) $154
    (c) $223
    (d) $500

61. In a previously undisturbed forest, periodic forest fires of moderate intensity are likely to
    (a) increase forest diversity.
    (b) decrease forest diversity.
    (c) decrease forest resiliency.
    (d) stabilize forest diversity.

Question 62 refers to the following figure, which shows total U.S. $PM_{2.5}$ emissions.

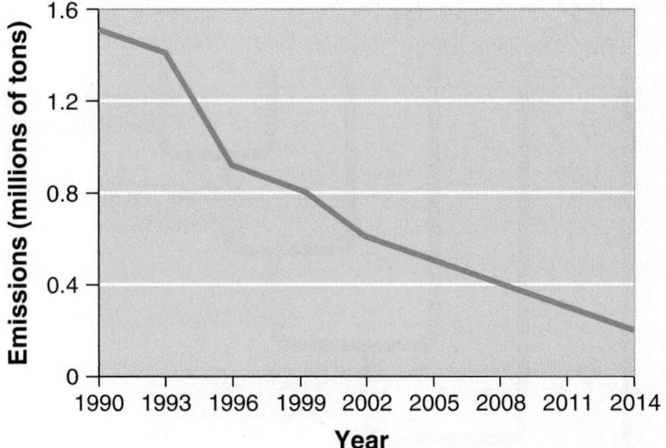

62. Between 1990 and 2014, what is the average percent reduction per year of $PM_{2.5}$ emissions?
    (a) 1.6 percent
    (b) 2.5 percent
    (c) 3.6 percent
    (d) 10.1 percent

63. Which is likely to increase the rate of saltwater intrusion?
    I. pumping freshwater from a well near a coastline
    II. precipitation from a massive hurricane
    III. a rise in sea level
    (a) I only
    (b) II only
    (c) I and II
    (d) I and III

64. Suppose that all plants in an ecosystem are capable of absorbing 10,000 J of solar energy per day, and they release 9,500 J of solar energy per day due to respiration. What is the net primary productivity of that system?
(a) 500 J
(b) 19,500 J
(c) 500 J per day
(d) 19,500 J per day

65. Including the cost of externalities in the price of a product will eventually result in
(a) increased supply and reduced demand of the product.
(b) reduced supply and increased demand of the product.
(c) increased supply and increased demand of the product.
(d) reduced supply and reduced demand of the product.

Question 66 refers to the following figure which is a phylogeny based on morphology.

Lamprey Sunfish Newt Lizard Bear Chimpanzee
Common mammal ancestor
Common amniote ancestor
Common tetrapod ancestor
Common jawed vertebrate ancestor
Common vertebrate ancestor

66. Which is TRUE?
I. Bears and newts share a common ancestor.
II. Lizards share greater morphological similarity to bears than to chimpanzees.
III. Sunfish and lampreys had a different common ancestor from lizards.
(a) I only
(b) II only
(c) I and II
(d) I and III

67. Given unlimited resources and no threat of predation, parasites, or pathogens, a population will exhibit
(a) carrying capacity.
(b) a uniform distribution.
(c) logistic growth.
(d) exponential growth.

68. Tectonic plates move because of convection cells within the
(a) mantle.
(b) subduction zone.
(c) lithosphere.
(d) inner core.

69. Which is concerned with the international trade of animals?
I. the Lacey Act
II. CITES
III. the Endangered Species Act
(a) I only
(b) II only
(c) III only
(d) I and II

70. If Earth were being warmed by elevated solar radiation and not by elevated CO_2, which would be TRUE?
(a) The average temperature over Central America would remain the same.
(b) The average temperature over Central America would increase.
(c) The average temperature over Iceland would decrease.
(d) The average temperature over Central America would decrease.

Question 71 refers to the following graph which shows U.S. per capita wheat flour use, 1964–2012.

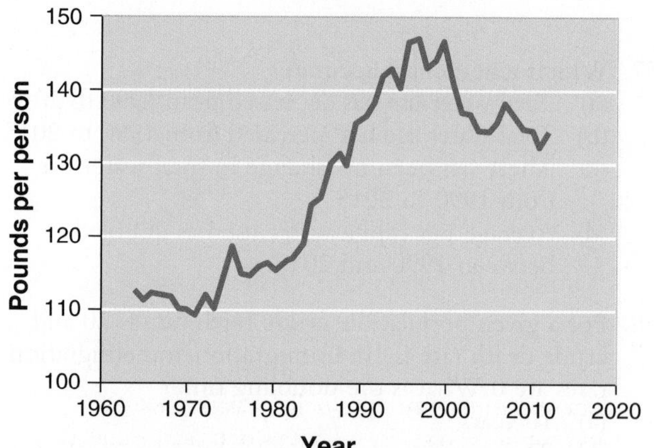

Data from: USDA, Economic Research Service calculations using data through the second quarter of 2011 from U.S. Department of Commerce, Bureau of the Census' Flour Milling Products (MQ311A) and U.S. Department of Commerce, Bureau of Economic Analysis' Foreign Trade Statistics.

71. Based only on the data provided in the graph, which of the following statements could be inferred?
(a) The amount of wheat flour used per person in the United States rose between 1972 and 1998.
(b) Population densities in the United States increased between 1972 and 2012.
(c) The average amount of wheat flour used per person in the United States has remained steady between 1968 and 2012.
(d) Population abundance in the United States increased between 1972 and 2012.

72. An increase in the average albedo of the Earth's surface would lead to
(a) more solar energy reflected by the atmosphere and a gradual cooling of Earth.
(b) more solar energy reflected from Earth's surface and a gradual cooling of Earth.
(c) less solar energy reflected by the atmosphere and a gradual warming of Earth.
(d) less solar energy reflected from Earth's surface and a gradual warming of Earth.

73. Radiocarbon dating is a process that determines the age of material based on
(a) conversion of an unstable isotope of carbon into a stable isotope of nitrogen.
(b) splitting ^{14}C isotopes into two separate atoms.
(c) conversion of a stable isotope of carbon into an unstable isotope of nitrogen.
(d) release of electrons from a carbon atom into a nitrogen atom.

74. What is the purpose of water impoundments at hydroelectric plants?
(a) to provide a constant source of water to drive the turbines
(b) to divert water away from the plant
(c) to prevent or minimize downstream water damage during large storm events
(d) to reduce rates of siltation

75. Many developing nations still rely on wood as a source of energy and heat. For this practice to be carbon neutral,
(a) any ash generated by the combustion of wood must be buried in a landfill or composted.
(b) all carbon released from the combustion of wood must be absorbed by plants.
(c) nations must start using a different, renewable energy source.
(d) trees must absorb more carbon than is emitted from the combustion of the wood.

76. Over thousands of years, rainfall erodes mountainsides that contain precious minerals such as gold. These minerals are gradually washed into streams and carried down into valleys between mountains. The most common method of mining for these minerals is
(a) open pit mining.
(b) placer mining.
(c) subsurface mining.
(d) mountaintop removal.

77. In recent decades, the increased severity (i.e., intensity, spread, and duration) of forest fires has been primarily caused by
(a) prescribed burning.
(b) fire suppression.
(c) clear-cutting.
(d) tragedy of the commons.

78. The strategy of planned obsolescence often leads to
(a) lower rates of recycling.
(b) greater amounts of MSW.
(c) more open-loop recycling.
(d) lower GDP.

79. Suppose that a specific plant requires 6,000 additional glucose molecules provided to each cell to double its productivity. How many additional CO_2 molecules will each cell need?
(a) 6,000
(b) 36,000
(c) 40,000
(d) 62,000

80. What is the primary cause of ocean acidification?
(a) absorption of CO_2 by the oceans
(b) coral bleaching
(c) acid mine drainage
(d) melting of Arctic ice sheets

Section 2: Free-Response Questions

Write your answer to each part clearly. Support your answers with relevant information and examples. Where calculations are required, show your work.

1. Many of the islands in the South Pacific Ocean are separated from each other by hundreds of miles and are at least 800 miles from mainland Australia. Early settlers on many of these islands founded colonies with little immigration or emigration until the advent of modern methods of transportation.
(a) Explain why the human population of a small South Pacific island might be at greater risk of genetic defects relative to the population of mainland Australia. (2 points)
(b) Define genetic drift and discuss why the original colonists of a small island would be more susceptible to this process than individuals in mainland Australia. (2 points)
(c) For a newly founded island, suppose that two children are born each year for every 10 individuals and one adult out of every 100 individuals dies each year. Assume there is no immigration or emigration. Calculate the population growth rate of the island. (2 points)
(d) Given the population growth rate calculated in (c), calculate the doubling time of the island population. (2 points)
(e) The island of New Caledonia is approximately the same distance from mainland Australia as Norfolk Island, but is approximately 500 times larger. Use the theory of island biogeography to explain how the number of species on these two islands will differ and why. (2 points)

2. In a series of studies, researchers explored the psychological effects of green space (i.e., areas with plants) in cities. They reported a significant, positive association between the amount of green space around schools and the ability for students to focus on coursework. In addition, they found that human populations living near green space have significantly fewer individuals diagnosed with depression and anxiety. The diversity of species in the green space also matters: One study reported that the number of people in an area who take medicine for depression is negatively correlated with the diversity of plants and birds in that same area.
   (a) Provide TWO additional positive externalities not mentioned in the report above and TWO negative externalities associated with green space in cities. (2 points)
   (b) List THREE principles of smart growth that involve the implementation, design, and maintenance of green spaces. Explain how green space contributes to each of those three principles. (3 points)
   (c) Describe a natural experiment to explore the relationship between green space around schools and student performance on standardized exams. Your description should contain the following:
      (i) a hypothesis for your experiment. (1 point)
      (ii) a description of the experimental procedure and the data to be collected. (2 points)
      (iii) a description of the results that would support your hypothesis. (2 points)

3. The city of Cape Town, South Africa, is home to 3.75 million people. In recent years, the city and surrounding countryside have experienced the worst drought in recorded history. Water reserves are generally replenished by rainfall during the months of July and August, but there has been less rainfall each year since 2014. At the same time, the population size of the city has grown by 2.5 percent each year for the past several decades. To address the water shortage, the government has limited the use of water by individual households. However, many critics suggest that such limitations will not significantly help the shortage since households are not necessarily the dominant consumer of water. Instead, they point out that most electricity generation in Cape Town is derived from coal-fired electricity generation plants, which use large quantities of water for cooling.
   (a) Describe how a coal-fired electricity generation plant generates electricity, including how water is used, recycled, and lost in the process. (2 points)
   (b) For every 1 kWh of electricity, electricity generation plants must heat 700 gallons of fresh water. Although the water is cooled and recycled, 5 percent is lost in the process. If an average home in Cape Town uses 650 kWh per month, how many gallons of water per home are lost each month? (2 points)
   (c) Name two alternative methods of electricity production that do not use water and explain how each generates electricity. (2 points)
   (d) To avoid excessive water use, many people have started using disposable paper plates instead of reusable ceramic plates. Using cradle-to-grave analysis, how might you determine whether reusable ceramic plates or disposable paper plates use less water? (4 points)

Cumulative AP® Environmental Science Practice Exam 2

Section 1: Multiple-Choice Questions

Choose the best answer for questions 1–80.

1. Which is the most abundant greenhouse gas?
   (a) CH_4
   (b) H_2O
   (c) CO_2
   (d) NO_2

2. Subsistence economies occur primarily in
   (a) developed nations.
   (b) urban areas.
   (c) suburban areas.
   (d) developing nations.

3. Why do chemical manufacturers test the effects of any new chemical on a bird, a mammal, a fish, and an invertebrate?
   (a) These animals tend to exhibit a more acute response than other types of animals.
   (b) These animals are most important to the environment.
   (c) These animals are thought to be the most sensitive species.
   (d) These animals are the most important to people.

Question 4 refers to the following figure.

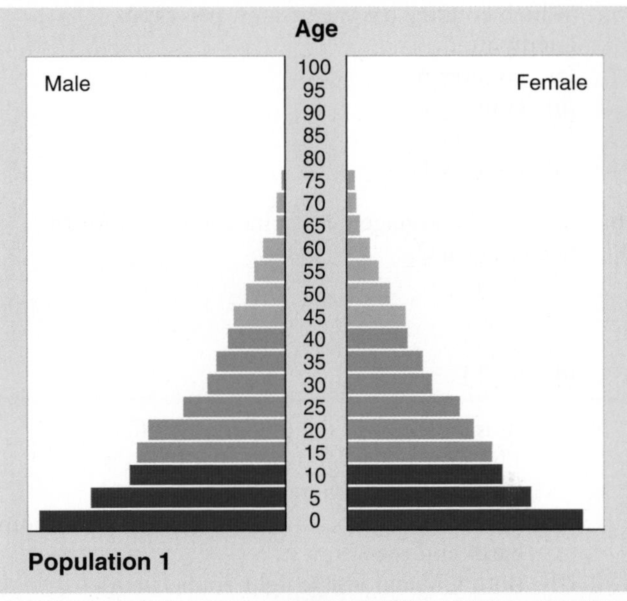

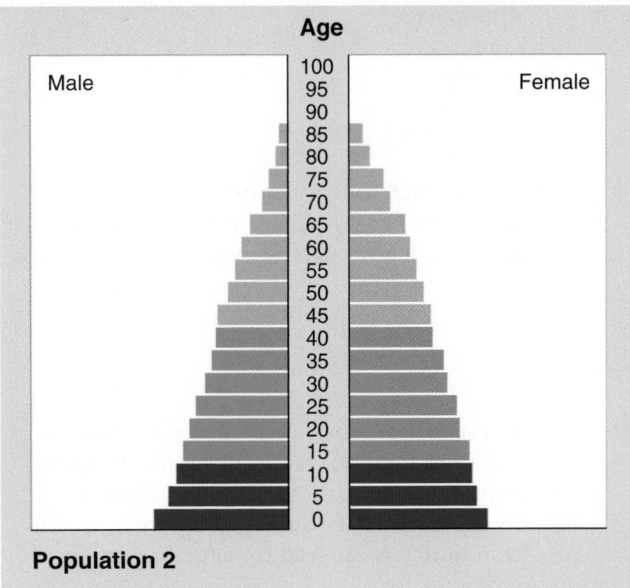

4. Which statement best describes the two populations depicted in the graph?
   (a) Both populations are growing and population 2 is experiencing faster growth than population 1.
   (b) Population 1 is growing and population 2 is stable.
   (c) Population 1 is growing and population 2 is shrinking.
   (d) Both populations are growing and population 1 is growing faster than population 2.

Questions 5 and 6 refer to the following table.

Country	Total annual energy use	Total population
Country A	100 million J	1 million people
Country B	100 million J	10 million people
Country C	100 million J	100 million people
Country D	10 million J	10 million people

5. Which country has the greatest per-capita energy use?
   (a) country A
   (b) country B
   (c) country C
   (d) country D

6. What is the average per-capita energy use for the five countries?
   (a) 2.5 J
   (b) 10 J
   (c) 22.6 J
   (d) 62.2 J

7. Which is a process in the greenhouse effect?
   I. The solar radiation that strikes Earth is converted into infrared radiation.
   II. Greenhouse gases absorb infrared radiation from Earth and then emit it.
   III. Both UV and visible light enter Earth's atmosphere.
   (a) I only
   (b) III only
   (c) I and III
   (d) I, II, and III

8. Last year, the total daily water use in Norway was 1.56×10^6 m^3. The population of Norway is 5.2×10^6 people. What was per capita daily water usage in Norway?
   (a) 150 L
   (b) 200 L
   (c) 250 L
   (d) 300 L

9. Artificial selection is best defined as a mechanism of
   (a) evolution that can lead to novel phenotypes.
   (b) recombination that can lead to evolution.
   (c) evolution that can lead to mutations.
   (d) evolution that can lead to natural selection.

Questions 10 and 11 refer to the following information.

For most plants and algae, chlorophyll *a* is the molecule that absorbs solar energy for use in photosynthesis. High amounts of chlorophyll *a* in water could indicate an increasing risk of algal blooms. In one study, researchers questioned how the addition of different nutrients, alone or in combination, altered the amount of chlorophyll *a* in the water of a lake. Their results are shown in the graph.

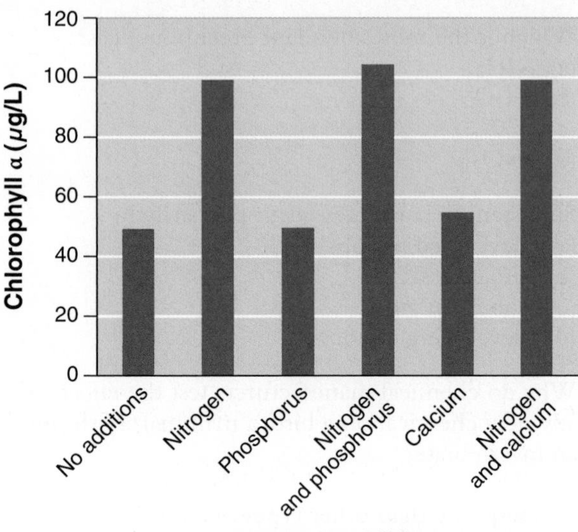

10. What is the limiting nutrient(s) for the lake that the researchers studied?
    (a) calcium
    (b) phosphorus
    (c) nitrogen and phosphorus
    (d) nitrogen

11. Which activity will likely lead to an algal bloom in this lake?
    (a) application of nitrogen fertilizer on surrounding farmlands
    (b) increased use of phosphorus-based detergents
    (c) adding calcium to the lake
    (d) greater rates of denitrification by aquatic bacteria

12. Approaches to pollution control that impose regulations, green taxes, and fines on industries are known as
    (a) ecocentric strategies.
    (b) incentive-based regulations.
    (c) command-and-control strategies.
    (d) international initiatives.

13. After the financial crisis of 2008, the city of Detroit lost most of its industry and many residents left. In the city's attempts to rejuvenate itself, managers focused on emphasizing the unique character of individual districts, such as Mexicantown, Corktown, and Chaldean town. According to the principles of smart growth, this strategy of development is known as
(a) establishing a sense of place.
(b) residential zoning.
(c) limiting urban growth boundaries.
(d) transit-oriented development.

14. Which would reduce the amount of transpiration that occurs in a forest?
(a) an increase in precipitation
(b) a decrease in the average number of leaf stomata
(c) a reduction in herbivory
(d) an increase in the rate of primary productivity

15. Which is likely a consequence of the 2001 Stockholm Convention?
(a) an increase in the number of species placed on the IUCN Red List
(b) an expansion of the EPA's recommendations for smart growth
(c) a sharp reduction in the number of hazardous chemicals used in both industry and farming
(d) a reduction in the number of genetically modified foods allowed to be sold

16. The addition of which material can increase the cation exchange capacity of soil?
(a) silt
(b) sand
(c) clay
(d) nutrients

17. What is the difference between modern carbon and fossil carbon?
(a) Modern carbon provides a more fuel-efficient energy source than fossil carbon.
(b) Modern carbon has cycled in and out of the atmosphere more recently.
(c) Modern carbon has been engineered to burn at a higher temperature than fossil carbon.
(d) Modern carbon rarely emits greenhouse gases, unlike fossil carbon.

18. Green taxes are best defined as
(a) taxes imposed on farmers.
(b) taxes that help offset environmental degradation.
(c) taxes imposed on eco-friendly products.
(d) taxes imposed on states that released historically large amounts of pollution.

19. Food scraps that are buried in a landfill often do not decompose because
(a) there are no detritivores.
(b) there is limited oxygen.
(c) there are few bacteria.
(d) there is limited carbon dioxide.

20. Which organization is responsible for preventing environmental degradation, enforcing environmental regulations, and collecting environmental data among nations?
(a) United Nations Environmental Programme
(b) United Nations Educational, Scientific, and Cultural Organization
(c) World Health Organization
(d) United Nations Development Programme

21. Development of residential and commercial areas often includes large areas of land covered in asphalt. What might be a consequence of increased asphalt cover?
I. reduced albedo
II. greater erosion of surrounding soil
III. an increase in trapped solar energy
(a) II only
(b) III only
(c) I and III
(d) I, II, and III

22. Which would NOT improve the efficiency of a vehicle powered by hydrogen fuel cells?
(a) reducing the energy input needed for the electrolysis of water
(b) reducing the energy input needed for transporting hydrogen gas
(c) collecting any water produced during electricity production
(d) reducing the weight of the vehicle

23. If there are 2.5×10^4 trees in a forest and selective logging removes 1.3×10^3 trees, how many trees are left in the forest?
(a) 2.35×10^3
(b) 2.37×10^3
(c) 2.37×10^4
(d) 2.90×10^4

24. Legal sewage dumping generally occurs during
(a) snowstorms.
(b) warm seasons.
(c) periods of heavy rain.
(d) high tide.

25. Suppose that you recently installed a 10,000-watt photovoltaic solar array on the roof of your house. One year after installing the array, you notice that your annual electricity bill decreased by $1,500. If electricity costs $0.15 per kWh, what is the capacity factor of your solar array?
(a) 11.5 percent
(b) 27.5 percent
(c) 40 percent
(d) 40.5 percent

26. Which statement is NOT correct concerning the intertropical convergence zone (ITCZ)?
   (a) Movement of the ITCZ generates wet and dry seasons in tropical biomes.
   (b) Adiabatic cooling generates clouds above the ITCZ.
   (c) The ITCZ causes trade winds along Earth's surface to move away toward the poles.
   (d) The ITCZ is generally characterized as a hot and moist region.

27. Which best explains why chlorofluorocarbon (CFC) in the stratosphere is harmful for life on Earth?
   (a) CFCs interact with solar radiation to create ultraviolet waves that cause skin cancer.
   (b) The reaction of ozone with CFCs reduces a valuable source of oxygen in the atmosphere.
   (c) CFCs occupy holes in the ozone layer.
   (d) CFCs break apart ozone, thus preventing ozone from absorbing UV radiation.

28. The Ring of Fire is an area of
   (a) increased tectonic activity that occurs along plate boundaries in the Pacific Ocean.
   (b) increased island formation that occurs along subduction zones in the Atlantic Ocean.
   (c) increased tectonic activity that occurs along metamorphic rock formations in the southern hemisphere.
   (d) increased island formation that occurs along divergent plate boundaries in the Pacific Ocean.

29. A NIMBY attitude often leads to an increase in
   I. health risks in poor areas.
   II. environmental protection.
   III. environmental justice.
   (a) I only
   (b) II only
   (c) I and II
   (d) I and III

Questions 30 and 31 refer to the following figure.

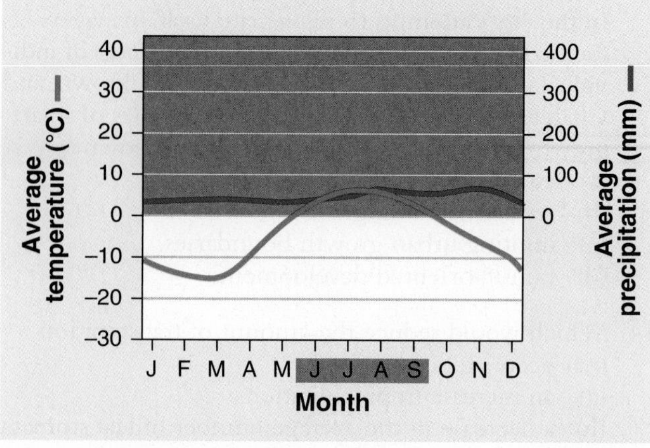

30. During what time of the year is peak productivity?
   (a) October to December
   (b) June to September
   (c) January to May
   (d) January to March

31. Which biome is most likely represented by the figure?
   (a) temperate seasonal forest
   (b) temperate grassland
   (c) seasonal rainforest
   (d) tundra

32. The highest concentration of man-made air pollution occurs over
   (a) Europe.
   (b) the United States.
   (c) the Sahara Desert.
   (d) eastern Asia.

33. In recent years, algal blooms have led to a decline in the population sizes of large fish that are commonly exploited by recreational fishing. Such declines are best described as a
   (a) a loss of intrinsic value.
   (b) a loss of regulating services.
   (c) a loss of instrumental value.
   (d) a loss of species richness.

34. In April 2009, a giant avalanche within Banff National Park in Alberta, Canada, buried and killed almost all of the caribou population living within the park and, as a result, decreased the population's genetic diversity. This represents
   (a) natural selection.
   (b) a bottleneck effect.
   (c) genetic drift.
   (d) a founder's effect.

35. When a population's carrying capacity is included in the intrinsic growth model, the model takes the form of
(a) an S-shaped growth curve.
(b) a type I survivorship curve.
(c) a J-shaped growth curve.
(d) a type II survivorship curve.

36. Three heavy metals commonly associated with water pollution are
(a) iron, arsenic, and potassium.
(b) lead, mercury, and arsenic.
(c) lead, iron, and calcium.
(d) mercury, nitrate, and sulfur.

37. Which is not an organic component of the carbon cycle?
(a) calcium carbonate
(b) dead tree leaves
(c) oil
(d) DNA

38. A group of finches reached the Galápagos Islands several million years ago. The island was previously not colonized by any birds. The initial population of finches survived and bred with low genetic variation. This is an example of
(a) a founder's effect.
(b) a bottleneck effect.
(c) natural selection.
(d) genetic drift.

39. Strontium-90 is one of the radioactive waste products from the decay of U-235 in nuclear reactors. It has a half-life of 30 years. After 90 years, what fraction of strontium-90 will remain in the waste?
(a) 1/4
(b) 1/6
(c) 1/8
(d) 1/16

40. For one island ecosystem, ecological efficiency between all adjacent trophic levels is 10 percent. Each year, the island produces about 30,000 kg of plant material that can be consumed by primary consumers. What is the expected mass of all secondary consumers on the island?
(a) 200 kg
(b) 300 kg
(c) 400 kg
(d) 4,000 kg

41. Which does NOT contribute to particulate matter pollution?
(a) diesel exhaust
(b) dust storms
(c) coal burning
(d) electrostatic precipitators

42. Teratogens are chemicals that
(a) interfere with the normal development of embryos.
(b) disrupt the nervous system of animals.
(c) cause uncontrolled cell growth.
(d) interfere with the normal functioning of hormones.

43. Which order shows increasing water-use efficiency?
(a) spray irrigation; furrow irrigation; drip irrigation
(b) furrow irrigation; spray irrigation; drip irrigation
(c) drip irrigation; spray irrigation; furrow irrigation
(d) spray irrigation; drip irrigation; furrow irrigation

Question 44 refers to the following information.

A forest manager is conducting a study to determine the maximum sustainable yield from a forest. To do this, the manager harvests a different biomass of trees from four forest plots. Each plot starts off with an identical composition and biomass of trees. After harvesting, the manager tracks the biomass of new tree growth during the following year, as shown in the table.

	Trees harvested (kg)	New tree growth (kg)
Plot 1	10,000	7,000
Plot 2	8,000	8,000
Plot 3	6,000	9,000
Plot 4	5,000	10,500

44. What is the maximum sustainable yield of the forest?
(a) 5,000
(b) 6,000
(c) 8,000
(d) 8,500

45. Renewable or potentially renewable energy sources include
I. gasohol.
II. wind.
III. manure.
(a) I only
(b) II only
(c) I and III
(d) II and III

46. Effective strategies to reduce total fertility rates include
I. enhanced educational opportunities for women.
II. encouraging gender equity.
III. reducing GDP.
(a) I only
(b) II only
(c) I and II
(d) I and III

47. Considering the economies of scale, which activity would be the most profitable?
(a) large-scale nonconventional farming
(b) large-scale conventional farming
(c) small-scale CAFOs
(d) small-scale intercropping

48. Suppose that you dig a hole in the ground with a shovel. After digging down 12 inches, you notice that water starts to form a small puddle at the bottom of your hole. Which best explains what happened?
(a) You have tapped into a confined aquifer.
(b) Atmospheric water has condensed inside the hole.
(c) You have dug below the water table.
(d) You have struck an artesian well.

49. An increase in the abundance of greenhouse gases in our atmosphere has led to
(a) an increase in carbon dioxide.
(b) increased albedo.
(c) a greater amount of Earth's energy retained by its atmosphere.
(d) more solar energy reflected by the Earth and released to space.

50. The average vehicle in the United States produces 9 kg of CO_2 emissions for each gallon of gasoline consumed, and the average car obtains 24 mpg. In 2011, individuals drove an average annual distance of 9,450 miles. In 2014, individuals drove an average annual distance of 9,700 miles. How much more CO_2 pollution did the average driver produce annually in 2014 relative to 2011?
(a) 81 kg
(b) 84 kg
(c) 90 kg
(d) 94 kg

51. A soil-and-clay cover placed over a filled sanitary landfill
I. reduces the amount of leachate generated by the landfill.
II. reduces the amount of methane production within the landfill.
III. provides a substrate for planting shallow-rooted vegetation.
(a) I only
(b) III only
(c) I and III
(d) I, II, and III

52. Suppose that you want to test whether playing classical music to gerbils has an effect on their ability to learn the quickest pathway to food in a maze. What might be the best treatment and control for this experiment?
(a) Treatment: gerbils exposed to classical music for 3 minutes; control: gerbils exposed to random noise for 3 minutes
(b) Treatment: gerbils exposed to silence for 3 minutes; control: gerbils exposed to random noise for 1 minute
(c) Treatment: gerbils exposed to random noise for 3 minutes; control: gerbils exposed to classical music for 3 minutes
(d) Treatment: gerbils exposed to classical music for 2 minutes; control: gerbils exposed to jazz for 2 minutes

53. Malnourishment can be the result of
I. undernutrition.
II. overnutrition.
III. food insecurity.
(a) I only
(b) I and II
(c) II and III
(d) I, II, and III

54. Which correctly shows an order of increasing energy density?
(a) charcoal, wood, uranium, corn ethanol
(b) wood, charcoal, corn ethanol, uranium
(c) uranium, corn ethanol, charcoal, wood
(d) corn ethanol, wood, uranium, charcoal

55. What objections did the United States raise regarding the Kyoto Protocol?
I. unfair treatment between developed and developing nations
II. uncertainty in the causes underlying climate change
III. unreachable CO_2 reduction targets in the United States
(a) I only
(b) II only
(c) I and II
(d) II and III

56. Which is an example of open-loop recycling?
I. shredding used plastic soda bottles and using the material to make polyester cloth
II. melting glass bottles into materials for stained glass windows
III. melting down used aluminum cans to make new cans
(a) I only
(b) III only
(c) I and II
(d) I, II, and III

57. For a given population, an average of two infants per family die before they reach adulthood. What would replacement level fertility be?
(a) 0
(b) 2
(c) 2.1
(d) 4

58. Which soil is most productive for growing plants?
(a) clay
(b) a mix of equal parts silt and clay
(c) a mix of silt, clay, and sand
(d) a mix of equal parts sand and silt

59. The construction of a roadway divides two populations of moose. The population on the northern side of the road has only a few members, whereas the population on the southern side of the road is much larger. The construction of a habitat corridor between these two populations is likely to
I. reduce inbreeding depression.
II. establish a metapopulation.
III. expand the fundamental niche of the moose.
(a) II only
(b) III only
(c) I and II
(d) I, II, and III

60. An ecosystem service is best described as
(a) any harmful process from an ecosystem.
(b) benefits humans receive from ecosystems.
(c) a process that provides aid to developing nations.
(d) a process that promotes sustainable development.

61. Which is a method of preventing a tragedy of the commons?
(a) reducing the cost of private transportation
(b) regulating the use of public property
(c) lowering taxes levied on the use of public land
(d) eliminating zoning laws

62. The abundance of fish along the western coasts of most continents is usually caused by
(a) upwelling.
(b) gyres.
(c) an El Niño-Southern Oscillation event.
(d) thermohaline circulation.

63. Which is essential for natural selection to occur?
(a) Genetic drift alters the frequency of genes.
(b) Most individuals can survive equally well in a population.
(c) Differences in phenotypes are linked to genotypes.
(d) Genetic variation is reduced.

64. A new coal-fired electricity plant is being constructed near a town that requires 90,000 MWh per month to supply all its 150,000 households. The plant will be able to supply 500 MW. What is the minimum capacity factor that the plant must maintain?
(a) 0.25
(b) 0.40
(c) 0.60
(d) 0.95

65. The processes that transfer nitrogen gas from the atmosphere into the biosphere and release nitrogen back into the atmosphere are known as
(a) fixation and denitrification.
(b) assimilation and mineralization.
(c) mineralization and nitrification.
(d) ammonification and assimilation.

66. Which of the following best explains why the Sahara Desert is so dry?
(a) It is in the area where dry air from the equatorial convection cells (Hadley cells) sinks to the Earth's surface.
(b) The Earth's rotation forces moisture away from the area toward the equator.
(c) The Earth's rotation forces moisture away from the area toward the poles.
(d) It is in the area where moist air from the equatorial convection cells (Hadley cells) rise away from the Earth's surface.

67. Which is a dominant source of lead pollution in the United States?
I. burning gasoline
II. acid mine drainage
III. corroding pipes
(a) I only
(b) II only
(c) I and II
(d) I and III

68. Insects that produce many larvae without any predator defenses are likely to exhibit
(a) type I survivorship.
(b) stable growth.
(c) type III survivorship.
(d) bottleneck effects.

69. Developers have recently proposed the construction of a new strip mall in a rural area that is a short driving distance to a nearby city. Projects such as this contribute to
(a) urban blight.
(b) smart growth.
(c) urban pollution.
(d) urban sprawl.

70. In 2015, the average residential household in the state of Louisiana used 1,250 kWh per month whereas the average residential household in the state of Hawaii used only 500 kWh per month. Residents of Hawaii pay an average of $0.35 per kWh whereas residents of Louisiana only pay $0.09 per kWh. What is the average annual electric bill for residents of each state?
(a) $112.50 for Louisiana residents; $175 for Hawaii residents
(b) $1,250 for Louisiana residents; $500 for Hawaii residents
(c) $1,350 for Louisiana residents; $2,100 for Hawaii residents
(d) $2,100 for Louisiana residents; $1,350 for Hawaii residents

71. In one lake, researchers estimate that there are 5×10^3 largemouth bass and 2×10^4 total fish. What percentage of the fish in the lake are largemouth bass?
(a) 10 percent
(b) 25 percent
(c) 50 percent
(d) 70 percent

72. A large increase in tipping fees at a landfill is likely to
I. reduce total MSW.
II. increase illegal dumping.
III. increase recycling.
(a) II only
(b) III only
(c) I and II
(d) I, II, and III

73. Where are CO levels likely to be the highest?
(a) in China, during the winter
(b) in Europe, during the winter
(c) in South Africa, during the winter
(d) in North America, during the summer

74. Thermal inversions typically occur within the
(a) stratosphere in valleys between mountains.
(b) troposphere on the East Coast of the United States.
(c) stratosphere in open plains.
(d) troposphere in valleys between mountains.

75. The IUCN Red List was designed to document
(a) endangered and threatened species.
(b) exotic species.
(c) invasive species.
(d) species that pose threats to ecosystem services.

76. Electricity-generating steam power plants need to dissipate excess heat. This is done by extracting water from a nearby lake or reservoir and returning it at a higher temperature. When the water returns to the lakes, it is much warmer than when it left. Which is likely to increase the risk of thermal shock for organisms in a lake near an electricity-generating plant?
(a) releasing water at the top of a lake instead of the bottom
(b) reducing the capacity factor of the electricity-generating plant
(c) an increase in the height of the cooling towers
(d) a decrease in lake volume

77. Carbon dioxide is
(a) essential to the persistence of life on Earth.
(b) responsible for all changes in Earth's climate.
(c) very stable and persists in the atmosphere for hundreds of years.
(d) the most abundant type of gas in the environment.

Question 78 refers to the following graph.

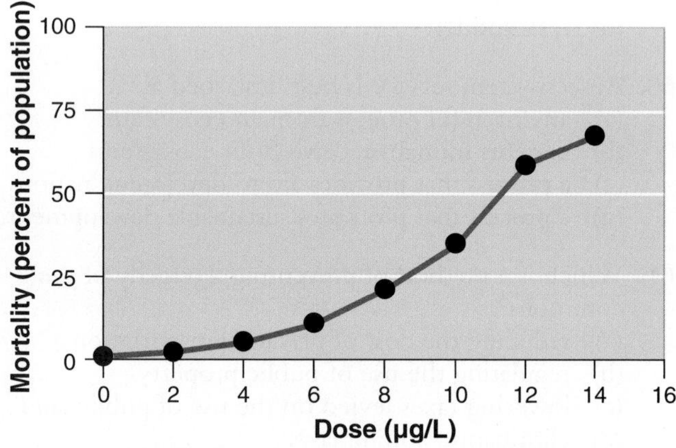

78. The graph shows the results of a dose-response study on rats exposed to a new pesticide. Based on the graph, what is the safe concentration regulatory agencies typically deem for rats and for humans?
(a) 1.1 μg/L for rats; 0.011 μg/L for humans
(b) 1.4 μg/L for rats; 0.0014 μg/L for humans
(c) 11 μg/L for rats; 0.011 μg/L for humans
(d) 14 μg/L for rats; 0.0014 μg/L for humans

79. To study the effects of a chemical contaminant on people, researchers can conduct studies that monitor people after they have been exposed to a chemical contaminant or studies that monitor people who might be exposed to a chemical contaminant. These types of studies are known as
(a) LD50 and ED50 studies.
(b) prospective and retrospective studies.
(c) chronic and acute studies.
(d) manipulative and natural studies.

80. Which fuel is least likely to be a source of lead emissions to the atmosphere?
(a) gasoline
(b) diesel
(c) firewood
(d) coal

Section 2: Free Response Questions

Write your answer to each part clearly. Support your answers with relevant information and examples. Where calculations are required, show your work.

1. Many clothes are now made from polyester thread derived from recycled plastic. When a piece of polyester cloth is washed, many small polyester fibers are loosened and flushed down the drain with wastewater. Researchers estimate that 70,000 microscopic polyester fibers are flushed with every wash cycle in the average washing machine. Wastewater treatment facilities can remove 95 percent of these fibers before the water exits the sanitation process. However the remaining fibers escape and enter natural aquatic systems where they are often mistaken for food by detritivores and herbivores. Many consumers appear to pass these fibers through their guts without harm, but some species show signs of digestive distress from ingesting these fibers. In addition, organic pesticides can bind to these fibers. Consequently, ingesting the fibers can cause pesticide absorption to occur.
(a) Define bioaccumulation and biomagnification, and discuss how they apply to the emerging problem of polyester thread contamination in aquatic systems. (2 points)
(b) In one city with an older sewage system, researchers notice that the concentration of polyester fibers in natural aquatic systems was higher immediately after a major rainfall. Explain why this might occur. (2 points)
(c) Describe a dose-response experiment that could determine the lethal concentration of polyester fibers for aquatic organisms. Along with your description of the experiment, sketch a graph that shows the predicted outcome and indicate the LD50 value. (2 points)
(d) Suppose that researchers genetically modified existing bacteria into a new strain that can digest plastic and they are considering introducing it into natural aquatic systems.
   i. Describe a strategy that researchers could implement to incorporate the precautionary principle of environmental policy. (2 points)
   ii. Describe a strategy that researchers could implement to incorporate the innocent-until-proven-guilty principle of environmental policy? (2 points)

2. In early 2017, residents of Delhi, India, witnessed record concentrations of smog that were hazardous to human health. Although this was not a new phenomenon for the city, pollution concentrations were so high that the governor shut down schools and businesses. Delhi was not alone; many areas of India that lie north of the country's vast mountain ranges experienced such levels of air pollution. Research indicated this smog was caused by emissions from the combustion of fossil fuels, as well as the burning of the previous year's rice crop residue. In addition, northern India's location adjacent to mountains often generates thermal inversions that exacerbate the pollution.
   (a) Describe how photochemical smog formed. (2 points)
   (b) Identify ONE primary pollutant and ONE secondary pollutant associated with this photochemical smog? (2 points)
   (c) Describe a thermal inversion and discuss how it can lead to higher concentrations of air pollution. (2 points)
   (d) Explain TWO benefits associated with burning fields that contain the remains of a harvested crop. (2 points)
   (e) Researchers have noticed that the pH of rainfall over northern India has dropped to 5.0 over the past decade. Discuss how this phenomenon is associated with photochemical smog. (2 points)

3. The Clean Water Act places regulations on development on or near existing wetlands. An environmental impact statement must be generated for any new development. In addition, development should actively avoid disturbing existing aquatic habitats. However, in some situations it is not possible to avoid disturbing existing ecosystems. In such cases, the law requires that a developer mitigate damage by creating new wetland habitat for every wetland destroyed. To assist with this, entities known as wetland mitigation banks create man-made wetlands in areas that have not been previously disturbed. These banks then sell individual wetlands to developers who use them to meet their legal obligations to mitigate any wetland damage they have created.
   (a) Identify TWO ecosystem services that a wetland provides. (2 points)
   (b) Explain the purpose of an environmental impact statement. (3 points)
   (c) To effectively mitigate wetland damage, it is most desirable for mitigation banks to create wetlands in areas that are protected from human development. Identify and describe THREE international categories of protected public lands. (3 points)
   (d) As an employee of a wetland mitigation bank, you are asked to monitor the health and quality of the new wetlands. Describe TWO environmental indicators you could use for this task. (2 points)
   (e) Although there are no streams entering or exiting the wetlands you are monitoring you notice that the water level rises and falls throughout the year. List TWO processes in the hydrological cycle that could contribute to this fluctuation. (2 points)

Appendix

Reading Graphs

Environmental scientists often use graphs to display the data they collect. Unlike a table that contains rows and columns of data, graphs help us visualize patterns and trends in the data and they communicate our results more clearly. Because graphs are such a fundamental tool of environmental science as well as most other sciences, we have designed this appendix to help you become familiar with the major types of graphs that environmental scientists use; we also discuss how to create each type of graph and how to interpret the data that are presented in the graphs.

Scientists use graphs to present data and ideas

A graph is a tool that allows scientists to visualize data or ideas. Organizing information in the form of a graph can help us understand relationships more clearly. Throughout your study of environmental science you will encounter many different types of graphs. In this section we will look at the most common types of graphs that environmental scientists use.

Scatter Plot Graphs

Although many of the graphs in this book may look different from each other, they all follow the same basic principles. Let's begin with an example in which researchers who were investigating a possible relationship between per capita income and total fertility rate plotted data points for 11 countries. We can examine this relationship by creating a *scatter plot graph*, as shown in **FIGURE A.1**. In the simplest form of a scatter plot graph, researchers look at two variables; they put the values of one variable on the x axis and the values of the other variable on the y axis. By convention, the place where the two axes converge in the bottom left corner, called the origin, represents a value of 0 for each variable. The units of measurement tend to get larger as we move from left to right on the x axis, and from bottom to top on the y axis. In our example, per capita income is on the x axis and total fertility rate is on the y axis. As you can see in the graph, as income increases the fertility rate decreases.

When two variables are graphed using a scatter plot, we can draw a line through the middle of the data points that describes the general trend of these data points—as you can see in Figure A.1. Because such a line is drawn in a way that fits the general trend of the data, we call it *the line of best fit*. The line of best fit allows us to visualize a general trend. In our graph, the addition of a line of best fit makes it easier to identify a trend in the data; as we move from low to high income we observe lower fertility. This is known as a negative relationship between the two variables because as one variable gets larger the other variable gets smaller. When graphing data using a scatter plot graph, the line of best fit may be either straight or curved.

Line Graphs

A *line graph* displays data that occur as a sequence of measurements over time or space. For example, scientists have estimated the number of humans living on Earth from 8,000 years ago to the present time.

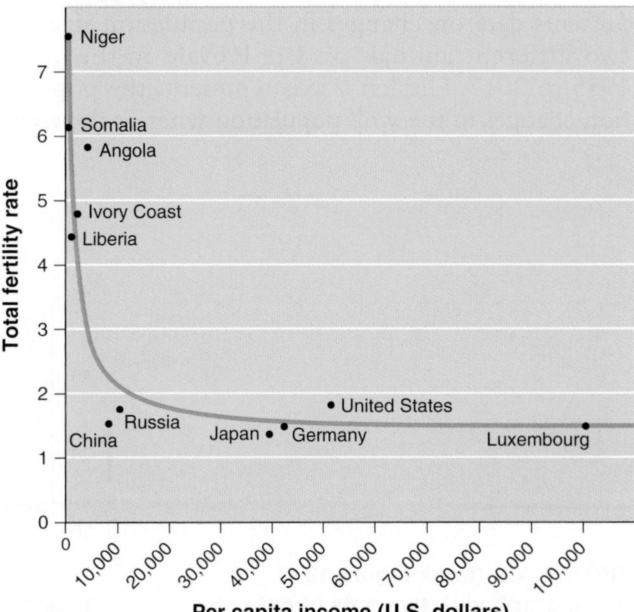

FIGURE A.1 (book Figure 23.2) Scatter plot graph. In this graph, we place data points that coincide with the income of different countries on the x axis and the corresponding fertility rate of each country on the y axis.

Using all of the available data points, a line graph can be used to connect each data point over time, as shown in **FIGURE A.2**. In contrast to a line of best fit that fits a straight or curved line through the middle of all data points, a line graph connects one data point to another, so it can be straight or curved, or it can move up and down as it follows the movement of the data points.

When a graph includes data points with a very large range of values, the size of the graph can become cumbersome. To keep the graph from becoming too large, we can use a break in the axis. For example in our graph of human population growth, we see that the size of the population varied little from 7000 BCE to 2 million BCE. Showing the data for those years would not provide much additional useful information but it would make the graph a lot wider. The break in the *x* axis between 7000 BCE and 2 million BCE, indicated by the double hatch marks, allows us to shorten the *x* axis. The double hatch marks indicate that we are condensing the middle part of the *x* axis.

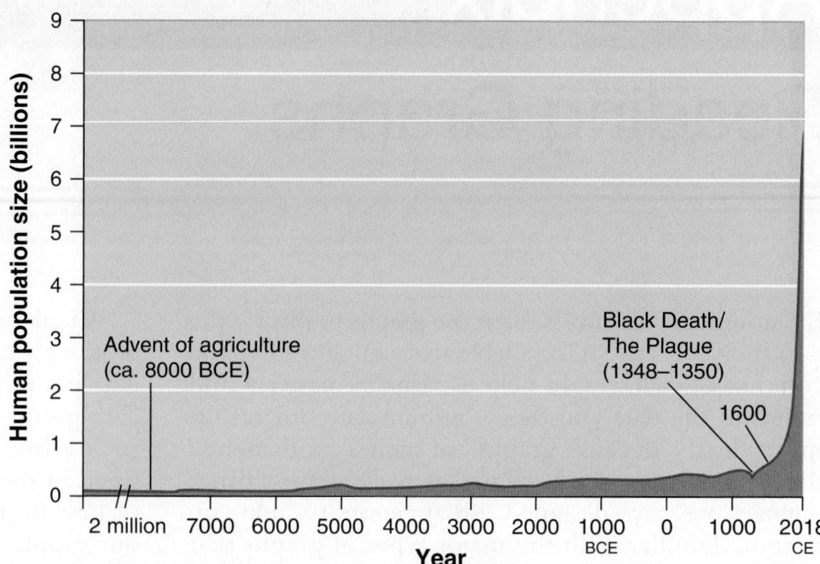

FIGURE A.2 (book Figure 22.1) Line graph. In this graph, a line is used to track the change in human population size over time. Note that this graph extends far back in time and there is only a small increase in population size from 2 million BCE to 7000 BCE, so a break in the *x* axis is used to allow a shorter axis. This allows us to focus on the period of rapid population growth that happened during the past 7,000 years.

Line graphs can also illustrate how several different variables change over time. When two variables contain different units or a different range of values, we can use two *y* axes. For example, **FIGURE A.3** presents data on changes in the population sizes of two different animals on Isle Royale in the years 1955 to 2017. The left *y* axis represents the population changes in the wolf population whereas the right *y* axis represents the population changes in the moose population during the same time span.

Bar Graphs

A *bar graph* plots numerical values that come from different categories. For example, in **FIGURE A.4** the *x* axis contains categories that represent regions of the world. The *y* axis represents a numerical value—the number of people infected with HIV. The visual impact of the different bar heights provides a

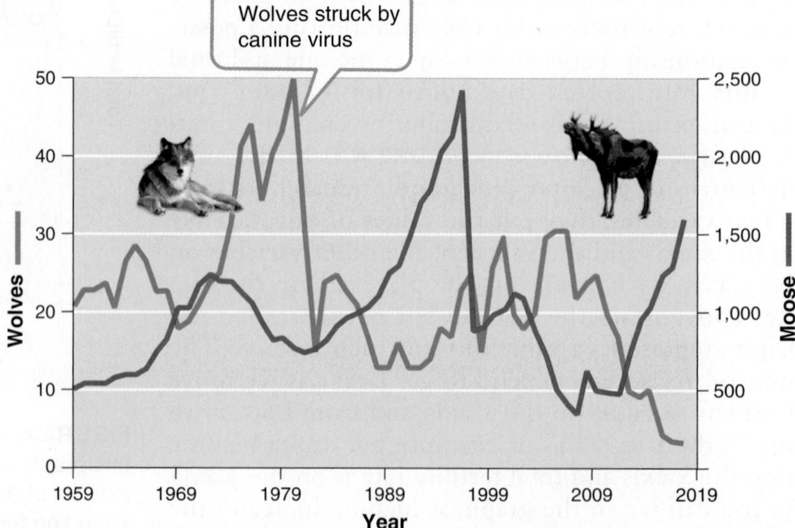

FIGURE A.3 (book Figure 19.6)
A line graph with two sets of data. By graphing changes in the populations of both wolves and moose, we can see that declines in wolves are associated with increases in moose.

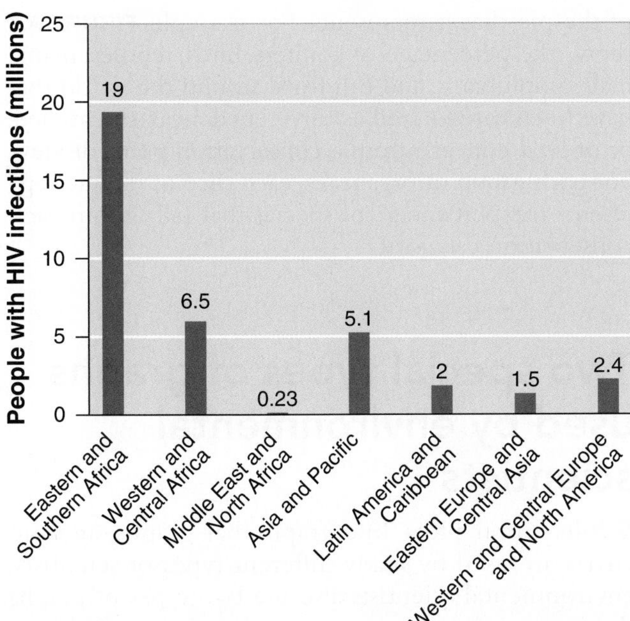

FIGURE A.4 (book Figure 22.7) Bar graph. When we have numerical values that come from different categories we can use a bar graph to plot data. In this example, we can plot the number of people infected with HIV in several areas around the world.

dramatic comparison of the incidence of HIV in different regions.

A bar graph is a very flexible tool and can be altered in several ways to accommodate data sets of different sizes or even several data sets that a researcher wishes to compare. When scientists measured the net primary productivity of different ecosystems, as shown in **FIGURE A.5**, they put the categories—various ecosystems—on the *y* axis and the plotted values—net primary productivity—on the *x* axis. This orientation makes it easier to accommodate the relatively large amount of text needed to name each ecosystem.

FIGURE A.6 on page APP-4 shows an example of a bar graph that presents two sets of data for each category. In this example, the bar graph is rotated such that the categories are on the *y* axis and the numeric data are on two *x* axes. The upper *x* axis represents total annual energy consumption of each country. The lower *x* axis plots per capita (per person) annual energy consumption. Notice how much information we can gather from this graph; we can compare the total annual energy consumption versus per capita annual energy consumption within each country and also compare the energy consumption among countries.

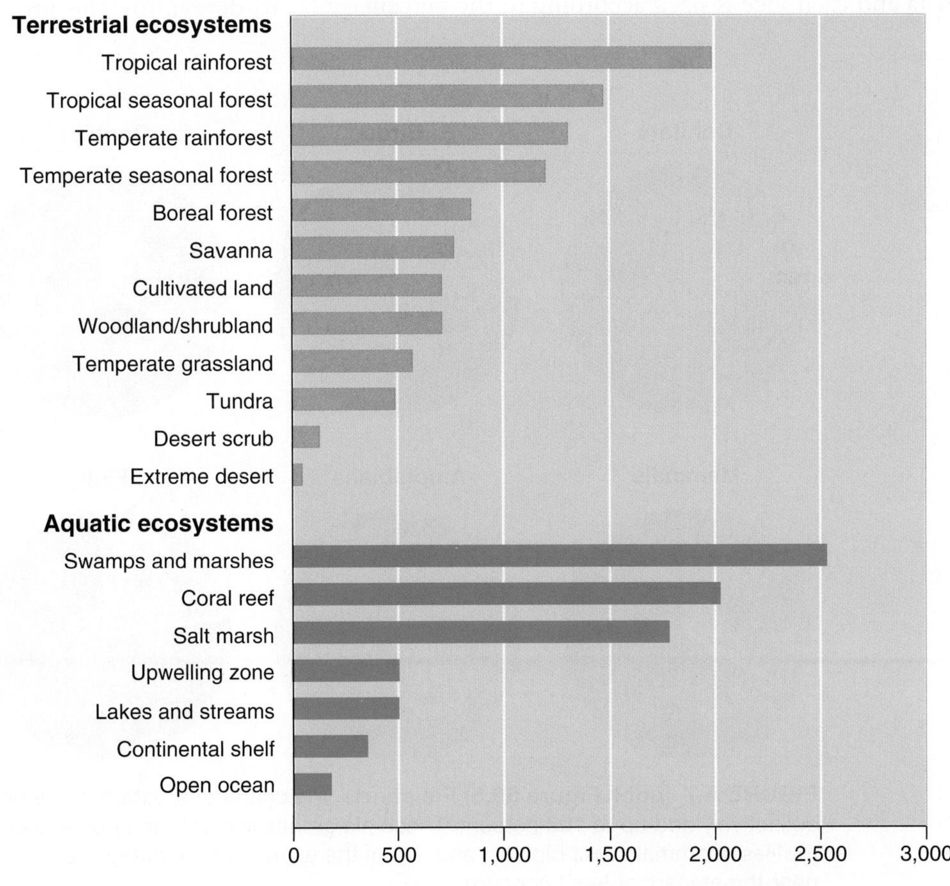

FIGURE A.5 (book Figure 6.8) A rotated bar graph. Bar graphs can place the categories on either the *x* axis, as in Figure A.4, or on the *y* axis, as in this figure that plots the net primary productivity of different ecosystems.

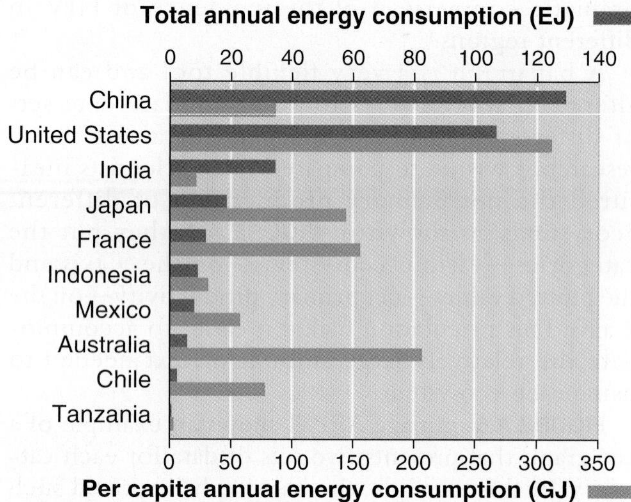

FIGURE A.6 (book Figure 34.2) A rotated bar graph with two sets of data. In this bar graph, we have nations as our categories and two sets of numerical data: the total energy consumed by each country and the per capita annual energy consumed.

Pie Charts

A *pie chart* is a graph represented by a circle with slices of various sizes representing categories within the whole pie. The entire pie represents 100 percent of the data and each slice is sized according to the percentage of the pie that it represents. For example, **FIGURE A.7** shows the percentage of conifers, birds, reptiles, mammals, amphibians, and fish from around the world that have been categorized as threatened, near-threatened, or of least concern from a conservation point of view. For each group of organisms, each slice of the pie represents the percentage of species that fall within each conservation category.

Two special types of graphs used by environmental scientists

While scatter plots, line graphs, bar graphs, and pie charts are used by many different types of scientists, environmental scientists also use two types of graphs that are not common in most other fields of science: *climate diagrams* and *age structure diagrams*. Although these two types of graphs are discussed within the text, we provide them here for review.

Climate Graphs

Climate diagrams are used to illustrate the annual patterns of temperature and precipitation that help to determine the productivity of biomes on Earth.

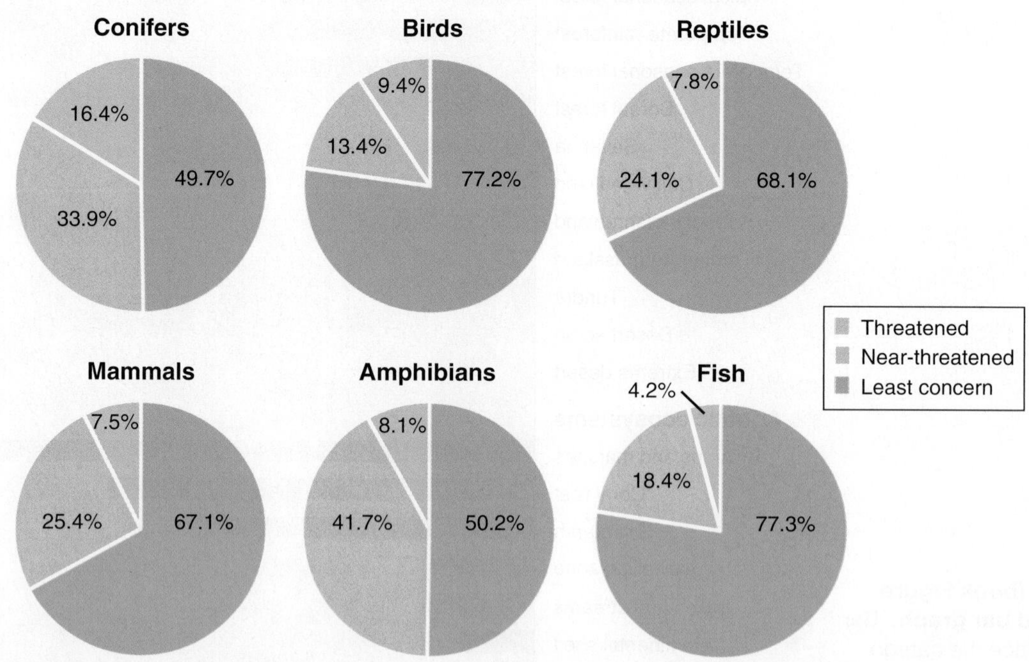

FIGURE A.7 (book Figure 59.5) Pie chart. Pie charts plot data that are percentages and collectively add up to 100 percent. These pie graphs illustrate the percentages of conifers, birds, reptiles, mammals, amphibians, and fish of the world that are categorized as either threatened, near-threatened, or least concern.

FIGURE A.8 shows two hypothetical biomes. By graphing the average monthly temperature and precipitation of a biome, we can see how conditions in a biome vary during a typical year. We can also observe the specific time period when the temperature is warm enough for plants to grow. In the biome illustrated in Figure A.8a, the growing season—indicated by the shaded region on the *x* axis—is mid-March through mid-October. In Figure A.8b, the growing season is mid-April through mid-September.

In addition to identifying the growing season, climate diagrams show the relationships among precipitation, temperature, and plant growth. In Figure A.8a, the precipitation line is above the temperature line in every month. This means that water supply exceeds demand, so plant growth is more constrained by temperature than by precipitation throughout the entire year. In Figure A.8b, the precipitation line intersects the temperature line. At this point, the amount of precipitation available to plants equals the amount of water lost by plants through evapotranspiration. When the precipitation line falls below the temperature line from May through September, water demand exceeds supply and plant growth will be constrained more by precipitation than by temperature.

Age Structure Diagrams

Age structure diagrams are visual representations of age distribution for both males and females in a country. **FIGURE A.9** on page APP-6 presents four examples. Each horizontal bar of the diagram represents a 5-year-age group and the length of a given bar represents the number of males or females in that age group.

While every nation has a unique age structure, we can group countries very broadly into three categories. Figure A.9a shows a country with many more young people than older people. The age structure diagram of a country with this population will be in the shape of a pyramid, with its widest part at the bottom, moving toward the smallest at the top. Age structure diagrams with this shape are typical of countries in the developing world.

A country with a smaller difference between the number of individuals in the younger and older age groups has an age structure diagram that looks more like a column. With fewer individuals in the younger age groups, we can deduce that the country has little or no population growth. Figure A.9b shows the age structure of people in the United States, which is similar to the age structure of people in Canada, Australia, Sweden, and many other developed countries. Panels (c) and (d) show countries with a proportionally larger number of older people. This age structure diagram resembles an inverted pyramid. Such a country has a decreasing number of males and females within each younger age range and that number will continue to shrink. Italy, Germany, Russia, and a few other developed countries display this pattern. In recent years China has also begun to show this pattern.

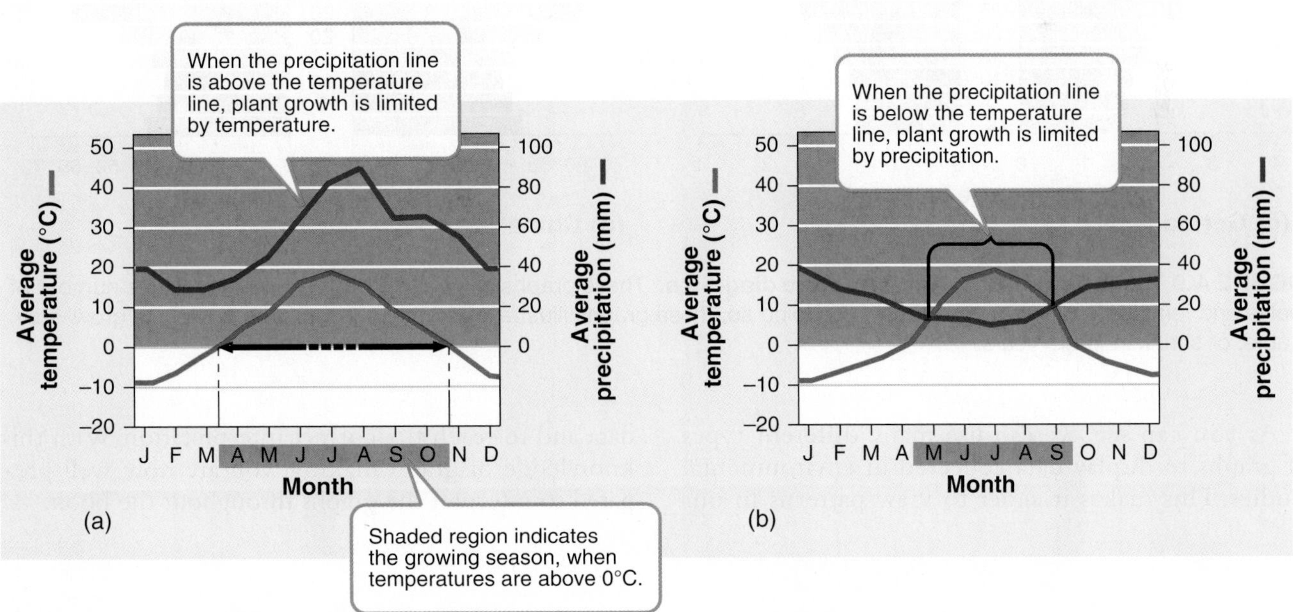

FIGURE A.8 (book Figure 12.4) Climate graph. Climate graphs are a special type of graph that plot monthly temperatures and monthly precipitation in a way that tells us whether plant growth is more limited by temperature or water. These diagrams help us understand the productivity of different biomes.

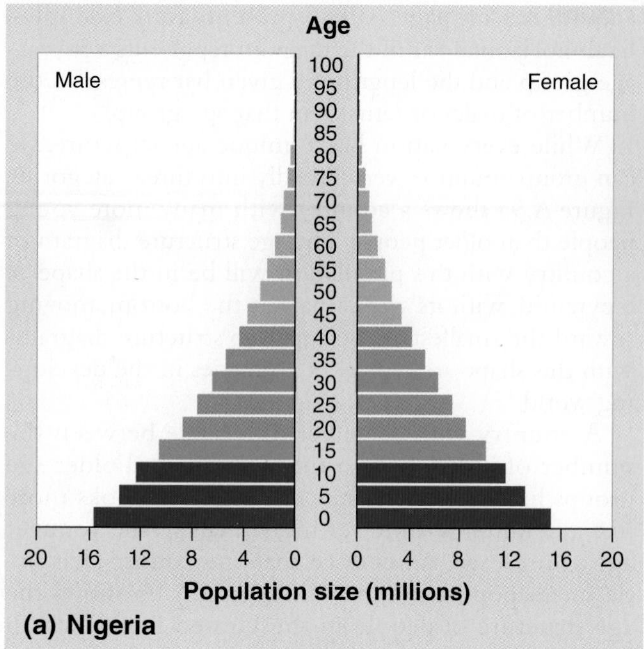

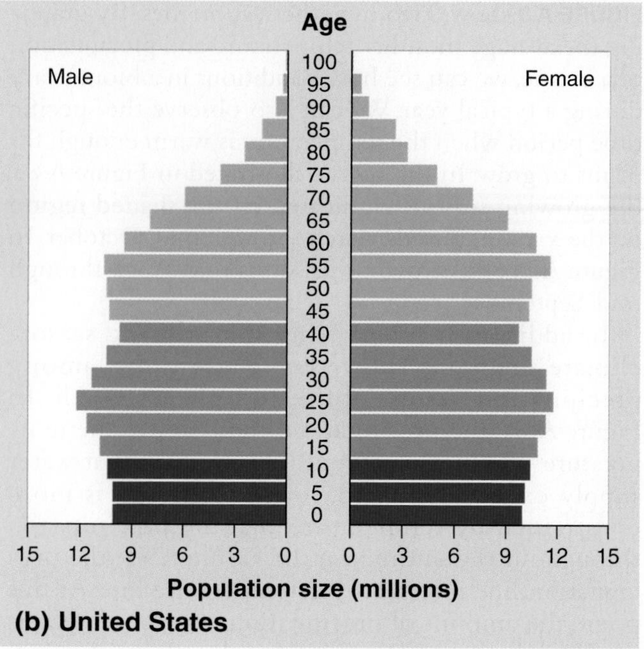

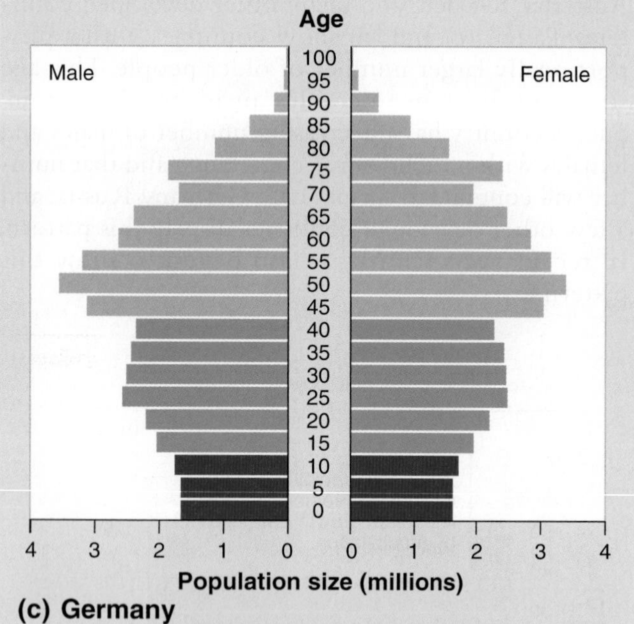

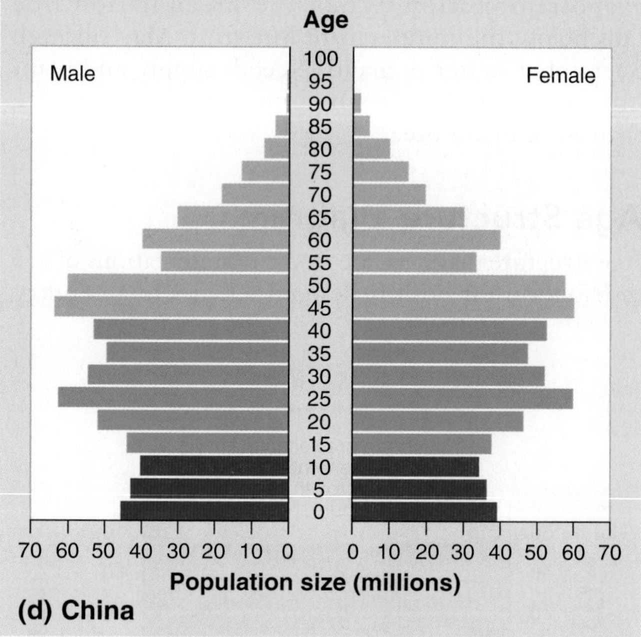

FIGURE A.9 (book Figure 22.8) Age-structure diagrams. These graphs allow us to understand the relative number of males and females in different age classes. In doing so, these graphs illustrate whether a population is likely to grow, stay stable, or shrink in future years.

As you can see, we can use many different types of graphs to display data collected in environmental studies. This makes it easier to view patterns in our data and to reach the correct interpretation. With this knowledge of graph making, you are now well prepared to interpret the graphs throughout the book.

Glossary / Glossário

GLOSSARY	GLOSSÁRIO
A	
abiotic Nonliving.	**abiótico** Sin vida.
accuracy How close a measured value is to the actual or true value.	**exactitud** Fidelidad que hay entre un valor medido y su valor real.
acid A substance that contributes hydrogen ions to a solution.	**ácido** Sustancia que aporta iones de hidrógeno a una solución.
acid deposition Acids deposited on Earth as rain and snow or as gases and particles that attach to the surfaces of plants, soil, and water.	**depósitos ácidos** Ácidos que se depositan sobre la Tierra en la forma de lluvia y nieve, o pueden ser gases y partículas que se adhieren a la superficie de plantas, suelos y agua.
acid precipitation Precipitation high in sulfuric acid and nitric acid from reactions between water vapor and sulfur and nitrogen oxides in the atmosphere. *Also known as* **acid rain.**	**precipitación ácida** Precipitación con una concentración alta de ácido sulfúrico y ácido nítrico producto de reacciones entre el vapor de agua y los óxidos de nitrógeno y azufre en la atmósfera. *También se conoce como* **lluvia ácida.**
acid rain *See* **acid precipitation.**	**lluvia ácida** *Ver* **acid precipitation.**
Acquired Immune Deficiency Syndrome (AIDS) An infectious disease caused by the human immunodeficiency virus (HIV).	**síndrome de inmunodeficiencia adquirida (SIDA)** Enfermedad infecciosa causada por el virus de la inmunodeficiencia humana (VIH).
active solar energy Energy captured from sunlight with advanced technologies.	**energía solar activa** Energía captada de la luz solar con tecnologías de punta.
acute disease A disease that rapidly impairs the functioning of an organism.	**enfermedad aguda** Enfermedad que rápidamente perjudica el funcionamiento de un organismo.
acute study An experiment that exposes organisms to an environmental hazard for a short duration.	**estudio agudo** Experimento que expone a los organismos a un peligro ambiental durante un período de tiempo breve.
adaptation A trait that improves an individual's fitness.	**adaptación** Característica que mejora la adecuación de un individuo.
adaptive management plan A plan that applies flexibility so that managers can modify it as changes occur.	**plan de manejo adaptativo plan de manejo adaptativo** Plan que presenta flexibilidad para que los directores puedan modificarlo a medida que ocurren cambios.
adiabatic cooling The cooling effect of reduced pressure on air as it rises higher in the atmosphere and expands.	**enfriamiento adiabático** Efecto enfriador al reducir la presión sobre el aire a medida que este asciende en la atmósfera y se expande.
adiabatic heating The heating effect of increased pressure on air as it sinks toward the surface of Earth and decreases in volume.	**calentamiento adiabático** Efecto calentador al aumentar la presión sobre el aire a medida que este desciende hacia la superficie de la Tierra y su volumen disminuye.
aerobic respiration The process by which cells convert glucose and oxygen into energy, carbon dioxide, and water.	**respiración aeróbica** Proceso mediante el cual las células convierten la glucosa y el oxígeno en energía, dióxido de carbono y agua.

affluence The state of having plentiful wealth including the possession of money, goods, or property.	**abundancia** Estado de contar con una plenitud de riqueza, inclusive poseer dinero, bienes o propiedades.
age structure A description of how many individuals fit into particular age categories in a population.	**estructura de edades** Descripción de cuántos individuos se ajustan a las distintas categorías de edad en una población.
age structure diagram A visual representation of the number of individuals within specific age groups for a country, typically expressed for males and females.	**diagrama de la estructura de edades** Representación visual de la cantidad de individuos dentro de grupos de edades específicos por país; característicamente se expresa por separado para varones y para mujeres.
agribusiness *See* **industrial agriculture.**	**agroindustria** *Ver* **industrial agriculture.**
agroforestry An agricultural technique in which trees and vegetables are intercropped.	**agrosilvicultura** Técnica agrícola en la que se combinan los cultivos de árboles y los de comestibles.
A horizon Frequently the top layer of soil, a zone of organic material and minerals that have been mixed together. *Also known as* **topsoil.**	**horizonte A** A menudo es la capa superficial del suelo, una zona de material orgánico y minerales que se han entremezclado. *También se conoce como* **manto.**
air pollution The introduction of chemicals, particulate matter, or microorganisms into the atmosphere at concentrations high enough to harm plants, animals, and materials such as buildings, or to alter ecosystems.	**contaminación aérea** La introducción de sustancias químicas, materias particuladas o microorganismos en la atmósfera a concentraciones que son lo suficientemente altas como para causar daño a las plantas, los animales o a materiales tales como edificios, o que alteran ecosistemas completos.
albedo The percentage of incoming sunlight reflected from a surface.	**albedo** Porcentaje de luz solar que llega y que se refleja a partir de una superficie.
algal bloom A rapid increase in the algal population of a waterway.	**eflorescencia de algas** Crecimiento acelerado de la población de algas en una vía navegable.
alien species *See* **exotic species.**	**especie introducida** *Ver* **exotic species.**
allergen A chemical that causes allergic reactions.	**alérgeno** Sustancia química que produce reacciones alérgicas.
allopatric speciation The process of speciation that occurs with geographic isolation.	**especiación alopátrida** Proceso de especiación por aislamiento geográfico.
ammonification The process by which fungal and bacterial decomposers break down the organic nitrogen found in dead bodies and waste products and convert it into inorganic ammonium (NH_4^+)	**amonificación** Proceso mediante el cual saprófitos fúngicos y bacterianos descomponen el nitrógeno orgánico que se encuentra en organismos muertos y en desechos, y lo convierte en amonio inorgánico (NH_4^+)
anaerobic respiration The process by which cells convert glucose into energy in the absence of oxygen.	**respiración anaeróbica** Proceso mediante el cual las células convierten glucosa en energía en la ausencia de oxígeno.
anemia A deficiency of iron.	**anemia** Deficiencia de hierro.
annual plant A plant that lives only one season.	**planta anual** Planta que vive solamente una estación.
anthropocentric worldview A worldview that focuses on human welfare and well-being.	**visión antropocéntrica del mundo** Visión del mundo que se concentra en la felicidad y el bienestar humanos.
anthropogenic Derived from human activities.	**antropogénico** Derivado de actividades humanas.
aphotic zone The deeper layer of ocean water that lacks sufficient sunlight for photosynthesis.	**zona afótica** Profundidad submarina en la que se carece de suficiente luz solar para que haya fotosíntesis.
aquaculture Farming aquatic organisms such as fish, shellfish, and seaweeds.	**acuacultura** Cultivo de organismos acuáticos tales como peces, mariscos y algas marinas.
aquatic biome An aquatic region characterized by a particular combination of salinity, depth, and water flow.	**biomedio acuático** Región acuática caracterizada por una combinación específica de salinidad, profundidad y flujo acuático.
aqueduct A canal or ditch used to carry water from one location to another.	**acueducto** Canal o trocha que se utiliza para transportar agua de una ubicación a otra.

aquifer A permeable layer of rock and sediment that contains groundwater.	**acuífero** Capa permeable de roca y sedimento que contiene aguas freáticas.
artesian well A well created by drilling a hole into a confined aquifer.	**pozo artesiano** Pozo que se crea taladrando un agujero en un acuífero cautivo.
asbestos A long, thin, fibrous silicate mineral with insulating properties, which can cause cancer when inhaled.	**asbesto** Mineral silicato fibroso, delgado y largo con propiedades aislantes, que puede causar cáncer si se lo inhala.
ash The residual nonorganic material that does not combust during incineration.	**ceniza** Material residual no orgánico que no se quema durante la incineración.
assimilation The process by which producers incorporate elements into their tissues.	**asimilación** Proceso mediante el cual los productores incorporan elementos en sus tejidos.
asthenosphere The layer of Earth located in the outer part of the mantle, composed of semi-molten rock.	**astenosfera** Capa de la Tierra ubicada en la parte exterior del manto. Está compuesta por piedra semifundida.
atmospheric convection current Global patterns of air movement that are initiated by the unequal heating of Earth.	**corriente de convección atmosférica** Patrones globales de movimientos de aire iniciados por diferencias en el calentamiento de la Tierra.
atom The smallest particle that can contain the chemical properties of an element.	**átomo** Partícula más pequeña que puede contener las propiedades químicas de un elemento.
atomic number The number of protons in the nucleus of a particular element.	**número atómico** Número de protones en el núcleo de un elemento dado.
autotroph *See* **producer.**	**autótrofo** *Ver* **producer.**

B

background extinction rate The average rate at which species become extinct over the long term.	**tasa de extensión en trasfondo** Tasa promedio en la que las especies se extinguen a largo plazo.
base A substance that contributes hydroxide ions to a solution.	**base** Sustancia que aporta iones de hidróxido a una solución.
base saturation The proportion of soil bases to soil acids, expressed as a percentage.	**saturación base** Proporción de bases en el suelo, comparadas con los ácidos en el suelo. Se expresa en forma de porcentaje.
becquerel (Bq) Unit that measures the rate at which a sample of radioactive material decays; 1 Bq = decay of 1 atom or nucleus per second.	**becquerelio (Bq)** Unidad que mide la tasa de desintegración de una muestra de un material radiactivo; 1 Bq = desintegración de 1 átomo o núcleo por segundo.
benthic zone The muddy bottom of a lake, pond, or ocean.	**zona béntica** Fondo de lodo de un lago, una laguna o un océano.
B horizon A soil horizon composed primarily of mineral material with very little organic matter.	**horizonte B** Horizonte del suelo compuesto principalmente de material mineral con muy poco material orgánico.
bioaccumulation An increased concentration of a chemical within an organism over time.	**bioacumulación** Mayor concentración de una sustancia química en un organismo con el pasar del tiempo.
biocentric worldview A worldview that holds that humans are just one of many species on Earth, all of which have equal intrinsic value.	**visión biocéntrica del mundo** Visión del mundo en la que los seres humanos son solo una de muchas especies en la Tierra, y todas y cada una de dichas especies tienen el mismo valor intrínseco.
biochemical oxygen demand (BOD) The amount of oxygen a quantity of water uses over a period of time at specific temperatures.	**demanda bioquímica de oxígeno** Volumen de oxígeno que una cantidad de agua utiliza durante un período de tiempo a temperaturas específicas.
biodiesel A diesel substitute produced by extracting and chemically altering oil from plants.	**biodiésel** Sustituto del diésel producido de aceite extraído de plantas, que luego se altera químicamente.

biodiversity The diversity of life forms in an environment.	**biodiversidad** Diversidad de las formas con vida en un entorno.
biodiversity hotspot An area that contains a high proportion of all the species found on Earth.	**punto caliente de biodiversidad** Zona que contiene una proporción alta de todas las especies que se encuentran en la Tierra.
biofuel Liquid fuel created from processed or refined biomass.	**biocombustible** Combustible líquido creado a partir de biomasa procesada o refinada.
biogeochemical cycle The movements of matter within and between ecosystems.	**ciclo biogeoquímico** Movimientos de materia dentro de ecosistemas y entre ellos.
biomagnification The increase in chemical concentration in animal tissues as the chemical moves up the food chain.	**biomagnificación** Aumento en la concentración química en tejidos animales a medida que la sustancia química asciende por la cadena trófica.
biomass The total mass of all living matter in a specific area.	**biomasa** Masa total de materia viviente en una zona específica.
biophilia Love of life.	**biofilia** Amor por la vida.
biosphere The region of our planet where life resides, the combination of all ecosystems on Earth.	**biosfera** Región de nuestro planeta en la que reside la vida, la combinación de todos los ecosistemas en la Tierra.
biosphere reserve Protected area consisting of zones that vary in the amount of permissible human impact.	**reserva de la biosfera** Zona protegida que consta de extensiones que varían en la cantidad de impacto humano que permiten.
biotic Living.	**biótico** Con vida.
bird flu A type of flu caused by the H5N1 virus.	**gripe aviaria** Tipo de influenza causada por el virus H5N1.
bitumen A degraded petroleum that forms when petroleum migrates to the surface of Earth and is modified by bacteria.	**betún** Petróleo degradado que se forma cuando el petróleo migra a la superficie de la Tierra y se ve modificado por bacterias.
boreal forest A forest biome made up primarily of coniferous evergreen trees that can tolerate cold winters and short growing seasons.	**selva boreal** Biomedio boscoso compuesto primordialmente por coníferos de hoja perenne capaces de tolerar inviernos fríos y temporadas de crecimiento breves.
bottleneck effect A reduction in the genetic diversity of a population caused by a reduction in its size.	**efecto cuello de botella** Reducción en la diversidad genética de una población atribuible a una reducción en su volumen.
bottom ash Residue collected at the bottom of the combustion chamber in a furnace.	**ceniza de fondo** Residuo que se recoge en el fondo de la cámara de combustión de una caldera.
broad-spectrum pesticide A pesticide that kills many different types of pest.	**plaguicida de espectro amplio** Plaguicida que mata muchos tipos diferentes de plagas.
brownfields Contaminated industrial or commercial sites that may require environmental cleanup before they can be redeveloped or expanded.	**zona contaminada** Sitios comerciales o industriales que podrían precisar una limpieza ambiental antes de que se puedan volver a urbanizar o se puedan ampliar.
brown smog *See* **photochemical smog.**	**esmog marrón** *Ver* **photochemical smog.**
bycatch The unintentional catch of nontarget species while fishing.	**pesca incidental** Captura no intencional de especies que no son las anheladas cuando se pesca.

C

capacity In reference to an electricity-generating plant, the maximum electrical output.	**capacidad** Al referirse a centrales generadores de electricidad, la producción eléctrica máxima.
capacity factor The fraction of time a power plant operates in a year.	**factor de capacidad** Fracción de tiempo que una central eléctrica funciona durante un año.
cap-and-trade An approach to controlling CO_2 emissions, where a cap places an upper limit on the amount of pollutant that can be emitted and trade allows companies to buy and sell allowances for a given amount of pollution.	**sistema de control e intercambio** Enfoque que propone el control de las emisiones de CO_2. El control consiste en limitar la cantidad máxima que se puede emitir de un agente contaminador y el intercambio les permite a las empresas comprar y vender cupos de una cantidad de contaminación.

capillary action A property of water that occurs when adhesion of water molecules to a surface is stronger than cohesion between the molecules.	**capilaridad** Propiedad del agua que ocurre cuando la adhesión de las moléculas de agua a la superficie es mayor que la cohesión entre las moléculas mismas.
carbohydrate A compound composed of carbon, hydrogen, and oxygen atoms.	**carbohidrato** Compuesto que consta de átomos de carbono, hidrógeno y oxígeno.
carbon cycle The movement of carbon around the biosphere.	**ciclo de carbono** Movimiento del carbono alrededor de la biosfera.
carbon neutral An activity that does not change atmospheric CO_2 concentrations.	**emisión neutra de carbono** Actividad que no cambia las concentraciones atmosféricas de CO_2.
carbon offsets Methods of promoting global CO_2 reduction that do not involve a direct reduction in the amount of CO_2 actually emitted by a company.	**mitigación de emisiones de carbono** Métodos para fomentar la reducción total de CO_2 que no implican una reducción directa de la cantidad de CO_2 que una empresa dada de hecho emana.
carbon sequestration An approach to stabilizing greenhouse gases by removing CO_2 from the atmosphere.	**captura de carbono** Enfoque para estabilizar la emisión de gases de efecto invernadero mediante la eliminación de CO_2 en la atmósfera.
carcinogen A chemical that causes cancer.	**carcinógeno** Sustancia química que causa cáncer.
carnivore A consumer that eats other consumers.	**carnívoro** Consumidor que se come otros consumidores.
carrying capacity (*K*) The limit of how many individuals in a population the environment can sustain.	**capacidad poblacional (*K*)** Límite del número de individuos de una población que el entorno puede sustentar.
cation exchange capacity (CEC) The ability of a particular soil to adsorb and release cations.	**capacidad de intercambio de cationes** Capacidad de un suelo dado de adsorber y de liberar cationes.
cell A highly organized living entity that consists of the four types of macromolecules and other substances in a watery solution, surrounded by a membrane.	**célula** Entidad viviente muy bien organizada que consta de cuatro tipos de macromoléculas y de otras sustancias en una solución acuosa, rodeadas por una membrana.
cellular respiration The process by which cells unlock the energy of chemical compounds.	**respiración celular** Proceso mediante el cual las células liberan la energía de compuestos químicos.
cellulosic ethanol An ethanol derived from cellulose, the cell wall material in plants.	**etanol celulósico** Etanol derivado de celulosa, el material de la pared celular en las plantas.
chemical energy Potential energy stored in chemical bonds.	**energía química** Energía potencial almacenada en los enlaces químicos.
chemical reaction A reaction that occurs when atoms separate from molecules or recombine with other molecules.	**reacción química** Reacción que ocurre cuando algunos átomos se separan de las moléculas o se recombinan con otras moléculas.
chemical weathering The breakdown of rocks and minerals by chemical reactions, the dissolving of chemical elements from rocks, or both.	**erosión química** Descomposición de rocas y minerales por reacciones químicas; disolución de elementos químicos de rocas, o ambas cosas.
chemosynthesis A process used by some bacteria in the ocean to generate energy with methane and hydrogen sulfide.	**quimiosíntesis** Proceso que utilizan algunas bacterias en el océano para generar energía valiéndose del metano y el ácido sulfhídrico.
child mortality The number of deaths of children under age 5 per 1,000 live births.	**mortalidad infantil** Cantidad de fallecimientos de niños menores de 5 años por millar de partos vivos.
C horizon The least-weathered soil horizon, which always occurs beneath the B horizon and is similar to the parent material.	**horizonte C** Horizonte del suelo menos erosionado, que siempre ocurre debajo del horizonte B y es parecido al material parental.
chronic disease A disease that slowly impairs the functioning of an organism.	**enfermedad crónica** Enfermedad que lentamente va perjudicando las funciones de un organismo.
chronic study An experiment that exposes organisms to an environmental hazard for a long duration.	**estudio crónico** Experimento que expone los organismos a un peligro ambiental durante un período de tiempo prolongado.

Clean Water Act Legislation that supports the "protection and propagation of fish, shellfish, and wildlife and recreation in and on the water" by maintaining and, when necessary, restoring the chemical, physical, and biological properties of surface waters.	**Ley de Agua Limpia** Legislación estadounidense que fomenta la "protección y propagación de peces, moluscos y crustáceos, y vida silvestre, así como recreación en o sobre el agua" mediante el mantenimiento y, cuando se precise, la restauración de las propiedades químicas, físicas y biológicas de las aguas superficiales.
clear-cutting A method of harvesting trees that involves removing all or almost all of the trees within an area.	**tala indiscriminada** Método de cosechar árboles que implica la tala de todos o casi todos los árboles dentro de una zona.
climate The average weather that occurs in a given region over a long period of time.	**clima** Promedio del estado del tiempo que ocurre en una región dada durante un período de tiempo prolongado.
climax community Historically described as the final stage of succession.	**comunidad clímax** Históricamente se le conoce como la etapa final de sucesión.
closed-loop recycling Recycling a product into the same product.	**reciclaje en circuito cerrado** Reciclaje de un producto para formar el mismo producto.
closed system A system in which matter and energy exchanges do not occur across boundaries.	**sistema cerrado** Sistema en el que no se dan intercambios de materia y energía a través de sus respectivas fronteras.
coal A solid fuel formed primarily from the remains of trees, ferns, and other plant materials preserved 280 million to 360 million years ago.	**carbón** Combustible sólido formado principalmente de los residuos de árboles, helechos y otros materiales vegetales preservados hace entre 280 y 360 millones de años.
cogeneration The use of a fuel to generate electricity and produce heat. *Also known as* **combined heat and power.**	**cogeneración** Uso de un combustible para generar electricidad y producir calor. *También se conoce como* **energía y calor combinados**.
combined cycle A power plant that uses both exhaust gases and steam turbines to generate electricity.	**ciclo combinado** Central eléctrica que utiliza tanto los gases de salida como turbinas de vapor para generar electricidad.
combined heat and power *See* **cogeneration.**	**energía y calor combinados** *Ver* **cogeneration.**
command-and-control approach A strategy for pollution control that involves regulations and enforcement mechanisms.	**enfoque de mando y control** Estrategia para el control de la contaminación que incluye reglamentaciones y mecanismos para verificar el acatamiento.
commensalism A relationship between species in which one species benefits and the other species is neither harmed nor helped.	**comensalismo** Relación entre especies en la que una especie se beneficia y a la otra especie no se la ha perjudicado ni ayudado.
commercial energy source An energy source that is bought and sold.	**fuente comercial de energía** Fuente de energía que se compra y se vende.
community All of the populations of organisms within a given area.	**comunidad** Poblaciones de todos los organismos que viven dentro de un área dada.
community ecology The study of interactions between species.	**ecología de la comunidad** Estudio de las interacciones entre las especies.
competition The struggle of individuals to obtain a shared limiting resource.	**competencia** Batalla entre individuos por obtener un recurso limitante compartido.
competitive exclusion principle The principle stating that two species competing for the same limiting resource cannot coexist.	**principio de la exclusión competitiva** Principio que explica que dos especies que compiten por el mismo recurso limitante no pueden coexistir.
composting Creation of organic matter (humus) by decomposition under controlled conditions to produce an organic-rich material that enhances soil structure, cation exchange capacity, and fertility.	**compostación** Creación de materia orgánica (humus) mediante la descomposición en condiciones controladas con el fin de producir un material de riqueza orgánica que mejora la estructura del suelo, aumenta la capacidad de intercambio de cationes y fomenta la fertilidad.
compound A molecule containing more than one element.	**compuesto** Molécula que contiene más de un elemento.

concentrated animal feeding operation (CAFO) A large indoor or outdoor structure designed for maximum output.	**operación concentrada para la alimentación de animales** Estructura grande interior o al aire libre que se ha diseñado para generar una producción máxima.
cone of depression An area lacking groundwater due to rapid withdrawal by a well.	**cono de depresión** Área que carece de aguas subterráneas debido a que estas han sido retiradas rápidamente por medio de un pozo.
confined aquifer An aquifer surrounded by a layer of impermeable rock or clay that impedes water flow.	**acuífero cautivo** Acuífero rodeado por una capa de roca o arcilla impermeable que impide el flujo de agua.
consumer An organism that is incapable of photosynthesis and must obtain its energy by consuming other organisms. *Also known as* **heterotroph.**	**consumidor** Organismo que es incapaz de hacer la fotosíntesis y debe obtener su energía consumiendo otros organismos. *También se conoce como* **heterótrofo.**
contaminated water Wastewater from toilets, kitchen sinks, and dishwashers.	**agua contaminada** Aguas negras provenientes de inodoros, lavamanos y lavaplatos.
contour plowing An agricultural technique in which plowing and harvesting are done parallel to the topographic contours of the land.	**arar en curvas de nivel** Técnica agrícola en la que se ara y se cosecha en paralelo a los contornos topográficos del terreno.
control group In a scientific investigation, a group that experiences exactly the same conditions as the experimental group, except for the single variable under study.	**grupo de control** En una investigación científica, grupo que experimenta exactamente las mismas condiciones que el grupo experimental, con la excepción de una variable que se está estudiando.
control rod A cylindrical device inserted between the fuel rods in a nuclear reactor to absorb excess neutrons and slow or stop the fission reaction.	**vara de control** Dispositivo cilíndrico que se inserta entre las barras de combustible en un reactor nuclear para absorber los neutrones que sobran y para enlentecer o detener las reacción de fisión.
Convention on Biological Diversity An international treaty to help protect biodiversity.	**Convenio sobre la Diversidad Biológica** Tratado internacional que busca proteger la biodiversidad.
Convention on International Trade in Endangered Species of Wild Fauna and Flora (CITES) A 1973 treaty formed to control the international trade of threatened plants and animals.	**Convenio sobre el Comercio Internacional de Especies Amenazadas de Fauna y Flora Silvestres (CITES)** Convenio promulgado en 1973 con el fin de controlar el comercio internacional de plantas y animales amenazados.
convergent plate boundary An area where plates move toward one another and collide.	**borde de placas convergentes** Zona en la que las placas tectónicas se acercan una a la otra y chocan una con otra.
coral bleaching A phenomenon in which algae inside corals die, causing the corals to turn white.	**blanqueamiento de corales** Fenómeno en el que las algas dentro de los corales fallecen, motivo por el cual los corales se ponen blancos.
coral reef The most diverse marine biome on Earth, found in warm, shallow waters beyond the shoreline.	**arrecife de coral** Biomedio más diverso de la Tierra, se encuentra en aguas tibias y de poca profundidad a cierta distancia del litoral.
core The innermost zone of Earth's interior, composed mostly of iron and nickel. It includes a liquid outer layer and a solid inner layer.	**núcleo interno** Zona interior de la Tierra, compuesta en su mayor parte de hierro y níquel. Incluye una capa externa líquida y una capa interna maciza.
Coriolis effect The deflection of an object's path due to the rotation of Earth.	**efecto Coriolis** Desviación de la trayectoria de un objeto debido a la rotación de la Tierra.
corridor Strips of natural habitat that connect populations.	**corredor** Franjas de hábitat natural que interconectan poblaciones.
covalent bond The bond formed when elements share electrons.	**enlace covalente** Enlace que se forma cuando los elementos comparten electrones.
cradle-to-grave analysis *See* **life-cycle analysis.**	**análisis de la cuna a la tumba** *Ver* **life-cycle analysis.**
crop rotation An agricultural technique in which crop species in a field are rotated from season to season.	**rotación de cultivos** Técnica agrícola en la que las especies de cultivos en un campo se rotan de estación a estación.

crude birth rate (CBR) The number of births per 1,000 individuals per year.	**tasa bruta de nacimientos** Cantidad de nacimientos por millar de individuos por año.
crude death rate (CDR) The number of deaths per 1,000 individuals per year.	**tasa bruta de fallecimientos** Cantidad de fallecimientos por millar de individuos por año.
crude oil Liquid petroleum removed from the ground.	**petróleo crudo** Petróleo líquido que ha sido extraído del suelo.
crust In geology, the chemically distinct outermost layer of the lithosphere.	**corteza** En geología, la capa más externa y distinguible en términos químicos de la litosfera.
crustal abundance The average concentration of an element in Earth's crust.	**abundancia en la corteza** Concentración promedio de un elemento en la corteza terrestre.
CTL (coal to liquid) The process of converting solid coal into liquid fuel.	**carbón a líquido** Proceso para convertir el carbón sólido en combustible líquido.
cultural eutrophication An increase in fertility in a body of water, the result of anthropogenic inputs of nutrients.	**eutrofización cultural** Aumento en la fertilidad de una extensión de agua como resultado del ingreso antropogenético de nutrientes.
curie A unit of measure for radiation; 1 curie = 37 billion decays per second.	**curio** Unidad con la que se mide la radiactividad; 1 curio = 37 mil millones de desintegraciones por segundo.

D

dam A barrier that runs across a river or stream to control the flow of water.	**presa** Barrera que atraviesa un río o arroyo. Se usa para controlar el flujo de agua.
dead zone When oxygen concentration become so low that it kills fish and other aquatic animals.	**zona muerta** En una extensión de agua, zona con una concentración de oxígeno extremadamente baja y muy poca vida.
decomposers Fungi and bacteria that convert organic matter into small elements and molecules that can be recycled back into the ecosystem.	**saprofitos** Hongos o bacterias que convierten la materia orgánica en elementos y moléculas diminutos que se pueden reciclar y devolver al ecosistema.
demographer A scientist in the field of demography.	**demógrafo** Científico que trabaja en el campo de la demografía.
demography The study of human populations and population trends.	**demografía** Estudio de las poblaciones humanas y sus tendencias poblacionales.
denitrification The conversion of nitrate (NO_3^-) in a series of steps into the gases nitrous oxide (N_2O) and, eventually, nitrogen gas (N_2), which is emitted into the atmosphere.	**desnitrificación** Conversión, mediante una serie de pasos, de nitrato (NO_3^-) en los gases óxido nitroso (N_2O) y, con el tiempo, nitrógeno gaseoso (N_2), que se emite a la atmósfera.
density-dependent factor A factor that influences an individual's probability of survival and reproduction in a manner that depends on the size of the population.	**factor densodependiente** Factor que influye en la probabilidad de supervivencia y reproducción de una manera que depende del volumen de la población.
density-independent factor A factor that has the same effect on an individual's probability of survival and the amount of reproduction at any population size.	**factor densoindependiente** Factor que tiene el mismo efecto en la probabilidad de supervivencia y cantidad de reproducción de un individuo en una población de cualquier volumen.
Department of Energy (DOE) The U.S. organization that advances the energy and economic security of the United States.	**Departamento de Energía (DOE)** Dependencia del gobierno federal estadounidense que fomenta la seguridad energética y económica de Estados Unidos.
dependent variable A variable that is dependent on other factors.	**variable dependiente** Variable que depende de otros factores.
desalination The process of removing the salt from salt water. *Also known* as **desalinization.**	**desalinización** Proceso mediante el cual se le quita la sal al agua salobre.
desertification The transformation of arable, productive land to desert or unproductive land due to climate change or destructive land use.	**desertificación** Transformación de terrenos arables y productivos a suelos no productivos debido a cambios climáticos o al uso destructivo del suelo.

detritivore An organism that specializes in breaking down dead tissues and waste products into smaller particles.	**detritívoro** Organismo que se especializa en descomponer tejidos muertos y desechos en partículas más pequeñas.
developed country A country with relatively high levels of industrialization and income.	**país desarrollado** País con niveles relativamente altos de industrialización e ingresos.
developing country A country with relatively low levels of industrialization and income.	**país en vías de desarrollo** País con niveles relativamente bajos de industrialización e ingresos.
development Improvement in human well-being through economic advancement.	**desarrollo** Mejoras en el bienestar humano mediante adelantos económicos.
die-off A rapid decline in a population due to death.	**mortandad** Disminución rápida en una población debido a muertes.
dike A structure built to prevent ocean waters from flooding adjacent land.	**dique** Estructura construida para prevenir que las aguas del océano inunden los terrenos contiguos.
disease Any impaired function of the body with a characteristic set of symptoms.	**enfermedad** Toda función corporal disminuida que presenta un conjunto característico de síntomas.
distillation A process of desalination in which water is boiled and the resulting steam is captured and condensed to yield pure water.	**destilación** Proceso de desalinización en el que el agua se hierve y el vapor resultante se capta y se condensa para producir agua pura.
distribution Areas of the world in which a species lives.	**distribución** Zonas del mundo en las cuales vive una especie.
disturbance An event, caused by physical, chemical, or biological agents, resulting in changes in population size or community composition.	**perturbación** Acontecimiento causado por agentes físicos, químicos o biológicos que produce cambios en el volumen de la población o en la composición de la comunidad.
divergent plate boundary An area beneath the ocean where tectonic plates move away from each other.	**borde de placas divergentes** Zona en la profundidad del océano en la que las placas tectónicas se separan y alejan una de la otra.
DNA (deoxyribonucleic acid) A nucleic acid, the genetic material that contains the code for reproducing the components of the next generation, and which organisms pass on to their offspring.	**ADN (ácido desoxirribonucleico)** Ácido nucleico, material genético que contiene el código para la reproducción de los componentes de la siguiente generación, y que los organismos pasan a sus descendientes.
dose-response study A study that exposes organisms to different amounts of a chemical and then observes a variety of possible responses, including mortality or changes in behavior or reproduction.	**estudio de dosis-respuesta** Estudio que expone organismos a diferentes cantidades de una sustancia química y luego observa una diversidad de respuestas posibles, las que incluyen la mortalidad o cambios en la conducta o reproducción.
doubling time The number of years it takes a population to double.	**tiempo de duplicación** Número de años que precisa una población para duplicarse.

E

earthquake The sudden movement of Earth's crust caused by a release of potential energy along a geologic fault and usually causing a vibration or trembling at Earth's surface.	**terremoto** El movimiento repentino de la corteza de la Tierra producido por la liberación de energía potencial a lo largo de una falla geológica y que por lo general produce una vibración o temblor en la superficie terrestre.
Ebola hemorrhagic fever An infectious disease with high death rates, caused by several species of the Ebola virus.	**fiebre hemorrágica del ébola** Enfermedad infecciosa con tasas de mortalidad altas, causada por varias especies del virus del Ébola.
ecocentric worldview A worldview that places equal value on all living organisms and the ecosystems in which they live.	**visión ecocéntrica del mundo** Visión que le da el mismo valor a todos los organismos vivos y los ecosistemas en los que residen.
ecological economics The study of economics as a component of ecological systems.	**economía ecológica** Estudio de economía como componente de sistemas ecológicos.

ecological efficiency The proportion of consumed energy that can be passed from one trophic level to another.	**eficiencia ecológica** Proporción de energía consumida que se puede pasar de un nivel trófico a otro.
ecological footprint A measure of how much an individual consumes, expressed in area of land.	**presencia ecológica** Medida de cuánto consume un individuo, expresada en extensión de tierra.
ecologically sustainable forestry An approach to removing trees from forests in ways that do not unduly affect the viability of other noncommercial trees.	**silvicultura sostenible ecológicamente** Enfoque de talar árboles en los bosques de maneras que no afecten indebidamente la viabilidad de los demás árboles.
ecological succession The predictable replacement of one group of species by another group of species over time.	**sucesión ecológica** Reemplazo predecible de un grupo de especies por otro grupo de especies a través del tiempo.
economics The study of how humans allocate scarce resources in the production, distribution, and consumption of goods and services.	**economía** Estudio de cómo los seres humanos reparten recursos escasos en la producción, la distribución y el consumo de bienes y servicios.
economies of scale The observation that average costs of production fall as output increases.	**economías de escala** Observación de que los costos promedio de la producción caen en la medida en que aumenta la producción.
ecosystem A particular location on Earth with interacting biotic and abiotic components.	**ecosistema** Ubicación particular en la Tierra que tiene componentes bióticos y abióticos que interactúan entre sí.
ecosystem engineer A keystone species that creates or maintains habitat for other species.	**ingeniero del ecosistema** Especie clave que crea o mantiene el hábitat para otras especies.
ecosystem services The processes by which life-supporting resources such as clean water, timber, fisheries, and agricultural crops are produced.	**servicios del ecosistema** Procesos mediante los cuales se producen los recursos que sustentan la vida, tales como el agua limpia, la madera, las pesquerías y la agricultura.
ED50 The effective dose of a chemical that causes 50 percent of the individuals in a dose-response study to display a harmful, but nonlethal, effect.	**ED50** Dosis efectiva de una sustancia química que causa que el 50 por ciento de los individuos en un estudio de dosis-respuesta presenten un efecto nocivo pero no letal.
edge habitat Habitat that occurs where two different communities come together, typically forming an abrupt transition, such as where a grassy field meets a forest.	**hábitat de borde** Hábitat que ocurre donde dos comunidades diferentes se unen. Característicamente forman una transición abrupta, como cuando un pastizal se topa con un bosque.
E horizon A zone of leaching, or eluviation, found in some acidic soils under the O horizon or, less often, the A horizon.	**horizonte E** Zona de lixiviación o eluviación que se encuentra en algunos suelos acidizados debajo del horizonte O o, con menor frecuencia, el horizonte A.
electrical grid A network of interconnected transmission lines that joins power plants together and links them with end users of electricity.	**red eléctrica** Red de líneas de transmisión interconectadas que unen varias centrales generadoras y las conectan con los usuarios finales de la electricidad.
electrolysis The application of an electric current to water molecules to split them into hydrogen and oxygen.	**electrólisis** Aplicación de una corriente eléctrica a moléculas de agua para separarlas en hidrógeno y oxígeno.
electromagnetic radiation A form of energy emitted by the Sun that includes, but is not limited to, visible light, ultraviolet light, and infrared energy.	**radiación electromagnética** Forma de energía emitida por el Sol que incluye, sin limitarse, luz visible, luz ultravioleta y energía infrarroja.
element A substance composed of atoms that cannot be broken down into smaller, simpler components.	**elemento** Sustancia compuesta por átomos que no se pueden desintegrar en componentes más pequeños y sencillos.
El Niño–Southern Oscillation (ENSO) A reversal of wind and water currents in the South Pacific.	**El Niño–oscilación austral** Inversión de las corrientes de viento y de agua en el Pacífico Sur.
emergent infectious disease An infectious disease that has not been previously described or has not been common for at least 20 years.	**enfermedad infecciosa emergente** Enfermedad infecciosa que no se ha descrito antes o que no ha sido común durante al menos 20 años.
emigration The movement of people out of a country or region.	**emigración** Desplazamiento de personas que salen de un país o región.

eminent domain A principle that grants government the power to acquire a property at fair market value even if the owner does not wish to sell it.	**derecho de expropiación** Principio según el cual se le otorga al gobierno el poder de adquirir un inmueble al valor del mercado incluso si el propietario no desea venderlo.
endangered species A species that is in danger of extinction within the foreseeable future throughout all or a significant portion of its range.	**especie en peligro de extinción** Especie que está en peligro de desaparecer dentro de un futuro predecible en la totalidad o en una parte significativa de su hábitat.
Endangered Species Act A 1973 U.S. act designed to protect species from extinction.	**Ley de Especies en Peligro de Extinción** Ley promulgada en 1973 en Estados Unidos mediante la cual se protegen de extinción ciertas especies.
endemic species A species that lives in a very small area of the world and nowhere else.	**especie endémica** Especie que vive en una zona muy reducida del mundo y en ningún otro sitio.
endocrine disruptor A chemical that interferes with the normal functioning of hormones in an animal's body.	**interruptor endocrino** Sustancia química que interfiere con el funcionamiento normal de las hormonas en el organismo de un animal.
energy The ability to do work or transfer heat.	**energía** Capacidad de hacer trabajo o transferir calor.
energy carrier Something that can move and deliver energy in a convenient, usable form to end users.	**transportador de electricidad** Algo que puede transportar y entregar electricidad de una manera que sea conveniente y fácil para los usuarios finales.
energy conservation Finding and implementing ways to use less energy.	**conservación de energía** Hallar y poner en práctica maneras de consumir menos energía.
energy efficiency The ratio of the amount of energy expended in the form you want to the total amount of energy that is introduced into the system.	**eficiencia energética** Cantidad de energía que se expende en la forma deseada, expresada como proporción de la cantidad de energía que se introduce en el sistema.
energy intensity The energy use per unit of gross domestic product.	**intensidad energética** La energía por unidad del producto bruto interno.
energy quality The ease with which an energy source can be used for work.	**calidad de la energía** Facilidad con la cual una fuente de energía se puede aprovechar para hacer trabajo.
energy subsidy The fossil fuel energy and human energy input per calorie of food produced.	**subsidio energético** La energía de combustibles fósiles y la energía humana consumidas por caloría de alimento producida.
entropy Randomness in a system.	**entropía** Aleatoriedad en un sistema.
environment The sum of all the conditions surrounding us that influence life.	**entorno** Suma de todas las condiciones circundantes que influyen en la vida.
environmental economics A subfield of economics that examines the costs and benefits of various policies and regulations that seek to regulate or limit air and water pollution and other causes of environmental degradation.	**economía ambiental** Subcampo de la economía en la que se estudian los costos y los beneficios de las normas y las reglamentaciones que regulan o limitan la contaminación del aire y del agua, así como otras causas de degradación ambiental.
environmental hazard Anything in the environment that can potentially cause harm.	**peligro ambiental** Toda cosa en el medioambiente que potencialmente pueda ser perjudicial.
environmental impact statement (EIS) A document outlining the scope and purpose of a development project, describing the environmental context, suggesting alternative approaches to the project, and analyzing the environmental impact of each alternative.	**declaración de impacto ambiental** Documento que detalla el alcance y el propósito de un proyecto de desarrollo. Describe el contexto ambiental, sugiere enfoques alternativos que podría tomar el proyecto, y analiza el impacto ambiental de cada alternativa.
environmental indicator An indicator that describes the current state of an environmental system.	**indicador ambiental** Indicador que describe el estado actual de un sistema ambiental.
environmentalism A social movement that seeks to protect the environment through lobbying, activism, and education.	**ecologismo** Movimiento social que busca la protección del medio ambiente mediante el cabildeo político, el activismo y la educación.

environmentalist A person who participates in environmentalism, a social movement that seeks to protect the environment through lobbying, activism, and education.	**ambientalista** Persona que participa en el ambientalismo, un movimiento social que aspira a proteger el medio ambiente por medio del cabildeo, el activismo y la educación.
environmental mitigation plan A plan that outlines how a developer will address concerns raised by a project's impact on the environment.	**plan de mitigación ambiental** Plan que detalla cómo el encargado tratará las inquietudes que surgen en torno al impacto ambiental de un proyecto.
Environmental Protection Agency (EPA) The U.S. organization that oversees all governmental efforts related to the environment, including science, research, assessment, and education.	**Agencia de Protección Ambiental (EPA)** Dependencia del gobierno federal estadounidense que mantiene un control de todos los esfuerzos del gobierno relacionados con el medio ambiente, incluidas las ciencias, las investigaciones, las valoraciones y la educación.
environmental science The field of study that looks at interactions among human systems and those found in nature.	**ciencia ambiental** Campo en el que se estudian las interacciones entre sistemas humanos y sistemas que se hallan en la naturaleza.
environmental studies The field of study that includes environmental science and additional subjects such as environmental policy, economics, literature, and ethics.	**estudios ambientales** Campo de estudio que incluye la ciencia ambiental y otros temas tales como política ambiental, economía, literatura y ética.
environmental worldview A worldview that encompasses how one thinks the world works; how one views one's role in the world; and what one believes to be proper environmental behavior.	**visión ambiental del mundo** Visión del mundo que abarca la manera como pensamos que funciona el mundo; cómo uno ve el papel que uno desempeña en el mundo; y lo que uno considera que es el comportamiento ecológico apropiado.
epicenter The exact point on the surface of Earth directly above the location where rock ruptures during an earthquake.	**epicentro** Sitio exacto en la superficie terrestre directamente encima de la ubicación en la cual hay una ruptura de la roca durante un terremoto.
epidemic A situation in which a pathogen causes a rapid increase in disease.	**epidemia** Situación en la que un patógeno causa la propagación rápida de una enfermedad.
erosion The physical removal of rock fragments from a landscape or ecosystem.	**erosión** Retiro físico de fragmentos de roca de un paisaje o ecosistema.
estuary An area along the coast where the fresh water of rivers mixes with salt water from the ocean.	**estuario** Zona en una costa donde el agua dulce fluvial se mezcla con el agua salada del océano.
ethanol Alcohol made by converting starches and sugars from plant material into alcohol and CO_2.	**etanol** Alcohol que se elabora convirtiendo almidones y azúcares de material vegetal en alcohol y CO_2.
eutrophic Describes a lake with a high level of productivity.	**eutrófico** Se dice de un lago con un gran nivel de productividad.
eutrophication A phenomenon in which a body of water becomes rich in nutrients.	**eutrofización** Fenómeno en el que una extensión de agua se enriquece de nutrientes.
evapotranspiration The combined amount of evaporation and transpiration.	**evapotranspiración** Cantidad combinada de evaporación y transpiración.
evolution A change in the genetic composition of a population over time.	**evolución** Cambio en la composición genética de una población a través del tiempo.
evolution by artificial selection The process in which humans determine which individuals breed, typically with a preconceived set of traits in mind.	**evolución por selección artificial** Proceso en el que los seres humanos determinan qué individuos se reproducen, característicamente teniendo en cuenta un conjunto de rasgos.
evolution by natural selection The process in which the environment determines which individuals survive and reproduce.	**evolución por selección natural** Proceso en el que el entorno determina qué individuos sobreviven y se reproducen.
exotic species A species living outside its historical range. *Also known as* **alien species.**	**especie exótica** Especie que vive por fuera de su hábitat histórico. *También se conoce como* **especie introducida**.

exponential growth model ($N_t = N_0 e^{rt}$) A growth model that estimates a population's future size (N_t) after a period of time (t), based on the intrinsic growth rate (r) and the number of reproducing individuals currently in the population (N_0).	**modelo de crecimiento exponencial ($N_t = N_0 e^{rt}$)** Modelo de crecimiento que estima el volumen futuro (N_t) de una población después de un período de tiempo (t), con base en la tasa de crecimiento intrínseco (r) y la cantidad de individuos que se pueden reproducir en la actualidad en la población (N_0).
externality The cost or benefit of a good or service that is not included in the purchase price of that good or service or otherwise accounted for.	**externalidad** Costo o beneficio de un bien o servicio que no se incluye en el precio de compra de dicho bien o servicio o de lo contrario contabilizado.
extinction The death of the last member of a species.	**extinción** Muerte del último integrante de una especie.
extrusive igneous rock Rock that forms when magma cools above the surface of Earth.	**roca ígnea extrusiva** Roca que se forma cuando el magma se enfría en la superficie de la corteza terrestre.
exurb An area similar to a suburb, but unconnected to any central city or densely populated area.	**exurbio** Zona parecida a un suburbio, pero desconectada de la ciudad central o de la zona de densidad poblacional alta.

F

family planning The practice of regulating the number or spacing of offspring through the use of birth control.	**planificación familiar** Práctica de regular el espaciamiento de la descendencia gracias al uso de un método de control de la natalidad.
famine The condition in which food insecurity is so extreme that large numbers of deaths occur in a given area over a relatively short period.	**hambruna** Condición en la que la inseguridad alimentaria es tan extrema que ocurren grandes cantidades de muertes en una zona dada o dentro de un período de tiempo relativamente breve.
fault A fracture in rock caused by a movement of Earth's crust.	**falla** Fractura en la roca causada por un movimiento de la corteza terrestre.
fault zone A large expanse of rock where a fault has occurred.	**zona de falla** Extensión amplia de roca en la cual ha ocurrido una falla.
fecal coliform bacteria A group of generally harmless microorganisms in human intestines that can serve as an indicator species for potentially harmful microorganisms associated with contaminated sewage.	**bacteria fecal coliforme** Grupo de microorganismos que en general no son nocivos en el intestino humano que pueden ser una especie indicadora de los microorganismos potencialmente dañinos que se asocian con aguas negras contaminadas.
Ferrell cell A convection current in the atmosphere that lies between Hadley cells and polar cells.	**célula de Ferrell** Corriente de convección en la atmósfera que yace entre las células de Hadley y las células polares.
first law of thermodynamics A physical law which states that energy can neither be created nor destroyed but can change from one form to another.	**primera ley de termodinámica** Ley física que declara que la energía no se puede ni crear ni destruir, pero sí se puede transformar de una forma a otra.
fishery A commercially harvestable population of fish within a particular ecological region.	**pesquería** Población de peces dentro de una región ecológica dada. Se puede cosechar con fines comerciales.
fishery collapse The decline of a fish population by 90 percent or more.	**colapso de pesquería** La caída en un 90 por ciento o más de una población de peces.
fish ladder A stair-like structure that allows migrating fish to get around a dam.	**escalera para peces** Estructura en forma de escalera que facilita la migración de los peces que ha sido por un embalse.
fission A nuclear reaction in which a neutron strikes a relatively large atomic nucleus, which then splits into two or more parts, releasing additional neutrons and energy in the form of heat.	**fisión** Reacción nuclear en la que un neutrón bombardea un núcleo atómico relativamente grande, lo cual produce su rotura en dos o más partes, liberando otros neutrones y energía en forma de calor.
fitness An individual's ability to survive and reproduce.	**aptitud** Capacidad que tiene un individuo para sobrevivir y reproducirse.

flex-fuel vehicle A vehicle that runs on either gasoline or a gasoline/ethanol mixture.	**vehículo de combustible flexible** Vehículo que funciona ya sea con gasolina o con una mezcla de gasolina y etanol.
floodplain The land adjacent to a river.	**terreno inundable** Terreno contiguo a un río.
fly ash The residue collected from the chimney or exhaust pipe of a furnace.	**ceniza voladora** Residuo que se recoge de la chimenea o del tubo de salida de una caldera.
food chain The sequence of consumption from producers through tertiary consumers.	**cadena trófica** Secuencia de consumo desde productores hasta consumidores terciarios.
food insecurity A condition in which people do not have adequate access to food.	**inseguridad alimentaria** Condición en la que las personas no cuentan con acceso adecuado a alimentos.
food security A condition in which people have access to sufficient, safe, and nutritious food that meets their dietary needs for an active and healthy life.	**seguridad alimentaria** Condición en la que las personas tienen acceso asegurado a alimentos nutritivos en cantidad suficiente para satisfacer las necesidades dietéticas que se precisan para llevar una vida activa y saludable.
food web A complex model of how energy and matter move between trophic levels.	**red trófica** Modelo complejo que explica cómo la energía y la materia se desplazan entre niveles tróficos.
forest Land dominated by trees and other woody vegetation and sometimes used for commercial logging.	**bosque** Terreno dominado por árboles y otra vegetación leñosa, que a veces se usa para la explotación forestal comercial.
fossil carbon Carbon in fossil fuels.	**carbono fósil** Carbono en combustibles fósiles.
fossil fuel A fuel derived from biological material that became fossilized millions of years ago.	**combustible fósil** Combustible derivado de materiales biológicos que se fosilizaron hace millones de años.
founder effect A change in the genetic composition of a population as a result of descending from a small number of colonizing individuals.	**efecto fundador** Cambio en la composición genética de una población por descender de un número reducido de individuos colonizadores.
fracking Hydraulic fracturing, a method of oil and gas extraction that uses high-pressure fluids to force open cracks in rocks deep underground.	**fraqueo** Fracturación hidráulica, método para la extracción de petróleo y gas que se vale de líquidos a gran presión que abren a la fuerza fisuras en rocas que se encuentran a grandes profundidades subterráneas.
fracture In geology, a crack that occurs in rock as it cools.	**fractura** En geología, fisura que ocurre en la roca cuando esta se enfría.
freshwater wetland An aquatic biome that is submerged or saturated by water for at least part of each year, but shallow enough to support emergent vegetation.	**humedales de agua dulce** Biomedio acuático sumergido o saturado de agua al menos una parte de cada año, pero suficientemente pando para dar sustento a vegetación emergente.
fuel cell An electrical-chemical device that converts fuel, such as hydrogen, into an electrical current.	**celda de combustible** Dispositivo electroquímico que convierte un combustible, por ejemplo, el hidrógeno, en una corriente eléctrica.
fuel rod A cylindrical tube that encloses nuclear fuel within a nuclear reactor.	**barra de combustible** Tubo cilíndrico que contiene el combustible nuclear dentro de un reactor nuclear.
fundamental niche The suite of abiotic conditions under which a species can survive, grow, and reproduce.	**nicho fundamental** Conjunto de condiciones abióticas bajo las cuales una especie puede sobrevivir, crecer y reproducirse.

G

gene A physical location on the chromosomes within each cell of an organism.	**gen** Ubicación física en los cromosomas dentro de cada célula de un organismo.
gene flow The process by which individuals move from one population to another and thereby alter the genetic composition of both populations.	**migración genética** Proceso mediante el cual los individuos se desplazan de una población a otra y así alteran la composición genética de ambas poblaciones.
genetic diversity A measure of the genetic variation among individuals in a population.	**diversidad genética** Medida de la variación genética entre individuos en una población.

genetically modified organism (GMO) An organism produced by copying genes from a species with a desirable trait and inserting them into another species.	**organismo transgénico** Organismo producido al copiar genes de una especie con un rasgo deseado e insertarlos en otra especie.
genetic drift A change in the genetic composition of a population over time as a result of random mating.	**deriva genética** Cambio en la composición genética de una población a través del tiempo producto del apareamiento aleatorio.
genotype The complete set of genes in an individual.	**genotipo** Conjunto completo de genes de un individuo.
genuine progress indicator (GPI) A measure of economic status that includes personal consumption, income distribution, levels of higher education, resource depletion, pollution, and the health of the population.	**indicador de avance genuino** Medición del estado económico que incluye el consumo personal, la distribución de ingresos, los niveles de educación superior, el agotamiento de recursos, la contaminación y el estado de salud de la población.
geographic isolation Physical separation of a group of individuals from others of the same species.	**aislamiento geográfico** Separación física de un grupo de individuos de otros de la misma especie.
geothermal energy Heat energy that comes from the natural radioactive decay of elements deep within Earth.	**energía geotérmica** Energía térmica que proviene de la descomposición natural radiactiva de elementos en lo profundo de la Tierra.
global change Change that occurs in the chemical, biological, and physical properties of the planet.	**cambio global** Cambio que se da en las propiedades químicas, biológicas y físicas del planeta.
global climate change Changes in the average weather that occurs in an area over a period of years or decades.	**cambio climático global** Cambios en el promedio del estado del tiempo que se presentan en una región durante un período de años o décadas.
global multidimensional poverty index (MPI) An internationally comparable measurement index that measures education, health, and living standards.	**Índice de pobreza multidimensional (IPM)** Índice de medición a escala global que mide y compara la educación, la salud y el nivel de vida.
global warming The warming of the oceans, land masses, and atmosphere of Earth.	**calentamiento global** Calentamiento de los océanos, las masas terrestres y la atmósfera de la Tierra.
gray smog *See* **sulfurous smog.**	**esmog gris** *Ver* **sulfurous smog.**
gray water Wastewater from baths, showers, bathroom sinks and washing machines.	**aguas grises** Aguas servidas provenientes de tinas, duchas, los lavabos y máquinas lavadoras de ropa.
greenhouse effect Absorption of infrared radiation by atmospheric gases and reradiation of the energy back toward Earth.	**efecto invernadero** Absorción de radiación infrarroja por los gases atmosféricos y reirradiación de la energía de vuelta a la Tierra.
Green Revolution A shift in agricultural practices in the twentieth century that included new management techniques, mechanization, fertilization, irrigation, and improved crop varieties, that resulted in increased food output.	**Revolución verde** Modificación en el cultivo agrícola que se dio en el siglo XX, la cual incluyó nuevas técnicas de gestión, mecanización, fertilización, riego, así como cultivos mejorados, con lo cual se obtuvo un aumento en la producción alimentaria.
green tax A tax placed on environmentally harmful activities or emissions in an attempt to internalize some of the externalities that may be involved in the life cycle of those activities or products.	**impuesto verde** Gravamen que se impone a las actividades o emisiones que pueden ser perjudiciales para el medio ambiente con el fin de tratar de internalizar algunas de las externalidades que pueden estar implícitas en el ciclo de vida de dichas actividades o productos.
greenhouse gases Gases in Earth's atmosphere that trap heat near the surface.	**gases de efecto invernadero** Gases de la atmósfera terrestre que atrapan el calor y lo mantienen cerca de la superficie.
greenhouse warming potential An estimate of how much a molecule of any compound can contribute to global warming over a period of 100 years relative to a molecule of CO_2.	**potencial de calentamiento por gases de efecto invernadero** Estimación de cuánto una molécula de cualquier compuesto puede contribuir al calentamiento global durante un lapso de 100 años con relación a una molécula de CO_2.
gross domestic product (GDP) A measure of the value of all products and services produced in 1 year in one country.	**producto bruto interno (PBI)** Medición del valor de todos los productos y servicios producidos en un país en un año.

gross primary productivity (GPP) The total amount of solar energy that producers in an ecosystem capture via photosynthesis over a given amount of time.	**productividad bruta primaria (PBP)** Monto total de energía solar que los productores de un ecosistema captan por fotosíntesis durante un lapso de tiempo dado.
ground source heat pump A technology that transfers heat from the ground to a building.	**bomba térmica de origen terrestre** Tecnología mediante la cual se transfiere el calor de la tierra al interior de un edificio.
groundwater recharge A process by which water percolates through the soil and works its way into an aquifer.	**recarga de aguas freáticas** Proceso mediante el cual el agua se filtra a través del suelo y avanza hasta llegar al acuífero.
gyre A large-scale pattern of water circulation that moves clockwise in the Northern Hemisphere and counterclockwise in the Southern Hemisphere.	**giro oceánico** Sistema a gran escala de la circulación del agua que se desplaza en el sentido de las manecillas del reloj en el hemisferio septentrional (Norte) y en el sentido contrario al de las manecillas del reloj en el hemisferio austral (Sur).

H

habitat An area where a particular species lives in nature.	**hábitat** Lugar en la naturaleza donde vive una especie en particular.
Hadley cell A convection current in the atmosphere that cycles between the equator and 30° N and 30° S.	**célula de Hadley** Corriente de convección en la atmósfera entre el ecuador y 30° N y 30° S.
half-life The time it takes for one-half of an original radioactive parent atom to decay.	**hemivida** Tiempo necesario para que se desintegre la mitad de un átomo radiactivo original.
hazardous waste Liquid, solid, gaseous, or sludge waste material that is harmful to humans, ecosystems, or materials	**residuos nocivos** Material líquido, sólido, gaseoso o fangoso que es perjudicial para los seres humanos, ecosistemas o materiales.
haze Reduced visibility.	**bruma** Visibilidad reducida.
herbicide A pesticide that targets plant species that compete with crops.	**herbicida** Plaguicida que se enfoca en especies de plantas que compiten con los cultivos.
herbivore A consumer that eats producers. *Also known as* **primary consumer.**	**herbívoro** Consumidor que come productores. *También se conoce como* **consumidor primario.**
herbivory An interaction in which an animal consumes a producer.	**herbivorio** Interacción en la que un animal consume un productor.
heterotroph *See* **consumer.**	**heterótrofo** *Ver* **consumer.**
Highway Trust Fund A U.S. federal fund that pays for the construction and maintenance of roads and highways.	**Fondo Fideicomisario para el Transporte** Fondo federal estadounidense que paga por la construcción y el mantenimiento de carreteras y autopistas.
horizon A horizontal layer in a soil defined by distinctive physical features such as texture and color.	**horizonte** Capa horizontal en un suelo que se define por sus rasgos físicos característicos tales como textura y color.
hot spot In geology, a place where molten material from Earth's mantle reaches the lithosphere.	**punto caliente** En geología, sitio en el que el material en estado líquido del manto terrestre alcanza la litosfera.
Hubbert curve A bell-shaped curve representing oil use and projecting both when world oil production will reach a maximum and when the world will run out of oil.	**curva de Hubbert** Curva en forma de campana que representa tanto el aprovechamiento de petróleo como proyecciones de cuándo la producción mundial de petróleo alcanzará un máximo y cuándo se agotará el petróleo en todo el mundo.
human capital Human knowledge and abilities.	**capital humano** Conocimientos y habilidades de los seres humanos.
human development index (HDI) A measurement index that combines three basic measures of human status: life expectancy; knowledge and education; and standard of living.	**índice de desarrollo humano** Índice de medición que combina tres mediciones básicas del estatus del ser humano: expectativa de vida, conocimientos y educación.

Human Immunodeficiency Virus (HIV) A type of virus that causes Acquired Immune Deficiency Syndrome (AIDS).	**virus de la inmunodeficiencia humana (VIH)** Tipo de virus que causa el síndrome de inmunodeficiencia adquirida (sida).
humus The most fully decomposed organic matter in the lowest section of the O horizon.	**humus** La sustancia orgánica en mayor descomposición en la parte más baja del horizonte O.
hydroelectricity Electricity generated by the kinetic energy of moving water.	**hidroelectricidad** Electricidad generada por la energía cinética del agua en movimiento.
hydrogen bond A weak chemical bond that forms when hydrogen atoms that are covalently bonded to one atom are attracted to another atom on another molecule.	**enlace de hidrógeno** Enlace químico débil que se forma cuando los átomos de hidrógeno que tienen un enlace covalente con un átomo son atraídos por otro átomo en otra molécula.
hydrologic cycle The movement of water through the biosphere.	**ciclo hidrológico** Movimiento del agua por la biosfera.
hydroponic agriculture The cultivation of plants in greenhouse conditions by immersing roots in a nutrient-rich solution.	**agricultura hidropónica** Cultivo de plantas en condiciones de invernadero mediante la inmersión de las raíces en una solución rica en nutrientes.
hypothesis A testable conjecture about how something works.	**hipótesis** Conjetura comprobable acerca de cómo funciona algo.
hypoxic Low in oxygen.	**hipóxico** De poco oxígeno.

I

igneous rock Rock formed directly from magma.	**roca ígnea** Roca que se formó directamente del magma.
immigration The movement of people into a country or region, from another country or region.	**inmigración** Desplazamiento de personas a un país o región, provenientes de otro país o región.
impermeable surface Pavement or buildings that do not allow water penetration.	**superficie impermeable** Pavimento o edificios que no permiten que el agua penetre.
inbreeding depression When individuals with similar genotypes—typically relatives—breed with each other and produce offspring that have an impaired ability to survive and reproduce.	**depresión por endogamia** Cuando individuos con genotipos parecidos — característicamente parientes — se aparean entre sí y producen descendientes a quienes les es imposible sobrevivir y reproducirse.
incentive-based approach A strategy for pollution control that constructs financial and other incentives for lowering emissions based on profits and benefits.	**enfoque basado en incentivos** Estrategia para controlar la contaminación que propone incentivos económicos y de otra índole por disminuir las emisiones. Se basa en utilidades y beneficios.
incineration The process of burning waste materials to reduce volume and mass, sometimes to generate electricity or heat.	**incineración** Proceso de quemar desechos para reducir su volumen y masa. A veces se usa para generar electricidad o calor.
independent variable A variable that is not dependent on other factors.	**variable independiente** Variable que no depende de otros factores.
indicator species A species that indicates whether or not disease-causing pathogens are likely to be present.	**especie indicadora** Especie que señala si es o no probable la presencia de patógenos que causan enfermedades.
individual transferable quota (ITQ) A fishery management program in which individual fishers are given a total allowable catch of fish in a season that they can either catch or sell.	**cuota transferible individual** Programa de gestión de pesquerías en el que a cada pescador individual se le otorga un volumen de pesca para la temporada. Dicha cuota se puede aprovechar pescándola o vendiéndola.
induced demand The phenomenon in which an increase in the supply of a good causes demand to grow.	**demanda inducida** Fenómeno en el que un aumento en la oferta de un bien da por resultado un aumento de la demanda.
industrial agriculture Agriculture that applies the techniques of mechanization and standardization to the production of food. *Also known as* **agribusiness.**	**agricultura industrial** Agricultura que aplica las técnicas de mecanización y normalización a la producción de alimentos. *También se conoce como* **agroindustria.**

industrial smog *See* **sulfurous smog.**	**esmog industrial** *Ver* **sulfurous smog.**
infant mortality The number of deaths of children under 1 year of age per 1,000 live births.	**mortalidad infantil** Número de muertes de niños menores de un año por millar de partos vivos.
infectious disease A disease caused by a pathogen.	**enfermedad infecciosa** Enfermedad causada por un patógeno.
infill Development that fills in vacant lots within existing communities rather than expanding into new land outside the city.	**aprovechamiento** Desarrollo de lotes baldíos dentro de comunidades establecidas en lugar de expandirse a nuevas tierras fuera de la ciudad.
innocent-until-proven-guilty principle A principle based on the belief that a potential hazard should not be considered an actual hazard until the scientific data definitively demonstrate that it actually causes harm.	**principio de ser inocente hasta que se compruebe la culpabilidad** Principio que se basa en la convicción de que un posible peligro no debe ser considerado un peligro real sino hasta que se cuente con datos científicos que demuestren de manera definitiva que es perjudicial.
inorganic compound A compound that does not contain the element carbon or contains carbon bound to elements other than hydrogen.	**compuesto inorgánico** Compuesto que no contiene el elemento carbono ni contiene carbono enlazado a elementos que no sean el hidrógeno.
inorganic fertilizer *See* **synthetic fertilizer.**	**fertilizante inorgánico** *Ver* **synthetic fertilizer.**
input An addition to a system.	**entrada** Algo que se le agrega al sistema.
insecticide A pesticide that targets species of insects and other invertebrates that consume crops.	**insecticida** Plaguicida que se enfoca en especies de insectos y otros invertebrados que consumen cultivos.
instrumental value Worth as an instrument or a tool that can be used to accomplish a goal.	**valor instrumental** Valor como instrumento o recurso que se puede emplear para lograr un cometido.
integrated pest management (IPM) An agricultural practice that uses a variety of techniques designed to minimize pesticide inputs.	**gestión integrada de plagas** Técnica agrícola que emplea una diversidad de técnicas diseñadas para minimizar el uso de plaguicidas.
integrated waste management An approach to waste disposal that employs several waste reduction, management, and disposal strategies in order to reduce the environmental impact of MSW.	**gestión integrada de desechos** Manera de enfocar la disposición de desechos que emplea varias estrategias para la reducción, la gestión y la eliminación de desechos con el fin de mitigar el impacto de los desechos sólidos municipales.
intercropping An agricultural method in which two or more crop species are planted in the same field at the same time to promote a synergistic interaction.	**cultivos asociados** Técnica agrícola en la que dos o más especies de cultivos se siembran en el mismo terreno con el fin de fomentar una interacción de sinergia.
intermediate disturbance hypothesis The hypothesis that ecosystems experiencing intermediate levels of disturbance are more diverse than those with high or low disturbance levels.	**hipótesis de la perturbación intermedia** Hipótesis que explica que los ecosistemas experimentan niveles intermedios de perturbaciones que son más diversos que aquellos con niveles de disturbios altos o bajos.
intertidal zone The narrow band of coastline between the levels of high tide and low tide.	**zona intermareal** Banda delgada en el litoral entre los niveles conocidos de marea alta y marea baja.
intertropical convergence zone (ITCZ) The latitude that receives the most intense sunlight, which causes the ascending branches of the two Hadley cells to converge.	**zona de convergencia intertropical** Latitud que recibe la luz solar más intensa, lo que lleva a que las ramas ascendentes de la dos células de Hadley se junten.
intrinsic growth rate (*r*) The maximum potential for growth of a population under ideal conditions with unlimited resources.	**tasa de crecimiento intrínseco (*r*)** Máximo potencial de crecimiento de una población en condiciones ideales con recursos ilimitados.
intrinsic value Value independent of any benefit to humans.	**valor intrínseco** Valor independiente de todo beneficio para los seres humanos.
intrusive igneous rock Igneous rock that forms when magma rises up and cools in a place underground.	**roca ígnea intrusiva** Roca ígnea que se forma cuando el magma asciende y se enfría en un sitio subterráneo.
invasive species A species that spreads rapidly across large areas and causes harm.	**especie invasora** Especie que se disemina rápidamente a través de zonas extensas y causa daño.

inversion layer The layer of warm air that traps emissions in a thermal inversion.	**capa de inversión** Capa de aire cálido que atrapa las emisiones en una inversión térmica.
ionic bond A chemical bond between two ions of opposite charges.	**enlace iónico** Enlace químico entre dos iones con cargas opuestas.
IPAT equation An equation used to estimate the impact of the human lifestyle on the environment: Impact = population × affluence × technology.	**ecuación IPAT** Ecuación que se usa para estimar el impacto del estilo de vida humano en el medio ambiente: Impacto = población × afluencia × tecnología.
isotopes Atoms of the same element with different numbers of neutrons.	**isótopos** Átomos del mismo elemento con diferentes cantidades de neutrones.
J	
joule (J) The amount of energy used when a 1-watt electrical device is turned on for 1 second.	**julio (J)** Cantidad de energía que se usa cuando se enciende un dispositivo eléctrico de 1 vatio por un segundo.
J-shaped curve The curve of the exponential growth model when graphed.	**curva con forma de J** Curva del modelo de crecimiento exponencial cuando se representa de manera gráfica.
K	
keystone species A species that that is not very abundant but has large effects on an ecological community.	**especie clave** Especie que eso no es muy abundante, pero tiene grandes efectos en una comunidad ecológica.
kinetic energy The energy of motion.	**energía cinética** Energía del movimiento.
***K*-selected species** A species with a low intrinsic growth rate that causes the population to increase slowly until it reaches carrying capacity.	**especie de selección *K*** Especie con una tasa de crecimiento intrínseco baja que es motivo de que la población crezca lentamente hasta que llegue a su capacidad poblacional.
Kyoto Protocol An international agreement that sets a goal for global emissions of greenhouse gases from all industrialized countries to be reduced by 5.2 percent below their 1990 levels by 2012.	**protocolo de Kyoto** Convenio internacional que fija una meta para el total de emisiones mundiales de gases de efecto invernadero provenientes de todos los países industrializados. Para el año 2012, dichos gases se habrían de reducir por 5.2 por ciento a los niveles que tenían en 1990.
L	
Lacey Act A U.S. act that prohibits interstate shipping of all illegally harvested plants and animals.	**Ley Lacey** Ley federal estadounidense que prohíbe el despacho interestatal de plantas y animales obtenidos de manera ilícita.
latent heat release The release of energy when water vapor in the atmosphere condenses into liquid water.	**liberación térmica latente** Liberación de energía cuando el vapor de agua en la atmósfera se condensa y forma agua líquida.
law of conservation of matter A law of nature stating that matter cannot be created or destroyed; it can only change form.	**ley de conservación de la materia** Ley natural que declara que la materia no se puede crear ni destruir; sólo puede cambiar de forma.
LD50 The lethal dose of a chemical that kills 50 percent of the individuals in a dose-response study.	**LD50** Dosis letal de una sustancia química que mata el 50 por ciento de los individuos en un estudio de dosis-respuesta.
leachate Liquid that contains elevated levels of pollutants as a result of having passed through municipal solid waste (MSW) or contaminated soil.	**lixiviado** Líquido que contiene niveles elevados de contaminantes como resultado de haber pasado por desechos sólidos municipales o por suelos contaminados.
leach field A component of a septic system, made up of underground pipes laid out below the surface of the ground.	**campo de lixiviación** Componente de un sistema séptico compuesto por tuberías subterráneas dispuestas en un diseño planificado.
leaching The transportation of dissolved molecules through the soil via groundwater.	**lixiviación** Transporte de moléculas disueltas a través de la tierra por medio de aguas freáticas.

leapfrogging The phenomenon of less developed countries using new technology without first using the precursor technology.	**brincos superadores** Fenómeno en países menos desarrollados que se valen de tecnologías nuevas sin tener que usar primero la tecnología precursora.
least concern species Species that are widespread and abundant.	**especies menos preocupantes** Especies generalizadas y abundantes.
levee An enlarged bank built up on each side of a river.	**dique de contención** Orilla agrandada que se amplía a cada costado de un río.
life-cycle analysis A systems tool that looks at the materials used and released throughout the lifetime of a product—from the procurement of raw materials through their manufacture, use, and disposal. *Also known as* **cradle-to-grave analysis.**	**análisis de ciclo de vida** Recurso en sistemas que estudia los materiales utilizados y emanados durante la vida completa de un producto (es decir, desde que se compra la materia prima hasta que se fabrica, utiliza y tira). *También se conoce como* **análisis de la cuna a la tumba.**
life expectancy The average number of years that an infant born in a particular year in a particular country can be expected to live, given the current average life span and death rate in that country.	**expectativa de vida** Cantidad promedio de años que se puede prever que viva un bebé nacido en un año dado y en un país específico, dada la media de vida o la tasa de fallecimientos en dicho país.
limiting nutrient A nutrient required for the growth of an organism but available in a lower quantity than other nutrients.	**nutriente limitante** Nutriente que se precisa para el crecimiento de un organismo, pero que está disponible en cantidad menor que otros nutrientes.
limiting resource A resource that a population cannot live without and that occurs in quantities lower than the population would require to increase in size.	**recurso limitante** Recurso sin el que una población no puede vivir y que existe en cantidades menores de las que la población precisaría para aumentar en volumen.
limnetic zone A zone of open water in lakes and ponds.	**zona limnética** Espacio de aguas abiertas en lagos y lagunas.
lipid A smaller organic biological molecule that does not mix with water.	**lípido** Molécula biológica orgánica pequeña que no se mezcla con agua.
lithosphere The outermost layer of Earth, including the mantle and crust.	**litosfera** Capa exterior de la Tierra que incluye el manto y la corteza.
littoral zone The shallow zone of soil and water in lakes and ponds where most algae and emergent plants grow.	**zona litoral** El espacio pando de tierra y agua en lagos y lagunas, en los que crece la mayoría de las algas y las plantas emergentes.
logistic growth model A growth model that describes a population whose growth is initially exponential, but slows as the population approaches the carrying capacity of the environment.	**modelo de crecimiento logístico** Modelo de crecimiento que describe una población cuyo crecimiento inicialmente es exponencial, pero que pierde velocidad a medida que se aproxima a la capacidad poblacional del entorno.
London-type smog *See* **sulfurous smog.**	**esmog tipo londinense** *Ver* **sulfurous smog.**
Los Angeles–type smog *See* **photochemical smog.**	**esmog tipo angelino o de Los Ángeles** *Ver* **photochemical smog.**
Lyme disease A disease caused by a bacterium (*Borrelia burgdorferi*) that is transmitted by ticks.	**la enfermedad de Lyme** es una enfermedad causada por una bacteria (*Borrelia burgdorferi*) transmitida por garrapatas.

M

macroevolution Evolution that gives rise to new species, genera, families, classes, or phyla.	**macroevolución** Evolución de la que surgen nuevas especies, géneros, familias, clases o filos.
macronutrient One of six key elements that organisms need in relatively large amounts: nitrogen, phosphorus, potassium, calcium, magnesium, and sulfur.	**macronutriente** Uno de seis elementos clave que los organismos necesitan en cantidades relativamente grandes: nitrógeno, fósforo, potasio, calcio, magnesio y azufre.
mad cow disease A disease in which prions mutate into deadly pathogens and slowly damage a cow's nervous system.	**enfermedad de las vacas locas** Enfermedad en la que priones mutan en patógenos mortales que lentamente perjudican el sistema nervioso de las vacas.

magma Molten rock.	**magma** Roca fundida.
malaria An infectious disease caused by one of several species of protists in the genus *Plasmodium*.	**malaria** Enfermedad infecciosa causada por una o varias especies de protistas del género *Plasmodium*.
malnourished Having a diet that lacks the correct balance of proteins, carbohydrates, vitamins, and minerals.	**desnutrido** Que cuenta con un régimen alimenticio que carece del equilibrio adecuado de proteínas, carbohidratos, vitaminas y minerales.
mangrove swamp A swamp that occurs along tropical and subtropical coasts, and contains salt-tolerant trees with roots submerged in water.	**manglar** Pantano que se da en las costas tropicales y subtropicales, que contiene árboles que toleran la sal y cuyas raíces permanecen sumergidas en el agua.
mantle The layer of Earth above the core, containing magma.	**manto** Capa de la Tierra que está sobre el núcleo interno y que contiene el magma.
manufactured capital All goods and infrastructure that humans produce.	**capital manufacturado** Todos los bienes y infraestructura que producen los seres humanos.
manure lagoon Human-made pond lined with rubber built to handle large quantities of manure produced by livestock.	**lago anaeróbico o de estiércol** Estanque artificial con forro de caucho o hule construido para manejar grandes cantidades de estiércol producido por ganado.
Marine Mammal Protection Act A 1972 U.S. act to protect declining populations of marine mammals.	**Ley de Protección de los Mamíferos Marinos** Ley federal estadounidense promulgada en 1972 con el fin de proteger las poblaciones de mamíferos marinos en disminución.
market failure When the economic system does not account for all costs.	**falla del mercado** Ocurre cuando el sistema económico no tiene en cuenta todos los costos.
mass A measurement of the amount of matter an object contains.	**masa** Medición de la cantidad de materia que contiene un objeto.
mass extinction A large extinction of species in a relatively short period of time.	**extinción masiva** Extinción terminal de una especie en un período de tiempo relativamente corto.
mass number A measurement of the total number of protons and neutrons in an element.	**número másico** Medición de la cantidad total de protones y neutrones que tiene un elemento.
matter Anything that occupies space and has mass.	**materia** Todo aquello que ocupa espacio y tiene masa.
maximum contaminant level (MCL) The standard for safe drinking water established by the EPA under the Safe Drinking Water Act.	**nivel máximo de contaminantes** La norma que define el agua potable, establecida por la EPA conforme a la Ley de Agua Potable Segura.
maximum sustainable yield (MSY) The maximum amount of a renewable resource that can be harvested without compromising the future availability of that resource.	**producción sostenida máxima** Cantidad máxima de un recurso renovable que se puede cosechar sin comprometer la disponibilidad futura de dicho recurso.
meat Livestock or poultry consumed as food.	**carne** Ganado o aves que se consumen como alimento.
mesotrophic Describes a lake with a moderate level of productivity.	**mesotrópico** Se dice de un lago con un nivel moderado de productividad.
metal An element with properties that allow it to conduct electricity and heat energy, and to perform other important functions.	**metal** Elemento con propiedades que le permiten conducir energía eléctrica y térmica, y realizar otras funciones importantes.
metamorphic rock Rock that forms when sedimentary rock, igneous rock, or other metamorphic rock is subjected to high temperature and pressure.	**roca metamórfica** Roca que se forma cuando una roca sedimentaria, una roca ígnea o cualquier otra roca metamórfica se somete a temperatura y presión alta.
metapopulation A group of spatially distinct populations that are connected by occasional movements of individuals between them.	**metapoblación** Grupo de poblaciones separadas espacialmente que están conectadas entre sí por desplazamientos ocasionales de individuos entre ellas.
microevolution Evolution below the species level.	**microevolución** Evolución por debajo del nivel de especie.
mineralization The process by which fungal and bacterial decomposers break down the organic matter found in dead bodies and waste products and convert it into inorganic compounds.	**mineralización** Proceso mediante el cual los saprofitos fúngicos y bacterianos desintegran la materia orgánica que se encuentra en los cuerpos muertos y los desechos, y la convierten en compuestos inorgánicos.

mine tailings Unwanted waste material created during mining including mineral and other residues that are left behind after the desired metal or ore is removed.	**relaves** Materiales de desecho que sobran del proceso de minería. Entre ellos hay minerales y otros residuos que quedan luego de que se extrae el metal deseado, o mena.
modern carbon Carbon in biomass that was recently in the atmosphere.	**carbono moderno** Carbono en forma de biomasa que hace poco estuvo en la atmósfera.
molecule A particle that contains more than one atom.	**molécula** Partícula que contiene más de un átomo.
monocropping An agricultural method that utilizes large plantings of a single species or variety.	**monocultivo** Método agrícola en el que se siembran grandes plantaciones de una sola especie o variedad.
mountaintop removal A mining technique in which the entire top of a mountain is removed with explosives.	**remoción de cumbres** Técnica de minería en la que la cumbre completa de una montaña se elimina con explosivos.
multiple-use lands A U.S. classification used to designate lands that may be used for recreation, grazing, timber harvesting, and mineral extraction.	**terrenos de usos múltiples** Clasificación que se usa en Estados Unidos para designar terrenos que se pueden usar para fines recreativos, de pastoreo, de silvicultura y para la extracción de minerales.
multi-use zoning A zoning classification that allows retail and high-density residential development to coexist in the same area.	**zonificación de múltiples usos** Clasificación territorial que autoriza que coexistan en la misma zona tiendas detallistas y urbanizaciones residenciales de alta densidad.
municipal solid waste (MSW) Refuse collected by municipalities from households, small businesses, and institutions.	**desechos sólidos municipales** Basura recogida por el municipio de los hogares, la pequeña empresa y las instituciones.
mutagen A type of carcinogen that causes damage to the genetic material of a cell.	**mutágeno** Tipo de agente carcinógeno que causa daños a los materiales genéticos de una célula.
mutation A random change in the genetic code produced by a mistake in the copying process.	**mutación** Cambio aleatorio en el código genético producido por un error en el proceso de copia.
mutualism An interaction between two species that increases the chances of survival or reproduction for both species.	**mutualismo** Interacción entre dos especies que aumenta las posibilidades de supervivencia o reproducción para ambas especies.

N

narrow spectrum pesticide *See* **selective pesticide.**	**pesticida estrecho del espectro** *Ver* **narrow spectrum pesticide.**
National Environmental Policy Act (NEPA) A 1969 U.S. federal act that mandates an environmental assessment of all projects involving federal money or federal permits.	**Ley Normativa Nacional de Integración Ambiental** Ley de EE. UU., promulgada en 1969, que obliga a realizar una evaluación ambiental en todo proyecto en el que se usen dineros federales o que precisen una autorización federal.
national wilderness area An area set aside with the intent of preserving a large tract of intact ecosystem or landscape.	**área silvestre nacional** Área que se ha separado con la intención de conservar un terreno de mucha extensión preservando intacto un ecosistema o paisaje.
national wildlife refuge A federal public land managed for the primary purpose of protecting wildlife.	**refugio silvestre nacional** Terreno de propiedad del gobierno federal administrado para el efecto principal de proteger la vida silvestre.
native species Species that live in their historical range, typically where they have lived for thousands or millions of years.	**especies nativas** Especies que viven dentro de sus hábitats históricos, generalmente donde se han arraigado durante miles o millones de años.
natural capital The resources of the planet, such as air, water, and minerals.	**capital natural** Recursos del planeta, tales como aire, agua y minerales.
natural experiment A natural event that acts as an experimental treatment in an ecosystem.	**experimento natural** Acontecimiento natural que hace las veces de tratamiento experimental en un ecosistema.
near-threatened species Species that are very likely to become threatened in the future.	**especies casi amenazadas** Especies que con mucha probabilidad van a verse amenazadas en un futuro.

negative feedback loop A feedback loop in which a system responds to a change by returning to its original state, or by decreasing the rate at which the change is occurring.	**circuito de realimentación negativa** Realimentación en la que un sistema responde a un cambio mediante el regreso a su estado original o mediante la disminución de la tasa a la cual sucede el cambio.
net migration rate The difference between immigration and emigration in a given year per 1,000 people in a country.	**tasa neta de migración** Diferencia entre inmigración y emigración en un año dado por cada millar de personas que viven en un país.
net primary productivity (NPP) The energy captured by producers in an ecosystem minus the energy producers respire.	**productividad primaria neta** Energía captada por productores en un ecosistema menos la energía que respiran los productores.
net removal The process of removing more than is replaced by growth, typically used when referring to carbon.	**eliminación neta** Proceso de eliminar más de lo que se reemplaza con crecimiento. Típicamente se usa el término al referirse al carbono.
neurotoxin A chemical that disrupts the nervous systems of animals.	**neurotoxina** Sustancia química que altera el sistema nervioso de los animales.
niche generalist A species that can live under a wide range of abiotic or biotic conditions.	**generalista del nicho** Especie que puede vivir bajo una amplia gama de condiciones abióticas o bióticas.
niche specialist A species that is specialized to live in a specific habitat or to feed on a small group of species.	**especialista del nicho** Especie especializada para vivir en un hábitat específico o para alimentarse de un grupo pequeño de especies.
nitrification The conversion of ammonia (NH_4^+) into nitrite (NO_2^-) and then into nitrate (NO_3^-).	**nitrificación** Conversión de amonio (NH_4^+) a nitrito (NO_2^-) y luego a nitrato (NO_3^-).
nitrogen cycle The movement of nitrogen around the biosphere.	**ciclo de nitrógeno** Movimiento del nitrógeno alrededor de la biosfera.
nitrogen fixation The process that converts nitrogen gas in the atmosphere (N_2) into forms of nitrogen that producers can use.	**fijación de nitrógeno** Proceso por el que el gas nitrógeno de la atmósfera (N_2) se convierte en formas de nitrógeno que los productores pueden utilizar.
nomadic grazing The feeding of herds of animals by moving them to seasonally productive feeding grounds, often over long distances.	**pastoreo nomádico** Alimentación de manadas de animales mudándolas cada temporada de un terreno de alimentación productiva a otro, a menudo cubriendo grandes distancias.
nondepletable An energy source that cannot be used up.	**inagotable** Fuente de energía que no se puede agotar ni consumir en su totalidad.
nonpersistent pesticide A pesticide that breaks down rapidly, usually in weeks or months.	**plaguicida no persistente** Plaguicida que se descompone rápidamente, generalmente en semanas o meses.
nonpoint source A diffuse area that produces pollution.	**fuente sin punto** Zona indefinida que produce contaminación.
nonrenewable energy resource An energy source with a finite supply, primarily the fossil fuels and nuclear fuels.	**recurso energético no renovable** Fuente de energía con un abastecimiento finito, primordialmente combustibles fósiles y combustibles nucleares.
no-observed-effect level (NOEL) The highest concentration of a chemical that causes no lethal or sublethal effects.	**nivel sin efecto observable (NOEL, por sus siglas en inglés que significan, No-observed-effect level)** La concentración más alta de un químico que no causa efectos letales o subletales.
no-till agriculture An agricultural method in which farmers do not turn the soil between seasons as a means of reducing topsoil erosion.	**agricultura sin arar** Método agrícola en el que los agricultores no voltean los suelos entre estaciones a fin de minimizar la erosión de la capa superior o manto.
nuclear fuel Fuel derived from radioactive materials that give off energy.	**combustible nuclear** Combustible derivado de materiales radiactivos que producen energía.
nuclear fusion A reaction that occurs when lighter nuclei are forced together to produce heavier nuclei.	**fusión nuclear** Reacción que ocurre cuando se amontonan núcleos ligeros para producir núcleos más pesados.
nucleic acid Organic compounds found in all living cells.	**ácido nucleico** Compuestos orgánicos que se encuentran en todas las células vivientes.

null hypothesis A prediction that there is no difference between the groups or conditions that are being compared.	**hipótesis nula** Predicción de que no hay diferencia entre los grupos o las condiciones, o declaración o concepto que puede ser falsificado o que se puede demostrar que es erróneo.

O

Occupational Safety and Health Administration (OSHA) An agency of the U.S. Department of Labor, responsible for the enforcement of health and safety regulations.	**Administración de Seguridad y Salud Ocupacional (OSHA)** Dependencia del gobierno federal estadounidense. Forma parte del Departamento del Trabajo y se responsabiliza de hacer cumplir las reglamentaciones de salud y seguridad.
ocean acidification An increase in the acidity of the oceans.	**acidificación de los océanos** El aumento en el nivel de acidez de los océanos.
O horizon The organic horizon at the surface of many soils, composed of organic detritus in various stages of decomposition.	**horizonte O** Horizonte orgánico en la superficie de muchos suelos, compuesto por desperdicios en diversas etapas de descomposición.
oil sands Slow-flowing, viscous deposits of bitumen mixed with sand, water, and clay.	**arenas petrolíferas** Depósitos bituminosos viscosos de flujo lento mezclados con arena, agua y arcilla.
oligotrophic Describes a lake with a low level of productivity.	**oligotrópico** Se dice de un lago con un nivel bajo de productividad.
open ocean Deep ocean water, located away from the shoreline where sunlight can no longer reach the ocean bottom.	**mar abierto** Aguas profundas del océano, ubicadas a una distancia del litoral donde la luz solar ya no puede llegar hasta el fondo.
open-loop recycling Recycling one product into a different product.	**reciclaje en circuito abierto** Reciclaje de un producto para terminar con otro producto diferente.
open-pit mining A mining technique that creates a large visible pit or hole in the ground.	**minería a cielo abierto** Técnica de minería en la que crea un tajo o agujero grande y visible en el suelo.
open system A system in which exchanges of matter or energy occur across system boundaries.	**sistema abierto** Sistema en el que los intercambios de materia o energía se dan a través de las fronteras del sistema.
ore A concentrated accumulation of minerals from which economically valuable materials can be extracted.	**mena** Acumulación concentrada de minerales de los que se pueden extraer materiales de valor económico.
organic agriculture The production of crops in a way that sustains or improves the soil without the use of synthetic pesticides or fertilizers.	**agricultura orgánica** Producción de cultivos sin usar plaguicidas ni fertilizantes sintéticos.
organic compound A compound that contains carbon-carbon and carbon-hydrogen bonds.	**compuesto orgánico** Compuesto que contiene enlaces de carbono con carbono y carbono con hidrógeno.
organic fertilizer Fertilizer composed of organic matter from plants and animals.	**fertilizante orgánico** Fertilizante hecho de materia orgánica tomada de plantas y animales.
output A loss from a system.	**salida** Pérdida en un sistema.
overnutrition Ingestion of too many calories and a lack of balance of foods and nutrients.	**sobrenutrición** Ingestión de demasiadas calorías y la falta de equilibrio entre alimentos y nutrientes.
overshoot When a population becomes larger than the environment's carrying capacity.	**extralimitación** Cuando una población crece más que la capacidad poblacional del entorno.
oxygenated fuel A fuel with oxygen as part of the molecule.	**combustible oxigenado** Combustible que incluye oxígeno como parte de la molécula.
ozone (O_3) A secondary pollutant made up of three oxygen atoms bound together.	**ozono (O_3)** Contaminante secundario compuesto por tres átomos de oxígeno ligados entre sí.

P

pandemic An epidemic that occurs over a large geographic region.	**pandemia** Epidemia que cuando ocurre abarca una región geográfica amplia.
parasitism An interaction in which one organism lives on or in another organism.	**parasitismo** Interacción en la que un organismo vive sobre o en otro organismo.

parasitoid A specialized type of predator that lays eggs inside other organisms—referred to as its host.	**parasitoide** Tipo especializado de depredador que pone huevos dentro de otros organismos, a los que se los designa huéspedes.
parent material The underlying rock material from which the inorganic components of a soil are derived.	**material parental** Material de rocoso subyacente a partir del cual se derivan los componentes inorgánicos del suelo.
Paris Climate Agreement A pledge by 195 countries to keep global warming less than 2 °C above pre-industrial levels. *Also known as* the **Paris Climate Accord**.	**Acuerdo de Paris sobre el cambio climático** Compromiso de 195 países de mantener el aumento de la temperatura global por debajo de 2 °C con respecto a los niveles preindustriales.
particles *See* **particulate matter.**	**partículas** *Ver* **particulate matter.**
particulate matter (PM) Solid or liquid particles suspended in air. *Also known as* **particulates; particles.**	**materia particulada** Partículas sólidas o líquidas suspendidas en el aire. *También se conoce como* **particulados; partículas.**
particulates *See* **particulate matter.**	**particulados** *Ver* **particulate matter.**
passive solar design Construction designed to take advantage of solar radiation without active technology.	**diseño solar pasivo** Construcción diseñada para aprovechar la radiación solar sin tecnología activa.
pathogen A parasite that causes disease in its host.	**patógenos** Parásitos que pueden enfermar a su huésped (u hospedador).
peak demand The greatest quantity of energy used at any one time.	**demanda pico** Cantidad máxima de energía que se utiliza simultáneamente.
peak oil The point at which half the total known oil supply is used up.	**pico petrolero** Punto en el cual se ha consumido la mitad del total conocido de las existencias de petróleo.
per capita Amount per each person in a country or unit of population.	**per cápita** Cantidad de algo por persona en un país o unidad de población.
perchlorates A group of harmful chemicals used for rocket fuel.	**percloratos** Grupo de sustancias químicas nocivas que se utilizan de combustible en cohetes.
perennial plant A plant that lives for multiple years.	**planta perenne** Planta que vive múltiples años.
periodic table A chart of all chemical elements currently known, organized by their properties.	**tabla periódica** Cuadro de todos los elementos químicos conocidos a la fecha, organizados según sus propiedades.
permafrost An impermeable, permanently frozen layer of soil.	**permafrost** Capa del suelo impermeable y permanentemente congelada.
persistence The length of time a chemical remains in the environment.	**persistencia** Tiempo que una sustancia química permanece en el medio ambiente.
persistent pesticide A pesticide that remains in the environment for a long time.	**plaguicida persistente** Plaguicida que permanece en el medio ambiente mucho tiempo.
pesticide A substance, either natural or synthetic, that kills or controls organisms that people consider pests.	**plaguicida** Sustancia, sea natural o sintética, que mata o controla organismos que las personas consideran plagas.
pesticide resistance A trait possessed by certain individuals that are exposed to a pesticide and survive.	**resistencia a plaguicidas** Rasgo de ciertos individuos que se exponen a un plaguicida y sobreviven.
pesticide treadmill A cycle of pesticide development, followed by pest resistance, followed by new pesticide development.	**círculo vicioso de plaguicidas** Ciclo de la creación de un plaguicida, seguida de resistencia al mismo de las plagas, seguida de la creación de un nuevo plaguicida.
petroleum A widely-used fossil fuel that occurs in underground deposits, composed of a liquid mixture of hydrocarbons, water, and sulfur.	**petróleo** Combustible fósil de uso extendido que se da en yacimientos subterráneos, compuesto por una mezcla líquida de hidrocarburos, agua y azufre.
pH The number that indicates the relative strength of acids and bases in a substance.	**pH** Cifra que indica la potencia relativa de ácidos y bases en una sustancia.
phenotype A set of traits expressed by an individual.	**fenotipo** Conjunto de rasgos que se manifiestan en un individuo.
phosphorus cycle The movement of phosphorus around the biosphere.	**ciclo de fósforo** Movimiento de fósforo alrededor de la biosfera.

photic zone The upper layer of ocean water in the ocean that receives enough sunlight for photosynthesis.	**zona fótica** Capa superior del agua del océano en la que el océano recibe suficiente luz solar para la fotosíntesis.
photochemical oxidant A class of air pollutants formed as a result of sunlight acting on compounds such as nitrogen oxides.	**oxidante fotoquímico** Clase de contaminantes aéreos que se forman como resultado de la acción de la luz solar sobre compuestos tales como óxidos de nitrógeno.
photochemical smog Smog that is dominated by oxidants such as ozone. *Also known as* **Los Angeles–type smog; brown smog.**	**esmog fotoquímico** Esmog dominado por oxidantes tales como ozono. *También se conoce como* **esmog angelino o de Los Ángeles** o como **esmog marrón.**
photon A massless packet of energy that carries electromagnetic radiation at the speed of light.	**fotón** Paquete de energía sin masa que transporta radiación electromagnética a la velocidad de la luz.
photosynthesis The process by which producers use solar energy to convert carbon dioxide and water into glucose.	**fotosíntesis** Proceso mediante el cual los productores aprovechan la energía solar para convertir el dióxido de carbono y el agua en glucosa.
photovoltaic solar cell A system of capturing energy from sunlight and converting it directly into electricity.	**celda solar fotovoltaica** Sistema para captar la energía del sol y convertirla directamente en electricidad.
phylogeny The branching pattern of evolutionary relationships.	**filogenia** Diseño de ramas de las interrelaciones evolucionarias.
physical weathering The mechanical breakdown of rocks and minerals.	**desgaste físico** Descomposición mecánica de rocas y minerales.
phytoplankton Floating algae.	**fitoplancton** Algas flotantes.
pioneer species A species that can colonize new areas rapidly and grow well in full sunshine.	**especie pionera** Especie que puede colonizar áreas nuevas rápidamente y crecer bien cuando hay plenitud de luz solar.
placer mining The process of looking for minerals, metals, and precious stones in river sediments.	**minería de arenal** Proceso mediante el cual se buscan minerales, metales y piedras preciosas en los sedimentos de ríos.
plague An infectious disease caused by a bacterium (*Yersinia pestis*) that is carried by fleas.	**peste** Enfermedad infecciosa causada por una bacteria (*Yersinia pestis*) transportada por pulgas.
planned obsolescence The process of designing a product so that it will need to be replaced within a few years.	**obsolescencia programada** Proceso de diseñar un producto de tal manera que tenga que ser reemplazado en pocos años.
plate tectonics The theory that the lithosphere of Earth is divided into plates, most of which are in constant motion.	**tectónica de placas** Teoría que explica que la litosfera de la Tierra está dividida en placas, la mayoría de las cuales está en constante movimiento.
point source A distinct location from which pollution is directly produced.	**fuente de punto** Ubicación definida en la que se produce la contaminación directamente.
polar cell A convection current in the atmosphere, formed by air that rises at 60° N and 60° S and sinks at the poles, 90° N and 90° S.	**célula polar** Corriente de convección en la atmósfera, formada por aire que asciende a 60° N y 60° S y se hunde en los polos, 90° N y 90° S.
polar molecule A molecule in which one side is more positive and the other side is more negative.	**molécula polar** Molécula en la que un lado es más positivo y el otro es más negativo.
polychlorinated biphenyls (PCBs) A group of industrial compounds used to manufacture plastics and insulate electrical transformers, and responsible for many environmental problems.	**bifenilos policlorados** Grupo de compuestos industriales que se usan para fabricar plásticos y aislar transformadores eléctricos. Son responsables de muchos problemas medioambientales.
population The individuals that belong to the same species and live in a given area at a particular time.	**población** Individuos que pertenecen a la misma especie y conviven en un área dada en un momento dado.
population density The number of individuals per unit area at a given time.	**densidad poblacional** Cantidad de individuos en un área dada en un momento dado.
population distribution A description of how individuals are distributed with respect to one another.	**distribución de la población** Descripción de cómo los individuos se distribuyen con respecto a los demás.

population ecology The study of factors that cause populations to increase or decrease.	**ecología poblacional** Estudio de los factores que llevan a que las poblaciones aumenten o disminuyan.
population growth models Mathematical equations that can be used to predict population size at any moment in time.	**modelos de crecimiento poblacional** Ecuaciones matemáticas que se pueden utilizar para predecir el volumen poblacional en un momento dado.
population growth rate The number of offspring an individual can produce in a given time period, minus the deaths of the individual or its offspring during the same period.	**tasa de crecimiento poblacional** La cantidad de descendientes que un individuo puede producir en un período de tiempo dado, menos los fallecimientos del individuo o su descendencia durante ese mismo período de tiempo.
population momentum Continued population growth after growth reduction measures have been implemented.	**momento poblacional** Continuación del crecimiento de la población después de implementarse medidas para reducir la población.
population pyramid An age structure diagram that is widest at the bottom and smallest at the top, typical of developing countries.	**pirámide poblacional** Diagrama de estructuras de edad que es más ancho en la parte inferior y más pequeño en la cima, aspecto característico de los países en vías de desarrollo.
population size (*N*) The total number of individuals within a defined area at a given time.	**volumen poblacional (*N*)** Número total de individuos dentro de un área definida en un momento dado.
positive feedback loop A feedback loop in which change in a system is amplified.	**circuito de realimentación positiva** Realimentación en la que se amplifica cualquier cambio en un sistema.
potential energy Stored energy that has not been released.	**energía potencial** Energía almacenada que no se ha liberado.
potentially renewable An energy source that can be regenerated indefinitely as long as it is not overharvested.	**potencialmente renovable** Fuente de energía que se puede regenerar indefinidamente siempre y cuando no se sobreaproveche.
power The rate at which work is done.	**potencia** Ritmo al cual se trabaja.
precautionary principle A principle based on the belief that action should be taken against a plausible environmental hazard.	**principio precaucionario** Principio basado en la convicción de que se debe poner en práctica alguna acción contra un peligro ambiental plausible.
precision How close the repeated measurements of a sample are to one another.	**precisión** Cercanía que existe entre las reiteradas mediciones de una muestra.
predation An interaction in which one animal typically kills and consumes another animal.	**depredación** Interacción en la que un animal característicamente mata y consume otro animal.
prescribed burn A fire deliberately set under controlled conditions in order to reduce the accumulation of dead biomass on a forest floor.	**quema prescrita** Incendio que se comienza a propósito y en condiciones controladas a fin de reducir la acumulación de biomasa muerta en un suelo forestal.
primary consumer *See* **herbivore.**	**consumidor primario** *Ver* **herbivore.**
primary pollutant A polluting compound that comes directly out of a smokestack, exhaust pipe, or natural emission source.	**contaminante primario** Compuesto contaminante que proviene directamente de una chimenea, un tubo de escape o una fuente de emisión natural.
primary succession Ecological succession occurring on surfaces that are initially devoid of soil.	**sucesión primaria** Sucesión ecológica que se da en las superficies que inicialmente carecen de tierra.
prion A small, beneficial protein that occasionally mutates into a pathogen.	**prión** Proteína pequeña y benigna que a veces muta en un patógeno.
producer An organism that uses the energy of the Sun to produce usable forms of energy. *Also known as* **autotroph.**	**productor** Organismo que aprovecha la energía del Sol para producir formas utilizables de energía. *También se conoce como* **autótrofo.**
profundal zone A region of water where sunlight does not reach, below the limnetic zone in very deep lakes.	**zona de profundidad** Región del agua a la que no llega la luz solar, debajo de la zona limnética en lagos muy profundos.
prospective study A study that monitors people who might become exposed to an environmental hazard, such as a harmful chemical in the future.	**estudio prospectivo** Estudio en el que se vigila a personas que en un futuro podrían quedar expuestas a un peligro medioambiental como las sustancias nocivas.

protein A critical component of living organisms made up of a long chain of nitrogen-containing organic molecules known as amino acids.	**proteína** Componente crítico de los organismos vivientes compuesto por una larga cadena de moléculas orgánicas que contienen nitrógeno conocidas como aminoácidos.
provision A good that humans can use directly.	**provisión** Bien que los seres humanos pueden aprovechar directamente.

R

radioactive decay The spontaneous release of material from the nucleus of radioactive isotopes.	**descomposición radiactiva** Liberación espontánea de material desde el núcleo de los isótopos radiactivos.
radioactive waste Nuclear fuel that can no longer produce enough heat to be useful in a power plant but continues to emit radioactivity.	**residuos radiactivos** Combustible nuclear que ya no puede producir suficiente calor para ser útil en una planta generadora pero que sigue emitiendo radiactividad.
rain shadow A region with dry conditions found on the leeward side of a mountain range as a result of humid winds from the ocean causing precipitation on the windward side.	**sombra pluviométrica** Región de condiciones secas que se encuentra en el lado de sotavento de una cordillera como resultado de los vientos húmedos que provienen del océano que producen precipitación en el lado de barlovento.
rangeland Grassland primarily used for grazing cattle.	**pastizal** Pastizales usados principalmente para pastoreo de ganado.
range of tolerance The limits to the abiotic conditions that a species can tolerate.	**gama de tolerancia** Límites de las condiciones abióticas que puede tolerar una especie.
REACH A 2007 agreement among the nations of the European Union about regulation of chemicals; the acronym stands for registration, evaluation, authorization, and restriction of chemicals.	**REACH** Convenio logrado en el año 2007 entre los países de la Unión Europea en el que se controlan las sustancias químicas. La sigla representa en inglés registro, evaluación, autorización y restricción de las sustancias químicas.
realized niche The range of abiotic and biotic conditions under which a species actually lives.	**nicho realizado** Gama de condiciones abióticas y bióticas bajo las cuales vive una especie.
recombination The genetic process by which one chromosome breaks off and attaches to another chromosome during reproductive cell division.	**recombinación** Proceso genético en el cual un cromosoma se separa y se adhiere a otro cromosoma durante la división reproductiva de células.
recycling The process by which materials destined to become municipal solid waste (MSW) are collected and converted into raw material that is then used to produce new objects.	**reciclaje** Proceso en el que se recogen los materiales destinados a convertirse en desechos sólidos municipales y se convierten en materia prima que luego se aprovecha para producir objetos nuevos.
Red List A list of worldwide threatened species.	**Lista Roja** Lista de las especies amenazadas en todo el mundo.
reduce, reuse, recycle A popular phrase promoting the idea of diverting materials from the waste stream. *Also known as* **the three Rs.**	**reducir, reutilizar, reciclar** Frase de gran acogida que fomenta el concepto de extraer materiales del flujo de desechos. *También se conoce como* **las tres erres.**
renewable In energy management, an energy source that is either potentially renewable or nondepletable.	**renovable** En la gestión de energía, fuente energética que tiene el potencial de ser renovable o inagotable.
replacement-level fertility The total fertility rate required to offset the average number of deaths in a population in order to maintain the current population size.	**fertilidad a nivel de reemplazo** Tasa de fertilidad total requerida para compensar la cantidad promedio de fallecimientos en una población a fin de mantener el volumen poblacional actual.
replication The data collection procedure of taking repeated measurements.	**replicación** Procedimiento de recolección de datos en el que se repiten las mediciones.
reproductive isolation The result of two populations within a species evolving separately to the point that they can no longer interbreed and produce viable offspring.	**aislamiento reproductor** Lo que resulta cuando dos poblaciones dentro de una especie evolucionan separadamente hasta el punto de que ya no pueden aparearse ni producir descendientes viables.

reserve In resource management, the known quantity of a resource that can be economically recovered.	**reserva** En la gestión de recursos, la cantidad conocida de un recurso que se puede recuperar en términos económicos.
reservoir The water body created by damming a river or stream.	**estanque** Extensión de agua creada cuando se le coloca una presa a un río o arroyo.
resilience The rate at which an ecosystem returns to its original state after a disturbance.	**resiliencia** Ritmo al cual un ecosistema regresa a su estado original después de una perturbación.
resistance A measure of how much a disturbance can affect flows of energy and matter in an ecosystem.	**resistencia** Medición del efecto que una perturbación puede tener respecto de los flujos de energía y materia en un ecosistema.
resource conservation ethic The belief that people should maximize use of resources, based on the greatest good for everyone.	**ética de conservación de recursos** Noción de que las personas deben aprovechar al máximo los recursos, basándose en el mayor bienestar para todos.
resource partitioning When two species divide a resource based on differences in their behavior or morphology.	**partición de recursos** Cuando dos especies se dividen un recurso según diferencias en su comportamiento o morfología.
restoration ecology The study and implementation of restoring damaged ecosystems.	**ecología de restauración** Estudio de ecosistemas perjudicados y de cómo restaurarlos.
retrospective study A study that monitors people who have been exposed to an environmental hazard such as a harmful chemical at some time in the past.	**estudio retrospectivo** Estudio en el que se vigila a personas que en algún momento en el pasado fueron expuestas a un peligro medioambiental como las sustancias nocivas.
reuse Using a product or material that was intended to be discarded.	**reutilizar** Hacer uso de algún producto o material que se iba a tirar.
reverse osmosis A process of desalination in which water is forced through a thin semipermeable membrane at high pressure.	**ósmosis inversa** Proceso de desalinización en el que el agua se pasa por una membrana delgada y semipermeable bajo gran presión.
Richter scale A scale that measures the largest ground movement that occurs during an earthquake.	**escala de Richter** Escala que mide los movimientos telúricos de mayor escala que ocurren durante un terremoto.
RNA (ribonucleic acid) A nucleic acid that translates the code stored in DNA, which makes possible the synthesis of proteins.	**ARN (ácido ribonucleico)** Ácido nucleico que traduce el código almacenado en el ADN, que posibilita la síntesis de proteínas.
rock cycle The geologic cycle governing the constant formation, alteration, and destruction of rock material that results from tectonics, weathering, and erosion, among other processes.	**ciclo rocoso** Ciclo geológico que dirige la constante formación, alteración y destrucción de material rocoso que es resultado de la tectónica, la desintegración y la erosión, entre otros procesos.
route of exposure The way in which an individual might come into contact with an environmental hazard such as a chemical.	**ruta de exposición** Manera en que un individuo puede entrar en contacto con un peligro medioambiental como un producto químico.
***r*-selected species** A species that has a high intrinsic growth rate, which often leads to population overshoots and die-offs.	**especies de selección** *r* Especie que tiene una alta tasa de crecimiento intrínseco, lo cual a menudo conduce a extralimitarse en la población y a mortandades.
runoff Water that moves across the land surface and into streams and rivers.	**escorrentía** Agua que se desplaza a través de la superficie terrestre y desemboca en ríos y arroyos.
run-of-the-river Hydroelectricity generation in which water is retained behind a low dam or no dam.	**flujo del agua** Generación hidroeléctrica en la que el agua se retiene detrás de una presa pequeña o se aprovecha sin presa.

S

Safe Drinking Water Act Legislation that sets the national standards for safe drinking water.	**Ley de Agua Potable Segura** Legislación federal estadounidense que fija las normas para el agua potable segura.

salinization A form of soil degradation that occurs when the small amount of salts in irrigation water becomes highly concentrated on the soil surface through evaporation.	**salinización** Forma de degradación del suelo que ocurre cuando una cantidad reducida de sales en el agua de riego termina con una concentración alta en la superficie del suelo por medio de la evaporación.
salt marsh A marsh containing nonwoody emergent vegetation, found along the coast in temperate climates.	**marisma salina** Marisma que contiene vegetación emergente no leñosa que se encuentra en el litoral en climas templados.
saltwater intrusion An infiltration of salt water in an area where groundwater pressure has been reduced from extensive drilling of wells.	**intrusión de agua salada** Infiltración de agua salada en una zona en la que la presión de las aguas freáticas se ha visto reducida por la extensa perforación de pozos.
sample size (*n*) The number of times a measurement is replicated in data collection.	**tamaño de la muestra (*n*)** Cantidad de veces que se replica una medición en una recolección de datos.
sanitary landfill An engineered ground facility designed to hold municipal solid waste (MSW) with as little contamination of the surrounding environment as possible.	**basurero sanitario** Instalación terrestre diseñada para retener desechos sólidos municipales con el mínimo posible de contaminación del entorno ambiental.
saturation point The maximum amount of water vapor in the air at a given temperature.	**punto de saturación** Cantidad máxima de vapor de agua en el aire a una temperatura dada.
scavenger An organism that consumes dead animals.	**carroñero** Organismo que consume animales muertos.
scientific method An objective method to explore the natural world, draw inferences from it, and predict the outcome of certain events, processes, or changes.	**método científico** Método objetivo para explorar el mundo natural, trazar inferencias del mismo y predecir el resultado de ciertos eventos, procesos o cambios.
seafloor spreading The formation of new ocean crust as a result of magma pushing upward and outward from Earth's mantle to the surface.	**ensanchamiento del fondo marino** Formación de una nueva corteza marítima como resultado de la presión hacia arriba y hacia fuera del magma sobre el manto terrestre hacia la superficie.
secondary consumer A carnivore that eats primary consumers.	**consumidor secundario** Carnívoro que se come consumidores primarios.
secondary pollutant A primary pollutant that has undergone transformation in the presence of sunlight, water, oxygen, or other compounds.	**contaminante secundario** Contaminante primario que se ha transformado ante la presencia de luz solar, agua, oxígeno u otros compuestos.
secondary succession The succession of plant life that occurs in areas that have been disturbed but have not lost their soil.	**sucesión secundaria** Sucesión de flora que ocurre en áreas que han sido perturbadas pero que no han perdido su suelo.
second law of thermodynamics The physical law stating that when energy is transformed, the quantity of energy remains the same, but its ability to do work diminishes.	**segunda ley de termodinámica** Ley física que manifiesta que cuando la energía se transforma, la cantidad de energía sigue igual pero su capacidad de hacer trabajos disminuye.
sedimentary rock Rock that forms when sediments such as muds, sands, or gravels are compressed by overlying sediments.	**roca sedimentaria** Roca que se forma cuando sedimentos tales como lodos, arenas o gravas se comprimen sobre sedimentos sobreyacentes.
seismic activity The frequency and intensity of earthquakes experienced over time.	**actividad sísmica** Frecuencia e intensidad de terremotos que ocurren en un lapso de tiempo.
selective cutting The method of harvesting trees that involves the removal of single trees or a relatively small number of trees from the larger forest.	**tala selectiva** Método de tala de árboles que implica retirar un solo árbol o una cantidad relativamente reducida de árboles entre los muchos que hay en un bosque o selva.
selective pesticide A pesticide that targets a narrow range of organisms. *Also known as* **narrow-spectrum pesticide.**	**plaguicida selectivo** Plaguicida que se enfoca en una gama estrecha de organismos. *También conocido como* **pesticida de espectro estrecho**.
sense of place The feeling that an area has a distinct and meaningful character.	**sentido de sitio** Sentido de que una zona tiene un carácter distintivo y significativo.
septage A layer of fairly clear water found in the middle of a septic tank.	**residuos sépticos** Capa de agua relativamente clara que se encuentra en el centro del tanque séptico.

septic system A relatively small and simple sewage treatment system, made up of a septic tank and a leach field, often used for homes in rural areas.	**sistema séptico** Sistema relativamente pequeño y sencillo para el tratamiento de aguas residuales. Consta de un tanque séptico y un campo de lixiviación. A menudo se utilizan en hogares en zonas rurales.
septic tank A large container that receives wastewater from a house as part of a septic system.	**tanque séptico** Recipiente grande que forma parte de un sistema séptico y que recibe aguas negras del hogar.
severe acute respiratory syndrome (SARS) A type of flu caused by a coronavirus.	**síndrome respiratorio agudo grave (SRAG)** Tipo de gripe causada por uno de los virus de la corona.
sex ratio The ratio of males to females in a population.	**proporción sexual** Proporción de machos a hembras en una población.
shifting agriculture An agricultural method in which land is cleared and used for a few years until the soil is depleted of nutrients.	**agricultura de desplazamiento** Método agrícola en el que el terreno se despeja y se usa unos pocos años hasta que se agoten los nutrientes del suelo.
sick building syndrome A buildup of toxic pollutants in an airtight space, seen in newer buildings.	**síndrome de edificio enfermo** Acumulación de contaminantes tóxicos en un espacio hermético. Se registra en los edificios más recientes.
siltation The accumulation of sediments, primarily silt, on the bottom of a reservoir.	**sedimentación** Acumulación de sedimentos, primordialmente cieno, en el fondo de un embalse.
siting The designation of a landfill location, typically through a regulatory process involving studies, written reports, and public hearings.	**ubicación** Designación del sitio donde se ubicará un basurero, por lo general mediante un proceso reglamentado que incluye estudios, informes escritos y audiencias públicas.
sludge Solid waste material from wastewater.	**fangos residuales** Material de desecho sólido que proviene de aguas negras.
smart grid An efficient, self-regulating electricity distribution network that accepts any source of electricity and distributes it automatically to end users.	**red inteligente** Red de distribución eléctrica eficiente, que se autorregula, y que recibe cualquier fuente de electricidad y la distribuye automáticamente a los usuarios finales.
smart growth A set of principles for community planning that focuses on strategies to encourage the development of sustainable, healthy communities.	**crecimiento inteligente** Conjunto de principios de planificación comunitaria que se concentra en estrategias que fomentan la creación de comunidades sostenibles y saludables.
smog A type of air pollution that is a mixture of oxidants and particulate matter.	**esmog** Tipo de contaminación del aire que es una mezcla de oxidantes y materias particuladas.
soil compaction A process where repeated trampling by humans, machinery, or animals causes a compaction of soil and a reduction in pore space.	**compactación del suelo** Proceso en el cual el impacto por el paso de humanos, maquinaria o animales causa compactación y reducción de porosidad en el suelo.
soil degradation The loss of some or all of a soil's ability to support plant growth.	**degradación de los suelos** Pérdida parcial o total de la capacidad que tiene el suelo para sustentar el crecimiento vegetal.
solubility How well a chemical dissolves in a liquid.	**solubilidad** La facilidad con que una sustancia química se disuelve en un líquido.
source reduction An approach to waste management that seeks to cut waste by reducing the use of potential waste materials in the early stages of design and manufacture.	**reducción de fuentes** Manera de enfocar la gestión de desechos mediante la cual se procura reducir el uso de materiales que posiblemente se tengan que desechar en las etapas iniciales de diseño y fabricación.
speciation The evolution of new species.	**especiación** Evolución de especies nuevas.
species A group of organisms that is distinct from other groups in its morphology (body form and structure), behavior, or biochemical properties.	**especie** Grupo de organismos que se distinguen de otros grupos en su morfología (forma y estructura del cuerpo), conducta o propiedades bioquímicas.
species diversity The number of species in a region or in a particular type of habitat.	**diversidad de especies** Número de especies en una región o en un tipo de hábitat en particular.
species evenness The relative proportion of individuals within the different species in a given area.	**uniformidad de especies** Proporción relativa de individuos dentro de especies diferentes en un área dada.

species richness The number of species in a given area.	**riqueza de especies** Cantidad de especies en un área dada.
spring A natural source of water formed when water from an aquifer percolates up to the ground surface.	**manantial** Fuente natural de agua que se forma cuando el agua de un acuífero se filtra hasta la superficie.
S-shaped curve The shape of the logistic growth model when graphed.	**curva en forma de S** Representación gráfica de la forma del modelo de crecimiento logístico.
stakeholder A person or organization with an interest in a particular place or issue.	**parte interesada** Persona u organización con un interés en un sitio o asunto específico.
standing crop The amount of biomass present in an ecosystem at a particular time.	**cultivo activo** Cantidad de biomasa presente en un ecosistema a una hora dada.
steady state A state in which inputs equal outputs, so that the system is not changing over time.	**régimen estacionario** Régimen en el que las entradas son equivalentes a las salidas, de manera que con el pasar del tiempo el sistema no cambia.
stewardship The careful and responsible management and care for Earth and its resources.	**conciencia fiduciaria** La gestión y cuidado esmerados y responsables de la Tierra y sus recursos.
Stockholm Convention A 2001 agreement among 127 nations concerning 12 chemicals to be banned, phased out, or reduced.	**Convenio de Estocolmo** Convenio suscrito en 2001 por 127 países referente al tratamiento de 12 sustancias químicas que se han de prohibir, eliminar de manera paulatina o reducir.
stratosphere The layer of the atmosphere above the troposphere, extending roughly 16 to 50 km (10–31 miles) above the surface of Earth.	**estratosfera** Capa de la atmósfera encima de la troposfera, que se extiende aproximadamente 16 a 50 km (de 10 a 31 millas) sobre la superficie terrestre.
strip mining The removal of strips of soil and rock to expose ore.	**minería a cielo abierto** Remoción de tajos de suelo y roca para dejar expuesto un mineral.
subduction The process of one crustal plate passing under another.	**subducción** Proceso en el que una placa de la corteza pasa debajo de otra.
sublethal effect The effect of an environmental hazard that is not lethal, but which may impair an organism's behavior, physiology, or reproduction.	**efecto subletal** Efecto de un peligro medioambiental que no es letal pero que puede perjudicar la conducta, fisiología o reproducción de un individuo.
subsistence energy source An energy source gathered by individuals for their own immediate needs.	**fuente de energía de subsistencia** Fuente energética recogida por individuos para satisfacer sus necesidades inmediatas.
subsurface mining Mining techniques used when the desired resource is more than 100 m (328 feet) below the surface of Earth.	**minería subsuperficial** Técnicas de minería que se emplean cuando el recurso deseado se encuentra a más de 100 m (328 pies) por debajo de la superficie terrestre.
subtropical desert A biome prevailing at approximately 30° N and 30° S, with hot temperatures, extremely dry conditions, and sparse vegetation.	**desierto subtropical** Biomedio predominante a aproximadamente 30° N y 30° S, con temperaturas calientes, condiciones extremadamente secas y vegetación escasa.
suburb An area surrounding a metropolitan center, with a comparatively low population density.	**suburbio** Zona circundante de un centro metropolitano, con una densidad poblacional comparativamente baja.
sulfur cycle The movement of sulfur around the biosphere.	**ciclo de azufre** Desplazamiento del azufre alrededor de la biosfera.
sulfurous smog Smog dominated by sulfur dioxide and sulfate compounds. *Also known as* **London-type smog; gray smog; industrial smog.**	**esmog sulfuroso** Esmog dominado por el dióxido de azufre y compuestos sulfatados. *También se conoce como* **esmog londinense; esmog gris; esmog industrial.**
Superfund Act The common name for the Comprehensive Environmental Response, Compensation, and Liability Act (CERCLA); a 1980 U.S. federal act that imposes a tax on the chemical and petroleum industries, funds the cleanup of abandoned and nonoperating hazardous waste sites, and authorizes the federal government to respond directly to the release or threatened release of substances that may pose a threat to human health or the environment.	**Ley** *Superfund* Nombre popular de la Ley Integral de Respuesta, Indemnización y Responsabilidad Medioambiental (CERCLA); ley federal estadounidense promulgada en 1980 que impone un gravamen a las industrias químicas y petroleras, fondos para la limpieza de sitios con desechos nocivos que se hayan abandonado o que hayan dejado de funcionar, y que autoriza al gobierno federal a responder directamente a la liberación o posible liberación de sustancias que podrían atentar contra la salud humana o contra el medioambiente.

surface tension A property of water that results from the cohesion of water molecules at the surface of a body of water and that creates a sort of skin on the water's surface.	**tensión superficial** Propiedad del agua que resulta de la cohesión de moléculas de agua en la superficie de una extensión de agua y que crea una especie de piel sobre la superficie del agua.
survivorship curve A graph that represents the distinct patterns of species survival as a function of age.	**curva de supervivencia** Representación gráfica de los distintos patrones de supervivencia de las especies como función de la edad.
sustainable agriculture Agriculture that fulfills the need for food and fiber while enhancing the quality of the soil, minimizing the use of nonrenewable resources, and allowing economic viability for the farmer.	**agricultura sostenible** Agricultura que satisface la necesidad de alimento y fibra al mismo tiempo que mejora la calidad del suelo, minimizando el uso de recursos no renovables y facilitando la viabilidad económica del agricultor.
sustainable development Development that balances current human well-being and economic advancement with resource management for the benefit of future generations.	**desarrollo sostenible** Desarrollo en el que se equilibran el bienestar y el progreso económico de los seres humanos de hoy con la gestión de los recursos por el bien de las generaciones futuras.
sustainability Living on Earth in a way that allows humans to use its resources without depriving future generations of those resources.	**sostenibilidad** Vivir en la Tierra de una manera que les permite a los humanos aprovechar sus recursos sin privar a generaciones futuras de tales recursos.
swine flu A type of flu caused by the H1N1 virus.	**gripe porcina** Tipo de gripe causada por el virus H1N1.
symbiotic relationship The relationship between two species that live in close association with each other.	**relación simbiótica** Relación entre dos especies que viven en asociación de proximidad una con la otra.
sympatric speciation The evolution of one species into two, without geographic isolation.	**especiación simpátrida** Evolución de una especie en dos, sin que medie aislamiento geográfico.
synergistic interaction A situation in which two risks together cause more harm than expected based on the separate effects of each risk alone.	**interacción sinérgica** Situación en la que dos riesgos se unen para producir más perjuicios que lo previsto con base en los efectos separados de cada riesgo por sí solo.
synthetic fertilizer Fertilizer produced commercially, normally with the use of fossil fuels. *Also known as* **inorganic fertilizer.**	**fertilizante sintético** Fertilizante producido de manera comercial, generalmente a partir de combustibles fósiles. *También se conoce como* **fertilizante inorgánico**.
systems analysis An analysis to determine inputs, outputs, and changes in a system under various conditions.	**análisis de sistemas** Análisis para determinar las entradas, las salidas y los cambios en un sistema en diversas condiciones.

T

technology transfer The phenomenon of less developed countries adopting technological innovations developed in wealthy countries.	**transferencia de tecnología** Fenómeno de los países menos desarrollados que adoptan innovaciones tecnológicas creadas en los países adinerados.
tectonic cycle The sum of the processes that build up and break down the lithosphere.	**ciclo tectónico** Suma de los procesos que componen y descomponen la litosfera.
temperate grassland/cold desert A biome characterized by cold, harsh winters, and hot, dry summers.	**pradera templada o desierto frío** Biomedio caracterizado por inviernos fríos y agrestes, y veranos calientes y secos.
temperate rainforest A coastal biome typified by moderate temperatures and high precipitation.	**bosque lluvioso templado** Biomedio litoral caracterizado por temperaturas moderadas y precipitación abundante.
temperate seasonal forest A biome with warm summers and cold winters with over 1 m (39 inches) of precipitation annually.	**bosque templado estacional** Biomedio con veranos cálidos e inviernos fríos con más de 1 metro (39 pulgadas) de precipitación al año.
temperature The measure of the average kinetic energy of a substance.	**temperatura** Medición de la energía cinética promedio de una sustancia.
teratogen A chemical that interferes with the normal development of embryos or fetuses.	**teratógeno** Sustancia química que interfiere con el desarrollo normal del embrión o feto.

terrestrial biome A geographic region categorized by a particular combination of average annual temperature, annual precipitation, and distinctive plant growth forms on land.	**biomedio terrestre** Región geográfica categorizada por una combinación dada de los promedios de temperatura anual y de precipitación anual, así como por las plantas terrestres características que se distinguen en la zona.
tertiary consumer A carnivore that eats secondary consumers.	**consumidor terciario** Carnívoro que come consumidores secundarios.
the three Rs *See* **reduce, reuse, recycle.**	**las tres erres** *Ver* **reduce, reuse, recycle.**
theory A hypothesis that has been repeatedly tested and confirmed by multiple groups of researchers and has reached wide acceptance.	**teoría** Hipótesis que múltiples grupos de investigadores han comprobado y confirmado en repetidas ocasiones y que ha logrado amplia aceptación.
theory of demographic transition The theory that as a country moves from a subsistence economy to industrialization and increased affluence it undergoes a predictable shift in population growth.	**teoría de transición demográfica** Teoría que explica que en la medida en que un país pasa de una economía de subsistencia a la industrialización y a una mayor abundancia, experimenta un desplazamiento predecible en su crecimiento poblacional.
theory of island biogeography A theory that demonstrates the dual importance of habitat size and distance in determining species richness.	**teoría de biogeografía de islas** Teoría que demuestra la importancia doble de la extensión del hábitat y la distancia para determinar su riqueza en especies.
thermal inversion A situation in which a relatively warm layer of air at mid-altitude covers a layer of cold, dense air below.	**inversión térmica** Situación en la que una capa de aire relativamente cálida a media altura cubre una capa de aire frío y denso más abajo.
thermal mass A property of a building material that allows it to maintain heat or cold.	**masa térmica** Propiedad de un material de construcción que le permite mantenerse caliente o frío.
thermal pollution Nonchemical water pollution that occurs when human activities cause a substantial change in the temperature of water.	**contaminación térmica** Contaminación del agua sin sustancias químicas que ocurre cuando ciertas actividades humanas causan un cambio sustancial en la temperatura del agua.
thermal shock A dramatic change in water temperature that can kill organisms.	**choque térmico** Cambio drástico en la temperatura del agua que puede matar organismos.
thermohaline circulation An oceanic circulation pattern that drives the mixing of surface water and deep water.	**circulación termohalina** Patrón de circulación oceánica en el que se mezclan las aguas superficiales con las aguas profundas.
threatened species According to the International Union for Conservation of Nature (ICUN), species that have a high risk of extinction in the future; according to U.S. legislation, any species that is likely to become an endangered species within the foreseeable future throughout all or a significant portion of its range.	**especies amenazadas** Según la Unión Internacional para al Conservación de la Naturaleza (ICUN), las especies que presentan un riesgo grave de extinción en el futuro; según la legislación estadounidense, toda especie que presenta una probabilidad de ser una especie en peligro de extinción dentro de un futuro previsible en todo o en una parte significativa de su hábitat.
tidal energy Energy that comes from the movement of water driven by the gravitational pull of the Moon.	**energía mareomotriz** Energía que se genera del movimiento del agua impulsado por la atracción gravitatoria de la Luna.
tiered rate system A billing system used by some electric companies in which customers pay higher rates as their use goes up.	**sistema tarifario escalonado** Sistema de facturación utilizado por algunas compañías eléctricas en la que los clientes pagan tarifas más altas a medida que aumenta su consumo de energía.
tiered water-pricing system A water allocation system that charges rates that increase with the amount of water consumed.	**sistema escalonado de tarifas de agua** Sistema en el que la tarifa por el reparto de agua aumenta según la cantidad de agua consumida.
tipping fee A fee charged for disposing of material in a landfill or incinerator.	**tarifa de carga** Cuota que se cobra por enviar material a un basurero o incinerador.
topsoil *See* **A horizon.**	**manto** *Ver* **A horizon.**

total fertility rate (TFR) An estimate of the average number of children that each woman in a population will bear throughout her childbearing years.	**tasa de fertilidad total** Estimación del número promedio de niños que cada mujer en una población dará a luz durante sus años en edad reproductiva.
toxicity Harm, illness, or death caused by chemical means through ingestion, inhalation, or absorption.	**toxicidad** Daño, enfermedad o muerte causados por la ingestión, inhalación o absorción de sustancias químicas.
tragedy of the commons The tendency of a shared, limited resource to become depleted if it is not regulated in some way.	**tragedia de los comunes** Tendencia de un recurso compartido y limitado si no está regulado de alguna manera.
transform fault boundary An area where tectonic plates move sideways past each other.	**borde de una falla de transformación** Zona en la que las placas tectónicas se desplazan lateralmente, una pasando la otra.
transit-oriented development (TOD) Development that attempts to focus dense residential and retail development around stops for public transportation, a component of smart growth.	**urbanización orientada por el tránsito** Desarrollo urbano que trata de concentrar la creación de zonas residenciales y de venta detallista alrededor de paradas del transporte público. Es uno de los componentes del crecimiento inteligente.
transpiration The release of water from leaves during photosynthesis.	**transpiración** Liberación de agua desde las hojas durante la fotosíntesis.
tree plantation A large area typically planted with a single rapidly growing tree species.	**plantación de árboles** Zona extensa en la que característicamente se han sembrado especies de árboles de crecimiento rápido.
triple bottom line An approach to sustainability that considers three factors—economic, environmental, and social—when making decisions about business, the economy, and development.	**triple fin de cuentas** Manera de enfocar la sostenibilidad en la que se tienen en cuenta tres factores: el económico, el medioambiental y el social, cuando se toman decisiones sobre los negocios, la economía y el desarrollo.
trophic levels The successive levels of organisms consuming one another.	**niveles tróficos** Niveles sucesivos de organismos que se consumen unos a otros.
trophic pyramid A representation of the distribution of biomass, numbers, or energy among trophic levels.	**pirámide trófica** Representación gráfica de la distribución de biomasa, cifras o energía entre los niveles tróficos.
tropical rainforest A warm and wet biome found between 20° N and 20° S of the equator, with little seasonal temperature variation and high precipitation.	**bosque lluvioso tropical** Biomedio cálido y húmedo que se encuentra entre 20° N y 20° S del ecuador, con poca variación estacional de temperatura y niveles altos de precipitación.
tropical seasonal forest/savanna A biome marked by warm temperatures and distinct wet and dry seasons.	**bosque/sabana tropical estacional** Biomedio que se distingue por sus temperaturas cálidas y sus estaciones de lluvia y sequía claramente distinguibles.
troposphere A layer of the atmosphere closest to the surface of Earth, extending up to approximately 16 km (10 miles).	**troposfera** Capa de la atmósfera contigua a la superficie de la Tierra que se extiende en forma ascendente unos 16 km (10 millas).
tuberculosis A highly contagious disease caused by the bacterium *Mycobacterium tuberculosis* that primarily infects the lungs.	**tuberculosis** Enfermedad muy contagiosa causada por la bacteria *Mycobacterium tuberculosis.* Primordialmente afecta los pulmones.
tundra A cold and treeless biome with low-growing vegetation.	**tundra** Biomedio frío y carente de vegetación arbórea con vegetación de poco crecimiento.
turbine A device that can be turned by water, steam, or wind to produce power.	**turbina** Dispositivo que puede ser girado por agua, vapor o viento para producir energía.
type I survivorship curve A pattern of survival over time in which there is high survival throughout most of the life span, but then individuals start to die in large numbers as they approach old age.	**curva de supervivencia tipo I** Patrón de supervivencia temporal en el que existe un nivel de supervivencia alto durante la mayor parte de la duración de la vida, pero luego los individuos comienzan a fallecer en gran volumen a medida que se aproximan a la vejez.

type II survivorship curve A pattern of survival over time in which there is a relatively constant decline in survivorship throughout most of the life span.	**curva de supervivencia tipo II** Patrón de supervivencia temporal en el que existe una disminución relativamente constante en la supervivencia durante la mayoría de la duración de la vida.
type III survivorship curve A pattern of survival over time in which there is low survivorship early in life with few individuals reaching adulthood.	**curva de supervivencia tipo III** Patrón de supervivencia temporal en el que existe un bajo nivel de supervivencia temprano en la vida y son pocos los individuos que llegan a la adultez.

U

uncertainty An estimate of how much a measured or calculated value differs from a true value.	**incertidumbre** Estimación de la diferencia que hay entre un valor medido o calculado y el valor real.
unconfined aquifer An aquifer made of porous rock covered by soil out of which water can easily flow.	**acuífero no cautivo** Acuífero hecho de roca porosa, cubierto por suelo del cual fácilmente puede fluir agua.
undernutrition The condition in which not enough calories are ingested to maintain health.	**subnutrición** Condición producto de no haber ingerido suficientes calorías para mantener el estado de salud.
United Nations (UN) A global institution dedicated to promoting dialogue among countries with the goal of maintaining world peace.	**Organización de Naciones Unidas (ONU)** Institución global dedicada a fomentar el diálogo entre países con el objetivo de mantener la paz en el mundo.
United Nations Development Programme (UNDP) An international program that works in 170 countries around the world to advocate change that will help people obtain a better life through development.	**Programa de Naciones Unidas para el Desarrollo (UNDP)** Programa internacional que funciona en 170 países del mundo con el fin de promover cambios que les ayudarán a las personas a lograr una mejor vida gracias al desarrollo.
United Nations Environment Programme (UNEP) A program of the United Nations responsible for gathering environmental information, conducting research, and assessing environmental problems.	**Programa de Naciones Unidas para el Medio Ambiente (UNEP)** Programa de Naciones Unidades responsable por recoger información, realizar investigaciones y evaluar problemas de índole ambiental y ecológica.
upwelling The upward movement of ocean water toward the surface as a result of diverging currents.	**surgencia** Movimiento hacia arriba del agua oceánica hacia la superficie como resultado de corrientes divergentes.
urban area An area that contains more than 385 people per square kilometer (1,000 people per square mile).	**zona urbana** Extensión que contiene más de 385 personas por kilómetro cuadrado (1,000 personas por milla cuadrada).
urban blight The degradation of the built and social environments of the city that often accompanies and accelerates migration to the suburbs.	**deterioro urbano** Degradación del entorno construido y del ambiente social de la ciudad que a menudo acompaña y acelera la migración hacia los suburbios.
urban growth boundary A restriction on development outside a designated area.	**límite del crecimiento urbano** Limitación de la urbanización por fuera de una zona designada.
urban sprawl Urbanized areas that spread into rural areas, removing clear boundaries between the two.	**expansión urbana** Zonas urbanizadas que se desparraman hacia las zonas rurales, eliminando los límites claros entre las dos.
valuation The practice of assigning monetary value to intangible benefits and natural capital.	**valoración** Asignación de un valor monetario a los beneficios intangibles y al capital natural.
volatile organic compound (VOC) An organic compound that evaporates at typical atmospheric temperatures.	**compuesto volátil orgánico** Compuesto orgánico que se evapora a las temperaturas atmosféricas características.

V

variable Any categories, conditions, factors, or traits that differ in the natural world or in experimental situations.	**variable** Cualquier categoría, condición, factor o rasgo que difiere en el mundo natural o en situaciones experimentales.
volcano A vent in the surface of Earth that emits ash, gases, or molten lava.	**volcán** Escape en la superficie terrestre por el que se emiten cenizas, gases o lava fundida.

W

waste Material outputs from a system that are not useful or consumed.	**desecho** Efluentes materiales de un sistema. Se trata de efluentes que ni son útiles ni se pueden consumir.
waste stream The flow of solid waste that is recycled, incinerated, placed in a solid waste landfill, or disposed of in another way.	**flujo de desechos** El flujo de desechos sólidos que se reciclan, incineran, tiran en un basurero o eliminan de alguna otra manera.
waste-to-energy A system in which heat generated by incineration is used as an energy source rather than released into the surrounding environment.	**desecho a energía** Sistema en el que se genera calor mediante incineración. Dicho calor se utiliza como fuente de energía en lugar de liberarlo en el medio ambiente.
wastewater Water produced by livestock operations and human activities, including human sewage from toilets and gray water from bathing and washing of clothes and dishes.	**aguas residuales** Aguas producidas por operaciones de ganado y actividades humanas, incluidas aguas negras humanas de inodoros y aguas grises de baños y del lavado de ropas y trastes.
water footprint The total daily per capita use of fresh water.	**consumo de agua** El uso total diario per cápita de agua dulce.
water impoundment The storage of water in a reservoir behind a dam.	**embalse de agua** Almacenamiento de agua en un reservorio, detrás de una presa.
waterlogging A form of soil degradation that occurs when soil remains under water for prolonged periods.	**saturación** Forma de degradación del suelo que ocurre cuando este permanece bajo agua por lapsos de tiempo prolongados.
water pollution The contamination of streams, rivers, lakes, oceans, or groundwater with substances produced through human activities.	**contaminación del agua** Contaminación de arroyos, ríos, lagos, océanos o aguas freáticas con sustancias producidas mediante actividades humanas.
water table The uppermost level at which the water in a given area fully saturates rock or soil.	**nivel freático** El nivel más alto en el que el agua satura totalmente la roca o el suelo.
watershed All land in a given landscape that drains into a particular stream, river, lake, or wetland.	**cuenca hidrográfica** Todo el terreno en un paisaje dado que desemboca en un arroyo, río, lago o humedal dado.
weather The short-term conditions of the atmosphere in a local area, which include temperature, humidity, clouds, precipitation, and wind speed.	**estado del tiempo** Condiciones de la atmósfera a corto plazo en una zona local; incluye la temperatura, la humedad, la nubosidad, la precipitación y la velocidad del viento.
well-being The status of being healthy, happy, and prosperous.	**bienestar** Estado de ser feliz, y de gozar de buena salud y prosperidad.
West Nile virus A virus that lives in hundreds of species of birds and is transmitted among birds by mosquitoes.	**virus del Nilo Occidental** Virus que vive en centenares de especies de aves y que es transmitido entre aves por mosquitos.
wind energy Energy generated from the kinetic energy of moving air.	**energía eólica** Energía generada a partir de la energía cinética del aire en movimiento.
wind turbine A turbine that converts wind energy into electricity.	**turbina aerogeneradora** Turbina que convierte la energía del viento en electricidad. *También se conoce como* **torre eólica.**
woodland/shrubland A biome characterized by hot, dry summers and mild, rainy winters.	**terreno de árboles y arbustos** Biomedio caracterizado por veranos calurosos y secos, e inviernos suaves y lluviosos.
World Bank A global institution that provides technical and financial assistance to developing countries with the objectives of reducing poverty and promoting growth, especially in the poorest countries.	**Bando Mundial** Institución de alcance mundial que ofrece asesoría técnica y asistencia económica a los países en vías de desarrollo, con los objetivos de reducir la pobreza y fomentar el crecimiento, sobre todo en los países menos adinerados.
World Health Organization (WHO) A global institution dedicated to the improvement of human health by monitoring and assessing health trends and providing medical advice to countries.	**Organización Mundial de la Salud (OMS)** Institución de alcance mundial que se dedica a mejorar la salud humana mediante el monitoreo y la evaluación de tendencias sanitarias. Además, ofrece consejos médicos a los países que los soliciten.

X	
xeriscaping A style of landscaping that removes water-intensive vegetation from lawns and replaces it with more water-efficient native landscaping.	**xeriscape o xeropaisajismo** Estilo de paisajismo que retira la vegetación que requiere de riegos suplementarios de los jardines y la sustituye con tipos de vegetación nativa cuyo consumo de agua es más eficiente.

Z	
Zika virus disease A disease caused by a pathogen that causes fetuses to be born with unusually small heads and damaged brains.	**enfermedad por el virus del Zika** Enfermedad causada por un patógeno que causa microcefalia (reducción del tamaño de la cabeza) y daños cerebrales en los fetos.
zoning A planning tool used to separate industry and business from residential neighborhoods.	**zonificación** Recurso de planificación que se utiliza para separar la industria y el comercio de los barrios residenciales. **horizonte A** A menudo es la capa superficial del suelo, una zona de material orgánico y minerales que se han entremezclado. *También se conoce como* **manto.**

Index

Note: Page numbers in **boldface** indicate definitions; those followed by f indicate figures; those followed by t indicate tables.

Math Review

Metric Prefixes A prefix in front of a unit of measure indicates a multiple or fraction of that unit. The table below lists the most frequently used prefixes of the metric system, which you should know for the AP® Environmental Science Exam.

	Prefix	Equivalence	Scientific notation	Abbreviation
Largest	mega	1,000,000	10^6	M
	kilo	1,000	10^3	k
	hecto	100	10^2	h
	deca	10	10^1	da
Base unit	e.g., meter, gram, liter, joule, watt	1	10^0	m, g, L, J, W
	deci	0.1	10^{-1}	d
	centi	0.01	10^{-2}	c
	milli	0.001	10^{-3}	m
Smallest	micro	0.000001	10^{-6}	μ

Scientific Notation *(See pages 200, 293, 349.)*

Use scientific notation when dealing with very large or very small numbers. Here are some examples of numbers written in scientific notation:

$34{,}670 = 3.467 \times 10^4$
$0.0053 = 5.3 \times 10^{-3}$
$259 = 2.59 \times 10^2$

The first part of this notation (e.g., 2.59) is called the digit term. The second part of the notation (e.g., 10^2) is called the exponential term. When adding or subtracting numbers that are written in scientific notation, first make sure that both numbers use the same exponent. Then add or subtract the digit terms and retain the original exponential term.

Example

$$1.23 \times 10^5 + 1.51 \times 10^6 = (0.123 + 1.51) \times 10^6 = 1.633 \times 10^6$$

When multiplying in scientific notation, multiply the two digit terms and then add the two exponential terms.

Example

$$(2.4 \times 10^2) \times (3.0 \times 10^4) = (2.4 \times 3.0) \times (10^{2+4}) = 7.2 \times 10^6$$

When dividing in scientific notation, divide the two digit terms and then subtract the two exponential terms.

Example

$$(3.6 \times 10^4) \div (1.2 \times 10^2) = (3.6 \div 1.2) \times (10^{4-2}) = 3.0 \times 10^2$$

Converting Units *(See pages 10, 13, 47, 49, 136, 349, 381, 422, 520, 532, 590.)*

To convert a value into different units, multiply by one or more ratios that are equal to 1, where the numerator and denominator are equivalent. The units in the numerator and denominator should cancel each other out until only the desired units are left.

Example

Convert 6.85 km into inches. (1 km = 1,000 m, 1 m = 3.28 feet, and 1 foot = 12 inches)

$$\frac{6.85\ \text{km} \times}{} \left| \frac{1{,}000\ \text{m} \times}{1\ \text{km}} \right| \frac{3.28\ \text{foot} \times}{1\ \text{m}} \left| \frac{12\ \text{inches}}{1\ \text{foot}} \right. = 269{,}616\ \text{inches}$$

Calculating Averages *(See pages 20, 136, 564.)*

To find the average, or mean, of a set of values, add all the given values and divide the total by the number of given values.

average (mean) $= (x + y + z) \div n$*, where* $n =$ *the number of given values.*

Example

The average (mean) of 107, 53, 100, 114, and 236 = (107 + 53 + 100 + 114 + 236) ÷ 5 = 610 ÷ 5 = 122

Percent of a Total Value *(See pages 91, 167, 584, 590, 658.)*

To calculate the percent of a total value, divide the subset value by the total value, then multiply by 100%.

percent of a total value = (subset value ÷ total value) × 100%

Example

A group of rabbits consists of 10 brown, 15 white, and 25 black rabbits.

The percentage of brown rabbits = [10 ÷ (10 + 15 + 25)] × 100% = (10 ÷ 50) × 100% = 0.2 × 100% = 20%

Percent Change *(See pages 349, 556, 706, 726.)*

To calculate percent change, take the difference between the final and initial values and divide this difference by the initial value. Then multiply by 100%.

percent change = [(final value − initial value) ÷ initial value] × 100%

Example

Annual U.S. meat consumption in 1960 = 60 kg per person

Annual U.S. meat consumption in 2012 = 75 kg per person

The percent change from 1960 to 2012 = [(75 kg − 60 kg) ÷ 60 kg] × 100% = (15 ÷ 60 kg) × 100% = 0.25 × 100 = 25%

Annual Rate of Change *(See pages 278, 556, 695.)*

To calculate the annual rate of change for a given measurement, first find the difference between the final value and the initial value. Then divide that difference by the total number of years considered, which is the difference between the final year and the initial year.

annual rate of change = (final value − initial value) ÷ (final year − initial year)

Example

Atmospheric CO_2 = 320 ppm in 1960

Atmospheric CO_2 = 390 ppm in 2010

Average annual rate of change = (390 ppm − 320 ppm) ÷ (2010 − 1960) = 70 ppm ÷ 50 years = 1.4 ppm/year

pH Scale *(See page 41.)*

The pH scale is logarithmic, meaning that each number on the scale changes by a factor of 10.